# Golf Turf Management

# Golf Turf Management

L.B. (Bert) McCarty
Department of Plant and Environmental Sciences
Clemson University
Clemson, South Carolina

CRC Press is an imprint of the
Taylor & Francis Group, an **informa** business

**Disclaimer:** All mentioned chemicals are for reference only and may not be available for turf use. They may be restricted by some state, province, or federal agencies; thus, be sure to check the current status of the pesticide being considered for use. Always read and follow the manufacturer's label as registered under the Federal Insecticide, Fungicide, and Rodenticide Act. Mention of a proprietary product does not constitute a guarantee or warranty of the product by the author or the publisher and does not imply approval to the exclusion of other products that also may be suitable.

CRC Press
Taylor & Francis Group
6000 Broken Sound Parkway NW, Suite 300
Boca Raton, FL 33487-2742



ISBN 13: 978-1-138-47638-7 (pbk)
ISBN 13: 978-1-138-58406-8 (hbk)



**Library of Congress Cataloging-in-Publication Data**

Names: McCarty, L. B. (Lambert Blanchard), 1958- author.
Title: Golf turf management / author: L.B. McCarty.
Description: Boca Raton : Taylor & Francis, 2018. | Includes bibliographical references and index.
Identifiers: LCCN 2017061730| ISBN 9781138476387 (pbk. : alk. paper) | ISBN 9781138584068 (hardback : alk. paper)
Subjects: LCSH: Turf management. | Turfgrasses. | Golf courses--Maintenance.
Classification: LCC SB433 .M33 2018 | DDC 635.9/642--dc23
LC record available at https://lccn.loc.gov/2017061730

**Visit the Taylor & Francis Web site at**
**http://www.taylorandfrancis.com**

**and the CRC Press Web site at**
**http://www.crcpress.com**

*To those golf course superintendents who grow and maintain the best playing conditions in the world and to the students who should relentlessly pursue knowledge in turf management, personnel skills, and human relationships.*

*To my parents, family, students, and colleagues, who have provided vast opportunities for me to learn and work.*

# Contents

## *SECTION IV Turfgrass Fertilization and Water Management*

## SECTION V Turfgrass Management Practices

## SECTION VI Pest Management and Control Practices

## SECTION VII Budgeting Practices

# Preface

Golf courses are continually increasing in number, as well as sophistication in terms of design, management, and increased scrutiny from the general public and regulatory agencies. Golf course management quality and intensity range from very low-maintained facilities to exquisite, highly maintained touring courses. Many resort courses rely on the tourist industry; however, on other courses, membership and daily fees are major sources of income. In these cases, the year-round conditions become very important.

This book is intended as a reference guide for golf course superintendents, assistants, club managers, green's committee members, students, and regulatory agencies in their efforts to grow and maintain some of the world's most prestigious courses. Authors with expertise in specific areas of turfgrass and environmental science have reviewed this book, making the information as complete and up to date as possible. However, management and pesticide recommendations are constantly being updated. New products, grasses, and management techniques continue to evolve, while older ones often disappear. Contact your state university turf specialist or county cooperative extension service office, and attend the various turfgrass field days and turfgrass associations' annual conferences and trade shows, for the latest recommendations.

The use of trade names in this publication is solely for the purpose of providing specific information. It is not a guarantee or warranty of the products named, and does not signify that they are approved to the exclusion of others of suitable composition.

The pesticide recommendations presented in this publication were current with state and federal regulations at the time of publication. The user is responsible for determining that the intended pesticide use is consistent with the directions on the label of the product being used. Use pesticides safely. Read and follow label directions because this constitutes federal law.

A new accompanying publication, *Advanced Turfgrass Management Laboratory Manual*, is a complete laboratory manual implementing many of the concepts covered in *Golf Turf Management*. Included in this manual are

- Review of Basic Mathematical and Chemistry Concepts—Autotutorial
- Turfgrass Anatomy, Morphology, and Identification
- Soil Chemical Properties
- Soil Physical Properties and Modification—Parts 1 and 2
- Basic Surveying Techniques and Determining Areas
- Seed Calculations
- Fertilizer Technology and Calculations
- Tissue and Soil Testing and Interpretation
- Irrigation Water—Part 1: Quality; Part 2: Application and Distribution
- Sprayer Calibration and Calculations
- Budgeting and Personnel Management
- Effective Presentations and Persuasive Public Speaking
- Field Trips or Guest Speakers

This publication should be considered by all turfgrass students and professionals.

# Acknowledgments

Any project of such magnitude is definitely the result of the efforts of many competent, dedicated professionals. The author wishes to express gratitude to the following reviewers: first edition: Ed Freeman, Professor Emeritus, University of Florida; and golf course superintendents Fred Biggers, Don Garrett, Chuck Green, Will Holroyd, and David Lowe; second edition: Michael Ventola, Sandhills Community College; John Wildmon, Lake City Community College; J. Phillip Thomas, Kirkwood Community College; Bob Emmons, State University of New York at Cobleskill; third edition: Dr. Jeff Higgins, Harrell's Fertilizer; Dr. Grady Miller, Dr. Dan Bowman, Dr. Jim Kerns, Dr. Rick Brandenburg, and Dr. Charles Peacock, North Carolina State University; Dr. Bruce Martin, Dr. Bob Cross, Dr. Ray Hubbard, Jr., Dr. Haibo Liu, and Dr. Philip Brown, Clemson University; Dr. Clint Waltz, University of Georgia; Dr. Cale Bigelow, Purdue University; fourth edition: Dr. Grady Miller, Dr. Jim Kerns, Dr. Rick Brandenburg, and Dr. Charles Peacock, North Carolina State University; Dr. Bruce Martin, Dr. "J.C." Chong, Dr. Bob Cross, Dr. Philip Brown, Dr. Bob Bellinger, and Bobby Kerr, Clemson University; Dr. Cale Bigelow, Purdue University; Dr. Todd Bunnell, Brightview Golf Management. Hopefully no one has been omitted, from this list; if so, I apologize for the oversight.

# About the Author

**Bert McCarty** is a professor of plant and environmental sciences specializing in turfgrass science and management at Clemson University in Clemson, South Carolina. A native of Batesburg, South Carolina, McCarty received a BS degree from Clemson University in agronomy and soils, an MS from North Carolina State University in crop science, and a PhD from Clemson University in plant physiology and plant pathology. Dr. McCarty spent nine years as a turfgrass specialist at the University of Florida in Gainesville. While there, he oversaw the design and construction of a state-of-the-art research and education turfgrass facility named "The Envirotron." He also was author or coauthor of the books *Best Management Practices for Florida Golf Courses*, *Weeds of Southern Turfgrasses*, and *Florida Lawn Handbook*. In 1996, he moved to Clemson University, where he is currently involved in research, extension, and teaching activities. He has published numerous articles dealing with all phases of turfgrass management and is in constant demand to conduct presentations, workshops, training sessions, and on-site visits concerning all phases of turfgrass establishment and management. He is currently coauthor of the books *Color Atlas of Turfgrass Weeds* (third edition); *Weed Control in Turf and Ornamentals*, *Southern Lawns*, *Managing Bermudagrass Turf*, and *Fundamentals of Turfgrass and Agricultural Chemistry*. He is coauthor for the GCSAA workshops Weed Control, Advanced Weed Management, Managing Bentgrass Golf Greens in Heat Stress Environments, and Managing Ultradwarf Bermudagrass Golf Greens. He also teaches an undergraduate/graduate course, Advanced Turfgrass Management. He is a fellow for the Crop Science Society of America and the American Society of Agronomy. He is also recipient of the Fred Grau Outstanding Turfgrass Scientist award and is active in a number of professional societies. He is an oft-requested speaker, teacher, and consultant.

# Commonly Used Abbreviations

| | |
|---|---|
| @ | at |
| Δ | change in |
| $\theta$ | soil water content |
| $\theta_{fc}$ | volumetric water content at field capacity |
| $\theta_g$ | gravimetric water content |
| $\theta_v$ | volumetric water content |
| $\theta_{wp}$ | volumetric water content at wilting point |
| $\mu$ | micro |
| $\pi$ | pi (~3.14) |
| $\rho_b$ | bulk density |
| $\rho_s$ | particle density |
| $\Psi$ or $\Psi_w$ | water potential |
| $\Psi_g$ | gravitational potential |
| $\Psi_o$ | osmotic potential |
| $\Psi_p$ | pressure (or matric) potential |
| $\Psi_t$ | total water potential |
| ~ | approximate |
| / | per |
| % | percent |
| A | cross section of column |
| ABA | abscisic acid |
| ac or a | acre |
| ac-ft | acre foot |
| ac-in | acre inch |
| adj SAR | adjusted sodium absorption ratio |
| AEP | air entry point |
| aka | also known as |
| ASTM | American Society for Testing and Materials |
| atm | atmosphere |
| ATP | adenosine triphosphate |
| B | boron |
| BMPs | best management practices |
| bu | bushel |
| C | centigrade or Celsius |
| $C_3$ | cool-season turfgrasses |
| $C_4$ | warm-season turfgrasses |
| $Ca^{+2}$ | calcium |
| $CaCO_3$ | calcium carbonate |
| $CaSO_4$ | calcium sulfate (aka gypsum) |
| cb | centibar |
| cc | cubic centimeters |
| CEC | cation exchange capacity |
| Cl | chlorine |
| $Cl^-$ | chloride |
| cm | centimeter |
| $cm^3$ | cubic centimeter |
| $CO_2$ | carbon dioxide |
| $CO_3^{-2}$ | carbonates |
| cos | cosine |
| Cu | copper |
| D | diameter of the particle or its pore |
| $D_w$ | equivalent depth of water or void ratio |
| dH | hydraulic head length (total) |
| dist | distance |
| dL | core length |
| dS | decisiemens |
| DU | distribution uniformity |
| $D_x$ | gradation index based on diameter of particles |
| EC | electrical conductivity |
| $EC_{dw}$ | electrical conductivity of a saturated paste extract by the plant grown |
| e.g. | for example |
| ESP | exchangeable sodium percentage |
| ET | evapotranspiration |
| etc. | etcetera |
| $ET_p$ | potential or pan evapotranspiration |
| F | Fahrenheit |
| $Fe^{+2}$, $Fe^{-3}$ | iron |
| FDR | frequency domain reflectometry |
| ft | foot |
| $ft^2$ | square foot |
| g | gram |
| GA | gibberellic acid |
| gpa | gallons per acre |
| gpm | gallons per minute |
| gal | gallon |
| ht | height |
| $H_2O$ | water |
| ha | hectare |
| $HCO_3$ | bicarbonates |
| Hg | mercury |
| HOC | height of cut |
| hr | hour |
| i.e. | that is |
| in. | inch |
| J | joule |
| $K^+$ | potassium |
| $K_2O$ | potash |
| K | temperature constant |
| $K_c$ | crop coefficient |
| kg | kilogram |
| kPa | kilopascal |
| $K_{sat}$ | saturated hydraulic conductivity |
| L | liter |
| lb | pound |
| m | meter |
| $m^2$ | square meter |
| mbar | millibar |
| MHz | megahertz |
| $Mg^{+2}$ | magnesium |
| mg | milligrams |
| mi | mile |
| mL | milliliter |
| mm | millimeter |
| mmhos | millimhos |
| Mo | molybdenum |

| | |
|---|---|
| Mn | manganese |
| mol | mole |
| MPa | megapascal |
| mph | miles per hour |
| N | nitrogen |
| $Na^+$ | sodium |
| NADPH | nicotinamide adenine dinucleotide phosphate |
| nm | nanometer |
| $NO_3^-$ | nitrate |
| noz | nozzles |
| $O_2$ | oxygen |
| OH | hydroxyl ion |
| OM | organic matter |
| P | phosphorus |
| PGR | plant growth regulator or retardant |
| pH | $-\log[H^+]$ |
| $P_2O_5$ | phosphate |
| PLS | pure live seed |
| ppm | parts per million |
| psi | pounds per square inch |
| pt | pint |
| PTO | power take-off |
| Q | quantity of water |
| r | radius |
| RH | relative humidity |
| RSC | residual sodium carbonate |
| S | sulfur |
| SAR | sodium adsorption ratio |
| sec | second |
| SMRC | soil moisture retention curve |
| $SO_4^{-2}$ | sulfate |
| TDR | time domain reflectometry |
| TDS | total dissolved salts |
| TSS | total suspended solids |
| U.S. | United States |
| U.S.DA | United States Department of Agriculture |
| U.S.GA | United States Golf Association |
| V | velocity of the falling particle |
| v | volume |
| $V_a$ | volume of air |
| $V_s$ | volume of solids |
| $V_t$ | total volume |
| $V_v$ | volume of voids |
| $V_w$ | volume of water |
| VWC | volumetric water content |
| WDPT | water drop penetration time |
| WHC | water-holding capacity |
| wt | weight |
| Zn | zinc |

# *Section I*

*Turfgrasses for Golf Courses*

# 1 Turfgrasses

## THE GOLF INDUSTRY

The game of golf originated in the mid-fourteenth century along the coasts of Scotland and Ireland where "linksland" was formed by the forces of nature. In 1522, during the reign of Mary, Queen of Scots, St. Andrews, the first permanent golf course, came into existence. St. Andrews is a linksland golf course with few artificial characteristics. Receding seas left sandy waste areas with ridges and furrows, while natural plateaus became greens and tees (Figure 1.1). Burrowing animals, along with wind and rain, formed bunkers.

In the United States, golf began in 1787 with the South Carolina Golf Club in Charleston, but was later formalized in 1884 with the opening of Oakhurst Golf Club in West Virginia. Currently, over 15,000 golf courses exist in the United States (75% public, 16% municipal) occupying > 2 million acres (~1 million ha) and serving about 26 million golfers (Figure 1.2). The average size of each 18-hole course—6,000 yards (5,486 m) in length—ranges between 120 and 240 acres (49 to 91 ha), with an average of 133 acres (54 ha) for 18-hole facilities and 62 acres (25 ha) for 9-hole courses. If water, buildings, and parking lots are included, total acreages increase to almost 200 acres (81 ha) for 18 holes (Table 1.1). "Executive" golf courses, usually consisting of only par 3 and par 4 holes, typically have a par between 58 and 64 with lengths between 3,500 and 5,000 yards (3,200 and 4,500 m). Greens and tees account for ~6% of this total, and on a typical 18-hole facility, greens occupy from 2.1 to 3.3 acres (0.85 to 1.3 ha). Green surrounds (collar, bunkers, and grassy surrounds) compose from 2.5 to 5.0 acres (1 to 2 ha), tees 0.4 to 3.0 acres (0.16 to 1.2 ha), fairways 30 to 60 acres (12 to 24 ha), and roughs 35 to 90 acres (14 to 36 ha). The wooded area for an 18-hole course averages 35 acres (14 ha), while 20 acres (8 ha) is average for a 9-hole course (Figure 1.3).

Golf courses in the United States provide a total economic impact of an estimated $69 billion each year (Figure 1.4). One in 75 U.S. jobs is impacted by the golf industry. Turfgrasses also provide soil erosion control, dust stabilization, heat dissipation, noise abatement, air pollution control, wildlife habitat, safety to competitive athletic participants, and increased property values, and are an integral component of the landscape (Table 1.2) (Figure 1.5). Turfgrasses provide many of these benefits due to their high number of plant shoots and roots: 49 to 85 billion shoots/acre (123 to 213 billion/ha), with up to 163 billion shoots/acre (408 billion/ha) for putting greens and a combined root weight of up to 14,363 pounds/acre (16,000 kg/ha) for a lawn. Due to this high shoot and root mass, turfgrasses are often used as filter strips for mining operations, animal production facilities, and agricultural croplands. Research also demonstrates bare ground loses almost 200 pounds soil/acre (224 kg/ha) during a 3-inch (7.6 cm) rainstorm; conversely, turfgrass-covered ground only loses between 9 and 54 pounds soil/acre, or 10 to 60 kg/ha (Gross et al. 1991).

### Golf Course Rating

Golf courses are rated to develop a handicap system, which is ideally equitable everywhere. This rating includes the length (yardage) of the course and measurement of difficulty. A course rating team reviews numerous variables of a course to rate its relative difficulty for all levels of golfers. The standards of yardage for par are defined in the USGA *Rules of Golf* (Table 1.3). A regulation 18-hole facility typically has a total par between 70 and 72 and between 6,000 and over 7,200 yards (~5,500 and 6,500 m) for men and 4,500 and 5,800 yards (4,100 and 5,300 m) for women. The average handicap index of the American golfer is 14.3 for men and 26.5 for women.

## TURFGRASS TAXONOMY

Living organisms are classified based on shared characteristics and natural relationships. The grass family, *Poaceae* (formerly known as *Gramineae*), includes six subfamilies and over 700 genera with more than 10,000 species. About 150 grass genera and 1,500 species are found growing in the United States; however, less than 25 species are important as turfgrasses. By having such a wide diversity of habitats, grasses show a considerable measure of ecological adaptation.

### Scientific Names

Living organisms are identified by a Latin binomial classification system. The first name is the Latin description for the genus while the second name represents the species. The taxonomic authority may be added to the Latin binomial following the species designation to indicate who first identified the species. For example, perennial ryegrass has the Latin binomial (also called the scientific name) of *Lolium perenne* L., with the L. serving as an abbreviation representing Carl Linnaeus, a botanist who first described this species.

Other variations of author citations occur for turfgrass taxonomy. For example, bermudagrass has the Latin binomial of *Cynodon dactylon* [L.] Pers., which indicates the botanist, Carl Linnaeus (or L.), first described the species but in a different genus or as a separate species or at a different rank. Later, another author, Christiaan Persoon (abbreviated as Pers.), moved it to the *Cynodon* genus and is considered the primary author while Linnaeus, in brackets, is the secondary author.

Other rules apply with multiple authorities. For example, with the Latin name of mascarenegrass, (*Zoysia tenuifolia*

**FIGURE 1.1** Golf originated in Scotland, where wind, rain and burrowing animals formed course features.

**FIGURE 1.3** In addition to its economic impact, golf provides many environmental benefits such as helping to control soil erosion as well as providing habitats for a diverse number of animals and plants.

**FIGURE 1.2** In the United States, over 16,000 golf courses serve about 26 million golfers.

**FIGURE 1.4** In the United States, 1 in 75 jobs is impacted by the golf industry.

Willd. ex Trin.), the term "ex" represents that Carl Ludwig Willdenow (abbreviated Willd.) first proposed the name, but that Carl Bernhard von Trinius (abbreviated Trin.) later provided the recognized valid description.

Further subdivisions are necessary when important differences exist within a species but not to the extent to

**TABLE 1.1**
**Typical Size for an 18- and a 9-Hole Golf Course**

| | 18 Holes (6,000 yards[a]) | | 9 Holes | |
|---|---|---|---|---|
| **Area** | **Size, Acres (ha)** | **Percentage of Size** | **Size, Acres (ha)** | **Percentage of Size** |
| Rough/out-of-play/water/woods | 147 (60) | 76 | 45 (18.2) | 60 |
| Fairways | 35 (14) | 18 | 25 (10) | 33 |
| Buildings/parking lots | 5 (2) | 2.6 | 3 (1.2) | 4 |
| Greens/greens surround | 3 (1.2) | 1.6 | 1.5 (0.61) | 2 |
| Tees | 3 (1.2) | 1.5 | 1 (0.4) | 1 |
| Total | 193 (78) | 100 | 70 to 80 (28 to 32) | 100 |

[a] 6,000 yards = 5,486 m.

**TABLE 1.2**
**Temperatures of Various Soil Surfaces with Air Temperatures of 90°F (33°C) and 45% Relative Humidity, Clemson, SC**

| Surface | Temperature, °F (°C) |
|---|---|
| Algae mat (wet) in golf green | 101 (38) |
| Asphalt | 140 (60) |
| Bare soil | 109 (43) |
| Baseball infield clay | 121 (49) |
| Bermudagrass (well watered) | 98 (37) |
| Bermudagrass (wilted) | 130 (54) |
| Bunker sand (white) | 102 (39) |
| Concrete slab | 119 (48) |
| Gravel roadbed | 109 (43) |
| Synthetic turf with crumb rubber | 160 (71) |
| Synthetic turf and no crumb rubber | |
| • Light-green color | 136 (58) |
| • Dark-green color | 155 (68) |
| Water (stream) | 82 (28) |

**FIGURE 1.5** Golf courses typically increase property values and convert unproductive land into highly desirable travel destinations and residential communities.

**TABLE 1.3**
**Yardage Guidelines for Par Ratings of Golf Courses and Playing Lengths for Normal and Championship Play (USGA)**

| | Yardage (Meter) Minimum and Maximum | |
|---|---|---|
| Par | Men | Women |
| 3 | ≤250 (229) | ≤210 (192) |
| 4 | 251 to 470 (229 to 430) | 211 to 400 (192.8 to 366) |
| 5 | 471 to 690 (431 to 631) | 401 to 575 (367 to 526) |
| 6 | >691 (632) | ≥575 (526) |
| Normal play | 6,200 to 6,600 (5,667 to 6,032) | 5,000 to 5,600 (4,520 to 5,118) |
| Championship play | 6,600 to 7,200 (6,032 to 6,580) | 5,800 to 6,400 (5,301 to 5,850) |

warrant separation into a different species. Cultivars, a contrived word meaning CULTIvated VARiety (abbreviated as cv.), represent a subdivision of cultivated plants and subspecies (ssp.). Variety (var.) and form (f.) describe further subdivisions in wild populations of plants. For example, the Latin designation for the Rebel cultivar of tall fescue is *Festuca* (or *Lolium*) *arundinacea* Schreb. cv. Rebel or *Festuca arundinacea* Schreb. "Rebel," while the subdivision of the perennial (creeping) biotype of the wild population of annual bluegrass is *Poa annua* L. ssp. *reptans* or *Poa annua* L. var. *reptans*.

Interspecific hybrids are designated by the names of the two parent species included in the Latin name separated by an "X." For example, Emerald zoysiagrass is referred to as *Zoysia japonica* X *Z. tenuifolia* since it is a selected hybrid between *Zoysia japonica* and *Zoysia tenuifolia*.

## Turf Quality

The overall quality of a turf stand is determined by its (1) visual and (2) functional quality components. For example, lawns are evaluated more on the visual quality of turf while roadsides are judged more on the functional quality of turf. Golf course turf quality is evaluated based on the combination of aesthetics and playability; thus, they require both excellent visual and functional qualities.

The most visible components of turf quality include density, color, growth habit, texture, smoothness, and uniformity (Figure 1.6). Density is a measure of the number of aerial shoots per unit area. Texture is a measure of the leaf blade width, with most golf course turf in the medium- to fine-texture range. Density and texture most often combine to determine the smoothness of a turf surface. Uniformity refers to the even appearance of the turf with regards to variation in turfgrass shoots and the presence of weeds. Smoothness is a surface feature of the turf affecting visual quality and playability; an example might be a poor surface formed from

**FIGURE 1.6** Besides color, turf quality includes density, growth habit, texture, smoothness, and uniformity.

using a dull mower, which shreds or frays plants. Growth habit describes the type of shoot growth, such as bunch-type, rhizomatous, and stoloniferous. Color is a measure of the light reflected by a turfgrass. Color ranges from light to very dark green and is often a useful indicator of the general condition of the plants. Unfortunately, color alone is often the visible component most used by golfers to judge turf quality.

In addition to color and other listed components of turfgrass quality, golf course turf is often rated by its rooting, recuperative ability, leaf growth rate, grain, resiliency, and putting speed (ball roll distance).

## ADAPTATION

### Climate

Climate (or environment) is the primary factor determining the region of adaptation of turfgrasses (Figure 1.7). Temperature and moisture are the major components of climate influencing adaptation (further discussed in Chapter 2). Light and wind are two additional climatic parameters influencing turfgrass adaptation. Grasses evolve in nature through natural selection or naturalization in order to adapt to specific regions. This naturalization may involve mutations (rapid genetic changes), physiological adjustments over long periods, or, most likely, some combination of numerous change processes.

Turfgrasses are divided into two groups based on their temperature requirements. Cool-season (or $C_3$) grasses, such as bentgrass, fine fescue, tall fescue, rough bluegrass (or *Poa trivialis*), and Kentucky bluegrass, are adapted and grown in cooler temperatures and subarctic regions of the world, with optimum growth at temperatures between 60°F and 75°F (15.5°C and 24°C) (Figure 1.7). Their peak growth periods occur in spring and fall. Conversely, their growth is slow during summer and winter. They often become dormant during these periods and can eventually die if management practices are not manipulated to better favor their growth and survival.

Warm-season (or $C_4$) grasses grow best at air temperatures between 80°F and 95°F (26°C and 35°C) and night temperatures in the high 60s°F (15.6s°C) to low 70s°F (21s°C). Warm-season grasses such as bahiagrass, bermudagrass, carpetgrass, centipedegrass, St. Augustinegrass, and zoysiagrass are better adapted in the southern (tropical and subtropical) regions but can grow further into the transition zone. The main flush of growth of warm-season grasses occurs during summer and will continue even at 55°F to 60°F (12.8°C to 15.6°C) but go dormant at temperatures less than 50°F (10°C). In spring when temperatures rise above 50°F (10°C), warm-season turfgrasses resume growth. The greatest stress period for warm-season grasses is winter due to potential low-temperature damage.

**FIGURE 1.7** Climate is the primary factor determining the adaptation range of turfgrasses.

### Climatic Zones

**Cool humid zone.** The cool humid zone consists of two areas: (1) a larger area in the north central to northeastern parts of the continental United States, and (2) a smaller region along the Pacific coast extending from Washington state through Oregon into northern California (Figure 1.7). It is characterized by cold winters and mild to hot summers. Annual rainfall in the eastern and central regions ranges from 20 to 45 inches (51 to 114 cm), while the Pacific coast region ranges from 16 to 150 inches (41 to 381 cm).

The primary turfgrasses in the cool humid area are the bentgrasses, bluegrasses, fescues, and ryegrasses. In the southern portion of the cool humid area, tall fescue, zoysiagrass, and cold hardy bermudagrasses are sometimes used.

**Cool arid zone.** The cool arid zone consists of two areas: (1) the Northern Great Plains to the east, and (2) the intermountain region (Rockies) to the west (Figure 1.7). Elevations in this zone vary from 1,000 feet (305 m) in the plains to more than 14,000 feet (4,267 m) in the western mountain sections. This zone experiences hot summers and cold winters, depending on elevation. As altitude increases, temperature decreases. Typically, a 1°F (~0.6°C) temperature change occurs with every 300-foot (91 m) altitude differential. Annual rainfall ranges from less than 10 inches (25 cm) in the intermountain plateaus to ~25 inches (64 cm) along the eastern side of the zone. The majority of this precipitation falls during spring and summer. Snow accumulations can be substantial, especially at higher elevations.

Turfgrasses used most often in the cool arid zone include Kentucky bluegrass, bentgrass, and fine fescues. In many areas, quality turf is achieved only with irrigation. In the Plains area without irrigation, drought-tolerant native species are commonly grown, including buffalograss, gramagrass, bluestem, and various wheatgrasses (Figure 1.8). Low temperatures throughout most of this region limit the use of most warm-season grasses. However, along the southern borders of Kansas and Oklahoma and the northern border of Texas, bermudagrass and zoysiagrass are used to a limited extent.

**Hot arid zone.** The hot arid zone is a semiarid to arid region extending in a wide belt from central Texas to California (Figure 1.7). Rainfall is less than 5 inches (13 cm) in desert parts of Arizona and California to about 20 inches (51 cm) in western Texas. Relative humidity tends to be low and, along with mostly clear skies, the region generally experiences cool nights and hot days.

**FIGURE 1.8** Low-maintenance grass use has increased on golf courses. Shown is crested wheatgrass.

Bermudagrass is the primary turfgrass used for golf courses in the hot arid zone. It is often overseeded or painted for winter color. Zoysiagrass is also used to a limited extent. Golf greens are either bermudagrass or bentgrass. Other grasses used on low-maintenance (nonirrigated) fairways and roughs include buffalograss, crested wheatgrass, gramagrass, and, sometimes, lovegrass. Kikuyugrass is sometimes used but generally does not produce a high-quality playing surface unless it is intensively managed. Without irrigation, high-quality turfgrass is difficult to achieve in the hot arid zone. Due to high evaporative rates and limited rainfall, salinity problems are also typical in this area.

**Warm humid and hot humid zones.** The warm humid zone runs from southern Virginia through the Carolinas to central Texas. It extends northward into southern Illinois and Oklahoma (Figure 1.7). The hot humid zone includes the coastal areas of the southern Atlantic states, all of Florida, and the Gulf coastal regions to the southern tip of Texas. Many of these areas are considered subtropical. Humidity is usually high during summer months with rainfall ranges from a low of 20 inches (51 cm) in central Texas and western Oklahoma to a high of about 70 inches (178 cm) along the Gulf coast and about 80 inches (208 cm) in the southwestern portion of the Appalachian Mountains.

The primary turfgrasses grown in these two regions include bermudagrass, zoysiagrass, bahiagrass, St. Augustinegrass, carpetgrass, and centipedegrass. Ryegrass is used for temporary winter overseeding. Kentucky bluegrass and tall fescue are used along the northern edge of the warm humid zone. Creeping bentgrass is often used on golf greens in the northern portions of the warm humid zone and on fairways at higher (e.g., >2,000 feet, 610 m) elevations. Seashore paspalum is also used along coastal areas in the hot humid zone, especially where poorer quality irrigation water is used.

**Transition zone.** An imaginary line passing east to west at 37° north latitude separates the temperate and subtropical zones in the eastern United States (Figure 1.7). This line marks the center of a 200-mile-wide (322 km) belt, called the transition zone, where cool-season and warm-season turfgrasses encounter the limits of their southern and northern adaptations, respectively. Many warm- and cool-season turfgrasses can be found growing together in the transition zone.

Turfgrasses often grown in the transition zone include bermudagrass, zoysiagrass, tall fescue, bentgrass, and Kentucky bluegrass. Some shade is generally needed in the transition zone to support permanent tall fescue and Kentucky bluegrass, while full-sun conditions are needed for bermudagrass. It is typically too cold for most cultivars of St. Augustinegrass, carpetgrass, centipedegrass, seashore paspalum, and bahiagrass and too hot for the fine fescues and perennial ryegrasses as a permanent turf cover, unless grown at higher (e.g., >2,000 feet, 610 m) elevations.

## CULTIVARS AND THEIR CHARACTERISTICS—WARM-SEASON GRASSES (TABLE 1.4)

### Bermudagrass

Bermudagrass is a long-lived perennial grass originating under subhumid, open, closely grazed rangelands characterized by hot, dry summers around the Indian Ocean ranging from Eastern Africa to the East Indies. Bermudagrass grows almost continuously across all continents and islands between the latitudes of 45°N and 45°S (Figure 1.9). It is believed to have been introduced to the United States from Africa in 1751 or earlier.

The *Cynodon* genus comprises nine species with chromosomal numbers ranging from 18 to 54. The four main turf-type bermudagrasses are (1) common bermudagrass (*Cynodon dactylon* (L.) Pers. var. *dactylon*), a tetraploid with 36 (predominantly) chromosomal numbers; (2) African bermudagrass (*C. transvaalensis* Burtt-Davy), a diploid with chromosome numbers of 18 (predominantly); (3) their interspecific hybrid, Magennis bermudagrass [*Cynodon* X *magennisii* Hurcombe (= *C. dactylon* X *C. transvaalensis*)], a naturally occurring triploid with 27 (predominantly) somatic chromosomes; and (4) Bradley bermudagrass (*C. incompletus* var. *hirautus*), an aneuploid with 18 somatic chromosomes. Hexaploid *Cynodons*, with 54 chromosomes, also exist but are rarely used.

**Advantages.** Bermudagrass produces a vigorous, deep-rooted, light to dark green, dense turf that is well adapted to most soils and climatic regions in hot and warm climates (Figure 1.10). Bermudagrass has excellent wear, drought, and salt tolerance and is a good choice for oceanfront property. It establishes rapidly, produces lateral stems, is competitive against weeds and, depending on the variety, is available as seed, sod, or sprigs (Figure 1.11).

**Limitations.** Bermudagrass has several cultural and pest problems that may restrict its use in certain situations. In most temperate areas, bermudagrasses become dormant (turn brown) with sufficient frost. Overseeding in fall with a temporary cool-season grass or artificial coloring are common practices to maintain year-round green color (Figure 1.12). Bermudagrasses are susceptible to several nematode, insect, and disease problems. Bermudagrass also has very poor shade tolerance and should not be grown underneath tree canopies

**TABLE 1.4**
**General Characteristics of Warm-Season Turfgrasses**

| Characteristic | Bahiagrass | Bermudagrass | Buffalograss | Carpetgrass | Centipedegrass | Kikuyugrass | Seashore Paspalum | St. Augustine-grass | Zoysiagrass |
|---|---|---|---|---|---|---|---|---|---|
| **Establishment** | | | | | | | | | |
| Planting method | Seed/veg. | Seed/veg. | Seed/veg. | Seed/veg. | Seed/veg. | Veg. | Seed/veg. | Veg. | Seed/veg. |
| Rate of coverage | Very slow | Very fast | Medium | Slow | Slow | Medium | Slow | Medium | Slow |
| Days to germination | 21 to 35 | 14 to 21 | 14 to 21 | 14 to 21 | 35 to 40 | N/A | N/A | N/A | 21 to 28 |
| Growth rate | Slow | Med/fast | Med/fast | Slow | Slow | Fast | Slow/med | Med/fast | Slow |
| Density | Poor/fair | Excellent | Fair | Fair | Good | Fair | Excellent | Excellent | Excellent |
| Leaf texture | Coarse | Fine/med | Fine/med | Medium | Medium | Medium | Fine/coarse | Coarse | Fine/med |
| Polystand compatibility | Medium | Medium | Poor | Poor | Poor | Poor | Poor | Poor | Poor |
| **Cultural Requirements** | | | | | | | | | |
| Maintenance intensity | Low | High | Low | Low | Low | Medium | Medium | Medium | Medium |
| Mowing height (inches) [fairways] | 3 to 4 | 0.5 to 1.5 | 0.5 to 3.0 | 1 to 2 | 1 to 2 | 0.5 | 0.75 to 1.5 | 2.5 to 4.0 | 0.75 to 2.0 |
| Mowing frequency (days) | 7 to 14 | 1 to 4 | 3 to 14 | 10 to 14 | 10 to 14 | 2 to 4 | 1 to 7 | 7 to 14 | 1 to 14 |
| Irrigation requirements | Low | Low | Low | Medium | Medium | Low | High | High | Medium |
| Evapotranspiration (ET) (mm/day) | 2.5 to 8.5 | 6 to 7 | <6 | 6 to 7 | 6 to 7 | 5.8 to 9.0 | 2.5 to 8.5 | 7 to 8.5 | 6 to 7 |
| Nitrogen needs | Low | High | Low | Low | Low | Medium | Low | Medium | Low to medium |
| **Performance** | | | | | | | | | |
| Heat tolerance | Excellent | Excellent | Excellent | Excellent | Excellent | Excellent | Excellent | Excellent | Excellent |
| Cold tolerance | Fair | Good | Fair/good | Poor/fair | Fair | Poor | Poor | Poor/fair | Fair/good |
| Killing temperature, °F | ~23 | ~19 | N/A | N/A | ~11 | N/A | ~19 | ~23 | ~6 |
| Drought resistance | Excellent | Excellent | Excellent | Poor | Poor | Good | Excellent | Good | Excellent |
| Shade tolerance | Fair | Poor | Poor | Fair | Fair | Poor | Fair/good | Good/v. good | Fair/good |
| Wear tolerance | Excellent | Excellent | Fair/good | Poor | Poor | Good | Good | Fair/good | Excellent |
| Recuperative potential | Fair | Excellent | Good | Poor | Poor | Good | Good | Good | Excellent |
| Low mowing tolerance | Poor | Excellent | Fair | Fair | Fair | Good | Good | Poor | Good/excellent |
| Salt tolerance | Poor | Good | Fair | Poor | Poor | Excellent | Excellent | Fair/good | Good |
| Soil pH range | 4.5 to 7.0 | 4.5 to 8.0 | 6.0 to 7.5 | 5.0 to 6.5 | 4.5 to 6.5 | N/A | 5.5 to 8.0 | 6.0 to 8.0 | 6.0 to 8.0 |
| Acidic soil tolerance (pH < 5.5) | Good | Good | Poor | Good | Excellent | N/A | Good | Fair | Fair |
| Thatch tendency | Low | High | Low | Low | Medium | High | Medium | High | High |
| Disease tendency | Low | Low/med | Low | Low | Low | Low | Medium | Medium | Low/med |
| Submersion tolerance | Fair/good | Excellent | Excellent | Excellent | Poor | Good | Good | Excellent | Excellent |

*Notes:* Cultivar differences may occur. N/A = not available; veg. = vegetative; med = medium.

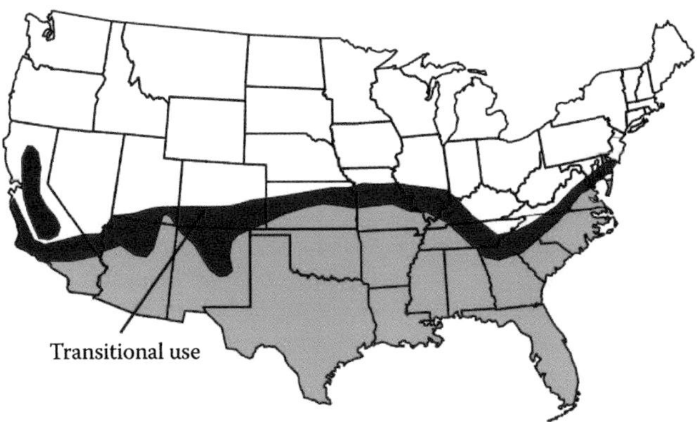


**FIGURE 1.9** Bermudagrass growing area in the United States.

**FIGURE 1.10** Bermudagrass produces a deep-rooted, dense turf adapted to most soils in hot and warm climates.

**FIGURE 1.11** Due to its density and recuperative ability, bermudagrass is often used for putting surfaces.

**FIGURE 1.12** Where bermudagrass goes dormant in winter, courses may overseed to provide green color.

**FIGURE 1.13** Bermudagrass inherently has poor shade tolerance.

or building overhangs (Figure 1.13). It spreads very rapidly by both above- (stolons) and belowground (rhizomes) runners, which are very difficult to maintain and control along golf greens, bunkers, flower beds, walks, and borders. Due to its rapid growth, thatch buildup and grain development can become problems. A reel mower should also be used to produce the highest possible quality turf stand. Bermudagrass is subject to periodic "winter kill" (Figure 1.14) when grown in the upper transitional zone where temperatures periodically drop below ~10°F (–12°C).

**Cultivars.** Prior to the mid-1940s, golf courses used common bermudagrass from tee to green. Common bermudagrass provided a coarse-textured, uneven, and thin putting surface. Many golf courses still had sand greens where a roller was used to smooth the ball line toward the cup and heavy oils or diesel fuel packed the sand. Common bermudagrass seed was not certified until 1963 in an attempt to rid seed from a tall, rapidly growing giant bermudagrass (*Cynodon dactylon* var. *aridus*). This common bermudagrass was also called "Arizona common" since the production fields were in Arizona. The first recognized turf-type bermudagrass cultivar in the United States was "St. Lucie," a fine-textured dwarf plant used in Florida lawns.

In 1947, the first recorded release of an improved vegetatively established bermudagrass occurred when "U-3" bermudagrass was provided by the United States Golf Association (USGA) in cooperation with the United States Department of Agriculture (USDA). It survived as far north as Cleveland, Ohio, and central Pennsylvania (40° north). "Ormond" and

**FIGURE 1.14** Periodic "winter-kill" can occur when bermudagrass is grown in the upper transitional zone.

"Texturf 10" were released after "U-3." During the 1950s and 1960s, Dr. Glen W. Burton of the USDA in Tifton, Georgia, released several interspecific hybrids, including "Tiflawn," "Tifgreen," "Tifway," and "Tifdwarf," which are still used. These are from various crosses of common bermudagrass (2n = 36) with African bermudagrass (2n = 18) to produce sterile triploid hybrid bermudagrass (2n = 27).

The first improved seeded bermudagrass was "Guymon," released in 1982 for its improved cold tolerance, darker green color, and lower seedhead counts compared to common bermudagrass. Until then, "Arizona common" or simply "common" was the only seeded bermudagrass (2n = 36) sold. "NuMex Sahara," the next improved seeded common bermudagrass, was released in 1987. Today, there are numerous bermudagrass cultivars used for turf (Table 1.5). Most seed is available with the hull removed (*hulled*) or with the hull remaining (*unhulled*). Hulled seed germinates faster but costs more; unhulled seed persists longer during unfavorable weather before germinating.

**Tifway (or Tifton 419).** Tifway is a dark-green bermudagrass released in 1960 with medium texture and high shoot density. It is the current industry standard and is used in warm-season areas for quality fairways, most roughs, sports fields, and lawns (Figure 1.15). It produces few seedheads, and those produced are very elongated and thin. It tends to green-up quicker in spring compared to common and forms a much denser, finer textured stand (Figure 1.16). It must be vegetatively established by sprigs, plugs, stolons, or sod and is prone to spring dead spot disease when overmanaged. Various off-types, mutants, and/or contaminants have plagued the industry in recent years.

**"Ultradwarf" bermudagrasses.** Ultradwarf bermudagrasses are low-growing grasses for closely mowed, highly maintained areas such as putting greens and lawn bowling greens. They are believed to be somatic mutations from Tifdwarf or Tifgreen, possess a low growth habit, and are frequently maintained at or below 1/8 inch (3.2 mm) mowing height (Table 1.5). They have good density with cold tolerance similar to Tifdwarf. They require relatively high fertility to maintain their appearance and vigor. As these are triploid hybrids, they produce no viable seed and must be vegetatively propagated. Due to tight density, these grasses are thatch prone (Figure 1.17); thus, they require periodic vertical mowing, grooming, and topdressing. They tend to be slower growing, develop a purplish color during cool (50s°F, 10s°C) temperatures, and often develop grain (Figure 1.18).

**TABLE 1.5**
**Characteristics of Popular Improved Bermudagrass Varieties**

| **Greens Grade Varieties** |
|---|
| These are used for golf greens, collars, approaches, and high-maintenance tees. They all are vegetatively propagated with no available seed: Aussie Green, Champion, Classic Dwarf, Cotton Creek Dwarf, Emerald Dwarf, FloraDwarf, Jensen Dwarf, G-12, Mini Verde, MS Supreme, Novatek, Quality Dwarf, Reesegrass, Sunday, Tifdwarf, TifEagle, Tifgreen (328) |
| **Vegetative Propagated Varieties for Fairways, Roughs, and Tees** |
| Baby Bermuda, Bulls-Eye (MS Choice), Cardinal, Celebration[a], Discovery, FloraTeX, GN-1, Guymon, Latitude 36[b], Midfield[b], Midiron[b], Midlawn[b], Midway[b], MS Express, MS Pride, NorthBridge[b], Ormond, Patriot[b], Plateau, Quickstand[b], Santa Ana, Shanghai, Sunturf, Texturf 10, TifGrand, Tiflawn (57), TifSport[a], TifTuf, Tifton 10 (or T-10), Tifway (419), Tifway II, Transcontinental, Windsor Green, Wintergreen |
| **Seeded Varieties for Fairways, Roughs, and Tees** |
| Ashmore, Astro, Barbados, Blackjack, Blue-Muda, Bradley, Burning Tree, Casino Royal, Cheyenne, Common, Contessa, DelSol, GoldGlove, Guymon, Hollywood, Jackpot, Kashmir, LaPaloma, LaPrima (a blend of certified bermudagrasses), Majestic, Mercury, Mirage, Mohawk, Northshore SLT, NuMex Sahara, Panama, Paradise, Primavera, Primo, Princess 77, Pyramid 2, Riviera[b], Royal Bengal, Sahara, Savannah, Shangri-La, Soliel, Sonesta, Southern Star, Sultan, Sunbird, Sundance II, Sundevil II, Sunsport, SunStar, Sultan, Sydney, Tif No. 1-4, Transcontinental, U-3, Veracruz, Yukon[b], Yuma |

*Note:* This list is subject to change. Refer to your state turfgrass specialist or www.ntep.org for the latest varieties and their comparisons.
[a] Improved shade tolerance.
[b] Improved cold tolerance.

**FIGURE 1.15** Tifway (aka, 419) bermudagrass is the industry standard for quality fairways and most roughs.

FIGURE 1.16 Tifway bermudagrass contaminated with lighter green, coarser textured common bermudagrass.

FIGURE 1.17 "Ultradwarf" bermudagrass golf green with finer texture and denser stands.

FIGURE 1.18 "Grain" associated with bermudagrass from leaf blades lying in alternating directions.

TifEagle is a 1997 release developed as an induced mutant by cobalt radiation from another bermudagrass. It resembles other "dwarfs" in its extremely fine leaf texture and excellent density, but appears to have the least problems with genetic instability. These grasses tend to become thatchy and produce a spongy putting surface if insufficient maintenance is provided. This is an important consideration before a course decides to utilize one of these cultivars.

## MUTANTS OR OFF-TYPES

In recent years, considerable attention has been given to the occurrence of mutants or off-types of grasses in previously pure bermudagrass stands (Figure 1.19). In addition to different color, texture, and density, off-types often have differential susceptibility to environmental stresses such as high temperature, humidity, and reduced sunlight intensity. Additional stresses such as low mowing height or verticutting often delay recovery. When they occur, off-types typically start becoming noticeable about seven years following planting. When over 50% of the surface becomes contaminated, it is almost impossible to provide a quality playing surface.

It is believed that off-types have arisen from the chance mutation of the parent material or possibly contamination through mechanical means, improper or insufficient control of an existing grass when converting cultivars, or encroachment from collars. Mutations are abrupt inheritable changes brought about by alterations in a gene or a chromosome, or by an increase in chromosome number. Aneuploidy—the presence of an abnormal number (extra or missing) of chromosomes in a cell—is believed to be the source of off-types, though it is not known what is causing this or why. The rate of mutation occurrence can be increased artificially, but results cannot be controlled. As they are usually recessive, mutations may be unexpressed for many generations.

FIGURE 1.19 Mutant or "off-type" patches of different colored and textured bermudagrasses.

Mutations are produced by internal disorders, such as inaccurate gene duplication, and by natural external forces, such as severe temperature changes and sunlight radiation. They are induced experimentally by use of cobalt radiation, x-rays, chemicals, and sudden temperature change. Natural mutations appear very infrequently, while artificial mutations occur more quickly. Tifway II, Tifgreen II, and TifEagle are induced mutations of original grasses created by exposing parent material to artificially high levels of radiation. Tifdwarf, FloraDwarf, and Pee Dee 102 bermudagrasses are believed to be natural mutants from Tifgreen bermudagrass.

Since Tifdwarf is a probable vegetative mutant from Tifgreen bermudagrass, a possibility exists that an original planting of Tifdwarf can undergo another mutation to produce a different grass. Champion, Mini Verde, MS Supreme, Sunday, and others are believed to be mutations of Tifdwarf bermudagrass. Mutations offer breeders new ways of introducing genetic variability into breeding lines, but also may cause existing materials to be somewhat unstable; thus, they may produce undesirable off-types after several years of growth.

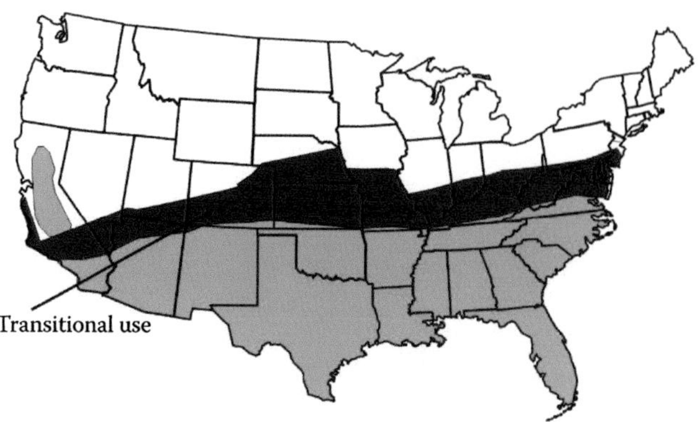


**FIGURE 1.21** Zoysiagrass growing area in the United States.

## Zoysiagrass

Zoysiagrasses, named after Karl von Zois, an eighteenth-century Austrian botanist, are warm-season grasses (2n = 40) native to the hot, humid Southeast Asia region including China, Korea, and Japan. They are becoming more important as golf course grasses due to their excellent wear tolerance, slow growth rate, improved winter hardiness, salt tolerance, unique green color during the summer, and golden color during winter months (Figure 1.20). Zoysiagrasses also have better shade tolerance than bermudagrass and lower fertility requirements (Table 1.5). Zoysiagrasses are well adapted for use on golf course fairways, tees, bunker faces, collars, and some greens in hot and warm climatic zones (Figure 1.21).

Zoysiagrasses as a whole are very slow-growing grasses; thus, they have slow recuperative potential. This slow lateral growth, however, can be advantageous as zoysiagrasses are often planted on the perimeter of sand traps and golf greens to greatly reduce or slow off-site lateral growth (Figure 1.22). They tend to develop thatch over time (Figure 1.23) and are susceptible to several diseases, most notably rusts and brown patch. Most improved zoysiagrasses have to be propagated vegetatively and are extremely slow in becoming established. Other limitations include slow recovery from damage; poor growth on acidic and compacted soils; higher water requirements than most bermudagrasses; susceptibility to bermudagrass invasion; and damage by nematodes, hunting billbugs, and several diseases. Zoysiagrass, being native to areas that naturally receive heavy yearly precipitation, also tends to be shallow rooted and is weakened when grown in soils low in K. For maximum quality, a reel mower should be used.

Several species and varieties of zoysiagrasses are used for golf courses (Table 1.6). These vary widely in color, texture, and establishment rate. This is not an exhaustive list; only the more common varieties will be discussed.

**FIGURE 1.20** Zoysiagrass fairway adjacent to a hybrid bermudagrass rough.

**FIGURE 1.22** Zoysiagrass planted around bunkers due to its naturally slow lateral growth.

**FIGURE 1.23** "Thatchy" zoysiagrass from maintenance neglect.

***Zoysia japonica* Steud.** This species, commonly called Japanese, Chinese, or Korean common lawngrass, was introduced into the United States as seed in 1930 from Kokai, North Korea. It has a coarse texture; hairy, light-green leaves; a faster growth rate; and the best cold tolerance of the zoysiagrasses. Although cold tolerant, it quickly loses color with slight frost. Hunting billbugs and nematodes cause considerable damage. *Zoysia japonica* periodically has commercially available seed. It can be used for lawns or general turf areas where convenience of establishment by seed is more important than quality. The optimum soil temperature range for seed germination is 70°F to 95°F (21°C to 35°C), with slowest germination occurring at the lower range.

Traditionally, Meyer zoysiagrass has been the most important vegetative *Z. japonica*. It is also called Z-52 or Amazoy; it is an improved selection of the *Zoysia japonica* released in 1951 and named after Frank N. Meyer, who was one of the first plant explorers to collect zoysiagrass seed in Korea in 1905. Meyer has a deep-green color and medium leaf texture (Figure 1.24). It spreads faster than many other varieties, although it produces few rhizomes. Meyer makes an excellent fairway once established. Meyer is less shade tolerant than Emerald but is the most cold-tolerant zoysiagrass. Hunting billbugs, brown patch disease, nematodes, and particularly mole crickets are potential problems with Meyer. Meyer is the zoysiagrass often advertised as the "miracle" or "wonder" grass in newspapers and magazines.

***Zoysia matrella* (L.) Merr.** Also called Manila grass since it was introduced from Manila, the Philippines, in 1912, this species produces a finer and denser turf stand than *Zoysia*

**FIGURE 1.24** Meyer zoysiagrass fairway.

**TABLE 1.6**
**Zoysiagrasses Used for Fine Turf Purposes**

| | Propagation Method | |
|---|---|---|
| **Zoysiagrass (Common Names)** | **Vegetative** | **Seeded** |
| *Zoysia japonica* (Japanese, Chinese, or Korean common) | Belair, BK-7, Carrizo, Chisholm, Crowne, El Toro, Empire, Empress, GNZ, Himeno, JaMur, Marion, Meyer (aka, Z52 and Amazoy), Miyako, Palisades, Serene, Southern Gem, Sunburst, UltimateFlora, VJ, Y2, Zoyboy | Cathay, Chinese Common, Compadre, Companion, Compatibility, J-36 and 37, Ming, SR 9150 and 9200, Sunrise, Sunstar, Traveler, W3-2, Zen 100-300, Zenith, Zoysia japonica |
| *Zoysia macrantha* Desv. (prickly couch) | Icon, Nara | – |
| *Zoysia matrella* (Manila grass) | Cashmere, Cavalier, Diamond, Facet, L1F, Matrella, M60, M85, Rollmaster, Omni, PristineFlora, Royal, Shadow Turf, Zeon, Zorro, ZOYSIU | – |
| *Zoysia japonica* x *Z. matrella* | Z-3 | – |
| *Zoysia tenuifolia* (Mascarene grass, templegrass, or Korean velvetgrass) | Common | – |
| *Z. japonica* x *Z. tenuifolia* | Emerald, Geo, Pristine | – |
| *Z. japonica* x (*Z. matrella* x *Z. tenuifolia*) | DeAnza, Victoria | – |
| Z. sinica | – | J-14 |
| Other cultivars (these are mostly selections from *Z. japonica*) | Links, Marquis | – |

*japonica* but is less winter hardy. Its leaf blades are thin, sharply pointed, and wiry. Manila grass resembles bermudagrass in texture, color, and quality and is recommended for a high-quality, high-maintenance turf where a slow rate of establishment is not a disadvantage. These typically have better shade (low light) tolerance compared to the japonicas. *Zoysia matrella* appears to be highly susceptible to damage by nematodes.

***Zoysia japonica* x *Z. matrella*.** The only reported cultivar of this cross is Z-3 in Hawaii. It is medium textured, apple-green in color, and forms a tight mat.

***Zoysia tenuifolia* Willd. ex Trin.** Also called Mascarene grass and Korean velvetgrass, this species is the finest textured, most dense, least winter-hardy zoysiagrass available. It is native to the Far East and was introduced from the Mascarene Islands. It has good wear tolerance but is extremely slow to spread and develops excessive thatch, giving it a puffy appearance. Since it also is the least cold-hardy zoysiagrass, *Zoysia tenuifolia* is best adapted only to tropical and subtropical climates.

***Zoysia japonica* x *Z. tenuifolia*. Emerald.** Emerald zoysia is a selected hybrid between *Zoysia japonica* and *Z. tenuifolia* released in 1955, although recent DNA fingerprinting suggests it is possibly a cross between *Z. matrella* and *Z. pacifica*. This hybrid combines the winter hardiness, color, and faster growth rate of *Z. japonica* with the fine texture and density of *Z. tenuifolia* (Figure 1.25). Emerald resembles Manila grass in color, texture, and density, but is faster spreading and has wider adaptation. Emerald zoysia is recommended for quality turf where time and money allow for an intermediate maintenance program. Emerald is among the most beautiful of the zoysiagrasses, but it also is subject to thatch buildup, puffiness, and to dollar and leaf spot and brown patch disease. Seed of Emerald zoysiagrass is very expensive and does not reproduce true to the variety. Other cultivars include Geo and Pristine and are noted for good shade tolerance and fine leaf texture.

**FIGURE 1.25** Emerald zoysiagrass is a hybrid combining the fine texture and density of *Zoysia tenuifolia* with the winter hardiness and faster growth of *Z. japonica*.

***Zoysia japonica* x (*Z. matrella* x *Z. tenuifolia*).** Two cultivars released in 1993 from the University of California are DeAnza and Victoria. These hybrids include El Toro and other experimental lines as parents, which provide them with faster lateral growth than most traditional cultivars. These grasses have been suggested to be classified as *Zoysia japonica*.

**Additional zoysiagrass cultivars.** Considerable research is currently being conducted on breeding and selecting additional zoysiagrass cultivars for the turfgrass sector. A new turf-type seeded zoysiagrass (*Z. sinica* Hance) also is being investigated for turfgrass potential, especially in low- to moderate-maintenance saline areas. J-14 is a commercial variety of *Z. sinica*. Other zoysiagrass species being investigated for turf use include *Z. macrostaycha* Franch. et Sav. and *Z. korenia*, a very salt-tolerant zoysiagrass species.

### Bahiagrass

Bahiagrass (*Paspalum notatum* Flugge.) is a very coarse-textured species originally from rangelands and forest fringes in Argentina and Brazil in South America. It was introduced into the United States in 1914 and forms a low to fairly dense turf. It has very tough leaves and spreads vegetatively by short. stout rhizomes and stolons (Figure 1.26). Its inflorescence bears two racemes as opposed to three to five for dallisgrass (*P. dilatatum*) or seven to nine for vaseygrass (*P. urvillei*). It is best adapted to the warm humid and hot humid zones. Bahiagrass has excellent drought tolerance, due in part to an extensive root system. It has fair shade and good wear tolerance; does particularly well on infertile, droughty sands; and seems affected little by nematodes (Table 1.4). Fertility requirements are minimal. A major drawback, however, is the constant production of tall (2 to 4 feet, 0.6 to 1.2 m), V-shaped seedheads continuously throughout summer months (Figure 1.27). Salt tolerance is poor and a high mowing height (2 to 3 inches, 5 to 7.6 cm) is required to maintain an acceptable stand. Bahiagrass also often expresses iron chlorosis in spring and fall months. Often used as a roadside grass because of its toughness, bahiagrass frequently escapes

**FIGURE 1.26** Bahiagrass in low-maintenance areas due to its tolerance of drought and nematodes.

**FIGURE 1.27** Bahiagrass production of V-shaped seedheads throughout summer.

**FIGURE 1.28** Centipedegrass is a crab-apple green, low-maintenance grass used for specialty areas.

into bermudagrass fairways and roughs. However, as water resources become more restrictive, fertilizer rates become reduced, and effective nematicides are lost, bahiagrass use may increase. Cultivars include Argentine, Common, Paraguay, Pensacola, RCP, Riba, and Wilmington, though breeding efforts are being made to produce shorter growing, turf-type cultivars.

### Carpetgrass

Common carpetgrass (*Axonopus affinis* Beauv.) and tropical carpetgrass (*A. compressus* [Swarty] Beauv.) originated from the eastern regions of the hot, humid, poorly drained soils of South Central American continents including the West Indies. These are coarse-textured species that form a fairly dense turf. Carpetgrass spreads by stolons and appears similar to St. Augustinegrass and centipedegrass except for its lighter green color and wavy leaf margin. It has limited cold, drought, and wear tolerance and only fair shade tolerance (Table 1.4). Carpetgrass produces tall, unsightly crabgrass-like seedheads throughout the summer, giving the turf a ragged, unmowed appearance. Leaf tips also easily discolor with cool weather. Carpetgrass is also susceptible to brown patch disease. Like centipedegrass, carpetgrass grows best on acidic (pH 5 to 6) soils. It does tolerate wet, poorly drained soils very well, and often becomes established in these locations before escaping into other sites. Chase is an improved cultivar.

### Centipedegrass

Centipedegrass [*Eremochloa ophiuroides* (Munro.) Hack.] originates from hot, humid, high-rainfall areas of China and forms a relatively dense turf (Figure 1.28). Seeds were found in the baggage of Frank N. Meyer, USDA plant explorer, who disappeared on his fourth trip to China in 1916. It spreads vegetatively by slow-growing stolons and is a diploid (2n = 18). It has medium leaf texture and a natural light-green (crab apple-green) color (Table 1.4). Stems may be red, yellow, or green. Due to exposed stolons and a shallow root system, centipedegrass has limited cold and drought tolerance. It is typically more cold tolerant than St. Augustinegrass and carpetgrass but less so than bermudagrasses and most zoysiagrasses. Shade tolerance is fair. Maintenance requirements in terms of fertility and mowing frequency are low.

Centipedegrass grows on a variety of soils but is best adapted to slightly acidic (pH 5 to 6) soils. It has poor salt tolerance and is susceptible to ring nematode damage. It also often experiences iron chlorosis during spring months. It is sometimes planted as golf course roughs due to its unique, lighter green color and coarser texture. However, in this situation, centipedegrass invariably receives excessive amounts of fertilizer and water, is mowed too short, and is often accidentally treated with sensitive herbicides. Results are often disappointing. Centipedegrass, however, does fairly well in shady areas that receive minimum play and maintenance. Cultivars include AU Centennial, Centi-Seed (common), Common, Covington, Hammock, Oaklawn, Pavilion, Santee, TennTurf, and TifBlair. The improved cultivars have better cold tolerance than common does. AU Centennial is only vegetatively propagated.

### Kikuyugrass

Kikuyugrass (*Cenchrus clandestinus* (Hochst ex. Chiov.) Morrone, formerly *Pennisetum clandestinum* Hochst ex. Chiov.) is a coarse- to medium-textured, lime-green, dioecious species originating from heavily grazed pastures in the hot, arid regions of the Kenyan highlands in eastern Africa (Figure 1.29). The common name is derived from a native tribe. It spreads by thick, leafy rhizomes, stolons, and seed, and has a very rapid growth rate under warm, moist environments. It is found primarily in California, certain islands of Hawaii, Australia, New Zealand, South Africa, Colombia, Mexico, South America, Spain, and other tropical areas of the world, where it is considered a weed as well as a turfgrass. It has a very aggressive growth habit, often invading and overtaking areas containing more desirable species, and was initially used in California for erosion control.

FIGURE 1.29 Kikuyugrass is a coarse- to medium-textured, lime-green, tough grass often considered a weed.

TABLE 1.7
**Advantages and Limitations of Using Kikuyugrass as a Fine Turf Species**

| Advantages | Limitations |
|---|---|
| • Excellent tolerance to traffic | • Medium to coarse texture |
| • Rapid recovery from injury | • Rapid thatch accumulation |
| • Good heat, drought, and salinity tolerance | • Natural lime-green color |
| • Tough, competitive grass for hitting golf shots | • Potential for mower scalping |
| • Good growth and color retention at temperatures from 50°F to 60°F (10°C to 15.6°C) | • Often has a silvery cast appearance from production of white filaments above the turf canopy |
| • Tolerant of a wide range of soil and water environments | • Frequent mowing, verticutting, and aerifying needed to minimize scalping and to control rapid growth |
| • Good pest resistance | • Damaged by many postemergence herbicides |
| • Low nutrient requirements, e.g., 2 to 3 pounds N/1,000 square feet (10 to 15 g N/m$^2$) yearly | • Other pests and their control are largely unknown |

Kikuyugrass forms a dense, tough sod when mowed close, although the production of white filaments often gives it a silvery cast (Figure 1.30). Its leaf width ranges from 1/8 to 1/4 inch (3.2 to 6.4 mm), similarly to *Zoysia japonica* and St. Augustinegrass. Due to its similar texture, growth habit, and thick stolons, kikuyugrass is often confused with St. Augustinegrass. It has good drought and wear tolerance but poor cold and limited shade tolerance. Compared to bermudagrass, kikuyugrass grows better under low light conditions and is not as prone to dormancy in cool (50°F to 60°F, 10°C to 15.6°C) weather. It usually outcompetes bermudagrass where climates are mild, hazy, and subtropical. Kikuyugrass can escape and become very problematic. Hosaka is a cultivar established mostly by sprigs. Whittet is a forage-type cultivar from which AZ-1 (or Arizona-1) was selected. AZ-1 is a less invasive, turf-type cultivar with good traffic tolerance. Kenda is an additional improved cultivar.

When mowed high (>1.5 inches, 3.8 cm), kikuyugrass forms an open growth habit due to the fleshy stolons and relatively wide leaf blade. When mowed at lower heights, stand density generally improves. Due to its rapid growth rate, kikuyugrass is thatch prone. Greens and tee encroachment can be problematic and, when left unmowed in roughs, the grass is very difficult to hit out of. Best growth often occurs in medium- to heavy-texture soils with a neutral to alkaline pH. Advantages and limitations of kikuyugrass are listed in Table 1.7.

### Seashore Paspalum

Seashore paspalum (*Paspalum vaginatum* Swartz.) is a sexually reproducing diploid (2n = 20) species characterized by a dense-growth habit with variable texture and relatively dark-green color (Figure 1.31). It originated from South Africa and, secondarily, from Argentina north into Brazil, where it is often found growing along seashores (hence, its common name) and in brackish water. In the United States, it grows

FIGURE 1.30 Kikuyugrass may appear silvery due to the production of white filaments.

FIGURE 1.31 Figure Seashore paspalum forms a light-green, dense turf tolerable to a wide range of water quality.

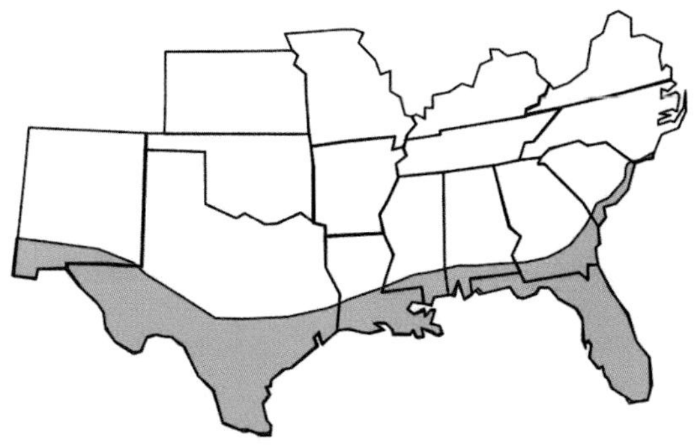

**FIGURE 1.32** Seashore paspalum growing area in the United States.

in coastal areas of the Southeast to Texas (Figure 1.32) and southward into Mexico and Argentina.

Seashore paspalum is best grown in high-draining sand content environments where salinity or poor water quality is an issue and inadequate quality water is available. Due to self-incompatibility, seed production is inhibited; thus, seashore paspalum is mostly propagated vegetatively. Traditional selections include Excaliber (formerly Adalayd) and Futurf. Newer releases include Aloha, AP-10, Boardwalk, FWY-1, Neptune Salam (fairways), Platinum TE (greens, tees, fairways), SeaDwarf (greens), SeaGreen (greens), Sea Isle I (tees and fairways), Sea Isle 2000 (greens), Sea Isle Supreme, SeaSpray (seeded), SeaStar, and SeaWay (fairways). Advantages and limitations of seashore paspalum are shown in Table 1.8.

### St. Augustinegrass

St. Augustinegrass [*Stenotaphrum secundatum* (Walt.) Kuntze] is believed to be native to open to lightly shaded, high-rainfall, and humid regions of coastal South and Central America including the West Indies (Figure 1.33). It spreads by thick, long stolons and forms a very coarse-textured turf of medium density. Due to the lack of cold tolerance, St. Augustinegrass use is restricted to coastal areas of the lower Atlantic states, the Gulf Coast regions, all of Florida, and parts of southwest and southern California. Its drought tolerance is fair to good but often develops chlorosis (yellowing) when grown on alkaline soils. Certain cultivars have good shade and salt tolerance. Chinch bugs, gray leaf spot disease, St. Augustinegrass decline (SAD) virus, and brown patch disease are problems. Mowing height ranges from 1.5 to 3 inches

**TABLE 1.8**
**Advantages and Limitations of Seashore Paspalum as a Golf Course Grass**

| Advantages | Limitations |
|---|---|
| • Possesses highest salinity tolerance of warm-season turfgrasses—withstanding short irrigation duration with seawater (~35,000 ppm salts); however, turf must be fully grown in and have excellent soil drainage available; with diligence plus right growing conditions, water with ~5,000 ppm total dissolved salts can be used | • Limited shade tolerance, requiring about six to eight hours of full sunlight daily for acceptable performance, similarly to bermudagrass<br>• Limited cold tolerance, restricting its geographic range to tropical areas, or coastal areas in subtropical zones |
| • Possesses a shiny, glassy, dark-green hue, similar to Kentucky bluegrass or perennial ryegrass, with similar striping capabilities<br>• Compared to bermudagrass, has better low-light and cool-temperature tolerance from shorter winter days or cloudy, rainy, foggy, or smoggy conditions | • Cannot be effectively and rapidly established with irrigation water high in salinity (>5,000 ppm total dissolved salts) due to suppression of root growth; fresh water needed during grow-in or extended time will be necessary; fresh water also needed periodically during prolonged droughts; Leaf tip burn can occur following irrigation with water with high-salt content |
| • Due to its good water logging or low $O_2$ tolerance, can be inundated for short intervals with minimal detrimental effects<br>• Requires 30% to 50% less N than bermudagrass; if fertilized similarly to bermudagrass, tends to thatch and then severely scalps | • Sandy profile with high percolation rates necessary for salt management; rivers and salt-laden estuaries as irrigation sources often contain high levels of silt and clay and, with prolonged use, may reduce infiltration rates of soils<br>• Irrigation heads must operate properly and uniform coverage is needed. Granular Ca is used to displace excess soil Na, which is then leached, requiring a soil profile with high percolation capabilities; aggressive deep-tine aerification needed in poorly draining soils; other nutrients, such as P, K, Mg, and Mn, also critical for good turf performance, requiring constant tissue and soil testing and regular nutrient spoon-feeding<br>• Thatch prone, so additional grooming, verticutting, and topdressing are needed; due to tougher leaves, mower bed-knives and reels dull more quickly, requiring constant sharpening<br>• Few pesticides labeled specifically for seashore paspalum; bermudagrass and other grassy weeds can be problematic. Weed control, especially bermudagrass suppression, involves using less N, using saline irrigation to selectively slow bermudagrass growth, and applying rock salt to infested areas; all planting stock should be certified bermudagrass free. |

**FIGURE 1.33** St. Augustinegrass forms a medium- to coarse-texture turf noted for its shade tolerance.

(3.8 to 7.6 cm). Its use on most golf courses is limited to shade areas or lower quality roughs. Cultivars include Amerishade, Bitterblue, Captiva, Classic, Delmar, DeltaShade, Emerald Blue, Floralawn, Floratam, Floratine, Gulf Star, Island, Jade, Levi, Mercedes, Palmetto, Raleigh, Raleigh-S, Sapphire, Seville, Sunclipse, Texas Common, Winchester, Woerner Classic, and 80-10.

## CULTIVARS AND THEIR CHARACTERISTICS—COOL-SEASON GRASSES (TABLE 1.9)

### Bentgrasses

Bentgrass (*Agrostis* sp.) is native to the cool, moist climate of central Europe and is adapted to cool, humid environments from the Northeast to the Northwest in the United States (Figure 1.7). Bentgrasses, however, are used outside this climatic region predominantly only on small, highly maintained areas such as golf greens. Bentgrass provides a year-round green surface; it does not require yearly overseeding or artificial coloring and transition in spring as bermudagrass does (Figure 1.34). Hot (temperatures >85°F, 29°C), humid summer conditions are unfavorable for its growth and development. Under such conditions it often becomes very shallow rooted due to the depletion of carbohydrate reserves and is very susceptible to stresses such as drought, traffic, and diseases, especially dollar spot and *Pythium* spp. (Figure 1.35). The microclimate surrounding the bentgrass grown in these areas is often altered to better favor its survival. This includes soil preparation, irrigation, surface and subsurface air circulation, shade, and other parameters. Due to its aggressive stoloniferous growth, bentgrass can become a serious weed in other cool-season turfgrasses such as fescue and Kentucky bluegrass. Various bentgrass varieties produce a broad range of colors, from greenish-yellow to dark green and dark blue-green.

**Varieties.** The genus *Agrostis* has ~125 species, which evolved from Europe and are adapted to the cooler regions of the world. Turf-type bentgrasses currently include creeping (*Agrostis palustris* subsp. *stolonifera* L.), colonial (*A. tenuis* Sibth., also listed as *A. capillaris* L.), redtop (*A. alba* L.), and velvet bentgrass (*A. canina* L.).

Creeping and colonial bentgrasses are the two most widely used species for golf course purposes (Table 1.10). Creeping bentgrass is a fine-textured, cool-season perennial with excellent cold tolerance and is stoloniferous—hence, the name "creeping." It has less wear tolerance than many other turfgrass species, poor soil compaction tolerance, and slow recuperative ability following damage or environmental stress. Creeping bentgrass also produces heavy thatch, which must be controlled. It is the primary bentgrass used for cool-season golf greens in the temperate and transition zones of the United States. Newer cultivars should be mowed lower, aerated, and topdressed more as they tend to be more aggressive in their growth habit. However, the fine-textured, aggressively growing newer varieties are more susceptible to thatch buildup, scalping, and "puffiness"; thus, they need more aggressive management. Velvet bentgrass is used for greens in cooler portions of the world. It produces one of the finest bladed, densest putting surfaces available and possesses good shade tolerance (Table 1.11). Imperfections such as ball marks and diseases, however, are easily seen with velvet bentgrass and it is very prone to thatch buildup, thus becoming puffy and scalping easily.

Colonial bentgrass, by contrast, has minimal creeping tendencies (Table 1.12) and is currently being used in the maritime regions of Canada and the Northeast, Great Lakes, and northwestern areas of the United States. It is being genetically improved for fairway use. Under close mowing, creeping bentgrass tends to form a fine-textured, dense, and low-growing turf, while colonial bentgrass forms an upright turf with more resistance to dollar spot disease and less N requirements. It is often mixed with fine fescues. Other bentgrass species currently being investigated for use on golf courses include Dryland (*Agrostis castellana*), Idaho (*A. idahoensis*), and Browntop (*A. capillaris*). For tees and fairways, blends are often used to broaden the genetic base, thereby reducing disease occurrence.

### Kentucky Bluegrass

Kentucky or smooth-stalked bluegrass (*Poa pratensis* L.) is the most widely used cool-season turfgrass (Figure 1.37) due to its high quality and ability to survive summer drought through dormancy. It is native to cool, open areas in Europe and was introduced in the United States during the 1600s. It possesses a vigorous rhizome system that forms a very dense cover, allowing it to recover from divots and traffic damage, and has fair to good drought tolerance. During extended heat and drought stress without irrigation, it will enter dormancy where the aboveground tissue ceases growth and can turn brown. With favorable growing conditions, new shoots are initiated from rhizomes and crowns. Kentucky bluegrass also has greater cold temperature tolerance than grasses such as perennial ryegrass or tall fescue (Table 1.9). Overall, shade tolerance is poor to fair, but breeding efforts are improving this as well as tolerance of low mowing height (less than

**TABLE 1.9**
**General Characteristics of Cool-Season Turfgrasses**

| Characteristic | Annual Bluegrass | Kentucky Bluegrass | Browntop | Colonial Bentgrass | Creeping Bentgrass | Perennial Ryegrass | Red Fescue | Chewings Fescue | Tall Fescue | Crested Wheatgrass |
|---|---|---|---|---|---|---|---|---|---|---|
| **Establishment** | | | | | | | | | | |
| Planting method | Seed | Seed/veg. | Seed/veg. | Seed/veg. | Seed/veg. | Seed | Seed | Seed | Seed/veg. | Seed |
| Rate of coverage | Very fast | Slow | Medium | Medium | Slow | Very fast | Medium | Very slow | Medium | Slow |
| Days to germination | N/A | 8 to 21 | 7 to 14 | 7 to 14 | 7 to 14 | 5 to 12 | 10 to 21 | 10 to 21 | 10 to 14 | N/A |
| Growth rate | Fast | Fast | Fast | Fast | Fast | Fast | Slow/med | Slow/med | Fast | Slow |
| Density | Excellent | Good | Excellent | Excellent | Excellent | Very good | Excellent | Very good | Good | Fair |
| Leaf texture | Fine | Medium | Fine | Fine | Very fine | Medium | Very fine | Very fine | Med/coarse | Medium |
| Polystand compatibility | Low | High | Medium | Medium | Medium | Medium | High | Medium | Medium | Low |
| **Cultural Requirements** | | | | | | | | | | |
| Maintenance intensity | High | Medium | Medium | Medium | High | Medium | Low | Low | Medium | Low |
| Mowing height (inches) [fairways] | 0.3 to 1.0 | 1.5 to 2.5 | 0.5 to 1.0 | 0.5 to 1.0 | 0.25 to 0.5 | 0.75 to 2.5 | 1.5 to 2.5 | 1.5 to 2.5 | 1.5 to 3 | 1.5 to 2.5 |
| Mowing frequency (days) | 1 to 3 | 5 to 7 | 3 | 1 to 3 | 1 to 3 | 5 to 7 | 7 | 7 | 7 to 14 | 7 to 14 |
| Irrigation requirements | High | Medium | Medium | Medium | High | Medium | Low | Low | Low | Low |
| Evapotranspiration (ET) (mm/day) | >10 | >10 | N/A | N/A | >10 | 8.5 to 10 | 7 to 8.5 | 7 to 8.5 | >10 | N/A |
| Nitrogen needs | Medium | Medium | Medium | Medium | High | Medium | Low | Low | Medium | Low |
| **Performance** | | | | | | | | | | |
| Heat tolerance | Poor | Fair | Fair/good | Poor/fair | Fair/good | Poor/fair | Fair | Fair | Good | Fair |
| Cold tolerance | Poor | Very good | Good | Good | Good | Fair | Good | Good | Good | Good |
| Killing temperature (°F) | N/A | N/A | N/A | N/A | N/A | N/A | N/A | N/A | N/A | N/A |
| Winter performance | Fair | Fair | Very good | Good | Fair | Excellent | Very good | Very good | Good | Fair |
| Drought resistance | Poor | Good | Poor | Poor | Poor | Good | Very good | Excellent | Excellent | Excellent |
| Shade tolerance | Poor | Fair | Fair/good | Fair/good | Fair/good | Fair | Excellent | Excellent | Good | Fair |
| Wear tolerance | Poor | Good | Fair | Fair | Fair | Excellent | Good | Fair | Good | Poor |
| Recuperative potential | Poor | Good | Fair | Fair | Good | Poor | Fair | Poor | Good | Poor |
| Low mowing tolerance | Excellent | Poor | Excellent | Excellent | Excellent | Very good | Very good | Good | Fair | Poor |
| Salt tolerance | Poor | Poor | Good | Fair | Poor/good | Good | Good | Good | Good | Good |
| Soil pH range | 5.5 to 6.5 | 6.0 to 7.0 | 5.0 to 6.0 | 5.5 to 6.7 | 5.5 to 6.5 | 6.0 to 7.0 | 5.5 to 6.5 | 5.5 to 6.5 | 4.7 to 8.5 | 6.0 to 8.0 |
| Thatch tendency | Medium | Medium | High | High | High | Low | Medium | Medium | Low | Low |
| Disease tendency | High | Medium | High | High | High | Med/high | Medium | Medium | Low | Low |
| Submersion tolerance | Fair | Good | N/A | N/A | Good | Fair | Fair | N/A | N/A | N/A |

*Note:* Cultivar differences may occur. N/A = not available; veg. = vegetative; med = medium.

1 inch, 2.5 cm). Under a high-maintenance program, thatch and diseases (e.g., summer patch, necrotic ring spot, rust, powdery mildew, leaf spot/blight) can be serious problems, and billbugs, grubs, and mites are troublesome insects.

Kentucky bluegrass seed, compared to ryegrass, can be slow to germinate and become established and has weaker seedlings. At warmer soil temperatures (>60°F, 15.6°C), germination occurs in about 10 days. Cooler temperatures (<55°F, 12.8°C) extends germination to 14 to 21 days. For good quality, density, and beauty, Kentucky bluegrass requires moderate to high fertility. Herbicide and plant growth regulator (PGR) tolerance can be marginal. "Merion" was the first improved Kentucky bluegrass developed, primarily for its resistance to leaf spot disease. Over 200 cultivars currently exist; consult your state turfgrass specialist and www.ntep.org for independent cultivar comparison.

**Hybrid bluegrass.** Crosses between Texas (*Poa arachnifera* Torr.) and Kentucky bluegrass have produced hybrid bluegrasses with characteristics of each. Texas bluegrass is indigenous to the prairie regions of the southwestern United States and is more drought and heat tolerant than Kentucky bluegrass. Like Kentucky bluegrass, the hybrids produce rhizomes and like Texas bluegrass, the hybrids possess better heat tolerance and near year-round green color. These are often designated as HT-KBG or heat-tolerant Kentucky bluegrass. Ameriblue, Bandera, Dura Blue, Fahrenheit 90, Fire and Ice, Longhorn, Reveille, Solar Green, SPF 30, Spitfire, Thermal Blue, and Thermal Blue Blaze are some releases.

FIGURE 1.34 Creeping bentgrass as a putting surface.

FIGURE 1.35 Creeping bentgrass grown outside its adaptation region with insufficient inputs.

Hybrid bluegrasses currently appear best for use on golf course roughs, especially in shade, as they perform best at higher mowing heights. Pest problems for hybrid bluegrass are similar to those for Kentucky bluegrass. They are typically sold as a seed mixture with tall fescue.

### Tall Fescue

Over 100 *Festuca* spp. are known. Generally, these are divided into two types: fine- or coarse-leaf fescues. Tall fescue (*Lolium arundinaceum* (Schreb.) Darbysh., formerly *Festuca arundinacea* Schreb.) is a coarse-textured, bunch-type species originally from southern Europe where summers are warmer and drier than in other European regions. Tall fescue forms a low-density turf with wider leaves than other cool-season turfgrasses (Figure 1.38). It establishes quickly, is deep-rooted, wear tolerant, has good heat and cold tolerance, is a slow thatch producer, and has fewer pest problems than most cool-season turfgrasses (Table 1.9). Although improvements have been made toward developing a rhizomatous cultivar, tall fescue has an overall bunch-type growth habit. Due to its clumping growth habit, tall fescue often requires reseeding every two to three years to maintain desirable turf density.

Tall fescue has very poor low-temperature hardiness, thereby limiting its use for those areas that experience severe winter temperatures. However, it has the best heat tolerance of the cool-season turfgrasses and is commonly grown in the transition zone, mid-Atlantic region, and the upper portion of the warm-humid zone. It also survives in arid climates when irrigated. Finer textured cultivars are available with good color and traffic tolerance. Shade tolerance is fair but the grass will tolerate infertile, saline, alkaline, wet, and dry soil conditions. It is not as susceptible to gray leaf spot as perennial ryegrass and has good herbicide tolerance.

Tall fescue germinates quickly but not as fast as perennial ryegrass and its texture is coarser than Kentucky bluegrass and perennial ryegrass. When overirrigated and/or -fertilized, tall fescue often becomes susceptible to leaf spot and brown patch diseases, as well as white grub damage. To maintain stand density and to avoid scalping, tall fescue is mowed at 3 inches (7.6 cm). Due to its coarse leaf texture, low shoot density, and clumpy growth habit, tall fescue can be considered a serious weed in finer textured monostand turfs such as Kentucky bluegrass. When seeded with Kentucky bluegrass, the seed mix on a weight basis should be at least 70% tall fescue. "Rebel" was the first introduced turf-type tall fescue and provided improved leaf width, stand density, color, and disease resistance. Over 200 cultivars are currently available with numerous new releases yearly. Grande II, Labrinth, Regiment II, Rhizing Star, RTF, and Tulsa II are initial

## TABLE 1.10
## Bentgrass Cultivars Used for Golf Course Purposes

**Creeping Bentgrass (*Agrostis palustris* subsp. *stolonifera* L.)**

Penn A and G Series (A-1; A-4; G-1; G-2; G-4; G-6), Alpha, Armor, Authority, AU Victory, Backspin, Bardot, Barifera, Barracuda, Benchmark DSR, Bengal, Brighton, Carman, Cato, Century, CEO, Cobra 1-2, Crenshaw, Crystal BlueLinks, CY-2, Declaration, Dominant (SR 1019 + SR 1020), Dominant Plus/Xtreme, Emerald, Flagstick, Focus, Grand Prix, 18th Green, Greenwich, HTM, Imperial, Independence, Inverness, Kingdom, Kingpin, L-93 XD, Lopez, LS-44, Luminary, Mackenzie, Mariner, Memorial, National, Nightlife, Ninety-Six Two, Piranha, Penncross, Penneagle II, Pennlinks II, Pin-Up, PLS, Princeville, Proclamation, ProCup, Prominent, Providence, Pure Distinction, Pure Select, Putter, Regent (Normark 101), Sandhill, Seaside II, Shark, Southshore, SR 1150, SR 7200, T-1, T-2, Tendenz, Tiger Shark, TourPro, Trueline, Tyee, V-8, Villa, Viper, 007, 13M[a]

**Colonial Bentgrass (*A. tenuis Sibthorp*)**

Alister, Allure, Astoria, Bardot, Barostis, Boral, Egmont, Exeter, Greentime, Glory, Heriot, Holfior, Revere, SR 7000/7100/7150, Tendenz, Teno, Tiger II, Tracenta, Vitek[b]

**Velvet Bentgrass (*A. canina subsp. canina* L.); Dryland or Highland Bentgrass (*A. castellana*); Browntop (*A. capillaris*); Redtop (*A. alba* L.)**

Acme (velvet), Albert Royal (browntop), Alister (browntop), Barracuda (redtop), Bavaria (velvet), BR 1518 (dryland), Egmont (browntop), Exeter (dryland), GoldStar, Greenwich (velvet), Highland (dryland), Jorvik (browntop), Kernwood (velvet), Kingston (velvet), Legendary (velvet), Leirin (browntop), Piper (velvet), Raritan (velvet), Reten (redtop), Rudiger (redtop), Sefton (browntop), SR 7200 (velvet), Streaker (redtop), Venus (velvet), Vesper (velvet), Villa (velvet)[c]

**Idaho Bentgrass (*Agrostis idahoensis* Nash)**

GolfStar, Spike[d]

*Note:* This list is subject to change as new varieties are continually being released and others discontinued. Check with your state turfgrass extension specialist, seed supplier, and www.ntep.org for the latest available and recommended cultivars for your specific area and needs.

[a] Penncross, released in 1954, is a polysynthetic variety between the accession numbers 10(37)4 (or Pennlu creeping bentgrass), 9(38)5, and 11(38)4. However, it lacks resistance to *Pythium* and *Rhizoctonia* diseases, has a shallow rooting system, becomes grainy, and tends to segregate into various patches in the fall. Improved varieties have better putting and color characteristics but still lack specific disease resistance, especially to dollar spot, and have limited heat tolerance. Blends show potential for overcoming some of these limitations. Improvements are being made on heat tolerance, putting quality (density and fine leaf blades), and summer rooting. Seaside and Mariner (a selection from Seaside) are noted for their salt tolerance.

[b] These are bunchgrasses that tend to grow more upright and require less water and fertility than creeping bentgrass. They generally do not tolerate close mowing like creeping bentgrass. They have been used mostly as a component of mixtures for fairways and general turfgrass areas in coastal regions of the Pacific Northwest, the Northeast, and northern Europe (summer temperatures less than 85°F, 29°C). Newer cultivars may prove better adapted to other regions and other uses.

[c] These produce some of the finest textured turfgrasses used for putting surfaces due to upright growth habit and density (Figure 1.36). However, due to lack of heat, disease, and traffic tolerance, they are used mostly in upper New England and the Pacific Northwest, New Zealand, Australia, and in other regions as part of blends. Velvet bentgrass is also noted for its shade tolerance (Table 1.11). Redtop establishes the most quickly of the *Agrostis* species and performs well on wet, poorly drained sites. Velvet bentgrass is sensitive to iron chlorosis on alkaline soils and requires routine topdressing and grooming due its thatch buildup nature.

[d] Idaho bentgrass has a fine texture, upright leaves, dark-green color, and noncreeping (bunch-type) growth habit. It is useful for winter overseeding or as a low-maintenance permanent turf in fairways or roughs.

## TABLE 1.11
## Advantages and Weaknesses of Using Velvet Bentgrass as a Fine Turf Species

| Advantages | Weaknesses |
|---|---|
| • Produces high-quality turf | • Susceptible to seeding *Pythium*—use treated seed |
| • Produces high density and fine leaf texture | • Susceptible to copper spot disease |
| • Disease resistance to dollar spot and brown patch | • Slower to repair damage |
| • Reduced water requirement: 40% to 60% ET | • Thatch control program necessary |
| • Excellent shade tolerance | • Susceptible to moss if overwatered |
| • Low fertility (postestablishment) | • Fewer herbicides registered for use |
| • Grows best in acidic soils | • Marginal salt tolerance |
| • Good wear tolerance | |

## TABLE 1.12
## Advantages and Limitations of Using Colonial Bentgrass as a Fine Turf Species

| Advantages | Limitations |
|---|---|
| • Low N requirement | • Brown patch susceptible; normally not lethal |
| • Snow mold resistance | • Reduced stolon and rhizome growth |
| • Dollar spot resistant against most strains | • Often seeded with chewings fescue for divot repair |
| • Less thatch compared to creeping bentgrass | • Lighter green color than creeping bentgrass |
| • Better wear tolerance than creeping bentgrass | • More sensitive to ethofumesate than creeping bentgrass |
| • Golf ball sits upright | • Fewer herbicides registered for use |
| • Low water use, ~50% ET | • Can be susceptible to take-all patch |
| • Provides good quality when blended with fine fescues | |

**FIGURE 1.36** Velvet bentgrass produces a finely textured putting surface.

**FIGURE 1.37** Kentucky bluegrass is the most widely used cool-season turfgrass.

releases with rhizomatous growth habits, while Pure Gold and Tomahawk GT are early members of a new generation of glyphosate-tolerant cultivars. Refer to your state turfgrass specialist, seed company representative, and www.ntep.org for comparison of cultivars.

### Perennial Ryegrass

Perennial ryegrass (*Lolium perenne* L.) is a short-lived perennial with a medium leaf texture and good shoot density originating from open to fringed forest regions of southern Europe and western Asia to northern Africa (Figure 1.39). It has a nonspreading, bunch-type growth habit, good wear resistance, finer leaf texture, good soil compaction tolerance, good drought tolerance, good herbicide tolerance, fast establishment rate, and bright-green color (Table 1.9). Perennial ryegrass tolerates lower mowing heights than Kentucky bluegrass or tall fescue. However, it does not tolerate low or high temperature extremes, has slow recovery from divots, and has limited shade tolerance. It has become very susceptible to gray leaf spot disease, which is expensive and troublesome to control. Perennial ryegrass also is susceptible to *Pythium* blight and has marginal cold tolerance. Perennial ryegrass is primarily used as a component in seed mixtures with other cool-season turfgrasses, such as Kentucky bluegrass, to broaden the resistance to various diseases and other pests. It is often used for overseeding purposes in bermudagrass and zoysiagrass for winter color (Figure 1.40). Tetraploid varieties are being developed for better drought tolerance, disease resistance, and winter performance. Refer to your state turfgrass specialist, seed company representative, and www.ntep.org for comparison of cultivars.

### Fine Fescue

Fine fescue is a general term used for several fine-leafed *Festuca* spp. (Figure 1.41). Originally from the cool, forested European Alps, fine fescues have delicate and wiry leaves

**FIGURE 1.38** Tall fescue as a low-maintenance, heat-tolerant, cool-season turfgrass.

**FIGURE 1.39** Perennial ryegrass fairway with desirable striping characteristics.

**FIGURE 1.40** Perennial ryegrass used for winter color in dormant bermudagrass.

**FIGURE 1.41** Fine fescue in low-maintenance, shade-prone areas.

that are usually less than 0.5 mm wide and have a clumping, bristle-like appearance. These normally are maintained at 2 to 3 inches (5 to 7.6 cm) in height and should only receive low annual fertility (Table 1.9). Traffic and mowing should be withheld during heat or drought stress as turf injury or death often occurs. Due to shallow root systems, fine fescues tend to go dormant and turn brown in heat and drought conditions. Red fescue (*F. rubra* L. spp. *rubra*, spp. *litoralis*, and ssp. *tricolphylla*) and spreading fescue (*F. pratensis* Huds.) have slow spreading rhizomes, while chewings fescue (*F. rubra* ssp. *commutata* Gaud.), sheep fescue (*F. ovina* L.), blue fescue (*F. ovina* ssp. *glauca*), and hard fescue (*F. longifolia* auct. non Thuill.; *F. brevipila* Tracey) have a bunch-type growth habit.

As a group, the fine fescues are noted for shade tolerance, winter hardiness, and adaptability to infertile and dry soil conditions (Table 1.13). They do not tolerate wet, poorly drained soils well, but are particularly adapted to dry, shady conditions as well as to low-maintenance situations. Fine fescues rarely are used alone but are usually mixed with other cool-season grasses such as Kentucky bluegrass, perennial ryegrass, and colonial bentgrass. They tend to eventually dominate the stand when grown in shady conditions. They are also used in blends for winter overseeding of warm-season turfgrasses. Seeds tend to germinate in 7 to 10 days under good environmental conditions. Aurora Gold is a glyphosate-tolerant hard fescue cultivar. Table 1.14 lists some of the currently available fine fescue cultivars.

**TABLE 1.13**
**Characteristics of Using Chewings and Creeping Red Fescue as Fine Turf Species**

| Chewings Fescue | Creeping Red Fescue |
|---|---|
| • Rapid germination, close to perennial ryegrass | • Good salt and shade tolerance |
| • Winter-active growth | • Resistance to rapid blight disease |
| • Better heat tolerance with improved cultivars | • Produces short rhizomes for wear repair |
| • Lower height of cut tolerance | • Tolerant of fairway mowing heights |
| • Shade tolerance | • Better heat tolerance with improved cultivars |
| • Good summer patch and leaf spot resistance | • Excellent fall, winter, and spring color |
| • Improved wear tolerance with improved cultivars | |

Hard fescue tends to have a deep-green color, whereas sheep fescues are bluish-green. Creeping red fescue performs well in shady conditions in humid regions, whereas chewings, hard, and sheep fescues grow well in full sun or partial shade, especially in the drier regions. These often are planted in shaded, low-traffic areas, poor soils, or in secondary roughs or out-of-play areas. If left unmowed, their leaf growth and seedheads often provide a low-maintenance, natural appearing area.

### Annual Bluegrass

Annual bluegrass (*Poa annua* L.) is a highly variable species originating from Europe, but it is widely distributed and found on every continent. It has traditionally been the most troublesome winter annual weed in most turfgrass situations (Figure 1.42). This is due to its prolific seed production, which disrupts the appearance and uniformity of the turf stand; its competitive growth habit; and poor heat and disease tolerance, which often leaves bare or thin areas in the turf. Additionally, the annual biotypes have a bunch-type growth habit of low density (Table 1.9). Selective herbicide control has also been marginal at best. Various forms or biotypes behave as true annuals while others have perenniated. Perennial or "creeping" biotypes (*P. annua* L. f. *reptans* (Hausskin.) T. Koyama) have been identified that have more desirable characteristics for putting surfaces. The perennial biotypes have a more prostrate, spreading growth habit with short stolons. These also have very high shoot density and much less seed production as the grass allocates more of its energy into vegetative growth. *Poa annua* (2n = 28) is believed to have derived from hybridization of two *Poa* ancestors, *P. supina* and *P. infirma*.

**TABLE 1.14**
**Fine Fescue Cultivars**

| Red Fescue | Chewings Fescue | Hard Fescue | Sheep Fescue | Others |
|---|---|---|---|---|
| Aberdeen, Arctared, Aruba, Audubon, Badger, Barcrown, Bargena series, Blanka, Beudin, Bodega, Boreal, Camilla, Cardinal series, Celestial, Cezanne, Chantilly, Cindy, Cindy Lou, Crossbow, Dawson Red, Edgewood, Ensylva, Epic, Eugene, Florensate, Florentine, Flyer, Fortitude, Foxfire, Foxy, Franklin, Frigg, Garance, Garnet, Gibraltar, Hector, Heidrun, Herald, Illahee, Inverness, Isolde, Jasper series, Jitka, Joppa, Lisamba, Logro, Lustrous, Medallion, Miramar, Mirka, Musette, Mystic, Nikolka, Pathfinder, Petruna, Navigator II, Pennlawn, Player, Rainier, Razor, Reptans, Revere, Rose, Rosecity, Rossinate, Ruby, Salem, Salsa, Shademark, Shademaster III, Shoreline, SR 5200 series, Sylvester, Terka, Tiara, Treasure, Valda, Victorka, Vista, Wendy Jean, Weston, Zamboni | Agram, Alice, Ambassador, Andulka, Aranka, Atlanta, Banner, Banner II, Barborka, Bargreen, Barlineus, Barnica, Bridgeport, Brittany, Calliope, Camaro, Carmen, Cascade, Castle series, Center, Checker, Citera, Compass II, Conductor, Culumbra, Darwin, Disnity, Dover, Eco, Enjoy, Fairmont, Garnet, Highlight, Hood, Intrigue 2, Jamestown series, K2, Klarka, Koket, Lacrosse, Lifalla, Longfellow series, Mary, Molinda, Monika, Musica, Nikky, Nimrod, Ombetta, Polaris, Protenza, Radar, Sandpiper, Scarlet, Shadow III, Southport, SR 5000 series, Tiffany, Treazure II, Trophy, Victory I-II, Waldorf, Windward, Wintergreen, Wrigley 2, Zodiac | Anvil, Attila, Aurora I-II, Aurora Gold, Barcrown, Bardur, Barreppo, Beacon, Berkshire, Biljart, Bonny Dunes blend, Brigade, Chariot, Clarinet, Defiant, Discovery, Dorotka, Durar, EcoStar, Eureka I-II, Firefly, Gladiator, Gotham, Hardtop, Harpoon, Heron, Jana, Jetty, Minimus, Minotaur, Moxie, Nordic, Osprey, Oxford, Predator, Reliant series, Rescue, Rescue 911, Resolute, Rhino, Ridu, Scaldis, Silvana, Spartan series, SR 3000 series, Stepanka, Stonehenge, Sword, Valda, Vernon, Viking, Waldina, Warwick | Azay, Azay Blue, Azure, Barok, Bighorn, Black, BlueMesa, Little Bighorn, Lucka, MX-86AE, MX-07, Quatro. | Arcta (Slender), Bighorn GT (Blue), Cindy Lou (Strong), Chorus (Strong), Diego, Fenway, Gordon, Libano (Slender), Marker (Slender), Minotaur (Blue), Navigator II (Strong), Perville, Raisa, Rose, Seabreeze GT (Slender), Sealink (Slender), SeaMist (Slender), Shademark, Silhouette, SR 3200 (Blue), SR 5000 series (Strong), Sunset, Trapeze, Weston, Wilma, Wrigley |

*Note:* Check with www.ntep.org, your seed supplier, and state turfgrass specialist for the latest recommended cultivar(s) for your particular area and needs.

**FIGURE 1.42** Annual bluegrass (*Poa annua*) can be a desirable putting surface or major weed.

Peterson and True Putt creeping bluegrasses are two of the first commercial improved perennial biotype bluegrasses for putting greens.

## Canada Bluegrass

Canada bluegrass (*Poa compressa* L.) is a bluish-green, weakly rhizomatous grass, originally from Europe, which forms a low-density turf of inferior quality. It tolerates droughty and infertile soils, low temperatures, and shady conditions in the cooler, humid areas of the United States and Canada better than Kentucky bluegrass. It is mostly used in colder climates where it is too dry to grow Kentucky bluegrass without irrigation. Its overall low quality is best suited for minimum-use or low-maintenance situations in cool temperate and subarctic climates. Improved cultivars include Canon, Reubens, and Talon.

### Supina Bluegrass

Supina bluegrass (*Poa supina* Schard.), a naturally apple-green grass native to the European Alps, can tolerate extreme shade (80% to 90%) and heavy traffic (Figure 1.43). It can be undesirable due to its light-green color, though darker-green ecotypes are being developed. It has high seed cost and a shallow root system. It is a common grass type used on golf courses in Europe and in the cooler regions of the United States. It displays good disease resistance but is susceptible to *Microdochium* patch (pink snow mold). Commercial varieties include Supra and Supranova.

### Other Cool-Season Turfgrasses

Numerous additional cool-season turfgrasses are available, with most playing only a very small role on golf courses. Many are used as blends with other cool-season grasses, especially for overseeding purposes or for very low-maintenance, nonirrigated, out-of-play areas. Several also can become very weedy species. Refer to the next section on native grasses and the chapter on overseeding for more information.

## NATIVE, LOW MAINTENANCE, AND ORNAMENTAL GRASSES

It has become popular to use native grasses in an attempt to reduce inputs such as water, fertilizer, and pesticides (Figure 1.44). These are typically incorporated outside the primary play areas as most native grasses provide loosely knitted, blue-green colored turf stands when grown naturally. When placed under higher maintenance such as periodic irrigation and fertilizing, many of these become weakened and thin, become disease susceptible, and face invasion from other grasses and weeds. If used, native grasses should be maintained as close to their natural adaptation as possible; mow these high (>2 inches, 5 cm), fertilize only lightly (0 to 2 pounds N/1,000 square feet annually, 0 to 98 kg N/ha), and only irrigate occasionally.

Several grasses native to the arid Great Plains regions of North America are rarely used as a turfgrass. Most of these originated as pasture grasses, forage, or for roadside soil stabilization. They are occasionally used for low-maintenance roughs and out-of-play areas on golf courses due to their distinct characteristics such as seedhead height or plant color;

**SELECTING A TURFGRASS FOR ROUGHS AND FAIRWAYS**

Availability of a reliable irrigation system specifically designed for roughs is an initial question, if expectations are high. In much of the northern United States and Canada, Kentucky bluegrass, bentgrass, and fine fescues are widely used. Tall fescue and perennial ryegrass are less winter hardy; thus, they are used more in the midcentral regions of the United States including the transition zone. Table 1.15 lists the advantages and limitations of each species. Cool-season grass cultivars constantly change as new selections are introduced and older ones become less important. Contact your state turfgrass specialist for the latest releases recommended in your area.

Tall fescue competes better with annual bluegrass but requires an extensive time to establish a good stand. Perennial ryegrass establishes quickly and is tolerant to annual bluegrass control herbicides. However, summertime stress along with *Pythium*, gray leaf spot, and dollar spot diseases can cause stand decline.

Current trends in cooler regions use predominantly pure perennial ryegrass stands in the upper transition zone and the cool semiarid zone, especially if the mowing height is maintained at or below 1 inch (2.5 cm) (Table 1.16). With pure ryegrass, however, gray leaf spot and winter-kill periodically thin or weaken the stand. Courses attempt to overcome this by adding a small amount (e.g., 20% to 30%) of Kentucky bluegrass, since it has better recuperative ability due to rhizomatous growth, or by switching to bentgrass or zoysiagrass in more southern areas. However, if Kentucky bluegrass is used, the mowing height should be maintained above one 1 (2.5 cm). In nonirrigated areas of the Midwest, low-maintenance grasses such as buffalograss or weeping alkalinegrass may be considered. These, however, should be viewed as low-maintenance grasses, which do not provide the color, density, and texture of irrigated Kentucky bluegrass, tall fescue, bentgrass, or ryegrass.

Blends of approximately 85% tall fescue plus 15% Kentucky bluegrass work reasonably well. Often, sod is used to re-establish highly visible areas such as near greens surrounds. For acceptable long-term rough turf, weed and insect control programs are necessary. Grub control products applied in early summer can provide season-long control. Fungicides also are being used more often in cool-season grass roughs and good fertility programs are needed to maintain thick, healthy grass. Finally, heavily trafficked areas, such as those near greens, need special attention. This includes spreading wear to prevent concentrated traffic damage, additional N for recovery, aggressive aeration, and possibly walk-mowing during periods of summer heat stress or when soils are saturated.

In the upper areas of cool regions, winter-kill periodically occurs to ryegrass and tall fescue; thus, Kentucky bluegrass is often used alone. However, the bluegrass should be mowed above 1 inch (2.5 cm) in these situations for best turf growth and hardiness. In addition, a blend of two or more bluegrass cultivars should be used to help in pest resistance.

Fine fescues are used, especially in shady areas or areas without irrigation, but these tend to establish slowly and are susceptible to several diseases.

In the southern United States, bermudagrass is used widely for fairways and roughs. Irrigated fairways use a hybrid bermudagrass while nonirrigated ones use an improved common bermudagrass. Zoysiagrass is being used more when cold temperature damage or shade is an issue with bermudagrass or if a different hue or textured grass is desired. Some

**TABLE 1.15**

**Region of Adaptation, Advantages, and Limitations of Various Turfgrasses Used for Golf Course Fairways and Roughs**

| Turfgrass | Region of Adaptation | Advantages | Limitations |
|---|---|---|---|
| Bentgrass (*Agrostis* spp.) | Most cool-region zones (temperate regions) if irrigation is available | • Tolerates close (<0.5 inch, 1.3 centimeters) mowing<br>• Produces dense stands from stolons, has excellent recuperative ability, and does not require yearly interseeding<br>• Stripes easily, good color contrast with darker colored bluegrass roughs<br>• Provides excellent playing surface<br>• Tolerates close (<0.5 inch, 1.3 cm) mowing | • Generally intolerant of heat and drought<br>• Maintenance expense due to susceptibility to diseases such as *Pythium,* dollar spot, and brown patch<br>• Must control thatch<br>• Susceptible to winter desiccation<br>• *Poa annua* infestations require lightweight mowers, clipping removal, aerification, and continued herbicide/PGR use |
| Bermudagrass (*Cynodon* spp.) | Tropical to subtropical regions into lower half of the transition zone, southern Illinois, Missouri, and Kansas | • Tolerates poor soils, hot climates<br>• Efficient water user<br>• Excellent wear tolerance<br>• Good pesticide options<br>• Rapid and cost effective to establish<br>• Low-cost option for fairways | • High N requirements<br>• Aggressive growth habit<br>• Shade intolerant<br>• Nematode susceptibility<br>• Cold sensitive in northern locations<br>• Dark tan/brown winter color; poor spring playability<br>• Poor spring transition when overseeded |
| Buffalograss (*Buchloe dactyloides*) | Drier regions of the Midwest such as Kansas and Nebraska | • Low maintenance in terms of water and nutrient needs and pest problems<br>• Tolerates hot, dry conditions<br>• Seed (burs) are available<br>• Infrequent mowing | • Produces a coarse–rough texture and somewhat thin density that is weed prone<br>• Produces unsightly seedheads (burs)<br>• Poor shade, traffic, and herbicide tolerance<br>• Disease susceptible in high humidity |
| Fine fescue (*Festuca* spp.) | Shady areas in upper transition zone and northward | • Withstands very low maintenance in terms of nutrient and water requirements<br>• Slow growth rate, requiring about half the mowings of taller grown grasses<br>• Alternative to Kentucky bluegrass in unirrigated, infrequently mowed roughs | • Goes dormant with warm summer temperatures<br>• All are bunch grasses except the rhizomatous creeping red fescue<br>• Susceptible to dollar spot and red thread<br>• Slow recuperative rates |
| Kentucky bluegrass (*Poa pratensis*) | Irrigated regions in the cool zones, Midwest throughout Canada into Alaska | • Excellent color, texture, and density<br>• Affordable seed<br>• Good recuperative ability from rhizomes<br>• Excellent cold tolerance | • Slow establishment from seed<br>• Shallow root system—thus, poor water use<br>• Poor tolerance to low (<1.5 inches) mowing heights<br>• Susceptible to diseases (e.g., summer patch)<br>• Poor shade tolerance |

(*Continued*)

**TABLE 1.15 (CONTINUED)**
**Region of Adaptation, Advantages, and Limitations of Various Turfgrasses Used for Golf Course Fairways and Roughs**

| Turfgrass | Region of Adaptation | Advantages | Limitations |
|---|---|---|---|
| Perennial ryegrass (*Lolium perenne*) | Lower temperate (cool, humid) zone into upper transition zone; also southern portions of Midwest | • Inexpensive seed that establishes rapidly | • Requires more frequent mowing than fine fescue or Kentucky bluegrass |
| | | • Has bright-green color, which stripes well | • Thins out and becomes "stemmy" at lower fertility levels |
| | | • Compatible with Kentucky bluegrass | • Highly disease susceptible to *Pythium*, red thread, and gray leaf spot |
| | | • Good wear and drought tolerance | • Slow to spread, slow recovery from divots |
| | | • Good tolerance to low mowing | • Usually requires yearly interseeding |
| | | • Good tolerance to ethofumesate and other products for *Poa annua* control | • Rapid spring and early summer growth |
| | | • Good playability | • Only fair cold tolerance, susceptible to ice damage in shaded, low-lying areas |
| Seashore paspalum (*Paspalum vaginatum*) | Tropical to subtropical regions up to coastal Carolina | • Excellent salt tolerance in good draining soils | • Limited cold and shade tolerance |
| | | | • Bermudagrass invasion |
| | | • Bright-green color, which stripes well | • Few herbicides labeled for use |
| | | • Stiff blades, so golf balls sit up | • Eventually becomes thatchy |
| | | • Low N requirement. | • Moderate to high disease potential |
| | | • Forms excellent density | • Mostly vegetatively established |
| Tall fescue (*Lolium arundinacea*) | Irrigated and nonirrigated lower maintenance courses throughout the temperate region | • Good heat drought tolerance | • Clumpy growth habit requires reseeding to maintain density |
| | | • Easily established by seeding | |
| | | • Compatible with Kentucky bluegrass and perennial ryegrass | • Intolerant to close (<2 inches, 5 cm) mowing |
| | | | • Slow to recover from damage |
| | | | • Requires more frequent mowing than other cool-season grasses |
| | | | • Susceptible to brown patch |
| Weeping alkaligrass (*Puccinella distans*) | High-sodium regions of the Midwest | • Excellent salt tolerance | • Produces open, coarse stands |
| | | • Good cold tolerance, fair heat tolerance | • Gray-green, tufted bunch-growth habit |
| | | • Low maintenance in terms of nutrient and water requirements | • Requires overseeding to maintain a uniform, dense stand |
| Wheatgrasses (*Agropyron* and *Pascopyrum* spp.) | Drier regions of Midwest and intermountain region of western United States | • Provides an alternative to Kentucky bluegrass in nonirrigated regions | • Forms coarse-textured, low-density turf stands |
| Zoysiagrass (*Zoysia* spp.) | Tropical and subtropical areas into the upper transition zone | • Can be low maintenance after establishment | • Expensive and slow to establish from sprigs or seed, slow to recover from divots |
| | | • Produces dense, weed-resistant turf, with good playability | • Traffic restrictions when dormant |
| | | | • Eventually becomes thatchy |
| | | • Best winter-hardy, warm-season grass | • Winter damage in wet areas |
| | | • Light tan color when dormant | • Bermudagrass invasion and susceptible to brown patch |

architects use other grasses such as centipedegrass or bahiagrass roughs with bermudagrass fairways to provide contrasting color and textural differences. However, in most cases, this is a maintenance mistake as many of the pesticides, especially herbicides, used on bermudagrass are invariably sprayed on these other turfgrasses, causing damage. Also, the higher N and watering requirements of bermudagrass usually weaken these other grasses. Seashore paspalum is used in barrier-island conditions where brackish irrigation water and sandy soils exist.

**TABLE 1.16**
**Common Seed Mixtures for Cool-Season Grasses within Three Major Climatic Zones**

| Fairways: Sun, with Higher Fertility | | Roughs | | | | |
|---|---|---|---|---|---|---|
| **Irrigated** | **Nonirrigated** | **Sun** | **Shade** | **Shade, Nonirrigated, Low Fertility** | **Grass Bunkers** | **Low-Maintenance Areas** |
| **Upper Portion of the Cool Humid Transition Zone and Cool Semiarid (Upper Great Plains) Areas** | | | | | | |
| **Mowing height > 1 inch, 2.5 cm:** 70%–80% Ky. bluegrass blend + 20%–30% p. ryegrass<br>**Mowing height < 1 inch, irrigated:** 100% p. ryegrass blend *or* 100% creeping bentgrass for higher budgets | 100% tall fescue *or* buffalograss *or* alkaligrass in salty environments | 60% Ky. blue blend + 20% chewings fescue + 20% p. ryegrass<br>or<br>90% tall fescue + 10% Ky. blue | 40% Ky. blue + 40% chewings fescue + 20% p. ryegrass<br>or<br>90% tall fescue + 10% Ky. blue | 20%–40% Ky. blue blend + 60-80% fine fescue blend<br>or<br>60% Ky. blue blend + 40% p. ryegrass | Fine fescue blend such as 80% hard fescue + 20% chewings fescue | Cool, humid areas: fine fescue blend such as 80% hard fescue + 20% chewings fescue<br>Cool, arid areas: 40% buffalograss + 40% blue grama + 20% hard fescue |
| **Lower Portion of the Cool Humid Transition Zone and Cool Semiarid (Lower Great Plains) Areas** | | | | | | |
| **Mowing height > 1 inch:** 10%–30% Ky. blue blend + 70%–90% p. ryegrass *or* 70%–90% tall fescue<br>**Mowing height < 1 inch, irrigated:** 100% p. ryegrass blend *or* 100% creeping bentgrass<br>**Gray leaf spot areas:** 100% bentgrass | 100% buffalograss *or* tall fescue *or* alkaligrass in salty environments | 90% tall fescue + 10% Ky. blue<br>or<br>60% Ky. blue + 20% p. ryegrass + 20% chewings fescue | 90% tall fescue + 10% Ky. blue<br>or<br>65% chewings fescue + Ky. blue + 15% p. ryegrass | 10%–20% Ky. blue + 40%–60% p. ryegrass + 30%–40% fine fescue<br>or<br>15% p. ryegrass + 50% tall fescue + 25% red fescue + 10% Ky. bluegrass | 100% tall fescue blend<br>or<br>80% hard fescue + 20% chewings fescue<br>or<br>80% tall fescue + 10% Ky. blue + 10% p. ryegrass | 100% tall fescue blend<br>or<br>80% hard fescue + 20% chewings fescue<br>or<br>10% p. ryegrass + 50% tall fescue + 30% red fescue + 10% Ky. bluegrass |
| **Cool Humid Pacific Coast Zone** | | | | | | |
| **Mowing height > 1 inch:** 10%–30% Ky. blue blend + 70%–90% p. ryegrass *or* tall fescue<br>**Mowing height < 1 inch, irrigated:** 100% p. ryegrass blend.<br>**Gray leaf spot areas:** 100% bentgrass | 100% tall fescue | 60% Ky. blue blend + 20% chewings fescue + 20% p. ryegrass | 40% Ky. blue + 40% chewings fescue + 20% p. ryegrass | 60%–80% fine fescue + 20%–40% colonial bentgrass<br>or<br>60% Ky. blue + 40% p. ryegrass | Fine fescue blend such as 80% hard fescue + 20% chewings fescue | Fine fescue blend such as 80% hard fescue + 20% chewings fescue<br>or<br>35% hard fescue + 20% red fescue + 20% p. ryegrass |

these help provide distinct differences from the desired playing surface. Blends are often planted. Weed control prior to planting, and for the first two years thereafter, is essential for success. Examples include the warm-season grasses buffalograss, gramagrass, switchgrass, bluestems, Indiangrass, and lovegrass (Table 1.15) and the cool-season grasses wheatgrass, timothy, redtop, orchardgrass, fine fescues, and alkaligrass.

Some courses have opted to incorporate extensive areas with minimum vegetation to mimic earlier times when most resources such as irrigation and fertilizers were very limited (Figure 1.45). Although inputs such as mowing, fertilization, and irrigation are obviously reduced or eliminated, dramatic increases in hand labor for weeding and maintaining area integrity often occur. Regardless, the uniqueness of these areas and potential reduced artificial inputs have other courses considering similar trends.

## Warm-Season Grasses

**Buffalograss.** Buffalograss [*Buchloe dactyloides* (Nutt.) Engelm.] is a low-maintenance, perennial warm-season species native to the arid and semiarid regions of the North American western Great Plains from southern Canada to the highlands of Central Mexico receiving low rainfall (10 to 30 inches yearly, 25 to 76 cm). It is named from the American buffalo (bison) for which it served as the primary grazing

**FIGURE 1.43** Supina bluegrass is an apple-green, extremely shade-tolerant cool-season grass.

**FIGURE 1.44** Native grasses are largely low maintenance and provide distinct color and textural characteristics.

**FIGURE 1.45** Selective areas with extensive natural areas and limited irrigation.

**FIGURE 1.46** Buffalograss is a dull green, low-maintenance, warm-season species native to the U.S. Great Plains.

species. It spreads by stolons and seed and forms a medium- to fine-textured, relatively thin turf with a soft blue-green color that resembles common bermudagrass (Figure 1.46) but provides a distinct color contrast to hybrid bermudagrass and zoysiagrass. The grayish-green color is due to the fine hairs covering the leaves. Curling of the leaf blades also is a distinctive feature. Ecotypes are adapted to either southern (hot) or northern (cold) areas; thus, the origin of improved cultivars should be considered before selecting them for turf purposes. Buffalograss is adapted for use on nonirrigated lawns, parks, golf course roughs, and roadsides in the warm zones of arid and semiarid regions.

Buffalograss is a dioecious species, meaning it produces separate male and female flowers and plants. The female flower is a short spike where the glumes, lemma, and palea form a bur-like enclosure for the seed. The male flower is more prominent as its seed stalk extends above the turf canopy. Advantages and limitations of buffalograss are listed in Table 1.15.

*Available varieties.* Vegetative varieties include Bonnie Brae, Buffalawn, Buffalo (609), Density, Eco, Legacy, Midget, Oasis, Prairie, Prestige, Scout, Sapphire, Stampede, Turffalo, and UC Verde. They have better density, are lower growing, and have more competitive turf-type characteristics compared to older varieties. These were mostly bred and perform quite well in the arid southwestern United States, but have shown varying potential to withstand humid conditions elsewhere. However, as newer cultivars are developed, a wider use range of buffalograss may occur. Seeded varieties include Bison, Bowie, Cody, Comanche, Frontier, Hays, Palmetto, Plains, Sharp's Improved, Sundancer, Tatanka, Texoka, and Topgun.

In general, cultivars that are diploids (have 20 chromosomes) and most tetraploids (40 chromosomes) lack cold tolerance and are better adapted to southern growing areas. Most intermediate (or pentaploid) types with 50 chromosomes and hexaploids (60 chromosomes) have better cold tolerance. Prairie and 609 are more susceptible to winter-kill, while Legacy, Prestige and Turffalo have better winter hardiness. The intermediate types

**TABLE 1.17**
**Commonly Used Ornamental Grasses and Sedges in Golf Course Landscapes**

| Latin Name | Common Name(s) | General Descriptions | Potential Uses |
|---|---|---|---|
| | | **Cool-Season Grasses** | |
| *Acorus* sp. | Sweet flag | • Cool season | • Good shade tolerance and for water gardens |
| *Calamagrostis* x *acutiflora 'Karl Foerster'* | Feather reedgrass | • Cool season; clump forming; erect; flowering in summer | • Provides excellent contrast to low shrubs; produces attractive bronze-colored inflorescence |
| *Deschampsia caespitosa* | Tufted hairgrass | • Cool season | • Low maintenance, use in mid- or foreground bed, plant in front of dark background; good shade tolerance |
| *Festuca glauca* (*F. amethystine, F. cinerea, F. ovina, F. ovina* var. *glauca*) | Blue fescue | • Cool season, clump forming, forms low-growing ground covering tufts | • Accent plant, foreground bed, ground cover |
| *Helictotrichon sempervirens* | Blue oatgrass | • Cool season, moderate size, clump forming | • Blue-gray foliage, tolerates dry sites, use in the foreground of beds, specimen, ground cover |
| *Leymus arenarius* Glaucus | Lymegrass, dunegrass | • Cool season, stoloniferous | • Useful in large, open areas as a groundcover or for erosion-control perennial borders, dune plantings |
| *Phalaris arundinacea* Picta | Ribbongrass, gardener's garters | • Cool season, stoloniferous | • Spreads rapidly, useful in masses, as a groundcover, or for erosion control in shaded areas |
| *Seslaria caerulea* | Blue moorgrass | • Cool season, clump forming, forms low-growing ground covering tufts | – |
| | | **Warm-Season Grasses** | |
| *Andropogon gerardii* | Big bluestem, turkeyfoot | • Warm season, tall (5 to 8 feet, 1.5 to 2.4 m), native to central United States | • Mass planting, mid- or background bed, screen, low maintenance; full sun to light shade |
| *Andropogon* or *Schizachyrium scoparium* | Little bluestem | • Warm season, low (1 to 3 feet, 0.3 to 0.9 m) growth habit, short rhizomes | • Good fall color, for dry, sand, and rocky soils in full sun; not for formal gardens, groundcover, mid- or foreground bed |
| *Andropogon virginicus* | Broomsedge | • Weak clump-forming warm-season perennial; native to United States. | • Blue-green spring and summer color, orange winter foliage, good for acidic, nutrient-poor soils; full sun |
| *Arundinaria pygmaea* | Dwarf bamboo | • Spreading perennial | • Good groundcover for shade |
| *Aruno donax* | Reedgrass, giant reed, giant cane, carrizo | • Warm season, tall, coarse textured, clump forming | • Good accent plant, requires space, good near water's edge or for poor soils |
| *Bouteloua curtipendula* | Side oats gramagrass | • Warm season, low (1 to 3 feet, 0.3 to 0.9 m) growth habit | • Good ornamental seedheads; good in dry soils and heavy foot traffic; full sun |
| *Carex* and *Cyperus* sp. | Sedge | • Warm season, clump forming, not cold or drought hardy | • Annuals for borders or mixed with container plants; good shade and moist site tolerance |
| *Chasmanthium latifolium* | Northern sea oat, wild oats, upland oats | • Warm season, clump forming, native to United States, heavy seed producer | • Handsome bamboo-like quality, specimen, moderately shade tolerant, produces glittering seedheads in fall |
| *Cortaderia selloana* | Pampasgrass | • Warm-season perennial, large, upright, clump forming | • Excellent specimen in larger areas; flowers late summer |
| *Eragrostis* spp. | Lovegrass | • Warm-season perennial, clump forming, thin bladed | • Excellent drought tolerance; for dry, sandy sites |
| *Miscanthus sinensis* | Miscanthus, Chinese silvergrass, eulalia grass | • Warm season, numerous varieties available; tall or short growing, clump forming, good foliage color | • Good cold tolerance, variegated cultivars available, good fall color, site in masses, as specimens in a hedge or screen, in containers |

*(Continued)*

**TABLE 1.17 (CONTINUED)**
**Commonly Used Ornamental Grasses and Sedges in Golf Course Landscapes**

| Latin name | Common Name(s) | General Descriptions | Potential Uses |
|---|---|---|---|
| *Panicum virgatum* | Switchgrass, panicgrass | • Warm season, tall (6 feet, 1.8 m), clump forming, native to United States | • Back of borders in sunny areas, perennial beds, erosion control, moist sites, mass plantings, not for refined landscapes; full sun to light shade |
| *Pennisetum* spp. | Fountaingrass | • Warm season, numerous varieties available, clump forming, purple flowers | • Good for low accent in annual or perennial borders or in containers, building foundations, use in masses, groups, or as single specimen |
| *Saccharum ravennae* | Ravennagrass, plumegrass | • Warm season, tall, clump forming | • Good accent plant, near water, tall screen or windbreak |
| *Sorghastrum nutans* | Indiangrass | • Warm season, tall (5 to 8 feet, 1.5 to 2.4 m) clump forming, native to United States | • Specimen or in masses, screen, windbreak; full sun to light shade |
| *Spartina* spp. | Prairie cordgrass | • Warm season, tall (5 to 8 feet, 1.5 to 2.4 m), rhizomatous, native to United States | • Large open areas as a groundcover or for erosion control; good for moist sites mid- or background bed; full sun |
| *Sporobolus heterolepis* | Prairie dropseed | • Warm season, low (1 to 3 feet, 0.3 to 0.9 m) growth habit, native to United States | • Good in dry, rocky sites; full sun |
| *Tripsacum dactyloides* | Eastern gamagrass, fakahatcheegrass | • Robust, warm season, native to United States | • Low salt tolerance; good in moist or dry sites; seed and rhizomatous |

tend to have a dwarf growth habit and less vigor. Tetraploid types are a lighter apple-green color, are more vigorous, and have more sod tensile strength than most diploids or hexaploids.

**Bluestems.** Big bluestem (*Andropogon* or *Schizachyrium gerardii* Vitman) grows up to 6 feet (1.8 m) high at maturity (Figure 1.47). It has strong, deep roots and short rhizomes (Table 1.17). Seeds are planted in midspring at 5 to 20 pounds/acre (5.6 to 22 kg/ha) of pure live seed. Seeds mature in late summer. Big bluestem will not tolerate continued close mowing. It is good as a strong accent plant and as a backdrop for shorter plants, and turns bright red in fall.

Little bluestem (*Andropogon scoparius* Michx.) reaches a mature height of about 4 feet (1.2 m). It is more drought resistant than big bluestem and should be used in sandy sites. It produces a noticeable crimson fall color and is planted in midspring at 8 to 12 pounds/acre (9 to 13 kg/ha).

Broomsedge (*Andropogon virginicus* L.) is sometimes a weed species, and is a bunchgrass that grows on low-fertility acidic (pH 4.0) soils (Figure 1.48). In contrast to the other bluestems, broomsedge has a shallow root system, seedheads are partially enclosed in the sheath, and mature plants have a distinct brownish-orange color, although some variants are silver. Seedheads may reach 4 feet (1.2 m) tall. It is planted in midspring at 8 to 12 pounds/acre (9 to 13 kg/ha). Other bluestems include sand bluestem (*Andropogon hallii* Hack.), yellow bluestem (*Bothriochloa ischaemum* Keng.), and Old World or Caucasian bluestem (*Bothriochloa caucasica* C. E. Hubb.).

**FIGURE 1.47** Bluestems as accent plants in low-maintenance areas.

**FIGURE 1.48** Broomsedge is a bunchgrass adapted to low-fertility, acidic soils.

**Cordgrass.** Prairie cordgrass (*Spartina pectinata* Link.) is a relatively tall-growing (4 to 6 feet, 1.2 to 1.8 m) grass grown for its long, graceful leaves and ability to grow in relatively moist sites, where it is often used to stabilize streambanks, shorelines, and other wet soil sites (Figure 1.49).

**Dropseed.** Prairie dropseed (*Sporobulus heterolepis* A. Gray) is a short grass 1 to 3 feet tall (0.3 to 0.9 m). It is an elegant growing grass with fountain-like emerald green leaves

**FIGURE 1.49** Prairie cordgrass or *Spartina* as accent plants in low-maintenance areas.

**FIGURE 1.50** Prairie dropseed is a short, clumping grass with fountain-like emerald-green leaves.

originating from a tight clump (Figure 1.50). It is seeded in fall or early spring at 8 to 12 pounds/acre (9 to 13 kg/ha).

**Grama grasses.** Sideoats grama (*Bouteloua curtipendula* Michx. Torr.) is a perennial, warm-season, bunch-type plant native to the North American Great Plains; it rarely exceeds 3 feet (0.9 m). It has a rolled vernation and long flower stalks with short, dangling purplish spikes (Figure 1.51). Seeds mature in late summer. Due to its relatively tall growth habit, sideoats grama is better suited to infrequently mowed sites. It is used for regrassing dry sites such as rangelands, erosion control along highways, and in roughs and out-of-play areas on golf courses. Seeding rates range from 8 to 12 pounds/acre (9 to 13 kg/ha) in midspring. It is often planted with other prairie grasses such as buffalograss, bluestem, lovegrass, dropseed, and blue grama.

Blue grama (*Bouteloua gracilis* (HBK) Lag. ex Steud.) is a warm-season grass with a relatively fine leaf texture, weak rhizomatous growth habit, good heat and cold tolerance, and low fertility and irrigation needs. It also has a low fertility

**FIGURE 1.51** Sideoats grama is a perennial, warm-season bunch-type grass native to the U.S. Great Plains.

requirement, requires infrequent mowing, germinates and establishes quickly, and typically has few insect and disease problems. Wear and shade tolerance are poor, however. The fine, curling basal leaves have a distinctive grayish-green color and a folded vernation. It is adapted to higher pH soils (6.5 to 8.5) of warmer arid to semiarid plains regions and has few pests. It is best suited to low-maintenance, nonuse sites such as out-of-play areas and roadsides; it is often planted with buffalograss. Seeding rate is 1 to 3 pounds/1,000 square feet (4.9 to 15 g/m$^2$).

**Lovegrass.** The lovegrasses (*Eragrostis* spp.) are distinct bunchgrasses noted chiefly for their ability to grow on low-fertility, low pH, sandy or rocky soils (Figure 1.52). They are best suited to the southern transition zone of the United States and similar climates in the world. Seeding rates are 1 (roadsides) to 10 (golf courses) pounds/acre (1.1 to 11 kg/ha) timed in late spring. These reach a height of 1 to 1.5 feet (0.3 to 0.5 m). These perennial grasses have an extensive but shallow fibrous root system and eventually form a bunch about 2 feet (0.6 m) in diameter. Examples include weeping lovegrass [*E. curvula* (Schrad.) Nees.]; sand lovegrass [*E. trichodes* (Nutt.) Wood]; Boer lovegrass (*E. chloromelas* Steud.); and Lehmann lovegrass (*E. lehmanniana* Nees.). Boer lovegrass is a long-lived perennial bunchgrass, less winter hardy but more drought tolerant than weeping lovegrass. Lehmann lovegrass has prostrate stems, which root and produce new plants at the nodes; thus, it creeps. It is the least winter hardy of the lovegrasses. Sand lovegrass tolerates sandy, droughty soils.

**Switchgrass.** Switchgrass (*Panicum virgatum* L.) is a perennial tall, sod-forming grass native to the Great Plains and most of the eastern United States that reaches 4 to 6 feet (1.2 to 1.8 m) in height (Figure 1.53). It is a coarse-stemmed plant spreading slowly by short rhizomes, which form dense bunches or colonies. It is seeded at 5 to 20 pounds/acre (5.6 to 22 kg/ha) pure live seed in spring and forms a distinct yellow fall color.

**Indiangrass.** Indiangrass [*Sorghastrum nutans* (L.) Nash] is another native grass from the prairies of the eastern Great Plains and eastern United States. It is a perennial bunchgrass that spreads by long rhizomes. It is deep-rooted and grows 3 to 6 feet (0.9 to 1.8 m) in height. It produces distinct yellow panicles (plumes), which are particularly showy in fall (Figure 1.54). It is planted in spring at 6 to 10 pounds/acre (6.7 to 11 kg/ha) pure live seed.

**FIGURE 1.52** Lovegrasses are bunchgrasses able to grow on low-fertility, low-pH, sandy, or rocky sites.

**FIGURE 1.53** Switchgrass is a perennial tall, sod-forming grass with a distinct yellow fall color.

## Cool-Season Grasses

**Crested wheatgrass.** Crested wheatgrass (*Agropyron cristatum* (L.) Gaertn.), also called fairway wheatgrass, is best adapted to cooler areas of the semiarid to arid regions. It is also a low-maintenance species requiring little supplemental irrigation and fertilization (Table 1.17). It has a medium to coarse (2 to 8 mm) leaf texture, dark- to bright-green color, a bunch-type growth habit, rolled vernation, membranous ligules, pubescent leaf, and slow recuperative potential (Figure 1.55). Crested wheatgrass begins vigorous growth in early spring, often goes dormant during summer drought, and resumes growth in the fall with precipitation. It is used on nonirrigated lawns, fairways, and roadsides. Cultivars include Ephraim, Fairway, Hycrest II, Parkway, RoadCrest, and Ruff. Western wheatgrass (*Pascopyrum smithii*) is similar except for its short rhizomes. Newer cultivars are noted for rhizomatous growth. Seeding rate is 5 pounds/1,000 square feet (24 g/m$^2$). Desert wheatgrass (*Agropyron desertorum* L.) and tall wheatgrass (*Agropyron elongatum* L.) also are being evaluated as potential turfgrasses, especially in saline environments.

**FIGURE 1.54** Indiangrass is a native U.S. bunchgrass producing distinct yellow panicles (plumes) in fall.

**Prairie junegrass.** Prairie junegrass or crested hairgrass [*Koeleria macrantha* (Ledeb.) Schultes] is a perennial, cool-season, native grass to the United States that is used as a low-maintenance turf in the northern Great Plains (Figure 1.56). It has a slow vertical growth rate, moderate drought tolerance, fair to good salt tolerance (4 to 8 dS/m), and is adapted to dry, sandy soil. Barleria and Barkoel are two commercially available cultivars.

**Alkaligrass.** Alkaligrass (*Puccinella distans* [L.] Parl.) or "weeping alkaligrass," is a bunch-type, cool-season perennial grass with exceptional tolerance to high sodium and high pH (7 to 8) soils. It possesses a gray-green color, has a membranous ligule, rolled vernation, boat-shaped leaf tips, and parallel light lines to the blade's midrib resembling a bluegrass (*Poa*) species. It grows best in the cool-arid regions and along roads that have been treated with salt for snow and ice removal. Alkaligrass is often found growing in low, salt-laden areas where sodium tends to accumulate. Plants are leafy and leaves are narrow with a natural dark-green color. Seeding rates are 1 to 3 pounds/1,000 square feet (49 to 146 kg/ha), down to 10 pounds/acre (11 kg/ha) when used in soil conservation mixtures. Seed germination requires 14 to 21 days, and is often blended with fine fescue and bluegrasses in higher maintained turf situations. It is often crowded out by tall fescue, smooth bromegrass, or Kentucky bluegrass on neutral to acidic soils. Cultivars available include the Fults (or Fultz) series, Chaplin, and Salty.

**Miscellaneous grasses.** Several other miscellaneous low-maintenance cool-season grasses include rough hairgrass (*Agrostis scabra*); crested dogtail (*Cynosurus cristatus*); orchardgrass (*Dactylis glomerata* L.); timothy (*Phleum pratense* L.); tufted hairgrass (*Deschampsia caespitosa* (L.) Beauv.); junegrass (*Koeleria gracilis* (L.) Pers.); alpine fescue (*Festuca ovina* var. *brachyphylla*); blue fescue (*Festuca ovina*

**FIGURE 1.55** Crested wheatgrass is a cool-season bunchgrass with dark- to bright-green color.

**FIGURE 1.56** Prairie junegrass is a perennial, cool-season, U.S. native grass grown in the northern Great Plains.

ssp. *glauca*); meadow fescue (*Festuca elatior* L. or *F. pratensis* Huds.); and sheep fescue (*Festuca ovina* L.). Numerous *Poa* species are also being screened and bred for fine turfgrass purposes and include alpine bluegrass (*Poa alpina* L.); bulbous bluegrass (*P. bublosa* L.); creeping bluegrass (*P. reptans*); fowl bluegrass (*Poa palustris*); Texas bluegrass (*P. arachnifera* Torr.); wood bluegrass (*P. nemoralis* L.); and Canby bluegrass (*P. canbyi* (Scribn.) Piper), as well as interspecific hybrids among different *Poa* species such as Kentucky bluegrass and Texas bluegrass.

**Sand dune grasses.** Northern oceanside courses usually stabilize sand dunes with the cool-season grass American beachgrass (*Ammophila breviligulata* Fernald), while southern oceanside courses often use the warm-season grass sea-oats (*Uniola paniculata* L.). Saltgrass (*Distichlis* spp.) is a warm-season, rhizomatous, sod-forming dioecious grass adapted to highly salt-laden soils and brackish marshes. Its use as a low-maintenance turf may increase as the use of poorer (salty) irrigation sources also increase.

### Ornamental Grasses

With the desire to add unique color and textural characteristics to the golf landscape, a variety of ornamental grasses, sedges, and rushes are available. The foliage color can range from green to variegated yellow or white to red, with textures varying from fine to heavy and coarse. These are used as specimen or accent plants or as mass plantings (Figure 1.57). The height of these plants ranges from about 2 (0.6 m) to more than 10 feet (3 m). Some are stoloniferous while others have clump growth habits. Refer to Table 1.17 for an overview of some of the available plants for various landscapes. New grasses and other landscape plants are constantly being introduced. Attend conferences, field days, visit websites, and ask for referrals to keep up with the latest introductions and their merits.

**FIGURE 1.57** Ornamental grasses add unique color and textural characteristics to the landscape. Shown is feather reedgrass (*Calamagrostis* x *acutiflora*).

# 2 Turfgrass Physiology and Environmental Stresses

## PLANT CHARACTERISTICS

Flowering plants (also called angiosperms) are divided into two subclasses named for the number of cotyledons, or food storage organs, possessed by their seed embryo (Table 2.1). Dicotyledons, also known as **dicots**, have two storage organs, while monocotyledons, also known as monocots, have one. Monocots typically have long, two-ranked, thin leaves with parallel veins and scattered bundles of vascular tissue (Figure 2.1). Their floral parts (sepals, petals, and sex organs) are arranged in threes. Dicots, however, have leaves of various shapes, often wider than long, with a vein network (Figure 2.2). They often have showy flowers, as well as a continuous cylinder of vascular tissue in the stem. Vascular tissue of the stem is also arranged in a ring of bundles surrounding a central pith. Their flowers are mostly arranged in fours or fives. Perennial dicots also produce bark through secondary thickening of cambium cells. Monocot leaves are usually attached directly to the stem, while dicot leaves usually have short stalks called petioles. Monocot plants also lack a cambium, and thus cannot develop secondary thickening like dicot plants can. As a result, monocot plants do not grow as tall or strong as dicot plants such as trees (palm trees and bamboo are notable monocot exceptions). All turfgrasses are monocots, while most broadleaf weeds are dicots.

Cells are the small units that make up all organisms and their parts. Cells that are grouped and perform the same functions are tissues. These various tissues are the components of the plant's systems of structures. Cells are able to divide or enlarge in meristematic regions, which increases the size of the plant.

The vascular system in plants consists of xylem and phloem. These are tube-like tissues that connect belowground with aboveground plant structures and serve as a pipeline through which water ($H_2O$), nutrients, and carbohydrates move to various portions of the plant. Xylem carries $H_2O$ and nutrients absorbed by the roots and plant growth regulators (hormones) produced in roots to the stems and leaves. The driving force for this movement is transpiration—the loss of $H_2O$ to the atmosphere through evaporation at the leaf surface. This loss creates a tension that "pulls" the $H_2O$ and dissolved nutrients in the xylem upward into the plant. Food and plant growth regulators manufactured by photosynthesis in leaves travel down in the plant via the phloem system. Monocot xylem and phloem tissue are primary, meaning they never increase and are not replaced after they form during the stem's initial development.

### SEED

Grasses are either annuals (live one year or less) or perennials (live more than two years). Turfgrasses are most often grown as perennials, meaning the plant lives and continually replaces all of its components during the course of two or more years. Turf management involves ensuring this replacement growth occurs at the rate needed by losses due to natural senescence, injury, or disease.

Most grasses are herbaceous (or nonwoody) plants. The seed of a grass is actually a fruit known as a caryopsis. The caryopsis contains the true seed, which typically has a relatively elaborate embryo situated at one end next to its food source, the endosperm. The embryonic shoot and root axes are found at the center of the scutellum. When germination begins, the caryopsis absorbs $H_2O$ and swells. If seed dries following this, viability is significantly lowered. The first indication of growth is appearance of the primary root, the radicle, which pushes through its protective organ, the coleorhiza. The developing shoot then expands within a protective modified leaf, the coleoptile, which splits through the caryopsis. The first leaf then elongates through a pore at the apex of the coleoptile. After entering light, new leaves develop from the apical meristem. Chlorophyll is then synthesized and photosynthesis begins.

The primary root lives only one to two months and is not the main absorbing system of grass plants. Water and nutrient absorption is performed by adventitious roots, which arise from the lowermost nodes of the first shoot and tillers. In bunchgrasses, adventitious roots usually originate from basal nodes of the main stem or from tillers near ground level. In sod-forming grasses, adventitious roots also may develop from nodes of stolons or rhizomes (Figure 2.3).

Requirements for germination vary between grasses, but all require $H_2O$, favorable temperatures, and $O_2$. Light is essential for seed of certain grass species (e.g., goosegrass, crabgrass, and bluegrass). Optimum germination of cool-season grasses occurs when air temperatures are between 60°F and 80°F (16°C and 27°C), while warm-season grasses require 70°F to 95°F (21°C to 35°C). Seed viability and longevity increase when stored in cool (<80°F, 27°C), dry (<20% relative humidity) conditions. Bermudagrass seed is generally short lived (less than or equal to one year), fescue and ryegrass seed are intermediate, and bentgrass and buffalograss seed are considered long lived (greater than two years).

Certain grass seeds have a hard, impermeable coat or hull for survival purposes. Germination generally increases if this

**TABLE 2.1**
**Distinguishing Characteristics of Monocots Compared to Dicots**

| Characteristic | Monocots | Dicots |
|---|---|---|
| Seed cotyledons | One | Two |
| Leaf venation | Parallel | Netted |
| Ligules | Present, rarely absent | Absent |
| Vascular bundles | Scattered | Distinct |
| Vascular tissue growth | Primary | Secondary |
| Meristems | Basal | Terminal |
| Root system | Fibrous without cambium layer | Tap root with a cambium layer |
| Flowers | Not showy | Showy |
| Flower parts | Group of threes | Group of fours or fives |
| Members | Grasses, sedges, and rushes | Most broadleaf plants |

**FIGURE 2.1** Grass leaves (monocots) typically have long thin leaves with parallel veins.

**FIGURE 2.2** Dicot leaves have various shapes, often wider than long, with a vein-like network.

coat is removed (scarified) before planting, as seen in bermudagrass, buffalograss, zoysiagrass, and bahiagrass.

## Crowns

Crowns are, arguably, the most important organ in grasses. The crown is the primary meristematic tissue or growth zone for cell division and enlargement of established plants and is located at the base of the plant near the soil surface where leaves, roots, and stems join (Figure 2.3). All new leaf, root, and stem growth originates at the crown. Crowns consist of a stem apex, unelongated (or highly compacted) internodes, axillary buds, and nodes, and they are generally very short (1/8 inch, 3.2 mm). Adventitious roots initiate from the lower nodes of crowns while lateral shoots and leaves initiate from the upper portion where the apical meristem is located. The crown is the primary meristematic tissue and as long as it is not damaged, grasses can recover from environmental stresses, dormancy, excessive defoliation, and pest damage. Healthy crowns are white and turgid (filled with $H_2O$) while damaged crowns turn brown and dry out.

Grasses are well adapted to mowing since leaf formation continues after each defoliation. This occurs because the primary growing point, the crown, is located at the base of the plant, close to the soil surface and below mower blades. Also, grass internodes do not normally elongate, except when flowering. Keeping crown tissue alive is a key to growing plants.

## Leaves

The organs of grasses are the shoots (stems plus leaves, referred to as primary shoot) and roots. As the plant develops from germinating seed, the shoot (or culm) becomes apparent. At this stage, the shoot consists of a series of concentric leaves, with the oldest on the outside and younger ones forming in the center, pushing upwards until they emerge.

As seedlings develop, emerging leaves show variations in structure, which become important for species identification. Leaf width, shape, hairs, vernation, collars, ligules, leaf tips, and so on all aid in the identification of turfgrasses.

Leaves consist of a sheath, blade, and ligule. The sheath is the lower portion of the leaf, which is wrapped around the

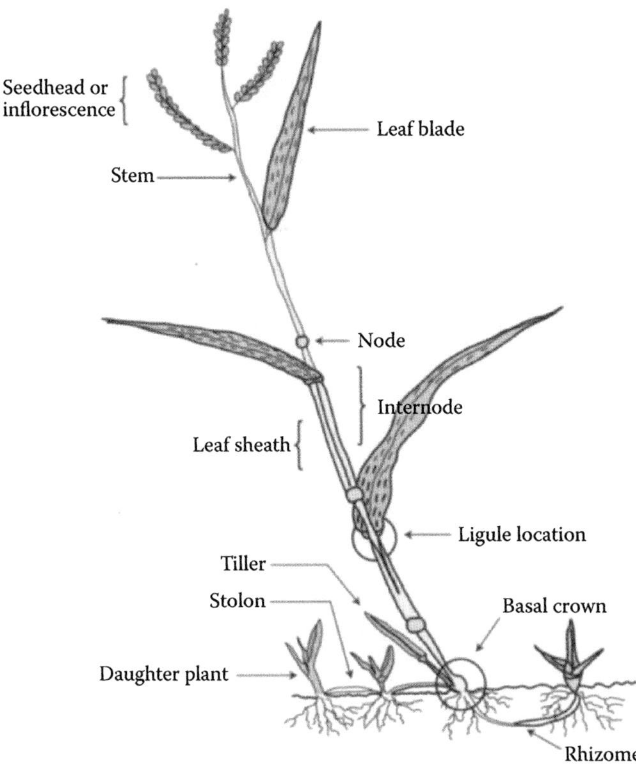


**FIGURE 2.3** Primary morphological parts of a sod-forming grass.

shoot above the node and from which the leaf bud emerges (Figure 2.3). Blades (also called lamina) are parallel veined and typically flat, long, and narrow. Leaves on mature plants are borne on the stem, alternately in two rows, one at each node. A large vein, the midrib, extends through the middle of the blade and lesser veins run parallel on each side of it. In some grasses, the edge of the blade feels rough (or scabrous) when rubbed. The blade may have hairs on either the upper or lower surface or both.

Vernation is the arrangement of the youngest leaf protruding from the sheath of an older leaf in the bud shoot and is either rolled or folded (Figure 2.4). Turfgrasses with rolled vernation include annual ryegrass, buffalograss, creeping bentgrass, and zoysiagrass. Bermudagrass, perennial ryegrass, bluegrass, and St. Augustinegrass have folded vernation.

Leaf tip shapes also aid in separating and identifying certain turfgrasses. Pointed leaf tips are the most common for turfgrasses (Figure 2.5). Examples include bahiagrass, ryegrass, fescue, zoysiagrass, and bermudagrass. Blunted or rounded leaf tips are associated with carpetgrass, centipedegrass, and St. Augustinegrass. Boat-shaped leaf tips are

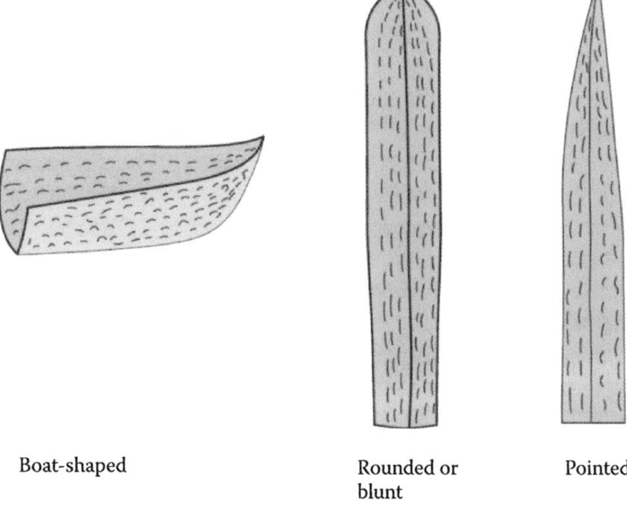


**FIGURE 2.5** Turfgrasses' leaf tip shapes typically are boat shaped, rounded, or pointed.

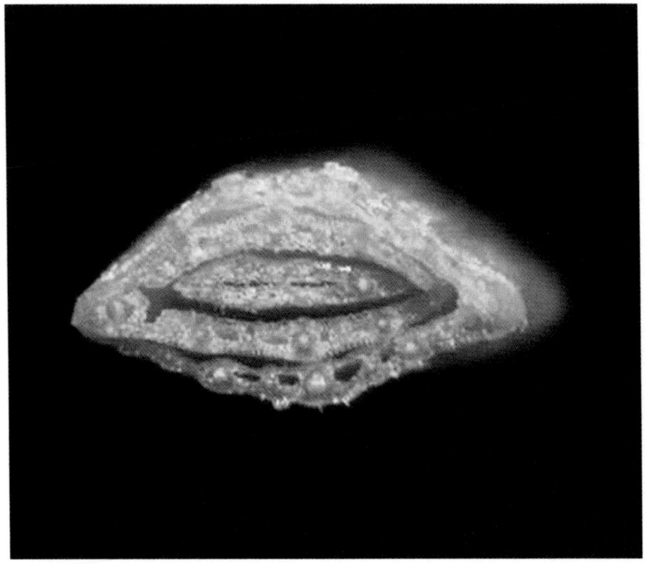

**FIGURE 2.4** Grass leaf vernation is either folded (left) or rolled (right).

associated with the bluegrasses and centipedegrass. Intervals between leaf junctions are internodes.

As leaves mature and die, they are replaced by new ones developing within the sheath of the next oldest leaf. A specialized intercalary (separated) meristem is located between two previously differentiated tissues of certain organs such as between the leaf blade and sheath or between a node and an internode. Intercalary meristematic tissue called leaf primordia located at the leaf base is the site of new leaf initiation. Consequently, new growth occurs at the base of the sheath just above the ligule, and is the youngest portion of the leaf, while the oldest portion is at the tip of the blade. Since the leaf tip is physiologically more mature, it is the first part to senesce when the leaf dies. The new leaf emerges at the top of the plant.

The collar region is where the leaf blade and sheath join, and contains the collar, auricles, and ligule. The ligule, one of the most important diagnostic features of grasses, is a projection or extension from the top of the leaf sheath where the leaf sheath and blade join. It can vary considerably in its texture, size, and shape (Figure 2.6). Most commonly, the ligule is either a membranous structure, a fringe of hairs, or absent. Auricles are appendages extending from the edge of the leaf where the blade and sheath meet (Figure 2.7). Auricles are claw-like, short, or absent. The development of these appendages marks the end of elongation growth, at which point the leaf has reached its final length. Meanwhile, the next leaf is moving up inside the previous leaf's sheath.

Photosynthesis, or food production, occurs in leaf cells with the presence of chlorophyll (Figure 2.8). The outer

**FIGURE 2.7** Auricles are appendages extending from the leaf and can be claw-like (shown), short, or absent.

**FIGURE 2.6** The ligule is a projection from the top of the leaf sheath and, if present, is often membranous (left) or hairy (right).

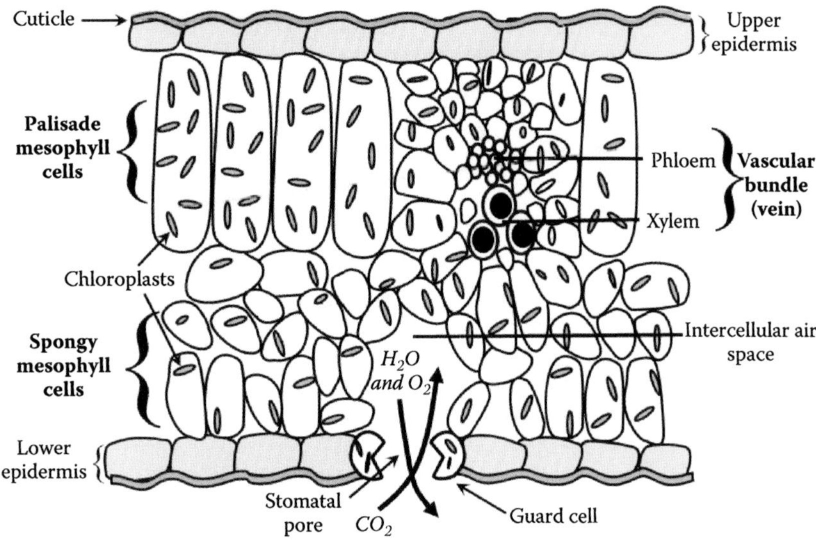


**FIGURE 2.8** Grass leaf cross section with chloroplast-containing mesophyll cells where photosynthesis occurs.

layer of cells in leaves is the epidermis and often is coated with a thin waterproof wax covering called the cuticle. The cuticle serves, in part, as a sealant to prevent excess $H_2O$ loss from the leaf. Openings in the leaf epidermis, called stomata, allow gas exchange in the plant. Most commonly, carbon dioxide ($CO_2$) is absorbed during photosynthesis, $O_2$ is exchanged during respiration, and $H_2O$ vapor is lost during transpiration. Stomata close at night since photosynthesis is not occurring. During periods of drought, stomata also close to prevent $H_2O$ vapor loss. This closure also stops photosynthesis since the necessary gases ($CO_2$) for this process cannot be absorbed.

The mesophyll consists of cells between the upper and lower epidermal layers. Except for veins, mesophyll cells contain chlorophyll. Typically, mesophyll cells located at the top are tall and cylindrical, forming the palisade layer. Cells below the palisade layer are irregular in shape and have large spaces between them. They form a spongy layer in which $CO_2$ and $O_2$ move.

Leaf number, size, and emergence rate are governed by numerous factors including the plant's genetic makeup and exposure to certain environmental parameters. Turfgrass leaf life expectancy is approximately one to two months. Turf managers, therefore, need to stimulate new leaf development to replace older, brown-colored (termed necrotic) dying leaves.

Temperature and light are the environmental factors with the greatest influence on leaf appearance, assuming other parameters such as soil moisture, soil $O_2$, and rootzone area are not limiting. Generally, warmer temperatures, higher light intensities, and longer day lengths promote leaf emergence, especially with warm-season grasses. If either becomes limiting (e.g., too high or low; too long or short), the corresponding leaf appearance rate is influenced. Nitrogen (N) is the nutrient with the greatest influence on leaf growth rate. Nitrogen application stimulates new, darker-green leaves but not the greening of older leaves. Increasing N fertilization also tends to increase leaf area.

Leaf size is also influenced by the environment. Again, temperature, light intensity, and nutrient supply are the most important factors. In general, leaves tend to be shorter and wider during cool temperatures, and longer and narrower when it is warm.

Reduced light intensity (e.g., shade) usually causes grass leaves to become wider and longer. Leaf thickness and corresponding weight decline; thus, under shade, grass leaves stretch and widen, but are low in weight and lack rigidity. The leaf under low-light conditions develops this larger surface area in an attempt to capture as much filtered sunlight as possible.

## Stems

Stems are the basic structural feature from which turfgrasses develop and are the connecting structures between leaves and roots. Stems consist of internodes spaced between nodes with attached leaves. Grasses are composed of three different types of stems: crowns, flowering culm (or seedhead), and lateral stems (rhizomes, stolons, and tillers). The jointed stem of a grass is distinctly divided into a series of nodes and internodes and is terminated by an apical bud. Internodes are usually cylindrical and hollow, while nodes are solid. Successive leaves are initiated by the swollen meristematic stem apex at the node, which in most species remains short during vegetative development, so leaves arise close together. Since they originate from the same point, the number of nodes and internodes equals the number of leaves.

Most species delay extended stem development until shortly before flowering. When stem development does occur, the structure is fairly long and narrow with elongated internodes separated by solid nodes. Nodes generally appear as slight swellings of the stem. Stems are either solid or hollow, but typically are hollow in turfgrasses. In solid stems, the central parenchyma provides storage for carbohydrates, which may be used later during plant development. Modified stems and leaves make up the inflorescence and fruits.

Alternately appearing on either side of the crown is a series of axillary buds that give rise to lateral stems, such as tillers, rhizomes, and stolons. Tillers, or primary lateral shoots, develop when lateral stems grow up within the leaf lying under the node. As the leaf sheath develops from the crown, it appears to wrap completely around the crown, including the axillary buds. Therefore, developing axillary buds must grow either through or within the leaf sheath. When these are retained within the surrounding leaf sheath and grow upright, tillering is referred to as intravaginal and produces a tufted or bunch-type (noncreeping) growth habit. Ryegrass, colonial bentgrass, and fescue have bunch-type growth habits, spread very slowly, and tend to grow in clumps or bunches.

Extravaginal tillering (or shoot development) occurs where the lateral stem elongates and penetrates through the side of the surrounding leaf sheath to produce a spreading or creeping growth habit. Extravaginal tillers form either rhizomes (belowground stems) or stolons (aboveground stems that possess fully developed leaves) and are referred to as secondary lateral shoots. Because they are stems, rhizomes and stolons possess nodes, internodes, and an apical bud. Both can root and produce new shoots at the nodes, permitting a single plant to spread over a wide region. Rhizomes and stolons serve as major storage areas for reserve carbohydrates and provide sod strength and recuperative ability. Creeping red fescue and Kentucky bluegrass are examples of rhizomatous turfgrasses.

Unlike roots, which grow by adding cells at their tips, rhizomes grow by intercalary meristematic activity in the vicinity of stem nodes. The resulting elongation of stem internodes is partly responsible for pushing the rhizome tip through soil. Rhizomes grow faster under long days, high light, and low N levels.

Stolons resemble rhizomes in that they have definite nodes, internodes, and nodal meristems from which secondary structures arise. They are more stem-like than rhizomes since they creep above ground and their leaves develop and function normally (Figure 2.3). St. Augustinegrass, creeping bentgrass, centipedegrass, and *Poa trivialis* are examples of stoloniferous turfgrasses. Some turfgrasses, such as bermudagrass and zoysiagrass, produce both rhizomes and stolons.

Tillering is favored by high light intensity, frequent mowing, and moderate N levels, and is generally greatest for cool-season grasses in the spring and fall (Table 2.2). Conversely, as light is reduced from 100% intensity, tiller production also declines. Shade, therefore, discourages tiller formation.

Tiller production also is influenced by mowing height and frequency. Physiological control of tillering is influenced by the concentration of the internal growth hormone, auxin. When sufficient concentrations of auxin are produced in the apical meristem and expanding leaves, tillering is inhibited, a process referred to as apical dominance. Tillering is stimulated when the auxin in shoot tips is removed by mowing. Thus, frequently mowing golf turfgrasses at less than or equal to 1 inch (2.5 cm), for example, encourages tiller production. However, if mowed less frequently above this height, auxin accumulates at the leaf tips, tillering is suppressed, and the plant uses its carbohydrate reserves to form new leaves at the expense of tiller production.

**TABLE 2.2**
**Parameters Influencing Tillering in Turfgrasses**

| Parameter | Tillering Response |
|---|---|
| Temperature | • Temperatures of 50°F to 60°F (10°C to 16°C) are optimum for cool-season grasses and of 80°F to 90°F (27°C to 32°C) for warm-season grasses. |
| Light intensity | • As light intensity is reduced (or shade increases), tillering is also reduced. |
| Nitrogen | • Moderate N levels encourage tillering while high levels may favor shoot growth over tillering and root growth. |
| Mowing | • Proper mowing height and frequency remove the tillering-inhibiting hormone, auxin; thus, tillering is promoted. |
| Day length | • Longer days encourage tillering. |
| Moisture | • Tillering decreases as soil moisture is reduced from field capacity. |

## Roots

Roots anchor plants to soil and take up $H_2O$ and nutrients. They also function in food storage and are the primary source of certain plant growth regulators, such as gibberellins and cytokinins.

Turfgrasses have fibrous, branched root systems, mostly located in the upper foot of soil (Figure 2.9). The primary (also called seminal) roots arise from the root tissues of the

**FIGURE 2.9** Turfgrass roots are fibrous and branched, mostly located in the upper portion of the soil.

seed embryo and generally persist for only a short time (one to two months) after germination. The primary root is the first structure to emerge from the embryo. Secondary (also called adventitious) roots arise two to three weeks after germination at the lower nodes just below the internodal intercalary meristem of young stems and comprise the major part of the permanent root system. The root system is progressively replaced by adventitious roots, which arise at nodes of creeping stems (stolons) and lower plant crowns, and from older roots.

The root cap, the tips of roots, is a thimble-shaped group of protective cells for the meristematic region located just behind it. The cap minimizes damage to this meristematic region as the root grows (or pushes) through soil. The meristematic region behind the root cap is the site of active cell growth. Here cells divide and elongate, pushing the root through soil. Root hairs are tiny projections on the outer surface of roots that increase the surface area of the root system. Plant $H_2O$ and nutrient uptake occur here. Unlike leaves, roots have a growing point at their tips that is capable of branching into soil air spaces at every opportunity. However, grass roots lack a cambium for secondary thickening.

The life expectancy for most turfgrass roots is less than one year. Roots of cool-season grasses grow prolifically in fall and winter months when soil temperatures are about 55°F (13°C). Roots begin to decline in the spring when flowering occurs and eventually die or are severely restricted during periods of summer heat and drought (Table 2.3). Creeping bentgrass, for example, ceases root initiation when soil temperatures exceed 75°F (24°C). Roots of cool-season grasses are smaller in diameter and grow shallower than warm-season grasses. When maintained as turf, cool-season grasses such as bentgrass, *Poa trivialis*, and ryegrass rarely have roots greater than 1 foot deep; most often, they are much shorter.

**TABLE 2.3**
**Primary Parameters Influencing Turfgrass Rooting**

| Parameter | Rooting Response |
|---|---|
| Soil temperature | • Optimum soil temperature for cool-season grasses is 50°F to 65°F (10°C to 16°C) and 75°F to 85°F (24°C to 29°C) for warm-season grasses. Root growth stops at soil temperatures of 45°F (7°C) for warm-season grasses and above 90°F (32°C) for cool-season grasses. |
| Soil pH | • Soil pH outside 5.0 to 8.0 may limit root growth. |
| Plant age | • Young plants generally produce the most roots. |
| Moisture | • Moisture extremes (too dry or wet) discourage rooting. |
| Soil compaction/ soil $O_2$ levels | • Rooting is reduced due to physical impedance from soil compaction and low soil $O_2$ levels. Anaerobic conditions (without $O_2$) may lead to toxic sulfide, $CO_2$, and methane buildup. |
| Mowing height | • Higher mowing heights usually promote deeper, more extensive rooting systems. Cool-season grass rooting is especially sensitive to mowing height, being restricted as this is decreased. |
| Nutrient availability | • This is promoted with moderate N and adequate levels of K, P, Fe, and Ca. High N levels decrease root growth, especially with cool-season grasses. Using a balance of ammonium and nitrate N sources generally provides best rooting. |
| Light intensity | • Shade (reduced light levels) discourages deeper rooting. |

Roots of warm-season grasses begin growing in spring when soil temperatures at 4 inches (10 cm) reach ~64°F (18°C). Tillers, from nodes of stolons and rhizomes, break dormancy and begin growth. Carbohydrate reserves are converted to soluble sugars and new leaves begin to appear.

Roots of warm-season grasses have a major flush of growth in late spring, summer, and early fall months when soil temperatures reach about 80°F (27°C). Roots tend to store few energy reserves during summer due to their repeated use for regrowth following mowing. During late fall, plant-soluble sugars are converted to starch granules and stored in roots, stolons, and rhizomes during winter months. In early spring, when warm-season turfgrasses are greening-up from winter dormancy, these starch energy reserves are converted to soluble sugars for growth, and are depleted. Root systems may experience severe die-back with the appearance of new leaves. This die-back, called "spring root decline," lasts one to three weeks, during which time plants are susceptible to environmental and chemical stresses or from competition from overseeded cool-season grasses. New roots are initiated from crown, rhizomes, and stolons.

Low soil temperature will adversely affect bermudagrass rooting even if air temperatures are high. These conditions often exist in early spring and fall months. Similar rooting decreases occur if light levels are reduced, presumably due to less carbohydrate assimilation with lower photosynthesis. Complete green-up requires two to six weeks, depending on the temperature (faster green-up occurs with warmer temperatures).

If temperatures remain favorable for shoot growth during the spring transition period, normal green-up and recovery can proceed. If these favorable growth conditions, however, are interrupted by a late freeze, the young, tender green leaves may be killed and new leaves then must be regrown. If several intermittent freezes occur, an enormous drain is placed on the root energy reserves, often resulting in thin turf stands.

Age also influences root growth patterns. Young plants produce maximum root growth. Young plant roots are generally white, small in diameter, fibrous, and multibranched. As plants mature, their roots change in color from white to brown, become larger in diameter, and become less efficient in nutrient and $H_2O$ absorption. Environmental and biotic extremes negatively impacting turf rooting are temperatures outside the optimum range, soil compaction, low (<5.0) soil pH, low soil $O_2$, drought, and excessive moisture (Table 2.3). Cultural practices restricting rooting include improper mowing practices (height and frequency), excessive N and thatch levels, phosphorus (P) and potassium (K) deficiencies, and improper herbicide use. In general, the combination of continuous high soil moisture (frequent, light irrigation), close mowing, and high N fertility discourages rooting of most turfgrasses.

Plant growth also is interrelated between roots and shoots. Roots depend on shoots (leaves and stems) for carbohydrates produced during photosynthesis. Meanwhile, shoots depend on roots for $H_2O$ and nutrients. Typically, shoots have priority over roots for available carbohydrates, while seed development has priority over shoots and roots. Root growth also does not occur until shoot and seed demands are satisfied. Shoot growth is favored over rooting when (1) soil temperatures are above the optimum for root growth, (2) turf is mowed at close mowing heights, (3) excessive N is applied, and (4) conditions of low-light intensities exist, which reduce plant photosynthesis and thereby limit the carbohydrates available for root development.

**FIGURE 2.10** Turfgrass seedhead or inflorescence where seeds are formed.

### Inflorescence

The seedhead or inflorescence of a turfgrass is the reproductive organ where seeds are formed (Figure 2.10). Unlike other organs, inflorescences are not present throughout the life of the plant. They originate when a grass plant enters the reproductive stage and an elongated stem from the apical meristem of the crown is produced. Flowers and seed appear at the top of this elongated stem, called the flowering culm. The unit of the grass inflorescence is composed of a group of subunits called spikelets (Figure 2.11). Spikelets are composed of two bracts or reduced leaves called glumes, which enclose one or more florets. Florets consist of a lemma, palea, and the enclosed flower. Mature florets harvested from the inflorescence of flowering grass plants are commonly referred to as grass seed. The axis or branch of the spikelet is referred to as the rachilla. The spikelets usually are in groups or clusters, which constitute the inflorescence.

Several spikelet cluster types exist, including raceme, spike, and panicle (Figure 2.12). The simplest is the raceme where spikelets are borne on individual stalks (or pedicels) on an unbranched main axis. St. Augustinegrass, bahiagrass, zoysiagrass, and centipedegrass have raceme inflorescences. The spike differs from the raceme, since a spike has sessile (without a stalk or pedicel) spikelets on the main axis while racemes have simple stalked spikelets. Wheatgrass and ryegrass have the spike form of inflorescence (Tables 2.4 and 2.5).

The panicle is the most common type of grass inflorescence. In this case, spikelets are similar to the raceme because they are attached to the main axis (or rachis) but are in a branched inflorescence. Racemes have simple stalked spikelets. Some panicles are tightly branched while some are multibranched, resembling the limbs and leaves of trees. Most turfgrass inflorescences have the panicle arrangement, including bluegrass, creeping bentgrass, and tall fescue.

Seedhead formation in most turf situations is undesirable because it disrupts the uniformity of the turf stand (Figure 2.13) and becomes a visibility hazard, such as with bahiagrass or tall fescue seedhead formations along roadsides.

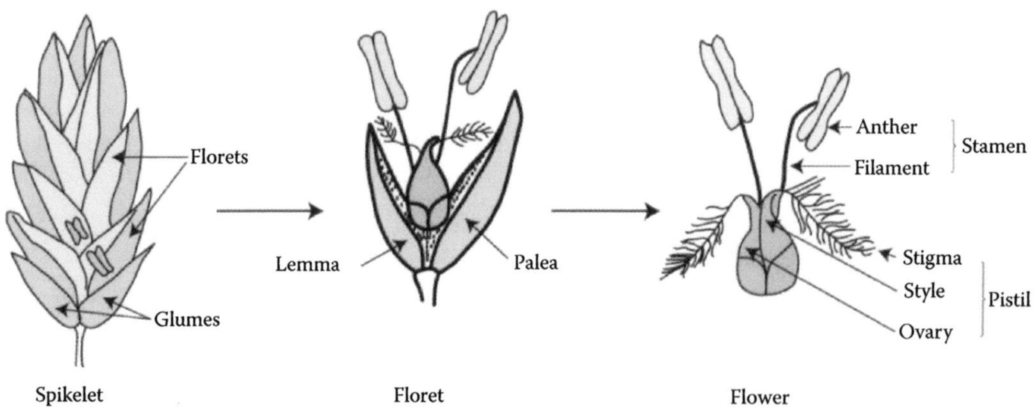


**FIGURE 2.11** The grass inflorescence is composed of subunits called spikelets composed of one or more florets which house flowers.

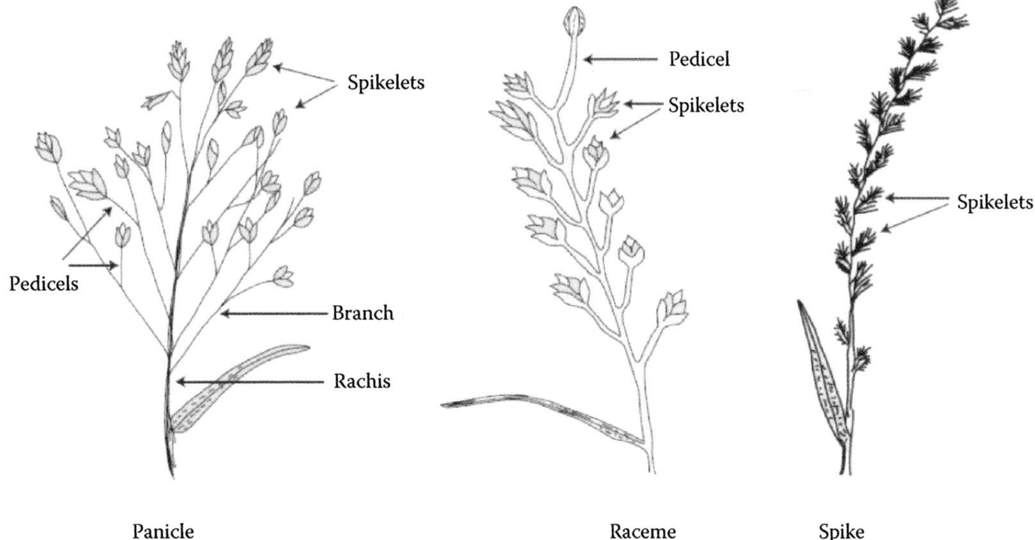


**FIGURE 2.12** Several grass spikelet clusters exist, including panicle, raceme, and spike.

In addition, seedhead formation occurs at the expense of vegetative growth since the plant uses its food and energy to form seedheads instead of producing leaves and roots. Therefore, flowering inhibits tiller production and occurs by using carbohydrate reserves located in roots.

Seedheads develop during the spring or summer, depending on the turfgrass species. For cool-season grasses, inflorescence development occurs predominantly in spring, while development occurs in the summer for warm-season species. Day length and temperature are the two major environmental factors regulating the timing of flowering for most turfgrass species. Long-day plants are those triggered to flower when a critical minimum day length is exceeded. They tend to flower in late spring through early summer when days are longest. Under shorter day lengths, long-day plants remain in the vegetative state. Short-day plants flower only when the day length is shorter than a critical maximum period. They tend to flower during the shorter days of late summer and fall. These plants remain in the vegetative state and do not flower when day length is long. Warm-season grasses tend to be short-day plants (bahiagrass being a notable exception) and flower when days are less than a critical length.

As mentioned, most cool-season grasses are long-day plants and flower only when the photoperiod exceeds a certain critical length. However, the effect of day length is highly modified by temperature. Cool-season grasses tend to have floral induction in fall in response to cooler (e.g., 32°F to 50°F, 0°C to 10°C) temperatures. Higher night temperatures (~54°F to 64°F, 12°C to 18°C), however, tend to delay or inhibit flower initiation in cool-season grasses. Conversely, short-day grasses (most warm-season grasses) often require night temperatures above 54°F to 61°F (12°C to 16°C) for floral initiation. Frequent mowing, adequate soil moisture, and increased N levels also discourage seedhead formation.

Day-neutral plants do not flower in response to day length and can flower as soon as they are mature. For example, annual bluegrass and, arguably, common bermudagrass are day-neutral plants and are insensitive to, or only weakly influenced by, photoperiod and produce seedheads over a wide range of photoperiods. Turfgrasses also develop seedheads when under environmental stresses such as drought (Figure 2.14).

## TURFGRASS PHYSIOLOGY

All plant physiological processes are interconnected as illustrated in Figure 2.15. The process of photosynthesis supplies plants with energy and biomass. During photosynthesis, leaves capture energy from the sunlight and store this energy in the chemical bonds of newly synthesized sugar molecules. Photosynthesis requires carbon dioxide ($CO_2$) from the atmosphere and water ($H_2O$) from the soil; the process eventually releases oxygen ($O_2$).

Overall chemical reaction of photosynthesis:

$$\underset{\text{carbon dioxide}}{6\,CO_2} + \underset{\text{water}}{6H_2O} \xrightarrow[\text{photosynthesis}]{\lambda(\text{light})} \underset{\text{oxygen}}{6O_2} + \underset{\text{carbohydrate}}{C_6H_{12}O_6}$$

Sugars (carbohydrates) formed in photosynthesis are transported throughout the plant, bringing carbon and energy to sites where they are needed. During respiration, the reversal of photosynthesis occurs; the energy contained in the photosynthetically produced sugars is transferred to adenosine triphosphate (ATP), a cellular "energy currency" that fuels many chemical reactions, including those involved in biosynthesis and mineral nutrient uptake. Sugars and mineral nutrients form the raw materials from which proteins, enzymes, nucleic acids, hormones, and all other chemical components of the plant are synthesized.

**TABLE 2.4**
**Distinguishing Morphological Characteristics of Warm-Season (or $C_4$) Turfgrasses**

| Turfgrass | Leaf Texture (width, mm) | Lateral Shoot | Leaf Vernation | Ligule Type (length, mm) | Auricles | Leaf Tip | Inflorescence | Additional Characteristics |
|---|---|---|---|---|---|---|---|---|
| Bahiagrass | Coarse (4–8) | V. short stolons, rhizomes | Folded or rolled | Membrane (1) | Absent | Pointed | Branched, spike-like racemes ("V" shaped) | • Tall, "V" shaped seedheads; wide leaf blade, hairy margins; reddish-purple stem base; drought and nematode resistant; low maintenance |
| Bermudagrass | Fine to coarse (1–3) | Stolons, rhizomes | Folded | Hair (1–3) | Absent | Pointed | Four or five digitate spike-like racemes | • Turf types with fine leaf texture and dense stand density; collar narrow, covered with hairs; some hairs on leaf surface; poor shade tolerance |
| Blue grama | Fine to medium (1–2) | Rhizomes | Rolled | Hair (0.1–0.5) | Absent | Pointed | Spike, spreading at maturity | • Blue-green color from hairy leaves; adapted to warmer regions of the arid transition zone; low maintenance |
| Buffalograss | Fine to medium (1–3) | Stolons | Rolled | Hair (0.5–1) | Absent | Pointed | Separate male and female flowers | • Collars hairy; blue-gray color due to leaf pubescence; drought resistant; poor shade tolerance; male flowers appear as curved branches at top of main stem; female flowers hard burs above leaf sheath |
| Carpetgrass | Medium to coarse (4–8) | Stolons | Folded | Fringe of hairs (1) | Absent | Rounded (blunt) | Raceme (usually three) | • Similar to St. Augustinegrass except for lighter green color leaves, which twist and have "waves" along their margins; seedheads resembling "crabgrass seedheads" with two to five spikes at their apex; tolerates wet conditions; collar narrow, continuous (indistinct) |
| Centipedegrass | Medium (3–5) | Stolons | Folded | Short membrane (0.5), hairs across top | Absent | Rounded (blunt) | Single spike-like solitary racemes | • Natural yellow-green color, medium textured leaves; hairs along edges; collar broad, much constricted, hairy; grows best at slightly acidic (5.0 to 6.0) soil pH |
| Kikuyugrass | Medium (4–5) | Stolons, rhizomes | Folded | Hair (2) | Absent | Pointed | Two to four spikelets | • Very tough grass with hairy leaves; poor shade tolerance; usually considered a weed; very difficult to eradicate once established; use cautiously |
| Seashore paspalum | Medium to coarse (3–8) | Stolons, rhizomes | Folded or rolled | Membrane (0.8–1.2) | Absent | Pointed | Spike-like racemes | • Good salt but poor cold tolerance; blue-green color; tolerates brackish water |
| St. Augustinegrass | Coarse (4–10) | Stolons | Folded | Hair (0.3) | Absent | Rounded, boat shaped | Single spike-like racemes | • Has long, thick stolons and wide leaf blades; cold tolerance poor to fair; good shade and salt tolerance |
| Zoysiagrass | Fine (2–3) to coarse (4–6) | Stolons, rhizomes | Rolled | Hair (0.2) | Absent | Pointed | Spike-like terminal racemes | • Stiff to the touch; collar covered with long hairs; hairy on leaf surface; dark-green, dense turf; good shade tolerance (cultivar dependent) |

**TABLE 2.5**
**Distinguishing Morphological Characteristics of Cool-Season (or $C_3$) Turfgrasses**

| Turfgrass | Leaf Texture (width, mm) | Lateral Shoot | Leaf Vernation | Ligule Type (length, mm) | Auricles | Leaf Tip | Inflorescence | Additional Characteristics |
|---|---|---|---|---|---|---|---|---|
| Annual bluegrass | Fine (0.5–2.2) | Tillers (annual biotype), short stolons (perennial biotype) | Folded | Membrane (1.5–2) | Absent | Boat shaped | Panicle | • Pale-green color; dull under leaf; prolific seedhead producer; transparent, parallel "light" lines on either side of the midrib; a major weed |
| Annual ryegrass | Medium to coarse (3–7) | Tillers | Rolled | Membrane (1–2) | Long, narrow, claw-like | Pointed | Spike | • Glossy under leaf; red stem base; lighter green color and coarser leaf texture vs. perennial ryegrass; seed (spikelets) awned |
| Colonial bentgrass | Fine (1–3) | Tillers, short stolons | Rolled | Membrane (0.4–1.2) | Absent | Pointed | Panicle | • Tufted (patch-like) growth habit; rhizomes and stolons either absent or short |
| Creeping bentgrass | Fine (1–3) | Stolons | Rolled | Membrane (1–3) | Absent | Pointed | Panicle | • Leaf blades rough along edges |
| Crested wheatgrass | Medium (2–5) | Tillers | Rolled | Membrane (0.5–1.5) | Claw-like | Pointed | Spike | • Prominent veins above and midrib below; blue-green color; adapted to cooler regions of arid transition zone |
| Fine fescue:<br>• Red fescue<br>• Chewings fescue | Fine (0.5–2) | Rhizomes, tillers | Folded | Membrane (0.2–0.5) | Absent | Pointed | Panicle | • Very fine, needlelike, leaf texture; red fescue with rhizomes, chewings without; additional fine fescues: hard, sheep, creeping, and meadow |
| Kentucky bluegrass | Medium (2–4) | Rhizomes, tillers | Folded | Membrane (0.5–2) | Absent | Boat shaped | Panicle | • Transparent, parallel "light" lines on either side of midrib |
| Orchardgrass | Coarse (2–10) | Tillers | Folded | Membrane very tall (3.5–7) | Absent | Pointed | Panicle | • Coarse leaf texture; pale blue-green color; prominent leaf midrib; seed contaminant in tall fescue |
| Perennial ryegrass | Medium (2–5) | Tillers | Folded | Membrane (0.5–1.5) | Short, nonclasping | Pointed | Spike | • Prominent veins and leaf midrib; very glossy under leaf; red stem base; seed (spikelets) unawned |
| Roughstalk bluegrass | Fine (1–6) | Short stolons | Folded | Membrane (2–6) | Absent | Boat shaped | Panicle | • Short creeping stolons; light-green leaf color; glossy under leaf; leaf margins rough |
| Tall fescue | Coarse (5–10) | Tillers, short rhizomes | Rolled | Membrane (0.2–0.8) | Rudimentary to absent | Pointed | Panicle | • Leaf blade edges rough; bunch-type growth; prominent midrib and veins; red stem base |
| Velvet bentgrass | Fine (<1) | Short stolons | Rolled | Membrane (0.4–0.8) | Absent | Pointed | Panicle | • Leaf blade edges rough |

FIGURE 2.13 In golf, seedhead formation is mostly undesirable as it disrupts turf stand uniformity.

FIGURE 2.14 Seedhead formation is influenced by environmental variables such as drought and day length.

## Photosynthesis

The reason why some plants, such as bentgrass, "prefer" cooler climates while others, such as bermudagrass, "prefer" warm climates was not fully understood until the 1950s and 1960s. During this time, researchers discovered the existence of two distinct photosynthetic pathways in plants. One, the Calvin–Benson or $C_3$ cycle, operates primarily in cool-season turfgrasses (Figure 2.16), while the other, the Hatch–Slack or $C_4$ cycle, operates primarily in warm-season turfgrasses (Figure 2.17).

Understanding how these two photosynthetic pathways differ in physiology and stress response allows superintendents to manipulate environmental conditions; this allows turfgrass species to grow in areas where they do not naturally occur.

The chemical reactions of photosynthesis can be divided into two phases. In the first phase, called the light reactions, the sun's light energy is captured and stored in the chemical bonds of two "energy transfer" molecules, ATP and NADPH. During this process, $H_2O$ is consumed and $O_2$ is released. The light reactions are identical in $C_3$ and $C_4$ turfgrasses. In the second phase, called the dark or carbon reactions, carbon from atmospheric $CO_2$ and energy from ATP and NADPH produced in the light reactions are used to form triose phosphates, the chemical precursors of simple sugars. The dark reactions proceed differently in $C_3$ and $C_4$ turfgrasses. Before examining this difference, leaf anatomy and photosynthesis common to all turfgrasses will be covered.

**Basic reaction of photosynthesis.** The simple reaction of photosynthesis shown earlier disguises the complex nature of the photosynthetic process, which involves numerous electron transfers, enzymatic reactions, and cellular compartments. On the left-hand side of the equation are the raw materials of photosynthesis: $CO_2$ and $H_2O$. On the right-hand side are the products: $O_2$ and a six-carbon sugar molecule such as glucose or fructose. The equilibrium constant for this equation

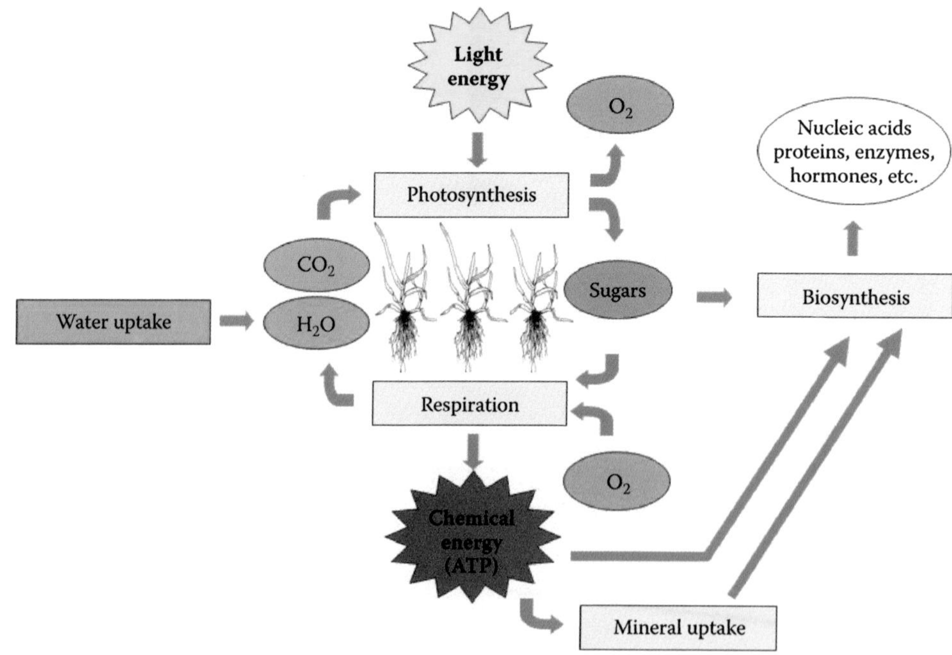


FIGURE 2.15 Illustration demonstrating how plant physiological processes are interrelated.

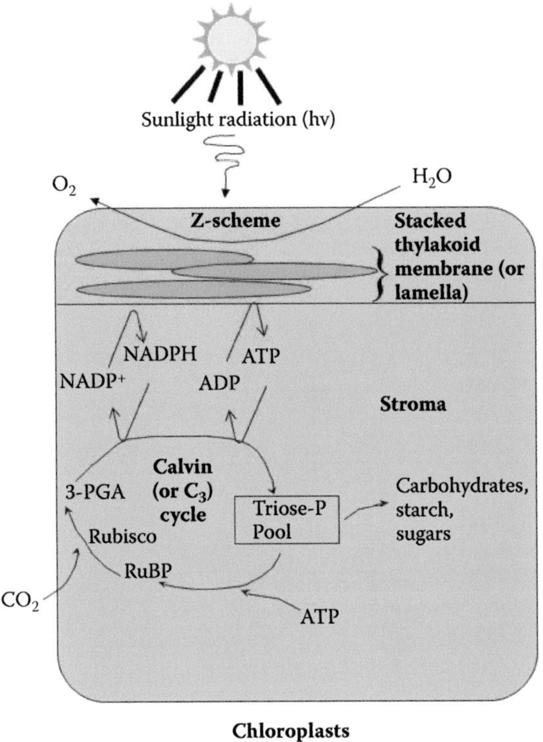


**FIGURE 2.16** Illustration of the Calvin (or $C_3$) cycle of photosynthesis in cool-season turfgrasses.

is approximately $10^{-500}$, meaning this reaction is impossible without a large input of energy. Light provides this energy.

**Nature of light.** Light is a form of electromagnetic radiation with the qualities of both a particle and a wave. One light particle is called a photon, and each photon contains a discrete quantity of energy called a quantum. The magnitude of a photon's energy is related to its wavelength, or the distance between adjacent wave peaks. The higher the wavelength of light is, the less energy is contained in its photons. The energy of blue light (430 nm) is "high," about 70 kcal/einstein, whereas the energy of red light (680 nm) is "low," about 40 kcal/einstein. Thus, x-rays, whose wavelengths are on the order of 0.1 nm, have a great deal of energy, while infrared light waves, whose wavelengths are on the order of $10^{12}$ nm, have relatively low energy (Figure 2.18) and are experienced by plants as heat.

The human eye can perceive light within the wavelength range of 400 nm (violet) to 700 nm (red). Photosynthesis uses light within the same range. For this reason, light between 400 and 700 nm is often referred to as photosynthetically active radiation, or PAR. Leaves can also absorb light at wavelengths above and below PAR, but this is not used for photosynthesis. High-energy, short-wavelength ultraviolet (UV) light can damage DNA, membranes, and other cellular components.

**Pigment composition.** Light energy is captured by pigment molecules embedded in the thylakoid membranes, including chlorophyll (chlorophyll a and b) and carotenoids. These pigment molecules absorb light in the 400 to 700 nm wavelength range in sufficient amounts to support photosynthesis. Chlorophyll molecules are green-colored pigments; carotenoids are light yellow to orange in color. Two kinds of carotenoids exist: red to orange carotenes and deeply yellow xanthophylls. Plants also contain other pigments such as the anthocyanins (red to blue colored). These additional pigments contribute to light harvesting in photosynthesis; provide various colors of flowers, fruits, and other plant parts; and possibly aid in plant defense mechanisms.

Each type of pigment in the thylakoid membrane absorbs a specific wavelength of light. Chlorophylls a and b absorb blue (430 nm) and red (680 nm) light and appear green in color. Their structure includes a tetrapyrrole ring (containing four pyrroles joined; see Figure 2.19) consisting of N, carbon (C), hydrogen (H), and oxygen ($O_2$) and containing conjugated unsaturated bonds with a magnesium (Mg) ion in the center and a long phytol tail. Carotenoids, including carotenes and xanthophylls, contain C, $O_2$, and H and include a long chain of

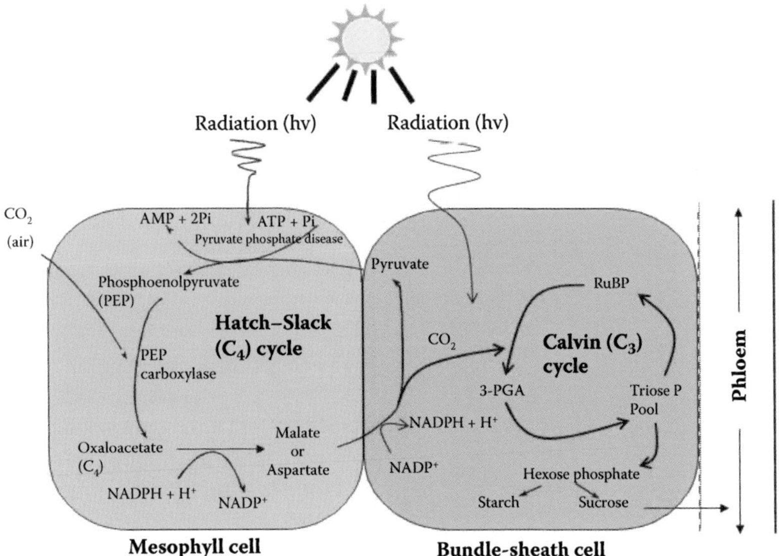


**FIGURE 2.17** Illustration of the Hatch–Slack (or $C_4$) cycle of photosynthesis in warm-season turfgrasses.

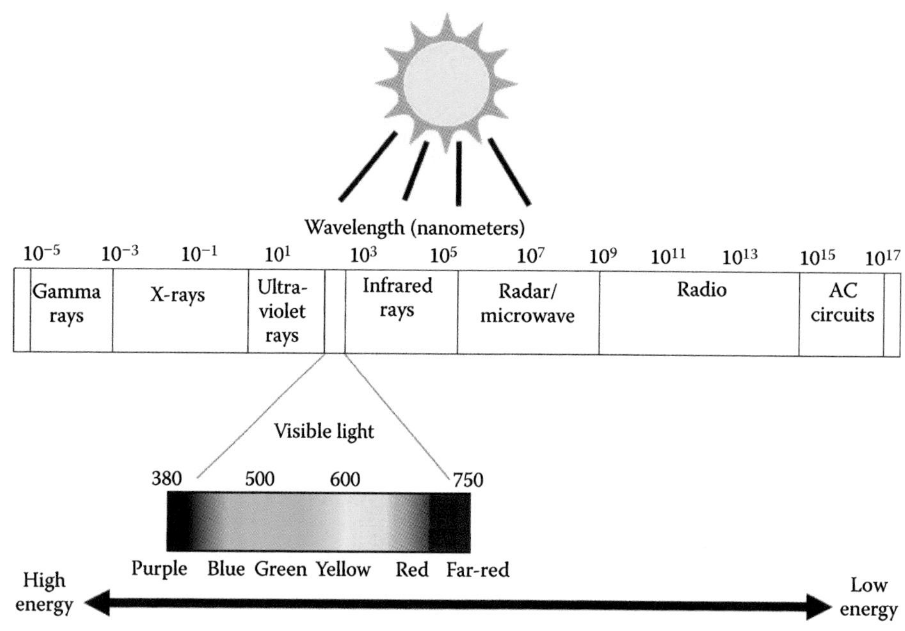


**FIGURE 2.18** Sunlight is electromagnetic radiation related to its various wavelengths. Photosynthesis occurs between 400 and 700 nanometers.

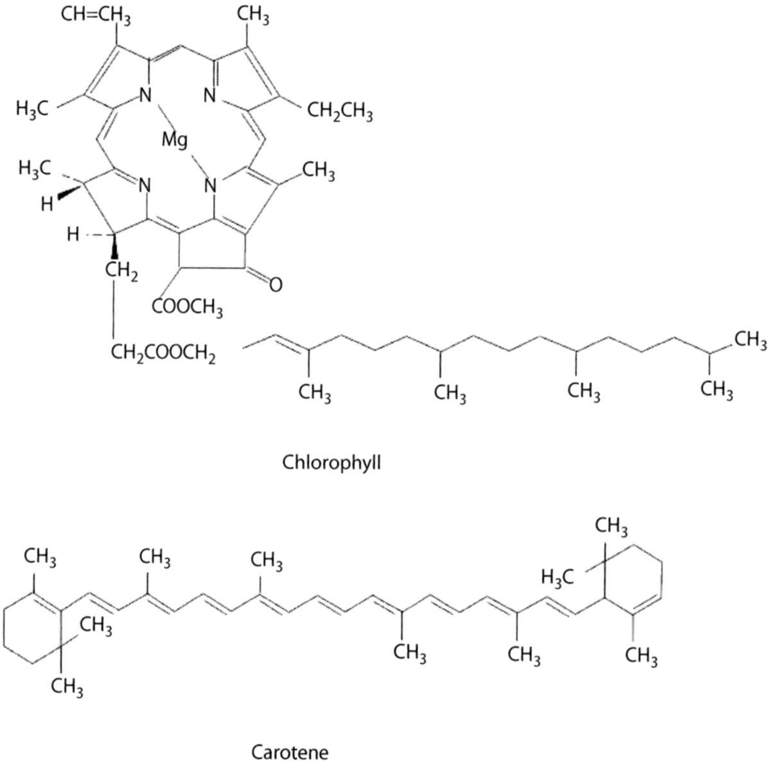

Chlorophyll

Carotene

**FIGURE 2.19** Green colored chlorophyll and orange colored carotene are two light-harvesting pigments.

conjugated carbon double bonds cyclized on either end. These unsaturated bonds allow the plant pigment molecules to capture and transfer light energy to chlorophyll without damage to the plant's cellular structure.

Besides capturing light energy, carotenoids also help in the dissipation of excess light energy at very high light intensities. By absorbing excess photons and dissipating their energy harmlessly as heat, carotenoids protect the rest of the cell from photo-oxidative damage.

### Light Reactions

As mentioned, photosynthesis is actually a combination of two separate but related processes—a light reaction and a dark (or carbon) reaction. Light is absorbed by pigments and its energy is converted into usable chemical energy (ATP and NADPH). In the dark or carbon reaction, these products of the light reaction are used to fix atmospheric $CO_2$ into carbohydrate (sugars). The use of energy derived from the absorption of light is possible because of a series of oxidation/reduction reactions carried out by chemicals arranged in two parts of what is termed the "light reactions."

Components of the light reactions (also called the electron transport system or "Z" scheme; Figure 2.20) are designated photosystem I (PSI) and photosystem II (PSII). These photosystems can be considered as light-driven electron pumps that work in two different, but overlapping, areas. PSII absorbs light in the red region (around 680 nm), while PSI absorbs light in the far-red region (around 700 nm). The PSII contains a different chlorophyll, designated $P_{680}$, as well as chlorophyll a (chl a) antenna and carotenoids. The PSI complex contains the following pigments: a special chl a or $P_{700}$ in the reaction center; chl a antenna; and the carotenoids β-carotene and xanthophyll. Photosystem II and PSI are arranged sequentially in the electron transport chain to transfer energy from $H_2O$ to $NADP^+$ (Figure 2.20). Electron transport involves several carrier molecules connecting the two photosystems. The electrons are finally passed to ferredoxin (Fd), a water-soluble iron-sulfur protein found on the stromal side of the thylakoid membrane. Ferredoxin donates the electron to $NADP^+$, reducing it to NADPH.

Overall, the electron transport system accomplishes the transfer of electrons (energy) from $H_2O$ to $NADP^+$ to form NADPH with the evolution of $O_2$ and the production of ATP as previously discussed. NADPH is vital for the reduction of $CO_2$, and ATP is an energy-carrying molecule used in many energy-requiring processes including various synthesis reactions (e.g., the synthesis of proteins). The formation of ATP is referred to as photophosphorylation.

**Practical implications of the light reactions.** Passing high-energy electrons through a series of electron carriers is not without risk. Cold temperatures, high light levels, and insufficient $H_2O$ may all cause more light energy to be absorbed than can be processed in photosynthesis. Under these conditions, the majority of the photosynthetic electron carriers become reduced and are unable to accept more electrons. Excited electrons may then be captured by $O_2$, converting it to one of several highly damaging forms, or a reactive oxygen species, which indiscriminately attack proteins, enzymes, membranes, and other cellular components.

This vulnerability of the light reactions is exploited by many herbicides. About half of all commercial herbicides directly target carriers of the photosynthetic electron transport chain, deliberately blocking electron flow and/or diverting electrons to $O_2$. Some herbicides, like atrazine, block electron flow between plastoquinone A ($Q_a$) and plastoquinone B ($Q_b$) in PSII. Others, like diquat, prevent the flow of electrons to ferredoxin in PSI. Still others compete with plastoquinone for binding to cytochrome $b_6f$.

## Carbon Reactions

The biosynthesis of carbohydrates is, with respect to the amounts produced, the most important biochemical process in the chloroplast. The following reaction shows the involvement of the products of the light reaction to convert atmospheric $CO_2$ to plant-usable carbohydrates (designated as $CH_2O$).

Carbon (or dark) chemical reaction:

$$CO_2 + 3ATP + 2NADPH + 2H^+ \rightarrow CH_2O + H_2O + 3ADP + 3P + 2NADP^+$$

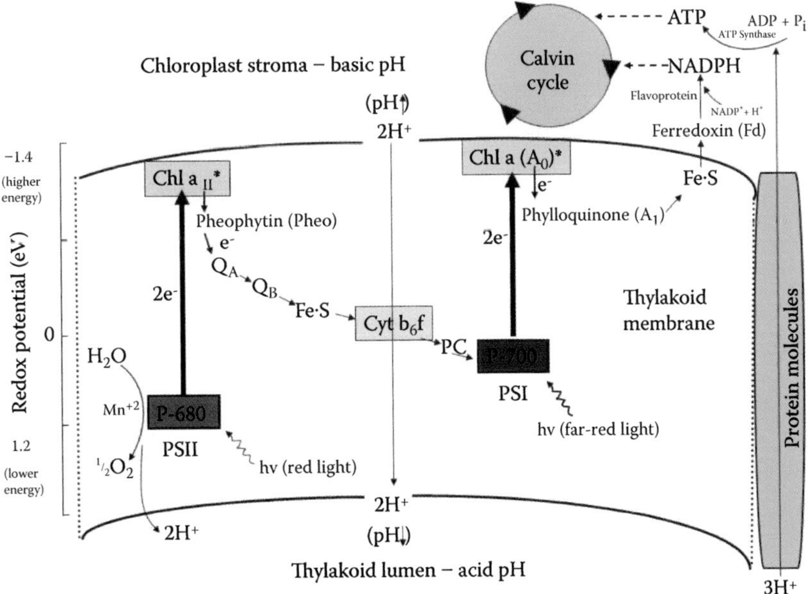


**FIGURE 2.20** Two light reactions (or "Z"-scheme) of photosynthesis occur with PSI harvesting far-red (700 nanometers) wavelengths while PSII harvests red light (680 nanometers). ATP and NADPH are two chemical energy forms produced from the Z-scheme used subsequently in the Calvin cycle.

These reactions were thought to be independent of light; therefore, they were previously referred to as dark reactions. However, since these reactions depend on products of light-regulation processes (ATP and NADPH), they are now referred to as the carbon reactions of photosynthesis.

Carbon dioxide is absorbed through leaf stomates. In addition to providing an entry of atmospheric $CO_2$ into the leaf, stomates release $H_2O$ vapor during transpiration. Normally, stomates are open during the day or in the light and closed during the night or dark in an attempt to conserve $H_2O$. The carbon in $CO_2$ is in a highly oxidized state, whereas carbon in a carbohydrate such as sucrose is in a highly reduced state. The electrons that have been transported to $NADP^+$, forming NADPH in the light reaction, are used to reduce the oxidation level of carbon during the dark reaction.

### $C_3$ (or Cool-Season) Plants

The first product formed from $CO_2$ fixation is a three-carbon sugar, 3-phosphoglycerate (3-PGA; Figure 2.16), which gives rise to the designation of $C_3$ or Calvin cycle. As shown in Figure 2.16, $CO_2$ is added to a receptor molecule, ribulose-bisphosphate (RuBP), a five-carbon sugar, producing two molecules of 3-PGA via the enzyme ribulose 1,5-bisphosphate carboxylase/oxygenase (designated as RuBP carboxylase/oxygenase, or RuBisCO). The Calvin cycle involves a number of other intermediate compounds (sugars). The result of this cycle is the production of a pool of triose phosphates. These three-carbon compounds are the precursors to six-carbon sugars including glucose and fructose. The net reaction consumes six $CO_2$ molecules, yielding a six-carbon sugar through the energy provided in the light reactions. These basic six-carbon sugars ($C_6H_{12}O_6$) are later used to synthesize larger carbohydrate molecules such as sucrose, starch, cellulose, and many others.

The sugar-phosphates are used in plant growth and development while the ADP and $NADP^+$ are cycled back to the light reaction to form additional ATP and NADPH. The reactions of the Calvin cycle are common to all photosynthetic plants and are the only known series of reactions that produce a net gain in carbohydrates.

Cool-season turfgrass species all are $C_3$ plants; thus, they are genetically adapted to cooler regions. As temperatures increase, the ratio of $CO_2$ to $O_2$ decreases. As a result, photorespiration increases relative to photosynthesis. When grown in high-temperature environment regions, $C_3$ plants weaken. Their microenvironments must be manipulated to relieve heat stress on $C_3$ plants or they may eventually die.

### $C_4$ (or Warm-Season) Plants

Until 1965, scientists thought the Calvin cycle was the only pathway of carbon fixation in higher plants. Research in sugarcane, a $C_4$ plant, revealed that instead of a three-carbon compound (3-PGA) being the first product of carbon fixation, a four-carbon compound, oxaloacetic acid (OAA), was the first product formed (Figure 2.17). This gave rise to the designation of the $C_4$ cycle or $C_4$ plants. This cycle allows these plants a more efficient method of fixing $CO_2$. Warm-season turfgrasses minimize photorespiration, but at the cost of additional ATP required for carbon fixation. Under high light and temperatures, this trade-off benefits the plant-photosynthetic rates in $C_4$ species without a loss in efficiency. However, under low-light conditions, the additional ATP cost for $C_4$ carbon fixation can slow growth.

Most (but not all) $C_4$ plants are monocots and presumably evolved from the tropics since high temperatures and light intensities are needed for their optimal performance. All warm-season turf species are $C_4$ plants; thus, they are adapted to warmer temperatures and higher light intensities (Table 2.6). Other $C_4$ grasses include dallisgrass, crabgrass, and goosegrass, all of which are common weeds in turf (Figure 2.21). These weeds, when present in $C_3$ turfgrasses such as fescue, ryegrass, or bluegrass, have the distinct physiological growth edge during warm summer months since they grow faster and use energy more efficiently under these conditions.

Carbon dioxide fixation in $C_4$ plants begins in the mesophyll cells (Figure 2.17), where it is combined with phosphoenolpyruvate (PEP) to produce oxaloacetate (OAA). The carboxylation enzyme is phosphoenolpyruvate carboxylase (designated as PEP carboxylase). The OAA is quickly converted into either malate or aspartate (both four-carbon compounds), which is then transported to cells surrounding the vascular bundles of the leaf, or bundle sheath cells. In the bundle sheath cells, $CO_2$ is cleaved from the four-carbon acids and combined with ribulose bisphosphate, producing 3-PGA and triose phosphates via the Calvin cycle. In $C_4$ plants, the four-carbon acid acts to shuttle and concentrate $CO_2$ in the bundle sheath cells. Identical to $C_3$ plants, NADPH and ATP from the light reaction are required for $CO_2$ fixation and functioning of the pathway. Thus, in $C_4$ plants, $CO_2$ fixation involves a combination of two carbon fixation pathways and seems much too complex to be efficient. However, consideration of this mechanism combined with a distinct leaf anatomy, described next, will illustrate a unique evolutionary development.

### Kranz Anatomy

It is important to note that the internal structural anatomy of a $C_4$ leaf is different from that of a $C_3$ plant. In $C_4$ plants, vascular tissues (phloem) are surrounded by bundle sheath cells, which also contain chloroplasts and are the site of the second $CO_2$ fixation via the Calvin cycle. These are surrounded by mesophyll cells that largely fill the leaf, resulting in little air space. Mesophyll cells are the site of the first $CO_2$ fixation and responsible for the transport of the four-carbon acid to the bundle sheath cells. The decarboxylation reaction in the bundle sheath creates a very high concentration of $CO_2$, referred to as Kranz (German for "wreath") anatomy. In contrast, $C_3$ plants do not contain chloroplasts in their bundle sheath cells; therefore, they are not able to concentrate $CO_2$ for maximum RuBisCO activity. Rather, in $C_3$ plants, there

**TABLE 2.6**
**Distinguishing Characteristics between $C_3$ (Cool-Season) and $C_4$ (Warm-Season) Turfgrasses**

| Characteristics | $C_3$ (cool season) | $C_4$ (warm season) |
|---|---|---|
| Leaf anatomy | • Flattened mesophyll allows large air spaces and less concentrated $CO_2$ levels; photosynthesis occurs in mesophyll cells. | • Mesophyll is arranged radially around bundle sheaths (Kranz- or halo-type anatomy); this close association allows $C_4$ plants to concentrate $CO_2$ in bundle sheath cells; photosynthesis occurs in mesophyll and bundle sheath cells. |
| $CO_2$-compensation concentration | • 30 to 70 ppm | • <10 ppm |
| Primary $CO_2$ receptor molecule | • RuBP | • PEP |
| First product of photosynthesis | • $C_3$ acids (PGA) | • $C_4$ acids (oxaloacetate) |
| Photosynthesis depressed by $O_2$ | • Yes | • No |
| $CO_2$ released in light (photorespiration) | • Yes | • No |
| Light levels required for photosynthesis | • 1/4 to 1/2 full sunlight | • Full sunlight (~2,000 μmol/m$^2$/second) |
| Net photosynthetic capacity | • Low to high | • High to very high |
| Water use efficiency (mg dry weight produced per g $H_2O$ lost) | • Low (1.05 to 2.22) | • High (2.85 to 4.00) |
| Dry-matter production capacity | • Medium | • High |
| Food storage compound | • Simple sugars (fructosan) | • Starch |
| Optimum photosynthesis temperatures | • 59°F to 77°F (15°C to 25°C) | • 80°F to 117°F (27°C to 47°C) |
| Optimum root growth temperatures | • 50°F to 65°F (10°C to 18°C) | • 75°F to 85°F (24°C to 29°C) |

**FIGURE 2.21** Warm-season (or $C_4$) grasses such as crabgrass have a distinct physiological advantage over cool-season (or $C_3$) grasses, such as tall fescue, during summer stress.

are large air spaces since the parenchyma cells are arranged into two distinct tissues: the palisade layer and the spongy parenchyma. Thus, in the $C_4$ leaf, the distance for $CO_2$ diffusion to the carboxylation sites is shorter compared to that in $C_3$ plants. The distance between the mesophyll cells and the bundle sheath cells is also shorter, reducing the distance for intercellular transport of malic and aspartic acids. Thus, the distinct anatomy of $C_4$ plants enables a warm-season grass to concentrate $CO_2$ in the bundle sheath cells at about a 12-fold increase compared to $C_3$ plants.

### Photorespiration

The enzyme responsible for $CO_2$ fixation in the Calvin cycle is RuBisCO, which can react with $O_2$ as well as with $CO_2$. High $O_2$ levels favor the oxygenase activity whereas high $CO_2$ levels favor the carboxylase activity. The oxygenase activity converts RuBP to 2-phosphoglycolate and 3-phosphoglycerate (3-PGA), rather than the two molecules of 3-PGA produced during carboxylase activity (the normal carboxylase activity of RuBP carboxylase/oxygenase). This oxygenase activity is termed photorespiration (Figure 2.22). It is not true respiration as in glycolysis or the Krebs cycle where carbon compounds are converted to $CO_2$, ATP, and carbon compounds with consumption of $O_2$. Therefore, photorespiration in $C_3$ plants is thought wasteful and reduces photosynthetic efficiency and plant productivity. Although photosynthetically inefficient, photorespiration maintains the flow of energy (NADPH and ATP) from the light reactions to the dark reactions to prevent damage to the photosynthetic apparatus of the light reactions in high-light environments and is necessary for nitrate assimilation. In $C_4$ plants, PEP carboxylase is not affected by $O_2$ as RuBisCO is in $C_3$ plants; thus, in $C_4$ plants, photorespiration is not significant.

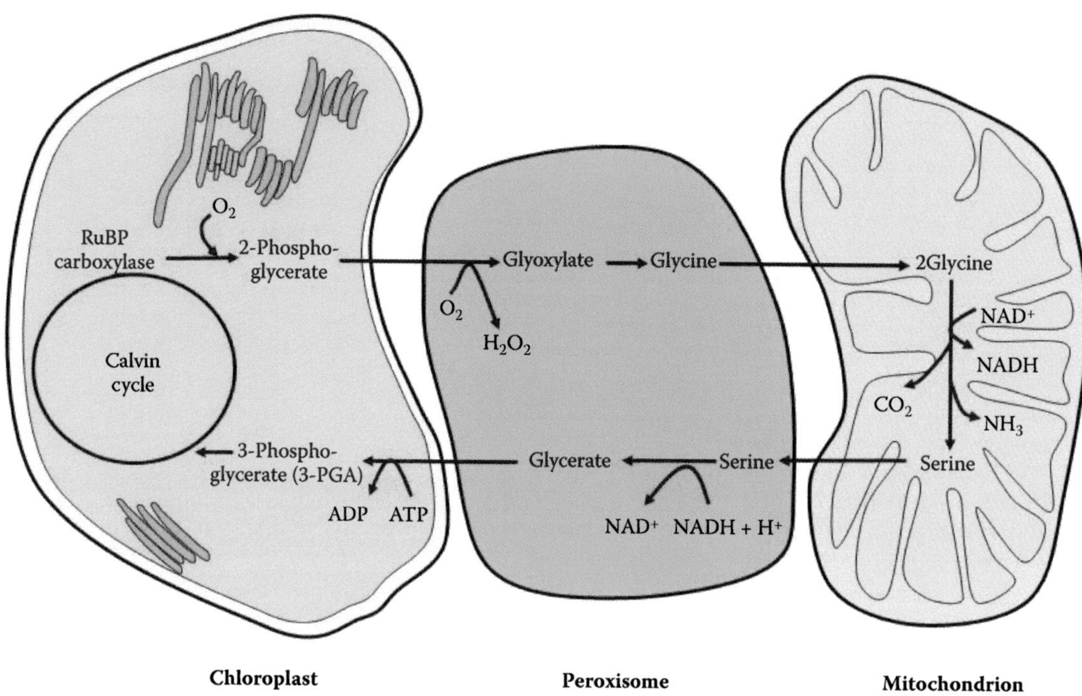


**FIGURE 2.22** Illustration of photorespiration where cool-season ($C_3$) plants close their stomates when stressed to conserve moisture. If this occurs for an extended period, insufficient $CO_2$ is captured, causing a deficient carbohydrate (energy) balance.

Photorespiration reaction in $C_3$ plants:

$$\begin{matrix} 3O_2 \\ 2ATP \\ H_2O \\ 2Fe^{+2} \end{matrix} + 2RuBP \xrightarrow{\text{Rubisco}} \text{3-phosphoglycerate}$$

$$+ \text{2-phosphoglycolate} + \begin{matrix} CO_2 \\ 2ADP \\ 3P_i \\ 2Fe^{+3} \end{matrix}$$

Photorespiration is also referred to as the $C_2$ cycle as two molecules of the 2-carbon compound, 2-phosphoglycolate, are converted to one molecule each of $CO_2$ and 3-PGA. These products are returned to the Calvin cycle to be converted back to RuBP via 3-PGA. For every two glycolate molecules passing through the photorespiration cycle, three $O_2$ are consumed, one $O_2$ is released, one ATP unit is consumed, and one PGA is made. Three times more $O_2$ is used in photorespiration than $CO_2$ is released.

Environmental parameters favoring photorespiration are bright, hot, dry days. Under these conditions, $C_3$ plants try to reduce transpirational $H_2O$ loss by closing their stomates. In turn, atmospheric $CO_2$ intake levels are reduced and $O_2$ levels begin to build up in the leaf from photosynthesis. Higher $O_2$ levels reduce the efficiency of $CO_2$ fixation by RuBisCO. Increasing temperatures increase photorespiration in $C_3$ plants, but since this process does not occur in $C_4$ plants, increasing temperatures normally only increase growth. As temperatures rise, $C_3$ plants gradually have decreased efficiencies due to faster $CO_2$ loss by photorespiration and may be up to 50% less efficient in carbon fixation compared to $C_4$ plants (Figure 2.23).

Since $C_3$ plants cannot concentrate $CO_2$ as $C_4$ plants do, their stomates must stay open longer to capture the necessary $CO_2$ levels to incorporate into carbohydrates. However, the more they stay open, the larger amount of $H_2O$ lost via transpiration. Thus, $C_4$ plants lose less $H_2O$ per unit of $CO_2$ fixed compared to $C_3$ plants and are more efficient at utilizing $H_2O$ and in photosynthesis. Cool-season turfgrass species all are $C_3$ plants; thus, they are genetically adapted to cooler regions. Since increasing temperatures result in decreasing ratios of $CO_2$ to $O_2$, photorespiration increases relative to photosynthesis under these conditions. When grown in high-temperature environment regions, $C_3$ plants are weakened

**FIGURE 2.23** Initial symptoms of photorespiration in bentgrass where damage to cell membranes has begun.

from photorespiration and prolonged transpiration. Their microenvironments must be manipulated to resemble cooler areas or they may eventually die. This is most applicable in turfgrass management scenarios where creeping bentgrass is grown in hot, humid climates.

Overall, energy in the form of ATP required to fix $CO_2$ into carbohydrates is initially higher in $C_4$ plants. Five molecules of ATP are required per $CO_2$ fixed in $C_4$ plants while only three molecules of ATP per $CO_2$ fixed are required in $C_3$ plants. However, additional ATP molecules are consumed in photorespiration in $C_3$ plants.

## $C_4$ Plants Compared to $C_3$ Plants

In addition to differences in carbon fixation efficiency, leaf anatomy, $H_2O$ usage, and energetics of photosynthesis, there are other differences between $C_4$ plants and $C_3$ plants. These physiological differences affect plant response to various environmental conditions and growth or productivity.

**Nitrogen use.** Not only are $C_4$ plants more efficient at utilizing $CO_2$ and $H_2O$ compared to $C_3$ plants, but they also utilize N more efficiently. A twofold greater dry matter production per unit of leaf N and nearly a twofold greater rate of photosynthesis often occur with $C_4$ plants compared to $C_3$ plants. The efficient utilization of N can be attributed to the "$CO_2$ trap" or $CO_2$ concentration or saturation found within $C_4$ plants. Warm-season plants possess an active isozyme of PEP carboxylase, thus allowing only 10% to 25% of the leaf N to be used in RuBisCO. Cool-season plants do not have a $CO_2$ trap; therefore, they invest 40% to 60% of their leaf N in RuBP carboxylase.

**Temperature tolerance.** In addition to greater efficiencies, $C_4$ plants have an ability to withstand certain other environmental pressures that $C_3$ plants cannot. High temperature is the primary stress $C_4$ plants are better adapted to. As mentioned before, $C_3$ plants do not photosynthesize as well as $C_4$ plants in warmer temperatures because of photorespiration. As the temperature becomes warmer, the reaction of RuBP carboxylase with $O_2$ occurs faster and the growth of $C_3$ plants is slowed. Eventually, heat stress in $C_3$ grasses occurs by damaging cell membranes, especially by denaturing proteins. However, as temperatures fall, $C_4$ plants do not photosynthesize as well as $C_3$ plants. This is perhaps related to the cold sensitivity of the enzyme pyruvate phosphate dikinase. This enzyme is needed to regenerate PEP from pyruvate, and its cold sensitivity would stop photosynthesis in $C_4$ plants (Figure 2.24).

**Light.** $C_4$ plants exhibit a nonsaturating growth curve at light intensities found in nature, meaning they require full sunlight for optimum photosynthesis. $C_3$ plants, however, are fully saturated at one-half full sunlight. At conditions beyond one-half full sunlight, photosynthesis decreases in $C_3$ plants because of photorespiration. Therefore, $C_4$ plants require full sunlight for maximum growth while $C_3$ grasses in general can photosynthesize at much lower light intensities. Overall, $C_4$ plants can utilize $H_2O$ and N more efficiently than $C_3$ plants. However, the additional energy cost of the $CO_2$ concentration mechanism makes $C_4$ plants less efficient in their utilization of light. This is one reason most shade-adapted turfgrasses are $C_3$ plants.

**FIGURE 2.24** Bermudagrass purpling from chilly night temperatures.

## RESPIRATION

Respiration is essentially the reversal of photosynthesis where the sugars and starches (carbohydrates or food) synthesized during photosynthesis are utilized to provide energy and metabolites for plant growth and maintenance. This energy from carbohydrates is converted to NADH and ultimately into ATP.

**Respiration reaction:**

$$C_6H_{12}O_6 + 6O_2 \rightarrow 6CO_2 + 6H_2O + \text{energy}\,(\text{ATP})$$

Plant and animal cells must respire to live; this takes place inside mitochondria. Mitochondria are membrane-enclosed organelles found throughout the cytosol of all plant cells. Plant respiration uses $O_2$ as a final electron acceptor through the first step of respiration called **glycolysis** (an anaerobic process located in the cytosol), which splits glucose to form two molecules of pyruvic acid. The second step of respiration, called the citric acid cycle (also called the Krebs cycle or TCA cycle), is an aerobic or $O_2$-requiring process located in the mitochondria where $CO_2$ and $H_2O$ are the final by-products, as well as energy in the form of ATP (Figure 2.25). Oxygen released from photosynthesis can be used directly in respiration while $CO_2$ released from respiration can be used in photosynthesis.

Carbohydrates, such as glucose, are relatively stable under cellular conditions, and will not break down at any appreciable rate. However, in the process of glycolysis ("sugar splitting" in Greek), carbohydrates are activated, can undergo energy-releasing reactions, and are eventually converted to pyruvate. The net gain for each glucose molecule consumed in glycolysis is two ATP units, two NADH molecules, two molecules of pyruvate, and 686 kilocalories of free energy released per molecule.

Glycolysis is essentially an inefficient process if considered as a free-standing oxidative pathway. Its end product, pyruvate, contains only slightly less energy than the starting carbohydrate material. Under aerobic conditions, the pyruvate from glycolysis is channeled to the mitochondria,

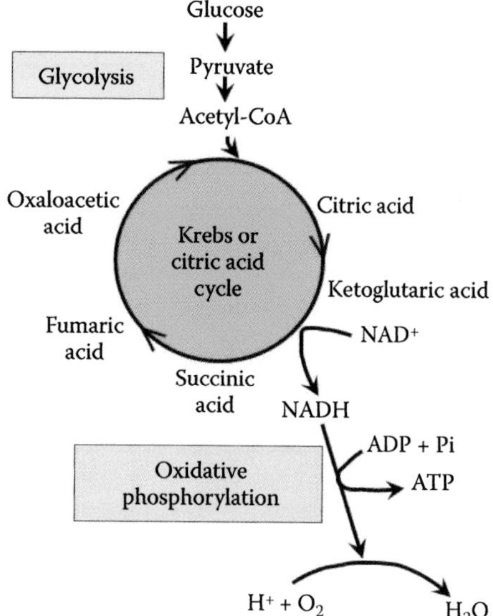


**FIGURE 2.25** Illustration of respiration where glucose is converted to usable plant energy with the eventual formation of NADH, ATP, and water.

where it is oxidized to form the NADH, $FADH_2$, ATP, and carbon intermediates used in other metabolic reactions such as the synthesis of amino acids necessary for protein synthesis. Such a process is referred to as the citric acid cycle, the tricarboxylic acid cycle (TCA cycle), or the Krebs cycle (named after Hans Krebs, the discoverer). Pyruvate is converted into acetyl-CoA, which is combined with a four-carbon compound (OAA) to form citric acid or citrate (hence, the other designated name). By the end of two cycle turns, acetyl-CoA is completely oxidized into $CO_2$ and $H_2O$, producing NADH and $FADH_2$. The high-energy compounds from the Krebs cycle (NADH and $FADH_2$) are converted to ATP through a series of $O_2$-dependent electron transfers. This mitochondrial process is the ATP-producing electron transport chain. Collectively, the electron transport chain and the synthesis of ATP are referred to the final step of respiration called oxidative phosphorylation. Glucose oxidation via glycolysis and the Krebs cycle in combination with mitochondrial activities yields a net sum of 36 molecules of ATP compared to only 2 molecules through glycolysis alone.

## Photosynthesis and Respiration Balance in Plants

In light, photosynthesis and respiration occur simultaneously. However, the photosynthetic rate usually occurs more rapidly, resulting in an accumulation of carbohydrates and a net positive release of $O_2$ by plants. In darkness, plants release $CO_2$ and energy from respiratory oxidation reactions. Between 25% and 75% of the daily carbohydrates produced by photosynthesis are consumed by respiration, depending on environmental conditions and the turfgrass species. When photosynthesis exceeds the capacity of chloroplasts to export triose phosphate sugars, glucose is formed and polymerized to form starch, which temporarily accumulates in the chloroplasts. When light dims and photosynthesis slows, as in the fall, this starch is hydrolyzed to glucose, synthesized to sucrose, and transported to storage sink organs—leaf sheaths, crowns, rhizomes, seeds, and stolons. When plants need energy for cell division or growth, starch is converted to glucose. In cool-season grasses, sucrose is synthesized into various fructose polymers (fructans) in vacuoles and stored temporarily in leaves or for longer periods in crowns, rhizomes, and stolons. Unlike starch, fructans are $H_2O$ soluble and help cool-season grasses tolerate freezing temperatures.

As mentioned, when photosynthesis exceeds respiration, an excess (or positive balance) of carbohydrates occurs and is stored as fructans (cool-season grasses) or starch (warm-season grasses). However, when carbohydrate production from photosynthesis is insufficient for plant functions, stored carbohydrates are used for growth until they are exhausted. Growth then declines, eventually weakening plants to the point of metabolic dysfunction, leading to plant death. This occurs, for example, when cool-season grasses are grown in warm or hot environments for extended periods (Figure 2.26). Eventually, respiration exceeds carbohydrate production by photosynthesis. For creeping bentgrass, this temperature point is 86°F (30°C). Storage carbohydrates are then used for plant growth, resulting in shallower roots, weaker stolons, and eventual plant death unless temperatures abate. Untimely N applications during this period also exacerbate carbohydrate consumption by stimulating shoot growth and respiration at the expense of root growth. Other examples of respiration exceeding photosynthesis are long exposures to shaded conditions and excessively low mowing heights. These two conditions reduce photosynthetic capacity while respiration continues, leading to carbohydrate depletion.

Sufficient levels of storage carbohydrates also are important for winter survival and subsequent spring regrowth or green-up of perennial grasses. Warm-season grasses, for example, become dormant (brown) in areas with sufficient frost and lose their capability to photosynthesize. Crowns and roots of these plants, however, remain alive and continue to respire, albeit at reduced levels. Therefore, proper turfgrass management practices maximizing summer and fall

**FIGURE 2.26** Advanced photorespiration in bentgrass where turf thins and algae invades with noticeably healthier turf in aerification holes.

carbohydrate production and storage minimize depletion of food reserves over the winter.

## GROWTH HORMONES

Unlike mammals, plants lack a nervous system and sensory organs that allow them, among other things, to respond to various environmental stimuli. Plants, however, employ chemicals called growth hormones, which are synthesized in one part of the plant, translocated to other parts, and, in low concentrations, cause various growth responses. A plant hormone is a chemical messenger involved in the regulation of plant metabolism, growth, and morphogenesis. Due to their low concentrations found in plants, discovery and understanding of most plant growth hormones have been delayed until the advent of highly sensitive scientific equipment. Growth-promoting hormones include auxins, gibberellins, and cytokinins, while growth-inhibiting hormones include ethylene and abscisic acid (Figure 2.27).

### GROWTH PROMOTERS

**Auxins.** The discovery of auxin (from the Greek *auxein*, meaning "to increase" or "to grow") can be traced to the observations of Darwin (around 1880), who was investigating the phototropic curvature of grass seedlings' coleoptiles. Went (1926) later found evidence leading to the physiological proof of auxin as a plant growth substance. It is produced in meristematic or other actively growing tissue, moves throughout the plant, and is capable, in minute concentrations, of affecting elongation in cells below this meristem or site of production. Indole-3-acetic acid (IAA) is the primary plant auxin. Other auxin-like compounds include phenylacetic acid, indole-3-butyric acid (IBA), naphthalene acetic acid (NAA), and 2,4-dichlorophenoxy acetic acid (2,4-D).

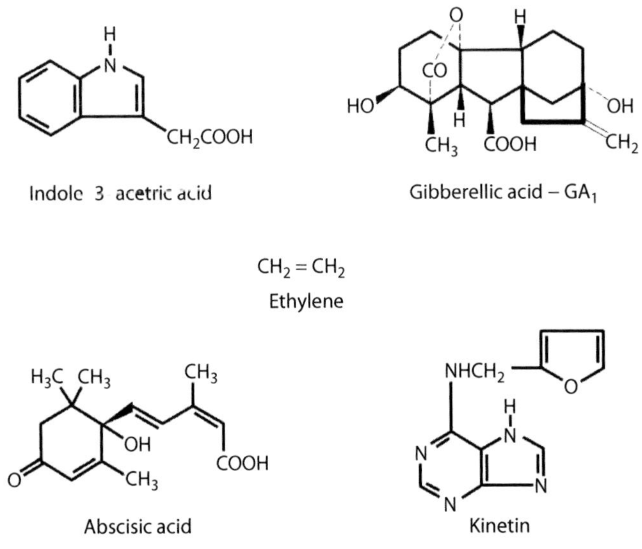

Indole 3 acetric acid

Gibberellic acid – $GA_1$

$CH_2 = CH_2$
Ethylene

Abscisic acid

Kinetin

FIGURE 2.27 Chemical structures of major plant growth hormones. Ethylene and abscisic acid are considered growth inhibitors; the others are growth promoters.

The primary influence of auxin is stimulation of cellular elongation (Table 2.7). Phototropism (light-induced bending) of stems is induced by a differential distribution of IAA within the shoot while auxins stimulate cell division and cell differentiation (e.g., into xylem and phloem tissue). Auxin also influences apical dominance by inhibiting lateral bud development and delaying leaf senescence and fruit ripening, but promoting flowering and fruit set. Low concentrations of auxin also promote root growth.

Excessive auxin levels can inhibit growth, depending on the tissue involved. For example, the amount of auxin needed to stimulate plant growth in a stem, making it bend toward light, will stop growth in a root, bending it away from light. Stolon shoot tip removal by mowing or slicing removes the auxin; thus, apical dominance is eliminated and additional stolon production is promoted from axillary buds. Excessive auxin levels may even kill the plant, as evidenced by the auxin-like herbicide, 2,4-D.

**Gibberellins.** Gibberellins were first discovered when rice plants were noticed to grow excessively tall when infected by the fungus *Gibberella fujikuroi*, which produced certain gibberellins. Japanese scientists isolated the substance and in 1954 and 1955, the structure of the biologically active gibberellin, $GA_1$, was established. Approximately 125 gibberellins have been identified and most plants contain complex mixtures of several. The most important gibberellins in agriculture are $GA_1$ and $GA_3$ and, to a lesser degree, $GA_4$ and $GA_7$ (Table 2.7). Gibberellins basically stimulate internodal elongation in intact plants. Recently, through genetic engineering, scientists have regulated a plant's height by controlling the level of bioactive gibberellin and other enzymes.

Gibberellins are synthesized in young leaves (as is auxin) and in developing roots and seeds via a branch of the terpenoid pathway. They promote both cell division and cellular elongation, and are also known to stimulate flowering by overcoming cold-temperature requirements needed by certain plants to flower (referred to as vernalization). In addition, gibberellins stimulate the need of some plants for long days, cause juvenile growth patterns in plants to mature, reverse dwarfism in certain plants, stimulate the release of certain hydrolytic enzymes in some seeds, and can cause parenthogenesis, which is seed production in some plants without pollination. Commercial formulations of gibberellic acids are sometimes used to aid bermudagrass color recovery after a mild cold stress period or light frost or to serve as antidotes when excessive GA-inhibiting plant-growth regulators are applied to turf.

Several plant-growth regulators retard shoot growth through the interference of gibberellin biosynthesis. Cell division and seedhead formation are not greatly influenced, but cell elongation is retarded (Figure 2.28). The results from using these growth retardants are miniature plants. Examples of gibberellin-inhibiting growth retardants include trinexapac-ethyl, prohexadione-Ca, paclobutrazol, and flurprimidol.

**Cytokinins.** Cytokinins are compounds that stimulate cytokinesis, or cell division (Table 2.7). These were discovered

**TABLE 2.7**
**Growth Hormones Influencing Plant Growth and Development**

| Hormone | Examples | Site of Production | Major Growth Influence |
|---|---|---|---|
| | | **Growth Promoters** | |
| Auxin | Indole-3-acetic acid (IAA), phenylacetic acid, naphthalene acetic acid (NAA), dichlorophenoxy acetic acid (2,4-D) | • Meristematic tissue, young leaves, developing fruits, and seed | • Stimulates cell elongation, division, and differentiation; promotes apical dominance, which inhibits lateral bud development; delays leaf senescence and fruit ripening |
| Gibberellins | $GA_1$, $GA_3$, $GA_4$, $GA_7$, and many others | • Meristematic tissue in young leaves, developing roots, and seeds | • Promotes cell division and cellular elongation |
| Cytokinins | Kinetin and other derivatives of adenine | • Root tips | • Stimulates cytokinesis (cell division) in shoot and leaf formation; slows plant aging and leaf senescence by delaying chlorophyll and protein degradation; may help plants overcome external stresses |
| | | **Growth Inhibitors** | |
| Ethylene | Ethylene gas ($C_2H_4$) | • Shoots | • Promotes plant aging and senescence; inhibits root growth, lateral bud development, and cell elongation |
| Abscisic acid | ABA | • Green tissue (chloroplasts) | • Inhibits coleoptile growth and promotes senescence and abscission; ABA levels increase when plants are exposed to stresses such as drought, flooding, starvation, injury, and salinity |

**FIGURE 2.28** Annual bluegrass stunted growth from increasing rates of a gibberellic acid inhibiting regulator (left to right, foreground) compared to untreated (background).

when studying the division of wounded potato tuber tissues. Most cytokinins are derivatives of adenine (a purine), one of the key bases found in nucleic acids. All naturally occurring cytokinins are derivatives of adenine. An important adenine-derived cytokinin is kinetin, which does not naturally occur but rather is formed when DNA is broken down. Zeatin is the naturally occurring cytokinin in plants.

Currently, it is believed the primary sites of cytokinin synthesis in plants are root tips that are then translocated to shoots through the vascular system. They are also synthesized in tissues with high meristematic activity (e.g., cambium, vegetative apices, and young leaves). Cytokinins are known to promote cell division and cell enlargement; delay plant aging and leaf senescence by slowing the rate of chlorophyll disappearance and protein degradation; replace the light requirements for certain processes, such as seed germination, pigment synthesis, and chloroplast development; stimulate lateral shoot (branch) development in some plants; and inhibit branch formation in roots. Exogenous applications of cytokinins (10 μmol) as zeatin riboside to bentgrass under heat stress can promote root growth and initiation. This alleviates leaf senescence and thereby improves turf quality. Exogenous application of certain chemicals, such as trinexapac-ethyl, has also shown an increase in cytokinin levels in plants and is believed to be involved in increasing stress tolerance.

## Growth Inhibitors

**Ethylene.** Ethylene is a gas discovered when oranges and bananas were shipped together and the bananas prematurely ripened. This was due to a gaseous substance, ethylene, emitted by the oranges. Around 1960, when it became possible to reliably measure small amounts of ethylene, it was demonstrated to have important physiological roles in plant growth and development. In addition to simulating ripening of fleshy fruits, ethylene is known to induce aging (senescence) in many plants (e.g., abscission of leaves and floral parts) (Table 2.7).

Ethylene also is known to inhibit root growth, inhibit lateral bud development, inhibit cell elongation, increase membrane permeability, and cause epinasty (downward curvature of leaves). Ethylene, however, enhances a cellular increase in width. The results are shorter but broader stems and petioles. Ethylene production by plants is high wherever auxin concentration is high. Therefore, shoots produce ethylene, whereas roots produce little. Ethylene gas is produced when the sulfur-containing amino acid, methionine, is degraded and is very mobile in plants by diffusion. Ethephon (e.g., Proxy) is a commercial plant growth regulator that causes ethylene production in plants. Its main use is to suppress *Poa annua* seedheads.

**Abscisic acid (ABA).** ABA was discovered around 1960 by two groups whose research was based on two different physiological functions: "abscisin," causing the abscission of cotton fruits; and "dormin," inducing the dormancy of buds in birch trees. In 1965 these substances were found to be the same compound and the term "abscisic acid" was adopted. ABA is present in all vascular plants, inhibits coleoptile growth, and promotes senescence and abscission by blocking DNA and certain protein synthesis (Table 2.7).

An increase in endogenous plant ABA levels occurs when plants are exposed to a variety of stress conditions such as flooding, drought, salinity, injury, and starvation. This is especially acute when plants wilt from drought stress. ABA is believed to be a signaling messenger for stomatal closure in response to drought or $H_2O$ stress. Increased ABA levels have also been associated with increased low-temperature tolerance in plants. The role ABA and ethylene play in turfgrass growth and development is still poorly understood.

### Biostimulants

Biostimulants are materials that promote plant growth when applied in small quantities. Research indicates biostimulants may increase photosynthesis rates, reduce turfgrass senescence, enhance seed germination and root growth, help sod establish faster, and increase salt tolerance and drought resistance after application. They often contain cytokinins as their principal ingredient and tend to help plants withstand harsh environments through enhanced cytokinesis (Table 2.7). They also exhibit other growth-regulating actions as found from auxins, gibberellin, and abscisic acid. Seaweed extract (or seaweed kelp), most commonly from the sea kelp species Ascophyllum nodosum, is a common commercial cytokinin-containing growth material used in turf. A commercially available synthetically produced cytokinin is benzyladenine (or BA).

Products with cytokinin-like properties also are available, such as the triazole systemic fungicides—propiconazole, triadimefon, and others. These fungicides inhibit sterol biosynthesis, cause an increase in ABA synthesis, and inhibit gibberellin synthesis. Sterols are part of cell membranes and help protect plants and promote growth. When sterol and gibberellin biosynthesis are inhibited, plant ABA levels rise, favoring plant stomatal closure and thereby creating an increase in plant $H_2O$ content. Cytokinin levels also tend to decrease when drought stress begins and ABA level increases, resulting in closed stomata. Further evidence suggests these triazole fungicides translocate to shoot tips and cause growth reduction of the foliage but not the roots. Therefore, enhanced root development occurs with reduced shoot growth and increased plant $H_2O$ content. Additionally, research shows triazole-treated plants not only inhibit reduced transpiration, but also, when plants are under moisture stress, increase yields, delay senescence, and increase chlorophyll levels as well as carbohydrate levels. These compounds also have protected plants from chilling temperatures, heat stress, and ozone exposure.

The levels of cytokinin in plants regulate, to an extent, the growth response. For example, large applications of cytokinin have been shown to inhibit root formation, but it is known that a small amount of cytokinin is necessary for the formation of rooting. Root growth, therefore, appears to require a certain cytokinin level and a favorable auxin-to-cytokinin ratio.

One of the greatest benefits turfgrass receives from biostimulant application is when the turf is being grown under a physiological stress. These stresses include cold (or chilling) temperatures, low soil nutrient levels (especially K and/or P), low mowing height stress, and periods of rapid growth where these and other resources become limiting. Other possible stresses that may be partially overcome by biostimulant applications include salinity and nematode exposure. Newly planted turfgrasses also may benefit from biostimulant use. Replicated field research of biostimulants on turfgrasses is limited; thus, caution should be used when applying these materials.

## ENVIRONMENTAL STRESSES

### Light Intensity (or Shade)

Light intensities also greatly influence turfgrass growth. Interestingly, less than 40% of the energy in sunlight is within the 400 and 700 nm wavelengths necessary for photosynthesis (Figure 2.18). Over 60% of sunlight is >700 nm (or infrared light) long and <1% between 200 and 400 nm (or ultraviolet light). Of the sunlight intercepted by leaves, only 1% to 5% is used for photosynthesis. Scientists measure and quantify the intensity of sunlight reaching the ground on a cloudless day at sea level as between 1,800 and 2,300 μmol/m²/second. Warm-season grasses have a saturation point for photosynthesis from 1,794 to 2,139 μmol/m²/second compared to the much lower value—534 to 1,072 μmol/m²/second—for cool-season grasses.

Warm-season plants (such as bermudagrass) grow best when exposed to full sunlight. At lower light intensities (<70% full sunlight), $C_4$ plants respond by developing narrow, elongated leaves, thin upright stems, elongated internodes, and weak rhizomes and stolons. For example, canopy height of bermudagrass under low light intensities increase over 100% compared to plants exposed to full sunlight, while cool-season grass leaf elongation is about 35% higher under reduced irradiance. Due to lower photosynthetic rates, total nonstructural carbohydrates are reduced, eventually causing turf stand thinning. When mowed excessively low under low light intensities (e.g., shaded golf greens), $C_4$ plants frequently thin and are eventually invaded by other pests such as algae, *Poa annua*, or goosegrass. The increase in shade tolerance of newer bermudagrasses is related to their ability to remain in a prostrate growth habit and not become stemmy when grown in shade, thus avoiding scalping when mowed.

In the field, only a small fraction of the light striking the upper leaves of a turf stand filters through to leaves below. The second layer of leaves receives about 10% of the light striking the top layer, and only 1% strikes the lower third layer. The arrangement and angle of the leaves also helps determine how much light will pass to lower leaf layers. The more upright the leaves, the more sunlight can reach the lower layers and the more efficiently the plants can utilize full sunlight. This explains why, during shorter light duration days (e.g., late summer and fall), $C_4$ plants become "stemmy." This is an attempt by the plant to allow lower leaves to capture as much sunlight as possible to make the plant as photosynthetically efficient as possible to store carbohydrates for upcoming winter months. This is also an attempt by the plant to enhance its chances of growing above the canopy and acquiring a greater share of unfiltered, photosynthesis-active light. During longer days (e.g., late spring and summer), $C_4$ grasses have more of a prostrate (or decumbent) growth habit. Lower leaves are not as important to the plant during these times since the longer days allow enough sunlight absorption by the upper leaves to adequately sustain growth.

**Trees.** In the natural environment, trees and grasses rarely are found growing together; the major exception is the oak savannas of the eastern fringes of the tall grass prairies. However, on golf courses, trees and other shade sources often occur together. Turfgrass health under trees depends on the intensity and quality of light passing through a tree's canopy. Turfgrasses under trees receive less intense and less effective wavelengths. Tall trees obviously intercept light first, leaving grasses only speckled flecks and filtered light to use for photosynthesis. Tree roots also provide competition to turfgrasses for moisture and nutrients.

Conifer (e.g., pines) and shade-intolerant trees tend to have more open growth habits, allowing more sunlight to penetrate. In some instances, conifer trees (e.g., pines) planted far enough apart will allow sufficient light through to support a somewhat thin stand of grass. Shade-tolerant trees, such as oaks, sycamores, and maples, however, have denser canopies of foliage, allowing little sunlight to penetrate. Deciduous trees also tend to selectively filter out the photosynthetically active wavelengths (blue-violet and orange-red wavelengths) necessary for grass to grow. This further adds to the problem of trying to grow grass underneath shade-tolerant deciduous trees.

**Light duration.** Minimum daily duration of light is necessary for all turfgrasses. Duration depends on the time of year (days are longest on June 21 in the northern hemisphere and shortest on December 21), latitude (the sun's angle relative to the earth's axis), and size and location of the shade source. Optimum growth of warm-season grasses occurs when daily solar radiation day length is greater than 13 hours. When solar radiation day length is reduced below 13 hours, growth is slowed regardless of whether adequate irrigation or N fertilizer is available. The opposite effect occurs with some cool-season grasses. For example, shorter day length periods, like those experienced in spring and fall, promote the greatest tillering of most cool-season grasses. Bermudagrass golf greens require eight hours minimum of full sunlight, year-round. When full sunlight hours are reduced below eight hours, bermudagrass, especially in shorter mowed areas such as greens, will gradually thin. Taller mowed bermudagrass, such as fairways and approaches, can withstand less sunlight compared to greens due to possessing higher leaf surface area for light capture from higher mowing heights (Table 2.8).

In winter, when days are shortest and the sun is lowest on the horizon, excessive shade from trees 30 to 50 feet (9 to 15 m) away from the greens often occurs (Figure 2.29). In the northern hemisphere, this is especially true for trees on the southeastern to southwestern direction of greens. As these trees mature and grow larger, golf courses must implement an ongoing aggressive selective limb pruning and possible tree removal program.

In addition to reducing light, shaded areas generally have (1) increased competition to the turf for nutrients and $H_2O$ by trees and shrubs, (2) increased humidity, (3) reduced or restricted air movement, (4) longer periods of dew and frost occurring during the morning before "burning-off," and (5) lower temperatures.

Greens experiencing other stresses, such as diseases, excessive soil compaction, poor drainage, and salinity stress, are especially low-light sensitive. Greens typically experience a gradual thinning, increased occurrence of algae, increased diseases, poor surface drainage, and slow recovery time from

**TABLE 2.8**
**Approximate Daily Light Intervals Necessary for Golf Course Rough Heights Unless Indicated Otherwise**

| Grass and Cultivar | Light (mol/m²/day) |
|---|---|
| TifEagle bermudagrass | 32.5 |
| Creeping bentgrass green | 32.0 |
| Tifway bermudagrass | 22.4 |
| TifGrand bermudagrass | 18.6 |
| Celebration bermudagrass | 19.5 |
| Argentine bahiagrass | 16.0 |
| Seadwarf seashore paspalum | 13.5 |
| Tifblair centipedegrass | 13.3 |
| Palisades zoysia (*japonica*) | 12.0 |
| Floratam St. Augustinegrass | 11.9 |
| Diamond zoysia (*matrella*) | 11.7 |
| Pristine zoysia (*matrella*) | 11.3 |
| Captiva St. Augustinegrass | 11.2 |
| Jamur zoysiagrass (*japonica*) | 10.5 |

**FIGURE 2.29** Bermudagrass damage from winter shade plus mower traffic.

surface damage. Optimum growth of $C_3$ plants also appears related to having good morning sunlight. By afternoon during summer in many places, cloud and haze build up to naturally reduce the quality and quantity of sunlight reaching the earth's surface. Therefore, reducing morning shade sources, thereby allowing the earliest sunlight, appears best for most $C_3$ grasses. Early sunlight also reduces the time needed for dew evaporation and frost melting. The opposite is true for $C_4$ plants, where afternoon sun is most important as sunlight intensity is greatest then.

Although $C_3$ plants, such as bentgrass, theoretically require less sunlight to grow than $C_4$ plants, the best bentgrass greens occur when grown in wide-open, shade-free areas. Other factors, such as reduced air movement (higher humidity) and longer drying periods created by shade sources often promote other life-threatening conditions such as diseases. Taller mowed cool-season turfgrasses require four to five hours of full sunlight daily or an entire day of filtered light.
Best management practices for shaded turf:

- Selectively remove and/or prune trees and shrubs.
- Prune shallow tree roots to reduce competition with turf for soil moisture and nutrients.
- Mow the turf high to allow maximum leaf surface to capture filtered sunlight for photosynthesis.
- Reduce rates of N, which encourages shoot growth at the expense of roots and also results in turf scalping.
- Use gibberellic acid-inhibiting plant growth regulators to prevent stem elongation, thus reducing scalping.

## Temperature Stress

Turfgrasses exhibit optimum growth within a certain range of temperatures (Table 2.9). Extremely warm or hot temperatures during late spring and/or summer can cause stress to cool-season turfgrasses, while cold temperatures during winter and/or spring months can cause stress to warm-season turfgrasses. It is often difficult to identify any one single factor predisposing turfgrasses to temperature stress. More commonly, it is a combination of parameters influencing their susceptibility. Injury from temperature stress often is a reflection of the accumulation of various stresses over an entire growing season, which ultimately leads to less hardy turfgrass plants.

### Low-Temperature Stress

Winter turfgrass injury due to low-temperature stress is commonly referred to as "winter-kill," which, in the simplest of terms, refers to the loss of turfgrass during the winter or early spring season (Figure 2.30).

The growth of warm-season turfgrasses slows significantly when the average daytime temperatures decrease to below 60°F (15.5°C), and leaf discoloration (onset of dormancy) begins when temperatures drop below 50°F (10°C). However, warm-season turfgrasses can continue to slowly grow with nighttime temperatures as low as 34°F (1.1°C) as long as daytime temperatures increase to near 70°F (21°C). Permanent turfgrass injury to warm-season turfgrasses often occurs if ambient temperatures drop rapidly to below 23°F (–5°C) or drop gradually to below 10°F (–12.2°C).

During the spring transition period, turfgrasses are very vulnerable to "low-temperature injury," due to the biological and physiological growth conditions of turfgrasses during this time (Figure 2.31). At ~60°F (15.5°C), lateral buds at the nodes of rhizomes and stolons begin to break dormancy, the stored carbohydrate or food reserves (starch) are converted to soluble sugars, and the growth or greening process begins. Simultaneously, there is a rapid die-back of old roots just prior to the production of new roots. This phase of turfgrass growth is very susceptible to low-temperature stress or injury. Also, areas planted (seeded, sprigged, or sodded) late the previous year are typically more susceptible to low temperatures since the growing time for adequate establishment and root development is insufficient.

Maximum cold temperature tolerance is usually obtained in early winter and declines throughout the winter as the grass is exposed to warmer temperatures, freeze–thaw cycles, anaerobic conditions, and crown hydration. For example, annual bluegrass that survives 15°F (–9°C) in January may be killed by that same temperature in March. Late winter and early spring are the most critical periods of susceptibility to winter injury. During this time, grass plants are beginning to grow and are losing carbohydrates and cold hardiness as crowns become hydrated. This, followed by a late and rapid cold snap, often causes the most damage.

Low-temperature or cold-weather damage to plants is a collective term used to describe several forms of injury. It is important to distinguish between these forms of low-temperature injury and have an understanding of the conditions promoting the specific types of injury. Rarely does just one variable cause low-temperature damage to turfgrasses. In general, the major types of low-temperature injury are caused by one or more of the following:

1. Crown hydration
2. Direct low-temperature exposure

**TABLE 2.9**
**Approximate Air and Soil Temperatures Affecting Turfgrass Growth and Development**

| | Cool-Season ($C_3$) Turfgrasses | | Warm-Season ($C_4$) Turfgrasses | |
|---|---|---|---|---|
| **Turfgrass Response** | **Air** | **Soil** | **Air** | **Soil** |
| | | **Temperature, °F (°C)** | | |
| Heat kill | 131 (55) | – | >140 (>60) | – |
| Maximum for shoot growth | 90 (32) | – | 120 (49) | – |
| Maximum for root growth | – | 70 to 75 (21 to 24) | – | – |
| Suggested for planting | 75 to 80 (24 to 27) | 60 to 70 (15.6 to 21) | 80 to 90 (27 to 32) | 70 to 80 (21 to 27) |
| Optimum for shoot growth | 59 to 77 (15 to 25) | – | 80 to 100 (27 to 38) | – |
| Optimum for root growth | – | 50 to 65 (10 to 18) | – | 75 to 85 (24 to 29) |
| Minimum for shoot growth | 40 (4.4) | – | 55 (13) | – |
| Minimum for root growth | – | 33 (0.6) | – | 35 to 50 (1.7 to 10) |
| Possible low-temperature kill | – | 20 (–6.7) | – | 25 (–4) |
| Possible chilling injury | – | – | 50 (10) | – |
| Trigger spring root decline | – | – | – | 64 (18) |
| | | **50% Stand Kill** | | |
| Alkaligrass | – | –6 to –17 (–21 to –27) | – | – |
| Perennial ryegrass | – | 5 to 23 (–15 to –5) | – | – |
| Fine fescue | – | –6 to –11 (–21 to –24) | – | – |
| Kentucky bluegrass | – | –6 to –22 (–21 to –30) | – | – |
| Annual bluegrass | – | –4 to 14 (–20 to –10) | – | – |
| Roughstalk bluegrass | – | –40 (–40) | – | – |
| Bentgrass | – | –31 (–35) | – | – |
| Bahiagrass | – | – | – | 23 (–5) |
| Bermudagrass | – | – | – | 18 to 23 (–8 to –5) |
| Buffalograss | – | – | – | –10 (–23) |
| Centipedegrass | – | – | – | 18 to 21 (–8 to –6) |
| Zoysiagrass | – | – | – | 6 to 12 (–14 to –11) |
| St. Augustinegrass | – | – | – | 23 (–5) |

*Note:* Soil temperatures are at a 4-inch (10-centimeter) depth. Data not available for all environmental parameters for all grasses.

**FIGURE 2.30** Low-temperature damage (or "winter-kill") to bermudagrass.

**FIGURE 2.31** Bermudagrass damage from a combination of low temperatures, traffic, and shade.

3. Desiccation
4. Traffic effects
5. Ice cover (suffocation or anoxia)
6. Frost heaving
7. Diseases

**Crown hydration.** Crown hydration occurs when turfgrass plants remain in or under constant moisture following a thaw. Consequently, these plants absorb high levels of $H_2O$. This is especially acute for turf located in standing $H_2O$. If temperatures then decrease below freezing, ice crystals form

**FIGURE 2.32** Zoysiagrass damage from ice crystals forming within plant cell walls (or intracellularly), rupturing them.

**FIGURE 2.33** Direct low temperature when turfgrass plants are exposed to a rapid decrease in temperatures below 23°F (–5°C).

within the plant cell walls (or intracellularly), rupturing them and thereby killing tissues (Figure 2.32). Intracellular freezing injury (aka, "direct" kill) is common in spring when plants become deacclimated, especially in low, poorly drained areas. Ice crystals can also form "extracellularly," or in intercellular spaces between protoplasts and the cell wall. As ice crystals form, the vapor pressure becomes lower than in the protoplasts and $H_2O$ is drawn from within cells. Protoplasts then shrink in size and the concentration of dissolved substances, such as salt ions and organic acids, increases and becomes toxic via osmotic (or moisture stress) damage. This is also known as "indirect" kill. Plants surviving winter are more successfully able to tolerate ice formation between cells (extracellularly) by having increased cell membrane fluidity (flexibility) that can tolerate freezing. Plants increase this fluidity by increasing the ratio of short chain unsaturated fatty acids (those containing double bonds) in membranes to saturated fatty acids (those containing single bonds). Having higher ratios of unsaturated fatty acids prevents membranes from gelling and freezing, allowing the protein components of membranes to alter their shape or change position during freezing without permanent damage. If temperatures remain low for a long period of time, cells may become dehydrated to a point where injury or death occurs. Extracellular freezing is most common during dry, cold weather on hillsides exposed to winds, or low-mowed golf greens on sand-based soils.

**Direct low-temperature exposure.** Another common type of low-temperature injury is direct low-temperature exposure. Basically, two types of injury are associated with direct low-temperature exposure: (1) lethal and (2) nonlethal. When turfgrass plants are exposed directly to a rapid decrease in temperature below 23°F (–5°C), injury can be lethal, resulting in significant stand loss (Figure 2.33). Turfgrass leaves initially appear water soaked, turn whitish-brown in color, and then turn dark brown. Low-temperature exposure causes plants to lose control of membrane function, resulting in solute leakage and, eventually, a water-soaked appearance. Damaged turfgrass tends to mat over the soil surface and emit a distinctly putrid (rotten) odor. Similar turf browning occurs to warm-season grasses in fall when temperatures drop to ~50°F (10°C) during periods of high light intensity. This type of injury, referred to as "chilling" injury, only results in the normal discoloration (browning) of turfgrass leaves without lethal injury to the plants.

Insufficiently hardened turfgrass plants have lush growth (or high moisture content) and thus become more susceptible to direct low-temperature injury. Hydrated, nonhardy cells are more susceptible to freeze damage than dehydrated, acclimated cells. Turfgrass plants become "hardened" by being exposed to gradual decreases in temperature and frosts prior to a drastic, sudden drop in temperature or a hard freeze. This hardening process (often referred to as osmotic adjustment) involves an environmentally induced 10% to 25% reduction in plant tissue water content and an accumulation of carbohydrate reserves. This reduces ice formation inside cells, preserving cell walls. Three to five light frosts prior to a sudden, hard freeze help increase soluble carbohydrate reserves, amino acids, membrane structure, and soluble protein constituents in turfgrass plants. Solute accumulation depresses the freezing point, enabling crown tissue to withstand low temperatures without severe protoplasmic membrane disruption. Although intercellular water freezes at 14°F to 23°F (–10°C to –5°C), concentrated solutes from osmotic adjustment may serve like antifreeze agents, lowering plant cell freezing to –4°F to –40°F (–20°F to –40°C), depending on the type and amount of solutes present. Turfgrasses not exposed to these preconditioning frosts have succulent crown tissues, which are highly susceptible to freeze damage.

Maximum low-temperature tolerance for warm-season turfgrasses is reached in early winter. Plant damage occurs most often in late winter or early spring, just prior to or during spring greening (green-up). This is mostly associated with widely fluctuating temperatures during this time in which plants lose their earlier cold-hardening effects. Wide temperature swings also may interrupt spring greening as warm temperatures favor turfgrass growth and occasional freezes kill the new leaves. Thus, new leaves must be reproduced, delaying the complete greening process in spring while placing a greater demand on carbohydrate reserves in the roots, rhizomes, and stolons (Figure 2.34). This repeated "die-back" and regeneration of new leaves depletes energy reserves,

**FIGURE 2.34** Bermudagrass spring damage from repetitive new leaf production followed by freezes.

reduces rooting, and makes the plant more susceptible to direct low-temperature injury.

With cool-season grasses, low-temperature damage often occurs following a short period of mild weather where thaw occurs, causing crown hydration. A subsequent late winter or early spring freeze often leads to severe turf loss, especially when turf surfaces are wet and the surface is composed predominantly of *Poa annua* compared to bentgrass.

An initial low mowing height (scalping height) in spring after the last expected frost date will help remove the brown and/or dead overlying turfgrass plant material damaged from low temperatures. By removing this turfgrass material, sunlight can reach the soil surface more effectively and warm the soil, thereby producing quicker green-up of the turfgrass.

**Desiccation.** Another very common form of low-temperature turfgrass injury is desiccation, caused by high winds in combination with low relative humidity. This desiccation or "drying out" occurs on turfgrass plants unprotected from wind with limited soil moisture due to either drought conditions or low temperatures (frozen soil moisture) (Figure 2.35). When temperatures rise above 32°F (0°C), turfgrass leaves lose $H_2O$ to the atmosphere due to the pressure gradient created from low-humidity conditions. Consequently, turfgrass crowns, nodes, and roots dry out, causing desiccation of these plant tissues and important plant proteins. Plant injury from desiccation can result in significant losses of turfgrass. It is important to maintain adequate soil moisture during these conditions, especially in spring when turfgrass roots are developing. If the area is exposed to high winds and not continuously covered by snow, artificial covers are often placed to help protect the turf from desiccation. Sand topdressing, wind breaks, and snow fences are additional ways to prevent turf desiccation.

**Traffic.** Traffic can play a significant role in low-temperature turfgrass injury caused by vehicular and/or foot traffic that occurs on frozen or slush-covered turfgrass (Figure 2.36). The most common type of low-temperature injury due to traffic is from frost damage where mechanical injury ruptures plant cells in the early morning before frost melts. Frozen turfgrass crowns are easily damaged from traffic, which should be minimized or diverted from frozen turfgrass until the soil and turfgrass plants have completely thawed. Lightly syringing frosted turfgrass areas prior to allowing traffic will aid in reducing traffic injury. Highly trafficked areas are also prone to high soil compaction. Generally, compacted soils reduce turfgrass rooting and overall plant health, resulting in weak turf that is more susceptible to low-temperature injury. Such areas should be heavily cultivated during summer months to improve soil physical properties.

**Ice cover.** Low-temperature injury from ice cover results from turfgrass plants being covered by ice for extended periods of time (greater than 50 days). Injury occurs from $O_2$ suffocation (termed anoxia) and toxic gas buildup (Figure 2.37). *Poa annua* greens, for example, can withstand ~50 to 60 days under ice; creeping bentgrass and Kentucky bluegrass may survive 150 days of coverage. This is possibly due to *Poa annua* being more sensitive to low $O_2$ levels than creeping bentgrass. Interestingly, excess thatch may also contribute to these as soil microbes deplete the limited supply of $O_2$ under ice as they slowly decompose the thatch layer. Even if the ice is removed from *Poa annua* greens after 45 days' exposure, turf damage is not lessened.

**FIGURE 2.35** Zoysiagrass winter damage amplified by drying winds.

**FIGURE 2.36** Low-temperature injury amplified by traffic on frozen turf.

**FIGURE 2.37** Low-temperature injury by extended ice cover, resulting in oxygen deficiency.

A key to alleviating turf damage from extended ice cover is physically disrupting or breaking the ice (Figure 2.38). This allows toxic gases to escape. A darkening agent such as compost, charcoal, or natural organic fertilizer may be used to absorb heat and create pores in the ice to allow for gas exchange. The melted ice must drain away or the turf remains susceptible to refreezing during the next cold snap. Other ways to minimize the risk of winter injury from cold temperatures or ice coverage include:

- Avoiding excessive fall N, which causes lush growth during the hardening process
- Providing surface and subsurface drainage, especially collar "dams" that prevent $H_2O$ from draining off greens' surfaces
- Raising the mowing height in fall to encourage carbohydrate production and storage
- Reducing shade, which weakens grass grow and also contributes to longer ice covering
- Avoiding excessive (>3%, by weight) thatch layering
- Using covers on chronically affected greens
- Reducing *Poa annua* populations in sites susceptible to low temperatures

**FIGURE 2.38** Physically disrupting an ice cover to allow toxic soil gases to escape.

**Frost heaving.** An uncommon type of low-temperature turfgrass injury is frost heaving, caused by the freezing and thawing cycles on wet soils where the soil surface moves due to ice formation. Heaving pushes turfgrass crowns and roots out of the soil, where they are exposed. Frost heaving mainly occurs on newly germinated seedlings.

**Diseases.** The interaction of diseases and low temperatures can combine to cause more winter-kill than either parameter alone. Presumably, the disease organism somehow predisposes the grass to winter-kill by reducing internal moisture supplies, altering cell wall components, elevating plant N levels, or producing or disrupting normal plant products, such as growth hormone production, which influence low-temperature tolerance. Spring dead spot on warm-season grasses and various snow molds on cool-season grasses are common examples of a disease/low-temperature interaction.

### Best Management Practices Minimizing Low-Temperature Stress

Several parameters influence the susceptibility of warm-season turfgrasses to low-temperature stress; however, only a few are controllable by turfgrass managers. Protecting turfgrass crowns from low temperatures is essential for survival as leaves, roots, and lateral shoots regenerate from them. Temperature, freeze and thaw frequency, and freeze and thaw rate are factors outside a turfgrass manager's control. Conversely, several parameters within the control of turfgrass managers include:

- Degree of shading
- Drainage
- Fertility
- Irrigation
- Mowing
- Pest control
- Soil cultivation
- Thatch level
- Traffic
- Turfgrass selection

**Shade.** Exposure of turfgrass plants to sunlight is critical for several reasons, such as food production (photosynthesis) and hardening. Prolonged shade reduces carbohydrate levels in turfgrass plants, produces weaker plants, and suppresses soil temperatures (Figure 2.39). Shaded areas also remain colder for prolonged periods due to minimal sunlight exposure and tend to maintain dew or ice coverage longer than exposed, sunny areas; this can promote diseases. In addition, competition from tree roots for nutrients and $H_2O$ weakens turfgrass plants. Bright sunlight exposure also signals plants to harden off during fall. Turfgrass plants are "hardened" by being exposed to light frosts prior to a heavy freeze. Northern or northeastern facing slopes receive less (or little) direct sunlight in the winter months and are more susceptible to low-temperature stress. Interestingly, due to tree root respiration

**FIGURE 2.39** Low-temperature damage amplified by shade, reducing the plant's ability to store carbohydrates.

that increases soil temperatures, shade can moderate light frost or freezes, especially on cool-season turfgrasses.

One method to determine and document the degree of shading on each golf hole is using time-lapse photography. This can be performed by setting up a camera on a tripod either at the teeing area or the putting green area and taking time-lapse photographs at one-hour intervals for the entire day to document a representative degree of shading for each hole. These photographs can be invaluable if or when a severe loss of turfgrass stand occurs due to low-temperature stress and/or data are needed to selectively remove some trees.

**Drainage.** Poor surface and/or subsurface drainage can result in direct low-temperature damage or injury to the crowns of turfgrass plants. The level of $H_2O$ or moisture content within turfgrass crowns is positively correlated to the degree of low-temperature injury. The higher the moisture content of turfgrass crowns is, the higher is the susceptibility to low-temperature damage. Turfgrass plants growing in areas with poor drainage (surface or subsurface), high compaction, or excessive irrigation are at a greater risk of low-temperature injury.

**Fertility.** Research indicates excessive late-season N fertilization can decrease the low-temperature tolerance of some turfgrasses. Late-fall fertilization typically coincides with the last fall mowing. Late-fall N fertilization promotes succulent or "lush" turfgrass growth and reduced root carbohydrate formation by forcing shoot growth over root or rhizome growth as well as encouraging diseases such as snow mold or spring dead spot. Turfgrass susceptibility to low-temperature stress can be reduced by applying a late-summer application of a 4-1-2 or 3-1-2 ratio of N-P-K fertilizer. Adequate levels of potassium (K) have been shown to improve low-temperature stress tolerance. Potassium plays a regulatory role in plant $H_2O$ relations, serving as cellular antifreeze; reduces the amount of free $H_2O$ in the plant cells; and lowers the cellular freezing point. Soil test results should indicate medium to high levels of K, while leaf tissue analysis should be at least 1.5% K. However, excessive levels of K have not been shown to increase winter hardiness, and the high salt content from these may cause crown desiccation. In terms of N, consider using soluble sources when growth ceases; slow-release sources are used four to five weeks prior to this.

**Irrigation.** Water is the primary requirement for turfgrass growth and survival. Irrigation should be utilized to "condition" the turfgrass. Several methods exist on scheduling irrigation; however, irrigation should be managed according to the exact amount of $H_2O$ needed at any given time or any given condition. Just as excessive moisture may increase low-temperature damage or injury, the lack of moisture can be equally damaging. A certain amount of $H_2O$ or moisture is needed for crown tissues to survive low-temperature stress. If a moisture deficiency occurs, crown tissues become desiccated; thus, they are severely weakened and more susceptible to low-temperature stress and injury. Maintaining soil moisture at or slightly below field capacity and preventing standing $H_2O$ helps keep crown tissue properly hydrated.

**Mowing.** Mowing is the most basic, yet most important, cultural turf management practice. Improper mowing decreases root and rhizome growth and depletes food reserves needed for the regeneration of shoot growth. If turfgrass plants are mowed too low in late summer, rooting will be discouraged and the accumulation of carbohydrates reduced, which can contribute to a greater susceptibility to low-temperature stress (Figure 2.40). The one-third rule of mowing height should be utilized to schedule mowing frequency. If mowing height is increased in late summer or early fall, the leaf area of the turfgrass is increased; therefore, photosynthesis and carbohydrate production are increased. In addition, by increasing the mowing height, turfgrass crowns and growing points are better insulated from low-temperature exposure.

**Pests.** Pests damaging to turfgrass roots, such as mole crickets, nematodes, spring dead spot, root rot diseases (*Pythium* species), patch diseases (*Rhizoctonia, Gauemannomyces*, and *Leptosphaeria* species), and white grubs, may decrease the tolerance of turfgrasses to low-temperature stress. In addition, improper application timing of certain pesticides affecting root growth (certain pre-emergence herbicides) may contribute to low-temperature stress and injury.

**FIGURE 2.40** Low-temperature damage by perpetual low mowing (scalping) and uncontrolled traffic.

**Soil cultivation.** Soil cultivation is accomplished by several means, such as aerification, spiking, and vertical mowing. These operations help relieve soil compaction, improve surface drainage, allow for deeper and faster root and water penetration, and reduce thatch accumulation. Applying topdressing after soil cultivation operations also improves soil structure. Additional solid-tine aerification may be justified in late fall to help drain low areas and aid in favorable gas exchange.

**Thatch level.** Thatch is the accumulation of organic material (turfgrass leaves, stems, stolons, clippings, etc.) within the top layer of the turfgrass biosphere. As thatch accumulates, the crowns, rhizomes, and stolons of the turfgrass tend to grow and develop within this thatch layer. These growing points then become elevated above the more insulating soil zone and therefore are more exposed. Frequent soil cultivation and topdressing are the most effective means of controlling and/or reducing thatch levels.

**Traffic.** Traffic (either vehicular or foot) can be very damaging to frozen turfgrass crowns, leading to the eventual death of plants (Figure 2.41). In addition, traffic leads to soil compaction that reduces soil $O_2$ levels (porosity) and $H_2O$ infiltration and percolation rates. Compacted soils may not drain excessive $H_2O$; as a result, they often remain wet and saturated, thereby increasing the risk of low-temperature stress and injury. Most often, these problems are observed on teeing areas, approach areas to tees and putting greens, and in golf cart traffic areas onto and in the fairways. Reducing, diverting, or withholding traffic can aid in relieving soil compaction and decreasing the chances of low-temperature stress and injury.

**Turfgrass selection.** Turfgrass species and varieties or cultivars within a particular species have varying degrees of low-temperature tolerance (Table 2.10). Generally, turfgrass cultivars possessing greater low-temperature tolerance tend to exhibit a deeper growing, denser rhizome network. Deeper growing roots and rhizomes are more insulated by being deeper in the soil profile; therefore, they are less exposed to low-temperature damage or injury. Stoloniferous grasses, such as St. Augustinegrass and centipedegrass, have growing points less protected from low temperatures. Extensive research and turfgrass breeding efforts are ongoing for improved low-temperature tolerance.

**FIGURE 2.41** Avoiding turf damage by delaying play due to frost.

**TABLE 2.10**
**Relative Low-Temperature Tolerance of Turfgrasses[a]**

| Low-Temperature Tolerance | Warm-Season Species | Cool-Season Species |
|---|---|---|
| Most tolerant of low temperatures ↑ | Buffalograss | Roughstalk bluegrass |
| | Zoysiagrass | Creeping bentgrass |
| | Common bermudagrass | Kentucky bluegrass |
| | Hybrid bermudagrass | Canada bluegrass |
| | Centipedegrass | Colonial bentgrass |
| | Bahiagrass | Annual bluegrass |
| | Carpetgrass | Fine fescues |
| | St. Augustinegrass | Tall fescue |
| | Seashore paspalum | Perennial ryegrass |
| ↓ Least tolerant of low temperatures | | Annual ryegrass |

[a] Cultivars within a species can vary.

**Turf covers.** Turf covers may help protect turfgrass crowns from direct low winter temperatures and desiccation (Figure 2.42). Snow is probably the best insulator and should remain as long as possible. Artificial covers that use an air layer to insulate the turf from extreme moisture and temperatures are the next best option (Figure 2.43). However, covers should not be used to prematurely green-up plants if temperatures are still likely to suddenly drop. For bermudagrass greens, covers should be considered if temperatures are forecasted for below 25°F (–4°C). If temperatures are forecasted for the midteens or lower, double covers or adding pine straw to covering provide additional insulation. In extremely cold areas, permeable insulating covers along with pine straw or other porous material are applied to act as an insulator. On top of the insulating material an impermeable cover is laid. This layering system is especially beneficial in preventing turf desiccation and delaying crown hydration. These are applied in late fall once plants have stopped growing. A preventive fungicide to protect against snow mold is applied prior to covering. In addition, various venting systems may also be installed to prevent anoxic conditions.

**Frequent soil sampling.** Soil samples should be taken on a frequent basis until a baseline for certain parameters is completely documented. These are utilized to determine chemical analysis (fertility levels), disease pathogens, nematodes, and physical property analysis (particle size, bulk density, porosity, water infiltration rates). All of the information garnered from these samples could be correlated to low-temperature stress tolerance or lack thereof.

**Documentation.** Documentation or recordkeeping is probably one of the most overlooked aspects of turfgrass management and one of the most important. Historical documentation of parameters such as shade patterns, wind movement,

**FIGURE 2.42** A cover that protects turfgrass crowns from direct low winter temperatures and desiccation.

**FIGURE 2.43** Green turf (left) where a cover was used during cold temperatures.

and drainage problems can be directly correlated to the risks associated with low-temperature stress and turfgrass damage or injury. These parameters should be documented for each golf hole and a proactive management plan initiated to improve problem areas. Other maintenance practices should be documented as well, such as irrigation amounts, mowing heights, mowing patterns, rainfall amounts, minimum and maximum temperatures, frost occurrences, length of frost, fertilizer applications, and pesticide applications. All of these and other parameters should be documented in detail and retained for later use, if needed.

## ICING GREENS FOR WINTER PROTECTION

"Icing" has been touted as a means of protecting warm-season golf greens against winter damage similarly to protecting citrus and berries in subtropical areas like Florida (Figure 2.44). The theory behind icing golf greens is to provide a thin ice cover that will hopefully moderate soil temperatures around 32°F (0°C).

**FIGURE 2.44** Icing of a green in an attempt to "protect" against winter damage. Little benefit occurs from this.

Two types of freeze events are common. Advective freezes occur when a dry (low dew point), cold air mass moves in quickly with high (>5 mph, 8 km/hour) wind, normally with cloud cover. Icing provides little to no protection from advective freezes as desiccation from winds is as damaging as low temperatures. Radiant freezes occur when heat is lost from soil and plants and radiates back to the sky. This occurs on calm, clear nights with no cloud cover. Icing of crops works best with radiant freezes. In many areas, advective freezes are much more common and damaging as they are associated with rapidly moving cold fronts.

When protecting against radiant freezes in citrus, $H_2O$ is constantly added (irrigated) lightly at 0.25 to over 0.40 inch/hour (6.4 to 10 mm) to take advantage of the slight heat released when water molecules turn into ice crystals. This is referred to as latent heat of fusion where 80 calories heat is released when 1 g of $H_2O$ freezes. This heat is retained in the canopy of citrus trees, providing a "blanket" for protection, keeping the plant surface near 32°F (0°C). Icing protects trunks of young citrus trees, but not the fruit. Conditions of low temperatures, low humidity, or high winds require more $H_2O$.

The heat of vaporization is the heat lost when liquid $H_2O$ changes to vapor. Evaporative cooling is avoided during the icing event as 1 gram of evaporating $H_2O$ removes ~600 calories of heat from the air, which is about 7.5 times more than the latent heat of fusion. Therefore, 7.5 times more $H_2O$ must be freezing than evaporating to maintain stable heat loss near 32°F (0°C) (and not sustainably drop). If adding this amount of $H_2O$ does not occur, then evaporation removes heat from the crop, causing damage. Increasing evaporation is why icing during advective (windy) freezes or when air temperatures are below 22°F (–5.6°C) is not very beneficial.

Ice is a poor insulator. Temperatures of ice-covered plants drop if freezing stops and evaporation starts, especially with low relative humidity. Since evaporation is increased by wind, speeds ≥5 mph (8 km/hour) greatly reduce effectiveness of sprinkler irrigation for freeze protection. Finally, in citrus, if the icing system is working properly, the ice will be clear in color. If ice has a milky, white appearance (presence of air bubbles), water is freezing before it strikes the plant, severely reducing its beneficial properties.

The major damage to turfgrass is when too little or too much moisture is in the plant during cold temperatures. Icing greens would probably work best for a one-time freeze. However, since this is rarely the case, icing would melt during the day, saturating the soil and thus causing plant crowns to excessively hydrate. When temperatures drop below freezing that night, intracellular ice formation would possibly occur, causing cellular damage and eventual crown death (Figure 2.45).

Just as excessive moisture may increase low-temperature damage or injury, the lack of moisture can be equally damaging. A certain amount of $H_2O$ or moisture is needed for crown tissues to survive low-temperature stress. If a moisture deficiency occurs, crown tissues become desiccated and thus have less buffering capacity against temperature swings. Maintaining soil moisture at or slightly below field capacity and preventing standing $H_2O$ help keep crown tissue properly hydrated.

**FIGURE 2.45** Damage to a golf green intentionally "iced" the previous winter.

### Diagnosing Low-Temperature Injury

Each year, prior to normal turfgrass green-up in spring, diagnostic measures should be taken by superintendents suspecting possible cold weather injury to determine the extent of the damage.

Superintendents should anticipate that certain areas of turfgrass may be more susceptible to low-temperature injury than other areas. Cultural factors contributing to low-temperature injury and increased susceptibility include poor drainage, soil compaction, excessive thatch, reduced light intensity (shade), north or northeastern facing slopes, excessive fall N fertilization, and low mowing heights. Any areas exhibiting one or more of these may be more susceptible to low-temperature injury.

If any of the aforementioned conditions are present, or have occurred, and low-temperature injury is suspected, utilize one of the following methods to determine the severity of the potential damage:

Method one

1. Remove the individual turfgrass plant crown near the soil surface.
2. Cut a slice through it with a sharp razor blade or knife.
3. Examine the crown under a 10× magnifying lens. If the crown tissue is firm and white in color with turgid cells, it should be healthy and has survived the low-temperature stress. Crowns exhibiting dark color or brownish color with a "mush" or soft appearance have been injured from low-temperature exposure.

Method two

1. Use a cup cutter to remove samples from the suspected damaged areas.
2. Label the samples appropriately.
3. Remove soil from turfgrass samples by washing with water.
4. Remove the "old" dormant top growth from the turfgrass samples.
5. Expose the crown, rhizomes, and stolons by pulling the sample apart.
6. Place the turfgrass sample in a clear, plastic Ziploc® bag along with a wet paper towel, but do not seal shut.
7. Place the bags under a grow light or in a southern facing window to create summer-like growing conditions.
8. Rewet the paper towel in the bag as needed to prevent the sample from drying out.
9. If healthy, greening of the surviving samples should occur within five to seven days.

Method three

1. Collect turfgrass plugs using a cup cutter from suspected low-temperature damage areas.
2. Place or plant these plugs in a suitable container of native soil with drainage holes, properly labeling them.
3. Place the containers in a greenhouse, or in a room beneath a heat lamp or grow light or, as a last resort, in a southern facing window.
4. Keep the turfgrass plugs adequately watered.
5. Note that turfgrass should initiate growth and greening within 7 to 10 days.
6. Assess the amount of greening after plugs have been grown for two to three weeks. Suspected areas with less than 50% greening should be considered extensively damaged from low-temperature exposure and

will probably require renovation. Less damaged areas may recover with proper management practices and extra attention.

7. Repeat this sampling procedure on a 14- to 21-day interval through the periods of potential cold weather injury.

Turfgrass samples should be taken during winter months to assay low-temperature damage. Samples exhibiting less than 50% greening represent areas with extensive damage and will probably require renovation.

## Heat Stress on Cool-Season Turfgrasses

In many areas, high-temperature/humidity stress is a concern for superintendents with cool-season turfgrasses present on the golf course. Generally, high-temperature stress is closely associated with $H_2O$ stress. A cool-season turfgrass may turn brown and enter a dormancy phase if rainfall or irrigation is not available during high-temperature stress periods of summer. This phase of dormancy occurs with cool-season turfgrasses as a means of survival during severe high-temperature and/or $H_2O$ stress. These dormant turfgrass plants will initiate growth from basal crowns and lateral stems when the stress is relieved.

The growth of cool-season turfgrasses slows significantly when the average daytime temperatures increase above 90°F (32°C). This is especially acute if night temperatures remain in the high 60s°F (15.5s°C) or above and if the relative humidity of air remains near saturation. However, cool-season turfgrasses will continue to grow with temperatures above 90°F (32°C) as long as nighttime temperatures fall below approximately 65°F (18°C) and adequate but not saturated soil moisture is available. Low nighttime temperatures provide turfgrass with a recovery time to mobilize stored carbohydrate reserves for energy production. Permanent turfgrass injury to cool-season turfgrasses often occurs if ambient temperatures exceed 100°F (38°C).

The heat load that leaves are exposed to in full sunlight is very high. Means of dissipating this is by reflection and refraction of long-wave radiation, by heat conduction and convection to cool air (termed sensible heat loss), and from evaporation (or latent) cooling from $H_2O$ loss. Sensible heat loss and evaporative heat loss are the most important means of regulating leaf temperatures. When transpiration is high, heat dissipation is sufficient. However, when plants close stomates to conserve $H_2O$, high leaf temperatures quickly develop as well as insufficient internal plant $CO_2$ levels for photosynthesis. Turf managers then often induce artificial conduction and convection (sensible) heat loss by installing fans.

In general, there are three basic types of high-temperature stress to cool-season turfgrasses: (1) heat exposure, (2) scalding, and (3) desiccation. As mentioned, each of these stresses has an association with $H_2O$ stress.

**Heat exposure.** Heat exposure can be lethal to cool-season turfgrasses when temperatures exceed the range for plant growth processes. As temperatures rise above optimum levels, plant photosynthesis decreases while respiration increases. Direct heat stress kill of the turfgrass can occur if the temperatures are well in excess of 100°F (38°C) as the exposure causes denaturation of proteins and cellular enzymes, rupturing of cell membranes, and cell protoplasm precipitation (or leaking). Heat exposure can cause gradual decline in cool-season turfgrass plants—termed indirect heat stress—by slowing the growth of roots and shoots when soil temperatures at 2 inches (5 cm) exceed 75°F (24°C), especially at or above 84°F (29°C). Indirect heat stresses occur with decreased photosynthesis, including the interruption of electron transport in photosystem II, reduction in RuBP carboxylase activity, decreased chlorophyll content, and leaf senescence from decreased cytokinin activity. This indirect stress leads to slow decline of cool-season grasses and may also lead to plant death if the stress is not relieved.

Photosynthesis produces carbohydrates that are consequently used in respiration to produce ATP for energy. When ambient temperatures rise above a specific level, $C_3$ plants cannot produce enough carbohydrates to fulfill the demand for respiration. This is especially important for low mowed turf, which removes large amounts of leaf area used for photosynthesis while respiration continues. This temperature is referred to as the temperature compensation point. At this temperature, the amount of $CO_2$ fixed by the dark reactions of photosynthesis is equal to the amount of $CO_2$ released by mitochondrial respiration. Prolonged exposure of temperatures in excess of the temperature compensation point leads to continuous depletion of carbohydrate reserves and eventual weakening of the turf. For creeping bentgrass golf greens, this temperature is 86°F or 30°C. Ultimately, this leads to exhaustion of carbohydrate reserves via respiration, eventually causing "summer bentgrass decline" (Figure 2.46). Reducing the soil temperature below 84°F (29°C) allows resumed bentgrass growth.

Summer decline symptoms in cool-season grasses include leaf chlorosis (yellowing), shoot density reduction, and root dieback. These symptoms are often confused with pathological problems. Summer decline symptoms, however, are not

**FIGURE 2.46** "Summer bentgrass decline" where extended heat (>86°F, 30°C) eventually led to exhaustion of carbohydrate reserves via respiration.

alleviated or reversed when treated with fungicides as turf quality and root growth continue to deteriorate following application.

Physiologically, free radicals often form inside cells during heat stress, including superoxide radical $(\cdot O_2^-)$, hydrogen peroxide ($H_2O_2$), hydroxyl free radical ($\cdot OH$), and singlet oxygen ($^1O_2$). Free radicals react with macromolecules such as unsaturated fatty acids and proteins, leading to oxidative damage to cell membranes (termed lipid peroxidation) and eventual turf decline. Antioxidants are used by plants to reduce the negative effects of free radicals, protecting plant tissue, especially during heat stress.

With reduced rooting from heat stress, turf plants may suffer from nutrient and moisture deficiencies. Cytokinin synthesis in roots also declines as roots die back. Reduced moisture levels reduce transpirational cooling, which lowers leaf temperatures 5°F to 18°F (3°C to 10°C) below air temperatures. Turf managers often attempt to alleviate these stresses by incorporating higher mowing heights, applying nutrients in liquid forms, and installing above- and belowground fans to help dissipate heat buildup.

**Scalding.** Scalding occurs when turfgrasses are in standing pools of $H_2O$ and intense sunlight creates a rise in the water's temperature. Consequently, the plants are "scorched" or scalded from the $H_2O$ temperature (Figure 2.47).

**FIGURE 2.47** Scald from standing water exposed to intense sunlight, leading to lethal water temperature levels.

**FIGURE 2.48** Turf desiccation where transpirational losses exceeded soil moisture levels.

**Desiccation.** Desiccation of turfgrass shoots can occur due to an imbalance in $H_2O$ relations between the plant and the atmosphere (Figure 2.48). Shoots can be desiccated even if soil moisture is adequate. This type of desiccation occurs when the rate of transpiration (loss of $H_2O$ to the atmosphere) of shoots exceeds the rate of $H_2O$ absorption from soil by roots. Warm, low-humidity days are a typical scenario for turf desiccation. Although soil moisture is adequate, it cannot provide the $H_2O$ necessary for the transpiration demand due to low atmospheric humidity levels. Desiccation can also occur if soil moisture levels are inadequate to provide roots with enough moisture to sustain the transpiration rates.

### Best Management Practices for Reducing Heat Stress of Cool-Season Turfgrasses

High-temperature stress to cool-season turfgrasses can be minimized by using cultural practices that promote maximum rooting and reduced direct heat stress. Root development is directly correlated to $H_2O$ absorption and uptake from the soil. If adequate soil moisture is available and plants can access this moisture, then the plant can maintain its ability to "cool" itself via transpiration. This is acutely important with cool-season turfgrasses associated with putting greens due to the low mowing heights. Root development is also associated with mowing height or shoot height. Turfgrasses mowed at a lower mowing height have a shorter or shallower root system. Therefore, it is most important to manage these closely mowed areas more intensely to promote maximum root growth and development. Five basic management practices help plants survive severe high-temperature stress and/or $H_2O$ stress:

1. Irrigation and fans
2. Mowing
3. Soil cultivation
4. Fertilization
5. Pest control

**Irrigation and fans.** Irrigation is a very important management practice affecting turfgrass survival against heat and moisture stress. In spring, when root growth rates are rapid, cool-season turfgrasses should be irrigated sparingly to encourage plants to root deeper in "search" of moisture. If plants are overwatered in spring, they develop short root systems, creating a "lazy" root system that is not required to grow and develop in order to search for deeper moisture. However, during heat stress periods, it is important to maintain adequate soil moisture to prevent any loss of turfgrass roots and also provide the necessary moisture to allow the transpirational cooling necessary for survival. Deep, infrequent watering practices should be used to minimize continuous saturated soil conditions (Figure 2.49). In addition to causing a poor soil atmosphere with limited soil $O_2$, saturated soils have a high heat capacity, not allowing rootzone temperature cooling during nighttime hours compared to drier soils.

**FIGURE 2.49** Hand-watering desiccated spots to provide transpiration cooling.

To maximize desirable heat dissipation on leaves, turf managers often combine light misting and the use of fans to create artificial air movement (Figure 2.50). This combination maximizes heat loss by reducing the boundary layer surrounding leaves. As a result, an evaporational cooling effect occurs from $H_2O$ loss of the applied moisture; also, conduction and convection (sensible heat loss) reduce acute heat accumulation on leaves. A reduction of 12°F occurs for surface temperatures when misting is combined with air movement from fans.

**Mowing.** Mowing height can make a significant difference in cool-season turfgrass survival during high-temperature stress, especially on cool-season turfgrass putting greens. A reduction in mowing height on a bentgrass putting green during high-temperature stress is often detrimental. The mowing heights of cool-season turfgrasses on putting greens should be increased during these stress periods to enhance root survival and transpiration rates.

**Soil cultivation.** Soil cultivation or aerification is considered to be a key management practice to the survival of cool-season turfgrasses in high temperatures. Soil cultivation is critical to the management of poor drainage, soil compaction, soil $O_2$ and $CO_2$ levels, and turfgrass growth. It is important to aerify cool-season turfgrasses in the fall and early in spring during their optimum growth period to stimulate shoot and root growth as much as possible. These core aerification events are supplemented with additional monthly venting events using solid-tine aerification, hydrojetting, slicing, and spiking. Cool-season turfgrasses need as deep a root system as possible going into the summer when high-temperature stress is expected. In addition, adequate drainage during these high-temperature periods is important to minimize disease pressure.

**FIGURE 2.50** Using fans to create artificial air movement to maximize heat dissipation and ET.

**Fertilization.** Proper fertilization is also important to minimize high-temperature stress. During early summer, prior to the onset of high temperatures, reducing the amount of N fertilization and increasing the amount of K fertilization on cool-season turfgrasses is generally recommended. This will promote more root growth versus shoot growth and improve stress tolerance. Heavy summer N fertility will allocate sugars and energy into shoot production, decreasing root carbohydrate reserves and growth. A fertilizer with an N-to-K ratio of 1:1 is often utilized prior to the onset of high temperatures. Slow-release N sources should be used cautiously to avoid excess N release under high temperatures. On putting greens, liquid applications are used to control N availability to plants during this time.

**Pest control.** Pest control is of paramount importance during high-temperature stress periods on turfgrasses. Disease pressures such as dollar spot, anthracnose, patch diseases, bacteria, crown rot, and *Pythium* can all be detrimental during summer. Preventative fungicide applications should be utilized where possible to minimize disease potential. In addition, any pesticide that may negatively affect root development, such as some herbicides, should be avoided during these stress periods.

## Drought Stress

In addition to having lower water-use rates than cool-season grasses, warm-season grasses also tend to be more drought tolerant (Table 2.11). This is partly due to differences in rooting depth; also, warm-season grasses possess $C_4$ photosynthesis, which makes them more water-use efficient. Several of the cool-season grasses, such as Kentucky bluegrass and creeping bentgrass, avoid severe drought stress by going dormant. Although they may appear dead, they will recover quickly when $H_2O$ is resupplied.

Established perennial species generally have better drought tolerance due to their deep rooting and rhizomatous growth. Annual species that are shallow rooted are more susceptible to drought conditions. Also, turfgrasses grown under arid conditions lose more $H_2O$ (~30%) than similar turf grown under humid conditions.

In general, a low water-use rate by a turfgrass is based on its ability to maintain dense shoot coverage, slow vertical leaf growth rate, narrow leaf width, horizontal leaf orientation, and color. As discussed, cultural practices and environmental parameters also have considerable influence on each of these contributing factors.

**TABLE 2.11**
**Relative Drought Resistance of Turfgrasses[a]**

| Relative Drought Resistance | Turfgrass |
|---|---|
| Excellent | • Bahiagrass<br>• Blue grama<br>• Buffalograss<br>• Common bermudagrass<br>• Wheatgrass<br>• Zoysiagrass |
| Very good | • Hybrid bermudagrass<br>• St. Augustinegrass |
| Good | • Canadian bluegrass<br>• Centipedegrass<br>• Fine fescue<br>• Kentucky bluegrass<br>• Seashore paspalum<br>• Tall fescue |
| Fair | • Perennial ryegrass |
| Poor | • Annual bluegrass<br>• Annual ryegrass<br>• Carpetgrass<br>• Colonial bentgrass<br>• Creeping bentgrass<br>• Roughstalk bluegrass |

[a] Listed alphabetically within each category.

## SALINITY

Saline $H_2O$ can cause stress and injury to plants by several means. Direct injury occurs with the accumulation of salts such as sodium, chloride, boron, and other ions in soil as well as ion accumulation in plants (Figure 2.51). These cause certain enzymes to be inactivated and also inhibit protein synthesis and eventually photosynthesis. Salts, however, reduce plant growth, mostly by their osmotic effect. The more salt in its rootzone, the harder a plant must work to take up $H_2O$. This direct osmotic stress causes $H_2O$ to move from the plant into the soil because of a salt concentration gradient. Salt stress is similar to drought stress in appearance, except the soil can still be moist. The initial response of turfgrass to salinity stress is reduced shoot growth. Leaf blades become narrower and stiff, and they can become darker green, or even blue-green, in color as the plant cannot take up enough $H_2O$. There also is a tendency for turf to wilt faster than normal from the osmotic stress. Root growth initially increases, but over time, it is reduced, presumably as a response to reduced shoot growth.

**FIGURE 2.51** Turf damage from salt exposure.

Shoot and root growths are reduced at high salt levels through both direct and indirect salt injury. Leaf tipburn and a general thinning of turf occur at higher salinity levels. Due to mowing, leaf tips containing high levels of sodium are removed in turf. However, on golf greens, such a small amount of leaf blade is removed during mowing that sodium accumulation may be more problematic. Young plants are more sensitive to salinity stress than mature ones. Salinity also reduces the number of seeds that germinate and delays establishment by sprigs. A common symptom of turf damaged by high-salinity soils is spotty stands of grass.

Plant nutrient deficiencies may also occur in saline soils due to the typically high levels of sodium and calcium, which reduce uptake of other nutrients by competition. The most common example of this is the antagonistic effects of sodium on plant uptake of calcium, K, magnesium, and ammonium nitrate. As excess sodium displaces calcium in root cell walls and membranes, these cells start to leak out their contents, especially K.

Plant tolerance to salt stress varies greatly. Some plants avoid salt stress by excluding salt absorption by roots; excluding salt from meristems, especially in shoots; extruding excess salts; sequestering (capturing) Na and Cl in vacuoles of root cells; or diluting absorbed salts. Other plants tolerate salt stress by adjusting their metabolism to withstand direct or indirect injury. Still others regulate their internal osmotic potential of tissue fluid to compensate for increases in substrate osmotic potentials, often using internal supplies of calcium to displace the sodium. In most cases, the mechanism of salt tolerance in plants is a combination of several methods. As a compensatory adaptive mechanism to nutrients and $H_2O$ stress under saline conditions, plants often increase root biomass to enlarge their $H_2O$ and nutrient absorbing area. Highest root weights occur at intermediate salt levels but decline at high levels. Ornamental plants tend to be much more sensitive to sodium, chlorine, and boron than turfgrasses.

### Turfgrass Salt Tolerance

Turfgrass species have been classified according to salt tolerance (Table 2.12). Most turfgrass comparisons are based on the salt levels that cause a 50% reduction in top or root growth. Only a few species grow well under saline conditions. Zoysiagrass, seashore paspalum, bermudagrass, and St. Augustinegrass are the best warm-season turf species to grow if irrigation is limited to saline $H_2O$; alkaligrass, tall fescue, and perennial ryegrass are the best cool-season turfgrasses. Creeping bentgrass/*Poa annua* mixed greens become difficult to maintain when irrigation $H_2O$ electrical conductivity ($EC_w$)

**TABLE 2.12**
**Relative Salt Tolerance of Turfgrass Species Grown in Solution**

| Salt Tolerance | Species (listed alphabetically) | $EC_w$ (or dS/m) | Parts per Million |
|---|---|---|---|
| Excellent | • Alkaligrass<br>• Seashore paspalum | >10 | 6,500 |
| Good | • Bermudagrass<br>• Buffalograss<br>• Kikuyugrass<br>• St. Augustinegrass<br>• Wheatgrass<br>• Zoysiagrass | 6–10 | 4,000–6,500 |
| Fair | • Annual ryegrass<br>• Bluegrama<br>• Chewings fescue<br>• Creeping bentgrass cv. Seaside, Mariner<br>• Orchardgrass<br>• Perennial ryegrass<br>• Smooth brome<br>• Tall fescue | 3–6 | 2,000–4,000 |
| Poor | • Bahiagrass<br>• Centipedegrass<br>• Colonial bentgrass<br>• Creeping bentgrass<br>• Creeping red fescue<br>• Kentucky bluegrass<br>• Meadow fescue<br>• *Poa annua*<br>• *Poa trivialis* | <3 | <2,000 |
| Very poor | • Bentgrass/*Poa annua* golf greens | 1.5–2.0 | 960–1,280 |

*Sources:* Modified from Harivandi, M. A. et al., 1992, Agronomy Monograph Series 32, 208–230. Madison, WI: ASA, CSSA, and SSSA; Kopec, D. M., 1996, *Golf Course Irrigation* 4(3):14–16; McCarty, L. B., ed., 2011, *Best Golf Course Management Practices* (3rd ed.). Upper Saddle River, NJ: Prentice-Hall, Inc.

nears 1.5 to 2.0 dS/m (soil $EC_e$ >3.0 dS/m), while bermudagrass greens show stress at a higher $EC_w$ of 4 to 15 (soil $EC_e$ 6 to 20 dS/m). Certain cultivars of creeping bentgrass, such as Seaside, Seaside II, and Mariner, can tolerate an $EC_w$ up to 6.0 dS/m, but have limited tolerance to heat stress. Under saline conditions, grasses require good draining soils and moist soil conditions to produce good quality turf; adequate leaching (whether it is from rainfall or excess saline irrigation); specific management techniques, such as the use of gypsum or acidified $H_2O$; and skilled turf managers.

Salinity values and the species rankings are relative to each other, meaning these can often shift in real agronomic conditions. For example, as moisture stress and temperatures increase and soil infiltration/percolation rates and atmospheric relative humidity decrease, salinity tolerance levels by all turfgrasses also decrease. This is compounded by surface compaction, reduced air movement, and high $H_2O$ pH and bicarbonate levels. Salt, drought, and temperature stresses are covered in more detail in subsequent chapters.

# Section II

## Soil Management Practices

# 3 Soil Chemical Properties

## HORIZONS

Soil consists of mineral particles, organic matter, water, air, and organisms that support plant life. Soils are formed over long periods of time through the weathering of rock minerals (in the case of mineral soils) and by the deposition of vegetative matter (in the case of organic soils). Soils vary widely in properties and in their suitability for various types of land use. The type of soil that forms is dependent on the parent material (type of rock minerals or sediments), climate (particularly rainfall and temperature), vegetation, landscape position or topography (consisting of elevation, slope, and aspect), and time in which these factors have been acting.

Every golf course superintendent should have an understanding and appreciation of soil chemical properties influencing his or her turf. These include soil organic matter, cation exchange capacity (CEC), soil pH, buffer capacity, base saturation, soil salinity, and available soil nutrients. These properties not only directly influence the turfgrass stand but also indirectly influence other soil characteristics such as nutrient and pesticide leaching, disease activity, and availability of native and applied nutrients.

Soil chemical and physical characteristics often change dramatically with soil depth. Distinct changes result in layers of soil being formed, which are referred to as horizons; the totality of these is called the soil profile (Figure 3.1). The properties of a soil are dependent on characteristics of both surface and subsurface horizons.

The A horizon is the uppermost layer of soil, typically several inches (~5 cm) to nearly a foot (0.3 m) thick, and is characterized by an accumulation of organic matter. Organic matter gives this "topsoil" its brown to black color and high amount of biological activity (fungi, bacteria, earthworms, insects, etc.). Most of a turfgrass root system resides in the A horizon.

Horizons below the A level are referred to as subsoil and contain declining biological activity with increasing depth into the profile. The B horizon is defined as the zone of clay accumulation that occurs below the A horizon, with some presence of turfgrass roots. The C horizon is found below the B horizon. Plant rooting can occur in the B and C horizons, depending on the suitability of their chemical and physical properties and their distance from the soil surface.

## CATION EXCHANGE

Plant nutrients in soil solution possess an electrical charge, either positive (called cations) or negative (called anions). Many soil nutrients occur naturally as complex, insoluble compounds that may not be readily available for plant use. Over time, as weathering occurs, these elements slowly enter the soil solution and become available to plants. Organic matter also may be degraded by the soil microbial population, releasing its nutrient constituents for possible plant uptake.

**Organic matter.** Organic matter is the resulting residue from the decay of plant material, mainly by bacteria and fungi. It is an intermediate phase of carbon between photosynthetically fixed carbon and completely oxidized carbon as $CO_2$. It consists of ~50% carbon by mass, 20% to 40% nonhumic substances (simple sugars, amino acids, polysaccharides, proteins, and lipids), and 60% to 80% humic substances (humus, humates) consisting of humic acids, fulvic acids, and humin. Organic material mostly has a positive effect on soil quality as it influences soil physical and chemical properties such as cation exchange, pH, and air–water exchange. For example, 2% organic matter by weight in the top 4 inches (10 cm) of soil equates to 460 pounds organic matter per 1,000 square feet (225 kg/100 $m^2$). Under normal conditions, it is estimated that 0.1% soil organic matter increases yearly, or about 35 pounds organic matter per 1,000 square feet (17 kg/100 $m^2$). For soils undergoing intense turf management practices, this increase is probably significantly higher.

The two properties accounting for soil reactivity are surface area and surface charge. Surface area and charge are largely a direct result of particle size and shape and are mostly influenced by soil organic matter content and clay-sized particles.

As soils are formed during the weathering processes, some minerals and organic matter are broken down to extremely small particles called colloids. Colloids have tremendous amounts of surface area; therefore, they have high impact on soil properties. As a result, colloids are primarily responsible for the chemical reactivity of soils. Colloids from clay and organic matter sources typically have a net negative (–) charge, developed during the formation process. These, therefore, can attract and hold (a process called ionic bonding) positively (+) charged particles, called cations (including many nutrients), similarly to opposite poles of a magnet attracting each other.

One of the most important parameters affecting cationic nutrient supply to the turfgrass root is cation exchange capacity (CEC), which refers to the quantity of negative charges on soil particles provided by clay and organic matter. The negative charges attract cations, hence the name "cation exchange capacity" (Figure 3.2). A low CEC value indicates fewer negatively charged exchange sites to bind with cations. Many essential plant nutrients exist in the soil as cations and are accumulated by the grass plant in this form. Potassium ($K^+$), calcium ($Ca^{+2}$), and magnesium ($Mg^{+2}$) can be found in relative abundance on the exchange sites. Iron ($Fe^{+2}$), zinc ($Zn^{+2}$), copper ($Cu^{+2}$), manganese ($Mn^{+2}$), and ammonium $\left(NH_4^+\right)$ occur to a lesser extent. Sodium ($Na^+$), which may cause

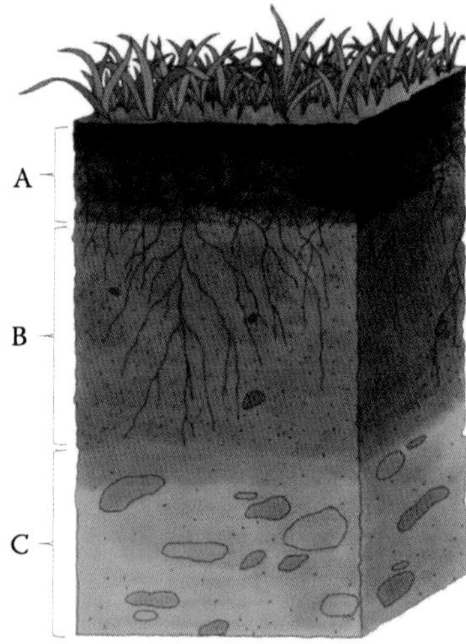


**FIGURE 3.1** Illustrations of common soil horizons with most biological activity in the A and B horizons.

severe problems in soils irrigated with poor-quality irrigation water, also occurs as a cation. Soils high in Na ions are dispersed, resisting water infiltration, while soils high in Ca ions are typically well aggregated, having high infiltration rates. Hydrogen ($H^+$) and aluminum ($Al^{+3}$) are the other predominant cations occupying soil exchange sites. These two cations are responsible for the detrimental effects on turfgrass health that occur in extremely acidic soils. The major ions influencing plants are listed in Table 3.1.

Cation exchange is a reversible process where cations in solution are exchanged (or replaced) with other cations on the soil's negatively charged solid phase, such as clay minerals or organic matter. Exchangeable cations are not readily leached unless they are replaced (or exchanged). Ions more strongly attracted to the cation exchange sites can replace the existing ones. For example, Ca can be exchanged for K or H, or vice versa. Also, this exchange may occur when the soil solution concentration is altered and is not in equilibrium with cations on the exchange sites. Only a small percentage of the essential plant nutrient cations ($K^+$, $Ca^{+2}$, $Mg^{+2}$, and $NH_4^+$) will be "loose" in the soil water and thus available for plant uptake. Therefore, CEC is important because it provides a reservoir of nutrients to replenish those removed from the soil water by plant uptake. Similarly, cations in soil water that are leached below the rootzone by excess rainfall or irrigation water are replaced by cations formerly bound to the CEC. Higher CEC values indicate the soil has more negative charges and thus will hold more nutrients. Low CEC soils hold fewer nutrients and thus require more frequent additions of nutrients such as K and Ca to remain sufficient.

Cations are attracted to the CEC site in different strengths. This determines which cations are more easily lost from the soil by leaching. Usually, a cation with a high charge of two or three (e.g., $Ca^{+2}$ or $Al^{+3}$) is preferentially held over cations with a lesser charge (e.g., $K^+$). Also, smaller ions have greater affinity than larger ones. Cations also are adsorbed and exchanged on a chemically equivalent basis. For example, one $K^+$ can replace one $Na^+$, while two $K^+$ cations are required to replace or exchange for one $Ca^{+2}$. The bonding strength and presence of exchangeable cations in fertile soils most often occur in the following sequence.

General order of bonding strength of exchangeable cations:

$$Al^{+3} > H^+ > Ca^{+2} > Mg^{+2} > NH_4^+ > K^+ = Na^+$$

Calcium ($Ca^{+2}$) is the most tightly held nutrient cation. Heavy applications of $Ca^{+2}$ may induce deficiencies of magnesium ($Mg^{+2}$) and potassium ($K^+$), in part due to enhanced leaching of these cations. The strong bonding of $Ca^{+2}$ is also the reason gypsum (calcium sulfate) applications are frequently used to displace (or exchange) sodium ($Na^+$) from salt-affected soils.

Cation exchange capacity is calculated by summing the exchangeable cations (CEC = $H^+ + Al^{+3} + Ca^{+2} + Mg^{+2} + K^+ + Na^+$) and is expressed in units of milliequivalents (meq) per 100 g of soil (meq/100 g) or centimoles of charge per kilogram of soil ($cmol_c$/kg). Numerically, meq/100 g and $cmol_c$/kg are the same. In the lab, these cations are extracted from soil by using an "exchange" cation such as $NH_4^+$, $Na^+$, or $H^+$ in sufficient strength to remove most of them. One milliequivalent of negative charge on a clay particle is neutralized by one milliequivalent of cation, but the cation weight may differ depending on the atomic weight of the cation.

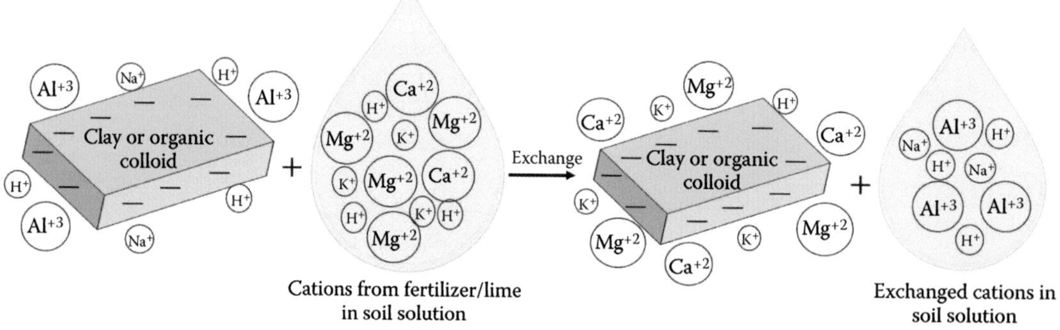


**FIGURE 3.2** Illustration demonstrating desirable soil cation exchange where cations from fertilizer or lime "exchange" with undesirable ones on clay or organic soil colloids.

**TABLE 3.1**
**Major Ions Influencing Plant Growth, Shown as Their Most Common Form in Soil Solutions**

| | |
|---|---|
| | **Major Exchangeable Cations** |
| $Ca^{+2}$, $Mg^{+2}$, $Na^{+}$, $K^{+}$, $NH_4^{+}$, $Al^{+3}$, $H^{+}$ | • Mostly as exchangeable cations in soils; readily manipulated by liming, acidification, or irrigation; $H^{+}$ concentrations determine active soil pH while $Al^{+3}$ is a portion of reserve soil acidity; $Na^{+}$ ions can disrupt soil physical characteristics by causing deflocculation; productive soils typically contain abundant exchangeable $Ca^{+2}$ |
| | **Major Anions** |
| $NO_3^-$, $SO_4^{-2}$, $Cl^-$, $HCO_3^-$, $CO_3^{-2}$, $SeO_4^{-2}$, $Br^-$, $I^-$ | • Weakly retained or repelled in soils and in much lower concentrations than the major cations; $SO_4^{-2}$ and $NO_3^-$ are important plant nutrients; $SO_4^{-2}$, $Cl^-$, and $HCO_3^-$ salts accumulate in saline soils; $HCO_3^-$ ions occur mostly when soil pH is >8.0 |
| | **Weakly Soluble Anions** |
| $H_2PO_4^-$, $HPO_4^{-2}$, $H_3BO_3$, $H_2BO_3^-$, $Si(OH)_4$, $MoO_4^{-2}$, $H_2S$, $HS^-$, $F^-$, $H_2AsO_4^-$, $CrO_4^{-2}$ | • Moderately to strongly retained by soils; soil pH greatly influences their retention; $MoO_4^{-2}$ and $Si(OH)_4$ are more soluble at high pH; phosphate is more soluble at slightly acid pH; borates are most soluble of this group |
| | **Aluminum and Transition Metals** |
| $Al^{+3}$, $AlOH^{+2}$, $Al(OH)_2^{+1}$, $Fe^{+2}$, $Fe(OH)_2^{+1}$, $Mn^{+2}$, $Cu^{+2}$, $Zn^{+2}$ | • Insoluble hydroxides that accumulate as silica and other weathered ions; Cu and Zn more soluble and increasingly available with increasing soil acidity; strongly complexed by organic matter |
| | **Toxic Ions** |
| $Al^{+3}$, $Cd^{+2}$, $Pb^{+2}$, $Hg^{+2}$ | • $Al^{+3}$ is a hazard to plants while the others are more of a concern with animals |
| | **Active in oxidation-reduction (redox) reactions** |
| Organic C $\rightarrow HCO_3^-$<br>$-NH_2 \rightarrow NO_3^-$<br>$-SH \rightarrow SO_4^{-2}$<br>$Fe^{+2} \rightarrow FeOOH$<br>$Mn^{+2} \rightarrow MnO_2$<br>Organic Se $\rightarrow SeO_4^{-2}$<br>Organic Hg $\rightarrow Hg^{+2}$ | • Changes in the oxidation state of soil carbon (C), N, and S compounds key components of soil biochemistry, with $O_2$ being the main electron acceptor; Fe(III), Mn(III-IV), nitrate, and sulfate are electron acceptors when soil $O_2$ levels are low, leading to anaerobic conditions and possible "black layer"' formation |

### Example

What would be the CEC by volume (milliequivalents/liter) of a sand with a bulk density of 100 pounds/cubic foot (or 1.6 g/mL) and a CEC value of 2 meq/100 g soil?

$$\frac{2\,\text{meq}}{100\,\text{g}} \times \frac{1.6\,\text{g}}{\text{mL}} \times \frac{1{,}000\,\text{mL}}{\text{L}} = \frac{32\,\text{meq}}{\text{L}}$$

## Soil Type

The CEC of a soil can be estimated from its texture due to the direct relationship between soil texture and clay mineral content and the indirect relationship of texture and soil organic matter (generally the sandier the soil is, the lower is its organic matter content). The CEC of soil organic matter ranges from 150 to 350 meq/100 g. Clay minerals may range from as low as 4 meq/100 g to as high as 150 meq/100 g (Table 3.2).

Organic matter, or humus, and clays have large CEC values and surface areas. Organic matter surfaces are negatively charged due to the following reaction with hydroxyl ions as soil pH is increased.

**TABLE 3.2**
**Cation Exchange Capacity (CEC) Examples of Various Soils**

| Soil Texture | CEC (meq/100 g) | Relative Level |
|---|---|---|
| Sand[a] | 0 to 6 | Very low |
| Sandy loam | 6 to 12 | Low |
| Loam | 12 to 30 | Medium |
| Silt loam | 10 to 50 | Medium |
| Clay loams | 4 to 30 | Medium |
| Clays | 18 to 150 | High |
| Organic matter/humus | 150 to 350 | Very high |

[a] A newly constructed USGA green typically contains 1% to 3% organic matter (by weight).

Cation exchange capacity of organic matter greatly increases as soil pH increases:

$$\text{COOH(organic matter carboxyl group)} + OH^- \text{ (hydroxyl ion; } \uparrow\text{pH)} \rightarrow COO^- \text{(cation exchange site)} + H_2O\text{(water)}$$

**Sands.** Sandy soils generally have low CEC values due to their relatively small surface areas, resulting in less exposed negative charges to attract or hold cations (Figure 3.3). Sandy soils also contain relatively little organic matter and clay, which increase soil CEC. Nutrients in low CEC soils are also more easily leached. This results in low CEC of soils that are relatively infertile and have low water-retention potential. Most of the essential plant nutrients must, therefore, be applied in a turf fertilizer program. These soils generally profit from additions of clay or organic matter to increase their CEC values. The primary factor affecting the amount of variable charge in the soil is pH. Increasing pH increases CEC. For example, increasing pH from 5.5 to 6.5 may double the CEC of a typical sandy soil when most of the CEC arises from organic matter. A pH of 6.5 is generally the highest pH recommended for most grasses. Maintaining soil pH near this level provides near-optimum CEC and P availability while providing acceptable conditions for micronutrient availability.

**Clays.** If clay is present, the type is important as the surface charge and expandability strongly influence CEC values. Due to their small size, clay particles expose a large external surface (Figure 3.4). In addition, certain clay particles (e.g., montmorillonite) have relatively extensive internal surfaces due to the plate-like crystal units that compose each particle. Clays that swell in the presence of water (Figure 3.5) increase their surface area and allow the replacement of cations in the internal surface. Much of the charge in montmorillonite and vermiculite is permanent and is located within the structure of the clay particle. This charge arises by a process called isomorphous substitution during the formation of the clay. Negative charges are created when cations of lesser charge are substituted for cations of higher charge. For example, $Al^{+3}$ ions can replace $Si^{+4}$ ions within clay sheets when they expand, while $Mg^{+2}$ and $Fe^{+2}$ or $Fe^{+3}$ can substitute for $Al^{+3}$ ions. This regulates the permanent charge of most soils and is influenced little by the soil solution. The lower charge but similarly sized replacement cations result in a net negative charge, which attracts other cations thereby increasing its CEC value. Montmorillonite and vermiculite clays, with high CEC and surface area, have greater nutrient-holding capacity than nonexpandable, lower CEC clays such as kaolinite, chlorite, or illite (Table 3.3).

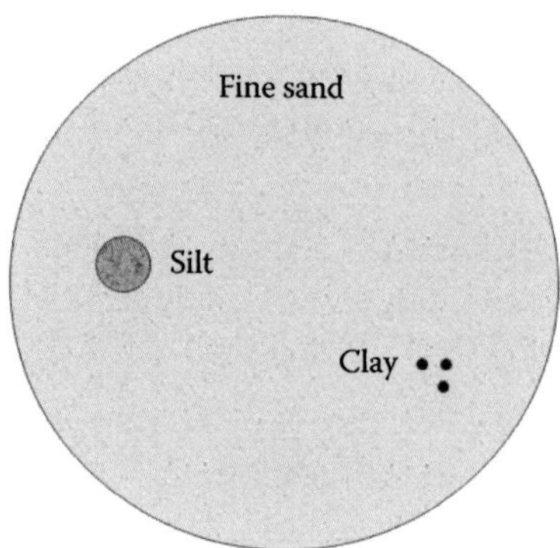


**FIGURE 3.4** Relative size of clay, silt, and fine sand particles. Due to their relatively large external surface areas, silt and clay particles typically have much higher CEC values compared to sand.

## Example

A fairway soil contains 40% clay and 5% humus. Determine the CEC of the soil, assuming the clay is kaolinite. From Table 3.3, kaolinite clay has a CEC range of 4 to 6 meq/100 g and humus 150 to 350 meq/100 g.

**Step 1:** Determine the CEC contribution of the kaolinite clay. With a CEC range of 4 to 6 meq/100 g, an average of 5 meq/100 g is used. The CEC of the clay contribution is determined by multiplying the percent clay (40%) by the CEC of pure kaolinite (5 meq/100 g):

$$\text{CEC contribution from clay} = 0.4 \times 5\ \text{meq/100 g} = 2\ \text{meq/100 g}$$

**Step 2:** Determine the CEC contribution of the humus. With a CEC range of 150 to 350 meq/100 g, an average of 250 meq/100 g is used. The CEC of the humus contribution is determined by

**FIGURE 3.3** Sandy soils' low CEC values are due to their small surface areas and low organic matter content.

FIGURE 3.5 A shrink–swell clay that swells when wet, increasing surface area to exchangeable cations.

**TABLE 3.3**
**Typical Characteristics and Cation Exchange Capacities (CECs) of Various Clay Minerals and Humus**

| Material | Structural Type | Permanent Charge | Expands in Water | Surface Area | CEC at pH 7 (meq/100 g) |
|---|---|---|---|---|---|
| Kaolinite | 1:1 | Very low | No | Extremely low | 4 to 6 |
| Chlorite | 2:1:1 (2:2) | Medium | No | Low | 20 to 30 |
| Mica (Illite) | 2:1 | High | Slightly | Low | 20 to 40 |
| Montmorillonite (Smectite) | 2:1 | Medium | Yes | High | 60 to 100 |
| Vermiculite | 2:1 | High | Limited | Medium | 100 to 150 |
| Humic acid | Organic | High | Slightly | High | 150 to 350 |

multiplying the percent humus (5%) by the CEC of pure humus (250 meq/100 g):

$$\text{CEC contribution from humus} = 0.05 \times 250\ \text{meq/100 g}$$
$$= 12.5\ \text{meq/100 g}$$

**Step 3:** To determine the CEC of the fairway soil, sum the contribution of each component.

$$\text{CEC of fairway soil} = 2\ \text{meq/100 g (from clay)}$$
$$+12.5\ \text{meq/100 g (from humus)}$$
$$= 14.5\ \text{meq/100 g}$$

## Increasing Soil CEC

The CEC of many soils also is pH dependent. With organic matter, most CEC sites are pH dependent with a 10% to 30% increase in CEC per pH unit. Maintaining a pH between 5.5 and 6.5 is the first step to increase CEC values in sandy soils. Adding organic matter or inorganic amendment prior to construction or by topdressing following establishment would be the next most efficient means of increasing sandy soil CEC values (Figure 3.6). Increasing CEC by adding clay or organic matter to soil, however, is difficult due to the large amount of material needed. For example, to increase the CEC of a soil 1 meq/100 g, it is necessary to increase the soil content of clay or organic matter by 10 tons/acre (22,417 kg/ha) of soil (assuming a CEC of 100 meq/100 g for the clay or organic matter and incorporation 6 inches, 15 cm, deep). Adding this much material would increase the clay or organic matter content of the soil 1%. In general, organic matter has a greater potential to increase CEC than clay. For example, a 5% content of organic matter by weight equals the CEC provided by a 30% content of an illitic clay. Although additions of clay and organic matter may increase the CEC of sandy soils, excessive increases in these materials may adversely affect the soil's physical properties such as infiltration and percolation rate.

**Amount of amendments to use.** The approximate amount of amendment needed to increase soil CEC can be calculated using the following equation:

$$A = \frac{I \times S}{C - I}$$

**FIGURE 3.6** Dark-green turf from increasing CEC of a sandy soil by adding montmorillonite clay. (Courtesy Todd Bunnell.)

where

A = dry weight of amendment required
I = amount of CEC increased desired (meq/100 g or $cmol_c/kg$)
S = dry weight of soil or sand being amended
C = CEC of amendment to be used (meq/100 g or $cmol_c/kg$)

In addition, if significant amounts are present, the water content by weight of the amendment is needed.

### Example

Suppose a 6,000-square-foot (55.7 $m^2$) green is being constructed with a 3 meq/100 g CEC and the superintendent wishes to increase it to 5 meq/100 g using zeolite (CEC of 120 meq/100 g and negligible water content). How much amendment is necessary for a 12-inch (0.3 m) deep rootzone? The sand to be amended weighs ~450,000 pounds (204,300 kg):

$$A = \frac{2.0\text{ meq/100 g} \times 450{,}000\text{ pounds}}{(120\text{ meq/100 g} - 2.0\text{ meq/100 g})} = \frac{900{,}000\text{ pounds}}{118}$$
$$= 7{,}627\text{ pounds (3,463 kg) of dry amendment required}$$

If the amendment contains appreciable amounts of water, the value of A should be corrected to account for this. Multiply A by one plus the decimal value of water content. For example, if the amendment (often organic in nature) contains 20% water on a dry weight basis, multiply A (7,627 pounds) by 1.20 to obtain the weight of moist amendment required (or 9,152 pounds, 4,155 kg, in this example).

## Base Saturation of the Cation Exchange Capacity or Percentage Cation Saturation

The CEC of the soil exchange complex (organic matter and clay) generally increases as the soil pH increases due to the dissociation of H ions from surface sites, as previously illustrated for COOH. The degree to which the exchange sites are saturated with basic cations such as $Ca^{+2}$, $Mg^{+2}$, $Na^+$, and $K^+$, as opposed to acid cations $H^+$ and $Al^{+3}$, is referred to as the base saturation of a soil (Table 3.4). In general, the pH and fertility of a soil increase as the percentage base saturation increases. Higher base saturation generally increases the ease with which cations are absorbed by plants because higher concentrations of cations are maintained in the soil solution. The greatest availability of most nutrients to plants is in the soil pH range of 6 to 7. In acidic soils (pH < 5.5), exchangeable $Al^{+3}$ may be present and, along with $Mg^{+2}$ and/or $H^+$ ions, can suppress plant uptake of $Ca^{+2}$ or $K^+$. However, toxic $Al^{+3}$ becomes less prevalent, and therefore less detrimental to the plant, as the base saturation increases. Maintaining soil pH between 5.8 and 7.0 is a much better way to avoid $Al^{+3}$ and $H^+$ toxicity than trying to attain 100% base saturation.

**TABLE 3.4**
**Desired Cation Exchange Capacity (CEC) and Percent Base Saturation Ranges and Ratios in Soils according to Base Cation Saturation Ratio (BCSR) Concept**

| Cation | Percent CEC | Percent Base Saturation |
|---|---|---|
| $Ca^{+2}$ | 60 to 70 | 50 to 85 |
| $Mg^{+2}$ | 10 to 20 | 18 to 22 |
| $K^+$ | 5 to 10 | 3 to 9 |
| $Na^+$ | 0 to 1[a] | 0 to 1 |
| $H^+$ | 0 to 10[a] | 0 |
| Ca:Mg | – | <6.5:1 |
| Ca:K | – | <13:1 |
| Mg:K | – | <2:1 |

[a] A Na+ or H+ percentage above 15% may indicate a nutrient imbalance or potential toxic level.

In order to achieve 100% base saturation in many soils, the pH must exceed 7.0, which in turn results in some micronutrient deficiencies.

Cation saturation is a useful measurement to manage soil $Na^+$ and can be utilized to determine soil $Mg^{+2}$ availability. When $Na^+$ exceeds 15% of the CEC, water and air infiltration into the soil may be reduced and poor growing conditions may result. To overcome this problem, $Ca^{+2}$ is added to replace the $Na^+$ on the exchange sites, most often through calcium sulfate ($CaSO_4$ or gypsum). Sodium in the soil water is then leached out of the rootzone by excess irrigation or rainfall. The amount of $Ca^{+2}$ needed to replace the $Na^+$ is based on the amount of exchangeable $Na^+$ as well as $Na^+$ saturation.

### Example

From the following soil test report data, calculate the percent base saturation: $Ca^{+2}$ = 20, $Mg^{+2}$ = 7.0, $K^+$ = 0.2, $Na^+$ = 0.3, $Al^{+3}$ = 7.0, and $H^+$ = 4.0 meq/100 g.

**Step 1:** CEC, the total of all cations present, is determined:

$$20 + 7.0 + 0.2 + 0.3 + 7.0 + 4.0 = 38.5\,\text{meq/100 g}$$

**Step 2:** Next, the base saturation is the sum of the basic cation $Ca^{+2}$, $Mg^{+2}$, $K^+$, and $Na^+$. Therefore, the calculation is the sum of the basic cations divided by the CEC and then multiplied by 100:

$$\text{Percent base saturation} = \frac{20.0 + 7.0 + 0.2 + 0.3}{38.5} \times 100$$
$$= 71.4\%$$

## Measuring Soil CEC

Cation exchange capacity can be estimated or measured by several different methods. Since the direct measurement of CEC is costly and time consuming, most soil-testing laboratories routinely estimate CEC by measuring the predominant extractable cations ($Ca^{+2}$, $Mg^{+2}$, $K^+$, $Na^+$) and estimating extractable acidity ($H^+$ and $Al^{+3}$) from soil pH and buffer pH.

Estimates will be erroneously high in two commonly occurring situations. First, CEC will be overestimated if a soil is sampled shortly after a heavy fertilizer or gypsum application. This error arises as the cations in the soil solution are misrepresented as exchangeable. These conditions may be found on a sand-based green with recent fertilization. Second, extraction of high-pH soils containing calcium carbonate (limestone) with an acid-extracting agent will also provide inflated CEC estimates. The acid extractant dissolves a portion of the limestone and considers the $Ca^{+2}$ removed as exchangeable. For golf greens, a target CEC is 80 to 160 meq/L (or 5 to 10 meq/100 g) (Shaw and Andrews 2001).

**Equivalent weight.** Cations are adsorbed and exchanged on a chemically equivalent weight (eq) basis. An equivalent weight (also called moles of ion charge) is the amount (or weight) of an ion needed to equal the amount (or weight) of another ion. Equivalent weights are calculated by dividing the atomic weight of a cation by its valence, regardless of the sign. For example, one equivalent of $Ca^{+2}$ can be calculated by dividing the atomic weight of Ca (40 g/mol) by the valence of Ca (+2) to obtain 20 g. Likewise, one equivalent of $K^+$ can be calculated by dividing the atomic weight of K (39 g/mol) by the valence of K (+1) to obtain 39 g. Therefore, for one equivalent of $Ca^{+2}$ to equal one equivalent of $K^+$, it requires only 20 g of Ca compared to 39 g of K. Calculating milliequivalent weight is the same, only the amounts are in milligrams instead of grams. For a further demonstration of equivalents, see the following example. Also, Table 3.5 lists equivalent and milliequivalent weights of common cations.

### Example

How many milliequivalents of $K^+$ are in 78 mg of $K^+$?

**Step 1:** Determine the mass of $K^+$ in 1 meq of $K^+$. The periodic table shows K has an atomic mass of 39 mg/mmol (millimole). The single positive sign on the cation ($K^+$) indicates K has a valence of one. With this information, milliequivalents can be calculated by dividing the atomic mass by the valence. For K, 1 meq of $K^+$ weighs 39 mg:

$$1\,\text{meq}\,K^+ = \frac{39\,\text{mg/mmol}}{1} = 39\,\text{mg}$$

**TABLE 3.5**
**Equivalent and Milliequivalent Weights of Common Soil Cations**

| Cation | Molecular Weight | Valence | Equivalent wt (g) | Milliequivalent wt (mg) |
|---|---|---|---|---|
| Hydrogen ($H^+$) | 1 | 1 | 1 | 1 |
| Aluminum ($Al^{+3}$) | 27 | 3 | 9 | 9 |
| Calcium ($Ca^{+2}$) | 40 | 2 | 20 | 20 |
| Magnesium ($Mg^{+2}$) | 24 | 2 | 12 | 12 |
| Potassium ($K^+$) | 39 | 1 | 39 | 39 |
| Sodium ($Na^+$) | 23 | 1 | 23 | 23 |

**Step 2:** To determine the number of milliequivalents $K^+$ in 78 mg of $K^+$, it is easiest to set up a ratio:

$$\frac{1\,\text{meq}\,K^+}{39\,\text{mg}} = \frac{X\,\text{meq}\,K^+}{78\,\text{mg}}$$

$$X\,\text{meq}\,K^+ = \frac{78\,\text{mg}}{39\,\text{mg}} = 2$$

One milliequivalent $K^+$ is 39 mg; therefore, 78 mg = 2 meq $K^+$.

The concept of cation exchange is based on moles of charge. One mole of an element is the quantity of it having a mass in grams equal to the atomic weight. For example, the atomic weight of Na is 23, so 1 mol of Na weighs 23 g. One mole of an element contains Avogadro's number ($6.022 \times 10^{23}$) of atoms of the element. Therefore, 1 mol of charge is $6.022 \times 10^{23}$ charges. So, if Na exists as a cation ($Na^+$), then a solution containing 1 mol of Na would contain $6.022 \times 10^{23}$ positive charges and weigh 23 g. In the case of a divalent cation (having two positive charges per mole) like $Ca^{+2}$, only half of its mass would be needed to equal the same amount of charge as a monovalent cation. Trivalent cations (e.g., $Al^{+3}$) would require a third as much. The next example determines the CEC of a given soil.

## Example

Calculate the CEC of a soil using the following information:

| Cation | Atomic wt (g/mol) | Amount on Exchange Sites (ppm or mg/kg) |
|---|---|---|
| Calcium ($Ca^{+2}$) | 40 | 4000 |
| Magnesium ($Mg^{+2}$) | 24 | 840 |
| Potassium ($K^+$) | 39 | 78 |
| Sodium ($Na^+$) | 23 | 69 |
| Aluminum ($Al^{+3}$) | 27 | 630 |
| Hydrogen ($H^+$) | 1 | 40 |

**Step 1:** Since CEC is generally expressed in milliequivalents/100 g soil, convert milligrams/kilogram soil by multiplying by 0.1 (100 g divided by 1,000 g/kg).

| Cation | Atomic Weight (g/mol) | Amount on Exchange Sites (mg/kg) | Amount on Exchange Sites (mg/100 g) |
|---|---|---|---|
| Calcium ($Ca^{+2}$) | 40 | 4000 | 400 |
| Magnesium ($Mg^{+2}$) | 24 | 840 | 84 |
| Potassium ($K^+$) | 39 | 78 | 7.8 |
| Sodium ($Na^+$) | 23 | 69 | 6.9 |
| Aluminum ($Al^{+3}$) | 27 | 630 | 63 |
| Hydrogen ($H^+$) | 1 | 40 | 4 |

**Step 2:** Now calculate the mass of milliequivalents for each cation. Recall, milliequivalents are equal to the atomic weight divided by the valence (e.g., $Ca^{+2}$ has a valence of two while $K^+$ has a valence of one). Therefore, 1 meq of $Ca^{+2}$ is equal to the atomic weight (40 g/mol) divided by the valence (2), and thus 1 meq of $Ca^{+2}$ has a mass of 20 mg.

| Cation | Atomic Weight (g/mol) | Amount on Exchange Sites (mg/100 g) | Mass of 1 meq of Cation (mg) |
|---|---|---|---|
| Calcium ($Ca^{+2}$) | 40 | 400 | 20 |
| Magnesium ($Mg^{+2}$) | 24 | 84 | 12 |
| Potassium ($K^+$) | 39 | 7.8 | 39 |
| Sodium ($Na^+$) | 23 | 6.9 | 23 |
| Aluminum ($Al^{+3}$) | 27 | 63 | 9 |
| Hydrogen ($H^+$) | 1 | 4 | 1 |

**Step 3:** To calculate the milliequivalents of cation on the exchange site, divide the amount on the site by the mass of 1 meq of the cation (e.g., for $Ca^{+2}$, divide 400 mg/100 g soil by 20 mg to find 20.0 meq $Ca^{+2}$ 100/g soil).

| Cation | Atomic Weight (g/mol) | Amount on Exchange Sites (mg/100 g) | Mass of 1 meq of Cation (mg) | Amount on Exchange Sites (meq/100 g) |
|---|---|---|---|---|
| Calcium ($Ca^{+2}$) | 40 | 400 | 20 | 20.0 |
| Magnesium ($Mg^{+2}$) | 24 | 84 | 12 | 7.0 |
| Potassium ($K^+$) | 39 | 7.8 | 39 | 0.2 |
| Sodium ($Na^+$) | 23 | 6.9 | 23 | 0.3 |
| Aluminum ($Al^{+3}$) | 27 | 63 | 9 | 7.0 |
| Hydrogen ($H^+$) | 1 | 4 | 1 | 4.0 |

**Step 4:** To calculate the CEC, sum the number of cations on the exchange site in milliequivalents/100 g soil [e.g., CEC = $Ca^{+2}$ (20.0) + $Mg^{+2}$ (7.0) + $K^+$ (0.2) + $Na^+$ (0.3) + $Al^{+3}$ (7.0) + $H^+$ (4.0) = 38.5 meq/100 g].

| Cation | Atomic Weight (g/mol) | Amount on Exchange Sites (mg/100 g) | Mass of 1 meq of Cation (mg) | Amount on Exchange Sites (meq/100 g) |
|---|---|---|---|---|
| Calcium ($Ca^{+2}$) | 40 | 400 | 20 | 20.0 |
| Magnesium ($Mg^{+2}$) | 24 | 84 | 12 | 7.0 |
| Potassium ($K^+$) | 39 | 7.8 | 39 | 0.2 |
| Sodium ($Na^+$) | 23 | 6.9 | 23 | 0.3 |
| Aluminum ($Al^{+3}$) | 27 | 63 | 9 | 7.0 |
| Hydrogen ($H^+$) | 1 | 4 | 1 | 4.0 |
| CEC = | | | | 38.5 |

### Example

**A.** A sample from an 85:15 (sand:sphagnum peat by volume) putting green mix is oxidized (burned) and determined to have 1% organic matter by weight. Calculate a CEC for the mixture; assume the CEC of the peat alone is 125 meq/100 g.

Due to the relatively low surface area and lack of surface negative charges to retain cations, the CEC contribution of the sand in the mix can be considered negligible. Therefore, the peat will be responsible for nutrient retention and thus the calculation is simple. CEC is determined by multiplying the amount, by weight, of organic matter (1%) by the CEC of the organic matter (125 meq/100 g):

$$\text{CEC of 85:15 green's mix} = 0.01 \times 125\,\text{meq/100 g} = 1.25\,\text{meq/100 g}$$

Thus, the CEC of the 85:15 mix is 1.25 meq/100 g.

**B.** Determine the potential CEC of the following golf green rootzone mix:

| Volume (%) | CEC (meq/100 g) | Bulk Density (g/mL) |
|---|---|---|
| 80% Sand | 2 | 1.6 |
| 10% Peat | 150 | 0.08 |
| 10% Zeolite | 100 | 0.80 |

CEC (by volume):

$$\text{Sand: } 0.80 \times \frac{2\,\text{meq}}{100\,\text{g}} \times \frac{1.6\,\text{g}}{\text{mL}} \times \frac{1000\,\text{mL}}{\text{L}} = \frac{25.6\,\text{meq}}{\text{L}}$$

$$\text{Peat: } 0.10 \times \frac{150\,\text{meq}}{100\,\text{g}} \times \frac{0.08\,\text{g}}{\text{mL}} \times \frac{1000\,\text{mL}}{\text{L}} = \frac{12.0\,\text{meq}}{\text{L}}$$

$$\text{Zeolite: } 0.10 \times \frac{100\,\text{meq}}{100\,\text{g}} \times \frac{0.80\,\text{g}}{\text{mL}} \times \frac{1000\,\text{mL}}{\text{L}} = \frac{80.0\,\text{meq}}{\text{L}}$$

Total CEC: 25.6 + 12.0 + 80 = 118 meq/L. To convert this to CEC by weight:

**Step 1:** The average density of the rootzone mix is determined by

$$\text{Sand: } 80\% \times \frac{1.6\,\text{g}}{\text{mL}} = \frac{1.280\,\text{g}}{\text{mL}}$$

$$\text{Peat: } 10\% \times \frac{0.08\,\text{g}}{\text{ml}} = \frac{0.008\,\text{g}}{\text{mL}}$$

$$\text{Zeolite: } 10\% \times \frac{0.80\,\text{g}}{\text{mL}} = \frac{0.08\,\text{g}}{\text{mL}}$$

Total: 1.28 + 0.008 + 0.08 g/mL = 1.368 g/mL.

**Step 2:** From the previous calculations, convert g/mL to meq/100 g soil:

$$\frac{118\,\text{meq}}{\text{L}} \times \frac{\text{L}}{1{,}000\,\text{mL}} \times \frac{\text{mL}}{1.368\,\text{g}} = \frac{86\,\text{meq}}{1{,}000\,\text{g}} \text{ or } \frac{8.6\,\text{meq}}{100\,\text{g}}$$

In summary, the type of cation bound to exchange sites on soil colloids is important in determining the fertility and pH of a soil. The greater the quantity of $Ca^{+2}$, $K^{+}$, $Mg^{+}$, $Fe^{+2}$, and other secondary plant minerals is, the more fertile is the soil. Also, the more $Al^{+3}$ and $H^{+}$ held by the soil, the more acidic it is.

## Anion Exchange Capacity

Several nutrients are present in the soil and absorbed by the plant as anions (negatively charged ions) rather than as cations (Table 3.1). These include nitrate $(NO_3^-)$, phosphate $(H_2PO_4^-)$, sulfate $(SO_4^{-2})$, and chloride ($Cl^-$). Carbonate $(CO_3^{-2})$ and bicarbonate $(HCO_3^-)$ may also be found in soil as a result of liming or the use of irrigation water high in these anions. Anion exchange capacity (AEC) is the sum of exchangeable anions a soil can adsorb, expressed as centimoles of charge per kilogram of soil ($cmol_c$/kg or meq/100 g). AEC reflects the amount of positive charge on the surface of organic matter and clay colloids. AEC only occurs to any appreciable extent when soil pH is below 5. Since surface soil pH is usually maintained at higher pH levels, little anion exchange occurs in the rootzone of soils supporting turfgrass growth. However, anion exchange may occur in the subsoil of some turf soils. The $Cl^-$, $SO_4^{-2}$, and $NO_3^-$ anions are the ones commonly exhibiting net anion repulsion; thus, they are retained less in soils.

Turf managers should remember that the form of N available to plants in warm and pH-neutral soils is primarily nitrate $(NO_3^-)$. Turfgrasses also can directly utilize ammonium nitrogen $(NH_4^+)$, but some ammonium nitrogen is readily converted to nitrate nitrogen by microorganisms in many soils via nitrification. Since little AEC exists in soils, nitrate $(NO_3^-)$ is not retained and leaches through the rootzone. Nitrate moved below the rootzone will leach to lower depths or will be denitrified. Other forms of N in the soil are generally held as insoluble organic matter compounds. Relatively small, frequent N applications are, therefore, generally more efficient for plant use compared to heavy, infrequent applications. Slow-release N sources also help to minimize the amount leached, because of the more gradual release of N into the soil solution.

Phosphate and sulfate anions are retained in soils by mechanisms other than anion exchange. Retention of these ions is favored by high iron and aluminum oxide content and low pH. The strength of the bonding between these anions and the soil, particularly in the case of phosphate, may be quite strong with little of each nutrient remaining available to plants. In addition, little downward movement of phosphate occurs in most soils, except in extremely sandy soils and with high application rates of phosphate. Sulfate is retained much more weakly than phosphate; therefore, retention of sulfate does not occur in the presence of phosphate in soils. Leaching of sulfate from soil layers associated with high levels of P readily occurs. Retention of sulfate, however, will occur when the leaching sulfate encounters layers of low P soil.

### Example

A soil analysis report indicates the following cations are present per 100 g soil:

| | | | |
|---|---|---|---|
| $Ca^{+2}$ | 5 meq | $H^+$ | 1 meq |
| $K^+$ | 6 meq | $NO_3^-$ | 4 meq |
| $Mg^{+2}$ | 2 meq | $Na^+$ | 0 meq |
| $Al^{+3}$ | 4 meq | | |

What is the CEC, the percent base saturation, and percent exchangeable acidity of this soil?

$$\text{CEC} = Ca^{+2} + Al^{+3} + K^+ + H^+ + Mg^{+2}$$

$$= (5+4+6+1+2) = 18\ \text{meq/100 g soil}$$

$$\%BS = \frac{Ca^{+2} + K^+ + Mg^{+2}}{\text{CEC}} \times 100\% = \frac{(5+6+2)}{18} \times 100\% = 72\%$$

$$\%EA = \frac{H^+ + Al^{+3}}{\text{CEC}} \times 100\% = \frac{1+4}{18} \times 100\% = 28\%$$

### Flocculation and Dispersion

Cations can also influence the interactions of soil colloids and soil structure. The term flocculation describes the attraction of soil colloids to one another (Figure 3.7). Clay particles have an electronegative charge on their surface and therefore have the ability to attract cations. When cations swarm to these negative charges, they provide bridging for clays to form structure. Cations, such as $Ca^{+2}$, $Fe^{+2}$, and $Al^{+3}$, with a high charge density and/or a small hydrated radius, promote flocculation or clay structure. These cations suppress the negative charge of the clays, allowing them to attract one another and thereby improving soil structure.

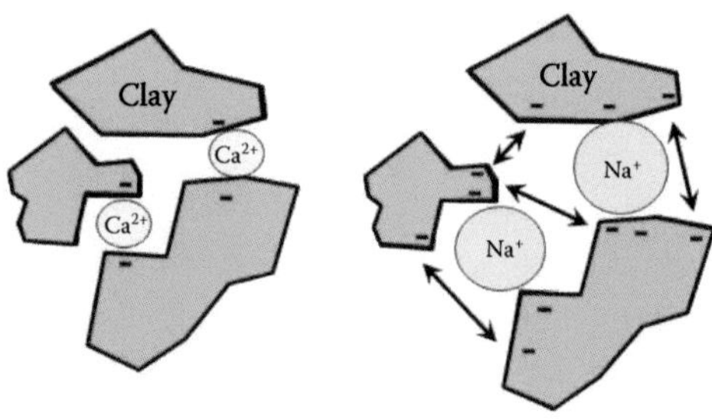


**FIGURE 3.7** Illustration of "deflocculation" (right) where sodium ions replace calcium, causing clay particles to disperse (or separate) and leading to poor soil structure, reduced drainage, and lower oxygenation.

Some cations disperse, rather than attract, clay particles. This process is referred to as dispersion or deflocculation. Dispersion is common in arid and semiarid regions where sodium ($Na^+$) accumulates in soils because of low rainfall. Sodium is responsible for dispersion since the large hydrated radius of a Na ion does not allow it to get close enough to the clay particle to completely neutralize the electronegative charge (Figure 3.7). Sodium is also monovalent (only reacts with one exchange site), thereby failing to create a bridge for clay attraction. The lack of colloid attraction degrades soil structure. Soils with poor structure fail to sustain plant growth effectively due to reduced drainage and oxygenation. Dispersed soils can be amended with applications of gypsum, also known as calcium sulfate ($CaSO_4$) (Figure 3.8). Gypsum improves flocculation by replacing the $Na^+$ ion with $Ca^{+2}$. Soil dispersion is generally not a problem in sand-based golf greens because of the absence of electrochemical charges on and the relatively uniform sizes of sand particles. However, $Na^+$ can cause other problems in sand-based soils, as discussed in the water quality chapter.

## SOIL ACIDITY

Soil pH is an important chemical property of soils with a substantial effect on plant nutrient availability, root function and health, pesticide effectiveness, and microbial activity. The actual concentration of $H^+$ and $OH^-$ ions is extremely low. A soil with a pH of 7.0 only has 0.0000001 mol ($10^{-7}$) of $H^+$ ions per liter of soil solution. These low values are cumbersome to work with, so a logarithmic scale is used. As a result, soil pH is defined as the negative logarithm of the $H^+$ activity or concentration:

$$\text{pH} = \frac{1}{\log[H^+]} = -\log[H^+]$$

where hydrogen ion activity [$H^+$] is in moles per liter.

A relationship exists between $H^+$ and $OH^-$ ion concentrations in water at 25°C.

Relationship between hydrogen and hydroxyl ion concentrations:

$$[H^+]\,(\text{hydrogen ion}) \times [OH^-]\,(\text{hydroxyl ion}) = 10^{-14}$$

For example, if the $H^+$ activity is 0.0001 mol/L ($1 \times 10^{-4}$ mol/L), then the negative log of this activity is 4. At a pH of 6, the $H^+$ activity is 0.000001 and, at a pH of 8, the $H^+$ activity is 0.00000001.

Hydroxyl ion ($OH^-$) is a base and counteracts acidity. In soil water, $H^+$ and $OH^-$ are inversely related (when one increases, the other decreases). When $H^+$ and $OH^-$ are equal, the pH is 7 and the solution is termed neutral—neither acidic nor basic. When $H^+$ exceeds $OH^-$, the pH is less than 7 and the solution is acidic. The lower the pH is, the greater is the acidity.

**FIGURE 3.8** Demonstration of how added gypsum (right) causes desirable soil flocculation, allowing better soil water penetration and distribution.

Conversely, when $H^+$ is less than $OH^-$, the solution is basic and the pH is greater than 7. The higher the pH is, the less acidic the solution is. Since the pH scale is logarithmic, a 10-fold increase in $H^+$ or $OH^-$ ions occurs when pH values change by one unit. A 100-fold change occurs when pH values change by two units and a 1,000-fold change occurs when pH values change by three units. For example, when pH changes from 6.0 to 5.0, a 10-fold increase in $H^+$ ion concentration occurs while concurrently, a 10-fold decrease in $OH^-$ ions occurs. A 100-fold increase in $H^+$ ions occurs for a pH change from 6.0 to 4.0 and a 1000-fold increase with a pH from 6.0 to 3.0. The pH of several common substances is listed in Table 3.6.

**TABLE 3.6**
**Typical pH Values for Several Common Substances**

| Substance | pH | Description |
|---|---|---|
| Lye (bleach) | 13.0 | Strong alkaline |
| Soap | 9.3 | ↑ |
| Antacid tablets | 9.4 | \| |
| Baking soda | 8.0 | \| |
| Seawater | 7.9 | ↓ |
| Human blood | 7.3 | Weak alkaline |
| Pure water | 7.0 | Neutral |
| Fresh milk | 6.7 | Weak acid |
| Natural rainfall | 5.6 | ↑ |
| Sour milk | 4.7 | \| |
| Beer | 4.4 | \| |
| Coffee | 4.2 | \| |
| Orange juice | 3.7 | \| |
| Vinegar | 2.9 | \| |
| Lemon juice | 2.4 | ↓ |
| Battery acid | 0.5 | Strong acid |

Buffer pH differs from soil pH as it estimates the amount of limestone required to increase soil pH. To determine this, acid soil is added to a buffered, high-pH solution, often the A–E (Adams–Evans) buffer, SMP (Shoemaker, McClean, and Pratt) buffer, or the Moore–Sikora buffer. The quantity of limestone required to achieve a target soil pH is estimated from the reduction in pH of the soil–buffer combination relative to the initial pH of the buffer alone. By calibrating pH changes in the buffered solution (most often pH of 8) that accompany the addition of known amounts of acid, the amount of lime required to bring a soil to a particular pH can be determined. Different laboratories, however, use different buffer solutions, so readings may not be universal.

A lime recommendation is based on a calculation on the difference between the original soil pH value and the lime index. A higher lime index (or buffer pH) indicates the soil is more susceptible to changes in pH, a common trend for many sand-based soils. Generally, soils containing clay or organic matter have a lower lime index (buffer pH), which indicates higher acid reserves in a soil and that larger quantities of lime will be required to raise its soil pH.

Salt pH testing, which predicts the pH of a soil following fertilization, is sometimes performed. This test is designed for agricultural soils depleted of salts following a winter thaw and generally is not applicable for turfgrass.

An acid is a substance that releases (or donates) $H^+$ ions (also called protons) to another substance. The extent to which an acid dissociates (converts to an $H^+$ ion and an accompanying anion) determines its strength. When saturated with $H^+$, a soil behaves as a weak acid.

Hydrochloric (HCl), sulfuric ($H_2SO_4$), and nitric ($HNO_3$) acids are "strong" acids, because they dissociate completely, meaning they readily release all of their $H^+$ ions into solution, thus dramatically reducing the solution pH. Weak acids, such as carbonic, boric, and acetic acids, dissociate only partially when placed in solution; therefore, they do not reduce the pH as much as strong acids do. Neutralization of an acid occurs when it reacts with a base (a source of hydroxyl or $OH^-$ ions) to form a salt and water. In turf, acid neutralization is usually achieved by using lime as the base.
Using lime to neutralize an acid:

$$AH^+\text{(potential acid)} + BOH\text{(base, e.g., lime)} \rightarrow BA\text{(salt)} + H_2O\text{ (water)}$$

For example:

$$2HCl\text{ (hydrocloric acid)} + Ca(OH)_2\text{(calcium hydroxide)} \rightarrow CaCl_2\text{(calcium chloride)} + 2H_2O\text{(water)}$$

At low pH values (<5.5), the solubility of some phytotoxic elements, such as $Al^{+3}$, increases. Sodium ions often are excessive when the soil pH is >8.5.

The $H^+$ ions liberated from exchange sites and the $H^+$ produced by hydrolysis of liberated $Al^{+3}$ ions lower soil pH to a range of about 4.5 to 5. Application of a liming material, however, produces hydroxyl ions ($OH^-$), which react with (neutralize) the liberated $H^+$ to produce water. The hydroxyl–Al [$Al(OH^{+2})$] polymer formed is irreversibly sorbed to the surface of inorganic soil colloids and, as a result, decreases CEC and a pH of ~5 to 6.5.

As soil pH increases, Al ions eventually precipitate, leaving negative charges that allow other available cations, such as $Ca^{+2}$, to be retained until they can be absorbed by plants. With this decrease in the degree of $Al^{+3}$ and $H^+$ saturation of the exchange complex, the degree of saturation with exchangeable bases increases. This occurs at a pH of ~6.5 to 8.5. Exchangeable bases such as $Ca^{+2}$ and $Mg^{+2}$ then dominate, while the $Al^{+3}$ ions form insoluble gibbsite and the $H^+$ ions react with hydroxyl ions ($OH^-$) to form water.

**Soil pH effects on plants.** The soil pH range for optimum turf growth is 5.5 to 6.5. Soil pH below and above this range may impact plants in several ways. First, and likely most important, is the abundance of toxic elements Al, H, and Mn in the soil solution. Aluminum becomes more prevalent in soils when pH is below 5.8; however, Al toxicity to turfgrass roots generally does not occur until pH falls below 5.5 and, more typically, not until around pH 5.0. Hydrogen ion damage of root systems typically does not occur until soil pH approaches 4.0. Root systems become short, thickened, and brown with damage from Al and/or H. The uptake of nutrients, particularly P, K, Mg, and Ca, and water is usually reduced by these toxicities. A decreased tolerance to environmental stresses, such as drought and heat, results, as does a decrease in the turf's recuperative potential. Manganese toxicity, which affects the aboveground portion of the plant, can occur at soil pH below 5.5, but is not common in turfgrasses. Conversely, as the soil pH becomes more basic, several nutrients, including Al, P, Fe, Mn, B, and Zn, will become less available and plant deficiency symptoms will appear.

Thatch accumulation and algae problems also tend to be associated with low soil pH (<5). Naturally present bacteria that decompose thatch are less prevalent and less effective at low soil pH. Bacterial activity is also important for the release of N from organic and some slow-release fertilizers; low pH slows the release of N from these materials.

### Liming Materials

A liming material by definition contains calcium (Ca) and neutralizes acidity. All liming material, whether it is the oxide (CaO), hydroxide [$Ca(OH)_2$], or carbonate ($CaCO_3$) form of Ca, reacts with $CO_2$ and water to yield the bicarbonate form [$Ca(HCO_3)_2$] when applied to an acidic soil. The Ca (or Mg) of the liming material displaces $H^+$ and $Al^{+3}$ ions from the soil exchange complex and places them in solution with a resulting increase in the base saturation. In addition to raising the pH of acidic soils, liming materials can contribute to the plant nutrients Ca and/or Mg, make P more available, improve microbial activity, and make K more efficient in plant nutrition. The general reaction of a liming material such as calcium carbonate ($CaCO_3$) being added to the soil can be written as
Two-step process when soil is limed with calcium carbonate ($CaCO_3$):

Step 1:

$$CaCO_3\text{(lime)} + H_2O \rightarrow Ca^{+2}\text{(calcium)} + 2OH^- + CO_2 \uparrow \text{(carbon dioxide)}$$

Step 2:

$$H^+\text{(hydrogen ion)} + OH^-\text{(hydroxyl ion)} \rightarrow H_2O\text{(pH} \uparrow\text{)}$$

$$Al^{+3} + 3OH^- \rightarrow Al(OH)_3\text{ gibbsite (pH} \uparrow\text{)}$$

As shown in the second reaction, H ions in the soil solution then combine with the hydroxyl ions ($OH^-$) to form water. This neutralization of the H ions results in an increased (or higher) soil pH.

Liming materials most commonly consist of Ca and calcium–magnesium carbonates (including ground, pelletized, and flowable limestone), calcium hydroxide, calcium oxide, marl,

**TABLE 3.7**
**Characteristics of Commonly Used Liming Materials**

| Source | Alternative Name(s) | Chemical Symbol | Characteristics |
|---|---|---|---|
| Calcium carbonates | Calcite, calcitic limestone, aglime, marl, chalk | $CaCO_3$ | • ~40% Ca when pure; often used when Mg is not needed; marl is lime deposits from the bottom of small freshwater ponds while chalk is soft limestone deposits in oceans |
| Calcium-magnesium carbonates | Dolomite | $CaMg(CO_3)_2$ | • ~50% calcium carbonate, 35% magnesium carbonate, and 15% soil and other impurities; used when Mg is needed; often pelletized to ease application |
| | Dolomitic limestone | $CaMg(CO_3)_2$ | • Has unequal proportions of Ca and magnesium carbonate with 21% to 30% Ca and 6% to 11% Mg; used when Mg is needed |
| Calcium oxide and/or magnesium oxide | Unslaked lime, burned lime, quicklime | CaO or CaO + MgO | • White powdery substance; most effective and quickest reacting liming material; high burn potential; caustic to handle; difficult to soil mix |
| Calcium hydroxide | Slaked lime, hydrated lime, builder's lime | $Ca(OH)_2$ | • White powdery substance similar to calcium oxide except moisture has been added; second most effective and quickest reacting material; however, has high burn potential; caustic to handle; difficult to soil mix; also used to make mortar |
| Slags | Blast-furnace slag, basic slag, electric-furnace slags | CaO plus impurities, mostly by calcium silicate ($CaSiO_3$) | • By-product of Fe manufacturing that is ground and screened; effective if available and economically priced; also contains P (2% to 17%), Mg (~3%), and Fe |

and slags (Table 3.7). Ground limestone is the most inexpensive source but is dusty and not as easily spread as the pelletized form.

Pelletized limestone is ground limestone (either calcitic or dolomitic) that is aggregated into larger particles to aid in spreading and reducing dust. These pellets disintegrate quickly when they become wet.

In addition, localized sources of other materials containing Ca and/or Mg are used as liming agents, including fly ash from coal-burning electrical power plants, ash from hardwoods, sludge from water-treatment plants, pulp mill lime, flue dust from cement manufacturing, and carbide lime. The relative purity or neutralizing value (also referred to as the calcium carbonate equivalent) is shown in Table 3.8. Application rates are adjusted according to the source used, as impurities such as clay often lower the neutralizing values of many limestone sources. A good lime source should have a neutralizing value of 70 or higher.

The higher the neutralizing value is, the less lime is required to change pH. For example, if burned lime had a neutralizing value of 160, 63 pounds of this material would produce the same effect on pH as 100 pounds of calcium carbonate. Table 3.9 lists various conversion factors for Ca and Mg sources used for liming.

**TABLE 3.8**
**Relative Neutralizing Values of Common Liming Materials Compared to Calcium Carbonate**

| Material | Relative Neutralizing Value |
|---|---|
| Calcium carbonate ($CaCO_3$) | 100 |
| Burned or quicklime (calcium oxide, CaO) | 150 to 175 |
| Hydrated lime (calcium hydroxide, slake lime, builder's lime, $Ca(OH)_2$) | 120 to 135 |
| Dolomitic lime ($CaMg(CO_3)_2$) | 95 to 108 |
| Burned oyster shells ($CaCO_3$ + impurities) | 90 to 110 |
| Calcitic lime ($CaCO_3$ + impurities) | 85 to 100 |
| Baked oyster shells ($CaCO_3$ + impurities) | 80 to 90 |
| Sugarbeet lime ($CaCO_3$ + 4% to 10% organic matter) | 80 to 90 |
| Marl ($CaCO_3$ + impurities) | 50 to 90 |
| Basic slag (calcium oxide, CaO + impurities, or $CaSiO_3$) | 50 to 70 |
| Cement kiln dusts | 40 to 100 |
| Power Plant Ash (CaO, $K_2O$) | 5 to 50 |
| Gypsum (calcium sulfate, $CaSO_4$) | 0 |

**TABLE 3.9**
**Conversion Factors for Various Calcium and Magnesium Sources Used for Liming**

| | | | | | | | | | |
|---|---|---|---|---|---|---|---|---|---|
| $CaCO_3$ | × | 0.56 | = | CaO | $MgCO_3$ | × | 0.48 | = | MgO |
| $CaCO_3$ | × | 0.40 | = | Ca | $MgCO_3$ | × | 0.29 | = | Mg |
| CaO | × | 0.71 | = | Ca | MgO | × | 0.60 | = | Mg |
| CaO | × | 1.78 | = | $CaCO_3$ | MgO | × | 2.09 | = | $MgCO_3$ |
| Ca | × | 1.40 | = | CaO | Mg | × | 1.66 | = | MgO |
| Ca | × | 2.50 | = | $CaCO_3$ | Mg | × | 3.48 | = | $MgCO_3$ |
| $CaCO_3$ | × | 0.84 | = | $MgCO_3$ | $MgCO_3$ | × | 1.19 | = | $CaCO_3$ |
| $CaMg(CO_3)_2$ | × | 0.22 | = | Ca | | | | | |
| $CaMg(CO_3)_2$ | × | 0.63 | = | CaO | | | | | |
| $CaMg(CO_3)_2$ | × | 1.09 | = | $CaCO_3$ | | | | | |

**Calcium and calcium-magnesium carbonates.** Calcium and calcium–magnesium carbonates are commonly referred to as calcitic and dolomitic limestone. Various forms of Ca and calcium-magnesium carbonates occur, including:

- **Calcite or calcitic limestone** (also known as calcium carbonate or $CaCO_3$) is composed primarily of crystalline calcium carbonate, containing ~40% Ca when pure and only trace amounts of $Mg^{+2}$. It is also called aglime.
- **Dolomite** [$CaMg(CO_3)_2$] has equal molecular proportions of calcium ($CaCO_3$) and magnesium carbonate ($MgCO_3$).
- **Dolomitic limestone** has unequal proportions of calcium and magnesium carbonate, with Ca dominant, from 21% to 30%, and normally $Mg^{+2}$ of 6 to 11%.

Limestone is most often obtained by open-pit mining. The neutralizing capability of most agricultural limestones containing Ca and calcium–magnesium carbonates is commonly between 90% and 98% because of associated impurities such as clay. Dolomitic lime is recommended for acidic soils low in Mg and for sandy soils of low nutrient-retention capacity.

**Calcium oxide and/or magnesium oxide (CaO, or CaO plus MgO).** This source is also known as *unslaked lime, burned lime*, or *quicklime*. Calcium oxide is produced by roasting calcitic or dolomitic limestone ($CaCO_3$, or $CaCO_3$ and $MgCO_3$) in an oven, thus driving off $CO_2$; this results in a white powder (CaO, or CaO and MgO), which is caustic and difficult to handle. It is the most effective and most quickly reacting of all liming materials. It has a neutralizing value of up to 179% compared with calcium carbonate (calcite). However, it has a high potential to injure (burn) plants and an increased cost, and is difficult to thoroughly mix with the soil without almost immediately caking. Calcium oxide is not widely used or recommended in turf management. It is often used in the building industry for various brick-laying mortars.

**Calcium hydroxide ($Ca(OH)_2$).** This source is also known as *slaked lime, hydrated lime*, or *builder's lime*, since it also is used in making mortar. It is produced by hydrating (adding water to) calcium oxide (CaO), and is the second most effective and quick reacting of liming materials, with a neutralizing value of up to 136%. Like calcium oxide, calcium hydroxide is a white, powdery substance, which is difficult and unpleasant to handle and has a high degree of burn potential to plants. Ammonia gas, which may injure grass, also can form if calcium hydroxide and ammonia-based N fertilizers are applied close (e.g., within two weeks) to each other.

**Marl.** Marl, also known as bog lime, is mined from shallow, natural deposits of soft calcium carbonates and is often quite moist and impure. Shell marl is bog lime or marl containing numerous shells. When present, marl may become mixed with surface organic soils or peat. These normally acidic organic soils thus become neutral or even alkaline due to the liming action of the marl. Many of the peat or muck sod farms in south Florida, for example, are on soils with marl intermixing. These soils are almost always low in Mg as well as K, P, Cu, and Zn. Marl, when mined for liming purposes, has a neutralizing value between 50% and 90%, depending on the amount of impurities (clay and organic matter) it contains. Oyster shells and other seashells are composed mostly of calcium carbonate and can also make acceptable liming materials when properly pulverized. Chalk is from soft calcitic limestone deposits in oceans.

**Slags.** Slags are the by-product of smelting operations and contain mostly calcium silicate ($CaSiO_3$). The Ca and Mg are present as silicates and not as carbonates or oxides as in limestone or quicklime. A slag is formed when metal is made. The slag is then ground, screened, and shipped for use as a liming material. Three important slags include blast-furnace slag, basic slag, and electric-furnace slag. Their neutralizing values range from 65% to 90%, and each has been shown to be effective liming material if locally available and economically priced. They often contain significant levels of phosphorus (2% to 17%), Mg (about 3%), and Fe.

**Liquid lime materials.** Liquid lime materials, which consist of finely ground limestone (50%), water (47%), clay (2%), and a dispersing agent (1%), are available. Since the particle size of the limestone is extremely small, the reaction rate is as rapid as a finely ground limestone. The liming material has a fineness of <60 mesh (or <250 μm). The advantage of using a liquid lime rather than a finely ground dry limestone is that liquid lime can be applied more uniformly and with less mess. Liquid lime materials are generally more costly than dry limestone, but their ease in utilization may justify the increased cost in some situations. Due to potential plant phytotoxicity, only small amounts of liquid lime can be applied at one time and the spray suspension may be abrasive to sprayer parts.

**Pelletized limestone.** This form of limestone consists of small particles compressed to form a granule. Upon application and wetting, the pellet is dispersed into smaller particles. The advantage of pelletized lime over standard limestone materials is minimum dust and ease of application. Complete dispersion of the individual particles is necessary for a pelletized lime to be as effective as standard finely ground limestone.

**Fineness and purity of liming material.** The fineness of a liming material will determine how fast it reacts with a soil to neutralize acidity. The finer the limestone is, the more soil surface it contacts and the more quickly it reacts. However, the cost of limestone increases with its fineness. Powder-like limestone also is difficult to handle and to apply uniformly. Although faster acting, the effects of fine limestone are not as long-lasting as for coarser materials since coarser materials dissolve more slowly. Therefore, most agricultural limestone contains both coarse and fine materials. In general, a pulverized limestone, which will pass a 10-mesh screen (at least 50% passing a 100-mesh screen), should provide desirable results, and yet be economical. Particles unable to pass an eight-mesh screen have almost no effect on soil pH.

**Incorporation and mixing.** Since limestone is relatively insoluble (only 1 pound, 454 kg, will dissolve in 500 gallons, 1,893 L, of water), the extent of contact between the soil and the limestone is important in determining how fast the limestone will react and increase soil pH. The best time to adjust soil pH is during construction. Uniformly mixed limestone maximizes the neutralization of soil acidity, whereas clumps of limestone react slowly. Limestone should be thoroughly mixed into the upper 6 to 8 inches (15 to 20 cm) of soil based on soil analysis of the same soil depth.

Once the turf is established, lime applications are applied to the soil surface and rates should be based on the soil analysis of the upper 4 inches (10 cm) of soil. Surface-applied limestone is in contact with very little soil; therefore, it reacts extremely slowly and the downward movement of neutralization is minimal. With established turf, soil pH should be monitored and adjusted (if necessary) annually, since correction with lime will be slow. Liming after core aerification uniformly mixes limestone into the soil on established turfgrass and increases the amount of neutralization occurring beyond the upper inch (2.5 cm) of soil.

It should be noted that limestone crushed for road construction is generally too coarse to be effective in lowering soil pH. Even if this material is applied in excessive amounts, it is usually inferior to a finer ground source in raising soil pH.

## Selecting a Liming Material

When choosing a liming material, the degree of fineness, neutralizing value, Mg content, plant-tissue burn potential, and moisture content should each be considered. In most areas, materials with the least cost per unit of neutralizing value should be considered, assuming the same degree exists for fineness of all materials. However, many soils are deficient in Mg; thus, dolomitic lime is often used. Turf managers also must carefully consider the burn potential of available materials, especially when irrigation is not readily available and/or temperatures are hot. In these cases, a more slowly reacting dolomitic limestone may still be the best choice.

Most lime sources contain a certain degree of moisture to minimize dust. In many states, a maximum moisture content of 10% is allowed to prevent consumers from paying for more moisture than needed to reduce dust.

Most commercially available limestone sources contain impurities, reducing their overall effectiveness and ability to be active in the soil. Limestone quality is determined by two factors: (1) fineness and (2) purity or calcium carbonate equivalent (CCE). These two factors combine to produce the effective calcium carbonate (ECC) rating:

$$\text{ECC (effective calcium carbonate)} = \text{CCE (calcium carbonate equivalent)} \times \text{fineness factor}$$

The fineness factor determines the activity and speed of reaction to neutralize acidity. Most states have laws requiring percentages of lime that must pass through a 20-mesh sieve (0.84 mm opening). Activity factor of limestone sizes is based on the fraction of the material effective over a two- to three-year period. Activity factors can be found in Table 3.10.

CCE measures the ability to neutralize acidity relative to an identical weight of pure $CaCO_3$. In most cases, natural limestone has CCE values less than 100% because of impurities. CCE is determined in laboratories by combining $CaCO_3$ with hydrochloric acid (HCl). The solution is then titrated with a base, sodium hydroxide (NaOH) to determine the CCE. Liming materials can have CCEs ranging from 50 to 150. A liming material with a CCE of 50% would have only one-half (0.5) of the neutralizing value of pure $CaCO_3$. In contrast, a liming material with a CCE of 150% would have 1.5 times the neutralizing value of pure $CaCO_3$.

**TABLE 3.10**
**Determining Limestone Activity over a Two- to Three-Year Period Based on Its Particle Size**

| Sieve Size | Particle Size (mm) | Activity Factor |
|---|---|---|
| Coarser than 8 mesh | >2.36 | 0 |
| 8 to 40 mesh | 0.425 to 2.36 | 25 |
| 40 to 60 mesh | 0.250 to 0.425 | 60 |
| Finer than 60 mesh | <0.250 | 100 |

*Note:* For example, according to this table, 60% of limestone passing through a 40-mesh sieve but not through a 60-mesh one would be active in increasing pH over two to three years.

### Example

A 100 g sample of limestone was shaken through a set of sieves. Five grams were retained on the 8-mesh sieve, 5 g were retained on the 40-mesh, 20 g were retained on the 60 mesh, and 70 g passed the 60 mesh. The limestone has a CCE of 95%. Find the ECC and lime required to satisfy a recommendation of 5,000 pounds lime/acre (560 g/m²).

**Step 1:** Determine the fineness factor:

| Sieve size | Retained (g) | × | Activity Factor | = | Percent Available for Reaction |
|---|---|---|---|---|---|
| Coarser than 8 mesh | 5 g | | 0 | | 0 |
| 8 to 40 Mesh | 5 g | | 0.25 | | 1.25 |
| 40 to 60 Mesh | 20 g | | 0.60 | | 12 |
| Finer than 60 mesh | 70 g | | 1.00 | | 70 |
| Sum | | | | ≤ | 83.25 |

In two to three years, approximately 83% of the liming material will be active.

**Step 2:** Find ECC:

Since the CCE was given, multiply the CCE and fineness factor to determine the ECC:

$$\begin{aligned}\%\,\text{CCE} &= \%\,\text{fineness factor} \times \text{ECC}\\ &= 95 \times 83.25\\ &= 79.1\end{aligned}$$

**Step 3:** Limestone recommendation:

The ECC was found to be 79.1%; therefore, 79% of the applied limestone will be active in neutralizing acidity. The limestone recommendation was 5,000 pounds/acre pure $CaCO_3$:

$$\frac{5{,}000\ \text{pounds}}{\text{acre}} \times \frac{1}{79\%} = 6{,}329\ \text{pounds/acre}\ (7{,}089\ \text{kg/ha})$$

One would need to apply 6,329 pounds/acre (709 g/m²) of the tested limestone to equivalently apply 5,000 pounds/acre (560 g/m²) of pure $CaCO_3$.

**Rates.** Rates of lime required to neutralize excessive acidity vary with the degree of soil weathering and soil texture (Table 3.11). Lime rate requirements depend on (1) the change of soil pH desired, (2) the buffering capacity of the soil, and (3) the chemical composition and fineness of the liming material to be used. High CEC soils, such as clay or muck soils, may require up to double the amount needed for soils of low CEC, such as sands. Sandy soils, however, often require more frequent applications than clay soils as the liming agents are more easily leached from the soil.

Recommended rates of lime should be determined by a soil testing lab and generally do not exceed 50 to 100 pounds/1,000 square feet (244 to 488 g/m²). If hydrated or burned lime is used, rates over 25 pounds/1,000 square feet (122 g/m²) should be avoided in one application. Rates exceeding this, or applications during periods of hot temperatures, can injure (burn) turfgrass. The turf should be dry at the time of application and should be immediately irrigated to wash all lime off the leaves and into the soil. If over 100 pounds/1,000 square feet (488 g/m²) are needed, splitting applications several months apart will improve the efficacy of lime use. Soil analysis will usually indicate the specific amount and type of lime needed to raise the pH of a specific soil type. Aerifying prior to lime application aids in placing the material into the rootzone.

Intensively weathered soils are often highly $Al^{+3}$ dominated with low $Ca^{+2}$ and $Mg^{+2}$. To replace (or exchange) this $Al^{+3}$, the amount of lime needed is commonly considered to be 1.5 times the amount of exchangeable $Al^{+3}$ listed on the soil test report.

The purpose for applying the lime will also influence the application rate used. As noted, relatively large amounts are needed if the desire is to increase soil pH. One ton or more may be needed per acre (224 g/m²) for this purpose. If the purpose of liming is to supply Ca or Mg to the soil as a nutrient, the amount applied is greatly reduced. If the desire is to supply these nutrients and not raise soil pH, then another source of Ca or Mg should be used. For example, gypsum and magnesium sulfate supply Ca and Mg, respectively, without affecting soil pH.

**TABLE 3.11**
**Examples of Rates of Dolomitic Limestone or Calcium Carbonate Required to Raise pH of Sand, Loam, Clay, or Muck Soil to 6.5**

| | Pounds Required per 1,000 square feet[a] | | |
|---|---|---|---|
| Current pH | Sand | Loam | Clay or Muck |
| 6.0 | 20 | 35 | 50 |
| 5.5 | 45 | 75 | 100 |
| 5.0 | 65 | 110 | 150 |
| 4.5 | 80 | 150 | 200 |
| 4.0 | 100 | 175 | 230 |

[a] Multiply values by 48.83 to obtain kilograms/hectare or by 4.88 to obtain grams/square meter.

### Example

1. A soil test report indicates acidity of 1 meq/100 g of $H^+$ and 3 meq/100 g of $Al^{+3}$. How many pounds of $CaCO_3$ are needed per acre to totally replace these? Molecular weights: Ca = 40, C = 12, O = 16, H = 1, Al = 27. It is assumed one acre that is 6 inches (15 cm) deep weighs 2,000,000 pounds (907,200 kg).

**Step 1:** The equivalent weights of $Ca^{+2}$, $Al^{+3}$, and $H^{+}$ need to be determined:

Equivalent weights = molecular weights ÷ valence charge

| | | |
|---|---|---|
| $Ca^{+2}$ | = 40 g | ÷ 2 (valence charge) = 20 g |
| $Al^{+3}$ | = 27 g | ÷ 3 = 9 g |
| $H^{+}$ | = 1 g | ÷ 1 = 1 g |

**Step 2:** Determine the total acidity of the soil composed of $H^{+}$ and $Al^{+3}$ ions from the soil test report:

$$\frac{3\text{ meq Al}^{+3}}{100\text{ g soil}} + \frac{1\text{ meq H}^{+}}{100\text{ g soil}} = \frac{4\text{ meq acid}}{100\text{ g soil}}$$

**Step 3:** Determine the total amount of $CaCO_3$ needed to neutralize 4 meq acid/100 g soil and convert to a per-acre basis.

$$\frac{4\text{ meq}}{100\text{ g soil}} \times \frac{1\text{ eq}}{1{,}000\text{ meq}} \times \frac{20\text{ g Ca}^{+2}}{\text{eq Ca}^{+2}} \times \frac{2{,}000{,}000\text{ pounds}}{\text{acre}}$$

$$\times \frac{1\text{ pound CaCO}_3}{40\%\text{ active ingredients}} = \frac{4{,}000\text{ pounds}}{\text{acre}}$$

2. You have the option to completely replace 5 meq/100 g of $Al^{+3}$ with either $MgCO_3$ or $CaCO_3$. Which would you choose if the least amount of product was desired (1 acre = 2,000,000 pounds soil, and equivalent weights of $Ca^{+2}$ = 20 g and $Mg^{+2}$ = 12 g)?
**Step 1:** Determine how much $CaCO_3$ is needed to replace 5 meq/100 g $Al^{+3}$:

$$\frac{5\text{ meq Al}^{+3}}{100\text{ g soil}} \times \frac{1\text{ eq}}{1{,}000\text{ meq}} \times \frac{20\text{ g Ca}^{+2}}{\text{eq Ca}^{+2}} \times \frac{2{,}000{,}000\text{ pounds}}{\text{acre}}$$

$$\times \frac{1\text{ pound CaCO}_3}{40\%\text{ active ingredients}} = \frac{5{,}000\text{ pounds CaCO}_3}{\text{acre}}$$

**Step 2:** Determine how much $MgCO_3$ is needed to replace 5 meq/100 g $Al^{+3}$:

$$\frac{5\text{ meq Al}^{+3}}{100\text{ g soil}} \times \frac{1\text{ eq}}{1{,}000\text{ meq}} \times \frac{12\text{ g Mg}^{+2}}{\text{eq Mg}^{+2}} \times \frac{2{,}000{,}000\text{ pounds}}{\text{acre}}$$

$$\times \frac{1\text{ pound MgCO}_3}{28.6\%\text{ active ingredients}} = \frac{4{,}196\text{ pounds CaCO}_3}{\text{acre}}$$

Therefore, 4,196 pounds/acre (470 g/m$^2$) of $MgCO_3$ is needed to neutralize 5 meq $Al^{+3}$/100 g soil compared to 5,000 pounds $CaCO_3$ (9,560 g/m$^2$).

Adding liming materials can raise the pH of a soil. However, over time, most soils eventually return to their original pH level. Therefore, long-term pH change often requires periodic liming. Using plants naturally adapted to the existing soil pH may be easier (and cheaper) in the long term, especially in out-of-play areas.

### Acidic Irrigation Water

With the increase in acid rain and acidic lakes, problems with excessively low (<5.6) pH irrigation water have also increased. This is most notable around heavy industrial areas, presumably due to industrial air and water pollution problems. Acidic irrigation water not only may create problems for plants, but also causes considerable corrosion to irrigation pumping and piping systems. A quality, powdered carbonate-containing material such as dolomitic lime is the material most often injected into an irrigation system to raise the pH of the irrigation water.

## SOIL ALKALINITY

Soil alkalinity occurs when an excess of $Ca^{+2}$, $Mg^{+2}$, or $Na^{+}$ ions is present. Moderate alkalinity occurs in the pH range of 7.5 to 8.5, with excessive alkalinity above these figures. A 2% to 3% by weight calcium carbonate content usually results in a soil pH range between 7.5 and 8.5. In the United States, soil alkalinity is generally of greatest concern in arid (dry) or semiarid western regions, where rainfall is minimal, or along coastal regions. Leaching of $Ca^{+2}$, $Mg^{+2}$, and $Na^{+}$ ions does not readily occur in these areas. However, due to deposits of calcium carbonates in the form of shell marl and/or underlying limestone, or to overliming or irrigation with water from limestone aquifers, other selected areas, such as coastal regions, also can have excessive soil pH values.

Deficiencies of several micronutrients necessary for plant growth may occur as a result of excessive soil alkalinity. Iron, Mn, Zn, B, and Cu all tend to be less available to plants under these conditions. Chlorotic, unhealthy-appearing plants, often resembling N deficiency, may result (Figure 3.9). Superintendents not experienced in managing grass grown under alkaline conditions may be confused by these symptoms.

### Reducing Soil pH

Reducing soil alkalinity often is a never-ending chore, especially if high pH water is used for irrigation. Turf managers must constantly regulate and adjust their management programs to compensate for alkalinity. The best materials for acidifying soil are elemental sulfur and aluminum sulfate (Table 3.12). Elemental sulfur (S) is usually the most efficient and practical of these sources. Sulfur-coated ureas and acid-forming fertilizers such as ammonium sulfate also are sometimes used to reduce soil pH due to the limited H ions produced via nitrification. However, they do not generally contain enough S or supply enough H to quickly correct the alkalinity. They can gradually reduce or regulate pH if used long term. Sulfuric acid or sulfur dioxide can be injected into irrigation water to reduce its pH. Materials such as calcium sulfate (gypsum or $CaSO_4$), copper sulfate ($CuSO_4$),

**FIGURE 3.9** Yellow, chlorotic turf from micronutrient deficiency, resembling lack of nitrogen.

**TABLE 3.12**
**Sulfur-Containing Compounds That Produce an Acidifying Effect**

| Material | S (%) | S Content (pounds/ton) | N (%) | Other (%) |
|---|---|---|---|---|
| Aluminum sulfate | 14 | 288 | 0 | 11(Al) |
| Ammonia-sulfur solution | 10 | 200 | 74 | |
| Ammonium bisulfite solution | 17 | 340 | 9 | |
| Ammonium polysulfide solution | 40 | 800 | 20 | |
| Ammonium sulfate | 24 | 484 | 21 | |
| Ammonium thiosulfate solution | 26 | 520 | 12 | |
| Aqua-sulfur solution | 5 | 100 | 20 | |
| Lime sulfur (dry) | 57 | 1140 | 0 | 43 (Ca) |
| Lime sulfur (solution) | 23 to 24 | 480 | 0 | 9 (Ca) |
| Sulfuric acid (100%) | 33 | 654 | 0 | |
| Sulfuric acid (66° Be = 93%) | 30 | 608 | 0 | |
| Sulfur (elemental) | 100 | 2000 | 0 | |
| Sulfur dioxide | 50 | 1000 | 0 | |

potassium sulfate ($K_2SO_4$), and magnesium sulfate (Epsom salt or $MgSO_4$) are neutral salts and have little acidity effect.

Table 3.13 indicates the amount of calcium carbonate neutralized by 100 pounds (454 kg) of various amendments and how to convert the weight of these amendments to pounds of S.

For example, 100 pounds (45.4 kg) ammonium sulfate are needed to neutralize approximately 45 pounds (20 kg) of calcium carbonate, while 6.9 pounds (3.1 kg) of the ammonium sulfate is required to equal the acidifying effects of 1 pound (0.454 kg) of elemental sulfur.

**Elemental sulfur.** Elemental (or granular) sulfur ranges from 99% to 20% purity or less (often 90% powdered S plus 10% bentonite clay), with the lower contents in low-grade deposits containing clay and other materials. It should not be used on putting greens for reducing soil pH. It is a yellow, inert, water-insoluble crystalline solid (Figure 3.10). Pelletized water-dispersible granules and porous granules of S (termed popcorn sulfur) formed from molten S and water are available. Applied elemental sulfur is oxidized by soil microorganisms (the prominent group being various *Thiobacillus* species) into sulfuric acid.

Conversion of elemental sulfur to sulfuric acid by soil microorganisms:

$$2S\,(\text{elemental sulfur}) + 3O_2\,(\text{oxygen}) + 2H_2O\,(\text{water}) \rightarrow$$
$$\textit{Thiobacillus}\text{ spp. } 2H_2SO_4\,(\text{sulfuric acid})$$

Sulfuric acid then reacts with insoluble limestone ($CaCO_3$) to form calcium sulfate ($CaSO_4$) of gypsum. The Ca in gypsum can then displace Na ions on soil exchange sites which are removed from the soil by leaching. Sulfuric acid is considered a "strong" acid since it readily dissociates, releasing its H ions that lower soil pH. The conversion of elemental sulfur to sulfuric acid, because it is a biological reaction involving

**TABLE 3.13**
**Pounds of Calcium Carbonate ($CaCO_3$) Neutralized by 100 Pounds of Amendment and Factor Needed to Calculate Pounds of Amendment Equivalent to Pounds of Sulfur**

| Amendment | Pounds $CaCO_3$ Neutralized by 100 pounds (45.4 kg) of Amendment | Factor to Calculate Pound Amendment Equivalent to Pounds Sulfur |
|---|---|---|
| Aluminum sulfate | 45 | 6.9 |
| Ammonium nitrate | 59 | 5.3 |
| Ammonium sulfate | 110 | 2.8 |
| Diammonium phosphate | 70 | 4.5 |
| Monoammonium phosphate | 65 | 4.8 |
| Sulfuric acid | 100 | 3.1 |
| Sulfur | 312 | 1.0 |
| Sulfur coated urea | 118 | 2.6 |
| Urea | 84 | 3.7 |
| Ureaformaldehyde | 68 | 4.6 |

*Note:* Multiply pounds $CaCO_3$ by 0.454 to obtain kilograms.

**FIGURE 3.10** Sulfur pellets that are heated and incorporated into irrigation water to reduce pH and subsequent bicarbonate concentrations.

microorganisms, increases with increasing temperature. This conversion does not occur readily when soil temperatures are below 40°F (4.4°C). A steady increase in oxidation occurs above 40°F (4.4°C); however, a *sharp* increase occurs at temperatures above 70°F (21°C). Smaller particle-sized material also reacts faster than larger particles.

Oxidation also is optimum when soil moisture is near field capacity. Excessive soil moisture, however, reduces the degree of soil aeration needed for oxidation; thus, S conversion is reduced.

Approximately one-third the quantity of S is required to reduce soil pH by one unit as the amount of lime needed to raise it the same amount (Tables 3.11 and 3.14). It would be difficult to decrease the pH of soils containing sufficient free $CaCO_3$ (calcium carbonate) (e.g., >8%) without adding excessive and, most often, impractical S amounts. A simple test for $CaCO_3$ is to place several drops of a liquid acid (such as vinegar) on the soil. If bubbling occurs, the acid is reacting with $CaCO_3$, releasing $CO_2$.

After application, elemental sulfur should be thoroughly mixed into the top 6 to 8 inches (15 to 20 cm) of soil. Optimum results follow thorough mixing either by a power rotor-tiller or by running a disk in several directions across the treated area. Adequate soil moisture (though not saturation) is necessary for this conversion to occur. Irrigation should therefore begin immediately after S incorporation and should be continued at regular intervals to maintain moisture levels near field capacity, as long as oxidation is desired. This process may take several months, especially if initial soil pH is high or if soil temperatures are cool.

**Established turf.** Once the turf is established, significantly reducing soil pH becomes much more difficult because of the

**TABLE 3.14**
**Approximate Amounts of Elemental Sulfur (99% Purity) per 1,000 square feet to Lower Soil pH to 6.5 in the Top 7 inches (17.8 cm) of Soils before Planting**

| | Pounds/1,000 square feet[a] | | |
|---|---|---|---|
| **Current pH** | **Sandy** | **Loam** | **Clay or Muck** |
| 8.5 | 35 to 50 | 50 to 60 | 60 to 70 |
| 8.0 | 25 to 30 | 30 to 35 | 40 to 50 |
| 7.5 | 10 to 15 | 15 to 20 | 20 to 25 |
| 7.0 | 2 to 3 | 3 to 5 | 5 to 8 |

*Note:* To determine application rates of other amendments, use the conversion factors in the last column of Table 3.13.

[a] Multiply by 43.5 to convert rates to pounds/acre or 4.88 to obtain grams/square meter.

plant-tissue burn that can occur when using high rates of acidifying materials. Since elemental sulfur is sparingly soluble, it also does not readily move down the soil profile, and tends to reduce soil pH (by an amount greater than desired) only in a thin zone near the application surface. This results in little benefit to the roots from the acidifying effect. Care must be taken not to create too toxic a surface acidic layer near the turf–plant crown surface. To minimize the injury potential and to facilitate sulfur's acidifying effects in the turfgrass rootzone, it is suggested that elemental sulfur application be in conjunction with turf coring or aerification.

If a soil is inherently alkaline, the turf manager has two methods of reducing pH once the site has been established in grass. One is to use ammonia-based fertilizers, such as ammonium nitrate or ammonium sulfate. Hydrogen ions produced during nitrification (the conversion of the ammonium to nitrite and nitrate) will help reduce soil pH.

The general ranking of soil acidification by using N fertilizers includes:

Ammonium sulfate > DAP = ammonium nitrate > ammonia = urea > calcium ammonium nitrate

The other procedure is to add small, frequent applications of S to the turf. Only elemental sulfur or S compounds oxidized to sulfate ion produce the desired acidification (Table 3.12). Up to 5 pounds of elemental S may be applied per 1,000 square feet (24 g/m$^2$) on taller maintained grasses, such as fairways or roughs. Applications to putting greens should be no greater than one-half pound elemental S/1,000 square feet (2.4 g/m$^2$). Applications should be spaced at least three to four weeks apart. Application during hot weather also should be avoided, as should application to wet leaf surfaces, and each application should be followed by irrigation. Total application of elemental sulfur should not exceed 10 pounds/1,000 square feet yearly (49 g/m$^2$). Excessive amounts of S and/or excessive watering (e.g., lack of soil $O_2$) can result in hydrogen sulfide ($H_2S$) formation, which reacts with trace metal ions and precipitates in soils as compounds such as $FeS_2$, instead of escaping as a gas. The formation of a "black layer" 1 to 2 inches (2.5 to 5 cm) below the putting green surface is evidence of this. Other toxic components associated with soils of low $O_2$ content include carbon dioxide ($CO_2$) and methane ($CH_4$). Proper aerification and irrigation management minimize the chance of such conditions occurring. Applications during winter may not produce the desired results until spring, when temperatures are warm enough to drive the reaction. Excessive application therefore may occur since immediate results are not seen and additional materials are applied in an attempt to speed up the reaction. Oftentimes in spring, following this scenario, the added S begins to oxidize at the same time and the pH in the top inch (2.5 cm) of soil drops.

**Acid injection.** Conditions that may warrant acid injection into the irrigation source include:

1. When total bicarbonate and carbonate levels exceed free Ca and Mg totals, such as soil residual sodium carbonate (RSC) values > 15%
2. When residual sodium carbonate (RSC) values exceed 1.25
3. When sodium adsorption ratios (SARs) exceed 6 meq/L

If the soil pH is raised by bicarbonate levels in the irrigation source, then a sulfur-injection system may be used to maintain or lower water pH. Reducing soil pH also helps dissolve insoluble carbonates (salts) into much more soluble bicarbonates to help flush and remove various salts from the soil surface. Reducing soil pH by dissolving carbonates into soluble bicarbonate:

$$CaCO_3(\text{insoluble calcium carbonate}) + H^+(\text{acid injection})$$
$$\leftrightarrow Ca^{+2}\downarrow(\text{soluble calcium}) + HCO_3^-\downarrow(\text{soluble bicarbonate})$$

Elemental sulfur, in its pure form, is insoluble and therefore cannot be applied alone with the irrigation water. However, a slurry of finely graded S with about 2% clay results in a suspension containing 40% to 60% S, and can generally be successfully used with the irrigation water. Pelletized forms (chips) of S also are available that must be heated prior to irrigation injection (Figure 3.10).

Sulfuric acid ($H_2SO_4$) also is sometimes added to the irrigation water. Its reaction time is much faster than that of elemental sulfur, it does not require a bacterial oxidation process like elemental sulfur, and it reacts with Ca in the soil to form gypsum in place. Sulfuric acid is combined with the water before it enters the pump station. The sulfuric acid is first diluted and then mixed with a large volume of water in a wet well, avoiding pressure injection that may be unsafe. Courses without a wet well may be restricted to an acidic fertilizer injection system using fertigation techniques. All equipment used to handle and apply sulfuric acid should be constructed of stainless steel or other noncorrosive materials. Workers also must be extremely careful when handling or applying this material, and should always wear protective clothing.

Another product often used as an acidifying solution is monocarbamide dihydrogensulfate (or *N*-phuric or pHairway®), which is formed from the reaction of urea and sulfuric acid. The acidic fertilizer or monocarbamide dihydrogensulfate should be injected into a plastic pipe rather than an iron irrigation line to minimize corrosion. The operator also should keep good records as dual application of an acidic fertilizer with a urea nitrogen source in this fertigation setup may lower the soil pH more than desired.

Phosphoric acid also is used for acidification of irrigation water. Although useful, at pH above 6.0, phosphoric acid may react with Ca and Mg ions in the water to form insoluble Ca or magnesium phosphate, which may clog or significantly interfere with irrigation head operation. Refer to Chapter 8 on water quantity and quality for more information on acid injection.

**Benefits of sulfur.** In addition to lowering soil pH, several additional benefits are associated with S applications. Sulfur is required as one of the 17 essential elements for plant growth, and is necessary for root growth, chlorophyll production, protein synthesis, and tissue development. Turfgrasses require almost as much S, for example, as they do P. For nutritional purposes, turfgrasses require approximately 1 pound S/1,000 square feet/year (4.9 g/m$^2$).

Sulfur also sometimes reduces the incidence of several plant diseases, especially "patch" diseases, and annual bluegrass (*Poa annua*) levels. Research indicates one-half pound elemental S per 1,000 square feet (2.4 g/m$^2$) on golf greens, and 2 pounds/1,000 square feet (9.8 g/m$^2$) on higher cut turfgrass, reduce several patch diseases. Rates of 3.5 pounds/1,000 square feet/year (17 g/m$^2$) have gradually reduced annual bluegrass stands. Superintendents should regularly test the soil to ensure soil pH is not drastically reduced after S application in a short period of time.

## SOIL OXYGEN

Adequate soil oxygen ($O_2$) is an essential part of encouraging desirable turf rooting, which in turn helps maintain healthy plants. Table 3.15 lists the major composition of clean, dry air near sea level. However, soil air differs from atmospheric air as carbon dioxide ($CO_2$) levels are higher, ranging from 0.03% to 21%; $O_2$ content is slightly lower, ranging from 0% to 21%; and water vapor content is much higher, often 80% to almost 100%. Soil $O_2$ is utilized by plant roots and microbial respiration, producing $CO_2$.

The composition of soil air is dependent on the amount of respiration and air exchange between the soil and atmosphere. When temperatures are higher (>70°F, 21°C), $O_2$ use by plant roots and soil microorganisms can deplete $O_2$ from soil water in 24 hours. As $O_2$ is consumed by this respiratory process, the concentration of $CO_2$ proportionally increases.

Ideally, soils should consist of roughly 50% pore space with half of this filled with gases (mostly, $O_2$). When this percentage drops to below about 20% due to either higher bulk densities or increased soil moisture, the soil becomes anaerobic (without $O_2$), and a gradual decline in turfgrass vigor and quality will follow. Plant root growth is restricted through soil compaction by reducing root water utilization, restricting nutrient uptake, lack of $O_2$, accumulation of $CO_2$ and other toxic gases (e.g., as methane ($CH_4$) and sulfides), and physically restricting root penetration. Accompanying this decline in plant growth (especially root growth) is the reduction of nutrient and water absorption and the formation of certain inorganic compounds toxic to plant growth, including organic gases such as methane and ethylene ($C_2H_4$); nitrogenous gases such as dinitrogen ($N_2$), nitrous oxide ($N_2O$), nitric oxide (NO), and ammonia ($NH_3$); and sulfur gases such as hydrogen sulfide ($H_2S$). As mentioned, through the processes of plant respiration and the decomposition by microorganisms of organic residues, $O_2$ is utilized and $CO_2$ is produced.

**TABLE 3.15**
**Composition of Clean, Dry Air near Sea Level**

| Component | Volume Content | |
|---|---|---|
| | Percent | Parts per Million |
| Nitrogen ($N_2$) | 78.09 | 780,900 |
| Oxygen ($O_2$) | 20.94 | 209,400 |
| Argon (Ar) | 0.93 | 9,300 |
| Carbon dioxide ($CO_2$) | 0.0332 | 332 |
| Methane ($CH_4$) | 0.00015 | 1.5 |
| Hydrogen ($H_2$) | 0.00005 | 0.5 |
| Nitrous oxide ($N_2O$) | 0.000033 | 0.33 |
| Carbon monoxide (CO) | 0.00001 | 0.1 |
| Ozone ($O_3$) | 0.000002 | 0.02 |
| Ammonia ($NH_3$) | 0.000001 | 0.01 |
| Nitrogen dioxide ($NO_2$) | 0.0000001 | 0.001 |
| Sulfur dioxide ($SO_2$) | 0.00000002 | 0.0002 |

The rate in which soil $O_2$ exchanges with atmospheric $O_2$ is the oxygen diffusion rate (or ODR). The ODR through a soil typically represents $O_2$ utilization by plant roots instead of the actual soil $O_2$ concentration level. Oxygen must diffuse through large soil pores and through the water layer surrounding roots before it reaches the root. Therefore, moist soils have a lower ODR than drier soils. When soils are saturated, ODR is limited to the solubility at a given temperature. Increasing soil organic matter provides an energy source for soil microorganisms, resulting in increased soil $CO_2$ as $O_2$ is consumed. Increased surface winds also increase soil $O_2$ exchange, as a pressure gradient is produced at the soil surface.

No single value exists as the critical value of soil $O_2$ for all plants in all situations. The critical soil ODR for root growth of many plants is approximately $20 \times 10^{-8}$ g/cm/minute (or 0.20 μg/cm/minute) but may be less for some grasses, such as creeping bentgrass ($5 \times 10^{-8}$ g/cm/minute), Kentucky bluegrass (5 to $9 \times 10^{-8}$ g/cm/minute), or common bermudagrass (15 to $20 \times 10^{-8}$ g/cm/minute). Values less than this are often seen following irrigation or rainfall, especially on compacted soils (Figure 3.11). Soils composed of large pores allow

**FIGURE 3.11** Turf damage from lack of sufficient soil oxygen levels.

**FIGURE 3.12** Increased turf rooting where deep tine aerification provides a desirable soil oxygen channel.

more rapid ODR than soils composed of smaller pores. For example, clay soils with small pores often have a slow ODR, essentially reducing the depth to which roots will penetrate. Kavanagh and Jelley (1981) reported $CO_2$ levels at 130 mm depth in golf greens as ranging from 0.5% to 2.2% for sandy soil greens and 3.2% to 4.7% in clay loam soil greens. They also noted that traffic and irrigation or rainfall events significantly influenced these values. Oxygen diffuses 10,000 times quicker through air than water; thus, a minimum continuity of air-filled porosity is required for roots and aerobic microorganisms to respire. Compaction and excessive soil moisture often limit diffusion, leaving the highest ODR near the soil surface and decreasing values with increasing soil depth.

If no mechanism exists to replenish $O_2$ and permit the escape of $CO_2$, plant growth will be reduced. Plants attempt to adapt to low soil $O_2$ levels by increasing surface rooting compared to deeper rooting. Soil aeration is the mechanism of gas exchange in soils that prevents $O_2$ deficiency and $CO_2$ toxicity (Figure 3.12).

### Reduced Soil Oxygen Levels

**ATP production.** Oxygen utilization by plant roots and soil microorganisms drastically increases with increasing temperature, and this can totally deplete soil $O_2$ within 24 hours. When soil $O_2$ is depleted, the Krebs (or TCA) cycle stops and

ATP can be produced only by fermenting pyruvate (from glycolysis) to lactate. Fermentation is very inefficient in ATP production, with only 2 mol produced per mole of hexose sugar respired, compared with 36 mol of ATP produced per mole of hexose in aerobic respiration. Thus, initial root injury by $O_2$-deficient soils is from a lack of ATP for other metabolic processes (Figure 3.13). Without adequate ATP, cellular pH levels are not maintained, causing membrane deterioration.

**Oxidation-reduction reactions.** Another important chemical characteristic of soils related to soil aeration is the oxidation and reduction (also called redox) states of the chemical elements in these soils. If a soil is well aerated (or oxidized), oxidized states such as ferric iron, manganic manganese, nitrate, and sulfate dominate. Soils with red, yellow, and reddish-brown colors usually indicate well-oxidized conditions. In poorly drained and poorly aerated soils, the reduced forms of such elements are found (e.g., ferrous iron [pyrite ($FeS_2$) and ferrous sulfide (FeS)], manganous, ammonium, and sulfides). Reduction occurs when $O_2$ is removed by organisms that can extract the $O_2$ component of oxidized elements, hydrogen is gained, or an electron is gained in a solution. The presence of these reduced forms is an indication of restricted drainage and poor aeration. Iron in its reduced ferrous form causes a gray-green color formation (gleying) of mineral soils (Figure 3.14). If present, Fe can quickly react with and neutralize sulfide ions to form ferrous (or iron) sulfide. Ferrous sulfide gives the characteristic black color (e.g., "black layer") of many anaerobic soils (Figure 3.15).

When wetland-type soils are drained or aerified, the ferrous sulfide and elemental sulfur in them are oxidized,

**FIGURE 3.13** Summer bentgrass rooting (left) following three annual aerifications compared to one (middle) or none (right).

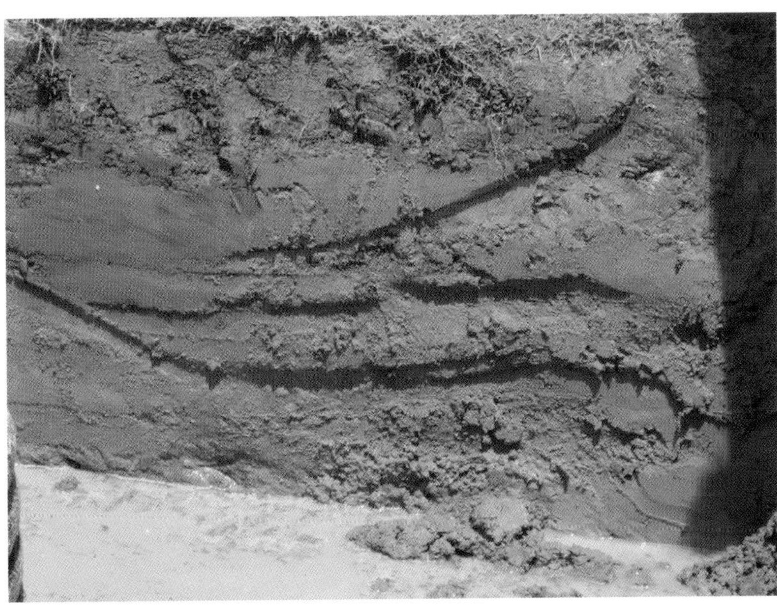

**FIGURE 3.14** Gray-colored "gumbo" clay from excessive saturation.

**FIGURE 3.15** "Black layer" formation in a golf green indicating anaerobic (insufficient oxygen) soils.

The lower the redox potential value is, the less soil $O_2$ is available. Oxidation-reduction (or redox potential) provides a measure of the tendency of a system to reduce or oxidize chemicals and is usually measured in volts or millivolts. If it is positive and high, strong oxidizing conditions exist. If it is low or even negative, elements are found in reduced forms indicating that $O_2$ is less available to plants.

In a well-drained or aerated soil, the redox potential is in the 400 to 700 mV range. As aeration is reduced, the redox potential declines to a level of about 300 to 350 mV. At lower redox potentials, $O_2$ dissolved in soil water is used by soil microbes. Combined $O_2$ in nitrates, sulfates, and ferric oxides is utilized for microbial metabolism, and these nutrients are changed into reduced forms as listed in Table 3.16. Under drastic waterlogged conditions, the redox potential value may be lowered to an extreme of –400 mV. Toxic compounds such as hydrogen sulfide and elemental sulfur then develop since anaerobic bacteria use the $O_2$ molecule in sulfate, leaving a reduced form of sulfide. The rotten egg smell of anaerobic soils results when sulfides are formed; these sulfides are highly toxic to plants. Other toxic substances may also form when compounds are reduced under waterlogged soils. For example, sulfide and methane may be produced from soil sugars found in decaying plants.

Soil sulfide and methane production from decaying plants:

$$SO_4^{-2}\text{(sulfate)} + 8e^-\text{(electrons)} + 9H^+ \rightarrow HS^-\text{(sulfides)} + 4H_2O\text{(water)}$$

$$CO_2\text{ (carbon dioxide)} + 8e^-\text{(electrons)} + 8H^+ \rightarrow CH_4\text{ (methane)} + 2H_2O\text{(water)}$$

Acid soils can intensify the adverse effects of poor aeration on several of these reduced forms of compounds such as iron and manganese.

ultimately forming sulfuric acid, causing soil pH to dramatically drop. Enormous amounts of lime are then required to neutralize this acidity.

Table 3.16 lists the oxidized and reduced forms of several important elements. The approximate range of redox potential values where soils are either oxidized or reduced are given and, if reduced, the relative ranking of this reduction.

**TABLE 3.16**
**Oxidized and Reduced Forms of Inorganic Redox Components under Low Soil Oxygen Levels**

| Element | Normal Form in Oxidized Soils | Form in Reduced (Anaerobic) Soils | Approx. Redox Values for Change of Forms (mV) | Bacteria Type and Conditions Present |
|---|---|---|---|---|
| Oxygen | $O_2$ | $H_2O$ (water) | 350 | Aerobic bacteria and conditions |
| Nitrogen | $NO_3^-$ (nitrate) | $SO_4^{-2}$ (ammonium)<br>$N_2O$ (nitrous oxide)<br>$N_2$ (nitrogen gas) | 250 | Facultative anaerobic bacteria and conditions |
| Manganese | $Mn^{+4}$ or $MnO_2$ (manganic oxides) | $Mn^{+2}$ (manganous oxides) | 225 | – |
| Iron[a] | $Fe^{+3}$ or $Fe_2O_3$ (ferric oxides) | $Fe^{+2}$ (ferrous oxides) | 120 | – |
| Sulfur | $NH_4^+$ (sulfate) | $S^{-2}$ (sulfide)<br>$H_2S$ (hydrogen sulfide) | –75 to –150 | Obligate anaerobic bacteria |
| Carbon | $CO_2$ (carbon dioxide) | $CH_4$ (methane) | –250 to –350 | – |

*Note:* The reduction reaction for iron is $Fe^{+3e-} \rightarrow Fe^{+2}$ (or, more specifically, $FeOOH + e^- + 3H^+ \rightarrow Fe^{+2} + 2H_2O$).

## SOIL TEMPERATURES

Soil temperature, along with water, influence the biological and chemical activity of a soil and the adaptability of plants that will grow in it. Temperature governs evaporation, aeration, and the types and rates of chemical reactions in the soil. As soils cool, their chemical and biological rates are proportionally slowed, often limiting the availability of necessary nutrients and slowing plant processes such as seed germination and root growth.

The sun's radiation is the primary heat source for most soils. Much of the heat energy from solar radiation can be transferred from one environmental feature to another by various processes, including evaporation, radiation, conduction, convection, and advection (Figure 3.16). Soil temperature at any given point depends on the ratio of energy being absorbed to that being lost. Heat lost by the earth (known as far-infrared radiation) is replaced by energy from the sun (in the form of incoming shortwave solar radiation). The equation demonstrating net energy ($R_N$) of incoming shortwave solar radiation ($R_S$) minus heat lost by the earth as reflected shortwave radiation ($R_R$) and longwave infrared radiation ($R_L$) is

$$R_N\,(\text{net energy}) = R_S\,(\text{shortwave radiation from sunlight})$$
$$-\,R_R\,(\text{reflected shortwave radiation})$$
$$-\,R_L\,(\text{reflected longwave infrared radiation})$$

The angle at which the sun's rays strike the soil surface also influences soil temperatures. Soil energy absorption (and resulting temperature) is greatest when the sun is directly overhead. As the sun's angle increases from overhead, a corresponding decrease in soil energy absorption occurs. For example, a 20° southern slope, a level soil, and a 20° northern slope in the northern hemisphere on the summer solstice (approximately June 21) will be in the proportion of 106:100:81.

Heat retention by a soil is primarily determined by its composition, color, moisture content, and cover. If soil moisture content is high, more heat is needed for temperature increases because the heat retention capacity of water is three to five times greater than for soil minerals. For example, only 0.2 of a calorie of heat energy is required to increase the temperature of 1 g of dry soil by 1°C compared to 1 calorie of heat energy required for each degree increase per gram of water. Conversely, the higher the soil moisture content is, the more heat is held by the soil. For superintendents, this is an important consideration. During summer stress periods on bentgrass greens, for example, a goal is to reduce the heat load in soil. Superintendents, therefore, should provide only enough soil moisture to prevent wilt, allowing evaporative cooling to reduce this heat load. Excessive moisture also promotes other problems such as algae and diseases. Tools such as surface and subsurface air units provide better control of potentially harmful excessive soil moisture and gases.

Darker color soils (e.g., high organic matter soils) absorb more heat than lighter color soils (e.g., sands). However, this increase in heat retention by certain soils may be offset by their higher moisture-holding capacities. Compacted soils also tend to hold more heat than loosened soil, but due to the compaction, these soils have more material to heat per volume and tend to be buffered in heat gain and loss compared to coarser materials.

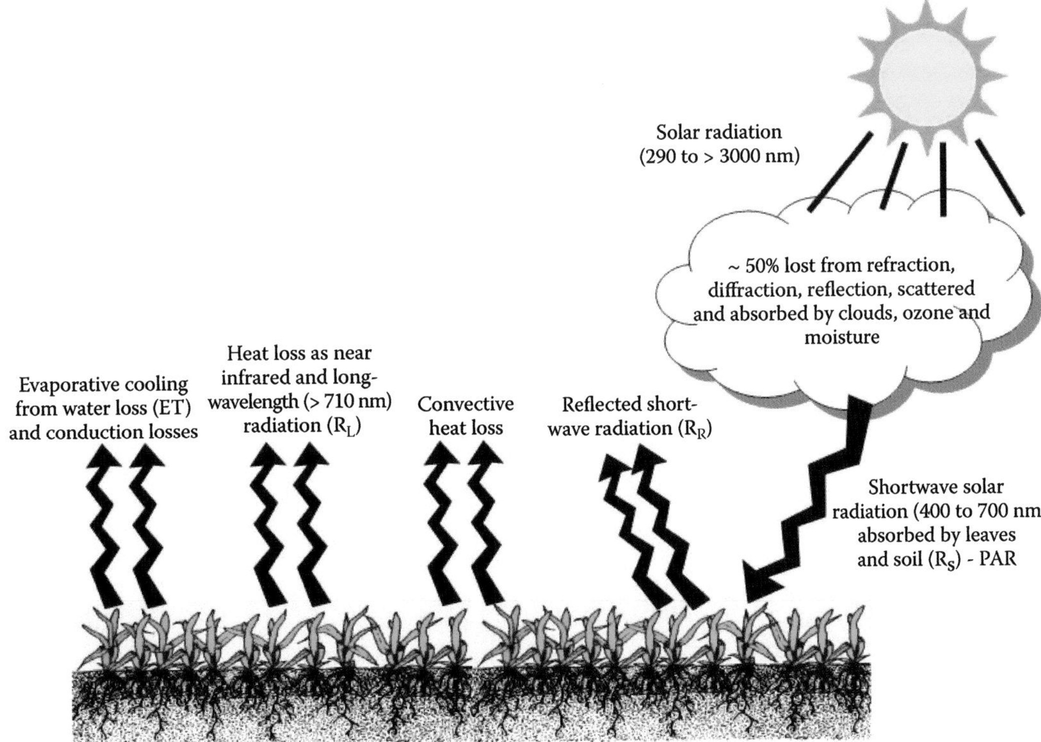


**FIGURE 3.16** Illustration of solar radiation inputs and losses on a turf stand.

Atmospheric temperature, like the soil, is greatly influenced by its moisture content. Evaporation of water absorbs relatively large amounts of heat (580 calories are absorbed to evaporate each gram of liquid water), leaving a cooler surface. Condensation of moisture, likewise, releases large amounts of heat back to the system (580 calories per gram of water). Conversely, when water is frozen, heat is given off as 60 calories per gram of water. When ice melts, 60 calories of heat are absorbed per gram of water.

Typically, due to heat flow being slower in soil than in the atmosphere, a lag occurs between maximum air temperatures and soil temperatures. Lag is also influenced by soil depth and time of day. The deeper the soil is, the longer a temperature change takes to occur; thus, there is a lag in maximum daily temperature and soil temperature and the temperature fluctuates less over time. Soil depths below about 12 inches (0.3 m) are seldom affected by daily temperature fluctuations. On cool nights, deeper soil layers do not cool as quickly as surface soil due to the insulating effect of the overlying soil.

Microenvironments also influence soil temperatures. For example, relatively flat soils will absorb less heat when the sun is low in the southern sky, such as during winter. However, south and southwestern facing slopes during winter will absorb more energy from the sun since these are at a more acute angle with the sun.

### Example

Consider two similar golf green soils; $Soil_1$ is wet with 35 g water/100 g soil, while $Soil_2$ is drier with only 15 g water/100 g soil. The following equation is used to calculate the weighted average heat capacity of a mixture of substances:

$$\text{Heat capacity (or } C) = \frac{C_1M_1 + C_2M_2}{M_1 + M_2}$$

where $C_1$ and $M_2$ are the heat capacity and mass of substance 1 (soil, in this case), and $C_2$ and $M_2$ are the heat capacity and mass of substance 2 (water, in this case):

$C_1$ = heat capacity of soil, 0.2 calorie/g of soil
$M_1$ = mass of soil, in this case, 1.0 g soil
$C_2$ = heat capacity of water, 1.0 calorie/g water
$M_2$ = mass of water or number of calories needed to raise the temperature of the water by 1°C

The amount of water present (g) is multiplied by 1.0 calorie/g. For example, for $Soil_1$, 35 g water/100 g of soil (or 0.35 g water/g soil) occurs. This is multiplied by 1.0 calorie/g, which is the number of calories needed to raise the temperature of 0.1 g of water by 1°C to achieve 0.35:

$$\text{Heat capacity Soil}_1 = \frac{C_1M_1 + C_2M_2}{M_1 + M_2} = \frac{(0.2\,\text{cal/g} \times 1\,\text{g}) + (1\,\text{cal/g} \times 0.35\,\text{g})}{1.0\,\text{g} + 0.35\,\text{g}} = 0.407\,\text{cal/g}$$

$$\text{Heat capacity Soil}_2 = \frac{C_1M_1 + C_2M_2}{M_1 + M_2} = \frac{(0.2\,\text{cal/g} \times 1\,\text{g}) + (1\,\text{cal/g} \times 0.15\,\text{g})}{1.0\,\text{g} + 0.15\,\text{g}} = 0.304\,\text{cal/g}$$

The wetter $Soil_1$ has a heat capacity of 0.407 calorie/g, where the drier $Soil_2$ has a heat capacity of 0.304 calorie/g. The wet soil, therefore, needs an additional 0.103 calorie (0.407 – 0.304) of solar radiation for every degree of temperature rise of the dry soil and will warm up and cool down more slowly than the dry soil.

## SOIL ORGANISMS

In addition to the mineral and organic composition of soils, a wide array of organisms also inhabit soils. They range in size from microscopic (bacteria, nematodes, algae, actinomycetes, and fungi) to groups visible to the naked eye (earthworms, insect larvae, mole crickets). Soil organisms make up less than 1% of the soil by volume, but are necessary to sustain plant life. Without a healthy and balanced complement of these organisms, many of the essential processes involved in plant growth could not be accomplished. Soil organisms also contribute, through their death, to soil nutrient levels. For example, 100 pounds (45 kg) of dead microbes provide 10 pounds (4.5 kg) N, 2 pounds (0.9 kg) P, and 1.5 pounds (0.7 kg) K. Soil organisms also fix atmospheric N, make many elements available, and recycle nutrients. For most soils, if conditions favor turf growth, it is highly likely there is an active soil organism population. Being mostly invisible, the science of soil microbiology is often perplexing and an area of numerous unproven products and practices.

Biologists divide the living world into two groups of organisms based on the types of living cells that compose each organism. Eukaryotes are composed of cells with a membrane-bound nucleus; these cells divide through mitosis and include higher animals and plants. Prokaryotes have cells with no nucleus. Instead, they have a single, circular DNA molecule (chromosome) not bound by a membrane. Their cells also divide by binary division instead of mitosis. Bacteria and mycoplasmas are important prokaryotes.

Organisms may also be further classified according to their carbon, $O_2$, and energy requirements. Autotrophs are organisms that obtain atmospheric carbon from $CO_2$ and methane ($CH_4$), inorganic nutrients from the soil solution, and their energy from the sun through photosynthesis to synthesize cellular components for growth, survival, and reproduction. Higher plants and algae are autotrophic. Some bacteria are also autotrophic since they use $CO_2$ and inorganic nutrients similar to the higher plants. However, these bacteria obtain energy from the oxidation of inorganic elements such as N or S. These autotrophic, nitrifying bacteria (*Nitrosomonas* spp. and *Nitrobacter* spp.) are essential to produce nitrate $\left(NO_3^-\right)$ for the growth of turfgrasses. Other autotrophic bacteria

(*Thiobacillus* spp.) convert elemental sulfur (S) to sulfuric acid ($H_2SO_4$), which lowers the soil pH.

Heterotrophs are consuming organisms, meaning they obtain carbon and energy by ingesting carbohydrates and other organic substances from organic materials in the soil such as humus, thatch, roots, and even other organisms. They may be subdivided as primary consumers that feed on dead organic matter and secondary consumers that are predators and parasites. Aerobes are organisms that require free $O_2$ in the air to perform respiration. Anaerobes can use the combined $O_2$ from various compounds such as nitrate or sulfate. Examples of heterotrophs include certain bacteria, fungi, protozoa, and animals.

### Bacteria

Bacteria are the most abundant of the soil organisms. For example, a single gram of soil (about the size of a raisin) contains more than 10 billion bacteria representing 10,000 different species. Most soil bacteria are very small (<1/1,000,000 inch, $<2.5 \times 10^{-5}$ mm), single-cell microorganisms without a nucleus or chlorophyll. These exist as mats, clumps, and filaments on and around soil particles and roots. Their shape ranges from spheres (cocci), ovals (spirals), or rods (bacilli) to branching filaments (actinomycetes). They are responsible for many key biochemical reactions necessary to support higher plants, including nitrification, N fixation, and S oxidation, which provide these nutrients to plants. Due to the abundance of roots and organic matter, bacteria are found in the highest numbers in grasslands. Heterotrophic bacteria decompose organic materials, as their primary function in ecosystems is to release nutrients for recycling to higher plants. Bacteria populations, like most soil organisms, fluctuate with the season. Numbers are generally highest in temperate regions in early summer and in fall. Bacteria are rapid multipliers, doubling in population in as little as 20 minutes. Cyanobacteria, also known as blue-green algae, contain chlorophyll and fix limited amounts of N under damp, wet conditions.

Bacteria can be divided into three basic groups based on differences in their cell walls. The first group, containing mycoplasmas and phytoplasmas, have no cell walls and are not normally found living freely in soil. The other two groups are separated based on the composition of their cell walls, which is determined by a test called the Gram stain. Gram-positive bacteria retain an applied purple dye and therefore have a purple color; Gram-negative bacteria do not retain this dye and remain red in color. Gram-positive bacteria, in general, have stress-tolerating, rigid cell walls and often dominate in extreme soil conditions (hot, cold, dry, etc.). Gram-negative bacteria are involved in N cycling, have relatively weak cell walls, and usually are the dominant bacteria under normal growing conditions.

On golf greens, two main bacteria genera occur: *Bacillus* and *Pseudomonas* (Figure 3.17). Other genera in lower occurrence include *Clavibacter, Flavobacterium, Microbacterium, Arthrobacter*, and *Xanthomonas*. Soil aeration largely determines the type of bacteria in soil. Where gaseous $O_2$ is available (e.g., aerated, oxidized soils), aerobic bacteria exist and use $O_2$ to oxidize or decompose organic matter. Bacteria also help decompose turf thatch by breaking down simple organic compounds such as plant exudates and other compounds. In the absence of gaseous $O_2$, anaerobic *Thiobacillus* bacteria dominate, utilizing compounds such as nitrates, sulfates, and iron (ferric) oxides in metabolism. As mentioned, the reduced forms of some elements, such as Fe and Mn, may be present in sufficiently high quantities to be toxic to higher plants. When sulfate is used rather than $O_2$, hydrogen sulfide ($H_2S$) is produced that can be toxic and smells like rotten eggs, possibly leading to "black layer" formation.

Great interest exists in some bacteria as possible biological control agents. These include *Bacillus popillae* and *B. thuringensis* for control of certain insects, and *Beauvaria* and *Pasteuria* species for biological control of certain nematodes and possibly soil insects.

Research on the use of bacteria as a possible biological control agent is growing. A biopesticide identifying a specific

**FIGURE 3.17** Bentgrass damage from harmful bacteria.

weed, insect, disease, or nematode pest that it can control must be registered as a biopesticide. Such materials are usually a very specific bacterium or fungus, and will be registered with the U.S. Environmental Protection Agency as to their specific pesticidal activity. Information about biopesticides is available on the EPA's website at www.epa.gov/pesticides/biopesticides/.

Biological control agents differ from "bioproducts" or biostimulants, which are general mixes containing a wide variety of growth-enhancing compounds including fertilizer nutrients, bacteria, fungi, vitamins, or hormones. Such products cannot claim a specific pesticidal activity and instead will have vague claims, such as "improves turf health."

### Fungi

Soil fungi are multicellular organisms without chlorophyll and are aerobic and heterotrophic in their nutritional requirements. Examples of fungi include single-celled yeasts, multicellular filamentous mildews, molds, rusts, smuts, and larger mushrooms. Although fungi can sometimes be pathogenic to cultivated plants, most present in the soil are actually beneficial due to their ability to decompose complex organic residues (cellulose, starch, gums, lignin, proteins, and sugars), help control other microbes such as nematodes and bacteria, and physically build soil structure. This leads to the formation of desirable humus and aggregate stabilization. Soil fungi, called mycorrhizae, are involved in an important symbiotic association with roots of higher plants to aid in improved plant nutrition and exchange, especially phosphorus (Figure 3.18). Endophytic fungi also are beneficial in turf for their ability to discourage certain insects.

### Actinomycetes

Developmentally, soil actinomycetes are a special type of primitive bacteria with some featured characteristics of fungi. Actinomycetes are single celled without chlorophyll, have branched mycelia, and are aerobic and heterotrophic in nature. They occur second only to bacteria in abundance in soils. They are of great importance in the decomposition of soil organic matter and the liberation of nutrients from this material. As with fungi, they are also able to degrade the more complex-resistant compounds such as cellulose, lignin, chitin, and phospholipids. The presence of actinomycetes in abundance in soils is an indication of their capacity to attack complex compounds. Actinomycetes, particularly *Streptomyces* spp., produce antibiotics and volatile substances that give good soil a rich, sweet smell. They, as well as most soil organisms, prefer a pH of ≥5, with most activity occurring between 6.5 and 8. If the pH is >5.5, bacteria and actinomycetes are favored, while fungi grow well over a wide range of soil pH but have less competition from bacteria and actinomycetes when pH is <5.5. Fungi most often dominate low pH soils.

### Algae

Algae are multicellular organisms with chlorophyll and are aerobic and autotrophic in their nutritional requirements. They thrive on or near the soil surface since most are able to photosynthesize; thus, they require sunlight to produce food. The predominant types of algae are cyanobacteria or blue-green algae (some of which also fix N), green algae, and diatoms. Green algae are evident when soil pH is low (<5.0), while blue-green algae prefer a soil pH between 7 and 8.5 and seldom are found when soil pH drops below 5.0.

### Protozoa

Protozoa are the simplest form of animal life in soil and are the most abundant of soil invertebrates. They are primitive,

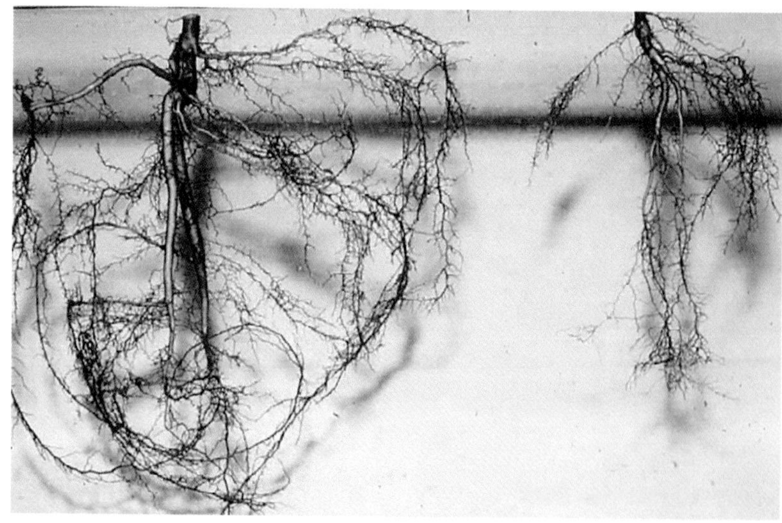

**FIGURE 3.18** Increased rooting (left) when soil fungi (mycorrhizae) have a symbiotic association with plants.

unicellular organisms and are generally found in the upper 6 inches (15 cm) of soil. Protozoa include organisms such as amoeba and paramecium. Their life cycle consists of an active stage and a resting (or cyst) stage that allows them to resist adverse environmental conditions and survive for many years.

Protozoa feed on organic matter and other microbes, and some are cannibalistic. Some amoeba and other ciliates can divide several times daily and may consume several thousand bacteria per division. Despite the fact that they are very abundant and common in soil, little is known regarding the function of protozoa. They are thought to be primarily involved in regulating bacterial and algal populations.

### Other Soil Organisms

Larger soil animals and insects inhabiting soils include nematodes, earthworms, grubs, slugs, protozoa, mole crickets, and mites. Most of them feed primarily on other soil animals, bacteria, fungi, actinomycetes, and plant debris; thus, they add to organic residue decomposition. However, certain organisms (e.g., nematodes, grubs, and mole crickets) feed directly on plant roots and can reduce the vigor of turfgrasses. Others, such as ants and termites, can substantially alter soil structure and till the soil, but also become a nuisance when in direct contact with humans or wooden structures. Large burrowing animals such as armadillos, moles, gophers, skunks, raccoons, and prairie dogs eat vegetation and soil insects and disrupt the turf surface through their digging activities. One approach of combining earthworms and beneficial bacteria is using red worms to decompose cow manure into usable compost. The manure is first sterilized to rid it of various weeds, insects, and diseases. Earthworms are then introduced, accelerating the decomposition and stabilization of the organic matter. The remaining compost is rich in nutrients and contains high organic matter. Worm Power is a commercial form of this technology.

### Abundance of Soil Organisms

A nonsterilized biologically active soil can contain as many as 45 quadrillion ($10^{15}$) microorganisms/1,000 square feet of turfgrass rootzone. These include up to $10^9$ bacteria, $10^8$ actinomycetes, $10^6$ fungi, and $10^5$ yeast and algae per gram of soil. Normal populations, however, can be deficient in sand-based rootzones like golf greens, especially those that have been recently sterilized (fumigated). Some time may be required for such systems to obtain a stable microbial population, presumably due to the time necessary for organic matter to build up from root and shoot turnover. Amendments, such as organic matter, added to sand-based rootzones shorten the time needed to stabilize the soil organism population. Until a stable population of organisms occurs, the lack of microbial activity to help minimize environmental extremes and soil pathogens may explain certain problems when growing-in new greens. Conditions favoring soil organisms include:

1. Available food source, most often an organic carbon source such as decaying turf plant parts or soil organic matter
2. Sufficient moisture, such as 50% to 100% field capacity
3. Aeration
4. Soil pH between 6 and 8
5. Soil temperatures from 50°F to 104°F (10°C to 40°C)
6. Inorganic nutrients such as N, P, K, S, and micronutrients

### Benefits of Soil Organisms

Functions/benefits of soil organisms include:

1. Soil organic matter/thatch decomposition
2. Nutrient transformation, such as N, P, S, Fe, and Mn transformation
3. Soil structure enhancement and stabilization
4. Increases in soil CEC (nutrient retention) and chelation
5. Degradation or transformation of pesticides
6. Nutrient source as organisms decompose
7. Pathogens, both beneficial and detrimental

Soil organisms are very important in the breakdown and recycling of plant and animal residues with a corresponding release of water, heat, $CO_2$, and nutrients such as N, P, and S. To thrive, soil organisms require sufficient soil temperature, a source of energy (most often organic carbon), nearly neutral soil pH, sufficient moisture and aeration, and essential inorganic nutrients and minor elements. An active soil microbial population decomposes grass clippings, dead roots, and other dead microbes. From this, nutrients are recycled back to inorganic, plant-available forms. For example, one year after adding plant residues to soil, 60% to 80% of its carbon is typically converted to carbon dioxide by soil organisms. Approximately 5% remains as either live soil organisms (or biomass); 5% as nonhumic compounds such as acids, polysaccharides, and polyuronides; and the remaining 10% to 30% as humic substances. Soil organisms also function in the formation of beneficial humus, improve soil physical properties such as improved aeration and better tilth, break down certain pesticides and a variety of environmental pollutants, and provide antagonistic reactions against plant root pathogens.

The amount of available N often controls this rate of organic matter decomposition as it is needed to build proteins in new bacterial and fungal colonies. The carbon:nitrogen ratio (or C:N ratio) indicates the N content in the microorganisms and in organic matter; a material low in N content is indicated by a wide C:N ratio. A C:N ratio of 10:1 is normally optimum for organisms involved in organic matter decomposition. Soil amendments with higher ratios than this (e.g., 74:1 for rice hulls, 80:1 for straw, 142:1 for fir bark, and 400:1 for sawdust) need supplemental N applications to support and encourage microbial activity and to provide N for higher plants; otherwise, the bacteria will consume the N, the higher plants will show N deficiency, and the high carbon containing materials will decompose slowly.

Soil organisms account for many of the biochemical reactions and changes, ranging from physical disintegration of large plant residues by insects and earthworms to the eventual complete decomposition of these residues by smaller organisms such as bacteria, fungi, and actinomycetes. The most abundant plant substrates for soil microorganisms are cellulose, hemicellulose, lignin, and humus (Table 3.17).

Accompanying these decaying processes is the release of several nutrient elements including N, P, and S from organic combination:

- **Cellulose** is a long-chain polymer of glucose molecules. Breakdown of soil cellulose is slow and is most often performed by fungi, bacteria, protozoa, and actinomycetes.
- **Hemicellulose** is a water-insoluble polysaccharide mostly made of hexose and pentose sugars not structurally related to cellulose. Hemicellulose is broken down faster than cellulose and can be performed by a wider range of microorganisms.
- **Lignin** is a complex polymer of subunits of a $C_6$–$C_3$ resinous material that impregnates cell walls as plants age. Lignin normally is found mixed with cellulose; thus, it is difficult to extract alone. As the plant lignin content rises, decomposition rates decrease. High lignin content in turfgrasses, such as zoysiagrass and bluegrass, tends to develop heavy thatch layers due to the relative inability of most soil microorganisms to break lignin down. Fungi in the basidiomycete family (e.g., mushrooms) are the main lignin decomposers since they are best suited for decomposition of hard, woody substrates (like tree bark).
- **Humus** is the final product of organic matter decomposition in soils and consists of numerous chemical substances such as water-soluble sugars and amino acids, and water-insoluble *fulvic acid, humic acid,* and *humin.* Humus is the relatively stable, dark-colored colloidal organic matter in soils. It is very important in improving soil nutrient exchange, soil tilth, and soil water-holding capacity. It is a product of microbial biosynthesis from organic residues; therefore, its susceptibility to further microbial decomposition is very low, and because of the advanced stage of decomposition, humus does not greatly stimulate increased microbial activity.

Carbohydrates, proteins, sugars, and organic acids are easily broken down by soil organisms while lignaceous materials, lipids, and soil humus are more resistant and therefore slower to degrade. Numerous heterotrophic bacteria, including those called ammonifiers, help in the breakdown and conversion of organic N from organic matter to ammonium $\left(NH_4^+\right)$. Autotrophic bacteria called nitrifiers then convert ammonium into nitrite $\left(NO_2^-\right)$ and nitrate $\left(NO_3^-\right)$ nitrogen. Nitrite nitrogen is toxic to higher plants, whereas ammonium and nitrate nitrogen are forms used by plants. *Nitrosomonas* spp. convert ammonium to nitrite while *Nitrobacter* spp. convert nitrite to nitrate (Figure 3.19). These bacteria are extremely difficult to purify and to enrich. Therefore, it is highly unlikely they will be found in various microbial products.

The biological fixing of N requires a considerable amount of energy. Sixteen ATPs are needed to make two molecules of ammonia from one dinitrogen. Therefore, sufficient carbon is necessary to supply this massive amount of energy. Sufficient $O_2$ and N levels are also needed for these bacteria to convert N.

Many soil organisms produce substances that are toxic, or antagonistic, to the growth of other soil organisms. An example involves certain soil fungi and bacteria that parasitize and kill nematodes. The antagonistic relationships help control or reduce plant root parasites. Other organisms compete very strongly for nutrients with pathogenic bacteria and fungi to inhibit their growth and thus reduce their damage to higher plants. Hyphae from mycorrhizal fungi can also compensate for the loss of roots to nematodes and diseases.

### Potential Harmful Effects of Certain Soil Organisms

In the absence of $O_2$, many microorganisms utilize oxidized forms of soil nutrients to conduct respiration rather than $O_2$. The resulting reduced forms of these nutrients [e.g., hydrogen sulfide ($H_2S$) and methane ($CH_4$)] may build up to toxic levels in plant roots if $O_2$ is not reintroduced into the root profile. This is one of the important reasons to aerify soils that tend to compact, such as in heavy traffic areas like golf greens and athletic fields.

Under very wet conditions, microorganisms convert nitrate-N to gaseous products in a process known as denitrification. While this may prevent leaching of excess nitrate-N to groundwater, it is an economical loss of N and could result

**TABLE 3.17**
**Relative Decomposition Rate of Various Organic Substances in Soils**

| Organic substance | Approx. Percentage of Total Organic Matter | Decomposition Rate |
|---|---|---|
| Carbohydrates, simple proteins, starches, sugars | 1 to 5 | Rapidly |
| Crude proteins | 5 to 20 | ↑ |
| Hemicellulose | 10 to 25 | \| |
| Cellulose | 30 to 50 | ↓ |
| Fats, waxes, lignins, lipids, humus | 10 to 30 | Very slow |

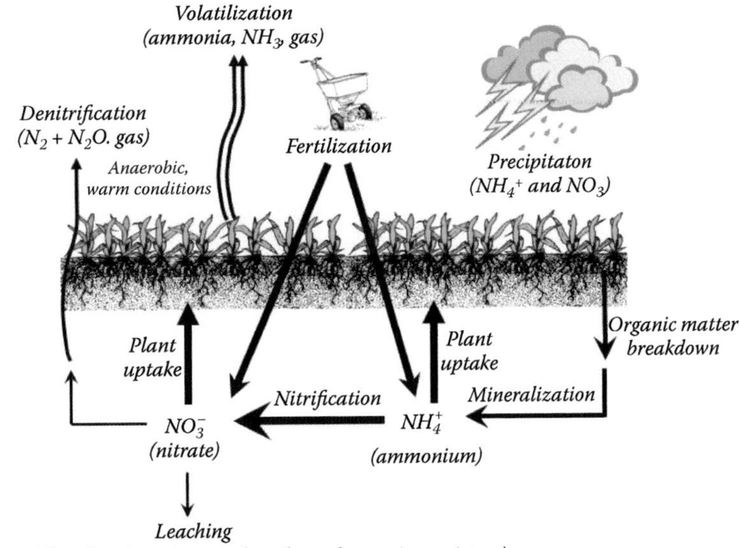


**FIGURE 3.19** Illustration of nitrogen fate when applied to turfgrass.

in N deficiency for plants. Sulfur may also be lost as gaseous $H_2S$ under wet conditions.

During the nitrification process, soil acidity is generated through the production of $H^+$ ions, and **lime** must be periodically applied to maintain the proper soil pH for the growth of turfgrasses.

**Pesticide failure.** In an effort to control harmful organisms in turfgrasses, repeated applications of certain pesticides may be used. With this continuous use, soil microorganisms begin to use certain pesticides rapidly as sources of energy and carbon to the extent that the efficacy of the pesticide is lost. This process is known as enhanced biodegradation and can result in performance failure in certain insecticides, fungicides, herbicides, and nematicides. Biodegradation is normally a slow transformation of complex organic chemicals by microorganisms into simple end products such as $CO_2$ and $H_2O$; enhanced biodegradation is a rapid breakdown or detoxification of a pesticide faster than normal by a biological agent in the soil. For example, when a golf course superintendent notices that a pesticide does not control pests as expected, enhanced biodegradation could be a possibility. Microorganisms, therefore, detoxify the pesticide before the chemical can control the pest.

Enhanced biodegradation of pesticides is responsible for the loss of efficacy and, ultimately, for the performance failures of certain pesticides in certain regions of the United States. Not using the product for two to three years and/or rotating to a product in another chemical class are potential solutions to this problem.

## INOCULATING SOIL WITH ORGANISMS AND OTHER PRODUCTS

The turf industry is famous for its spectacular claims of unproven products. These magical growth additives usually have colorful descriptive terms like "root biostimulants," "all natural," "organic," "conditioner," "biologicals," "super," "hormones," and "secret" to describe them. Although some of these prove partially beneficial, caution should be heeded before treating a whole golf course. Ask the sales representative for research data/results from your state land grant university. If a product is used, treat a small area first, such as the practice or nursery green or driving area, before treating large areas. With the exception of establishing new greens from seed on pure sand, soil scientists generally agree that adding exogenous microbes is not needed. Even if adding a microbe would be beneficial, knowing which one to add is puzzling. Also, it is challenging to change a soil's microbial population by adding a small number of bacterial species via a microbial inoculant as soils typically have an innate population of 10 billion bacteria per gram of soil. Trying to determine which organisms are actually beneficial to the turfgrass biosphere also is highly debatable and largely unknown. If unknown inoculants are added, these could actually compete with the naturally occurring beneficial microbes present; therefore, they may be more harmful than beneficial. For example, it is estimated ~$10^{15}$ bacteria occur per 1,000 square feet (93 $m^2$), 3 inches (7.6 cm) deep. If one gallon of a commercial product containing $10^6$ bacteria per milliliter is applied on this area, only ~0.0004% population alteration would occur.

### Microbial Inoculants

In an attempt to boost soil microbial populations, various inoculants are available. These are applied frequently, often nightly, many times through irrigation systems. Questions associated with this topic include (Zuberer 2012):

1. Are turfgrass soils lacking in soil microbes?
2. Do microbes need to be added?
3. Should the microbes be fed in order to feed the plant?

Questions often asked concerning the benefits of inoculating (adding) soil organisms (mostly bacteria) on golf greens are often in reference to (1) disease suppression, (2) thatch/OM decomposition, (3) improved soil structure, (4) enhanced soil fertility, and (5) soil health or sustainability.

Keeping the inoculum alive and stable is a major manufacturing, distribution, and storage challenge, as these organisms are living, need constant nutrition and moisture, and have natural fluctuations in consistency and populations. Dehydration is one method to overcome some of these problems, but this technology has not been perfected for all organisms.

If applied during daylight, many of the microbes are sensitive to ultraviolet light and to drying conditions. Furthermore, applied inoculants become exposed to existing, adapted organisms; thus, they must be competitive, and therefore are often applied daily to try to produce a stable population. Without direct contact with soil particles and, especially, organic matter, most bacteria quickly die. Continual applications of these organisms probably offer a much better chance of survival and beneficial results compared to a one-time application. If perceived benefits are seen, ascertain whether or not these are a result of the additive or from a change to favorable weather, nutrients added to the mix, or other management practices. Be sure to save untreated areas for a comparison, preferably three or four sites, with and without the additive. These should be monitored for a substantial period of time to achieve consistent results.

Soil microbiology is a relatively new and unexplored science in turfgrass management. Much is to be learned on what organisms are present, their particular roles, their interactions with other organisms, and various turf responses to added inoculants. Much of the data published so far have been positive, but some negative data have seldom been published. Furthermore, much of the research has been on field crops where yield rather than aesthetics is the measure of success. With so much uncertainty, use these products with care and objectively to determine if or to what extent they may be beneficial to your particular operation.

### Compost Tea

Compost tea is a generic term for the official name, Actively Aerated Compost Tea (AACT), where oxygen is added into water to promote aerobic microorganisms such as various bacteria, fungi, actinomycetes, mycorrhizae, and protozoa. The liquid is applied to foliage or soil to boost plant health and growth. Common brewing ingredients include a food source such as composted fruit, fish, oatmeal, alfalfa meal, soybean meal, manure, and animal feed and an energy source such as sugar, corn syrup, maple syrup, and molasses. Additional items are often used, such as humic acid, seaweed kelp, yucca, forest soil, and vermicompost (worm castings). Air is inserted into water and temperatures maintained above 60°F (15.6°C), to facilitate microorganisms. Many additional components and brewing techniques are available.

Compost teas are claimed to provide many benefits, such as healthier plants, some nutrition, humic acid, improved CEC, and certain pathogen suppression. They typically benefit soils by improving soil CEC values and water-holding capacity, and providing some nutrition. These benefits typically are more evident as soils are sandier in their physical composition. Regular applications are necessary to sustain such benefits. Compost bins and brewers are becoming more popular on golf courses to promote healthier soil and plants as well as to allow custom tea brewing and provide more environmentally friendly conditions.

### Biostimulants

Biostimulants are used to encompass non-nutritional growth-promoting substances such as microbes, plant-growth hormones, soil conditioners, and microbe energy sources. Although not marketed as fertilizers or pesticides, these products are often touted to improve turfgrass health, vigor, and overall quality, especially turf under environmental or cultural stress. Rarely does a biostimulant elicit a noticeable growth response in healthy turf. In some instances, biostimulants will provide an erratic growth response but usually only if the plant is exposed to a stressful environment such as cold weather, heat stress, nematodes, dollar spot disease, or moisture extremes, or with young seedlings growing in sand-based (e.g., low N and carbon-containing) rootzones.

Biostimulants are believed, at least in part, to promote the plant production of antioxidants, which react with toxic $O_2$ radicals that form when plants are under stress. Toxic $O_2$ radicals are often referred to as free radicals and include the superoxide $\left(O_2^-\right)$, singlet oxygen ($^1O_2$), hydrogen peroxide ($H_2O_2$), and hydroxyl ($OH^-$) radical. If they remain unchanged, $O_2$ radicals cause pigment breakdown (or bleaching), causing damage to plant cell walls, mitochondria, and chloroplasts and leading to inefficient photosynthesis, eventual cell death, and, finally, plant death.

The benefits of most biostimulants are difficult to measure as a wide array of ingredients, from "seaweed" to "cultural living microorganisms" to various "natural" or "organic" compounds, are shown on the label (Table 3.18). In an attempt to produce a more consistent growth response, manufacturers often fortify biostimulants with various nutrients such as N, Fe, and S. This then becomes very confusing to the turf manager trying to determine whether the biostimulant or nutrient component of the mixture produces the growth response.

Most biostimulants contain at least one growth hormone—often cytokinin (for rooting) or gibberellic acid (for shoot growth). However, if an antigibberellic acid plant growth retardant is being regularly used (such as trinexapac-ethyl, flurprimidol, prohexadione-Ca, or paclobutrazol), adding a biostimulant containing gibberellic acid may negate or diminish the desired effects of the PGRs.

Hormones are well known to promote certain plant responses. However, an overabundance or imbalance of hormones also has the potential to inhibit the same growth responses. Typically, certain environmental and cultural stresses can limit the plant's ability to produce the specific

**TABLE 3.18**
**Common Ingredients Listed on the Labels of Biostimulants**

| | | |
|---|---|---|
| • Activated nutrients | • Fungi | • Organic chelates |
| • Active humic acid | • Gibberellic acid | • Peptides |
| • Amides | • Growth stimulators | • PGRs |
| • Amino acids | • Humic substances | • Plant extracts |
| • Antioxidants | • Humic/fulvic acids | • Plant hormones |
| • Bacteria | • Hydrated organic proteins | • Plant nutrients |
| • B-12 vitamins | • Intermediate metabolites | • Polysaccharides |
| • Carbohydrates | • Invert sugars | • Proteins |
| • Carbon-rich organics | • Kelp extract | • Scientifically balanced formulation |
| • Cellulose fiber | • Lignin | • Sea kelp |
| • Chelated micronutrients | • Manure extract | • Seaweed |
| • Chelates | • Metabolites | • Secondary nutrients |
| • Chemical activators | • Micronutrients | • Simple sugars |
| • Complex sugars | • Minerals | • Soil conditioners |
| • Cultured living microorganisms | • Monosaccharides | • Sugar acid chelates |
| • Cyanobacteria | • Mycorrhizae | • Vitamins |
| • Cytokinin | • Natural wetting agents | • Wetting agents |
| • Disaccharides | • N-fixing bacteria | • Yeast |
| • Enzymes | • Non-ionic wetting agents | • Yucca extract wetting agent |
| • Fermentation materials | • Nutrient broth | |

hormones it needs for normal growth and development. However, little is known about when and what stressful conditions require specific hormones, and their balance for maximum benefit. For example, scientists postulate that when plants are under stress, they produce greater levels of the hormone ethylene, which signals the initiation of leaf senescence and conservation of energy reserves. Plants stop growing and the levels of growth hormones, cytokinins and auxin, drop. Biostimulants may change this hormonal balance to favor cytokinins and auxin over ethylene enough so that antioxidant production can continue and the plant can "protect" itself during periods of stress when it normally begins to shut down. Most times, this is a hit-or-miss proposition. Turf managers, therefore, should test these products on a small scale before using them indiscriminately. Always leave an untreated area in these tests so that side-by-side comparisons can be made, and also apply an area with just the fertilizer component in the biostimulant to indicate if it is just a turfgrass fertility response or a true biostimulant reaction.

**Sucrose addition.** Another contemporary practice in the turf industry is artificially adding a sucrose (carbon) source to golf greens with the idea that added carbon provides energy for soil microbes. This increased activity, in turn, aids in thatch and organic matter decomposition or indirectly feeds plants. Sucrose sources include molasses, honey, syrup, and sugar. Research, however, indicates little influence of these sources on the number of bacteria or fungi and only extremely high rates (60 to 600 pounds/acre, 67 to 670 kg/ha) elicit increased respiratory response. At these high rates, soil organisms become so active they can quickly deplete the soil $O_2$ levels.

# 4 Soil Physical and Drainage Properties

## SOIL PHYSICAL PROPERTIES

Soil is a mixture of mineral matter and organic sources. Ideally, mineral soils consist of minerals (~50% by volume), water (~25%), and air (~25%). Some soils also contain a small amount (<5%) of organic matter (Figure 4.1). In turf, as greens age, it is not unusual to find these percentages similar to 60% solids, 25% water, and 15% air. Excessive organic matter (>4% by weight, 15% to 20% by volume), compaction, and lack of aeration (porosity) are contributing factors to why aging golf greens fail.

Minimal soil physical measurements needed to evaluate potential components of a rootzone are

1. Particle size analysis (texture)
2. Bulk density and porosity (total, capillary, and noncapillary)
3. Infiltration and saturated hydraulic conductivity (or Ksat)
4. Water-holding capacity
5. Soil moisture retention curves including soil depth

### SOIL PARTICLE ANALYSIS

Many of a soil's physical properties are influenced by the size distribution of its particles. For the purposes of quantifying and describing soil texture, soil mineral particles are subdivided into three fractions based on the average diameter of the particle: sand, silt, and clay. The relative proportion of these fractions in a soil determines its texture. According to the U.S. Department of Agriculture (USDA) classification system, particles greater than 2 mm in average diameter are classified as gravel. Sand particles are between 0.05 and 2.0 mm, silt between 0.002 and 0.05 mm, and clay less than 0.002 mm (Table 4.1). Subdivisions within the sand fraction include: very coarse, coarse, medium, fine, and very fine. The U.S. Golf Association (USGA) utilizes the USDA system with a slight modification in the classification of very fine sand (USGA, 0.05 to 0.15 mm versus USDA, 0.05 to 0.10 mm) and fine sand (USGA, 0.15 to 0.25 mm versus USDA, 0.10 to 0.25 mm).

Particle-size analysis provides a general description of physical and textural soil properties and is the basis for assigning the soil's textural class name (e.g., sand, sandy loam, clay, etc.). Once the percentages of sand, silt, and clay in a soil have been identified, the soil's textural class can be determined by using the USDA textural triangle (Figure 4.2). The USDA textural triangle is made up of 12 soil textural classes. To determine soil textural class from sieve and sedimentation analysis, first find the percent of sand along the base of the triangle and follow the corresponding diagonal line up and to the left. Then find the percent of clay on the left leg of the triangle and draw a horizontal line toward the right leg of the triangle. The intersection of the sand and clay percentage lines indicates the textural class of the soil. Some familiar soil textural names are loamy sand, sandy loam, loam, and silt loam.

#### Example

Determine the soil classification name of a sample with a particle-size distribution of 40% sand, 40% silt, and 20% clay.

From the soil textural triangle, the 40% sand and 20% clay lines intersect at the center of the *loam* textural classification.

General descriptions for soil classification include:

1. A "sand" soil is high (85% to 100%) in sand content and contains small amounts (0% to 15%) of silt and clay.
2. As the silt or clay content increases, the soil becomes either "loamy sand" or "sandy loam." A "loamy sand" is 70% to 90% sand, 0% to 30% silt, and 10% to 30% clay. A "sandy loam" is 45% to 85% sand, 0% to 50% silt, and 15% to 55% clay. A "loam" soil contains 40% sand, 40% silt, and 20% clay.
3. A "clay" soil is mostly (~60%) clay, with ~20% each of sand and silt.

The term "loam" implies the soil possesses properties of all three particle fractions (sand, silt, and clay), but none to a dominant degree. If one particular particle size sufficiently alters the properties of the soil, then a modifying term is used (e.g., clay loam, silt loam, or sandy loam). Loam-textured soils are generally considered to possess the best overall physical properties for plant growth.

The sand fraction consists mostly of primary minerals such as quartz, feldspars, micas, and other weather-resistant minerals. Silts are basically weathered sands and primary minerals plus additional minerals susceptible to weathering. Silt particles feel smooth like powder. Clay particles tend to be plate-like rather than spherical like many sand particles, and they consist mostly of secondary or clay minerals like kaolinite, montmorillonite, vermiculite, illite, and sesquioxides (iron and aluminum oxides). Surface areas of clay particles are many times greater than those of sand or silt; thus, they adsorb much more water. Conversely, when certain clay soils are dried, very hard soil clods often form due to the enormous area of contact between the plate-shaped clay particles.

Some sand companies provide particle-size distribution information, but in many cases the analysis is based on engineering

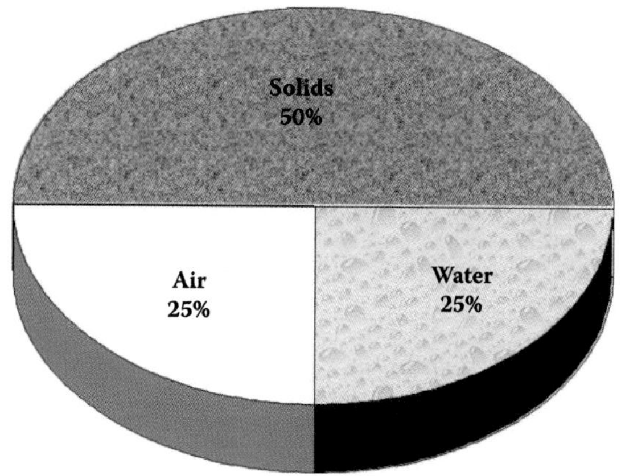


FIGURE 4.1 Soil is ideally composed of ~50% minerals, ~25% water, and ~25% air. Some soils also contain a small amount (<5%) organic matter.

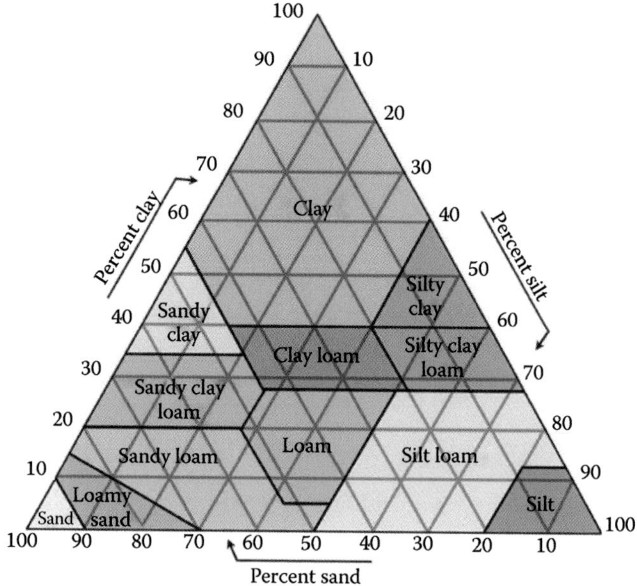


FIGURE 4.2 Textural triangle is used to classify a soil based on its percent clay, silt, and sand content.

criteria rather than the USDA sieve sizes. The construction industry often defines sand broadly by particle-size distribution as follows:

**Concrete sand** is typically composed of a broad distribution of sand-sized particles and some fine gravel-sized particles.

**Mason's sand** is similar to concrete sand but without the fine gravel.

**Dune sand** is deposited by wind or water; thus, it has a narrow particle-size range.

**River sand** is from river bottom deposits and can vary in particle-size uniformity from very uniform to highly variable, and often contains higher percentages of silt and clay.

## Determining Particle Size

Particle-size analysis is based on sieving and sedimentation methods. To determine particle size, a sample of soil is broken up and particles larger than silt (e.g., gravel and sand) are separated into their various size groups, as outlined in Table 4.1, by the use of sieves (Figure 4.3). The weight of each group is found to determine a percentage of total sample weight. Due to their light weights and extremely small sizes, silt and clay percentages are determined based on the rate of settling of these two separate soils in a suspension. This method involves the proportionality of clay and silt settling rates relative to their size (particle diameter). The larger the particle (e.g., sand or gravel) is, the quicker it will settle in a suspension (e.g., water) solution. Conversely, the smaller the particles (e.g., silt and clay) are, the more slowly settling will occur. This is referred to as Stokes's law. Sand generally takes only several minutes or less to settle, while silt takes several hours and clay requires up to several days (Table 4.1).

**TABLE 4.1**
**Particle-Size Classifications as Determined by the U.S. Department of Agriculture (USDA) and the U.S. Golf Association (USGA)**

| Textural Name | | Particle-Size Range (mm) USDA | USGA | U.S. Standard (Sieve Number) | Sieve Opening (mm) | Number of Particles per Gram | Objects with Similar Relative Sizes | Typical Settling Velocity |
|---|---|---|---|---|---|---|---|---|
| Gravel | Gravel | > 4.00 | > 3.40 | 4 | 4.76 | <2 | Medicine ball | 20 cm/second |
| | Very fine gravel | 2.00 to 4.00 | 2.00 to 3.40 | 10 | 2.00 | 11 | Basketball | 3 cm/second |
| Sand | Very coarse sand | 1.00 to 2.00 | 1.00 to 2.00 | 18 | 1.00 | 90 | Soccer ball | 1 cm/second |
| | Coarse sand | 0.50 to 1.00 | 0.50 to 1.00 | 35 | 0.50 | 720 | Tennis ball | 13 cm/minute |
| | Medium sand | 0.25 to 0.50 | 0.25 to 0.50 | 60 | 0.25 | 5,700 | Golf ball | 3 cm/minute |
| | Fine sand | 0.10 to 0.25 | 0.15 to 0.25 | 140 | 0.10 | 46,000 | Play marble | 31 cm/hour |
| | Very fine sand | 0.05 to 0.10 | 0.05 to 0.15 | 270 | 0.05 | 722,000 | Match head | 6 cm/hour |
| Silt | – | 0.002 to 0.05 | 0.002 to 0.05 | – | – | 5,776,000 | Sesame seeds | 1.3 mm/hour |
| Clay | – | < 0.002 | < 0.002 | – | – | 90,260,853,000 | Table salt | <1.3 mm/hour |

**FIGURE 4.3** Sieves physically separate gravel and sand particles into their variously sized groups.

Another method of measuring particle-size distribution in suspension is based on the rate of settling by monitoring changes in the specific gravity of the suspension. A hydrometer is a device used to measure suspension density at various times, thus reflecting the amount of particles that remain in suspension after a certain settling time. A hydrometer with a Bouyoucos scale in grams per liter (g/L) is used to determine the amount of soil in suspension (Figure 4.4). The greater the density of a suspension is, the greater is the buoyant force on the hydrometer and the higher the reading. As particles settle out of the suspension, density decreases and a lower reading is obtained. Since temperature influences the settling rate, a temperature correction must be made if the suspension temperature differs from the temperature at which the hydrometer is calibrated.

## Example

From the following data, calculate the texture of a 50 g soil sample:

| **Hydrometer Reading (temperature corrected)** | **g/L** |
|---|---|
| 40 seconds | 31 |
| ~7 hours | 16 |

1. $\% \text{ Silt and clay} = \dfrac{\text{40-second reading}}{\text{dry weight soil (g)}} \times 100$

$$= \frac{31}{50} \times 100$$

$$= 62\%$$

(After 40 seconds, the sand has settled out leaving just silt and clay in suspension.)

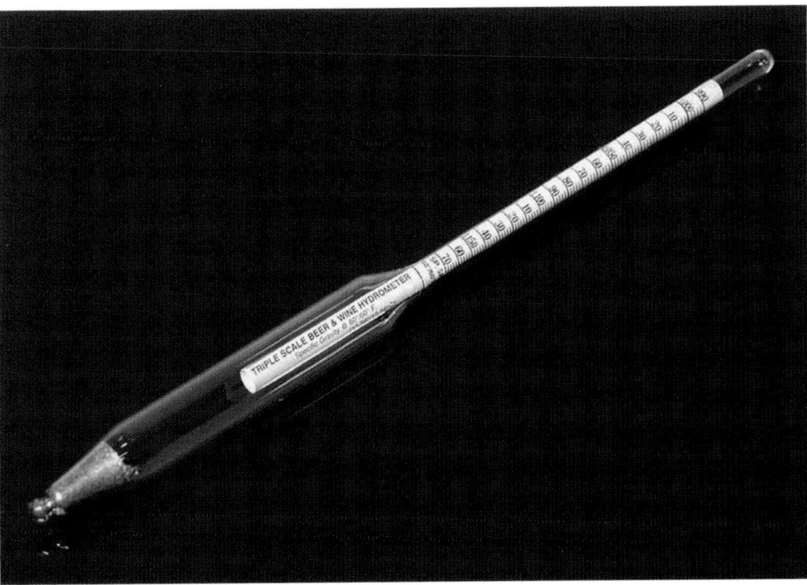


**FIGURE 4.4** A hydrometer measuring a suspension's density over time. The greater the density (more soil) is, the greater the is buoyant force on the hydrometer and the higher the reading is.

2. $\% \text{ Clay} = \dfrac{\text{7-hour reading}}{\text{dry weight soil (g)}} \times 100$

$= \dfrac{16}{50} \times 100$

$= 32\%$

(After ~7 hours, the sand and silt have settled out, leaving just clay in suspension.)

3. $\% \text{ Silt} = \%(\text{silt} + \text{clay}) - \% \text{ clay}$

$= 62\% - 32\%$

$= 30\%$

4. $\% \text{ Sand} = 100\% - \% \text{ silt} + \text{clay}$

$= 100\% - 62\%$

$= 38\%$

This soil (38% sand + 30% silt + 32% clay) is classified by the textural triangle as a *clay loam*.

## Particle Shape and Soil Structure

Particle shape may also influence whether a particular soil is acceptable for the construction of golf greens and sports fields. Nearly round sand particles can result in excessive percolation rates and less stable soil surfaces. Particles that are too angular may provide excessive bulk density that leads to unacceptable soil compaction. Soil structure is the grouping of individual soil particles into aggregates, which are formed when soil particles adhere to each other and behave as a single soil unit. Aeration, hydraulic conductivity, bulk density, and porosity are soil properties highly influenced by structure. Soil structure, like texture, may vary with depth in the soil, as well as across the landscape. However, unlike texture, soil structure may be altered. Destruction of soil structure is generally detrimental and may occur rapidly with poor management practices. By comparison, improving soil structure is a slow and difficult process.

Soils where structure is absent are often referred to as "structureless." This term denotes a condition where individual soil particles are not arranged into aggregates. Sand-textured soils with low organic matter contents or excessive sodium ions are often essentially structureless. Some subsoil horizons may also be structureless.

## Soil Particle and Bulk Density

Two important measurements of soils include particle density and bulk density. Particle density is the average density of soil particles and is defined as mass (or weight) of dry soil per unit volume of soil solids, not including pore volumes occupied by air or water. Generally, mineral soils (sands, silts, clays) have higher particle densities than organic matter. Particle density can vary considerably, but for most mineral soils, this range has a narrow limit of 2.60 to 2.75 g/cm$^3$ (or g/cc or g/mL). An average mineral soil (minerals and organic matter) has a particle density of 2.65 g/cm$^3$. For comparison, water, concrete, steel, and lead have densities of 1.0, 2.4, 7.7, and 11.3 g/cm$^3$, respectively, while organic matter has a lower particle density of ~1.1 g/cm$^3$.

Bulk density is a measure of soil compaction (or density) and is defined as the mass (weight) of dry soil per given unit volume, including both solids and pores occupied by air and water:

$$\text{Bulk density} = \frac{\text{weight of dry soil (g)}}{\text{total soil volume (cm}^3\text{)}}$$

Unlike particle density, bulk density is an indicator of pore space volume in addition to soil solids. Since most soils are about half solids and half pore space by volume, bulk densities tend to be about half of the particle density. A typical acre of soil 6 inches deep weighs ~2,000,000 pounds, assuming an average bulk density of 1.5 g/cm$^3$.

Bulk density (or mass) of a soil is related to its density (or weight) per volume:

$$\overbrace{6 \text{ inches} \times \frac{1 \text{ foot deep}}{12 \text{ inches}} \times \frac{43{,}560 \text{ square feet}}{\text{acre}}}^{\text{Volume}}$$

$$\times \overbrace{\frac{1.5 \text{ g}}{\text{cm}^3} \times \frac{1 \text{ pound}}{454 \text{ g}} \times \frac{28{,}320 \text{ cm}^3}{\text{cubic feet}}}^{\text{Density}}$$

$$= \overbrace{\frac{2{,}037{,}917 \text{ pounds}}{\text{acre}}}^{\text{Mass}}$$

Bulk densities of soils generally range from 1.0 to 1.9 g/cm$^3$ (Table 4.2). Clay, clay loam, and silt loam soils normally range from 1.0 to 1.6 g/cm$^3$, while sands and sandy loams normally range between 1.2 and 1.8 g/cm$^3$. Organic soils have extremely low bulk densities (0.2 to 1.1 g/cm$^3$) due to low particle densities and large amounts of pore space. Highly compacted soils may have bulk densities as high as 2.0 g/cm$^3$ or more, as they have less large pore space and thus higher bulk densities (Figure 4.5). This results in slower infiltration and percolation. Noncompacted sandy soils have low bulk densities due to a high percentage of macropores and can have infiltration and percolation rates as high as 4 feet/hour (1.3 m), while compacted clay loams with more micropores have significantly lower rates, often less than 1 inch/hour (2.5 cm). Modern golf green and sports field designs integrate compaction-resistant soil mixes of predominantly

**TABLE 4.2**
**Typical Bulk Density and Total Porosity Values for Various Soil Textural Classes**

| Soil Textural Class | Bulk Density (g/cm$^3$, Pounds/cubic foot) | Porosity (%) |
|---|---|---|
| Sands or compact clay | 1.4 to 1.8 (87 to 112) | 32 to 47 (low) |
| Loam | 1.2 to 1.6 (75 to 100) | 39 to 55 (medium) |
| Loose silt loams or clay | 1.0 to 1.4 (62 to 87) | 47 to 62 (high) |
| Organic soils | 0.2 to 1.1 (12 to 62) | 62 to 92 (very high) |

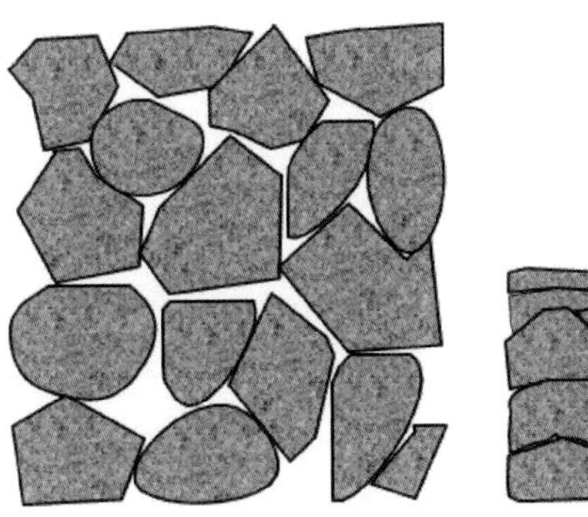


**FIGURE 4.5** Bulk density measures the compaction or compressibility of a soil based on the formula mass of soil per unit volume.

medium- and fine-sized sands with organic matter and other materials. These mixes withstand compaction while retaining enough moisture to maintain plant survival. The bulk density range for golf greens is between 1.25 and 1.55 g/cm³, with an optimum level of 1.40 g/cm³.

### Example

1. What is the bulk density of a soil sample that weighs 120 g and occupies a volume of 75 cm³?

$$\text{bulk density}(\rho_b) = \frac{\text{dry weight (g)}}{\text{volume of soil (cm}^3)} = \frac{120\text{ g}}{75\text{ cm}^3} = 1.60\text{ g/cm}^3$$

2. What is the volume (m³) of 500 kg of soil with a bulk density of 1.30 g/cm³?

$$500\text{ kg} \times \frac{1000\text{ g}}{\text{kg}} \times \frac{\text{cm}^3}{1.30\text{ g}} \times \frac{\text{m}^3}{(100{,}000\text{ cm})^3} = 0.3846\text{ m}^3$$

## Soil Porosity

Soil porosity, pore space, or void space is the percentage of total soil volume not occupied by solid particles, or the percentage of total soil volume occupied by air and water. In dry soils, pores are mostly filled with air. In moist soils, pore spaces contain both air and water. The arrangement and size of solid particles in soil largely determine pore space. If solid particles lie close together, as in many fine sands or compacted subsoils, porosity is low. If they are arranged in porous aggregates, as is often found in medium-texture soils high in organic matter, pore space per unit volume will be high. Organic matter increases soil porosity by promoting soil particle aggregation and improving soil structure. Porosity can be determined from bulk density by the equation

$$\text{Porosity}(\%) = 100 - \left(\frac{\text{bulk density}}{\text{particle density}} \times 100\right)$$

Generally, particle density can be assumed to be 2.65 g/cm³ and bulk density is determined on an undisturbed soil core. Sandy soils generally have a total pore space between 32% and 47%, while finer textured soils vary in total pore space from 47% to 62% (Table 4.2).

### Example

A soil core 5 cm in diameter and 6 cm in height weighs 220 g when collected and 190 g when oven-dried ($M_s$). Following grinding, a soil-water volume was 171 mL after being poured into 100 mL of water. Calculated soil bulk density ($\rho_b$), particle density ($\rho_S$), and percent pore space ($f_t$) (1 mL water = 1 cm³ water):

Core volume ($V_t$)
$$= \pi r^2 h = (3.14)(2.5\text{ cm})^2(6\text{ cm}) = 117.8\text{ cm}^3$$

Volume water displaced by soil ($V_s$)
$$= 171\text{ mL} - 100\text{ mL} = 71\text{ mL or } 71\text{ cm}^3$$

Bulk density ($\rho_b$)
$$= \frac{M_s}{V_t} = \frac{190\text{ g}}{117.8\text{ cm}^3} = 1.61\text{ g cm}^{-3}$$

Particle density ($\rho_s$)
$$= \frac{M_s}{V_s} = \frac{190\text{ g}}{71\text{ cm}^3} = 2.67\text{ g cm}^{-3}$$

Total porosity ($f_t$)
$$= \left(1 - \frac{\text{bulk density}, \rho_b}{\text{particle density}, \rho_s}\right) \times 100 = \left(1 - \frac{1.61\text{ g/cm}^1}{2.67\text{ g/cm}^1}\right) \times 100 = 40\%$$

Small pores, called capillaries or micropores, hold water against the pull of gravity and are responsible for a soil's water-holding capacity. Larger noncapillaries or macropores drain rapidly and are typically air filled and contribute to drainage characteristics of a soil. The relative proportion of macro- and microporosity is primarily determined by soil texture and structure. Sandy soils typically have a high proportion of macropores because sand particles, often spherical as well as large, cannot pack closely together; thus, the pore sizes tend to be large. Clay and silt particles, however, pack together tightly because these particles are very small and plate-like (flat).

The suggested porosity range for golf greens is 35% to 55% total pore space with an optimum range of 40% to 55% by volume. Capillary porosity is usually between 15% and 25% and noncapillary porosity between 15% and 35%. Ideally, capillary

**TABLE 4.3**
**Plant-Available Moisture Range for Various Textural Classes and at Field Capacity (One-Third Bars) and Permanent Wilting Point (15 Bars)**

| | In Water/Foot of Soil | | Percent Moisture Content by Volume | |
|---|---|---|---|---|
| **Textural Class** | **Total** | **Available** | **One-Third Bars** | **15 Bars** |
| Sand | 0.6 to 1.8 | 0.4 to 1.0 | 7 | 3 |
| Sand loam | 1.8 to 2.7 | 0.9 to 1.3 | 22 | 9 |
| Loam | 2.7 to 4.0 | 1.3 to 2.0 | 28 | 15 |
| Silt loam | 4.0 to 4.7 | 1.8 to 2.4 | 32 | 14 |
| Clay loam | 4.2 to 4.9 | 1.8 to 2.1 | 33 | 21 |
| Clay | 4.5 to 4.9 | 1.8 to 1.9 | 40 | 29 |
| Organic soils | – | 2.0 to 3.0 | – | – |

*Note:* Multiply values by 8.33 to obtain centimeters/meter. Multiply bars by 100 to obtain kilopascals (kPa), by 0.001 to obtain joules/kilogram (J/kg), or by 1,020 to obtain grams/cubic centimeter.

and noncapillary pore space should be equal at 25% of the total soil volume. The minimum air-filled porosity at which soils will support good turfgrass growth is between 10% and 15%. Lower porosity values indicate a clay soil and/or excessive soil compaction. These porosity ranges are for a rootzone mix that has been compacted, allowed to percolate water for 24 hours, and then exposed to a 40 cm tension (or suction).

Water-retention capacity at 40 cm tension for oven-dry soils typically ranges from 12% to 25% by weight, with 15% (1.5 $cm^3$ water held per 10 cm soil) being optimum. Table 4.3 lists plant-available moisture for various textural classes. This table also presents typical values of available moisture for various soil types. This may be estimated in the field by applying a known amount of water to the soil when the profile water content is near the permanent wilting point, observing the volume of soil wetted, and calculating the volume of water stored per unit volume of soil.

## SOIL–WATER RELATIONSHIPS

### Describing Soil Moisture

Several values are used to describe soil moisture, including saturation, field capacity, and permanent wilting point. Soil is completely saturated when all the pores are filled with water, such as during or immediately following a heavy rain. Water then occupies about 50% of the total soil volume. The addition of more water results in surface runoff and puddling (termed gravitational water), which is unusable by plants.

After drainage has removed water from the macropores and the two forces of gravity and capillary tension become equalized, the soil is at field capacity (which is equivalent to water-holding capacity), with water normally occupying 20% to 35% of the total volume (Figure 4.6). Any water in excess of field capacity will drain due to gravitational pull. Since having saturated soil for an extended period is undesirable, the primary function of drainage is removal of this excess gravitational water from soil.

In constructed greens, field capacity is determined by applying a force of 40 cm of tension to the soil to simulate the gravitational force on a rootzone 12 to 16 inches (30 to 40 cm) thick. In native soils, field capacity is usually determined at 0.33 bars [33 kilopascals (kPa)]. The higher the clay and organic matter content of the soil are, the greater the water content at field capacity is. Sand-textured soils may have as little as 7% moisture at field capacity, whereas clay soils may have as much as 40% moisture at field capacity (Table 4.3).

Soil does not stay at field capacity very long. Evaporation of water from the soil surface and soil water absorption by plant roots deplete water from the larger micropores, and soil begins to dry out. Forces of soil adsorption and capillarity pull at water molecules and hold them (Figure 4.7). As particle and pore size decrease, these combined forces strengthen. At some point, roots can no longer take up water from the smallest pores as it is held too tightly, and the plant starts to wilt. A soil is considered to be at the permanent wilting point when its moisture content yields a severely wilted plant that is unable to recover, even after irrigation. This usually occurs when soil moisture is around 10% to 15% of the total volume. Water held at a tension of 15 bars (1,500 kPa) is often considered the lower limit of available water. In sandy soils, the amount of water held is small in proportion to the total, but in clayey soils it can be a large percentage of the total soil water content (Table 4.3). These values are estimates, and actual values depend on the soil texture and structure, and the type of plant growing in the soil.

Volumetric water content is the most common measure used to express the quantity of water in a soil. Volumetric water content is defined as the volume of water per unit volume of soil. This value is usually expressed in inches of water per inch (cm/cm) of soil (Table 4.3). For example, a volumetric water content of 10% refers to 1 inch of water in 10 inches of soil. The units of depth may be used to describe volumetric measurements since cross-sectional area is the same for both water volume and total soil volume. This is a convenient unit

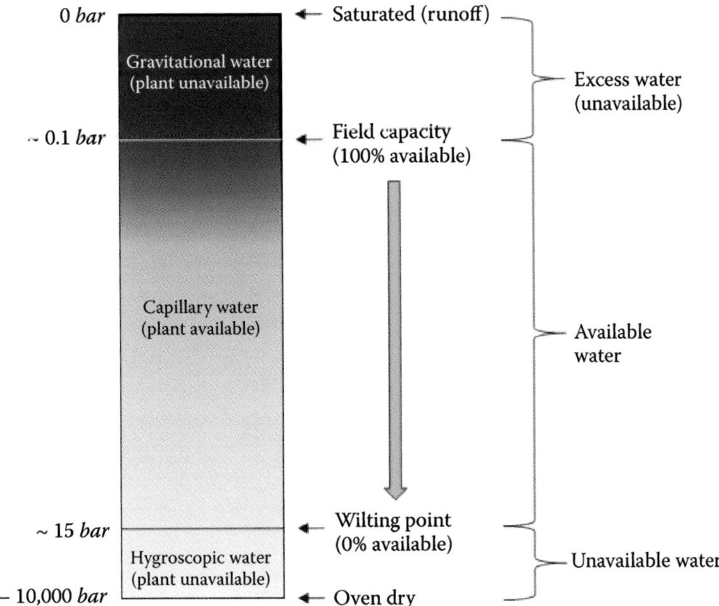


**FIGURE 4.6** Graph illustrating various moisture levels in soil. Capillary water is plant available while gravitational and hygroscopic water are not.

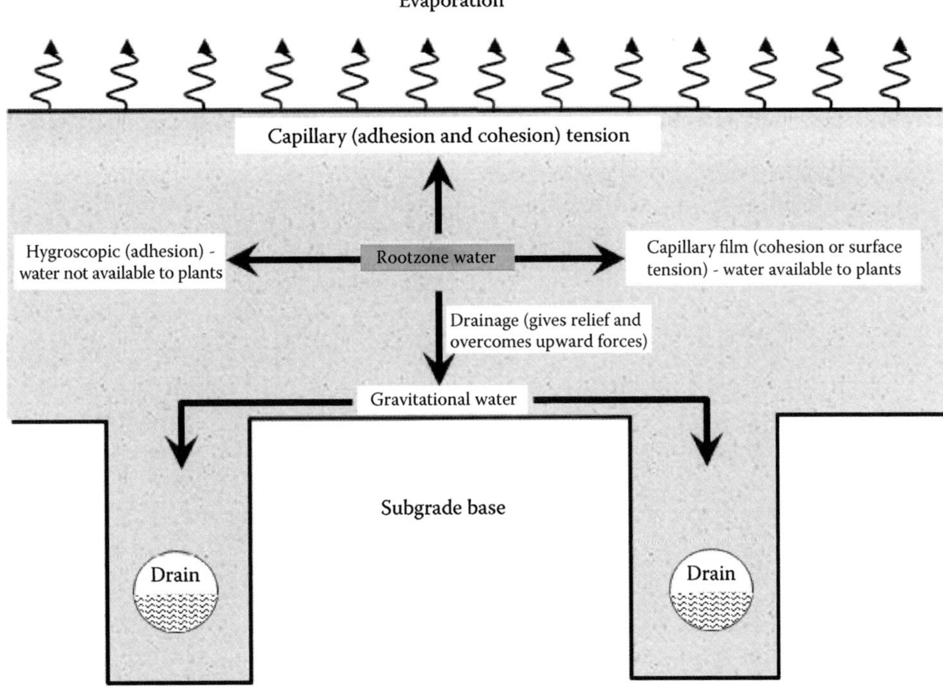


**FIGURE 4.7** Avenues of water movement in soil including unavailable hygroscopic (adhesion) and gravitation water, and available capillary (adhesion and cohesion) tension and capillary film water.

because rainfall, irrigation, and turf water use are expressed in the same units.

Soil scientists measure the amount of water left in the soil at various tensions (or pressures) to create a moisture release (or retention) curve. Different soils have different-shaped moisture release curves with different heights of perched water tables. The shape and characteristics of this curve reflect the particle-size distribution of a soil and degree of soil compaction. Moisture release curves indicate water movement and amounts at various tensions, allowing scientists to predict soil water behavior at various depths and drying points. Refer to McCarty, Hubbard, and Quisenberry (2016) for additional information on these topics.

Plant-available water. The amount of water held in the soil between field capacity and the permanent wilting point is termed plant-available water (Figure 4.6). Soils with a high percentage of silt have the greatest plant-available water content—as much as 25% or 3 inches/foot soil (25 cm/m). Sand-textured soils have less available water than silt-textured soils because the total amount of water they hold is less. Plant-available water content of clay-textured soils is also less than that of silty soils because much of the water in a clay soil is held too tightly to be used by plants.

If the available water content of the soil and the depth of turfgrass rooting are known, then the amount of water the plant has access to can be estimated by the following:

$$\text{Total water available} = \frac{\text{rooting depth} \times \text{\% volumetric water content between field capacity and writing point}}{100}$$

### Example

If a bentgrass green has an average rooting depth of 2 inches (5 cm), and the soil it is grown on has a field capacity of 21.4% and a permanent wilting point of 14.6%, how much water is available from the soil?

$$\text{Total water available}$$
$$= \frac{\text{rooting depth} \times \text{\% volumetric water content between field capacity and wilting point}}{100}$$
$$= \frac{\text{rooting depth} \times (\text{field capacity} - \text{permanent wilting point})}{100}$$
$$= \frac{2\text{ inches} \times (21.4 - 14.6)}{100}$$
$$= 0.136\text{ inch}$$

Therefore, 0.136 inch (3.45 mm) of water is available to the grass from the top 2 inches (5 cm) of soil. If water use by the turf is 0.10 inch (2.5 mm) per day, then daily irrigation is needed so that moisture stress of the plant does not occur.

## Infiltration and Percolation Rates

Infiltration rate is the speed at which water enters the soil surface. Percolation rate is the speed at which water moves downward through the soil profile. Infiltration determines whether applied water enters soil, runs off (if sloped), or puddles. The rate at which water enters soil is in part dependent on soil texture and structure and the impact of these factors on soil porosity. Soils with a high proportion of noncapillary (or large) pores have high infiltration rates. Coarse, sandy soils have high infiltration rates because the large sand particles result in an abundance of noncapillary pores. Finer textured soils (having more silt and/or clay) may also have high infiltration rates if good structure results in the presence of large pores between structural aggregates. More often, however, fine-textured soils have slow infiltration rates because most of the pores are small and accept water slowly (Figure 4.8). The infiltration rate of a soil is also highest when it is dry. As a soil becomes wetter, infiltration rates decrease.

On an established turfgrass site, an infiltrometer is often used to determine the soil's hydraulic conductivity (Figure 4.9). Water is added to the single- or double-ring infiltrometer and, after a period of time, the depth of water absorbed is measured. Hydraulic conductivity of established turf sites can also be determined by extracting intact soil cores, taking them to the laboratory, subjecting them to a hydraulic head, and using Darcy's law to determine saturated hydraulic conductivity (or Ksat).

Saturated hydraulic conductivity is defined as the proportional relationship of water flow through a saturated soil in response to a given difference in head (pressure). Large pores allow for high saturated hydraulic conductivities. Sandy soils, therefore, generally have much higher saturated hydraulic conductivities than clay-type soils. Hydraulic conductivity for rootzone mixes can be determined in the laboratory (Figure 4.10). Combinations of sand, soil, and/or organic

**FIGURE 4.8** Slower infiltration and drainage rates are often associated with fine-texture soils.

**FIGURE 4.9** Measuring a soil's ability to accept and percolate water using a double ring infiltrometer.

**FIGURE 4.10** Laboratory measurement of hydraulic conductivity of water flow through a saturated soil.

material (usually peat) can be mixed in various ratios for testing purposes. The USGA guidelines for rootzone mixes recommend a saturated hydraulic conductivity between 6 and 24 inches (15 and 61 cm) per hour. The saturated hydraulic conductivity of rootzone mixes will decrease over time due to compaction and organic matter accumulation.

Infiltration and percolation rates of golf greens are critical as they determine playability after rainfall. Although relatively high initial percolation rates can be achieved on sand-based greens, percolation rates usually decrease over time. Surface soil compaction from player traffic and maintenance machinery also decreases infiltration rates. Macropores may be blocked by silt and clay, which may be inadvertently added in soil amendments and irrigation water, or by wind-blown soil and dust. Sand-sized particles from sandstone are not stable and may break down with traffic. Excessive accumulation of soil organic matter in sand-based greens also decreases macroporosity and the percolation rate (Figure 4.11). However, because a rapid percolation rate is generally correlated with a low water-holding capacity, a balance between percolation and water-holding capacity is needed.

In fairways, roughs, and tees, infiltration rates not only determine playability after rainfall, but also dictate the rate at which irrigation can be supplied. The application rate of irrigation sprinklers must be less than the soil's infiltration rate so that irrigation water does not run off.

**FIGURE 4.11** Excessive soil organic matter decreases soil porosity and percolation.

## Water Movement in Soil

Soil water movement is largely dependent on the balance of forces that it is under: (1) water being pulled downward by gravity (or hydraulic gradient), (2) water sticking to itself due to hydrogen bonding (called surface tension or cohesion), and (3) water sticking to other surfaces, such as soil particles (referred to as adhesion, adsorption, or hygroscopic moisture) (Figure 4.7). This relationship is described by Darcy's law.

**Gravity (or hydraulic gradient).** Gravity is the constant downward-pulling force that moves water through soil. As the soil profile depth increases, so does the height (elevation) of the water column being pulled down and the weight (or pressure) of water. Gravity limits the height to which water can rise by capillary tension, and limits the thickness of water film held by tension around each soil particle.

**Surface tension (or cohesion).** Due to their hydrogen-bonding characteristic, water molecules are attracted to each other in all directions and are attracted much more to each other than to adjacent air molecules. Surface tension is created when water meets air, causing water molecules to become round and bead up. The smaller the drop of water is, the stronger the surface tension becomes, and the more difficult it becomes to break this tension. The surface of water can be bent slightly by gently touching it. However, if the force applied is too great, the surface will break. Wet soils have less cohesive tension than dry soils.

**Water adhesion (or adsorption) to soil particles.** As soil particle and pore sizes decrease, an increase in attraction (or adhesion) occurs between soil particles and soil water. In most instances, water adheres to soil particles very strongly, much more so than to other water molecules. This adhesive force "pulls" or holds water against the force of gravity. It can, in fact, pull water in all directions away from a water source in soil, independent of gravity (Figure 4.12). If a dry column of soil is placed in contact with free water, moisture will rise into the soil. However, capillary water movement is generally limited (rarely more than 4 inches, 100 mm) from its source, and is very slow. Furthermore, the smaller the soil pore spaces are, the slower that water will move laterally (sideways) and the more tightly the water is held. Soils with a larger pore size (e.g., coarse-textured sand) hold less moisture at a given potential than soils with a smaller pore size (e.g., silt or clay).

**FIGURE 4.12** Demonstration of a sand's adhesive ability to pull water up against the force of gravity. (Courtesy Todd Bunnell.)

## Capillarity

Capillary tension is the combined force of surface tension and adhesion that retains water in small soil pores against the pull of gravity. This retained water held in the soil against gravity is collectively referred to as capillary water. It includes the film of water left around soil particles and water in capillary soil pores after gravity water has drained. Once water molecules wet a particle, they seek another dry surface on which to cling. As water moves, it pulls additional water along with it. This pulling action produces a negative pressure or vacuum.

A similar reaction can be seen with a sponge and water. When dry, contact with water (capillary tension) causes a sponge to wet-up both upwards and sideways. Likewise, if the sponge is highly saturated, some water will initially drip out of it, but a certain amount will be held. This is capillarity and adsorption forces causing the sponge to hold water and is, in a way, very similar to soils. The extent to which capillary action works depends on the size of spaces or channels formed when soil particles pack together. The smaller the channels are (e.g., more compacted or finer textured soil), the greater the capillary action is.

Surface tension is weaker than adhesion and is the first force to break under the pull of gravity. Therefore, surface tension is the limiting factor in the amount of water a pore can hold. Water held by adhesive forces requires greater gravitational forces to remove it. In a saturated soil, all the pore spaces are filled with water. Additional water has nowhere to go, and gravity pulls water downward, causing the soil to drain. The larger pores eventually empty out and drainage stops. Water in the small (or capillary) pores is retained by capillary forces, allowing it to be used by plants.

## Water Potential ($\psi_w$)

In order for water to move in soil, work must be done on it by the previously discussed forces. To predict its movement in soil, the energy potential of water is considered. Water always moves from a point of higher amount (or total potential or energy) to a point of less amount (or lower total potential). Total water potential ($\Psi_w$) is the sum of gravitational potential ($\Psi_g$), pressure (or turgor or matric) potential ($\Psi_p$), and osmotic (or solute) potential ($\Psi_o$). Other forces act on water, but are almost always insignificant.

$$\Psi_w = \Psi_g + \Psi_p + \Psi_o$$

**Gravitational potential ($\psi_g$).** This is determined by water's elevation relative to an arbitrary reference level. Water above the reference elevation has positive (+) gravitational potential. Water below the reference elevation has negative (–) gravitational potential. Gravitational potential is independent of the chemical and pressure conditions of water. It depends solely on relative elevation.

**Pressure potential ($\psi_p$).** This is determined by the comparison to water at atmospheric pressure (e.g., at a free-water surface). Soil water at a hydrostatic pressure greater than atmospheric has a positive pressure potential. For example, water below a free-water surface (such as a groundwater table) has a positive pressure potential. This relates to the increasing pressure felt by divers as they descend. Water at the free-water surface has zero pressure potential, while at hydrostatic pressure less than atmospheric (in other words, under suction or tension), a negative pressure potential occurs. Negative pressure potential is often termed capillary or matric potential (sometimes designated as $\Psi_m$). This is seen above the free-water surface where water is held in micropores by capillary and adsorptive forces.

**Osmotic (or solute) potential ($\psi_o$).** This is created by the presence of solutes (or salts) in the water solution. The more solutes present, the more their molecules (or ions) are attracted to water. Due to solute presence, a greater amount of tension or suction (or work) is required by plant roots to extract water from soil. Soil salts and fertilizer (such as nitrogen or potassium) are common sources of these solute salts.

### Example

A soil has a water table 50 cm deep and has its reference point at the soil surface. The quantity of solutes in the soil is negligible. Calculate $\Psi_w$, $\Psi_o$, $\Psi_p$, and $\Psi_g$ in this system.

**Step 1:** To determine $\Psi_w$, we know the water table is at 50 cm depth. Since the water table is at the interface between saturation and unsaturation, it is assumed that $\Psi_p$ is zero—the soil is so saturated that water is no longer attracted to the soil surface and no weight of water is above this point to cause a pressure potential.

**Step 2:** Since $\Psi_p$ is 0 and $\Psi_o$ is negligible, the only potential working at this point is $\Psi_g$; since the reference point is at the soil surface and the water table is 50 cm below the soil surface, $\Psi_g$ is therefore –50 cm.

**Step 3:** If the information is now entered into the equation: $\Psi_w = \Psi_g + \Psi_p + \Psi_o$, it is possible to calculate total potential ($\Psi_w$):

$$\Psi_w = (-50) + 0 + 0$$

Therefore, the total potential in the system is –50 cm.

## Pore Size

The rate of water movement down through a soil is strongly affected by the size of pore through which it must move. This rate is approximately proportional to the square of the pore diameter or four times its radius squared. As the soil water content decreases due to drainage (gravity pull), the rate of drainage decreases at an exponential rate since the remaining water is held in the smaller, often narrower pores. Once the soil is no longer saturated, water is prevented from moving downward at the rate of saturated hydraulic conductivity. This occurs even though large pores exist in the soil below the draining front. Water is prevented from filling these larger pores because it is held in the smaller pores above by adhesion, and by the surface tension at the ends of the smaller pores. If the soil becomes saturated, water will quickly enter and fill these large pores and move downward, again at the rate of saturated hydraulic conductivity.

### Perched Water Table

A perched water table (or zone of free water) is a zone of saturated soil just above the interface between a finer textured soil over a coarser textured soil. This is often a desirable condition, created by design for golf greens and sports fields, where finer sand is placed over a coarser aggregate layer such as gravel (Figure 4.13). Water will not drain from the upper finer textured soil until a sufficient zone of saturation develops above the coarser soil (Figure 4.14). At that point, the weight of the water cannot be contained by the capillary retention forces, and water starts flowing downward. A perched water table (or saturated zone) will extend from the top of the coarser layer up to the air entry point.

**FIGURE 4.14** Perched water table formed between two dissimilar soil layers. A zone of saturation forms until sufficient weight of water breaks the capillary forces between the two layers (WSU).

This saturated zone develops as the adhesive force of the finer diameter soil particles on water is greater than the drainage force due to gravity. The finer textured and/or more compacted a soil becomes, or the greater the difference in particle-size distribution between the soils, the harder it is for water to cross over from one to the other, and the higher the perched water table will be. Conversely, a coarser soil or two soils with similar particle size distributions will have a shallower perched water table. If the topsoil depth is not greater than the depth of the perched water table, the whole soil profile will remain saturated and will not drain. The upper portion of this saturated zone of free water is known as the capillary fringe.

By creating an adequate perched water table, fast-draining, low-compacted sand can be used as a rootzone. When the correct depth of top sands is placed over the correctly sized gravel layer, the perched water table provides a reservoir of water for grass to use, but the complete profile still drains sufficiently. Drainage flow will continue from the sand into the gravel due to the combined forces of gravity and adhesive forces at the contact points between the sand and gravel particles. This "flow" continues until the surface tension in the pores of sand and the adhesive forces of sand particles in contact with the gravel equal those forces (gravity and adhesion) pulling the water down. At this point, equilibrium is reached and a saturated perched water table exists above the sand/gravel interface. Optimally, this zone of saturated soil should extend no more than about 6 inches (15 cm) up from the interface. Unless additional water is added to the system above, no further water moves downward out of the perched water table.

**FIGURE 4.13** A perched water table created by placing a sand-based rootzone over a coarser gravel layer.

### Choosing Sands for Golf Greens

Several factors should be considered when choosing a sand for golf green construction, including the climate (especially rainfall amount and intensity), locally available sand, type of turfgrass being grown, irrigation water quality, amount of play, and golfer expectations. These considerations help determine an acceptable range of saturated hydraulic conductivity and water-holding capacity for the rootzone mix. Unfortunately, it is impossible to optimize both hydraulic conductivity and water-holding capacity because, as one increases the other decreases. A rootzone mix with a high hydraulic conductivity will drain rapidly after heavy rains, allowing play to resume quickly, and will provide a firm playing surface. However, moisture- and nutrient-holding capacity will be low, necessitating frequent irrigation and fertilizing. Low hydraulic conductivity mixes will hold more moisture and nutrients, but playing conditions after a heavy rain may be delayed compared with conditions having a high hydraulic conductivity. Rootzone mix characteristics can be altered by adjusting the sizing of the sands or by the addition of soil amendment such as peat.

## DRAINAGE

Water management is the primary key to success for most commercial turfgrass facilities. Soil serves as the storehouse for water used for plant growth that must be readily available to satisfy the demand created by transpiration. Being able to apply water when needed (irrigation) and being able to expediently remove excess water (drainage) ensure good turfgrass growth and prevent prolonged play delay. Improper or inadequate drainage is the most common agronomic problem cited by golf course superintendents and sports field managers. As with many topics in turfgrass management, drainage is a subject widely misunderstood, full of myths, nonscientifically based practices, and unproven materials and products.

### Drainage Methods

Two primary forms of drainage are utilized in turfgrass facilities: surface and subsurface:

1. In surface drainage, land surfaces are reshaped, sloped, and smoothed as needed to eliminate ponding and to induce gravitational flow overland to an outlet. Diverting and excluding water from an area often involve diversion ditches, swales, and floodways.
2. With subsurface drainage, soils may be modified to induce surface water infiltration and percolation through the rootzone to buried drains that collect and transport excess soil water to an outlet. The drop in pressure (or water potential) due to outlet discharge induces excess soil water flow into the drains. Subsurface drainage may also involve interceptor drains oriented perpendicularly to the direction of groundwater flow.

A combination of surface and subsurface drainage is often required to quickly remove water from the soil surface to minimize delays in play, avoid excessive compaction, and allow maintenance practices to continue (Figure 4.15). First, soil water movement and the factors influencing this movement will be discussed.

### Surface Drainage

Surface drainage is often a missing component in the design of modern golf courses. Some of the major problems of poor playability and performance of facilities are caused by insufficient surface drainage, especially when the rootzone has poor internal drainage properties. Almost all long-term successful turfgrass facilities have adequate surface slope (grade) to remove excess surface water. Surface drainage uses the potential energy existing due to elevation change to provide a hydraulic

**FIGURE 4.15** Surface and subsurface drainage are often needed to expediently remove water from the soil surface.

gradient. The surface drainage system creates a water-free surface by moving surface water to an outlet at a lower elevation. For native soil–constructed (e.g., push-up) golf greens characterized by low infiltration and poor internal drainage from high silt and clay content of the soil, surface drainage represents the only effective method for removal of excess surface water. For optimal surface drainage, these greens essentially need to be dome shaped, often referred to as "turtle-backed."

As water enters the soil, pores (large and small) near the soil surface fill first. When pores become full, gravity begins to pull water downward. Water on the soil surface will puddle (or pond) if the water application rate exceeds the amount that gravity can pull further down the profile. Once soil saturation is reached in shallow golf green or sports field profiles, the rate of water entering the soil is dependent on the rate at which the subsoil can remove it. If water sits or ponds on the surface, the whole topsoil is saturated. This is most common in surface depressions and on flat surfaces. If play commences while soil is saturated, moisture acts like a lubricant, allowing the soil particles to compress and causing compaction. Turf plants and roots are easily damaged when soils are saturated. In addition, saturated soils contain less oxygen, thus encouraging anaerobic conditions that lead to root loss and possible buildup of toxic gases such as carbon dioxide and methane, as well as substances such as iron and aluminum oxides, the chief causes of black layer.

A major advantage of good surface drainage is the capability to remove large volumes of water. This capability is especially important during heavy rainfall events. For example, a 1-inch rainfall across 1 acre equals 27,154 gallons (25 mm over 0.40 ha = 102,870 L).

**Slopes.** The slope at which a particular surface should be constructed is determined by several variables. Slopes up to 3% (1:33) are acceptable for soils with poor infiltration rates. A minimum of a 1% slope (1:100) is almost always necessary for proper surface drainage, except with extensively modified rootzones and subsurface drainage such as USGA or California-style constructed greens. For these modern greens, the surface slope surrounding the cup should typically be no more than 3% for bermudagrass or ryegrass and no more than 2% for bentgrass greens or ball roll distances may become excessive. For most nonmodified soils, a 1.5% (1:66) to 2.5% (1:40) slope is usually adequate.

The following equation calculates the velocity of water across a bare surface as influenced by the surface slope and depth of ponded water or rainfall:

$$V = 0.35 \times D^{0.67} \times S^{0.5}$$

where

V = velocity, inches/second
D = water depth (inches)
S = slope

### Example

1. A 1-inch (25 mm) rainfall onto saturated soil with a 1% slope would yield:

$$V = 0.35 \times 1^{0.67} \times 0.01^{0.5}$$

$= 0.035$ inch(0.9 mm) water movement over a bare surface per second

2. A similar rainfall on a 2% slope would yield:

$$V = 0.35 \times 1^{0.67} \times 0.02^{0.5}$$

$= 0.05$ inch(1.3 mm) water movement over a bare surface per second

3. On a 3% slope, velocity increases to:

$$V = 0.35 \times 1^{0.67} \times 0.03^{0.5}$$

$= 0.06$ inch(1.5 mm) water movement over a bare surface per second

These examples demonstrate the large amount of water surface drained by properly designed and constructed slopes. Insufficient slope means water must be drained through soil infiltration, which can be too slow to be sufficient.

The length of slope becomes important as areas at the bottoms of long slopes (about 75 yards or 69 m) remain wet for longer periods than areas further up the slope; thus, they become subject to wear and compaction. Such areas are often found at the intercept of surface drainage from the fairway and front of golf greens. Golf course fairways should be designed so that surface drainage is toward the outside of the fairway, rather than down the slope toward the green (Figure 4.16). A maximum distance for surface drainage is approximately 150 feet (46 m).

**Existing surface depressions.** Once depressions form, several means of correction are used. Small (shallow) depressions can gradually be brought up to grade by frequent, light topdressing (adding sand). Topdressing should never be so heavy as to smother the grass and should only be performed when the grass is actively growing and not under any environmental or agronomic stresses. However, topdressing and turfgrass repairs can cause dissimilar layering or stratification problems if finer soil is used as a topdressing material. A 1/8 inch (3.2 mm) layer of fine soil (such as silt or clay) is enough to disrupt normal water flow in an otherwise ideal soil profile. Successive layering further slows drainage to the point at which it may stop.

For deeper depressions, topdressing may take too long or may require excessive amounts of fill to bring the area up to grade. The existing turf should then be cut (lifted), the area brought up to grade with soil, and turf relaid. Additional topdressing will probably be needed thereafter to help smooth the surface and fill in any seams between sod pieces.

## Subsurface Drainage

Subsurface drainage involves water movement through a soil profile and often includes the installation of subsurface drains to remove excess water that can create undesirable (e.g., saturated) growing conditions. Water available to plants is held in soil by capillarity, while excess water flows by gravity into drains. This lowers the groundwater level below the rootzone

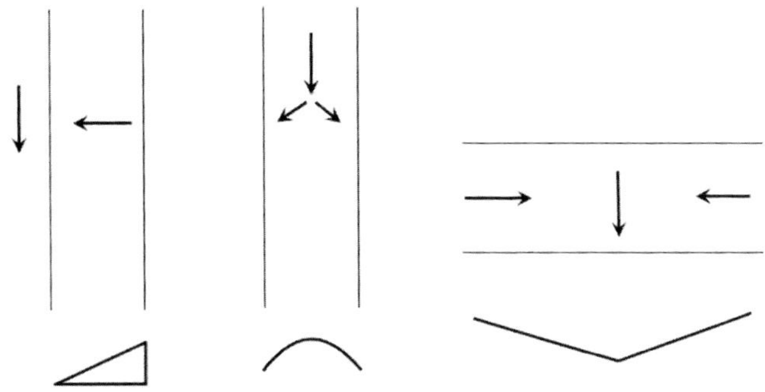

**FIGURE 4.16** Golf course fairway design where surface drainage is directed toward the outside.

of plants. The movement of water into drains for turf facilities is influenced primarily by

1. Soil permeability. This includes soil horizontal and vertical water permeability.
2. Drain spacing. This is most often calculated using Hooghoudt's equation.
3. Depth of drain. Drain depth and spacing are interrelated. As the depth of the drain increases, generally so does the spacing distance needed between drain lines.
4. Drain size. This is more correctly described as the ability of the drain to lower the water potential sufficiently to promote water movement to and out of the drain.

**Soil modification to improve permeability.** Soil modification to enhance internal soil moisture percolation is a common practice in the turfgrass industry. However, several misconceptions exist regarding soil modification to improve permeability. One such misconception is manifested in the practice of applying a 2- to 6-inch (5 to 15 cm) layer of sand over a native soil with little or no surface slope provided and no subsurface drain lines installed. This is often referred to as the "bathtub" effect, where the finer textured native soil will not adequately drain and the area holds water like a bathtub (Figure 4.17). Heavy rainfall then causes saturation of the added sand layer and surface water accumulates, causing poor playing conditions. This is why most heavily used turf areas need 10 to 12 inches (25 to 31 cm) of modified topsoil and properly spaced drain lines to lower this excess surface moisture further down in the soil profile. The drains act similarly to a drain in a bathtub, providing a means of water removal.

Another misconception is that an inch or so of a coarse sand can be tilled into the top 3 to 6 inches (7.6 to 15 cm) of native soil to enhance internal percolation. Unfortunately, this practice is rarely successful. Sand often has a wide range

**FIGURE 4.17** Placing a shallow layer of sand over heavier soil rarely improves drainage.

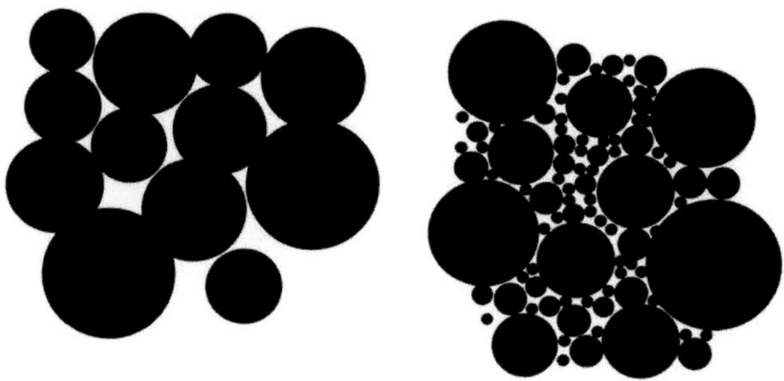

**FIGURE 4.18** Illustration of how mixing coarser textured sand with finer soil (right) "clogs" internal pores.

of particle sizes. This variety in particle size allows smaller silt and clay particles to become dispersed among the larger sand particles, effectively reducing the pore space for water to percolate. Similarly, adding sand to native soil, which often has a high degree of silt and/or clay, can "clog" these larger internal sand pores, again reducing internal percolation (Figure 4.18). Lastly, trying to uniformly "mix" the surface applied sand with the underlying soil is virtually impossible with a tractor-mounted roto-tiller. These machines will not provide the blended soil mix desired. Proper mixing requires "off-site" machine blending (Figure 4.19).

Table 4.4 demonstrates the results of blending high-quality (USGA specified) sand into a native Cecil clay soil. The sand–clay blend was performed "off-site" in a laboratory, providing a very uniform distribution of sand and soil in the various ratios. As shown in Table 4.4, adding just 10% clay soil to this sand reduced its hydraulic conductivity by almost 85% (from 58 inches/hour to 9 inches/hour, 148 to 23 cm/hour). Conductivity values quickly dropped as the clay soil content increased; for example, with a 50:50 blend, the hydraulic conductivity was less than 0.2 inch/hour (0.5 cm/hour), totally unacceptable by today's standards.

**FIGURE 4.19** Undesirable, nonuniform "on-site" mixing of peat into sand by roto-tilling.

**TABLE 4.4**
**Hydraulic Conductivity of a USGA Medium Sand and a Cecil Clay Soil, Alone and in Various Combinations**

| | Hydraulic Conductivity ($K_{sat}$) | |
|---|---|---|
| **Sand:Soil Ratio** | **(Inches/Hour)** | **(cm/Hour)** |
| 0:100 | 0.07 | 0.18 |
| 10:90 | 0.05 | 0.13 |
| 20:80 | 0.03 | 0.06 |
| 30:70 | 0.09 | 0.22 |
| 40:60 | 0.13 | 0.33 |
| 50:50 | 0.15 | 0.39 |
| 60:40 | 0.19 | 0.47 |
| 70:30 | 1.89 | 4.80 |
| 80:20 | 3.24 | 8.23 |
| 90:10 | 9.01 | 22.89 |
| 100:0 | 58.1 | 147.6 |

Furthermore, adding 20% sand to the soil reduced drainage more than 50% compared to straight (100%) soil. This again represents small soil particles "clogging" the larger pores between sand particles.

The following equation provides a guideline for using a suitable sand with a soil of known mechanical composition to create a rootzone with the desired drainage rate:

$$A = \frac{[R - B]}{[C - R]} \times 100$$

where

A = weight of sand to add to 100 weight units of the original soil

B = of original soil in the desired particle-size range (e.g., 0.125 to 0.5 mm)

C = of desired particle-size range (e.g., 0.125 to 0.5 mm) in the sand used as an amendment
R = of desired particle-size range (e.g., 0.125 to 0.5 mm) sand in the final mix

### Example

Assume the particle-size distribution (%) and bulk density values are found in the sand and soil sources in Table 4.5.

If a 9 inches/hour (23 cm/hour) percolation rate is desired for this sand–soil rootzone, the R value would be 90% as determined from Table 4.5 in the desired particle-size range of 0.125 to 0.5 mm.

$$A = \frac{[R-B]}{[C-R]} \times 100 = \frac{[90-35]}{[65-90]} \times 100 = 220$$

Therefore, 220 tons of sand per 100 tons of soil would be required to raise the percentage of soil particles between 0.125 and 0.5 mm to 90% in the final mix.

**Note:** Values generated are absolute values regardless of their positive or negative signs, as the actual calculation value in the previous example is –220.

If mixed on a volume basis (such as with off-site blending) instead of a weight basis, one must find the volumetric ratio of sand to soil using the equation: volume = mass/density. The bulk density of sand in this example is 1.65 g/cm³ and soil is 1.35 g/cm³, giving:

$$\text{Volume(ratio)} = \frac{V_{sand}}{V_{soil}} = \frac{220}{1.65} \div \frac{100}{1.35} = \frac{220}{100} \times \frac{1.35}{1.65} = 1.8$$

Therefore, 1.8 unit volumes of this particular sand are needed per one unit volume of this soil to achieve the desired ratio of 220 tons of sand per 100 tons of soil.

**Calculating volume to volume (v/v) ratios.** If one wishes to determine the outcome of mixing sand with topsoil on a volume to volume (v/v) ratio basis, the following calculations can be performed for a sand to soil ratio mix:

$$\text{New \% particle size} = \frac{[\text{sand fraction \%} \times \text{ratio sand}] + [\text{soil fraction \%} \times \text{ratio soil}]}{\text{total sand} + \text{soil ratio}}$$

### Example

Calculate the new percent particle size in the 0.5 to 0.25 mm range from the sand/soil ratio listed in Table 4.5 in a 3:1 ratio.

$$\text{New \% particle size} = \frac{[\text{sand fraction \%} \times \text{ratio sand}] + [\text{soil fraction \%} \times \text{ratio soil}]}{\text{total sand} + \text{soil ratio}} = \frac{[44 \times 3] + [15 \times 1]}{3+1} = 37$$

The following example demonstrates how to determine the new particle-size distribution obtained by tilling a known volume of sand into native soil.

### Example

1,500 tons (1,814 metric tons) of sand with a bulk density of 1.65 g/cm³ is tilled into the top 5 inches (13 cm) of native soil 1.7 acres (0.7 ha) in area. Calculate the predicted new particle-size distribution percentages and bulk densities.

**TABLE 4.5**
**Calculated Values of Various v/v Ratios of Sand to Soil from Known Particle-Size Distribution and Bulk Density Values**

| Soil Type | Percent Particle-Size Distribution (mm) | | | | | | | Bulk Density |
|---|---|---|---|---|---|---|---|---|
| | 2 to 1 | 1 to 0.5 | 0.5 to 0.25 | 0.25 to 0.125 | 0.125 to 0.05 | 0.05 to 0.002 | <0.002 | (g/cm³) |
| **Known Values** | | | | | | | | |
| Sand | 3 | 32 | 44 | 21 | 0 | 0 | 0 | 1.65 |
| Soil | 3 | 17 | 15 | 20 | 5 | 29 | 11 | 1.35 |
| **Calculated Values of Various Sand:Soil Ratios** | | | | | | | | |
| 1:1 | 3 | 24.5 | 29.5 | 20.5 | 2.5 | 14.5 | 5.5 | 1.50 |
| 2:1 | 3 | 27 | 34 | 21 | 1.7 | 9.7 | 3.7 | 1.55 |
| 3:1 | 3 | 28 | 37 | 21 | 1.3 | 7.3 | 2.8 | 1.58 |
| 9:1 | 3 | 30.5 | 41 | 21 | 0.5 | 2.9 | 1.1 | 1.62 |

First, determine the depth of 1,500 tons of sand over the 1.7 acre:

$$\frac{1{,}500\text{ tons}}{1.7\text{ acres}} \times \frac{2{,}000\text{ pounds}}{\text{ton}} \times \frac{454\text{ g}}{\text{pounds}} \times \frac{1\text{ cm}^3}{1.65\text{ g}} \times \frac{1\text{ acre}}{43{,}560\text{ square feet}}$$

$$\times \frac{1\text{ square foot}}{929\text{ cm}^2} \times \frac{1\text{ inch}}{2.54\text{ cm}} = 4.7\text{ inches deep}$$

Since 4.7 inches (12 cm) of sand in depth is to be tilled into the top 5 inches (13 cm) of soil, this can be approximated using a 1:1 ratio as presented in Table 4.5.

Although this equation helps predict projected particle-size distribution and bulk density values of two known sand/soil sources, it cannot be reliably used to predict percolation rates. For example, with the same sand and soil from the previous example in a 1:1 ratio, the following calculations could be performed to predict a percolation rate for the mix:

$$\text{Predicted percolation rate} = \frac{[\text{perc. rate sand} \times \text{ratio sand}] + [\text{perc. rate soil} \times \text{ratio soil}]}{\text{total sand} + \text{soil ratio}}$$

$$= \frac{[58 \times 1] + [0.07 \times 1]}{1+1}$$

$$= 29\text{ inches/hour}(74\text{ cm})$$

However, when actual samples are mixed in a 1:1 ratio, the percolation rate is only 0.15 inch/hour (0.4 cm/hour) (Table 4.4). The small amount of fine-textured clay in the soil mix is sufficient to "clog" the pores in the sand, thus reducing the actual percolation. This demonstrates the importance of actually measuring particle-size distribution, bulk density, and hydraulic conductivity (percolation rate) of the various soil/sand mix being considered as well as the ratios of each.

Table 4.6 reflects the percent (by volume) change when a known amount of amendment is mixed into a soil. For example, if a contractor places 2 inches (5 cm) of an amendment on the existing soil surface and roto-tills this 6 inches (15 cm) deep, the theoretical percent volume this added amendment occupies is 25%.

**TABLE 4.6**
**Amount (Inches) of Surface-Applied Amendment Converted into Percent Volume**

| Amount Amendment Mixed into Upper Soil | Inches of Soil | = | Percent by Volume |
|---|---|---|---|
| 0.5 | 2 | | 20 |
| 0.5 | 3 | | 14 |
| 0.5 | 4 | | 11 |
| 0.5 | 6 | | 8 |
| 1 | 2 | | 33 |
| 1 | 3 | | 25 |
| 1 | 4 | | 20 |
| 1 | 6 | | 14 |
| 1.5 | 2 | | 43 |
| 1.5 | 3 | | 33 |
| 1.5 | 4 | | 27 |
| 1.5 | 6 | | 20 |
| 2 | 2 | | 50 |
| 2 | 3 | | 40 |
| 2 | 4 | | 33 |
| 2 | 6 | | 25 |

## Lateral Soil Water Movement

Lateral (sideways) water movement in a soil is influenced or restricted by three factors:

1. Depth (hydraulic head) of the saturated free-water zone in the topsoil
2. Hydraulic conductivity of the rootzone soil
3. Slope of the subgrade or base

As a soil absorbs more water, a saturated zone eventually develops and reaches the subsoil base. Until this saturated zone reaches the subsoil base and a buildup of "free water" occurs, little water will move laterally (sideways). This saturated zone of free water is the only water moved sideways by gravity. The smaller the soil pore space is, the more slowly water will move laterally.

At the top of this saturated section of the profile, a zone (called the capillary fringe) exists where water is held in the pores and will not move laterally out of this zone into a drain or a hole. Water will not enter a hole dug into this capillary fringe zone even though the surrounding soil is saturated, since the surface tension and adhesive forces of soil particles in the capillary fringe zone are greater than the saturated soil. If a hole is dug that penetrates below the capillary fringe, it will fill with water, but only to the level of the top of the free-water zone. This miniature drainage occurs as there is a sufficient "head" of water above it "pushing" the water out of the pores. This helps illustrate the principle that the shallower the topsoil (rootzone mix) is, the closer drains need to be spaced, since only water from the free-water zone will move into subsoil drains. Drain pipes lower the depth of free water and the capillary fringe moves down at the same rate. As water moves into the drain pipes, the height of the free-water zone diminishes, as does the gradient from the drain to the top of the free-water zone at the midpoint between the drains. As this gradient diminishes, so does the rate of water movement into the drains. This occurs because the component of gravity that pulls water laterally decreases as the height of the free-water zone decreases.

Lateral water movement ceases when the free-water zone is removed. This occurs even if the capillary fringe is still saturated. Hooghoudt's equation (discussed later) is used to calculate the rate at which the saturated free-water zone of the topsoil will drain at the midpoint between two drains (the slowest draining point).

With USGA-specified greens, the subgrade should be sloped to conform to the general slope of the finished surface grade. Although the surface may consist of complex contouring, the subgrade need only consist of an inclined plane with straight, evenly spaced contours sufficient to avoid low spots within the subgrade.

**Gravel size and shape.** In USGA specification profiles, the height of the perched water table is also determined by the number of gravel particles in the gravel layer in contact with the sand above. As the gravel size becomes smaller or contains many fine particles, there is more contact with the sand above, a shallower perched water table develops, and more water flows downward across these contacts (Figure 4.20). In addition, if the gravel particle shape becomes flatter and narrower, it is able to pack closer together, lie more horizontally, and thus create a large surface area in contact with the sand. Gravel rounder in shape will have only a small point of contact with the sand and less water will flow downward across these contacts, creating a higher (or deeper) perched water table.

In the laboratory, the correct gravel size in relation to the average size of sand particles the gravel will contact can be determined. The properties used in specifying materials used in construction include: D-value (diameter value), coefficient of uniformity (also called the gradation index), and percentage of particles within a desirable size range. The $D_x$ value is the sieve opening size through which x percent of particles in a sample pass. Two D-values are commonly used, with the larger value indicating the general coarseness of the sample while the lower D-value often reflects the largest makeup of particles in a sand-based rootzone mix.

$D_{85}$ is defined as the particle diameter size below which 85% of soil particles (by weight) are smaller. $D_{15}$ is the particle diameter size below which 15% of soil particles (by weight) are smaller. In sand-over-gravel profiles, the largest 15% of the sand particles "bridge" with the smallest 15% of the gravel particles. This bridging is caused by the irregular shape of particles, friction between particles, and the weight of the material above, all working to "lock" the smaller particles into voids in the gravel (Figure 4.21).

Once particles have locked together and "bridged" above the voids, this prevents further significant particle movement, while maintaining adequate permeability. For bridging to occur, the USGA recommends gravel $D_{15}$ be less than or equal to eight times the sand rootzone $D_{85}$. To maintain sufficient permeability across this gravel/rootzone interface, gravel $D_{15}$ should be greater than or equal to five times the sand rootzone $D_{15}$.

The uniformity coefficient describes the uniformity of the distribution of particle sizes, helping indicate the potential for interpacking. Interpacking of sands with a large gradation index reduces total pore space and reduces hydraulic conductivity. The gravel should have a uniformity coefficient (gravel $D_{90}$/gravel $D_{15}$) less than or equal to 3.0. Lastly, gravel should totally pass through a 0.5-inch (12 mm) sieve, with less than 10% passing a 0.08-inch (2 mm) sieve (no. 10) and greater than 5% passing a 0.04-inch (1 mm) sieve (no. 18) (Table 4.7).

Sand meeting the USGA specs will have a $D_{85}$ between 0.4 and 0.7 mm (0.016 and 0.03 inches); in other words, 15% of the sand particles will be larger than this size. Using the bridging factor, gravel $D_{15}$ (smallest 15% diameter of the gravel) should not be larger than five times sand $D_{85}$ (0.4 to 0.7 mm), or 2 to

**FIGURE 4.20** Smaller diameter gravel increases contact with the sand, allowing a perched water table to develop.

**FIGURE 4.21** When properly sized sand is matched with gravel, "bridging" occurs, allowing water to move but not sand.

3.5 mm (five times 0.4 to 0.7 mm). If the gravel is too coarse, the bridging factor will be too high, and if the sand is very dry, it may migrate into the gravel. When sand and gravel are matched to bridge, sand will not migrate into the gravel voids, even though many of these gravel voids are larger than 0.4 to 0.7 mm.

### Example

The following information was determined on a potential rootzone mix and gravel sample. Can this rootzone mix meet the bridging criteria to prevent migration into the gravel?

**Step 1:** For bridging to occur, $D_{15}$ (gravel) $\leq 8 \times D_{85}$ (rootzone). From the data in the table, a regression equation is developed to fit the line or curve to determine $D_{15}$ and $D_{85}$. For sands, this curve is often sigmoidally shaped while it is linear or quadratic in nature for gravel. For $D_{15}$ (gravel), this is 2.95 mm while for the $D_{85}$ (rootzone), it is 0.84 mm. Therefore, $5 \times 0.84 = 4.20$ mm and 2.95 mm is $\leq 4.20$ mm. Bridging of the coarsest 15% rootzone particles should occur with the finest 15% gravel particles. This gravel is considered compatible for bridging with the sand.

Again, a regression equation is needed to determine $D_{15}$ for the rootzone, which is 0.22 mm. From this, the $D_{15}$ (gravel) is $\geq 5 \times D_{15}$ (rootzone), which is $5 \times 0.22 = 1.10$ mm. Therefore, adequate permeability should occur with this gravel and rootzone sand.

**TABLE 4.7**
**Size Recommendations by the USGA for Gravel When an Intermediate Sand ("Choker") Layer Is or Is Not Used in Golf Green Construction**

| Performance Factors | Recommendation[a] |
|---|---|
| | **Gravel Size without Intermediate Layer** |
| Bridging factor | • $D_{15}$ of gravel $\leq$ eight times the $D_{85}$ of the rootzone |
| Permeability factor | • $D_{15}$ of the gravel $\geq$ five times the $D_{15}$ of the rootzone |
| Uniformity factor | • $D_{90}/D_{15}$ ratio of gravel $\leq 3.0$ |
| | • No particles of gravel greater than 12 mm |
| | • Not more than 10% of gravel diameter less than 2 mm |
| | • Not more than 5% of gravel diameter less than 1 mm |
| | **Gravel Size with Intermediate Layer** |
| Gravel size | • Not more than 10% of particles greater than 12 mm (1/2 inch) |
| | • At least 65% of particles between 6 and 9 mm (1/4 and 1/8 inches) |
| | • Not more than 10% of particles less than 2 mm |
| Intermediate (choker) layer material size | • At least 90% of particles between 1 and 4 mm. |

[a] $D_{15}$, gravel = particle diameter below which 15% of gravel particles (by weight) are smaller. $D_{85}$, rootzone = particle diameter below which 85% of rootzone particles (by weight) are smaller.

| | Particle Diameter (%) | | | | | | | | |
|---|---|---|---|---|---|---|---|---|---|
| | | | | | Sand | | | | |
| | Gravel | | Fine Gravel | | V. Coarse | Coarse | Medium | Fine | V. Fine |
| | mm | | | | | | | | |
| Sample | 6.3 | 4.75 | 3.35 | 2.0 | 1.0 | 0.50 | 0.25 | 0.15 | 0.05 |
| | Gravel | | | | | | | | |
| % Retained (>D) | 33.2 | 21.9 | 26.1 | 14.8 | 3.2 | 0.8 | 0.0 | 0.0 | 0.0 |
| % Passed (<D) | 66.8 | 44.9 | 18.8 | 4.0 | 0.8 | 0.0 | 0.0 | 0.0 | 0.0 |
| | Rootzone Mix | | | | | | | | |
| % Retained (>D) | 0.0 | 0.0 | 0.0 | 0.3 | 5.5 | 36.2 | 38.9 | 12.7 | 3.3 |
| % Passed (<D) | 0.0 | 0.0 | 0.0 | 99.7 | 94.2 | 58.0 | 19.1 | 6.4 | 3.1 |

*Note:* For adequate permeability $D_{15}$ (gravel) $\geq 5 \times D_{15}$ (rootzone)

**Step 2:** For uniformity coefficient of gravel, $D_{90}$ (gravel)/$D_{15}$ (gravel) $\leq$ 3.0. From the regression equation, $D_{90}$ (gravel) = 7.30 mm and $D_{15}$ (gravel) = 2.95 mm. Therefore, $D_{90}$ (gravel)/$D_{15}$ (gravel) = 7.30/2.95 = 2.47, which is less than 3.0. Thus, the uniformity of coefficient of this gravel passes the criteria in Table 4.7.

**Slope.** Lateral water movement in soil is generally limited in distance and time. However, gravity is able to "pull" water down a sloped base (subgrade). The steeper the subgrade slope is, the greater the effect of gravity is. Generally, water will move laterally (sideways) along the subgrade's surface in direct proportion to the subgrade's slope. For example, if the slope is 2% (1:50), water will move laterally 2% (or 1/50) as fast as it will move downward based on its hydraulic conductivity. If the hydraulic conductivity of a soil is 15 inches/hour (381 mm), the maximum rate at which water would move laterally due to gravity would be 2% × 15 = 0.3 inch (7.6 mm) per hour.

In addition, for water to continuously drain (move) down the subgrade, water must be removed from the end or low point of the subgrade with drains or ditches. If this water is not removed, an equilibrium will be reached, often resulting in ponding of water on the surface and excessively wet conditions at the end of a slope or against an impermeable obstruction such as a wall. Such conditions frequently occur when water drains down banks or hills onto a flatter playing surface (Figure 4.22). In this case, disposal of water from the bottom of the slope can be achieved by (1) installing a cutoff surface drain at the top of the hill to collect water before it reaches the hill, (2) constructing a terrace to move the water gently across or around the perimeter of the hill, or (3) most commonly, placing an interceptor drain at the bottom of the hill.

## Drain Lines

Subsurface drain lines are designed to function as open channels, meaning the water flow through the pipe is from the influence

**FIGURE 4.22** A "seep" where surface and subsurface water drains down an incline onto a flatter surface.

**FIGURE 4.23** Demonstration of water movement toward a subsurface drainage line (WSU).

of gravity due to the slope or grade of the pipe, not from pressure pushing water through the pipe (Figure 4.23). If the subsurface drainage pipe tries to convey more water than it was designed for, it will first fill to capacity and then become pressurized along some portion of its length. When perforated pipe become pressurized, the water escapes through the inlet holes of the pipe. This pressure on the water in the drain line trench can create flow to the surrounding soil, causing the soil to become saturated. This saturated soil can become flowable and easily erodes. When under extreme pressure, drainage pipe can erode out of its trench, requiring extensive repair. This is especially true near the outlet of a long run of pipe at a steeper slope. This pipe pressure can also create a floating "lens" of water between the turf and soil. The turf and thatch layer essentially floats off the soil surface, creating play and maintenance problems. The key is to size the drainage pipe properly, taking into account the expected flows, so that pipe behaves as an open channel.

Two parameters largely determine the rate at which water is removed by a drain: namely, (1) depth and (2) spacing of drain lines. In addition, slopes of drain lines in the trenches also affect drainage capacity. Generally, the deeper the drain lines and closer their spacing, the quicker and more effectively soil moisture is removed. However, water cannot enter a drain any faster than the soil around it can conduct water into it. Optimum depth and spacing are directly related to the permeability of the soil. Since golf greens have a relatively shallow rootzone (approximately 1 foot, 30 cm) of highly permeable soil (sand) and need to quickly and completely remove surface water so that play can resume, their optimum drain spacings are much narrower (closer) than in most unmodified soil situations.

As mentioned, the closer drains are together, the faster a profile will drain. Also, as the free-water depth in the soil profile decreases, so too does the gravitational gradient. A deeper topsoil has a greater storage space in the profile for the free-water zone. Therefore, in shallower soils, the rate of drainage and soil water storage capacity decrease and drains need to be spaced closer together. Golf green drainage lines should be spaced so water will not have to travel more than 10 feet to reach any individual line. If the golf course is situated on an area with a high water table, it may be necessary to place larger drainage lines deeper into the subgrade to lower the water table and handle the increased internal flow of water.

Calculating drainage line spacing can be done with a modification of Hooghoudt's equation. In this equation, the drain discharge is assumed to equal the incoming rainfall or irrigation, and the water table midway between drains is maintained at a steady height above the drain level. Water enters the soil more rapidly nearer drains than midway between them. The equation takes into account both horizontal flow and radial flow caused by the convergence of flow lines over the drains (Figure 4.24). In shallow topsoil, widely spaced drains only remove water from a very small area immediately adjacent to the drains and do not adequately drain the topsoil between them.

Hooghoudt's equation:

$$S = \sqrt{\frac{4Kh^2}{v}}$$

where

$S$ = drain line spacing (inches); the units used for $h$ must be the same as those used for $S$

$K$ = saturated hydraulic conductivity (inches/hour) of the soil

$h$ = height of the free-water zone, also known as the water table, midway between the two drains (inches)

v = drain discharge rate, assumed to equal irrigation or rainfall rate (inches/hour). Normally, the anticipated maximum rainfall or irrigation event rate is used here.

Two calculations are suggested; one involves the assumption of a worst-case scenario where the free-water zone extends to the surface or the total depth of the topsoil, resulting in ponding. The other calculation is performed with the free-water zone lowered by 2 or 3 inches. This gives an indication of how quickly water can be removed from the top of the profile. Often this removal is slow if drains alone are being relied on.

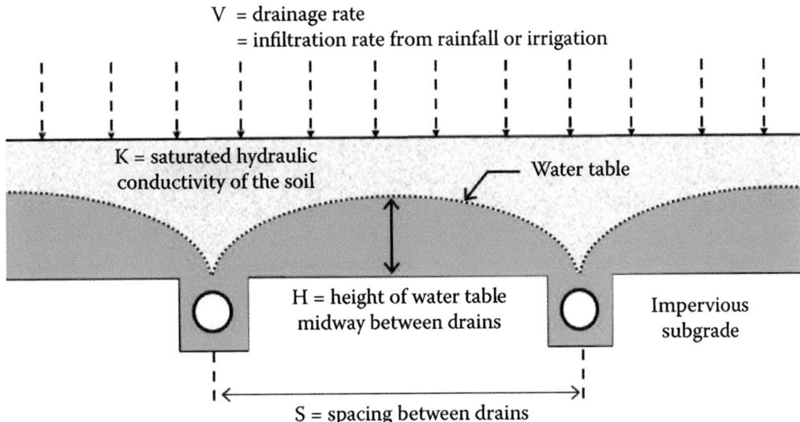


**FIGURE 4.24** Variables used in Hooghoudt's equation to determine drain line spacing.

Since $S$ and $h$ are squared in Hooghoudt's equation, varying them will change the drainage rate by the square of the magnitude of drainage in distance apart or height, respectively. In other words, if drain spacing ($S$) is halved, or depth of the saturated zone ($h$) is doubled, the effective drainage rate of drains increases fourfold. Conversely, shallower topsoil and more widely spaced drains decrease soil water drainage rates exponentially.

Two points are illustrated by Hooghoudt's equation:

1. As the allowable free water (water table) depth decreases (e.g., the shallower the topsoil), the gravitational gradient decreases, resulting in decreased drainage, and the closer drain lines need to be spaced. Conversely, the deeper the topsoil is, the greater the storage space in the profile for the free-water zone should be and the further apart drain lines may be spaced.
2. The closer the drains are, the faster a profile will drain.

**Note:** Hooghoudt's equation becomes inaccurate when drain spacing ($S$) approaches the same value as the height of the saturated free-water zone ($h$). However, this does not occur often in turfgrass facilities since relatively shallow topsoils are used. Hooghoudt's equation cannot be used in a two-tier soil profile with a sand rootzone over a gravel bed (USGA-specified green). The gravel bed allows rapid vertical rootzone drainage and movement to the drain pipes, and soil water movement is most influenced by the saturated hydraulic conductivity of the rootzone. For Hooghoudt's equation to be accurate, the soil must be uniform in hydraulic conductivity and must have an impervious layer located below the soil and the drain.

### Example

If the hydraulic conductivity of a loam soil is 12 inches/hour (30 cm/hour), the height from the drain line to the soil surface is 18 inches (46 cm), and an anticipated rainfall event is 1 inch (25 mm) per hour, determine the drain line spacing required to prevent ponding.

$$S = \sqrt{\frac{4Kh^2}{\text{V}}}$$

$$= \frac{4(12\ \text{inches/hour})\,(18\ \text{inches})^2}{1\ \text{inch/hour}}$$

$$\text{S} = 124\ \text{inches}(10.3\ \text{feet})$$

**Calculating necessary soil hydraulic conductivity.** Hooghoudt's equation also can be rearranged to calculate the desired hydraulic conductivity (percolation) for a given drain line spacing:

$$K = \frac{S^2\text{v}}{4h^2}$$

### Example

1. If an area has a proposed drain spacing of 10 feet (120 inches, 305 cm) between drain lines, a 10-inch (25.4 cm) deep rootzone above the drains, a 1 inch/hour (2.54 cm/hour) anticipated rainfall rate, and the free-water zone extends to the surface, as might occur after prolonged rain, determine the necessary hydraulic conductivity (inches/hour) of the soil.

$$K = \frac{S^2\text{v}}{4h^2}$$

$$= \frac{(120\ \text{inches})^2(1\ \text{inch/hour})}{4(10\ \text{inches})}$$

$$= \frac{14{,}400\ \text{cubic inches/hour}}{400\ \text{square inches}}$$

$$= 36\ \text{inches/hour}(91\ \text{cm/hour})$$

2. If the same area had a 12-inch (or 1 foot, 0.3 m) deep rootzone instead of 10 inches (25 cm), what would be the necessary hydraulic conductivity of the soil?

$$K = \frac{S^2 v}{4h^2}$$
$$= \frac{(120 \text{ inches})^2 (1 \text{ inch/hour})}{4(12 \text{ inches})^2}$$
$$= \frac{14{,}400 \text{ cubic inches/hour}}{576 \text{ square inches}}$$
$$= 25 \text{ inches/hour} (64 \text{ cm/hour})$$

3. If the same 10-inch (25 cm) rootzone area had an anticipated maximum rainfall of 0.5 inch/hour (13 mm) instead of 1 inch/hour (25 mm), what would be the necessary hydraulic conductivity of the soil?

$$K = \frac{S^2 v}{4h^2}$$
$$= \frac{(120 \text{ inches})^2 (0.5 \text{ inch/hour})}{4(10 \text{ inches})^2}$$
$$= \frac{7{,}000 \text{ cubic inches/hour}}{400 \text{ square inches}}$$
$$= 18 \text{ inches/hour} (46 \text{ cm/hour})$$

**Calculating drainage rates.** Hooghoudt's equation can also be rearranged to calculate the drainage rate between subsoil drains.

$$v = \frac{4Kh^2}{S^2}$$

where v = drainage rate (inches/hour) of the saturated free-water zone at the midpoint between drains.

### Example

1. A soil has a hydraulic conductivity rate of 2 inches/hour (5 cm/hour), a saturated depth midway between the drains of 10 inches (or 0.83 foot), and drains spaced 10 feet (3 m) apart. Determine the drainage rate at the midpoint between the drain lines.

$$v = \frac{4Kh^2}{S^2}$$
$$= \frac{4(2 \text{ inches/hour})(0.83 \text{ foot})^2}{(10 \text{ feet})^2}$$
$$= 0.055 \text{ inch/hour} (1.4 \text{ mm/hour})$$

2. If the soil's depth in the preceding example is increased to 12 inches (1 foot), determine the new drainage rate at the midpoint between the drain tiles.

$$v = \frac{4Kh^2}{S^2}$$
$$= \frac{4(2 \text{ inches/hour})(1 \text{ foot})^2}{(10 \text{ feet})^2}$$
$$= 0.08 \text{ inch/hour} (2 \text{ mm/hour})$$

3. If the same soil has drain tile spaced at 15 feet (4.5 m) instead of 10 feet (3 m), what will be the resulting drainage rate?

$$v = \frac{4Kh^2}{S^2}$$
$$= \frac{4(2 \text{ inches/hour})(1 \text{ foot})^2}{(15 \text{ feet})^2}$$
$$= 0.036 \text{ inches/hour} (0.9 \text{ mm/hour})$$

Increasing tile spacing from 10 to 15 feet (3 to 4.5 m) decreases the drainage rate from 0.08 to 0.036 inch/hour(2 to 0.9 mm/hour).

4. Now determine the drainage rate for the above example if tile lines are spaced 5 feet (1.5 m) apart.

$$v = \frac{4Kh^2}{S^2}$$
$$= \frac{4(2 \text{ inches/hour})(1 \text{ foot})^2}{(5 \text{ feet})^2}$$
$$= 0.32 \text{ inch/hour} (8 \text{ mm/hour})$$

Decreasing tile spacing from 10 to 5 feet (3 to 1.5 m) increases the drainage rate from 0.08 to 0.32 inch/hour (2 to 8 mm/hour).

These examples illustrate that, the closer the drain tiles are or the deeper the saturated rootzone is, the faster a profile will drain. Specifically, if the drain spacings are halved, drainage increases fourfold. Similarly, as soil depth is doubled, drainage increases fourfold.

**Determining drain line discharge rates.** If the length of the drain line is known (Figure 4.25), then the total amount of water expected to drain from a particular area following a known amount of rainfall or irrigation can be determined from the following equation, modified from Darcy's and Hooghoudt's equations:

$$Q = \frac{2Kh^2 w}{S}$$

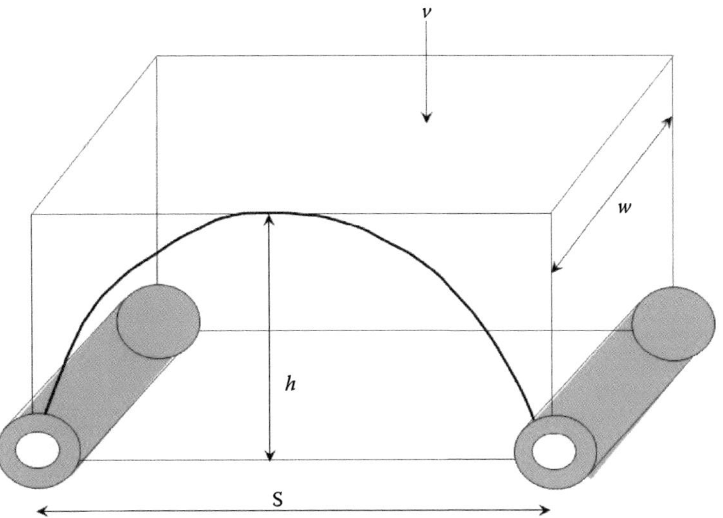


**FIGURE 4.25** Variables used to determine drain line discharge rates.

where $Q$ = discharge rate of water through a drain and $w$ = length of the drain line (Figure 4.25).

### Example

Determine the volume of water flowing from an area with a drain spacing of 10 feet (3 m), drain lines 12.5 feet (3.8 m) long in a 1-inch (2.54 cm) rainfall event, a rootzone hydraulic conductivity of 16 inches/hour (41 cm/hour) and a 10-inch (25 cm) rootzone depth (water has a volume of 0.00434 gallon/cubic inch, 1 mL/cm$^3$).

$$Q = \frac{2Kh^2w}{S}$$

$$= \frac{2(16\text{ inches/hour})(10\text{ inches})^2(150\text{ inches})}{120\text{ inches}}$$

$$= 4{,}000\text{ cubic inches/hour}$$

$$= \frac{4{,}000\text{ cubic inches}}{\text{hour}} \times \frac{0.00434\text{ gallon}}{\text{cubic inch}}$$

$$= 17.4\text{ gallons/hour}$$

Therefore, drain lines should be selected that can remove at least 18 gallons/hour (6 L/hour). The rational method is another means of calculating water discharge from an area based on rainfall intensity, drainage areas, and a rational runoff coefficient.

## Multiple Layered Systems

In USGA-designed turf systems, two- or three-layer (-tier) systems are used to allow turf managers to take advantage of a perched water table. When determining drain spacing for such designs, two methods are used by soil scientists. The first and simplest is used when an intermediate (choker) layer is not present by inserting the height and percolation values of the gravel layer in Hooghoudt's equation in place of the topsoil mix (Figure 4.24).

The second method is used when multiple layers are present and takes into account the depth and hydraulic conductivity of the rootzone and intermediate layer. This method determines the drainage rate as regulated by the saturated hydraulic conductivity of the rootzone and the intermediate layer. The downward permeability is calculated using Luthin's equation (Figure 4.26):

$$\text{Apparent permeability} = \frac{[h_1 + h_2 + \ldots h_n]}{[(h_1/K_1) + (h_2/K_2) + \ldots (h_n/K_n)]}$$

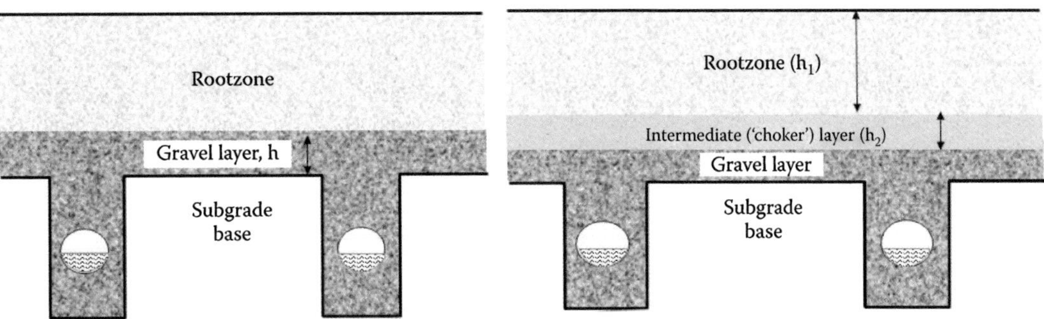


**FIGURE 4.26** Variables used to determine drain line spacing when an intermediate ("choker") layer is present.

where

$n$ = the total number of layers
$h_1$ = depth of layer 1
$h_2$ = depth of layer 2
$h_n$ = depth of layer $n$
$K_1$ = saturated hydraulic conductivity of layer 1
$K_2$ = saturated hydraulic conductivity of layer 2
$K_n$ = saturated hydraulic conductivity of layer $n$

The generated apparent permeability value is incorporated into Hooghoudt's equation as the overall $K$ value. Each method assumes the topsoil mix has an infiltration and percolation rate greater than any anticipated rainfall event.

### Example

If a 12-inch (30 cm) topsoil has a percolation rate of 13 inches/hour (33 cm/hour), a 2-inch (5 cm) intermediate (choker) layer has a percolation rate of 20 inches/hour (51 cm/hour), a 4-inch (10 cm) underlying gravel layer has a percolation rate of 500 inches/hour (1,270 cm/hour), and an anticipated rainfall maximum of 2 inches/hour (5 cm/hour), what drain line spacing should be used according to each of the previous methods?

When the intermediate layer is not installed, the gravel values are used in Hooghoudt's equation, as follows:

$$S = \sqrt{\frac{4Kh^2}{v}}$$

$$= \sqrt{\frac{4(500\text{ inches/hour})(4\text{ inches})^2}{2\text{ inches/hour}}}$$

$$= 126\text{ inches (10.5 feet)}$$

In this example, sufficiently sized drain lines should be spaced 10.5 feet (3 m) apart.

If an intermediate (choker) layer is used, a permeability value is first generated from both the rootzone and intermediate layer values using Luthin's equation. This apparent permeability value is then used in Hooghoudt's equation.

Apparent permeability

$$= \frac{[h_1 + h_2 + \ldots h_n]}{[(h_1/K_1) + (h_2/K_2) + \ldots (h_n/K_n)]}$$

$$= \frac{[12\text{ inches} + 2\text{ inches}]}{[(12\text{ inches} \div 13\text{ inches/hour}) + (2\text{ inches} \div 20\text{ inches/hour})]}$$

$$= \frac{14\text{ inches}}{(0.92 + 0.1\text{hour}} = 13.7\text{ inches/hour}$$

Therefore, water flowing at a maximum rate of 13.7 inches/hour (35 cm/hour) could enter a gravel layer typically 4 inches (10 cm) deep with an elevated percolation rate.

The 13.7 inches/hour value and 14 inches total are then inserted into Hooghoudt's equation as follows:

$$S = \sqrt{\frac{4Kh^2}{v}}$$

$$S = \sqrt{\frac{4(13.7\text{ inches/hour})(14\text{ inches})^2}{2\text{ inches/hour}}}$$

$$= 73\text{ inches(6.1 feet)}$$

As demonstrated, Luthin's equation typically yields more conservative values in terms of drain line spacing recommendations. This further demonstrates the economic savings of using properly sized rootzone and gravel layers, rather than an intermediate layer. Not only are there savings from deleting the intermediate layer, but also wider drain line spacings are typically allowed.

**Determining drain size and length.** Sizing drain pipe for a particular area requires a considerable amount of information, including proposed drain depth, slope, width, length, and spacing; average rainfall event (inflow rates); soil type; area to be drained; and surface slope. First, the amount of water to drain following a rainfall event needs to be determined.

### Example

Calculate the volume of water to be drained following an anticipated rainfall event of 2 inches/hour (50 mm/hour), using a 2-inch (50 mm) pipe with 1% slope and a drain spacing of 10 feet (3 m) [1 cubic foot (0.028 $m^3$) = 7.5 gallons (28 L)]. Every foot (0.3 m) of trench should collect:

$$10\text{ feet} \times 1\text{ foot} \times 2\text{ feet} \times 1\text{ foot/12 inches}$$

$$= 1.7\text{ cubic feet/hour} \times 7.5\text{ galons/cubic foot}$$

$$=\sim 13\text{ gallons/hour or } 0.21\text{ gallons/minute(0.79 L/minute)}$$

The next question is how long of a run this 2-inch pipe can handle. Each linear foot of pipe should collect 0.21 gallon/minute (0.79 L/minute) and the maximum drainage rate the pipe can handle is 7.9 gallons/minute (30 L/minute) (this is from the manufacturer's specification for 2-inch (50 mm) pipe with 1% slope). Therefore, the 2-inch pipe's effective length is 7.9 gallons/minute ÷ 0.21 gallon/minute = 38 feet (11.4 m). A collector (lateral) drain would be needed after a maximum 2-inch (5 cm) pipe run of 38 feet (11.4 m).

In a similar scenario, a 4-inch (10 cm) diameter drain pipe is used instead of 2-inch (5 cm) pipe. According to the manufacturer's specifications, the 4-inch pipe has a maximum flow rate on a 1% slope of 0.85 gallon/second (or 51 gallons/minute). Therefore, the 4-inch pipe's effective length would be 51 gallons/minute ÷ 0.21 gallon/minute = 242 feet (72.6 m) compared to 38 feet (11.4 m) for the 2-inch (5 cm) diameter pipe. A 4-inch

(10 cm) drain pipe is the current standard for most golf greens, and it is more than sufficient to handle most rainfall events. Lateral lines are typically increased to 6 inches in diameter to handle the total output of a draining green.

## Drain Line Types

A wide array of drain line types, sizes, and configurations is available. A common misconception is that all of these work equally well. Unless the drainage line provides lower water potential than the surrounding soil and the pipe is laid on grade in the bottom of the trench, water will not efficiently enter or move down it. A suitable outlet is also needed to remove drained water.

**Corrugated pipe.** In the past, drainage lines were built from agricultural clay tile, concrete, or flexible corrugated plastic. Today, 4-inch (10 cm) diameter corrugated high-density polyethylene (HDPE) pipe with perforations (slits or holes) is the industry standard, due to its ease of installation and relatively low cost (Figure 4.27). The perforations in HDPE pipe are laid facing downward on a bed of gravel to prevent clogging by downward-migrating soil particles. Single-wall corrugated pipe is flexible and easier to work with but is not as strong as double-wall pipe. Installers recommend a minimum of 20 inches (51 cm) of backfill to prevent crushing of single-wall pipe. Double-wall corrugated pipe can be safely buried with a minimum of 12 inches (30 cm) of backfill. Double-wall drain lines have smooth inner walls, which provide more efficient water removal than lines with corrugated inner walls but then have a quicker discharge rate.

Water enters the bottom of the pipe(s) through slits or holes fabricated by the manufacturer (Figure 4.28). Water moves down through the medium and stops at the bottom of the trench. As the water level rises in the trench, it moves into the pipe through the holes and then moves down the pipe. An alternative to corrugated HDPE pipe is rigid PVC pipe with two rows of holes drilled adjacent to each other over the length of the pipe for drainage water to enter. PVC pipe is generally more expensive but is able to withstand heavier traffic pressure than flexible HDPE pipe.

**Strip drains.** Strip drains exist where a narrow (2 to 4 inches [5 to 10 cm] wide) trench is dug, a 1-inch (2.5 cm) wide perforated drain (or "strip") sleeve is installed, and trenches are backfilled with sand (Figure 4.29). Drain types used include 1-inch (2.5 cm) wide cloth-wrapped waffle- or honeycomb-shaped drain sleeves, vertically stacked small diameter pipes (composite drains), or other similar narrow sleeved material. Drains are placed at the bottom of the trench and extend about halfway to the soil surface (Figure 4.30). Water enters the lowest pipe of a composite drain first. Once this pipe is filled, water will enter and flow through the next lowest, and so on. Narrower trenches are used for this stacked pipe, reducing the costs of trenching and fill material. However, the smaller diameter pipe means greater surface area is contacting the water; thus, there is more friction loss and lower water-carrying capacity. This makes stacked pipe less efficient for water removal and more prone to clogging than single, larger tile lines.

Strip drains allow an increase in water infiltration rates into the soil surface without complete renovation of the native soil profile. However, in many cases the trenches filled with gravel and sand on grade actually work as "dry wells," where they lower the water table, removing surface water. The drain lines are not directly involved in this water removal. Drain lines must be installed deep enough to avoid disruption or displacement by heavy equipment.

A secondary tier of sand-only slit drains may also be installed at 0.5- to 1-foot (0.15 to 0.3 m) spacing at right angles to the mini-pipe slit drains to provide a more intensive means for water interception. In this system, the sand-filled trenches drain away some surface water without it having to traverse across the field.

**Slit drains.** A modification of the slit drain is a **French drain**, which involves a trench 4 to 8 inches (10 to 20 cm) wide dug on a 1% to 3% slope, with a drainage pipe laid at its bottom and backfilled with sand (Figure 4.31).

**FIGURE 4.27** High-density polyethylene (HDPE) pipe used for subsurface drainage.

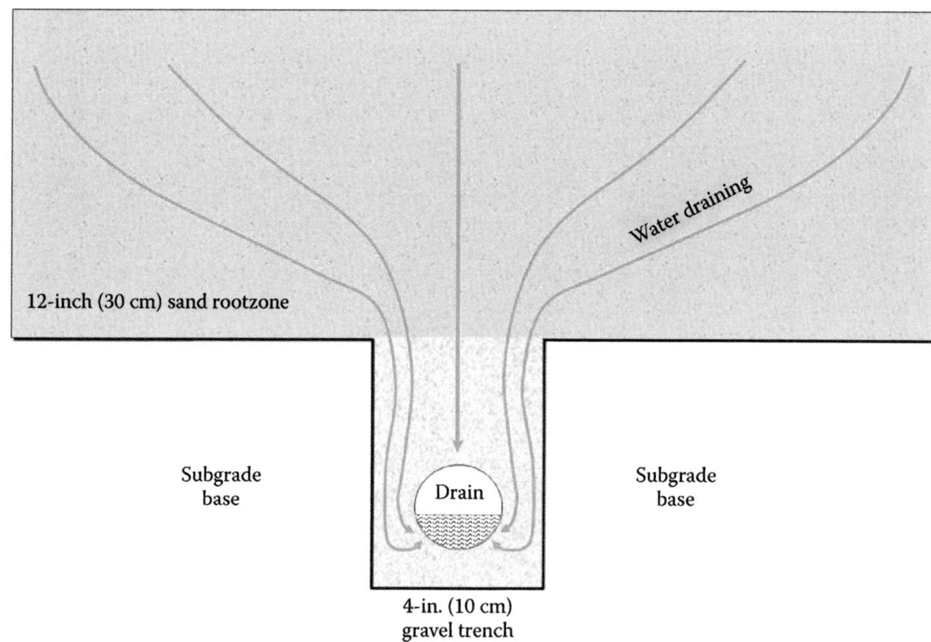


**FIGURE 4.28** Illustration of soil water movement toward a subsurface drain line.

**FIGURE 4.29** Strip drain where smaller diameter pipes are stacked, allowing narrower trenches to be dug.

This drainage system provides a wider trench than slit drains, which extends life expectancy and potentially drains a larger area. Additional fill material is necessary for this system and more surface area disruption occurs. However, many facilities have ready access to the trenching equipment and the wider trench is usually easier to work with. In heavy soils, the wider trench filled with sand easily desiccates, shows nutrient deficiencies, and may become more susceptible to low-temperature damage (Figure 4.32).

Advantages of slit and strip drains include installation with minimal surface disturbance and the need for less labor, as these steps are largely mechanized. They also provide drainage to poorly designed and constructed facilities at an attractive cost compared to complete soil profile renovation (Figure 4.33). Sufficient surface slopes must be present for lateral water movement to these drains. Gravel and possibly small pipe in the bottom of the slits also are needed to expedite water removal from the site. Traffic from play, soil migration, erosion, and aerification can also cause glazing of silt and/or clay over the sand, reducing water infiltration. Topdressing with finer sand or soil used to fill the trench can also clog pore spaces and disrupt water infiltration. Heavy equipment can crush or disturb the integrity of the drain. Smaller diameter (1 inch, 2.5 cm) pipes used in the strip drains are also prone to eventual clogging by downward-migrating soil particles. These drains typically require closer spacings than traditional lateral drains; thus, more pipe junctions are needed.

However, the complex design of the drains and golf course personnel turnover in management often lead to eventual disturbances of the drain's integrity. Due to these inherent limitations, strip or slit drains should be viewed as a technique for removing the symptoms of waterlogging rather than as a technique for curing the problem. They also have a finite life span and often are a means of providing a temporary fix that will have to be performed repeatedly to succeed continually. They generally are not viewed as a substitute for sound surface and

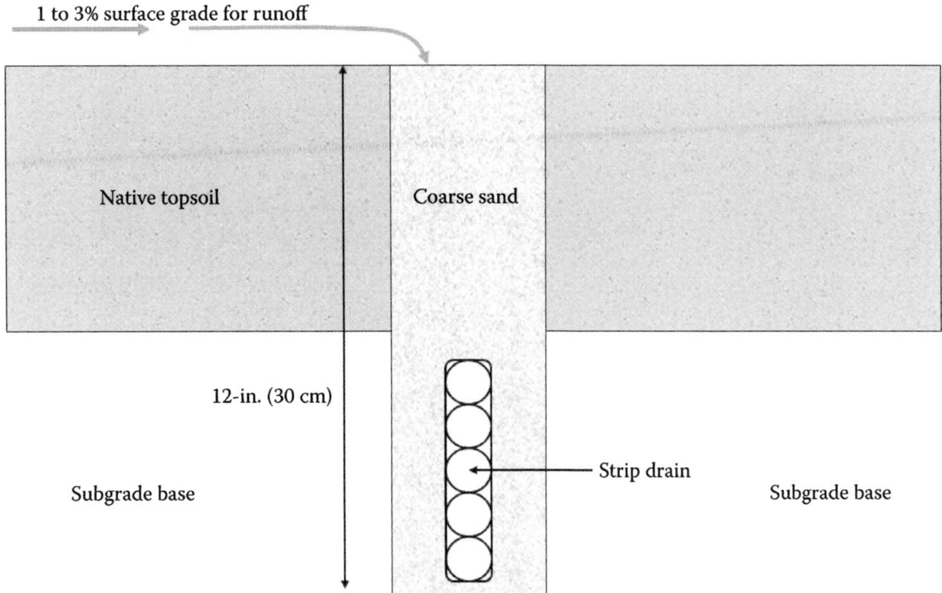


**FIGURE 4.30** Illustration how strip drains intercept surface runoff.

**FIGURE 4.31** Slit drains where a trench is dug, drainage pipe installed on a grade, and backfilled with sand.

subsurface drainage planning and installation, which should occur prior to construction.

**Cell system.** A modification of the sand-slit drainage method is installing a narrow, 3-inch (7.6 cm) wide trench and inserting a 2-inch (5 cm) diameter perforated, corrugated pipe that is often wrapped with geotextile fabric. The trench is then backfilled with sand (Figure 4.34). Line spacings are often 6 to 8 feet (1.8 to 2.4 m) on sand-based greens or contoured areas or narrower spacings (5 to 6 feet, 3.7 to 1.8 m) in heavier soils. The pipe should be on a 0.5% to 1.0% slope and slits are often "recharged" every couple of years by removing the turf above the slit, and new sand is inserted to the depth of organic matter accumulation.

### Additional Drainage Considerations

**Filter cloth.** Nylon-netted filter drainage sleeves, which are wrapped around perforated pipe to prevent soil particles from impeding water flow into drainage lines, are also available. There is much debate over the probability of these nets/sleeves clogging over time from soil particle movement toward the drain. If excessive (>5% to 10%) silt and clay are present in the topsoil, these drainage sleeves may clog. In this situation, filter cloth should be considered to line the drainage ditch but should not be physically wrapped around the individual drain lines. It is also believed that this cloth can become clogged from bioproducts from algae and other organisms that may colonize the perpetually wet cloth.

**FIGURE 4.32** Turf desiccation and nutrient deficiency over sand-filled slit drain trenches.

**FIGURE 4.33** Narrow slits backfilled with sand to temporarily provide surface drainage.

**FIGURE 4.34** Cell system drainage where small-diameter drains are installed in narrow trenches.

A suitable alternative is a knitted or nonwoven geotextile fabric.

**Drainage line patterns.** Typically, a gridiron or herringbone pattern is used for drainage line arrangement (Figure 4.35). The drainage pattern should be designed so that drain lines are placed nearly perpendicular to the slope and rotated downhill as required to drain. However, any pattern is acceptable as long as each line has a continuous downward slope and water in golf greens does not have to travel more than 10 feet (3 m) to a drainage line. An additional lateral drain line is placed at the furthest downslope location of the green, adjacent to the perimeter of the green. This perimeter drain (referred to as a "smile" drain) helps avoid wet areas where the modified greens sand meets the native soil.

**Drainage line trenches.** Trenches where golf green drainage lines are to be laid should be cut a minimum of 6 to 8 inches (15 to 20 cm) in depth into the subgrade and 5 to 6 inches (12.7 to 15 cm) in width (Figure 4.36). In native soil, 3- to 4-foot (0.9 to 1.2 m) deep drain lines are sufficient. Lines less than 2 feet (0.6 m) deep become subject to damage or disruption by heavy machinery or excessive traffic.

The bottom of the trench should be a minimum of 2 inches (5 cm) wider than the outside diameter of the pipe. Trenches up to 12 inches (30.5 cm) wide have been utilized. However, more gravel is needed to fill the wider trenches, which increases cost. Normally, a drainage line trench should be no more than twice the width of the drain pipe. A 5- to 6-inch (12.7 to 15 cm) wide "U"-shaped trench will allow

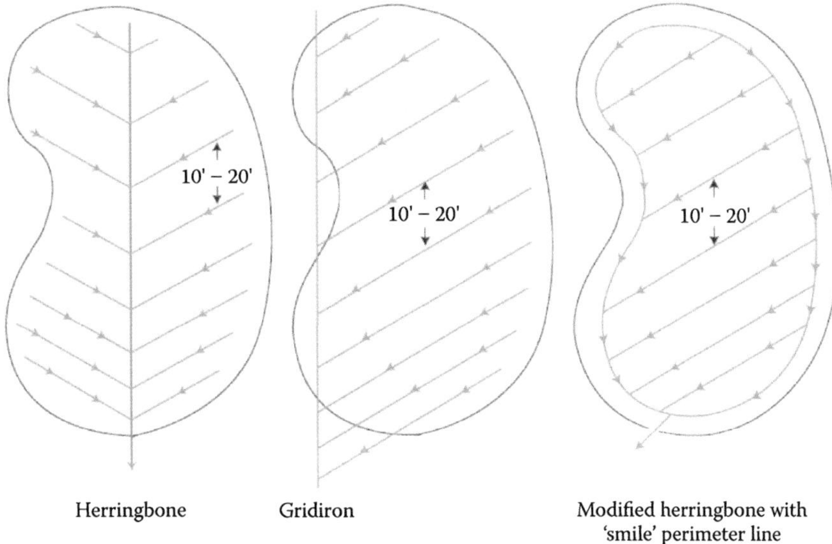


**FIGURE 4.35** Common drainage tile patterns used on golf greens where water will not have to travel more than 10 feet (3 m) before encountering a drain.

**FIGURE 4.36** Installing drainage lines on a golf green and backfilling with gravel.

for a half- to 1-inch (1.27 to 2.54 cm) bed of gravel to be placed around (below, above, and on either side of) a four-inch (10 cm) diameter drain line to reduce washing of subgrade soil into the drain line. The soil displaced by digging the trench should be removed or placed between drainage lines to provide a slight slope toward the trench and then compacted.

Prior to digging trenches, the area should be surveyed. Proposed trench lines should be staked and labeled with the desired depth of cut. Drain lines should not be placed any deeper than necessary to obtain the desired slope. Trenches should have a minimum downward slope of 0.5% (1 foot of drop for every 200 feet, 0.3 m per 60 m) and a maximum slope of 4% (1 foot of drop for every 25 feet, 0.3 m per 7.5 m). Slopes of 1% to 2% (1 foot of drop for every 100 feet, 0.3 m per 30 m, to one foot of drop for every 50 feet, 0.3 m per 15 m, respectively) are ideal. Drain lines with slopes of 0.5% or less are difficult to properly grade, install, and maintain due to the slight elevation changes and slow flow rates. Drain lines with slopes greater than 4% will lose lateral drainage capability. Steeper slopes also require greater elevation changes within the drain line and a lower outlet point.

When establishing the subgrade of a drain system, it is best to start at the outlet and establish the grade of the main collector line. After establishing this main line grade, the grade of each lateral can be determined. Care must be taken to ensure the drainage trench; drain lines always slope downward to avoid any entrapment or collection of water along the drainage lines. If a section of pipe is lower than the section closer to the outlet, water will pool in the lower section. This causes any sediment in the water to settle and collect in the bottom of the pipe, eventually clogging and slowing drainage. Grades of all main and lateral drainage lines should be checked with a level prior to backfilling.

Drain line ends and joints should be covered with asphalt paper, fiberglass composition, or plastic spacers or caps to prevent gravel and/or soil from entering the drain lines. It is common for the main drain lines into which the lateral feeder lines flow to have their upper end extended to the soil surface and properly capped. If these lines become clogged or contaminated from gravel, soil, or some other material, the cap can then be removed and lines flushed. This can greatly extend the useful life of the drainage system without the need of disturbing the playing surfaces to clean individual lines. After the drain line is laid, trenches should be backfilled with gravel and care taken not to displace the slope or joint covers of the drain tiles. An alternative design element is to have the main lateral line daylight into a catch basin. This allows for periodic flushing of the line and also allows air entry into the line for more efficient drainage.

## FAIRWAY DRAINAGE

Surface drainage, as discussed previously, is the first and quickest means of removing excess surface moisture. However, in areas that do not surface drain sufficiently, subsurface drainage is often used, which involves an underground system of perforated pipes. When a soil's water table approaches the soil surface, then waterlogged conditions occur. The underground system of perforated pipes intercepts this groundwater as the rising water enters pipes from below. The pipe system then routes this water to an outlet and maintains the water table at a reasonable depth below the soil surface.

### Subsurface Drainage Design

Subsurface drain lines can be installed as either a singular system or as composite systems (Figure 4.37). A singular system consists of an array of individual drain lines, each emptying into an outlet. Composite systems consist of laterals connected to a common main line.

Similarly to golf greens, fairway subsurface drainage design can have a variety of patterns such as gridiron, herringbone, or random. Gridiron and herringbone patterns are used to drain larger areas; random drains are used when small localized areas need drainage while areas in between are satisfactorily drained. A gridiron system is used to drain an area

**FIGURE 4.37** Installing drainage lines on a fairway.

with a uniform slope in one direction and a herringbone system is generally best used to drain an area with a swale near the center.

With each design, the main drainage lines should generally follow natural valleys and be perpendicular to the contours. Lateral drain lines are often laid across the slope with a gentle downward grade of 0.5% to 2.0%. These drains intercept subsurface interflow that usually moves perpendicular to the contours. Lateral drains should maintain a sufficient uniform grade while keeping the laterals at a consistent depth from the soil surface. Lateral lines typically are from 2 to 2.5 feet (0.6 to 0.8 m) deep. Spacing varies from as much as 25 inches (64 cm) up to 30 feet (9 m) on highly permeable sandy soils to as little as 15 feet (4.6 m) on less permeable soils such as clays and silt loams. Hooghoudt's equation, as discussed previously, can be used to determine the drain tile spacing or hydraulic conductivity needed for a particular drain spacing design. Modifications of Hooghoudt's equation are available for designing fairway subsurface drainage systems where a drainage coefficient is used to estimate water loss from a soil profile and is then multiplied by the area and converted to the desired units. This provides the outflow volume of drainage, which allows one to then choose the appropriately sized pipe to carry this flow using a drainage pipe capacity chart.

## Interceptor Drains

Surface drainage from areas adjacent to golf course fairways, such as parking lots, hills, or adjacent fairways, often becomes problematic. Water infiltrating into the soil can either continue to move downward to eventually recharge the groundwater or can move laterally through the soil down a hill and is referred to as interflow (Figure 4.38). Interflow is the major source of water for stream and pond recharge during periods between rains and slows considerably near the bottom of a hill. Wetter soils near the hillside base often occur and result in a **seep**. Attempts to drain seeps by installing subsurface drainage typically fail since the source of the seeping water remains unchecked. Usually this water is easily collected by installing surface cutoff (or interceptor) drains to collect the water at the bottom (or "toe") of slopes, prior to entering the playing surface, or by diversion using surface terraces (or swales) (Figure 4.39). Interceptor drains consist of a gravel- or coarse sand-filled trench cut along the contour and perpendicular to the overland flow.

Sloping water tables are found in slightly rolling, hilly, or mountainous areas. The free groundwater in these areas will flow in the direction of the slope, usually along an underlying impervious soil layer. Precipitation on the soil surface percolates downward until it encounters this impervious layer and then flows laterally over this layer. The most likely place for a water table (seep) to appear at the soil surface is near the intersection of a steep and a flatter slope. This is a common problem on golf courses, such as when an elevated green meets the surrounding land area. Wet seep areas are also common on approaches where the fairway slopes downhill toward the green, which is slightly elevated. Here the approach may be wet from irrigation water being retained in the green base material, and a seep may occur in the same approach area from a surfacing water table on the fairway side. Interceptor drains are placed in these situations where the free groundwater of the hill meets the flat area to intercept the water flowing on the slowly permeable subsoil layer before it appears on the soil surface.

Determining placement of an interceptor drain can best be performed by digging test holes or miniature wells (called piezometers) when most of the area is dry enough to use, but the seep area is still wet. Piezometers are small-diameter pipes driven into the subsoil so that no leakage occurs around the pipes and water entrance is only from the open bottom. This indicates hydrostatic pressure of groundwater at the specific point in the soil. The piezometers should extend, in a grid pattern, upslope from the seep area to a depth of

**FIGURE 4.38** Uncontrolled water flow across a fairway during intense rainfall.

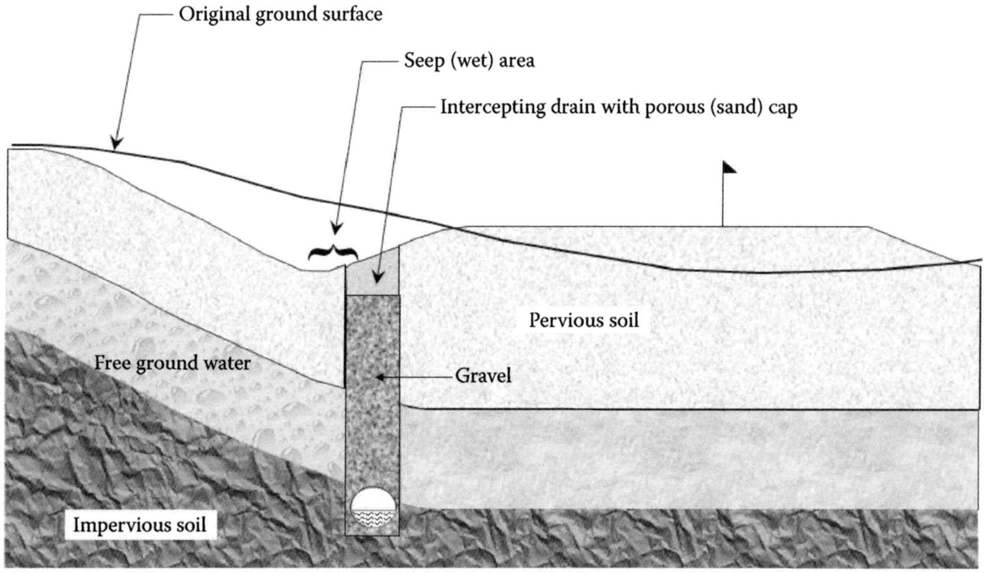


**FIGURE 4.39** Illustrating an intercept drain to prevent a seep area forming at the bottom of a grade.

2 to 3 feet (0.6 to 0.9 m). By observing the water level in the piezometer holes 24 hours after they have been dug, the depth to the water table or water flowing over the impervious layer in the ground can be determined. The trench should be dug to ~2.5 feet (0.75 m) deep to extend below the water table. To facilitate drainage, the trench should be backfilled to the depth of the water table with gravel. If the water table intersects the soil surface, additional drains may be necessary. If not, additional interceptor drains may be needed further down the slope.

The bottom of drain trenches should be uniform in slope to prevent depressions and should have a minimum slope of 2% (1:50) if a pipe is not placed at the bottom of the trench. Placement of a pipe in the trench allows grades as low as 0.5% (1:200). This allows quick removal of surface water and helps prevent ponding, wheel depressions, and trash accumulation. Fairway drains should be spaced no more than about 45 to 50 yards (41 to 46 m) apart. Mowable drains or graded drains are ideal to minimize maintenance requirements and to facilitate play.

### Springs

Springs are weak points in the soil strata where groundwater is under sufficient pressure to allow surfacing of the water. Springs are drained by placing a perforated drain pipe directly in the actual spring head to a depth of about 2.5 feet (0.76 m) and 5 to 10 feet (1.5 to 3.0 m) beyond it and filled with gravel to facilitate water entry into the drain.

### Outlets

Water intercepted by surface and subsurface drainage designs requires a suitable outlet to discharge this flow—typically into channels, streams, or lakes (Figure 4.40). If the outlet is inadequate, the effectiveness of the entire drainage system can be reduced. Outlet types include the classical outlet or extension of the subsurface drainage pipe to the discharge location, pumped outlets, siphon outlets, dry wells and subsurface reservoirs, and wetlands.

With classical outlets, the location of the drainage pipe outlet must be at the low point of the drainage system. Efficient drainage system design requires identifying the outlet location for an area and then extending the drainage system array upslope from this location. An adequate slope must also occur along the entire run of the system, along with adequate soil cover as a protection from crushing; excessively deep excavations are avoided.

The drain outlet often is the weakest portion of a drainage system since it is exposed and subject to damage or clogging. To prevent this, extending the tile or plastic tubing directly to the discharge point should be avoided. A section of non-perforated plastic or metal pipe 10 to 15 feet (3 to 4.5 m) in length is used to carry the water from the point where sufficient soil cover is available to the discharge to avoid crushing that may occur if insufficient cover is present to protect the pipe. A concrete collar is placed at this intersection of pipes to prevent pipe displacement. The outlet pipe should be the same size as or larger than the main discharge line at the collar and should discharge at least 1 foot (0.3 m) above the normal water level in the receiving waterway. If flooding periodically occurs, the outlet pipe should be equipped with a floodgate to prevent water backing up into the pipe. Otherwise, the outlet pipe should be covered with a wire mesh to prevent animals from entering it.

Sometimes a pump and siphon outlet is necessary if a gravity outlet is unavailable or the area to be drained is completely contained with a large depressed area. A pumped outlet consists of an automatically controlled pump with float switches set to start and stop levels and placed within a

**FIGURE 4.40** Outlet of a subsurface drainage system over rocks.

small sump to provide some degree of active water storage. A siphon outlet is when the entire drainage system is located in a depression and a sump contains a nonperforated, 2-inch (5 cm) siphon tube leading to a remote discharge location. As long as the entrance and exit of the siphon pipe remain under water, the tube can convey water across higher elevations than the location of the sump or relief point. This system w best for relatively flat areas that are connected to an irrigation line so that it can be primed and occasionally flushed.

**Dry wells.** Dry wells are holes dug into the ground at the end of a drain line that are used to receive normal drainage water from relatively small areas (Figure 4.41). They are used when discharge locations are too far to trench and pipe. Dry wells are usually buried beneath the soil surface at the lowest point of elevation of the surrounding area. They usually consist of a surface inlet emptying below into a bed of gravel or rock. Sufficient gravel or rock along with sufficient depth and width of the dry well will be needed for the surface water percolation. Stormwater wetlands are constructed systems designed to mitigate downstream impacts of stormwater quantity and quality by temporarily storing drainage waters in shallow pools and marshes. Drainage design specialists consider these and other options when planning stormwater and normal surface and subsurface drainage systems.

High water table. High water tables are often adjacent bodies of water and can become a problem if the table reaches

**FIGURE 4.41** Use of a dry well to receive normal drainage water from a relatively small area.

within 12 to 16 inches of the surface. These cause chronic wet areas that encourage anaerobic conditions, promote water-loving plants such as sedges, and reduce playability of the area (Figure 4.42). Installing piezometers indicates if a high water table exists, as water will seep into the bottom of the piezometer hole. If a high water table exists, the adjacent body of water will need to be lowered, the saturated water table lowered using subsurface drains, or the wet area elevated.

To lower a high water table, subsurface drains will need to be placed into the soil 2.5 to 4 feet (0.76 to 2.2 m) deep. A suitable outlet also is needed for the subsurface drain lines and, if not available, the water should be collected and pumped to one.

## Sand Capping

"Sand capping" is a procedure where various depths of sand are added to low lying, poorly drained areas in an attempt to allow quicker play and maintenance to commence following rain (Figure 4.43). Sand capping basically raises the turf above a site with a high water table or insufficient elevation differences between drain tile and an adjacent outlet. Depths used depend on budget constraints, field expectations, anticipated annual rainfall, rootzone mixture used, if a gravel bed is used, if drain lines are installed, and if so, their spacing. Typically, these depths range from 6 to 12 inches (15 to 30 cm). Although several means of determining appropriate sand capping depth can be used,

**FIGURE 4.42** Silt and clay deposit adjacent to a pond, reducing playability and increasing maintenance of the area.

**FIGURE 4.43** "Sand-capping" a fairway to allow quicker play following a rain event.

optimum rootzone depth combines maximum perched water table depth (air entry) with adequate depth of drained sand for root growth. From soil moisture retention curves, two commonly used methods are

1. Aeration porosity values are determined and rootzone depth (tension) is at
   a. 10% Aeration (air-filled) porosity (minimum)
   b. 15% Aeration porosity (better)
   c. 35% Aeration porosity (best)
2. Following the depth of air entry point (top of perched water table) of the sand:
   a. To this value, add 4 inches (10 cm) for golf greens
   b. To this value, add 6 inches (15 cm) for fairways

Consult a qualified soil scientist to run soil moisture retention curves for the sands being considered and to discuss how to appropriately interpret the results. Refer to McCarty, Hubbard, and Quisenberry (2006) for additional information on this and other topics.

# Section III

## Golf Course Construction and Establishment Practices

# 5 Golf Course Construction and Renovation

## ENVIRONMENTAL GOLF COURSE DESIGN

The construction or renovation of a golf course is a tremendous undertaking, requiring the expenditure of significant time and capital. Over the years, however, far greater sums will be spent on upkeep and maintenance, especially if the course was initially designed and/or constructed poorly. It is best to do the job correctly initially to ensure the course will not close in the future for renovation(s) to correct earlier mistakes.

Certain construction practices are expensive and necessary, but others are not needed and inevitably drive up the total cost. The difference between an average and a good golf course is often not so much a question of money as it is a question of time, dedication, and attention to detail in the design and construction phases.

Regardless of how the golf course construction or renovation is initiated, the overall objective remains the same—construct features (e.g., greens, tees, bunkers) that will distinguish the golf course, make it a challenging yet enjoyable experience to play, and allow it to be maintained in proper playing conditions at a competitive cost consistent with the expected return on the investment.

### Preplanning—Team Approach

The most effective approach to ensure a successful project completion is to form a team of experts. This team should include a golf course architect, engineer(s), landscape architect, water resource specialist, environmental specialist, and other consultants as dictated by the peculiarities of the golf course and the watershed in which it is sited. This team then works with the developer, community representatives, environmental groups, and regulatory agencies to determine the project goals.

### Property Selection

A critical consideration influencing all elements of construction is property selection. Instead of being liabilities, golf courses have transformed barren and unproductive land sites such as reclaimed strip mines, quarries, and sanitary landfills into community assets (Figure 5.1). Many of today's golf courses are the principal part of residential or resort developments with their own priorities for property selection.

Several principal considerations help determine whether a piece of property can be utilized for a golf course site: (1) economics, (2) acreage, (3) topography, (4) soil, (5) vegetation, (6) environment, (7) water, and (8) utilities.

**Economics.** In simplest terms, supply and demand should dictate if a particular site is economically feasible for a golf course. Market information to verify the need for a golf course, including such things as demographics, potential golfing demand, and existing golf accessibility, should be considered during this evaluation process. In addition, costs associated with land purchase and construction must be considered and entered into the economic equation to determine if a piece of property will provide an adequate return on the investment.

**Acreage.** Generally speaking, the more property available for a course, the better. An 18-hole golf course and practice facility generally requires a minimum of 150 acres (61 ha). Additional land is needed if extensive wetlands, steep topography, or other features preclude these areas from being utilized in the scheme of the golf course. The configuration of the property also is important for the routing of the golf course, as well as siting the practice facilities, clubhouse, parking lot, maintenance facility, and possibly swimming and tennis facilities. Given less property to work with, the golf course architect will have to "force" the golf holes on the acreage.

Housing development also is a reality surrounding many courses, and a balance on land allocation between the course and development is necessary. Many courses have been built with housing development as the primary consideration.

**Topography.** Topography is a very integral part of property selection for a golf course site because it gives the golf course its character by determining property access, traffic routing (flow), and the layouts of holes. It also contributes to the construction costs as large-scale earth moving is expensive and should be held to a minimum (Figure 5.2). If a golf course has more elevation change than 150 feet (45 m) between the highest and lowest point, it will be very difficult to walk. In addition, maximum visibility should be provided from the tee to the landing area, and from the landing area to the green, to ensure player safety as well as speed of play.

**Soil.** Soil and the related aspect of drainage are the next most important factors associated with property selection. Coarser textured soils (loam and sandy soils) are preferred for a golf course site over finer textured soils (clay). Fine-textured soils have inherent problems, making it difficult to maintain high-quality turfgrass. These soils are very pronc to soil compaction, especially under intense traffic, and soil compaction restricts and/or impedes water infiltration and internal drainage. Poor drainage delays golfers from playing following rain,

**FIGURE 5.1** Golf courses often turn unproductive land such as landfills (shown) into community assets.

**FIGURE 5.2** Shaping of a constructed golf course to minimize unnecessary elevation changes.

which reduces cash flow. Finer textured soils are also slow to thaw after being frozen, increasing the chances of low-temperature stress and injury and, again, delaying play. Generally, as soils compact and drain poorly, there is an increased need for cart paths. On a parcel of 150 acres (61 ha) or more, one expects to encounter a variety of soil types that create either problems or opportunities for the golf course architect and superintendent (Figure 5.3).

**Vegetation and wildlife habitat.** Native vegetation adds immensely to the beauty, character, and playability of a golf course. Generally, some select trees and shrubs have to be removed during construction. This can be an expensive undertaking depending on the density of the trees and shrubs. Laws in most regions require an environmental impact study to be conducted and approved prior to any large-scale tree or dirt removal. After approval, trees may be removed and sold to help offset the clearing costs, or burned where allowed.

**Environment.** In today's regulated world, even if all of the property criteria previously listed were ideal, it still may be impossible to build a golf course due to environmental restrictions. The most common environmental considerations or restrictions are the presence of wetland areas, floodplains, or endangered plants or animals (Figure 5.4).

Just as restrictions on the filling of wetlands have been imposed because of their ecological impact in the recharging of the groundwater supplies, other ecological and biological considerations may restrict the site as a golf course. Floodplains cannot be graded unless the work is determined not to increase the downstream flow of water. This means a golf course site located on a floodplain is subject to flood damage. Property considered a wildlife habitat also may have restrictions placed on its use by environmental agencies, especially those areas with "endangered" species. Property deemed to have archaeological or historical significance can also be deemed sensitive. Antigrowth factions frequently use environmental issues to challenge the use of a piece of property for a golf course. It is highly advisable and often required to perform a plant and animal inventory

**FIGURE 5.3** A variety of soils and topography are often encountered when constructing a golf course.

**FIGURE 5.4** Protecting endangered animals during golf course design and construction.

of the property prior to construction to track changes before and after the course is built. If planned correctly, animal populations and diversity often increase with the presence of a golf facility.

## Wetlands

A wetland is often defined as any area where the water table rises to within 6 inches (15 cm) of the surface for more than two weeks annually. Wetlands include natural bodies of water as well as marshes, bogs, and seasonally flooded lands (Figure 5.5). By definition, many wetlands may not be detectable to the untrained eye. Wetlands must accurately be located by a specialist to ensure the golf course site can be utilized.

In the United States, such areas are subject to the jurisdiction of the Army Corps of Engineers. Wetlands cannot be filled under present statutes, although in some states fills may be allowed if new wetlands are created elsewhere.

Natural and constructed wetlands and riparian zones may be incorporated into golf course design. These features, when fully integrated into the water management strategy, act as natural filters to remove nutrients, pesticides, and suspended particulates (e.g., soil, microorganisms) from runoff water. Many courses in coastal areas are fortunate to have an abundance of natural wetlands and riparian zones (Figure 5.6).

**FIGURE 5.5** Protecting wetlands during golf course construction and maintenance.

**FIGURE 5.6** Golf courses designed to protect natural wetlands and riparian zones.

These geographic features provide a critical wildlife habitat. In turn, the wildlife and the wetlands significantly contribute to the aesthetics of the course itself.

The key is to create a hydrologic design that begins with irrigation and rainfall; follows runoff from tees, fairways, and greens; and treats this runoff with vegetative filter strips, wetlands, and riparian zones. Collection of this treated runoff into ponds and lakes may provide a much-needed source of clean irrigation water for reuse on the course.

Wetlands can be located almost anywhere on a golf course. This utility facilitates the integration of a challenging hole design with water and chemical management. Wetlands can be incorporated into streams by adding control structures. Blocking the entire stream is a reasonable alternative only in low-order streams. This approach is usually not cost effective and is particularly vulnerable during high flow and flooding. An alternative would be to provide another channel for high-flow periods. This would preserve the integrity of the wetland during intense storm events and flooding.

*Riparian wetlands* are those adjacent to flooding streams. These wetlands periodically receive flood waters and, in natural systems, may be seen as bottomland hardwood forests. Forested riparian zones adjacent to small creeks and drainage ditches are extremely useful for water and chemical management. For example, runoff from a playing surface may be directed via drain pipe or shallow depressions to a riparian ditch. This ditch, with a gentle-to-steep slope depending on the terrain, then empties into a wetland immediately upstream from a lake or larger order stream. When the complete golf course hydrologic plan is designed, the use of several small wetlands instead of a few larger wetlands should be considered. There are several advantages to locating several small wetlands in the upper reaches of the golf course (but not in the streams themselves) rather than fewer larger wetlands in the lower reaches. A particularly useful design might be the construction of multiple small wetlands in the landscape to intercept small streams and drainage tiles prior to reaching the stream.

The size and shape of the wetland should be dictated by other phonographic features, such as slope. Short, wide wetlands might be appropriate for intercepting diffuse overland flow in areas with gentle slopes; long, narrow wetlands might be more appropriate for ditches, swales, and streams in areas with steeper slopes. In extreme cases of the latter, terraced wetlands placed into the watershed in a stair-step style are most appropriate.

### Audubon International Cooperative Sanctuary Program

Audubon International began its Audubon Cooperative Sanctuary Program (ACSP) in 1991 to "educate people about environmental stewardship and motivate them to take action in their daily lives which will enhance and protect wildlife and their habitats and conserve natural resources." The ACSP has devised four programs tailored for homeowners, businesses, schools, and golf courses.

A golf course can become certified in the ACSP in six areas:

1. Environmental planning
2. Wildlife and habitat management
3. Chemical use reduction and safety
4. Water conservation
5. Water quality management
6. Outreach and education

If a course successfully becomes certified in all six categories, it receives a Certified Audubon Cooperative Sanctuary designation (Figure 5.7).

Once a course is certified, it must be recertified after two years. This involves completing a case study, site assessment, and environmental planning. For further information on this sanctuary program, contact Audubon International at

Audubon International
46 Rarick Road
Selkirk, NY 12158
Phone: (518) 767-9051
Fax: (518) 767-9076
www.audubonintl.org

**Water.** Water is a key component of turfgrass growth and maintenance, and its accessibility and quality are critical in the property selection for a golf course site. A typical 18-hole golf course requires between 4,000 and 10,000 gallons/acre (37,000 to 93,500 L/ha) for turfgrass per day during summer. Sources include streams, creeks, rivers, ponds, lakes, drainage canals, wells, municipalities, utility companies, or a combination of these (Figure 5.8). Water quality biologists advise no more than 5% of the low-water flow be removed, or damage to fish and plants may result. Therefore, it usually is necessary to locate an underground supply of clean, salt-free water to supply irrigation. The use of effluent water (treated wastewater) from development and/or municipality sources is becoming increasingly common. If effluent water is considered, ensure an adequate supply is available and note if the course is required to use (or store) a certain amount on a daily basis.

Baseline data should be generated prior to any construction on all water sources entering and exiting the property. With the criticism that golf courses are major sources of environmental pollutants, this baseline data are necessary to indicate otherwise. Data should include coliform bacteria counts, $O_2$, pH, phosphate, nitrates, heavy metals, pesticides, temperatures at various depths, total suspended solids, and any other suspected pollutants.

**Utilities.** Electrical and fuel requirements are substantial for the construction and maintenance of a golf course. The availability of and/or distance to these sources should

**FIGURE 5.7** Audubon International currently provides golf course certification in several areas.

**FIGURE 5.8** Sufficient water quality and quantity are key requirements for golf courses.

be determined for a particular piece of property. The costs associated with connecting electrical lines can be substantial, especially if these costs cannot be offset by a surrounding development.

## Problem Avoidance

During the preconstruction conference, the following items should be individually addressed to avoid future confusion and resulting problems:

1. Review details of agreement.
2. Review plans and specifications.
3. Determine who will act as representative and agent to make decisions for contractor, owner, and golf course architect.
4. Locate all property lines and define limits of work.
5. Determine and locate all utilities on the site.
6. Define types of work and schedules of other contractors to be working on the site.
7. Determine schedule for electrical power to be installed to irrigation pumps.
8. Determine water source for irrigation system.
9. Inform all parties of the delays and destruction that weather may cause.
10. Submit materials for approval.
11. Set procedure for final inspection.
12. Determine when owner's responsibility to accept and maintain the golf course begins.
13. Schedule construction meetings in advance.

## Architectural Design

The selection of a golf course architect is one of the most important aspects in the successful construction of a golf course. The golf course architect's primary task is to route the golf holes to take greatest advantage of a given piece of property's assets and blend them with the natural surroundings. One, therefore, should spend considerable time in evaluating and selecting a golf course architect and not make hasty decisions. A team of qualified professionals led by the architect should be assembled to perform a thorough site analysis to address any complex planning, design, or construction issues.

**Golf course design.** The architectural plan usually revolves around using the natural topography to make interesting golf holes (Figure 5.9). The first step in this process is to look on topographical maps for suitable putting green sites and fairway landing areas. When evaluating a site, the architect emphasizes contour lines of the topographic map to optimize the land's natural features. For instance, if the site is relatively flat, the architect focuses on areas where contour lines are close together to maximize as many of the limited natural features as possible. In contrast, if the property is hilly and/or steep, the architect looks for widely spaced contour lines that indicate flatter areas for possible landing sites. A slope of more than 5% on a putting green or 10% on a fairway is considered excessively steep, and it is extremely difficult to blend landing areas with adjacent slopes through grade leveling. Putting green sites generally are located on natural plateaus, at a natural saddle in the ground, at the crest of a knoll, in a punch bowl or valley, or carved out of a slope. Once a green site is discovered, a "natural" golf hole may be possible by tracing back along a visible line of approach to a suitable point for either the tees of a shorter hole or a landing area of a longer hole. Obviously, not every hole will feature a natural green site or natural landing area. The idea is to discover as many "natural" holes as possible with at least one natural feature to distinguish each from the others (Figure 5.10). Usually, however, the "natural" golf holes are disjointed and occur randomly across a piece of property. The key is to connect the golf holes into suitable loops of either 9 or 18 holes.

**Clubhouse site selection.** Once the golf holes have been looped together, the architect must select an appropriate clubhouse location site. Several factors should be considered in this site selection: (1) access to the golf course; (2) room for facility expansion, if needed; and (3) the ability to accommodate at least two starting golf holes, two finishing holes, a practice range, a parking lot, an entrance road, and access to

**FIGURE 5.9** A golf hole adjacent to an interesting natural feature such as a waterfall.

**FIGURE 5.10** Incorporating natural scenery in a golf hole design.

fuel and power supplies. Other amenities such as swimming pools, tennis courts, and health clubs also may be desired and will need to be accounted for. Most often, the best clubhouse sites are near the middle or eastern half of the property due to golfer problems associated with the rising and setting sun.

**Irrigation design.** Another aspect of golf course design requiring special attention is irrigation. An irrigation system must provide sufficient water to sustain turfgrass during periods of drought. In temperate climates, the design may be as simple as a single irrigation line running down the middle of each fairway with regularly spaced sprinkler heads. However, in steeper terrain or more sophisticated golf course layouts, proper irrigation design is essential to ensure the golf course is evenly and accurately watered. In arid or subtropical climates, the irrigation design must allow all turfgrass areas to receive adequate coverage (Figure 5.11).

Irrigation design, as well as sprinkler head selection and location, depends on a number of local parameters. Undulating terrain, varying soil types, prevailing wind patterns, tree or building location, and south-facing slopes are all-important microenvironment considerations influencing irrigation systems. A typical fairway irrigation system is designed with two rows of heads spaced ~65 feet (20 m) apart to form an equilateral triangle that provides head-to-head coverage. Swing joints are used under each irrigation head (Figure 5.12). Isolation valves also should be strategically placed throughout the golf course to allow one section of the system to be turned off while repairs or additions are made on another section. Other valves needed on some golf courses are drain valves and pressure reduction valves. Drain valves are installed on low spots of the course to purge the water system during freezing temperatures. In areas of extreme topographical

**FIGURE 5.11** Extended drought often indicates the quality of an irrigation system. (Courtesy USGA Agronomists.)

**FIGURE 5.12** Using swing joints with irrigation heads to avoid pipe breakage from traffic.

elevation change, pressure reduction valves may be necessary to counter the added line pressure caused by gravity.

Modern irrigation systems also integrate radio controllers, lightning protection, and computers to allow maximum watering efficiency, pump power, and system longevity (Figure 5.13). Satellite controllers that control a "zone" or group of sprinklers are often connected to these main computers to allow microcontrol of irrigation needs. Other water-saving techniques include integrating soil moisture sensors and weather stations into the design. These monitor moisture

**FIGURE 5.13** Modern irrigation systems integrate radio controllers, lightning protection, and computers for maximum watering efficiency, pump power, and system longevity.

levels, wind speeds, rainfall, and evapotranspiration rates to closely match turf water needs in relation to soil moisture levels and weather conditions. Low-volume, low-pressure heads that use about one-half the pressure and water volume of larger heads are being implemented, as well as using partial turning heads on the perimeter of the property. With the ever-increasing technology and sophistication of products, it is wise to hire an irrigation designer that specializes in this field.

### Developing the Design

The sum of all reconnaissance and analysis should be a series of maps illustrating existing roads and property boundaries; water sources for both irrigation and consumption; topography; sensitive wildlife habitat; potentially high erosion areas (steep slopes); wetlands and required buffer areas; drainage patterns including floodplains, vegetative cover, historical or archaeological sites, rights-of-way or easements; utilities including power and sewer; scenic views and vistas; adjacent land uses; and other information critical to planning and designing the golf course. It is critical, at this point, to be intimately familiar with all local and state rules and regulations governing the construction and management of the golf course. This will help establish realistic goals and produce the most efficient planning and design. In particular, this approach will avoid costly revisions and delays during the review, permit, and construction process.

A well-balanced landscape design for the golf course results in a mix of shrubs, trees, grassy areas, and water features that sustain and encourage wildlife and plant diversity. This design should balance the correction of poor drainage and erosion with the need to maintain wetland habitats.

Wetland habitats are particularly critical, not only for wildlife but also for water treatment, processing, and storage. Whenever possible, golf courses should be designed so irrigation and stormwater runoff move from the edges into the middle of the course (Figure 5.14). Drainage ditches should be bisected by small swales or even natural or constructed wetlands. These geographic features slow down water and allow for assimilation of both nutrients and pesticides by vegetation. These geographic features can be picturesque playing hazards that make the course more challenging.

When considering enhancing wildlife and their habitats, the core requirements include:

- Providing adequate space
- Providing appropriate and sufficient food sources such as feeder boxes, food gardens (plots), or landscapes (Figure 5.15)
- Cover or protection such as nesting boxes, natural (or no-mow) areas, wildflowers and native grasses
- Sufficient quality water planted with wildlife-friendly plants, sized and shaped properly, and surrounded with "no spray or treatment" buffers or zones (Figure 5.16)

In many locations, water quantity and quality may be the limiting factors in golf course development. A wise water-use plan might include the recapture and reuse of irrigation water as well as the use of secondary treated effluent water from a municipality or surrounding housing development for irrigation. Wetlands are extremely important for both their water-holding and water-purification capacity. These features will increase water recapture by the golf course as well as help alleviate fears by homeowners over the use of treated wastewater (effluent) on the course.

**FIGURE 5.14** Golf course design so irrigation and stormwater runoff is centralized.

**FIGURE 5.15** Golf courses provide desirable habitats for a number of animals and plants.

**FIGURE 5.16** Golf courses often provide extensive wetland protection with nontreated buffer zones.

## CONSTRUCTION

Good golf courses are the result of (1) careful, detailed, and organized design and planning; (2) good communication among all parties; and (3) attention to construction details and management. This occurs from detailed contracts and specifications.

For those undertaking new construction, it is important to have an overall understanding of the steps and processes required to build a golf course. Construction is not a simple series of distinctive steps, but instead is a combination of overlapping steps performed on each successive golf hole (Table 5.1).

The procedures outlined in Table 5.1 are followed on each golf hole, but the holes are not normally constructed individually or numerically. The most effective construction operation maximizes the management and overlapping of these steps, creating the most efficient use of equipment and other resources. The length of time to complete these steps typically ranges from 150 days to two years.

Scheduling of the construction process is very important, with construction and shaping of features and turfgrass establishment ideally based on the optimum date for establishing the turfgrasses in a given climate. It is very important to recognize that it may require a month or longer to "grass" all 18 golf holes; therefore, the target date for planting turfgrasses should consider this. In addition, sufficient time should be allowed for delays due to inclement weather conditions, especially if the golf course is being built on poorly drained soil or potentially highly erodible soil sites.

### Construction Documents

Once the planning and design processes have been completed, the construction phase of development is initiated. The environmental issues concerning construction will have been addressed during the design of the golf course. The construction documents will vary depending upon the architect and local regulations, but typically include:

1. Only qualified contractors experienced in the unique needs of golf course construction will be used.
2. Staking plans to locate the key points of the golf course (tees, landing areas, and greens) in the field for review and construction. This should emphasize protecting water sources, reducing wildlife and plant disruption, minimizing topsoil loss, and avoiding environmentally sensitive areas.
3. Erosion control and stormwater management plans show the location of features and methods of controlling stormwater and erosion on disturbed areas of the site during construction. Sediment traps and basins are constructed and established with plant material to provide soil stabilization.
4. Clearing plans indicate the limits of clearing necessary for golf course construction. Specimen trees to be saved or areas of vegetation to be preserved will be shown on this plan or designated in the field.
5. Grading and drainage plans outline the construction of the golf course and the earth work necessary to create features and produce the proper drainage. Long or steep slopes are broken with diversions to keep runoff velocities low and divert storm water runoff (Figure 5.17).
6. Golf green plans provide details for the construction of each green complex.
7. Construction details and sections show how the features (trees, bunkers, mounding, ponds, etc.) are to be constructed in conjunction with the grading and drainage plan.

## TABLE 5.1
## A General Outline of Steps Involved in Constructing a Golf Course

**Preplanning Phase**

1. Verify the need for a golf course, select a property, assess the suitability of the site, and establish project goals.
   a. Secure funding/financing.
2. Assemble a team of professionals to do a thorough site analysis and address planning, design, and construction issues.
3. Begin preparation of the architectural design (golf course, clubhouse, irrigation).

**Planning, Phase I**

1. Initiate permitting and engineering, especially concerning environmental issues, land use, and construction regulations.
2. Meet with appropriate regulatory agencies, environmental groups, and concerned citizens to review and receive input on the initial design and construction plan. The overall design is then adjusted based on these inputs and submitted for approval.
3. Hire a qualified superintendent to help administer construction bidding and site inspection.
4. Complete an environmental audit on plant and animal species present and determine baseline water quality of all water entering and exiting the property prior to any construction.

**Planning, Phase II**

1. Routing plan
2. Strategy plan
3. Strategic narrative
4. Scope of work
5. Construction narrative
6. Estimated preconstruction costs
   a. Golf course–related engineering costs
   b. Design costs
   c. Travel costs
   d. Soil- and water-testing costs
   e. Surveying costs
   f. Utility sources and costs
7. Estimated hard construction costs and confirmations
   a. Mobilization
   b. Clearing costs
   c. Grading costs
   d. Excavation costs
   e. Shaping costs
   f. Drainage costs
   g. Irrigation costs
   h. Feature construction costs (putting surfaces, tees, bunkers, etc.)
   i. Abutments and retaining walls
   j. Water feature costs
   k. Lake construction costs (lining, if required; water level controls; out fall structures, if required)
   l. Cart path costs
   m. Bunker sand costs
   n. Soil amendment costs
   o. Topsoil management costs
   p. Finishing costs
   q. Grassing costs
   r. Landscaping costs
8. Estimated postconstruction costs
9. Complete project budget definition

*(Continued)*

## TABLE 5.1 (CONTINUED)
## A General Outline of Steps Involved in Constructing a Golf Course

10. Complete construction schedule
11. Color rendering

**Working Drawings, Phase III**

1. Staking and clearing plan to locate key points such as tees, landing areas, and greens
2. Erosion and sediment control plan (with outside engineer) of disturbed sites during construction
3. Clearing plan to indicate clearing necessary for construction and specimen trees or areas of vegetation to be preserved
4. Grading and drainage plan to indicate overall grading and elevations for the necessary earthwork
   a. Storm drainage in concert with project engineer
   b. Specific golf course drainage
      i. Water or drainage retention facilities (dams, spillways)
      ii. Surface storm drainage (catch basins, ditches)
      iii. Subsurface storm drainage (including corrugated metal pipe and polyethylene tile)
5. Grassing plan of specific turfgrasses
   a. Types
   b. Planting rates
   c. Soil amendments
   d. Application windows
6. Irrigation (to be provided by outside contractor)
   a. Design and materials
   b. Field direction
   c. Pump station
   d. Control system
   e. Volume delivery service
7. Cut and fill plan
8. Cart path plan
9. Putting surface/green area shape plans to indicate details on construction of greens complex to include putting surface, surrounding features, and subsurface drainage
10. Special water features
11. Abutment and retaining wall plans
12. Shelter locations
13. Clubhouse area site plan (with land planner)
14. Maintenance area site plan (with land planner)
15. Complete practice facility plan
16. Putting green plan
17. Landscape considerations
   a. Tree and shrub location for retention
   b. Conservation or natural areas
18. All specifications for above, outlining the methods and details of construction

**Bid/Negotiation, Phase IV**

1. Complete bid package
   a. Working drawings (above)
   b. Technical specifications
   c. General conditions
   d. Bid for contract
   e. Construction scope of work
   f. Special instructions
2. Contractor qualification
3. Prebid conference
4. Postbid analysis

*(Continued)*

**TABLE 5.1 (CONTINUED)**
**A General Outline of Steps Involved in Constructing a Golf Course**

5. Price negotiation
6. Selection consultation and logic
7. Construction agreement document

**Construction Progress, Phase V**

1. Specific approvals of levels of construction completion
   a. Verification of centerlines and perimeters
   b. Clearing approval
   c. Rough grading and grubbing of roots, stumps, rocks, and stones approval
   d. Storm drainage and erosion control approval
   e. Golf drainage installation approval
   f. All landform approvals
   g. Green subgrade approvals
   h. Irrigation head placement approval
   i. Irrigation wet well and slab installation approval
      i. Pump station installation approval
   j. General irrigation
      i. Installation observation
      ii. Controller and quick coupler location approval
   k. Temporary golf course drainage installation approval
   l. Permanent golf course drainage installation approval
   m. Tee construction subgrade approval
   n. Final tee installation approval
   o. Bunker drainage installation approval
   p. Bunker sand approval
   q. Putting surface materials confirmation (by USGA lab report)
   r. Cart path location approval
      i. Bridges
      ii. Shelters, bathrooms, drinking fountains
   s. Abutment location approval
   t. Abutment material approval
   u. Soil management approval
   v. Final grade approval
   w. Seedbed preparation approval
      i. Redistribution of topsoil
      ii. Rootzone soil modifications (amendments, greens mix)
      iii. Soil pH and fertility adjustments (based on soil tests)
      iv. Fumigation of rootzone mix (pest control for greens, tees)
      v. Final grading
   x. Pregrassing finish approval
   y. Final grassing lines approval
   z. Sod, seed, sprig installation recommendations and approval
      i. Grow-in
      ii. Irrigation
      iii. Fertilization
      iv. Mowing
         aa. Water feature approval
2. Notice of substantial completion
3. Draw request verification and recommendations
4. Completion of final punch list
5. Execution of final punch list or automatic retention
6. Change order verification and negotiation
7. Verification and monitoring of schedule progress
8. Field staking of landscaping

**Postconstruction, Phase VI**

1. Grow-in budget and consultation

8. Irrigation plans and details provide information for the type of irrigation system and pump station to be installed for the golf course.
9. Grassing plans indicate the areas where specific turfgrasses and, in some cases, ornamental grasses are to be planted on the golf course.
10. Landscape plans serve as a guideline of where plant material is to be installed to enhance the golf course design. As a part of this plan, conservation areas can be established throughout the golf course.
11. Specifications and bid documents outline the methods and details of construction for the course completion.

### Permitting and Engineering

Permitting and engineering for a golf course construction project can require considerable time, frustration, and expense. Permits should be obtained as soon as possible prior to the initiation of the detailed design. Approvals will be needed before the architect spends considerable time designing the golf course, assuming there is accessibility to the entire site. Wetlands and floodplains must be identified and approval sought for their use in the golf course layout. Other possible needed permits include damming of rivers or streams where water flow may be interrupted, land clearing, vegetation burning, proposed water use source, and use of electricity or other power sources.

### Meeting with Local Groups

Either before or immediately after the permitting and engineering phase, the project team should meet with all regulatory agencies, environmental groups, and interested citizens to review and modify the design, if needed, based on their input (Figure 5.18). Inputs and concerns by these groups should be implemented whenever possible to minimize future criticism and to project a neighborhood-friendly image.

### Construction Process

The golf course superintendent should be hired early in the design and construction process. The superintendent will inspect the construction process daily and serve as the on-site representative for the owner and the architect. During the construction process, site visits are made by the golf course architect, accompanied at times by other members of the consultant team, to inspect the work and see that the intended level of design and quality in the course is being accomplished. These visits ensure that close interaction between the architect, design team, and construction team will ultimately produce the distinctive features and character of the golf course. These visits also provide the opportunity to monitor the controls and management techniques in place for environmental protection.

**FIGURE 5.17** Incorporating a swell or diversion above a green to reroute storm water runoff away from it.

**FIGURE 5.18** Meeting with governmental agencies and concerned groups during a golf course design phase.

## Site Preparation

There are several steps needed in preparing a site as a golf course, including (1) surveying and staking, (2) identifying and developing the power and water sources, (3) land clearing, (4) rough grading, (5) soil cultivation, and (6) shaping of features such as greens, tees, and bunkers.

**Surveying and staking.** The first step of site preparation involves property surveying and staking including identifying sensitive areas such as wetlands and floodplains and not disturbing them during construction. Other reference points such as fence rows, large trees, trails, and streams also are incorporated into the initial base map (Figure 5.19). From the base map, the final routing and general size and location of greens, tees, fairways, natural hazards, and site characteristics are initially identified. Additional drawing sheets are developed to identify the anticipated clearing, thinning, grubbing, and disposal operations.

The next base sheet identifies the overall grading design, including existing and proposed contours and subsurface drainage. Additional base sheets are developed to indicate irrigation, trees, utilities, shelters, planting areas, cart paths, roadways, and so on.

Based on the architect's staking plan, a "centerline" is cleared and staked off down the middle of each golf hole. Stakes along the centerline are usually placed on 100-foot (30 m) intervals from the back of the rear tee, through the center of each fairway, to the center of the putting green (Figure 5.20). Other stakes may be placed to mark property lines and other features such as bunkers, ponds, and so

**FIGURE 5.19** Identifying "keep" features (such as trees) before golf course clearing.

**FIGURE 5.20** Stakes identifying the centerline and perimeters of a proposed golf hole.

on. Once the centerline has been staked for each hole and agreed upon by all concerned parties, the surveyor establishes a permanent bench mark (or elevation point) to serve as a reference for all subsequent elevations and elevation changes.

As stated, the bench mark usually is located in the center of the golf course construction site; however, uneven terrain and/or unavoidable obstacles may necessitate several permanent bench marks across the site. Once established and identified, the perimeter of each hole is staked at 25- to 100-foot (7.5 to 30 m) intervals using the centerline stakes plus any elevation differences from the bench mark. These perimeter stakes identify and outline the shape of the fairway to indicate the initial ground surface contours according to the architect's plans. All stakes should be appropriately marked (tees, greens, perimeters, etc.) to preclude errors during the land clearing operations, especially if the rough grading plan requires extensive earth moving.

**Erosion control and environmental protection.** Most jurisdictions have regulations to minimize soil erosion and siltation on a construction site. The project's civil engineer usually prepares plans that include control devices such as sediment basins, silt fences, hay bales, and other structures conforming to current guidelines and regulations (Figure 5.21). These control and/or protection measures are installed before the clearing and grading operations begin.

**Identify and develop utility sources.** Power and water accessibility to the construction site is imperative in developing, building, and maintaining a golf course. The irrigation pumping station will require a power source such as electricity or fuel. If the power source is a considerable distance from the site, then connection costs will substantially increase.

The same is true for an available water source on site. Water is essential to the establishment and maintenance of turfgrass. The preferred water source is one located within the confines of the golf course property. Sources for irrigation water

**FIGURE 5.21** A comprehensive soil erosion plan is necessary before construction begins on a golf course.

include: (1) subsurface water such as wells and springs; (2) surface water such as ponds, lakes, or streams; and (3) processed water such as gray water (or effluent), municipal sources, or possibly desalinized water. In recent years, effluent wastewater use as an irrigation source has increased, especially in arid climates where water is extremely limited. Golf courses are ideal candidates for effluent water use as irrigation and will play an increasing role in recycling, conservation, and treatment of this most vital natural resource. Often, holding ponds are needed if several wells are drilled or if a certain amount of effluent water must be accepted by the course, regardless of need. Time must be allowed for the proper permitting of needed utilities, especially if well drilling, pond construction, or access to a river or stream is required.

**Land clearing.** Before initiating land clearing, the golf course architect and the developer/owner should agree on the overall golf course design and layout. During surveying and staking, particular attention should be given to "specimen" trees and vegetation that add to the natural beauty of the site and environment. These should be identified and incorporated when possible into the strategic layout of the golf course as the construction process continues. The land-clearing operation should be conducted in various phases or stages to prevent any irreparable damage to the site during clearing.

The first stage of land clearing involves tree removal along each hole's centerline stakes. Final tree selection and thinning is then completed. Grass requires light for optimum growth, yet tree selection is very important in the overall golf design of a hole. The architect should mark all trees to be preserved, removing all not needed to help speed play and to readily allow movement of construction equipment and future turf-maintenance equipment. Pruning lower limbs of remaining trees also should be considered to further facilitate equipment movement, play speed, sunlight penetration, and air circulation. Although owners and members desire large trees near greens and tees for aesthetics, shade and turfgrasses do not mix. Grass on greens and tees requires a minimum of eight hours of full sunlight year-round. Shade is most pronounced during late fall and winter months, when days are shorter and the sun is lower on the horizon. Shade patterns during these times should be closely considered when deciding on tree removal. If usable timber exists from this tree removal, the owner may wish to salvage and sell it.

Generally, the contractor is provided a clearing plan from the architect that outlines the clearing width. This width varies from 20 to 50 feet (6 to 15 m) on each side of the centerline depending on the hole's length and shape. The resulting vegetation is stockpiled in the center of each hole, typically windrowed, and burned or removed from the construction site (Figure 5.22).

After the initial clearing, the architect walks each golf hole, visualizes how each fits in with the terrain, and estimates how it will play (Figure 5.23). The architect then initiates the second stage of land clearing by specifically marking the boundary lines for each golf hole. This stage of land clearing is intended to open up each golf hole enough so the architect can further visualize the playing characteristics of each hole. Architects are typically conservative during this second stage of clearing so as not to remove too many trees that may strategically affect the way the hole plays. The remaining trees should be fenced off to prevent soil compaction and accidental contact with construction equipment. Whenever possible, leave trees and underbrush in out-of-play areas undisturbed to provide wildlife habitat.

The final stage of land clearing involves the removal of stumps, roots, rocks, and stones. The objective of this final stage is to leave the site free of rocks and organic debris to a depth of 12 to 15 inches (30 to 38 cm), thereby preventing interference with subsequent construction activities. This process is sometimes referred to as *grubbing*. This very important operation should not bury any tree stumps, branches, or other vegetation, as these will decompose over time and possibly leave depressions that may collect water and produce uneven playing or riding surfaces. From this decomposition process, soil microorganisms may also initiate fairy ring disease development, which disrupts the turfgrass surface uniformity later. Stumps are removed with a large hydraulic backhoe or with a bulldozer and a clearing blade to push out the biggest

**FIGURE 5.22** Stockpiled vegetation from initial clearing of a new golf hole.

**FIGURE 5.23** Initial clearing of a golf hole.

material. Large debris removal is followed by a bulldozer with a root rake and finished with a tractor-pulled rake or a large root disk. Depending on the site, it may be necessary to excavate a nonplay area to bury these stumps and roots, especially if burn permits are unavailable. Small rocks in the upper 2 inches (5 cm) are removed with rock-picking machines. Every effort should be made to remove or dispose of debris and to save the topsoil.

**Rough grading.** Rough grading is the site preparation stage where all major contours and features of the golf course subgrade are set to shape the future final product within 6 inches (15 cm) of the final grade. Rough grading is sometimes referred to as heavy earth moving since it involves bulldozers, pans, power shovels, dredges, sand pumps, and other large equipment to reposition soil around the golf course site (Figure 5.24). During this phase of construction, the sculptural aspects of the design begin to take shape. The proposed subgrade contour is finalized, and it is common to have the land forms shaped several times to achieve the exact look the architect desires (Figure 5.25). Some architects provide detailed grading plans while others prefer to "work on site" with specialized equipment operators (or shapers) that contour the subgrade as the golf course is constructed. Shapers often make or break an architect's design and are key personnel when bringing out the final beauty and playability of a hole. Rough grading and the installation of major drainage overlap somewhat during the construction process and usually are simultaneously completed.

While forming the subgrade, topsoil is often removed or stripped from high areas of the subgrade and stockpiled for later use during construction. This topsoil is then respread over depressed or low areas of the fairways and roughs as needed to maintain the proper elevations. In the past, grading stakes and strings were utilized to aid in achieving the proper grade levels; however, laser grading has largely replaced these.

Initial shaping of the subgrade contours involves the placement of fixed grading stakes referenced to the centerline and permanent bench mark elevations. The bulldozer and/or pan operators then follow these routes of elevation to their indicated depths as outlined by the architect's plans. In areas

**FIGURE 5.24** Rough graded golf hole.

**FIGURE 5.25** Shaping the final grade of a golf hole according to the architect's design.

where topsoil has been excavated and removed, the subgrade should be established at an appropriate level below the finished grade to compensate for adding topsoil back to the grade. These stakes remain in place until major grading is completed and topsoil is placed.

It is also quite common to encounter rocks during the rough grading process. Loose rocks and large stones should be excavated and/or removed. Rocks and stones can be sold to local landscapers, disposed of by burying them in nonplay areas, or possibly utilized in the construction process for stabilizing banks, ditches, or streams.

**Soil cultivation.** Construction sites often become highly compacted due to traffic from heavy construction equipment operation, site deliveries, and/or site cleanup operations. Soil cultivation may be necessary on these areas to alleviate any soil compaction problems prior to "grassing" the golf course. Soil cultivation can be accomplished utilizing several different types of equipment, including chisel plows, subsoilers, disks, harrows, or plows.

## DRAINAGE

It has been stated that successful golf course architecture is nothing more than making drainage features appear pleasing to the human eye! Good drainage is essential to the health of the grass and its ability to tolerate concentrated traffic from play and maintenance equipment use. If drainage is not sufficient, it will be difficult to maintain the turfgrass without causing physical damage, and a finely groomed playing surface will be elusive to achieve, regardless of budget. As previously stated, without good drainage, golfers are delayed from playing; thus, potential revenue is lost (Figure 5.26).

The stormwater drainage system must be designed to conform with local regulations regarding upstream and downstream watersheds. Nonconformance to these plans without written review and approval can result in civil and criminal actions against all parties involved with the project. Golf courses, especially around residential or resort developments, should pay close attention to drainage and drainage

**FIGURE 5.26** Without sufficient drainage, golf is delayed following heavy rainfall, losing potential revenue.

requirements since the amount of surface runoff is increased with storms. Ponds, grass swales, and wetlands are incorporated into this water management plan to reduce runoff, create natural habitats, and help filter sediments and nutrients from the water. Because, invariably, 50- to 100-year rainfall maximums will occur within several years of opening, they should be anticipated. The stormwater drainage system should account for these.

**Water or drainage retention facilities.** Lakes, ponds, dams, spillways, wetlands, and other water-retention facilities are often integral parts of the stormwater management system and an irrigation source for the course (Figure 5.27). A civil engineer prepares construction drawings that are in conformance with the regulatory standards for dams, overflow structures, and inlet and outlet structures. Water-retention facilities should be shaped and graded concurrently with other rough grading operations. Retention structures are normally created at the low end of the golf course or throughout the course to collect the surface runoff water during a storm and is then released at a naturally occurring rate prior to construction. These water-retention facilities are usually lined with some type of impervious material such as polyethylene, clay soil, or bentonite, according to design specifications.

### Surface Storm Drainage

Surface drainage is probably the most important component of the stormwater drainage system since it is the most rapid and easiest means of removing excess water from the golf course. Surface drainage of fairways and roughs includes the design and installation of culverts, channels, diversions, surface inlets, catch basins, and interceptors. The initial surface storm drainage design ensures these surface drainage structures are sized properly to handle the anticipated storm flow. Rain on a heavily textured soil will yield more runoff compared with a lightly textured soil. Correspondingly, a compacted soil will also yield more runoff compared with an uncompacted soil. Surface drainage can be accomplished by providing contours to

**FIGURE 5.27** Incorporating water features into a golf hole.

**FIGURE 5.28** Runoff onto greens or into bunkers significantly increases maintenance costs.

fairways and primary roughs to divert the surface runoff water away from the golf course. Runoff should be directed away from features such as greens and bunkers, where it will cause immense maintenance problems (Figure 5.28). Any surface with less than 1% slope is likely to shed water too slowly; 2% or 3% is ideal. Surface stormwater drainage work could include grading and shaping drainage swales, establishing drainage ditches, stabilizing stream banks, and constructing overflow and drainage structures connecting water-retention facilities. Water in active play areas generally should not be allowed to drain more than 150 feet (45 m) before intercepting a drain or before it is out of play. Shaping is performed skillfully to direct this water and to hide catch basins or drop inlets.

The first step in surface drainage design is to determine peak runoff rate for the specific site. Several methods exist to calculate peak runoff rates. A commonly used method is the *rational method*, which simply determines peak runoff rate from knowing the expected rainfall intensity for a given period (e.g., inches/hour) and area (acres) of the watershed involved, and using an appropriate coefficient value based on the soil and land use of the watershed area. An experienced drainage specialist is needed to determine and design appropriate surface drainage for these. Peak runoff rates are commonly used to design channels, diversions, surface inlets, storm sewers, and culverts.

Once rough grading and surface drainage are completed, any specialized construction, such as stone or wood walls, is installed. These walls not only stabilize banks and slopes that otherwise would collapse, but also add a distinct look to many facilities.

### Subsurface Storm Drainage

Subsurface drainage is merely the management of stormwater infiltration, percolation, and movement through the soil profile. Subsurface drainage is needed where natural low areas occur, or where water is slow in drainage (Figure 5.29).

**FIGURE 5.29** Subsurface drainage is needed where natural low areas occur.

**FIGURE 5.30** Installing subsurface drain lines on an insufficiently draining green.

Subsurface drainage is imperative where the subgrade is a clay soil or if the soil has an impermeable layer (e.g., hardpan); otherwise, the area could remain excessively wet for several days after heavy rainfall.

The most common means of subsurface drainage involves surface catch basins connected to a network of underlying drainage lines (tile or plastic pipe) (Figure 5.30). The drainage lines intercept the water as it moves through the soil profile and then route it to an outlet located several feet lower than the existing grade. Subsurface drainage is commonly used in greens, bunkers, some tees, and low-lying fairways.

The vertical elevations of stormwater drainage lines are expressed as invert elevations (or vertical elevation measured at the bottom of the interior of the drainage pipe or drainage structure) and directly relate to the final elevations of the playing surfaces. Any unaccounted changes to the invert elevations could seriously hamper the golf course's playability and maintenance. Depending on the location, underground piping may be performed before major earth-moving occurs.

**Subsurface drainage outlet.** The first phase of installing subsurface storm water drainage involves locating an adequate outlet for the drainage water. Typically, drainage lines are routed into nearby water-retention facilities such as ditches, ponds, larger drainage lines, catch basins, French drains, or other sources. The discharge lines should be nonperforated (pipe or tile) and should not be laid straight down a steep slope, but instead have a gradual slope to reduce the flow rate of the drainage water from the area. In some cases, a suitable discharge source may not be readily available. If this is the case, a sump and pump may be required.

**Sump and pump drainage.** A sump is considered a pit or reservoir serving as a drain receptacle for liquids (water). This sump is commonly in the form of a tank or several concrete rings placed on top of each other and enclosed with a cover (Figure 5.31). A low-lift pump (or sump pump) is then installed inside the sump at the lowest point with a float-activated switch so the water level may be controlled within specified limits. Once a predetermined amount of water is allowed to drain into the sump, the discharge water is then pumped up to an appropriate discharge area. Sumps should be located away from the fairway and in areas receiving little to no traffic. Covering the main drainage line outlet of the sump with a mesh wire screen is also advisable to prevent animals from entering and possibly causing damage.

**Subsurface drainage line spacing.** If the golf course is situated on an area with a high water table, it may be necessary to place larger drainage lines deeper into the subgrade to lower the water table and handle the increased internal flow of water. Calculating the precise drainage line spacing needed can be done with a modification of Hooghoudt's equation (see Chapter 4).

**Drainage line types.** In the past, drainage lines consisted of agricultural clay tile, concrete, or flexible corrugated plastic. Today, 2- to 4-inch (5 to 10 cm) diameter flexible corrugated pipe with perforations (slits) is utilized due to ease of installation and relatively low cost. Main drain lines are bigger—typically, 6 to 8 inches (15 to 20 cm) or larger in diameter. The perforations in the plastic pipe should be laid facing downward on a bed of gravel to prevent them from clogging by downward-migrating soil particles.

**Drainage line patterns.** Typically, a gridiron or herringbone design pattern is used for drainage line arrangement. The drainage pattern is placed diagonally to the grade. However, any pattern is acceptable as long as each line has a continuous downward slope and water does not have to travel more than 150 feet (45 m) to intercept a drainage line.

**Drainage line trenches.** Trenches in which the drainage lines are to be laid should be a minimum of 6 to 8 inches (15 to 20 cm) in depth into the subgrade and 5 to 6 inches (13 to 15 cm) in width. Trenches up to 12 inches (30.5 cm) wide have been utilized; however, the wider the trench is, the greater is the amount of gravel needed to fill the trench; this elevates costs. Normally, a drainage line trench should be no more than twice the size of the drainage line. A 5- to 6-inch (13 to 15 cm) wide "U"-shaped trench will allow for a 0.5- to 1-inch (1.3 to 2.5 cm) bed of gravel to be placed around (below, above, and on either side) a 4-inch (10 cm) diameter drainage line. The soil created from digging the trench should

**FIGURE 5.31** Sump and pump pit where accumulated water is pumped to an appropriate discharge site.

either be removed or placed between drainage lines to provide a slight contour before being compacted.

Prior to digging of trenches, the area should be surveyed and staked with proper labeling for the desired depth of cut. Drainage lines should not be placed any deeper than necessary to obtain the desired amount of slope (Figure 5.32). Grade stakes should be placed in order to achieve a minimum downward slope of 0.5% (1 foot of drop for every 200 feet, 0.3 m in 60 m). Ideally, a slope of 1% to 2% (1 foot of drop for every 100 feet, 0.3 m in 30 m, and 1 foot of drop for every 50 feet, 0.3 m in 15 m, respectively) would be preferred, with a maximum slope of 4% (1 foot of drop for every 25 feet, 0.3 m in 7.5 m). Drainage lines with 0.5% slope or less are difficult to properly grade, install, and maintain due to the slight elevation changes and slow flow rate. Slopes of greater than 4% will lose lateral drainage capability and also require greater elevation changes within the drainage line, as well as a lower outlet point.

Care must be taken to ensure the drainage trench and drainage lines are always sloping downward to preclude any entrapment or collection of water along the drainage lines. Drainage lines should be placed diagonally to the slope and not at right angles. All main and lateral drainage lines should be checked with a level prior to backfilling to ensure the grade provides adequate drainage. Drainage line ends or joints should be covered with asphalt paper, fiberglass composition, and plastic spacers or caps to prevent gravel and/or soil from entering the drainage lines. It is recommended the

**FIGURE 5.32** Using a laser guided trencher to install drainage lines.

main drainage lines into which the lateral feeder lines flow have their upper end extended to the soil surface and properly capped. In the future, if these lines become clogged or contaminated from gravel, soil, or some other material, the cap can then be removed and the line or lines flushed. This will greatly extend the useful life of the drainage system without the need of disturbing the playing surfaces to clean individual lines.

**French drains.** French or slit drains are a modification of subsurface drainage lines. French drains involve digging a drainage trench and then backfilling with either gravel and/or coarse sand. The top 2 to 4 inches (5 to 10 cm) of the French drain trench should be filled with coarse sand to serve as a growing medium for the turfgrass. This sand should be compacted prior to installing sod to decrease the amount of settling that will occur over time. If sod is installed, its soil should be removed by washing prior to installation to minimize any negative impact from introducing dissimilar soil from the sod onto the sand medium. French drains have the advantage of being easily installed at a fraction of the cost of tile lines.

### Installation of Irrigation System

An irrigation system provides sufficient water to sustain turfgrass growth during drought periods and is the most important component in maintaining proper turfgrass conditions for optimum play (Figure 5.33). In certain climates, this may require only a simple irrigation design with a single irrigation line running down the middle of each fairway. However, in most climates, the irrigation system is more sophisticated so as to cover all areas intended for turfgrass. On modern golf courses, the irrigation system can cost as much as one-fourth of the total construction budget. In addition, many golf courses today are utilizing a variety of turfgrass species for greens, tees, fairways, and roughs. Therefore, a carefully designed irrigation system is required to allow an adequate supply of water for each of these turfgrasses according to its needs and to be able to efficiently and accurately distribute the water to these areas.

The installation of the irrigation system normally begins after the shaping of features within a few inches of design-finish grade and installation of major drainage have been completed, often done as a set of two to three golf holes together. To maximize efficiency and workmanship, irrigation installation usually begins at a logical starting point such as the pumping station and radiates outward, or starts at the furthest point and works back toward the pump station. The system must be in place and operational before planting or establishing turfgrass begins. The irrigation system can also be used to settle the greens mix or rootzone mix prior to planting, and to settle backfilled areas during the construction process.

A quality irrigation system is dependent on accurate design plans and specifications (Figure 5.34). The first step in installation involves the designer laying out or staking the complete irrigation system. Staking should be accurately established for the entire irrigation system including the main lines, sprinkler heads, valves, satellite controllers, and other key features. This staking should be done exactly according to the design specifications to ensure sprinkler head spacing and coverage are adequate. Individual sprinkler head staking is needed to ensure the design specifications accurately fit the actual topography. If conditions dictate, this is the time to make modifications to the irrigation design and system by adding lines and/or sprinkler heads to ensure adequate coverage and performance. The stakes also serve as a guide or map for the trenching operation.

The trenching route should follow the design layout as marked by the staking operation. The depth and width of the trenches should be determined by the pipe size, soil depth

**FIGURE 5.33** A properly designed and working irrigation system is the lifeline of a golf course.

**FIGURE 5.34** Improperly designed and operating irrigation system.

needed to prevent freezing, and slope needed to provide adequate drainage. The completed trenches should be smooth, free of rocks, and on the specified grade level.

After trenching, irrigation lines or pipe is installed. Installation of irrigation pipe normally begins at the water source with the main line(s) installed first. The control, drain, and gate valves are then installed in coordination with the lateral irrigation lines. Next, the control lines are placed in the trenches along with the appropriate wiring (Figure 5.35). The trenches are then partially backfilled with either sand or other fine-textured soil free of rocks and/or stones. It is generally advised that joints and fittings associated with the main line(s) not be backfilled initially so they can be checked for leakage and easily repaired during the initial start-up of the irrigation system.

After the irrigation pipe has been installed, the system is connected to the water source (pumping station) and backfilled with sufficient soil to prevent movement. The pump station is the heart of an irrigation system and is generally delivered preassembled, placed on the pump pad, and wired. The system should then be activated and "flushed" by passing sufficient water through all pipe, fittings, valves, and sprinkler head risers to ensure no soil, pipe chips, or other debris is present in the lines.

After the system has been completely flushed, sprinkler heads are installed, sometimes during the flushing operation. Sprinkler heads are normally installed several inches above the projected rough grade level through the use of "swing" joints and are left to operate in this position until after the turfgrass has become established (Figure 5.36). Sprinkler heads

**FIGURE 5.35** Wiring used during irrigation system installation.

**FIGURE 5.36** Leaving sprinkler heads above grade until turf grow-in occurs.

are then lowered to the permanent finished grade after the soil has settled and the turfgrass has become established. The system is then pressure tested to ensure it can operate above the normal operating pressure. The system is also checked for any leakage at joints at this time prior to backfilling. This is most often done by pressurizing the system to about 120 psi (827 kPa) and allowing it to sit for 24 hours. If the pressure drops more than several pounds, leaks are likely and must be detected and fixed, followed by another 24-hour pressure test until the pressure stabilizes.

After successful pressure testing, irrigation lines are then backfilled to a level slightly above the original soil level to compensate for settling and compaction over the irrigation trenches. The irrigation system is then ready to be properly adjusted for achieving the accurate pressure setting and water distribution. A final "as built" site plan for the irrigation system is recommended for future reference by the golf course superintendent to maintain, winterize, repair, or make modifications.

## Seedbed Preparation

**Redistribution of topsoil.** Whether the area is to be established by sprigging, sodding, or seeding, soil preparation for each is the same. After final subgrade shaping, features construction, and irrigation installation, the stockpiled topsoil can then be respread. Due to the potential damage created from equipment operation, this phase is sometimes completed before irrigation is installed. Particular attention is needed in preserving the final shaping contours and grade when redepositing topsoil.

The topsoil is spread evenly over all prepared subgrade areas except the greens, tees, and bunkers. The generally accepted standard requires a minimum of 6 inches (15 cm) of topsoil over all fairways, tee slopes, and green surroundings, and a minimum of 4 inches (10 cm) over the roughs. If the topsoil is imported, the golf course architect must specify its structure. Once the topsoil has been evenly spread, seedbed preparation begins by disking or harrowing the soil to a depth of 6 inches (15 cm). The soil is then raked to a depth of 4 inches (10 cm) to remove any rocks and/or stones larger than 0.5 inch (1.3 cm) in diameter. The upper few inches of soil are then ready for any necessary amendments or modifications.

**Rootzone soil modifications.** Sometimes the soil(s) must be modified to ensure adequate and uniform soil moisture, drainage, and nutrient exchange. For economic reasons, the modification of fairway soils is performed much less frequently than on greens or tees. Determining which type of soil amendment and how much to add is crucial, since the use of an improper soil amendment can forever diminish the quality of the turfgrass site.

**Sand soil types.** In coarse-textured sandy soils, adding the proper soil amendment(s) is recommended to increase the soil's water- and nutrient-holding capacities. In these coarse-textured soils, drainage is normally not a problem, but adequately holding soil moisture and nutrients potentially is. In addition, sandy soil modified with an amendment discourages pests such as nematodes. In sandy soils, the addition of a good topsoil generally provides the desired results at the most economical price. However, the topsoil source must be acceptable to prevent later drainage problems resulting from excessive silt, clay, or pests.

Other soil amendments used in sandy soils include peat sources, rotted sawdust, rice hulls, calcined clays, and colloidal phosphate. These amendments improve soil aeration, increase nutrient- and water-holding capacity, and encourage microbial activity. Peat sources are generally added at a rate of 45 to 90 cubic yards/acre (85 to 170 $m^3$/ha).

Other soil amendments are added according to the type of material used and the degree of soil modification required. These amendments are often incorporated at a rate of 10% to 15% by volume in the top 4 to 6 inches (10 to 15 cm) of the

**TABLE 5.2**
**Soil Amendment Volumes (Cubic Yards) for Modifying Existing Depths**

| Desired % of Amended Soil | Modification Depth (inches)[a] | | | | | | | | |
|---|---|---|---|---|---|---|---|---|---|
| | 4 | 5 | 6 | 7 | 8 | 9 | 10 | 11 | 12 |
| | Cubic Yards[b] | | | | | | | | |
| 5 | 0.61 | 0.77 | 0.93 | 1.08 | 1.23 | 1.39 | 1.55 | 1.71 | 1.86 |
| 10 | 1.23 | 1.54 | 1.85 | 2.16 | 2.47 | 2.78 | 3.09 | 3.40 | 3.71 |
| 15 | 1.85 | 2.32 | 2.78 | 3.24 | 3.70 | 4.17 | 4.63 | 5.09 | 5.55 |
| 20 | 2.47 | 3.09 | 3.71 | 4.32 | 4.94 | 5.55 | 6.17 | 6.79 | 7.40 |
| 25 | 3.08 | 3.86 | 4.63 | 5.40 | 6.17 | 6.95 | 7.72 | 8.49 | 9.27 |
| 30 | 3.70 | 4.64 | 5.56 | 6.48 | 7.41 | 8.33 | 9.26 | 11.18 | 12.10 |
| 35 | 4.32 | 5.40 | 6.48 | 7.57 | 8.64 | 9.72 | 10.80 | 11.87 | 12.95 |
| 40 | 4.94 | 6.18 | 7.41 | 8.64 | 9.88 | 11.13 | 12.34 | 13.61 | 14.81 |
| 45 | 5.55 | 6.95 | 8.33 | 9.72 | 11.10 | 12.52 | 13.88 | 15.17 | 16.67 |
| 50 | 6.17 | 7.72 | 9.26 | 10.80 | 12.34 | 13.88 | 15.42 | 16.96 | 18.50 |
| 55 | 6.78 | 8.49 | 10.19 | 11.88 | 13.58 | 15.28 | 16.98 | 18.67 | 20.37 |
| 60 | 7.41 | 9.26 | 11.11 | 12.86 | 14.81 | 16.67 | 18.52 | 20.37 | 22.22 |
| 70 | 8.64 | 10.80 | 12.96 | 15.11 | 17.28 | 19.44 | 21.60 | 23.77 | 25.92 |
| 80 | 9.88 | 12.35 | 14.81 | 17.28 | 19.76 | 22.22 | 24.69 | 27.16 | 29.63 |

[a] Multiply inches by 2.54 to obtain centimeters.
[b] Multiply cubic yards by 0.7645 to obtain cubic meters.

seedbed. Table 5.2 lists the volumes of soil amendments (cubic yards to add to 1 cubic yard of soil) to provide the desired percentages of various treatment or incorporation depths. Refer to the section on "soils" for additional information regarding soil amendments and their characteristics.

### Example

Determine how much soil amendment volume (cubic yards) is needed to modify the top 6 inches (15 cm) with 15% amendment by volume.

From Table 5.2, the intersection of 15% desired amended soil to a depth of 6 inches (15 cm) indicates that 2.78 cubic yards (2.1 $m^3$) of soil amendment is needed per cubic yard (0.76 $m^3$) of soil.

**Clay soil types.** Fine-textured, native clay soils also may be modified to facilitate internal drainage. The modification of fine-textured, heavy soils such as clay can be both an agronomic and financial challenge. These fine-textured soils are prone to compaction and generally require substantial amounts of soil amendment for any significant improvement to be observed. A properly sized sand source is generally the most popular and recommended amendment for improving clay-type soils. However, as discussed in Chapter 4, not all sands are created equal. Some types are better suited for making concrete while others are better for soil modification. Do not use any sand without careful investigation into its usefulness for this application. It is advisable to consult a competent soil-testing laboratory well in advance of the soil modification step.

**Modification depth.** The soil should be modified or amended to a minimum depth of 6 inches (15 cm), with 10 to 12 inches (25 to 30.5 cm) being the ideal depth. "On-site" mixing of the soil amendment will require extensive harrowing, disking, and rototilling to ensure as uniform an incorporation of the amendment as possible. These areas should be mixed or incorporated a minimum of four times (passes) in different directions. Pockets of "pure" soil amendment or soil should be avoided to prevent poor draining areas or excessive drying.

**Soil pH and fertility adjustments.** Soil test results provide the basis for soil pH and fertility adjustments prior to turfgrass establishment. These adjustments can be made during the rootzone modification stage while mixing the soil amendments, or afterwards by incorporating the necessary soil pH or fertility products. A better distribution of these materials normally can be accomplished by incorporating these during the soil modification stage.

As a general rule, the soil pH should be between 5.5 and 7.0. Agricultural limestone (calcium carbonate) or dolomitic limestone is applied to soils with a low pH. Sulfur is utilized to adjust soils with a high pH. Gypsum is sometimes applied to compensate for high-salt soils. The exact amount of these materials must be determined by a soil test. Refer to the chapter on soil chemical properties for additional information.

The addition of starter fertilizer into the soil is required in almost all golf course sites. The specific nutrient application rate should be based on soil test recommendations. The fertilizer is normally applied for broadcast just prior to turfgrass planting or establishment and incorporated into the top 4 inches (10 cm) of the soil rootzone. Research studies indicate a fertilizer with a 1-2-2 ratio of N-P-K is most beneficial during the turfgrass establishment phase. However, the specific recommendations for adjusting soil pH and soil fertility levels should be based on the type(s) of turfgrass to be established and the specific requirements.

## Fumigation of Rootzone Mix

Fumigating the rootzone soil mix for greens, and sometimes tees, is recommended just prior to the final grading to control disease organisms, insects, nematodes, weed seeds, and perennial weeds (Figure 5.37). The more common fumigant materials include methyl bromide, metam-sodium, or dazomet, although methyl bromide is being phased out in the United States. Fumigation requires special application equipment and is toxic; therefore, it should only be applied by properly certified and trained professionals. Fumigation treatments normally require about 48 hours of exposure to achieve adequate results. Another 48 hours or so are normally needed to allow the rootzone mix to "air out" after this gas treatment has been applied and the plastic covers are removed.

Fairways should be periodically tilled before establishment to reduce weeds or else sprayed with a nonselective herbicide prior to planting. A minimum of three applications of a nonselective herbicide will be necessary to control existing perennial grassy weeds such as common bermudagrass, nutsedge, kikuyugrass, quackgrass, or torpedograss.

## Final Grading

Final grading of the areas to be planted with turfgrass consists of smoothing the seedbed to remove any depressions or rough areas, breaking up large clods of dirt, removing rocks larger than 2 inches (5 cm), removing roots and stems, smoothing low and high spots, and conditioning the seedbed for planting (Figure 5.38). For small areas, such as tees and greens, and for working and smoothing around all sprinkler heads, hand- or mechanical raking is utilized. Larger areas such as fairways require grading and fitting tools such as box blades, wire mats, boarded mats, plank drags, and other specialized harrows and/or equipment. This consists of disking the soil to

**FIGURE 5.37** Fumigating is the best assurance of controlling pests. (Courtesy Fred Yelverton.)

**FIGURE 5.38** Final graded turf area ready for planting.

break up the clods and float up any rocks; removing rocks with a rock rake; using a landscape rake to windrow this debris, which will be picked up by a pan or front-end loader; and finally smoothing the surface with a land leveler, box scraper, or finishing rake (Figure 5.39). These operations should be performed when the soil is moist (not wet).

A vibratory roller or similar roller normally provides the firming needed to settle an area before turfgrass establishment. It is key to remember this final grading process is the last chance to smooth and firm the seedbed prior to turfgrass establishment. Once the turfgrass is established, any undulations create an uneven playing surface and will remain until corrective measures are implemented. The final grading process or finished grade and last 24-hour irrigation check should be approved by the golf course architect before "grassing" begins, as this is the last chance to make major modifications or repairs before planting.

### Turfgrass Establishment

The specification of different turfgrasses for use on the golf course is the golf course architect's last responsibility, but one not to be taken lightly considering the amount of money to be spent on future turf maintenance. The threat of erosion during establishment also must be considered or much money and time and many delays could occur from even one severe thunderstorm (Figure 5.40).

Once the final grading process is complete, the area must be planted as quickly as possible before rain, wind, traffic, or other conditions damage the prepared seedbed. If the seedbed becomes damaged, the surface must be carefully reshaped and the seedbed again prepared for "grassing."

The surface soil condition in terms of firmness is also important. If too compacted, seeds will have a difficult time germinating and becoming established, and the soil will

**FIGURE 5.39** Mechanical devices often used during final grading.

**FIGURE 5.40** Erosion control means used before and during turfgrass establishment.

require additional cultivation. If the soil is too wet, it could rut during planting and mowing, causing a very uneven and bumpy playing and maintenance surface. If it's too dry and powdery, obtaining a proper finishing grade is again difficult and planting will be delayed until a more favorable soil moisture level is achieved.

The type of turfgrass specified will largely determine the method of planting or establishment. Turfgrasses are established by either seeding or by vegetative propagation (refer to the chapter on turfgrass establishment for more specific information).

**Planting sequence.** Golf greens, collars, and slope areas just within the throw of the greens sprinklers are normally the first areas planted. After planting by hand-sprigging, seeding, or sodding, these areas are lightly rolled to smooth the surface and to ensure good soil contact.

Tees, bunker edges and outslopes, catch basins, pond and ditch banks, and other areas prone to severe erosion also are planted early in the sequence. Often, these areas are sodded to reduce erosion. Fairways are planted next, followed lastly by the rough area. These are sodded or planted mechanically by seeding or sprigging, with close attention given to minimizing rutting.

Seeded areas are then mulched to minimize wind and water erosion and to help conserve soil moisture. Mulching materials include straw, wood, or paper fiber or a synthetic plastic coating. A tacking material such as an asphalt binder is often used to prevent wind and water movement of the seed and mulching material (Figure 5.41).

## Accessibility, Bridge, and Cart Path Construction

Construction of access roads, service roads, bridges, and cart paths normally commences during or immediately after the turfgrass has been planted (Figure 5.42). The initial construction steps, such as grading and subgrade preparation, are coordinated with these corresponding construction phases of the golf course. The installation of a gravel layer over

**FIGURE 5.41** Hydromulched hillside to reduce erosion potential.

**FIGURE 5.42** Bridge construction during turf establishment.

**FIGURE 5.43** Cart path installation.

the compacted subgrade for these accessories is sometimes completed during the final grading process and then surface graded either immediately or after turfgrass planting.

Cart paths, when properly designed, installed, and maintained, provide traffic flow, prevent damage to the turf from excessive traffic, allow rapid transport for maintenance personnel and equipment, and produce a professional-appearing facility without worn, bare areas from unrestricted cart traffic (Figure 5.43). In areas where vehicular traffic should be restricted to the cart paths, a low-face curbing, fence, rope, or other means should be installed to prevent unbounded access. Cart paths are typically constructed with: (1) a local aggregate over a compacted subgrade, (2) asphalt, or (3) concrete. All cart paths need to be designed and constructed with the considerations of water flow and retention, climate extremes, dust/dirt movement, and sufficient width. Cart paths should be a minimum of 6 feet (1.8 m) wide—preferably, 8 feet (2.4 m). This provides sufficient width for two carts to pass without leaving the path and allows the access of larger maintenance equipment to the course.

## PUTTING GREEN CONSTRUCTION

Golf greens typically experience heavy use throughout the year. Public courses may receive as many as 400 rounds of golf played per day, or over 100,000 rounds of golf per year. Although putting greens represent ~2% of the total course area, ~50% of the game is actually played on them. This concentrated traffic combined with daily mowing almost guarantees a problem with soil compaction, especially if the greens are poorly or incorrectly constructed with inferior or unproven materials (Figure 5.44).

A golf green must rapidly accept and drain away excess water and yet retain enough moisture to avoid excessive, frequent irrigation. Rootzones constructed from the native soil frequently turn into mudbaths. The objective of the following section is to discuss the proper sequence of decision making and construction processes needed to create a high-quality putting green.

## INITIAL PLANNING

The most important phase of green construction is the initial planning. A golf course planning committee should include the club president, greens committee chairperson, golf course architect, club pro, building contractor, and, most importantly, golf course superintendent. Communication is the key to success. Several key topics requiring early priority are discussed in the following sections.

### Location

Locating a green is almost as much an art as a science. Nonagronomic considerations provide the aesthetic background and challenge for the golfer. Natural features, such as water, topography, scenic views, and natural hazards, such as trees, are incorporated into the location, shape, and size of a green. Agronomic inputs, such as type of soil, proximity to irrigation, and competition from trees, should also be part of this decision. Even the best designed hole will be limited by parameters affecting the health of the turf.

### Drainage

Being able to control soil moisture is the key parameter in the success or failure of a golf green (Figure 5.45). Greens, bunkers, and surrounding mounds should be located and shaped so surface water from heavy rains will drain quickly away from, rather than onto, the green. Locating greens in consistently wet areas, such as along river beds, floodplains, or marshes, must also be carefully planned to allow for adequate subsurface drainage. A current trend in the industry is to build greens elevated above the fairway. Elevation encourages surface and subsurface drainage, helps prevent runoff from

**FIGURE 5.44** Damaged turf from excessive/uncontrolled traffic.

**FIGURE 5.45** Precisely controlling soil moisture largely determines the success or failure of a golf green.

adjacent slopes onto the green's surface, and adds character to the hole.

Greens generally require extensive drainage systems, with drain lines placed every 10 to 20 feet (3 to 6 m) across the entire green. An extra drain line (often referred to as a "smile" line) along the front edge of the green should be installed to handle the large volumes of water collecting at this traditional low spot.

## Shade

Shade affects turfgrass quality and health by reducing the amount of carbohydrates produced through photosynthesis. Without adequate carbohydrate reserves, the turfgrass plant is more susceptible to stress and considerably less able to recover from frequent mowing. Sunlight also helps dry greens after a heavy dew or rain and discourages algae and moss buildup, as well as fungal diseases.

Planning should consider that the sun is lowest on the horizon during fall and winter months. Trees to the east, south, and southwest of greens will cast longer shadows during cooler months and may cause shade problems even though trees may be some distance from the green. Shade also may increase the potential for cold damage to grasses, since shade-covered greens stay colder longer than greens in full sunlight. It is suggested during the planning stage that summer (June) and winter (December) shade patterns for proposed golf green sites be sketched every two hours from 8 p.m. until dark. These sketches will indicate which trees may need to be removed, or, if trees cannot be removed, whether the putting green may need to be relocated. If tree removal is not possible, light can be increased by pruning, and air movement improved with fans.

**Adjacent holes, houses, and roads.** Another location consideration is a green's relationship to adjacent golf holes, housing developments, highways, and other high-population areas. Many times a green can be strategically placed to guide players away from these areas. However, tree barriers, shrub lines, and nets are sometimes required to protect nearby personnel and property.

**Size and slope.** The size of a golf green should be large enough to allow for adequate selection of pin placement but not so large as to become a maintenance burden. Smaller sized greens will readily show the effects from traffic concentration, while larger ones increase maintenance costs. In general, golf greens range from 5,000 to more than 7,500 square feet (465 to 697 m$^2$), averaging about 6,000 square feet (557 m$^2$). Greens on par-3 holes are usually larger, whereas longer holes have smaller putting surfaces, therefore requiring a more accurate shot. Often a municipal or public course has a minimum size of 6,500 square feet (604 m$^2$); resort courses have a minimum size of 5,000 square feet (465 m$^2$) while private course golf green size varies with each hole.

In order to provide challenge and interest to players, a good putting green design should incorporate unique characteristics in size, shape, contour, and location of bunkers or other hazards. The placement of bunkers and shaping of contours surrounding a green should prevent concentrated traffic in any one area, and also allow adequate room for efficient turfgrass maintenance practices. The outline of the green should avoid any sharp turns, as the resulting pressure exerted by mowers will cause compacted, worn areas. Severe contours or mounds also are not necessary to produce a good test of putting. Instead, they limit cup placement and produce droughty, easily scalped areas (Figure 5.46).

The newer "ultradwarf" grasses on greens allow closer mowing heights than older cultivars. Sharp contours or ridge lines are easily scalped or gouged by low-mowing machines. In addition, surface slopes in excess of 2.5% to 3% in combination with these "ultradwarf" grasses reduce the usable hole locations and provide more penal play characteristics.

## Profile

The modern putting green consists of two to four distinct components, including (from top to bottom) the rootzone medium, choker sand layer (optional), gravel layer, and drain lines. The rootzone medium is the finest textured, the choker layer is intermediate textured, and the gravel layer is the coarsest textured component. This profile creates a perched water table in the finer textured layer, since water will not move (or percolate) readily from the small pores of the finer textured layer into the large pores of the coarser layer unless the finer layer is saturated with water to some depth (Figure 5.47). An example of this principle involves placing a saturated sponge on top of a bed of gravel, coarse sand, or another material. The water will stay in the sponge due to the differential particle size between it and the coarser material (called granular discontinuity). However, if additional water is added to the sponge, the water's weight will eventually break the tension between the two materials and water will start flowing. There are several successful putting green construction systems, each using some or all of these components.

**USGA specifications.** The best known and most widely used system is a tiered or layered system adopted by the U.S. Golf Association (USGA) (Figure 5.48). In this system, 12 to 14 inches (30.5 to 35.6 cm) of rootzone medium overlay a 2- to 4-inch (5 to 10 cm) coarse sand layer (choker), which in turn covers a 4-inch (10 cm) layer of gravel. Drainage is provided by drain lines cut into subsoil at 15- to 20-foot (4.5 to 6 m) spacings. The gravel blanket helps move water rapidly to the drainage lines and out of the green while the choker layer prevents migration of sand into the gravel layer. The physical textural difference between the gravel and rootzone mix creates a capillary break, where water will not move freely into the gravel unless the rootzone mix above it is saturated.

USGA greens, if constructed properly, have a history of providing many years of satisfactory service. However, appropriate

**FIGURE 5.46** Severe contours or mounds limit cup placement and often produce droughty, easily scalped areas.

FIGURE 5.47 A perched water table formation when a finer textured soil is placed over a coarser textured one (WSU).

FIGURE 5.48 Cross section of a USGA-adopted golf green soil profile with a sand "choker" layer.

sands and gravel may be difficult and expensive to obtain, and the expertise and care required in construction are demanding.

**Hybrid greens.** A modification to the USGA system allows the intermediate choker layer to be eliminated (Figure 5.49). This deletion depends on whether very specific criteria, as determined by laboratory analyses, are met by the rootzone medium and the gravel (see Chapter 4). If the gravel is too coarse or the rootzone medium too fine, problems may arise when sand from the rootzone migrates into and clogs the coarse gravel layer.

**Sand greens.** Figure 5.50 offers a one-tier profile with simplified construction standards for pure sand golf greens (often referred to as the "California method"). It consists of 12 to 14 inches (30.5 to 35.6 cm) of an appropriate rootzone sand overlaying the native soil. Drain lines are trenched into the subgrade and backfilled with gravel. Unlike the previous two profiles, the 4-inch (10 cm) gravel layer is deleted, as is the 2- to 4-inch (5 to 10 cm) choker layer. This type of green is simple and relatively inexpensive to construct. It can perform quite well if the native soil underlying the rootzone is either impermeable, or a layer of plastic (e.g., six mil) is placed on the subsoil before adding the rootzone mix to prevent the downward movement of water. However, if the native soil readily drains, moisture will be sucked out of the rootzone medium and the green will be extremely droughty and difficult to manage. Research also indicates that, for sand greens to drain as fast as two-tier greens, sand greens must percolate at least 20 inches/hour (51 cm/hour) greater than percolation rates in two-tier greens. Sand greens are most difficult to grow-in as they tend to remain dry and lack nutrient-holding capabilities. Drainage line spacing also should be based on the permeability of the rootzone sand, average rainfall rate, and the amount of water to be removed or retained. This drain spacing is much more critical for water removal than in a two-tier profile; a qualified laboratory should be consulted to make this determination. The planning committee should consider all three green profiles and weigh the benefits of each against its negative points. Generally, better results can be expected from those greens depicted in Figures 5.48 and 5.49; however, under financial restraints, Figure 5.50 may be a suitable alternative.

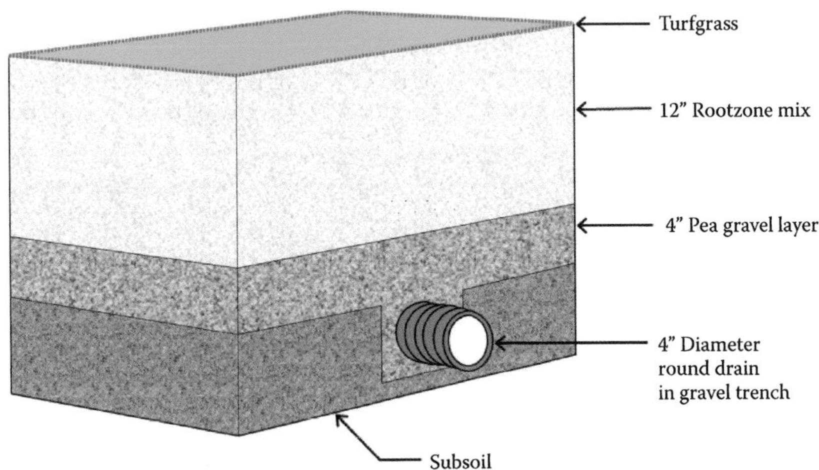


FIGURE 5.49 Illustration of a USGA-adopted golf green soil profile without the sand "choker" layer.

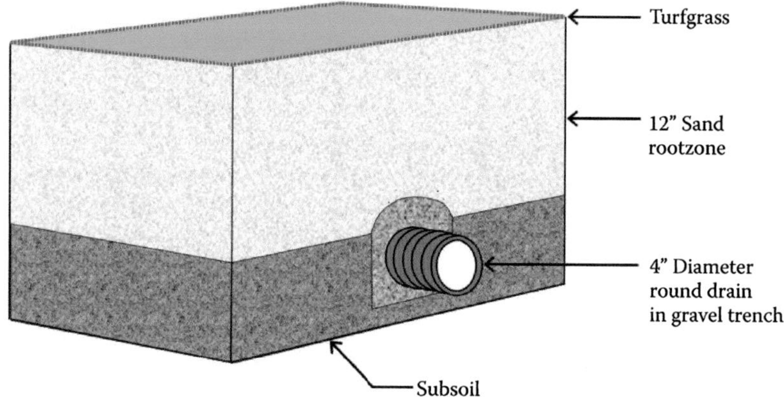


**FIGURE 5.50** Illustration of a "sand" or "California" golf green profile without the gravel sublayer.

**Native soil greens.** Despite the advances of putting green construction, a high interest remains in building native topsoil-containing greens, mainly for financial reasons. In temperate areas, for example, where play only occurs a limited amount of time annually, it may be difficult to justify the considerably increased funds for a modified soil profile green. If native soil is used, however, one must recognize that it will not drain as well internally; thus, sufficient (e.g., 1% to 3%) surface drainage must be included in the design. Also, soil compaction is the other major culprit of using most native soils. Compaction is minimized during construction by keeping large, heavy machines off the greens. Also, compaction may require additional coring (aerification) and topdressing once the greens are established to help combat drainage problems. The advantages of native soil greens include: (1) they are cheaper to build, (2) they hold water and nutrients much more efficiently than sand-based soils, and (3) they are less likely to have drastic changes in soil chemical properties (e.g., soil pH). However, be prepared to accept some risk of failure with them since they vary from a proven standard. Visit similar examples of the type of construction and materials being considered. No substitute for proper materials and construction methods exists.

**Tax depreciation.** The Internal Revenue Service (IRS) in the United States has ruled the cost of building or rebuilding "modern" golf greens is depreciable over 15 years (20 years under the Alternative Depreciation System). However, "push-up" or native soil greens without internal tile drainage systems, along with the cost of general earthmoving, grading, or initial shaping of all golf greens, are nondepreciable. "Push-up" greens are considered undistinguishable from land, have an unlimited useful life, and therefore are not depreciable. "Modern" greens typically have lost drainage effectiveness after 20 years and their tile systems are replaced; thus, they have a determinable useful life.

## ROOTZONES

Green construction is a time-consuming and costly procedure when building a golf course because of the extensive excavation and restructuring of the area and the expensive materials used. After the design of the greens has been agreed upon, construction steps involve:

1. Professional surveying and staking of the green
2. Shaping and compacting the subgrade and grading the surrounding area
3. Installing subsurface drainage
4. Placing the gravel layer, if used
5. Installing the choker layer, if used
6. Off-site mixing of rootzone mix and its subsequent placement
7. Installing the irrigation system
8. Sterilizing the soil (sometimes, optional)
9. Settling and finishing grading prior to planting

Shortcuts usually result in long-term dissatisfaction by the golfers and increased maintenance for the superintendent.

### Surveying and Initial Staking

The architect provides a plan drawn to scale detailing the slope and shape of the intended green. A competent, licensed surveyor is then responsible for ensuring this plan is transferred to the field. A permanent bench mark (permanent elevation point) must first be established from which all subsequent elevations are made. Bench marks are usually centered in the golf course construction site, but uneven terrain or unavoidable obstacles may necessitate the use of several bench marks throughout the course. Once the bench mark is established, the perimeter of the putting green is staked at intervals of every 15 to 25 feet (4.5 to 7.5 m) or at every change in direction or elevation (Figure 5.51). The purpose of the perimeter stakes is to identify the outline of the green and provide initial surface contours according to the architect's drawings. These stakes should be properly coded (e.g., colored) to minimize errors.

### The Subgrade

**Contours.** Final subgrade contours should closely reflect the contours of the surface. Consequently, successful green

**FIGURE 5.51** Staked perimeter of a golf green indicating the depths of layers to be added.

construction starts with a properly planned and constructed subgrade (Figure 5.52). Internal drainage follows the contours of the subgrade. Under normal circumstances, subgrade contours should not be sloped exclusively toward the front of the green since this will cause the front edge to be extremely wet. A soggy turf exposed to concentrated foot traffic quickly becomes worn and thin. It is better to have the green's slopes draining away from high-traffic areas and also from any side facing the cart path's entrance and exit.

Depending on the green design and elevation of the site, the subgrade will be built into the existing grade or cut into the subsoil. If the grade is to be cut into the subsoil, the stripped topsoil may be stockpiled for future construction, such as mounds adjacent to the green, or distributed over the fairway and rough. Usually, greens built into the existing grade are elevated, requiring outside fill material for the subgrade. Heavier soils, such as clays, are desirable for the subgrade since these are easily compacted to form a firm base that does not readily shift or settle. In either case, the subgrade must be compacted to prevent future settling that might create depressions or pockets of poor drainage or, in the event of a higher grade, droughty areas. This is accomplished with a power-driven vertical compactor (modified jackhammer), a vibratory plate, or with a water-filled mechanical roller operated in several directions across the subgrade.

The subgrade is constructed 14 to 18 inches (36 to 46 cm) below the planned surface, and should look like the finished green, but at a lower elevation. Contours of the subgrade should match those of the surface to within a tolerance of 1 inch (2.5 cm). The gravel layer must conform to the finished surface grade even if the subgrade does not. Initial shaping of subgrade contours involves placement of fixed grade stakes that are referenced to the permanent bench mark. The bulldozer operator then follows these premarked stakes to the depths indicated. Once the initial grade is established, it should be resurveyed and then inspected by the architect to ensure that settled contour elevations match original specifications.

A uniform subgrade, or uniform depth of green, is critical since soil and water physics that dictate the amount of water retained in a soil profile are inversely proportional to its depth. This means the deeper a soil profile is, the less water the top surface will hold. Uneven soil profile depths will have areas that remain excessively dry (high spots), while others will remain soggy (low spots). This greatly increases costs later as

**FIGURE 5.52** The final subgrade should closely reflect the contours of the surface.

the superintendent struggles to maintain uniform soil moisture, usually by using extensive hand-watering (Figure 5.53).

The finished subgrade should be smooth; free of any pockets, rocks, or tire tracks; and firm enough to support construction equipment to prevent settling later. Any plants growing in the subgrade should be removed or killed before applying gravel or sand layers.

**Drainage installation.** Putting green rootzones are formulated to drain quickly and allow play to be resumed shortly after heavy rain or irrigation. However, installation of a well-designed drainage system is critical for water removal from the subgrade, especially if the native soil is a clay or has an impermeable layer (Figure 5.54). Without drainage, the green could remain excessively wet and unplayable for several days after heavy rain.

**Outlet location.** The first task in drainage installation is locating an adequately sized outlet area for the water. Typically, drain lines are routed into nearby ditches, ponds, retention areas, larger drain lines, existing French drains in the fairways, or nearby out-of-play grass areas. Discharge lines are normally nonperforated pipe and should be laid across, rather than down, a steep slope to reduce the flow rate from the green. In some cases, a suitable discharge area may not be readily available and a sump and pump may be required (see earlier discussion on this).

**Spacing.** Drain lines should be spaced 10 to 20 feet (3 to 6 m) apart. If the golf green is in an area with a high water table, it may be necessary to place larger drain lines deeper into the subgrade to lower the water table and handle the increased drainage.

**Design.** Typically, drainage lines are installed diagonally to the grade in a gridiron or herringbone pattern. However, any arrangement is acceptable as long as each line has a continuous downward slope and water does not have to travel more than 10 feet (3 m) to a drain line. Greens with slopes greater than 2% or having surface water runoff from higher surroundings should have an interceptor drain line that rings the perimeter of the green, especially in the front or lowest areas (Figure 5.55).

Herringbone designs are generally the most popular and are well suited for irregularly shaped or relatively large turf areas

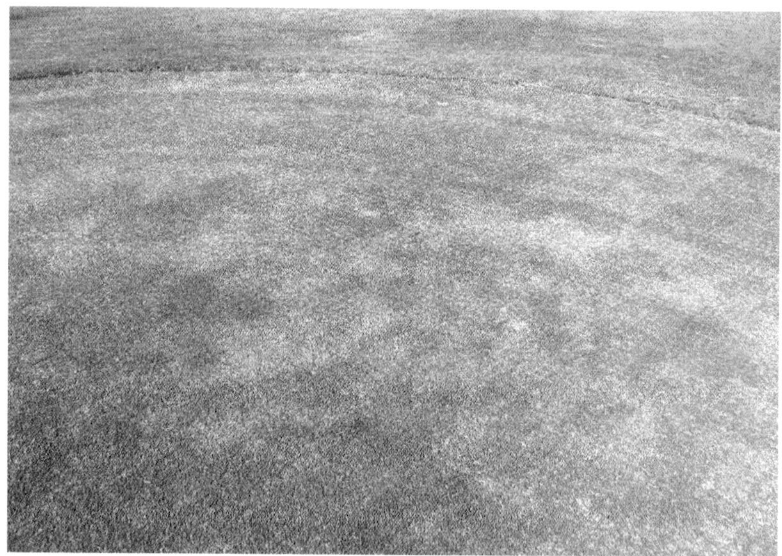

**FIGURE 5.53** Varying turf quality due to excessive deep soil profile, which holds less moisture near the surface.

**FIGURE 5.54** Subsurface drainage installation of a renovated golf green.

**FIGURE 5.55** Using a perimeter or "smile" drain to remove moisture in the front or lowest areas of greens.

due to the numerous lateral drain lines. However, herringbone systems are complicated to install and the pipes may be difficult to locate once installed. If slit drainage is needed later, cutting the slits at 90° angles to the lateral lines becomes difficult.

**Type.** Two- to four-inch (5 to 10 cm) diameter corrugated, flexible, plastic pipe with slits is widely used because it is easy to install and inexpensive. Always place the slits of the plastic pipe face-down on the gravel bed to prevent clogging of drain lines with soil migrating downward from the rootzone. Nylon drain sleeves that wrap around the line are available. However, if silt and clay exist in the rootzone, these may plug the filters and ultimately restrict drainage. Another popular design is to place a fabric along the perimeter of the tile ditch, fill to grade with gravel, and place the edges of the fabric over the drainage ditch (Figure 5.56). Other pipe or tile designs are also available; however, little research exists on the total benefits of these. Refer to Chapter 4 for additional information on determining if other tile types may be successfully used.

**FIGURE 5.56** Placing a fabric over the drainage ditch to minimize soil contamination in it.

FIGURE 5.57 Using flat drainage pipe instead of traditional round pipe to save on installation costs.

An alternative design involves using flat drainage pipe instead of the traditional round pipe (Figure 5.57). The flat pipe is laid directly on the subgrade base and is not cut into the subgrade as with round pipe. Pea gravel is then placed around the flat pipe. The flat pipe still must be on a downward grade to facilitate drainage. This technique is cheaper as drainage ditches are not needed and less gravel is required to surround the flat pipe. Limited research suggests this pipe design is beneficial; however, because it is such a new technique, this construction design has not been proven for all situations and environmental conditions.

**Installation.** Drain lines are laid in trenches dug into the subgrade 6 to 8 inches (15 to 20 cm) deep and 6 to 8 inches (15 o 20 cm) wide. Wider trenches are sometimes used, but this means more gravel is required to fill the trench at higher cost. Normally, the trench width and depth should be no greater than twice the diameter of the drain line. Soil (or spoil) dug from the trenches should be removed or spread between the drain lines and then compacted to provide a slight crown. A one-inch (2.5 cm) bed of pea gravel should be placed in the bottom of the trenches before the drain line is laid. Once drain tile is installed, the trenches should be filled with gravel (Figure 5.58). Care should be taken not to contaminate the gravel with surrounding native soil or drainage may be sacrificed.

**Slopes.** Before excavation, the trenches should be surveyed and staked with the desired depth of cut clearly marked. Drains should be placed only as deep as necessary to obtain the desired slope. Stakes should be marked to give drain lines a minimum downward slope of 0.5% (or 1 foot/200 feet, 0.3 m/60 m), an ideal slope of 1% to 2% (or 1 foot/100 feet to 1 foot/50 feet, 0.3 m/30 to 15 m), and a maximum slope of 3% to 4% (or 1 foot/33 feet to 1 foot/25 feet, 0.3 m/9.9 to 7.5 m). Care must be taken to ensure the trench and drain line are always sloping downward so pockets of standing water do not develop. These lines should be placed diagonally to the slope of the green and not at right angles. All main and lateral lines should be double-checked with a level prior to backfilling to ensure the grade provides the desired drainage. Joints connecting drain lines should be covered with tape, asphalt paper, fiberglass composition, plastic spacers, or covers to prevent gravel and sand from entering the line.

It is recommended that the main drain line have its upper end extended to the soil surface and capped (Figure 5.59). If this line becomes clogged with soil in the future, the cap can be removed and the line periodically flushed. This greatly

FIGURE 5.58 Installing gravel in drainage trenches and as the golf green bed.

**FIGURE 5.59** Daylighting and capping the main drainage pipe for future cleanout capability.

extends the useful life of the drainage system and reduces the need to disturb the playing surface to clean the lines.

**Grade stakes.** After the drain lines are installed, the surveyor should place grade stakes into the subgrade (Figure 5.60). With allowance for the depth needed to drive the stakes into the ground, each stake should be marked at 4 inches (10 cm), 6 to 8 inches (15 to 20 cm), and 18 to 20 inches (46 to 51 cm) above the subgrade, respectively. These markings correspond to the intended depth of the gravel, coarse sand layer (if used), and rootzone layer required in the green profile. Stakes should be placed at frequent spacing throughout the putting green site to indicate any changes in elevation or contouring of the surface.

**Gravel and coarse sand layers.** After the drains are installed, the packed subgrade should be covered with 4 inches (10 cm) of "pea" gravel (1/4 to 2/5 inch in diameter, 6.4 to 10 mm) with it conforming to the proposed final surface grade of the green to a tolerance of plus/minus 1 inch (2.5 cm) (Figure 5.61).

The gravel layer serves several purposes:

1. It is very porous and allows water to rapidly move laterally to the drain lines.
2. It separates the subgrade from the rootzone and prevents the subgrade soil from extracting water from the rootzone.
3. It impedes salt movement from the subsoil into the rootzone. Near coastal areas, where the water table may be contaminated by seawater intrusion, salts can move to the soil surface during periods of hot, dry weather.
4. It helps prevent an excessively wet rootzone due to a rising water table.
5. The interface between the gravel and the sand or rootzone mix above temporarily creates the perched water table, which increases the water-holding capacity of the rootzone mix.

**FIGURE 5.60** Grade stakes indicating the depth of rootzone mix to be installed.

**FIGURE 5.61** "Pea" gravel (~0.25 to 0.4 inch, 6.4 to 10 mm) is often used in golf greens.

Successful greens have been constructed without the four-inch (10 cm) pea gravel layer. However, it is critical that drainage trenches are backfilled with some gravel. If this strategy is chosen, the parent subgrade soil must be compacted and/or a six-mil plastic layer used to separate the rootzone medium from the subsoil and from the collar. It is recommended that courses with adequate financial resources not eliminate the gravel layer since it increases the probability of success.

**Gravel size.** The physical properties of the rootzone mix will determine the size of gravel to be used. In theory, the diameter of the gravel should be five to seven times the diameter of the sand used to construct the rootzone. This will permit stable "bridging" between the sand and gravel and prevent migration of smaller particles from the rootzone into the gravel. For example, if the sand used to construct the rootzone is approximately 1 mm (1/24 inch) in diameter, then 6 mm (1/4 inch) pea gravel is used.

If the proper-sized "pea" gravel is not available, then a 2- to 4-inch (5 to 10 cm) layer of coarse sand (1 to 4 mm) is placed on top of the gravel layer. This coarse sand layer is commonly referred to as the "choker" layer, which acts as a barrier to prevent soil particles from the rootzone from migrating downward into the gravel. It is best to install the coarse sand layer manually to prevent mixing with or into the gravel bed. Sand should be delivered and dumped on the outside perimeter of the green and moved into place in wheelbarrows on a plywood board path. If 1/4 to 2/5 inch (6.4 to 10 mm) of pea gravel is available and the rootzone particle size conforms to those limits previously discussed, then the choker layer may not be necessary. Normally, it is cheaper and easier to use properly sized pea gravel alone compared to using coarser sized gravel plus a choker layer since the choker layer must be evenly spread by hand instead of using a machine. The operator of a tractor or bulldozer will have difficulty spreading the choker layer uniformly and a chance exists the heavy equipment may crush the underlying drain lines. Refer to Chapter 4 for additional information on determining choker layer needs.

**Stone type.** Superintendents should carefully choose their gravel. Several types of stone are sold for use in drainage—crushed limestone, crushed granite, and river rock or gravel (mainly quartz). Other local sources of stone may be available. Granite and quartz gravels are best since they are strong and less likely to be crushed. Softer gravels, such as limestone (calcium carbonate), may break down over time due to the weight of the overlying soil and to chemical reactions with acidic water. Gravel suspected of being soft should be tested by a soils laboratory using the LA abrasion test (ASTM procedure C-131) and values should exceed 40. Gravel sources other than granite and quartz should also be analyzed to determine weathering stability using the sulfate soundness test (ASTM procedure C-88). Weight loss should be less than 12%.

**Alternatives to the choker layer.** A topic currently under debate is whether it is possible to substitute a nonbiodegradable woven filter (geotextile) fabric for the coarse-sand choker layer to prevent the downward migration of fine particles (Figure 5.62). Earlier versions of these fabrics commonly clogged within a year or two of installation. As newer and improved materials are developed, this clogging problem may be eliminated. Recent research suggests a benefit of using or substituting geotextile fabric for the coarse-sand choker layer. An appropriate selected material should maintain acceptable water infiltration and percolation rates, and not clog.

## Rootzone Mix Selection

In earlier times, the most common rootzone material for green construction was simply native soil. A bulldozer operator would "push up" the surrounding soil to a final grade, followed by grass planting. These greens performed adequately as long as traffic was light and the soil was not excessively wet.

**FIGURE 5.62** Geotextile fabric to prevent downward migration of soil particles into the gravel layer.

As golf became more popular and courses received additional play, many of these greens declined or failed. They became seriously compacted, drained poorly, became algae infested, and were more susceptible to damaging outbreaks of disease, particularly *Pythium*. An extensive survey of "push-up" greens revealed the poorest turf was associated with heavier loams and clay soils, while the healthiest turf was usually growing on sands or sandy soils. Today this seems obvious, since it is well known that sands resist compaction, maintain good drainage, and promote deep rooting, but at the time it was a revelation. This insight led to recommendations that putting greens should be constructed using sands or sandy soils as the primary ingredient.

Numerous refinements have been made to the sand-based rootzone over the past several decades, and modern recommendations for rootzone materials are considerably more specific and detailed. Because the success or failure of a putting green often hinges on the performance of the turfgrass root system, experts agree that choosing the rootzone mix is the most important decision when constructing golf greens.

### Sand Sources

**Particle size.** The successful use of sand for greens construction depends primarily on three factors: (1) average particle size, (2) uniformity of particles, and (3) correct mixing of the sand with amendments. All sands are not created equal. Highly uniform sands are well suited for constructing golf greens while less uniform sands are better for making concrete or providing a stable road bed.

Uniform sands are characterized as having most of the individual particles similar in size, which is termed a narrow particle-size distribution. This is important, since like-sized particles do not interpack, and result in good and stable porosity in the soil (Figure 4.18). By contrast, nonuniform sands have particles ranging from very coarse to very fine in size, and these can interpack and have higher bulk density (or compaction). Intermediate particles fill the spaces (pores) between the largest particles, smaller particles fill the spaces between the intermediate particles, and silt and clay can fill any remaining spaces. The net result is a dense sand with reduced pore space, smaller average pores, and a tendency to compact.

Chapter 4 lists the U.S. Department of Agriculture (USDA) particle-size classification for those materials of general interest for building the rootzone of desirable putting greens. Some sand companies provide particle-size distribution, but in many cases the analysis is based on engineering criteria, not the USDA sieve sizes.

**Soil testing.** Materials being considered for rootzone construction should be submitted to a soil laboratory for testing. Only a reputable, reliable soil-testing facility should be used to determine the best rootzone mixture for a particular location. Before choosing a soil-testing laboratory, consider the experience and reputation of the laboratory and the sample turnover rate it can provide. If the laboratory personnel do not have turf experience, then they probably do not understand the specific goals of putting green construction.

Laboratory testing of rootzone materials is important to help formulate the final mix, as well as for quality control once a mix has been chosen. Although they may look identical, sands can change dramatically depending on where a supplier happens to be digging. Sometimes particle size varies with the depth in the pit, or between adjacent pits. It is the superintendent's responsibility to make sure all materials delivered for greens construction meet the original specifications. Random samples of the rootzone components and proposed mixes should be taken and submitted to the soils laboratory to ensure that physical properties do not change. This is where quick turnaround of the samples by the laboratory becomes important.

Soil-testing laboratories usually need a minimum of 2 gallons (7.6 L) of sand and 1 gallon (3.8 L) each of organic matter, soil, and gravel. Once the samples are received, the lab will analyze the physical characteristics of each component to determine the best proportions for the rootzone mix. Included in the analyses are saturated hydraulic conductivity (infiltration and percolation rate), particle-size distribution, total pore space, moisture retention, bulk density, and mineral derivation.

**Shape.** This indicates the predominant shape of sand particles in a mix. Sand shape varies with weathering, chemical properties, and geographic location. The four basic shapes are round, angular, subangular, and crushed. Crushed, sharp sand particles tend to damage turf roots and do not pack properly. Very round sand acts like ball bearings and will shift under foot or machine traffic. It also tends to be somewhat droughty. Sands that are flat may pack excessively. Angular or subangular sands are preferred because they form a more stable matrix, are more resistant to compaction, and provide desirable capillary pore space for drainage and root growth.

**Composition.** Quartz silica sand is preferred for every golf course use (bunkers, greens, tees, and fairways) because it is very resistant to breakdown and retains its original shape. The chosen sand should contain 95% or greater quartz silica. Manufactured sands generally have poor quality and undesirable chemical content, while calcareous sands are soft, unstable, and have a high pH, which can reduce the availability of some micronutrients (e.g., Fe) to grass.

**Particle and bulk density.** Particle density is defined as the mass (or weight) of dry soil per unit volume of the soil solids (excluding pores). If one could melt the soil into a solid mass of known volume, this would be its density. An average value of 2.65 g/cm$^3$ has been found for most mineral soils, and is the standard value used by soil scientists and soil laboratories in calculating other soil properties. Particle density decreases if peat or inorganic amendments are added to the soil.

Bulk density, defined as the mass (or weight) of dry soil per actual volume of the soil, is the more important parameter for golf course superintendents. Bulk density, unlike particle density, includes the pore space volume as well as the volume of soil solids. Since most soils are about half solids and half pore space, bulk densities tend to be about half the particle density. The higher the bulk density of a soil is, the greater is the compaction. The bulk density range for golf greens is between 1.25 and 1.55 g/cm$^3$, with a lower limit of 1.20 g/cm$^3$, an upper limit of 1.60 g/cm$^3$, and an optimum level of 1.40 g/cm$^3$. Incorporating organic matter, such as peat, is one means of reducing the bulk density of a golf green rootzone.

Sand size and the uniformity of the sands influence bulk density. Widely graded sands tend to pack tightly, producing higher densities. More uniform sands tend to pack less (or not at all), resulting in lower bulk densities.

**Soil porosity. Soil porosity** or **total pore space** is the fraction of soil volume not occupied by solid particles. Porosity is important, since it is the pores between solid particles that hold both air and water. The arrangement and size of particles largely determines pore space. Optimum porosity for golf greens ranges from 40% to 55% by volume, evenly divided between smaller capillary (water filled) and larger noncapillary (air filled) pore space (Table 5.3).

**TABLE 5.3**
**Suggested Porosity of Golf Green Soils**

| Reference | Porosity (cm$^3$/100 cm$^3$ or Percent by Volume) | | |
|---|---|---|---|
| | Capillary | Noncapillary (Aeration Porosity) | Total |
| USGA (or two-tier) greens | 15 to 25 | 15 to 30 | 35 to 55 |
| California (sand or one-tier) greens | 10 to 20 | 15 to 30 | 35 to 55 |

Smaller capillary pores largely determine the amount of water held by soil, while larger noncapillary pores determine air content. Noncapillary pores also control how quickly water and air move through a soil. If capillary pores predominate, moisture-holding capacity of the soil will be high, but water and air movement may be inhibited. If noncapillary pores predominate, excessive drainage and high aeration result at the expense of adequate moisture-holding capacity. Also, as bulk density increases, total soil porosity decreases, and vice versa. For example, a rootzone with a bulk density of 1.55 g/cm$^3$ and a particle density of 2.65 g/cm$^3$ has a total porosity of 42%. However, if the bulk density (compaction) increases to 1.65 g/cm$^3$, total porosity would be reduced to 38%.

**Particle-size analysis.** The mineral fraction of soil is composed of sand, silt, and clay. The relative proportion of these in a soil determines its texture. A particle-size analysis provides a general description of physical soil properties to soil scientists, and is the basis for assigning the textural class name, such as sand, sandy loam, clay, and so on, to the soil sample. Once the percentage of sand, silt, and clay has been determined, the specific textural class of the soil can be determined from the U.S. Department of Agriculture's textural triangle (Chapter 4). Native soils used for push-up type golf greens should fall in the sand or loamy sand textural classes. If a modern sand rootzone is being constructed, a pure sand free of silt and clay should be used.

**Sand specification.** The particle-size distribution should be determined for all sands being considered for the rootzone. Values from the analyses can then be compared to the specifications listed in Table 5.4, which summarizes recommendations from several different sources.

Although experts may disagree on precisely which sands are best suited for golf greens, three general recommendations should be met:

1. The sand should be free of silt and clay (Figure 5.63). If present, silt should not exceed 5% (clay 3%) by volume. Larger amounts of silt and clay will reduce infiltration and percolation. Riverbed sands or other sedimentary type soil or peat are often unacceptable due to their high clay or silt content. Sometimes these sands are washed to remove the silt and clay, but all materials being considered should be tested before use.

**TABLE 5.4**
**Suggested Specifications for Sandy Soils Used for Turfgrass Rootzones by Various References**

| Textural Name | Particle Size | ASTM Mesh | Approx. Saturated Hydraulic Conductivity | United States Golf Association (USGA) | | Univ. of California (sand only) | Australian Turfgrass Research Institute | |
|---|---|---|---|---|---|---|---|---|
| | Millimeters | | Inches/Hour | | | % | | |
| Clay | <0.002 | – | – | ≤3 | | | 5 to 10 | |
| Silt | 0.002 to 0.05 | – | <5 | ≤5 | ≤10 | 0 to 8 | | <30 |
| Very fine sand | 0.05 to 0.15 | 270 | 5 | <5 | | | – | |
| Fine sand | 0.15 to 0.25 | 100 | 18 | ≤20 | | | | |
| Medium sand | 0.25 to 0.5 | 60 | 59 | ≥35 (75 ideal) | ≥60 | 82 to 100 | | 60 to 70 |
| Coarse sand | 0.5 to 1.0 | 35 | 217 | <45 | | | | 10 to 30 |
| Very coarse sand | 1.0 to 2.0 | 18 | >217 | <7 | ≤10 | 0 to 10 | | <5 |
| Gravel | >2 | 10 | – | <3 | | | | 0 |

*Note:* Multiple inches per hour by 2.54 to obtain centimeters per hour.

**FIGURE 5.63** Sand used for a golf green rootzone mix.

2. The sand should also be free of very coarse sand and gravel. If present, very coarse sand should not exceed 7% (gravel 3%) by volume. If these limits are exceeded,
   a. Large particles may cut or bruise the stolons/rhizomes of the turfgrass.
   b. Large particles tend to accumulate at the soil surface, resulting in hard greens.
   c. Large particles at the surface may dull mower blades.
   d. Large particles make cup-setting and core aerification difficult.
   e. The soil may not hold adequate water or nutrients.
3. The sand should have a particle-size distribution with the majority (>80%) of the particles falling in the fine, medium, and coarse (0.1 to 1.0 mm) sand fractions. Within this range, the medium-sized particles (0.25 to 0.5 mm) should comprise at least 50% to 70%.

**Infiltration and percolation rates.** Even though the particle-size distribution of a sand falls within the ranges listed in Table 5.4, the sand could have unacceptable infiltration or water-retention values. Therefore, it is essential that the soils lab perform a compacted infiltration test (also referred to as hydraulic conductivity or "perc" rate) on all sands before use. This will help eliminate questionable sands that might create problems later. Refer to Chapter 4 for details on this procedure.

For most putting greens, the initial percolation rate should be 10 to 15 inches (25 to 38 cm) per hour. This rate will decline over time by ~33%, but it should still be well above the minimum of 4 inches (10 cm) per hour. Bermudagrass greens with adequate surface slope can have slightly lower percolation rates of 6 to 10 inches (15 to 38 cm) per hour. High or "accelerated" percolation rates of 12 to 14 inches/hour (30 to 36 cm/hour) may be appropriate for bentgrass/*Poa* courses at some locations to handle heavy rains, or if irrigation water quality is poor or cool-season turfgrasses are being grown outside their range of adaptation. In this case, the sand should contain a minimum of 65% coarse (0.5 mm) and medium (0.25 mm) size particles. Even more desirable would be a sand comprising 75% medium sand, and the majority of the remaining 25% comprising coarse sand. Alternatively, a minimum amount of organic matter should be added to the rootzone mix.

However, accelerated (e.g., >16 inches/hour, 41 cm/hour) draining rootzones often require a longer period of time for full maturity due to excessive moisture applied and lower cation (nutrient) exchange capacity. These also tend to remain firm longer after grow-in and are more difficult to manage due to their low water- and nutrient-holding capacities. Sand (or California-style or one-tier system) greens also require a rootzone permeability about 20 inches/hour (51 cm/hour) greater than a two-tier (USGA) green to provide a similar drainage rate with rates ranging from 15 to 50 inches/hour (38 to 127 cm/hour).

### Soil Amendments

**Organic amendments.** The addition of well-decomposed organic amendment enhances soil structure by improving soil aggregation, nutrient retention, and the water-holding capacity of sands. These are typically inexpensive and somewhat short-lived. However, there are differing opinions regarding the use of organic amendments for golf greens.

Ideally, organic amendments being considered would contain high organic content, low ash content, uniform fiber sizes, and low to medium pH ranges. Organic amendments commonly used in rootzone mixes include peats, composts, and humates such as those listed in Table 5.5. These improve germination and establishment by increasing moisture- and nutrient-holding capacity. However, peats are not all the same; they vary in their botanical origins and quality. Table 5.6 lists specific details of the major organic sources used to amend turf-grown soils.

Of organic matter, 3% to 5% is adequate for sand-based rootzone media with 10% of soil organic matter for soil-based rootzones. Another common belief is that the thatch/mat layer that typically develops several years after establishment will substitute against adding organic amendments during construction. Thatch, however, is not considered organic matter. Until it is decomposed, it will not return any nutrients to the soil but will actually tie them up.

A drawback of organic materials is ensuring that well or properly decomposed sources are used. Their influence may be short-lived as soil microoganisms naturally decompose them. Many materials are not sized for easy, uniform application. They also tend to be regional in nature and some (e.g., peat moss) are extremely difficult to rewet once dry. Their cleanliness (lacking toxic resins, etc.) and high C-to-N ratios should also be checked before use.

### Characteristics of Organic Amendments

**Percent organic matter.** Percent organic matter is determined by a loss on ignition. A peat should have a minimum organic matter content of 85%. Compost, by nature, has a lower organic matter content than peat with a 60% organic matter content as a minimum.

**Fiber content.** Fibrous peats act like sponges that hold water and are characteristic of younger peats such as sphagnum moss peat. Older peats are less fibrous as they decompose; thus, they have less water-holding capacity. Fiber sizes should also be uniform. Due to a lack of reliable testing methods on fiber content, this is not routinely performed.

**Density.** Density is the weight of peat per unit volume and is influenced by the age, texture, and moisture content of the peat. A lightweight (low density) peat requires more peat on a volume basis to produce desirable physical soil properties in a mix.

**Carbon-to-nitrogen ratio.** Soil microorganisms use the carbon in organic matter as a food (or energy) source. However, as these soil organisms multiply, the N in the organic matter source may be insufficient to support them. The organisms then use soil N sources for this food source that otherwise would be used by the turfgrass plant. Additional amounts of N can typically be added to the rootzone mix to compensate for this loss and will be mandatory if a sawdust organic or poor-quality compost source is used.

**Ash content.** Ash is what remains after the organic carbon of organic matter content is burned off and is calculated by subtracting the organic matter content from 100%. Lower ash content reflects a purer organic matter content composition of the material.

**Moisture content.** Dry peats often are hydrophobic (water-hating); thus, they are difficult to re-wet once they have dried out. Wet peats, meanwhile, tend to be lumpy. A moisture content between 35% and 55% is characteristic of most high-quality peats.

#### Peat

Peat is a generic term for partially decomposed plant material formed in bogs under cool and moist conditions (Figure 5.64). It is also the most commonly used organic amendment for putting green soil mixes. Most commercial peats are mined in Canada, North Dakota, Minnesota, and Europe, and since they are derived from different plant materials decomposing under different environmental conditions, these products vary considerably in pH, water retention, organic content, ash and fiber content, and level of decomposition. Peats are broadly classified into moss peat, reed-sedge peat, and peat humus. Moss peats are composed of sphagnum, hypnum, and other mosses. Reed-sedge peat is formed from reeds, sedges, cattails, marsh grasses, and other plants. Peat humus

**TABLE 5.5**
**Qualitative Comparison of Soil Organic Amendments Used to Modify Golf Green Rootzones**

| Soil Amendment | pH | Cation Exchange Capacity | Water-Holding Capacity | Durability (Years) |
|---|---|---|---|---|
| Peat humus | Acid | Good | Good | 5+ |
| Reed-sedge peat | Acid | Good | Good | 4 to 5 |
| Peat moss | Acid | Fair | Excellent | 1 to 3 |
| Rice hulls | Acid | Fair | Poor | 1 to 3 |
| Ground fir bark | Acid | Fair | Fair | 5 |
| Lignified wood waste | Acid | Poor–fair | Good | 8+ |
| Sawdust | Acid | Fair–good | Fair–good | 1+ |
| Sphagnum moss peat | Acid | Good | Excellent | 1 to 3 |

**TABLE 5.6**
**Chemical and Physical Properties of Organic Materials Used in Turf**

| Organic Sources | pH | OM (%) | CEC (Meq/100 g) | EC (dS/m) | Bulk Density (g/cm³) | C:N Ratio | Nutrient Release | Microbial Activities | Hazardous Risks | Blending with Sand |
|---|---|---|---|---|---|---|---|---|---|---|
| Reed-sedge peat | 5.1 to 7.5 | >85 | >140 | 1.0 to 1.5 | 0.19 | 23:1 | Slow | Low | Low | Excellent |
| Reed-sedge peat (lower quality) | 4 to 5 | 30 to 80 | 100 to 150 | 2.4 | 0.072 to 0.112 | 40:1 | Slow | Very low | Low | Excellent |
| Sphagnum peat | 3 to 4 | >95 | 100 to 200 | 2.0 to 3.0 | 0.08 to 0.16 | 47:1 | Slow | Very low | Low | Good |
| Hypnum moss peat | 5 to 7 | <80 | 140 | 0.7 | 0.16 to 0.29 | 35:1 | Slow to mod.[a] | Very low | Low | Good |
| Peat humus | 5.0 to 7.5 | <50 | 1 to 120 | — | 0.32 to 0.64 | 48:1 | Mod. | Low | Low | Excellent |
| Rice hulls | 6.4 | 72 | 60 | 0.95 | 0.50 | 19 to 50:1 | Slow | Mod. to high | Low | Good |
| Cotton burr compost | 5 to 7.5 | 38 | 200 | 1.7 | 1.20 | 8 to 22:1 | High | High | Mod. | Poor |
| Composted bark | 4 to 5.5 | 64 | 50 | 2.5 | 0.40 | 115:1 | Slow | Mod. | Mod. | Poor |
| Manure | 6 to 8 | 4 to 36 | 75 to 100 | 7.4 | 0.70 | 12:1 | Mod. to high | Very high | Low | Poor |
| Biosolid | 6.6 to 7.2 | 24 | 80 | 2.2 to 3.5 | 1.40 | 20 to 100:1 | Mod. | Mod. | High | Good |
| **Desired range** | **6.0 to 7.5** | **>80** | **>100** | **<1** | **<0.5** | **<30:1** | **Slow** | **Mod. or lower** | **Low** | **Good or better** |

*Source:* Modified from Li, D., 2004, *Turfgrass Trends* 60(4):58–61.
[a] mod. = moderate.

is decomposed to the point where the original plant materials are not recognizable. Peats used to modify sands should be high in organic content and low in ash. Fibrous peats are preferred over sedimentary and woody-type peats. Amounts of peat used range from 5% to 20% by volume. Table 5.7 lists specific characteristics of some commonly used peats for modifying rootzone mixtures.

## Humus

Humus is synonymously used with soil organic matter and is the relatively stable, dark-colored colloidal organic matter in soils that contains no recognizable plant parts. It improves soil nutrient exchange, soil tilth, and soil moisture-holding capacity. Chemically, humus is a very complex and diverse group of molecules in the final stage of decomposition, including

**FIGURE 5.64** Peat is a commonly used organic soil amendment for greens construction.

nonhumic biochemical substances such as water-soluble sugars, amino acids, fats, and water-insoluble humic substances, which are unclear molecular structures with high molecular weights and of brown and other dark colors. Humic substances are further divided into *humins, humic acids*, and *fulvic acids* in accordance to their solubility in acids and alkali. Humate (or humin) is partially decomposed humus, or soil organic matter, and is defined as the salt of the humic acid. It has cation exchange sites that bind protons ($H^+$), Ca, Na, Fe, K, Cu, Al, and Mg. Humin is not soluble in water or alkali. Under acid soil conditions, most of the humate exchange sites will be occupied by protons. To take advantage of the extremely high cation exchange potential of humate, the pH of the soil must be raised to remove protons from the exchange sites and replace them with nutrient cations.

Humic acid is commercially produced by adding a dilute (~2%) alkali, usually sodium hydroxide, to a humus-bearing material, usually leonardite or possibly coal, lignite, or

**TABLE 5.7**
**Characteristics of Commonly Used Peats for Soil Modification**

| Type | Composition | Color | Level of Decomposition | Remarks |
|---|---|---|---|---|
| Peat humus (cultivated peat, black peat) | Nonfibrous; fine textured | Dark brown to black | Advanced | • Well-decomposed form of OM, thus very stable; longest durability of peats (~5 years); low to intermediate water-holding capacity (three to six times its weight); low C:N ratio; due to fine texture and higher density, often used in topdressing mixes has a moist density of 30 to 40 pounds/cubic foot (481 to 641 g/m³) and 15 to 20 pounds OM/ cubic foot (240 to 320 g/m³) |
| Reed-sedge peat (lowmoor peat) | Semifibrous; fine textured | Reddish brown to dark brown | Partially to substantially | • Originate from reeds and sedges with good (85% to 95%) OM; older than moss peats and therefore more stable; finer in texture (denser), and therefore used often for topdressing; good water-holding capacity (four to seven times its weight); intermediate in durability; usually high in exchangeable Ca; somewhat low (23)C:N ratio; moist density of 15 to 18 pounds/cubic foot (240 to 288 g/m³) and 6 to 9 pounds OM/cubic foot (96 to 144 g/m³); CEC = 118 meq/100 g; pH of 6.6 |
| Hypnum moss peat (peat moss, sphagnum peat, highmoor peat) | Fibrous | Tan to brown | Partially | • Difficult to mix into soil; usually quite acidic and low in Ca; high in water-holding capacity and intermediate in durability |
| Sphagnum moss peat (top moss) | Fluffy, fibrous young residue of moss | Yellow to tan | Fresh | • Young peats harvested from surfaces of bogs, dried, and screened; lightweight (low density), thus requiring more on a volume basis needed in a mixture; 6 to 9 pounds/cubic foot (96 to 144 g/m³) moist density and 3 to 5 pounds OM/ cubic foot (48 to 80 g/m³); high (e.g., >95%) OM and fibrous; high (65) C:N ratio; high water-holding capacity (10 to 14 times its weight); CEC = 75 meq/100 g and pH = 4.3; difficult to mix into soil; decomposes faster than peats; used more as a surface cover and for packing boxes |
| Sedimentary peat | Nonfibrous | Brown to black | – | • Contains silt and ash; typically hard and lumpy; low water-holding capacity; least desirable of the peats; use with caution |

*Note:* OM = organic matter.

well-composed peat. Leonardite is a lignitic–organic material related to coal. The sodium hydroxide separates the humus from the alkali-insoluble plant residues present in the leonardite. Acid is added to this humus extraction that produces lignin, humic acid, and fulvic acid. About 50% of the leachate is insoluble precipitate of lignin, which is the portion of plants that is difficult to decompose. The other 50% of the leachate is the water-soluble portion—40% of this being humic acid and the other 10% fulvic acid. The exchange sites of humic acid are filled predominantly with protons ($H^+$ ions)—hence, the name—while fulvic acid contains much of the biostimulant-like materials. These acids have a minimum effect on soil pH because the acids are insoluble in water.

In high-sodium (or sodic) soils, humic acid dissolves to form a black organic crust called black alkali on the soil surface, restricting infiltration. Algae also often forms on this black alkali to further restrict water infiltration.

As mentioned earlier, when the predominant exchange sites contain cations other than H, such as Ca, Na, Fe, K, Cu, Al, and Mg, the material is called humate. Humates vary widely depending on the source, and commercial sources often contain fulvic and humic acids, but are still referred to as humates. The general rule of thumb is that the higher the organic matter and humic acid content is, the stronger the soil reaction will be and, therefore, the better the humate. Granular humates have been used to increase the microbial population of soil, provide a slow release of humic substances, and sustain fertility. Liquid products may affect the plant directly, perhaps by enhancing rapid foliar uptake of liquid fertilizers and micronutrients.

Determining which source of humate to buy, or whether to buy at all, is somewhat confusing. When considering granular sources, use the humate with the highest content of humic acids, the highest organic matter content, and the highest carbon and lowest ash content. Also consider whether the humate matrix is sand or clay, since clay is undesirable on putting greens.

As with many things in life, a small quantity of humates may be beneficial, especially in sand-dominated greens low in organic matter, but excessive amounts may inhibit growth. Always follow the manufacturer's rate and timing recommendations and ask to see a detailed analysis of the product and independent test results.

A charcoal-containing material used for carbon sequestration is *biochar* and it is being investigated as a possible soil amendment. Wood-based biochar generally provides the most consistent results, such as increased nutrient retention, moisture holding, and, possibly, improved soil aeration. Additional research is needed to fully explore its long-term benefit as a soil amendment.

### Sawdust

Sawdust is sometimes used as an organic amendment. Fresh sawdust, however, can have adverse effects by causing severe N deficiencies. The C-to-N ratio of sawdust can be as high as 400. Carbon-to-nitrogen ratios above ~20 cause a nutrient imbalance that results in soil microorganisms rapidly using up most or all of the available N. Some sawdusts also contain excessive salts. If used, sawdust should be well composted, be free of toxic materials, and have some N fertilizer added for plant use.

### Muck Soil

Muck soils are available in certain areas and appear similar to peat with their dark color. However, muck soils are usually a silt loam soil with only 20% to 60% organic matter compared to 60% to 95% organic matter for peats. Due to their silty loam nature, muck soils should not normally be used in golf greens due to their tendency to compact and reduce drainage.

### Composts

Composts are organic residues that have been mixed, piled, and moistened to allow thermophilic decomposition until the original organic materials have been substantially altered or decomposed. The use of composts for sand-based rootzone mixes has become popular as regional composting facilities are built. Quality composts have high organic matter content (>60%), are free from heavy metal and chemical contaminants and disease organisms, and are low in silt and clay.

Composts used in sand-based rootzone mixes should be of the highest quality (not from yard waste), preferably aerobically composted in a closed vessel system to ensure complete digestion. These are referred to as biosolid (or sludge) composts. Composts should then be ground and screened through a 1/4 (preferably) to 1/2 inch (6.4 to 13 mm) screen. Composts are often rich in N and P, resulting in rapid and relatively easier grow-in. They also display suppression of certain diseases such as *Pythium* root rot, brown patch, and dollar spot. Carbon-to-nitrogen ratios vary but tend to be low in composts. Their pH values also vary but tend to be high. Composts being considered should have an excellent record of being used as a plant-growing medium and not be contaminated with heavy metals, plastic film, glass, or pesticides such as herbicides or defoliants (as with cotton burs). If a record is not available, a bioassay should be performed to ensure they do not contain any contaminants such as heavy metals, plastic film, glass, or toxins, and are low in silt and/or clay contents. Recently, manure composts have been developed using red worms to decompose the organic matter compared to soil organisms as with traditional composts. This quicker process produces a consistent, granulated material that is easier to work with.

### Other Sources

Other organic materials have been used as components of a rootzone mix. These include rice hulls, composted bark, and various animal and vegetable by-products. However, availability and product consistency may vary by location. When considering organic sources, ensure the product is (1) finely shredded (e.g., screened to 1/4 to 1/2 inch, 6.4 to 13 mm) to achieve the best possible mixing; (2) very low in silt, clay, and salt; (3) well decomposed; and (4) free of toxic chemicals.

Organic amendments should persist in the soil for at least two years. Due to poor drainage and the introduction of pests, organic materials such as animal manure and mushroom compost generally are not acceptable amendments for golf green soil mixes.

### Inorganic Soil Amendments

A number of inorganic amendments (Table 5.8) are marketed as rootzone amendments, and may merit consideration if they are readily available, meet the infiltration and percolation specifications, and are affordable. These are mined, highly porous materials that have been processed, heated, and sized (Figure 5.65). Bulk densities are typically below 1 g/cm$^3$. Inorganic soil amendments do not promote microbial growth the way organic materials do, but may improve either water-holding capacity or aeration of the sand due to their high internal pore space. Since they do not decompose, inorganic amendments usually persist. Also, inorganic soil amendments tend to displace sand on a one-to-one basis, while organic amendments do not. Less total material is needed when using inorganic amendments compared to organic.

#### Example

If a 90-to-10 sand-to-amendment ratio is desired, 100 cubic yards (77 m$^3$) of organic amendment are necessary for each 1,000 cubic yards (765 m$^3$) of sand since organic amendments do not appreciably displace (add to the total volume or bulk of) the sand. Inorganic amendments, however, do displace the sand; thus, only 900 cubic yards (689 m$^3$) of sand are needed with the 100 cubic yards (77 m$^3$) of inorganic amendment to achieve a total volume of a 90-to-10 mixture.

Some unstable inorganic materials, however, may crush into finer particles when subjected to compaction. Inorganic amendments also may retain and release water tightly and retain nutrients less efficiently than organic amendments. They are more difficult for grow-ins and are costlier than most organic amended sands. A ranking of inorganic amendments according to turfgrass establishment from best to worst is

$$\text{Peat} > \text{clinoptilolite zeolite} = \text{porous ceramic} \geq \text{calcined diatomaceous earth} = 100\% \text{ sand}$$

**Calcined clay (also called porous ceramics).** Calcined clay is formed when clays such as illites, montmorillonite, and attapulgite are fired to about 1,200°F (650°C), crushed, and then screened. By firing (heating) these materials, they expand, increasing air-filled porosity, permeability, cation exchange capacity (nutrient retention), and water-holding capacity. This firing or heating process is often referred to as "calcining." Firing of calcined clays is similar to that used for ceramics, so they are often referred to as "porous ceramics." After firing, calcined clay particles become stable, maintaining their original shape and hardness, even when wet. Due to their porosity, calcined clays are also used as absorbents in cat litter and for grease and oil spills. However, cat litter and industrial absorbents are not manufactured to the same hardness specifications as agricultural products and should not be used in rootzone mixes. Calcined clays have an average bulk density of 0.6 g/cm$^3$ and a wide range of particle sizes. *Turface, Ag-Sorb (Oil Dri), Profile, Pro's Choice*, and *Profile Fine* are commercial examples of porous ceramic clay sources.

**Vitrified clays** are similar except they are heated at higher (~2000°F, 1093°C) temperatures; thus, they vitrify (become

**TABLE 5.8**
**Comparison of Miscellaneous Inorganic Soil Amendments Used in Golf Green Construction**

| Soil Amendment | pH | Cation Exchange Capacity | Water-Holding Ability | Durability (Years) | Relative Cost |
|---|---|---|---|---|---|
| Porous ceramics | | | | | |
| • Calcined clay | Slightly acidic | Mod.[a] | High | 10+ | Mod.–high |
| • Calcined diatomaceous earth | Slightly acidic | Mod. | High | 10+ | Mod.–high |
| • Vitrified clay | Slight acidic | Mod. | mod. | 10+ | Mod.–high |
| Clinoptilolite/zeolite | Neutral | High | High | 10+ | Mod.–high |
| Peat | Acidic | High | High | <5 | Low |
| Perlite | Neutral | Low | mod. | 10+ | Mod.–high |
| Pumice | Neutral | Low | Low | 10+ | Mod.–high |
| Sand | Neutral | Low | Low | 10+ | Low |
| Vermiculite | Neutral | Low | Low–mod. | 10+ | Mod.–high |
| Colloidal phosphate | Neutral | Mod. | mod. | 10+ | Mod.–high |

[a] mod. = moderate.

**FIGURE 5.65** Inorganic amendments for golf greens are typically processed, heated, and sized porous materials.

glassy). This results in coarser textured, shale-based material with high particle integrity, making them hard and stable. However, vitrified clays tend to absorb less water, behave similarly to porous sand, and are redder in color. Choose a particle size compatible with the sand to be used. *Greenschoice, Fieldschoice*, and *Diamond Pro* are commercial examples of vitrified clays.

**Calcined diatomaceous earth.** Diatomaceous earth is a natural material composed of the microscopic silicon skeletons of diatoms. Diatoms are single-celled ocean organisms whose cell walls consist of interlocking parts and valves containing approximately 85% to 95% silica. The skeletons of diatoms have a high degree of internal pore structure and large water-holding capacities. Clay binders (<5%) may or may not be used with diatomaceous earth. The products can be extruded or formed into pellets, and then fired to form rigid, highly porous particles compatible with many sands. These products have a bulk density ranging from ~0.5 to 0.9 g/cm$^3$, and have a moderate cation-exchange capacity. Like calcined clays, these materials are used to increase air-filled porosity, permeability, cation exchange, and water-holding capacity. Commercial sources include *PSA*, a diatomaceous earth; *Axis*, a nonfired calcined diatomaceous earth composed of poorly crystalline silica (silicon dioxide); and *Isolite*, an extruded diatomaceous earth.

**Perlite.** Perlite is a light, expanded, porous material produced by heating volcanic rock. It is resistant to weathering but is fragile and may be easily broken by compaction; thus, it is not normally recommended for greens. Bulk density ranges from 0.10 to 0.14 g/cm$^3$. Particle size will influence performance. Finer materials can increase porosity and available soil water. A coarser size will also increase porosity but reduces available water.

**Pumice.** Pumice is another volcanic rock product containing approximately 70% silica. Pumice, like perlite, is available in various grades and also may be unstable. Finer grades increase soil water availability while coarser grades increase porosity.

**Zeolite/clinoptilolite.** Zeolites are natural minerals noted for their very high cation-exchange capacity and low bulk densities. They have found widespread use in industry to remove environmental pollutants and as filtering and purifying agents, and have been introduced to the turfgrass market. Zeolites are mixed with sand primarily to increase cation-exchange capacity to improve nutrient retention, but in some cases may also improve aeration, drainage, and water-holding capacity. Due to their ability to bind cations (especially K), zeolite-based amendments may reduce nutrient leaching from porous sand rootzones. Additionally, they may be formulated to slowly release K and/or ammonium to the turf, and thereby help control nutrition levels. Commercial sources include *Clinolite, Ecolite, EcoSand, EcoSand X, Zeoponix*, and *Zeopro. Zeopro* is a zeolite with its cation exchange sites fortified with nutrients.

**Colloidal phosphate.** Colloidal phosphate is formed when hard-rock and land-pebble phosphates are washed in preparation for market. Soft phosphate, a by-product of this process, is washed into holding ponds. This soft phosphate is recovered and marketed under the name colloidal phosphate and contains 18% to 23% $P_2O_5$ (7% to 9% P). Colloidal phosphate increases the cation-exchange capacity of a sand and increases the small pore space. It also decreases hydraulic conductivity and improves the wettability of hydrophobic sands. Due to this decrease in hydraulic conductivity, colloidal phosphate should be used to amend topsoil but not golf-green sand. A suggested mix with sand is 5% to 8% colloidal phosphate by volume. Order dry material.

**Native soil.** Many areas have desirable local soil sources that, when used in small amounts, provide desirable water- and cation-exchange capacities. A 5% to 10% soil addition often provides these desirable characteristics at economic prices. Care, however, must be used when choosing a soil source since excessive silt and clay may be present. A loamy sand or sandy loam material is best. Representative soil samples along with the sand source should be analyzed by a reputable soil laboratory to determine their compatibilities.

Research and experience show many of these inorganic soil amendments can increase porosity and water retention of sand mixtures as well as decrease bulk density. However, research also suggests too much of several of these may limit the amount of available water to plants. Another concern is the long-term stability of the material once traffic, compaction, and excessive rainfall are introduced on the amended turf. A soil-testing laboratory should be consulted before using any soil amendment for golf-green construction.

## ROOTZONE INSTALLATION

### Off-Site Mixing

After formulating a soil mix, the next step is to uniformly blend the components. All rootzone mixing should be completed off site. The use of commercial blending equipment is strongly suggested (Figure 5.66). When mixed on site (e.g., soil components placed on top of each other and rototilled in), the individual components are often poorly distributed, resulting in localized areas of wet or dry spots (Figure 5.67). Most rototillers only penetrate the top 6 to 8 inches (15 to 20 cm) of soil, leaving the lower 6 inches (15 cm) unamended. For smaller jobs, mixing can be accomplished by tumbling the sand plus amendments in a concrete mixer or by spreading the measured quantities on a hard, smooth surface (such as pavement) and then moving the pile repeatedly in several directions using a front-end loader or tractor.

Once the mixing operation is underway, random samples should be obtained and checked by the laboratory to ensure specifications are being met. Typically, a small amount of mix (25 to 50 tons, 23 to 25 metric tons) should be initially blended by the machine and the mix tested for its suitability. Afterwards, the rootzone mix is blended in increments of 500 to 1,000 tons (450 to 900 metric tons) and tested to ensure quality control.

**FIGURE 5.66** "Off-site" blending of sand and amendments for golf green rootzone.

**FIGURE 5.67** Poor turf from "on-site" rototilling amendments into a green's rootzone.

Each sample is compared to the initial sample to check if the physical specifications are within preapproved tolerance levels. If this cannot be performed, then every truckload of each component utilized in the rootzone mix should be checked at delivery to ensure specifications are met. To construct a sampling tube, use a PVC pipe about 50 inches (127 cm) long, cut at a 45° angle at one end. From a stockpile, take random samples from the top to the bottom and around the circumference of the pile. Brush away the outer 6 inches (15 cm) of the pile and push the sampling tube as far as possible into the stockpile. Tap the sample with a rubber mallet into a clean bucket, thoroughly mixing the various samples into one.

**Moisture.** Sand should be periodically moistened during the mixing operation to facilitate uniform distribution. Adequate moisture also provides the desired "bridging" between the rootzone mixture and underlying gravel layer. Fibrous organic matter should also be moistened to prevent it from clinging to wet sand. Incorporate a starter fertilizer and/or lime as needed.

**Starter fertilizer.** Soil test results should be used as the basis for determining the amount of fertilizer or lime needed. Phosphorus and K are essential nutrients often low in high-sand-content greens. When soil test results are unavailable, a complete fertilizer such as 10-20-20 should be added at 1 to 2 pounds/cubic yard (594 to 1,188 g/m$^3$) of mix. An organic fertilizer such as processed sewage sludge should also be added to high-sand-content greens at 1 pound of a 6-2-0 or equivalent per cubic yard (594 g/m$^3$) of mix. These should be performed during off-site mixing to encourage uniform distribution throughout the soil profile. Some prefer to mix fertilizers at the surface and rototill twice into the top 6 to 8 inches (15 to 20 cm). In this case, 20 pounds of 10-20-20 (10 kg/100 m$^2$) or equivalent and 20 pounds of 6-2-0 (10 kg/100 m$^2$) or equivalent are added per 1,000 square feet and rototilled twice. A granular micronutrient package and a humate or carbohydrate soil conditioner also may be added according to label directions. These become especially important if sand-only (e.g., no peat) greens are constructed.

**Settling and sloping greens.** Once the 4-inch (10 cm) gravel or 2- to 4-inch (5 to 10 cm) choker layer (if used) is evenly distributed, the rootzone mix is spread on top to a depth of 14 to 15 inches. This will settle to a final depth of ~12 inches (30 cm). Depending on the amount and type of organic amendment, ~20% settling of the original mix can be expected. Severely sloping greens typically have excessively wet conditions in low areas, while dry conditions occur in elevated areas. Increasing the rootzone depth at the lower ends of sloping greens reduces (lowers) soil moisture; shallower depths of rootzone mix on elevated areas increase soil moisture in the upper layer.

**Installation.** The rootzone mix is transported to the edge of the green and then unloaded. To prevent disturbance and tire rutting, the soil at the edge of the green should be dry and firm, or plywood sheets should be placed on the soil surface to prevent compaction from the delivery trucks. This becomes critical if the soil surrounding the work site is wet or loose.

A small crawler tractor with a blade is used for pushing and positioning the rootzone mix to a rough grade (Figure 5.68). The tractor should always be operated with its weight

**FIGURE 5.68** Distributing and positioning the rootzone mix on a golf green.

on the distributed rootzone mix and not directly on top of the underlying gravel. This minimizes the possibility of crushing or displacing the underlying drain and gravel layer. Grade stakes placed at 10- to 15-foot (3 to 4.5 m) intervals should be used as a guide in spreading the rootzone mix to ensure that final contours are developed. Once the initial rough grading is completed, irrigation installation and soil sterilization should follow. After the irrigation system is installed, the entire green should be settled and firmed by thorough wetting. This wetting will also check the effectiveness of the drainage system.

## Building Materials

To estimate costs, the USGA Green Section has tabulated the quantities of materials required per 1,000 square feet of putting surface (Table 5.9). In general, 1 ton (907 kg) of sand will cover ~20 square feet (1.9 m$^2$) to a depth of 12 inches (30 cm) (refer to the calculation section at the end of this chapter). It is recommended that an additional two-year supply of rootzone mix be purchased for future topdressing and minor repairs. If a sand dissimilar to that originally used for construction is applied as topdressing, the chance of introducing excessive amounts of very fine sand, silt, and clay is increased. Fine-textured materials could clog soil macropores, resulting in reduced infiltration and percolation and eventual failure of the green. Other costs involve comparing amendments such as organic versus inorganic materials, transportation costs, a 10% to 20% "waste" factor, laboratory testing, blending costs, and sales tax.

**TABLE 5.9**
**Estimated Materials Needed to Build 1,000 square feet of Putting Green Surface**

| Material | Depth, Inches (cm) | Material Required per 1,000 square feet |
|---|---|---|
| Gravel | 4 (10) | 12 Cubic yards (10 m$^3$/100 m$^2$)) |
| Coarse sand (choker) layer | 2 to 4 (5 to 10) | 6 to 12 Cubic yards (5 to 10 m$^3$/100 m$^2$) |
| Rootzone mixture | 12 (30) | 37 Cubic yards (31 m$^3$/100 m$^2$) |
| Tile | – | ~100 Linear feet (30 m) |

**Irrigation installation.** A green normally has an underground, automatic irrigation system installed at the perimeter of the collar (Figure 5.69). Individual greens usually require four to six irrigation heads with spacing determined by the size and shape of the green and the infiltration rate of the soil. Normal head spacing is approximately 60 feet (18 m), varying from 45 to 90 feet (14 to 27 m). Having operational control of each head is preferred over systems that provide total green or zone irrigation control. Individual head control increases irrigation flexibility by allowing for wind correction, watering localized dry spots, and other special local needs. One or two quick-coupler hose-end outlets should also be installed around each green to facilitate syringing, watering localized dry spots, and for hydroaerifying or emergency irrigation. Irrigation line trenches are installed along the perimeter of the green—never across or through it.

Irrigation heads need to be strategically placed to minimize the amount of water applied to bunkers. Constant watering of bunkers results in erosion, soggy shots for the players, and algae and weed encroachment. Separate irrigation lines (or dual heads) should be provided for slopes and areas surrounding the greens. Normally, the grass species or cultivar used here is different from what is used on the green; thus, it has different water requirements. The native soils in these areas also are often heavier and drain more slowly compared to the modified greens mix. Most native soils also hold more water and do not need to be irrigated as frequently as the well-drained green. Water from fairway irrigation also should just reach the perimeter of the green, but not be on it.

The irrigation pipe bed should be fully settled before planting to prevent future uneven surfaces around the green. Soil from the irrigation line should not be allowed to contaminate the green's soil profile. During establishment, sprinkler heads are often placed 3 to 4 inches (7.6 to 10 cm) above the

**FIGURE 5.69** Properly designed and installed irrigation system around a golf green.

finish grade for easy marking. These are lowered after turf is matured to the finish grade. Swing joints, a system of fittings that permits irrigation head adjustment without raising or lowering the irrigation pipe, are used for each head to prevent irrigation head or line damage from foot, vehicular, or maintenance traffic. Swing joints also readily allow for head adjustment once the turf is established and the soil settles.

### Soil Sterilization

Soil sterilization is the next step. Sterilization eliminates most weeds, insects, and nematodes in the rootzone mix. This is truly a case of "an ounce of prevention is worth a pound of cure," since the cost of controlling pests, especially nematodes and weeds, in a nonsterilized green can be considerable.

The most common soil fumigant is metam sodium. Metam sodium does not absolutely require a cover, but is more effective if one is used. Additionally, approximately two weeks are generally required to air the metam sodium-treated soil before planting. A combination of glyphosate plus fluazifop plus triclopyr ester applied prior to metam sodium use improves long-term control if switching turfgrasses and control of nematodes is not a concern. It is suggested that superintendents unfamiliar with soil sterilants contract with a custom applicator.

### Final Grading

Once fumigation is completed, the final grade should be rechecked using a level or transit to ensure it conforms to the original specifications. Any final grading should be performed manually using shovels, push boards, and/or laser-guided drags (Figure 5.70). Once the final grade is established, the green is ready for firming. A small crawler tractor, a mechanically powered 1-ton roller, a vibratory settler, a drag with weight, or a tractor with wide tires is operated back and forth in all directions until the entire surface has been compacted (Figure 5.71). The profile can be encouraged to settle and become firm by irrigating deeply enough so water continuously flows out of the drain pipe. This usually requires up to one hour of irrigation the first time. This heavy irrigation should also be repeated the day prior to planting to ensure sufficient soil moisture for seedlings or sprigs. The rootzone is deemed firm when foot or tire prints are less than 1/4 inch (6.4 mm) deep. If they are deeper than this, additional irrigation and rolling are needed. The outer perimeter of new greens often does not settle sufficiently, and this turf is often higher than the remainder of the green, which can lead to scalping.

Once a compacted rootzone is achieved, the surface should be smoothed by hand raking, although certain types of mechanical sand rakes are acceptable as long as the final grade is not significantly disturbed (Figure 5.72). The area is now ready to be planted.

### Additional Information

**Collars.** The collar region should be constructed using the same techniques as were used for the green since they receive similar traffic. This should reduce problems and improve turfgrass quality. The same grass should be used for both the green and the collar. This helps minimize encroachment of coarser fairway grasses into putting greens. Planting an additional strip of a slow-growing grass, such as zoysiagrass, between the collar and fairway grasses in order to provide an additional buffer to minimize encroachment should be considered. Generally, since collars are mowed higher than the greens, their maintenance levels are somewhat lower.

**Excessive drying.** Collars sometimes have a problem with being droughty. This occurs from the adjacent native soil, which is usually finer in texture and has greater tension for water in the coarser rootzone mix of the collar. This can be prevented by ringing the collar with an impermeable barrier or by grading the collar's perimeter to a slope of 60% to 80% (minimum 3:1) (Figure 5.73). In the "ringing" process, a strip of polyethylene sheeting is inserted between the outer soil and the sandy rootzone mix to act as a vertical barrier, preventing lateral transfer of water into the adjacent soil (Figure 5.58). This polyethylene barrier should be positioned on the outside perimeter of the collar's modified rootzone mix before the mix is placed. Extending

**FIGURE 5.70** Final grading of a golf green prior to planting.

**FIGURE 5.71** Firming or settling the rootzone by repeatedly driving over it.

**FIGURE 5.72** Using a drag to smooth a green's surface prior to planting.

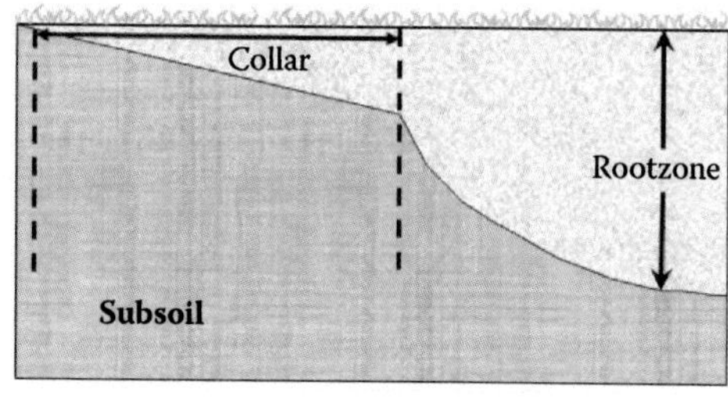


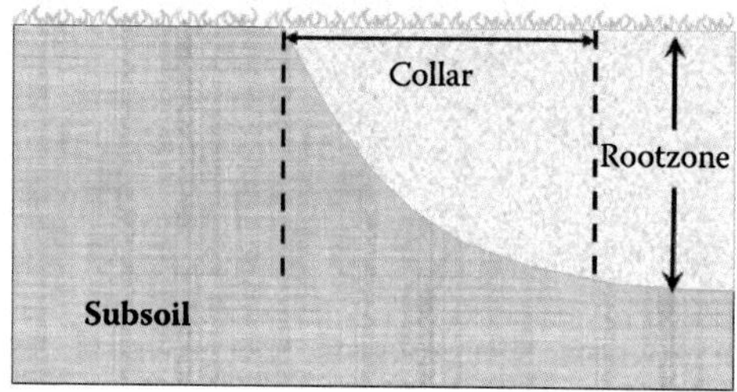


**FIGURE 5.73** Illustration of a desirable collar profile (bottom) versus one that is too shallow (top).

the polyethylene 4 to 6 inches (10 to 15 cm) above the soil during construction will minimize soil erosion into the green construction site. The width of the collar should be a minimum of 3 feet (0.9 m) and, ideally, 5 feet (1.5 m).

## CALCULATIONS

### Estimating Tons of Material for a Given Area

In order to estimate the necessary amount of material to cover a given area, the following assumption is made: 1 cubic yard of sand, choker sand, or pea gravel weighs ~2,600 pounds (or 1.3 tons).

| | | | | |
|---|---|---|---|---|
| 1 cubic yard ($yd^3$) | = | 27 cubic feet | | |
| Cubic feet ($ft^3$) | = | area (square feet) | × | desired depth (feet) |
| Number of tons | = | number of cubic yards | × | 1.3 |

#### Example

To determine the tons of mix required for a 5,000 square-foot green area 1 foot deep,

**Step 1:** Determine the three-dimensional area of green:

$$\text{Square feet} \times \text{depth(1 foot)} = \text{cubic feet of mix}$$

$$5{,}000 \text{ square feet} \times 1 \text{ foot} = 5{,}000 \text{ cubic feet}$$

**Step 2:** Determine the necessary amount of sand needed:

$$\text{Cubic feet of mix} \div 27 = \text{cubic yards of mix}$$

$$5{,}000 \text{ square feet} \div 27 = 85 \text{ cubic yards}$$

**Step 3:** Convert this volume to tons:

$$\text{Cubic feet of mix} \times 1.3 = \text{tons of mix}$$

$$185 \text{ cubic yards} \times 1.3 =\sim 240 \text{ tons}$$

### Estimating Volumes of Material for a Given Area

To estimate fill dirt or topsoil volumes needed to cover a given area, use the following (this assumes the units are the same for all variables and the thickness and shape of the layer is reasonably uniform):

$$\text{Volume} = \text{surface area or cross-sectional area} \times \text{depth}$$

#### Example

To determine topsoil amount (cubic yards ) needed to spread a 1-inch (2.54 cm) layer over an area of 9,500 square feet,

**Step 1:** Convert all values to a common unit: 1 inch = 0.08333 foot
**Step 2:** Compute the volume: 0.08333 foot × 9,500 square feet = 791.6 cubic feet
**Step 3:** Convert to desired units (cubic yards): 791.6 cubic feet ÷ 27 = 29.3 cubic yards

## CONSTRUCTION AND SHAPING OF FEATURES

Detailing of the features, such as tees and bunkers, is the cornerstone of golf course architecture. Interestingly, the putting greens and tees account for less than 5% of the total golf course area, yet play a role in 75% of all golf strokes.

### Teeing areas

Due to concentrated use, tees receive more traffic per square area than any other part of most golf courses (Figure 5.74). Tees should be well drained, nearly flat, with a 1% slope away from the golfer's entrance side and a 1% slope to its back, and free of any depressions that could hold water (Figure 5.75). Teeing areas are often constructed like putting greens, with the same kind of drainage system, layers of the subgrade, and greens mix or rootzone mix for the surface. Use 6 to 8 inches (15 to 20 cm) of an amended sand with drain tile installed at junctions between tees or

**FIGURE 5.74** Practice tees often receive more traffic per square area than any other parts of courses.

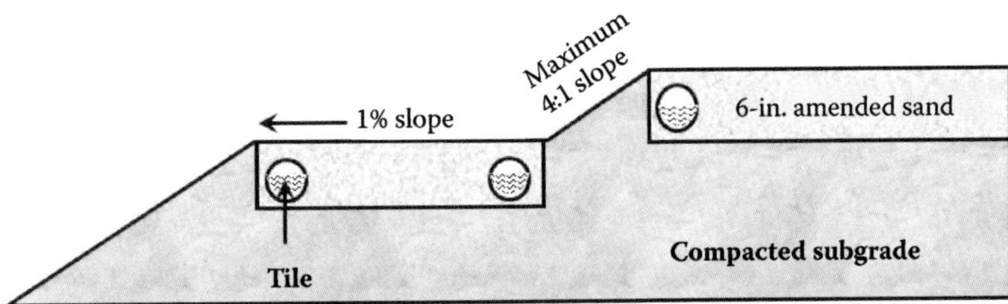


**FIGURE 5.75** Illustration of a tee with a layer of amended sand, a surface slope, and possible tile installation.

throughout on exceptionally large tees. If the rootzone depth sand-based tees are 6 to 8 inches (15 to 20 cm) deep, the drain line trench should be excavated deeper with adequate gravel envelope over the crown of the pipe. A 7:2:1 or 8:1:1 sand to organic matter, such as peat, to soil ratio is a popular mixture for tees. After the tee is rough graded, tile should be installed, especially when tees are constructed at various elevations and connected. If the subsoil of a teeing area is not modified, sufficient surface contouring will be necessary for surface water removal.

Courses cannot have enough tee area, especially on par-3 holes (Figure 5.76). Tee surface area should be one to two times larger than the green surface area for a particular hole. Another specific method to determine tee size is to allow 150 square feet (9.3 m$^2$) of usable tee footage for every 1,000 rounds of golf played yearly on par-4 or -5 holes. On par-3

**FIGURE 5.76** Sufficient teeing areas are necessary to help distribute wear.

holes and the first and tenth holes, where irons or mulligans are used, 200 square feet (18.6 $m^2$) of usable tee footage for every 1,000 rounds played yearly is suggested.

Another common mistake when designing tees and greens is allowing trees to remain, especially those located to the east, south, or southwestern sides (Figure 5.77). Shade from these growing trees increases over time, often ruining even the best constructed turf area. Eight hours of sunlight should be received by tees daily. Multiple entrance and exit points should also be provided. Restriction points like steps, planter boxes, severe slopes, etc. can cause concentrated traffic patterns that easily wear. Steep slopes on tee sides also can become severe hazards to golfers slipping trying to enter or exit the tee. Also, on steep tee slopes, mowers slip when operating, thus tearing or shearing the turf, and spray equipment may easily topple when loads shift. Tee slopes should be ≤18° (or 3:1).

### Bunkers

Bunkers are the most common hazards incorporated into golf courses to force strategic play (Figure 5.78). Bunkers constitute about 3% of the average U.S. golf course acreage, yet they often receive more than 80% of golfer complaints, and up to 25% of a maintenance staff's time is spent maintaining them. Depth of sand, "fried-egg" lies (sand too soft), lack of consistency, and varying moisture levels between bunkers are common complaints by players concerning modern bunkers. Bunkers come in all shapes and sizes and a good golf course architect should blend the golfer's skill level with maintenance needed in the design (Figure 5.79). Today's players generally expect bunkers to be firm, uniform, and consistent between holes. Bunkers also need to have a clear delineation of perimeters, including an appropriate lip to separate sand from turf areas, and should be free of weeds.

**FIGURE 5.77** Due to heavy play, tees require adequate sunlight to recover.

**FIGURE 5.78** Bunkers are the most common hazards incorporated into golf courses to force strategic play.

**FIGURE 5.79** Bunkers come in many shapes and sizes, blending the golfer's skill level with maintenance needs.

Proper bunker construction is important to avoid costly maintenance and/or unsightly problems later. For aesthetics, a steep, flashed-sand bunker facing is often used. However, these are extremely prone to erosion from wind and water. Disproportional amounts of labor are then needed to shovel the sand back on the facing, often causing "fried-egg" lies until the pushed back sand has time to settle. Heavy rain events also tend to cause contamination of the sand with underlying silt and clay, quickly reducing drainage and causing color differences and organic matter to float to the surface. Flat bunkers, meanwhile, have little exposed sand faces and rarely have sand displaced by heavy rain. Flat bunkers also typically provide a more consistent playing surface compared to steep, flashed-faced bunkers. Stacks or layers of artificial turf are a current means of reducing bunker face maintenance while retaining the natural look of traditional bunkers.

Mounds and contours surrounding bunkers also influence play and maintenance. Large, steep bunker mounds and bunkers with intricate edges and numerous fingers (capes) might be more eye-catching, but they are considerably more expensive to maintain due to labor required for hand-raking and trimming (Figure 5.80). Several additional ways to reduce bunker maintenance costs include:

- Avoid liners such as soft fabric that are easily snagged by mechanical rakes.
- Provide multiple access points for mechanical rakes.
- Reduce hand-raking by matching the turning radius of mechanical rakes to the size of the bunker.
- Reduce washing by having a maximum 25% slope.
- Avoid outside surface drainage into bunkers.
- Install internal tile drainage including a "smile" or perimeter drain, especially in low or wet sides.
- Capes (or fingers) should be wide enough to allow trim mowers down-and-back motion.
- Install a sprinkler system that irrigates only the bunker sand and not the surrounding area, allowing better moisture control of the sand.

**Drainage.** Contouring to divert surface drainage away from bunkers is the first drainage consideration. Excessive surface

**FIGURE 5.80** Bunker fingers or capes are considerably more expensive to maintain due to hand labor.

drainage from adjacent areas often causes washouts and trash collection in bunkers. Mounding or intercept drains are means to avoid this.

During construction, particular attention should be given to the overall topography around bunkers. Failure to direct surface runoff water away or around bunkers will lead to extensive erosion problems along with the possible contamination of the bunkers with unsightly clay or other native soils (Figure 5.81). Bunker erosion and soil contamination require costly maintenance and/or replacement measures and invariably occur just prior to important tournaments and other events.

Excessive surface runoff flow onto or over bunkers should be addressed during construction. Small diversion trenches or swells and/or interceptor drainage lines should be installed at the interface of the bunker lip (or edge) on the higher grade. These should be extended (or "wrapped") around the bunker's edge, preventing higher ground surface runoff from reaching the sand.

The subgrade of bunkers is firmly compacted or, if consisting of unstable soil, the subgrade may be covered by a layer of clay similar to a pond liner, a geotextile membrane, or some other stabilizing or sealant treatment (Figure 5.82). Sufficient sand must be continuously maintained over these liners to prevent snagging by players and maintenance equipment, leading to extensive hand-raking by maintenance personnel, an expensive adventure. Bunker interiors should have between 2% and 3% slope to ensure drainage. Drainage lines should be placed in larger bunkers and typically require a herringbone arrangement like putting greens (Figure 5.83), especially in cape and bay bunkers. The main drain line follows the valley of the bunker floor while all perforated drainage pipe is encased in a gravel envelope no less than 3 inches (7.6 cm) above the crown of the pipe to protect it from

**FIGURE 5.81** Failure to divert surface water from bunkers, plus inadequate internal drainage.

**FIGURE 5.82** Using a sealant to help stabilize a bunker's subgrade.

**FIGURE 5.83** Subsurface drainage in bunkers often resembles those of greens.

crushing. This means that for a 4-inch drainage pipe, the trench should be at least 6 inches (15 cm) wide and 8 inches (20 cm) deep. A perimeter (also called "smile") drain should be placed at the interface between steep slopes and inside the bunker to intercept the increased water drainage from the slope. Drainage lines are connected to an adequate outlet. A 2-inch (5 cm) gravel blanket is placed over the bunker floor. Washed pea gravel or crushed stone 1/4 to 3/8 inch (6.4 to 9.5 mm) deep is best.

**Liners.** Liners between the soil and the bunker sand have evolved from using sod, to soft or fabric liners, to the more durable, composite liners of today. These help reduce washouts and prevent "bleeding" of underlying soil with the bunker sand. Liners can be perforated polyethylene, plastic, rubber, filter cloth, mesh, synthetic fabric, and/or net (Figure 5.84). Geotextile fabric liners, which are stapled to the base of the bunker, have been in use for the past several decades. Liners should be placed in the bottom of sand bunkers over the gravel blanket prior to the addition of 4 to 6 inches (10 to 15 cm) of compacted sand with 2 to 3 inches (5 to 7.5 cm) on bunker faces. Bunker liners can become damaged from the use of powered sand rakes if the sand depth is not adequate. Though better than not having one, fabric liners often suffer snagging from mechanical rakes or golfers, staples being displaced by frost heaving, and loss of drainage due to algae formation, silt and clay deposits from the irrigation system, or from surface drainage from surrounding areas into the bunkers.

In recent years, more durable liners have been introduced and are gaining favorable use. These typically consist of a porous hard base material such as gravel, porous asphalt and concrete, or other similar material that is then sealed by special polymers (Figure 5.85). These have sufficient porosity between the base materials to allow drainage but are somewhat rigidly bound by the polymer to help reduce washouts and soil contamination. Another liner option consists of artificial-type turf dyed white (Figure 5.86). Although the newer, more durable liners are improvements, costs have to be

**FIGURE 5.84** A variety of liners such as the shown cloth help reduce washing of bunker sand.

**FIGURE 5.85** Using a special porous polymer as a bunker liner.

**FIGURE 5.86** Using white artificial turf as a bunker liner.

carefully weighed with these newer technologies to other liner options, as well as the shrink–swell tendencies of the soil they are constructed on. Durable liners may crack or otherwise be displaced if the underlying soil has a high shrink–swell tendency.

**Sand.** Sand used in bunkers should drain well and play firm. The sand selected should be determined based on the size of the particles, purity, shape, composition, and color. Bunkers need sufficient sand depths to prevent player injuries. This typically involves 4 to 6 inches (10 to 15 cm) of sand at the base of a bunker and 2 to 4 inches (5 to 10 cm) of sand across the bunker faces. Angular sand helps reduce compaction (firm) and drainage issues. Lastly, bunkers should be maintained to minimize the potential for a "fried egg" or for a ball to become completely buried in the sand.

No "official" specifications exist dealing with properly selected bunker sand. Bunker sands generally fall into two broad categories: (1) native sands mined and screened to achieve the proper particle size distribution and (2) manufactured or crushed sands. which are mined, mechanically crushed, and then sometimes screened. Due to the crushing process, manufactured sand typically is twice the cost of native sands. Factors that help determine the quality of a bunker sand include:

- Particle size
- Particle shape and penetrometer values
- Crusting potential
- Hardness and chemical reaction (pH)
- Drainage via hydraulic conductivity,
- Color
- Playing quality

With regard to particle size, an initial set of specifications was provided by the USGA where 100% of the sand should fall in the range of 0.25 to 1.00 mm. This was later refined to suggest ≥75% of the sand particles should fall in the 0.25 to 0.50 mm range. However, other guidelines have been proposed since sands with these specific USGA textures are very difficult to find and very costly to make (Tables 5.10 and 5.11). More than 65% of the bunker sand should fall in the coarse- to medium-sand range (0.5 to 0.25 mm). Sands in this range usually sift through the grass and down to the soil surface after being blasted out of the bunker. Smaller particles may create drainage problems when blasted onto greens, while silt and clay (particles < 0.05 mm) should be minimized in the sand to prevent surface crusting. Sand particles exceeding 1 mm tend to remain on the putting surface and interfere with mowing operations and putting.

If available, golfers typically prefer white-colored sand. However, the decision on color should not supersede the size and drainage quality of the sand in question. Crusting of bunkers occurs when poor-quality sand (excessive silt and clay) is used (Figure 5.87). This involves formation of a thin layer (1/8 to 1/4 inch, 3.2 to 6.4 mm) of dried, stiff surface sand causing a golf club to excessively bounce off the sand surface, severely hampering a proper follow-through swing. To prevent this, sands with excess (>3%) silt and clay particles, calcareous sand, and sands with less than 20 inches/hour (51 cm/hour) hydraulic conductivity should be avoided. As calcareous sands weather (break down, mostly from acidic rainfall or irrigation water), calcium carbonates dissolve, acting as weak cementing agents, thereby forming a crusted restrictive drainage layer. To avoid this problem, a silica (quartz)-based sand is preferred. Bunker contamination also occurs over time from washouts, dust and organic debris blown in, silt-contaminated irrigation water, and other debris.

Buried balls ("fried-egg" lie) occur when sand is too soft. To test this, the bunker sand is subjected to a penetrometer test (Figure 5.88). Bunker sand with a penetrometer reading <1.8 kg/cm$^2$ (or 0.18 MPa) is too soft. Readings between 1.8 and 2.4 kg/cm$^2$ are acceptable with readings >2.4 kg/cm$^2$ being desirable. Sand particles may also be angular, subangular, subrounded, or rounded in shape, depending upon the sharpness of the edges and corners of the particles. Desirable bunker sand shape is angular with a low degree of sphericity ("roundness"), or in other words, flatter particles. Highly angular sands tend to pack tightly (soft) and may injure (abrade) turf roots. Highly rounded (spherical) sands usually do not adequately pack, remaining loose and unstable, increasing buried lies. Desirable sands with sharp angles resist movement under impact from a golf ball, and thus contain fewer buried balls. A handheld penetrometer is used to measure a sand's ability to resist compaction (Table 5.11).

Finally, if several sands meet these criteria and remain as possible candidates, courses often set up a demonstration bunker site with each candidate sand. Players can then compare how these play and thus help make the final decision on what bunker sand is selected.

The bunker sand should be inspected and approved for cleanliness and freedom of weeds at its origin. Subsequently, each truckload of sand should also be checked upon arrival before being placed into the bunkers. The sand is then introduced or placed into the bunkers and spread by various means. A small tractor plus box blade, small bulldozer, mechanical rakes, or raking by hand are all acceptable means of spreading the sand. A "gunite" machine has also been utilized to

**TABLE 5.10**
**Sand Particle Size for Bunkers**

| Size (mm) | Percent |
|---|---|
| 0.05 to 0.1 (Very fine sand) | <5 |
| 0.1 to 1.0 (Fine + medium + coarse sand) | 78 to 100 |
| 1.0 to 2.0 (Very coarse sand) | <15 |
| >2.0 (Fine gravel) | <2 |

*Source:* Brown, K. W., and J. C. Thomas, 1986, *Golf Course Management* 54:64–70.

**TABLE 5.11**
**Sand Criteria for Bunkers**

| Test | Good | Fair | Poor |
|---|---|---|---|
| Silt and clay | <3% | 3% | >3% |
| Saturated hydraulic conductivity ($K_{sat}$) (inches/hour) | >20 | 20 | <20 |
| Ball penetration (MPa) (packing tendency) | >0.24 (low) | 0.18 to 0.24 (slight to moderate) | <0.18 (high) |
| Crusting | None | Slight | Predominant |
| Setup (thick crust formation or adherence) | None | Slight | Predominant |
| Shape | Angular | Subangular | Round |

*Source:* Brown, K. W., and J. C. Thomas, 1986, *Golf Course Management* 54:64–70.

FIGURE 5.87 Crusting of bunker sand from contaminated silt and clay.

FIGURE 5.88 Penetrometer used to test a bunker sand's firmness to avoid buried ("fried-egg") balls.

spread the sand. This machine is capable of receiving bulk sand and, under high pressure, delivers the sand into the bunker with sufficient force so the sand is compacted and efficiently spread evenly at the same time.

## RENOVATION

Over time, due to increased play, competition, tree growth, player demands, and better grasses and maintenance equipment, many golf courses face the critical financial decision of renovation. Also, many older courses were not built using today's standards or materials, which are designed to handle heavy traffic or play (Figure 5.89). The life expectancy of the various components of a golf course depend, among other things, on the environment, quality of original construction, subsequent maintenance practices, playing condition standards, previous upgrades, and budgets. Improved golf equipment and better conditioned players have forced many older and shorter courses to renovate. Older courses also typically had 8,000 to 10,000 rounds of golf played yearly compared to 30,000 to 40,000 rounds currently played on many courses. These older courses often fall prey to excessive compaction and steadily decline, especially on greens and tees. Older tees also are often inadequately sized and need enlarging.

Webster defines renovation as "...a restoration to an earlier condition or to impart new vigor to; to revive." In some instances, renovation projects can be compared to plastic surgery, where dramatic results and improvements are obvious; in others, however, it can resemble malpractice! In reality, renovation projects are like plastic surgery because the renovation process leaves some scars to the golf course that will require time to heal (mature). Therefore, one cannot fail to appreciate the importance of communication before and during any renovation project. Not only do the positive aspects of the project need to be communicated, but also the negatives and risks associated with the renovation. Computer-generated three-dimensional images of the proposed renovations are also available and are often used to convince membership to support the proposed changes.

There are three different types of golf course renovations: (1) restoration to the original design and form, (2) renovation or improvement of the design to update and/or enlarge areas more like a modern golf course, and (3) complete redesign.

### RESTORATION

There are very few golf course managers and/or members who would admit they have deliberately or by neglect allowed their golf course to change for the worse. Restoration projects

**FIGURE 5.89** Golf courses are often periodically renovated to modernize materials and course design.

usually are focused on the preservation of a "classic" design such as the work of notable golf course architects. It is imperative to select a golf course architect with an appreciation of the original golf course architect's work and who understands the design principles.

## Renovation

Many older golf courses were constructed without standards or were not designed for the level or talent of play they receive today. The rationale for renovating a golf course arises from many considerations ranging from speed of play to the need for additional golf holes. Some of the reasons for considering renovation include:

- Lengthening the golf course or improving the strategy of play
- Expanding the number of golf holes or adding a short game practice area
- Correcting initial poor construction and/or drainage (Figure 5.90)
- Expanding the size or recapturing the original size of tees, greens, or ponds
- Releveling, adding, or reconfiguring tees
- Speeding or facilitating play by installing new cart paths
- Alleviating soil compaction problems on tees and/or greens
- Upgrading or changing the turfgrass species and/or variety
- Reestablishing turfgrass due to loss from some form or combination of stresses (e.g., winter-kill)
- Converting from a mixture of turfgrass species with contaminants and/or "off-types" to a monoculture or pure stand of a turfgrass species
- Rebuilding bunkers with proper sand and drainage
- Removing or replanting mature trees that unfairly restrict play or limit turfgrass growth

**FIGURE 5.90** Renovating bunkers to improve drainage, appearances, and playability.

Renovation can extend from a single tee or green to an entire golf course. Comparing the economics of the renovation and the long-term implications are needed before making a decision (Table 5.12). Golf courses that have undergone properly planned renovations rarely regret the money or time invested.

When renovating, a qualified golf course architect should be hired to review and make suggestions on course changes. A master plan will then be developed by the architect that outlines these potential changes and upgrades and is then presented to the greens committee, board, and, finally, to the general membership. This master plan allows a course to make renovation changes in phases as funds become available.

## Redesign

Many golf courses could be dramatically improved for the enjoyment of members through a redesign of the course. However, if a golf course is committed to redesign, the focus should be on the improvement of the golf course's strategic interest and not merely on its appearance. It is important to note that with any form of golf course renovation, careful attention should be given to the existing facilities, such as water lines, power lines, sewer lines, easements, and drainage outlets.

### Rebuilding Greens

Golf greens, like many things in life, can deteriorate and become unsuitable as a playing surface, especially when proper agronomic practices are not performed during the life of the green. Many reasons contribute to the decision to rebuild greens; reducing internal drainage and the desire to change grasses are the most common (Figure 5.91). Testing the profile for adequate drainage is a key component of the decision of rebuilding. Undisturbed soil cores of the whole rootzone profile can be extracted using PVC pipe and sent to a laboratory to specifically quantify the profile drainage. Other ways of achieving this, or at least providing preliminary indication of insufficient drainage, include measuring surface drainage using a nondestructible double-ring infiltrometer. This can be performed on multiple sites to provide an overall picture of how greens are accepting and subsequently draining water through the rootzone. Another method to ascertain if the gravel layer and drain lines are working satisfactorily is to dig a test pit on the high side of a green where the turf and rootzone are removed, leaving the gravel bed. Water is then introduced into the gravel to indicate its ability to function as designed, as well as check water flow through the drainage system. Other components of greens that often change over time and should be periodically checked include organic matter content at various depths; topdressing layering; sand, silt, clay, and gravel content; pH; bulk density; and total porosity including capillary and noncapillary.

Another common but unrecognized problem is varying depths of rootzones used in the initial construction of the green. Rootzone depth typically is designated to be 12 inches (30 cm) ± 1 inch (2.54 cm) in order to have uniform moisture-holding capacity throughout the profile. If the profile is deeper than 12 inches (30 cm), these areas will tend to stay dry while shallower profiles will stay wet. This uneven soil moisture not only complicates turf uniformity but also prevents the surface from holding shots consistently.

### Regrassing or Resurfacing Greens

If the original soil mix is acceptable and adequate subsurface drainage is present, courses wanting to change putting surface grass or improve upper soil drainage often can strip the existing grass, refill the sand to bring the surface back to grade, fumigate, and replant. Steps include:

1. Cut the existing turf as deep as possible with a sod cutter and remove the grass by hand or by scraping it off with a front-end loader. This extraction depth is typically 2 to 2.5 inches (5 to 6.4 cm) or to the depth of the existing thatch/organic layering.
2. If an excessive thatch and organic matter layer is present, remove an additional 2 to 4 inches (2.5 to 5 cm) of rootzone material. If not removed, these layers disrupt proper drainage. If deep layering is suspected, aggressive core aerification can also be performed to help break these up.
3. Bring the subsoil back to grade using suitable laboratory-tested sand.
4. To mix the new sand and existing subsoil, rototill as deep as possible and in multiple directions. If this is not performed adequately, layers develop that disrupt proper drainage.
5. Following rototilling, the surface is roughly shaped and packed using a water-filled roller or wide tractor tires making multiple passes. A rule of thumb is to firm and float the seedbed until an average-size person will leave no deeper than a 1/4 inch (6.4 mm) footprint when walking on it.
6. Fumigate with a tarp to assure the original grass is eradicated and to kill nematodes, noxious weeds, and diseases. Once the grass is planted, if any original grass is not eradicated, contaminated surfaces eventually occur.
7. The area is then final-shaped, recompacted, and then planted (Figure 5.92).

This renovation method generally requires four to eight weeks to complete. Warm-season grasses should be planted in early summer; planting cool-season grasses in late summer allows maximum growth before seasonal stresses occur.

### Tees

Tees often become unlevel over time. To correct this, they are often stripped, soil mix replaced to grade, laser-leveled, fumigated, and regressed. Smaller tees often need enlarging or additional tee surface area added. Surface and subsurface drainage may need to be upgraded or added as well as a pure and/or

**TABLE 5.12**
**Typical Life Expectancy for Various Golf Course Components**

| Component | Warm Climates | Cool Climates |
|---|---|---|
| | Years | |
| **Greens/collars/surrounds rebuilding** | | |
| • USGA/California/sand based | 12 to 25+ | 20 to 35+ |
| • Pushup/native soil | 25 to 50+ | 25 to 50+ |
| • Tees/surrounds rebuilding | 12 to 15 | 15 to 20 |
| **Bunkers** | | |
| • Rebuild (+ surrounds) | 10 to 15 | 15 to 20 |
| • Replace sand | 5 to 8 | 8 to 12 |
| • Replace drainage pipes/liners | 5 to 8 | 8 to 12 |
| **Regrassing** | | |
| • Greens—USGA/California/Sand based | 8 to 12 | 12 to 15 |
| • Greens—Pushup/native soil | 8 to 12 | 12 to 15 |
| • Tees | 8 to 12 | 12 to 15 |
| • Fairways | 10 to 15 | 12 to 15 |
| • Approaches | 10 to 15 | 12 to 15 |
| • Collection areas | 10 to 15 | 12 to 15 |
| • Intermediate roughs | 10 to 15 | 12 to 15 |
| • Roughs | 15 to 20 | 20 to 25 |
| • Native roughs/natural areas | 5 to 20 | 15 to 25 |
| **Bunkers** | | |
| • Rebuild (+ surrounds) | 10 to 15 | 15 to 20 |
| • Replace sand | 5 to 8 | 8 to 12 |
| • Replace drainage pipes/liners | 5 to 8 | 8 to 12 |
| **Irrigation system replacements** | | |
| • Controllers/wiring/sprinkler heads/swing joints | 8 to 10 | 10 to 12 |
| • Piping/valves fittings—PVC | 20 to 25 | 27 to 30 |
| • Piping/valves fittings—HDPE | 40 to 45+ | 45 to 50+ |
| • Pump stations, fertigation pumps/injectors | 8 to 10 | 12 to 15 |
| **Drainage (fairways/roughs) pipe replacement** | | |
| • Corrugated metal pipe | 15 to 20 | 25 to 30 |
| • Corrugated single-wall pipe | 15 to 20 | 25 to 30 |
| • PVC/double-wall pipe | 20 to 30 | 25 to 35 |
| **Cart path replacement** | | |
| • Asphalt | 7 to 10 | 12 to 15 |
| • Concrete | 15 to 20 | 15 to 20 |
| **Irrigation lakes/ponds** | | |
| • Dredging | 10 to 15 | 15 to 20 |
| • Lake liner replacement | 15 to 20 | 20 to 25 |
| • Rebuild lake/pond edges | 10 to 20 | 15 to 25 |
| • Pine straw/wood mulch replacement | 0.5 to 1 | 1 |
| **Driving range** | | |
| • Relevel/regrass practice tee | 7 to 10 | 8 to 12 |
| • Rebuild practice tee | 10 to 12 | 12 to 15 |
| • Rebuild target greens/target bunkers | 10 to 15 | 12 to 15 |
| **Tunnels** | | |
| • Replacement | 20 to 30 | 25 to 30 |
| • Sumps/pipes/grates/lights/mirrors/path replacement | 10 to 15 | 15 to 20 |

*Source:* Modified from Buchen, T., 2017, *Golf Course Management* 85(8):48–56.

**FIGURE 5.91** Stripping existing turf when renovating golf greens.

**FIGURE 5.92** Planting (sodding) a newly renovated green with zoysiagrass.

improved turf cultivar introduced. Tree removal also is often necessary as shade and heavy tee use often leave bare surfaces.

### Fairways and Roughs

Replacing outdated and contaminated cultivars is common on many fairways, approaches, collection areas, and roughs. Some courses also sand-cap low-lying fairways to improve drainage, firm playing conditions, and reduce time needed to resume maintenance practices. Soil moisture release curves are used to scientifically indicate the capping depth needed for a particular sand and site. If costs prevent total sand-capping, aggressive sand topdressing can be implemented to gradually reach these goals.

### Bunkers

Bunkers are a common source of complaints from golfers. Inconsistent firmness, moisture levels, depths, and soil contamination top this list. Recent improvements in bunker liners have virtually eliminated contamination and erosion issues and thus a common reason for renovation. Drainage and bunker facing irrigation often need installation or replacement. Turf facing the green adjacent bunkers also is often slowly altered over time from constant blasting of sand from them. Installing 14-gage trace wires around bunker lips is a popular way to identify original design and elevations

### Other Areas

Common components such as irrigation and drainage systems, cart paths, irrigation lakes/ponds as well as bridges, retaining walls, tunnels, and driving ranges also become worn and periodically need replacing and/or upgrading (Table 5.12). Courses may also add components such as a short game practice area if space allows as these have become more popular. Table 5.12 lists the typical life expectancies of these for courses located in warm or cool environments. This table also lists several construction methods or components so courses can decide if a particular option is more cost effective over the long run compared to one that may need replacement more frequently.

# 6 Turfgrass Establishment and "Grow-in"

## TURFGRASS SELECTION

The modern golfer demands and expects high-quality turfgrasses and near-perfect playing conditions. This level of quality, in turn, requires a higher intensity of turfgrass maintenance with sufficient personnel and budgets, and the demand for high quality typically starts before a course opens. Following golf course construction, everyone wants "instant" turfgrass so the golf course can open as soon as possible. Golfers are eager to play. Owners, developers, and/or financial partners want to start receiving a return on their investment while the general contractor worries about soil erosion from wind or water, hoping they will not have to reshape the course. These expectations place great pressure on the golf course superintendent and/or general contractor. Newly established turfgrass is very vulnerable to the elements and must receive adequate time to develop and sufficiently mature to withstand the wear and stresses associated with golf play.

Frequently, the question, "How fast can it be done?" is asked rather than, "How long will it take?" Neither can be answered with accuracy because of weather uncertainties and varying environmental conditions. However, turfgrass managers can plan ahead and optimize the timing of turfgrass establishment to improve the chances of providing a quality turfgrass playing surface as quickly as possible.

One of the most important decisions in regard to establishment is the selection of turfgrass species and cultivars for a golf course. A basic understanding of the local environmental patterns, cultural requirements, and species adaptations is necessary to make these decisions (Figure 6.1). Other questions to address include the level of resources (e.g., time and money) allocated for the maintenance of the turfgrass. Refer to Chapter 1 to determine the options in selecting the appropriate turfgrass species or cultivar for your specific area and use.

### Plant Material Quality

The components of turfgrass quality reflect its genetic makeup, environmental influences on the turfgrass plant, and expressed characteristics from biochemical and physiological processes. The characteristics of turfgrass quality have traditionally been established by the personal preferences and needs of the end user. In addition, turfgrass quality characteristics vary with seed versus vegetative materials.

Numerous varieties of a specific turfgrass species typically exist. According to the Plant Variety Protection Act of 1970, a variety must be new, distinct, uniform, and stable (DUS). However, a new variety does not necessarily have to be "better" or an improvement on an older one, even though this is largely understood. The label will indicate "PVP number XYZ, approved (or pending), unauthorized propagation prohibited."

PVP is a form of patent on seed-propagated plants and provides seed companies a return on their investment. In the United States, PVP species since 1994 are protected for 20 years (18-year protection was offered prior to 1994). The distinguishing characteristic must be stable and observable for this 20-year period of protection.

In the application to the USDA-ARS PVP office, five exhibits (checklists) must be completed:

1. Origin and breeding history of the variety
2. Statement of distinctness, showing how the variety differs from all other varieties in that species
3. Objective description of the variety, describing crucial characteristics of the new variety as compared to a control or standard variety
4. Additional data, such as tables, figures, photographs, molecular analysis, and so on, that are useful in describing and distinguishing the variety
5. Description on how ownership of the variety was obtained

Once a certificate of protection is issued, the protected variety can be sold or advertised for sale by variety name only by the owner of the certificate or with the owner's permission. The vegetative plants produced following planting are also protected and may not be increased without the owner's permission. Superintendents, however, can use the nursery green, for example, to repair regular greens but cannot legally use the turf from the nursery to grass another golf course.

Turfgrass seed quality is determined by the seed production and processing practices. The two basic components analyzed for turfgrass seed quality are seed purity and seed germination. These can be analyzed by any reputable seed laboratory. *Seed purity* is the percentage of pure seed in a bag compared to other crop or weed seed and inert materials, while *seed germination* indicates the expected germination percentage. Turfgrass seed certification programs also exist to guarantee the "trueness to type" from a genetic perspective.

To ensure quality, one should personally inspect all turfgrass materials (seed and vegetative) from which planting stock will be utilized. It is imperative to utilize the same source of turfgrass planting stock over time to avoid contamination of the initial planting stock with "off-types" or mutations. This includes not only material from the same grower, but also the same fields or blocks, as variations often occur between them. Most states have turfgrass certification programs providing a paper trail to ensure quality, but these are not standardized or monitored across state lines. Use reputable

**FIGURE 6.1** Turfgrass selection involves understanding local environmental patterns, cultural requirements, and species adaptation to them.

producers and sellers because, once the turf is established, it becomes very expensive, difficult, and time consuming to convert grasses. Proceed with extreme caution on lowest bid or lowest cost sources.

## METHODS OF ESTABLISHMENT

As discussed, turfgrasses can be established by two basic means: (1) seeding or (2) vegetative. These are influenced by having proper soil preparation, soil moisture, balanced nutrients, adequate light, and favorable temperatures.

### SEEDING

Seeding normally is cheaper compared to vegetative planting methods and reduces the potential of introducing soil layering; however, the establishment time with seeding exceeds the time from vegetative means, and the area may be more prone to erosion. Seeded varieties exist for both cool- and warm-season turfgrasses. Most cool-season turfgrasses such as bluegrass, bentgrass, fescues, and ryegrasses can be established by seeding. However, only a limited number of warm-season golf course turfgrasses can be established by seeding. Bahiagrass, buffalograss, carpetgrass, centipedegrass, common bermudagrass, St. Augustinegrass, and zoysiagrass are warm-season grasses that can be established by seeding, but the time required to achieve an acceptable playing surface is much longer compared to vegetative propagation of these species. In addition, most varieties of St. Augustinegrass and zoysiagrass do not produce many viable seeds, or seeds often do not remain "true to type" once they germinate; therefore, these turfgrasses are normally established by vegetative means.

**Seed labels.** The Federal Seed Act and state laws regulate the sale, transportation, and distribution of seed. Up to four paper trail steps exist when new seeds are being developed and produced for consumers. These steps help ensure that the intended purity and identity of the plant are maintained and available.

The first step involves breeder seed, which is produced under the direct supervision of the breeder. In most cases, the same species must not have previously been grown in the same field for five years and must be separated from any other variety of this species by 900 feet (275 m). This helps ensure purity. From breeder seed, foundation seed is produced as a primary source of a genetically identified variety from which all subsequent increases are made. Foundation seed is developed and produced under supervision of the breeder, the foundation seed agency, a private association or cooperation of seed growers, private business, or the agricultural experiment station. As the field grows in, it is inspected by both the breeder and certification agency for any off-types or weeds. Foundation seed is then released for surveillance to other growers, whose subsequent second-generation seeds are referred to as registered seed.

Registered seed is the progeny of foundation seed used to maintain satisfactory genetic identity and purity, and has been approved and certified by the seed certification agency according to the rules defined by the Association of Official Seed Analysts. Registered seed goes to those farmers who grow seed for the retail market. This final product is checked by the certification agency, who certifies varietal characteristics. Such a seed can then be marketed as certified seed—seed that is genetically pure. The color of the certified seed tag is blue, the registered seed tag color is purple, and the breeder and foundation seed tag is normally white. Superintendents should always buy and plant certified seed as this is the only means of ensuring seed purity.

### STEPS TO INCREASE NEED AND MAINTAIN PURITY

Normally, seed laws and regulations require the following to be indicated on the labels (Table 6.1):

1. Species, variety, strain
2. Purity
3. Germination
4. Weeds, including noxious weeds

**TABLE 6.1**
**Components of a Typical Seed Label**

| | |
|---|---|
| Seed kind | Creeping bentgrass |
| Variety | T-93 |
| Lot number | XYZ-3-111 |
| Pure seed | 99.31% |
| Crop seed | 00.00% |
| Inert matter | 00.69% |
| Weed seed | 00.00% |
| Noxious weeds | None found |
| Origin | Oregon |
| Germination | 93% |
| Test date | 12/20 |
| AMS (agricultural marketing number) | 50 |
| Net weight | 25 pounds (11.4 kg) |
| Manufacturer's code | ABC 123 |
| Seed company address | Best Seed Company<br>1234 Seed Road<br>Anywhere, OR 12345 |

Two basic components analyzed for turfgrass seed quality are seed purity and seed germination. These can be analyzed by any reputable seed laboratory. Seed purity is the percentage of pure seed in a bag compared to other crop or weed seed and inert materials, while seed germination indicates the expected germination percentage. Turfgrass seed certification programs also exist to guarantee the "trueness to type" from a genetic perspective.

When seed is obtained, it will not be 100% pure or have 100% germination. It will contain some inert matter (e.g., chaff, dirt, sand, broken seed, etc.), weed seed, and other crop seed. Its germination is also usually less than 100%. Therefore, 1 pound of seed will not result in 1 pound of pure, viable seed.

For seed to germinate, it requires moisture, suitable temperature, and air. The seed absorbs water and, if other conditions are suitable, cells of the embryo become active. Cell division starts and seed germinates. Some seed may be viable but dormant. Dormancy, a resting stage of the seed, may be caused by a seedcoat impervious to water; therefore, the seed does not germinate. Soaking or scratching (also called scarifying) the seedcoat may be beneficial to promote germination. Potassium nitrate ($KNO_3$) as a dilute solution (0.1% to 1.0%) is often used in the soaking process to help loosen or dissolve the seedcoat, enhancing germination. Often a 0.2% solution of $KNO_3$ is used by dissolving 2 g of laboratory grade $KNO_3$ in 1,000 mL of distilled water and seed is wet. Prechilling seed also may enhance germination of many species. Seed is placed in moist substrata at about 40°F (5°C) for five days or longer. Other physiological and biochemical reasons also may exist for this dormancy. The seed may carry inhibitors that have to undergo chemical modification before the embryo is released to sprout. Seed testing laboratories grow seed under favorable conditions to measure germination and list this fact on the seed label.

Germination is the quality of seed most likely to change. Germination of seed stored in high moisture and high temperatures typically drops in a few weeks. Best storage conditions are cool and dry places that are free of rodents such as rats. Time also reduces germination. Seed older than one year generally loses a significant percentage of germination and should be replaced.

Other things being equal, the heavier the seed of a grass is, the better the germination is and more vigorous the young seedlings are. This weight is due to more food in the endosperm for the embryo to draw from. However, the larger the seed is, the more that is needed to achieve the same number of seedlings in a given area.

**Pure live seed.** *Pure live seed* (or PLS) is often used to express the quality of seed, although it is not normally shown on the label. Pure live seed is the percentage of pure seed that will germinate. This value is determined by multiplying percentage of pure seed by percentage of germination and dividing by 100.

### Example

From the label example in Table 6.1, the bentgrass has 99% purity and 93% germination. What is the pure live seed?

$$\text{Pure live seed} = \frac{\text{germination} \times \text{purity}}{100}$$
$$= \frac{99 \times 93}{100}$$
$$= 92\%$$

This means 92% of the package contents consist of pure seed that can germinate and produce plants.

**Seed number.** When seeding to establish turf, the number of seedlings per unit area (such as per square inch) is an important indicator of the amount of seed actually required. Since seed size varies among turfgrass species, the exact number of seedlings per unit area varies with the species and the intent of the seeding. When establishing a new area, the ideal number of seedlings per square inch varies from about 8 to 18 (1.2 to 2.8/cm$^2$), with 10/square inch (or about 1,500 seedlings/square foot, 1.5 per cm$^2$) being a good medium range.

### Example

This example involves a mixture of bentgrass and *Poa trivialis* seed with the following information:

| Grass | Percent Purity | Percent Germination | Seeds/Pound (kg) |
|---|---|---|---|
| Creeping bentgrass | 95 | 90 | 6,000,000 (2,724,000) |
| *Poa trivialis* | 90 | 85 | 2,500,000 (1,135,000) |

If a seeding mixture contains 60% creeping bentgrass and 40% *Poa trivialis* seed, what is the maximum seedling density per square inch resulting from a seeding rate of 8 pounds seed/1,000 square feet?

**Step 1:** Determine the percentage of seed weight of each grass in the 8-pound seeding rate.

$$\text{Bentgrass: } 8 \text{ pound seed} \times 60\% = 4.8 \text{ pounds bentgrass in this 8-pound seeding rate}$$

$$\textit{Poa trivalis}\text{: } 8 \text{ pound seed} \times 40\% = 3.2 \text{ pounds } \textit{Poa trivalis} \text{ in this 8-pound seeding rate}$$

**Step 2:** Find the total PLS for each grass:

$$\text{PLS: \% purity} \times \text{\% germination} \times \text{seeds/pound} \times \text{seeding rate}$$

$$\text{Bentgrass: } 0.95 \times 0.90 \times 6{,}000{,}000 \times 4.8 \text{ pounds} = 24{,}624{,}000 \text{ PLS total}$$

$$\textit{Poa trivialis}\text{: } 0.90 \times 0.85 \times 2{,}500{,}000 \times 3.2 \text{ pounds} = 6{,}120{,}000 \text{ PLS total}$$

**Step 3:** Convert this total seed-to-seedling density per square inch, remembering the initial seeding rate was based on 1,000 square feet:

$$\text{Bentgrass: } \frac{24{,}624{,}000 \text{ PLS}}{1{,}000 \text{ square feet}} \times \frac{1 \text{ square inch}}{144 \text{ square inches}} = 171 \text{ bentgrass seedlings/in}^2$$

$$\textit{Poa trivialis}\text{: } \frac{6{,}120{,}000 \text{ PLS}}{1{,}000 \text{ square feet}} \times \frac{1 \text{ square inch}}{144 \text{ square inches}} = 43 \ \textit{Poa trivialis} \text{ seedlings/square inch}$$

**Seeding rates.** Turfgrass seeding rates are influenced by the seed size, costs, and growth habit. A general guide is to apply a sufficient number of pure live seed to develop 1,000 to 2,000 seedlings per square foot (10,760 to 21,500 per m$^2$). In general, smaller sized seed such as buffalograss, bentgrass, roughstalk bluegrass, and centipedegrass have lower seeding rates (based on weight) and are more expensive. Turfgrasses with a creeping growth habit also are normally seeded at lower rates since the developing stolons or rhizomes aid in establishment and coverage. Table 6.2 provides the seeding rates, approximate number of seeds per pound, and the optimum time of the year for seeding for most available turfgrasses. Table 6.3 lists similar planting information for native and prairie grasses and selected legumes.

### Example

How many pounds of common bermudagrass seed (2,000,000 seeds/pound), 93% purity and 88% germination, would be needed to achieve 1,500 PLS per square foot over a 5,500 square foot area?

**Step 1:** Determine the total number of seedlings required:

$$\frac{1{,}500 \text{ seedlings}}{\text{square foot}} \times 5{,}500 \text{ square feet} = 8{,}250{,}000 \text{ total seedlings over the whole area}$$

**Step 2:** Determine the PLS per pound of seed: % germination × purity × seed no./pound

$$0.88 \times 0.93 \times 2{,}000{,}000 = 1{,}636{,}800 \text{ PLS per pound}$$

**Step 3:** Divide the total seedlings needed by the PLS per pound:

$$8{,}250{,}000 \div 1{,}636{,}800 = 5 \text{ pounds seed needed to provide } 1{,}500 \text{ PLS/square foot over a } 5{,}500 \text{ square foot area } (4.4 \text{ g/m}^2)$$

### Example

An assistant superintendent calls his local seed supplier and requests a particular cultivar of perennial ryegrass for overseeding. The supplier says three lots are available, depending on age. Determine which of the following seed lots is considered the "best buy."

| | Lot A | Lot B | Lot C |
|---|---|---|---|
| Germination | 95% | 75% | 90% |
| Purity | 65% | 80% | 95% |
| Cost (per pound) | $2.00 | $1.75 | $2.10 |

**Step 1:** Calculate PLS for each:

$$\frac{\text{\% germination}}{100} \times \text{\% purity}$$

A = 62%
B = 60%
C = 86%

**Step 2:** Calculate cost per pound of PLS using the following:

$$\frac{\text{cost/pound}}{\text{PLS}} \times 100$$

A = \$3.24
B = \$2.87
C = \$2.46

Although Lot C costs more per pound of seed, when this cost is considered per pound of PLS, it actually is the "best buy."

**TABLE 6.2**
**Seeding Rates, Approximate Seed per Pound, and Optimum Time of Year for Planting Various Turfgrasses for Golf Course Fairways**

| Turfgrass | Seeding Rate | | Approx. Seeds/Pound[b] | Optimum Seeding Time | Optimum Germination Temperatures °F[c] | Days to Germination |
|---|---|---|---|---|---|---|
| | Pounds/1,000 Square Feet | Pounds/Acre | | | | |
| Annual bluegrass | 3 to 7 | 130 to 300 | 2,250,000 | Late summer into early fall | 68 to 86 | N/A |
| Annual ryegrass | 5 to 9[a] | 220 to 400 | 225,00 | Fall | 59 to 77 | 3 to 7 |
| Bahiagrass | 3 to 8 | 130 to 350 | 170,000 | Spring | 68 to 95 | 7 to 21 |
| Bermudagrass | | | | | | |
| • Hulled | 1 to 2 | 45 to 90 | 2,000,000 | Spring into early summer | 68 to 95 | 10 to 20 |
| • Unhulled | 4 to 8 | 175 to 350 | 2,000,000 | Spring into early summer | 68 to 95 | 14 to 21 |
| Blue grama | 1 to 3 | 45 to 130 | 900,000 | Spring | 68 to 86 | 14 |
| Buffalograss | | | | | | |
| • Hulled | 0.5 to 2 | 22 to 90 | 200,000 | Spring | 68 to 95 | 1 to 7 |
| • Burs | 1 to 2 | 45 to 90 | 50,000 | Spring | 68 to 95 | 7 to 28 |
| Carpetgrass | 3 to 4 | 130 to 175 | 1,300,000 | Spring | 68 to 95 | N/A |
| Centipedegrass | 0.25 to 1 | 11 to 45 | 400,000 | Spring | 68 to 95 | 10 to 20 |
| Colonial bentgrass | 1 to 2[a] | 45 to 90[a] | 7,000,000 | Fall | 50 to 86 | 6 to 12 |
| Creeping bentgrass | 0.5 to 1[a] | 22 to 45[a] | 6,000,000 | Fall | 59 to 86 | 4 to 12 |
| Crested wheatgrass | 3 to 5 | 130 to 220 | 320,000 | Late summer into early fall | 59 to 86 | 14 |
| Fine fescue—hard | 4 to 5 | 175 to 220 | 350,000 | Fall | 59 to 77 | 10 to 20 |
| Fine fescue—chewings/red | 4 to 5 | 175 to 220 | 375,000 | Fall | 59 to 77 | 5 to 10 |
| Intermediate ryegrass | 6 to 10 | 260 to 450 | 250,000 | Late summer into early fall | 59 to 77 | 3 to 7 |
| Kentucky bluegrass | 2 to 3[a] | 90 to 130 | 2,250,000 | Late summer into early fall | 59 to 77 | 6 to 30 |
| Kikuyugrass | 0.5 | 25 | N/A | Spring | 68 to 86 | 3 to 10 |
| Lovegrass | 0.25 to 0.5 | 11 to 22 | 1,500,000 | Spring | 58 to 95 | 14 |
| Perennial ryegrass | 6 to 10[a] | 260 to 450[a] | 250,000 | Late summer through early fall | 59 to 77 | 3 to 7 |
| Redtop | 0.5 to 1 | 22 to 45 | 5,000,000 | Fall | 68 to 86 | 4 to 10 |
| Roughstalk bluegrass (*Poa trivialis*) | 1 to 2[a] | 45 to 90 | 2,500,000 | Late summer through early fall | 68 to 86 | 10 to 21 |
| St. Augustinegrass | | | | | | |
| • Hulled | 0.5 to 0.75 | 22 to 33 | 380,000 | Spring | N/A | N/A |
| • Unhulled | 0.5 to 0.75 | 22 to 33 | 450,000 | Spring | N/A | N/A |
| Tall fescue | 5 to 8 | 220 to 350 | 250,000 | Late summer through early fall | 59 to 86 | 6 to 12 |
| Velvet bentgrass | 0.5 to 1 | 22 to 45 | 8,000,000 | Fall | 68 to 86 | N/A |
| Zoysiagrass | 1 to 2 | 45 to 90 | 1,000,000 | Spring | 68 to 95 | 10 to 14 |

[a] Seeding rates for overseeding purposes are higher than listed.
[b] Seed counts per pound can vary significantly among cultivars of a species.
[c] Optimum germination temperatures are listed by the Association of Official Seed Analysts. The lower value indicates a continuous 16-hour temperature exposure time, while the higher value indicates a continuous eight-hour temperature exposure time needed for germination at that particular temperature. C = (F-32) × 5/9; multiply pounds/1,000 square feet by 4.88 to obtain grams/square meter; multiply pounds/acre by 1.12 to obtain kilograms/hectare or by 0.112 to obtain grams/square meter.

**Optimum seeding time.** Since superintendents have no real means of warming soil, adjusting the planting time according to the proper season is the only means of influencing soil temperatures. The optimum time of year for seeding varies with the turfgrass species (Table 6.2). These recommended seeding times correspond with the optimum temperatures required for maximum germination and growth. The optimum seeding time for cool-season turfgrasses is in the fall when soil temperatures are between 60°F and 70°F (15.6°C and 21°C) at a depth of 4 inches (10 cm). Little or only sporadic seed germination will occur when soil temperatures are below 50°F (10°C). Seeding should cease ~30 days before the first expected killing frost. As a second choice, cool-season turfgrasses may be seeded in the early spring before soil temperatures increase above the required growth range. However, seeding these turfgrasses during spring usually results in immature seedlings during the summer heat stress and drought periods, which transcends into a higher susceptibility to diseases such as brown patch and leaf spot. Annual weeds can also be a problem if thin turfgrass occurs from the late seeding in spring.

**TABLE 6.3**
**Seeding Rates, Approximate Seed per Pound, and Optimum Time of Year for Planting Various Native and Prairie Grasses and Legumes for Golf Course Roughs and Out-of-Play Areas**

| Plants | Seeding Rate (pounds/acre) | Approx. Number of Seeds/Pound | Optimum Seeding Time | Minimum Rainfall (inches) | Preplanting Treatment | Days to Germination |
|---|---|---|---|---|---|---|
| | | | **Warm-Season Grasses** | | | |
| Broomsedge | 8 to 12 PLS[a] | 800,000 | Spring | 30 | Soak in water for 24 hours | 14 |
| Little bluestem | 5 to 40 PLS | 260,000 | Spring | 16 | – | 14 |
| Big bluestem | 10 to 60 PLS | 150,000 | Spring | 20 | – | 14 |
| Blue grama | 2 to 20 PLS | 900,000 | Spring | 10 | – | 14 |
| Sideoats grama | 5 to 40 PLS | 180,000 | Spring | 17 | – | 14 |
| Switchgrass | 4 to 30 PLS | 400,000 | Spring | 16 | – | 14 |
| Weeping lovegrass | 4 to 40 PLS | 1,500,000 | Spring | 15 | – | 14 |
| Japanese millet | 20 to 30 | 115,000 | Spring | – | – | 10 |
| Browntop millet | 10 to 20 | 142,000 | Spring | – | – | 7 |
| | | | **Cool-Season Grasses** | | | |
| Smooth bromegrass | 7 to 40 PLS | 140,000 | Spring | 16 | – | 14 |
| Crested wheatgrass | 10 to 30 PLS | 320,000 | Fall or early spring | 9 | – | 14 |
| Western wheatgrass | 10 to 40 PLS | 110,000 | Spring | 14 | – | 28 |
| Indian ricegrass | 7 to 30 PLS | 140,000 | Fall | 6 | – | 42 |
| Reed canary grass | 3 to 10 PLS | 530,000 | Spring or fall | Moist areas | – | 21 |
| Alkaligrass | 20 to 80 PLS | 1,900,000 | Spring or fall | – | – | 28 |
| | | | **Legumes** | | | |
| Lespedeza sp. | 25 to 60 | 90,000 | Spring or fall | – | Inoculate | 21 |
| Crown vetch | 20 to 25 | 110,000 | Spring or fall | – | Scarify and inoculate | 14 |
| Birdsfoot trefoil | 6 to 12 | 375,000 | Spring or fall | – | Scarify and inoculate | 12 |
| Strawberry clover | 5 to 10 | 300,000 | Spring or fall | – | Scarify and inoculate | 7 |
| Crimson clover | 15 to 30 | 140,000 | Spring or fall | – | Scarify and inoculate | 7 |
| Red clover | 6 to 12 | 275,000 | Spring or fall | – | Scarify and inoculate | 7 |
| White clover | 6 to 12 | 800,000 | Spring or fall | – | Scarify and inoculate | 7 |

[a] PLS = pure live seed = purity × germination. Multiply pounds/1,000 square feet by 4.88 to obtain grams/square meter; multiply pounds/acre by 1.12 to obtain kilograms/hectare or by 0.112 to obtain grams/square meter; multiply inches by 2.54 to obtain cemtimeters.

Warm-season turfgrasses are normally seeded in late spring or early summer when soil temperatures are between 70°F and 80°F (21°C and 27°C). Early or premature seeding of warm-season turfgrasses will delay their germination until soil temperatures reach the required germination range and they may lose some of their viability. Planting too late may not allow sufficient plant maturity before cooler fall temperatures and shorter day lengths slow growth.

## Seeding Methods

**Seeding.** Turfgrass seeds are generally planted using either a mechanical seeder (also called a slit seeder), gravity flow drop seeder, rotary (or centrifugal) spreader, or hydroseeder. After planting, seeds are lightly incorporated into the top one-fourth inch (6.4 mm) of the seedbed and the seedbed is firmed with a roller.

Mechanical or slit seeders are used on relatively flat surfaces (four-to-one slope is the maximum limit). These make a small furrow, drop seed into this furrow, cover the seed with approximately 1/4 inch (6.4 mm) of rootzone mix, and firm the seedbed by rolling (Figure 6.2). Many of these mechanical seeders are retrofitted with bins or hoppers to simultaneously apply starter fertilizers or lime.

Drop or rotary spreaders are utilized to broadcast-apply seed over a large area (Figure 6.3). However, the degree of postplanting maintenance is more extensive to achieve optimum results from these types of seeders. To ensure adequate seed distribution and coverage when utilizing these methods, the quantity of seed to be applied should be halved and distributed in two different directions at right angles to each other. Seeds are often mixed 50:50 with a granular product such as fine sand or organic fertilizer for easier and more even distribution. A drop spreader is often used to outline or highlight certain areas, thus minimizing cross-contamination of different turfgrass varieties. Once seeds have been broadcast or seeded, the area must be covered lightly with soil via topdressing to prevent them from desiccation or drying out.

**FIGURE 6.2** Mechanical slit seeder, which makes small furrows, drops and covers seed into them, and firms the seedbed by rolling.

**FIGURE 6.3** A drop seeder for seeding large areas. The foam markers help avoid skips and overlaps.

The area should then be lightly rolled to ensure adequate seed-to-soil contact for optimum germination. Alternatively, newly seeded areas are immediately pressed via tractor tires or mechanical bunker rake tires running back and forth until complete coverage is achieved (Figure 6.4).

This helps firm the seedbed surface besides providing excellent seed-to-soil contact. Extra attention to surface firming is often necessary for the 2- to 3-foot (0.6 to 0.9 m) outside perimeter of a green that is more difficult to reach by certain machines. Extra tracking or rolling during establishment often saves time and work later when trying to smooth this transition from collar to putting green.

**Hydroseeding and mulching.** Hydroseeding (hydraulic seeding) or hydromulching is used to seed areas that are severely sloping, susceptible to wind or water erosion, and/or relatively inaccessible (Figure 6.5). In some cases, hydroseeding can reduce the establishment time needed for a turf species. Also, if green mulch is used, a favorable illusion is created for a more finished look.

The process of hydroseeding consists of mixing seed, mulch, fertilizer, soil activators, and glues, and then spraying this mixture onto the soil. Hydroseeding uses water as a carrier source for seed, where both are applied under pressure (hydraulic). Hydromulching mixes seed with mulch and fertilizer and is often mixed with asphalt spray or synthetic binder to reduce wind and water erosion. If a binder is not used, the mulch is often crimped or pressed into the soil by a specialized machine. Typically, a mulch such as hay, straw, wood, paper, or fiber is applied to reduce soil erosion and help maintain good soil moisture. A rate of 1.5 to 2 tons of hay or straw mulch per acre (or one to three bales/1,000 square feet, 3,360 to 4,480 kg/ha) with 50 to 90 gallons of binder per acre

**FIGURE 6.4** Pressing a newly seeded golf green using bunker rake tires.

**FIGURE 6.5** Hydroseeding severely sloping and/or inaccessible areas.

(468 to 842 L/ha), depending on the slope, is most often used with turfgrass seeding. Ideally, after mulching, equal proportions of soil and mulch should be visible as one looks directly down on the seeded surface. The mulch is left to rot when used on fairways and roughs but is removed on greens. A leaf rake is delicately used to remove about half the mulch on greens when seedlings reach 2 to 3 inches (5 to 7.6 cm).

Wood cellulose mulch is similarly applied as hydromulching and has less chance of weed seed contamination and no need for crimping or tacking to stabilize it. Application rate with wood cellulose mulch is between 1,500 and 3,000 pounds/acre (1,680 and 3,360 kg/ha) in 3,000 to 10,000 gallons ($1.1 \times 10^4$ to $3.79 \times 10^4$ L) of water. Steeper slopes require the higher rates.

In the hydroseeding mix, polymer tackifiers (polyacrylamide, or PAM) are also used to help retain moisture in the mulch material, keep it in place as a tackifier, and prevent "splashing" onto off-seeding sites such as metal fences, roadways, or carpaths. Polymer tackifiers thicken water to a syrup-like consistency, enhancing their secure placement. PAM use rates are 1 to 1.5 pounds/1,000 square feet (0.5 to 0.7 kg/100 $m^2$). Steeper slopes require the higher rate.

Cross-contamination of grass varieties is a potential problem with hydroseeding. Wind and equipment operator skills determine whether this problem occurs.

**Seedbed preparation.** Proper seedbed preparation is essential for the successful establishment of turfgrasses. Refer to the chapter on golf course construction for the proper procedures to prepare a seedbed for planting. Other considerations when seeding cool-season grass greens include:

- Although grown in, seeded areas generally remain somewhat immature the first year.
- Slowly reduce mowing heights so as to ensure that this does not stress the young plants.
- Young plants easily wear. Consider rolling part of the time versus mowing only.

## Vegetative Propagation

Vegetative propagation involves planting a portion of the turfgrass plant (rhizome, stolon) and relying on this portion to form a mat of turfgrass. When properly timed and maintained, vegetative propagation normally produces an acceptable turfgrass playing surface faster than if the same turfgrass species was seeded. Turfgrass species commonly established by vegetative means include bahiagrass, centipedegrass, creeping bentgrass, hybrid bermudagrasses, St. Augustinegrass, velvet bentgrass, and zoysiagrass.

Four basic methods of vegetative turfgrass propagation exist:

1. Sprigging
2. Stolonizing
3. Plugging
4. Sodding

Sprigging and stolonizing are two means of vegetative propagation where the soil is not transferred. In contrast, plugging and sodding both transfer soil as part of the establishment unit, although "washed" sod is available without soil.

**Sprigging.** Sprigging utilizes both rhizomes and stolons as vegetative sources of turfgrass planting material. Sprigging is simply the process of broadcast planting these rhizomes and stolons (sprigs) or planting them in narrowly spaced furrows (Figure 6.6). These rhizomes and stolons have little to no soil associated with them. Sprigging is considered to be the most economical means of vegetative turfgrass establishment.

Sprigs can be purchased by the bushel or sod can be cut and shredded apart into sprigs. Other harvesting methods include roto-tilling of sod into sprigs and then raking them up. Tall grown bermudagrass is sometimes verticut to remove the rhizomes and stolons. This method generally requires a higher planting or sprigging rate since the roots and crowns of the turfgrass are missing. Refer to Table 6.4 for the vegetative planting rates of sprigs and/or plugs for the various turfgrasses.

One inconsistency within the turfgrass industry is the definition of a *bushel* for determining the planting rate of sprigs. A bushel of hybrid bermudagrass was originally determined by the amount of stolons and rhizomes required to plant a specific area. However, the introduction of lower growing varieties such as "Tifdwarf" necessitated a different planting rate than the taller growing "Tif" varieties. A bushel of turfgrass sprigs generally involves measuring the number of sprigs harvested by vertically removing all plant parts (rhizomes, stolons, stems, and leaves) to a soil depth of 2 inches (5 cm) from an area of 1 $m^2$ or from 1 square yard. A square yard of bermudagrass or zoysiagrass sod will yield 2,000 to 4,000 sprigs. One square yard of St. Augustinegrass and centipedegrass will only yield 500 and 1,000 sprigs, respectively. The approximate number of stolons per bushel for bermudagrass is 2,000; for St. Augustinegrass it is 500, and zoysiagrass is 3,000. One square yard of sod yields 1,296 1-square-inch (6.5 $cm^2$) plugs, 324 2-square-inch (13 $cm^2$) plugs, or 81 4-square-inch (26 $cm^2$) plugs.

Other definitions of a bushel also exist. For example, a "U.S. bushel," as defined by the National Institute of Standards and Technology, is the volume within 1.24 cubic feet (0.038 $m^3$). A "Georgia bushel" is about five times the volume of a U.S. bushel, and a "Texas bushel" is approximately twice the volume of a U.S. bushel. Knowing the exact definition and expectation of a bushel is necessary prior to signing a contract because the planting rate using Texas bushels is approximately one-third of that for a Georgia bushel.

Ideally, sprigs should average at least 6 inches (15 cm) in length, be planted with 8 to 12 live sprigs/square foot (85 to 130/$m^2$) and contain at least two vegetative nodes with only a few green leaves. If sprigs contain excessive leaves, they tend to dry out more rapidly—often within hours after harvest, leaving an aesthetically displeasing brown turfgrass cover. Excessive leaves also increase transportation and storage problems associated with sprigs. Turfgrass sprigs often appear brown and lifeless to those unfamiliar with the sprigging process; however, these brown sprigs soon are rejuvenated and grow if properly planted and cared for.

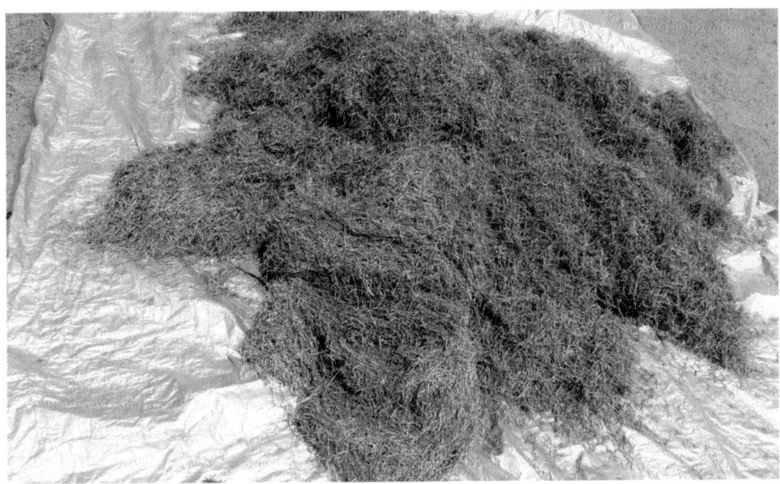

**FIGURE 6.6** Sprigs (rhizomes and stolons) used for planting.

**TABLE 6.4**
**Sprigging Rates for Various Grasses and Uses**

| Area | Variety | Planting Rate (One U.S. Bushel from 1 Cubic Yard of Sod or Equal to 1.24 cubic feet, 0.038 m³) |
|---|---|---|
| Putting greens | Bermudagrass | 15 to 30 bushels/1,000 square feet |
| | Bentgrass | 8 to 12 bushels/1,000 square feet |
| | Seashore paspalum | 8 to 15 bushels/1,000 square feet |
| | Zoysiagrass | 14 to 18 bushels/1,000 square feet |
| Tees | Bermudagrass | 10 to 20 bushels/1,000 square feet |
| | Bentgrass | 6 to 10 bushels/1,000 square feet |
| | Kikuyugrass | 8 to 14 bushels/1,000 square feet |
| | Seashore paspalum | 7 to 12 bushels/1,000 square feet |
| | Zoysiagrass | 9 to 16 bushels/1,000 square feet |
| Fairways/primary rough | Bermudagrass | 400 to 800 bushels/acre |
| | Kikuyugrass | 250 to 350 bushels/acre |
| | Seashore paspalum | 200 to 300 bushels/acre |
| | St. Augustinegrass | 175 to 300 bushels/acre |
| | Zoysiagrass | 300 to 400 bushels/acre |
| | Zoysiagrass | 40 to 50 square feet/100 square feet of nursery sod; plugs or sprigs placed 2 inches apart in 6-inch rows/1,000 square feet |
| Roughs | Bermudagrass | 200 to 400 bushels/acre |

*Note:* One bushel/1,000 square feet = 0.038 $m^3$/100 $m^3$. One bushel/acre = 0.086 $m^3$/ha.

Turfgrass sprigs need to be freshly harvested for best survival. If sprigs are allowed to dry out or overheat, their survival rate greatly diminishes. Best survival is by planting sprigs within 48 hours of harvest. Sprigs not planted within two days may experience more damage from respiration and appear moldy and smell musty. To increase survival, "turn" or rotate the sprigs if possible to allow air flow through them, keep them moist (not wet), and place them in shade to minimize heat buildup. Sprigs also can be covered with a wet tarp or cloth or placed under a mist system to extend their time of survival. Ideally, sprigs should be stored in a layer not more than 8 inches (20 cm) in depth.

Two primary methods used for sprigging are (1) broadcast sprigging, and (2) row planting. These methods are normally a single-pass, one-step operation, best suited for larger areas such as fairways since the equipment is large and bulky. Soil should be moist (not saturated) prior to planting to prevent root tip burn when contacting dry, hot soil. Sprigs should be watered immediately after planting, and watered several times daily for several weeks until the turfgrass establishes a root system. Do not sprig more area than can be immediately watered. Generally, turfgrass sprigs will cover and become established within 6 to 16 weeks after planting. The length of time for establishment depends on the sprigging rate, time of the year of planting, and maintenance practices implemented after planting. For example, 75 to 120 days are necessary for Tifdwarf bermudagrass to grow in and develop an acceptable playing surface, while 50 to 80 days are needed with the ultradwarf cultivars.

**Broadcast sprigging.** Broadcast sprigging involves spreading sprigs over the area (like mulch) either by hand or by specialized mechanical equipment (Figure 6.7). Sprigging rates for various turfgrasses are listed in Table 6.4. After spreading, sprigs are then cut (or pressed) into the soil with a light disk, covered with 1/2 inch (1.3 cm) of topdressing, and then rolled to firm the seedbed to ensure sprig-to-soil contact (Figure 6.8). Sprigs are then immediately watered. This method of sprigging provides very fast coverage and turfgrass establishment. However, since sprigs are planted at a shallow depth, they are susceptible to drying out. Light, frequent irrigation (four to six times daily) should be provided until the turfgrass roots become established (two to four weeks).

**Row planting.** Row planting involves a one-step, one-machine operation that opens a furrow or slit in the seedbed, drops sprigs into these furrows, presses the furrow or slits together, and rolls to smooth the surface and ensure good sprig-to-soil contact. Row planting is generally performed on 4- to 6-inch (10 to 15 cm) centers for even distribution and faster establishment (Figure 6.9). Little to no soil preparation is required for successful row planting, and in many cases golf play can resume immediately after planting.

**No-till sprigging.** A more contemporary means of sprigging is referred to as no-till. Minimum soil preparation is involved, reducing establishment costs. Typically, an area is treated one to three times with a nonselective herbicide such as glyphosate to kill existing vegetation. Sprigs are then hand-broadcasted and typically pressed or pushed in with a disk composed of circular blades. Variations of this process exist, such as using topdressing sand to cover the sprigs instead of pressing them in. There are several considerations in no-till sprigging, including, most importantly, ensuring complete kill of the existing vegetation prior to sprigging. Bermudagrass, for example, requires at least three applications of glyphosate for control. Existing soil characteristics also are not altered

**FIGURE 6.7** Broadcast sprigging by hand to ensure even plant material distribution and coverage.

**FIGURE 6.8** Pressing in a newly hand-sprigged golf green with rolling blades.

**FIGURE 6.9** A mechanical sprigger, which chops and plants sprigs from big roll sod.

by no-till plantings. If existing drainage, for example, is inadequate prior to no-till planting, it will remain that way after planting. Also, if excessive thatch/mat/organic layering exists, this is unaltered by no-till planting. If this layer is excessive, it is highly advisable to remove (strip) it prior to planting, adding sufficient sand to bring the grade to the desired level and then hand-sprigging.

**Stolonizing.** Stolonizing is similar to sprigging; however, only stolons are utilized as vegetative establishment units. Stolons are broadcast-spread across the prepared seedbed, covered lightly with soil and/or topdressing, and then rolled to firm the seedbed to ensure good stolon-to-soil contact. Stolonizing is most commonly practiced with creeping bentgrass and hybrid bermudagrasses. The planting rate for stolonizing is greater (~10%) than with sprigging due to greater mortality of stolons and lack of other vegetative parts such as rhizomes to become established. Stolons can also be planted or applied via hydromulching equipment as previously described. Hydrostolonizing is utilized on more severely sloped areas and requires less labor and time. However, establishment success of hydrostolonizing is less than with more conventional stolonizing methods.

**Plugging.** Plugging involves vegetative propagation of turfgrasses with plugs or small pieces of sod. Plugging is utilized in establishment of less aggressive turfgrass species such as centipedegrass, St. Augustinegrass, and zoysiagrass. Since soil accompanies plugging, it is safer than sprigging or stolonizing in terms of survival due to heat and drought stresses. Plugs are obtained from a nursery utilizing a specialized harvesting machine that cuts, divides, and separates sod plugs into either 1- to 4-inch (2.5 to 10 cm) diameter circular plugs or 2- to 4-inch (2.5 to 10 cm) square plugs. Plugs are then mechanically planted into the seedbed on 6- to 12-inch (15 to 30 cm) centers every 6 to 12 inches (15 to 30 cm) in a row (Figure 6.10). Obviously, the shorter the distance between two plugs is, the faster the turfgrass is established. After plugging, the area should be rolled to firm the seedbed and irrigated for more favorable establishment conditions.

**Sodding.** Sodding provides "instant greening" and excellent soil erosion control. However, it is initially more expensive and has the possibility of introducing different soil types or pests, such as nematodes, weeds, or fire ants, from the harvested field into the rootzone. As a minimum specification, sodding should be utilized for areas prone to soil erosion, such as steep slopes and areas surrounding bunkers, tees, and greens (Figure 6.11).

Large rolls of sod reduce labor and installation time, as well as the number of seams (Figure 6.12). Sod "washed" free of soil and/or grown on a custom growing medium (e.g., USGA-specified greens mix) also provides turfgrass managers with more options. On greens, for the first two years following sodding, four to six aerifications are needed, especially if thatch layering has formed. During grow-in on the sod farm, topdressing should be performed for greens-grown grasses to provide surface smoothness, reducing mower scalping.

Before installing sod, the seedbed should be watered. Rooting rate and establishment will be significantly increased if the underlying soil is moist and cool, and therefore receptive for rooting to occur. The sod should be installed soon, preferably within 48 hours, after harvest. When sod remains stacked or rolled on pallets for longer periods, respiration of the turfgrass plants creates heat that can dry out and injure the grass.

Start laying the sod along the straightest line possible. Edges of the sod should butt against each other tightly without stretching. Avoid gaps or overlaps by using machetes or sharp knives to trim around corners and edges. Joints between the sod should be staggered in a brick-like fashion so none of the edges of adjacent pieces of sod are parallel (Figure 6.13). It is important to handle the sod with care to avoid tearing or stretching. If sod is stretched, it will shrink as it dries out, leaving gaps between the edges. On slopes and bunker faces, place the sod across the slope and anchor the pieces with sod staples or wooden pegs until the roots have

**FIGURE 6.10** Plugging a golf green is time consuming and often slower to cover.

**FIGURE 6.11** Sodding erosion-prone areas.

**FIGURE 6.12** Sodded highly erodible areas such as a hillside.

**FIGURE 6.13** Sod pieces staggered in a brick-like fashion.

**FIGURE 6.14** Rolling newly laid sod to encourage root-to-soil contact and to smooth the surface.

become established. To help avoid indentations or air pockets, refrain from walking on the newly laid sod. Sometimes, the sod is installed by using plywood boards to ensure it is evenly installed. This also ensures good sod-to-soil contact.

After installation, the sod should be rolled to ensure evenness and smoothness and immediately watered heavily to wet the entire depth of the sod and top portion of the rootzone. Topdressing should be implemented to fill in creases and low pieces and to help conserve moisture (Figure 6.14). Although sodding produces an almost instant green-appearing turf, it should be allowed to knit down (or root) before it is subject to traffic and play. Four weeks is generally the minimum time necessary for this during periods of active turfgrass growth—longer when the turfgrass is not actively growing.

## "GROW-IN"

Immediately after seeding or vegetative propagation, turfgrass requires special attention and maintenance to ensure survival and healthy growth. This "grow-in" phase is the most overlooked expense of any golf course construction or renovation project. The "grow-in" period is the time between seeding or vegetative propagation (grassing) until the course opens for play. This phase can vary from a short time period such as 3 months to as long as 18 months depending on the location, turfgrass type, method of turfgrass planting, and local climate. Most courses should budget and plan for an average grow-in period of at least six months after planting starts. Labor is generally the greatest expense, but fertilizer and water are important needs during this period. It is important to recognize that the cost of "grow-in" cannot be offset by income since the golf course is not open for play.

### Seeding Establishment

Successful seeding establishment of turfgrass involves seed germination, initiation of roots, and lateral shoot development into mature plants. This involves imbibition (or uptake) of water into the seed and, in turn, primary root and shoot growth are initiated from which plants develop. Temperature, oxygen, light exposure, and water are all essential environmental parameters required for successful germination. However, sometimes successful seed germination does not occur even with favorable environmental conditions. Several parameters may influence successful germination of seed under favorable conditions, such as (1) presence of inhibitors, (2) an impermeable seed coat, (3) seed planted too deeply, (4) damaged seed, (5) injury from fertilizers and/or pesticides, (6) damage by insects or diseases, or (7) crusted soil surfaces. Successful establishment from seed germination can be enhanced with some turfgrass types by using pregerminated seeds, especially if temperatures are cool. Seeds can be pregerminated by placing them in a moist environment, such as a barrel of warm water, at favorable germination temperatures. These pregerminated seeds must be planted immediately to prevent injury to the primary root and shoot systems.

**Irrigation.** Proper water management is critical for the first several weeks after turfgrass planting. Irrigation or watering is essential to the germination process and survival of the turfgrass seedlings by preventing drought conditions and moisture stress. Irrigation or watering should be conducted during the predawn hours while moisture or dew is present. This prevents extended "free" water or moisture periods, which may enhance seedling diseases such as *Pythium* and *Fusarium* (Figure 6.15). The areas should be irrigated immediately after seeding with enough water to "moisten" the seed to ensure germination. Seeds must remain moist until they germinate and emerge (7 to 30 days). The seedbed should also remain moist with frequent, light irrigation several times daily (four to eight times per day). The first watering is commonly between 7 and 8 a.m. and the last between 6 and 7 p.m. The majority of watering is from 10 a.m. to 6 p.m., with no irrigation at night to minimize disease occurrence. Avoid excessive watering, as this can cause puddling of water or create soil erosion and "washes" that concentrate seed into areas or drifts. If this occurs, use a stiff-bristled brush or broom to redistribute the seed. Irrigation heads should also be frequently inspected to ensure proper working order and complete coverage. As seeds germinate and seedlings emerge, frequency of watering should be decreased and amount of water applied should be increased.

**FIGURE 6.15** *Pythium* on newly emerged bentgrass seedlings due to unseasonably warm temperatures.

**Fertilization.** Proper fertilization is essential for plant growth and development after seed germination. Prior to planting, a starter fertilizer with a 1-2-2 ratio of N-P-K or similar should be incorporated into the soil mix (Figure 6.16). The nitrogen (N) source should be slow release and applied at the equivalent of 1 pound N/1,000 square feet (4.9 g N/m$^2$). This should provide between 1.5 and 2 pounds P/1,000 square feet (7.3 to 9.8 g P/m$^2$). A granular micronutrient package and a humate or carbohydrate soil conditioner also may be added according to label recommendations. Conditioners are especially beneficial if pure sand greens are used.

After planting, the new seedlings should be fertilized approximately two to four weeks after seeding, or as a general rule, after the second mowing. A complete fertilizer (one containing N-P-K nutrients) designed specifically for use on turfgrass should be applied at an equivalent rate of 1 pound of N/1,000 square feet (4.9 g N/m$^2$) of turfgrass. Preferably, a slow-release N source should be a component of the N-P-K fertilizer. Cool-season grasses are fertilized weekly with 1/2 pound N/1,000 square feet (2.4 g N/m$^2$) on greens and tees, while fairways and roughs are fertilized every two weeks until establishment with 3/4 to 1 pound N/1,000 square feet (3.7 to 4.9 g N/m$^2$). A complete fertilizer is normally used for each of these fairway applications. For bermudagrass, 1 pound N/1,000 square feet (4.9 g N/m$^2$) applied weekly provides the quickest establishment.

Potassium (K) and phosphorus (P) should be added with each alternate N application in a balanced ratio until establishment. If sulfur (S) deficiencies are suspected, use potassium sulfate as the K source to supply the needed S. Supplemental liquid K and iron (such as a 0-0-28 plus Fe) should be applied every two to three weeks if color wanes or growth slows. Once fully established, the fertilizer program should then be converted to a less aggressive program (refer to the chapter on cultural practices). If cool-season grasses are seeded in spring, the aggressive fertilizer program should commence

**FIGURE 6.16** Applying a starter fertilizer and lime prior to planting a green.

until normal summer temperatures are anticipated. At this time, summer color is best maintained with soluble fertilizer sources. Monthly summer application of a 0-0-28 or 0-0-50 at 3/4 pound K/1,000 square feet (3.7 g N/m$^2$) will help harden grass for summer stress. Liquid Fe and/or a micronutrient spray supplements the K as needed for summer color.

**Soil amendments and conditioners.** Other soil amendments, such as lime, gypsum, or S, also are often needed in certain areas. Soil testing should be performed to determine these needs. Most of these should be incorporated before turf establishment. In acidic clay soils, for example, lime is often needed from 1 to 3 tons/acre (2.2 to 6.7 tons [metric]/ha). Commonly, 1.5 tons/acre (3.4 tons [metric] per ha) are initially applied and incorporated, followed by another ton/acre the following fall or early spring. Gypsum supplies calcium and helps dissociate sodic (sodium) soils. If needed, 300 to 400 pounds gypsum/acre (38 to 50 g/m$^2$) preplant are often suggested by soil testing. Sulfur helps promote grow-in and reduce soil pH in alkaline areas. High pH (≥8.0) fairways may need 150 to 200 pounds/acre (17 to 22 g/m$^2$) of a granular dispersible S, preplant. Greens also may need S prior to planting and also frequently receive additional S after planting via fertigation and/or with S-containing (e.g., potassium sulfate) or coated fertilizer products.

Biostimulants are a class of plant growth regulators (enhancers) that help turf growth under certain conditions. Their benefits are most noticeable when turf is grown on very sandy soils, such as golf greens, although benefits are not always experienced. Biostimulants containing seaweed extracts are composed of a variety of cytokinins, humates, and other materials. Cytokinins benefit root growth and positive results may be seen when used.

**Mowing.** Mowing is the most common turfgrass maintenance operation and is potentially one of the most damaging. A common mistake is waiting too long to mow the first time. Proper mowing promotes lateral growth and plant maturity. Mowing of the newly established turfgrasses should begin as soon as the turfgrass reaches 130% of the desired turfgrass mowing height. For sprigged bermudagrass greens, begin mowing when the grass reaches 2/3 to 3/4 inch (1.7 to 1.9 cm). Slowly lower this to 1/2 inch (1.3 cm) until complete coverage is achieved and then slowly lower it to the normal playing height. Mow fairways when the bermudagrass reaches 1.25 to 1.5 inches (3.2 to 3.8 cm) and bluegrass/ryegrass when they reach 1.5 to 2 inches (3.8 to 5 cm).

Bentgrass seedlings should be mowed within two to three weeks of germination, initially between 3/8 and 1/2 inch (9.5 to 12.7 mm). Lower the mowing height between 1/16 and 1/32 inches/week (1.6 to 0.8 mm) until the final desired cut is reached. It is imperative to use sharp blades to prevent turfgrass seedlings from being torn or "pulled up" during mowing. It is advisable to use walking reel mowers on greens and tees until complete coverage of turfgrass is achieved. Do not mow when the turfgrass is wet, and remove the clippings if their moisture levels promote clumping, thereby subsequently shading the newly established turfgrass. Solid front rollers should be used the first several months, as grooved rollers are often too damaging and can dig into the exposed soil surface.

**Pest control.** When establishing new turfgrass areas, pest problems such as diseases, insects, nematodes, and weeds must be minimized. This is best accomplished utilizing a preventative pest management strategy coupled with a routine pest monitoring or "scouting" schedule. Many pest problems can be alleviated by fumigating putting greens prior to planting to control pests and using clean soil, seed, sprigs, and sod.

As previously mentioned, irrigation or watering is required to achieve seed germination and turfgrass establishment; however, the amount and timing of its application can prevent or contribute to the development of pest problems. Most fungal disease pathogens require "free water" or very high humidity to initiate the disease infection process. Disease development is dependent on humidity and temperature. Irrigating in the evening before dew forms or in the morning after dew evaporates extends the dew or moisture period. Newly planted turfgrass seedlings, therefore, should be watered when dew or moisture is already present, such as in the predawn hours.

To reduce potential disease development to newly planted turfgrass seedlings, it is recommended to use seed treated with a fungicide. If untreated seed is used, then a fungicide should be broadcast-applied to the newly planted areas to minimize seedling diseases.

In addition to being unsightly, weeds compete with turfgrasses for light, soil nutrients, soil moisture, and space. Weeds can also play host to other pests such as diseases, insects, and nematodes; therefore, it is important to control these during turfgrass establishment.

Delay any herbicide applications as long as possible to allow the turfgrass to become well established. Hand-pulling or "roughing" should be conducted if only a few weeds are present; however, if many weeds emerge, the use of selective postemergence-applied herbicides may be required. These herbicides should be used judiciously only after careful herbicide selection.

For most new greens, the surface is not firm enough to support a self-propelled sprayer without ruts. For the first six to eight weeks after planting, a walk-type boom or back-pack sprayer system should be used.

### Vegetative Propagation Establishment

Turfgrass establishment from vegetative propagation requires favorable environmental conditions similar to those used for seeding. Temperature, oxygen, light exposure, and water are essential to the successful growth and development of vegetative materials. Bermudagrass is ideally sprigged between late spring through midsummer while cool-season grasses are stolonized in late summer. Proper cultural and turfgrass management practices also are needed to achieve successful establishment of a newly planted area. Under ideal growing conditions, sprigged areas require 6 to 10 weeks for complete establishment (Figure 6.17).

**Irrigation.** Proper irrigation or watering is one of the most important steps in achieving successful turfgrass establishment. Turfgrass establishment from sprigging, stolonizing, or plugging requires constant moisture due to the lack of soil associated with these vegetative materials. In contrast, sod

**FIGURE 6.17** A newly sprigged bermudagrass green (left) compared to five weeks later.

requires the least amount of water because soil associated with sod provides some moisture.

It is critical for newly sprigged, stolonized, or plugged turfgrasses to be irrigated or watered immediately after planting to avoid desiccation or drying out (Figure 6.18). Timing is not as critical with plugging due to some soil associated with the plugs. Plugging establishment is still best if irrigation occurs as soon as possible. Irrigation or watering should continue on a frequent basis to maintain a moist soil surface for the vegetative plantings. Frequent surface irrigation should be continued for a two- to three-week period or until establishment occurs.

**Fertilization.** Prior to planting, a starter fertilizer should be applied to the site to encourage good rooting and quicker turf cover. A starter fertilizer with a 1-2-2 ratio of N-P-K (such as 10-20-20) should be used with a slow-release N source. This ratio of fertilizer should be applied at the equivalent rate of 1 pound N/1,000 square feet (5 g N/m$^2$). After planting, it is recommended to fertilize with an equivalent rate of 1 pound water-soluble N/1,000 square feet of turfgrass (5 g N/m$^2$) every five to seven days for greens and tees and every 10 to 14 days on fairways or roughs. This can be split into 1/2 pound N/1,000 square feet (2.4 g N/m$^2$) twice weekly. Zoysiagrass, however, should receive this N rate every 30 days. With bermudagrass, each application of a quick-release fertilizer such as ammonium sulfate (21-0-0) or ammonium nitrate (45-0-0) should be alternated with a 1-2-2 ratio fertilizer (such as 5-10-10, or equivalent) at an equivalent rate of 3/4 to 1 pound of N/1,000 square feet (3.7 to 4.9 g N/m$^2$). Potassium nitrate is often successful for bentgrass. Lighter rates, 0.3 to 0.6 pound N/1,000 square feet (1.5 to 3.0 g N/m$^2$), applied every five days are generally more successful than heavier N rates applied less frequently. If additional color is needed or if growth slows, supplement these fertilizations as needed with a liquid N application at a rate of 0.2 pound N/1,000 square feet (1 g N/m$^2$). Micronutrient sprays of Fe, magnesium (Mg), and manganese (Mn) are applied as needed to aid in turf grow-in. Each granular fertilization should be irrigated or watered in immediately to avoid foliar turfgrass burn. In high sand content greens, expect to need 8 to 12 pounds N/1,000 square feet (39 to 59 g N/m$^2$) during the grow-in process.

**Mowing.** Mowing should be initiated after stolons have reached a growing height or length at about 2 inches (5 cm) (fairways) to ~0.180 inch (4.5 mm) (greens/tees). Mowing is recommended during midday, when the turfgrass is dry. Turfgrass clippings should be returned to the soil surface for the first couple of mowings to promote rooting of any stolons that may have been cut during mowing. The initial mowing height on fairways should be 2 inches (5 cm) or greater and then reduced over time as the turfgrass becomes established. Use only sharpened mower blades as dull blades shred and pull sprigs from soil. Mowing frequency should be two to three times weekly so that only small amounts of leaf tissue are removed during any one mowing event. On greens, over the first 30 days, gradually reduce the mowing height to about 0.140 inch (3.5 mm). Within about 45 days of planting, the mowing height can gradually be lowered to 1/8 inch (3.2 mm).

**FIGURE 6.18** New turf areas require sufficient irrigation to remain wet but not saturated during grow-in.

**Pest control.** Vegetative planting operations disturb the soil, which can promote germination and establishment of weeds during grow-in. In addition, an ample supply of moisture and fertilizer provides optimum growing conditions for weeds as well. Preemergence control of annual grass weeds such as crabgrass and goosegrass without impeding new bermudagrass growth can be obtained by applying an herbicide containing oxadiazon (e.g., Ronstar). Oxadiazon products should be applied immediately after vegetative planting at the recommended labeled use rate. The use of selective postemergence herbicides should be delayed for at least three weeks after vegetative planting.

**Topdressing.** Frequent topdressing during turfgrass establishment from vegetative planting of putting greens is recommended to achieve a smooth playing surface and enhance turfgrass establishment. The topdressing amount and frequency is dependent upon the existing smoothness of the surface. A light rate of 0.5 to 0.7 cubic foot per 1,000 square feet (1.5 to 2.1 $m^3/m^2$) is a good starting point for this. Use heavier rates if needed to fill in depressions. This should commence four to six weeks after seeding and three to four weeks after sprigging (about 80% soil coverage). Topdressing also serves as a "covering" for vegetatively planted sprigs, stolons, or plugs, thereby enhancing their establishment. Sodded areas should also be rolled throughout the grow-in period to push roots into the soil and to smooth the surface to prevent mower scalping. Weekly rolling should be performed until the eventual permanent mowing height is achieved.

A regular topdressing program should commence once 90% to 100% soil coverage is obtained. This includes regular light topdressing application on a weekly or biweekly basis when the grass is rapidly growing. Use consistent, good-quality topdressing sand that matches the greens construction media.

**Surface firmness.** When greens are planted and maintained at today's low mowing heights, a certain period of time is required to develop enough organic matter "pad" or "cushion" in the upper rootzone to provide surface resiliency. Complaints are often initially heard about hard greens that do not hold approach shots well. One to two full growing seasons are typically required to develop a mature surface. Aerification helps to "soften" greens and should commence and continue once the grass is sufficiently mature to withstand this process. The first several aerifications should be with 3/8 inch (9.5 mm) on 2-inch (5 cm) centers. One-inch (2.5 cm) centers can be used thereafter as greens become firmer and have more surface stability.

**Interseeding.** Interseeding is introducing a new cultivar into an existing one using seed, most often with bentgrasses. This is performed to help blend segregated patches, provide better fall and spring color and growth, promote active growth for improved turf density, provide damage repair for ballmarks, help prevent *Poa annua* invasion in ballmarks, and improve overall wear and traffic tolerance. Probably most important, interseeding allows the introduction of new and better varieties and genetics into an older cultivar for better rooting, putting performance, summer stress tolerance, and disease resistance. Interseeding is obviously more economical than renovation/resodding. However, it is not a one-time event, as interseeding only once annually typically provides only up to 15% success. Therefore, small quantities of seed are planted frequently until the desired success occurs. Two to eight interseedings during the growing season are not unusual, nor are multiple years of interseeding.

When converting, seedlings will need room in the existing turf for establishment. This is often performed by verticutting, aerifying, spiking, and the possibility of using plant growth regulators. With a continued interseeding program, increases in the new cultivar often occur within two years. However, with fewer interseeding events, three to five years may be required to experience this increase. Seeding in late spring through late summer increases germination due to warm soil temperatures. However, if nighttime temperatures remain over 82°F (28°C), temperatures are too high to realistically expect tender bentgrass seedlings to survive.

Commercial slit seeders are available for interseeding that are basically verticutting machines that also drop seed in the verticut slits. This should be performed in multiple directions to minimize seed rows from forming. Seeding following spiking is another popular means of interseeding. Depending on the density of existing turf, bentgrass seeding rates of one-fourth to 1/2 pound/1,000 square feet (~1 to 2 $g/m^2$) is performed about every two weeks in summer. Once the seed is applied, light topdressing is followed by light dragging or brushing of the green to help move seed into the spiker slits and to improve seed-to-soil contact. Additional seeding should follow normal aerifications, realizing that more coring holes provide a better (more successful) seedbed preparation.

## Budgeting

Budgeting during grow-in and for the first year after establishment is difficult depending on the time of year for planting, grasses grown, quality of course design and construction, unexpected weather phenomena, and so on. Other unexpected problems such as poor drainage sites, low areas, and additional tree removal/stump grinding are further examples. These problems should be expected, projected, and budgeted for so that enough funds are available to finish a quality job and to meet the anticipated high expectations of new members and players. Funds often become limited at this time and corners are cut. However, most of these occurrences are only one-time events and will not be dealt with again except on a limited basis. The following list includes predominant situations often occurring during this period:

1. **Labor**—In addition to the needed full-time superintendent, assistant superintendent, mechanic, irrigation technician, spray technician, and normal full-time or part-time employees, additional temporary employees may be necessary to finish unexpected special projects without sacrificing routine maintenance needs.
2. **Fertilizer and soil amendments**—During grow-in, much higher than normal fertilizer and soil amendments are used to hasten establishment. Also during the first year, an additional 15% or so above normal maintenance needs will be necessary to finish establishment in difficult areas. Uneven coverage and color may occur if amendments or fertilizers are not uniformly applied.
3. **Tree removal and replacement**—Despite best efforts, some unexpected tree die-out can be anticipated during grow-in. This also includes stump removal. Additional trees restricting air movement or creating excessive shade not identified in construction or grow-in may need to be removed. Some of these trees may need to be replaced in accordance with the course's master plan.
4. **Low spots and poor drainage areas**—Low spots that drain poorly invariably happen as settling occurs in fairways, tees, and collars. In most cases, low spots can be corrected by additional sand topdressing. If it is too severe, sod in these areas should be removed, the areas brought up to grade with soil or sand, and then resodded. Also, despite the best drawings, additional drainage will be needed in certain areas as settling occurs or wet springs surface. Sand, gravel pipe, and outlets will be needed to correct this. Other needed items are a backhoe and trencher and hand labor to install this equipment. Equipment rental may be an option for this equipment.
5. **Additional irrigation heads**—Due to poor or irregular pressure, shifting winds, or mounds that receive poor coverage, some additional irrigation heads may be necessary for complete course coverage, as are pipe and possibly control valves. Buy additional heads before the course is opened to have a stock on hand.
6. **Erosion control/sodding**—In 2003, under the U.S.EPA's Clean Water Act, the National Pollutant Discharge Elimination System (NPDES) storm water regulations went into effect. These regulations influence construction activity that disturbs five or more acres (as outlined under phase I of the regulations) or five acres or less (phase II) and are aimed at preventing sediments from migrating into waterways and disturbing endangered species habitats. To obtain a construction permit, NPDES requires a storm water pollution prevention plan that details pollution prevention BMPs slated for the proposed project. A notice of intent is submitted to the local administrative agency at least two days prior to breaking ground, and it includes certification that none of the construction activities will impact endangered or threatened species or critical habitat. Before disturbing the site, proper sediment control devices need to be in place, including silt fencing, ryegrass seed for winter stabilization (200 pounds/acre, 22g/m$^2$, on average), mulch, jute mat, hay bales, straw, soil, gravel, sand wood fiber filtration tubes, and sod. Additional sod should also be anticipated to help control erosion and to repair areas that die. Two to four truckloads often are needed for this.

Other miscellaneous items should also be anticipated during grow-in and first-year maintenance. These include extra gas and oil for equipment, soil testing, ornamentals for landscaping, bunker trim/spread, irrigation head leveling, and chemicals for weed control. During renovation, player interest and satisfaction can be maintained if temporary greens are needed by using a larger golf cup such as a large coffee or soup cup until renovations are completed (Figure 6.19).

**FIGURE 6.19** An oversized golf hole cup used during renovations to keep players engaged during the process.

# *Section IV*

## *Turfgrass Fertilization and Water Management*

# 7 Plant Nutrition and Turf Fertilizers

## NUTRIENT UPTAKE

Proper fertilization is essential for turfgrasses to sustain desirable color, growth, density, and vigor; to better resist diseases, weeds, and insects; and to provide satisfactory playing conditions. Turfgrass plants require 17 elements divided into two categories: **macronutrients** and **micronutrients** (Table 7.1). Macronutrients typically are found at concentrations greater than 500 to 1,000 ppm in plant tissue, while micronutrients are generally found at concentrations less than 100 ppm. All of these elements are equally essential for plant growth, regardless of amounts required, and nothing can substitute for them.

Macronutrients can further be subdivided into primary nutrients (nitrogen, phosphorus, and potassium) and secondary nutrients (calcium, magnesium, and sulfur). Carbon (C), hydrogen (H), and oxygen (O) are macronutrients obtained from air and water, and they are the building blocks for the process of photosynthesis. These three nutrients, when combined in the presence of chlorophyll and light through the process of photosynthesis, form carbohydrates, the sugars used to provide plant growth. Plants, therefore, do not obtain their food through soil; rather, they manufacture it via photosynthesis, utilizing the raw materials—mineral nutrients from the soil—as components of this food.

Carbon dioxide is absorbed by plants through leaf stomata, and water is absorbed through roots. Therefore, fertilizer practices affecting root growth and function, as well as stomata opening and closing, indirectly influence a plant's ability to produce food via photosynthesis.

Simple carbohydrates produced from photosynthesis are used to generate more complex compounds such as starch and amino acids that require other elements in addition to C, H, and O. The remaining essential elements are mainly absorbed into plants in the soil solution through roots. Understanding how these processes interact and are influenced by soil chemical properties is important to turf managers who strive to optimize their fertilization programs.

Large amounts of the nutrient anions are stored in soil organic matter. As organic matter decomposes, these nutrients are released. Up to 80% of soil S, 95% of soil nitrogen (N), and over 50% of total soil phosphorus (P) are stored in soil organic matter. Other anions, such as boron (B), molybdenum (Mo), and iron (Fe) and aluminum (Al) oxide reserves, are also held in organic matter.

Roots are the principal means by which nutrients and water enter plants. The root system is usually very large and extensive, allowing plants to make contact with a tremendous volume of soil. Root hairs greatly increase the surface area of roots and are the principal site of nutrient and water uptake.

The absorption (transfer from the soil solution into the root cells) of a nutrient element is primarily an "active" process and, as a result of a metabolic process, roots take in the nutrient element. For the element to be absorbed by plant roots, it must be dissolved in soil water.

Nutrient elements exist as either anions (negatively charged) or cations (positively charged) in the soil solution. As roots contact moist soil, anions and cations move by diffusion (concentration gradient) from soil water (higher nutrient concentration) into the outer free cellular space of the root (lower nutrient concentration). Many nutrients are not readily soluble in soil water. Therefore, before plant uptake occurs, these nutrients must be dissolved. This occurs by the creation of an acidic environment around the roots through carbonic acid formation from released $CO_2$ by the plant and water from the soil solution. Carbonic acid lowers soil pH, which, in turn, releases $H^+$ ions that can be "exchanged" with various soil clays and organic matter for cations such as K, Ca, and Mg.

Since active absorption is the primary mechanism for element uptake by plants, roots must be growing in an aerobic medium—that is, $O_2$ must be present. If $O_2$ is not present, respiration will cease and element absorption will not occur. Under these conditions, plants may develop nutrient deficiency symptoms. Any soil condition that affects soil aeration, such as compaction or continuous moisture saturation, will negatively influence root growth and nutrient uptake.

Temperature also influences nutrient uptake. Higher temperatures usually mean increased root growth; therefore, the plant encounters more soil nutrients. Warmer temperatures also increase leaf transpiration, which provides the "pull" for water (and dissolved nutrient) uptake.

Adequate soil nutrient levels also are needed for uptake by plants. The amount of an element in the soil solution is affected by the total concentration of the element in the soil, the soil pH, the concentration of other elements in the soil and soil solution, the extent of biological activity, and soil temperature.

### Mineral Mobility and Deficiencies

Nutrients are mobile if the plant can transport them from one tissue to another, and they tend to show deficiency symptoms in older tissue first, as plants will withdraw ("scavenge") these to support new growth. Mobile elements include N, P, K, Mg, and Zn, while S has limited mobility. Immobile elements show deficiencies in new growth as they are not transferred from older growth if external supplies are inadequate. Immobile elements also must be continuously applied if plants are foliar fed. Immobile elements include Ca, Fe, B, Mn, and Cu.

**TABLE 7.1**
**Elements, Their Most Common Available Forms for Plant Uptake, and Primary Functions in Turfgrass Growth**

| How Obtained | Element (Chemical Symbol) | Most Commonly Used Form(s) | Function in Plant Growth |
|---|---|---|---|
| | | | **Macronutrients** |
| From air and water | Oxygen (O)<br>Carbon (C)<br>Hydrogen (H) | $O_2$<br>$CO_2$<br>$H_2O$ | • Through photosynthesis, these elements are converted to carbohydrates, and finally into amino acids, proteins, protoplasm, enzymes, and lipids. |
| Primarily from fertilization | Nitrogen (N) | $NO_3^-$ (nitrate)<br>$NH_4^+$ (ammonium) | • Plant mobile element used in formation of amino acids, enzymes, proteins, nucleic acids, and chlorophyll. N generally increases color and shoot growth. Excessive amounts generally reduce heat, cold, and drought hardiness; disease and nematode resistance; wear tolerance; and root growth. |
| | Phosphorus (P) | $H_2PO_4^-$<br>$HPO_4^{-2}$ (phosphates) | • Plant mobile element that is a constituent of phospholipids and nucleic acids. Involved in a carbohydrate transport system that moves energy to all parts of the plant for vital growth processes. This function in root development is most vital; also hastens plant maturity and is needed for glycolysis, amino acid metabolism, fat metabolism, S metabolism, biological oxidation, and photosynthesis. In addition, P influences maturation, establishment, and seed production. |
| | Potassium (K) | $K^+$ | • Plant mobile element used in large quantities, second only to N; helps control and regulate various minerals; adjustment of stomatal movements and water relation; promotion of meristematic tissue and rooting; activation of various enzymes; synthesis of proteins; and carbenoxolone metabolism. Helps increase heat, cold, and drought hardiness; wear tolerance; and increases disease and nematode resistance. |
| | | | **Secondary Nutrients** |
| In some fertilizer formulations; available in most soils, and/or as part of conditioners such as lime, dolomitic lime, and gypsum | Calcium (Ca) | $Ca^{+2}$ | • Plant immobile element required for cell division (mitosis); cell membrane permeability; activates certain enzymes; provides chromosome stability and structure; enhances carbohydrate formation and translocation, and increases protein content of mitochondria; influences absorption of other plant nutrients, and also strongly influences soil pH and can improve soil structure, water retention, and infiltration by displacing Na ions. |
| | Magnesium (Mg) | $Mg^{+2}$ | • Plant mobile element that is a component of chlorophyll; assists in the stabilization of ribosome particles, and activates several plant enzyme systems such as carbohydrate and phosphate metabolism and cell respiration. |
| | Sulfur (S) | $SO_4^{-2}$ | • Partially a plant mobile element required for the synthesis of S-containing amino acids—cystine, cysteine, and methionine; required for protein synthesis and activation of certain enzymes and hormone constituents. |
| | | | **Micronutrients** |
| Most premium fertilizers | Iron (Fe) | $Fe^{+2}$ (ferrous)<br>$Fe^{+3}$ (ferric)<br>$Fe(OH)_2^+$ | • Plant immobile element necessary for chlorophyll, heme, and cytochrome production, and in ferredoxin, which participates in cellular respiratory (oxidation-reduction or electron-transfer reactions) mechanism; an essential component of Fe enzymes and carriers. Generally increases color, shoot, and root growth. |
| | Manganese (Mn) | $Mn^{+2}$, organic salts | • Plant immobile element that activates certain enzymes; needed in photosystem II of photosynthesis; connected with carbohydrate (N) metabolism, chlorophyll synthesis, oxidation-reduction process, phosphorylation reaction, and the citric acid (or tricarboxylic acid) cycle. |
| | Copper (Cu) | $Cu^{+2}$ (cupric)<br>$Cu^+$ (cuprous) | • Plant immobile element connected with oxidation-reduction or electron-transfer enzymes; found in cytochrome oxidase that is essential for plant (carbohydrate) metabolism, and is used for production of the enzyme polyphenol oxidase and to catalyze plant metabolism. |
| | Chlorine (Cl) | $Cl^-$ (chloride) | • Plant immobile element possibly required for photosynthesis of isolated chloroplasts and as a bromide substitute; believed to influence osmotic pressure and balance cell cationic charges; also affects root growth. |

*(Continued)*

**TABLE 7.1 (CONTINUED)**
**Elements, Their Most Common Available Forms for Plant Uptake, and Primary Functions in Turfgrass Growth**

| How Obtained | Element (Chemical Symbol) | Most Commonly Used Form(s) | Function in Plant Growth |
|---|---|---|---|
| | Zinc (Zn) | $Zn^{+2}$<br>$Zn(OH)^{+}$ | • Plant mobile elemental component of the enzyme dehydrogenase, which is needed for RNA and cytoplasmic ribosomes in cells, proteinases, peptidases, and indole-3-acetic acid (auxin) synthesis; involved in conversion of ammonium to amino N; necessary for chlorophyll production and promoting seed maturation and production. |
| | Boron (B) | $H_3BO_3$ (boric acid)<br>$HBO_3^{-2}$<br>$B_4O_7^{-2}$<br>$BO_3^{-3}$ | • Plant immobile element that facilitates sugar transport through membranes; involved in auxin metabolism in root elongation, protein, and phosphate utilization; influences cell division (growth) by control of polysaccharide formation. A nonmetal, one of two nonmetallic micronutrients called **metalloids.** |
| | Molybdenum (Mo) | $MoO_4^{-2}$ (molybdate) | • Required for the assimilation and reduction processes in N fixation to produce amino acids and proteins. |
| | Sodium (Na) | $Na^{+}$ | • Regulates stomatal opening and nitrate reductase. Toxic levels are generally more of a concern than deficiencies. |
| | Nickel (Ni) | $Ni^{+2}$ | • Most noted as a component of the enzyme urease, which is needed for the conversion of urea to ammonia in plant tissue. |

Plant deficiencies for nutrients often occur due to external conditions that prevent their uptake rather than their actually being lacking in the soil. Soil and tissue testing, therefore, should be used to ascertain if sufficient supplies are in the soil and if plant tissues are able to obtain them. Restricted root and tissue growth from improper soil pH, inadequate or excessive moisture, or temperatures outside the optimum growth range are several common reasons nutrient deficiency symptoms in plant tissue may occur, although tests indicate adequate soil nutrient levels. A common question in the industry is, "Why not feed plants enzymes, amino acids, proteins, vitamins, carbon, and organic acids rather than traditional fertilizer?" However, since plants make these and all other compounds from the essential elements, they would use extraordinary energy to degrade these additives into their individual components and then require additional energy to reassimilate them back into useable forms.

## PRIMARY NUTRIENTS AND FERTILIZERS

Nitrogen, P, and K receive a great amount of attention because they are typically deficient in soils and must be regularly applied. These elements are required in greatest amounts, and are therefore referred to as primary (or essential) nutrients or elements. The numerical designation on a fertilizer bag refers to the percentage of N, phosphate (expressed as $P_2O_5$), and potash (expressed as $K_2O$). Thus, a bag of 10-10-10 fertilizer would contain 10% N, 10% available phosphate ($P_2O_5$), and 10% potash ($K_2O$). The influences of increasing levels of essential nutrients on turfgrasses are listed in Table 7.2.

### Nitrogen

Nitrogen (N) is the key element due to its influence on color, growth rate, density, pest occurrence, and stress tolerance. Turfgrasses consist of between 20 and 60 g N/kg plant tissue (or 2% to 6% of the total dry matter), while soil organic matter typically contains about 5% N. In plants, N is used primarily for chlorophyll production, plant proteins, and nucleic acids. It is the most applied element and is required in larger quantities

**TABLE 7.2**
**Influence of Increasing Levels of Primary and Secondary Nutrients on Several Turfgrass Plant Responses**

| | Increasing Levels of | | | | | | |
|---|---|---|---|---|---|---|---|
| **Turf Response** | N | P | K | Fe | S | Ca | Mg |
| Turf color | ↑ | | ↑ | ↑ | ↑ | | ↑ |
| Shoot growth and density | ↑↓ | ↑ | | ↑ | ↑ | | ↑ |
| Rooting | ↓ | ↑ | ↑ | ↑ | ↑ | ↑ | |
| Carbohydrate formation | ↓↑ | | | ↑ | ↑ | ↑ | |
| Recuperative ability | ↓↑ | | | | | | |
| Heat, cold, and drought tolerance | ↓ | ↑ | ↑ | ↑ | | | |
| Wear tolerance | ↓ | | ↑ | | | | |
| Nematode tolerance | ↓ | | ↑ | | | | |
| Disease tolerance | ↓↑ | | ↑ | | ↑ | ↑ | |

*Notes:* ↑ = Ample supply of the nutrient usually increases the specific turf plant response; ↓ = excessive nutrient level usually decreases the specific turf plant response; ↑↓ = adequate nutrient levels increase the specific turf plant response while excessive amounts decrease the response.

**FIGURE 7.1** Nitrogen deficiency causes plants to grow poorly, be spindly, and lose color.

than any other element except C, H, and O. Nitrogen deficiency causes plants to grow poorly, be spindly, lose color, and be stunted (Figure 7.1). Problems, however, may develop if excessive N is used (Table 7.2). Excessive N increases shoot growth and selective diseases, as well as lowers stress tolerance to heat, cold, drought, and traffic. Most importantly, root and lateral shoot growth also are reduced.

**Origins and losses.** Turfgrasses may obtain N from organic matter decomposition and, to a small degree, from air oxidized by lightning and dispersed by rainfall. In soil, ammonium $\left(NH_4^+\right)$, nitrate $\left(NO_3^-\right)$, and nitrite $\left(NO_2^-\right)$ forms are the most important compounds and originate either from aerobic decomposition of organic matter or from the addition of commercial fertilizers. *Ammonium* $\left(NH_4^+\right)$ and *nitrate* $\left(NO_3^-\right)$ are the only forms of N absorbed or utilized by turfgrass plants. No matter what the N source applied is (e.g., manure, crop residues, organic matter, or commercial fertilizer), it has to be converted to one of these two forms for plant use.

*Mineralization* is the overall process where organic matter, organic fertilizers, and some slow-release fertilizers are broken down or transformed by soil microorganisms to provide available ammonium and nitrate forms for plants. Mineralization is a three-step process involving aminization, ammonification, and nitrification. Aminization and ammonification are steps of mineralization in which proteins, amines, and amino acids (usually from organic matter or humus) are converted to ammonium $\left(NH_4^+\right)$, a N source utilized by plants.

**Mineralization (Aminization and Ammonification):**

$$R\text{-}NH_2\text{(organic N)} + H_2O\text{ (water)} \xrightarrow{\text{microorganisms}} NH_4^+\text{(ammonium N)} + R\text{-OH(hydroxyl group)} + 275\text{ kJ energy}$$

Ammonium N $\left(NH_4^+\right)$ is then absorbed by plants or is further transformed to nitrate $\left(NO_3^-\right)$. Ammonium N is the preferred N source since extra energy (20 ATPs vs. 5 ATPs) from photosynthesis and respiration is required to convert nitrate into usable forms $\left(NH_4^+\right)$ by plants and because of its smaller likelihood of losses by leaching and denitrification (Figure 3.19, Chapter 3). In soils, ammonium N transformation to nitrate N is referred to as nitrification.

## Nitrification

$$2NH_4^+\text{(ammonium N)} + 3O_2 \xrightarrow[\text{warm temperatures}]{\text{Nitrosomonas bacteria}} 2NO_2^-\text{(nitrite N)} + 4H^+ + 2H_2O \xrightarrow[+O_2]{\text{Nitrobacteria}} 2NO_3^-\text{ (nitrate N)} + 2H_2O + \text{energy}$$

Nitrification is dependent on environmental conditions favoring soil microbiological activity. Warm temperatures, adequate soil moisture, and soil $O_2$ are necessary for this activity. However, nitrification does not readily occur under extreme temperatures (e.g., below freezing or above 105°F, 40.6°C), in saturated or poorly aerated soil, in excessively dry soil, or in low pH (<4.8) soil. Under these unfavorable conditions, microorganisms do not perform nitrification, and ammonium may accumulate.

Nitrate N $\left(NO_3^-\right)$ is readily soluble in water and may be repelled by negatively charged exchange sites of the soil components. Therefore, unless grasses rapidly utilize this form, it may be lost through leaching if excessive moisture is applied. This may be especially true during winter months when grass is not actively growing. In addition to nitrate and water, hydrogen ions ($H^+$) also are produced during nitrification, and a reduction in soil pH may be observed. This reduction is especially acute when a high rate of N is applied on sandy soils low in Ca. These soils are poorly buffered against pH changes induced through the acidifying effect of nitrification.

Besides leaching and clipping removal, additional avenues of N loss are through denitrification and volatilization. Denitrification is the conversion of nitrate N under anaerobic conditions to gaseous N ($N_2$) that can result in atmospheric losses. Certain anaerobic soil organisms can obtain $O_2$ from nitrates. They also can obtain $O_2$ from nitrites in waterlogged soils with a subsequent release of nitrous oxide and N gas.

Low soil $O_2$ levels and/or high soil moisture, alkaline (high pH) soils, and high temperatures favor denitrification. Applied N can be lost at the rate of 10% to 30% by denitrification in soils that arc compacted, waterlogged, or have an especially high pH (>7.5).

Volatilization is the conversion of ammonium N $\left(NH_4^+\right)$ to ammonia gas $\left(NH_3^-\right)$ that escapes to the atmosphere. If ammonium N comes in direct contact with free calcium carbonate in the soil, ammonium bicarbonate will be formed. This generally is of practical significance if the pH is >7 and the fertilizer is applied to bare soil. Ammonium bicarbonate is a relatively unstable compound. Upon exposure to the sun, it decomposes into ammonia ($NH_3$), $CO_2$, and $H_2O$.

Volatilization of ammonia N can usually be avoided by incorporating an ammonium N fertilizer source into soil. Ammonium N fertilizer also should not be applied immediately after lime application. Furthermore, irrigating with approximately 1/4 inch (0.6 cm) after fertilizer application will help eliminate this potential N loss.

**Turf color, growth,** and **density.** When plants are deficient in N, leaf color is initially an overall pale yellow-green color from **chlorosis.** Chlorosis reflects a reduction in chlorophyll production. Nitrogen is a part of the chlorophyll molecule, so it is essential in its manufacture. Since N is a mobile element, chlorosis usually appears first on the lower (older) leaves before eventually changing to yellow as the deficiency symptoms progress (often called "firing" for yellowing and senescence). In addition, plant growth rate and density may slowly decrease, resulting in weak turf that has difficulty recovering from damage.

Improper N fertilization can have an undesirable effect on grass rooting. Turfgrasses, in general, use carbohydrates stored in their roots to support shoot growth. If heavy amounts of N are used, excessive shoot growth occurs at the expense of roots. These roots, therefore, may not have enough recovery time to replenish their carbohydrates before being forced to support excessive shoot growth when N is reapplied. Turfgrass maintained at low N levels has up to twice as much root growth as plants maintained at high N levels.

In addition to forcing excessive shoot growth at the expense of root growth, physiological changes such as cell wall thinning, succulent tissue growth, and reduced root carbohydrate levels can also occur with excessive N use. Accordingly, increased stress susceptibility makes turfgrasses less hardy.

## Nitrogen Carriers

Turfgrass fertilizer carriers are available in an array of forms, such as granules, pellets, liquids (or solutions), powders, and suspensions. Granules are the most popular and range in size from coarse (1 to 3 mm in diameter) to fine, green's grade (<1 mm).

Atmospheric N is the original source of all N used by plants. When synthesized by man, over 90% of all N fertilizers are initially produced by reacting atmospheric N ($N_2$) and hydrogen gas ($H_2$) to form ammonia $\left(2NH_3^-\right)$ (called the *Haber process*). Large amounts of energy are required in the form of temperature and pressure for this process (Figure 7.2).

Gaseous ammonia N, also called anhydrous ammonia, is colorless and lighter than air but becomes a liquid when compressed and cooled. It contains 82% N by weight. From the basic ammonia compound, many different N-containing fertilizer compounds are manufactured. It may be liquefied to form anhydrous ammonia, dissolved in water to form N solution (called aqua), or formulated into other inorganic fertilizers.

Chemically, N sources are classified as either (1) quick release (also called soluble source, quickly available, readily available, fast acting, and other terms) forms that are water soluble and available as liquids or granules, or (2) slow release (also called water insoluble, controlled release, slow acting, delayed release, and others) forms, which are either coated products or controlled-release reacted products, available as granules, powders, or suspensions. Table 7.3 lists the most widely used N-containing turfgrass fertilizers and their characteristics; Table 7.4 lists the general characteristics of various N sources.

Nitrogen release (or availability) to plants from a particular carrier is facilitated by one of several mechanisms. The two most common release mechanisms include hydrolysis (dissolving of fertilizer with water) or microbial (Table 7.5). Microbial release (through *mineralization*) is very environmentally dependent. Extreme weather conditions (too hot,

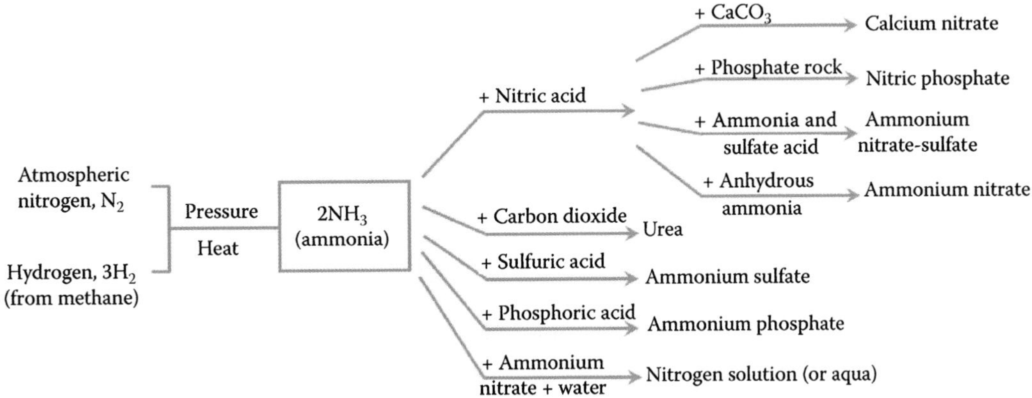


**FIGURE 7.2** Illustration of ammonia being produced and subsequently combined with other compounds to form commonly used fertilizers.

**TABLE 7.3**
**Primary Nitrogen Sources and Characteristics Used in Turf Fertilizers**

| Nutrient Source [Formula] | Approximate Nutrient Content ($N–P_2O_5–K_2O$) | Water-Insoluble N (%) | Salt Index Hazard | Comment (*Commercial Examples*) |
|---|---|---|---|---|
| | | | **Soluble or Inorganic Solids** | |
| Ammonium nitrate [NH4NO3] | 34-0-0 | 0 | High | • Water soluble; half N is ammonium form, other is the nitrate form; high burn potential; potential fire and explosive hazard; very hygroscopic (water loving) unless coated; medium acidifying effect. |
| Ammonium sulfate [(NH4)2SO4] | 21-0-0 | 0 | High | • Water soluble; contains 24% S; highest acidifying effect of listed sources. |
| Calcium nitrate [$Ca(NO_3)_2$] | 15-0-0 | 0 | High | • Very hygroscopic unless in airtight containers; contains 19% Ca and 1.5% Mg; fast acting with high burn potential; N release is not temperature dependent; used on sodic soils to displace Na; has alkalinity effect. |
| Monoammonium phosphate (MAP) [$(NH_4)HPO_4$] | 10-50-0 | 0 | Low | • Soluble N and P source; preferred to DAP when applied to alkaline soils; medium acidifying effect. |
| Diammonium phosphate (DAP) [$(NH_4)_2HPO_4$] | 18-46-0 | 0 | Low | • Soluble N and P source; ammonia losses can occur on alkaline soils; medium acidifying effect. |
| Potassium nitrate [$KNO_3$] | 13-0-44 | 0 | High | • Also known as saltpeter or nitre. Water-soluble K source with supplemental N; low chloride; has alkalinity effect. |
| Nitrate of soda (sodium nitrate) [$NaNO_3$] | 16-0-0 | 0 | Very high | • Water soluble; has highest burn potential of all materials; has alkalinity effect; imported from Chile. |
| Urea [$CO(NH_2)_2$] | 46-0-0 | 0 | High | • Contains 12% to 20% S; water soluble; rapid release; may volatilize if surface applied, especially under alkaline conditions; may leach rapidly if rainfall occurs immediately after application; medium acidifying effect; contains highest N% of all solid fertilizers. |
| | | | **Slow, Controlled-Release, or Water-Insoluble Solids** | |
| Sulfur-coated urea (or SCU) [$CO(NH_2)_2$ + S] | [31 to 37]-0-0 | 0 | Low | • Contains 10% to 20% S; slowly soluble; N release rate depends on temperature and coating integrity; has minor acidifying effect. *SCU*. |
| Polymer sulfur-coated urea (or PCSCU) | [39 to 43]-0-0 | 0 | Low | • Slowly soluble; N release rate is dependent on temperature and coating thickness; low acidifying effect; *Poly-Plus, Poly-S, TriKote, Poly-X Pro, XCU, NS-52*. |
| Methylene ureas (or MUs) | 40-0-0 | 14.5 | Low | • Sprayable powder, dispersible granulars, and suspension formulations (18-0-0) are available; *Nutralene; MU40; MUtech, METH-EX40; Chip; Contec DG, Coron, Nutri DG, NOVEX*. |
| Urea formaldehyde (or UF) [$CO(NH_2)_2 CH_2]_nCO(NH_2)_2$ | [18 to 38]-0-0 | ~27 | Low | • Slowly soluble; N release rate is dependent on temperature and formulation; low acidifying effect. Sprayable powder formulations are available; *Nitroform; MethEx38; UF38; Powder Blue, Blue Chip*. |
| Isobutylidene diurea (or IBDU) [$[CO(NH_2)_2]_2C_4H_8$] | 31-0-0 | 27 | Low | • Slowly soluble; N release is not temperature dependent but depends on moisture availability and particle size; low acidifying effect; sprayable powder formulations are available; *Slo-Release; IBDU, ParEx*. |
| Reactive layer coating (RLC)/resin-coated urea (RCU) | [37 to 44]-0-0 | 0 | Low | • Slowly soluble; has various thickness of resin or polymer coating; Coating thickness and soil temperatures control N release rate; *Polyon, Duration, ESN, Nutricote, Multicote, Osmocote, GoldCotes, Extend, V-Cote, SurfCote*. |
| | | | **Natural Organics** | |
| Milorganite | [5 to 6]-2-0 | 5.5 | Low | • Activated sewage sludge; N release rate increases with higher temperatures; contains micronutrients, especially Fe; low acidifying effect. |
| Actinite, Earthworks, Harmony, Hynite, Nature Safe, Ringer Turf, Sustane + others | [5 to 18]-[1 to 6]-[5 to 6] | Variable | Low | • N is from urea, methylene ureas, ammoniacal sources, and hydrolyzed poultry feather meal, sewage sludge, bone meal, leather tankage, fish meal, and blood; low acidifying effect. |

(*Continued*)

**TABLE 7.3 (CONTINUED)**
**Primary Nitrogen Sources and Characteristics Used in Turf Fertilizers**

| Nutrient Source [Formula] | Approximate Nutrient Content ($N–P_2O_5–K_2O$) | Water-Insoluble N (%) | Salt Index Hazard | Comment (*Commercial Examples*) |
|---|---|---|---|---|
| | | | **Solutions** | |
| Methylol urea + urea (formaldehyde + urea) | 30-0-0 | 30 | Low | • RESI-GROW GP-4340 (30-0-0); GR-4341 (30-0-2), and GR-4318; FLUF; Form-U-Sol; Homogesol-27; Slo-Release; Flormolene; CoRoN (28-0-0); Nitro-26 (26-0-0). |
| Methylene ureas + urea | 28-0-0 | 28 | Low | • Also formulated as a suspension as 18-0-0 with water-insoluble N. |
| Methylene ureas + triazone | 30-0-0 | 18 | Low | • A ~60% combination of MU and triazone and 40% urea; *Nitamin.* |
| Triazones + urea (formaldehyde + ammonia + urea) | [28 to 41]-0-0 | 28 to 41 | Low | • Considered a slow-realease N source; *N-Sure; Trisert; Formolene-Plus.* |
| Ammoniated solutions | [37 to 41]-0-0 | – | – | • – |
| Ammonium polyphosphate | 10-34-0 | 10 | Low | • – |
| Ammonium sulfate | 21-0-0 | 0 | High | • Contains 24% S; has acidifying effect. |
| Ammonium thiosulfate | 12-0-0 | 12 | High | • Contains 26% S. |
| Ammonium nitrate | 20-0-0 | 0 | High | • Often designated as AN-20. |
| Aqueous (or aqua) ammonia | [20 to 23]-0-0 | 0 | High | • Anhydrous ammonia + water. |
| Urea-ammonium nitrate | [28 to 32]-0-0 | 0 | High | • Used in colder climates where salting (precipitating) may occur; N content is ammoniacal N and nitrate-N; designated as UAN. |
| Calcium ammonium nitrate | 17-0-0 | 0 | High | • Clear solution, containing 8% Ca; often designated as CAN-17. |
| Potassium nitrate | 13-0-40 | 0 | High | • Has alkalinity effect. |
| Urea solution | [20 to 32]-0-0 | 0 | High | • Generally, most effective foliar N source. |
| Urea sulfuric acid | [10 to 28]-0-0 | 0 | – | • Contains 9% to 18% S. |
| | | | **Suspensions** | |
| Methylene ureas (or MUs) | 18-0-0 | 5 | 24 (Low) | Soluble granular formulas remain in suspension with mild agitation; *Flotec SG.* |

**TABLE 7.4**
**Advantages and Disadvantages of Various Nitrogen Fertilizer Sources**

| Advantages | Disadvantages |
|---|---|
| **Soluble (Liquid), Quick-Release Sources** | |
| • Rapid initial color and growth response<br>• Low cost per unit of N<br>• High in total N<br>• Odorless<br>• Versatile as a granular or as a liquid<br>• Maintain satisfactory N levels if applied frequently in small amounts<br>• Minimum temperature dependence for availability | • High potential for foliar burn, especially at higher application rates and temperatures<br>• Potential of undesirable growth surge<br>• Relatively short residual plant response; therefore, frequent applications needed<br>• Greater N loss potential to volatility, leaching, and runoff<br>• Often difficult to handle |
| **Synthetic Slow-Release Sources** | |
| • Slow-release liquid sources available<br>• High in total N<br>• Low potential for foliar burn<br>• Potential slow-release rates over extended time; product dependent | • High cost per unit of N<br>• Release rates depend on temperature, moisture, and soil microorganisms<br>• Due to slower release rates, more prone to mower removal |
| **Organic Slow-Release Sources** | |
| • Controlled release rates extending color response<br>• Low potential for foliar burn<br>• Other nutrients (e.g., P, K, and micronutrients) often included<br>• Viewed as a more "environmentally friendly" source<br>• Darker color may help warm soils | • Often have inconsistent release rates<br>• May have unpleasant odor<br>• High cost per unit of N<br>• May have high salt or metal content (e.g., chicken manure) |

**TABLE 7.5**
**Nitrogen Release Mechanisms for Various Granular Fertilizers**

| Release Mechanism | Comments |
|---|---|
| Water soluble (rapid hydrolysis) | • Water required to dissolve particle (hydrolysis) |
| Slow dissolution (slow hydrolysis) | • Primarily particle-size dependent; also influenced by soil moisture level, temperature, and sometimes pH |
| Osmosis across a barrier or layer | • Primarily particle-coating thickness and soil temperature dependent; also influenced by chemical nature, integrity, particle size, soil moisture, and microbial breakdown of any wax coating |
| Microbial degradation | • Primarily soil-temperature dependent; release is limited by cool temperatures, large particle size, low pH, and low soil $O_2$; simple organic sources release N before complex ones |

cold, or dry) may reduce or delay N release from microbial degradation.

Turfgrass response from N sources is influenced by application rate, particle size, solubility, temperature, and coating alterations. Response time, therefore, can be adjusted from a few weeks to several months in duration by manipulating these parameters. Table 7.6 lists the release means of major turfgrass fertilizers.

A *performance index number* (PIN) quantitatively and qualitatively rates fertilizers containing insoluble N. Ideally, this numerical system gives turf managers a more thorough understanding of the rating of the insoluble N fertilizers and helps predict how long a fertilizer will continue releasing N. The PIN measures percentages of (1) particle dispersion, (2) particle integrity, and (3) N activity index. Each of the three percentages is on a scale of 100; a perfect PIN rating is 300, while the lowest acceptable PIN value is 235. Particle dispersion is expressed as percentage of sample passing the dispersion test. A higher dispersion is desirable for water-insoluble N to assure the nutrient is delivered to the turf area. Minimum particle distribution is 80%. Particle integrity is the measure of how hard the fertilizer granule is. Integrity is important to assure durability through production and shipping to application. Minimum acceptable integrity is 90%. The nitrogen activity index is the percentage of water-insoluble N soluble in hot water with a minimum percentile of 40.

## Quick Release (Water Soluble)

**Urea.** Urea (~46% N) is one of today's most widely used N sources due to its relatively low cost and completely soluble nature. Once applied, urea is broken down into ammonium carbonate by the enzyme *urease* that is present in plant tissue and organic matter. Ammonium is a cation (positively charged) and is attracted to negatively charged clay particles, root hairs, and organic matter. Direct applications of urea to alkaline (high pH) surfaces can result in conversion of the ammonium carbonate to ammonia and $CO_2$, resulting in excess loss. This loss can be avoided by irrigating after application to incorporate the N.

A means of delaying the conversion of ammonium to nitrate is N stabilizers. Two types are nitrification inhibitors and urease inhibitors. Nitrapyrin (*N-Serve*), encapsulated nitrapyrin (*Instinct*), and dicyandiamide (DCD or *Didin*) are nitrification inhibitors that delay the conversion of ammonium to nitrite during nitrification. Urease inhibitors such as NBPT [N-(*n*-butyl) thiophosphoric triamide] and PPD [phenyl-phosphoro-diamidate] slow the activity of urease. *UMAXX* (47-0-0) and *UFLEXX* (46-0-0) are commercial stabilized urea products

**TABLE 7.6**
**Release Means for the Major Turfgrass Fertilizers**

| Nitrogen Source | Hydrolysis (moisture) | Microbial Dependent[a] | Coating Thickness/Durability |
|---|---|---|---|
| Isobutylidene diurea (IBDU) | Yes | – | – |
| Organic sources | – | Yes | – |
| Milorganite | – | Yes | – |
| Urea formaldehyde (UF) | – | Yes | – |
| Poly (or resin)-coated urea (PCU) | – | Yes | Yes |
| Methylene urea (MU) | – | Yes | Yes |
| Sulfur-coated urea (SCU) | Yes | – | Yes |
| Ammonium nitrate/sulfate | Yes | – | – |
| Potassium nitrate | Yes | – | – |
| Urea | Yes | – | – |
| Mono- or diammonium phosphate (MAP or DAP) | Yes | – | – |
| Calcium nitrate | Yes | – | – |

[a] Microbial N release generally increases with increasing temperatures, soil moisture, $O_2$ levels, and near-neutral soil pH.

that inhibit or delay nitrification utilizing DCD and urease (NBPT) activity. UFLEXX delays nitrification for six to eight weeks, while UMAXX delays this for 12 to 16 weeks. *Agrotain* (NBPT), *Stay-N*, and *NZone* are commercial urease inhibitors. *SuperN* is a liquid mixture of DCD and NBPT, while *HYDREXX* is a granular mixture of them.

**Ammonium sulfate/nitrate.** Ammonium sulfate/nitrate, ammonium phosphate, potassium nitrate, and calcium nitrate are other commonly used water-soluble N sources collectively referred to as inorganic salts. Once these fertilizers solubilize in soil, ammonium ions can be adsorbed by negatively charged clay or organic matter. As with urea, soil nitrobacteria convert this ammonium to nitrate, which is the main form available to plants. Unlike ammonium sulfate and phosphate, potassium nitrate and calcium nitrate fertilizers do not need to undergo conversion by nitrobacteria since their N source is already in nitrate forms. Ammonium phosphates and potassium nitrate are the primary sources of P and K, respectively, in inorganic salts, while calcium nitrate is often used on acidic soils.

## Slow-Release Nitrogen Sources

In an attempt to overcome some disadvantages of soluble N sources, fertilizer manufacturers have developed an array of slow- or controlled-release N sources (Figure 7.3). Controlled-release fertilizers provide plant nutrition over a very specific time period under a certain set of environmental conditions. Slow-release fertilizers have a less defined period of release. To qualify as a slow-release fertilizer, at least 15% of the nutrient must be slowly released over a period of time. Currently, several means are used to achieve slow-release characteristics:

1. Compounds with limited water solubility
2. Coating with a water-soluble barrier
3. Coating with a nonsoluble barrier where nutrients are released through osmosis or diffusion
4. Chemically reacted urea to make it more slowly available
5. Adding an inhibitor to slow N transformation, which extends its availability (also known as N stabilizers, as previously discussed)

Slow-release N sources generally provide a more uniform growth response and longer residual plant response. They also have less potential for N loss and allow a higher application rate than readily soluble sources. In addition, their burn potentials are lower because of their low salt index values. The application rate at which these sources release N may vary with fertilizer timing, source, temperature, moisture, pH, and particle size. Urea is often the background fertilizer used for many slow-release N sources.

Drawbacks of slow-release N sources include high per-unit cost and slow initial plant response. Some sources also are not adaptable to liquid application systems. General characteristics of several slow-release N sources are listed in Table 7.7.

### Controlled-Release Urea Reaction Synthetic Nitrogen Sources (Noncoated)

These products are slow- or controlled release by being formulated to have limited water solubility. They generally can be made in smaller particle sizes than coated products and thus have less mower pickup at lower mowing heights. They also are homogenous in terms of their having the same nutrient composition throughout the particle and are not release dependent on the integrity of a coating. The noncoated urea reaction fertilizers are made by reacting urea with either formaldehyde or isobutyraldehyde. Ureaform (UF) and methylene urea (MU) are fertilizers produced by reacting urea with formaldehyde. The composition and therefore the N release characteristics of methylene ureas are controlled by these ratios of short- to long-chain polymers. Isobutylidene diurea (IBDU) is produced from reacting urea with isobutyraldehyde. Microbial activity is required for decomposition and release of N and costs are higher than those for soluble sources.

*Isobutylidene diurea* (IBDU). IBDU is formed by reacting isobutyraldehyde with urea in an acid solution to form a material composed of a single molecule containing 31% N, 90% of which is water insoluble. In the presence of water, IBDU hydrolyzes back to urea and butyric acid. IBDU's N release rate is predominantly affected by soil moisture and particle size and is not as dependent on temperature. IBDU, therefore, releases more quickly in cool weather compared to slowly available N fertilizers dependent on the temperature activity of microbes. Higher soil moisture and smaller particle size result in a faster release rate. IBDU is used as a N source alone, or in homogenous fertilizers and fertilizer blends. It is sometimes blended with water-soluble N sources to overcome the initial delay in color response by IBDU alone.

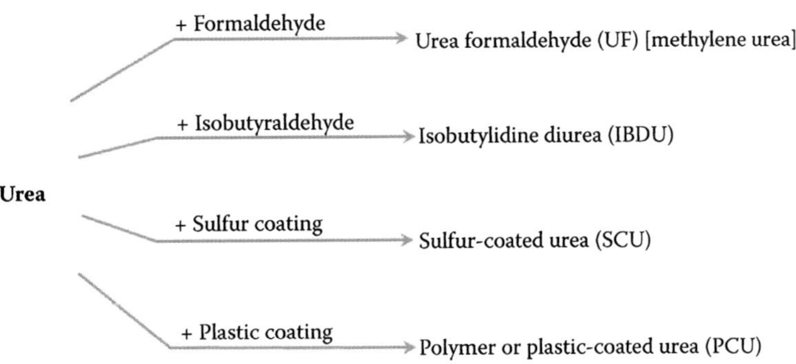


**FIGURE 7.3** Illustration of how several slow-release nitrogen fertilizers are formulated from urea.

**TABLE 7.7**
**General Characteristics of Several Slow-Release Nitrogen Sources**

| Response | Sulfur-Coated Urea (SCU) | Polymer-Coated Urea (PCU) | Urea Formaldehyde (UF) | Methylene Urea (MU) | Isobutylidene Diurea (IBDU) | Natural Organic |
|---|---|---|---|---|---|---|
| Approximate percent N | 37% N | 37% to 44% N | 38% N | 40% N | 31% N | 2% to 10% N |
| Immediately available N | 3.9% N | 0% N | 11% N | 26% N | 3.1% N | 10% N |
| Release time | 6 to 8 weeks | 8 to 36 weeks | 2 years | 8 to 12 weeks | 12 to 16 weeks | 8 to 52 weeks |
| Release mechanism | Coating breakdown | Osmosis | Microbial | Microbial | Hydrolysis | Microbial |
| Release requirements | Moisture | Moisture | Moisture and soil temperature | Moisture and soil temperature | Moisture | Moisture and soil temperature |
| Best response season | All seasons | Summer | Summer | Summer | Spring and fall | Summer |
| Initial response | Medium | Slow | Medium slow | Medium | Medium slow | Slow |
| Residual effect | Extended | Extended | Extended | Extended | Extended | Extended |
| Water solubility | Low | Low | Medium low | Medium | Medium low | Low |
| Foliar burn potential | Low | Low | Low | Low | Low | Low |
| Soil temperature release dependence | Low | Medium | High | Medium | Low | High |

Although IBDU's N release is independent of microbial activity and therefore more readily available during cool weather, this is somewhat dependent on its particle size, with finer particles providing a greater surface area and faster rate of hydrolysis (N release mechanism).

*Crotonylidene diurea* (CDU). Crotonylidene diurea is a white powder formed by reacting urea with crotonaldehyde or acetaldehyde. N release rate is slow as the material dissolves (hydrolyzes) in water to gradually decompose to form urea and crotonaldehyde. Like IBDU, CDU release rate depends on water availability and increases with smaller particle size. *Crotodur* and *CDU* are commercial products used in Europe and Japan.

*Ureaformaldehyde* (UF). Ureaformaldehyde is a generic designation for several methylene urea (MU) polymers made from reacting urea with formaldehyde to first form monomethylol urea. As the reaction continues, soluble methylene urea and ureaformaldehyde are produced (Figure 7.3). They are available as both granular and liquid forms. These products have varying lengths of polymers of methylene urea, depending on the proportion of ureas and formaldehyde in the initial reaction. These vary from short-chained water-soluble molecules to long-chained, highly water-insoluble molecules to provide controlled N release. Ureaform is the least water-soluble MU intermediate, while methylene diurea/methylene dimethylene (MDU/DMTU) is most water soluble. Nitrogen in ureaformaldehyde sources is classified as

1. **Cold water (72°F, 22°C) soluble nitrogen** (or **CWSN**)—This is readily available to plants and can appear as liquid formulations. It is made up of "free" urea or short polymer chains such as MU, MDU, and DMTU.
2. **Water-insoluble nitrogen** (or **WIN**)—This is slowly available to plant, and WIN values indicate the percentages of the following:
   a. **Cold-water-insoluble nitrogen** (or **CWIN** at 72°F, 22°C)—Plant availability is over several weeks by microbial degradation. Ureaform's 35% total N is at least 60% CWIN.
   b. **Hot-water-insoluble nitrogen** (or **HWIN** at 212°F, 100°C)—Plant availability is over many months by slow microbial degradation. It is the N fraction most resistant to microbial decomposition. Ureaform's HWIN is at least 40%.

The overall N availability is described by an *activity index*, which compares the proportion of hot-water-insoluble N with cold-water-insoluble N. The higher the activity index value is, the lower is the amount of HWIN present. By altering the proportion of urea and formaldehyde in the initial reaction, activity indexes are produced to meet the needs of turfgrass species at different geographical locations. As polymer lengths and number of longer polymers increase, solubility decreases, resulting in slower N release. These increase in length from methylene diurea (most soluble), dimethylene triurea, trimethylene tetraurea, and tetramethylene pentaurea to pentamethylene hexaurea (least soluble).

1. Urea + formaldehyde → methylol urea (MU) + $H_2O$ (one- to four-week release rate)
2. MU + urea → methylene diurea (MDU) + $H_2O$ (three to six weeks)
3. MDU + MU → dimethylene triurea (DMTU) + $H_2O$ (five to nine weeks)
4. DMTU + MU → trimethylene tetraurea (TMTU) + $H_2O$ (7 to 13 weeks)
5. TMTU + MU → tetramethylene pentaurea (TMPU) + $H_2O$ (10 to 16 weeks)
6. TMPU + MU → pentamethylene hexaurea (PMHU) + $H_2O$ (≥16 weeks)

Ureaform fertilizers contain 38% N with at least 60% of the total N as CWIN, and less than 15% of the total N as CWSN; they are commercially available as *Nitroform, Hydroform, Granuform, Ureaform, MethEx38, UF38,* and *Blue-Chip.* These sources are for soil applications only and not as liquids, except *Powder Blue*, which can be applied as a suspension, but is more costly than soluble N fertilizers. *Formolene Plus* (30-0-0) and *CoRon* (28-0-0) are UF solutions that contain water-soluble methylene ureas and some free urea. A recent line of shorter chain soluble UF fertilizers is *Triaform.*

A combination methylene diurea/dimethylene triurea (abbreviated MDU/DMTU) product containing two and three urea chain materials, respectively, is available as *Contec DG* and *Nutri DG.* These are granular products containing at least 40% N with less than 25% CWIN. They are formulated as dispersible granulars that quickly dissolve in the presence of water, reducing mower and traffic pickup losses.

All UF products depend on microbial breakdown for N availability. Therefore, environmental conditions favoring microbial activity (e.g., warm temperatures, >55°F, 13°C; neutral soil pH; and adequate soil moisture and $O_2$) promote N release. Conversely, low temperatures, acidic soils, and low soil $O_2$ inhibit N release from UF. Quickly available sources are usually applied alone or in combination with ureaform fertilizers during cool periods.

As with any N source, UF losses by mower pickup can be significant, especially immediately after application. To avoid this problem, turf should be allowed to dry for several days after application before mowing. Alternatively, clipping collection buckets can be removed to allow clippings and fertilizer granulars to return to the soil surface. The size guide numbers (SGNs) for greens, fairways, and general purpose are 100, 125, and 215, respectively.

A similar product to UF is *Nutralene*, a soluble methylene urea product containing ~40% N with 14% N being water insoluble. It is more readily available than UF because of its short-chained methylation of 1.7 to 1 urea to formaldehyde ratio. Its mechanism of N release depends on both microbial activity and hydrolysis. Its N release rate, therefore, is faster than for UF products, but its long-term fertilizer effects on the turf are shortened. Nutralene performs better during cooler temperatures than many fertilizers due to its quicker release during these conditions. Other short-chain aqueous solutions of methylol urea and free urea are *RESI-GROW, GP-4340 (30-0-0)*, and *GP-4341 (30-0-2).* Additional methylene urea products include *MU 40, Nutralene, Chips, Isotek* (IBDU + MU), *NX-Pro, Nitroform, NOVEX, MethEx-40*, and *MUtech. MESA* is a line of ammonium sulfate fertilizers combined with *MethEx-40* methylene urea. The MU is released by microbial activity and ammonium sulfate by hydrolysis.

Another flowable UF source, *FLUF*, contains 18% N, of which 20% to 25% of the total N is water insoluble. The initial response of FLUF is generally slower than urea but has less foliar burning potential. *Slo-Release, Homogesol-27*, and *RESI-GROW 4318* are other liquid UF sources containing water-soluble and water-insoluble methylene ureas.

**Coated, slow-release nitrogen sources.** Coated N fertilizers consist of urea or other soluble sources coated with a semipermeable barrier such as S, resins, plastics, and waxes (Figure 7.3). Nitrogen release rate is slow since the coating prevents wetting of the soluble N source. Release rates are dependent on coating degradation or the physical integrity of the coating. Nitrogen release occurs with diffusion of water into the granule, resulting in cracks forming in the coating (sulfur-coated urea [SCU] and polymer-coated sulfur-coated urea [PCSCU]) or by osmotic diffusion of the N source, such as urea, out of the granule. Examples include sulfur-coated urea, polymer-sulfur-coated urea, polymer-coated fertilizers, resin-coated urea, and urea coated with magnesium ammonium phosphate.

*Sulfur-coated urea and polymer-coated sulfur-coated urea.* Sulfur-coated ureas are particles of urea and other soluble N sources such as ammonium and nitrate inorganic salts, coated with a layer of S plus a sealant. Sulfur is most often used since it is relatively cost effective and is needed by plants. Sulfur-coated ureas are formulated by moving granulated or prilled preheated urea pellets through a stream of molten S using a rotating drum. It is then coated (sealed) with a microcrystalline wax to protect the surface from microbial degradation, to strengthen the S shell, and to decrease the initial rate of urea release. The product is then cooled and a diatomaceous earth or vermiculite clay conditioner is applied to further reduce cracking and to promote sealant stickiness. Urea is released through cracks, pinholes, and imperfections naturally occurring in the coating as particles cool. Water enters these cracks, followed by a rapid release of the dissolved urea from the core of the particle. If a wax sealant is used, N release then becomes temperature and moisture dependent as microbial activity is required to degrade the sealant.

The slow-release properties of SCU are from the variability in coating thicknesses of the individual particles and the various delay times associated with their moisture absorption. This release typically occurs over a 6- to 16-week period. SCU granules also are large in size; thus, they become subjected to damage during transportation, blending, and application, or by weight from mower reels, rollers, or wheels. Therefore, handling SCU should be kept to a minimum and drop spreaders avoided when making applications. Microprill and polycoating technologies have reduced these problems.

Rate of urea diffusion from SCU depends on microorganism activity, particle size, and, as previously discussed, coating thickness and integrity. To characterize the initial release rate of sulfur-coated products, a seven-day dissolution rate test is used. In a laboratory, a predetermined amount of sulfur-coated urea is immersed in distilled water at 100°F (38°C) for seven days. The amount of urea entering the solution is measured and is expressed as percentage of the total urea content. For example, SCU-30 is a product with 30% of the urea released at the end of the seven-day dissolution test. A seven-day dissolution rate of 25% to 35% is best for normal use rates on turfgrasses.

SCU applied during winter months may produce a mottled turfgrass appearance. The intensity of this mottled appearance is correlated with coating thickness and granule size. It normally dissipates two to four weeks after appearing, depending on the N application rate and weather conditions.

Sulfur-coated urea contains 28% to 32% N and 6% to 18% S by weight and can reduce soil pH slightly by reacting with water to form sulfuric acid. The sulfur coating also is a S source for plants. Sulfur-coated urea tends to be a lower cost, slow-release N source when compared to other coated materials. Leaching and volatilization losses generally are low, assuming excessive moisture is not applied. *SCU* is a commercial source.

Instead of using microcrystalline wax as with SCU, polymer sealants are more commonly used today and are referred to as polymer-coated sulfur-coated urea (PCSCU). With PCSCU, urea is first coated with S to form one layer and then coated with a polymer that further protects the nutrients. This, in combination with the S layer, determines the rate of release. Unlike SCU, these particles are not coated with wax. The release mechanism is the same for PCSCU as with SCU. It is controlled by water entry through coating, followed by swelling, and eventual bursting of the coating. An advantage is better resistance to abrasion than SCU. Dust problems when handling the material are also minimal, as are leaching potentials. *Poly-S, TriKote, Poly Plus, Poly NS-52, X-Cote,* and *XCU* are commercially available dual-coated fertilizer sources using this technology. Polymer-coated sulfur-coated products contain 38% to 43% N.

**Polymer- or plastic-coated urea.** A relatively new but similar technology to the resin-coating process involves coating a soluble N source, such as urea, nitrate, ammonium, or K, with alkyd resin, polyurethane, or polyolefin. Polymer-coated fertilizers rely on osmosis rather than coating imperfections, like SCU, to release N. Once inside the coated granules, water dissolves some of the solid fertilizer. As the fertilizer particle dissolves from absorbed water, internal pressure causes the gradual diffusion or release of the urea, or the urea is forced out through pores. This continues until all fertilizer has dissolved and diffused out. Nutrient release, therefore, is due to controlled diffusion through the coating and is fairly constant over time, and all the fertilizer is released from the coated granules. Release rates generally vary from 70 to 270 days depending on the thickness of the coating and dissolution of water into the granule. Higher temperatures also increase release rate and decrease longevity. Polymer coating also is more expensive than sulfur coating but generally is harder and will withstand abrasion better than the SCUs and PCSCUs. *ESN* is a commercial source.

*Reactive layer coating* (RLC) is a second method of polymer coating. Two coats of resin are applied with the first reacting with the urea and the second coating reacting with the first to form a harder coating. The polymer coating acts as a semipermeable membrane through which water moves from a dilute solution to a concentrate solution within the coating (osmosis). As the internal osmotic pressure increases, the concentration nutrient is then diffused out across the polymer coating.

Different coating thicknesses are used to obtain different release rates, depending on the soluble substrates used and water solubility. As the polymer coating thickness is increased, the release rate is reduced or becomes slower, and the final N analysis of the fertilizer is lower due to increased coating weight. Soil temperature is the most influential environmental parameter affecting nutrient release. As temperature rises, the nutrient release rate increases since more nutrient is dissolved in solution at higher temperatures. The rate of release is then controlled by the coating thickness. Polymer (often polyurethane) coated products include *Polyon, Duration, Nutricote, Extend, V-Cote, Osmocote, Multicote*, and *SurfCote*.

Additional coating materials are constantly being evaluated for their nutrient controlled-release properties. Examples include humic acid, humin, and fulvic acid particle coating. Fulvic acids are lowest in molecular weight and have good solubility and are often used for foliar application. Humic acids have higher molecular weights with moderate solubility and thus are more efficient as a soil application. Humins have even higher molecular weights and thus are applied to the soil. These are intended more to add organic carbon into the turf system versus providing controlled or slow nutrient release. They also increase cation exchange capacity (CEC) levels and have chelating characteristics.

### Natural Organic Nitrogen Sources

Natural organic N sources usually involve various levels of plant composted or waste (either human or animal) materials (Figure 7.4). Manure, sewage sludge, poultry litter, meat-processing waste, fish or feather meal, bone meal, humates, and composted plant residues are traditionally used natural organic N sources (Table 7.8). Nutrient contents are typically low—1% to 10% N, 0.2% to 6% P, and 0.5% to 6% K—and are not always in the ratios required by plants. Once applied, all organic N forms must be converted first to the ammonium form $\left(NH_4^+\right)$ and possibly to the nitrate $\left(NO_3^-\right)$ form by soil microorganisms prior to plant use. This requires appropriate moisture and temperatures, which are not always predictable. Advantages of these include low burn potential due to limited water-insoluble N, little effect on pH, and low leaching losses. Other advantages include a variety of nutrients included in addition to N, and certain organic sources that may possibly improve the physical condition of soils, especially sandy ones. In addition, depending on the local source, natural organic N sources may readily be available at competitive prices.

Recently, some natural organic N sources have been inoculated with various microorganisms or included with readily available organic materials to stimulate natural microbial population. In addition to claims of providing N, products claim to offer better nutrient retention; suppression of pests such as nematodes, soil insects, and diseases by the inoculated organisms; enhanced thatch decomposition by inoculation with bacteria or by stimulation of earthworm activity; and improved soil chemical and physical properties. These claims are rarely based on scientific research and results are often erratic.

**FIGURE 7.4** Numerous natural organic nitrogen fertilizer sources exist. These vary in their nutrient content, release rates, and plant burn potential.

**TABLE 7.8**
**Approximate Amounts of Macronutrients in Common Organic Fertilizer Sources**

| Nutrient Source | Nitrogen (N) | Phosphate ($P_2O_5$) | Potash ($K_2O$) |
|---|---|---|---|
| | Percent | | |
| | **Animal by-Products** | | |
| Dried blood | 13 | 2 | 1 |
| Blood meal | 10 to 14 | 1 to 2 | 0.6 to 1.0 |
| Bone meal, steamed | 3 | 25 | 0 |
| Dried fish meal | 10 | 7 | 3 |
| Tankage, animal | 7 | 10 | 1 |
| | **Excreta** | | |
| Cattle manure | 2 to 5 | 1.5 | 2 |
| Feather meal | 15 | – | – |
| Guano, bat | 8.5 | 5 | 1.5 |
| Horse manure | 2 to 8 | 1 to 3 | 2 to 7 |
| Poultry manure | 5 to 15 | 3 | 1.5 |
| Sewage sludge, activated | 6 | 3 | 0.5 |
| Sewage sludge, dried | 0.1 to 17 | 0.1 to 14 | 0.1 to 2.5 |
| Swine manure | 7 | 4 | 6 |
| | **Plant Residues** | | |
| Alfalfa meal | 2-3 | 0.5 to 1 | 1 to 3 |
| Corn gluten | 9 | 0.5 | – |
| Cottonseed meal | 6-7 | 0.4 to 3 | 1-2 |
| Garbage tankage | 2.5 | 3 | 1 |
| Linseed meal | 5.5 | 2 | 1.5 |
| Rapeseed meal | 5.5 | 2.5 | 1.5 |
| Rice-hull/turkey litter | 1.75 | 2.65 | 1.85 |
| Seaweed (kelp) | 0.20 | 0.10 | 0.60 |
| Soybean meal | 7 | 1.5 | 2.5 |
| Tobacco stems | 2 | 0.5 | 6 |

Low N release during cool weather due to reduced microbial activity and low N content should be considered prior to using organic N sources. This can result in large amounts of material having to be applied to supply a desirable N rate. Other considerations include greater costs per pound of nutrient than with soluble sources and that natural organic N sources may be difficult to store and to uniformly apply due to small particle size or dust. Depending on the source, some natural organic products produce objectionable odor after application and contain undesirable salts, heavy metals, and weed seeds. In general, natural organic sources such as manures and composted crop residues should not be used on golf greens because of potential soil drainage hindrance from the large amounts of material applied and from inorganic ash such as sand, silt, and clay remaining after the organic matter is combusted.

*Milorganite*, a product from the Milwaukee Sewage Commission, is the most popular commercial organic N source used on fine turf (Figure 7.5). It is an activated sewage sludge produced when raw sewage is inoculated with microorganisms, aerated to promote flocculation, filtered, dried, grounded, screened, and sterilized. Milorganite contains ~5% to 6% N, of which 92% is water-soluble 2% $P_2O_5$, 1.2% Ca, and 4% Fe. Milorganite is characterized by having a low burn potential on turfgrass leaves, and its uniform release of N over a three- to four-week period has a minimum effect on soil pH and salinity. It also has low leaching and volatilization losses; is a source of Fe, Cu, Ca, and Zn; and, due to its dark color, serves as a soil warmer during cool weather.

Disadvantages of Milorganite include being relatively costly per pound of N, having poor winter response since microorganisms are required for N utilization, and having relatively short residual N response. Milorganite also does not contain K. *Huactinite* is another sewage-based fertilizer, as are various other locally available products.

Other commercial slow-release natural organic N fertilizers from animal by-products include *Nature Safe, Sustane, Ringer, Worm Power*, and *Earthworks*. These contain various amounts and combinations of composted poultry litter, blood, manure, bone meal, wheat germ, and other additives, and also are often inoculated with various microorganisms and fortified with other N sources.

### Salt Index

Fertilizers behave just like any salt, such as table salt, by attracting water; thus, they pull moisture from plants they contact. If the fertilizers are applied at excessive rates or when days are hot and dry, plants can suffer fertilizer salt burn (Figure 7.6). The salt index measures the relative tendency of a fertilizer to increase the osmotic pressure of the soil solution compared to the increase by an equal weight of the reference material, sodium nitrate (Table 7.3).

Fertilizers vary in their capability to attract plant water, thereby causing burn. This can be minimized by using fertilizers with low salt indexes on days that are hot and/or by applying fertilizer on dry turfgrass leaves followed by immediate irrigation. If a soil test indicates soil salinity problems, a low salt index fertilizer should be considered. Another fertilizer burn involves animal urine (Figure 7.7). This often goes undetected until the turfgrass is severely damaged. Symptoms are typically a dark ring pattern with brown centers and urine smell that is detectable for several days.

## Phosphorus

Phosphorus (P), the second-most essential element for plant growth, is involved in the transfer of energy as the organic compound, adenosine triphosphate (ATP), during metabolic processes. It also is an essential component of DNA, RNA, and cellular membranes. Phosphorus content may range

**FIGURE 7.5** The most popular commercial organic nitrogen source on fine turf is Milorganite.

**FIGURE 7.6** Fertilizers can burn turf when applied at excessive rates or when conditions are hot and dry.

**FIGURE 7.7** Animal urine damage typically is circular and often has a darker perimeter ring of turf.

between 0.10% and 1.00% by weight, with sufficiency values from 0.20% to 0.40% in newly mature leaf tissue. Phosphorus is considered deficient when tissue levels are below 0.20% and excessive above 1%. The highest concentration of P is in new leaves and their growing point, but is readily mobile in plants.

Since P is fairly mobile within plants, deficiency symptoms initially occur in older (lowermost) tissue. Symptoms of P deficiency include slow growth, as well as weak and stunted plants possessing dark-green lower, older leaves from excess production of the plant pigment, anthocyanin. These older leaves eventually show a dull blue-green color with a reddish-purple pigmentation along the leaf blade margins due to sugar accumulation (Figure 7.8). Eventually, leaf tips turn reddish, which may then develop in streaks down the blade.

Phosphorus deficiency symptoms normally occur when root growth of turf plants is restricted, and when soil temperatures and $O_2$ levels decrease. Since roots are a principal site for energy (carbohydrate) storage, they are very dependent on adequate P levels (Figure 7.9). Early spring and fall are two seasons in which root growth is slowed; hence, P in the soil is not readily encountered. Similarly, P deficiencies often occur during turfgrass establishment, resulting from the initial restricted rooting of new seedlings. Establishment in sand soils also may have P deficiencies. Research indicates a 1:2:1 or 1:2:2 ratio of N-P-K as best for turfgrass establishment.

Phosphorus absorption by plants is largely due to the very slowly soluble orthophosphate ions, $H_2PO_4^-$ and $HPO_4^{-2}$, that are present in the soil solution. $H_2PO_4^{-2}$ ions dominate in acid

**FIGURE 7.8** Reddish-purple colored turf indicating phosphorus deficiency.

**FIGURE 7.9** Creeping bentgrass with complete fertilization (left), minus nitrogen (middle), and minus phosphorus (right).

soils. As soil pH increases, $HPO_4^{-2}$ ions dominate until conditions become alkaline, when $PO_4^{-3}$ ions then dominate. $H_2PO_4^-$ and $HPO_4^{-2}$ ions are both found at intermediate pH levels. When calcium carbonate $(CaCO_3^-)$ dominates at higher pH, P availability may be reduced. Heavier P fertilizer application under high pH conditions may be required to increase its availability. Maximum availability of soil P occurs between pH 6.5 and 7.5.

**Phosphorus forms at various soil pH levels:**

$$H_2PO_4^- \text{ (acid soils, pH} < 6.0) \leftrightarrow H_2O + HPO_4^{-2} \text{(neutral soils, pH} \sim 7.0) \leftrightarrow H_2O + PO_4^{-3} \text{(alkaline soil, pH} > 8.0)$$

The first P fertilizer was manufactured in England around 1830 from treating ground bones with sulfuric acid ($H_2SO_4$) to increase the solubility (or availability) of the P. Today, the most commonly used P fertilizers in turf include superphosphate, triple (or treble) superphosphate, and monoammonium and diammonium phosphate (MAP and DAP, respectively) (Table 7.9). Superphosphate consists of calcium phosphate and gypsum. It is produced by reacting rock phosphate with sulfuric acid ($H_2SO_4$) or phosphoric acid ($H_3PO_4$). Triple superphosphate is calcium phosphate formed when rock phosphate is treated with phosphoric acid; ammonium phosphates are produced by reacting anhydrous ammonia with phosphoric acid. Ammonium polyphosphates are used primarily in fluid fertilizer and are formed by reacting ammonia with superphosphoric acid. If irrigation water with appreciable Ca

**TABLE 7.9**
**Primary Phosphorus Sources and Characteristics Used in Turf Fertilizers**

| Source | Approximate Nutrient Content ($N-P_2O_5-K_2O$) | Salt Index Hazard | Comment |
|---|---|---|---|
| Monoammonium phosphate (MAP) [$(NH_4)H_2PO_4$] | 10-50-0 | Low | • Soluble P source, also provides N; preferred to DAP on alkaline soils; contains 1% to 2% sulfate ($SO_4^-$); medium acidifying effect. |
| Diammonium phosphate (DAP) [$(NH_4)_2HPO_4$] | 18-46-0 | Moderate | • Soluble P source; contains higher N than MAP; also reduces soil pH; preferred to MAP on acidic soils; ammonia losses on alkaline soils can occur; medium acidifying effect. |
| Superphosphate [$Ca(H_2PO_4)_2 + CaSO_4$] | 0-20-0 | Low | • Soluble P source, often used in mixed fertilizers; contains Ca (18% to 21%) and S (12%) as gypsum; neutral acidifying effect. |
| Triple (or treble) superphosphate [$Ca(H_2PO_4)_2H_2O$] | 0-46-0 | Low | • Concentrated P source containing Ca (13%); neutral acidifying effect. |
| Ammonium polyphosphate [$NH_4H_2PO_4$ + $(NH_4)_3HP_2O_7$)] | 10-34-0 | Low | • Liquid P and N solution source; used in fluid fertilizers; other grades also available. |
| Phosphoric acid [$H_3PO_4$] | 0-55-0 | High | • Very caustic; primarily used to make triple superphosphate, ammonium phosphate, and in fluid fertilizer. |
| Milorganite | 6-2-0 | Low | • Activated sewage sludge; contains micronutrients, especially Fe; low acidifying effect. |
| Colloidal phosphate | 0-8-0 | Very low | • Contains 20% Ca; low P availability; best used as a powder on acid soils; neutral acidifying effect. |

*Note:* Generally, the higher the salt index/unit of nutrient is, the higher the burn potential of the particular fertilizer material is.

content (>50 ppm Ca) is used, insoluble calcium phosphate may form. This may clog the irrigation system or reduce P availability. Injection of sulfuric acid or urea-sulfuric acid helps prevent or dissolve calcium phosphate precipitates that may form.

Other minor use P sources include bone meal, basic slag, urea-ammonium phosphate, magnesium ammonium polyphosphate (8-40-0), nitric phosphate, and calcium metaphosphate (or apatite). Bone meal contains 15% to 34% $P_2O_5$ plus ~4% N and is an animal bone by-product. Basic slag is a steel manufacturing by-product containing 10% to 18% $P_2O_5$ and 12% N. It also contains some Ca and should be finely ground before use. Phosphoric acid is a liquid containing 53% $P_2O_5$. Liquid forms of ammonium phosphate commonly used are 10-34-0 and 11-37-0 and are formed by reacting ammonia with phosphoric acid. Nitrogen, phosphate, and sulfuric acid can be added to make the desired grade.

The available P content in fertilizers is expressed as $P_2O_5$. To determine the actual P content from the percent of $P_2O_5$ in a fertilizer, the following conversion is needed:

$$\text{Percent of } P_2O_5 \text{ as expressed on fertilizer bag} \times 0.44$$
$$= \text{percent of P in fertilizer bag}$$

Phosphorus does not move or leach readily in soils due to its low solubility in the soil solution; therefore, P applications are not needed as regularly as N applications. A soil test is probably the best indicator of P levels in a soil. Indiscriminate P applications can form unhealthy, high levels. Iron deficiencies, for example, often result from high P-containing and/or alkaline soils. Phosphorus is most readily available to plants with the soil pH range of 5.5 to 6.5. At low pH (<5.0), soils containing Fe and Al form an insoluble complex with P, resulting in neither nutrient being easily available to the grass.

In alkaline soils (pH > 7.5), Ca forms insoluble complexes with P to render it unavailable as dicalcium phosphate [$Ca_2HPO_4$]. Soil pH adjustment may be necessary to prevent the formation of these complexes as applied P fertilizer is rendered unavailable. Interestingly, research indicates an increased uptake of P by plants when the ammonium form of N is applied with phosphate fertilizer.

## Potassium

Potassium (K) is an essential element not normally associated with a prominent visual response in terms of shoot color, density, or growth. Potassium is often called the "health" element since an ample supply increases the plant's tolerance to stresses such as cold, heat, drought, diseases, and wear (Table 7.2). Potassium is directly involved in maintaining the water status of the plant, the turgor pressure of its cells, and the opening and closing of its stomata. As its concentration increases in plants, cell walls thicken, tissue water content decreases, and plants become more turgid due to the regulation of stomatal opening. This is due to K providing much of the osmotic pressure necessary to pull water into plant roots, thereby improving the plant's drought tolerance. Thicker cell walls also resist pest pressure better. Potassium is required by more than 60 enzymes for activation; however, it is not synthesized into compounds such as chlorophyll, proteins, and nucleic acids as are N and P.

The critical level of K in plants is about fourfold that of P and almost the same as N. Dry matter of leaf tissue consists of 1.0% to 5.0% K. Sufficient values range from 1.5% to 3.0% in recently matured leaf tissue. Potassium deficiency occurs when tissue levels are less than 1.0%, and levels greater that 3.0% are considered excessive. However, most plants can absorb more K than needed; this is often referred to as luxury consumption as research indicates little growth benefits above about 2 pounds $K_2O$/1,000 square feet (10 g K/m$^2$) yearly. Also, an inverse relationship exists between K, Mg, and Ca in plants. As K levels increase, Mg deficiencies are the first to show, while Ca deficiencies occur at higher concentrations. In saline soils, an inverse relationship can occur where Ca, Mg, or Na ions compete with K for plant uptake.

Deficiencies may occur in soils low in micas, soils low in clay (fewer exchange sites), and soils high in sand, which are prone to leaching. Potassium deficiency symptoms include interveinal yellowing of older leaves, and the rolling and burning of the leaf tip (Figure 7.10). Leaf veins finally appear yellow and margins appear scorched (called necrosis). The turf stand will appear thin with spindly growth of individual plants (Figure 7.11). Potassium is a mobile element within plants; thus, it can be translocated to younger meristematic tissues from older leaves if a shortage occurs, while necrotic spots may form a unique pattern on the leaf margin in certain plants. Although similar in appearance, plant salinity damage differs by affecting newer leaves compared to older leaves with K deficiency.

**FIGURE 7.11** Potassium deficiency on creeping bentgrass.

Potassium fertilizer often is referred to as "potash." Early settlers coined the name after producing potassium carbonate needed for making soap by evaporating water filtered through wood ashes. The ash-like residue in the large iron pots was called "pot-ash," and this process was the first U.S. registered patent.

**FIGURE 7.10** Potassium deficiency on bermudagrass.

**TABLE 7.10**
**Primary Potassium Sources and Characteristics Used in Turf Fertilizers**

| Source | Approximate Nutrient ($N\text{-}P_2O_5\text{-}K_2O$) | Salt Index Hazard | Comment |
|---|---|---|---|
| Muriate of potash (potassium chloride) [KCl] | 0-0-[50 to 63] | High | • Most common K source; water-soluble granular; high burn potential; contains ~46% Cl; neutral acidifying effect. |
| Sulfate of potash (potassium sulfate) [$K_2SO_4$] | 0-0-[40 to 54] | Moderate | • Soluble granular source containing 17% S; used instead of potassium chloride where chlorine is not desirable, to reduce foliage burn potential, and to provide S; neutral acidifying effect; made by reacting KCl with sulfuric acid. |
| Potassium-magnesium sulfate (or K-Mag or Sul-Po-Mag) [$K_2SO_4 \cdot 2MgSO_4$] | 0-0-[18 to 22] | Moderate | • Contains 11% Mg and 23% S; also known as langbeinite; neutral acidifying effect. |
| Potassium nitrate [$KNO_3$] | 13-0-[37 to 45] | High | • Also known as saltpeter or nitre; produced by reacting KCl with nitric acid ($HNO_3$); K source with supplemental N; water soluble; low chloride, fire hazard; has alkalinity effect; used as a foliar source. |
| Potassium carbonate [$K_2CO_3$] | 0-0-[34 to 48] | – | • Granular and liquid forms available. |
| Potassium thiosulfate [$K_2S_2O_3$] | 0-0-25 | High | • Clear liquid containing 17% S; neutral to basic pH. |

Muriate of potash (potassium chloride) is the most often used K-containing fertilizer (Table 7.10) and originates from K salt deposits that have been mined and processed. These deposits developed where seawater once occupied the surface. Magnesium salts also are often found concurrently with K salts. As seawater evaporated, K salts crystallized to become beds of potassium chloride. A white soluble grade is used as a fluid source. Potassium sulfate forms when potassium chloride is reacted with sulfuric acid, and potassium nitrate forms when it is reacted with nitric acid. These forms are used to reduce the salt index compared to potassium chloride and are also carriers of S and N, respectively. Other K-containing fertilizers include:

- Potassium bicarbonate ($KHCO_3$): 0-0-46, soluble
- Potassium carbonate ($K_2CO_3$): 0-0-[34 to 48], granular and liquid
- Potassium hydroxide (KOH): 0-0-83, liquid
- Potassium polysulfide ($ks_x$): 0-0-22-23S, liquid
- Various potassium phosphates

The soluble K content in a fertilizer is expressed as $K_2O$. To determine the quantity of actual K supplied in a fertilizer source, the following conversion is used:

$$\text{Percent of } K_2O \text{ as expressed on fertilizer bag} \times 0.83 = \text{percent of K in the fertilizer bag}$$

The available form for plant use is the potassium ion ($K^+$), absorbed primarily from the soil solution. Other forms exist, but most are unavailable for plant use. Potassium is not readily held in sandy soils (low CEC soils) and can be lost by leaching; thus, tissue testing should be used along with soil testing to indicate if sufficient K levels are being absorbed by plants. This problem is not always appreciated, especially when the grass is subjected to heavy rainfall or watering.

Competition exists for plant entrance between K with Ca and Mg. Soils high in either or both of these elements will need additional K fertilization in order to satisfy plant needs. In sandy soils, or where turf clippings are not returned, a 2:1 or 1:1 ratio of N to K may be required to maintain adequate K supply. Frequent, light K treatments with these ratios should be considered with each N application. With fertigation, potassium nitrate, potassium sulfate, and potassium chloride have been successfully used.

## SECONDARY PLANT NUTRIENTS

Secondary elements consist of Ca, Mg, and S. They are required by plants in similar amounts as the primary nutrients. However, since these elements are less frequently deficient, they are classified as secondary nutrients.

### Calcium

*Calcium* (Ca), *magnesium* (Mg), and *sulfur* (S) are elements required in almost the same quantities as P. Calcium functions include strengthening cell walls to prevent their collapse; enhancing cell division; encouraging plant growth, protein synthesis, and carbohydrate movement; and balancing cell acidity (Table 7.1). Calcium also improves root formation and growth.

Plants only use the exchangeable calcium ion, $Ca^{+2}$. Deficiencies occur most often in sandy soils with low cation exchange capacity, extremely acidic (<5.0 pH) soils, soils saturated with Na, or extremely dry soils. Long term use of effluent water high in Na or Al may also induce Ca deficiency as can heavy applications of ammonium, K, or Mg fertilizers.

**TABLE 7.11**
**Primary Calcium and Magnesium Sources and Characteristics Used in Turf Fertilizers**

| Source | Approximate Nutrient Content ($N-P_2O_5-K_2O$) | Comment |
|---|---|---|
| Gypsum (calcium sulfate) [$CaSO_4{\cdot}2H_2O$] | 0-0-0 | • Contains 23% Ca + 19%S; neutral effect on soil pH; used to displace soil Na. Anhydrite sources do not contain water, contain 29% Ca; dissolves slower and does not prevent surface crusting associated with high salt or bicarbonate levels. |
| Calcium nitrate [$Ca(NO_3)_2$] | 16-0-0 | • Very hygroscopic; high burn potential; contains 24% Ca and 1.5% Mg; fast acting; N release is not temperature dependent; used on sodic soils to displace Na. |
| Dolomitic limestone [$CaMg(CO_3)_2$] | 0-0-0 | • Used to increase soil pH; contains 22% Ca and 13% Mg; very slowly available; low salt hazard. |
| Superphosphate [$Ca(H_2PO_4)_2 + CaSO_4$] | 0-20-0 | • Soluble P source often used in mixed fertilizers; contains Ca (18% to 21%) and S (12%) as gypsum. |
| Triple (or treble) superphosphate [$Ca(H_2PO_4)_2H_2O$] | 0-46-0 | • Concentrated P source + ~13% Ca. |
| Calcitic limestone (calcium carbonate) ($CaCO_3$) | 0-0-0 | • Contains 40% Ca; used to raise soil pH; very low salt hazard. |
| Magnesium sulfate (or Epsom salt) [$MgSO_4$] | 0-0-0 | • Contains 13% to 23% S and 10% to 17% Mg; water soluble; neutral salt with little effect on soil pH. |
| Potassium magnesium sulfate (or Sul-Po-Mag or K-Mag) [$K_2SO_4{\cdot}2MgSO_4$] | 0-0-[18 to 22] | • Contains 23% S and 11% Mg; water soluble. |
| Magnesium oxide (or Magnesia) [MgO] | 0-0-0 | • Contains 55% Mg; high salt hazard. |

Deficiency symptoms include distorted (e.g., twisted or deformed) young leaves that turn reddish-brown along their margins before becoming rose-red, finally leading to leaf tips and margins that wither and die. Roots also are short and bunched. Excessive Ca may tie up other soil nutrients, especially K and Mg, and less so P, Mn, Fe, Zn, and B.

Calcium is an immobile nutrient within plants. It does not move from older leaves to new ones and must be continuously supplied or young leaves will show deficiency symptoms early. It also does not translocate appreciably in the plant when foliar fertilizer is applied. Calcium is usually added in a liming program, by irrigation with high Ca-content water, or naturally through high pH soils. Commercial sources of Ca include calcitic and dolomitic limestone, gypsum, superphosphates, calcium nitrate, shells, slags, and water-treatment residue (Table 7.11). Calcium chloride and calcium nitrate are often used for foliar applications. When lime is not needed, gypsum or some other Ca source should be considered.

## Magnesium

Magnesium is essential for chlorophyll production in plants as chlorophyll molecules contain ~7% (Table 7.1). Magnesium also is essential for many of the energy reactions such as sugar formation. It acts as a carrier of P and regulates uptake of other plant nutrients. Deficiencies occur mostly in acidic sandy soils with low CEC or soils of extremely high pH, especially when clippings are continuously removed. Deficiencies can occur in soils with less than 40 pounds/acre (4.5 g/m$^2$) of Mehlich-I extractable Mg. Magnesium generally constitutes about 4% to 20% of a soil's CEC, compared with up to 80% for Ca and 5% for K. High Ca and K levels tend to reduce Mg uptake.

Magnesium is a mobile element in the plant and is easily translocated from older (bottommost) to younger plant parts as needed. Symptoms of deficiency include a general loss of green color starting at the bottom (older) leaves. Interveinal chlorosis then occurs; veins remain green, and older leaf margins turn a blotchy cherry-red with a striped appearance of light yellow or white between parallel veins. Necrosis eventually develops. Interveinal chlorosis can be caused by other deficiencies, but happens more so in neutral-to-alkaline soils. Therefore, a simple soil pH test can help narrow the potential deficiency problem to Mg. Magnesium deficiency can occur when (1) the soil pH is less than 5.5; (2) an imbalance occurs among the other major cations, $K^+$ and $Ca^{+2}$; and (3) plants are under stress for an extended period such as soils too wet, dry, or cold, and when root growth is limited. Sources of Mg include dolomitic limestone, sulfates of potash and magnesium, magnesium sulfate (Epsom salt), oxide, and chelates (Table 7.11). When lime is not needed, magnesium sulfate should be considered. A practical test to determine if Mg deficiency is occurring is to apply 1 pound (454 g) of Epsom salt in 3 to 4 gallons of water over 1,000 square feet (12 to 16.3 L/m$^2$). If the turf greens up within 24 hours or so, Mg deficiency should be suspected.

**Balance between potassium, calcium, and magnesium.** Soils should have a balance of these three major cations and if this balance is not maintained, the availability and uptake of one or more of them can be limited. Soils have a good balance if the total cation exchange capacity is about 60% to 80% Ca, 10% to 15% Mg, and 2% to 5% K.

## Sulfur

Sulfur is a constituent of three amino acids: cystine, cysteine, and methionine (Table 7.1). It is used as a building block of proteins and chlorophyll, and also reduces diseases. The acidifying effect of S also may increase the availability of essential elements such as Fe, Mn, Zn, and P, as well as help reclaim sodic soil. Sulfur content in leaf tissue ranges from 0.15% to 0.50% of the dry weight. Most plants use comparable S quantities as P and Mg.

The sulfate anion $\left(SO_4^{-2}\right)$ is the primary available form found in soil solution. Like nitrate, the sulfate ion can leach from soil. Deficiencies may occur when soil organic matter is low, grass clippings are removed, excessive watering occurs, and nonirrigated (no sulfate from water) and sandy soils predominate. Deficiency symptoms resemble N deficiency and include an initial light yellow-green color, with yellowing being most pronounced in younger leaves, since S has limited mobility in plants. Nitrogen deficiency is usually worse in older leaves while younger leaves are greener since N is plant mobile. Older leaves become pale, then turn yellowish-green in interveinal areas. Leaf tips are scorched along the margins. Roots tend to be longer than normal and stems become woody. Tissue testing is often necessary to distinguish between S and N deficiencies. Grass grown in sandy soils has been shown to respond to S applications.

Over 90% of available S exists in the organic matter, which has a ~10:1 N-to-S ratio. Deficiencies may occur when the N-to-S ratio is greater than 20:1 or at high soil pH (>7.0). Sulfur may be precipitated as calcium sulfate ($CaSO_4$ or gypsum), while at lower pH levels (<4.0), the sulfate anion may be adsorbed by Al- and/or Fe-oxides. Turf clippings with a high N-to-S ratio (≥20 to 1) decompose slowly and may slow thatch biodegradation as microorganisms require S to decompose plant residues.

Sulfur deficiencies have become more acute in recent years. This is probably due to less air pollution, a previously important source of S. Sulfur also was supplied as a contaminant in some fertilizer sources, such as superphosphate. However, many new high-analysis fertilizers frequently do not contain appreciable S. Less S is used today in fungicides and insecticides, and its primary soil source, organic matter, may be reduced over time.

Sulfur sources include gypsum, elemental S, ferrous sulfate, liquid form ammonium thiosulfate, potassium magnesium sulfate (or K-Mag or Sul-Po-Mag), ammonium sulfate, sulfur-coated urea, and potassium sulfate (Table 7.12).

In poorly drained, waterlogged soils where soil $O_2$ is exhausted, $SO_4^{-2}$ and sulfur-containing organic matter can be reduced by sulfate-reducing bacteria to toxic hydrogen sulfide ($H_2S$). Excessive applications of elemental S to golf greens also may encourage hydrogen sulfide buildup. In addition, insoluble sulfides may form by reacting with soil Fe or Mn.

Turf soils containing toxic levels of hydrogen sulfide or iron sulfate are acidic and commonly form a "black layer" several inches below the soil surface. They typically are characterized by the distinct hydrogen sulfide smell (e.g., a sewer or rotten egg). Low soil $O_2$ also has reduced states of Mn, Cu,

**TABLE 7.12**
**Primary Sulfur Sources and Characteristics Used for Turf Fertilizer**

| | Approximate Nutrient Content | | |
|---|---|---|---|
| **Sulfur (S) Carriers [Formulas]** | **Sulfur (S)** | **Others** | **Comment** |
| Gypsum (calcium sulfate) [$CaSO_4 \cdot 2H_2O$] | 16 to 24 | 20% Ca | • Granular source; commonly used on sodic soil. |
| Elemental sulfur [S] | Up to 99 | – | • Oxidized by soil microbes to sulfuric acid; lowers pH; foliage burn potential; available as a granular or suspension; slow acting; eye irritant. |
| Epsom salt (magnesium sulfate) [$MgSO_4$] | 13 to 23 | 10% to 17% Mg | • Granular and liquid forms available; neutral salt with little acidifying effect. |
| Ferrous sulfate [$FeSO_4 \cdot 7H_2O$] | 19 | 21% Fe | • Water soluble; usually applied foliarly. |
| Ferrous ammonium sulfate [$(NH_4)_2FeSO_4 \cdot 6H_2O$] | 15 | 14% Fe + 7% N | • Water soluble; usually applied foliarly. |
| Ammonium thiosulfate [$(NH_4)_2S_2O_3 + H_2O$] | 26 | 12% N | • Most widely used S source in clear liquid fertilizers; foliage burn potential. |
| Potassium thiosulfate [$K_2S_2O_3$] | 17 | 25% $K_2O$ | • Liquid form. |
| Potassium magnesium sulfate (or K-mag or Sul-Po-Mag) [$K_2SO_4 \cdot 2MgSO_4$] | 22 | 22% $K_2O$ + 11% Mg | • Commonly used on alkaline soils. |
| Ammonium sulfate [$(NH_4)_2SO_4$] | 24 | 21% N | • Water-soluble granular; high acidifying potential. |
| Sulfur-coated urea (or SCU) [$CO(NH_2)_2 + S$] | 10 to 20 | 32% N | • Slowly soluble. |
| Superphosphate [$Ca(H_2PO_4)_2 + CaSO_4$] | 12 | 20% $P_2O_5$ + 20% Ca | • Soluble P; contains gypsum (calcium sulfate); little acidifying effect. |
| Potassium sulfate (or sulfate of potash) [$K_2SO_4$] | 17 | 50% $K_2O$ | • Granular and liquid forms. |
| Sulfuric acid [$H_2SO_4$] | 20 to 33 | – | • Liquid injection in irrigation water; corrosive and caustic. |
| Urea-sulfuric acid [$CO(NH_2)_2 \cdot H_2SO_4$] | 9 to 18 | 10% to 28% N | • Liquid injection in irrigation water; corrosive and caustic. |

and Fe and results in gray- and blue-colored subsoils. This often occurs in poorly drained, anaerobic soils.

## MICRONUTRIENTS

Micronutrients, as previously discussed, are essential elements needed in relatively small (e.g., <50 ppm) amounts. These include boron (B), chlorine (Cl), copper (Cu), iron (Fe), manganese (Mn), molybdenum (Mo), zinc (Zn), and nickel (Ni). Many soils supply sufficient levels of micronutrients to the point they do not need to be added. In other cases, enough micronutrients are supplied in fertilizers as impurities. Deficiencies in micronutrients can become a problem, especially in areas with sandy and peat or muck soils, pockets of high pH and P-containing soil, and poor drainage or periods of extended, heavy rainfall (Figure 7.12).

Soil pH probably influences plants most by affecting the availability of important nutrients. For example, at lower pH values (<5), aluminum (Al), Fe, and Mn are highly soluble and may be present in sufficient quantities to actually be toxic to plants. High levels of Al can also reduce plant uptake of P, Ca, Mg, and Fe. At higher pH values (>7.0), nutrients such as Fe, Mn, Cu, and Zn are less soluble and, therefore, relatively unavailable for plant uptake, although molybdenum (Mo) availability actually increases at high pH (Table 7.13). The availability of P and B also may be hindered by pH values greater than 7.0.

A balance of micronutrients is particularly important because many plant functions require more than one element. Regular soil and tissue testing is the best preventive approach to solving many of the nutrient deficiency problems. Iron and Mn are two of the most common micronutrient deficiencies turf managers experience. However, if excessive or indiscriminate amounts of micronutrients are applied or soil pH is excessively low, toxicity to the plant can occur (Figure 7.13). An example involves growing turf on old vegetable or fruit

**FIGURE 7.13** Iron phytotoxicity typically begins as blackened turf followed by browning.

**FIGURE 7.12** Iron deficiency is typically characterized by a sporadic yellow, chlorotic appearance.

**TABLE 7.13**
**Absorption Rate and Mobility of Foliar Applied Materials in Plants**

| Foliar Absorption Rate | | | Internal Plant Mobility | | |
|---|---|---|---|---|---|
| **Quickly** | **Moderate** | **Slow** | **Readily** | **Partially** | **Immobile** |
| Potassium | Boron | Copper | Phosphorus | Boron | Calcium |
| Urea | Calcium | Iron | Potassium | Copper | Iron |
| Zinc | Sulfate | Magnesium | Sulfate | Manganese | Magnesium |
| | Phosphorus | | Urea | Molybdenum | |
| | Manganese | | Sugars | Zinc | |
| | Molybdenum | | Amino acids | | |

**FIGURE 7.14** Iron-deficient turf leaf (left) showing characteristic interveinal chlorosis.

production fields that were often sprayed with fungicides containing Cu, Zn, and/or S. Because of their relative immobility (with the exception of S) in soils, residues have become toxic to the turfgrasses in some cases.

Micronutrient deficiency symptoms can easily be confused with pest occurrence or other stresses (Figure 7.14). Micronutrient deficiency problems, however, usually are more localized and appear as irregular spots or in circular patterns. General nutrient deficiency symptoms are listed in Table 7.14; specific ones are outlined in Table 7.15. Table 7.16 offers a starting guideline for spot treating of micronutrients when sprayed on the foliage to the drip point. Sometimes, high-surface-area glasslike beads, called frits, are used for B, Cu, Zn, and other micronutrient applications. As the fritted materials weather in soil, they slowly release their micronutrients, avoiding some of the precipitation and sorption problems that might otherwise occur, especially in alkaline soils.

## Sources

**Chelates.** Chelates, chelating agents, or sequestering agents are cyclic structures of a normally nonsoluble divalent metal atom and an organic component that, when held together, become soluble in water. The chelation process allows nutrients to move through the soil solution to plants without being tied up with other soil chemicals. However, the activity of the metallic ion decreases in the aqueous solution. For example, in the absence of chelation and the addition of an inorganic ion such as ferric sulfate to calcareous soil, most of the Fe becomes unavailable by reacting with hydroxide ions (–OH) to form iron hydroxide (FeOOH). However, if the Fe is chelated (from the Greek *chele,* meaning claw), it remains available for plant uptake as shown:

$$\text{FeOOH (unavailable)} + \text{chelate}^{-3} + H_2O \leftrightarrow \text{Fe chelate(available)} + 3OH^-$$

Commercially available sequestered metallic ions are Fe, Cu, Zn, and Mn. Nutrient content is typically 5% to 14%. Organic compounds that have the ability to chelate or sequester these metallic ions include ethylenediaminetetraacetic acid (EDTA), diethylenetriaminepentaacetic acid (DTPA), cyclohexanediaminetetraacetic acid (CDTA), ethylenediaminedi (*o*-hydroxyphenylacetic acid) (EDDHA), citrate, and gluconate. Urea and humic acids also are used as chelating agents. Their stability fluctuates with various soil pH levels. For example, EDDHA is best for soil pH from 4 to 9, EDTA at soil pH <6.3, and DTPA at soil pH <7.5. Generally, the stability of metal chelates is greater at near neutral versus low or high pH levels.

**Iron oxides.** Although one of the highest percentage Fe sources (~69% to 77% Fe), iron oxides are nearly plant unavailable at soil pH > 6.0. Since Fe deficiency is less often at lower pH due to Fe being more soluble (available), iron oxides are not as beneficial. These also stain surfaces like concrete.

**Iron sucrate.** Iron sucrate is produced by blending iron oxide with sugars (often molasses) to form an iron-containing organic complex. This form is less water soluble and thus less prone to staining than iron oxide. It also may be more available (soluble) on high-pH soils.

**Iron sulfate.** Iron sulfate is a commonly used Fe fertilizer often containing ferrous iron that is generally more plant

**TABLE 7.14**
**General Plant Nutrient Deficiency Symptoms**

| Symptom | Deficient Nutrient |
|---|---|
| **Leaf Chlorosis Evident** | |
| 1. Entire leaf is chlorotic | |
| a. Only older leaves are chlorotic, then necrotic, then drop; growth noticeably slows; symptoms develop uniformly over the whole turfgrass area. | Nitrogen (N) |
| b. Initially, young leaves turn a light yellow-green color; eventually, all leaves are affected; appears similar to N deficiency symptoms. | Sulfur (S) |
| 2. Interveinal leaf chlorosis | |
| a. Only older and recently mature leaves show symptoms; leaf veins remain green; older leaf margins turn cherry-red. | Magnesium (Mg) |
| b. Only young leaves show interveinal chlorosis; leaves finally turn white; usually only patches of turf are affected (Figure 7.14). | Iron (Fe) |
| 1. Tan or gray necrotic spots also develop in chlorotic areas; leaf veins remain dark-green to olive color. | Manganese (Mn) |
| 2. Leaf tips remain green, then turn bluish, wither, and die. | Copper (Cu) |
| 3. Young leaves are dwarfed, dark, and desiccated-looking; shortened internodes; plants appear rosette. | Zinc (Zn) |
| **Leaf Chlorosis not Dominant** | |
| 1. Symptoms appear at plant base on older growth | |
| a. All leaves appear dark green, then possibly yellow; growth stunted; purple coloring in older (bottommost) leaves from increased anthocyanin development; leaf tips turn reddish. | Phosphorous (P) |
| b. Margins and tips of older leaves chlorotic, then scorched; small, whitish spots may be scattered over older leaves; turf stand composed of thin, spindly individual plants. | Potassium (K) |
| 2. Symptoms on new plant growth | |
| a. Terminal buds die, resulting in dwarf (rosette) plants. | Boron (B) |
| b. Margins of young leaves do not form; young leaves twisted or deformed; these do not or only partially unfold; young leaf margins turn reddish brown; roots are short and bunched; leaf tips and margins eventually wither and die. | Calcium (Ca) |

available than ferric iron, although this may be short in duration. Like most granular iron sources, iron sulfate can also stain concrete.

**Natural organics.** Iron humate, biosolids, and compost are examples of natural organics containing iron, either naturally or added. These generally are more plant available than other forms but due to low concentrations, large amounts of products may be needed to get an Fe response. These, however, tend to stain concrete less.

## Nonessential Elements

Sodium (Na), aluminum (Al), arsenic (As), and silicon (Si) are nonessential elements for turfgrass growth and development. These, in general, become toxic when levels are excessive and should not be applied in supplemental fertilizers, although calcium silicate use may suppress certain diseases (e.g., dollar spot) through the physical effects of silica deposits in leaves and through enhanced biochemical defense mechanisms.

## Fertilizer Formulations

### Granular Fertilizers

Granular fertilizers are manufactured when dry materials are mixed with water or with acids to form a slurry. The slurry is produced in a heated, revolving drum where differently sized particles form as the product dries. To prevent water adsorption from air and to prevent caking, materials like diatomaceous earth or kaolinitic clay are added to produce a granular coating. Oil may also be added to reduce dust. The final product should contain hard granules with uniform size and composition and have good storage and handling qualities.

The size guide number (SGN) and the uniformity index (UI) are two parameters used to characterize the particle size of dry fertilizer products. SGN is the median particle size (mm) of a product times 100 (Figure 7.15). A fertilizer with a large SGN will have a greater particle size than one with a smaller SCN. For example, a granular fertilizer with a 270 SGN would have a median particle size of 2.7 mm. Greens-grade fertilizer has a range from 80 to 100 SGN, fairway fertilizer ranges from 120 to 150 SGN, and high cut turf from 210 to 240 SGN.

The UI measures the size uniformity of a fertilizer source and is the particle size at which 95% of the material is retained, divided by the particle size at which 10% of the material is retained, multiplied by 100. The closer the UI is to 100, the closer the particles are to the same size. Most dry fertilizers have a UI between 20 and 50.

When two or more dry fertilizer materials are physically mixed, this is referred to as **bulk blending** (Figure 7.16). A fertilizer plant receives the individual fertilizer products from a basic producer, stores them, and blends them as needed in a mixing device. Materials more commonly used to make bulk blends are ammonium nitrate and sulfate, urea, diammonium phosphate, monoammonium phosphate, K carriers, and elemental S. Blends are typically available in bulk, minibulk, or bags. If materials of different sizes are used in bulk blending,

**TABLE 7.15**
**Micronutrient Forms, Deficiencies, and Sources for Turf Managers**

| Nutrient | Deficiency Occurrence | Deficiency Symptoms | Fertilizer Sources |
|---|---|---|---|
| Iron (Fe) | Oxidized ($Fe^{+3}$) form is found in aerobic soils while reduced ($Fe^{+2}$) form is in anaerobic soils. Iron levels in soils often are sufficient; however, soil conditions often render them unavailable. Deficiency occurs with excessive soil pH (>7.0), Ca, Zn, Mn, P, Cu, and bicarbonate ($HCO_3$) levels in irrigation water; poor rooting, poor soil drainage and thus low soil $O_2$ levels; and cold soils. Often seen in summer on wet soils following rainfall. At low soil pH, P can combine with Fe to form insoluble (unavailable) iron phosphate while at high pH, excessive P uptake by plants may inactivate absorbed Fe. For each increase in pH, there is a 100-fold decrease in soluble $Fe^{+2}$. A plant tissue ratio of P-to-Fe at 29:1 also provides healthy turf while a P-to-Fe ratio of 40:1 often expresses Fe deficiency. Deficiency symptoms are most severe during warm days/cool nights (e.g., early spring and fall) when root growth is insufficient to support shoot growth. | Chlorosis resembles N deficiency, except Fe chlorosis is interveinal (e.g., between leaf veins) and first occurs in youngest leaves since Fe is immobile within plants. N deficiency causes the entire leaf, including veins, to simultaneously yellow. Fe-deficient leaves finally turn white. Fe chlorosis tends to be in random scattered spots, creating a mottled appearance, and appears more severe when closely mowed. N deficiency develops uniformly over a large area and appears unaffected by mowing. Management practices to reduce deficiency are meant to lower the soil pH, improve drainage, reduce P fertilization, and use one of the listed foliar sprays. Excessive Fe levels can induce Mn deficiency and leaf blackening. Anaerobic, acidic, poorly drained soils can produce toxicity levels. Soil and tissue tests are often poor indicators for Fe deficiency. Apply a solution of 2 ounces (57 g) iron sulfate in 3 gallons (11 L) of water and spray the turf to the drip point to indicate if Fe deficiency is occurring (Figure 7.18). | • Ferrous sulfate [$FeSO_4 \cdot 7H_2O$] (19% to 21% Fe and 19% S); usually foliarly applied; low acidifying effect; water soluble<br>• Ferric sulfate [$Fe_2(SO_4)_2 \cdot 4H_2O$] 23% Fe; water soluble<br>• Ferrous ammonium sulfate [$FeSO_4 \cdot (NH_4)_2SO_4$] (5% to 14% Fe, 16% S, and 7% N); usually foliarly applied; also provides some N; medium acidifying effect; water soluble<br>• Chelated iron [Fe salts of -EDTA, -HEDTA, -EDDHA, or -DTPA] (5% to 14% Fe); longer greening effect than the other Fe sources; low acidifying effect; water soluble<br>• Iron frits (14% Fe)<br>• Ferrous oxide [FeO] 77% Fe. Ferric oxide [$Fe_2O_3$] 69% Fe; better for low pH (<6.0) soils<br>• Ferrous ammonium phosphate [$Fe(NH_4)PO_4 \cdot H2O$] 29% Fe, water soluble<br>• Iron humate, 27% Fe, from water purification<br>• Iron sucrate (granusol), 50% Fe reacted with molasses; less sidewalk staining |
| Zinc (Zn) | Alkaline soils decrease solubility and availability, as does excessive soil Cu, Fe, and Mn and soil moisture, N, and phosphates. Zn solubility increases 100-fold for each decreased pH unit. Above pH 7.7, Zn becomes $Zn(OH)^+$. Lower light intensities reduce root uptake. Sands (low organic matter content) also are likely to have lower Zn levels than clays. Some mine spoils and municipal wastes may contain high levels of Zn. | Interveinal chlorosis in both younger and some older leaves; mottled-chlorotic leaves, rolled and thin leaf blades; stunted, shortened internode growth; dark, desiccated-looking leaves (starting with the youngest ones); leaves finally turn white in appearance. Excessive Zn may inhibit root and rhizome development and induce Fe and Mg deficiencies. | • Zinc sulfate [$ZnSO_4 \cdot H_2O$] (35% Zn and 12% S); water soluble, foliarly applied; zinc chelate [NaZnEDTA] (9% to 14% Zn); foliarly applied; zinc oxide [ZnO] (78% Zn); water soluble, foliarly applied; zinc frits (4% to 7% Zn); zinc carbonate [$ZnCO_3$], 53% Zn; soil-applied zinc chloride [$ZnCl_2$], 48% Zn, water soluble; zinc nitrate [$(Zn(NO_3)_2 \cdot 6H_2O$] (22% Zn, 9% N). Certain fungicides (e.g., chlorothalonil-Zn) |
| Manganese (Mn) | Deficiencies occur in sand, peat, and muck soils (insoluble complexes are formed), alkaline soils high in Ca (for each increase in pH, there is a 100-fold decrease in soluble $Mn^{+2}$); at low temperatures; and in poor drainage. Excess Fe, Cu, Zn, K, and Na in low CEC soils can reduce Mn adsorption. Fe-to-Mn ratio in leaf tissue should be at least 2:1. Adjusting soil pH to below 7.0 usually reduces Mn deficiencies. | Yellowing (chlorosis) between veins (interveinal) of youngest leaves, with veins remaining dark green to olive green in color since Mn is an immobile element within plants; small, distinct necrotic leaf spots develop on older leaves; leaf tips may turn gray to white, droop, and wither. On closely mowed turf, mottled or blotchy appearance develops, with little or no response to N occurring. Toxicity can occur with anaerobic soils with pH < 4.8. Excessive Mn levels can induce Fe, Ca, or Mg deficiencies. Apply 1 ounce (28 g) of manganese sulfate (tecmangam) in 2 gallons (7.6 L) water to the point of drip for diagnosis purposes. | • Manganese sulfate [$MnSO_4 \cdot H_2O$] (26% to 28% Mn and 13%S); foliarly applied; manganese oxide [MnO] (33% to 77% Mn); mancozeb fungicides (16% Mn and 2% Zn); water soluble; manganese chelates [$Na_2Mn$ EDTA] (5% to 12% Mn); water soluble; tecmangam (20% to 25% Mn); manganese frits (3% to 6% Mn); manganese chloride [$MnCl_2$] (17% Mn); water soluble |

*(Continued)*

**TABLE 7.15 (CONTINUED)**
**Micronutrient Forms, Deficiencies, and Sources for Turf Managers**

| Nutrient | Deficiency Occurrence | Deficiency Symptoms | Fertilizer Sources |
|---|---|---|---|
| Copper (Cu) | Deficiency mostly occurs in sand, peat, muck, and high (>5%) organic soils due to tight binding properties of these for Cu. Excess Fe, N, P, and Zn, highly leached soils, and high soil pH encourage deficiency. Toxic levels can result from excess sewage sludge applications, use of poultry manures, copper sulfate, and Cu-containing pesticides such as Bordeau mixture. Liming to pH 7.0 is often the simplest means of overcoming Cu phytotoxicity. Reducing N fertilization may also help. | Deficiencies are rare. Deficiency symptoms include yellowing and chlorosis of younger leaf margins; leaf tips initially turn bluish, wither, and droop, eventually turn yellow, and die; youngest leaves become light green, and necrotic; plant dwarfing with inward rolling of leaves that develop a blue-green appearance; symptoms progress from the leaf tips to the base of the plant. Toxicity symptoms of excessive levels include reduced shoot vigor, poorly developed and discolored root systems, and leaf chlorosis resembling Fe deficiency. Excessive Cu levels suppress uptake of Fe, Mn, Zn, and Mo. | • Copper sulfate [$CuSO_4 \cdot 5H_2O$] (13 to 53% Cu and 13% S); foliarly or soil applied; copper oxide [CuO] (40% Cu); foliarly or soil applied; copper chelates [CuEDTA] (9% to 13% Cu); foliarly applied; copper chloride [$CuCl_2$], 47% Cu); foliarly or soil applied |
| Boron (B) | Organic matter is the principal source of B; availability increases with decreasing soil pH; deficiencies are most common in high (>6.5) pH, leached, or very dry, sandy soils. Ca also decreases translocation of B in plants. Liming acidic soils frequently causes a B deficiency. Excessive levels (>6 ppm) may occur with high-B–containing irrigation water, arid and semiarid soils, and some composts. | Thickening, curling, and chlorotic leaves develop on dwarf (rosette) plants; chlorotic streaks develop in the interveinal areas. Leaf tips turn pale green. Plants develop a "bronze" tint. B is immobile within the plant; symptoms, therefore, first appear in meristematic tissues and young leaves. Deficiencies are infrequent in turf. | • Borax (sodium borate) [$Na_2B_4O_7 \cdot 10H_2O$] (11% B and 9% Na); boric acid [$H_3BO_3$] 10% to 17% B; fertilizer borate [sodium tetraborate, $Na_2B_4O_7 \cdot 5H_2O$, 14% to 21% B; Sodium pentaborate, $Na_2B_{10}O_{16} \cdot 5H_2O$, 18% B]; solubor [$Na_2B_4O_7 \cdot 4H_2O + Na_2B_{10}O_{16} \cdot 10H_2O$] (20% soluble B); ammonium pentaborate [$NH_4B_5O_6{\cdot}4H_2O$] (20% B) |
| Molybdenum (Mo) | Availability increases with increasing soil pH; deficiencies are most common in acidic sands or highly weathered soils; excess Cu, Fe, Mn, or sulfate may reduce Mo utilization by plants. Deficiencies often occur in ironstone soils of Australia, New Zealand, and Holland. | Resembles mild N deficiency with pale yellow-green stunted plants; mottled yellowing of interveinal areas then appears in older leaves. Deficiencies are rare; lime acid soils. | • Ammonium molybdate [$(NH_4)_2MoO_4$] (54% Mo); liquid; sodium molybdate [$Na_2MoO_4 \cdot H_2O$] (40% Mo); water soluble; molybdenum trioxide [$MoO_3$] (66% Mo); molybdic oxide (47% Mo); molybdenum frits (fritted glass), 2% to 3% Mo |
| Chlorine (Cl) | Less available in alkaline soils, or soils high in $NO_3^-$ and $SO_4^{-2}$; very mobile in acid-to-neutral soils. Toxic levels reduce water availability to plants; cause premature leaf yellowing, leaf tip, and margin burning; and leaf bronzing and abscission. | Chlorosis of younger leaves and wilting of plants; not mobile within plants and accumulates in older parts. Deficiencies are rare. Cl is most commonly applied in large quantities along with the K source in fertilizers. | • Ammonium chloride [$NH_4CL$] (66% Cl, 25% N); acid-forming fertilizer; calcium chloride [$CaCl_2$] (65% Cl); magnesium chloride [$MgCl_2$] (74% Cl); potassium chloride [KCl] (47% Cl, 60% K); sodium chloride [NaCl] (60% Cl, 40% Na) |

segregation or separation of these may occur, causing uneven distribution of nutrients when spread.

Micronutrients can be added in bulk blending either by spraying a solution of the elements on the blend during mixing or by adding a granular form with a similar size as other components to the blend. Pesticides may also be added by spraying (impregnating) the bulk blends.

Homogeneous dry products contain granules or prills, with each having the same analysis. Ammonia, sulfuric and phosphoric acid, and additional raw materials are used. These may be applied directly to the soil or used in blends (Figure 7.17).

## Liquid Fertilizers

Foliar liquid fertilization (commonly referred to as foliar feeding) involves the use of a soluble nutrient form for plants. This concept is over 100 years old and provides quicker utilization of nutrients. However, the response is often temporary. Due to the small amounts required, micronutrient applications have traditionally been the most prominent use of foliar sprays (Figure 7.18). A difficulty in applying sufficient amounts of macronutrients such as N, P, and K without leaf burn also has been a problem. Other advantages and disadvantages of foliar liquid fertilization are listed in Table 7.4.

**TABLE 7.16**
**Solution Used to Spot Treat for Micronutrient Deficiencies Applied in 1 to 3 Gallons Water per 1,000 square feet (4 to 12 L/100 m²)**

| Micronutrient | Fertilizer Source | Rate: Fluid Ounces/Gallon/1,000 square feet (mL/L) | Rate: Pounds Element/1,000 square feet (kg/ha) |
|---|---|---|---|
| Iron (Fe) | Iron sulfate (20% Fe) | 2 (16) | 0.025 (1.2) |
| Manganese (Mn) | Manganese sulfate (27% Mn) | 0.75 (6) | 0.0125 (1.2) |
| Zinc (Zn) | Zinc sulfate (35% Zn) | 0.5 (4) | 0.010 (0.5) |
| Copper (Cu) | Copper sulfate (25% Cu) | 0.2 (1.6) | 0.003 (0.15) |
| Boron (B) | Boric acid (17% B) | 0.2 (1.6) | 0.002 (0.05) |
| Molybdenum (Mo) | Sodium molybdate (47% Mo) | 0.04 (0.3) | 0.001 |

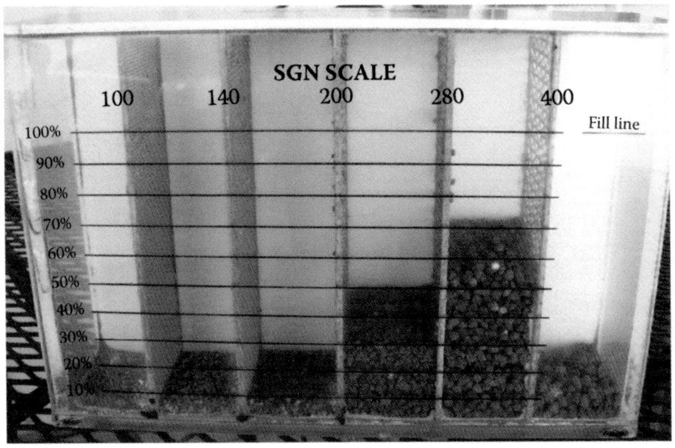

**FIGURE 7.15** Using sieves to determine the size guide number of granular fertilizer particles.

**Soluble sources.** Soluble or quickly available N sources are supplied in solid form including pure fine granules, crystals, or powders, mixed with water, resulting in an expedient response in terms of shoot growth and greening (Table 7.3). This occurs approximately 2 days after application, peaking in 7 to 10 days, and tapering off to original levels in two to four weeks, depending on the application rate and subsequent amount of water applied. The N in soluble sources is in either the ammonium or nitrate form (Figure 7.19). The ammonium N form (e.g., ammonium nitrate, ammonium sulfate) is prone to volatilization but less prone to leaching. The nitrate N form (e.g., potassium nitrate and calcium nitrate) is more prone to leaching.

Soluble (liquid) N sources have salt-like characteristics. They dissolve readily in water to form cations and anions. The greater availability of these ions corresponds with a greater burn potential of fertilizer (Figure 7.20). Burn potential can be lowered by making applications only to dry turfgrass surfaces and when air temperatures are <80°F (27°C). Watering-in soluble N immediately following application further reduces the chance of burning plant tissue. Other disadvantages of using soluble N sources can be minimized by frequently applying small amounts. Rates at or below 1/2 pound N/1,000 square feet (2.4 g N/m²) will minimize these problems but will increase application frequency and treatment costs.

*Fluid fertilizers*, also called liquid fertilizers, involve solutions, suspensions, and slurries and have been greatly expanded in use by the commercial lawn care industry. Liquid fertilizers are produced by neutralizing phosphoric acid with ammonia. Soluble N sources include various urea, methylol

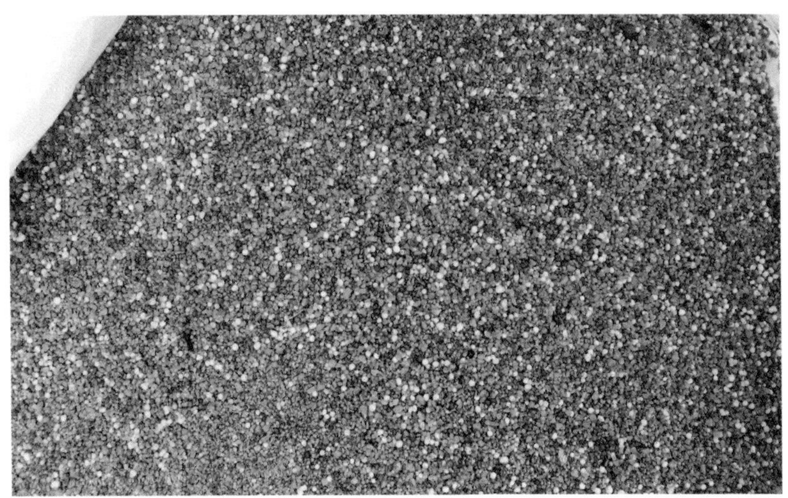

**FIGURE 7.16** Bulk blended fertilizer where two or more dry fertilizer materials are physically mixed.

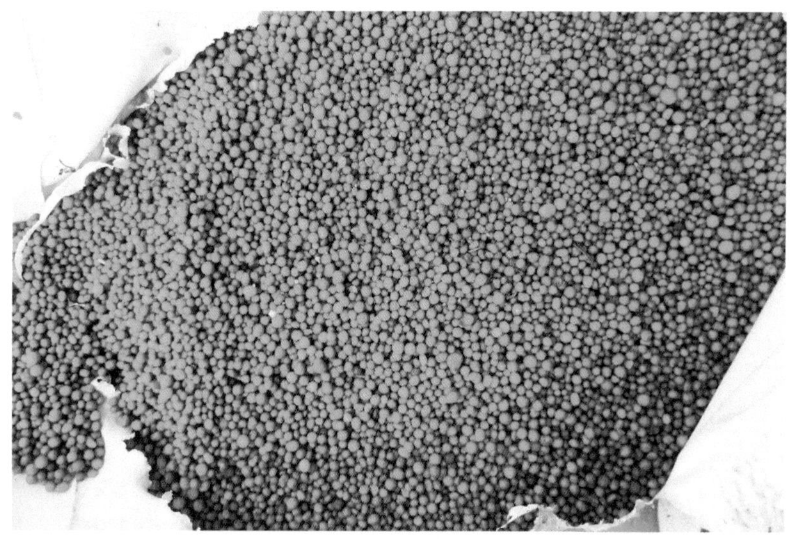

**FIGURE 7.17** Homogeneous fertilizer where each granule or prill contains a similar analysis.

**FIGURE 7.18** Green turf from an iron-containing fertilizer compared to the untreated foreground.

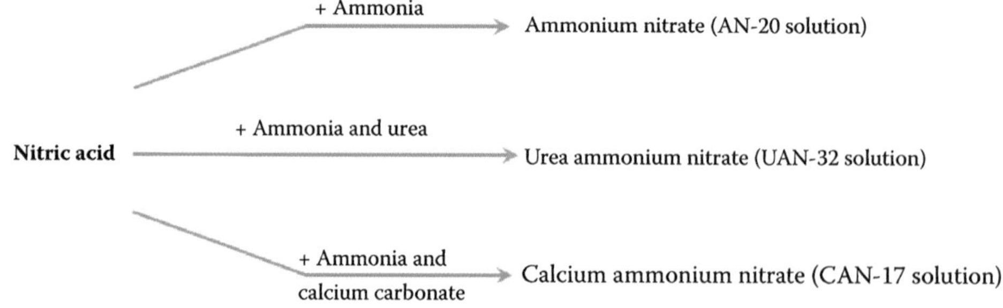


**FIGURE 7.19** Illustration on manufacturing ammonium nitrate liquid sources by adding various components to nitric acid.

urea, methylene urea combination, and urea/triazone products, which fully dissolve to form clear fluids.

*Suspensions* involve methylene urea totally or only partially dissolved. The undissolved portion may remain in suspension or a suspending agent, such as clay, may help stabilize the suspension. With slurries, the undissolved material settles and requires constant agitation to remain in solution. Powdered slow-release fertilizers form slurries while soluble granules tend to remain in solution. Suspension fertilizer's advantages over clear liquid are (a) higher analysis grades can be produced, (b) costs are lower because fewer pure products can be used, (c) larger quantities of micronutrients can be suspended, and (d) powdered forms of pesticides can normally be suspended and used.

**FIGURE 7.20** Soluble (liquid) fertilizers can burn turf if applied at excessive rates during stressful conditions.

Improving the efficiency of foliar nutrient application includes:

- Apply in early norming or evening hours when moisture is present on leaves.
- Apply when air temperatures are <85°F (29°C).
- Apply when relative humidity is high (>70%).
- Use spray volumes at or below 40 gallons per acre (3.7 L/100 m$^2$).
- Apply with a quality adjuvant.

*Foliar feeding* involves using low fertilizer rates (e.g., 1/8 pound N or Fe/1,000 square feet, 0.6 g/m$^2$) in low water application volumes (less than 1 gallon per 1,000 square feet, <4 L/100 m$^2$) compared to higher application volumes (more than 1 gallon/1,000 square feet, 4 L/100 m$^2$) for liquid fertilization (Figure 7.21). Spoon-feeding is often used synonymously as foliar feeding but refers to low rates (<0.4 pound N/1,000 square feet, <2 g/m$^2$) of fertilizer applied as soil or foliar treatments, as well as from granulars or liquids. Low nutrient and spray volumes are used to minimize costs and to supplement the normal fertilization program with nutrients absorbed directly by turfgrass leaves. Urea is a popular N form used for foliar application with lower rates (e.g., 0.10 pound N/1,000 square feet, 5 kg/ha) absorbed more completely than higher (e.g., 1/4 pound N/1,000 square feet, 12 kg/ha) ones. Foliar fertilization is often most effective (best taken up by plants) when root growth is slowed or restricted and when a quick color response or growth spurt is needed. Maximum foliar absorption into plants occurs when leaves are young and well hydrated, and stomates are fully open (such as during sunlight) and no moisture stress is occurring. Most absorption occurs within four hours following application.

*Liquid fertilization* involves using higher spray volumes (e.g., 3 to 5 gallons per 1,000 square feet, 12 to 20 L/m$^2$), to

**FIGURE 7.21** Foliar feeding turf using low fertilizer rates in low water application volumes.

wash the fertilizer off the leaves, increasing root absorption. With liquid fertilization, fertilizers and pesticides often are applied together.

The liquid application form in foliar feeding enters plants directly by penetrating leaf cuticles (via micropores) or through stomata, thereafter entering cells. Uptake is via diffusion (passive uptake) or active transport (requiring energy). Foliar applications provide quicker utilization of nutrients than through soil treatment. Research also indicates the physical form of the nutrient, dry or fluid, has no measurable effect on its agronomic properties, such as total amount of plant growth. Generally, cations (e.g., ammonium based, $K^+$) and neutral compounds (e.g., urea) are most easily absorbed by active transport than negatively charged anions such as nitrate and phosphate. Anions are repelled by the slightly negatively charged turf shoots. Once absorbed, nutrients move up in plants via xylem tissue or up or down in phloem tissue.

Because of its water solubility, urea is the most widely used fertilizer material and is often mixed with ammonium nitrate or potassium nitrate (Table 7.3). A higher analysis liquid can be formed by such combinations than if each individual component is dissolved by itself. Liquid urea is characterized by a quick response in terms of turf color and a medium- to high-burn potential. Low rates, applied more frequently, are required to promote even turf growth and color and to minimize burn potential when using liquid urea. Other quickly available liquid N carriers include ammonium sulfate, ammonium polyphosphate, ammonium thiosulfate, calcium nitrate, and potassium nitrate. Not all nutrients are readily absorbed by leaves. Table 7.17 lists the ones that are readily absorbed.

*Manufacturing of Liquid Fertilizers*

In the United States, greater than 50% of all fluid-mixed fertilizers are clear liquid solutions. Mixtures applied in liquid have a typical analysis of approximately 9% N, 3.9% P, and 7.5% K.

There are two general methods of manufacturing liquid-mixed fertilizers. The simplest method is known as the batch or suspension process, and simply consists of dissolving the correct proportions of the solid plant-food carriers, such as ammonium phosphates, urea, or potassium chloride, in water to produce the desired grade of final product. The weighed constituents are dissolved in the proper amount of water with a suitable mixing device. The solution may be heated to aid in the dissolving process since many dry fertilizer products absorb heat from water when mixed. The relatively higher cost of raw materials generally limits this to small operations or to companies engaged in manufacturing specialty grades. Solution grade forms of these products are used.

The second and most widely used method is based on the neutralization of phosphoric acid with ammonia to produce ammonium polyphosphates. Considerable heat is generated and the solution becomes hot during this process. Anhydrous or aqueous ammonia, or ammonia-ammonium nitrate, or ammonia-urea type N solutions are reacted with phosphoric acid solutions followed by the addition of solid sources of N and/or potash. Aqueous ammonia also is reacted with sulfate and elemental sulfur to form 12-0-0-26S ammonium thiosulfate. It can be used in a wide variety of N-P-K-S formulations and is essentially noncorrosive. The density of most common liquid mixtures will approximate 10 pounds/gallon (1.2 kg/L).

**Slow-release sources.** Several new liquid-based materials have better slow-release characteristics. These allow heavier rates to be applied less frequently without undesirable surges in growth or color and minimize turf foliar burn potential. They are often formed by altering the ratio of urea to formaldehyde to produce more water-insoluble N. The water-soluble N of these products contains unreacted urea, methylol urea, and methylene ureas. They are more expensive to produce but have less fertilizer burn potential, making them safer to use.

**Note:** Available commercial fertilizer sources constantly change in the marketplace as new formulations and delivery systems evolve. The commercial products, manufacturers, and distributors listed are not exclusive but are included for example only and do not constitute a warranty or guarantee.

**TABLE 7.17**
**Nutrient Forms Readily Absorbed by Leaves**

| N | P | K | S | Micronutrients |
|---|---|---|---|---|
| Urea | $P_2O_5$ (low solubility) | $K^+$ | $SO_4^{-2}$ | Fe |
| Ammonia | $H_2PO_4$ | | | B |
| Nitrate | Organic phosphates | | | Zn |
| Methylene urea (short chain) | | | | |
| Amino acids | | | | |

*Note:* Calcium, manganese, and magnesium also are absorbed by leaves.

## SOIL NUTRIENT ANALYSIS

A variety of analytical tools are available to assist turfgrass managers in developing an effective, efficient, and environmentally responsible nutrient management program for every area of the golf course. Test results indicate what is available to plants via a soil test and what is actually in plants by a leaf tissue test. The needs for conducting either of these tests vary depending upon management intensity, quality expectations, environmental conditions, turfgrass species, and geographic location. Soil testing measures an index of the amount of plant nutrient of available nutrient in the soil, and this is then correlated with the probability of a fertilizer response through soil test calibration.

### SOIL TESTING

Soil testing is one of the fundamental practices of turfgrass management. Native soil "push-up" greens should be tested at least annually while sand-based greens are tested two or more times annually. When sampling at initial establishment, a depth of 6 inches (15 cm) should be used. When sampling mature turf, soil should be sampled to a depth of 2 to 4 inches (5 to 10 cm) or to the depth where the majority of plant roots are located.

Following sampling, three basic steps are involved in developing a reliable soil test:

1. **Nutrient extraction.** Using specially designed extracting chemicals, this simulates a plant's ability to remove nutrients from the soil.
2. **Calibration.** This determines the meaning of values generated by the soil test extraction.
3. **Interpretation.** The information from the first two steps is interpreted and integrated into recommendations on the quantity of a particular fertilizer element needed.

Soil analysis provides information on relative levels of nutrients, organic matter, pH, buffer pH, soluble salts, and CEC (Table 7.18). However, significant variations often occur in reported values and nutrient recommendations between testing laboratories. This is primarily due to geographical differences in soil types, nutrient extraction solutions and analysis techniques, as well as philosophical differences in data interpretation. The ranking of the nutrient level, however, should be somewhat similar regardless of the extractant used. Table 7.19 lists the most commonly performed soil tests for turfgrass managers.

Distilled water is the simplest chemical extractant and is used for assessing soluble salts, S in certain soils, and soil pH. For most nutrients, additional chemicals are used to "unlock" or extract them from various soils. Soils vary on their acidity, origin, and texture in differing regions of the country and certain chemicals are more efficient for a particular soil. Weak acids (e.g., acetic, citric, lactic), diluted stronger acids (e.g., hydrochloric, sulfuric, nitric), double acids (two different acids combined), bicarbonates (e.g., ammonium, sodium), salt solution, specific ions (e.g., fluoride or ammonium), chelating agents and various combinations of these are used, depending on the soil in question. The appropriate extraction technique allows the most effective means of removing the available nutrients from the soil. If an inappropriate extraction method is used, some nutrients will not be removed and erroneously low readings will occur. The reverse may also happen when the wrong extraction method can cause higher nutrient readings than the actual available concentration in the soil. In addition, different testing laboratories use different report forms and different units to report results. Unfamiliar equations and chemical symbols also can add confusion to soil, tissue, and water tests.

**TABLE 7.18**
**Soil Testing Option Based on the Desired Objectives**

| Objective | Soil Test |
|---|---|
| Soil reaction and lime requirements | • Water, salt, and buffer pH |
| Nutrient element status | • *Major elements*: P, K, Ca, Mg, $NO_3$, $SO_4$<br>• *Micronutrients:* B, Cl, Cu, Fe, Mo, Ms, Zn |
| Toxic elements | • *Other elements:* Al, Na<br>• *Trace elements and heavy metals:* As, Cd, Co, Cr, Cu, Mn, Pb, Ni |
| Total salts | • Soluble salts |
| Physical and chemical characteristics | • Organic matter content; CEC |
| Mechanical analysis | • Soil texture, bulk density, porosity |
| Drainage potential | • Saturated hydraulic conductivity ($K_{sat}$) |
| Soil moisture characteristics | • Soil moisture retention curves |

In general, in low (<7) pH soils, the Mehlich I, II, and III; Bray 1 and 2; Morgan; and neutral ammonium acetate (pH 7) extractants are used. In high (>7) pH soils, the Olsen and buffered ammonium acetate (pH 8.5) methods are used with water and 0.01 *M* calcium carbonate methods as alternatives.

A turf manager should select a particular laboratory, preferably an accredited one that has experience analyzing samples from turfgrass sites, and stick with it. The chances of a lab switching analytical methods are minimal, and nutrient recommendations will be more consistent from year to year. For example, laboratories with only limited experience with turfgrass samples tend to overestimate P requirements and underestimate K needs. Since no single soil test analysis is applicable to all soils, it is best to use a lab located in your general geographic area.

## Soil Analysis Report

Two major philosophies for interpreting soil test results currently exist. They are the sufficiency level of available nutrients concept (SLAN) and the basic cation saturation ratio (BCSR), or nutrient maintenance concept. Although both philosophies measure the nutrient status of soil, they differ in how analytical results are interpreted and expressed. This can be confusing if the laboratory uses a combination of both concepts to make nutrient recommendations. Consequently, if a golf course manager regularly switches laboratories or is not familiar with reading and interpreting a soil test report, the information can be confusing, and inconsistency in a fertilizer program may result.

**Sufficiency level of available nutrient (SLAN) concept.** The SLAN philosophy assesses the actual levels of plant-available macronutrients, Ca, Mg, K, P, S, and micronutrients, Fe, Mn, Zn, Cu, and B in a soil sample and its ability to supply nutrients to plants. The SLAN attempts to correlate turf response to extractable soil nutrients and is based on the traditional concept that as the soil test index level for a given nutrient increases, the turfgrass plant response to the nutrient will increase up to a given critical level or threshold. The quantities of individual nutrients are typically expressed in terms of "low," "medium," "high," or "very high" (Tables 7.20 and 7.21). Based on this method, for example, if a soil test index indicates P is low, then a P application will probably improve plant growth. Conversely, if the soil test indicates P is present at "high" or "very high" levels, then a plant response is not expected and further applications on "very high" testing soils may even lead to other nutrient imbalances. Specific laboratory values for these criteria have been established by decades of field and greenhouse soil test calibration research on various crops over various soil types.

**Basic cation saturation ratio (BCSR) concept.** This philosophy is based on the concept of an "ideal" ratio of "basic"

**TABLE 7.19**
**Extraction Techniques Used to Determine Soil Nutrient Levels for Various Locations and Soil Types**

| Extraction Techniques | Primary Nutrients(s) Extracted | Comments |
|---|---|---|
| | **Single Acid** | |
| Ammonium acetate ($NH_4OAc$), pH 7.0 | K, Ca, Mg | • Widely used for CEC and K; in U.S. Northeast, used at pH 4.8 to extract Ca and Mg; not for calcareous soils (sands) |
| | **Double Acids** | |
| Mehlich I (also called the dilute double acid method; 0.05 *N* HCl + 0.025 *N* $H_2SO_4$) | P, K, Ca, micronutrients (Cu, Mn, B, Zn), Na, CEC, and base saturation | • Used in acidic clayey soils of U.S. Southeast with low CEC (<10 meq/100 g) and low (<5%) OM content for relatively soluble CaP, FeP, and AlP; may overestimate available P in calcareous soils |
| Mehlich II | P, K, Ca | • Extracts relatively soluble CaP, FeP, and AlP; some organic P; superior on volcanic ash or loess-derived soils |
| Mehlich III, pH < 2.0 (double acid + chelating agent) | P, K, Ca, Mg, micronutrients | • Universally used extractant; also used for cations (Cu, Fe, Mn, Zn) in many acidic, noncalcareous, and volcanic soils |
| Morgan (sodium acetate + acetic acid) | P, Ca, K, Mg | • Used in acidic U.S. northeastern and northwestern soil with CEC < 10 meq/100 g; extracts P dissolved by $CO_2$ |
| Bray P1 (ammonium fluoride + hydrochloric acid) | P | • Extracts P as relatively soluble CaP, FeP, and AlP when soil pH is <6.8 as well as some organic P |
| Ammonium acetate + acetic acid | S | • Used to extract sulfate (S) |
| | **Bicarbonate Based** | |
| Olsen's (0.5 *M* sodium bicarbonate, $NaHCO_3$), pH 8.5 | P, K | • Extracts P as CaP and some FeP; better choice for calcareous soils with pH is >6.8 than acid extractants (e.g., Bray 1) |
| Ammonium bicarbonate ($NH_4HCO_3$) or sodium bicarbonate ($NaHCO_3$) | Various cations | • Used in central western United States on calcareous soils for cation extraction |
| Ammonium chloride (0.5 *M* $NH_4Cl$) | Exchangeable basic cations | • Used on calcareous soils for CEC determination |
| Sodium acetate (0.4 *M* NaOAc) + sodium chloride (0.1 *M* NaCl) | Exchangeable basic cations | • Used on calcareous soils for CEC determination |
| | **Other** | |
| DTPA-TEA | Fe, Mn, Zn, Cu | • Chelating agent widely used for these nutrients |
| Hot water | S | • Used in arid and semiarid regions |
| Glass electrode pH meter on a 1:1, 1:2, or 1:5 (v/v) basis of soil:deionized water | Soil pH | • Soil pH; values must be correlated to compare with saturated paste extract values |
| Buffer solution, pH 8.0, + pH meter | Soil buffer pH | • Buffer pH (exchangeable or potential acidity) |
| Saturated paste extract | EC, SAR, B | • Electrical conductivity (EC); sodium absorption ratio (SAR) and B levels |
| Soil:water dilution on 1:2 or 1:5 (v/v) basis | Total soluble salts | • Generally not as accurate as saturated paste extraction |

*Note:* After extraction, nutrient concentrations are often determined by inductively coupled plasma (ICP) emission spectrometry.

**TABLE 7.20**
**Extractable Phosphorus, Potassium, Calcium, and Magnesium by Soil Test Ratings**

| Soil Test Index Rating | Phosphorus | Potassium | Calcium | Magnesium |
|---|---|---|---|---|
| | Pounds Element/Acre[a] | | | |
| Very low | 0 to 10 | 0 to 24 | 0 to 200 | 0 to 10 |
| Low | 11 to 30 | 25 to 70 | 201 to 400 | 11 to 32 |
| Medium | 31 to 60 | 71 to 156 | 401 to 800 | 33 to 60 |
| High | 61 to 120 | 157 to 235 | 801 to 2000 | >60 |
| Very high | >121 | >235 | >2000 | >60 |

*Source:* Clemson University Soil Testing Laboratory.
[a] Multiply values by 1.12 to obtain kilograms/hectare.

**TABLE 7.21**
**Predicting Plant Response to Nutrient Application Using Soil Test Index Values**

| Soil Test Index | | Probable Plant Response to Nutrient Applications | | | | |
|---|---|---|---|---|---|---|
| **Range** | **Rating** | **Phosphorus** | **Potassium** | **Manganese** | **Zinc** | **Copper** |
| 0 to 10 | Very low | High | High | High | High | High |
| 11 to 25 | Low | High | High | High | High | High |
| 26 to 50 | Medium | Medium/low | None | None | None | None |
| 51 to 100 | High | Low/none | None | None | None | None |
| >100 | Very high | None | None | None | None | None |

*Source:* North Carolina Department of Agriculture, Agronomic Division.

cations (primarily Ca, Mg, and K) for the given soil or crop occupying the majority of cation exchange sites in soil. This approach requires an accurate measurement of a soil's CEC. Soil testing laboratories often suggest that a balanced "ideal" soil will have a total base saturation of approximately 80%. This will comprise 65% of the exchange sites being occupied by Ca, 10% Mg, 5% K, less than 2% sodium, and 3% trace cations, with the remainder being H (10% or less). These percentages are a 13:2:1 ratio of Ca:Mg:K. When these soil nutrient levels are "balanced," "ideal" nutrient ratios for plant growth have been suggested: Ca:Mg of 6:1; Ca:K of 15:1; Ca:H of 3:1; and, Mg:K of 2:1. Once the cations are in "balance," these soils are purported to possess better aggregation, which improves water infiltration and percolation.

The primary focus of BCSR is on "fertilizing the soil" rather than "fertilizing the plant" as with the SLAN concept. With BCSR, annual soil tests are conducted to assess the presence of various nutrients on the cation exchange sites and normally recommend large quantities of Ca, Mg, and K, which, in turn, may significantly raise soil pH. Adjustments are suggested whenever nutrient ratios become out of "balance." The concept appears to work for many weathered, low-pH soils. However, applying additional Ca and Mg to naturally high pH soils where calcareous sands are used for putting greens, and/or soil tests indicate 90% Ca in soil, appears unnecessary. High Ca and Mg rates may reduce the availability of P and other critical cations and may lead to unnecessary fertilizer applications. This, in turn, often causes nutrient deficiencies and an undesirable pH. Unfortunately, research in this area of soil science is limited and the long-term benefit of fertilizing according to the BCSR philosophy is largely unproven. Additionally, many soil testing methods on calcareous soils often overestimate the true CEC due to the abundance of free Ca solubilized by some nutrient-extracting solutions.

**Test results.** Soil test reports often contain both types of data. SLAN data include extractable nutrient levels using the appropriate extraction technique and results are reported as parts per million (ppm = mg/kg), pounds nutrient/acre, or milliequivalent nutrient/100 g soil (meq/100 g). Nutrients routinely reported by SLAN include P, K, Ca, Mg, S, Fe, Mn, Zn, Cu, and Na. BCSR information provided on a routine soil test report includes CEC and ratios of the cations, Ca:Mg, Ca:K, and Mg:K.

When reported as parts per million, nutrient levels can be converted to pounds/acre. An acre of mineral soil, 6 to 7 inches (15 to 18 cm) deep, weighs approximately 2 million pounds (900,000 kg). To convert parts per million to approximate pounds/acre, simply multiply by two based on the following assumption:

$$\text{Parts per million} = \text{milligrams/kilograms}$$

$$\frac{\text{milligrams}}{\text{kilograms}} \times \frac{\text{pounds}}{454{,}000\,\text{mg}} \times \frac{0.454\,\text{kg}}{\text{pounds}} \times \frac{2{,}000.000\,\text{pounds}}{\text{acre (6 inches deep)}} \approx 2$$

$$\text{kilograms/hectare} = \text{milligrams/kilogram (or parts per million)} \times 2.25$$

Soil cations, such as Ca, Mg, K, and H, are expressed by their relative ability to displace other cations. For example, 1 meq of K is able to displace exactly 1 meq of Mg. The unit of measure milliequivalent/100 g serves this purpose. Cation exchange capacity and the total amounts of individual cations may be expressed using these units. To convert milliequivalents per 100 g of soil to pounds nutrient/acre furrow slice, the equivalent weight of the cation is used (refer to Chapter 3):

$$\frac{X\ \text{meq}}{100\,\text{g}} \times \text{cation equivalent weight (g)} \times \frac{1\ \text{eq}}{1{,}000\ \text{meq}} \times \frac{2{,}000{,}000\ \text{pounds}}{\text{acre}} = \frac{\text{pounds nutrient}}{\text{acre}}$$

### Example

Convert 3 meq $Ca^{+2}$ per 100 g soil to pounds $Ca^{+2}$ per acre.

$$\frac{3\ \text{meq}}{100\,\text{g}} \times \frac{40\,\text{g}\ Ca^{+2}}{2(\text{valance no.})} \times \frac{1\ \text{eq}}{1000\ \text{meq}} \times \frac{2{,}000{,}000\,\text{pounds}}{\text{acre}} = \frac{1{,}200\ \text{pounds}\ Ca^{+2}}{\text{acre furrow slice}}$$

Table 7.21 cross lists the amount of an element extracted from the soil related to its fertility rating index. Tables 7.22

through 7.25 list the recommended yearly amounts of phosphate ($P_2O_5$) and potash ($K_2O$) based on the P and K currently extractable from the soil. Table 7.26 summarizes recommended ranges for nutrient levels using various extractant procedures.

## Interpreting a Soil Analysis Report

A soil analysis report supplies a wealth of information concerning the nutritional status of a soil and may aid in the early detection of problems limiting turfgrass growth or desirable color. This is particularly important during the initial establishment or renovation of a turfgrass site. A typical soil analysis supplies information relative to CEC, soil acidity (pH), lime requirements, and soil P, K, Ca, and Mg status. Nitrogen content is rarely determined because of its dynamic status due to environmental conditions and microbial transformations. Additional information can be requested from lab reports such as soil organic matter content, soluble salts, and irrigation water analysis. Table 7.27 lists the results from a typical soil and plant tissue report. Recommendations, in terms of lime and nutrients, are usually listed at the end of the report.

**Various ratios of elements.** Ratios of various elements can be important for certain soil chemical reactions and influence nutrient availability. One of the most important ratios in turf management is the carbon-to-nitrogen (C:N) ratio of organic amendments like peat moss, composts, or other organic materials used for improving soil structure. The C:N ratio will influence organic matter decomposition and is an indicator of N availability. Generally, a C:N ratio between 10 and 12:1 for most organic amendments is desirable. Ratios greater than 20:1 may have an inefficient breakdown of organic matter due to the lack of N necessary to sustain soil organisms responsible for decomposition. When excessive carbon is present, soil microorganisms will immobilize all available N for their own metabolic activity. This condition is evident through chlorosis or yellowing—symptoms consistent with a N deficiency. Certain sawdust sources have a C:N ratio as high as 400:1. Turf managers who use these as soil amendments should add some readily available N, like urea, to the mixture.

**TABLE 7.22**
**Recommended Fertilizer Amounts for Bermudagrass and Zoysiagrass Golf Course Fairways Based on Existing Extractable Soil P and (K) Levels[a]**

| | Soil K Levels | | | | |
|---|---|---|---|---|---|
| Soil P Levels | Low | Medium | Sufficient | High | Excessive |
| | Pounds of N-$P_2O_5$-$K_2O$ Recommended per acre | | | | |
| Low | 120-100-80 | 120-100-60 | 120–100-30 | 120-100-0 | 120-100-0 |
| Medium | 120-40-80 | 120-40-60 | 120–40-30 | 120-40-0 | 120-40-0 |
| Sufficient | 120-40-80 | 120-40-60 | 120-40-30 | 120-40-0 | 120-40-0 |
| High | 120-0-80 | 120-0-60 | 120–0-30 | 120-0-0 | 120-0-0 |
| Excessive | 120-0-80 | 120-0-40 | 120–0-30 | 120-0-0 | 120-0-0 |

*Source:* Clemson University Soil Testing Laboratory.

[a] Apply one-half of the N with phosphate and potash in spring and in late summer. Apply additional N as needed in midsummer at 60 pounds/acre/application. Multiply values by 1.12 to obtain kilograms/hectare.

**TABLE 7.23**
**Recommended Fertilizer Amounts for Bermudagrass Golf Course Greens and Tees Based on Existing Extractable Soil P and K Levels[a]**

| | Soil K Levels | | | | |
|---|---|---|---|---|---|
| Soil P Levels | Low | Medium | Sufficient | High | Excessive |
| | Pounds of N-$P_2O_5$-$K_2O$ Recommended per 1,000 square feet | | | | |
| Low | 9.2-4.6-4.6 | 9.2-4.6-2.3 | 9.2-4.6-1.4 | 9.2-4.6-0.7 | 9.2-4.6-0 |
| Medium | 9.2-2.3-4.6 | 9.2-2.3-2.3 | 9.2-2.3-1.4 | 9.2-2.3-0.7 | 9.2-2.3-0 |
| Sufficient | 9.2-1.1-4.6 | 9.2-1.1-2.3 | 9.2-1.1-1.4 | 9.2-1.1-0.7 | 9.2-1.1-0 |
| High | 9.2-0-4.6 | 9.2-0-2.3 | 9.2-0-1.4 | 9.2-0-0.7 | 9.2-0-0 |
| Excessive | 9.2-0-4.6 | 9.2-0-2.3 | 9.2-0-1.4 | 9.2-0-0.7 | 9.2-0-0 |

*Source:* Clemson University Soil Testing Laboratory.

[a] Nitrogen applications should be alternated with complete fertilizers and modified to maintain desired growth and color. Multiply values by 49 to obtain grams/square meter.

**TABLE 7.24**
**Recommended Fertilizer Amounts for Bentgrass Golf Course Greens and Tees Based on Existing Extractable Soil P and K Levels[a]**

| Soil P Levels | Soil K Levels Low | Medium | Sufficient | High | Excessive |
|---|---|---|---|---|---|
| | Pounds of N-$P_2O_5$-$K_2O$ Recommended per 1,000 square feet | | | | |
| Low | 6.9-4.6-4.6 | 6.9-4.6-2.3 | 6.9-4.6-1.4 | 6.9-4.6-0.7 | 6.9-4.6-0 |
| Medium | 6.9-2.3-4.6 | 6.9-2.3-2.3 | 6.9-2.3-1.4 | 6.9-2.3-0.7 | 6.9-2.3-0 |
| Sufficient | 6.9-1.1-4.6 | 6.9-1.1-2.3 | 6.9-1.1-1.4 | 6.9-1.1-0.7 | 6.9-1.1-0 |
| High | 6.9-0-4.6 | 6.9-0-2.3 | 6.9-0-1.4 | 6.9-0-0.7 | 6.9-0-0 |
| Excessive | 6.9-0-4.6 | 6.9-0-2.3 | 6.9-0-1.4 | 6.9-0-0.7 | 6.9-0-0 |

*Source:* Clemson University Soil Testing Laboratory.

[a] Nitrogen applications should be alternated with complete fertilizers and modified to maintain desired growth and color. Multiply values by 49 to obtain grams/square meter.

**TABLE 7.25**
**Recommended Fertilizer Amounts for Kentucky Bluegrass, Ryegrass, and/or Fescue Golf Course Fairways Based on Existing Extractable Soil P and K Levels[a]**

| Soil P Levels | Soil K Levels Low | Medium | Sufficient | High | Excessive |
|---|---|---|---|---|---|
| | Pounds of N-$P_2O_5$-$K_2O$ Recommended per Acre | | | | |
| Low | 160-80-80 | 160-80-80 | 160-80-60 | 160-80-40 | 160-80-0 |
| Medium | 160-40-80 | 160-40-80 | 160-40-60 | 160-40-40 | 160-40-0 |
| Sufficient | 160-40-80 | 160-40-80 | 160-40-60 | 160-40-40 | 160-40-0 |
| High | 160-0-80 | 160-0-80 | 160-0-60 | 160-0-40 | 160-0-0 |
| Excessive | 160-0-80 | 160-0-80 | 160-0-60 | 160-0-40 | 160-0-0 |

*Source:* Clemson University Soil Testing Laboratory.

[a] Multiply values by 1.12 to obtain kilograms/hectare.

A soil test report utilizing SLAN can be used to determine the approximate amount of nutrient needed to be added to raise it into the medium range. From a SLAN soil test report, subtract the nutrient value listed (as parts per million) from the desired (or medium) range listed in Table 7.26. If the value is negative, supplemental fertilization is not needed. However, if this value is positive, multiply it by two to convert it to pounds nutrient/acre. Finally, divide the resulting value by 43.56 to indicate the pounds nutrient/1,000 square feet needed to be added to correct the deficiency.

### Example

If a SLAN-based soil test report indicates 30 ppm K, is additional K needed? If so, how much per 1,000 square feet is needed?

**Step 1:** Subtract the reported value from the desired (or medium) value listed in Table 7.26:

50ppm (from Table 7.26) − 30ppm (from the soil test report) = 20ppm K deficit

**Step 2:** Multiply the deficit value by 2 to determine its pounds/acre value (or by 2.25 to determine kg/ha).

20ppm (from step 1) × 2 = 40 founds K/acre (or 45kg K/ha) needed to correct the deficit

**Step 3:** To determine the approximate amount of K needed per 1,000 square feet, divide the step 2 value by 43.56 (or divide kg/ha by 100 to obtain g/m$^2$).

40pounds K/acre needed ÷ 43.56 = 0.92 pound K/1,000 square feet(or 4.5g K/m$^2$) needed correct the deficit

**TABLE 7.26**
**Relative Response Range of Soil Elements Analyzed by Indicated Extractant Techniques[a]**

| Element (abbreviation) | Comments | | |
|---|---|---|---|
| Nitrogen/organic matter (N/OM) | Typical levels are ≤5%. Due to changing levels in soils, N availability is hard to predict. Often, percent OM serves as a reserve for N. Labs, therefore, list an estimated N release figure based on the percent OM present. OM content is determined by loss on ignition. Nitrate-N ($NO_3$-N) levels using potassium chloride extraction are normal between 10 and 25 ppm. | | |
| Phosphorus (P) | P availability is greatest between a soil pH of 5.5 and 6.5. Desired P values for various extractant procedures include: | | |
| | Extraction technique | Desired P range (ppm) | |
| | Morgan | 10 to 20 | |
| | Bray P1 | 20 to 40 | |
| | Bray P2 | 40 to 60 | |
| | Mehlich I | 15 to 30 | |
| | Mehlich III | 25 to 55 | |
| | Olsen | 12 to 28 | |
| | Ammonium bicarbonate-DTPA | 8 to 45 | |
| Potassium (K) | Generally, higher K levels are required in high clay or organic matter containing soils or soils with high levels of Mg. Sandy soils require more frequent, light K applications versus heavier ones. A fertilizer ratio of N to K should be 3:2 or 1:1. Due to luxury consumption and leaching loss, levels above medium as reported by soil testing are mostly unnecessary. Desired K values for various extractant procedures include: | | |
| | Extraction technique | Desired K range (ppm) | |
| | **Mehlich 1** | | |
| | Sands/most soils | 50 to 100 | |
| | Fine textured (>35% clay) | 90 to 200 | |
| | **Mehlich 3** | | |
| | Sands/most soils | 50 to 116 | |
| | Fine textured (>35% clay) | 75 to 175 | |
| | **1 *M* ammonium acetate (pH 7.0)** | | |
| | Sands/most soils | 75 to 175 | |
| | Fine textured (>35% clay) | 100 to 235 | |
| | Olsen's | 155 to 312 | |
| | Morgan's | 120 to 174 | |
| Calcium (Ca) | Extraction technique | Desired range (ppm) | |
| | Mehlich I | 200 to 350 | |
| | Mehlich III | 500 to 750 | |
| | Ammonium acetate, pH 7.0 | 500 to 750 | |
| | Morgan | 500 to 750 | |
| | Saturate paste extract | 100 to 150 | |
| Magnesium (Mg) | Extraction technique | Mg in sands (<4 meq/100 g CEC) | Mg in clays (>4 meq/100 g CEC) |
| | | Parts per million | |
| | Mehlich I | 30 to 60 | 50 to 100 |
| | Mehlich III | 60 to 120 | 70 to 140 |
| | Ammonium acetate, pH 7.0 | 100 to 200 | 140 to 250 |
| | Morgan | >100 | >100 |
| | Saturated paste extract | 12 to 120 ppm | – |
| | **Comments**. With most soils, liming with dolomite to ensure an adequate soil pH will provide adequate Ca and Mg. Deficiencies are more common in soils containing sand, acid, and/or low organic matter. Calcareous sands and most soils in the Midwestern states will not require Ca additions. Use gypsum (calcium sulfate) if Ca is needed when soil pH is high. Consider using magnesium oxide, magnesium sulfate (Epsom salt), or sulfate of potash-magnesia (Sul-Po-Mag) if soil tests are low in Mg and lime is not required. Apply magnesium sulfate (Epsom salt) to test for Mg deficiency. Guidelines for Mg:K and Ca:K ratios based on saturation percentages on the soil CEC include: | | |

*(Continued)*

**TABLE 7.26 (CONTINUED)**
**Relative Response Range of Soil Elements Analyzed by Indicated Extractant Techniques[a]**

| Element (abbreviation) | Comments | | |
|---|---|---|---|
| BCSR Guidelines for Ca, K and Mg | Cations | Ratio | Possible deficiency |
| | Ca:K | <10:1 | • Ca deficiency may occur |
| | | <30:1 | • K deficiency may occur |
| | Mg:K | <2:1 | • Mg deficiency may occur |
| | | >10:1 | • K deficiency may occur |
| | Ca:Mg | >3:1 | • Ca deficiency may occur |
| | | >8:1 | • Mg deficiency may occur |
| Soil pH | Soil pH should ideally be between 5.5 and 6.5. Soil pH < 5.5 becomes highly acidic and can produce toxic elements. Soil pH > 7.0 often limits the availability of many minor elements. For most elements, maximum availability occurs between a pH of 5.5 and 6.0. | | |
| Cation exchange capacity | CEC between 5 and 35 meq/100 g is desired. CEC measures a soil's ability to hold the cations Ca, Mg, K, H, and Na. Increasing CEC generally occurs with increasing soil OM or clay content. Generally, the higher the CEC value is, the more productive the soil is. BCSR suggested range of a soil's CEC is 65% to 75% Ca, 12% to 18% Mg, and 3% to 5% K. | | |
| Percent base saturation | This is the proportion of CEC occupied by the cations Ca, Mg, K, H, and Na. Ideally, 80% or more base saturation is from Ca, Mg, and K ions. | | |
| Iron (Fe) | Soil levels should be between 12 and 25 ppm when using Mehlich extraction and 5 to 15 ppm using DTPA extraction. Soil pH and relative levels of other elements such as P are important when interpreting Fe soil tests. Generally, Fe becomes less available in alkaline or extremely acidic soils, cool soil temperatures, or when roots become dysfunctional due to heat or diseases, and soils with excessive P or moisture levels. See Cu for additional information. | | |
| Manganese (Mn) | Soil levels of Mn should be between 2 and 10 ppm when using Mehlich or DTPA extraction. Levels where a plant response to applied Mn may occur include: 3 to 5, 5 to 7, and 7 to 9 ppm for mineral or organic soils with pH 5.5 to 6.0, 6.0 to 6.5, and 6.5 to 7.0, respectively. Deficiencies are more prone to be present on coarse, sandy, acid soils receiving excessive water. See Cu for additional information. | | |
| Zinc (Zn) | Soil levels of Zn should be between 1 and 3 ppm using Mehlich extraction and 0.8 and 1.5 ppm using DTPA extraction. Plant response to applied Zn may occur at: 0.5, 0.5 to 1.0, 1 to 3 ppm for soils with pH 5.5 to 6.0, 6.0 to 6.5, and 6.5 to 7.0, respectively. Interactions with P and soil pH can alter needed application rates. See Cu for additional information. | | |
| Copper (Cu) | Plant response to applied Cu may occur at 0.1 to 0.3, 0.3 to 0.5, and 0.5 ppm for mineral soils with pH 5.5 to 6.0, 6.0 to 6.5, and 6.5 to 7.0, respectively. Deficiencies can occur on alkaline soils, high organic matter (peat and muck) soils, soils heavily fertilized with N, P, and Zn, and when flatwood soils are first cultivated. Toxic conditions may exist when Cu levels exceed 2 to 3, 3 to 5, and 5 ppm in mineral soils with pH of 5.5 to 6.0, 6.0 to 6.5, and 6.5 to 7.0, respectively. Additional levels used by many laboratories for micronutrient availability include: | | |

| Micronutrient | Extraction Technique: DTPA — Desired Range (ppm) | Mehlich III | Ammonium Bicarbonate-DTPA |
|---|---|---|---|
| Iron | 2.6 to 5.0 | 50 to 100 | 3.1 to 5.0 |
| Manganese | 1 to 2 | – | 0.6 to 1.0 |
| • Sand soils | – | 4.0 to 6.0 | – |
| • Clay soils | – | 8.0 to 12.0 | – |
| Zinc | 0.6 to 1.0 | 1.1 to 2.0 | 1.0 to 1.5 |
| Copper | 0.2 to 0.4 | 0.3 to 2.5 | 0.3 to 0.5 |

| Element (abbreviation) | Comments |
|---|---|
| Boron (B) | Soil B levels should be between 1 and 2 ppm when using Mehlich extraction, 0.5 and 1.2 ppm using the barium chloride–hot water extraction, and 0.2 and 0.7 ppm using saturated paste extraction. Boron deficiencies occur more commonly on sandy, low organic matter soils and alkaline soils. Boron is most soluble (available) under acidic soil conditions. |
| Sulfur (S) | Soil S levels, like N, are dependent on soil OM levels, are erratic to measure, and results are often meaningless. Soils that are low in OM, are well drained, have low CEC values, and are fertilized with excessive N can develop low S levels. Acceptable S ranges are 10 to 20 ppm for $Ca(H_2PO_4)_2$, 30 to 60 ppm for ammonium acetate (pH 7.0), and 15 to 40 ppm for Mehlich III extraction. Sulfate-sulfur ($SO_4$-S) level using ammonium acetate extraction is 10 to 15 ppm. Foliar application of magnesium sulfate (Epsom salt) indicates if S deficiencies exist by greening up within 48 hours after application. |

*Sources:* Modified from Carrow, R. N. et al., 2001, *Turfgrass soil fertility and chemical properties*. Chelsea, MI: Ann Arbor Press; Jones, J. B. et al., 1991, *Plant analysis handbook*. Athens, GA: Micro-Macro Publishing, Inc.; McCarty, L. B., ed., 2011, *Best golf course management practices* (3rd ed.). Upper Saddle River, NJ: Prentice-Hall, Inc.

[a] Refer to the specific soil testing facility and report to determine which nutrient extractant procedure was used and what the generated values actually represent. Multiply ppm values by 2 to obtain pounds nutrient/acre (pounds/acre) and by 2.25 to obtain kg/ha ~7 inches (18 cm) deep.

**TABLE 7.27**
**An Example of a Soil and Tissue Analysis Laboratory Report**

Grower:______________________
Received:__________ Processed:____________________
Sample ID:_________ Grass:________________________
Phone:_______ Fax:_______ E-mail:____________

| *Element* | *Plant Tissue* | *Soil Levels (pounds/acre)* | *Test Ratings* — *P = Plant*, *S = Soil* — *Deficient* / *Low* / *Medium* / *High* |
|---|---|---|---|
| Nitrogen (N) | 3.13% | – | PPPPPPPPPPPPPPPPPPPPPPPPP |
| Phosphorus (P) | 0.32% | 129 | PPPPPPPPPPPPPPPPPPPPPPPPPPPP<br>SSSSSSSSSSSSSSSSSSSSSSSSSSSSSSSSSSSS |
| Potassium (K) | 1.61% | 111 | PPPPPPPPP<br>SSSSSSSSSSSSSSSSSSSSSSSSS |
| Calcium (Ca) | 0.47% | 584 | PPPPPPPPPPPPPPPPPPPPPPPPPPPPPPPP<br>SSSSSSSSSSSSSSSSSSSSSSSSSS |
| Magnesium (Mg) | 0.35% | 196 | PPPPPPPPPPPPPPPPPPPPPPPPPPPPPP<br>SSSSSSSSSSSSSSSSSSSSSSSSSSSSSSSSSSSS |
| Sulfur (S) | 0.12% | 46 | PPPPPPPPP<br>SSSSSSSSSSSSSSSSSSSSSSSSSSSS |
| Boron (B) | 5 ppm | 0.20 | PPPPPPPPPPPPPPPPPPPPPPPPPPP<br>SSSSSSSSSSS |
| Copper (Cu) | 7 ppm | 0.80 | PPPPPPPPPPPPPPPPPPPPPPPPPPP<br>SSSSSSSSSSSSSSSSSSSSSSSSSSS |
| Iron (Fe) | 106 ppm | 103 | PPPPPPPPPPPPPPPPPPPPPPPPPPPPPP<br>SSSSSSSSSSSSSSSSSSSSSSSSSSSSSSSSSSSS |
| Manganese (Mn) | 38 ppm | 10 | PPPPPPPPPPPPPPPPPPPPPPPPPPPPPP<br>SSSSSSSSSSSS |
| Zinc (Zn) | 24 ppm | 4.0 | PPPPPPPPPPPPPPPPPPPPPPP<br>SSSSSSSSSSSSSSSSSSSSSSSSSSSS |
| Sodium (Na) | – | 11 ppm | SSS |
| Soluble salts | – | 0.080 mmho/cm | SSS |

| | | | | | |
|---|---|---|---|---|---|
| Soil pH (or active acidity): | 6.7 | | | | |
| Buffer pH (or lime index): | 7.9 | % K: | 2.5 | % H: | 66.7 |
| Organic Matter: | 1.1% | % Ca: | 24.0 | % Na: | 0.8 |
| Calculated CEC: | 6.0 meq/100 g | % Mg: | 6.5 | | |

Lime and fertilizer recommendations:
1. Lime (pounds/1,000 ft$^2$) 3. Phosphate (pounds $P_2O_5$/1,000 square feet)____________ 5. Other
2. Nitrogen (pounds N/1,000 square feet)_________4. Potash (pounds $K_2O$/1,000 square feet)______________

To convert tissue nutrient analysis from percent to grams/kilogram, multiply percent values by 10.
To convert micronutrient tissue analysis from parts per million to milligrams/kilogram, multiply parts per million by 1.
To convert pounds nutrient/acre for a 6-inch furrow slice to milligrams/kilogram (or parts per million), multiply pounds/acre by 2.
To convert pounds nutrient/acre for a 6-inch furrow to grams/square meter, multiply pounds/acre by 2.25.
To convert soil parts per million P to pounds $P_2O_5$/ac, multiply parts per million P by 4.6.
To convert soil parts per million K to pounds $K_2O$/acre, multiply parts per million K by 2.4.
To convert from pounds nutrient/acre to parts per million, divide pounds nutrient/acre by 2.
Cation exchange capacity is determined by summing the major cations:
CEC (meq/100 g) = $Ca^{+2} + Al^{+3} + K^{+} + H^{+} + Mg^{+2}$
Percent base saturation is determined by:

$$\% \text{ BS} = \frac{\text{meq of } Ca^{+2}, Mg^{+2}, K^{+}, \text{and } Na^{+}}{\text{CEC}} \times 100$$

**FIGURE 7.22** Leaf tissue analysis is a means to determine if sufficient nutrients are present in plants.

## LEAF TISSUE ANALYSIS

Leaf tissue analysis, in addition to a soil analysis, is another tool turfgrass managers can use to help develop or improve nutrient applications (Figure 7.22). It is most useful for intensively managed areas like golf greens and/or where the rootzone was constructed using a sand-based system having a relatively low nutrient-holding capacity. A tissue test determines if hidden nutrient deficiencies exist and determines how closely nutrients absorbed by the turf correlate with the soil test value. Unlike an index value generated with soil testing, tissue analysis reflects the actual concentration of nutrients in the sample.

Primary and secondary nutrients occur in relatively large quantities within plants; their concentrations are usually expressed in grams of the element per kilogram of plant dry weight or percentages. Micronutrients occur in relatively small quantities; their concentrations are usually expressed in milligrams of the element per kilogram of plant dry weight (also parts per million). Table 7.28 provides general guidelines for appropriate nutrient concentrations in leaf tissue for the common turfgrasses used on golf courses.

## FERTILIZER CALCULATIONS

The standard method for expressing fertilizer nutrient concentrations is "percentage by weight," whether the fertilizer is in liquid or solid form. For liquids, the weight must be determined before the amount of nutrients in solution can be known.

The percentages of most fertilizer nutrients, such as nitrogen (N), are expressed on an elemental basis. The percentages of phosphorus (P) and potassium (K) are expressed as the compounds phosphate ($P_2O_5$) and potash ($K_2O$), respectively. For example, a 5-5-10 fertilizer has a N-P-K ratio of 1-1-2. This 5-5-10 contains 5% N, 5% phosphate, and 10% potash. A 100-pound bag would contain 5 pounds N, 5 pounds $P_2O_5$, and 10 pounds $K_2O$. To convert $P_2O_5$ to elemental P and $K_2O$ to elemental K, the following conversions are used:

$P_2O_5$ contains 44% actual phosphorus (P)
$K_2O$ contains 83% actual potassium (K)

Turfgrass fertilizer programs are typically constructed around a target annual N requirement (e.g., 5 pounds N/1,000 square feet annually). Individual application rates are typically expressed as pounds N/1,000 square feet or per acre.

### Calculating Nutrient Percentages

To determine the percentage of a particular nutrient in a fertilizer product, the fertilizer chemical (molecular) formula and nutrient (atomic) weights are needed. For example, to determine the percentage of actual N and P in diammonium

**TABLE 7.28**
**Specific Ranges for Nutrients from Various Turfgrass Tissue Analysis**

| | | Bermudagrass | | | | | |
|---|---|---|---|---|---|---|---|
| | | Greens/Tees | | | Fairways | | |
| Element | | Low | Desired | High | Low | Desired | High |
| | | Percent | | | | | |
| Primary nutrients | N | 3.50 to 3.99 | 4.00 to 4.50 | >6.00 | 2.50 to 2.99 | 3.25 to 4.00 | >5.00 |
| | P | 0.15 to 0.24 | 0.25 to 0.35 | >0.60 | 0.12 to 0.14 | 0.25 to 0.35 | >0.50 |
| | K | 1.00 to 1.49 | 1.50 to 2.00 | >4.00 | 0.70 to 0.99 | 1.45 to 1.75 | >4.00 |
| Secondary nutrients | Ca | 0.30 to 0.49 | 0.50 to 0.60 | >1.00 | 0.30 to 0.45 | 0.45 to 0.50 | >1.00 |
| | Mg | 0.10 to 0.30 | 0.30 to 0.40 | >0.40 | 0.10 to 0.12 | 0.25 to 0.35 | >0.50 |
| | S | 0.15 to 0.50 | 0.50 to 0.60 | >0.60 | 0.12 to 0.14 | 0.45 to 0.50 | >0.50 |
| Micronutrients | | Parts Per Million | | | | | |
| | Al | – | <1,500 | – | – | <1,500 | >1,500 |
| | B | 4 to 5 | 15 to 20 | >30 | 4 to 5 | 10 to 20 | >30 |
| | Cu | 3 to 4 | 10 to 20 | >50 | 3 to 4 | 10 to 15 | >50 |
| | Fe | 40 to 49 | 300 to 400 | >400 | 40 to 49 | 250 to 450 | >450 |
| | Mn | 16 to 24 | 80 to 100 | >300 | 16 to 24 | 80 to 100 | >300 |
| | Zn | 15 to 40 | 40 to 80 | >250 | 15 to 19 | 40 to 80 | >250 |

| | | Creeping Bentgrass | | | | Perennial Ryegrass | | |
|---|---|---|---|---|---|---|---|---|
| | | Low | Desired | | High | Low | Desired | High |
| Element | | | Spring | Summer | | | | |
| | | Percent | | | | | | |
| Primary nutrients | N | <4.5 | 5.00 to 6.00 | 4.00 to 5.25 | >6.00 | 4.00 to 4.49 | 4.50 to 5.00 | >5.00 |
| | P | <0.3 | 0.35 to 0.40 | 0.35 to 0.60 | >0.60 | 0.30 to 0.34 | 0.35 to 0.40 | >0.40 |
| | K | 1.8 to 2.1 | 2.00 to 3.00 | 2.25 to 3.00 | >3.00 | 0.70 to 1.99 | 2.00 to 2.50 | >2.50 |
| Secondary nutrients | Ca | <0.45 | 0.45 to 0.60 | 0.45 to 0.60 | >0.75 | 0.20 to 0.24 | 0.25 to 0.30 | >0.30 |
| | Mg | <0.25 | 0.30 to 0.40 | 0.30 to 0.40 | >0.40 | 0.13 to 0.15 | 0.16 to 0.20 | >0.20 |
| | S | <0.30 | 0.50 to 0.60 | 0.50 to 0.60 | >0.70 | 0.22 to 0.26 | 0.27 to 0.32 | >0.32 |
| Micronutrients | | Parts Per Million | | | | | | |
| | Al | – | <1,500 | <1,500 | >1,500 | – | – | – |
| | B | <8 | 15 to 25 | 15 to 25 | >25 | <9.0 | 9 to 17 | >17 |
| | Cu | <8 | 10 to 20 | 10 to 15 | >30 | 4 to 5 | 6 to 7 | >8 |
| | Fe | <100 | 300 to 450 | 300 to 450 | >450 | <40 | 40 to 60 | >60 |
| | Mn | <50 | 90 to 130 | 90 to 130 | >150 | <2.0 | 2 to 10 | >10 |
| | Zn | <25 | 50 to 80 | 50 to 80 | >80 | 10 to 13 | 14 to 20 | >20 |

| | | Kentucky Bluegrass | | | Tall Fescue | | |
|---|---|---|---|---|---|---|---|
| Element | | Low | Desired | High | Low | Desired | High |
| | | Percent | | | | | |
| Primary nutrients | N | 2.01 to 2.59 | 2.60 to 3.50 | >3.50 | 2.50 to 3.39 | 3.40 to 3.80 | >3.80 |
| | P | 0.18 to 0.27 | 0.28 to 0.40 | >0.40 | 0.24 to 0.33 | 0.34 to 0.45 | >0.45 |
| | K | 1.50 to 1.99 | 2.00 to 3.00 | >3.00 | 2.20 to 2.99 | 3.00 to 4.00 | >4.00 |

*(Continued)*

## TABLE 7.28 (CONTINUED)
## Specific Ranges for Nutrients from Various Turfgrass Tissue Analysis

| | | | | |
|---|---|---|---|---|
| | | **Grams per Kilogram** | | |
| Secondary nutrients | Ca | 5 to 12.5 | 5 to 12.5 | |
| | Mg | 2 to 6 | 2 to 6 | |
| | S | 2 to 4.5 | 2 to 4.5 | |
| Micronutrients | | **Milligrams/Kilogram or Parts per Million** | | |
| | Fe | 35 to 100 mg/kg | 35 to 100 | |
| | Mn | 25 to 150 | 25 to 150 | |
| | Zn | 20 to 55 | 20 to 55 | |
| | Cu | 5 to 20 | 5 to 20 | |
| | B | 10 to 60 | 10 to 60 | |
| | Mo | 2 to 8 | 2 to 8 | |
| | | **St. Augustinegrass** | **Seashore Paspalum** | **Zoysiagrass** |
| **Element** | | **Desired** | **Desired** | **Desired** |
| | | **Percent** | | |
| Primary nutrients | N | 1.90 to 3.00 | 2.80 to 3.50 | 2.70 to 3.50 |
| | P | 0.20 to 0.50 | 0.30 to 0.60 | 0.11 to 0.39 |
| | K | 2.50 to 4.00 | 2.00 to 4.00 | 0.38 to 1.51 |
| Secondary nutrients | Ca | 0.30 to 0.50 | 0.25 to 1.50 | 0.31 to 0.54 |
| | Mg | 0.15 to 0.25 | 0.25 to 0.60 | 0.11 to 0.25 |
| | S | 0.20 to 0.45 | 0.20 to 0.60 | 0.20 to 0.45 |
| Micronutrients | | **Milligrams/Kilogram or Parts per Million** | | |
| | Fe | 50 to 300 | 50 to 500 | 35 to 100 |
| | Mn | 40 to 250 | 50 to 300 | 25 to 150 |
| | Zn | 20 to 100 | 20 to 250 | 20 to 55 |
| | Cu | 10 to 20 | 5 to 50 | 5 to 20 |
| | B | 5 to 10 | 5 to 60 | 10 to 60 |
| | Mo | 2 to 8 | 0.5 to 1.0 | 2 to 8 |

*Sources:* Modified from Carrow, R. N. et al., 2001, *Turfgrass soil fertility and chemical properties*. Chelsea, MI: Ann Arbor Press; Jones, J. B. et al., 1991, *Plant analysis handbook*. Athens, GA: Micro-Macro Publishing, Inc.; McCarty, L. B., ed., 2011, *Best golf course management practices* (3rd ed.). Upper Saddle River, NJ: Prentice-Hall, Inc.

phosphate (DAP), one needs to know the fertilizer chemical formula [$(NH_4)_2HPO_4$] and atomic weights of N (14), H (1), P (31), and $O_2$ (17). Thus, the total atomic mass for DAP is equivalent to 2[14 + (4 × 1)] + 1 + 31 + (4 × 17) = 136. The percentages of N and P are then calculated by adding up the total atomic weight of each element in DAP and determining what percentage of this total is N and P.

Table 7.29 provides conversion of percentage of an element in a fertilizer.

$$\%\,N = \frac{2N}{(NH_4)_2HPO_4} \times 100 = \frac{2 \times 14}{2[14 + (4 \times 1) + 1 + 31 + (4 \times 17)]} \times 100 = 21\%\,N$$

$$\%\,P = \frac{2N}{(NH_4)_2HPO_4} \times 100 = \frac{31}{2[14 + (4 \times 1) + 1 + 31 + (4 \times 17)]} \times 100 = 23\%\,P$$

**TABLE 7.29**
**Conversion of Percentage of an Element in a Fertilizer**

| A | B | A to B | B to A |
|---|---|---|---|
| Ammonia ($NH_3$) | Nitrogen (N) | 0.8224 | 1.2159 |
| Nitrate ($NO_3$) | Nitrogen (N) | 0.2259 | 4.4266 |
| Ammonium nitrate ($NH_4NO_3$) | Nitrogen (N) | 0.3500 | 2.8572 |
| Ammonium sulfate [$(NH_4)_2SO_4$] | Nitrogen (N) | 0.2120 | 4.7168 |
| Calcium ammonium nitrate solution (CAN 17) [$Ca(NO_3)_2 \cdot NH_4NO_3$] | Nitrogen (N) | 0.1707 | 5.8572 |
| Potassium nitrate ($KNO_3$) | Nitrogen (N) | 0.1386 | 7.2176 |
| Sodium nitrate ($NaNO_3$) | Nitrogen (N) | 0.1648 | 6.0679 |
| Monoammonium phosphate (MAP) [$NH_4H_2PO_4$] | Nitrogen (N) | 0.1218 | 8.2118 |
| Diammonium phosphate (DAP) [$(NH_4)_2HPO_4$] | Nitrogen (N) | 0.2121 | 4.7138 |
| Urea [$CO(NH_2)_2$] | Nitrogen (N) | 0.4665 | 2.1437 |
| Phosphoric acid ($P_2O_5$) | Phosphorus (P) | 0.4364 | 2.2914 |
| Phosphate ($PO_4$) | Phosphorus (P) | 0.3261 | 3.0662 |
| Monoammonium phosphate (MAP) [$NH_4H_2PO_4$] | Phosphoric acid ($P_2O_5$) | 0.6170 | 1.6207 |
| Diammonium phosphate (DAP) [$(NH_4)_2HPO_4$] | Phosphoric acid ($P_2O_5$) | 0.5374 | 1.8607 |
| Potash ($K_2O$) | Potassium (K) | 0.8301 | 1.2046 |
| Muriate of potash (KCl) | Potash ($K_2O$) | 0.6317 | 1.5828 |
| Potassium nitrate ($KNO_3$) | Potash ($K_2O$) | 0.4658 | 2.1466 |
| Potassium sulfate ($K_2SO_4$) | Potash ($K_2O$) | 0.5405 | 1.8499 |
| Potassium carbonate ($K_2CO_3$) | Potash ($K_2O$) | 0.6816 | 1.4672 |
| Gypsum ($CaSO_4 \cdot 2H_2O$) | Calcium (Ca) | 0.2326 | 4.3000 |
| Gypsum ($CaSO_4 \cdot 2H_2O$) | Calcium sulfate ($CaSO_4$) | 0.7997 | 1.2647 |
| Gypsum ($CaSO_4 \cdot 2H_2O$) | Calcium oxide (CaO) | 0.3257 | 3.0702 |
| Calcium oxide (CaO) | Calcium (Ca) | 0.7147 | 1.3992 |
| Calcium carbonate ($CaCO_3$) | Calcium (Ca) | 0.4004 | 2.4973 |
| Calcium carbonate ($CaCO_3$) | Calcium oxide (CaO) | 0.5604 | 1.7848 |
| Magnesium oxide (MgO) | Magnesium (Mg) | 0.6032 | 1.6579 |
| Magnesium sulfate ($MgSO_4$) | Magnesium (Mg) | 0.2020 | 4.9501 |
| Epsom salt ($MgSO_4 \cdot 7H_2O$) | Magnesium (Mg) | 0.0987 | 10.1350 |
| Sulfate ($SO_4$) | Sulfur (S) | 0.3333 | 3.0000 |
| Ammonium sulfate [$(NH_4)_2SO_4$] | Sulfur (S) | 0.2426 | 4.1211 |
| Gypsum ($CaSO_4 \cdot 2H_2O$) | Sulfur (S) | 0.1860 | 5.3750 |
| Magnesium sulfate ($MgSO_4$) | Sulfur (S) | 0.3190 | 3.1350 |
| Potassium sulfate ($K_2SO_4$) | Sulfur (S) | 0.1837 | 5.4438 |
| Sulfuric acid ($H_2SO_4$) | Sulfur (S) | 0.3269 | 3.0587 |
| Borax ($Na_2B_4O_7 \cdot 10H_2O$) | Boron (B) | 0.1134 | 8.8129 |
| Boron trioxide ($B_2O_3$) | Boron (B) | 0.3107 | 3.2181 |
| Copper sulfate ($CuSO_4$) | Copper (Cu) | 0.3981 | 2.5119 |
| Copper sulfate ($CuSO_4 \cdot 5H_2O$) | Copper (Cu) | 0.2545 | 3.9293 |
| Ferric sulfate [$(Fe_2(SO_4)_3$] | Iron (Fe) | 0.2793 | 3.5804 |
| Ferrous sulfate ($FeSO_4 \cdot H_2O$) | Iron (Fe) | 0.3150 | 3.1746 |
| Ferrous sulfate ($FeSO_4 \cdot 7H_2O$) | Iron (Fe) | 0.2009 | 4.9776 |
| Manganese sulfate ($MnSO_4$) | Manganese (Mn) | 0.3638 | 2.7486 |
| Manganese sulfate ($MnSO_4 \cdot 4H_2O$) | Manganese (Mn) | 0.2463 | 4.0602 |
| Sodium molybdate ($Na_2MoO_4 \cdot 2H_2O$) | Molybdenum (Mo) | 0.3965 | 2.5218 |
| Sodium nitrate ($NaNO_3$) | Sodium (Na) | 0.2705 | 3.6970 |
| Sodium chloride (NaCl) | Sodium (Na) | 0.3934 | 2.5417 |
| Zinc oxide (ZnO) | Zinc (Zn) | 0.8034 | 1.2447 |
| Zinc sulfate ($ZnSO_4$) | Zinc (Zn) | 0.4050 | 2.4693 |
| Zinc sulfate ($ZnSO_4 \cdot H_2O$) | Zinc (Zn) | 0.3643 | 2.7450 |

*Notes:* Multiply "A to B" to determine the amount of B that constitutes A, and "B to A" to find the equivalent of material B in terms of A. For example, to convert the amount of P in a 16-25-0, multiple 25% $P_2O_5$ by 0.4364 to obtain 10.9% P. To determine the percent MgO in an 11% Mg fertilizer, multiply 11 by 1.6579 to obtain 18.2% MgO.

## Granular Fertilizers

To find the amount of material needed to supply a certain amount of nutrient to a given area, divide the target application rate needed (e.g., 1 pound N/1,000 square feet) by the decimal fraction of the percent nutrient in the fertilizer source (e.g., 21% = 0.21):

$$\frac{\text{Rate of nutrient desired in a given area}}{\text{percent nutrient in fertilizer}} \times \text{total area to be treated}$$

### Example

A superintendent wishes to supply 1 pound N/1,000 square feet using a slow-release 20-5-10 fertilizer. How much 20-5-10 fertilizer is needed per 1,000 square feet to provide this?

$$\frac{\text{Rate of nutrient wanted}}{\text{percent nutrient in fertilizer}} = \frac{\text{1 pound N/1,000 square feet}}{0.2}$$

$$= \text{5 pounds 20-5-10 needed per 1,000 square feet}$$

The amount of $P_2O_5$ and $K_2O$ in this 5 pounds of 20-5-10 includes:

$P_2O_5$: 5 pounds fertilizer × 0.05 = 0.25 pound $P_2O_5$
$K_2O$: 5 pounds fertilizer × 0.10 = 0.5 pound $K_2O$

The amount of actual P and K in these 5 pounds includes:

P: 0.25 pound $P_2O_5$ × 0.44 = 0.11 pound P
K: 0.5 pound $K_2O$ × 0.83 = 0.41 pound K

## Liquid Fertilizers

To use liquid fertilizer, first find the weight of the liquid, usually listed on the product label as the weight of a single gallon. This is then multiplied by the percentages of the nutrient in the gallon.

### Example

Determine how much N, $P_2O_5$, and $K_2O$ are in 1 gallon of 12-4-8 liquid fertilizer weighing 10 pounds/gallon. How much of this liquid fertilizer is needed to apply 0.2 pound N/1,000 square feet?

**Step 1:** Determine actual weight of each nutrient per gallon of fertilizer by multiplying fertilizer weight by the decimal fractions of the nutrient percentages:

$$\text{N: 0.12(or 12\%)N} \times \text{10 pounds} = \text{1.2 pounds N/gallon}$$

$$P_2O_5\text{: 0.04(or 4\%)}P_2O_5 \times \text{10 pounds} = \text{0.4 pound } P_2O_5\text{/gallon}$$

$$\text{actual P: } 0.4 \times 0.44 = \text{0.176 pound P/gallon}$$

$$K_2O\text{: 0.08(or 8\%)}K_2O \times \text{10 pounds} = \text{0.8 pound } K_2O\text{/gallon}$$

$$\text{actual K: } 0.8 \times 0.83 = \text{0.664 pound K/gallon}$$

**Step 2:** Determine how much fertilizer should be applied by dividing the desired rate of 0.2 pound N/1,000 square feet by the actual amount of N per gallon of fertilizer:

$$\frac{\text{0.2 pound N/1,000 square feet}}{\text{1.2 pounds N/gal}} = \text{0.167 gallon required per}$$

1,000 square feet to deliver 0.2 pound N/1,000 square feet

### Example

1. How many pounds of an 18-6-12 granular fertilizer are required to fertilize a 7,500 square foot green at the rate of 0.75 pound N/1,000 square feet?
   **Step 1:** Determine how many pounds of the 18-6-12 fertilizer are needed to obtain 1 pound N. Since 18 is the desired percentage, it is divided into 1:

$$\frac{\text{0.75 pound N/1,000 square feet}}{\text{0.18 (or 18\%) N in the fertilizer}} = \text{4.2 lb of 18-6-12 gallon required per}$$

1,000 square feet to deliver 0.75 pound N/1,000 square feet

This means 4.2 pounds 18-6-12 contain 3/4 pound N. This may be checked by multiplying the answer, 4.2, by the decimal fraction of the percent N, 18% or 0.18:

$$\text{4.2 pounds} \times 0.18 = \text{0.75 pound N}$$

**Step 2:** Since 4.2 pounds 18-6-12 provides 3/4 pound N, multiply this value by the size of the area being treated. In this case, 7.5 units of 1,000 are in 7,500 square feet:

$$\text{4.2 pounds} \times 7.5 = \text{31.5 pound of 18-6-12 fertilizer}$$

provides $\frac{3}{4}$ pound N/1,000 square feet over a

7,500 square foot golf green

Another way to set this up involves:

$$\frac{4.2\text{ pounds}}{1{,}000\text{ square feet}} \times \frac{7{,}500\text{ square feet}}{\text{green}} = \frac{31.5\text{ pounds}}{\text{green}}$$

In summary, 31.5 pounds of 18-6-12 will provide 0.75 pound N on 7,500 square feet.

2. How many pounds of a 20-2-8 fertilizer are needed to apply 1 pound N/1,000 square feet on a 2.3 acre fairway?

**Step 1:** First, determine how many square feet are in 2.3 acres. One acre has 43,560 square feet:

$$43{,}560\text{ square feet/acre} \times 2.3\text{ acre}$$
$$= 100{,}188\text{ square feet}$$

**Step 2:** Next, determine how many pounds of 20-2-8 will provide 1 pound N/1,000 square feet:

$$\frac{1}{0.2} = 5\text{ pounds of 20-2-8}$$

**Step 3:** Finally, determine the total amount of fertilizer needed over the 2.3 acres:

5 pounds needed/1,000 square feet × 100.118 (1,000s) in 2.3 acres = 500 pounds total 20-2-8 needed to provide 1 pound N/1,000 square feet over 2.3 acres

3. How much of a 15-6-9 liquid fertilizer is required to apply 0.4 of a pound N/1,000 square feet on a 6,000 square foot green?

**Step 1:** The weight of the fertilizer is needed. The fertilizer label states 1 gallon of this liquid fertilizer weighs 11.2 pounds. If 1 gallon of fertilizer weighs 11.2 pounds and is 15% N, then 1 gallon contains 1.68 pounds N.

$$15\%\text{N} \times 11.2\text{ pounds} = 1.68\text{ pounds N}$$

**Step 2:** Since 1 gallon contains 1.68 pounds N, then 0.238 gallons or 30.4 ounces contain 0.4 pound N:

$$\frac{0.4\text{ pound N}}{1{,}000\text{ square feet}} \times \frac{1\text{ gallon fertilizer}}{1.68\text{ pounds N}}$$
$$= \frac{0.238\text{ gallon fertilizer}}{1{,}000\text{ square feet}} \times \frac{128\text{ ounces}}{\text{gallon}}$$
$$= \frac{30.4\text{ ounces}}{1{,}000\text{ square feet}}$$

**Step 3:** The green is 6,000 (6 × 1,000) square feet, so the total amount of liquid fertilizer needed is 6 × 0.238 gallon, or about 1.4 gallons to supply 0.4 pound of N per 1,000 square feet over 6,000 square feet.

4. If all other factors are equal, which of the following is the "best buy"?

| | Fertilizer A | Fertilizer B | Fertilizer C |
|---|---|---|---|
| Analysis: | 12-4-8 | 24-8-16 | 36-12-24 |
| Cost/ton: | $120 | $216 | $338 |

In order to compare the costs, first determine the cost of a unit of N for each fertilizer:

**Fertilizer A** is a 12% (0.12) N fertilizer. In 1 ton (2,000 pounds), there are 240 pounds N (0.12 × 2,000). If 240 pounds cost $120, then 1 pound costs:

$$\frac{\$120}{240\text{ pounds}} = \$0.50\text{ per pound of N}$$

**Fertilizer B** is a 24% (0.24) N fertilizer. In 1 ton, there are 480 pounds N (0.24 × 2,000). If 480 pounds cost $216, then 1 pound costs:

$$\frac{\$216}{480\text{ pounds}} = \$0.45\text{ per pound of N}$$

**Fertilizer C** is a 36% (0.36) N fertilizer. In 1 ton, there are 720 pounds N (0.36 × 2,000). If 720 pounds cost $338, then 1 pound costs:

$$\frac{\$338}{720\text{ pounds}} = \$0.47\text{ per pound of N}$$

Therefore, in this example, **fertilizer B,** at $0.45/pound of N, is the best buy.

5. How much of the following fertilizer material is required to apply a 3-1-2 ratio of N-P-K at a rate of 1 pound N/1,000 square feet on a 7,000 square foot green?
   A. 38-0-0 (urea formaldehyde)
   B. 0-20-0 (superphosphate)
   C. 0-0-50 (potassium sulfate)

A. Since urea formaldehyde is 38% (0.38) N, 1 ÷ 0.38 or 2.6 pounds of UF contain 1 pound N:

$$2.6\text{ pounds (UF)} \times 7\text{ (from 7,000 square feet)}$$
$$= 18.2\text{ total pounds}$$

B. The ratio of nutrients in this problem is 3-1-2. This means that, for every three units of N, one unit of $P_2O_5$ is needed, or for every unit of N, one-third unit of $P_2O_5$ is needed. Since the rate of N application is

1 pound/1,000 square feet, then the rate of $P_2O_5$ application is 0.33 pound/1,000 square feet. Superphosphate is 20% (or 0.20) $P_2O_5$:

$$\frac{1}{0.20} = 5 \text{ pounds of 0-20-0 will contain 1 pound of } P_2O_5$$

One-third pound of $P_2O_5$/1,000 square feet is needed, so:

$$0.33 \text{ pound} \times 5 = 1.65 \text{ pounds of 0-20-0 provides } 0.33 \text{ pounds } P_2O_5 \text{ per 1,000 square feet}$$

$$1.65 \text{ pounds} \times 7 \text{ (from 7,000 square feet)} = 11.5 \text{ total pounds 0-20-0 required to apply 0.33 pound } P_2O_5/1,000 \text{ square feet on the 7,000 square foot green}$$

C. The ratio of nutrients is 3-1-2. This means that, for every three units of N, two units of $K_2O$ are needed, or for every unit of N, two-thirds unit of $K_2O$ is needed. Since the rate of N application is 1 pound/1,000 square feet, the rate of $K_2O$ application is 2/3 pound/1,000 square feet. Potassium sulfate is 50% (or 0.50) $K_2O$:

$$\frac{1}{0.50} = 2 \text{ pounds of 0-0-50 will contain 1 pound of } K_2O$$

Since 2/3 pound of $K_2O$ per 1,000 square feet is needed, 2/3 (or 0.67) × 2 = 1.34 pounds 0-0-50 will provide 0.67 pound $K_2O$/1,000 square feet. Since our green is 7,000 square feet, multiply by 7:

$$1.34 \text{ pounds} \times 7 \text{ (from 7,000 square feet)} = 9.4 \text{ total pounds of 0-0-50 required to apply 0.67 pound } K_2O/1,000 \text{ square feet on the green}$$

The answer to this problem is therefore: 18.2 pounds 38-0-0, 11.5 pounds 0-20-0, and 9.4 pounds 0-0-50.

## FERTILIZER INJECTION (OR FERTIGATION) FORMULAS

For stationary irrigation systems, the formulas in the following examples are used to determine various calculations needed for fertigation.

### Example

**A.** $$\text{Hours of injection operation} = \frac{\text{acres irrigated} \times \text{pounds fertilizer to apply}}{\text{\% active ingredients of fertilizer} \times \text{pounds/gallon of fertilizer} \times \text{gallons per hour injected}}$$

A golf course sprinkler system covers 22 acres per setting. The injector pump delivers 50 gallons per hour (gph). The superintendent wants to apply 60 pounds N/acre using a 30-0-0 fertilizer weighing 10 pounds/gallon. How long should the injection pump be turned on?

$$\text{Hours of injection operation} = \frac{\text{acres irrigated} \times \text{pounds fertilizer to apply}}{\text{\% active ingredients of fertilizer} \times \text{pounds/gallon of fertilizer} \times \text{gallons per hour injected}}$$

$$= \frac{22 \times 60}{0.30 \times 10 \times 50} = 8.8 \text{ hours or 8 hours and 48 minutes}$$

**B.** $$\text{Gallons per hour to inject} = \frac{\text{acres irrigated} \times \text{pounds fertilizer to apply}}{\text{\% active ingredients of fertilizer} \times \text{pounds/gallon of fertilizer} \times \text{hours of injection}}$$

A superintendent irrigates 12 acres of fairways in six hours. He wishes to apply 60 pounds N in that six-hour period. He is using a 32% N source that weighs 10 pounds/gallon. How many gallons per hour should the injector pump deliver?

$$\text{Gallons per hour to inject} = \frac{12 \times 60}{0.32 \times 10 \times 6} = 37.5 \text{ gallons per hour to inject}$$

Other useful chemical injection formulas are included in the following examples.

**C.** $$\text{Gallons per hour to inject} = \frac{\text{acres irrigated} \times \text{pints/acre}}{\text{hours of injection} \times 8 \text{ pints/gallon}}$$

A superintendent wishes to inject a soil wetting agent into his chemigation system at the application rate of 6 pints/acre. He plans on diluting this rate into 100 gallons. His irrigation system covers 10 acres in four hours.

**Step 1:** The dilution rate must be determined. The total amount of wetting agent needed for 10 acres is

$$6 \text{ pints/acre} \times 10 \text{ acres} = 60 \text{ pints}$$

**Step 2:** Determine the total diluted rate needed for the 10 acres:

$$\frac{100 \text{ gallons}}{4 \text{ hours}} = 25 \text{ gallons or } 200 \text{ pints } (25 \text{ gallons} \times 8 \text{ pints/gallon})$$

**Step 3:** Determine how many gallons per hour the injection pump delivers to apply 200 pints/acre of diluted chemicals in this four-hour set:

$$\frac{10 \text{ acres} \times 200 \text{ pints/acre}}{4 \text{ hours} \times 8 \text{ pints/gallon}} = 62.5 \text{ gallons/hour}$$

**D.** $$\text{Hours of injection} = \frac{\text{acres irrigated} \times \text{pints/acre}}{\text{gallons/hour injected} \times 8 \text{ pints/gallon}}$$

A golf course green irrigation system is designed to cover 2.3 acres. A chemigation injection pump is set to deliver 2 gallons per hour and the superintendent wishes to apply a fungicide at 10 pints/acre. For how long should the injection pump be turned on?

$$\text{hours of injection} = \frac{\text{acres irrigated} \times \text{pints/acre}}{\text{gallons per hour injected} \times 8 \text{ pints/gallon}} = \frac{2.3 \text{ acres} \times 10 \text{ pints/acre}}{2 \text{ gallons/hour} \times 8 \text{ pints/gallon}}$$
$$= 1.4 \text{ hours or 1 hour and 24 minutes}$$

To determine the amount of nutrient(s) being applied in an effluent irrigation source or from fertigation, use the following formula:

**E.** $$\frac{\text{pounds nutrients applied}}{\text{acre}} = \text{parts per million of element in water} \times \text{acre-inches water applied} \times 0.226464$$

If 6 acre-inches of effluent water containing 150 ppm N are applied monthly, how many pounds N are applied per acre?

$$\frac{\text{pounds nutrients applied}}{\text{acre}} = \text{parts per million of element in water} \times \text{acre-inches water applied} \times 0.226464$$
$$= 150 \times 6 \text{ acre-inches} \times 0.226464 = 204 \text{ pounds N/ac}$$

## Parts per Million

In some instances, fertilizers and pesticides are applied in dilutions as parts per million. In this case, the active ingredient (percent active ingredient) of the material must be known. Several methods are then used to determine the amount of material needed in a given amount of water.

### Example

$$\textbf{A.}\ \text{fluid ounces to use per tank} = \frac{\text{parts per million desired} \times \text{gallons/tank} \times 8.34 \times 128}{1{,}000{,}000 \times \text{pounds active ingredient/gallon}}$$

or

$$\text{pounds to use per tank} = \frac{\text{parts per million desired} \times \text{gal/tank} \times 8.34}{1{,}000{,}000 \times \text{percent active ingredient}}$$

or

$$\text{parts per million} = \frac{\text{pounds to use/tank} \times 1{,}000{,}000 \times \text{percent active ingredient}}{\text{gallons/tank} \times 8.34}$$

A liquid fertilizer is to be applied at 50 ppm. How many fluid ounces of a 20-0-0 material is needed in a 2-gallon tank? This fertilizer weighs 10.5 pounds/gallon.

**Step 1:** First determine the pounds of active ingredient fertilizer per gallon of material:

$$\frac{10.5\ \text{pounds}}{\text{gallon}} \times 20\%\ \text{N (or 0.20)} = \frac{2.1\ \text{pounds active ingredient N}}{\text{gallon}}$$

**Step 2:** Next, insert the information into the formula:

$$\text{fluid ounces to use per tank} = \frac{\text{parts per million desired} \times \text{gallons/tank} \times 8.34 \times 128}{1{,}000{,}000 \times \text{pounds active ingredient/gallon}}$$

$$= \frac{50\ \text{ppm} \times 2\ \text{gallons/tank} \times 8.34 \times 128}{1{,}000{,}000 \times 2.1\ \text{pounds active ingredient N/gallon}} = 0.051\ \text{oz}$$

Another method is based on the conversion of 1 ounce per 100 gallons equals 78 ppm [1 ounce = 0.0078 gallon; (0.0078 ÷ 100) × 1 million = 78 ppm]. By using this, one determines the parts per million of an element in a fertilizer by multiplying the percent of the element by 78. The answer will be the ppm of the element per ounce of the fertilizer in 100 gallons of water:

$$\textbf{B.}\ \text{Fluid ounces to use per 100-gallon tank} = \frac{\text{parts per million desired}}{\text{percent active ingredient of fertilizer} \times 78}$$

A weight relationship can also be used to determine parts per million by using the following proportion:

$$\textbf{C.}\ \text{Parts per million} = \frac{\text{weight (e.g., pounds) to use} \times 1{,}000{,}000}{\text{weight of tank mixture (e.g., pounds)}}$$

or

$$\text{weight of material to be used} = \frac{\text{parts per million desired} \times \text{weight of tank mixture}}{1{,}000{,}000}$$

How much of a 20-20-20 fertilizer should be used to obtain a 200 ppm N solution in a 100-gallon tank?

$$\text{Fluid ounces to use per 100-gallon tank} = \frac{\text{parts per million desired}}{\text{percent active ingredient of fertilizer} \times 78}$$

$$= \frac{200\ \text{ppm}}{0.20 \times 78} = 12.8\ \text{ounces}$$

If an insecticide is added at 20.5 ounces to 50 gallons of water, what is the resulting parts per million?

$$\text{Parts per million} = \frac{\text{weight (e.g., pounds) to use} \times 1{,}000{,}000}{\text{weight of tank mixture (e.g., pounds)}}$$

$$= \frac{20.5\ \text{ounces} \times 1{,}000{,}000}{6{,}672\ \text{ounces (from 50 gallons} \times 8.43\ \text{pounds/gallon} \times 16\ \text{ounces/pound)}} = 3{,}073$$

Remember, this last equation is based solely on weight and does not account for active ingredients. Also, the weight of the material to be used must be in the same unit as the weight of the tank mixture.

Systems often have flowing water and inject a mixture at the proper rate to maintain a desired concentration level of a chemical. The following determines the injection rate necessary to maintain the desired concentration of a chemical:

**D.**

$$\text{Injection rate (gallons per minute)} = \frac{\text{parts per million desired} \times \text{water supply flow rate (gallons per minute)} \times 83}{\left[\text{percent active ingredients fertilizer} \times \text{specific weight of stock Fe solution (pounds/gallon)} \times 10{,}000\right] - \left[\text{parts per million desired} \times \text{specific weight of stock solution mix (pounds/gallon)}\right]}$$

Iron is to be injected to provide 10 ppm of free Fe into an irrigation system that has a flow rate of 500 gallons per minute. The Fe stock solution contains 5% free Fe and has a specific weight of 9.3 pounds/gallon.

$$\text{Injection rate(gallons per minute)}$$

$$= \frac{\text{part per million} \times \text{gallons per minute} \times 8.34}{[\%\ \text{active ingredients fertilizer} \times \text{pounds/gallon} \times 10{,}000] - [\text{parts per million} \times \text{pounds/gallon}]}$$

$$= \frac{10\ \text{parts per million} \times 500\ \text{gallons per minute} \times 8.3}{[5 \times 9.3 \times 10{,}000] - [10 \times 9.3]} = 0.09$$

Therefore, the injector should be set to deliver 0.09 gallon per minute of stock solution into the irrigation system to maintain an injected free Fe level of 10 ppm.

### Traveling Gun Irrigation Systems

For traveling gun irrigation systems, the injection rate is calculated from:

$$\text{injection rate} = \frac{\text{speed of traveling gun(feet/hour)} \times \text{distance between travel lanes(feet)} \times \text{fertilizer application rate(pounds/acre)} \times 100}{\%\ \text{active ingredients of fertilizer} \times \text{weight of fertilizer(pounds/gallon)} \times 43{,}560}$$

#### Example

Twenty pounds N are to be applied per acre to improve roughs irrigated with a traveling gun system. Travel lane spacings are 200 feet and the gun travels at 7 feet per minute (420 feet/hour). The fertilizer applied is ammonium nitrate (21% N; 10.73 pounds/gallon) at a desired rate of 20 pounds/acre. The required injection rate is calculated as:

$$\text{injection rate}$$

$$= \frac{\text{speed of traveling gun(feet/hour)} \times \text{distance between travel lanes(feet)} \times \text{fertilizer application rate(pounds/acre)} \times 100}{\text{percent active ingredients of fertilizer} \times \text{weight of fertilizer(pounds/gallon)} \times 43{,}560}$$

$$= \frac{420\ \text{feet/hour} \times 200\ \text{feet} \times 20\ \text{pounds/acre} \times 100}{21\% \times 10.73\ \text{pounds/gallon} \times 43{,}560} = 17.1\ \text{gallons per hour}$$

Therefore, to apply a fertilizer rate of 20 pounds N/acre, 17.1 gallons of ammonium nitrate would be injected per hour of traveling gun operation.

# 8 Water Management

## SOIL MOISTURE

Water is the primary material required for the growth and survival of turfgrasses; yet, at the same time, it is the most limiting. Plants consist of cells that are containers of water, and they maintain turgor (are rigid) when these cells are filled with water. Plants typically contain between 75% and 85% water by weight. Cells collapse when they lose water and, if enough cells lose their turgidity, leaves roll and turn a blue-green color, while stems droop. The plant is then considered wilted (Figures 8.1 and 8.2). Plants begin to die if the water content drops to 60% to 65% within a short period of time. Water acts as a buffer in plants against extreme temperature fluctuations, ensuring these processes occur rather slowly.

Unfortunately, rainfall does not occur frequently enough in most cases to provide adequate water to sustain turfgrasses, especially with the limited root systems associated with most golf course turf areas and soils used with low water-holding capacity. This situation is further intensified by warm weather and the high aesthetic demands of players. To ensure efficient watering, golf courses require well-designed irrigation systems based on soil infiltration rates, soil water-holding capacity, plant water-use requirements, depth of rootzone, conveyance losses from the surrounding area, and desired level of turfgrass appearance and performance (Figure 8.2).

### How Turfgrasses Obtain and Lose Soil Water

Water movement through soil into plants and loss through transpiration is somewhat synonymous to a wick-filled lantern. The wick, with its bottom dipped into a container filled with fuel, constantly transmits liquid from bottom to top where the burning fire consumes it. This is performed under the influence of physical forces imposed on the passive wick by the conditions at its two ends.

Roots absorb water from soil and transport it to the shoots through the xylem. To do this, several rules about water need reviewing. Water flows from a position of high energy to a position of lower energy. This situation is described in terms of *water potential* and expressed as bars and megapascals. In the case of roots absorbing water, soil water at field capacity is at a fairly high energy level. By contrast, the water inside a root is relatively low energy. Thus, there is a natural tendency for water to flow from the soil (high energy) into the root (low energy).

The second rule about water is that it "sticks" to itself. When water is sucked through a straw, it is pulled up against the force of gravity in a continuous column, due to the vacuum, or negative pressure, created by sucking. All the water behind the leading edge "sticks" to the water in front of it, and is pulled along. This ability to pull long columns of water up, against gravity, is fundamental for water to move to the top of plants.

The pathway for water loss from the leaf to the atmosphere includes the air space inside the leaf, the stomatal pore, the thin layer of still air adjacent to the leaf surface (referred to as the boundary layer), and the bulk atmosphere (Chapter 2). The concentration of water vapor will be essentially saturated inside the leaf air space, somewhat lower in the boundary layer, and usually much lower in the atmosphere. Water will diffuse (run downhill) from inside the leaf (high energy), through the boundary layer (lower energy), and into the bulk atmosphere (lowest energy). The rate of water vapor diffusion is regulated by the resistance encountered at the stomatal pore and the boundary layer. When stomata are open and the boundary layer is thin, as during windy conditions, water loss will be high. By comparison, water loss will be low when stomata are closed and the boundary layer is thick due to very still air.

With continued transpiration and in the absence of rainfall or irrigation, soil-available water will be nearly depleted and may approach the permanent wilting point. Water remaining in soil will be at a much lower energy level than the water at field capacity, and it may even be at a lower energy level than the water in the root. Under such conditions, a "downhill" path for water to flow into the root would no longer exist, and the plant would be unable to absorb soil water quickly, no matter how hard it "pulled." Water would still be transpired, but would not be replaced from the soil, resulting in wilt.

### Water Potential

As mentioned, when transpiration occurs, a gradient is formed from the leaves down through the plant to the roots, and water is literally "pulled" from the soil through the stem and eventually up to the leaves and atmosphere. Generally, this gradient becomes steeper from the soil through the plant's vascular system to the leaf stomata and into the atmosphere. This gradient (or energy level) is mathematically described as water potential and is expressed in negative values (Chapter 4). The more negative the value is, the lower is the water potential and the greater is the chance that water will flow in that direction. The chemical potential of water within this system is lower than that of pure free water due to dissolved solutes and pressure or gradient created by transpiration. This difference in water potential (or degree of "suction") provides the energy for water movement and largely determines the rate of water loss. In soils, water-potential gradients occur when one part of a soil is drier than another. A good example of this in turfgrass systems is the formation of dry spots on the edges of putting greens as moisture is pulled from a moist putting green soil (higher potential) to a dry neighboring native soil (lower potential).

Classic soil water potentials have been defined as 0 bar (or 0 MPa) when the soil is saturated and up to –15 bar (–1.5 MPa)

**FIGURE 8.1** Plant response to drought stress by rolling leaves (left) to reduce exposed surface area.

**FIGURE 8.2** Sufficient water is necessary for desired turfgrass performance and appearance.

when soils are so dry that plants are considered to be permanently wilted. For plant water potential, 0 bar represents fully turgid plants and ranges to –20 bars (–2.0 MPa) for severely wilted ones. In the atmosphere, 0 bar represents water-saturated air, or 100% relative humidity, while –1,000 bars (–100 MPa) is a very low relative humidity (e.g., arid). Due to this gradient, water will move from a site of high water potential (e.g., 0 bar) in soil to one of lower potential (e.g., negative value) in air. Table 8.1 indicates the relative humidity of air as a function of water potential.

The difference in soil water content between field capacity (–0.33 bar for loam or –0.1 bar for sand) and permanent wilting point (–15 bar or –1.5 MPa) is the amount of available water (Figure 4.6, in Chapter 4). For example, consider a soil with a field capacity of 35% volumetric water potential and a permanent wilting point of 15%. The difference, 20%, is the amount of water, expressed as a percent of total volume, that is potentially available to the plant. This can be used to calculate the amount of available water in a given rootzone. However, one cannot normally allow the soil to even dry out close to the permanent wilting point. Doing so, and making even the slightest error, can result in plant death, and the turf begins to display symptoms of drought stress much earlier, which usually results in immediate action.

### Example

A soil solution has a $\Psi_w$ of –0.3 MPa (or –3 bars) and the root cell has a $\Psi_w$ of –0.6 MPa (or –6 bars). Since the $\Psi_w$ of the root is less than that of the soil (–6 versus –3 bars), water can move from the soil into the root. If the $\Psi_w$ in

**TABLE 8.1**
**Water Potential Values in Relation to the Relative Humidity of Air[a]**

| | Water Potential | | |
|---|---|---|---|
| **Percent Relative Humidity** | **Bars** | **MPa** | **kPa** |
| 100 | 0 | 0 | 0 |
| 99.9 | –0.3 | –0.03 | –30 |
| 99 | –3.2 | –0.32 | –320 |
| 98 | –6.3 | –0.63 | –630 |
| 95 | –16 | –1.6 | –1,600 |
| 90 | –33 | –3.3 | –3,300 |
| 75 | –89 | –8.9 | –8,900 |
| 50 | –216 | –21.6 | –21,600 |
| 20 | –500 | –50.0 | –50,000 |
| 10 | –718 | –71.8 | –71,800 |

[a] The more negative the water potential values are, the greater affinity air has for water vapor in plant leaves.

leaves is –0.8 MPa (–8 bars), water will move from the roots through the crown and leaf sheaths into leaf blades. If the relative humidity of the atmosphere is 50%, this is equivalent to about $\Psi_w$ = –21.6 MPa (–216 bars). This means the force drawing water from the grass leaves is $\Psi_w$ = –0.8 MPa – (–21.6 MPa) = 20.4 MPa or 204 bars or 2,999 psi, a truly awesome transpirational force.

As air temperature rises, its water-holding capacity sharply increases. Thus, if a plant abruptly heats up, the relative humidity inside its leaf air space dramatically drops. As a result, water more readily evaporates from the leaf air space into the atmosphere, resulting in an increased water potential gradient (Table 8.1) and more transpiration.

## TURFGRASS EVAPOTRANSPIRATION RATES

Plants absorb water from the soil and lose water to the atmosphere. Only about 5% of all water consumed by turf is used in photosynthesis, carbohydrate synthesis, and other metabolic reactions. About 95% of this water is lost as vapor from the leaves, to the atmosphere, by the process of transpiration. Water is also lost by evaporation from soil and leaf surfaces. Evaporation is typically much lower than transpirational losses in a mature turf. The combined total of water lost through transpiration and evaporation is termed *evapotranspiration* (or ET). Evapotranspiration is usually expressed as inches or millimeters per day, week, or month. Since ET is the total water lost from the turf system, it represents the water demand, or the total amount to be replaced to maintain a healthy turf. Environmental parameters largely controlling ET are light intensity and duration, relative humidity, wind velocity, and temperature. Increasing solar radiation, temperature, and wind increases ET, while increasing relative humidity and cloud cover decrease ET. Other parameters affecting ET to a lesser extent include soil-water content, turf-root system development, inherent turf water needs and dehydration avoidance mechanisms, and turf cultural practices.

Although it might seem like transpiration is just a waste of water, it is in fact critically important as it cools the leaf. If not for transpirational cooling, a leaf could reach 120°F (49°C) or higher during midsummer, a lethal temperature for most plants. Fortunately, transpiration keeps leaves much cooler, usually below 90°F (32°C), due to the latent heat of vaporization for water, or the large amount of energy needed to convert 1 g of liquid water to 1 g of water vapor via evaporation (Figure 8.3). For example, for every calorie of solar energy absorbed by the plant, 1 g of turfgrass tissue (mainly water) will increase by nearly 2°F. Ten calories of solar energy could warm a gram of turf tissue by about 18°F. However, it takes a lot of energy (539 calories) to evaporate 1 g of water. By transpiring only 1 g of water, a turf plant loses enough energy to cool 539 g of plant tissue by ~2°F. Multiply this by the millions of grams of water a turfgrass area loses daily and the incredible cooling capacity of transpiration becomes evident. Humans use a similar process when perspiration evaporates, cooling their bodies.

**FIGURE 8.3** Energy is required to convert liquid water to vapor, leaving a cooler turf surface behind.

**TABLE 8.2**
**General Mean Summer Turfgrass Evapotranspiration Rates**

| | Summer ET Rates | | | |
|---|---|---|---|---|
| **Turfgrass** | **Inches/Day** | **Millimeters/Day** | **Inches/Week** | **Centimeters/Week** |
| Bahiagrass | 0.25 | 6.2 | 1.75 | 4.4 |
| Bermudagrass | 0.12 to 0.30 | 3.1 to 8.7 | 0.84 to 2.10 | 2.1 to 5.3 |
| Bluegrama | 0.23 | 5.70 | 1.60 | 4.1 |
| Buffalograss | 0.20 to 0.30 | 5.3 to 7.3 | 1.40 to 2.10 | 3.6 to 5.3 |
| Centipedegrass | 0.15 to 0.33 | 3.8 to 8.5 | 1.05 to 2.31 | 2.7 to 5.9 |
| Creeping bentgrass | 0.19 to 0.39 | 5.0 to 9.7 | 1.33 to 2.73 | 3.4 to 6.9 |
| Kentucky bluegrass | 0.15 to 0.26 | 3.7 to 6.6 | 1.05 to 1.82 | 2.7 to 4.6 |
| Perennial ryegrass | 0.15 to 0.44 | 3.7 to 11.2 | 1.05 to 3.08 | 2.7 to 7.8 |
| Seashore paspalum | 0.25 to 0.31 | 6.2 to 8.1 | 1.75 to 2.17 | 4.4 to 5.5 |
| St. Augustinegrass | 0.13 to 0.37 | 3.3 to 9.6 | 0.91 to 2.59 | 2.3 to 6.6 |
| Tall fescue | 0.15 to 0.50 | 3.6 to 12.6 | 1.05 to 3.50 | 2.7 to 8.9 |
| Zoysiagrass | 0.14 to 0.30 | 3.5 to 7.6 | 0.98 to 2.10 | 2.5 to 5.3 |

*Sources:* Modified from Beard, J. B., 1985, In *Turfgrass water conversation. Proc. Symposium ASPA*, San Antonio, TX. 15–16 Feb. 1983, ed. V. B. Youngner and S. T. Cockerham, 47–60. Oakland, CA: Cooperative Extension University of California; Carrow, R. N., Crop Science, 35, 1685–1690, 1995a; McCarty, L. B., ed. *Best golf course management practices* (3rd ed.). Upper Saddle River, NJ: Prentice-Hall, Inc.2011.

*Notes:* Low values within a range represent humid conditions; high values are for arid conditions. ET rates during nonsummer months generally are much lower.

Transpiration is also directly involved in mineral nutrition, both by causing soluble nutrients to be drawn to the roots along with soil water and by moving nutrients and certain hormones from roots to shoots. It is the diffusion of water through the stomata that creates a lower pressure potential, allowing water to overcome gravitational potential and move upward from the soil, into plant roots, and upward in the plant. When the transpiration stream is lacking, as when plants are grown in a saturated atmosphere (100% RH, $\Psi_t \sim 0$ MPa), nutrient deficiency symptoms, especially for N and Fe, often develop or "wet wilt" occurs, where plants collapse in the presence of water when evaporation exceeds root water uptake.

**Turfgrass water-use rates.** Water-use rates are usually expressed in inches or centimeters of water lost per day or per week. In general, warm-season grasses use less water due to their greater resistance to water stress than the cool-season grasses do (Table 8.2). This ranges between 35% and 50% less water required to maintain desirable warm-season grass color compared to that of cool-season grasses. For example, bermudagrass ET is between 0.3 and 0.9 cm of water per day, while tall fescue water use ranges from 0.4 to 1.3 cm per day. Lower values are associated with cooler or more humid regions of the United States, while higher values are typical of warm, arid regions. Tall fescue has the highest potential ET rates, but avoids drought stress due to its deep and extensive root system and ability to go dormant for short periods without lethal consequences.

## TURFGRASS WATER USE

### Cultural Practices

Turfgrass cultural practices influence water-use rates and efficiency. Mowing, fertilization, and irrigation are the primary cultural practices superintendents can use to control water loss and to encourage conservation.

**Mowing.** Managers should normally mow turf at the higher end of the optimum mowing height range. Taller mowed turf allows development of more efficient water-using plants. Higher mowed turf results in a deeper and denser root system that can extract water from a larger volume of soil. Using sharpened mower blades provides a "clean" cut, minimizing the amount of exposed leaf surface area and thereby reducing water loss via ET.

**Nutrition.** Nitrogen influences turf color, shoot-and-root growth, and water use. As a general rule, N should be applied at levels to maintain good color, moderate growth, and high recuperative ability. Moderate shoot growth usually equates to maximum root growth, and thus maximum water availability. Excessive N promotes shoot growth at the expense of root growth and can adversely affect water extraction.

Potassium is often referred to as the "stress" nutrient, since plants suffering from K deficiency are more susceptible to drought, cold, and other possible stresses. Fertilizer applications to a K-deficient soil may help the turfgrass plant maintain turgor and avoid wilt. Excessive N may override the beneficial effects of K.

Foliar Fe applications also have been shown to increase turfgrass rooting under certain environmental conditions. Liquid Ca applications also improve the drought tolerance of certain turfgrass species.

**Soil compaction.** Compaction decreases total pore space in a soil, mainly by eliminating macropores. This reduces soil aeration and may retard root growth and function. Turfgrasses may wilt in compacted, poorly aerated soils even when adequate moisture is present as impaired roots are unable to absorb water. Cultivating or coring compacted soils will improve aeration and enhance rooting and turf quality.

**Rooting depth.** Mowing, nutrition, and soil compaction all directly influence turfgrass water use. Other important parameters affecting available water include depth, activity, and spatial distribution of the root system. Very deep root systems will have access to available water much more than shallow ones.

### Example

1. A bermudagrass root system has a depth of 15 inches (38 cm) in a soil with 20% volumetric water content. How much available water to the turf and approximately how long (days) can the plant can go without irrigation at 0.25 in ET daily?

$$15\text{ inches} \times 20\% = 3\text{ inches (7.6 cm) available water}$$

$$3\text{ inches available water} \times \frac{\text{day}}{0.25\text{ inches ET}} = \sim 12\text{ days}$$

2. Determine days between irrigation for a bentgrass fairway with a root system 6 inches (15 cm) deep in spring (daily ET = 0.15 inch, 0.25 mm) and the same area in summer with a root system only 2 inches (5 cm) deep and daily ET = 0.30 inch (7.4 mm).
   Bentgrass (spring):

$$6\text{ inches} \times 20\% = 1.2\text{ inches (3 cm) available water}$$

$$1.2\text{ inches available water} \times \frac{\text{day}}{0.15\text{ inches ET}} = \sim 12\text{ days}$$

Bentgrass summer:

$$2\text{ inches} \times 20\% = 0.4\text{ inch (10 mm) available water}$$

$$0.4\text{ inch available water} \times \frac{\text{day}}{0.30\text{ inch ET}} = \sim 1.3\text{ days}$$

During spring, approximately eight days can pass between irrigations, while in summer, less than two days can. Rooting depth is clearly a dynamic characteristic of turfgrasses, changing seasonally and with specific turfgrass uses and species.

**Watering practices.** Irrigation practices, in terms of amount and frequency, can significantly increase drought tolerance by conditioning the turf. Allowing mild drought stress (soil drying) between watering promotes drought resistance without damage to the turf. Mildly stressed turf slows shoot growth and promotes root growth.

## Determining When to Irrigate

There are a number of methods used to determine how much water a turf requires at any given time, under any given environment. Several are indirect and base their estimates on measuring soil moisture. Others simulate evapotranspiration from the canopy.

**Visual symptoms.** A simple method used to determine when to irrigate is when visual symptoms of moisture stress are evident. Moisture-stressed grass appears blue-green or grayish-green in color (Figure 8.4), recuperates slowly (more than one minute) after walking or driving across it ("footprinting") (Figure 8.5), or wilts continuously. These symptoms occur when plant moisture is insufficient to maintain turgor due to more water being lost than being absorbed. As a result, the plant rolls its leaves to minimize exposed leaf surface and wilts to conserve moisture. This method is best used for low-maintenance turf such as golf course roughs.

**Evaporatory pans.** Another method of irrigation scheduling is the use of evaporatory pans. A U.S. Weather Service class A evaporatory pan is 122 cm in diameter, 25 cm deep, and is supported 15 cm above the ground. Evaporatory pans are filled with water and placed in a representative location, where water loss is measured over time. The amount of water evaporating from the pan correlates to that lost by evapotranspiration (ET).

The water amount lost through evaporation correlates with turfgrass ET, but is not exactly the same; turfgrasses use less water than the amount evaporated from the pan. A crop coefficient ($K_c$) value is needed to adjust this correlation (Table 8.3). Warm-season grasses use 55% to 65%, and cool-season grasses use 65% to 90%, of pan evaporation.

Calculating turfgrass $ET_p$ rates:

$$ET_p = K_c \times \text{pan evaporation}$$

### Example

If the evaporative pan shows a 1-inch (2.5 cm) water loss, how much would bermudagrass and bentgrass lose?

FIGURE 8.4 On golf greens, early symptoms of drought include turf turning a blue- or grayish-green color.

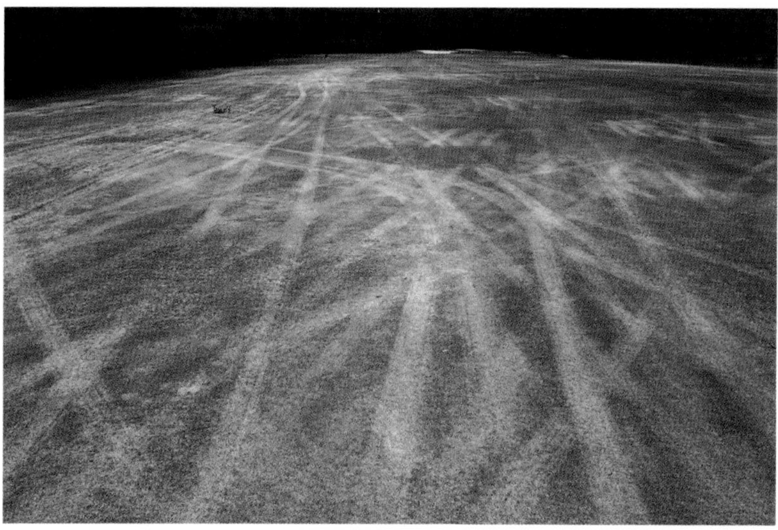

**FIGURE 8.5** Turf not rebounding from foot or vehicular traffic due to drought.

**TABLE 8.3**
**Crop Coefficient ($K_c$) Values for a Class A Evaporative Pan or the Penman–Monteith Equation**

| Grass | Class A Evaporative Pan $K_c$ Values | Penman–Monteith $K_c$ Values |
|---|---|---|
| Bermudagrass | 0.55 to 0.65 | 0.70 to 0.80 |
| Tall Fescue | 0.65 to 0.75 | 0.75 to 0.95 |
| Perennial Ryegrass | 0.65 to 0.75 | 0.80 to 1.0 |
| Kentucky Bluegrass | 0.70 to 0.80 | 0.85 to 1.0 |
| Creeping Bentgrass | 0.75 to 0.90 | 0.95 to 1.0 |

For bermudagrass,

$$ET_p = K_c \times \text{pan evaporation} = 1\ \text{inch} \times 0.6 = 0.60\ \text{inch}$$

For bentgrass,

$$ET_p = K_c \times \text{pan evaporation} = 1\ \text{inch} \times 0.85 = 0.85\ \text{inch}$$

Therefore, using the species-dependent corrective values, bermudagrass would actually have lost ~0.60 inch (1.5 cm), while bentgrass would have lost ~0.85 inch (2.2 cm).

**Atmometer.** The atmometer (also referred to as the "ET gauge" or "Bellani plate") also can be used to estimate evaporative demand (Figure 8.6). This relatively inexpensive device consists of a water reservoir connected to a porous plate covered by green fabric designed to simulate a leaf surface. Water from the reservoir is wicked through the plate to the fabric, where it evaporates. The drop in the reservoir is then easily measured on a daily basis, much like checking a rain gauge. Rates of water loss are directly related to weather conditions, especially temperature, wind, and humidity, and have been found to correlate very well with turfgrass water demand. Atmometers may be an attractive alternative to the costlier weather station-based system while still supplying similar information.

The atmometer should be located in a sunny, turf-covered area representative of the majority of the golf course. Additional units may be necessary for varying microclimates such as shady, windy, or stagnant areas, and irrigation rates should be adjusted accordingly.

## SOIL MOISTURE MEASURING DEVICES

Soil moisture measuring devices have been developed with the goal of indicating how much moisture is available to plants. Soil moisture is measured in two distinctly different methods: *quantitatively* (or *volumetric*), which is the actual amount of moisture in the soil, and *qualitatively* (or *tensiometric*), which is how tightly water is held by soil. Though numerous means of measuring these exist, the more common ones including gravimetric water content, time domain reflectometry (TDR), tensiometers, and frequency domain reflectomety (FDR) (or hand-push) probes.

The water content of different soils varies due to large differences between soils in their total particle surface areas. For example, moisture levels at field capacity for sands may be as low as 7%, whereas clays may have as much as 40% moisture content at field capacity. In another example, the permanent wilting point volumetric water content may range from 1% to 2% for sandy soils to 25% or 30% for clay (finer textured)

**FIGURE 8.6** An atmometer is a water-containing vessel that measures daily moisture loss via ET.

soils. This variation demonstrates that a measure of soil water (volumetric) content does not necessarily indicate the amount of water available to plants. A better indicator of a plant's soil-water availability is the energy status of water (called tensiometric or water potential), which measures the relative amount of work (or energy) needed to remove a unit of water from a particular soil.

## Quantitative Methods

Quantitative methods for measuring soil moisture include gravimetric sampling, neutron probe (or scatter), and dielectric constant (TDR and FDR) probes. The most accurate is the gravimetric water content method, where a volume of soil is weighed, dried, and then reweighed (Table 8.4). The impracticality of this method and the expense (>$5,000) for neutron probes have led to the development of other techniques.

Dielectric constant methods measure the soil's ability to transmit electricity in the form of moderate- to high-frequency electromagnetic waves or pulse with the value increasing as the water content of the soil increases (Figure 8.7). The permittivity constant for air is approximately 1; dry soil between 3 and 5; and about 80 for water. Values are related through calibration to known soil moisture content determined using either a neutron probe or the gravimetric sampling technique. The equipment consists of an electronic meter connected to from two to four rods placed into the ground. The instrument sends an electrical signal through the soil and the rods serve as the transmitter and receivers. TDR and FDR probes are currently the most commonly used dielectric devices. Although these devices are able to detect the amount of moisture in the soil, they do not determine how much of it is available to plants.

## Qualitative Methods

These methods measure how tightly soil moisture is held by soil particles but do not directly measure the quantity of water contained in it. As tension increases, water extraction

**TABLE 8.4**
**Comparison of Common Techniques of Measuring Soil Moisture**

| Technique | Measurable Range | Advantages | Disadvantages |
|---|---|---|---|
| **Gravimetric water content.** Measures soil moisture by weighing-drying-reweighing | Full range of water content (%) | • Simple equipment needs<br>• Highly accurate<br>• Easy interpretation | • Destructive sampling<br>• Labor intensive<br>• Collection, transport, and time restraints |
| **Time domain reflectometry** (TDR). Measures time for a high-frequency electromagnetic wave to travel using soil medium as a dielectric; moisture slows this down | Up to 50% volumetric water content (0.50 kg water $kg^{-1}$ soil) | • Accurate<br>• Minimal soil disturbance<br>• Soil-specific calibration is optional<br>• Relatively insensitive to temperature<br>• Also estimates, with limited accuracy, soil EC. | • Expensive<br>• Accuracy decreases in high-saline (>25 dS $m^{-1}$) conditions or heavy clay soils<br>• Relatively small sensing volume (about 1 inch, 2.5 cm, radius around probe) |
| **Frequency domain reflectometry** (FDR) or **hand-push probes**. Measures the change in a lower frequency of a capacitor using soil medium as a dielectric. | Up to 70% volumetric water content (0.70 kg water $kg^{-1}$ soil) or –7.0 MPa | • Relatively inexpensive<br>• Can be automated with irrigation<br>• Stable in different soil types and over a large range of moisture contents | • Needs soil-specific calibration for accuracy<br>• Samples small volume of soil (about 4 inches, 10 cm, radius around probe)<br>• Sensitive to soil air gaps, saline soils, and temperature |
| **Tensiometers**. Measure how tightly (the tension) water is held by soil | 0 to –0.08 MPa or 0 to –80 kPa | • Direct readout of soil water potential (or tension)<br>• Inexpensive<br>• Can be automated with irrigation<br>• Relatively reliable<br>• Good accuracy<br>• Unaffected by soil salinity | • Soil moisture retention curve needed to relate to soil water content<br>• Samples a small area near cup, thus needing multiple samples in larger areas<br>• Does not measure soil salinity content<br>• Exposed gauges; sensitive to disturbance and soil air gaps |

becomes more difficult for the plant. Tensiometers and porous blocks (i.e., gypsum, ceramic, nylon, and fiberglass) are qualitative methods (Table 8.3).

### Calibrating Soil Moisture Devices

When using any soil moisture measuring device for irrigation purposes, three critical soil moisture levels need to be quantified: (1) gravitational water, (2) field capacity, and (3) wilting point (Figure 4.6, Chapter 4). Moisture levels typically vary with soil type and uniformity and readings vary between soil measuring devices. Therefore, it is highly recommended that soil moisture measuring devices be calibrated for the particular soil under consideration.

The amount of moisture between field capacity and wilting is that available for plant use. Gravitational water is typically that in the soil following heavy rainfall and puddle disappearance. About 24 hours later, a reading should approximate field capacity. Field capacity is also the amount of soil moisture present about one hour following heavy irrigation. Multiple readings should be taken over the whole area and the probe length should extend to just below the average rooting depth. Readings should be taken at least twice daily until significant plant wilting is observed. Typically, multiple readings are taken in morning to determine an average moisture content for a green. This is repeated in the afternoon. The difference between morning and afternoon readings is the percent of volumetric water content (VWC) used for the day; based on the weather forecast, the amount of irrigation needed to replenish the moisture loss can be calculated.

#### Example

Your golf greens have significant wilt at 7% VWC and you use a 2% buffer to avoid catastrophic wilting. Therefore, the target value for the end of the day is 9% VWC. Tomorrow's predicted water use is 3%. How much water needs to be added? On average, your irrigation increases VWC 1% for every 1.25 minutes of run time.

$$3\%\ \text{VWC needed} \times \frac{1.25\ \text{minutes}}{1\%\ \text{VWC added}} = 3.75\ \text{minutes}$$

#### Example

From the use of a handheld TDR probe, determine a soil moisture management program including when to irrigate and how much water is needed to return the total moisture level to field capacity for two sands with $\theta_{fc}$ = 0.32 and 0.22 cm and $\theta_{wp}$ = 0.02 and 0.01 for sands 1 and 2, respectively. The TDR probe measures moisture in the top 10 cm (4 inches) of the soil profile.

**FIGURE 8.7** The turfgrass industry is a leader using moisture measuring probes for irrigation schedules.

**Step 1:** Determine available water for each sand using the following equation:

$$D_e = \text{soil depth}\ (\theta_{fc} - \theta_{wp})$$

where
$D_e$ = equivalent depth of available water in the top 10 cm (4 inches)
$\theta_{fc}$ = volumetric water content at field capacity
$\theta_{wp}$ = volumetric water content at wilting point

For sand 1: $D_e = \text{soil depth}\ (\theta_{fc} - \theta_{wp})$

$$= 10(0.32 - 0.02) = 3.0\ \text{cm}\ (1.8\ \text{inches})$$

For sand 2: $D_e = \text{soil depth}\ (\theta_{fc} - \theta_{wp})$

$$= 10(0.22 - 0.01) = 2.1\ \text{cm}\ (0.83\ \text{inch})$$

**Step 2:** If the effective rootzone is 10 cm (4 inches) deep and the turfgrass being used has an average ET rate of 0.2 inch/day (0.5 cm/day), the days between watering for each sand would be

$$\text{Sand 1: } 3.0\ \text{cm rootzone moisture} \times \frac{1\ \text{day}}{0.5\ \text{cm moisture used}} = 6\ \text{days}$$

$$\text{Sand 2: } 2.1\ \text{cm rootzone moisture} \times \frac{1\ \text{day}}{0.5\ \text{cm moisture used}} = 4.2\ \text{days}$$

Therefore, for sand 1, 3.0 cm of water would be needed every six days; sand 2 would need 2.1 cm every 4.2 days to return each to field capacity.

### Predictive Models or Evapotranspiration Feedback

Predictive models, such as the modified Penman–Monteith ET (also known as FAO 56) model based on weather station data and soil types, also are available. These are often referred to as irrigation or ET controllers in the industry and estimate or predict ET of the turf. They are relatively accurate and applicable, especially as long-term predictors of yearly turf water requirements. Models, however, are only as effective as the amount of data collected and the number of assumptions made. Weather data such as solar radiation, air and soil temperature, relative humidity, and wind speed are incorporated into certain model formulae, and soil moisture content is estimated. Accessible weather data, as well as specialized computer equipment and programs, must be available (Figure 8.8).

Evapotranspiration feedback strategies are also used to schedule irrigation. Weather station or evaporative pan data can be used to calculate water use. This value is referred to as potential ET ($ET_p$) and is used as a reference point. Actual turf water use usually is not quite as high as $ET_p$, so crop coefficient ($K_c$) is used to convert $ET_p$ to actual turf ET (as discussed previously on evaporative pans). Crop coefficients are fairly constant for a given species, but vary considerably between species (Table 8.2). For example, the $K_c$ of bermudagrass is about 0.75. This means bermudagrass will use about 75% as much water as is predicted from using environmental data to calculate $ET_p$. If environmental data indicate the theoretical reference crop used 2.2 inches (5.6 cm) of water for a given week in the summer, multiplying 2.2 by 0.75 indicates 1.65 inches (4.2 cm) of water is actually used by bermudagrass. Most cool-season grasses have a $K_c$ of ~0.85, indicating that they actually require 1.87 inches (4.75 cm) of water in the previous situation. These calculated water use rates are the "feedback" used to determine irrigation rates. Using the site information and weather data, ET feedback controllers run a "water balance" that keeps track of how much water is in the soil. Controllers then adjust the run timers (or amount) of water applied to the turf. These can be specified for specific irrigation heads used for watering different species of turf, soil types, or turf uses (greens, tees, fairways, etc.). Additionally, wireless soil moisture sensors can be placed at different locations on the golf course to monitor soil moisture levels, providing another tool for the golf course superintendent to accurately monitor field capacity and wilting points.

**FIGURE 8.8** On-site weather stations provide site-specific data to aid in irrigation scheduling.

These technologies increase the precision and conservation of irrigation resources.

## IRRIGATION STRATEGIES

With potential shortages of irrigation water, it is in the best interest of a golf course to conserve water whenever possible and to design irrigation programs that provide quality turf with minimum water use (Figure 8.9). Irrigating too heavily not only wastes valuable water, but also invites the potential for increased disease incidence, turf thinning, shallow rooting, reduced stress tolerance, and increased soil compaction and turf wear. Inefficient use of electricity and excessive wear and tear on the irrigation pumps and total system also are reasons to maximize water use. Annual water costs for courses located in deserts in the United States can be over $350,000.

Playing conditions also are influenced by watering practices. Overwatered golf courses tend to play much longer and have slower putting greens. Conversely, drier turf results in quicker putting surfaces and more bounce and roll—in effect, shortening the course. However, allowing turf to dry excessively increases the risk of losing turf from moisture stress and causes a reduction in aesthetic quality. Many courses also are restricted to the amount of water they can use and may be mandated to irrigate based on ET data, soil moisture levels, or other water need indicators. A major misunderstanding is not realizing that an irrigation system supplements rainfall, but does not replace it.

Steps in formulating an irrigation strategy include:

1. Calibrate an irrigation system's output and distribution uniformity (or DU).
2. Determine daily ET rates or soil moisture status by one of the methods discussed. A reasonable estimate of daily summer mean ET rates for various grasses is provided in Table 8.4.
3. Accurately track daily rainfall and ET rates so that a water budget can be set up and followed.
4. When irrigation is needed, use the appropriate crop coefficient percent (0.75 to 0.85) of daily ET rate and incorporate DU of the irrigation system as shown later.
5. Make adjustments for rainfall, varying microclimates, and forecasted weather.

### IRRIGATION SYSTEM CALIBRATION

The first step in irrigation scheduling is to determine how much water the irrigation system applies, expressed as inches per hour. This information is central to water management. The easiest and most common way to determine application rate is by "canning" the turf area. For small areas, a dozen or so empty tin cans are placed in a grid system across the turf with the location of each catch-can recorded (Figure 8.10). It is important that cans be the same size and fairly tall; soup or vegetable cans work well. The irrigation system is then activated for a timed period, usually 15 to 30 minutes, to let the cans collect 1/4 to 1/2 inch (6.4 to 13 mm) of water. The average amount of water in each can is then measured with a ruler and adjusted to the amount of water caught per hour. These cans are all emptied into a single can and water depth is measured with a ruler. Depth is then divided by the total number of cans to determine the average depth per can. This value must be divided by the time period to calculate application rate.

#### Example

Assume 12 cans were used to collect irrigation for a 30-minute period. The total depth of all cans was 4.4 inches (11 cm). What is the average amount applied per hour?

Dividing 4.4 inches (11 cm) by 12 gives 0.37 inch/can (0.94 cm). Now multiply the average depth—0.37 inch

**FIGURE 8.9** Irrigation design where water is conserved by treating fairways separately from roughs.

**FIGURE 8.10** "Canning" a turf area to quantify the distribution uniformity of an irrigation system.

(0.94 cm) per one-half hour—by two to calculate the application rate of 0.74 inch (1.9 cm) per hour.

The canning method also helps indicate the DU of the irrigation system, which is the ratio of underwatered area to the average applied within the sprinkler coverage area. Distribution uniformity is determined by identifying the volume applied to the driest 25% of the test area and dividing it by the mean depth of water in all cans. The equation of DU involves:

$$\text{Distribution uniformity (DU)} = \frac{\text{average } \textbf{least} \text{ amount of water depth collected in 25\% of all cans}}{\text{average depth of water collected for all cans}}$$

Typical DU values range from 55% to 80%, even if rainfall is not 100% uniform. The uniformity of water distribution is the weakest part of an irrigation system. The lower the value is, the less uniformity with which an irrigation system applies water; thus, more water and energy requirements are needed. Obtaining 80% DU is considered excellent (achievable), 70% as good (minimum), and 55% or less as poor. Means of improving existing DU values include (1) changing sprinklers and worn sprinkler nozzles; (2) pressure changes (increases); (3) changing sprinkler spacing, normally making them closer; and (4) maintaining level irrigation heads. The Center for Irrigation Technology is a resource for testing and selecting irrigation heads (www.fresnostate.edu/jcast/cit/index.html).

## Example

1. Determine the DU of the following conditions. A can test was performed with 20 cans evenly spaced 5 feet (1.5 m) apart in a grid system. After a 15-minute run cycle, the average depth in the five **least** filled cans was 0.2 inch (0.5 cm). The average depth measure in all cans was 0.33 inch (0.84 cm). The irrigation rate is then adjusted from inches per 15 minutes to inches/hour by multiplying the 0.33 inch (0.84 cm) and 0.2 inch (0.5 cm) depths by four to achieve 1.32 inches/hour (3.36 cm/hour) and 0.8 inch/hour (2.0 cm/hour), respectively. The DU value is then determined:

$$\text{Distribution uniformity (DU)} = \frac{\text{average } \textbf{least} \text{ amount of water depth collected in 25\% of all cans}}{\text{average depth of water collected for all cans}}$$

$$= \frac{0.8 \text{ inch}}{1.32 \text{ inches}}$$

$$= 0.61 \text{ (or 61\%)}$$

2. How much water would be needed to apply 0.5 inch (1.3 cm) over the entire area?

$$\frac{0.5 \text{ inch}}{0.61 \text{DU}} = 0.82 \text{ inch (2.1 cm) of irrigation needed}$$

to apply at least 0.5 inch (1.3 cm) over the area

3. How long would the irrigation system need to run to apply 0.82 inch (2.1 cm)?

From this information, it was determined the irrigation system delivered 1.32 inches/hour (3.4 cm/hour); therefore,

$$0.82 \text{ inch} \times \frac{1 \text{ hour}}{1.32 \text{ inches}} \times \frac{60 \text{ minutes}}{\text{hour}} = 45 \text{ minutes}$$

Irrigation system calibration, but not DU, can also be determined by knowing the amount (gallons) of water applied per irrigation head, the sprinkler spacing (feet), and by using one of the formulas listed in Table 8.5. Different formulas are needed depending upon irrigation head arrangement (square spacing, triangular spacing, or single-row design).

**TABLE 8.5**
**Irrigation Application Rates per Head Based on Head Spacing Pattern**

**Square Spacing Head Design**

$$\frac{96.3 \times \text{gallons/min applied per full circle head}}{(\text{sprinkler spacing, feet})^2} = \text{Inches/hour}$$

**Triangular Spacing Head Design**

$$\frac{96.3 \times \text{gallons/min applied per full circle head}}{(\text{sprinkler spacing, feet})^2 \times 0.866} = \text{Inches/hour}$$

**Single Row Spacing Head Design**

$$\frac{96.3 \times \text{gallons/min applied per full circle head}}{\text{Sprinkler throw diameter (feet)}^2 \times 0.80 \times \text{sprinkler spacing (feet)}} = \text{Inches/hour}$$

### Example

Determine inches of water applied per hour for an irrigation system designed with triangular spaced heads 50 feet apart that apply 30 gallons of water per minute per head.

From Table 8.5, the amount (inches) of water being applied per hour is determined as

In water applied/hour

$$= \frac{96.3 \times \text{gallons/minute per head}}{(\text{sprinkler spacing, feet})^2 \times 0.866}$$

$$= \frac{96.3 \times 30 \text{ gallons/minute applied per head}}{(50 \text{ feet})^2 \times 0.866}$$

$$= 1.33 \text{ inches/hour } (3.4 \text{ cm/hour})$$

**Determining irrigation rates and frequency.** In addition to the application rate and uniformity, the turf manager should know how much water the turf is using. This can be determined using reference ET from a weather station/computer system plus a crop coefficient specific for the turf species from data in Table 8.2, or with data from an atmometer or other devices as previously discussed. Historical weather information may also provide reasonable estimates of average water use. Managers also need to know where the roots are in the soil profile and approximately how much available water is held by the soil.

As previously noted, evaporation during hot, windy, and dry periods can reduce irrigation efficiency. Superintendents can avoid this by irrigating early in the morning before the temperature rises and humidity drops. Early morning irrigation also removes dew from the leaves and helps prevent diseases favored by irrigating in the evening.

**Water budgeting.** Budgeting water is analogous to handling money in a checking account (Figure 8.11). There are inputs (deposits), outputs (withdrawals), and a certain amount of water in the soil (standing balance). The flow of water (money) into and out of the "checking account" (rootzone) is simply followed over time. If the roots penetrate 12 inches (30 cm), the checking account is the water held in 12 inches (30 cm) of soil. If the roots penetrate only 2 inches (5 cm), the checking account is

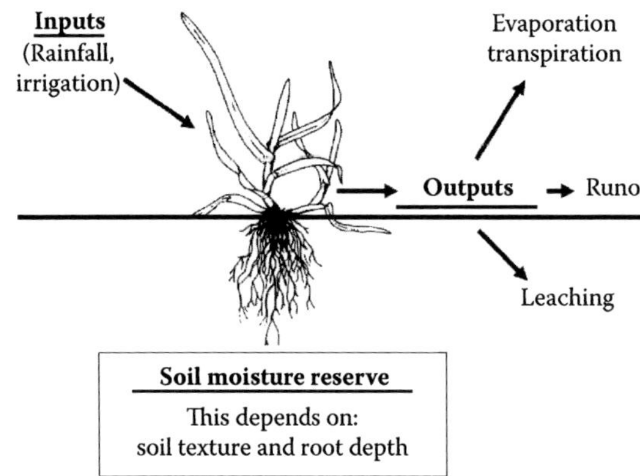


**FIGURE 8.11** Water budgeting mimics a checking account with various moisture inputs, outputs, and standing balances.

considerably smaller. Irrigation, therefore, should be applied to wet the area occupied by the roots (rootzone).

### Example

A silt-loam soil at field capacity has ~2.0 inches (5 cm) of water per foot of soil. A 12-inch (30 cm) deep root system growing in this soil will have access to 2.0 inches (5 cm) of available water. Determine the amount of turf water use and whether or not the turf should go another day without irrigation.

Weather station data and a predictive model estimate that, over a six-day period, 1.8 inches (4.6 cm) of water was used by the theoretical reference crop. Correcting this reference value using a $K_c$ of 0.7 for bermudagrass estimates the turf actually uses about 1.3 inches of water (1.8 × 0.7 = 1.26 inches, 3.2 cm). Subtracting this from the original 2.0 inches (5 cm) of available water indicates ~0.7 inch (1.8 cm) of water left in the soil. The turf should not go another day before watering since it is never a good idea to deplete most of the available water. Approximately 1.5 inches (3.8 cm) of irrigation should be applied to replace the 1.3 inches (3.3 cm) lost from the system. The soil is returned to field capacity without irrigating excessively and wasting water.

For most turfgrass examples, the amount of water at wilting point is negligible. Turfgrass rooting depth should be used instead of soil rootzone depth since most moisture obtained by plants in a reasonable time frame will be in the rooting depth and not below it. The usable soil moisture availability, divided by the high average daily requirement (ET values), provides irrigation frequency in days.

Determining approximate intervals (in days) between irrigation cycles:

Irrigation interval (days)

$$= \frac{\text{soil water content at field capacity} \times \text{rooting depth (inches)}}{\text{daily ET rate (inches/day)}}$$

### Example

Determine the time between irrigation cycles for a sand soil with a volumetric water content of 15% at field capacity, a rooting depth of 4 inches (10 cm), and a summer daily ET rate of 0.20 inch/day (5 mm/day):

$$\frac{0.15 \times 4 \text{ inches}}{0.20 \text{ inch/day}} = 3 \text{ days between irrigation cycles,}$$

which brings the soil back to field capacity

If rainfall occurs and it is more than the amount of water depleted during the period (1.3 inches, 3.3 cm), the rootzone is returned to field capacity and any excess is ignored since it will drain and not be stored in the rootzone. If it rains less than actual ET, the running deficit is calculated over several days, and irrigation is scheduled when ET has depleted the soil moisture to slightly more than 50% of the 0.6 inch (0.15 × 4 inches) of available water. A good rain gauge is needed to keep track of precipitation, and it is a good idea to use automatic pump shutdown switches to prevent irrigation after a significant precipitation. Conversion factors in Table 8.6 indicate gallonage required to apply certain amounts of irrigation.

With information on ET rates and sprinkler calibration available, each sprinkler's run time can be calculated by dividing the daily ET rate by the sprinkler output. For example, if the day's ET rate is 0.3 inch (7.6 mm) and the sprinkler output is 0.01 inch/minute (0.25 mm/minute), the irrigation time needed would be 30 minutes. However, this is adjusted according to the appropriate crop coefficient (e.g., 0.85 for bentgrass); therefore, 30 minutes is multiplied by 0.85 to give 25 minutes of run time needed. Distribution uniformity considerations should then be incorporated to ensure enough water is being applied uniformly across the turf area.

### Example

1. If 46 acres (18.6 ha) of turf were to receive 1 inch (2.5 cm) of water, what is the total amount of water, in gallons, needed?

   From Table 8.6, 1 acre-inch of water equals 27,154 gallons; thus,

   $$27{,}154 \text{ gallons} \times 46 \text{ acres} = 1{,}249{,}084 \text{ total}$$

   gallons water needed (4.7 million L)

2. If water costs are $0.03/cubic foot of water, what is the total cost of this volume? From Table 8.6, 1 cubic foot of water equals 7.48 gallons; thus,

   $$\frac{1 \text{ cubic foot}}{7.48 \text{ gallons}} \times 1{,}249{,}084 \text{ gallons total}$$

   $$\times \frac{\$0.03}{\text{cubic foot}} = \$5{,}010$$

**Time of day to irrigate.** As discussed earlier, water loss rates decrease with reduced solar radiation, little wind, high relative humidity, and low air temperatures. Superintendents can take advantage of these factors by irrigating when conditions do not favor excessive evaporation. Generally, irrigation should occur in early morning hours before air temperatures rise and relative humidity drops. Irrigating at this time also removes dew from leaf blades and allows sufficient time for infiltration into the soil but not enough to encourage disease development.

A problem with this timing is that golfers often begin play early in the morning since it is generally cooler. Therefore, superintendents may have to water at night. On most summer days, afternoon irrigation is not suggested unless lowering canopy temperature is important, fertilizer or pesticide application must be irrigated-in, or overseeding and turf establishment are being conducted. Watering efficiency may be reduced somewhat by mid-day irrigation. In addition, mid-day irrigation may result in compaction problems from concentrated play that normally occurs then. Therefore, superintendents should preferably irrigate in early morning, secondly at night, and least desirably, during the day.

## Irrigation Area Priorities

Golf course roughs comprise about 50% of a golf course's total acreage but they are not considered a primary playing area. Greens, meanwhile, typically comprise less than 5% of a course's total acreage but are considered the primary playing area. During periods of drought, obviously, most water savings could be realized by restricting irrigation to only priority play areas. The following is a general priority listing (from highest to lowest priority) applicable to most golf areas:

**TABLE 8.6**
**Conversions and Calculations for Determining Turfgrass Irrigation Needs**

| | | | | | |
|---|---|---|---|---|---|
| 1 acre-inch (water needed to cover 1 | = | 27,154 gallons | 1 acre-foot (water needed to | = | 325,851 gallons |
| acre to 1-inch depth) | = | 43,560 cubic inches | cover 1 acre to 1-foot depth) | = | 43,560 cubic feet |
| | = | 3,630 cubic feet | | = | 75,271,581 cubic inches |
| 1 inch/1,000 square feet | = | 623.33 gallons | 7.48 gallons | = | 1 cubic foot |
| | = | 83.33 cubic feet | | = | 1,728 cubic inches |
| 1 gallon | = | 0.134 cubic foot | 1 cubic foot | = | 7.4805 gallons |
| | = | 231 cubic inches | 1 psi | = | 2.31 feet of head |
| | = | 8.34 pounds water | 1 foot of head | = | 0.433 psi |
| 1 pound of water | = | 0.1199 gallon | 1 million gallons | = | 3.07 acre-feet |
| | = | 27.7 cubic inches | | | |

1. Greens
2. Greens surround
3. Tees
4. Fairway landing area
5. Remaining area of fairways
6. Primary roughs and trees
7. Secondary rough
8. Out-of-play areas

In the United States, water sources for golf course irrigation include: open water (lakes, ponds, etc.) (52%); on-site wells (46%); rivers, streams, creeks, canals (17%); municipal water systems (14%); and recycled water (12%).

### Example

Water use engineers employed at a municipality require a golf course to justify its water use permit in terms of total amount of water requested and how this value was determined (patterned after Green 2005).

A. Determine average yearly ET rate from one of the methods listed previously. In this example, 56.37 inches (4.7 feet, 1.4 m) is used.
B. Determine normal yearly precipitation rate. In this example, 10.67 inches (27 cm) is used.
C. Area of irrigated turfgrass. In this example, 110 acres is used (3.1 acres for greens, 3.7 for tees, 43.7 for fairways, and 59.5 for roughs),
D. Determine the irrigation efficiency (DU). In this example, 70% is used.

| | Greens | Tees | Fairways | Roughs |
|---|---|---|---|---|
| E. Turf area (acres) | 3.1 | 3.7 | 43.7 | 59.5 |
| F. Turfgrass | Bentgrass | Bermuda overseeded October–May | Bermuda overseeded October–May | Bermuda |
| G. $K_c$ (crop coefficient) | 0.8 | 0.75 | 0.75 | 0.65 |
| H. Turf water use [A × G] (which is ET × $K_c$) | 45.1 | 42.3 | 42.3 | 36.6 |
| I. 25% precipitation (inches): [B × 0.25][a] | 2.7 | 2.7 | 2.7 | 2.7 |
| J. Water use adjusted for 25% precipitation (inches): [H – I] | 42.4 | 39.6 | 39.6 | 33.9 |
| K. Irrigation water use (inches): [J/D] | 60.6 | 56.6 | 56.6 | 48.4 |
| L. K converted to feet: [K/12] (12 inches = 1 foot) | 5.1 | 4.7 | 4.7 | 4.0 |
| M. Annual irrigation use (acre-feet): [E × L] | 15.8 | 17.4 | 205.4 | 238.0 |
| N. Annual irrigation water use: [sum of M for all turf areas] | 477 acre-feet (or 155,430,927 gallons) | | | |

[a] Water use regulators often use a precipitation efficiency adjustment value to reflect the amount (percentage) of usable precipitation by plants. Rainfall is often at inefficient amounts (too high or low) or at the wrong agronomic time.

In this example, to compare calculated annual irrigation use to the overall formula, ET × area, the following was determined:

O. ET × 110 acres: A (feet) × C (total turfgrass area) or 4.7 feet × 110 acres = 517 acre-feet predicted by the simple formula.
P. Calculation efficiency for water budget: N/O × 100 or 477 acre-feet ÷ 517 acre-feet × 100 = 92%. This value indicates the simple formula of ET × area overestimated water needs by 8% compared to the preceding water budgeting process.

## WATER CONSERVATION

Daily water conservation practices integrate many of the previously mentioned practices and technologies. Using computerized irrigation systems to better pinpoint irrigation needs for various soil types or turfgrass use, utilizing weather stations to determine daily ET rates, installing soil moisture sensors to monitor soil moisture levels, and using automatic pump shutdown switches when significant rainfall occurs are examples of water conservation techniques (Figure 8.12).

A holistic approach to water conservation is required. If not, turf water conservation will probably be mandated by governing bodies and may include (a) changing the grass species, (b) allowing only native grasses and xeriscape designs, (c) reducing the area of irrigated turf, or (d) improving (updating and expanding) current irrigation designs to become more efficient. For example, the common types of Kentucky bluegrass typically have less drought tolerance than the compact type cultivars. The following is an example of questions to answer when developing best management practices (BMPs) for water for a particular golf facility (modified from the Georgia Golf Course Superintendent's Association; http://www.ggcsa.com/-best-management-practices-for-water-conservation).

Steps to develop BMPs for turfgrass water conservation include:

1. Site assessment for greens, tees, fairways, roughs, landscapes, and club grounds, including:
   a. Area size involved (square feet, acres, square meters, or hectares)
   b. Turf (plant) species involved
   c. General factors such as mowing height, soil type, special technology, other pertinent information
   d. Irrigation audit:
      - Pump station—year, type, pump size(s), gallonage, safety features, condition, maintenance schedule, other information
      - Controls—year, system type, number of field controllers, condition, other information
      - Irrigation system—year, type, valves, output and distribution efficiency (DU) for greens, tees, fairways, roughs, plus other information
2. Overall water needs
   a. Metering—number of meters, location(s), other information
   b. Recordkeeping—yearly usage, scheduling, other information (attach one year of records)

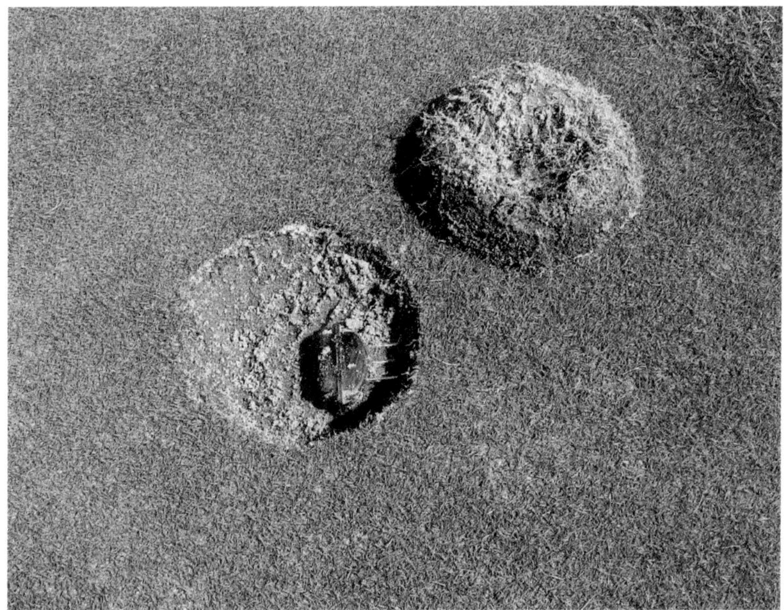

**FIGURE 8.12** Installed soil moisture probes pinpoint moisture levels throughout a golf course.

   c. Water testing—schedule, other information (attach most recent tests)
   d. Reservoir—size, type of water, source of water, other information
   e. Alternative water sources (yes or no); if yes, explain
   f. Future needs—explain in detail
3. Best management practices and current conservation measures
   a. Current irrigation control/costs—for pump station, controllers/computer, irrigation system components (sprinklers, pipe, valves, fittings, etc.), preventative maintenance of all these, other components
   b. Staffing control/maintenance costs—supervisor time, irrigation technician time, other assistance time (include diagnosis, repairs, recordkeeping, inventory, scheduling, etc.)
   c. Scouting costs—daily scouting time (explain)
   d. Hand-watering costs—daily hand-watering time (explain)
   e. Night watering capability—explain how this reduces loss and reduces disease occurrence
   f. Rain, leak loss costs
   g. Traffic controls/costs—daily traffic control time (explain)
   h. Management for water conservation (describe each):
      - Mowing heights
      - Soil cultivation (number of times yearly for greens, tees, and fairways)
      - Evapotranspiration utilization—list source for monitoring weather data to schedule irrigation events based on ET values
      - Landscape material selection explanation
      - Natural areas
      - Fertilization—yearly rates, slow versus quick release, stress nutrient use
      - Pest management (explain IPM programs)
      - Wetting agent use (explain products, timings, etc.)
      - Soil moisture sensors—calibrating and determining thresholds (saturation, field capacity, wilting point)
      - Recordkeeping:
         - Scouting labor hours and costs
         - Hand watering hours and costs
         - Irrigation repair hours and costs
         - Repair parts costs
         - Water usage weekly, monthly, and yearly
         - Water quality tests
         - Pesticide and fertilizer applications (in relation to irrigation)
         - Other methods
   j. Irrigation methods—combination of plant based, soil based, atmosphere based, and budget report
   k. Goal setting—for example:
      - Benefits of golf course and turf—that is, economic contributor, carbon dioxide exchange for oxygen, temperature moderation, erosion control, water filtering for improved water quality, wildlife sanctuary, recreational benefits, community outreach (i.e., First Tee programs), others
      - Publish this best management plan for use at club—articles in the club newsletter or on its web page explaining proper water use and efforts towards water conservation
   m. During drought, display water conservation plans (posters) in the pro shop and locker rooms and to patrons for use at home
4. Water conservation plan
   a. List reasons for water conservation, for example:

- Proper water management dictates that overwatering is unacceptable
- Economic considerations that inefficient watering costs money
- Depleted water supplies and reduced water quality
- Other reasons

b. List measures implemented by the course to reduce drought effects, for example:
   - Raise mowing heights where possible
   - Reduce or stop mowing nonirrigated areas
   - Increase hand-watering and wetting agent use
   - Improve uniformity by improving pressure regulations, levelling heads, etc.
   - Other reasons

c. Irrigation upgrades implemented by the course for increased water conservation
   - List possible options and costs

d. List and describe actual plans for water conservation at various mandated drought levels

5. Attachments, for example:
   a. Pump station records:
      - Most recent water quality test results
      - Man-hour records
      - Budgets
      - Repair records
      - Copies of publications

## HYDROPHOBIC SOILS AND THEIR MANAGEMENT

### Hydrophobic Soils

Hydrophobic (or "water-hating") soils such as those associated with localized dry spots, occur as organic matter decomposes and humic and fulvic acids (nonpolar) produced eventually coat individual sand grain particles (Figure 8.13). Sands are more prone to develop water repellency than finer soil textural classes due to the low or smaller surface area to volume ratio of sand particles. Certain fairy ring fungi also produce a mat of belowground hyphae that often become hydrophobic. These acids have extruding nonpolar ends that repel water particles, much like wax, leading to the dry spots. These conditions can be so severe that normal irrigation is often ineffective in restoring adequate soil moisture. Repeated wet and dry soil cycles aggravate hydrophobic soils. Other potential sources of organic acids that coat soil particles include exudates from turfgrass roots, lipids from decomposing organic matter, surface waxes from cuticles of turfgrass plant leaves, and fungal or soil microbial by-products.

Hydrophobic soils can cause problems on golf courses (especially golf greens) and other turf areas, in nurseries and greenhouses, and in open fields. Localized dry spots tend to be a surface phenomenon, in the top 2 inches (5 cm), but can occur up to 6 inches (15 cm) deep. Problems with hydrophobic soils are also commonly associated with citrus production areas, where mine spoils have been deposited, and with burned-over forestland and grassland.

A soil water repellency water drop penetration test is used to measure how hydrophobic a soil may be. Droplets of water are placed every inch down a soil core and the time required for the droplet to penetrate the soil reflects the soil's degree of repellency (Table 8.7).

### Adjuvants

An adjuvant is a spray additive that helps modify the surface properties of liquids to enhance their performance and handling. "Adjuvant" is a broad term and includes surfactants, wetting agents, crop oils, crop oil concentrates, activators, antifoaming agents, detergents, drift control agents,

**FIGURE 8.13** Localized dry spots develop as organic matter decomposes, coating sand particles with various "water-hating" (or hydrophobic) acids.

**TABLE 8.7**
**Degree of Soil Hydrophobicity (or Repellency) Based on the Water Drop Penetration Test**

| Water Drop Penetration Time (seconds) | Degree of Repellency |
|---|---|
| 0 to 5 | None |
| 5 to 60 | Slight |
| 60 to 600 | Moderate to high |
| 600 to 3,600 | Severe |
| >3,600 | Extreme |

emulsifiers, fertilizers, spreaders, sticking agents, dispersing agents, penetrants, pH modifiers, and compatibility agents.

**Surfactants.** Surfactants are adjuvants that produce physical or chemical changes at the interface of a liquid and another liquid, solid, or gas. These typically lower the surface tension of a liquid, allowing easier spreading, and lower the interfacial tension between two liquids. Since this occurs at the surface, the term "surfactant" is short for SURFace ACTive AgeNT. These facilitate emulsifying, dispersal, wetting, spreading, sticking, penetrating, or other surface-modifying properties of liquids into plants and soil. Surfactants are widely used in everyday life in medicines, medical care, fire extinguishers, paints, inks, adhesives, waxes, laxatives, hair conditioners, and agriculture. Surfactants include emulsifiers, detergents, dispersants, penetrants, soaps, spreaders, stickers, and wetting agents.

To understand how surfactants work, it helps to understand how water works. Each water molecule is bipolar, meaning it has a negative and a positive charge, similarly to a magnet. When several water molecules come into contact with each other, these positive and negative forces attract each other. This attraction of water molecules for each other is termed *cohesion*. The molecules on the surface of a water droplet are held together with more force than those of the interior water molecules. This causes surface tension, which causes the droplet to behave as if a thin, flexible film covered its surface, tending to keep the water molecules apart from other substances; this can prevent many things from going into solution and getting wet. This surface tension is the tendency of the water surface molecules to be attracted toward the center of the liquid, causing a water droplet with a dense, elastic membrane around it. Wetting agents help break this surface tension, so the water droplets break down, allowing dispersal. *Adhesion*, the attraction of water molecules to other substances, is the force causing water molecules to adhere to other objects, such as soil particles.

The effects of these forces can be illustrated by placing a drop of water on a napkin and another drop on a piece of waxed paper or a newly waxed vehicle. On the napkin, the force of adhesion between the water molecules and the paper molecules is greater than the force of cohesion that holds the water molecules together. As a result, the water droplet spreads out and soaks into the paper. Certain organic substances such as wax, however, do not have an adhesive force for water. On the waxed paper, therefore, the water "beads up"—that is, the droplet remains intact. The water molecules are not attracted to the wax that coats the paper's surface; instead, the water molecules cohere to each other. When the adhesive forces between water molecules and an object are weaker than the cohesive forces between water molecules, the surface repels water and is said to be hydrophobic (Figure 8.14).

Surfactants are composed of two parts: a water-soluble end that is polar or hydrophilic, meaning it is attracted to water, and an oil-soluble hydrocarbon chain that is lipophilic or nonpolar, meaning it is attracted to oil and not water. Water forms bonds with polar molecules but does not bond to nonpolar molecules and is repelled by them. Chemists manipulate the ratio of the hydrophilic (polar) portion of the molecule to the lipophilic (nonpolar) to produce different surfactants, with different molecular weights, and different characteristics. Thousands of potential combinations exist—thus, the reason for the hundreds of surfactants available. These components of a surfactant molecule help break water surface tension, allowing the solution to be more evenly dispersed on a surface and to reach its target. Two major types of surfactants are emulsifiers and wetting agents.

**Wetting agents.** Wetting agents are a type of surfactant that reduces interfacial tensions and causes a liquid to spread more easily over, or to penetrate, a solid surface, thus increasing the contact with treated surfaces (Figure 8.15). They can reduce this surface tension by 50% to 60% or more. In plant and soil sciences, wetting agents have a number of uses, including reducing soil hydrophobicity (i.e., localized dry spots), reducing dew and frost formation, firming bunker faces and sand, improving irrigation efficiency, reducing vehicle path dust, improving soil water infiltration, improving pesticide efficacy, and others. All uses require watering-in.

Wetting agents are classified based on how they ionize or separate into charged particles in water. Four types of wetting agents are

1. **Anionic**—negatively charged. These are often used for dispersion of clays in wettable dry granules as well as detergents, and degreasers. They may burn plants.
2. **Cationic**—positively charged, often used as biocides (disinfectants), soaps, shampoos, and fabric

**FIGURE 8.14** Water droplets on thatch indicating its hydrophobicity.

FIGURE 8.15 Water treated with a wetting agent (right), which readily enters soil compared to none (left).

softeners. These are strongly adsorbed to soil particles with high plant burn potential and are rarely used.

3. **Amphoteric**—charge is pH dependent of the solution and is seldom used.
4. **Nonionic**—neutrally charged. Also referred to as polyoxyethylene (POE) or alkylphenol ethoxylate (APE). Often used to enhance water movement into soil and are the preferred wetting agents in turf.

Nonionic surfactants do not ionize and thus remain uncharged. These are the most commonly used types of surfactants and are compatible with most pesticides. They are unaffected by water containing high levels of Ca, Mg, or ferric ions. They also can be used in strong acid solutions. Anionic wetting agents ionize with water to form a negative charge, while cationic ones ionize with water to form a positive charge. Anionic wetting agents may deleteriously impact soil structure (negative soil charges repel the negatively charged anionic wetting agents) and are often phytotoxic to plants. Amphoteric surfactants can be either anionic or cationic depending on the acidity of the solution. Cationic materials are strongly adsorbed to soil particles and may become ineffective. If used in hard water, anionic and cationic surfactants can cause an insoluble precipitate or foam to form. These are only occasionally used. Soaps and detergents are types of surfactants, but typically are anionic and react with salts in hard water and form a precipitate (scum) or foam, or are phytotoxic to plants.

Within the nonionic surfactant chemistry, two main groups of wetting agents are currently available: soil penetrants and water retainers.

1. **Soil penetrants.** These are often characterized as "water-moving" chemistry, classified by having ethylene oxide terminal functional groups. Ethylene oxide groups are hydrophilic, being able to attract or disperse water molecules. Soil-penetrating wetting agents generally are used to reduce water repellency. They thus increase water infiltration and percolation into the rootzone, providing more uniform soil moisture distribution within the profile leading to "fast and firm" playing conditions. Soil penetrants:
   a. Change tension
   b. Improve soil water infiltration
   c. Reduce water repellency
2. **Soil water retainers.** These are often characterized as "water-holding" chemistry, containing propylene oxide terminal functional groups. Propylene oxide groups are hydrophobic and thus repel water molecules. These are used where moisture retention is needed, especially sand-based rootzones with little organic matter and high infiltration and percolation rates. These are especially useful to help retain moisture during drought periods. Soil water retainers:
   a. Change tension
   b. Restore soil wettability
   c. Correct and/or avoid soil water repellency

Often, similar chemistry is used as soil penetrants and retainers; however, higher rates are needed to obtain retainer results. Higher rates obviously translate to higher costs. Others take advantage of both types of wetting agents as many newer commercial products are blends of each.

When a wetting agent is applied, the nonpolar ends react (or align) with the nonpolar ("water-hating") ends of the acid-coated sands. The polar ("water-loving") ends of the wetting agent then are exposed outward and can attract water, restoring wettability (Figure 8.16). Wetting agents, however, do not substantially remove the hydrophobic acid coating and do not aid in decomposing thatch, alleviating black layer, or reducing soil compaction. For most products, to minimize phytotoxicity, irrigation after wetting agent application is critical, as well as not treating when temperatures are extreme. When soil organic matter content exceeds ~3.0% (by weight) in the top 2 inches (5 cm) of soil, this organic matter may dry down more slowly when treated with a wetting agent. Increasing use rates above label recommendations generally does not increase the longevity or effectiveness of products and increases the chance of plant damage (Table 8.8). Wetting agents can improve the efficiency of irrigation and when water repellent soil conditions occur, wetting agent use may improve root growth and survival. Soil organisms naturally breaking down the materials eventually occurs, causing the need for repeat applications for extended results. When treating golf greens, it generally is best to treat the whole green versus just treating hot spots. Wetting agents do not solve a subsurface drainage problem but may help leach salts from the rootzone in certain situations.

In general, studies have shown that the extent of improvement in infiltration rate is affected by the type of wetting agent used, its dilution, previous use of wetting agents on the soil, and the water content of the soil at the time water is applied

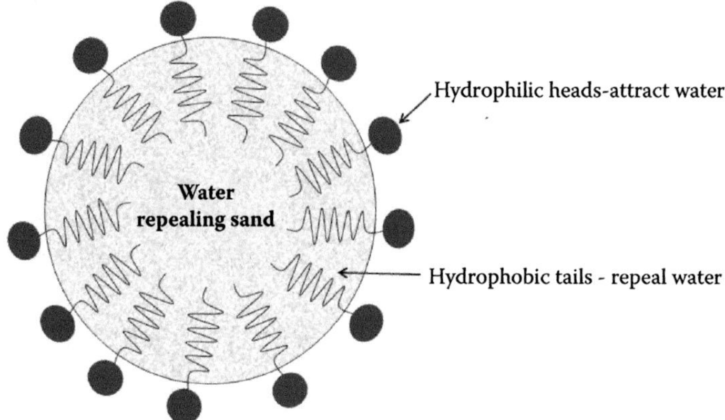


**FIGURE 8.16** Illustrating how wetting agent molecules align their hydrophilic heads to attract water to otherwise water-repealing sand.

**TABLE 8.8**
**Wetting Agent Use and Soil Water Repellency Development**

- On golf greens, soil water repellency often develops between 8 and 18 months after construction.
- The cause of repellency—organic coating of sand particles and thatch—cannot be practically prevented or removed.
- This tends to be a surface phenomenon in the top 2 inches (5 cm) and the degree of repellency varies across a green.
- When timing wetting agent use, early (preventative) applications typically provide better results than curative one. Reapplications are typically 30 days apart throughout summer.
- Soil water repellency is not associated with any particular turfgrass management practice.
- Incorporation of fine-texture materials (e.g., silt and clay) reduces or eliminates water repellency.
- Reducing soil organic matter below 3% (by weight) in the top 2 inches (5 cm) of soil helps reduce water repellency.
- Due to varying water repellency across a golf green, when treating with a wetting agent, it is best to treat the whole surface versus spot treating.
- Wetting agent use typically improves root growth and survival.
- Wetting agent use typically reduces the amount of irrigation required and/or improves its efficiency.
- No one wetting agent works universally under all environmental conditions and regions.

(Figure 8.17). Several studies have shown that the infiltration rate of a hydrophobic soil, once it has been wetted, remains higher than it was before it was wetted, even if it is allowed to dry out again. Applying wetting agents often reduces the severity of the condition, but best use is in combination with coring to allow water to pass through the hydrophobic surface layer. Also, keeping the soil moist seems to be the best defense against the development of dry spots, as allowing the soil to dry out intensifies the problem. For maximum efficiency, if your goal is to rewet a dry, hydrophobic soil, the area should be aerified, followed by prewetting the area, applying the wetting agent or soil surfactant, and then watering it rapidly and liberally into the soil profile. This should be performed early in the growing season before moisture stress causes severely hydrophobic soils.

## IRRIGATION WATER QUALITY

Golf courses increasingly face the use of poorer quality irrigation sources. Fortunately for turfgrass managers, with proper precautions and altered management practices, poor quality irrigation water may be used safely in most situations.

Successful irrigation management requires regular monitoring of both soil and water chemistry, especially salt content. Observations that suggest water quality problems include (1) soil water acceptance has changed (reduced) from irrigation or rainfall, (2) soils feel moist but plants still appear stressed by drought, (3) soil surface slick spots and wet areas, (4) soil surface crusting and flaking, and (5) wet soil surface but dry soil beneath. Table 8.9 lists the tests needed concerning soil and water quality.

Routine water tests are performed seasonally and soil tests twice yearly unless problems develop or water sources change, at which point additional testing is needed until a comfort level is reached. Specific ion toxicity, suspended solids, and soil nutrient imbalances are associated with effluent water use.

## SALTS

A salt is a combination of positively charged (cations) and negatively charged (anions) ions. Cations include Ca, Mg, Na, ammonium, and K; anions include carbonates, bicarbonate, nitrate, sulfate, chloride, and boron (Table 8.10). Table salt (sodium chloride) is also found in some soils. Insoluble salts (e.g., gypsum and lime) occur, but excessive soluble salts are the primary ones that may impede plant growth rather than the insoluble ones. High amounts of soluble salts in the soil solution reduce water availability, causing the turfgrass to

**FIGURE 8.17** Turf treated with a wetting agent (right) compared to none (left).

**TABLE 8.9**
**Chemical Tests Used to Determine the Quality of Irrigation Water**

| Test | Description |
|---|---|
| Water-soluble salts (or salinity drought hazard) | • Total salt content as measured by the electrical conductivity ($EC_w$) or total dissolved salts (TDS) of water; excessive salts produce plant physiological drought |
| Sodium status | • Soil Na level proportionally to Ca and Mg ions; Na measured by SAR, ESP, or adjusted SAR (adj. SAR); SAR also used to assess the Na levels of water; excessive Na causes soil structure deterioration |
| Alkalinity | • Bicarbonates and carbonates as measured by RSC; these tie up Ca in soil, not allowing it to displace Na ions |
| Acidity (pH) | • Measures acidity and corresponding lime requirements |
| Specific ions' toxicity | • Toxic ion levels, especially boron, chloride, fluoride, sulfate, and nitrate-N |
| Suspended solids | • Measured by TSS |
| Soil nutrient imbalance | • Based on (a) sufficiency levels of available nutrients and cation ratio, (b) CEC, (c) percent base saturation, and (d) percent organic matter |

**TABLE 8.10**
**Salts and Their Ionic Components Commonly Found in Soil and Irrigation Water**

| | | Fertilizer (soluble) Salts | |
|---|---|---|---|
| **Cations** | **Anions** | **Soluble Salts** | |
| Calcium ($Ca^{+2}$) | Chloride ($Cl^-$) | Sodium chloride (NaCl) | Potassium nitrate ($KNO_3$) |
| Magnesium ($Mg^{+2}$) | Sulfate ($SO_4^{-2}$) | Calcium chloride (CaCl) | Potassium chloride (KCl) |
| Sodium ($Na^+$) | Nitrate ($NO_3^-$) | Epsom salt ($MgSO_4$) | Potassium sulfate ($K_2SO_4$) |
| Potassium ($K^+$) | Borate ($BO_3^{-3}$) | Ammonium nitrate [$(NH_4)_2NO_3$] | Urea [$CO(NH_2)_2$] |
| Ammonium ($NH_4^+$) | Carbonate ($CO_3^{-2}$) | Ammonium sulfate [$(NH_4)_2SO_4$] | Sodium sulfate ($Na_2SO_4$) |
| | Bicarbonate ($HCO_3^-$) | Magnesium chloride ($MgCl_2$) | Sodium bicarbonate ($NaHCO_3$) |
| | | **Insoluble Salts** | |
| | | Limestone ($CaCO_3$) | Gypsum ($CaSO_4$) |

be prone to drought stress and stunting its growth, as well as typically causing a darker green color from more plant cells per unit area of stunted leaves. This is the most common salt problem involved with turfgrasses. Other problems with excessive salts are accumulation of potentially toxic ions such as boron and chloride and nutritional imbalances.

Soluble salts originate from the weathering of primary rocks and minerals with oceans becoming the eventual reservoir of these as water moves through the hydrological cycle. The following equation determines the amount of salt applied when irrigating with saline water:

$$\text{Pounds salt applied/acre} = \text{irrigation water salinity level (ppm or mg L}^{-1}) \times 2.72 \text{ million pounds (weight of water per acre-foot)} \times \text{acre-feet water applied}$$

#### Example

How much salt is applied per 1,000 square feet if 1 inch (2.5 cm) of water with salinity levels of 1.0 dS/m (~640 ppm) is used? One inch = 0.083 foot.

$$\begin{aligned}
&\text{Pounds salt applied acre} \\
&= \text{irrigation water salinity level (ppm or mgL}^{-1}) \\
&\times \frac{2.72 \text{ million pounds}}{\text{per acre-foot}} \times \text{acre feet water applied} \\
&= 640 \text{ ppm} \times \frac{2.72 \text{ million pounds}}{\text{acre-foot water}} \times 0.083 \text{ acre-foot} \\
&= 144 \text{ pounds salt applied/acre}
\end{aligned}$$

To determine the amount of salt applied per 1,000 square feet, divide 144 pounds salt/acre by 43.56. For this example, 3.3 pounds of salt are applied per 1,000 square feet when 1 inch of irrigation water with a salinity level of 640 ppm is used.

### Measuring and Classifying Irrigation Salinity

Four major units are used to express the concentration of constituents in water: (1) parts per million (ppm), milligrams per liter (mg/L), milliequivalents per liter (meq/L), and millimoles per liter (mmol/L).

Parts per million is one part salt to 1 million parts water, or milligram of salt per kilogram of solution. Since a kilogram of water equals 1 L, parts per million (ppm) is interchangeable with milligrams per liter (mg/L). The term ppm is used to report low concentrations in irrigation water of constituents such as Fe, manganese, boron, nitrates, and nitrate-nitrogen. Grains per gallon is an English system of units used mainly today to report hardness of water. To convert grains per gallon to parts per million, multiply this by 17.2.

Milliequivalents per liter is the most meaningful means of reporting the major chemical constituents of water. Salts are combinations of cations and anions in definite weight ratios. These weight ratios are based on the atomic weight of each constituent ion and upon its electrical charge (valence). An equivalent weight of an ion is its atomic weight divided by its valence. Therefore, milliequivalents per liter is a measurement of charge concentration per liter.

Unfortunately, no one method is universally used by various laboratories; thus, one needs to know how to convert from one unit of measurement to another. With solutions, parts per million, milligrams per liter, and milligrams per kilogram can be used interchangeably. To convert parts per million to milliequivalents/liter (or vice versa), the equivalent weight of the ion is needed (Table 8.11). The following calculations can then be used for converting one unit of measurement to another:

$$\text{Parts per million} = \text{equivalent weight} \times \text{millieqiuvalents/liter}$$

$$\text{Milliequivalents/liter} = \text{parts per million} \div \text{eqiuvalent weight}$$

$$\text{Milliequivalents/liter} = \text{molecular weight} \div \text{total valence number}$$

Salinity hazard is determined by measuring the ability of water to conduct an electrical current because increasing amounts of salt in water increases its conductivity. Salty water is a good conductor of electrical current, whereas pure water is a relatively poor conductor. Water salinity is expressed in two different ways, either as *electrical conductivity of the water* ($EC_w$) or *total dissolved salts* (TDS) (also reported as total soluble salts, TSS). There are several units commonly used to express $EC_w$: deciSiemens per meter (dS/m), millimhos per centimeter (mmhos/cm) (mhos = reciprocal of ohm), or micromhos per centimeter (μmhos/cm). The relationship between these units is (Table 8.12):

$$\begin{aligned}
1 \text{ dS/m} &= 1 \text{ mS/cm} = 0.1 \text{ S/m} = 1 \text{ mmhos/cm} \\
&= 1{,}000\ \mu\text{mhos/cm} = 640 \text{ ppm TDS}
\end{aligned}$$

Total dissolved salts are expressed in parts per million or milligrams per liter and are generally not measured directly, but rather are calculated from an $EC_w$ measurement.

$$\text{TDS (mg/L or ppm)} = EC_w \text{ (mmhos/cm or dS/m)} \times 640$$

Most laboratories use this relationship to calculate TDS from $EC_w$, but some multiply the amount by 700.

#### Example

1. An irrigation source has an $EC_w$ of 0.53 millimetermhos/cm. What would the $EC_w$ be in decisiemens/meter, micrometermhos/centimeter, and TDS parts per million?
   a. Since 1 dS/m = 1 mmhos/cm, then 0.53 dS/m = 0.53 mmhos/cm

**TABLE 8.11**
**Laboratory Analysis to Determine Water Quality and Factors for Converting Ion Concentration Reported in Parts per Million or Milligrams per Liter to Moles per Liter or Milliequivalents per Liter**

| Analysis | Reporting Symbol | Reporting Unit | Molecular Weight (g) | Milliequivalent Weight (mg/meq) |
|---|---|---|---|---|
| Electrical conductivity | $EC_w$ | mmhos/cm | – | – |
| Calcium | $Ca^{+2}$ | meq/L | 40 | 20 |
| Magnesium | $Mg^{+2}$ | meq/L | 24.3 | 12.2 |
| Sodium | $Na^+$ | meq/L | 23 | 23 |
| Carbonate | $CO_3^{-2}$ | meq/L | 60 | 30 |
| Bicarbonate | $HCO_3^-$ | meq/L | 61 | 61 |
| Chloride | $Cl^-$ | meq/L | 35.4 | 35.4 |
| Sulfate | $SO_4^{-2}$ | meq/L | 96 | 48 |
| Boron | B | mg/L | 10.8 | 10.8 |
| Nitrate-nitrogen | $NO_3$-N | mg/L | 14 | 14 |
| Acidity | pH | pH | – | – |
| SAR | – | meq/L | – | – |
| Potassium | $K^+$ | meq/L | 39.1 | 39.1 |
| Lithium | $Li^+$ | mg/L | 7 | 7 |
| Iron | $Fe^{+2 \text{ or } +3}$ | mg/L | 55.8 | 27.9 or 18.6 |
| Ammonium-nitrogen | $NH_4$-N | mg/L | 14 | 14 |
| Phosphate phosphorus | $PO_4$-P | mg/L | 31 | varies |

**Conversion Values between Milligrams/Liter and Milliequivalents/Liter for Various Water Constituents**

| Constituents | Multiply by the Following to Convert Milligrams/Liter (or parts per million) to Milliequivalents per Liter | Multiply by the Following to Convert Milliequivalents Per Liter to Milligrams/Liter (or parts per million) |
|---|---|---|
| Sodium ($Na^+$) | 0.043 | 23 |
| Magnesium ($Mg^{+2}$) | 0.083 | 12 |
| Calcium ($Ca^{+2}$) | 0.05 | 20 |
| Chloride ($Cl^-$) | 0.028 | 36 |
| Sulfate ($SO_4^{-2}$) | 0.021 | 48 |
| Bicarbonate ($HCO_3^-$) | 0.016 | 61 |
| Carbonate ($CO_3^{-2}$) | 0.033 | 30 |
| Alkalinity ($CaCO_3$) | 0.02 | 50 |

**TABLE 8.12**
**Conversion Factors for Irrigation Water Quality Reports**

| Component | To Convert | Multiply by | To Obtain |
|---|---|---|---|
| Water nutrient | mg/L or ppm | 1.0 | mg/kg |
| Total dissolved salts (TDS) | ppm or mg/L | 0.0016 | $EC_w$ (mmhos/cm or dS/m) |
| Total dissolved salts | ppm or mg/L | 2.72 | Pounds salts applied/acre-foot irrigation |
| Water salinity hazard | dS/m | 1.0 | 1 μmhos/cm |
| Water salinity hazard | 1 mmhos/cm | 1000 | 1 μmhos/cm |
| Water salinity hazard | $EC_w$ (dS/m or mmhos/cm) | 640 | TDS (mg/L or ppm) |
| Water salinity hazard | TDS | 0.0016 | $EC_w$ (dS/m or mmhos/cm) |
| Water $NO_3$-N, $SO_4$-S, B | ppm | 0.23 | Pounds/acre-inch water applied |
| Irrigation water | Acre-inch | 27,150 | Gallon of water |

b. Since 1 mmhos/cm = 1,000 μmhos/cm, then

$$\frac{0.53 \text{ mmhos}}{\text{cm}} \times \frac{1{,}000\ \mu\text{mhos}}{\text{cm}} \times \frac{\text{cm}}{1 \text{ mmhos}} = \frac{530\ \mu\text{mhos}}{\text{cm}}$$

c. To convert millimeterhos(mmhos)/centimeter to parts per million, multiply by 640:

$$\frac{0.53 \text{ mmhos}}{\text{cm}} \times 640 = 339 \text{ ppm TDS}$$

2. The salt content of a water sample is 1,121 mg/L TDS. What is the salt content in decisiemens/meter and micrometerhos(μmhos)/centimeter? (Table 8.12).
   a. To convert TDS (milligrams/liter or parts per million) to decisiemens/meter, divide by 640:

$$1{,}121 \text{ mg/L} \div 640 = 1.75 \text{ ds/m}$$

   b. To convert decisiemens/meter (or mmhos/centimeter) to μmhos/centimeter, multiply by 1,000:

$$1.75 \text{ dS/m (or mmhos/cm)} \times 1{,}000 = 1{,}750\ \mu\text{mhos/cm}$$

3. Convert 100 ppm Ca to milliequivalents/liter. The equivalent weight of $Ca^{+2}$ is 20.

$$100 \text{ ppm Ca} \div 20 = 5 \text{ meq/L of Ca}$$

Water sample salinities are often compared to those of seawater with an average $EC_w$ of 54 dS/m or about 34,500 ppm dissolved salts, Na levels of 457 meq/L, and *sodium adsorption ratio* (SAR) of 57.4 meq/L.

Irrigation water is classified based on the salinity hazard, which considers the potential for damaging plants and the level of management needed for utilization as an irrigation source (Table 8.13). Generally, higher salinity levels can be used on sandy soils where salts can be flushed compared to similar values on poorly draining clay soils that may cause problems. Under typical summer stress, $EC_w$ of turfgrass irrigation should ideally not exceed 1.25 dS/m soluble salts. Salinity levels above 3.0 dS/m are unsuitable for any length as an irrigation source.

## Water Sodium Hazard

The primary cause of sodic or saline–sodic soil is using high sodium ($Na^+$) content irrigation water. While $EC_w$ is an assessment of all soluble salts in a water sample, sodium hazard (termed sodic or saline–sodic soil) accounts for sodium's specific detrimental effects on soil physical properties. The potential for irrigation water to have poor infiltration properties or sodium hazards is assessed by determining the SAR and the electrical conductivity of the water ($EC_w$). The sodium adsorption ratio relates the concentration of Na to the concentration of Ca and Mg. Calcium and Mg counter the negative effects of Na on soil structure. The higher the Na level is in relation to Ca and Mg, the higher is the SAR, the poorer is the water infiltration, and the problems with soil deflocculation (dispersal) increase. Calcium will hold soil together (or flocculate), while Na pushes (or disperses) soil particles apart. Dispersed soil has poor structure with a consistency similar to fine powder. The dispersed soil readily crusts and possesses water infiltration and permeability problems. SAR is defined as

$$\text{SAR} = \frac{Na^+}{\sqrt{\dfrac{Ca^2 + Mg^{+2}}{2}}} \qquad \text{SAR} = \frac{Na^+}{\sqrt{Ca^2 + Mg^{+2}}}$$

Ion concentrations on the preceding equation on the left are expressed in milliequivalents per liter while the equation on the right is expressed in millimoles per liter (mmol/L).

### Example

A water sample test reports 1,000 mg/L $Na^+$, 200 mg/L $Ca^{+2}$, and 100 mg/L $Mg^{+2}$. Find the SAR value in milliequivalents/liter.

**Step 1:** Calculate the number of milliequivalents per liter for each ion (or refer to Table 8.11):

$$Na^+\text{: } 1{,}000 \text{ mg/L} \div 23 \text{ mg/meq} = 43.5 \text{ meq/L}$$

$$Ca^{+2}\text{: } 200 \text{ mg/L} \div 20 \text{ mg/meq} = 10 \text{ meq/L}$$

$$Mg^{+2}\text{: } 100 \text{ mg/L} \div 12.2 \text{ mg/meq} = 8.2 \text{ meq/L}$$

**Step 2:** Place these values into the SAR equation as

$$\text{SAR} = \frac{Na^+}{\sqrt{\dfrac{Ca^2 + Mg^{+2}}{2}}} = \frac{43.5}{\sqrt{\dfrac{10 + 8.2}{2}}} = 14.4 \text{ meq/L}$$

From Table 8.13, this value is classified as having increasing problems.

### Example

A water analysis indicates a Na concentration of 85 meq/L, a Ca concentration of 33.3 meq/L, and a Mg concentration of 7.1 meq/L. What is the SAR value for this water?

$$\text{SAR} = \frac{Na^+}{\sqrt{\dfrac{Ca^2 + Mg^{+2}}{2}}} = \frac{85}{\sqrt{\dfrac{33.3 + 7.1}{2}}} = 18.9 \text{ meq/L}$$

From Table 8.13, this value is borderline as a severe problem.

**TABLE 8.13**
**General Guidelines for Irrigation Water Quality Concerning Total Salinity, Sodium Hazard, And Ion Toxicity**

| Item | Units | Minor Problems | Increasing Problems | Severe Problems |
|---|---|---|---|---|
| **Water Salinity Hazard (Influences Water Availability to Plants)** | | | | |
| $EC_w$ (electrical conductivity from saturated paste extract) | mmhos/cm or dS/m | <0.75 | 0.75 to 3.0 | >3.0 |
| SAR (sodium adsorption ratio) | meq/L | <10 | 10 to 18 | >18 |
| | mg/L or ppm | <70 | 70 to 200 | >200 |
| TDS | mg/L or ppm | <500 | 500 to 2,000 | >2,000 |
| **Soil and Water Sodium/Ion Hazard (Influences Soil Water Infiltration and Soil Structure Properties)** | | | | |
| SAR | meq/L | <10 | 10 to 18 | >18 |
| | mg/L or ppm | <70 | 70 to 200 | >200 |
| Adjusted SAR | meq/L | <6.0 | 6.0 to 9.0 | >9.0 |
| $EC_e$ (1:1 saturated paste extract) | dS/m or mmhos/cm | <1.5 | 1.6 to 3.9 | >4.0 |
| $EC_e$ (1:2 dilution) | dS/m or mmhos/cm | <0.80 | 0.80 to 2.4 | >2.4 |
| TDS | mg/L or ppm | <2,500 | 2,500 to 7,500 | >7,500 |
| ESP (exchangeable sodium percentage) | % | <13 | 13 to 15 | >15 |
| pH | 1 to 14 | 6.0 to 7.0 | 7.0 to 8.0 | >8.0 |
| Bicarbonate $(HCO_3^-)$ | mg/L or ppm | <120 | 120 to 180 | >180 |
| | meq/L | <1.5 | 1.5 to 3.0 | >3.0 |
| Carbonate $(CO_3^{-2})$ | mg/L or ppm | <15 | 15 to 50 | >50 |
| | meq/L | <0.5 | 0.5 to 1.65 | >1.65 |
| RSC (residual sodium carbonate) | meq/L | <1.25 | 1.25 to 2.50 | >2.50 |
| TSS (total suspended solids) | mg/L or ppm | <5 | 5 to 10 | >10 |
| Water hardness (Ca + Mg) | mg/L or ppm | <200 | 200 to 400 | >400 |
| Calcium carbonate ($CaCO_3$) or alkalinity | mg/L or ppm | <50 | 50 to 300 | >300 |
| Calcium (Ca) | mg/L or ppm | <25 | 25 to 250 | >250 |
| Magnesium (Mg) | mg/L or ppm | <20 | 20 to 40 | >40 |
| Chloride ($Cl^-$) | meq/L | <2 | 2 to 10 | >10 |
| | mg/L or ppm | <70 | 70 to 350 | >350 |
| Boron (B) | mg/L or ppm | <1 | 1 to 2 | >2 |
| Organic matter | % | <1 | 1 to 3 | >3 |
| **Foliar Ion Toxicity for Turfgrass** | | | | |
| Calcium (Ca) | mg/L or ppm | <60 | 60 to 100 | >100 |
| | meq/L | <3 | 3 to 6 | >6 |
| Fluoride (F) | mg/L or ppm | <0.25 | 0.25 to 0.5 | >0.5 |
| Iron ($Fe^{+2}$ or $Fe^{+3}$) | mg/L or ppm | <0.3 | 0.3 to 5.0 | >5.0 |
| Magnesium (Mg) | mg/L or ppm | <25 | 25 to 50 | >50 |
| | meq/L | <3 | 3 to 6 | >6 |
| Nitrogen (N) | mg/L or ppm | <11 | 11 to 23 | >23 |
| Phosphorus (P) | mg/L or ppm | <0.40 | 0.40 to 1.0 | >1.0 |
| Potassium (K) | mg/L or ppm | <20 | 20 to 50 | >50 |
| Sodium (Na) | meq/L | <3 | >3 | – |
| | ppm or mg/L | <70 | 70 to 200 | >200 |
| Sulfates $(SO_4^{-2})$ | mg/L or ppm | <100 | 100 to 200 | >200 |
| **Soil Ion Hazard for Sensitive Ornamentals** | | | | |
| Boron (B) | meq/L | <1.0 | 1.0 to 3.0 | >3.0 |
| | mg/L or ppm | <1.0 | 1.0 to 3.0 | >3.0 |
| Chloride ($Cl^-$) | meq/L | <3 | 3 to 10 | >10 |
| | mg/L or ppm | <100 | 100 to 300 | >300 |
| Sodium (Na) | mg/L or ppm | <70 | 70 to 200 | >200 |
| | SAR (meq/L) | <3.0 | 3 to 9 | >9 |

*(Continued)*

**TABLE 8.13 (CONTINUED)**
**General Guidelines for Irrigation Water Quality Concerning Total Salinity, Sodium Hazard, And Ion Toxicity**

| Item | Units | Minor Problems | Increasing Problems | Severe Problems |
|---|---|---|---|---|
| | **Foliar Ion Toxicity for Sensitive Ornamentals** | | | |
| Ammonium-N ($NH_4$-N) | mg/L or ppm | <5 | 5 to 30 | >30 |
| Bicarbonate $(HCO_3^-)$ | meq/L | <1.5 | 1.5 to 8.5 | >8.5 |
| Unsightly foliar deposits | mg/L or ppm | <90 | 90 to 500 | >500 |
| Chloride ($Cl^-$) | mg/L or ppm | – | 50 | >50 |
| | meq/L | <3 | >3 | – |
| Residual chlorine ($Cl_2$) | mg/L or ppm | <1.0 | 1 to 5 | >5 |
| Sodium (Na) | meq/L | <3 | >3 | – |
| | mg/L or ppm | <70 | >70 | – |
| Manganese (Mn) | mg/L or ppm | <0.2 | – | >0.2 |
| Nitrate-N ($NO_3$-N) | mg/L or ppm | <50 | 50 to 100 | >100 |
| Nitrate $(NO_3^-)$ | mg/L or ppm | <45 | 45 to 150 | >150 |
| Copper (Cu) | mg/L or ppm | <0.2 | – | >5 |
| Fluoride (F) | mg/L or ppm | <1.0 | 1.0 | >1.0 |
| Zinc (Zn) | mg/L or ppm | <2.0 | – | >2.0 |
| pH | 1 to 14 | <8.4 | – | >9.0 |
| | **Potential Algal Blooms in Irrigation Storage** | | | |
| Phosphate ($PO_4$-P) | mg/L or ppm | <1.2 | 1.2 to 2.4 | >2.4 |

*Sources:* Modified from Ayers, R. S., and D. W. Westcot, 1985, Irrigation and drainage paper 29. Rome, Italy: Food and Agriculture Organization of the United Nations. http://www.fao.org; Harivandi, M. A. 2007, *Golf Course Management* 75(6):98–108; Huck, M., 2000, *USGA Green Section Record* 39(5):1–5.; McCarty, L. B., ed., 2011, *Best golf course management practices* (3rd ed.). Upper Saddle River, NJ: Prentice-Hall, Inc.

Since salts and Na do not act independently, the effect of Na on soil particle dispersion, and thus permeability, is counteracted by high concentrations of soluble salts (measured as $EC_w$) in the irrigation water. The effects of high SAR on irrigation water infiltration are dependent on the electrical conductivity of the water. Soil water infiltration rates generally increase with higher soil salinity and decrease with lower salinity or higher soil Na SAR values. For a given SAR, the lower the $EC_w$ is, the greater the dispersion or poorer infiltration properties are; the higher the $EC_w$ is, the better the infiltration is (Figure 8.18). For example, irrigation water with a SAR = 15 has poor infiltration properties with $EC_w$ of 0.5 dS/m but good infiltration properties with an $EC_w$ = 3.0 dS/m. When the $EC_w$ = 0.5 dS/m or less, the water has very few minerals to flocculate soil particles. Thus, irrigating with this pure water strips minerals from cation exchange capacity (CEC) sites, causing dispersed particles to settle closely next to each other. The result is a compacted soil surface that forms a thin crust layer impeding water flow into the soil. Problems can develop quickly when the $EC_w$ = 0.2 dS/m. In the case of pure water, the problem exists regardless of the SAR value since very few minerals are present to begin with. A rule of thumb is that if the SAR is more than 10 times greater than the $EC_w$, then poor water infiltration is likely to occur.

General guidelines for precautions and management of irrigation water with various SAR values and an $EC_w \geq$ 1.0 dS/m are provided in Table 8.14. Clay-textured soils can have structural permeability problems if a water SAR > 9 meq/L or 2.0 dS/m is used over an extended period that reduces infiltration, percolation, and drainage, often causing low soil $O_2$ problems. In the earlier example where the water sample had a SAR of 14.4 meq/L, problems could occur if this water source was used long term on finer textured soils.

### Example

A superintendent has two water sources to choose from based on their sodium hazard.

Sample 1. SAR = 5.0 and EC = 0.5.
Sample 2. SAR = 5.0 and EC = 1.5.

From Figure 8.18, which one is more suitable?

Sample 1. Water infiltration problems may occur, especially on finer textured clay or silt-based soils.
Sample 2. This sample is less likely to cause soil water infiltration problems.

Some labs report adjusted SAR values instead of SAR. The adjusted SAR includes the added effects of the precipitation or dissolution of Ca in soils and is related to carbonate $(CO_3^{-2})$ and bicarbonate $(HCO_3^-)$ concentrations. Bicarbonates can

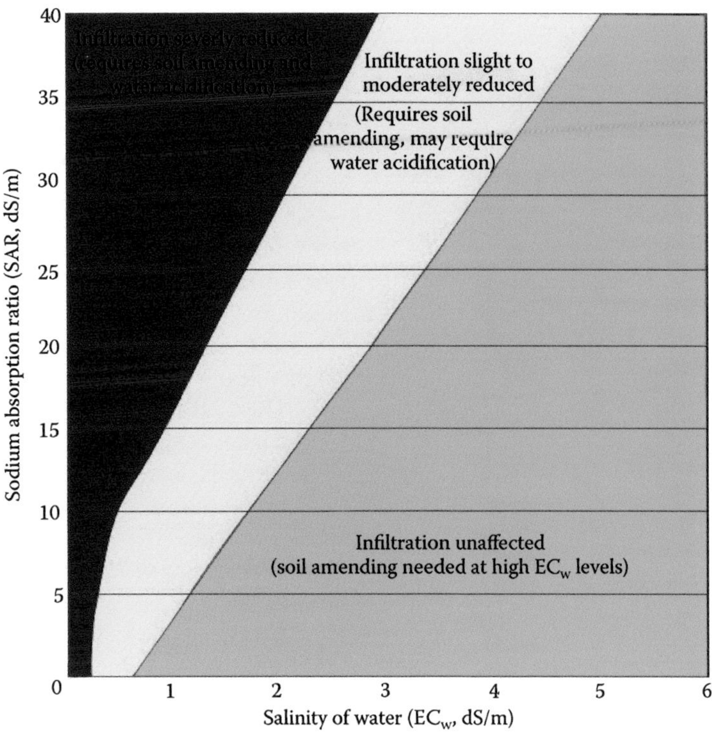


**FIGURE 8.18** Chart illustrating that soil water infiltration is influenced by its salinity and sodium absorption ratio.

**TABLE 8.14**
**SAR Values, Categories, and Precautions for Irrigation Sources with $EC_w \geq 1$ dS/m**

| SAR or adj SAR (meq/L) | Category | Precaution |
|---|---|---|
| 0 to 10 | Low-sodium water | • Little danger from structure deterioration to almost all soils. For ornamentals, water SAR values should be <10. |
| 10 to 18 | Medium-sodium water | • Problems on fine-textured soils and Na-sensitive plants, especially under low-leaching conditions. Soils should have good permeability. |
| 18 to 26 | High-sodium water | • Problems on Na accumulation on most soils. Good salt-tolerant plants are required along with special management, such as good drainage and the use of gypsum, and leaching. Generally, high and very high EC water should not be used for irrigating turfgrasses long term. |
| >26 | Very high sodium water | • Unsatisfactory except with high salinity ($EC_w$ > 2.0 dS/m), high Ca levels, and the use of gypsum. |

interact with soil Ca and Mg to precipitate out lime ($CaCO_3$) or magnesium carbonate ($MgCO_3$), leaving Na to dominate and causing an increase in sodium hazard. The adjusted SAR is a good predictor of what the SAR of a soil will be after several years of irrigation without any soil amendments. Adjusted SAs or residual sodium carbonate (RSC) values are best used when high levels of bicarbonate, $HCO_3^-$ (>2 meq/L or 120 mg/L), and carbonate, $CO_3^{-2}$ (>0.5 meq/L or 15 mg/L) occur.

## Soil Sodium Permeability Hazard

Although high Na levels in irrigation water can be directly toxic to plants (especially ornamentals), the most deleterious effect is on soil structure. Since sodium ions ($Na^+$) are monovalent (have only one positive charge), two Na ions are needed to displace divalent (two positively charged) ions such as calcium ($Ca^{+2}$) or magnesium ($Mg^{+2}$). This concern is greater on fine-textured soils such as clays and silt loams.

High soil Na causes finer textured soil clays and organic matter to crust badly, swell, separate, or disperse (termed *deflocculation*) to where aggregates break down into smaller units and smaller clay minerals and organic particles plug soil pores, reducing water infiltration and soil aeration (Figure 8.19). Soil then seals and becomes hard and compacted, reducing soil water and oxygen movement. Swelling occurs from Na ions replacing Ca and Mg cations in double layering of clays.

**Assessing soil salinity.** Saline soils are classified based on two criteria: (1) the total soluble salt or salinity content based on electrical conductivity of a saturated extract ($EC_e$), and (2) exchangeable sodium percentage (or sodium adsorption ratio). Additional information is also often used, such as carbonate content and potential toxic ions. It should be noted

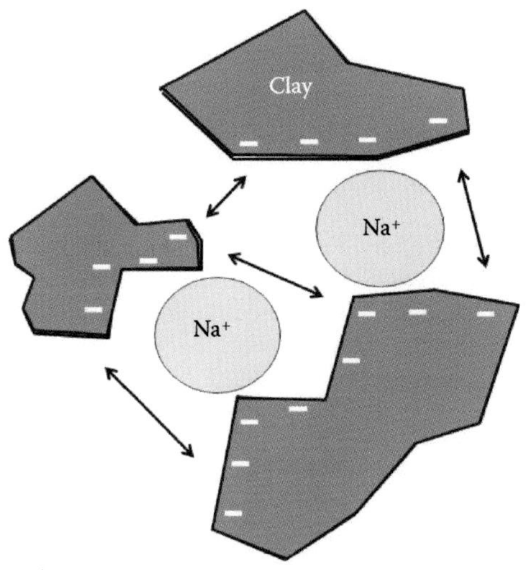


**FIGURE 8.19** Sodium ions can replace calcium and magnesium cations of certain clays, resulting in soil deflocculation (or breakdown).

that the saturated paste extract method is necessary for determining soil Na, Ca, and Mg for the SAR equation rather than other soil test extractant methods.

Soluble salts are measured in soils by the same basic method as used for water samples. A conductivity instrument measures electrical conductivity (EC) in an extract either from a saturated paste (preferred method) or from a soil:water dilution. As total salt concentration increases, $EC_e$ also increases. The SAR is a calculated value from a saturated paste extract sample based on milliequivalents/liter of Ca, Mg, and Na. The saturated paste extract is the most precise method to determine soil $EC_e$, SAR, and boron levels. A soil sample is brought just to the point of saturation using the irrigation source, allowing it to equilibrate for several hours, and then is subjected to vacuuming to extract the soil solution through filter paper. Spectrophotometers and other analytical equipment are then used to quantify the soil solution. Using the saturated paste extract, soils with $EC_e$ readings < 1.5 dS/m are considered to have low salt levels (Table 8.13). Soils with EC readings of 1.6 to 3.9 dS/m have medium levels. When soil readings are above 4.0 dS/m, soils are considered to have high salt levels and only salt-tolerant turfgrasses normally survive. $EC_e$ can be estimated by multiplying $EC_w$ by 1.5.

Soil water dilution ratios are either a 1:2 dilution (one part dry soil to two parts water) or a 1:5 dilution (one part soil to five parts water). Electrical conductivity readings from these two methods are not comparable, so the method used must be known in order to interpret the EC reading. Soil testing laboratories frequently use a 1:2 dilution method because it is more rapid than obtaining a saturated paste extract. The EC of a 1:2 extract is on average one-fifth the EC of a saturated paste extract on sand-based greens. To estimate the EC of a saturated paste from a 1:2 extract, multiply the EC of the 1:2 extract by five.

**Assessing soils for sodium problems.** *Sodicity* refers to high concentration of sodium ($Na^+$) while *salinity* refers to high concentrations of total salts including NaCl, $Ca^{+2}$, $Mg^{+2}$, and $SO_4^{-2}$. Salt-affected soil can be classified as *saline, sodic,* and *saline–sodic* soils (Table 8.15). Saline soils are the most common type of salt-affected soil and the easiest to reclaim (Figure 8.20). These soils are plagued by high levels of soluble salts, primarily chloride ($Cl^-$), sulfate $\left(NO_3^-\right)$, and sometimes nitrate $\left(SO_4^{-2}\right)$. Soluble salts are in such quantities they interfere with plant growth. Salts of low solubility, such as calcium sulfate (or gypsum, $CaSO_4$) and calcium carbonate ($CaCO_3$), may also be present. Because exchangeable Na is not a problem, saline soils are usually flocculated with good water permeability (Table 8.15). Saline problems generally occur when (1) insufficient rainfall leaches salts through the soil profile, (2) drainage is impaired, or (3) water contains high levels of salts. The total salt soil content is measured in terms of electrical conductivity of a saturated paste extract ($EC_e$).

Sodic soils, or soil structure deteriorated soils, have high levels of exchangeable Na, low total soluble salt content, $HCO_3 > 120$ mg/L or $CO_3^{-2} > 15$ mg/L, and these soils tend to disperse, reducing water infiltration and interfering with plant growth. Sodic soils also have a pH between 8.5 and 10 and are often called black alkali soils because the organic matter in the soil tends to disperse, creating black-colored puddles (Figure 8.21). Calcium and Mg ions in sodic soils tend to form insoluble calcitic lime, leaving low soluble Ca and Mg levels to displace Na ions, allowing the Na problems. The high Na concentration of a sodic soil not only injures plants directly but also degrades the soil structure, termed, "sodium hazard." Sodic soil cannot be improved by leaching the Na from the soil profile alone. Soil amendments are required to replace the Na in the soil in conjunction with leaching with acidified water. The presence of free lime (carbonates) in the soil allows the widest selection of amendments. To test for free lime, place a few drops of dilute acid or vinegar on the soil. Bubbling or fizzing indicates the presence of free lime or carbonates. Saline–sodic soils contain both high soluble salts and high exchangeable Na and interfere with plant growth. Like sodic soils, they are best reclaimed by adding a Ca-containing amendment and then leaching to remove excess soil Na ions.

Two laboratory measurements are used to assess whether soils contain excessive Na levels and if poor drainage and aeration are likely to occur. These measures are the *exchangeable sodium percentage* (ESP) and the SAR. The ESP identifies the degree or portion of the soil CEC occupied or saturated by Na, reported as milliequivalents/100 g or centimoles/kilogram, and is calculated as follows:

$$ESP = \frac{\text{exchangeable sodium (meq/L)}}{\text{cation exchange capacity (meq/L)}} \times 100$$

Unlike SAR, ESP does not consider the quantity of Ca and Mg ions relative to Na ions present.

### Example

A soil test indicates the $Na^+$ content of a soil is 6.9 meq/100 g and the CEC of the soil is 17.3 meq/100 g. Find the ESP of this soil.

**TABLE 8.15**
**U.S. Salinity Laboratory Classification of Soil Based on Salinity[a], Sodium[b], and SAR Determined in a Saturated Paste Extract**

| Classification | Total Soluble Salts (ECe; dS/mc | Sodium Status | | Soil pH (1–14) | Comments |
|---|---|---|---|---|---|
| | | ESP | SAR | | |
| Normal soil | <4 | <15 | <12 | 5.5 to 7.5 | – |
| Saline soil (high salt content) | ≥4 | <15 | <12 | <8.5 | • Nonsodic soil containing high total soluble salts, reducing plant growth. The exchange complex is usually dominated by $Ca^{+2}$ and $Mg^{+2}$; thus, soil pH is usually below 8.5. A white salt crust, referred to as "white alkali," forms on the soil surface as it dries and salts are deposited. Soil permeability or hydraulic conductivity is not adversely affected by adsorbed Na but high soil osmotic potential (wilt) from salts causes physiological drought (reduced plant water uptake). From this, turfgrass stands often have a bluish-green color and a spotty growth. Roots are often black and deteriorated and ornamental plants have leaf tip or margin burn. Saline soils are best reclaimed by aeration followed by leaching excess salts with low-SAR water. |
| Sodic soil (high sodium content) | <4 | ≥15 | ≥12 | >8.5 | • Sodic soils are nonsaline, but contain sufficiently high levels of exchangeable Na, and soil pH is generally 8.5 to 10 due to the hydrolysis of sodium carbonate. Referred to as "black alkali" due to black puddles of water, like oil, from dispersed soil organic matter that crusts after drying. Plant roots are often black, thin, and spindly, lacking root hairs, causing shoot drought stress symptoms. Sodic soils do not form a white salt crust on the soil surface, but clay particles are dispersed in these soils due to high levels of Na and low levels of Ca and Mg, resulting in a putty-like sticky consistency. Structureless soils result with low water infiltration, nutritional disorders, compaction risk, and air permeability that few plants can tolerate. Sodic soil is reclaimed by aeration followed by replacing Na with another cation, usually Ca in the form of gypsum ($CaSO_4$) or by liming plus adding a S compound such as sulfuric acid or iron sulfate. The Na must then be leached from the soil profile with excess water. |
| Saline-sodic soil (high salt and sodium content) | ≥4 | ≥15 | ≥12 | ≤8.5 | • Contain high total salts ($EC_e \geq 4.0$ dS/m) and high exchangeable Na (ESP ≥ 15% or SAR ≥ 12). These soils have characteristics intermediate between those of saline and sodic soils. Like saline soil, soil pH is usually below 8.5. The soils resemble saline soils if the soluble salts are not leached. With leaching, these soils become sodic (Na based), unless Ca and Mg are applied. Plants wilt and turn bluish-green due to high soil osmotic potential from soil Na. Reclaiming a saline-sodic soil is similar to reclaiming a sodic soil where Ca-containing amendments replace the Na, forming sodium sulfate, which is then leached from the soil profile. Other acid-forming amendments may be added if free lime is present to create soluble Ca in the soil. |

[a] $EC_e$ = electrical conductivity of a saturated soil paste extract.
[b] ESP = % of soil cation exchange capacity occupied by Na.
[c] $EC_e$ values are based on saturated extracts.

$$\begin{aligned} \text{ESP} &= \frac{\text{exchangeable sodium (meq/100 g)}}{\text{cation exchange capacity (meq/100 g)}} \times 100 \\ &= \frac{6.9\ \text{meq/100 g}}{17.3\ \text{meq/100 g}} \\ &= 40\% \end{aligned}$$

Soil SAR is a second, more easily measured property, analogous to the irrigation water SAR discussed earlier, which considers Ca and Mg ion content in the soil. Soil SAR is calculated from soil-test extractable levels of Na, Ca, and Mg (expressed in milliequivalents/100 g or millimoles/liter).

ESP indicates the probability a soil will disperse, thereby reducing the permeability of soil to water and air. In the environment, salts and Na do not act independently. High soluble-salt concentration can negate the soil particle dispersal (thus, impermeability) from the effects of Na. Table 8.15 lists the combined effects of salinity (EC) and soluble salt contents (ESP).

**FIGURE 8.20** Thin, white crusted turf with poor drainage from excessive soluble salts (salinity).

**FIGURE 8.21** Black puddles from soil organic matter dispersing in sodic soils.

Usually, few or only minor problems occur when ESP values are less than 13% to 15% (Table 8.15). An ESP > 15% or a soil SAR > 13 indicates a sodic soil where Na causes soil colloids to disperse and plug the soil's drainage pores, thereby reducing the permeability of the soil to water and air. Sodic soils become saturated with Na ions compared to Ca and Mg ions, especially if bicarbonate ions are present. Symptoms of reduced permeability include waterlogging, reduced infiltration rates, crusting, compaction, disease occurrence, weed invasion, and poor aeration. Sodic soils often have considerable clay that is sticky due to the Na. ESP and SAR are related and can be estimated by

$$ESP = \frac{1.475 \times SAR}{1 + (0.0147 \times SAR)}$$

## WATER CARBONATES AND PH

*Bicarbonate* $\left(HCO_3^-\right)$ and, to a lesser extent, *carbonate* $\left(CO_3^{-2}\right)$ are found in high-pH water. Although soil pH can be raised by high bicarbonate levels in water, the potential negative impact on soil permeability is more of a concern.

The primary source of carbonates and bicarbonates in soils is carbonic acid ($H_2CO_3$), which forms when carbon dioxide ($CO_2$) from microbial and root respiration reacts with water. Carbonic acid is a very weak, noncorrosive acid, and it is a chief constituent in carbonated beverages. In higher pH (6 to 9) soils, the abundance of hydroxyl ($OH^-$) ions reacts with the carbonic acid to initially form bicarbonate $\left(HCO_3^-\right)$ and, then, at pH > 9, carbonate $\left(CO_3^{-2}\right)$ ions.

**Step 1:**

$$H_2CO_3\ \text{(carbonic acid)} + OH^-\ \text{(pH 6 to 9)} \leftrightarrow HCO_3^-\ \text{(bicarbonate)} + H_2O\ \text{(water)}$$

**Step 2:**

$$HCO_3^-\ \text{(bicarbonate)} + OH^-\ \text{(pH > 9)} \leftrightarrow CO_3^{-2}\ \text{(bicarbonate)} + H_2O\ \text{(water)}$$

As the concentrations of bicarbonates and carbonates increase, more hydroxyl ions ($OH^-$) are formed with a corresponding reduction of hydrogen ions ($H^+$), causing an increase in pH.

When water containing $HCO_3^-$ dries at the soil surface, insoluble Ca and magnesium carbonates (calcitic lime) are formed. Since Ca and Mg are no longer dissolved, they do not counteract the effects of Na, and problems related to high ESP may occur. This results in an increase in the SAR and soil pH and a decrease in water quality and soil infiltration. White lime deposits may also become visible on turf leaves during hot, dry periods as bicarbonates are deposited during evaporation (Figure 8.22).

Two measurements are used for assessing the carbonate level of irrigation water: the direct measurement of carbonate and bicarbonate and the *residual sodium carbonate* equation. It is not the absolute levels of bicarbonates or carbonates present in the irrigation water that is most important, but rather their relative concentrations compared to Ca, Mg, and Na levels. RSC specifically measures the presence of excess carbonate $\left(CO_3^{-2}\right)$ and bicarbonate $HCO_3^-$ content over calcium ($Ca^{+2}$) and magnesium ($Mg^{+2}$) ions expressed as milliequivalents per liter.

Residual sodium carbonate equation:

$$RSC = \left(CO_3^{-2} + HCO_3^-\right) - (Ca^{+2} + Mg^{+2})$$

Assessment for poor water infiltration due to high carbonates and low Ca and Mg as determined by the RSC equation is listed in Table 8.13.

A negative RSC indicates Ca and Mg ions exceeding carbonates and bicarbonates; thus, Na buildup is unlikely. Many water test reports list RSC values as zero, even if the calculated value is less than zero. A positive RSC value indicates Na accumulation is possible because the water adds more carbonates than the divalent cations, $Mg^{+2}$ and $Ca^{+2}$, to the soil. Calcium then is lost from the soil by the chemical reaction:

$$CO_3^{-2}\text{(carbonate)} + Ca^{+2}\ \text{(soil calcium)} \rightarrow$$
$$CaCo_3\ \text{(insoluble calcitic lime)}$$

Under these conditions, gypsum or sulfuric acid water treatment may be necessary. Values greater than 1.5 meq/L may justify irrigation acid injection. Acid injection changes the carbonates and bicarbonates to carbon dioxide and water, allowing Ca or Mg to counter excess Na on the CEC sites. Normally, if irrigation RSC values are high but SAR values are low (<6 meq/L), acid injection is unnecessary since insufficient Na is present to cause a problem. This also is true in areas of high rainfall where Na is readily leached out of the soil profile. If hazardous RSC (high Na, >100 ppm) water is repeatedly used, the soil becomes alkaline and is likely to become sodic (structureless) over time. The seal can be broken up by a combination of cultivation (aerifying, spiking, slicing) and use of acidic fertilizers or elemental S. Gypsum can be applied directly through irrigation systems or directly on the soil to reduce Na accumulation on CEC sites.

### Example

A water analysis indicates 3 mg/L $CO_3^{-2}$, 128 mg/L $HCO_3^-$, 21 mg/L $Ca^{+2}$, and 10 mg/L $Mg^{+2}$. Determine the RSC in milliequivalents/litter.

**Step 1:** Convert milligrams/liter to milliequivalents/litter (Table 8.11):

$$\text{meq/L} = \frac{\text{mg/L}}{\text{equivalent weight}}$$

$$CO_3^{-2} = \frac{3}{30} = 0.1$$

$$Ca^{+2} = \frac{21}{20} = 1.05$$

$$Mg^{+2} = \frac{10}{12} = 0.83$$

$$HCO_3^- = \frac{128}{61} = 2.1$$

**Step 2:** Insert the values into the RSC equation:

**FIGURE 8.22** Lime deposits on tree trunks from irrigation water containing high bicarbonate levels.

$$RSC = (CO_3^{-2} + HCO_3^-) - (Ca^{2+} + Mg^{+2}) = (0.1 + 2.1) - (1.05 + 0.83) = 0.32 \text{ meq/L}$$

From Table 8.13, this level of RSC should pose little problem.

### Example

Determine RSC from the following water sample: bicarbonates $(HCO_3^-)$ = 4 meq/L, carbonates $(CO_3^{-2})$ = 1 meq/L, calcium ($Ca^{+2}$) = 1 meq/L.

$$RSC = (CO_3^{-2} + HCO_3^-) - (Ca^{+2} + Mg^{+2}) = (4 + 1) - (2 + 1) = 2 \text{ meq/L}$$

From Table 8.13, this RSC value could cause severe problems.

Bicarbonate levels alone are sometimes used to assess potential limitations of an irrigation water source (Table 8.13). Water containing 2 to 3 meq/L of bicarbonates can be managed by applying ammoniacal fertilizer as part of a regular fertilizer program to help reduce soil pH. Adding sufficient acid to decrease water to about pH 7.0 changes a positive RSC value to a negative one. Water with greater than 4 meq/L bicarbonates will need to be acidified. Blending poor-quality water with better quality water and applying soil amendments such as gypsum or S also are means to help manage bicarbonate problems. The negative effects on soil infiltration of bicarbonate and carbonate are negated by high levels of Ca and Mg. Bicarbonate and carbonate are good indicators of hazard when irrigation water Ca and Mg concentrations are low, but the RSC equation should be utilized when water Ca and Mg are high.

## pH

Although the pH of the water is seldom a problem in itself, very high or low pH may suggest other chemical constituents are present. Continued use of high bicarbonate and carbonate water leads to a high soil pH. High pH can induce Fe, Mn, and, to a lesser extent, Zn deficiencies by rendering these micronutrients unavailable to turfgrass roots. Unfortunately, simply adding these micronutrients in fertilizers is sometimes ineffective since these elements quickly become unavailable in high-pH soils. Using chelates and foliar applications helps avoid interactions between micronutrients and high-pH soils.

When Na is the predominant cation in the soil, sodium bicarbonate and sodium carbonate form causing high soil pH (pH around 10). Since these ions are water soluble and tend to ionize, high levels of bicarbonate and carbonate are encouraged. However, when Ca predominates, usually insoluble calcium carbonate forms and, unlike sodium carbonate, does not ionize to form more carbonate ions. Therefore, the soil pH generally stabilizes around 8.0.

## Alkalinity

Total alkalinity is the measure (as milligrams/liter or parts per million) of the amount of alkaline buffers (carbonates, bicarbonates, and hydroxides) in water. Alkalinity is not the same as pH because water does not have to be strongly basic (high pH) to have high alkalinity. Alkaline substances buffer the water against sudden changes in pH. Total alkalinity is a measure of how much acid is required to lower the pH of the water sample to the bicarbonate endpoint. In other words, this is how much acid it takes to lower the pH to the point where sufficient acid is present to potentially convert all of the bicarbonate $(HCO_3^-)$ to carbonic acid ($H_2CO_3$) and eventually to carbon dioxide ($CO_2$) and water.

$$H^+ \text{(from acid)} + HCO_3^- \text{(bicarbonates)} \leftrightarrow H_2CO_3 \text{ (carbonicacid)} \leftrightarrow CO_2\uparrow \text{(carbon dioxide)} + H_2O \text{ (water)}$$

The following example emphasizes the importance of *both* alkalinity and pH when adjusting the pH of water.

### Example

Superintendent A has a water with a pH of 9.3 and an alkalinity of 87 mg/L $HCO_3^-$ (total carbonates or TC = 1.42 meq/L). To reduce the pH of this water to 5.8, 15.8 fluid ounces of 35% sulfuric acid per 1,000 gallons of water (123 mL/L) are required. In contrast, Superintendent B has a water with a pH of 8.3 and an alkalinity of 378 mg/L $HCO_3^-$ (TC = 6.20 meq/L). To reduce this water to a pH of 5.8, it takes 68.6 fluid ounces of 35% sulfuric acid per 1,000 gallons of water (536 mL/L). Even though Superintendent B's water is one pH unit lower than Superintendent A's, it takes more than four times more acid to lower the pH to 5.8 due to the differences in alkalinity.

Sulfuric ($H_2SO_4$), phosphoric ($H_3PO_4$), nitric ($HNO_3$), or citric ($H_3C_6H_5O_7$) acid is commonly injected into irrigation water to neutralize water alkalinity. Acid injection will be discussed later in this chapter.

## MANAGING POOR-QUALITY WATER-USE SITES

Managing salinity, sodicity, and alkalinity problems requires constant attention (Table 8.16). Management practices that aid in remedying these problems include:

1. Site assessment to determine which, if any, water and soil treatments are best
2. Utilizing salt-tolerant grasses—warm-season turfgrasses generally are less salt sensitive compared to

**TABLE 8.16**
**Water and Soil Salinity Problems with Potential Management Solutions**

| Soil Salinity Problem | Potential Solutions |
|---|---|
| Total irrigation salt content (EC) | • Leaching; blending water sources; increase drainage and aeration; use salt-tolerant varieties |
| Soil SAR/adj. SAR | • Apply Ca amendment; apply S alone (in calcareous soils) or lime (in acidic soils); blending water sources; acid or S irrigation injection in severe cases |
| ESP | • Apply Ca amendment such as gypsum; apply S alone (in calcareous soils) or lime (in acidic soils); or S irrigation injection in severe cases |
| RSC | • Irrigation acid injection; S generator; S application in calcareous soils; blending water sources |
| Soil infiltration/permeability ($EC_w$ plus SAR) | • Gypsum additions to either (a) low $EC_w$ plus low SAR water or (b) low to moderate $EC_w$ plus high SAR water; blending water sources |
| Specific ion toxicity | • Establish tolerant varieties (especially ornamentals); blending water sources |
| Total suspended solids | • Irrigation line filtration; use of settling ponds |
| Nutrient imbalances | • Adjusting fertility programs |

cool-season turfgrasses, while most ornamentals are more salt sensitive
3. Diluting or blending poor-quality water with good-quality water
4. Flushing (leaching) excess salts by applying extra water
5. Modifying soils with various amendments to replace and leach Na from the soil
6. Amending irrigation water to correct Na and bicarbonate problems
7. Enhancing soil drainage by using sands and installing subsurface tile drainage, plus intensive cultivation to enhance infiltration, percolation, and drainage of salt-laden water (see earlier)
8. Using cytokinin and Fe-containing biostimulants as salt-stressed plants often exhibit low cytokinin activity, as well as using wetting agents and appropriate fertilizers
9. Raising the mowing height to promote more stress-tolerant plants
10. Routine use of wetting agents to help maintain good water infiltration and percolation to flush salts and Na below the rootzone

These practices center around maintaining the salt balance where the quantity of salt leaving an area has to be equal to, or greater than, the quantity of salt entering the area. As a rule of thumb, when using recycled irrigation water or water high in salts, 20% additional water will be required compared to using better quality water.

### Blending Water Sources for Reducing Salinity

High-salinity water that is unacceptable for use can be made suitable as an irrigation source by diluting it with nonsaline water. Enough nonsaline water must be available to create a mixed water of acceptable quality (e.g., not making a less saline water that is still unacceptable). The quality of a poor water source should improve proportionally to the mixing ratio with better quality water. For example, a water source with an $EC_w$ = 5 dS/m mixed equally with a source with an $EC_w$ = 1 dS/m should reduce salinity in the blend to ~3 dS/m. A chemical analysis of the blend should be performed to confirm this. The salinity of the mixture can be calculated with this equation:

$$EC_w(\text{blend}) = \frac{\text{volume (water A)} \times EC_w\ \text{(water A)} + \text{volume (water B)} \times EC_w\ \text{(water B)}}{\text{volume (water A)} + \text{volume (water B)}}$$

#### Example

Two water sources are available for irrigation. One has an $EC_w$ of 3.0 dS/m and the other, 0.6 dS/m. The water will be blended in equal amounts. What would the resulting $EC_w$ of the blended water be?

$$\begin{aligned} EC_w(\text{blend}) &= \frac{\text{volume (water A)} \times EC_w\ \text{(water A)} + \text{volume (water B)} \times EC_w\ \text{(water B)}}{\text{volume (water A)} + \text{volume (water B)}} \\ &= \frac{[1\ \text{gallon} \times 3.0\ \text{dS/m}] + [1\ \text{gallon} \times 0.6\ \text{dS/m}]}{1\ \text{gallon} + 1\ \text{gallon}} \\ &= \frac{3.6\ \text{dS/m}}{2} \\ &= 1.8\ \text{dS/m} \end{aligned}$$

Mixing of irrigation sources can occur in irrigation ponds or within the irrigation system itself. When mixing water sources in irrigation ponds, the nonsaline water should be added immediately prior to being used so as to reduce evaporative losses. Evaporation of surface water is not only an inefficient use of water, but it also increases the salinity of the water remaining in the pond. If blending is not an option, alternating irrigation events with effluent followed by fresh water helps leach salts.

### Leaching or Flushing Soils to Remove Salts

Salt buildup from salt-laden irrigation water occurs when rainfall is low and evaporative demand is high. As water evaporates from the soil surface, salt deposits are left behind. Applying water in an amount greater than ET to cause the applied water to flow (or leach) through the rootzone and wash away salts is the goal of leaching salt-laden soil. Steps involved when leaching or flushing soils to remove salts include:

1. Aerify to break hardpans or organic zone surface tension.
2. Add green's grade gypsum at 7 to 12 pounds/1,000 square feet (3.4 to 5.9 kg/100 $m^2$).
3. Start flushing, usually for about an hour. This is based on irrigation heads applying about an inch/hour (2.54 cm) and a sand-based rootzone holding ~1 to 1.5 inches (2.54 to 3.8 cm) water at saturation. If puddling occurs, stop, allow it to percolate, and then resume.
4. Assess flushing length of time by measuring $EC_w$ of discharged water with a portable meter. Once readings are stabilized or fall below preset thresholds, stop.
5. Afterwards, N and K may need to be added as they are commonly stripped by flushing.

Frequent flushing of the soil with good-quality irrigation water or rainfall is the best method of preventing excessive salt accumulation. Unfortunately, low-salinity irrigation sources are not always available and frequently saline irrigation water must be used to manage soil salinity. However, as long as the salinity of the irrigation water is acceptable, it can be used to leach accumulated salts from the turf rootzone. The goal is to maintain a soil salinity level that is not increased through salts added by irrigation and yet can support turfgrass growth. The use of soil amendments, such as gypsum, should be considered in conjunction with leaching irrigation applications in saline–sodic soils.

If saline water is used to reduce the salt level of the soil, irrigation must be applied at rates exceeding evapotranspiration to leach (or flush) excess salts out of the rootzone. Leaching of soluble salts in the soil solution is much more rapid and easier than removing Na on the CEC sites of sodic soils. On sodic soils, the Na is chemically bonded and must be replaced by Ca, Mg, or another cation before the Na can be leached from the soil solution. Soluble salts are already in the soil solution and thus are more easily leached. To determine the amount of excess water required to leach salt below the rootzone, the leaching requirement or fraction equation is often used, though other formulas and techniques are available.

Leaching requirement is the amount of extra water needed to leach salts from the rootzone and is defined as

$$\text{Leaching requirement (\% above ET to leach salts)} = \frac{EC_w}{5(EC_e) - ECd_w} \times 100\%$$

$EC_w$ is the electrical conductivity of the irrigation water and $EC_e$ is the salt-tolerance level of the grass.

#### Example

An irrigation water source has a salinity level of 2 dS/m. The turfgrass being grown has a tolerance of 4 dS/m. What would be the recommended amount of water needed to leach salt from the rootzone?

**Step 1:** Determine the leaching requirement for this sample and turfgrass.

$$\begin{aligned}\text{Leaching requirement} &= \frac{EC_w}{5(EC_e) - ECd_w} \times 100\% \\ &= \frac{2\text{ dS/M}}{5(4) - 2} \times 100\% \\ &= 11\%\end{aligned}$$

**Step 2:** The leaching requirement indicates 11% additional water above that normally applied based on ET rates would be needed to leach the salt from the soil. Therefore, for example, if 0.5 inch of water is needed to replace ET loses, then an additional 11% above 0.5 inch (0.5% × 11%), or 0.055 in would be required for a total of 0.56 inch to maintain a particular salinity level. One would then need to account for the distribution uniformity of the particular irrigation system to determine the total amount of irrigation needed to apply 0.56 inch. Table 8.17 lists these irrigation guidelines for leaching salts from soil with saline water.

As the irrigation water becomes saltier or the soil heavier, the leaching requirement becomes larger, meaning more water must be added for leaching to avoid salt accumulation. A guideline is for about 70% of the total soluble salts to be removed by leaching, 3 inches (7.6 cm) of water is needed per 12 inches (30 cm) of soil depth of a sandy soil, 6 inches (15 cm) of water per 12 inches (30 cm) of a medium-textured loam soil, and 9 inches (23 cm) of water per 12 inches (30 cm) of a clay (fine-textured) soil. Leaching $Na^+$ also removes nutrients such as K, Mg and others. If puddling occurs, it generally is better to have periodic leaching events (i.e., two to four times monthly at 0.2 to 0.4 inch, 0.5 to 1.0 cm, per application) compared to heavier, infrequent events (i.e., once monthly).

If saline water is the only source of water available for irrigation, it is helpful to predict how the leaching fraction of known irrigation water salinity will influence soil salinity over an extended period of time (Table 8.17). Applying a leaching fraction of 10% will lead to an $EC_e$ of $EC_w \times 2.1$, 15% to 20% will lead to an $EC_e$ of $EC_w \times 1.5$, and 30% will lead to an $EC_e = EC_w$. Finally, plants tolerate higher soil salinity levels if water stress is avoided by maintaining soil moisture. Adequate surface moisture also prevents capillary rise of subsurface water and salts.

**TABLE 8.17**
**Irrigation Guidelines for Leaching Salts from Soil with Saline Water**

| Irrigation Water $EC_w$ (dS/m) | Maximum Plant $EC_{dw}$ Tolerance Level, Measured by Saturated Soil Paste Extract (dS/m) | | |
|---|---|---|---|
| | 4 (Low) | 8 (Medium) | 16 (High) |
| | Water, Inches, to Replace Weekly ET Losses and Provide Adequate Rootzone Leaching[a] | | |
| 0.00 | 1.5 | 1.5 | 1.5 |
| 1.00 | 2.0 | 1.7 | 1.6 |
| 2.00 | 3.0 | 2.0 | 1.7 |
| 3.00 | 6.0 | 2.4 | 1.8 |

[a] Multiply inches by 2.54 to convert to cetimeters.

### Good Soil Percolation and Drainage

As previously mentioned, leaching works only with soils possessing good drainage. If compacted zones or abrupt changes in soil texture exist, less leaching occurs as water movement through the soil is reduced (Figure 8.23). Soil drainage can be improved through modifying rootzones, increased deep tine aerification, and the use of drain lines. Drain lines, spaced no more than 20 feet (6.1 m) apart, are used on golf greens for this purpose. Aerification also initiates deep root development prior to summer heat and increases tolerance of salt stress by reducing soil compaction and disrupting soil layering. Native (or pushup) greens often fail when effluent water is used, unless these techniques are aggressively incorporated.

For fairways, deep aerification has become standard on effluent-using courses to increase soil drainage and provide deep channels for incorporation of soil amendments. On tees and greens, deep aerification at least once yearly is typical along with supplemental monthly venting by spiking, slicing, quadratining, hydrojetting, or other techniques. For soils with limited infiltration properties, pulse irrigation is more effective where water is applied and allowed to infiltrate before reapplying.

### Counteracting Soil Sodium with Soil Amendments

Several soil amendments are used to replace Na in sodic soils in conjunction with leaching to remove salts from the rootzone. The amendments counteract Na by providing Ca either directly (contain Ca) or indirectly (provide an acid to dissolve calcium carbonate present in the soil, thus, releasing available Ca). Calcium arising from the soil amendments reacts with soil Na to displace it from the cation exchange sites on clay and organic matter particles. The released Na can then be leached out of the soil profile.

The presence of free lime (Ca) can be determined by adding a few drops of muriatic or sulfuric acid to a spoonful of soil. Bubbling or fizzing indicates presence of carbonate or bicarbonate. If these are present (soil contains lime), the acid-forming amendments in Table 8.18 can be used. If lime is absent, use Ca-containing amendments listed in Table 8.18.

Amendments used for the treatment of clay-textured sodic soils include gypsum, S, sulfuric or N-phuric acids, lime sulfur, ferric sulfate, calcium chloride, calcium nitrate, and calcium carbonate (Table 8.18). Due to their expense, calcium chloride and calcium nitrate are not widely used. Sulfuric acid is dangerous to handle and can be corrosive to some types of equipment. Ferric sulfate and lime sulfur also are usually too expensive for practical applications. Ground limestone is effective on acid soils, but due to its low solubility, its usefulness drops in high pH soils—which describes most sodic soils. Even though gypsum supplies Ca, it is a neutral salt and does not appreciably affect soil pH. Thus, for several reasons, gypsum is the material most often used for reclaiming sodic soils (Figure 8.24).

**Gypsum.** Gypsum is low to moderately soluble in water (2.41 g/L) and supplies soluble Ca to replace Na, as shown:

Reaction when using gypsum to reduce soil salinity:

$$2\text{Na}X\ (\text{soil sodium}) + \text{CaSO}_4\ (\text{gypsum}) \rightarrow$$
$$CaX_2\ (\text{soil calcium}) + \text{Na}_2\text{SO}_4\downarrow$$
$$(\text{sodium sulfate, removed by leaching})$$

The letter $X$ in this reaction indicates the exchange site for cations on the soil colloid. Gypsum, by mass action, drives Na off the soil exchange complex and replaces it with Ca. It leaves sodium sulfate ($NaSO_4$), which is soluble, and readily leaches downward with percolating water. The $Ca^{+2}$ in gypsum also will bind to CEC sites and help flocculate the otherwise dispersed (or deflocculated) soil. Calcium should be raised by 1 to 4 meq/L in the irrigation water in order to increase the $EC_w$ > 0.50 and experience a significant effect on water infiltration. The gypsum can also be land applied if it cannot be injected into the irrigation system. Gypsum applications to soil with pH values < 5.5 also often have a rapid green-up due to the more favorable conditions to stimulate nitrosomonas and nitrobacter bacteria to convert ammonia to nitrate.

**FIGURE 8.23** Replaced turf due to salinity buildup from insufficiently draining soil.

**TABLE 8.18**
**Relative Equivalent Amounts of Several Soil Amendments in Relation to Pure Gypsum or Elemental S**

| Amendment | Tons[a] of Amendment Equivalent to 1 ton Pure Gypsum | Tons[a] of Amendment Equivalent to 1 ton Pure Sulfur | Soil or Water Additive |
|---|---|---|---|
| | Calcium containing | | |
| Gypsum ($CaSO_4 \cdot 2H_2O$) (23% Ca) | 1.00 | 5.38 | Soil and water |
| • Use dihydrate for solutions | | | |
| • Use anhydrite for soil | | | |
| Calcium carbonate ($CaCO_3$) | 0.58 | 3.11 | Soil and water |
| Calcium chloride ($CaCl_2 \cdot H_2O$) | 0.86 | 4.62 | Soil and water |
| Calcium nitrate [$Ca(NO_3)_2 \cdot H_2O$] | 1.06 | 5.69 | Soil and water |
| Lime sulfur ($CaS_x$; 9% Ca and 24% S) | 0.68 | 4.19 | Soil and water |
| | Acid Forming[b] | | |
| Aluminum sulfate [$Al_2(SO_4)_3 \cdot 18H_2O$] | 1.29 | 6.94 | Soil |
| Ammonium polysulfate [$(NH_4)_2S_x$] | 0.37 | 1.95 | Soil |
| Ferric sulfate [$Fe_2(SO_4)_3 \cdot 9H_2O$] | 1.09 | 5.85 | Soil |
| Lime sulfur ($CaS_x$; 9% Ca and 24% S) | 0.78 | 4.19 | Soil and water |
| Elemental sulfur (S) | 0.19 | 1.00 | Soil |
| Sulfuric acid ($H_2SO_4$) (concentrated) | 0.61 | 3.20 | Soil and water |

[a] 1 ton = 2,000 pounds (907 kg).
[b] All S forms require lime to form gypsum and to help regulate soil pH levels.

**FIGURE 8.24** Gypsum (calcium sulfate) is the most commonly used amendment to reclaim sodic soils.

Gypsum is often needed when the following situations occur:

- Sodic or saline-sodic soils (ESP >15%) exist.
- Water with a SAR >10 causes poor water infiltration and sealing.
- Water with $EC_w$ <0.25 dS/m is very pure ("soft") and lacks sufficient Ca or Mg.

*Gypsum applications.* Gypsum is a by-product of phosphorus mining. Pure gypsum contains 26% Ca and 21% S; however, little of this is sold. Most commercial sources contain 50% to 90% $CaSO_4$ with approximately 23% Ca. If gypsum contains impurities or is wet, it will contain less Ca and/or S and larger quantities will be necessary. For comparison, dolomitic limestone contains 21% to 30% Ca, calcitic limestone, 32% Ca, and calcium nitrate, 19% Ca.

The effectiveness of gypsum increases with fineness. Gypsum used on turfgrass should be fine enough so at least 80% passes a U.S. standard no. 8 sieve. Finer, pulverized gypsum (like limestone) reacts more quickly with soil but becomes difficult to apply. Larger particles not able to pass a no. 8 sieve are too slow to dissolve, rendering them relatively ineffective. Finer particles will also give a better uniformity than large particles.

Irrigation is needed to dissolve gypsum. Several irrigations usually are required to dissolve gypsum and leach Na. Generally, if the Na problem is slight, passage of 1 foot (0.3 m) of water through the soil is sufficient to leach out the salt. Two feet (0.6 m) of water are needed on moderate Na problem soils while 3 feet or more are needed on severe Na soils.

*Gypsum rates.* When applied to reclaim high-Na soils, the amount of gypsum required depends upon the Na concentration of the soil, determined by a soil test and by soil texture. Suggested rates are listed in Table 8.19. The objective is to achieve ESP values below 10% on fine-textured soils and below 20% on coarser textured soils. These values reflect gypsum use on nonestablished (or preplant) soils. Rates should not exceed 5 tons/acre (11 metric tons/ha), per application. Rates higher than this should be split, with successive applications not made until sufficient time for some leaching has occurred. Additional needs should then be verified by a second soil test. The gypsum is applied and deeply plowed or tilled or injected prior to grass establishment.

On established grass, gypsum is added at rates ranging from 5 to 10 pounds/1,000 square feet (25 to 49 g/m$^2$) and applied monthly when needed. Light, frequent applications are more effective than heavy, less frequent ones, especially in sandy soils. Gypsum is slow to react and does not normally burn foliage; however, one should still apply it only during

**TABLE 8.19**
**Preplant Gypsum Amounts Required per 12 Inches (30 cm) of Soil Depth as Related to Soil Texture and Sodium Percentage**

| Soil Texture | Clay Content (%) | Exchangeable Sodium Percentage | | | | |
|---|---|---|---|---|---|---|
| | | 15 | 20 | 30 | 40 | 50 |
| | | Gypsum (tons/ac)[a] | | | | |
| Coarse (sands) | 0 to 15 | 2 | 3 | 5 | 7 | 9 |
| Medium (loams) | 15 to 55 | 3 | 5 | 8 | 11 | 14 |
| Fine (clays) | >55 | 4 | 6 | 10 | 14 | 18 |

[a] Multiply tons (2,000 pounds) per acre by 2.241 to obtain tons (metric) per hectare.

mild temperatures (e.g., ≤80°F, 27°C). Due to its low water solubility, time will be required before gypsum will disappear from the soil surface.

### Example

A soil with CEC of 20 meq/100 g contains 15% ESP, which needs to be reduced to 10% (a 5% total reduction). Determine the amount of gypsum needed per 6-inch-acre slice.

**Step 1:** Determine the milliequivalent of $Na^+$ that needs to be replaced and equivalent $CaSO_4$ per 100 g of soil. Note: 1 meq of $Na^+$ needs to be replaced by 1 meq $Ca^{+2}$ or 1 meq $Na^+$ = 1 meq $Ca^{+2}$.

$$5\% \times 20 \text{ meq CEC/100 g} = 1 \text{ meq Na}^+\text{/100 g}$$

$$\text{and 1 meq CaSO}_4\ 2\text{H}_2\text{O/100 g}$$

**Step 2:** Determine equivalent weight of gypsum.

$$1 \text{ meq CaSO}_4 \cdot 2\text{H}_2\text{O/100 g} \times 86 \text{ mg CaSO}_4 \cdot 2\text{H}_2\text{O/meq} = 86 \text{ mg CaSO}_4 \cdot 2\text{H}_2\text{O/100 g}$$

86 mg/meq is the molecular weight of $CaSO_4 \cdot 2H_2O$ (172) divided by two (the valence number of Ca in relation to Na) to obtain equivalent weights.

**Step 3:** Convert equivalent weights (mg/100 g) to pounds needed per 6-inch furrow slice.

$$86 \text{ mg CaSO}_4 \cdot 2\text{H}_2\text{O/100 g} \times 20 = 1{,}720 \text{ pounds CaSO}_4 \cdot 2\text{H}_2\text{O needed/6 inch-acre furrow slice depth to reduce ESP to 10\%.}$$

Twenty is a conversion factor to convert equivalent weight (milligrams/100 g) to pounds/6 inch-acre furrow slice from the following:

$$\frac{X \text{ mg sample}}{100 \text{ g}} \times \frac{2{,}000{,}000 \text{ pounds soil}}{6 \text{ inch-acre furrow slice}} \times \frac{1 \text{ g}}{1{,}000 \text{ mg}} \times \frac{\text{pounds ion}}{454 \text{ g ion}} \times \frac{454 \text{ g soil}}{\text{pounds soil}} = \frac{20 \text{ pounds}}{6 \text{ inch-acre furrow slice}}$$

Another way to determine the amount of gypsum needed involves:

**Step 1:** Determine the same as preceding step 1.
**Step 2:** Recognize 1 meq of $Na^+$ needs to be replaced by 1 meq $Ca^{+2}$ or 1 meq $Na^+$ = 1 meq $Ca^{+2}$.
**Step 3:** Determine the percent Ca contained in gypsum or calcium sulfate ($CaSO_4 \cdot 2H_2O$).

$$\frac{\text{Ca(g)}}{\text{CaSO}_4 \cdot 2\text{H}_2\text{O(g)}} = \frac{40\text{(g)Ca}^{+2}}{172 \text{ g CaSO}_4 \cdot 2\text{H}_2\text{O}}$$

$$= 23.2\% \text{ Ca in CaSO}_4 \cdot 2\text{H}_2\text{O}$$

Therefore, 1 pound $CaSO_4 \cdot 2H_2O$ contains 0.232 pound $Ca^{+2}$.

**Step 4:** Determine the pounds gypsum required to apply 1 meq $Ca^{+2}$/100 g soil.

$$\frac{1 \text{ meq Ca}^{+2}}{100 \text{ g soil}} \times \frac{1 \text{ eq Ca}^{+2}}{1{,}000 \text{ meq Ca}^{+2}} \times \frac{20 \text{ g Ca}^{+2}}{1 \text{ eq Ca}^{+2}} \times \frac{1 \text{ pound Ca}^{+2}}{454 \text{ g Ca}^{+2}} \times \frac{1 \text{ pound CaSO}_4 \cdot 2\text{H}_2\text{O}}{0.232 \text{ pound Ca}^{+2}} \times \frac{454 \text{ g soil}}{1 \text{ pound soil}} \times \frac{2{,}000{,}000 \text{ pounds}}{1 \text{ acre soil}} \times \frac{20 \text{ pounds}}{6 \text{ inch-acre furrow slice}}$$

$$= 1{,}720 \text{ pounds gypsum (CaSO}_4 \cdot 2\text{H2O) needed per 6 inch-acre furrow slice depth to reduce ESP to 10\%.}$$

Monthly gypsum applications are typically needed and are more efficiently utilized when made after monthly spiking or quadratining, followed by a heavy leaching cycle.

**Sulfur and other acid-forming amendments.** Elemental S and other acid-forming amendments and fertilizers (Table 8.20) may also be used to provide soluble Ca by lowering soil pH and dissolving calcium carbonate precipitated in the soil. If soil pH is only slightly elevated, routine applications of acid-forming N fertilizers containing or generating ammonium $\left(NH_4^+\right)$ may be sufficient to maintain soil pH. Ammonium sulfate is about three times more acidic than other commonly used N sources and is widely used as a primary N source where irrigation sources have moderate levels of bicarbonate. However, when irrigation sources have severe bicarbonate problems, elemental S is the most frequently utilized acid-forming amendment. Sulfur is more available and less costly than other amendments, such as aluminium sulfate. Extreme care should be exercised when using S to lower soil pH. Elemental S has a high potential to burn plant tissue and can lower pH excessively if used unwisely.

Sulfur furnishes Ca indirectly in a three-step process:

**Step 1:** S must first be oxidized by soil *Thiobacillus* bacteria to sulfuric acid.

$$2S\text{ (sulfur)} + 3O_2\text{ (oxygen)} + 2H_2O\text{ (water)} \rightarrow 2H_2SO_4\text{ (sulfuric acid)}$$

**Step 2:** Sulfuric acid then reacts with lime in the soil to produce gypsum, which is removed by leaching.

$$CaCO3\text{ (lime)} + H_2SO_4\text{ (sulfuric acid)} \rightarrow CaSO_4\text{ (gypsum)} + H_2O\text{ (water)} + CO_2\uparrow\text{ (carbon dioxide)}$$

**Step 3:** Gypsum then reacts with Na ions to produce soluble sodium sulfate.

$$2NaX\text{ (sodic soil)} + CaSO_4\text{ (gypsum)} \leftrightarrow CaX\text{ (calcium soil)} + Na_2SO_4\downarrow\text{ (sodium sulfate)}$$

The letter $X$ in this reaction indicates the exchange site for cations on soil clay and organic matter.

In soils lacking free lime, the reactions are

**Step 1:** Sulfuric acid is produced by the oxidation of S by soil bacteria.

$$2S + 3O_2 + 2H_2O \rightarrow 2H_2SO_4\text{ (sulfuric acid)}$$

**Step 2:** Sulfuric acid produced then reacts directly with Na ions on the cation exchange sites.

$$2NaX + H_2SO_4 \leftrightarrow 2HX \rightarrow Na_2SO_4\downarrow$$

The letter $X$ in this reaction indicates the exchange site for cations on soil clay and organic matter.

Sulfur use on sodic or saline–sodic soil not containing calcium carbonate causes excessively low pH. Under these conditions, lime should be periodically applied to the soil surface to provide Ca and the subsequent surface 1-inch (2.5 cm) and 4-inch (10 cm) zones pH monitored.

*Managing sulfur applications.* Sulfur-oxidizing bacteria are most active in wet, warm, well-aerated soil. Little activity occurs when soil temperatures are below 40°F (4.4°C). Sulfur application, therefore, should be avoided during cooler fall and winter months.

Due to its low solubility, when S is applied to the surface, it is slow to move into the soil. Therefore, large decreases in pH may occur in the thatch layer and immediate soil surface with little initial impact on the soil pH of the rootzone. Application rates also must be minimal to avoid damage to the crowns of the turfgrass plant. Rates applied to bermudagrass at fairway or rough height may be as high as 5 pounds/1,000 square feet (24 g/m$^2$). Applications to greens should not exceed 0.5 pound/1,000 square feet (2.4 g/m$^2$). Total annual applications should not exceed 10 pounds/1,000 square feet (49 g/m$^2$) on fairways. Sufficient irrigation water should be

**TABLE 8.20**
**Changes in Water Calcium and Electrical Conductivity due to Gypsum Addition**

| Gypsum Rate, Pounds/1,000 gallons | Gypsum Added in Acre-Feet of Water (pounds/acre-foot) | Ca Added (meq/L) | Increase in $EC_w$ (dS/m) |
|---|---|---|---|
| 0.72 | 234 | 1.0 | 0.075 |
| 1.44 | 468 | 2.0 | 0.150 |
| 2.16 | 702 | 3.0 | 0.225 |
| 2.88 | 936 | 4.0 | 0.300 |
| 3.60 | 1,170 | 5.0 | 0.375 |
| 4.32 | 1,404 | 6.0 | 0.450 |
| 5.04 | 1,638 | 7.0 | 0.525 |

*Notes:* Multiply pounds/1,000 gallons by 0.12 to obtain grams/1,000 liters. Multiply pounds/acre-foot by 0.3682 to obtain grams/cubic meter.

applied immediately after each application to wash the S from the turfgrass leaves to avoid leaf burn. It is wise to have the soil pH checked before reapplication of S to avoid overacidification. Sulfur application coincident with core aerification minimizes the potential for tissue burn and accelerates the acidification of the rootzone.

Commercial S ranges in purity from 50% to 99%. The value of S for reclamation depends on its purity and fineness. Like gypsum, the finer the material is, the faster it reacts in soil.

### Example

A water sample contains 1,800 ppm sodium (Na). If 7.5 acre-inches of water are applied per month, (a) how many pounds Na are being applied to the soil each year, (b) how much Ca would be needed per year to displace this Na, (c) how much gypsum ($CaSO_4$) would be needed to provide this Ca amount, and (d) how much calcium carbonate $\left(CaCO_3^-\right)$ would be needed to provide this Ca amount?

a.

$$\frac{1{,}800 \text{ mg Na}}{\text{L}} \times \frac{1 \text{ pound soil}}{454{,}000 \text{ mg}} \times \frac{7.5 \text{ acre-inches water}}{\text{month}}$$

$$\times \frac{12 \text{ months}}{\text{year}} \times \frac{27{,}154 \text{ gallons}}{\text{acre-inch}} \times \frac{3.785 \text{ L}}{\text{gallons}}$$

$$= 36{,}670 \text{ pounds Na yearly}$$

b.

$$36{,}670 \text{ pounds Na} \times \frac{454 \text{ g Na}}{\text{pounds Na}} \times \frac{1 \text{ eq Na}}{23 \text{ g Na}}$$

$$\times \frac{1 \text{ eq Ca}}{1 \text{ eq Na}} \times \frac{20 \text{ g Ca}}{1 \text{ eq Ca}} \times \frac{1 \text{ pound Ca}}{454 \text{ g Ca}}$$

$$= 31{,}887 \text{ pounds Ca}$$

c. **Step 1:** The molecular weight of gypsum $\left(CaSO_4^-\right)$ must be determined: Ca = 40 g, S = 32 g, O = 16 g.

$$CaSO_4 = 40 + 32 + (4 \times 16) = 136 \text{ g}$$

**Step 2:** One of two means of obtaining the desired amount of $CaSO_4$ can be used. The first sets up a ratio.

$$\frac{X \text{ pounds } CaSO_4}{31{,}887 \text{ pounds Ca}} = \frac{136 \text{ pounds } CaSO_4}{40 \text{ pounds Ca}} X$$

$$= 108{,}416 \text{ pounds or 54 tons}$$

d. If calcium carbonate ($CaCO_3$) was being used to displace the Na, how much would be needed?

**Step 1:** Determine the molecular weight of $CaCO_3$: Ca = 40 g, C = 12 g, O = 16 g.

$$CaCO_3 : 40 + 12 + (3 \times 16) = 100 \text{ g}$$

**Step 2:** The following second method is used to determine the rate of $CaCO_3^-$ needed to neutralize the Na:

$$\frac{31{,}887 \text{ pounds } Ca^{+2}}{\text{ac}} \times \frac{100 \text{ pounds } CaCO_3}{\text{eq } Ca^{+2}}$$

$$\times \frac{1 \text{ eq } Ca^{+2}}{40 \text{ pounds } Ca^{+2}} = 79{,}718 \text{ pounds or 40 tons}$$

### Amending Irrigation Water

**Irrigation water high in sodium.** Mixing high-SAR water with water low in both Ca and $EC_w$ does not reduce the sodium hazard of the mixture, because the SAR is generally not changed appreciably, but the $EC_w$ is reduced. Recall that high-SAR water with low $EC_w$ has worse infiltration properties than high-SAR water with high $EC_w$ (Table 8.13).

Adding gypsum to water with high SAR is one method of increasing the suitability of the water by displacing Na, adding plant-available Mg and C, and decreasing pH, which then increases P availability. Gypsum also decreases the SAR and increases the $EC_w$, which increases the infiltration properties of the water. The impacts of gypsum on Ca and $EC_w$ are listed in Table 8.20. The SAR and $EC_w$ for the water sample should be recalculated using the changes indicated. Adjust the $EC_w$ and SAR of the water with gypsum to produce water that has sufficient $EC_w$ and SAR to be considered acceptable as assessed by the criteria in Table 8.13.

Several forms of gypsum are available, including natural dihydrate gypsum and the natural anhydrite form. The dihydrate form dissolves more quickly because of attached water molecules. The finest, most soluble grade available should be used.

A gypsum machine with agitators and mixing tanks inject a slurry of suspended (not dissolved) particles through a filter into the water. Near 100% dissolution of the gypsum should occur within a few minutes. If a filter is not used, hard particles could be abrasive to the irrigation emitters.

The suggested injection gypsum rate range is between 2 and 2.5 meq/L. This must be continuously used if Na levels are excessive in the irrigation water source. Also, if the Na levels are severe, land application of gypsum will still probably be required. However, high salinity concentrations cannot be easily overcome. Dilution of high-salinity water with fresh water is probably the most practical solution. When total salinity is high, Fe plus a cytokinin as a foliar treatment is often beneficial since salt-stressed plants exhibit low cytokinin activity. Another method to determine the amount of 100% gypsum required per acre-feet of irrigation water is by multiplying the RSC value by 234 (or RSC × 86 = kilograms of gypsum/1,000 $m^3$ of water).

### Example

A superintendent wishes to know how much gypsum (75% pure) is needed to reduce the following water analysis SAR from 10 to 5 meq/L and how much the $EC_w$ will increase after adding this. Water analysis reveals 1.0 meq/L Ca, 7.5 meq/L Na, and 0.1 meq/L Mg.

**Step 1:** Solve the SAR equation for soluble Ca needed at the target SAR value.

$$SAR = \frac{Na^{+}}{\sqrt{\frac{Ca^{2} + Mg^{+2}}{2}}}$$

Insert the target SAR (5.0 in this example) and the concentrations of Na and Mg into the equation:

$$50 = \frac{7.5}{\sqrt{\frac{Ca^{2} + 0.1}{2}}}$$

$Ca^{+2}$ = 4.4 meq/L.

**Step 2:** Determine the amount of Ca needed.

$$[\text{Target Ca} - \text{initial Ca in water}] : [4.4 - 1.0] = 3.4 \text{ meq/L Ca.}$$

This is the amount of Ca to be added.

**Step 3:** Since gypsum is being used, the total amount needed is determined.

Conversion factor: 20 pounds of pure gypsum (23% Ca, $CaSO_4{\cdot}H_2O$) is needed to add 1 meq/L of Ca to 1 acre-inch of water.

$$[\text{Ca desired} \times 20 \times 100] \div \% \text{ gypsum purity}$$

$$= \text{pounds gypsum/acre-inch}$$

$$(3.4 \times 20 \times 100) \div 75\% = 91 \text{pounds}$$

$$75\% \text{ pure gypsum/acre-inch} (34 \text{ kg/m}^3)$$

$$\text{to lower SAR to } 5.0$$

**Step 4:** To calculate the increase in EC of water following this gypsum addition, convert pounds/acre-inch to parts per million. To do this, multiply by 4.4; to convert parts per million to EC, divide by 640. (91 × 4.4) ÷ 640 = 0.63 dS/m (or mmhos/cm). This is how much the conductivity of the irrigation water will increase from the gypsum addition, assuming the gypsum is completely soluble.

## Acid Injection

Acidification of irrigation water is effective in solving some problems but not others. Situations where water acidification may be beneficial include:

1. Water with a high (>1.25 meq/L) RSC: Lower water pH from acid injection will neutralize carbonates and bicarbonates in the water, making the water safe to use without the risk of damaging soil structure.
2. Water with a high SAR (>6 meq/L) and low bicarbonates/carbonates: One method of coping with high-SAR water is by adding S to the soil. Both acidification and S burners provide this. Adding S, however, does not lower the SAR of the water; it just makes the soil more manageable. The S reacts with insoluble calcitic lime in the soil to form gypsum, which in turns replaces Na ions on the soil CEC, which then leaches as sodium sulfate ($Na_2SO_4$), a soluble salt.
3. Soil ESP > 15%, which is Na-affected soil possessing poor infiltration and with free calcite (insoluble lime) present: Calcite (white crystals of calcium carbonate) forms at the soil surface in areas of low rainfall and high rates of evapotranspiration (e.g., arid regions). Calcite ($CaCO_3 \cdot MgCO_3$) deposits coat soil and sand particles, reducing soil water infiltration. The addition of acid helps dissolve calcite crystals. The goal is to lower the pH of the irrigation water to below 7.0 to promote calcite dissolving.
4. Water with a low RSC (low free Ca and Mg totals) and high bicarbonate content: In arid regions (<20 inches, 51 cm, rainfall per year), with fine-textured soil and soil Na levels that are moderate or high, it is often beneficial to acidify the water to lower soil pH and to maintain it at an acceptable level. Soils irrigated with poor-quality water typically have higher incidence of localized dry spots, standing water (poor infiltration), and poor fertilizer response, especially to Fe, S, manganese, phosphorous, and zinc.

High-bicarbonate–containing water may require acidification (via injection into the irrigation system) with sulfuric, N-phuric (a type of urea-sulfuric acid), or phosphoric acids or using S burners (or generators). To reduce handling and corrosion problems, sulfuric acid is often mixed with urea, commonly referred to as N-phuric acid. Water acidification is a two-step process that forms carbon dioxide ($CO_2$) and water ($H_2O$) from reacting an acid, such as sulfuric acid ($H_2SO_4$), with bicarbonate $\left(HCO_3^-\right)$ as shown earlier.

Other acids injected into irrigation water to neutralize water alkalinity include nitric ($HNO_3$) or citric ($H_3C_6H_5O_7$) acid (Table 8.21). When deciding on which acid is best for your situation, evaluate (1) safety and ease of use, (2) the relative cost of the acid, (3) plant nutrient being injected with the acid (how much N, P, or S will be injected into irrigation water), and (4) availability of the acid.

**Safety.** Citric acid, 75% phosphoric acid ($H_3PO_4$), and 35% sulfuric acid ($H_2SO_4$) are relatively safe to work with as compared to the 67% nitric acid (Table 8.21). Nitric acid is very caustic and can cause serious injury to exposed tissue, especially eyes and lungs. Avoid skin and eye exposure when handling any acid. Acid-resistant eyewear, gloves, and apron should be worn. Acids are corrosive and can damage clothing that is not immediately rinsed. Also, when mixing acid stock solutions, always add acid to a larger volume of water to create the stock. Never add water to concentrated acid.

**TABLE 8.21**
**Acids Commonly Used to Acidify Irrigation Water and Their Properties**

| Acid | Formulation and Density (d) or Formula Weight (FW) | Acid Needed for each meq/L of Alkalinity to Obtain a Water pH of ~5.8[a] | Nutrient Amount Provided by 1 Fluid Ounce of acid/1,000 Gallons Water[b] | Relative Safety[c] |
|---|---|---|---|---|
| Citric acid ($H_3C_6H_5O_7$) | 99.5% (w:w) granular FW = 192.1 | 9.1 ounces/1,000 gallons | None | • Can cause minor skin and eye irritation |
| | 50% (w:w) liquid d = 1.21 | 14.5 fluid ounces/1,000 gallons | None | • Can cause minor skin and eye irritation |
| Nitric acid ($HNO_3$) | 67% (w:w) liquid d = 1.42 | 6.6 fluid ounces/1,000 gallons | 1.64 ppm N | • Use extreme caution; very caustic and dangerous; avoid contact with fumes as well as acid |
| Phosphoric acid ($H_3PO_4$) | 75% (w:w) liquid d = 1.58 | 8.1 fluid ounces/1,000 gallons | 2.88 ppm P | • Slightly caustic; can cause skin and eye irritation as well as damage clothing |
| Sulfuric acid ($H_2SO_4$) | 35% (w:w) liquid d = 1.26 | 11.0 fluid ounces/1,000 gallons | 1.14 ppm S | • Slightly caustic; can cause skin and eye irritation as well as damage clothing |

[a] Add this amount for each milliequivalent/liter of alkalinity present. For example, if a water report indicates an alkalinity of 3 meq/L and sulfuric acid is used, add 33 fluid ounces of 35% sulfuric acid per 1,000 gallons of water (11 fluid ounces/milliequivalent/liter × 3 meq/L = 33 fluid ounces). Calculations are based on the following dissociation values: 2.07 meq H+ per 3 mg equivalent wt. of $H_3C_6H_5O_7$, 1 meq H+ per 1 mg equivalent wt. of $HNO_3$, 1.02 meq H+ per 3 mg equivalent wt. of $H_3PO_4$, and 1 meq H+ per 1 mg equivalent wt. of $H_2SO_4$.

[b] In this example, the acid would supply 38 ppm S at each irrigation (33 fluid ounces × 1.14 ppm S/fluid ounces = 33 ppm S).

[c] Use caution with **ALL** acids. Wear eye protection, acid-resistant gloves, and an acid-resistant apron when handling any acid.

**Cost.** In general, sulfuric and nitric acids are less expensive than phosphoric and citric acids. Citric acid is somewhat cost prohibitive for uses other than neutralization of water used in pesticide sprays and in fertilizer stock solutions. Citric acid is ideal as an acidifier for nutrient stock solutions and pesticide solutions, as it is much less likely to react with fertilizer salts or pesticides than the other three acids. However, the cost of citric acid makes it less desirable as an acidifier for large volumes of water used for irrigation and fertilization.

For acidification of irrigation water, sulfuric acid is recommended over nitric acid based on its relative safety. It is also recommended over phosphoric acid for very alkaline water due to the higher cost of phosphoric acid and the possibility of overapplication of P. As a general rule of thumb, for water acidification, multiply RSC by 133 to determine the amount of 100% pure sulfuric acid to add per acre-foot of irrigation water applied. To accurately predict the amount of acid required to acidify to a given pH, both the pH and alkalinity of the irrigation water must be known. Table 8.21 outlines initial amounts of acid to inject to lower the pH to approximately 5.8. However, this table is only taking alkalinity into account (and does not account for the starting pH of a water sample), so fine-tuning the amount of acid added will be necessary.

Another method to determine the amount of acid required to neutralize the alkalinity of water is by using the following equation:

$$A \times B \times C = \text{ounces of acid/1,000 gallons (mL/1,000 L)}$$
$$\text{water to lower the pH to 6.4}$$

where

A = cofactor determined by the existing water pH, which is listed in the following table

B = sum of the bicarbonate plus carbonate expressed as milliequivalents per liter

C = cofactor depending on the type of acid used

| Water pH | A value | Water pH | A value | Acid source | C value |
|---|---|---|---|---|---|
| 6.7 | 0.249 | 7.7 | 0.475 | 75% phosphoric | 10.6 |
| 6.9 | 0.342 | 7.9 | 0.484 | 85% phosphoric | 8.74 |
| 7.1 | 0.400 | 8.1 | 0.490 | 93% sulfuric | 3.72 |
| 7.3 | 0.437 | 8.3 | 0.494 | 61% nitric | 15.60 |
| 7.5 | 0.460 | 8.5 | 0.496 | | |

### Example

If the pH of a water source is 7.5, the carbonate + bicarbonate level is 3.4 meq/L, and 85% phosphoric acid is being considered. How much of 85% phosphoric acid is needed to reduce the water pH to 6.4?

$$A \times B \times C = \text{ounces of acid/1,000 gallons (mL/1,000 L)}$$
$$\text{water to lower the pH to 6.4}$$

$$0.46 \times 3.4 \times 8.74 = 13.7 \text{ ounces of 85\% phosphoric acid needed per 1,000 gallons (107 mL/1,000 L)}$$

**Sulfur "burners."** Other acidifying units (often called S generators) dissolve S chips or flakes (Chapter 3) into stored irrigation water to form sulfite ($SO_3$). The sulfite then reacts to form sulfurous acid and sulfuric acid, which then has the same effect as acid injection (Figure 8.25). The generator consists of a S chip storage hopper, oxidizing chamber, blower, and absorption tower. Pure elemental S chips or flakes are combusted in the oxidizing chamber to form sulfur dioxide ($SO_2$) gas. When $SO_2$ gas mixes with water, sulfurous acid $\left(H_2SO_3^-\right)$ is formed. Sulfurous acid is mild, only slightly corrosive, and easy to handle. In comparison, sulfuric acid ($H_2SO_4$) is an extremely strong and corrosive acid that is difficult to handle. Sulfurous acid improves water quality by lowering water pH and neutralizing bicarbonates and carbonates. It reduces pH by dissociating into hydrogen ions and sulphite. Hydrogen ions reduce water pH.

Bicarbonates and carbonates often form lime (calcium or magnesium carbonate), thereby increasing soil pH and tending to reduce soil percolation and drainage. These are neutralized by the sulfurous acid to form sulfite, carbon dioxide, and water.

The $CO_2$ gas escapes to the air. Since the S produced is in the $H_2SO_x$ form, *Thiobacillus* bacteria are not needed for oxidization as when elemental S is used, and this reaction is not dependent on soil temperatures.

Sulfur treatment will reduce water pH, bicarbonates, and carbonates. This treatment, however, will not in itself correct water Na problems. Sodium-rich water is usually injected with soluble gypsum and/or gypsum is added to the soil surface. Sulfur-treated water, however, helps maintain soluble Ca and Mg ions and reduce SAR values, thereby countering the detrimental effects of Na ions in the water.

Normally, a desirable soil pH range for turfgrasses is 5.5 to 7.0 and, for irrigation water, 6.0 to 8.0. The amount of acid added to water is most simply measured by pH probes. One probe is placed where water enters the pump station and another is placed downstream from the pump station. The amount of acid needed is based on the pH difference measured at the two probes. Other S-containing amendments are available (Table 8.18) and react similarly to sulfuric acid but react more slowly due to their lower solubility.

**FIGURE 8.25** Introducing sulfur to lower the pH of irrigation water.

## Water Hardness

Natural water, passing through rocks and soil, dissolves small quantities of chlorides, sulfates, and bicarbonates of calcium ($Ca^{+2}$), magnesium ($Mg^{+2}$), and iron ($Fe^{+2}$). Water containing such compounds is called **hard water** because soap has a "hard" time lathering in it. Ordinary soap is a Na or potassium salt of certain high-molecular-weight acids. When soap is added to hard water, the Ca, Mg, and ferrous Fe(II) salts of these acids react. This leads to the formation of insoluble metallic soaps that precipitate as a greasy scum on tubs or clothes. Until all of these Ca and Mg ions from the water are used up, additional soap will be needed to form lather.

$$\begin{aligned} &Ca^{+2}(\text{hardness ion} + CH_3(CH_2)_{10}COO^-(\text{soap}) \\ &\rightarrow [CH_3(CH_2)_{10}COO^-]_2CA^{+2} \downarrow \\ &(\text{soap scum;insoluble precipitate}) \end{aligned}$$

Hard water is undesirable due to the stains it causes on clothes, leather, and paper products during the manufacturing process. In addition, when exposed to heat, as in pipes and boilers, the bicarbonates decompose, forming carbonate precipitates. These precipitates build up, preventing good contact between the water and pipes. Therefore, the boiler pipes may overheat and fail under pressure. In turf, precipitates in hard water may clog irrigation heads and valves. In addition, excessive $Ca^{+2}$ ions may displace other desirable cations such as $K^+$ or $Fe^{+2}$ due to an increase in bicarbonates and water pH.

**Classifying water hardness.** Agriculture laboratories determine water hardness as the total lime-neutralizing value based on the following equation:

$$\text{Hardness} = (\text{percent Ca} \times 2.5) + (\text{percent Mg} \times 4.2)$$

or

$$\text{Hardness} = (\text{percent Ca} \times 2.5) + (\text{percent Mg} \times 4.2) + (\text{percent Fe} \times 1.8)$$

Interpretations for the total lime-neutralizing value are listed in Table 8.13 with <200 ppm having minor problems, 200 to 400 ppm having increasing problems, and >400 ppm having severe problems. Industrial water hardness is measured as parts per million or milligrams per liter of $CaCO_3$ regardless of what other minerals are present. Industrial water having a hardness of <50 ppm is considered "soft," 50 to 150 ppm considered "medium hard," and 100 to 300 ppm considered "hard."

### Example

The following water quality analysis report was generated for potential water sources for a golf course:

| Sample No. | Na | Ca | Mg | P | K | Cl | Carbonate ($CO_3$) | Bicarbonate ($HCO_3$) | Nitrate ($NO_3$) | Sulfate ($SO_4$) | TDS | pH | Conductivity (mmhos/cm) | SAR |
|---|---|---|---|---|---|---|---|---|---|---|---|---|---|---|
| | Parts per Million | | | | | | | | | | | | | |
| 1 | 1,800 | 131 | 190 | 1.55 | 73 | 6,800 | 117 | 610 | 6 | 86 | 9,100 | 8.4 | 8.0 | – |

1. Find the missing SAR value.
   **Step 1:** The units must be converted to milliequivalents per liter: Na (77), Ca (6.6), Mg (15.8), P (1.55), K (73), $CO_3$ (3.9), $HCO_3$ (9.8).
   **Step 2:** Insert the values into the SAR equation:

$$\begin{aligned} SAR &= \frac{Na^+}{\sqrt{\frac{Ca^2 + Mg^{+2}}{2}}} \\ &= \frac{77}{\sqrt{\frac{6.6+15.8}{2}}} \\ &= 23\ \text{meq/L} \end{aligned}$$

2. Calculate the RSC value:

$$\begin{aligned} RSC\ (\text{meq/L}) &= \left(CO_3 + HCO_3^-\right) - (Ca^{+2} + Mg^{+2}) \\ &= (3.9+9.8) - (6.6+15.8) = 8.7\ \text{meq/L} \end{aligned}$$

It is not always desirable for water to be completely free from minerals that make it hard, because very soft water is likely to be corrosive, especially if the pH value is very low. Highly alkaline water usually will leave a rustation on well screens and pipes. Acid water will be corrosive. Carbonic acid is a major source of low pH. Good water should be nearly neutral in pH.

One could add drops of an acid (like vinegar) to a dry soil sample and if fizz occurs (indicating $CO_2$ release), the water pH should be buffered to less than 7.0. These calcareous soils, however, can be acid treated (by irrigation injection) and the Ca released can help flocculate clay particles. A medium to high free-lime level on a soil test report also indicates soil calcium carbonate levels.

**Softening hard water.** Hardness of water is classified as (1) *temporary* (or *bicarbonate*) or (2) *permanent* (or *noncarbonate*). Temporary hard water is fresh water containing $Ca^{+2}$ and bicarbonate $\left(HCO_3^-\right)$ and can be softened by boiling it. However, in industry, temporary hard water is not softened by boiling due to the cost of fuel required. Chemical methods are used instead, such as the addition of ammonia ($NH_3$) or slaked lime [$Ca(OH)_2$]:

$$Ca^{+2} + 2HCO_3^- + 2NH_3(\text{ammonia}) \rightarrow CaCO_3 \downarrow + 2NH_4^+ + CO_3^{-2}$$

$$Ca^{+2} + 2HCO_3^- + Ca(OH)_2(\text{slacked lime}) \rightarrow 2CaCO_3 \downarrow + 2H_2O$$

Water purity can also be achieved by a combination of techniques such as deionization, removal of dissolved organic materials with activated carbon (or charcoal), and reverse osmosis to remove salts. Deionization involves removing ions with an ion-exchange resin. In this process, hard water containing $Na^+$, $Ca^{+2}$, $Cl^-$, and $SO_4^{-2}$ is added to two columns, the first containing $-SO_3^-$ groups bonded with hydrogen ($H^+$) ions. In this, $Na^+$ and $Ca^{+2}$ cations in the hard water are exchanged for the $H^+$ ions to form $-SO_3Na$, $-SO_3Ca^{+2}$, and an excess of hydrogen ions. In the second column, a hydroxide-containing anion-exchange resin such as $-N(CH_3)_3OH$ exchanges the $OH^-$ with the $SO_4^{-2}$ and $Cl^-$, leaving a number of $OH^-$ anions in solution. The $H^+$ from the first column combines with the $OH^-$ in the second column to form water. This formation also buffers the pH of the solution toward 7.0, thereby reducing bicarbonate $\left(CO_3^{-2}\right)$ and allowing certain nutrients to become more available. The cations on the cation-exchange resin can be replaced (or recharged) with hydrogen ions through passing dilute sulfuric acid through the column. The anions on the anion-exchange resin can be replaced (recharged) by hydroxide ions by passing a solution of sodium hydroxide (NaOH) through the column.

### Soft (Ultrapure) Water

Irrigation water is often considered "soft" or "ultrapure" if it has a low electrical conductivity ($EC_w$) of <0.50 dS/m, especially when ≤0.2 dS/m and bicarbonates are >120 ppm or if Ca levels are <20 ppm (1 meq/L). Prolonged use of this water may dilute and/or leach Ca and salts from the soil surface, reducing their soil stabilizing influence, resulting in soil surface crusting and reduced infiltration as clay particles disperse following drying, clogging smaller pore spaces. As a general rule of thumb to prevent soil dispersal, the irrigation water should have a minimum Ca content of 20 ppm (1 meq/L) and $EC_w$ of at least 0.5 dS/m. Soft water also may be corrosive to metal pipes, especially if the pH is low. Symptoms include more than normal runoff on sloped areas, reduced infiltration rates, and slower drainage of low areas. Plant nutrient response, especially to N, is generally lower and slower than normal and a distinct light green turf color often occurs.

Remedies for soft water include increasing salinity at the soil surface by applying gypsum, calcium chloride, phosphogypsum, or a S source plus lime to the soil. If soft water is a problem, these amendments should help increase low water infiltration. The second remedy is to add or increase the salinity concentration of the irrigation water above 0.50 dS/m. This is most often performed by injecting a Ca-containing source such as calcium sulfate (Epsom salt), calcium carbonate, or calcium chloride. To calculate the amount of amendment needed to add to irrigation water, the following formula is used:

$$[\text{target EC} - \text{current EC}] \times \frac{640 \text{ ppm salt}}{\text{EC water}} \times \frac{2.72 \text{ million pounds}}{\text{acre-feet water}} \div \%\text{ purity of material} = \frac{\text{pounds amendment}}{\text{acre-feet water}}$$

#### Example

If an irrigation water source has an EC of 0.1 mmhos/cm and it is desired to increase this to 0.50 mmhos/cm, how much gypsum with 75% purity is needed for each acre-foot of applied irrigation water?

**Step 1:** Determine how much the EC of the irrigation water needs to increase.

$$\text{Target EC} - \text{current EC} = 0.50 \text{ mmhos/cm} - 0.10 \text{ mmhos/cm} = 0.40 \text{ mmhos/cm}$$

**Step 2:** Determine the rate of pure gypsum needed.

$$0.4 \times 640 \times 2.72 = 696 \text{ pounds } 100\% \text{ pure gypsum added per acre-foot irrigation to increase EC from 0.1 to 0.5 mmhos/cm}$$

**Step 3:** Calculate the gypsum rate needed based on its purity.

$$\frac{696 \text{ pounds}}{75\% \text{ purity}} \times 100\% = 928 \text{ pounds of } 75\% \text{ pure gypsum applied per acre-foot irrigation water}$$

## EFFLUENT WASTEWATER USE

Use of effluent or recycled wastewater for irrigation purposes is a possible alternative water source for many turf growers. Irrigation with effluent water is an old practice used by the Greeks in Athens that was also used on U.S. golf courses as early as 1932. Turf often is a suitable commodity to use effluent water since it absorbs large amounts of N and other nutrients found in reclaimed water. This reduces the chances of groundwater contamination from effluent water use. Since turf is a perennial crop, a continuous supply is needed. Turf is often located near metropolitan areas; therefore, conveyance costs often are reduced. Certain states already have mandated that, if available, golf courses must use reclaimed water for irrigation. With this in mind, golf facilities should prepare to use reclaimed water or possibly face refused water rights (Figure 8.26). Since turf is not a food crop, potential health problems arising from reclaimed water are reduced.

Conditions to consider before using reclaimed water for golf course irrigation include:

- Does the reclaimed water have adequate quality and remain consistent over time?

**FIGURE 8.26** Sign indicating the golf course is being irrigated with reclaimed water.

- Are heavy metals present?
- Is it available free or at a reasonable cost?
- Does a predetermined amount have to be accepted daily regardless of need?
- Who is responsible for building and maintaining storage facilities when irrigation is not needed?
- Does it possess any detrimental health effects?
- Is it potentially injurious to plant life, fish, and wildlife?
- Will it cause any adverse effects on downstream water rights?

### Levels of Treated Water

The usability of effluent water on turf depends on the physical, chemical, and microbiological quality of the water. The terms potable water and domestic water are often used synonymously to describe drinking water; effluent-, reclaimed-, gray-, recycled-, or wastewater are terms used to describe water that has gone through one cycle of domestic use. Most effluent water from treatment plants is approximately 99.9% water and 0.06% solids (or sludge) (Table 8.22). Ideally, effluent or wastewater used for turf irrigation should principally come from an urban area without significant industrial input. This should guard against the possibility of excessive heavy metal content from industry. The treatment level of wastewater also should be at least **secondary.**

**Primary water.** Primary treatment begins with preliminary operations such as screening and sedimentation that remove organic and inorganic solids. After screening and possible grinding of debris, dense materials such as sand and stones are allowed to settle in a grit chamber. This material is normally washed and also is used as landfill.

Undissolved suspended matter is then removed in a second settling tank or a primary clarifier. Settled material forms a mass of raw sludge that is concentrated and used as landfill.

The remaining liquid in the settling tank is called primary effluent and may be chlorinated to destroy bacteria and reduce odor before it is used. Primary sedimentation removes approximately 60% to 70% of the suspended solids and 25% to 40% of the *biochemical oxygen demand* (BOD).

**Secondary water.** Nonchlorinated primary effluent water may be further treated in a biological process to break down complex organic matter during secondary treatment. Up to 90% of organic matter is removed by trickling the effluent through large cylindrical vats containing bacterial colonies that digest the complex and simple organic compounds. Water

**TABLE 8.22**
**General Characteristics of Effluent Water**

| Treatment Level | Characteristics |
|---|---|
| Primary | • Contains ≤50% of origin solids |
| | • Contains no large debris |
| | • Can have a bad odor |
| | • Contains <23 coliform bacteria per 100 mL |
| | • Not to be used for turf irrigation |
| Secondary | • >90% solids removed |
| | • Coliform bacteria count < 23/100 mL |
| | • Has a slight odor |
| | • Can be used to irrigate turf (no human contact) |
| Tertiary (or advanced) | • >99% solids removed |
| | • No or low odor |
| | • Highly purified |
| | • Coliform bacteria count < 2.2/100 mL |
| | • Can be used to irrigate turf |
| | • >1 mg/L biological oxygen demand (BOD) |
| | • Typically contains 2 to 10 mg N/L, 0.1 to 1.0 mg P/L, and 0 mg K/L |

is then chlorinated so that the coliform bacterial count is less than 23/100 mL of water and is the principal source of water for agricultural irrigation purposes.

**Tertiary water.** Although secondary water is usable for irrigation, many water plants continue treatment to the final stage, known as advanced or tertiary treatment, which involves using a charcoal or polymer bed for chemical coagulation and flocculation, sedimentation, filtration, or adsorption of compounds. It removes nonbiodegradable organic pollutants and most nutrients, such as N and P. It has no foul odor and contains a coliform bacterial count of less than 2.2/100 mL. This process can provide highly purified water and is similar to potable water treatment. It is the most commonly used effluent source on golf courses. In some instances, *quaternary treatment* is used in which UV light, ozone, or additional chlorination are used to further purify the water.

## Characteristics of Effluent Water

Effluent water has three major categories of characteristics that are modified during use: (1) biological composition, (2) organic composition, and (3) dissolved inorganic salts. Although biological composition of effluent water is of great concern because of pathogenic bacteria and viruses, chlorinated waters are not released for irrigation without prior approval of public health officials.

For irrigation purposes, the organic portion of the effluent water is generally of minimal consequence. The most influential characteristic of effluent water irrigation is the resulting higher salt load. Samples should be analyzed before use for

1. pH
2. Bicarbonate concentration
3. Dissolved solids
4. Salt and sodium hazard (permeability)
5. Heavy metal or toxic ion concentration

Some of these may accumulate in the soil or on leaves at levels injurious to plants.

**Total suspended solids.** Soluble inorganic materials such as clay, silt, and other soil constituents, as well as organic matter such as plant material, algae and bacteria, accumulate in secondary-treated water. Most tertiary-treated effluent water has negligible levels of suspended solids and is not a problem. Total suspended solid (TSS) level requirements have been set between 5 and 10 mg/L for most states. Continuous use of high TSS-containing water may eventually clog surface pores that inhibit water infiltration into the soil profile. Golf greens and tees to be watered with effluent should be constructed with high sand content and have infiltration and percolation rates at the higher rate end of the USGA specified range. Filtration may be needed to prevent clogging of sprinkler heads and valves. For example, drip irrigation emitters are prone to clogging when irrigation water has >50 mg/L (ppm) suspended solids. Increased aerification practices also may be necessary if TSS levels are excessive.

### Example

Determine the amount of TSS applied with each acre-foot (0.3 m) of irrigation water containing 19 ppm TSS.

**Step 1:** Multiply 19 ppm × 2.72 (an acre-foot of water weighs 2.72 million pounds) to determine the pounds of solids applied per acre-foot (0.3 m) of water.

$$19\,\text{ppm} \times 2.72 = 52\,\text{pounds/acre}\,(58\,\text{kg/ha})$$
$$\text{or}\ 1.19\,\text{pounds/1,000 square feet}\,(0.6\,\text{kg/100 m}^2)$$

**Step 2:** Is this a problem? Not likely, as only 52 pounds (58 kg/ha) are being applied over an acre.

**Turbidity** is also reported by many labs that analyze water quality. Turbidity measures the transmission of light through water. It is most affected by particulate matter suspended in the water. Unfortunately, standard guidelines on acceptable turbidity thresholds have not been developed.

**Salts and soil permeability.** Water management and adequate drainage are keys to successful effluent use in turf management. Generally, water going through one cycle of average home use accumulates approximately 300 ppm (or 0.48 dS/m) of *total dissolved salts* (TDS). The ability to leach out these salts is necessary to prevent accumulation of toxic concentrates. High TDS-containing effluent water also has a tendency to clog soil pores and coat the land surface. Coarse-textured soils, such as sandy loams, with a moderate permeability that can infiltrate approximately 2 inches/day (5 cm/day) or more on an intermittent basis is best for the use of wastewater. Soils with a hard pan, clay pan, or underlying rock may create a perched water table that promotes surface accumulation of salts and heavy metals. The TDS (meq/L) can be estimated by multiplying $EC_w$ (dS/m) by 10. For example, water having $EC_w$ of 2.50 dS/m contains ~25.0 meq/L of dissolved salts.

Initial tests should be conducted to show the SAR or ESP, which are indexes of the effect of Na in reducing soil permeability (the rate at which water passes into and through soil) based on the ratio of Na to Ca and Mg. A SAR of 10 or less is considered desirable. If greater than 10, soil permeability problems may occur. Gypsum or gypsite could be incorporated into the management program, or alternate application of fresh water with effluent water should be followed to help leach out Na. Additional tests include adjusted SAR, which accounts for adjustments in the influence of $HCO_3^-$ and the RSC value using Ca, Mg, $HCO_3$, and $CO_3$ concentrations.

Other potential problems with effluent irrigation include exposure of golf course maintenance equipment and golf carts to salt when these are driven through depressions holding standing effluent water. Internal switches and the underside bodies of carts are susceptible to rusting from salinity, as are seals and bearings associated with maintenance equipment. Using cart paths and frequently rinsing the equipment with fresh water will minimize these problems. Wetter playing conditions as a result of leaching programs will periodically occur. Areas damaged by salinity stress will require repair by supplemental seeding or sodding. Courses that overseed in winter also require higher seeding rates to compensate for seedling mortality from salt stress.

### Calculating Fertilizer Amounts from Effluent Irrigation

Most effluent irrigation sources contain a certain amount of nutrients. To compensate for these in a yearly fertilization program, the pounds/acre applied from irrigation water need to be determined. The following formula is one method of determining this. Another method was discussed previously in the nutrient content section of this chapter. Table 8.23 is useful in determining the amount of dissolved nutrients and salts applied per acre-foot of irrigation water.

#### Example

1. If irrigation water contains 5 ppm nitrate ($NO_3$-N), determine the amount of N applied per acre-foot.
   From Table 8.23, to convert pounds of N applied per acre-foot of water containing 5 ppm of $NO_3$-N, multiply 5 ppm × 2.72 = 13.6 pounds $NO_3$-N.
2. If 8 ppm of K is present in the irrigation water, determine how much potash ($K_2O$) is applied.
   From Table 8.23, 8 × 3.26 = 26 pounds of potash ($K_2O$) applied per acre-foot of water containing 8 ppm K.
3. How many pounds K and $K_2O$ are applied per acre-foot of irrigation water containing 10 ppm K?

**Step 1:** From Table 8.23,

$$10\,\text{ppm K} \times 3.26 = 32.6\,\text{pounds K applied/acre-foot of irrigation}\ (12\,\text{g/m}^3)$$

**Step 2:**

$$32.6\,\text{pounds K} \div 0.83 = 39.3\,\text{pounds } K_2O \text{ applied/acre-foot irrigation } (14.5\,\text{g/m}^3) \text{ containing 10 ppm K}$$

#### Example

If 7.8 acres of tees are irrigated an average of 4.5 acre-inches (or 0.375 acre-foot) every month containing 6 ppm $NO_3$, how much N is being applied?

**Step 1:** The amount of N in nitrate ($NO_3$) must be determined: molecular weight N = 14 g.

$$NO_3 = 14 + (3 \times 16) = 62\,\text{g}$$

$$\text{Percent N} = (14 \div 62) \times 100 = 22.6\%$$

From the analysis, 6 ppm $NO_3$ is in the water; thus, this is multiplied by 22.6% to obtain ppm N.

$$6\,\text{ppm} \times 22.6\% = 1.36\,\text{ppm N applied in 6 ppm } NO_3$$

**TABLE 8.23**
**Converting Nutrient Amounts [(meq/L or ppm (mg/L)] in a Water Analysis Report to Pounds Nutrient Applied per Acre-Foot of Irrigation Water (A × B × C = D)**

| Ion | Analysis Report (A) | Conversion Factor (B) | Equivalent Weight (C) | Pounds of Material Applied per Acre-Foot of Water (D) | Material |
|---|---|---|---|---|---|
| Ca | meq/l | 2.72 | 20.04 | 54.5 | Calcium |
| Mg | meq/l | 2.72 | 12.15 | 33.0 | Magnesium |
| Na | meq/l | 2.72 | 23.00 | 62.6 | Sodium |
| K | ppm | 3.26 | | 3.26 | Potash ($K_2O$) |
| $CO_3$ | meq/l | 2.72 | 30.00 | 81.6 | Carbonate |
| $HCO_3$ | meq/l | 2.72 | 61.02 | 166.0 | Bicarbonate |
| Cl | meq/l | 2.72 | 35.45 | 96.5 | Chloride |
| P | ppm | 6.22 | | 6.22 | Phosphate ($P_2O_5$) |
| $NO_3$ | ppm | 0.61 | | 0.61 | Nitrate |
| $NO_3$-N | ppm | 2.72 | | 2.72 | Nitrate-N |
| $SO_4$ | ppm | 0.90 | | 0.90 | Sulfur |

**Step 2:** Now determine how much N is being applied each month.

$$1.4\text{ ppm N}\times\frac{0.375\text{ acre-foot}}{\text{month}}\times 7.8\text{ acres}\times\frac{2.72\text{ pounds}}{\text{acre-foot}}$$
$$=10.8\text{ pounds N applied monthly over 7.8 acres}$$

**or**

From Table 8.23,

$$6\text{ ppm NO}_3\times 0.61\times\frac{0.375\text{ acre-foot}}{\text{month}}\times 7.8\text{ acres}$$
$$=10.7\text{ pounds N applied monthly over 7.8 acres}$$

## Toxic Ions

Irrigation water quality is also influenced by other specific ions such as Na, B, and chloride. Most irrigation water sources contain low levels of a variety of elements. Normally these pose minimal problems but can increase under conditions of inadequate leaching with quality water, poor soil permeability, and during periods of high evaporation (Tables 8.13 and 8.24).

**Sodium.** Sodium (Na) is of primary concern because it is often found in the largest amount. It is an antagonistic ion that displaces Ca in cell membranes, causes leaking of potassium, and can limit the availability of Fe and manganese in soils. Sodium toxicity appears as marginal scorch of older leaves. Leaf tissue content of 0.25% to 0.50% Na by weight can cause toxicity to sensitive ornamental plants. A water Na content of >3 meq/L (>70 ppm) can damage the foliage of sensitive ornamental plants. To convert milliequivalents/liter Na to parts per million, multiply the milliequivalents/liter value by 23.

**Sulfur.** Effluent water also often has an unusually high level of sulfate $\left(SO_4^{-2}\right)$. To convert parts per million sulfate to pounds $SO_4$–S per acre-foot of applied water, multiply sulfate parts per million by 0.90 (Table 8.23). Irrigation water, for example, with 100 ppm $SO_4^{-2}$ provides 2.1 pounds S/1,000 square feet (10 kg/100 $m^2$) with each foot of water applied. Sulfur can react with Fe and Mn under anaerobic conditions to create FeS and MnS, which then contribute to poor drainage and possible black layer.

Lime can be added to the soil to transform sulfate to insoluble gypsum ($CaSO_4$), which prevents the S from being reduced to FeS and MnS. Calcium carbonate applied at 10 pounds/1,000 square feet (49 kg/100 m) provides 3.8 pounds Ca (1.7 kg), which can react with 9.1 pounds (4.1 kg) $SO_4^{-2}$, or 3 pounds (0.4 kg) S to form gypsum.

**Iron.** Long-term use of water containing high amounts of Fe may cause

- Precipitation of P and Mo, causing deficiencies
- Staining of sidewalks, buildings, and equipment exposed to the water
- Plugging of irrigation components by anaerobic iron sludge deposits
- Manganese deficiency or, to a lesser extent, Zn and Cu deficiencies
- Possibility of clogging small irrigation heads if >1.5 ppm Fe is in the water

**Boron.** Boron (B or $BO_3^{-1}$) in irrigation water is rarely a problem with turfgrasses because B accumulates in leaf tips that are removed by regular mowing. However, other landscape plants may be more sensitive to B levels. Boron toxicity symptoms typically show first on older leaf tips and edges as yellowing, spotting, or drying of leaf tissue. The yellowing or spotting is sometimes followed by drying that progresses from near the tip along the leaf edges and toward the center between the veins (termed interveinal chlorosis). Landscape plants may be damaged if irrigation water has B contents above about 2 meq/L, or if leaf boron contents exceed 250 to 300 ppm (dry weight). High B levels are more commonly associated with alkaline soil formations in areas or times of low rainfall.

**Chloride.** High concentrations of chloride, sulfate, and bicarbonate ions also can cause specific ornamental plant injury under certain soil conditions. Chloride ($Cl^-$) is not adsorbed by soils but moves readily with the soil water. It inhibits plant water uptake as a salt; thus, it reduces nutrient uptake. It is absorbed by plant roots and accumulates in the leaves. Accumulation of chloride in sensitive plants of 0.30% to 1.00% by dry weight can lead to leaf burn and desiccation. Chloride toxicity symptoms include leaf burn or drying that typically occurs initially at the leaf tip of older leaves and progresses back along the edges as severity increases. Leaf drop and defoliation also occur with excessive leaf burn. Mowing of turf removes damaged leaf tips. Table 8.13 provides general ranges of elements and some expected results at various concentrations. A water chloride content of ≥100 mg/L can damage the foliage of sensitive ornamental plants. Chloride concentrations of 350 mg/L or higher in soil can cause damage. Multiply milliequivalents per liter of chloride by 36 to obtain parts per million. Water containing 1 meq/L chloride will deposit approximately 100 pounds (37 $g/m^3$) $Cl^-$ per acre-foot of water applied.

**Chlorine.** Chlorine (Cl) is the element and chloride ($Cl^-$) is the ion. Chlorine ($Cl_2$) toxicity normally occurs if reclaimed sewage water containing excessive disinfectants is used for irrigation. If stored between treatment and application in a holding pond, much of the free chlorine in water will dissipate as gas since it is very unstable. Foliage damage can occur to ornamentals if residual chlorine levels exceed 1 to 5 mg/L (or parts per million).

**Fluoride.** Fluoride (F) may also be another ion of concern to ornamental plants. Fluorine (F) is the element and fluoride ($F^-$) is the ion. Severe leaf tip burn and scorch can occur on ornamental plants exposed to water containing high levels of fluoride (>1 mg/L).

**Heavy metals.** Heavy-metal concentrations usually are not a problem with urban effluent water sources but are potentially major concerns with certain industrial effluent sources.

**TABLE 8.24**
**Potential Trace Element Tolerances for Irrigation Water**

| Element | Continuous Use (ppm) | Short-term Use on Fine-Textured Soils (ppm) | Comment |
|---|---|---|---|
| Aluminium (Al) | 1.0 to 5.0 | 20 | • Can cause nonproductivity in acid soils (pH < 5.5), but most alkaline soils (pH > 7.0) will precipitate the ion and eliminate any toxicity. |
| Arsenic (As) | 0.1 | 10 | • Toxicity to plants varies widely, ranging from 12 mg/L for sudangrass to <0.05 mg/L for rice. |
| Beryllium (Be) | 0.1 | 1.0 | • Toxicity to plants varies widely, ranging from 5 mg/L for kale to 0.5 mg/L for bush beans. |
| Boron (B) | 0.75 | 2.0 | • Accumulates in leaf tips. Sensitive landscape plants can be damaged if >2 mg/L B is in irrigation water. |
| Cadmium (Cd) | 0.01 | 0.05 | • Toxic to beans, beets, and turnips at concentrations as low as 0.1 mg/L. |
| Chlorine (Cl) | 10 | – | • Accumulates in leaf tips. Landscape plants can be especially sensitive. |
| Chromium (Cr) | 0.1 | 20 | • Not generally recognized as an essential growth element. |
| Cobalt (Co) | 0.05 | 10 | • Toxic to tomato plants at 0.1 mg/L in nutrient solutions. Tends to be inactivated by neutral and alkaline soils. |
| Copper (Cu) | 0.2 | 5 | • Toxic to a number of plants at 0.1 to 1.0 mg/L in nutrient solutions. |
| Fluoride (F) | 1.0 | ? | • Inactivated by neutral and alkaline soils. |
| Iron (Fe) | 5.0 | ? | • Not toxic to plants in aerated soils, but can add to soil acidification and in available P and Mo. Overhead irrigation may discolor plants, sidewalks, equipment, and buildings. |
| Lead (Pb) | 5.0 | 20 | • Can inhibit plant cell growth at very high concentrations. |
| Lithium (Li) | 2.5 | 5.0 | • Tolerated by most crops up to 5 mg/L; mobile in soil. Acts similarly to B. |
| Manganese (Mn) | 0.2 | 20 | • Toxic to a number of plant roots, but usually only in acidic, poorly drained soils. Keep pH 6.0 to 7.0 and provide good drainage. |
| Molybdenum (Mo) | 0.01 | 0.05 | • Not toxic to plants at normal concentrations. Can be toxic to livestock if forage has high concentrations. |
| Nickel (Ni) | 0.2 | 2 | • Toxic to certain plants at 0.5 to 1.0 mg/L; reduced toxicity at neutral or alkaline pH. |
| Selenium (Se) | 0.02 | 0.05 | • Toxic to plants at concentrations as low as 0.025 mg/L. |
| Tin (Sn) | ? | ? | • Effectively excluded by plants; specific tolerance is unknown. |
| Tungsten (W) | ? | ? | • See remarks for tin. |
| Vanadium (V) | 0.1 | 10 | • Toxic to many plants at relatively low concentrations. |
| Zinc (Zn) | 2.0 | 10 | • Toxic to many plants in widely varying concentrations; reduced toxicity at pH > 6.0 and in fine-textured or organic soils. |

*Sources:* Modified from Ayers, R. S., and D. W. Westcot. 1985. Irrigation and drainage paper 29. Rome, Italy: Food and Agriculture Organization of the United Nations. http://www.fao.org; McCarty, L. B., ed. 2011. *Best golf course management practices* (3rd ed.). Upper Saddle River, NJ: Prentice-Hall, Inc.

This is why industrial and mining effluent water sources are not generally recommended for turf irrigation.

Several trace elements in domestic effluent water could be present in potentially toxic amounts under certain conditions. Therefore, they should be periodically monitored. These include chlorine, boron, cadmium, copper, nickel, and zinc. The National Academy of Sciences recommends irrigation water should contain no more than 0.01 mg/L of cadmium, 0.2 mg/L of copper, 0.2 mg/L of nickel, and 2.0 mg/L of zinc.

### Managing Toxic Elements

Managing toxic ions is difficult in established landscapes. If a problem develops, replacing plants sensitive to particular elements is probably the easiest and cheapest means of overcoming it. Blending better quality water with that containing the toxic ions and applying additional water to leach the ions are additional means of managing toxic elements.

### Water Storage Facilities

A seasonal problem for turf managers who use wastewater is that most contracts require that a specific amount be accepted daily, regardless of weather conditions. In other words, a preset level of wastewater must be accepted per day, whether or not it is needed. Storage capability, therefore, is a major requirement when using effluent water and must be adequate to store enough water for the maximum days of nonirrigation (usually a minimum of three to five).

Storage ponds can be a source of algae, weeds, odors, and health problems if not properly designed and maintained. In fact, effluent water stored in ponds with a 1.1 ppm N may have excessive algae and aquatic plant growth. If water must be stored, an enclosed tank is preferred as this eliminates sunlight exposure and reduces algae formation. However, due to settling of suspended matter, storage tanks need periodic emptying and cleaning. Storage ponds generally are acceptable for wastewater storage as long as the storage amount does not impair the pond's ability to function as a stormwater management system. Generally, storage ponds do not have to be lined, although lining (minimum lake lining thickness of 40 mil) allows easier maintenance and cleaning following drawdowns. Storage ponds should also be at least 10 feet (3 m) deep, with good aeration, and have a three-to-one bankside slope to minimize aquatic weed problems. The deeper the pond is, the better, since this reduces sunlight penetration, keeps the water cooler (cool water holds 40% more oxygen than warm water), and helps manage algae better (Figure 8.27).

Adequate circulation and aeration are needed for odor and algae control; therefore, include electrical service in the course design. Aeration helps attack the cause(s) of poor water quality while other methods, such as dyes and algaecides, only treat the symptoms—algae. Artificial aeration helps provide oxygen when natural wind and wave action cannot provide sufficient levels. Fountains, intake-exit piping, air injection, waterfalls, and artificial wetlands are other means to provide pond circulation and aeration. Aerators need to be matched to the size of the pond or lake. If the aerator cannot move sufficient amounts of water, then disappointing benefits will be seen from it. Normally, with open ponds, high-quality filtration systems will be needed due to algae blooms to prevent clogging of irrigation heads and valves as well as golf-green rootzones. Generally, due to substantial labor and chemical costs, long-term costs associated with covered tanks are less than those for open ponds.

A week's supply of water (assuming 1.5 inches, 3.8 cm, of water applied on 100 acres, 40.5 ha, of turf) would be 4 million gallons (15 million L), or 535,000 cubic feet (15,133 $m^3$), translating into a lake that is 10 feet (3 m) deep measuring approximately 180 × 300 feet (1,528 $m^3$) in size. Several smaller ponds may fit a golf course layout better than one large lake.

## Example

1. How many gallons of water are in a storage pond with a surface area of 380,000 square feet and an average depth of 12 feet?

   **Step 1:** Determine the total acres of surface area of the pond.

$$\frac{380{,}000\text{ square feet}}{\text{pond}} \times \frac{1\text{ acre}}{43{,}560\text{ square feet}} = 8.724\text{ total acres}$$

   **Step 2:** Determine the total acre-feet of the pond.

$$8.724\text{ acres} \times 12\text{ feet} = 105\text{ acre-feet}$$

   **Step 3:** Convert acre-feet to gallons. One acre-foot equals 325,851 gallons; thus,

$$105\text{ acre-feet} \times \frac{325{,}851\text{ gallons}}{\text{acre-feet}} = 34{,}214{,}355\text{ gallons}$$

   Therefore, 34,214,355 gallons of water are in a pond of this size.

2. If 83 acres of turf are irrigated from this storage pond, how many inches of water could be applied? (Evaporation losses are not considered in this problem.)

   **Step 1:** Determine the amount of water needed to apply 1 acre-inch to 83 acres. Each acre-inch contains 27,154 gallons of water.

$$83\text{ acres} \times \frac{27{,}154\text{ gallons}}{\text{acre-inch}} = 2{,}253{,}782\text{ gallons}$$

   **Step 2:** Divide total capacity of the pond (34,214,355 gallons) by the amount of water needed over 83 acres.

**FIGURE 8.27** A major consideration when using reclaimed water is storage during periods of low demand.

$$34{,}214{,}355 \text{ gallons} \times \frac{\text{acre-inch}}{2{,}253{,}782 \text{ gallons}}$$
$$= 15.2 \text{ inches water available for 83 acres of turf}$$

## Golf Greens

In light of previously discussed advantages and disadvantages of using effluent water for golf course irrigation, it is suggested only tertiary treated wastewater be used on golf greens, especially those planted with cool-season turfgrasses. Turf mowed excessively low, such as golf greens, is constantly on the management edge in terms of maintaining a healthy, acceptable playing surface. Although primary and secondary wastewater impurities are low, golf greens do not need added stress in terms of salinity and salts. Salinity can be extremely detrimental to golf greens due to their relatively shallow and weak root systems. Continued use of wastewater with low to moderate TSS may, in time, reduce water infiltration and percolation to the point of reducing turf quality. For these and other reasons, primary and secondary reclaimed water sources are not recommended for golf greens.

If a turf manager must use one of these wastewater sources for irrigating golf greens, several prerequisites exist for success. The golf green should have excellent—more than 6 inches (15 cm)/hour—infiltration and percolation rates to prevent salinity buildup. A superintendent must also be allowed to frequently aerify, spike, and slice the soil surface to minimize crusting and algae development. Regular flushing with a freshwater source also is necessary to remove salts.

Higher mowed turf, such as fairways and roughs, is better able to tolerate higher salinity and TSS levels. Secondary wastewater, therefore, may be used successfully on such areas, assuming cultural practices such as aerifying and spiking are allowed. Exposure and environmental concerns, however, must be addressed before using this source on these areas.

## Irrigation System Design

Corrosion of metallic parts and plugging of nozzle orifices are two potential problems in irrigation components when wastewater is used. Chlorides and ammonia are corrosive components of many wastewater sources. Chlorides damage brass irrigation valves and fittings, as well as galvanized pipe and fittings. Ammonia can be corrosive to copper pipe with concentrations of N as low as 1.5 mg/L. Combinations of chlorides and ammonia substantially increase the chances of corrosion.

Nozzle orifice clogging is another potential problem when using wastewater. This usually is not a problem with secondary treated effluent if relatively large irrigation nozzle sizes are used. Normally, the high pressure used to irrigate serves as a self-cleaning mechanism.

Clogging problems may occur from algae growing in the nutrient-rich wastewater while in storage ponds or in the piping system. A filtration system should be provided to minimize algae introduction into pipes, and valves should be designed to handle wastewater. This is especially important if secondary treated effluent water is used.

Aboveground spigots, hose bibs, quick-coupling valves, and so on are not allowed when using effluent water. These must be in belowground boxes that are locked and clearly marked.

**Cross-contamination prevention.** The distribution system includes pipelines, pumping stations, and storage facilities. A cross connection between a potable water system and any nonpotable water source must be avoided. Ways of preventing this include:

- Irrigation lines should be tagged, colored, and designed so unauthorized persons cannot operate them. Most effluent lines and values are colored purple (Figure 8.28).
- Separation of domestic pipelines from reclaimed pipelines is necessary to reduce the possibility of cross contamination. This usually involves 10 feet

**FIGURE 8.28** Purple-colored pipe indicating reclaimed water is being used.

(3 m) horizontal and 1 foot (0.3 m) vertical separation where these lines cross.
- Easy connection to effluent water couplers should be prevented.
- Hose bibs are usually allowed if located in below-ground lockable vaults or if a special tool is necessary to access the effluent water.
- Backflow protection should be implemented.
- Minimum pipe and joint leakage is necessary.

### Other Information and Suggestions

- Bentgrass declines at 1.5 to 2.0 dS/m $EC_w$ and bermudagrass declines at 4 to 15 $EC_w$, but this may vary depending on soil factors, temperature, irrigation, cultivar, and so on.
- Juvenile plants are more sensitive to salt injury. Increasing overseeding rates 10% to 20% may be necessary to produce acceptable quality.
- pH values for most urban effluent water range between 6 and 8.
- Nematode concentrations have not been found to increase with effluent water use compared to freshwater use.
- Use of salt-tolerant grasses, such as hybrid bermudagrass or seashore paspalum, should be considered before application.
- Effluent water should not be sprayed on domestic water wells or reservoirs, or near drinking fountains or eating areas. Drinking fountains need self-closing covers.
- Irrigation systems should have proper filters to catch any damaging solids.
- Communicate to all persons about the procedure. Establish and maintain good relations with all involved governmental agencies.
- All irrigation should be timed to minimize public contact and to allow ample opportunity for land to dry out before it is reused.
- A backup system is critical for greens and tees, as some treatment plants shut down periodically for maintenance.
- Low-pressure sensors should be installed in the event of pressure drop.
- A 50- to 100-foot (15 to 30 m) vegetative buffer zone is often required between the edge of the spray and the nearest dwelling.
- Chlorine levels of 0.5 mg/L or higher are generally required in effluent water to reduce odors, slime, and bacterial growth.

## WATER TREATMENT OPTIONS

### Desalination

Desalination is the process of taking salt out of salt water. Several means are available to provide this, including (1) distillation, (2) reverse osmosis, (3) ion exchange, and (4) electrodialysis. Distillation and reverse osmosis remove water from solution while ion exchange and electrodialysis remove salt from the solution. Regardless of which desalination method is used, costs will depend on how much salt is in the source water and how clean the treated water must be. Fresh water typically has up to 500 ppm of total dissolved solids and usually does not require desalination. Brackish water contains 500 to 35,000 ppm and is the source of water most often used in desalination. Seawater contains more than 35,000 ppm of salt, which makes it the most costly source of water to treat.

A by-product of desalination is *brine,* a concentrated salt solution from each of these desalination processes that can be discharged into deep saline aquifers or into water with a higher salt content. Brine can also be mixed with treated effluent and sprayed over golf course rough areas.

**Distillation.** Distillation is the oldest form of desalination. Simply, distillation is where water is boiled and produces steam. The steam cools and condenses back into water, which is then collected and used. The salt does not vaporize, forming brine in the original container, so the distilled water is fresh and without salt. The distillation process requires high energy and thus is rarely used to commercially purify golf course irrigation water.

**Reverse osmosis.** Reverse osmosis is the desalination method gaining the most popularity. Osmosis occurs when fresh water and salt water are on opposite sides of a semipermeable membrane. Salt water has a higher osmotic pressure and, to balance this pressure, fresh water flows through the membrane and dilutes the salt water until the pressures of both are equal.

In reverse osmosis, a high-pressure pump applies more pressure on the salt water than the osmotic pressure of sea water. The water in the salt solution side flows through the membrane, leaving salt (brine) behind (Figure 8.29). Reverse osmosis is the desalination process many municipalities and some individual golf courses are starting to use. Although start-up costs are still high—often $1,000,000—technological advances have reduced energy, maintenance, and operation costs of this process. Reverse osmosis plants now require lower dosages of antiscaling chemicals; incorporate a membrane self-cleaning process (cross flow), greatly extending their useful life; and use membranes that can operate at lower pressures, thereby reducing energy costs (Figure 8.30).

**Ion exchange.** Ion exchange is a chemical process similar to the water softening processes. Salt is composed of positively charged sodium ($Na^+$) ions and negatively charged chloride ($Cl^-$) ions. An ion exchange uses resins to break up the salt by attracting the salt ions. A saline solution flows through a bed of granulated zeolite or an ion-exchange resin. The ions in solution become attached to the material in the bed and displace ions of the same sign. One resin attracts the chloride ions and releases hydroxyl ($OH^-$) ions. A second resin attracts Na ions, releasing hydrogen ions. In addition to the clean water that remains, the hydroxyl ($OH^-$) and hydrogen ($H^+$) ions combine to form a small amount of water. Brine is

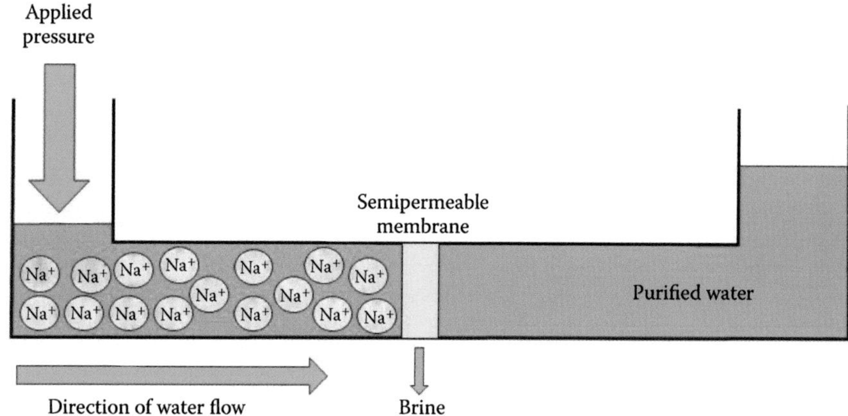


FIGURE 8.29 Illustration how reverse osmosis screens out sodium ions using a semipermeable membrane under pressure. Salt containing brine is generated.

FIGURE 8.30 Commercial reverse osmosis system enabling a golf course to irrigate with poorer quality water.

the by-product. The resins must be periodically regenerated or "recharged" by washing them with acids or bases.

**Electrodialysis.** Electrodialysis is similar to reverse osmosis because both use membranes. Electrodialysis is an ion exchange process where salt is removed by separating the ions using an electrical field instead of pressure as with reverse osmosis. In electrodialysis, source water flows into a unit with two membranes where an electric current causes the ions to separate. When a direct current is passed through the solution, the ions move through the membranes toward the electrodes of opposite charge, thereby depleting the salt in the center compartment. The positive Na ions go through one membrane into one compartment, while the negative chloride ions flow through a different membrane into another compartment. The end product remaining in the unit is fresh water.

### Additional Water Treatment Options

Turfgrass managers are constantly being inundated with certain products claiming to solve their specific problems. These in-line water-conditioning devices (aka, physical water conditioners) are touted to lessen the impact of reduced quality irrigation water on soil and plants. In most cases, water quality will not be improved unless something is added or removed from it. For most of these devices, little credible evidence exists in reputable scientific literature to support the claims and, in most cases, there is no scientific merit, based on current chemical knowledge, to believe they even can work. This certainly does not mean they do not work, but anyone contemplating the purchase of such devices should insist on data from credible sources; have a sound, solid performance guarantee; and, preferably, have a demonstration on his or her own course and unique situation. A website that addresses these and other water-treatment devices is www.chem1.com/CQ/gallery.html.

**Depressurizing water for scale control or catalytic devices.** Supposedly, when water is subjected to rapid reductions in pressure and turbulent flow, dissolved carbon dioxide escapes as a gas, rendering the solution more alkaline, and thereby causing dissolved calcium carbonates ("scale") to precipitate out and not be able to form scale deposits. However, this reaction requires heat, rather than pressure, to proceed. Also, it is not clear what happens to the carbon dioxide gas, or what prevents it from redissolving. The turbulent flow of water may also be over an alloy core composed of certain metals (catalysts) in order to change the configuration of minerals in the water so they no longer accumulate or bind to surfaces or form calcium carbonate (calcite) crystals. Neither process has been proven to perform as advertised.

**Magnetic (or electromagnetic) water treatment (MWT).** A long history exists of promoting magnets or electromagnetic devices from either the outside or inside of a pipe to alleviate the "hardness" of mineral-saturated water, thereby controlling the deposition of scale in various devices. The magnets supposedly cause carbonate salts, which ordinarily form scale deposits, to precipitate as small particles within the water. This is supposed to occur from a change in the morphology of calcium carbonate crystals formed upon evaporation of a solution exposed to a magnetic field. Specifically, the magnetic field deflects the ions of opposite charge to move in opposite directions. The resulting collisions then form a microscopic nucleus that causes calcium carbonate to precipitate out before forming scale.

Although polar, water has no paramagnetic properties because the partial charges produced between the $O_2$ and $H^+$ atoms in water and between larger ions in solution are too strong to undergo any significant deflection as they pass by an ordinary magnet. Most ionic motion in water is due to random thermal motions; thus, only a very small percentage of this motion would be nonrandom. Magnetic treatments also are erroneously claimed to reduce the surface tension of water or to break hydrogen bonds, allowing water molecule clusters to become individual water molecules. These individual water molecules are now supposedly able to dissolve calcium carbonate lime deposits. A variation of an electromagnetic device is one that implants "resonance frequencies" into water, which alters or restructures the water (or its electron orbit) so as to change its energy state. Again, little if any scientific evidence exists to support these devices.

**Electrical impulses or electrostatic precipitators.** Other similar devices claim to negatively charge the ions in water by producing high-frequency electrical impulses through the water. The charge can be directly applied or used as a hydroelectrical dynamic charge from water flowing along a metal pipe. This then supposedly converts poorly soluble bicarbonate crystals into a soluble powdered carbonate form that easily washes away. However, ions in water are always equally balanced in electric charge and cannot be altered by electrochemical processes. Ions cannot be injected into water using simple devices and, even if they could, they would exist for less than a second. Hydrogen bonds in water also cannot be permanently broken to make "wetter water" and to enable fertilizer to break up into smaller groupings and interface with smaller water groupings. Hydrogen bonds are constantly being broken apart and re-formed naturally, and cannot be enhanced by such simple devices. Similar erroneous claims also are often made with a variety of far-infrared or other electrically charged light-energy signals.

**Ozone injection and oxygen treatment.** Ozone ($O_3$) is found commonly in nature. It is formed following lightning strikes, or when an electrical discharge creates a spark. The ozone layer in the upper atmosphere provides a protective screen against dangerous solar radiation. Ozone is artificially created when air, dry air, or oxygen is drawn into ozone generators and is charged with high voltage. As the air is drawn through the ozone generator, the high voltage splits some oxygen molecules into oxygen atoms. Some of these atoms then quickly react with oxygen molecules to form ozone: $(O_1) + (O_2) = (O_3)$. Ozone is second only to fluorine as the most powerful oxidant. In wastewater treatment, ozone is used to inactivate and oxidize organic metals and most organisms faster than chlorine. Ozone also functions as a microflocculating agent to "polish" the water and improve clarity (clarifying Fe, S, and Mn). Soluble ferrous Fe, which is not filterable, is oxidized (the process of adding oxygen) to insoluble ferric iron, thus making it easily filtered. Bacterial iron can be killed through oxidation and by denying its food source (ferris iron). A similar process occurs for Mn. Odorous hydrogen sulfide, which is not filterable, is quickly converted to elemental S, which can be easily filtered. Again, the bacterial form is killed and filtered.

In treated turf water, ozone supposedly creates hydrogen peroxide and nitric acid, which are supposed to increase oxygen levels in water. Ozone is viewed more as a replacement or substitute for chlorine when purifying water for human contact. Its effect on water quality is mainly for Fe, S, and Mn, rather than for Na, other salts, bicarbonates, pH, or other problems with water quality.

**Vibrators and oscillators.** Several ring-shaped devices that are available are placed around irrigation pipes on the principle of oscillations (vibrations), which are then retained in the water. This supposedly softens the water, controls scale and corrosion, alters the chemistry of dissolved ions, etc.

## POND MANAGEMENT

The following discusses the various uses of water on a golf course (e.g., irrigation, hazards, and aesthetics). Attention will be given to sources of water used in irrigation and subsequent management of possible runoff, as well as the management of both existing and man-made bodies of water found on the property (Figure 8.31).

**Lakes/ponds/streams.** Flood control of existing water bodies and those constructed on the course is a very important issue to be addressed. All drainage structures should be maintained so as to ensure the proper control of excess water. In part, these aspects of management are built into the golf

**FIGURE 8.31** Ponds, lakes, and streams are desirable course features if constructed and maintained properly.

course at the design and construction phases. However, as a golf course ages, erosion should be curbed and drains often need some attention. A small sedimentation or settling pond upstream may be necessary to reduce erosion, especially if significant development or logging occurs upstream of the main golf course lakes or ponds. In most situations, lakes and ponds on the course will be used to collect runoff and will be able to hold nearly all input. It is the runoff into creeks and streams that is of most concern. The planting of aquatic macrophytes in these areas may help slow flow and filter contaminants. Lakes and ponds should maintain a minimum depth of 9 feet (3 m) at the deepest point to allow some sunlight penetration along the shoreline, but prevent total bottom rooted plants plus provide cooler waters to mix with warmer surface waters (Figure 8.32).

**Pond regions.** Ponds can be divided into various regions or areas, including the littoral region and open-water regions. The shore area or interface zone between the land of the drainage basin and open water is the *littoral region* (Figure 8.33). This is generally where most sunlight will penetrate to the bottom of the pond, often forming a ring of plants around the shore area according to water depth and wave action. The littoral region extends from the shoreline to the point where sunlight fails to penetrate to the pond's bottom; thus, aquatic plants no longer grow. Typical plants growing in this ring-like region are cattails or rushes, water lily or lotus, and submersed plants such as pondweeds. The littoral region is also the area in which over 90% of all algal species are found attached to macrophytes. These contribute significantly to the pond's productivity and provide habitats for microflora, as well as larger invertebrates and vertebrates. Typically, the littoral region is the most challenging region of a pond to manage. Shallow ponds (less than 6 feet, 2 m) are an even greater challenge to manage as the entire depth has sunlight penetration, producing warm temperatures and abundant plant and algae growth.

Beyond the littoral region is the *open water region* (also called the limnetic zone), an area where an upper layer of

**FIGURE 8.32** Dredging an unsightly, silted-in pond.

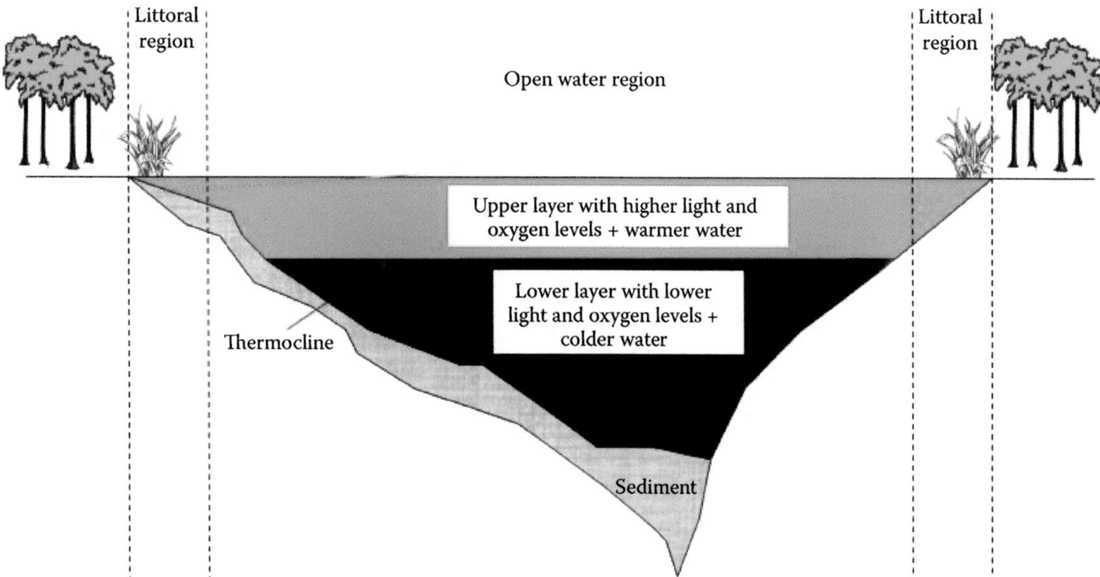


**FIGURE 8.33** Illustration of typical layers of water in a pond.

warm, well-lit, and relatively high $O_2$ levels and a lower layer of cooler, poorly lit, and lower $O_2$ water levels occur. The water in ponds gets colder as depth increases. Plants and animals in the open water region include water lilies, submerged pondweed, planktonic algae, zooplankton, invertebrates, and fish. Typically, this region is the easiest to manage.

**Pond food chain.** The food chain in a pond directly influences the amount of $O_2$ produced in the pond. As mentioned, phytoplankton (or "green") algae form the base of the food chain in a pond system. These are microscopic plants living suspending in water. Through photosynthesis, phytoplanktonic algae provide needed $O_2$ for fish and other aquatic organisms. Oxygen also is supplied by wave and wind action and by rainfall. Next higher in the food chain are zooplankton—tiny insect-like animals floating in the water and feeding on phytoplankton. Bream then feed on zooplankton and other aquatic insects. The top of the food chain in most ponds are bass. Bass feed on bream and help control their populations. In fact, ponds lacking a satisfactory bass population will be overcrowded with small bream. However, overstocking or using the wrong kind of fish could cause long-term fish kill. Consult a fisheries scientist before stocking any fish to determine how much and which kind(s) of fish are needed for a particular site.

Large-celled blue-green algae are not desirable for most ponds (Figure 8.34). They often predominate in warm, stagnant, nutrient-rich water and are a primary cause of pond scum. If these bloom, light often becomes limited and they will all die off at once. When this happens, the bacterial decomposition of the dead algae cells can use up all of the dissolved $O_2$, resulting in a fish kill. Adding to nutrients in a green algae bloom are calm winds, warm temperatures, and sunlight. They also can float to the surface during the day for sunlight, and then sink into nutrient-rich bottom waters to access N and P.

**FIGURE 8.34** Mostly undesirable algae due to stagnant, insufficient water movement.

## Managing Lakes/Ponds/Streams

Managing a pond involves controlling or manipulating three significant parameters:

1. Light and temperature
2. Nutrients
3. Oxygen ($O_2$)

Sunlight is the primary source of energy in pond dynamics. It fuels photosynthesis, which in turn regulates $O_2$ levels, as well as warms the water. As temperatures rise, the pond's surface warms, becoming lighter (less dense). *Thermal stratification* begins, allowing the water to "layer," with warm water being separated from the deeper, cooler water. Upper levels of ponds may become much warmer (termed epilimnion layer) than the cold-water bottom (termed hypolimnion). In small ponds (e.g., <15 acres, 6 ha), the warm layer extends down to about 6 feet (2 m) where a thermocline, the dividing line between the two layers, occurs. As water temperatures increase, the ability of water to hold $O_2$ decreases.

For example, water at 80°F (27°C) holds less than 40% of $O_2$ than water at 52°F (11°C) holds. Most of the $O_2$-producing phytoplankton live in the warmer upper layer; thus, most of the $O_2$ for fish to breathe also resides here. During stratification, lower waters are not exposed to photosynthesis and wave action—the two main means of dissolved $O_2$ in a pond. Little $O_2$ occurs below this warm layer. As a result, it does not support aquatic life very well, sulfides are converted to hydrogen-sulfide, insoluble Fe is converted to soluble forms, suspended solids increase as decomposition of waste materials on the pond bottom decreases, and this then becomes a less than ideal source of irrigation water for turf (Figure 8.35).

The benthic zone is the bottom of a pond that is normally nutrient enriched and $O_2$ starved. Undecomposed debris accumulates on the bottom as sludge and gradually decreases the depth of the pond. This sediment or sludge buildup can accumulate at a rate of 1 to 5 inches (2.5 to 12 cm) annually in temperate climates and up to 8 inches (16 cm) annually in tropical regions. For example, if a 3-inch (7 cm) accumulation occurs per year, a one surface acre (4,000 m$^2$) pond loses about 80,000 gallons or 300 m$^3$ of water storage capacity.

Thermal stratification must be interrupted or eutrophication will occur, where excess dissolved nutrients remain with insufficient dissolved $O_2$. "Turnover" of a pond often occurs when the surface is cooled from rain or cold temperatures. The temperature of the top layer then equals that of the bottom layer and the two water layers will mix. As the two layers mix, $O_2$ diffuses from the top of the pond to the bottom, leaving insufficient $O_2$ on the top to support fish life. Turnover of ponds may also occur as aquatic plants receive insufficient sunlight for normal photosynthesis during cloudy weather and dissolved $O_2$ becomes deficient in the system. Dissolved $O_2$ ranges from 0 ppm (no $O_2$) to ~9 ppm (the saturation point in water during typical summer temperatures). The normal $O_2$ content in a healthy pond ranges from ~5 to 9 ppm, day or night, based on normal photosynthesis and respiration of zooplankton. Oxygen levels are generally safe when all regions of a pond have at least 4 ppm (or mg/L) (Table 8.25). If a harmful algae bloom occurs, then the dissolved $O_2$ may range as high as 12 to 25 ppm in the day but drop to 0 to 4 ppm at night due to algal nighttime respiration. With depleted $O_2$, a cold rain may mix the warm and cold layers, breaking down the thermocline and depleting the pond of what little dissolved $O_2$ remains in the upper levels. Less desirable odoriferous anaerobic bacteria then take over. Without $O_2$, microorganisms, invertebrates, and fish die. Oxygen levels also tend to decline over time as ponds lose some of their holding capacity when sludge or sediment buildup occurs.

**TABLE 8.25**
**Desired Levels of Quality for Open Bodies of Water**

| Characteristic | Desired Level |
|---|---|
| Dissolved $O_2$ (prior to sunrise) | >4 mg/L (or ppm) |
| Biological $O_2$ demand (or BOD) | <5 mg/L |
| pH | 6 to 9 |
| Alkalinity | >50 mg/L |
| Chlorophyll | <2 mg/L |
| Suspended solids | <5 mg/L |
| Fecal coliform | <200 per 100 mL (no human contact if >400/100 mL) |
| Total nitrogen | <5 mg/L |
| Total phosphorus | <0.05 mg/L (<0.1 mg/L to prevent algae blooms) |

Oxygen depletion is from many situations. Most commonly, this occurs (1) during predawn hours, (2) on cloudy and windless days, (3) on hot and humid days, (4) when water nutrient content is high, and (5) upon extensive death and

**FIGURE 8.35** Phytoplankton on a pond's surface, which restricts oxygen levels and sunlight penetration.

subsequent decomposition of aquatic plants. Algae occurrence often indicates poor-quality water. Planktonic algae are single-cell plants at or near the surface and typically light green in color, creating the common "pea soup" appearance. The bloom of these algae typically occurs on cloudy, hot days or during predawn hours, and they consume all the available $O_2$ in the water. Fish kills and massive algae die-off follow. Cold weather, lack of sunlight, or a chemical application may cause this algae die-off. The water then quickly changes from the light green to a brown color.

> **WARNING SIGNS OF OXYGEN ($O_2$) DEPLETION IN PONDS**
>
> 1. Fish swim at the surface, gulping for air—especially in late night or early morning.
> 2. Water rapidly turns from light green to dark brown to black in color.
> 3. A rotten or putrid odor is detected.
> 4. Sudden algal death occurs.
>
> These conditions are most likely after extended hot, cloudy weather is interrupted by summer rain.

Ponds often naturally turn over in fall when temperatures begin to cool. The major difference between a fall turnover and a summer turnover is the cooler water temperatures. Cold water will hold more $O_2$ than warm water, so when a pond turns over in the fall, more $O_2$ is available in the epilimnion layer, and fish do not normally die.

Cultural eutrophication such as sewage disposal, land drainage, and fertilization may result in excess nutrient enrichment in ponds. A direct correlation exists with the level of available nutrients and the populations of algae, especially large-celled blue-green algae, and aquatic weeds (Figure 8.36). High N and P cause algal blooms, aquatic weeds follow, and water turbidity, pH, alkalinity, and dissolved $O_2$ are adversely affected in eutrophic ponds. Eventually, the bloom will become light limited, and the blue-green algae will all die off at once. The bacterial decomposition of the dead algae can then use up all of the dissolved $O_2$, resulting in fish kill. Vegetative life and sediment at the lake bottom are the primary sources of nutrients.

The second most common source of nutrients is runoff from surrounding turf, roads, farms, and other outlying areas and through inlet waters. Inlet waters can originate from effluent sewage, wastewater treatment facilities, and leaching from septic systems. These waters often have minimal $O_2$ and are high in P. A 30-foot (10 m) band (buffer) or "no fertilizer- or pesticide-treated zone" is suggested to be implemented to help eliminate nutrient runoff into the water. Using slow-release fertilizers also helps minimize runoff as well as a slightly elevated swale or berm around the pond. Balancing a pond's ecosystem must provide fewer destructive primary producers such as periphytic algae and manageable aquatic vegetation. Biodiversity of plants and animals is best for golf course ponds, resulting in as many levels of the food chain as possible. Strategically placed wetlands also capture unwanted nutrients prior to their entering ponds.

**Pond monitoring.** Monitoring ponds should consist of measuring turbidity, total ammoniacal N, P, temperature, pH, and dissolved $O_2$ (Table 8.25). Pond nutrients are most prolific during the summer, leading to foul odors. Aerobic bacteria digest algae and other nutrients seven times faster than anaerobic bacteria. When aerobic bacteria run out of $O_2$, the less desirable anaerobic bacteria take over.

**Aeration.** Adequate circulation and aeration are needed for odor and algae control; therefore, electrical service should be provided to all open water bodies in the course design. Aeration helps attack the cause(s) of poor water quality while other methods such as dyes and algaecides treat the symptoms—algae. Artificial aeration helps provide $O_2$ when natural wind and wave action cannot provide sufficient

**FIGURE 8.36** Algae- and aquatic weed-infested pond.

**FIGURE 8.37** Numerous pond aerators introduce oxygen into water and provide an attractive display.

levels. Aerators introduce $O_2$ into a larger volume of water, while fountains create an attractive display or spray pattern (Figure 8.37). Aerators use propellers to move a large volume of water and introduce a larger volume of $O_2$. In fountains, a centrifugal pump usually creates an attractive display, although volume is sacrificed to obtain the necessary pressure to achieve these heights and shapes.

Aerators need to be matched for the size of the pond or lake. A 1-acre (0.4 ha) pond 3 feet (1 m) deep contains about 1 million gallons (3.8 million L) of water. For aeration, 800 to 1,400 (3,000 to 5,300 L) gallons per minute (gpm) should be propelled into the atmosphere for a water surface area 0.5 to 1.5 acres (0.2 to 0.6 ha) in size. For smaller, 0.1- to 0.5-acre (0.04 to 0.2 ha) areas, 350 to 500 gpm are acceptable. If the lake is bigger, multiple smaller units generally distribute $O_2$ more evenly than one unit of larger horsepower. If the aerator cannot move sufficient amounts of water, then disappointing results will be observed.

**Types of aerators.** Many types of aerators are available, including:

1. **Floating fountains (also called display fountains).** These electrically charged floating devices with a motor or pump/motor send jets or fountains of water into the atmosphere, which dissolve $O_2$ and, through wave action following surface impact, support aerobic bacteria (Figure 8.38). Although spectacular, fountains normally do not move the volume of water that true aerators do.
2. **Aerating fountains.** This combination of an aerator and a fountain generally creates a spray pattern shaped like the flair of a trumpet or tuba (Figure 8.39).
3. **Floating surface aerators.** These devices, similarly to the floating fountains, do not create a fountain-type spray pattern when this is not desired. They agitate the surface in the form of a bubbling or boiling effect and are best in water less than 15 feet (5 m) deep.
4. **Aspirating aerator.** This device does not create a fountain pattern, but rather blows oxygenated bubbles directionally into the water body, deep below the surface (Figure 8.40). The units create artificial currents in long, narrow ponds or canals. The pond's bottom needs to be equipment-friendly, so as not to upset and mix the silt and any sludge throughout the pond and "islands" minimized to prevent normal channel flow. Aspirating aerators can be either floating or fixed-base mounted and are good for water bodies 3 to 12 feet (1 to 4 m) deep.
5. **Pond bottom aerator.** This completely submerged unit uses a propeller water transport system where bottom water is circulated to the surface. This cooler, denser water absorbs $O_2$ as it spreads across the pond and returns through the stratified layers. Also, as green algae from the bottoms become suspended near the surface, they can photosynthesize, adding $O_2$ back into the water. These devices help green algae survive, keeping fish, dissolved $O_2$, and pH in balance. Near-laminar-flow devices is another name for these machines.
6. **Diffused air or compressed air systems.** These systems involve a shore-mounted compressor that pumps $O_2$ through tubing into the pond. It requires quality weighted tubing and self-cleaning diffusers but offers superior aeration, circulation, and destratification. These are the most unobtrusive aerators and are most effective in water depths of 15 feet (5 m) or deeper.

Surface spray aerators can pump water from lower parts of the pond to break through the thermocline and into the air, where it can pick up $O_2$. Short, wide, full-volume water pumped at lower speeds maximizes recycling of water and

**FIGURE 8.38** Floating (also called display) pond fountain.

**FIGURE 8.39** A trumpet or tuba-shaped pond fountain.

**FIGURE 8.40** Aspirating aerators, which introduce oxygenated bubbles deep below the surface of the water.

injection of $O_2$ into water. Running fountain surface sprayers and subsurface aerators is maximized at night for $O_2$ benefits. Oxygen levels are typically lowest just before dawn. Increasing the $O_2$ content in lakes can help reduce nutrient levels, limit thermal stratification, and decrease algae growth.

**Beneficial algae.** The general category of predominantly beneficial green algae found growing on or attached to submerged structures in fresh water is *periphytic algae*. These can outcompete problematic blue-green and filamentous algae but require a submerged surface area to live. The usage of

**FIGURE 8.41** Water enhancement/protection by maintaining an adjacent nontreated vegetative buffer.

barley straw to control algae and clarify ponds was developed in England in the 1990s; however, mixed results have occurred since its effect on planktonic algae is less than on filamentous algae. Algae growth is believed to be inhibited through a chemical exuded from the decomposing barley straw; by a metabolic product produced by fungi that decomposes the barley straw, preventing algae growth; or by the barley tying up excessive nitrates $\left(NO_3^-\right)$. As a starting use rate, barley straw bales are applied at 225 pounds/acre (252 kg/ha) in shallow (4 to 5 feet, 1.2 to 1.5 m) ponds or a 2- × 4-foot (0.6 × 1.2 m) bale/5,000 $m^2$ of water, which is repeated twice yearly. Other structures used to encourage beneficial algae are rocks, rooted plants, Christmas trees, and other organic materials. Commercial mats are also available that provide enormous surface area for these beneficial algae.

Use of a filter system also is suggested to reduce algae that may enter the sprinkler system and clog nozzles. Even distribution by the irrigation system also helps prevent buildup of any harmful substances.

**Pond enhancements.** Prevention of algal blooms often requires controlling nutrients in the pond. If not prevented, restoring ponds after an algal bloom can be costly and time consuming. Ways to enhance pond health, attract wildlife, and prevent algal blooms include:

1. Have a minimum pond basin depth of 9 feet (3 m). This depth allows sunlight and discourages bottom-rooted plants and also provides cooler subsurface water to mix with surface waters.
2. Create islands and shallow water areas by adding rocks and logs. These help protect and provide nesting sites for various small creatures and create basking sites for turtles.
3. Provide shelter for fish by sinking trees.
4. Provide food and safe access to wildlife by planting a vegetative buffer around the edge of waterways and ponds (Figure 8.41). These plants also help filter runoff to improve water quality.
5. Plant emergent vegetation to provide food and cover for various birds, mammals, amphibians, and insects.
6. Plant submergent vegetation to provide cover for various aquatic life and to help maintain healthy $O_2$ levels.
7. Have a no-treatment buffer zone around the pond to act as a filter for surface-applied material.

**Herbicides.** Herbicides often are the fastest way to control weeds, but may not provide long-term control. Decaying weeds from herbicide control often contribute to $O_2$ depletion, as the bacteria that decompose weeds are $O_2$ consumers, leaving insufficient levels for fish. If aquatic herbicides are used, treat in early spring when water temperatures are cooler, thereby holding more $O_2$. If ponds need to be treated in summer, treat one quarter of the pond at a time. If the whole pond is treated in summer, dissolved $O_2$ levels are reduced by the decomposing bacteria, causing fish kills.

Additional water treatment means to reduce aquatic plants and algae are (1) ozone injection; (2) adding beneficial bacteria to lower nutrient levels, especially in effluent water; (3) stocking grass carp such as the White Amur or Triploid Grass Carp; (4) using dyes to alter the wavelength reaching aquatic plants, thus reducing their photosynthetic capability; and (5) dredging sediment.

# *Section V*

## *Turfgrass Management Practices*

# 9 Turf Cultural Practices

Golf courses require daily cultural practices to maintain a quality playing surface and an aesthetically pleasing turf. Frequent maintenance practices such as mowing, watering, fertilizing, and aerifying are necessary for this, and a well-defined and conceived short- and long-term maintenance plan often separates courses from local competitors. Courses typically begin to slowly decline when necessary practices such as aerifying or topdressing are curtailed or stopped.

Excessive organic matter and lack of aeration (sufficient porosity) are two contributing factors why aging golf greens fail. If left unchecked, organic matter increase is from <1% initially to 4% or 5% by weight (15% to 20% by volume) in the top 2 inches (5 cm) after several years. Reductions in saturated hydraulic conductivity along with reduced soil oxygen ($O_2$) levels are two negative results from this accumulation. Summer root decline often follows, adding to the decaying organic matter in the soil. For example, a USGA green one year after establishment has 20% aeration porosity and 25% capillary porosity. After six years, aeration porosity drops to 15% while capillary porosity increases to 30%. After 19 years, aeration porosity is between 5% and 10% and capillary porosity at 40% (Habeck and Christians, 2000). This indicates that very little $O_2$ is left in soil while excessive water-holding capacity occurs, encouraging puffiness, scalping, wet wilt, algae, moss, various diseases, carbon dioxide ($CO_2$) buildup, and eventually overall turf decline. Cultural practices such as vertical mowing, grooming, aerifying, and topdressing are necessary to keep the organic matter buildup in check (Table 9.1). This chapter outlines contemporary cultural practices and considerations for timings and frequencies superintendents can use when developing their master maintenance program.

## MOWING

Mowing is the most basic, yet most important cultural practice contributing to a desirable turf (Figure 9.1). Other cultural practices and many aspects of turf quality are affected by mowing, including density, texture, color, root development, and wear tolerance. Failure to mow properly usually weakens turf, resulting in poor density and quality.

Turfgrasses can be mowed relatively close to the ground due to their terminal growing point (crown) being located at or just below the soil surface (Figure 9.2). Regrowth from cell division and elongation takes place from growing points located below the height of the mower blade. In contrast, upright growing dicot plants have their meristematic tissue (growth points) at the top or tip of their stems. Consequently, mowing removes this growing point and many upright dicot weeds are eliminated since they do not have enough recuperative potential to regrow.

Turfgrass growth habit is also affected by mowing. *Tillering* is the development of primary lateral shoots from vegetative axillary buds on the crown. Mowing increases shoot density by promoting the growth of new tillers. As a result, mowing decreases root and rhizome growth because food reserves, following mowing, are utilized for new shoot tissue development at the expense of root and rhizome growth. Improper mowing frequency and radical height reduction magnify this problem. If the correct mowing frequency is followed, then the turf does not experience a stress period from the immediate loss of top growth and can recover more quickly. Infrequent mowing results in alternating cycles of elevated crowns followed by scalping, resulting in a further depletion of food reserves.

### Mowing Height

Mowing heights for golf course turf are mainly governed by the grass variety and intended use (Figure 9.3). For example, golf greens are generally mowed below 0.25 inch (6.4 mm) to provide the smooth, fast, and consistent playing surface golfers' desire. The turf species being mowed also influences height. For example, bentgrass withstands a shorter height than bluegrass. Mowing height also is influenced by cultivar (e.g., TifEagle bermudagrass tolerates shorter mowing than Tifgreen bermudagrass), while the newer ultradwarf creeping bentgrasses tolerate closer mowing than Penncross bentgrass. Other factors influencing mowing height include mowing frequency, shade- versus full-sun grown turf, mowing equipment, time of year, root growth, and moisture or temperature stress. Typical mowing heights for each grass species and use are listed in Table 9.2.

Mowing height (or height of cut, abbreviated HOC) refers to height of top growth immediately after the grass is cut. Determining HOC can be misleading to inexperienced mower operators. Often height is adjusted and checked on a level surface such as a work bench or roadway and is referred to as **bench setting** or **bench height**. However, mower wheels actually ride on top of grass shoots, causing the cutting units to be raised higher than the bench setting. Conversely, when a mower is operated on soft ground or when a thick, spongy thatch layer is present, the mower cuts lower than the bench setting, often resulting in undesirable scalping (Figure 9.4).

**Variables influencing mowing height.** Many factors influence the mowing height of grasses. Shoot tissue is the site of photosynthesis; thus, any leaf removal strongly influences the physiological and developmental condition of the turfgrass plant. If grass is mowed too low or too infrequently, excessive shoot tissue is removed and crown damage can occur. This reduces green plant tissue available for photosynthesis and results in scalped, off-color turf.

**TABLE 9.1**
**Target Ranges for Various Soil Physical and Organic Properties of Established Golf Greens Depending on Their Description**

| | Green Description | | | |
|---|---|---|---|---|
| | | | Well-Drained | |
| **Characteristic** | **Native Soil** | **Modified Soil** | **0 to 4 inches** | **4 to 12 inches** |
| Infiltration rate, inches/hour (cm/hour) | ≥2 (5) | ≥4 (10) | ≥6 (15) | ≥10 (25) |
| Noncapillary porosity (%) | ≥12 | ≥14 | ~20 | ≥20 |
| Capillary porosity (%) | 15 to 30 | <30 | 15 to 25 | <20 |
| Bulk density (g/cm$^3$) | 1.35 to 1.45 | 1.35 to 1.45 | 1.35 to 1.45 | 1.40 to 1.50 |
| Water retention (%) | 10 to 25 | <25 | 10 to 20 | 10 to 20 |
| | | **Organic content (%) at Various Depths** | | |
| • 0.25 to 1 inch (0.6 to 2.5 cm) | 1.5 to 2.5 | 1.5 to 3.0 | 1.5 to 2.5 | 0.1 to 1.0 |
| • 1 to 2 inches (2.5 to 5 cm) | 1.0 to 2.0 | 1.0 to 2.0 | 1.0 to 2.0 | 0.1 to 1.0 |
| • 2 to 3 inches (5 to 7.6 cm) | 0.5 to 2.0 | 0.5 to 2.0 | 0.5 to 2.0 | 0.1 to 1.0 |
| • 3 to 4 inches (7.6 to 10 cm) | 0.5 to 2.0 | 0.5 to 2.0 | 0.5 to 1.5 | 0.1 to 1.0 |

*Source:* http://www.istrc.com.

**FIGURE 9.1** Mowing is the most basic, yet important mechanical turfgrass practice.

**Root-to-shoot ratio.** Plants mowed too low also require a substantial amount of recovery time so roots can provide the food needed for shoot tissue production for future photosynthesis. Turfgrasses have a ratio of shoot-to-root tissue that is optimum to sustain turf growth. If mowed too low at one time, an imbalanced ratio occurs with more roots available than the plant physiologically requires (Figure 9.3). This excessive root mass is then sloughed off. Until the plant has time to regenerate new shoot tissue, it will become weaker and more susceptible to environmental and biotic stresses. Root growth is least affected when no more than 30% to 40% of the leaf area is removed at any one time.

**Root growth.** There is also a direct relationship between mowing height and rooting depth. As the mowing height is reduced, a corresponding reduction in rooting depth occurs (Figure 9.3). This again ties into the physiological aspect that less rooting is needed to support less top growth when the mowing height is lowered. Less rooting following a lower mowing height is the reason why golf greens need to be watered frequently and why fertilizer is frequently applied since the shallower roots have a decreased depth of soil from which to obtain these.

**Shade.** Other influences on mowing height and frequency include shade, type of mower being used, season of the year, and environmental stresses imposed upon the grass. Under shady conditions, grass leaves grow more upright in order to capture as much of the filtered sunlight for photosynthesis as possible. Mowing height, therefore, for grasses grown under these conditions needs to be raised at least 30%, and preferably 50%. If mowed continuously short, grasses grown under shaded conditions will gradually thin due to their inability to capture sufficient sunlight for photosynthesis (Figure 9.5).

**Mower type.** Mowing height is also influenced by the mower type being used. Rotary and flail mowers cut best

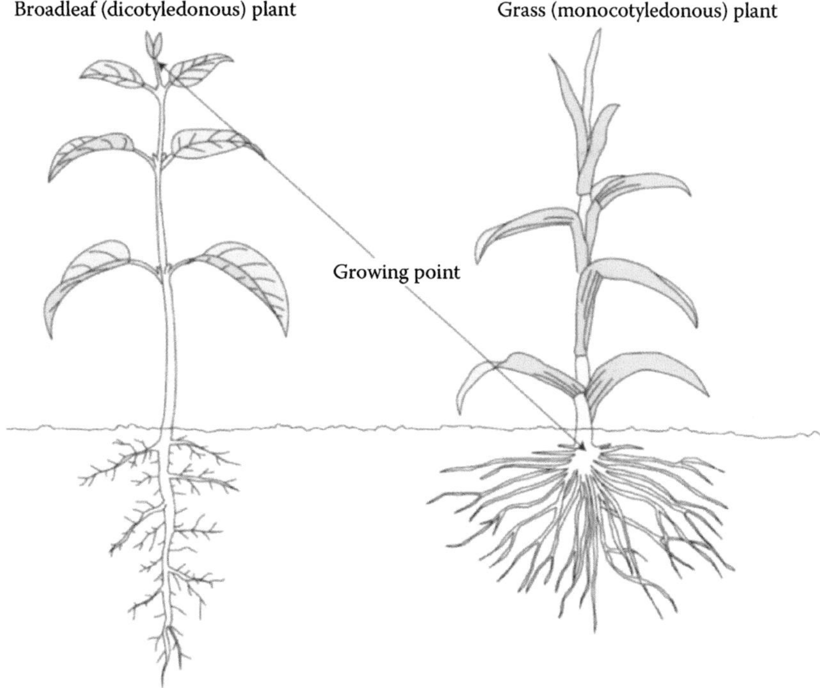


**FIGURE 9.2** Due to ground level growing points, grasses can be mowed lower than most broadleaf plants.

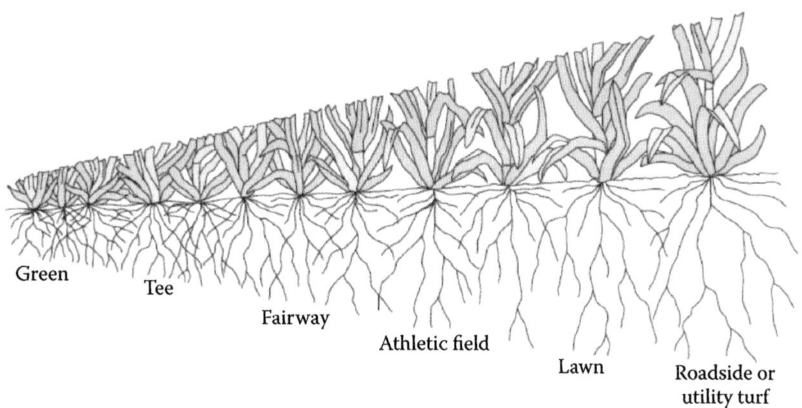


**FIGURE 9.3** A direct correlation generally exists between mowing heights and rooting depths.

at heights above 1 1/2 inches (3.8 cm). Conversely, reel mowers cut best at heights below one and a half inches (3.8 cm). Reel mowers are used on low-cut golf course play areas such as greens, tees, and fairways while rotary and flail mowers are mainly used in roughs and out-of-play areas.

**Season.** Mowing height may also be influenced by the season of year or by environmental stresses imposed upon the turf. In spring through midsummer, days become longer; therefore, warm-season turfgrasses tend to have a more prostrate (decumbent or spreading) growth habit. During this time, they can be mowed closer than during other portions of the year without serious consequences, as surface leaves capture sufficient sunlight to support the plants' photosynthetic needs. Close mowing in early spring also helps control thatch, increase turf density, remove excess residues or dead leaf tissue, and promote earlier green-up. Starting in midsummer, however, days become shorter and warm-season grasses start growing more upright to allow lower leaves to intercept more sunlight for photosynthesis.

Cool-season grasses tend to have a more upright growth habit during long summer days and are healthier if the mowing height is raised to maximize leaf area available for photosynthesis. Higher mowing height at this time also increases turf rooting; therefore, it reduces watering needs and stresses imposed by increased heat, moisture, or nematode activity. In fall, mowing height should be raised for warm-season grasses to reduce the chance of low-temperature damage during winter and to provide a cushion for grass crowns in winter when warm-season grasses are dormant. Cool-season grasses, meanwhile, can be mowed closer in fall when temperatures are cool. Their mowing heights, however, should be raised

**TABLE 9.2**
**Typical Mowing Heights and Frequencies for Turfgrass Species and Their Use**

| Species | Mowing Height (inches)[a] | | Mowings Per Week | |
|---|---|---|---|---|
| **Fairways** | | | | |
| Creeping or colonial bentgrass | 3/8 to 5/8 | | 3 to 5 | |
| Common bermudagrass | 3/4 to 7/8 | | 2 to 4 | |
| Hybrid bermudagrass | 3/8 to 3/8 | | 3 to 5 | |
| Kentucky bluegrass | 1 to 1 1/2 | | 1 to 2 | |
| Kikuyugrass | 1/2 to 3/4 | | 3 to 5 | |
| Perennial ryegrass, fine fescue | 1/2 to 7/8 | | 2 to 4 | |
| Seashore paspalum | 3/8 to 5/8 | | 2 to 5 | |
| Tall fescue | 1 1/4 to 2 1/2 | | 1 to 2 | |
| Zoysiagrass | 1/2 to 3/4 | | 2 to 3 | |
| **Roughs** | **Intermediate** | **Primary** | **Intermediate** | **Primary** |
| Common bermudagrass | 3/4 to 1 1/4 | 1 1/2 to 3 | 2 to 3 | 1 to 2 |
| Carpetgrass, centipedegrass | 1.0 to 2 | 2 to 3 | 2 to 3 | 1 to 2 |
| Hybrid bermudagrass | 3/4 to 1 1/4 | 1 1/2 to 2 | 3 to 4 | 1 to 3 |
| Kentucky bluegrass | 1 1/4 to 2 | 2 to 4 | 2 to 3 | 1 to 2 |
| Kikuyugrass | 1 to 1 1/2 | 1 1/2 to 2 | 2 to 3 | 1 to 3 |
| Ryegrass | 1 to 1 3/4 | 1 1/2 to 3 | 2 to 4 | 1 to 2 |
| Seashore paspalum | 3/4 to 1 1/4 | 1 1/2 to 2 | 2 to 3 | 1 to 2 |
| Tall fescue | 1 1/4 to 2 | 2 1/2 to 4 | 2 to 3 | 1 to 2 |
| St. Augustinegrass, bahiagrass[b] | – | 2 to 3 | – | 1 |
| Wheatgrass | – | 2 1/2 to 4 | – | 1 |
| Zoysiagrass | 3/4 to 1 1/2 | 1 1/2 to 2 | 2 to 3 | 1 to 2 |

[a] Multiply by 25.4 to obtain millimeters or by 2.54 to obtain centimeters.

[b] Long-term mowing heights below 3 inches (7.6 cm) may weaken certain cultivars of St. Augustinegrass, wheatgrass, Kentucky bluegrass, tall fescue, and bahiagrass.

**FIGURE 9.4** Scalping from mowers sinking in spongy thatch layers, causing lower cuts than the bench setting.

**FIGURE 9.5** Mowing shade-grown grass high allows maximum sunlight absorption for photosynthesis.

just prior to winter to improve winter color and hasten turf green-up the following spring.

## Mowing Frequency

Besides using properly maintained equipment, the most important aspect of mowing management is the frequency of cut—the more often, the better. Also, the shorter a grass is maintained, the more frequently it must be mowed. Mowing frequency often is a compromise between what is best for the turf and what is desired or practical. For example, daily mowing would be best for most turf, whether it is on a golf green, tee, fairway, or even a sports field. However, this is impractical except for smaller, highly maintained areas such as greens.

The growth rate of the grass should determine the frequency of cut. Growth rate is influenced primarily by the amount and source of nitrogen (N) fertilizer applied and by season or temperature. With warm-season grasses, higher levels of either result in faster top growth, thereby requiring an increased mowing frequency. By raising the mowing height, the frequency of cut is reduced, which helps compensate for faster growing turf. When considering cool-season grasses during summer in warm climates, higher temperatures do not result in faster growth rates. Depending on the species, peak growth occurs around 70°F (21°C) for cool-season grasses. Extreme temperatures above this will slow growth, thereby reducing the need for mowing.

**One-third rule.** The traditional rule is to mow often enough so as not to remove more than one-third of top growth at any one time (Figure 9.6). Removing more top growth decreases the recuperative ability of plants due to extensive loss of the leaf area needed for photosynthesis. This reduction in

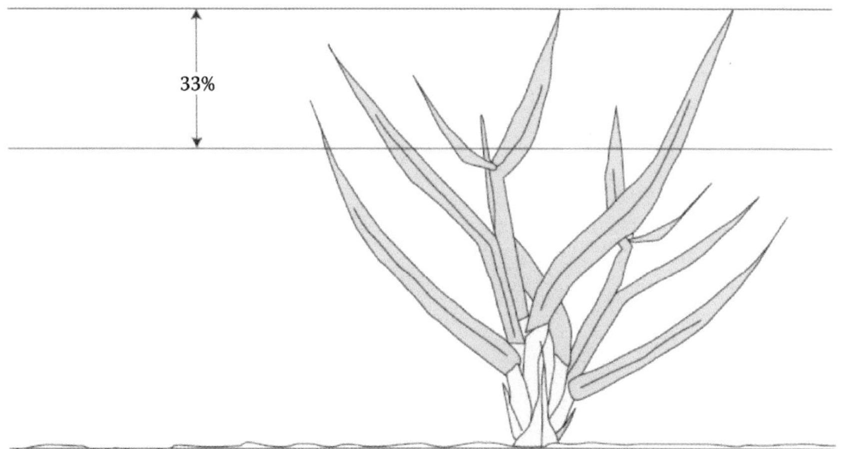


**FIGURE 9.6** Ideally, mowing should be no closer than 33% below the intended height.

**TABLE 9.3**
**Typical Mowing Frequencies during Periods of Active Growth Needed for a Given Height to Remove One-Third of the Leaf Surface**

| Desired Height (inches)[a] | Height Reached above the Desired Cutting Height before Mowing | Approximate Frequency (days) |
|---|---|---|
| ≤1/4 | 3/8 | Daily |
| 1/2 | 3/4 | 2 to 3 |
| 1 | 1 1/2 | 4 to 5 |
| 2 | 3 | 7 to 10 |
| 3 | 4 1/2 | 10 to 14 |

[a] Multiply inches by 25.4 to obtain millimeters.

photosynthesis (food production) can result in weakening or death of a large portion of the root system since carbohydrates in roots are then used to restore new shoot tissue. Consequently, root growth may stop for a period following severe defoliation since the regeneration of new leaves (shoots) takes priority over sustaining roots for food reserves. However, when only one-third or less of shoot growth is removed during one mowing, enough carbohydrates are available to simultaneously sustain shoot and root growth without significantly reducing either. Table 9.3 lists typical mowing heights and the resulting mowing frequency needed during active growing periods to maintain turf within the one-third rule of thumb. A golf course rough maintained at 3 inches (7.6 cm) needs to be mowed infrequently (once every 10 to 14 days) compared to a golf green mowed at 1/8 inch (3.2 mm), which requires daily mowing to prevent more than one-third of the height (0.04 inch, 1.0 mm) from being removed at any one period. Again, frequency is dependent on the grass growth rate.

**Scalping.** Scalping occurs when the turf has grown above the one-third maintenance height and is then mowed. Excessive growth most often is from high N fertilization rates, environmental conditions that favor rapid growth, and/or weather conditions that prevent normal mowing frequency (Figure 9.7). If turf becomes excessively tall, it should not immediately be mowed at the desired height of cut. The resulting severe scalping may stop root growth for extensive periods. Also, since scalping removes the majority of plant leaf tissue, sunlight more readily reaches the soil surface and weeds typically become more problematic.

Mowing is performed two to seven times weekly, with higher quality turf requiring the daily mowing frequency (Table 9.3). As grain development can eventually form, it is recommended to vary mowing direction and incorporate grooming as necessary. Hydraulically driven units are now the norm to prevent turf gouging and slipping that often occurs when mowing wet turf.

## Mowing Equipment

Mowing equipment has continued to increase in sophistication since beginning with a scythe. The first reel mower was developed in 1830 by Edwin Budding, a textile engineer who adapted the rotary shears used to cut carpet nap. Early mowers relied on hand, steam, or animal power to operate. In the early 1900s,

**FIGURE 9.7** Scalping (or browning) of turf from mowing too low or infrequently for the intended height.

gasoline-powered units became available, followed by diesel-powered units. Today there is a vast array of mower types, levels of sophistication, and costs available. When choosing a particular model, several considerations should initially be examined:

1. **Turf type and/or use.** Several types of mowers are available and used by golf course superintendents. These include *reel*, *rotary*, and *flail mowers*. Fine-textured turf species such as bentgrass, bermudagrass, or zoysiagrass should be mowed with a reel mower. This mower provides the finest, closest cut and is mandatory for high-maintenance areas such as golf greens, tees, and fairways. Turf species with wider leaf textures such as St. Augustinegrass or centipedegrass, or upright growing species such as bahiagrass, Kentucky bluegrass, or tall fescue, are usually mowed with a rotary mower. Flail mowers are often used in low-maintenance areas that do not demand the highest quality cut.
2. **The terrain to be cut.** A smooth and level terrain is best for reel mowers. A rough, hilly, swampy, or wooded terrain may be more suited for a rotary or flail mower.
3. **The size of the area.** With the ever-increasing costs of labor and fuel, it is more practical to buy the largest/fastest mower available to perform the job in the shortest amount of time. There are, of course, limits to this. For example, a weight limit or size limit restricts the type of mower usable in high-maintenance areas such as golf greens or those areas requiring trimming types of mowers (e.g., around bunkers).
4. **Simplicity of design, durability, and maneuverability.** Usually the simpler the mower design is, the easier it is to operate, adjust, and maintain. However, mowers need to be versatile enough to perform the job, such as hydraulically operating individual cutting units near stationary or inaccessible objects. Likewise, mowers should be well braced and ruggedly built. Maintenance should be easily and routinely performed in order to minimize the power and time required to operate the equipment. Machines with easily accessible lubrication points, belt adjustments, bearings, chains, and shafts will likely be better maintained by employees compared to designs that are not or that require extensive effort to reach.
5. **Engine types.** Diesel engines provide better fuel efficiency than gasoline engines, run cooler during hot weather, require no regular tune-ups, and generally have greater life expectancy. Diesel fuel, however, may be more expensive, and diesel engines generally weigh more, are more expensive to purchase, and require special maintenance personnel or equipment for repair.

Recently, due to noise levels associated with fuel-driven engines, battery-powered mowers have become available. Most units have sufficient power to mow at least 20 typically sized golf greens on one charge. Maintenance requirements with these also are minimal.

Equipment care is almost as important as initially choosing the right machine. Routine maintenance such as lubrication, oil changes, blade sharpening, tune-ups, belt adjustments, changing bearings, and proper cleaning are important in extending the useful life of equipment as well as lowering operating costs. Adequate, accurate records need to be maintained and observed to help pinpoint the costs of operation and to justify purchases of new equipment. In addition, proper storage facilities should be available to minimize the exposure of equipment to weather, to prevent accidents, and to maintain security. When a job is finished, the unit should be properly cleaned, rinsed, maintained, and then stored in a clean, dry, and secure area.

**Reel mowers.** Reel mowers consist of blades attached to a cylinder known as a reel. As this cylinder rotates, grass leaves are pushed against a sharp, stationary bed knife and clipped. A properly adjusted reel mower cuts grass as cleanly as a sharp pair of scissors and exhibits better mowing quality than other types of mowers (Figure 9.8). Reel mowers also require less power, consume less fuel and, therefore, are more efficient to operate than rotary or flail mowers. Reel mowers use up to 50% less fuel than rotary mowers when used at the same mowing speed.

The quality of cut for reel mowers is determined by mowing height, number of blades on the reel, roller type, rotational speed of the reel, and forward speed of the mower. Increasing the number of blades and sections of a mower typically improves uniformity and quality of cut, especially if the mower reels can flex or float over uneven terrain. At mowing heights more than 1 1/2 inches (3.8 cm), a reel has four or five blades; at heights of 1/2 to 1 1/2 inches (1.3 to 3.8 cm), a reel typically has five to eight blades. In order to use the same mower at a lower height of cut, the reel would have to be powered to revolve at a higher rate of speed than at a higher cutting height. A wavy or rippled appearance of the turf would develop. Nine or more blades per reel or a hydraulically powered reel are required at mowing heights below 1/2 inch (1.3 cm) to provide a smooth cut. Golf green mowers typically have 11 to 15 blades on a relatively small-diameter reel (less space between blades) in order to provide the desired smooth surface needed for putting.

The wavy or rippled appearance associated with an improper mower setting or height is from exceeding the designed "clip" of the blade (Figure 9.9). The *clip* is defined as the horizontal forward distance covered between the time each reel blade engages the bed knife, which depends on the number of blades and rotational speed of the reel, as well as the forward speed of the mower. As mentioned, the greater the number of blades on a reel is, the shorter is the distance the mower travels before the reel again engages the bed knife and the smoother the resulting cut will be. A faster rotating reel will produce a shorter clip length than a slower reel with the same number of blades. Usually, the most uniform cut occurs when the clip equals the mowing height. Mathematically, the frequency of clip can be calculated as

$$\text{Frequency of clip} = \frac{\text{forward speed (inches/minute)}}{\text{blade number} \times \text{reel rounds/minute}}$$

**FIGURE 9.8** A reel mower where rotating blades "cut" when contacting a stationary blade (or "bed knife").

**FIGURE 9.9** Wavy pattern from excessive mower speed for the mower blade rotation.

### Example

Determine the frequency of clip for a walk mower with a forward speed of 3.8 mph equipped with a 15-blade reel turning at 2,200 rounds per minute mowing at 0.125 inch (3.2 mm). Is this deemed "acceptable" (1 mph = 1,056 inches/minute)?

$$\text{Frequency of clip} = \frac{\text{forward speed (inches/minute)}}{\text{blade number} \times \text{reel rounds/min}}$$

$$= \frac{4{,}013 \text{ inches/minute}}{15 \times 2{,}200/\text{minute}}$$

$$= 0.122 \text{ inch}$$

Since the frequency of clip (0.122 inch) is similar to the mowing height (0.125 inch), a uniform cut should occur.

Ground-driven reel mowers have a constant clip and can only provide a smooth cut at or above a certain height; otherwise, ripples will result when grass is mowed below that height. This rippling also occurs when the forward speed of the mower is excessive. Conversely, if the clip is appreciably shorter than the mowing height, air movement generated by the reel prevents proper contact of grass leaves to the reel blades and results in a ragged, nonuniform cut regardless of ground speed. Hydraulic-driven reels provide a variable clip and usually extend the mowing height range that a ground-driven reel mower normally provides.

Adjusting the distance between reel blades and the bed knife, as well as sharpening blades by the standard process of backlapping, is used to provide a desired cut. Computerized spin grinding and running mowers without reel-to-bed knife contact are recent means of sharpening and operating blades

**FIGURE 9.10** Computerized reel mower blade and bed knife sharpener.

(Figure 9.10). Backlapping involves applying an emery powder/liquid soap slurry to the reel as it is rotating backward to help maintain a sharp cut between grindings. All reel blades should be aligned equally with the bed knife along its entire length. A reel properly adjusted to the bed knife will cleanly cut a piece of newsprint paper along the entire length of the reel without binding.

### BACKLAPPING REELS

**Backlapping** is a means to hone and sharpen reel mowers. This involves rotating the reels backwards or in reverse against the bed knife while a fluid-dispersed grinding compound is applied. The steps involved in this process include:

**Clean the reel and bed knife.** Rinse or blow off the reel and bed knife to remove any debris.

**Inspect the reel and bed knife.** Release the bed knife from the reel and feel along the knife to make sure the surface is even and not damaged. Grind (lathe) or replace as needed to ensure the surface is even and true. Spin the reel to see if it turns evenly and smoothly. Grab and move the reel in multiple directions to see if any give is present. If so, bearings may be worn or the main shaft bent or distorted. Repair or replace as needed.

**Grease reel bearings and rollers.** Use a high-quality, high-impact, water-resistant grease lubricant. If the new grease squeezes out water, more frequent greasing may be needed.

**Adjust the reel.** Adjust the reel so that it almost touches the bed knife. Place a piece of newspaper between the reel and bed knife and adjust the reel so that a slight tug occurs as the paper is pulled out. Work the paper up and down the length of the bed knife so that this tugging is uniform (parallel) across all reels along the bed knife.

**Backlap the reel.** Spin the reel backwards. It should not touch the bed knife. Apply 120-grit compound across the bed knife using a long-handled paintbrush so a grinding noise occurs. Continue applying the grit on the brush until it is used up (Figure 9.11). If the bed knife and reel are sharp, the job is completed. If not, adjust one end of the reel just enough to hear the grinding noise again and then repeat the same increment on the other end. Reapply grinding compound and continue this process until the blades are sharp. Zero metal-to-metal contact should result. If the bed knife does not become sharp after about 15 minutes, it probably needs to be sharpened (ground) on a lathe machine (Figure 9.12). A higher grit compound such as 180 can also be used to try to sharpen the bed knife.

**Clean the mowing unit.** Wash the reel and bed knife thoroughly to remove all grinding compound. Leave the backlapping machine running during this process to enhance cleaning. Once washing is completed, the reel should turn freely with no metal-to-metal contact. Adjust the height of cut and test the mower on a nursery or practice green for quality of cut.

**FIGURE 9.11** Honing reel mower blades by "backlapping" with an applied grit material.

**FIGURE 9.12** "Grinding" or sharpening the bed knife of a reel mower.

During the mowing operation, tight turns should be avoided or, at least, performed slowly. When turning, the mower strikes turf and has a tendency to dip and dig into it. Fast, tight turns often cause scalping or severe grass defoliation.

Except during adjustment or sharpening, reel mowers should only be operated when in contact with grass. The dew on and moisture within grass acts as a lubricant for the metal-to-metal contact between reel blades and the bed knife. If the reel is operated during transport or over nonvegetative surfaces such as roadways or sidewalks, the heat expansion of metal may result in severe wear or warping of reels and bed knives.

Reel mowers do have some disadvantages. Most notably, they are unable to aesthetically mow grass maintained above approximately 1 1/2 inches (3.8 cm) and cut coarse-textured turf. Similarly, tall seedheads, weeds, and tough seed stalks are not cut efficiently with reel mowers. Reel mowers, especially hydraulically driven ones, are more expensive than other mowers and usually require a higher level of skill to operate and maintain. Hydraulic leaks that normally kill the grass also are a concern (Figure 9.13).

**Front rollers.** An array of front-mounted mechanical rollers also is available for reel mowers. These often help provide a balance between healthy turf from higher mowing heights but

**FIGURE 9.13** Damaged turf from hydraulic fluid leak from a riding mower.

still have adequate putting surfaces. Several types of rollers are available, depending on the density and expectations of the turf.

1. **Grooved or slotted rollers.** Grooved rollers have alternating smooth and indented (grooved or slotted) surfaces (Figure 9.14). The indentations provide multiple contact points that allow maximum penetration of the turf.
2. **Swaged roller.** Swaged rollers allow only the outer edge of the roller to make contact with the turf. The roller penetrates the turf less than grooved (or slotted) rollers do.
3. **Full or smooth rollers.** Full rollers have a smooth surface with no indentations, thus penetrating the turf least. They are best used when mowing stiffer turf that tends to stand up following mowing and to smooth the putting surface.

Thatch or grain-prone turfgrasses generally have the greatest need for rollers to allow penetration. If they are

**FIGURE 9.14** "Grooved" (or slotted) rollers on mowers to allow maximum turf penetration.

mowing wet turf, rollers may require a roller-scraper or cleaning-brush option to prevent sand and clippings from building up on them.

In addition to rollers, a wide array of turf preparation devices is available. These help stand the turf erect for better mowing and aid in thatch/mat management and surface compaction. These devices include:

1. **Combs.** Combs are mounted behind rollers and raise grass blades to assist in cutting and preventing grain development. They are the least aggressive options among the turf preparation devices.
2. **Brushes.** Rotating or fixed brushes also help stand up the turf prior to mowing for a smooth cut and to help prevent grain formation (Figure 9.15). Brushes can be mounted in front of or behind the roller.
3. **Turf groomers.** Groomers are miniature verticutting devices with vertical rotating blades that stand grass blades up and cut grass stolons and stems (Figure 9.16). Turf groomers are the most aggressive of the preparation options and help in thatch/mat/organic matter control and allow faster greens speeds without lowering the height.
4. **Dethatching reels.** These are similar to turf groomers but have larger diameters and are more aggressive in severing stolons and removing thatch.
5. **Sweeper.** This attachment collects cores from aerifying undulating greens and uneven terrains.
6. **Spikers.** A spiker has a rotating blade with attached daggers to pierce the turf surface, creating small holes. These holes reduce surface compaction, allowing better water and $O_2$ penetration and quicker recovery compared to traditional core aerification.

**FIGURE 9.15** Fixed brush in front of a mower intended to stand up leaf blades prior to cutting.

**FIGURE 9.16** Turf "groomer" stands grass blades up for better cutting of stolons and stems.

**Rotary mowers.** Two impact-type cutting mowers are rotary and flail mowers. Rotary mowers have blades horizontally mounted to a vertical shaft that cuts grass by impact at a high rate of speed. The key to success with rotary mowers is to maintain a sharp, balanced blade. Rotary mowers cut grass similarly to a machete; as long as the blade is sharp and balanced, the quality of cut is acceptable. A dull mower blade shreds leaf blades instead of cutting them and leaf tips become jagged and frayed (Figure 9.17). When leaf tissue is mutilated from the use of an unsharpened rotary blade, wounds heal slowly and greater water losses occur through evaporation since the leaf area exposed to the environment is increased. Mutilated tissue that heals slower also provides invasion points for diseases. If rotary blades are nicked from hitting hard objects, they should be ground or filed to restore the original sharp cutting edges.

Rotary mowers have the advantage of being relatively inexpensive and more versatile than reel mowers. They can be used to cut very tall grass, coarse-textured grass, tough seedhead stalks, and weeds where reel mowers cannot. Rotary mowers are also more maneuverable, making it easier to trim around trees and buildings than with reel mowers. Rotary mowers generally have lower initial costs and simpler maintenance requirements. For example, replacement blades for rotary mowers are generally much cheaper compared to the cost of a reel mower blade.

Disadvantages of rotary mowers include their inability to provide a quality cut at heights lower than approximately 1 to 1 1/2 inches (2.5 to 3.8 cm). Newer, faster spinning blade units, however, are becoming available that allow lower mowing heights without sacrificing turf quality. Rotary mowers are dangerous, since hands or feet can accidentally be placed under the mowing deck while the blade is operating. The high speed in which blades rotate also can be dangerous if objects such as rocks or tree limbs are encountered. Rotary mowers are not usually designed to follow the surface contour as precisely as a reel mower. Therefore, at close mowing heights, the rotary mower is more likely to scalp turf as it travels across small mounds or ridges that often compose the turf surface. Individual floating mower decks help minimize scalping.

**Flail mowers.** Flail mowers are another impact-type cutting unit involving a number of small L- or Y-shaped blades (knives) loosely attached to a horizontal shaft. As the shaft rotates, the knives are held out by centrifugal force. Cut debris from flail mowers is recut until it is small enough to escape the close clearance between the knives and mower housing.

Advantages of flail mowers include their ability to cut tall grass into finely ground mulch and the ability of each blade to recoil without damage to the mower or creating a dangerous projectile if it strikes a hard object such as a rock. This is especially important when operating near bystanders. Flail mowers also avoid the windrowing associated with rotary units and generally can handle wet grass better since their discharge area runs the entire width of the machine.

Disadvantages include the flail mower's inability to provide a close, quality turf surface as does a reel mower and the difficulty in sharpening the small, numerous knives. Most units are bulky and heavy, often necessitating a designated tractor for smoother operation and handling. Flail mowers are most often used on low-maintenance utility turf mowed infrequently without a high aesthetic requirement, such as golf course roughs or out-of-play areas.

## Mowing Pattern

Mowing patterns imposed by operators can influence aesthetic as well as functional characteristics of a turf surface. For example, as turf is mowed in alternating directions, light- and dark-green strips occur, resulting from differing light reflections based on direction of grass growth. Double-cutting at right angles produces a checkerboard appearance (Figure 9.18).

**FIGURE 9.17** Shredded leaf tips from using dull mower blades.

**FIGURE 9.18** A checkerboard appearing turf from mowing at right angles.

On greens, repetitive, identical long-term mowing directions should not be used even though this may produce the alternating color differences. Mowing turf repeatedly in the same direction results in grass leaning or growing in the direction in which it is cut. This horizontal orientation of grass foliage in one direction is called *grain* (Figure 9.19). Grain results in an uneven cut, a streaked appearance, and a poor quality putting surface on golf greens. The ball tends to follow the grain and, when different grain is encountered, the ball reacts by bouncing, having its path slightly altered, or changing its rolling speed. Golf green mowing patterns or directions should be changed daily and cleanup laps reversed or skipped on a routine basis. Often a rotating clock pattern is followed for mowing directions and is changed daily. Similarly, fairways should be mowed side to side, and diagonally as well as longitudinally, to minimize wear, compaction, and grain development. A single mowing pattern can be maintained on fairways and tees for a period of time (e.g., tournament) for desirable aesthetics and later returned to a normal rotation of mowing direction to prevent excessive grain development.

Mowing continually in the same direction has the additional detriment of repeatedly scalping the same high spots as well as increasing compaction and rutting by mower wheels. Turning the mower at the same location and in the same direction also encourages severe damage from wear and soil compaction.

### Walking and Triplex Mowers

When triplex mowers were introduced in 1968, the potential cost and time savings provided were welcomed by many golf courses (Figure 9.20). However, over time, wear patterns and

**FIGURE 9.19** Golf green grain from leaf blades lying down in alternating directions.

**FIGURE 9.20** Triplex mowers save time and operating costs.

**FIGURE 9.21** Due to the centrifugal force exerted by triplex mowers, wear patterns can form.

stress-related problems on greens began to appear and many courses switched back to walking units.

The most common wear pattern problem with triplex mowers is termed "triplex ring" from tires following the same pattern during each cleanup lap around the perimeter of the green (Figure 9.21). The centrifugal force exerted on the mower tires and double cutting each day in these areas, due to the space required to raise and lower the mower attachments, contribute to turf decline in these areas. Skipping the cleanup lap every other day, slowing the triplex mower speed, avoiding triplexing when soil is saturated, using a walking mower to perform the cleanup pass, and double core aerating these areas help reduce this problem.

Another concern with triplex mowers is hydraulic leaks (Figure 9.13). Good preventative maintenance, leak detection devices, or the use of dyes in the hydraulic fluid help lessen this danger. Other challenges with triplex units are providing a true uniform mowing height over the course of mowing 18 greens and eventually changing (softening) the original shape of greens by avoiding sharp turns in the initial green's design.

With walking units, increased costs and staff limitations are considerations. Triplex mowers typically save considerable labor as three to four extra employees and sets of transportation equipment are needed to walk-mow 18 greens. One compromise is to walk-mow during weekdays and triplex mow on weekends. This avoids excessive employee overtime that can present a real obstacle with walking units.

## Grass Clippings

Clippings are often composed of 3% to 5% N on a dry weight basis and also contain significant amounts of phosphorus (P) and potassium (K). Clippings, therefore, serve as an important source of nutrients. If clippings are removed, additional fertilization is required to compensate for those nutrients. Removing clippings also poses a disposal problem to many superintendents since most municipal landfills no longer accept them, or if they do, an additional fee is assessed. Emptying the catcher or raking the clippings also requires additional time and labor. Under normal conditions, clippings should be allowed to fall back to the turf. Only when clippings are so heavy they smother the grass or interfere with the playing surface, such as golf greens, should they be removed. Often, plant growth regulators (PGRs) are used to manage excessive clipping yields during periods of rapid turf growth. These not only improve turf quality, but also can reduce the amount of time and labor needed for mowing events as further dragging or blowing of clippings is not needed.

By following the one-third rule on mowing frequency, heavy amounts of clippings are not deposited at one time. Soil organisms that naturally break down cellulose and lignin in grass clippings then have sufficient time to decompose them before accumulation problems develop. However, if excessive amounts are produced from mowing, such as following heavy N fertilization or from excessive scalping, natural decomposition may not be sufficient. A mat problem may develop under these conditions.

Clippings collected from golf greens should be disposed of properly to prevent undesirable odors near the playing area and to prevent fire hazards when clipping piles accumulate. Compost piles can be developed by alternating layers of clippings with a mixture of soil and N fertilizer. When composted, clippings can then be used as ground mulch in flower beds or inaccessible mowing areas. If not composted, clippings should be dispersed so that piles are not allowed to form.

## AERIFICATION

Turf on heavy use areas such as golf course greens, sports fields, and other high-traffic areas often deteriorates due to compacted soil, thatch development, and excessive use. Unlike annual row crops that are periodically tilled to correct these problems, turf managers cannot provide such physical disturbances without destroying the playing surface. Soil-related problems are usually confined to the upper 3 inches (7.6 cm) of the turf profile. Once formed, they may not be completely corrective, especially where improper site preparation occurred prior to establishment. However, a number of mechanical devices that provide soil cultivation with minimum turf surface disturbance have been developed. Cultivation is accomplished by core aerification, vertical mowing, spiking, slicing, and topdressing.

Aerification or "coring" is the removal of small soil cores or plugs of soil with grass from the turf surface, leaving a hole in the sod (Figure 9.22). Holes are normally 1/4 to 3/4 inch (6.4 to 19 mm) in diameter, with their depth and distance apart depending upon the type of machine used, forward speed, degree of soil compaction, and level of soil moisture present. Traditional aerifying machines penetrate the upper 2 to 4 inches (5 to 10 cm) of soil surface with cores spaced anywhere from 2 to 6 inches (5 to 15 cm) to the center. "Deep-tine" aerification equipment provides options for creating holes to depths of greater than 10 inches (25 cm) and diameters ranging from 1/8 to 1 inch (3.2 to 25 mm). Turf generally responds best when cores are close together and deep. This pattern removes more soil, exposes more surface area for water and nutrient intake, and alleviates compaction better than wider or shallower tine spacing and depth. Advantages and disadvantages associated with coring include the following.

### Benefits of Coring

1. Relieves surface and subsurface soil compaction
2. Allows deeper, more efficient penetration of water, air, topdressing sand, fertilizer, lime, and pesticides into the rootzone
3. Allows for penetration of $O_2$ into the soil for plant roots (Figure 9.23)
4. Allows for atmospheric release of toxic gases (e.g., $CO_2$, methane, carbon monoxide) from the rootzone, minimizing the occurrence of *black layer*

**FIGURE 9.22** Aerifying a golf course fairway where soil cores are removed.

**FIGURE 9.23** Brown, weak bentgrass between healthy plants in aerification holes.

**FIGURE 9.24** Excessive organic matter accumulation indicating insufficient cultural practices.

5. Improves surface drainage to help dry out saturated soils and prevent formation of puddles
6. Improves penetration of water into dry or hydrophobic soils (e.g., relieves localized dry spots)
7. Penetrates through soil layers that develop from topdressing with dissimilar materials. These form stratified layers where water will not drain out of the finer layer until it is fully saturated. A 1/8 inch (3.2 mm) layer of fine soil is sufficient to alter drainage in an otherwise perfect soil profile
8. Aids in soil replacement when combined with topdressing
9. Encourages organic matter/mat/thatch control by stimulating those environmental conditions that promote healthy soil microorganism activity that naturally decomposes the components attributed to thatch development (Figure 9.24)
10. Increases rooting by constructing a rootzone more conducive for active root growth (Figure 9.25)
11. Improves turf resiliency and soil cushioning
12. Improves seedbed preparation for renovation, overseeding, and interseeding

### Disadvantages of Coring

1. Temporary disruption of or damage to the playing surfaces

**FIGURE 9.25** Deeper rooting in aerification channels, which contain more oxygen than soil between them.

2. Increased surface desiccation as roots and soil surfaces are exposed
3. Temporarily providing a habitat in which cutworms, mole crickets, and other insects can reside

Generally, the benefits of aerification far outweigh any detrimental effects. Turf managers must decide which option is best to solve the existing problem.

**Soil compaction.** The primary goal of core aerification is to relieve soil compaction. However, for this to successfully occur, cores (soil) must be removed to reduce the amount of mass (or weight) of soil in a given area (or volume). Compacted soil occurs when mineral particles have been pressed close together from excessive or concentrated traffic, especially when soil is wet. Soil compaction reduces $O_2$ (porosity) levels in the soil. A soil should ideally be composed of between 25% and 40% air on a volume basis, but compacted soils have as little as 5%. Root function decreases under compaction due to the lack of $O_2$ needed for respiration and due to buildup of toxic gases such as $CO_2$, methane, or sulfides. Roots may also be unable to physically penetrate soil during growth when it is tightly compacted. New roots are often abundant along the sides of the aerifying holes, indicating the need for increased soil $O_2$ and reduced compaction.

Soil aerification also helps reduce thatch by providing a better microenvironment for soil organisms that naturally decompose thatch. The aerification process also brings up a small amount of soil to the surface, thereby acting similarly to a thin topdressing. Thatch accumulates more quickly on compacted soils than on well-aerated soils.

Compacted soil surfaces also reduce water infiltration and percolation rates. Dry soils in compacted areas are difficult to rewet and conditions such as localized dry spots often develop, especially on high sand content areas. However, compacted, saturated soils may not drain excessive water and often turn into mud with continued use. Such soils often remain wet for extended periods of time and become covered with an undesirable layer of algae or moss. The success of highly maintained turf areas is dependent on the superintendent having control over soil moisture, through adequate drainage and irrigation.

The best methods for preventing compaction are to build the greens and tees with a predominantly (>80%) medium to coarse sandy soil with minimum (<5%) content of silt and clay, and also surface grade for proper surface drainage. Compaction is much more likely on fine-textured clayey soils than on a coarser, sandy soil.

Reducing or redirecting the concentration of traffic is the next method of relieving soil compaction (Figure 9.26). For example, the correct placement of cart paths and sidewalks is imperative. Cart paths should normally be a minimum of 8 feet (2.4 m) wide to allow adequate passage for two-way traffic as well as larger maintenance vehicles such as tractors and mowers. Barriers such as curbs or railroad cross-ties should be used adjacent to high-traffic areas such as tees and greens to prevent carts from leaving the path.

Traffic should also be minimized or prevented when soil is wet. Water acts as a lubricant and traffic during these periods further aggravates soil compaction, reducing turfgrass growth and vigor. Traffic should be regulated after heavy rains, as well as after any mowing with large, heavy units. Wide turf tires should be used on all equipment to help distribute the weight of such vehicles over a larger area than allowed by regular tires.

Core aerification will usually soften hard, compacted turf surfaces. This is especially true when the spacing between holes does not exceed 2 inches (5 cm). Coring is most effective when soils are moist to facilitate penetration of tines,

**FIGURE 9.26** Reducing or redirecting concentrated traffic on wet soil to avoid unwanted compaction.

but should never be performed when soils are saturated, as rutting and additional compaction may occur.

### Types of Cultivation Methods

Many types of core aerifiers or cultivators are available. Equipment is classified as core cultivation or aerification, solid tine cultivation, slicing or spiking, water or air injection, and deep subsurface cultivation. Most aerifiers fall into one of two categories: vertical- or circular-motion (or disc-type) units. Minimal surface disruption accompanies vertical-motion core cultivators, and they are the preferred choice on closely mowed turf surfaces such as golf greens (Figure 9.27). Vertical units have the drawback of being relatively slow due to the synchronization of vertical and forward operations. However, improvement in speed and ease of operation have been achieved in recent years.

Circular-motion or disc-type cultivators have tines or spoons mounted on a drum or metal discs (Figure 9.28). Hollow tines or spoons are forced into soil as the drum or discs turn in a circular motion. Hollow drum units are available that remove extracted cores from the soil surface, while other units do not deposit cores back onto the soil surface. Circular-motion cultivators are preferred for aerifying large areas since the rotating units can cover more ground in a given time period than vertical-motion cultivators. However, circular-motion cultivators disrupt the turf surface more and do not penetrate as deeply as vertical-motion cultivators. Weights are often placed on top of circular-motion cultivators to increase penetration depth.

**FIGURE 9.27** Minimum turf disruption from a vertical-motion core aerifier.

FIGURE 9.28 Circular motion aerifiers have tines or spoons mounted on a drum or metal discs.

**Core removal.** Superintendents often consider whether or not to remove cores produced from aerifying from the playing surface (Figure 9.29). Cores must be removed from golf or bowling greens—and often from tees—due to the intensive maintenance of these areas. However, for turf areas other than golf or bowling greens, the decision involves a number of factors. First and foremost, the availability of equipment and labor should be considered. Core harvesting machines or large tractor-mounted blowers are needed to remove cores from large areas like fairways and roughs. Cores do not have to be removed from fairways and roughs if thatch control, temporary compaction reduction, or air and chemical entry are desired and the underlying soil is acceptable. If the rootzone mixture (soil) present is acceptable, then the cores should be broken up by lightly verticutting or dragging the area with a mat, brush, or piece of carpet. The remaining debris may then be blown off or picked up with a follow-up mowing (Figure 9.30). Before soil cores are matted, they should be allowed to dry sufficiently so they can easily crumble between the fingers. If cores are too dry when matted, they are hard and not easily broken up. If they are too wet, they tend to smear and be aesthetically undesirable. The positive influence of hollow-tine aeration on surface water penetration is from five to eight weeks using greater than or equal to 1/2 inch (1.2 cm) diameter holes.

**Hollow versus solid tines.** Solid tines are sometimes used for aerifying instead of hollow-cone tines. Creating holes by forcing solid tines into the turf is called *shatter-coring*. Solid tines do not remove soil cores; rather, they divot the soil surface and may compact soil along the sides and bottoms of the holes more severely than hollow tines. Uplifting and jarring soil using solid tines is often claimed to improve soil aeration. However, unless the soil's bulk density is reduced by removing the soil cores, areas receiving solid-tine aerification will

FIGURE 9.29 Mechanically removing aerification cores to provide a smoother playing surface.

FIGURE 9.30 Removing, by blowing, aerification cores off the playing surface.

only experience temporary benefits and the process must be repeated more frequently.

Solid tines, however, do not disrupt the playing surface as much as hollow-tine cultivation (Figure 9.31). This is an advantage during extreme weather conditions when the grass has ceased or reduced its growth rate. Using solid tines in overseeded turf or bentgrass greens temporarily reduces compaction and softens the greens with minimum disruption of the putting surface. Labor savings provide another advantage of solid tines.

Disadvantages of solid-tine cultivation include only providing short-term relief of compacted soil and an increased potential to develop a hard pan. The positive influence of solid-tine cultivation on water penetrating turf surface is also short lived—about three weeks. Therefore, a three-week schedule of using solid-tine cultivation would be necessary. Varying the depth of cultivation, allowing the soil to dry before aerifying, and the use of small-diameter (1/4 to 3/8 inch, 6.4 to 9.5 mm) solid tines are methods to help prevent hard pan development.

### Frequency and Timing of Cultivation

On fairways, at least one core aerification, plus verticutting, has proven extremely beneficial with additional events implemented in areas with longer growing seasons. A variety of

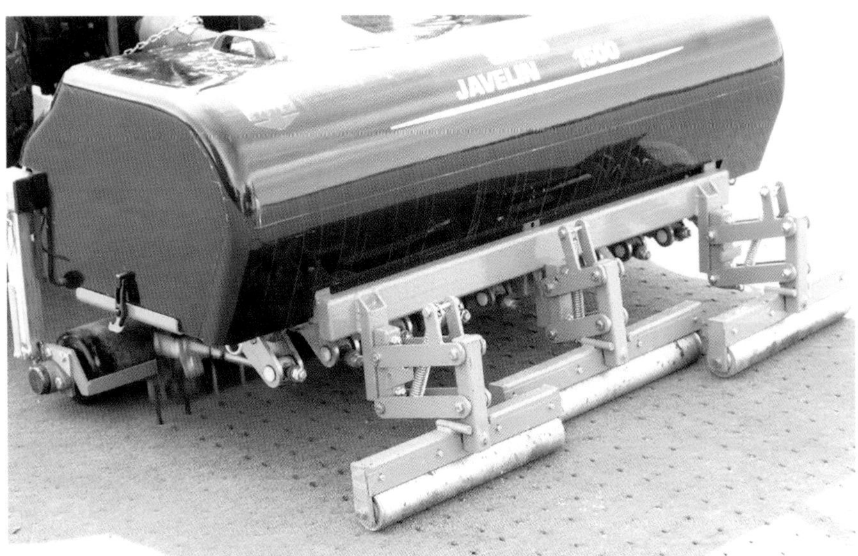


FIGURE 9.31 Solid-tine aerifier, which does not physically remove a soil core.

**TABLE 9.4**
**Aerification Tine Size Diameter and Hole Spacing Effects on the Turf Surface Area Displacement[a]**

| Tine Diameter (inch) | Tine Hole Spacing (inch) | No. Holes/ Square Foot | Surface Area Impacted per Tine (square inch) | Surface Area Displacement (%) | No. to Impact 20% Surface Area |
|---|---|---|---|---|---|
| 1/4 (0.25) | 1 × 1 | 144 | 0.049 | 4.9 | 4 |
| | 1 1/4 × 1 1/4 | 92 | | 3.1 | 6.5 |
| | 1 × 2 | 72 | | 2.5 | 8 |
| | 2 × 2 | 36 | | 1.2 | 16.7 |
| | 2 1/2 × 2 1/2 | 23 | | 0.8 | 25 |
| 3/8 (0.375) | 1 × 1 | 144 | 0.110 | 11 | 1.8 |
| | 1 1/4 × 1 1/4 | 92 | | 7.1 | 2.8 |
| | 1 × 2 | 72 | | 5.5 | 3.6 |
| | 2 × 2 | 36 | | 2.76 | 7 |
| 1/2 (0.5) | 1 × 1 | 144 | 0.196 | 19.6 | 1 |
| | 1 × 2 | 72 | | 9.8 | 2 |
| | 1 × 1 1/2 | 64 | | 8.7 | 2.3 |
| | 1 1/2 × 2 | 48 | | 6.5 | 3.1 |
| | 2 × 2 | 36 | | 4.9 | 4 |
| | 2 1/2 × 2 1/2 | 23 | | 3.1 | 6.5 |
| 5/8 (0.625) | 1 × 1 | 144 | 0.307 | 30.7 | 0.7 |
| | 1 × 2 | 72 | | 15.3 | 1.3 |
| | 1 × 1 1/2 | 64 | | 13.6 | 1.5 |
| | 2 × 2 | 36 | | 7.7 | 2.6 |
| | 2 1/2 × 2 1/2 | 23 | | 4.9 | 4 |
| | 5 × 5 | 5.8 | | 1.2 | 0.8 |
| 3/4 (0.75) | 2 1/2 × 2 1/2 | 23 | 0.44 | 7.1 | 2.8 |
| | 5 × 5 | 5.8 | | 1.8 | 11 |
| 1 (1.0) | 5 × 5 | 5.8 | 0.79 | 3.16 | 6.5 |

[a] Multiply inches by 25.4 to obtain millimeters or 2.54 to obtain centimeters; multiply square inches by 6.45 to obtain square centimeters.

front and rear rollers, brushes, groomers, and vertical mowing accessories are also available to improve the quality of cut and to help reduce or prevent grain or thatch development. These help promote stand density and upright shoot growth to provide better ball lie. The organic debris produced can be removed with sweeping, blowing, and/or vacuuming operations. Aggressive verticutting should be performed in early to midsummer (for warm-season grasses) and late summer or early spring (for cool-season grasses) when the grass is actively growing.

The frequency of core cultivation should be based on the intensity of traffic to which the turf is exposed. Other factors include soil makeup, hardness of the soil surface, drainage capability, and degree of compaction. Areas receiving intense, daily traffic, such as golf greens, approaches, landing areas, aprons, and tees, annually require a minimum of three core aerifications. Additional aerifications may be needed on exceptionally small greens where traffic is more concentrated, areas consisting of heavy soils high in silt and/or clay that do not drain well, areas with localized dry spots, or soils exposed to saline or effluent water use.

On greens, another means to determine how much to aerify each year is based on the amount of turf surface impacted by aerification. A 15% to 25% surface area being impacting on an annual basis is a reasonable routine goal for many well-established golf courses. A percentage basis is sometimes easier for nonagronomists to understand instead of trying to comprehend tine diameters and spacing. Neglected courses may need a more aggressive aerification program. As outlined in Table 9.4, using various tine sizes and hole spacings will determine the surface area of turf impacted. For example, a fourfold increase in surface area impacted occurs when using a 1/2 inch (1.3 cm) tine instead of 1/4 inch (0.6 cm) tines. An approximately 50% increase in surface area impacted occurs when using 5/8 inch (1.6 cm) tines instead of a 1/2 inch (1.3 cm) tine. Additionally, changing tine spacing can increase the amount of surface area impacted. For example, changing tine spacing from 2 × 2 inches (5 × 5 cm) to 1 × 2 inches (2.5 to 5 cm) (with 3/8 inch diameter tines, 1 cm) increases the surface area impacted by 100%.

### Additional Aerification Developments

**Needle tine.** Small-diameter, solid, "needle"- or "pencil"-tine aerifiers, ranging in size from 3/16 to 5/8 inch (0.48 to 1.6 cm) and up to 12 inches (30 cm) deep, can make an enormous number of tiny aeration channels in the soil profile (Figure 9.32).

**FIGURE 9.32** Various sizes, depths, and configurations of some available aerification tines.

These disrupt hardpans or layers that have formed, and provide passageways for soil water and air exchange, thereby aiding drainage and irrigation efficiency. The small-diameter tines are much less disruptive to the playing surface, allowing play immediately following their use. Since these machines place little stress on the turf, they are especially useful on cool-season grass greens during summer stress and on warm-season grass greens in late winter.

**Deep tine.** Several developments in technology provide turf managers with a wider choice of aerification strategies. One involves deep-tine cultivators that are able to extract 3/4 to 1 inch (1.9 to 2.5 cm) diameter cores to a depth of 8 to 12 inches, or 20 to 30 cm (Figure 9.33). Deep-tine units enable the superintendent to relieve the soil compaction layer (e.g., hardpan) that develops when traditionally used aerifiers constantly penetrate to shallower depths (e.g., 3 inches, 7.6 cm). Soil profiles consisting of many undesirable layers that develop from using different materials for topdressing are also penetrated. This enhances water penetration, soil aeration, and rooting. The soil profile of an undesirable green can also be improved by topdressing following deep aerification with a desirable rootzone mix.

**Deep drill.** Another development is the deep-drill aerifier. Drill bits of varying lengths and diameter are drilled into the turf, leaving a small casting of soil on the surface around each hole. This soil is usually then matted back into the turf. The biggest advantage of the deep-drill aerifier is the ability to provide a deep hole with the least disruption of the playing surface. Also, "drill-n-fill" units are now available that refill the drilled hole with a desired soil amendment such as sand (Figure 9.34). These units, however, are relatively slow and are generally more expensive to operate since a high degree of mechanization and numerous drill bits are needed to operate the machine. Additionally, since cores are not

**FIGURE 9.33** Deep-tine aerifier that penetrates 8 to 12 inches (20 to 30 cm) deep.

**FIGURE 9.34** Deep-drill aerifier, which bores or drills a core 8 to 12 inches (20 to 30 cm) deep.

physically extracted, the soil brought to the surface is difficult to remove.

**High-pressure water, sand, and air injection.** A novel aerification technique consists of using high-pressure water injection (Figure 9.35). Fine streams of high-velocity water (up to 5,000 psi, 34,450 kPa) are injected with minimum surface disruption occurring. Play, therefore, is not disrupted from high water pressure aerification holes as when traditional machines are used. High-pressure units also help wet hydrophobic soils such as localized dry spots, can also inject wetting agents, and distribute surface organic matter throughout the soil profile (Figure 9.36). Disadvantages include the initial high cost, the need for a water source at all aerification sites with certain units, and reduced effectiveness on heavy soils where the high-pressure water stream cannot adequately penetrate. Thatch control is also minimal, and sand cannot be incorporated back into the profile since the holes produced are generally not large enough. The hole spacing and penetration depth is, however, adjustable through multiple pulses by leaving the units in the raised position, changing the nozzle spacing, or by varying speed. Frequency of water injection on greens depends on the level of stress to be alleviated with a typical treatment on a three- to four-week basis. Water-injection cultivation should be viewed as a supplement to traditional core aerification and not as a replacement.

Other aerification developments include sand being forced into the soil using high pressure (e.g., Dry-Jet), and air injection tubes that insert high-pressure air at the bottom of an aerification tine's down stroke or through ports along the tine's surface (Figure 9.37).

Ways to enhance turf recovery following aerification include:

- Aerify only when the turf is naturally actively growing. For cool-season grasses, this is in spring and fall when soil temperatures are between 60°F and 65°F

**FIGURE 9.35** Hydraulic injection aerifier where high-pressure streams of water penetrate the turf surface.

**FIGURE 9.36** Soil profile from a hydraulic injection aerifier showing parabola-shaped downward movement of sand and organic matter.

**FIGURE 9.37** Air injection aerifier where high-pressure streams of air penetrate the turf surface.

(15°C and 18°C). For warm-season grasses, this is between late spring through summer when soil temperatures are above 75°F (24°C).

- Fertilize three to five days before or after core aerification, typically with 1/2 pound N/1,000 square feet (0.56 kg N/ha) for cool-season grasses and three-fourths to 1 pound N/1,000 square feet (0.84 and 1.1 kg N/ha) for warm-season grasses.
- Irrigate one to two nights prior to aerification to field capacity.
- Make clean aerification holes by replacing worn tines and setting the machine so that straight (perpendicular) motion occurs with the turf surface.
- Following aerification, clean the aerified surface completely of aerification cores and debris. Core harvesting equipment followed by blowers is the means to produce desired results.
- Topdress with dry sand on dry turf so as to fill holes completely. Dry sand applied to dry turf moves into holes most efficiently—filling holes flush with the turf surface to best "smooth" the playing surface. Sand in holes typically settles after several days, necessitating a follow-up lighter topdressing.
- Incorporate the sand by brushing or dragging. If performed too aggressively, the topdressing sand may bruise the turf, causing off-coloring until the turf grows sufficiently enough so that this is mowed off. Forced-air topdressing incorporation is becoming more popular due to better (more efficient) sand incorporation and less turf damage during stressful periods.

- Roll following sand incorporation to help smooth surface irregularities. This also helps prevent damage from raised tufts of grass following mowing.
- Roll for several days before mowing to avoid mower and reel damage and removal of excess sand.
- Irrigate and fertilize to stimulate turf recovery. The dry sand applied can deplete turf plants of moisture and fertilizer stimulates new growth. Hand-water and irrigate as needed for several days along with adding 1/4 to 1/2 pound of soluble N/1,000 square feet to aid in recovery.

### Spiking and Slicing

Two additional cultural practices available to help relieve surface compaction promote better water penetration and aeration, stimulate new shoot and root development, sever stolons and rhizomes, and help incorporate topdressing are spiking and slicing. These are generally pull-type, nonpowered units consisting of a series of blades mounted on a horizontal shaft. A slicer has thin, V-shaped knives bolted at intervals to the perimeter of metal wheels that cut into the soil (Figure 9.38). Turf is sliced with narrow slits about 1/4 inch (6.4 mm) wide and 2 to 4 inches (5 to 10 cm) deep. Slicing can be performed much faster than coring and does not interfere with the turf surface since soil cores are not removed; thus, no cleanup is necessary after the operation. Slicing is also performed on fairways and other large, high-traffic areas during midsummer stress periods when coring may be too injurious or disruptive. Slicing is an excellent tool to prevent surface crusting and algae development and can be used during periods of heavy rainfall when other aerification techniques cannot. Weekly use is recommended, especially during stressful growing conditions. Slicing, however, is less effective than coring and is most effective when used in conjunction with coring. As with coring, slicing is best accomplished on moist soils to facilitate penetration.

A spiker has solid tines mounted on a horizontal shaft (Figure 9.39). It provides an effect similar to that of a slicer, but the penetration is limited to approximately 1 inch (2.5 cm). The distance between perforations along the turf's surface is also shorter. For these reasons and since spiking

**FIGURE 9.38** A slicer with V-shaped blades that cut through the turf.

**FIGURE 9.39** Solid-tine spiker, which vertically penetrates turf and soil.

causes less surface disruption than coring, this is primarily practiced on greens and tees. Both a slicer and spiker help (1) break up soil surface crusting, (2) break up algae layers, and (3) improve water penetration and aeration. Solid tines are associated with a spiker, and holes are punched by forcing soil downward and laterally. This causes some compaction at the bottom and along the sides of the holes. Since only minor disruptions of soil surfaces occur, spiking and slicing can be performed more often (e.g., every 7 to 14 days) than with core aerification (e.g., every four to eight weeks).

## VERTICAL MOWING AND GROOMING

A vertical mower is a powered unit with a series of knives vertically mounted on a horizontal shaft. The shaft rotates at high speeds and the blades slice into the turf and rip out thatch and other debris (Figure 9.40). Depending on the task, the shaft can be raised or lowered to cut shallowly or more deeply into the turf. Vertical blade thickness varies between 1/32 and 1/4 inch (0.8 to 6.4 mm), according to use. Golf greens require thinner blades to prevent excessive surface damage, while fairways require heavier, thicker blades to obtain desired results.

### DEPTH

Different objectives can be met with vertical mowing depending on the depth of penetration and blade spacing. Grain is reduced when knives are set shallow enough to just nick the surface of the turf. Shallow vertical mowing also is used to break up cores following aerification, thereby providing a topdressing effect. Deeper penetration of knives stimulates new growth of creeping species when stolons and rhizomes are severed and also removes accumulated thatch. Seedbed preparation prior to overseeding also is accomplished by deep vertical mowing.

When dethatching is the objective, thatch depth will determine the depth of blades. The bottom of the thatch layer should be reached by vertical mowing, and the soil surface beneath the thatch layer should preferably be sliced. However, there is a limit to the depth at which blades should be set or excessive removal of turf roots, rhizomes, stolons, and leaf surface may occur. For example, blades should be set at a depth to just cut stolons and no deeper if new growth stimulation is the objective. Vertical blade spacing for thatch removal should be between 1 and 2 inches (2.5 to 5 cm). This range provides maximum thatch removal with minimal turf damage.

Deep vertical mower penetration requires the use of a heavy-duty machine that can penetrate 2 to 3 inches (5 to 7.6 cm). Deep vertical mowing grooves the turf surface, so topdressing is often required to smooth the surface and cover exposed stolons (Figure 9.41). Shallow-rooted or immature turf can be severely damaged or torn out by deep vertical mowing. Preliminary testing at the site to be verticut should be done by hand-pulling to measure whether favorable rooting of the grass exists. Irrigation and topdressing should follow such deep vertical mowing to prevent quick desiccation of exposed roots, rhizomes, and stolons, as well as to help smooth the turf surface and encourage turf recovery.

### FREQUENCY

The rate of thatch accumulation dictates the frequency of vertical mowing. Vertical mowing should begin once the thatch layer on golf greens exceeds 1/4 to 1/2 inch (6.4 to 13 mm). This layer can be periodically checked when cups are changed or at any time by using a knife to slice a plug from the green. Areas prone to thatching accumulation may require heavy vertical mowing several times per year. For warm-season grasses, the first is during mid- to late spring when the grass is actively growing. This removes thatch and encourages turf spread by slicing stolons and by warming the soil surface more quickly than if the thatch is allowed to remain. The second vertical mowing should be timed two to four weeks before the anticipated fall overseeding. This discourages late-season grass growth that can compete with the overseeded grasses, and exposes the soil surface so that grass seed can reach the soil better and have optimum germination. However, fall vertical mowing will result in a degree of surface damage that may not heal until the overseeding has time

**FIGURE 9.40** Debris from vertical mowing (verticutting) that removes thatch and slices stolons.

**FIGURE 9.41** Aggressive vertical mowing of a golf green.

to become established. Cool-season grass greens are vertically mowed in fall and again in spring when temperatures are mild, the grass is actively growing, and ~30 days prior to typical weather extremes (e.g., hot or cold). However, newer cultivars tend to form a tighter stand surface than older ones; thus, vertical mowing may be required more often with them.

Soil and thatch should be dry when deep vertical mowing is performed or turfgrass injury will be more extensive since moist conditions encourage excessive plant material to be removed. Following verticutting, debris should be disposed and the area immediately irrigated. Approximately five to seven days following heavy vertical mowing, N should be applied to encourage rapid recovery.

If the thatch layer has become excessive (more than 2 inches, 5 cm), it may become uncontrollable through vertical mowing (Figure 9.42). In such extreme conditions, the grass and thatch layer need to be removed with a sod cutter. Soil must be added to level the area and then reestablished. This problem can be best avoided by verticutting and topdressing frequently enough (e.g., every four weeks during the growing season) to keep the thatch under control. Additionally, judicious use of N fertilizer and pesticides will sometimes slow rapid and extreme thatch accumulations.

**FIGURE 9.42** Excessive thatch layer.

Interchangeable vertical mower units are now available for many of today's greens mowers. This equipment allows for frequent vertical mowing and simultaneous debris collection. Vertical blades on greens mowers should be set to only nick the surface of the turf. By conducting frequent, light vertical mowing, the severe vertical mowing needed for renovation may be avoided (Table 9.5). Large turf areas such as fairways are vertically mowed by using units powered via power take-off (PTO). Such units have heavily reinforced construction and large, thick (approximately 1/4 inch, 6.4 mm) blades that are able to penetrate the soil surface.

A variation of verticutting is fraze mowing. This unit can have tungsten blades, which strip off the existing thatch layer evenly across the unit (Figure 9.43), or be retrofitted with verticutting blades (Figure 9.44). If the blades are set to penetrate the soil deeper, they are also used to smooth uneven playing surfaces such as tees. This unit also can remove/reduce weeds such as crabgrass, goosegrass, and annual bluegrass by removing their seed from the soil. It requires considerable power to operate and a companion tractor and trailer to catch the stripped soil and debris, as well as a local disposal site. It

**TABLE 9.5**
**Turf Surface Area Impacted by Vertical Mowing Blade Widths[a]**

| Vertical Mower Blade Width (inch) | Spacing (inch) | Surface Area Impacted (%) | Vertical Mowings to Impact 25% of Surface Area |
|---|---|---|---|
| 5/64 | 1/2 | 15.6 | 1.6 |
| | 1 | 7.8 | 3.2 |
| 9/64 | 1/2 | 28 | 0.9 |
| | 1 | 14.1 | 1.8 |

[a] Multiply inches by 25.4 to convert to millimeters.

**FIGURE 9.43** Fraze mower retrofitted with blades to strip existing thatch and perform minor grading.

**FIGURE 9.44** Fraze mower retrofitted with verticutting blades to remove thatch.

is relatively slow to operate and, depending on how aggressively (low) the unit is set, considerable time will be needed for full turf recovery. The fraze-mowed turf also needs to be rhizomatous to recover; otherwise, reseeding is needed following its use.

## Grooming and Conditioning

The grooming mower (also called turf conditioner) is a variation in vertical mowing. Grooming keeps greens smooth and quick by reducing grain and removing excessive top growth. In front of the reel-cutting unit is an attached miniature vertical mower (often referred to as vertical grooming) with typical blade spacing of less than 1/4 inch (6.4 mm) set 1/16 to 1/8 inch (1.6 to 3.2 mm) below the height of cut that rotates through slots in the front roller (Figure 9.16). In front of the vertical groomer, a roller is mounted to increase turf penetration in dense turfgrasses and to improve the height of cut. A solid or full roller minimally penetrates, and performs best on stiffer turfgrasses that quickly rebound after the roller goes over them. A swage roller has thickened outer edges where the center of the roller does not touch the grass. It therefore penetrates the grass surface very little. Grooved rollers provide maximum turf penetration due to the many grooves or slits in them.

Each time turf is mowed with this unit, the turf is lightly vertically mowed (or groomed or conditioned). This unit improves the playing surface by standing up leaf blades before mowing, thereby removing much of the surface grain. New shoot development is also stimulated by slicing stolons and removing thatch near the surface. Frequent grooming, along with timely topdressing and aerification, helps eliminate the need for traditionally performed turf renovation by severe vertical mowing.

**FIGURE 9.45** Various grooming and conditioning attachments for golf green mowers.

Other groomers use a rotary brush that operates in the opposite direction of the mower blade. The brush stands the grass up prior to mowing, reducing grain and providing a smooth surface. Combs are also mounted behind the front roller to lift up (or comb) the grass to reduce grain and improve cutting (Figure 9.45).

## TOPDRESSING

Topdressing adds a thin layer of soil or sand to the turf surface, which is then incorporated into the turf by dragging or brushing (Figure 9.46). Benefits of a proper topdressing program include:

1. Increasing thatch decomposition
2. Providing a "truer" and smoother playing surface
3. Reducing grain
4. Enhancing turf recovery from injury
5. Encouraging a denser and finer textured turf
6. Enhancing overseeding
7. Modifying existing soil

Likewise, detriments of topdressing can occur if proven tips are not followed. These include:

1. Avoid topdressing during stressful (excessively dry and heat-stressed) conditions.
2. Use quality sand in the proper percentage range, without other issues such as weed seeds.
3. Topdress at a rate so as to avoid overbrushing afterwards, especially stressed greens.
4. Have a set of (old) reels and bed knives used exclusively following topdressing.
5. Several days are required for improved putting conditions following topdressing.

On newly vegetatively established turf, topdressing partially covers and stabilizes newly planted material, smooths gaps from sodding, and minimizes turfgrass desiccation. Topdressing is performed on established turf to smooth the playing surface, control thatch and grain, promote recovery from injury, and possibly change the physical characteristics of the underlying soil. Unfortunately, in recent years, many

**FIGURE 9.46** "Topdressing" or applying a thin layer of soil or sand to the turf surface.

superintendents have reduced the number of coring and topdressing procedures due to member complaints of disrupting play. These procedures, however, are sound, fundamental agronomic practices necessary to maintain an optimal putting surface. If eliminated, putting green quality will diminish over time. An effective topdressing program requires:

1. Selecting an appropriate topdressing material to avoid abrupt changes (e.g., layers) between the original soil and the topdressing material
2. Being consistent in application frequency and rate to help avoid alternating layers of sand and organic matter from forming
3. Adjusting the schedule to best suit a particular site or growing situation so as not to apply more than the turf can receive

### Topdressing Materials

When the underlying soil of the playing surface is unsatisfactory, a decision will be needed on whether to rebuild the facility or try to slowly change its composition through aggressive coring and topdressing (Figure 9.47). If the soil problem is considered severe, then the superintendent's Greens Committee will probably be disappointed with the coring and topdressing approach and should consider reconstruction.

If a topdressing program is chosen to try to eventually improve the soil makeup, then the next question is what material to use. This is one of the superintendent's most important long-term agronomic decisions. Using undesirable materials can be disastrous and ruin the integrity of many initially well-built facilities.

Many native soil playing surfaces ("pushup greens") are predominantly made up of fine-textured soils high in clay and/or silt. The thought is to introduce a coarser soil texture, most notably sand, to improve water percolation and aeration. A general rule of thumb is to never topdress with a sand that drains more slowly or is finer in texture than the existing rootzone mix. Current trends involve frequently topdressing with 80% or more of pure fine-to-coarse (0.15 to 0.75 mm) sand (industry designated "65" sand). Preferably, a minimum of 60% should be in the medium sand range (0.25 to 0.5 mm), with no more than 20% below 0.25 mm. This size sand is usually coarse enough to provide the desirable effects of changing the soil constituents and yet fine enough to be easily worked into the turf surface—but not so fine as to seal the surface and impede air and water movement. Also, sand particles larger than 1.0 mm in diameter are not likely to work themselves down into the turfgrass canopy. Therefore, they are picked up by mowers, causing equipment wear. A competent soil-testing laboratory should be consulted to test the sands in question before attempting to slowly change the rootzone of the green. When topdressing to fill aerification holes, a slightly coarser sand is often used. A minimum 60% medium sand (0.25 to 0.5 mm) is used (industry designated "35" sand).

When new greens are constructed, stockpiling enough rootzone mix to cover two to five years of routine topdressing is highly recommended to prevent the introduction of dissimilar soil into the green. The only difference in the stockpiled material and regular soil mix used in construction may be the absence of organic matter in this topdressing material. Enough organic matter is usually produced by normal plant growth for future needs.

One of the most commonly observed problems with improper topdressing is the formation of various alternating layers of soil (stratification) that arise from using different topdressing materials over time (Figure 9.48). Differences in textural characteristics between sands and organic matter layers result in poor root growth caused by physical barriers, lack of $O_2$, entrapment of toxic gases, microperched water tables, and dry zones. Only a 1/8 inch (3.2 mm) layer of different soil textures may disrupt normal soil water and gas movement. Once these layers are allowed to form, aggressive vertical mowing and coring are required to correct the problem, short of reconstruction. Aerification holes should extend at least 1 inch (2.5 cm) below the depth of the deepest layer. The use of one of the new deep-tine or deep-drill aerifiers often

**FIGURE 9.47** Deciding which topdressing sand to use is a key, long-term agronomic decision for golf courses.

**FIGURE 9.48** Undesirable layering from using different topdressing materials over time.

is required to reach these greater depths. Shallow spiking or coring above the layering is of little benefit.

Dry topdressing material penetrates core aerification holes better than damp or wet topdressing material. Topdressing should then be matted in by blowing or dragging a steel mat, brush, or piece of carpet over the areas in several directions to evenly distribute the material (Figure 9.49). This area should then be watered immediately to reduce soil drying and to provide further settling of newly topdressed soil. With the introduction of deep core aerifiers, changing the underlying soil characteristics may be expanded. Deep coring once per year, followed by heavy topdressing with a desirable sand, should be practiced to improve poor draining greens. Between these corings, conventional aerification and topdressing should still be performed. Over several years, progress can be made in improving the soil characteristics of the playing area by this technique, assuming that drain lines are installed beneath the surface.

### Topdressing Amounts and Frequency

**Rates.** The frequency and rate of topdressing depend on the objective. For greens, a suggested yearly total amount ranges from 25 to 45 cubic feet sand/1,000 square feet (0.75 to 1.3 $m^3$/100 $m^2$). This, along with aerifying, grooming and vertical mowing, helps prevent organic matter content from exceeding 3% (by weight). For routine topdressing, this is generally applied as light (e.g., 1/2 to 1 cubic foot/1,000 square feet, 0.015 to 0.03 $m^3$/100 $m^2$), frequent (e.g., every one to three weeks) events during periods of active turf growth to minimize sand layer development and to prevent smothering.

Following coring and heavy verticutting, moderate to heavy topdressing is used to help smooth the surface, fill coring holes, and cover exposed roots resulting from these processes. Irregular playing surfaces, refilling aerification holes, or soil profile renovation will require frequent and relatively heavy topdressing with rates ranging from 1/8 to 1/4 inch

**FIGURE 9.49** "Brushing" or incorporating in sand following topdressing.

**TABLE 9.6**
**Approximate Soil Volumes Needed to Topdress Golf Greens to Various Depths**

| | Depth | | Soil volume | | |
|---|---|---|---|---|---|
| **Relative Rate** | **Inches** | **Millimeters** | **Cubic feet/1,000 square feet** | **Cubic yards/5,000 square feet** | **Cubic meters/100 square meters** |
| Light | 1/167 (0.006) | 0.15 | 1/2 | 0.09 | 0.015 |
| | 1/83 (0.012) | 0.30 | 1 | 0.19 | 0.030 |
| | 1/55 (0.018) | 0.45 | 1 1/2 | 0.28 | 0.046 |
| Medium | 1/42 (0.024) | 0.61 | 2 | 0.37 | 0.061 |
| | 1/28 (0.036) | 0.91 | 3 | 0.56 | 0.091 |
| | 1/20 (0.048) | 1.22 | 4 | 0.74 | 0.122 |
| Heavy | 1/10 (0.10) | 2.54 | 8 | 1.5 | 0.244 |
| | 1/8 (0.12) | 3.05 | 10 | 1.9 | 0.30 |
| | 1/6 (0.17) | 4.23 | 14.2 | 2.3 | 0.38 |
| Extremely heavy | 1/4 (0.25) | 6.35 | 21 | 3.9 | 0.64 |
| | 1/2 (0.50) | 12.7 | 42 | 7.7 | 1.28 |

(3.2 to 6.4 mm) depth (~10 to 21 cubic feet soil/1,000 square feet, 0.3 to 0.64 $m^3/100\ m^2$) (Table 9.6), except if the turf has a limited capacity to absorb the material, as grass smothering would result and development of excessive layers is likely to occur. Excessive amounts also cause damage to mower blades and bed knives. Using old reels and bed knives is prudent following heavier topdressing. Greens should also be healthy enough to handle topdressing and brushing. Greens suffering from heat or drought stress or recovering from disease should be healthy and actively growing prior to undergoing stress from a topdressing event.

Specific soil volumes needed for a given desired depth can be determined by the following. It is normally easier to convert the desired depth in inches to feet.

| | | | | |
|---|---|---|---|---|
| Inches | × | 25.4 | = | Millimeters |
| Inches/1,000 square feet | × | 83.3 | = | Cubic feet/1,000 square feet |
| Cubic feet/1,000 square feet | × | 27 | = | Cubic yards/1,000 square feet |
| Cubic feet/1,000 square feet | × | 0.185 | = | Cubic yards/5,000 square feet |
| Cubic feet/1,000 square feet | × | 0.030463 | = | Cubic meters/100 square meter |
| Cubic yards/1,000 square feet | × | 0.822 | = | $m^3/100\ m^2$ |

### Example

To topdress a 6,000-square-foot green with 1/8 inch depth topdressing, the following calculations are needed:

**Step 1:** One-eighth in topdressing depth is converted to feet by dividing by 12 since there are 12 inches/foot.

$$1/8 \text{ inch} \div 12 \text{ inches/foot} = 0.011 \text{ foot depth of topdressing desired}$$

**Step 2:** Next, find the total volume, in cubic feet, of the area to be topdressed.

$$\text{Green size (6,000 square feet)} \times \text{topdressing depth (0.011 foot)} = 66 \text{ cubic feet}$$

**Step 3:** Finally, convert 66 cubic feet to cubic yards by dividing with 27 cubic feet/cubic yards:

$$66 \text{ cubic feet} \times \frac{\text{cubic yards}}{27 \text{ cubic feet}} = 2.4 \text{ cubic yards}$$

of soil volume needed to topdress 6,000 square feet green to a depth of 1/8 inch

**Aerified green.** The topdressing needs of an aerified green are greater than those for greens not aerified due to the filling of aeration holes. It generally is easier to first determine total volume of cored holes for 1 square foot and then convert this to the total surface area of the green being topdressed.

### Example

**Step 1:** Determine the total volume of all core holes for 1 square foot. A 1/2 inch (1.3 cm) diameter core is extracted to a depth of

3 inches (7.6 cm) on 2 inches (5 cm) centers, producing:

$$\frac{144\text{ inches}}{\text{square feet}} \times \frac{\text{cores}}{2 \times 2\text{ inches}} = \frac{36\text{ cores}}{\text{square feet}}$$

The volume of a core is [3.14 × (radius)$^2$] × height = [3.14 × (0.25 inches)$^2$] × 3 inches = 0.59 cubic inches per core.

**Step 2:** Determine the volume of cores per square foot:

$$\frac{0.59\text{ cubic inches}}{\text{core}} \times \frac{36\text{ cores}}{\text{square feet}} = \frac{21.2\text{ cubic inches}}{\text{square feet}}$$

**Step 3:** Determine the total amount of topdressing material by multiplying the amount per square foot by the area of the green. If a green is 6,000 square feet in area, then:

$$\frac{21.2\text{ cubic inches}}{\text{square feet}} \times \frac{6{,}000\text{ square feet}}{\text{green}}$$

$$= \frac{127{,}200\text{ cubic inches topdressing}}{\text{green}}$$

Since 1,728 cubic inches are in each cubic foot, this can be converted by:

$$\frac{127{,}200\text{ cubic inches}}{\text{green}} \times \frac{\text{cubic feet}}{1{,}728\text{ cubic inches}}$$

$$= \frac{73.6\text{ cubic feet topdressing}}{\text{green}}$$

Since 27 cubic feet are in each cubic yard, this can be converted by

$$\frac{73.6\text{ cubic feet topdressing}}{\text{green}} \times \frac{\text{cubic yards}}{27\text{ cubic feet}}$$

$$= \frac{2.73\text{ cubic yards topdressing}}{\text{green}}$$

Therefore, 2.73 cubic yards (2 m$^3$) of topdressing material is needed per 6,000 square feet (170 m$^3$) of green to fill 1/2 inch (13 mm) diameter holes 3 inches deep on 2-inch (5 cm) spacing.

**Frequency.** If the objective of topdressing is to change the characteristics of the underlying soil, then a heavier topdressing program following numerous core removal operations is required. However, following a rigorous coring and topdressing program, adequate modification of underlying soil still may require several years to accomplish.

If thatch control is the main objective of topdressing, the amount and frequency are governed by the rate of thatch accumulation. Thatch layering between 1/4 and 1/2 inch (0.64 and 1.3 cm) on golf greens is desirable. This relatively thin thatch layer cushions (holds) the approaching golf shot better and also provides a certain amount of protection of grass crowns from traffic. However, once this thickness is exceeded, frequent topdressing along with possible coring and verticutting are necessary. A suggested amount of topdressing when thatch is not excessive (1/4 to 1/2 inch, 6.4 to 13 mm) is approximately 1 cubic foot/1,000 square feet (0.03 m$^3$/100 m$^2$). If this relatively light rate does not adequately enhance the decomposition of the thatch layer, then the frequency of application and topdressing rate should be increased.

If the objective of topdressing is just to provide routine smoothing of the playing surface, then light, frequent topdressings are suggested. Matting or brushing the green following topdressing results in the material being dragged into low spots. Surface irregularities of the green are reduced and the area is somewhat leveled. Topdressing with 1/4 to 1/2 a cubic yard/5,000 square feet (0.04 to 0.08 m$^3$/100 m$^2$) of green surface every two to four weeks provides a smoother, truer playing surface. Light topdressing is also performed ~10 days prior to major club tournaments to increase the speed of greens and provide a smoother putting surface. Frequent, light topdressing should also be applied on new greens to cover stolons and to smooth the surface. This should be performed every two to four weeks until complete coverage or the desired smoothness is achieved.

**Fairways.** Topdressing fairways is also becoming more popular among superintendents (Figure 9.50). A fairway topdressing goal is to eventually develop a 4- to 6-inch (10 to 15 cm) homogenous combination of sand and organic matter. This practice helps provide firmer fairways for better ball roll and playability, extends play in spring and fall, promotes drier fairways from increased thatch decomposition, and has less traffic restrictions, especially after rain events. Fairways are typically topdressed monthly during the grass's growing season. Topdressing rates range from 4 to 6 tons/acre (9 to 22 metric tons/ha), and sand is most often applied using spinner-type spreaders. Annual depths of 1/2 to 3/4 inch (13 to 19 mm) are typical goals to eventually reach a depth of 4 to 6 inches (10 to 15 cm). Heavier, less frequent topdressing may lead to layering and golf shots resembling bunker play instead of fairways. Greens' grade quality topdressing sands for fairways are not normally necessary. Often a locally available mason's sand is sufficient if not contaminated by large particles or rocks. For courses that cannot afford "wall-to-wall" fairway topdressing, consider topdressing high-use areas such as landing areas, greens approaches, and collars.

**FIGURE 9.50** Topdressing fairways with sand for thatch decomposition and earlier play following rainfall.

## FERTILIZER PROGRAMS

Turfgrasses require regular fertilization to promote recovery from traffic, provide a pleasing green color, and replace nutrients lost by clipping removal. Furthermore, many courses are located in areas with extended growing seasons, high annual rainfall, and predominantly sandy soils that increase nutrient use and the need for a well-planned fertilization program.

For these and other reasons, no one single fertilization schedule or program will sufficiently meet the needs of every golf course or even every area on a golf course. In addition, excessive quality expectations and playing standards are often demanded by professional players. Club members, in turn, place much undue pressure on their superintendents to duplicate conditions viewed on television. This often results in lush, unhealthy plant growth that can drive up maintenance costs and potentially waste natural resources.

Questions to consider when developing a turfgrass fertility program include:

1. What are the aesthetic quality expectations of the area being maintained?
2. What are the use or playing requirements and recovery needs of this area?
3. What is the current nutritional status of the soil and plant?
4. What fertilizer source (carriers) and rates will best meet the aforementioned goals?
5. What fertilizer rate, application timing, and frequency should be used?
6. What type of fertilizer equipment or delivery method will be used to provide these results?

Steps when developing a fertility program include:

1. Determine annual nutrient needs depending on turf species, desired quality, budgets, length of growing season, clipping removal, soil types, irrigations/rainfall amounts and frequency, etc.
2. Schedule approximate application dates and rates.
3. Select appropriate fertilizer carrier(s).
4. Determine other nutrient and pH needs in addition to N by soil tests.
5. Consider other aspects such as price, availability, quality, ease of application, fuel costs, labor, etc.
6. Plan the program by developing spreadsheets indicating application dates, rates, carriers, etc.
7. Adjust the program as needed due to weather, pest outbreaks, traffic, etc.

### TIMING

Timing of nutrient applications is based predominantly on the growth cycle of the grass. Greatest nutrient needs are when shoots and roots are actively developing. Table 9.7 lists growth temperatures for cool- and warm-season turfgrasses. If temperatures are outside the growth range of the grass, fertilizer applications will be less efficiently utilized.

With cool-season grasses, the majority of the annual N fertilizer (~60% to 75%) should be applied in fall. This application

**TABLE 9.7**
**Air Temperatures and Soil Temperatures at 4 Inches (10 cm) Affecting Turfgrass Shoot and Root Growth**

| | Shoot Growth | | Root Growth | |
|---|---|---|---|---|
| **Turfgrass** | **Minimum** | **Optimum** | **Minimum** | **Optimum** |
| | °F (°C) | | | |
| Warm-season grasses | 55 (13) | 80 to 95 (27 to 35) | 50 to 60 (10 to 16) | 75 to 85 (24 to 29) |
| Cool-season grasses | 40 (4) | 60 to 75 (16 to 24) | 33 (1) | 50 to 65 (10 to 18) |

begins in late summer/early fall as temperatures begin to cool, encouraging renewed growth. A final fertilization application, high in K, just after the turf has stopped growing in the late fall and prior to freezes, improves future turf quality. In early spring, light applications of fertilizer should commence and be stopped or severely curtailed prior to the elevated temperatures of summer.

Fertilizer applications of warm-season grasses should begin during spring "green-up" after the last killing frost. This generally continues through the summer and early fall, with the last significant application timed ~30 days before the first historical fall killing frost.

Use of the turf site also influences N fertilization. Recovery from foot and vehicular traffic, soil compaction, and maintenance practices may or may not coincide with the natural growth cycles of the turfgrasses. Recovery from these activities may necessitate nutrient use outside the optimum growth temperatures of the grass. However, limits exist on which grasses may be forced to grow from N fertilization during periods of naturally occurring slowed growth. For example, excessive N fertilization of a cool-season grass during summer heat stress may weaken or even permanently damage the grass. Conversely, fertilizing a warm-season turfgrass aggressively in early spring or late fall may weaken or reduce the cold hardiness of these species.

### Nitrogen Rates

A general yearly range of N needs for bermudagrass golf greens is from 6 to 18 pounds N/1,000 square feet (29 to 88 g N/m$^2$). Seashore paspalum, zoysiagrass, bentgrass, and bentgrass/*Poa annua* greens annually require 3 to 8 pounds N/1,000 square feet (15 to 39 g N/m$^2$). Courses with sufficient resources, low cation exchange capacity (CEC) soils, longer growing seasons, excessive traffic, pest problems, and elevated player expectations would use the higher rate range. Those courses interested in maintaining a less intensive playing surface, or those with limited labor and financial resources, should use N rates in the lower range or apply very small doses frequently to ensure sufficient nutrient availability. Exceptions to these ranges may occur depending on various conditions or needs. For example, courses recovering from excessive traffic, pest or low-temperature stresses, or that are establishing new greens may require higher (~25% more) N than listed until their grass is sufficiently dense and mature.

Fairways and roughs are not normally managed as intensively as greens (Table 9.8). These areas are fertilized with 2 to 4 pounds N/1,000 square feet (1 to 2 kg N/100 m$^2$) annually with 3 to 6 pounds N/1,000 square feet (1.5 to 3.0 kg n/100 m$^2$) for longer growing season fairways. This translates to 1/2 to 1 pound N/1,000 square feet (1.12 kg N/ha) per actively growing month. This is often supplemented by spoon feeding with fertigation or broadcast soluble N spray applications.

### Frequency

The percentages of N fertilizer applied to warm- and cool-season turfgrasses during the year are listed in Table 9.9. In general, to maintain optimum color and density during periods of active growth, highly maintained bermudagrass

**TABLE 9.8**
**Typical Nitrogen Application Rates, Frequency, and Yearly Total Amounts for Golf Course Fairways and Roughs**

| | Fairways | | | Roughs | | |
|---|---|---|---|---|---|---|
| Grass | Single Rate (pounds N/acre)[a] | Frequency (weeks)[b] | Yearly Total (pounds N/acre) | Single Rate (pounds N/acre) | Frequency (weeks) | Yearly Total (pounds N/acre) |
| Annual bluegrass | 25 to 45 | 4 to 12 | 90 to 180 | 25 to 45 | 4 to 12 | 45 to 90 |
| Bentgrass | 25 to 45 | 4 to 12 | 90 to 180 | 25 to 45 | 4 to 12 | 45 to 90 |
| Bermudagrass | 25 to 45 | 4 to 16 | 45 to 180 | 45 | 8 to 16 | 0 to 90 |
| Fine fescue | 25 to 45 | 4 to 12 | 45 to 135 | 25 to 45 | 8 to 16 | 45 to 90 |
| Kentucky bluegrass | 25 to 45 | 4 to 12 | 60 to 180 | 25 to 45 | 4 to 12 | 45 to 90 |
| Ryegrass | 25 to 45 | 4 to 12 | 45 to 135 | 25 to 45 | 4 to 12 | 45 to 90 |
| Seashore paspalum | 25 to 45 | 4 to 12 | 90 to 180 | 25 to 45 | 4 to 12 | 45 to 90 |
| Zoysiagrass | 45 | 4 to 16 | 45 to 135 | 45 | 8 to 16 | 0 to 90 |

[a] Rates listed are typical for single applications. Single application rates, frequency, and yearly total N rates vary considerably between individual golf facilities, geographical regions, and desired results. Lower rates are acceptable for most golf facilities, especially for those turf areas not irrigated. Slow-release or controlled-release fertilizer sources should be considered with higher rates. Soluble sources should be used at lower rates and applied more frequently. Multiply pounds/acre by 1.12 to obtain kilograms/hectare.

[b] Frequency represents fertilizing only during periods of active growth for each respective grass. The highest application rates should be reserved for optimum growth seasons of each grass. Typically, this is late summer through early spring months for cool-season grasses and late spring through late summer/early fall for warm-season grasses. Fertilizer applications should be minimized or possibly eliminated during periods of natural slowed growth or where soils freeze.

**TABLE 9.9**
**Percentages of Nitrogen Fertilizer Applied to Turfgrass during the Year**

| | Cool-Season Grasses | | Warm-Season Grasses | |
|---|---|---|---|---|
| Season | Greens/Tees | Other | Overseeded | Nonoverseeded |
| Fall | 40 to 75 | 45 to 75 | 15 | 10 to 15 |
| Winter | 0 to 30[a] | 0 to 45 | 15 | 0 |
| Spring | 10 to 20 | 10 to 20 | 25 | 35 |
| Summer | 10 to 20 | 0 to 10 | 45 | 50 to 55 |

[a] Nitrogen use stops during winter in areas where soils routinely freeze.

golf greens need approximately 1/2 pound soluble N/1,000 square feet (2.5 g N/m$^2$) every 7 to 14 days. For those courses without these resources and for those with lower quality expectations, adequate bermudagrass can be maintained with 1/2 pound N/1,000 square feet (2.5 g N/m$^2$) applied every 14 to 21 days during periods of active growth. On intensively maintained courses, higher rates (e.g., 1 pound N/1,000 square feet every 7 to 14 days, 5 g N/m$^2$) may be necessary to encourage quicker turf recovery during times of heavy play, after aerification, or following spring transition from overseeding.

## Nitrogen Sources

The source of N used to fertilize golf courses affects the amount safely applied at one time. Usually, a combination of quick-release (water-soluble) and slow-release (water-insoluble) N sources is recommended to provide uniform grass growth. During cold temperatures, water-soluble sources or isobutylidenediurea (IBDU) provide the fastest turf response as they do not rely on soil microorganisms for N conversion and release. IBDU's release rate, however, is dramatically lower in high-pH soils like calcareous sands, and the desired response might not be achieved in a timely manner.

Another consideration involving N sources includes economics. Slow-release and natural organic sources cost considerably more compared to soluble sources. However, slow-release sources need to be applied less often and a labor savings may be achieved. Additionally, many water-soluble sources pose a salinity hazard, and may provide a desirable or undesirable soil acidifying effect. This is particularly true for the ammonium-containing fertilizers like ammonium sulfate and ammonium phosphate. Frequently applying small amounts of N (e.g., 1/8 to 1/4 pound N/1,000 square feet, 0.6 to 1.2 g N/m$^2$) as a liquid solution is preferred since this produces a higher quality turf, avoids growth flushes, and minimizes leaching and salt-accumulation potentials. In most cases a high-quality turfgrass can be maintained for a 90-day period without flushes of growth or major variations in color when slow-release N sources are used.

**Liquid fertilization.** Spoon (also called foliar) feeding during summer months with liquid fertilizer programs is often used on putting greens and tees with high sand content to help maintain desirable turf color, minimize growth surges to enhance ball roll, minimize thatch/mat accumulation, reduce undesirable stress responses, discourage *Poa annua*, and provide a continuous supply of elements such as N or K that are often easily leached from sand-based rootzones. Nitrogen rates for foliar feeding programs typically range from 0.05 to 1/4 pound N/1,000 square feet (1/4 to 1.2 g/m$^2$) on 7- to 14-day intervals. Phosphorus (as $P_2O_5$) is applied at about one-third the rate of N while K rates will be equal to half the rate of N. Other nutrients such as Ca, Mg, or Fe also may be added by foliar feeding to help regulate and maintain desirable turf color. Ideally, these nutrients should be supplied in 2 to 5 gallons of water/1,000 square feet (8 to 20 L/100 m$^2$) to ensure application uniformity and minimize turf burn.

Ammonium sulfate sources should be considered during periods of slowed or stressful turf growth. Plants eventually convert all N sources applied to ammonium $(NH_4^+)$, requiring extra energy to do so, often during times when grass energy reserves are at a minimum.

## Other Nutrients

**Potassium.** Potassium (K) is often referred to as the "health" element and, without a relatively available supply, turfgrasses will be more susceptible to environmental and pest stresses. Root growth is also related to K availability. Excessive K, however, may increase soil salinity levels, especially if the leaching capability is limited. Potassium also competes for occupancy on the soil cation exchange sites and may displace other essential elements like Ca or Mg. The following are guidelines for K fertilization (Carrow et al., 2001):

1. For native or unamended soils, use soil testing to determine available soil K levels and base K fertilization needs. On intensively managed turf, a high K soil test level should be maintained. On moderately to lower maintained turf, a medium K soil test level is normally sufficient.
2. On heavily leached sands:
   a. Apply a 1:1.5 ratio of N to potash ($K_2O$) when annual N rates are 1 to 3 pounds N/1,000 square feet (4.9 to 15 g/m$^2$).
   b. When annual N rates are 3 to 6 pounds /1,000 square feet (15 to 29 g/m$^2$), use a 1:1 N to $K_2O$ ratio.
   c. When annual N rates are more than 6 pounds / 1,000 square feet (29 g/m2), use a 1:0.75 or 1:0.5 N to K2O ratio to avoid the potential of salt buildup.
   d. In summer, apply one-fourth to one-half a pound $K_2O$/1,000 square feet (1.2 to 2.5 g/m$^2$) every two to six weeks.

These guidelines, in part, provide sufficient K for various N levels, but also help avoid salt injury exhibited as plant burn, wilting, and scalping. Using potassium sulfate

($K_2SO_4$) instead of potassium chloride (KCl) during summer stress further helps reduce fertilizer salt stress on cool-season grasses. In high-sodium environments, a 1:1 N to $K_2O$ ratio is suggested as excessive Na ions may displace some $K^+$ ions on low CEC sites, resulting in K loss by leaching. Excessive K applications can lead to deficiencies in Ca and Mg. In sandy soils, "low" or "very low" K levels are common. K soil test levels cannot be substantially increased unless the soil contains considerable organic matter. Tissue analysis provides a better indicator of K deficiency in sandy soils, with a value of 1% or less indicating deficiency.

**Phosphorus.** Soil phosphorus (P) levels tend not to fluctuate as readily as N or K. Soil-test results should be used to determine the amount needed for a particular turf area. Usually 4 pounds or less P/1,000 square feet (20 g P/m$^2$) are needed yearly. Phosphorus is generally not very water soluble; therefore, if needed, its efficiency is increased if applications follow core cultivation. This allows the material to be placed more directly in the rootzone where roots can access the nutrient. Over time, P levels may actually increase near the soil surface, especially where clippings are returned during mowing.

During establishment, P needs are often greatest due to the shallow and low-density root systems associated with young seedlings or recently installed sod. Phosphorus-level needs based on soil tests should be made prior to planting followed by tilling to the anticipated rooting depth of the turfgrass is an efficient means of utilizing this nutrient.

Phosphorus deficiency symptoms appear as a purple discoloration of the youngest leaf blades. However, this can be easily confused with purpling during cool periods when anthocyanins are exposed. Applying a small area or strip with P fertilizer is an easy way to determine if insufficient nutrition is the cause of purpling.

**Micronutrients.** Regular soil and tissue testing is the best preventative approach to solving many of the micronutrient deficiency problems. Iron and manganese are the most common micronutrient deficiencies turf managers experience. However, if excessive or indiscriminate amounts of micronutrients are applied or soil pH is excessively low, toxicity to the plant can occur.

Golf greens, due to their high sand content, typically have a low CEC (e.g., ≤5.0 meq/100 g). Exchange capacity, along with soil pH, are key components of a soil test report important to understanding micronutrient availability. By default, soils low in pH are saturated with hydrogen ions and are low in Ca, Mg, K, and other cations necessary for healthy plant growth. Thus, low-CEC and low-pH sand soils often experience Mg deficiency symptoms, especially during initial turfgrass establishment and where the rootzone has low organic matter. Also, sites receiving high K and/or Ca can induce Mg deficiency. Conversely, soils low in CEC but high in soil pH (>7.0) may experience an Fe or other micronutrient deficiency. This is amplified on sites receiving high bicarbonate-containing irrigation water and cultivars naturally sensitive to Fe chlorosis. Tissue testing and experimenting with small areas of a green by applying the suspected deficient nutrient are the best methods to determine the cause of the problem.

### Tees

Tees, like greens, should be fertilized sufficiently to sustain vigorous recuperative growth, but not to the point where wear tolerance is sacrificed. Often this is associated with succulent leaf growth or increased disease incidences. Tees, in general, are maintained almost as intensively as golf greens. This is especially true for sand-based tees and heavily used par-3 tees receiving excessive traffic and significant divoting. For par-3 tees, especially undersized, small tees, the fertilization program should range between three-fourths and equal to that used for greens. For most par-4 and par-5 tees, the fertilization program can be reduced to approximately half that used for golf greens. Potassium applications should be approximately half of N applications except where clippings are removed or when sand-based tees are constructed. In such cases, K application rates should equal those of N.

## WINTER COLOR

Warm-season grasses' growth stops when temperatures drop below 60°F (15.5°C), and discoloration (browning) can be expected if temperatures drop below 50°F (10°C) for an extended period. Warm-season grasses, therefore, are often overseeded with cool-season grasses or artificially colored during the fall and winter to provide a green playing surface with desirable color and better traffic tolerance (Figure 9.51). In addition, golf course fairways sometimes are overseeded to clearly mark suggested landing areas for golfers. Overseeding is an important economic aspect of golf because many resort courses enjoy their heaviest play from tourists during the fall, winter, and spring months, and overseeding helps attract golfers. Overseeding also improves winter and spring play conditions.

### Overseeding Myths

Although there are reasons or benefits associated with overseeding, some myths and drawbacks about it exist as well. One myth is that overseeding improves course drainage. During winter, evapotranspiration (ET) rates are low and rainfall is usually adequate. Therefore, overseeding does not normally significantly help "dry out" soil. Others contend that overseeding significantly protects the bermudagrass from winter-kill and will not have any negative impact on spring bermudagrass transition. Overseeding grasses compete very aggressively with bermudagrass until air temperature consistently reaches the high 80s°F (27s°C). This competition typically delays total green-up and fill-in of the bermudagrass and, if the overseeding quickly dies, the bermudagrass stand generally is thin (Figure 9.52).

If an area is overseeded for only one year, it will normally take at least two additional years for all of the overseeding seed to totally germinate. Therefore, courses will have recurring sporadic overseeding for up to three years after the last

**FIGURE 9.51** Overseeding during fall and winter to provide desirable color and better traffic tolerance.

**FIGURE 9.52** "Spring transition" issues when overseeded grass dies exposing the permanent grass.

overseeding year. Escaped clumps (often referred to as "renegade" ryegrass) of overseeding from the intended seeded areas are very unsightly, difficult to mow, difficult to eradicate, and often live into midsummer.

Overseeding is also believed to reduce weed population due to the competition. This is rarely the case and, in fact, overseeding usually slowly increases weed pressure. This is especially true for annual bluegrass. This seed is then easily tracked onto green and tee surfaces, thereby increasing its occurrence in these areas. The overseeding process of light, frequent watering and fertilizer application plus soil surface compaction also provides ideal growing conditions for annual bluegrass.

A good uniform irrigation system is also required for desirable overseeding. If not available, sporadic overseeding stands usually result. Overseeding is a disruptive process where golf courses typically have to close to prepare the seedbed, plant the seed, and then provide continuous watering to establish the stand. Finally, overseeding can be expensive as additional costs are incurred for seed; labor; extra watering needs; specialized equipment needs such as drop spreaders, vertical mower reels, and fairway vacuums; mowing expenses for fuel, labor, and equipment wear-and-tear; extra fertilizer and herbicide costs; use of plant growth regulators (PGRs); increased *Poa annua* infestations; and possible downtime for the course to actually seed.

Optimum overseeding performance is a sequential procedure requiring proper seedbed preparation and timing. Proper winter management and spring turf species transition also ensure optimum playing conditions during tournament play periods. Luck, mostly in terms of favorable weather, is as important as sound agronomic procedures.

### Timing

To maintain consistent turf quality, gradual transitions from the permanent grass to overseeded grass species in the fall and back to the permanent grass in spring are necessary.

Seeding too early can result in excessive permanent turf competition and the increased likelihood of turf failure due to diseases such as *Pythium* blight. Seeding too late in fall may result in unacceptable, weak, delayed, or reduced seed germination because of low temperatures.

There are several available means for determining when to overseed. A general indicator of optimum overseeding time is when late summer/early fall night temperatures are consistently around the 50°F (10°C) range. Other timing indicators include overseeding when soil temperatures at a 4-inch (10 cm) depth are in the mid-70s°F (14s°C), or the average mid-day air temperatures remain in the low 70s°F. Cool-season grass seed germination is favored by temperatures between 50°F and 70°F. Overseeding also should be timed at least 20 to 30 days before the first expected killing frost. This timing minimizes bermudagrass competition, and optimizes seed germination and establishment of the overseeded grasses. It also will reduce seedling diseases. Figure 9.53 indicates typical overseeding timings for various regions of the United States.

Other factors, such as tournament play and golf course usage, may dictate that seeding dates be altered. For example, light frosts can occur in subtropical areas such as southern Florida and California, but deep freezes are rare. Bermudagrass normally does not completely go dormant in these areas. Soil temperatures in these subtropical areas remain in the 80°F (27°C) range through November; thus, temperature-based indicators are not always applicable. Overseeding timing then becomes a management decision based primarily on the timing of play intensity, customer demands, availability of seed, and desired level of aesthetics. In general, overseeding is conducted prior to increased winter use in these areas, which typically occurs from late October through early December.

## Seedbed Preparation

Seedbed preparation planning should take place several months before the actual overseeding operation to ensure a seed source and a seed type. Purchasing of seed and its timely delivery also will allow flexibility for overseeding scheduling if done in advance. A healthy bermudagrass base is needed to withstand harsh cultural practices and the turf competition associated with overseeding. This can only be accomplished by careful management of bermudagrass throughout the year.

**Greens and tees.** Proper seedbed preparation ensures seedling roots will be in contact with the soil, which reduces susceptibility to drought and temperature stress (Figure 9.54).

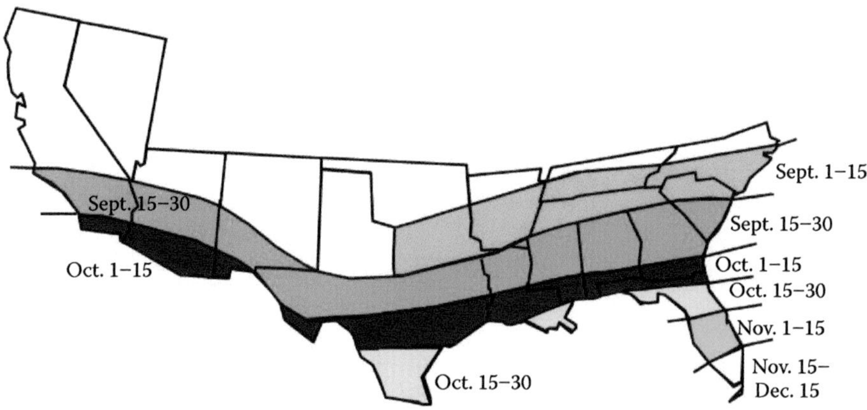


**FIGURE 9.53** Typical overseeding timings for the southern United States.

**FIGURE 9.54** Extensive verticutting prior to overseeding to ensure desirable seed-to-soil contact.

Thatch greater than half an inch (1.3 cm) associated with a bermudagrass or zoysiagrass base prevents this good seed-to-soil contact and therefore should be reduced before overseeding. Proper seedbed preparation to minimize thatch and allow optimal germination and growth includes the following procedures:

Three to six weeks prior to overseeding:

1. Nitrogen fertilization should be reduced or completely stopped three to four weeks prior to overseeding to minimize competitive permanent grass growth. Excessive growth at the time of overseeding will provide competition for the germinating seed and also may predispose the grass to winter injury.
2. Cultivate the soil by coring four to six weeks prior to overseeding to alleviate soil compaction and to open the turf. Allow the cores to dry and pulverize them by verticutting, power raking, or dragging. Coring is performed in advance of the actual overseeding date to allow coring holes to heal over, thereby preventing a speckled growth pattern of winter grass. The overseeded grasses typically grow faster with a darker green color in and around core holes than between them.
3. Following coring, verticut in several directions to reduce thatch and to open the soil surface to allow better soil-to-seed contact (Figure 9.55). Verticutting is often carried out in conjunction with mower scalping. Verticut debris should be removed by raking, blowing, or mowing with a catcher basket attachment. The depth of verticutting depends primarily on the depth of thatch. Thicker thatch layering requires deeper verticutting. Other parameters, such as algae formation, however, may dictate how severe this verticutting is. If algae is a problem, light, frequent grooming may substitute for verticutting greens.
4. Topdress with approximately half a cubic yard/1,000 square feet (0.015 $m^3$/100 $m^2$) following removal of the verticut debris in order to provide a smooth seedbed and to minimize the effects of the remaining thatch. Most consistent results occur when the seedbed is topdressed before and immediately after overseeding to provide desirable soil contact. Use a desirable topdressing mix (e.g., contains <10% of particles smaller than 0.1 mm and none above 1 mm in diameter).
5. Apply phosphate ($P_2O_5$) and potash ($K_2O$) at a suggested rate of 10 pounds of 0-9-27 (or equivalent) per 1,000 square feet (49 g/$m^2$), or at the rate suggested by a soil fertility test. Phosphorus and K will enhance overseeding rooting without promoting excessive bermudagrass or zoysiagrass top growth.

Ten to fourteen days prior to overseeding:

1. Reduce the mowing height and verticut lightly in two directions (just touching the soil surface) to open the turf and allow seed to fall into the turf canopy. This practice provides good seed-to-soil contact, prevents wind and water from carrying seed away, and provides sufficient time for verticut slits to recuperate before overseeding.
2. Use a power sweeper, a mower with basket attachments, or a blower to remove remaining debris, and then thoroughly irrigate. On sloped areas, if possible, discontinue mowing two or three days prior to overseeding, or raise the mowing height 1/4 or 5/16 inch (6.4 to 7.9 mm) one week prior to overseeding to provide an upright framework of grass to prevent

**FIGURE 9.55** Properly exposed seedbed prior to winter overseeding.

seed washing. Spiking or slicing in several directions just prior to seeding also enhances seed-to-soil contact. Care, however, might be taken to evenly distribute seed without concentrating it in spiked holes to prevent a speckled appearance (Figure 9.56).

Overseeding with small-seeded species such as fine fescue or *Poa trivialis* often does not require as extensive seedbed preparation as that for larger seeded grasses like ryegrass. However, some seedbed preparation is needed for all species, with the two steps just listed being the minimum requirements for all overseeding. A greater thatch layer and using larger seeded species require more seedbed preparation.

### Fairways

Fairways require an aesthetically pleasing appearance but are not prepared as intensely for overseeding as are greens and tees (Figure 9.57). However, preparation is needed to provide for uniform germination and overseeded grass establishment. Brushing with an industrial street brush throughout summer stands leaves up and provides a true cut. Continuous brushing does not leave grooves like verticutting can. Mow the base grass very close (e.g., 3/8 to 1/2 inch, 9.5 to 13 mm) just prior (two to three days) to overseeding and pick up clippings and debris. Brush again, if necessary. This also will open up the turf and will allow seed to reach the soil surface, provided thatch is not a problem. If thatch is a problem, coring should be performed several weeks in advance of overseeding. Cores can be broken up by dragging a piece of chain-link fence or a flexible metal mat over the surface after it has dried. Soil from this coring also provides a good germination media for the overseeded grass. Excessive fairway thatch layering, however, may require verticutting in conjunction to close mowing and coring prior to overseeding. Perform verticutting in several directions to minimize lines of overseeding from verticutting grooves (Figure 9.58).

**FIGURE 9.56** Speckled appearance from overseeding seed accumulating in spiking or aerification holes.

**FIGURE 9.57** Overseeded fairway for winter color.

**FIGURE 9.58** Insufficient removal of thatch prior to overseeding, resulting in sporadic germination.

### Additional Considerations

With more complaints from dust generated from aggressive verticutting and scalping and fears of permanent damage to radical practices on the turfgrass, as well as increasing disposal fees from the organic debris generated by these operations, golf courses have searched for alternatives. These range from using a variety of herbicides and PGRs to reduce turf growth, to much less aggressive vertical mowing and scalping as well as completely getting away from these practices by fairway topdressing only. Two schools of thought exist on verticutting depth. The traditional thought is to verticut deep enough to reduce surface thatch, making a seedbed that would allow maximum seed-to-soil contact. Another school of thought is to only verticut with blades no deeper than 1/4 inch (6.4 mm) so as not to cut stolons and injure the plant crowns. Both methods can work but an herbicide such as triclopyr, diquat, or pelargonic acid or a PGR such as trinexapc-ethyl is needed to slow bermudagrass growth. These chemicals are typically applied three to five days prior to seeding. Follow-up light scalping will also probably be needed with this method to remove excessive shoots.

With topdressing only, it is applied approximately one-half inch (6.4 mm) deep or about 40 tons of sand per acre (90 metric tons/ha) followed by smoothing with a drag mat and seeding ryegrass. Though more expensive upfront than other overseeding options, the sand topdressing method eliminates the aggressive mechanical practices, saving fuel, labor, equipment wear, and debris disposal fees. If used long enough, sand topdressing also helps provide firmer playing conditions and additional ball roll distances.

### Annual Bluegrass Weed Control

Annual bluegrass (*Poa annua* L.) is an undesirable bunch-type winter annual weed because of its clumping growth habit, its early spring die-back, and its prolific seedhead production (Figure 9.59). Control of annual bluegrass with traditional herbicides is difficult due to the potential injury of germinating overseeded grasses.

**FIGURE 9.59** Annual bluegrass (*Poa annua*) infestation in overseeded grass.

**Golf greens.** Several herbicide options are available for preemergence control of annual bluegrass in overseeded bermudagrass golf greens. Currently, pronamide, dithiopyr, foramsulfuron, trifloxysulfuron, and rimsulfuron are used prior in overseeding for *Poa annua* control. Refer to the weed control chapter for detailed information on these and other products for preemergence and postemergence *Poa* control.

Charcoal has been used in conjunction with pronamide when pronamide was applied less than 60 days prior to overseeding to negate its effect on the germination of overseeded grasses. Activated charcoal should be applied at 2 to 5 pounds/1,000 square feet (9.8 to 24 g/m$^2$). At least 14 days should be allowed between herbicide and charcoal applications, while overseeding should occur no sooner than seven days following charcoal application. There is an inherent chance the charcoal application may not totally work due to the difficulty of application. Charcoal also might negate the herbicidal effect of the pronamide on annual bluegrass germination.

**Fairways.** In addition to those products listed for annual bluegrass control in overseeded golf greens, several others can be used in overseeded fairways. Ethofumesate (Prograss 1.5 L) is used for preemergence and/or postemergence control of annual bluegrass in dormant bermudagrass tees, fairways, and roughs overseeded only with perennial ryegrass. In addition, bensulide, prodiamine, oxadiazon, benefin, and possibly others can be used for selective *Poa annua* control in overseeded grasses.

## Overseeding Grass Selection

The primary grasses used for overseeding are perennial ryegrass, annual ryegrass, intermediate ryegrass, roughstalk bluegrass or *Poa trivialis*, creeping bentgrass, and fine (chewings, creeping red, or hard) fescue. Improved cultivars of perennial ryegrass seeded alone, or in mixtures with bentgrass, roughstalk bluegrass, or fine fescue, are the grasses and mixtures most widely used on golf greens and tees. Each grass has advantages and disadvantages. Table 9.10 illustrates and describes each of these.

**Perennial ryegrass.** Perennial ryegrass has traditionally been the preferred grass used for most overseeding purposes. It is noted for its fast germination (typically five to seven days), finer leaf texture, darker green color, and seedling vigor, as well as for having better disease and traffic resistance than annual ryegrass (Figure 9.60). Perennial ryegrass also provides a highly desirable striped appearance when mowed in alternating directions. Many improved turf-type varieties recently have been released that provide fine putting surfaces if seeded at high rates. However, many new varieties have improved heat tolerance; thus, they may be more competitive during spring transition than the traditionally used annual (Italian) ryegrass. Poor (slow) putting surfaces also occur in fall until mowing heights are reduced three to five weeks following germination. If seeded alone, a blend of at least two (preferably three or four) improved turf-type cultivars should be used to help ensure good performance over a wide range of conditions. Mixtures of perennial ryegrass with creeping bentgrass, fine fescue, and *Poa trivialis* are often used on golf greens. Intermediate ryegrass cultivars (crosses between perennial and annual ryegrasses) are available to lessen transition issues including Midpoint and Transaction.

**Annual (Italian) ryegrass.** Annual ryegrass (*Lolium multiflorum*) has lost some of its importance as an overseeding grass in recent years since it produces a coarser, more open turf than many newer turf-type perennial ryegrasses and is extremely susceptible to *Pythium* diseases. Annual ryegrass also has poor heat and cold tolerance and often experiences early spring death that may result in poor seasonal transition. Like annual bluegrass, annual ryegrass quickly dies when a few warm days occur in early spring. This quick death may result in thin spots where the bermudagrass has not had time to fully green-up and cover any weak areas. However, annual ryegrass germinates quickly and is acceptable on fairways and other general use areas where color and appearance are not of greatest concern, or when budget constraints are important. A few improved turf-type cultivars of annual ryegrass include Attitude and Quickdraw.

***Poa trivialis* (roughstalk bluegrass).** *Poa trivialis* (often called roughstalk bluegrass, rough bluegrass, or "*Poa triv*") is native to wet, shaded environments of northern Europe. It has a fine texture, good density, and it is more cold tolerant than ryegrass. This is due, in part, to its high seed count, at approximately eight to one, compared to perennial ryegrass. This gives it a greater density in the number of seeds per square area than ryegrass. It is easier to establish in fall compared to ryegrass since radical increases in mowing heights are not necessary and golfers like it due to the minimum seedbed preparation needed for establishment. Roughstalk bluegrass also tolerates poorly drained soils and has good shade tolerance. These characteristics make it desirable as the sole component or part of a blend with perennial ryegrass or bentgrass (Figure 9.60). However, *Poa trivialis* is susceptible to diseases, especially dollar spot. It has poor wear tolerance and a natural yellow-green color. Hot spots or dry areas may develop and require daily hand-syringing. *Poa trivialis* also tends to die out earlier in the spring than most other overseeding grasses due to its low heat tolerance. This low heat tolerance normally is desirable since the spring transition is quicker and smoother. However, extended warm weather into the fall and early warm weather in the spring may prematurely weaken it. Numerous varieties of *Poa trivialis* are currently being developed and released. Refer to Chapter 1 for a listing.

## Blends and Mixtures

Blends (two or more cultivars of the same grass species) of perennial ryegrass, or mixtures (two or more cultivars of different grass species) of bentgrass and *Poa trivialis* (roughstalk bluegrass), or mixtures of perennial ryegrass and fine fescue are commonly utilized as a measure of protection against disease and environmental stresses. They are also used to enhance fine texture and smoothness. Mixtures typically consist of 70% to 90% perennial ryegrass, 20% to 30% fine fescue, and 20% to 30% *Poa trivialis*. If spring transition is a problem, reducing or eliminating the ryegrass component may be considered. An 80% fine fescue and a 20% roughstalk

**TABLE 9.10**
**Characteristics of Selected Grasses Used for Winter Overseeding**

| Grass | Green Color | Texture | Density | Establishment Rate (weeks) | Winter Performance | Spring Performance | Spring Transition | Shade Tolerance | Putting Quality | Competitiveness with Poa | Disease Resistance |
|---|---|---|---|---|---|---|---|---|---|---|---|
| Perennial ryegrass (turf-type) | Dark | Med.–fine (2–4 mm) | Good | Fast (3–4) | Good–excellent | Good–excellent | Slow | Fair–good | Fair–excellent | Very | Fair–good |
| Annual ryegrass | Light | Coarse (3–5 mm) | Poor | Very fast (2–3) | Fair–good | Poor | Med. | Fair | Poor | Very | Poor |
| Intermediate ryegrass | Intermediate | Coarse–med. (3–4 mm) | Fair | Fast (2–4) | Fair | Fair | Med. | Fair | Poor–fair | Very | Fair |
| Fine fescue | Light–dark | Very fine (1–2 mm) | Excellent | Slow (4–6) | Fair | Good | Fast | Very good | Good | Not very | Fair |
| Roughstalk bluegrass | Light–intermediate | Fine (1–4 mm) | Excellent | Slow (4–6) | Fair | Excellent | Fast | Good | Excellent | Not very | Poor–fair |
| Creeping bentgrass | Intermediate | Very fine (1–2 mm) | Excellent | Very slow (6–8) | Fair | Fair | Slow | Fair | Excellent | Not very | Fair |

**FIGURE 9.60** Overseeded roughstalk bluegrass (*Poa trivialis*) green and perennial ryegrass fairway.

bluegrass (by seed weight) mixture will provide a quicker, smoother spring transition. Roughstalk bluegrass also is typically mixed with perennial ryegrass and bentgrass to enhance its performance and to provide a more desirable spring transition. A typical seed-weight formula is 85% perennial ryegrass plus 15% bluegrass, or 60% bentgrass plus 40% bluegrass. Since no one grass provides all optimum characteristics necessary for overseeding, turf managers often choose different mixtures or blends to suit their needs.

**Seed quality.** Only certified (blue tag) seed should be used when selecting species and cultivars for overseeding. Maximum percent purity and acceptable germination should be specified at the time of purchase. Seed also should be certified free of hard-to-control weeds such as annual bluegrass.

Fungicide-treated seed should be requested to reduce potential seedling loss due to *Pythium* and *Rhizoctonia* diseases. Metalaxyl (Apron) or etridiazole (Koban) are fungicides commonly used for this purpose. Bentgrass used for overseeding is normally not treated with a fungicide. Extra seed (approximately 10%) also should be purchased for repairing small areas that may be lost from pests, weather, or traffic. All seed should be protected from rodents and stored in a cool, dry place to retain seed viability.

**Seeding rate.** The grasses being used and the desired turf density will dictate seeding rate. Small seeded grasses, such as bentgrass and roughstalk bluegrass, can be seeded at lower rates when compared to large seeded grasses such as annual and perennial ryegrass. They will provide a comparable number of plants per given area. Areas receiving heavy traffic require a higher seeding rate than those areas where winter color is the main objective. Table 9.11 lists suggested overseeding rates. If either of these high- or low-seeding

**TABLE 9.11**
**Overseeding Rates for Various Grasses and Mixtures (by Seed Weight) Used on Golf Courses[a]**

| Grass | Greens | Tees | Collar/Aprons | Fairways |
|---|---|---|---|---|
| | Pounds/1,000 square foot | | | Pounds/Acre |
| Perennial ryegrass | 25 to 40 | 15 to 20 | 10 to 20 | 250 to 850 |
| Italian (annual) ryegrass | 35 to 50 | 15 to 25 | 15 to 25 | 250 to 500 |
| Chewings (fine) fescue | 25 to 30 | 10 to 20 | 5 to 20 | – |
| Bentgrass | 2 to 5 | 2 to 3 | 2 | – |
| *Poa trivialis* | 6 to 12 | 5 to 7 | 4 to 7 | – |
| **Enhanced performance and better spring transition:** 75% perennial ryegrass + 25% chewings fescue | 30 to 40 | 10 to 20 | 10 to 20 | 150 to 250 |
| **Enhanced performance and better spring transition:** 75% perennial ryegrass + 25% *P. trivialis* | 30 to 40 | 10 to 20 | 10 to 20 | 150 to 250 |
| **Better performance in shady or wet areas:** 60% bentgrass + 40% *P. trivialis* | 5 to 7 | 3 to 4 | 2 to 4 | – |
| **Typical mixture:** 60% perennial ryegrass + 25% chewings fescue + 15% *P. trivialis* (widely adaptable to poorly drained or shady areas) | 25 to 30 | 10 to 15 | 10 to 15 | 150 to 250 |
| **Enhanced fall establishment and better spring transition:** 80% chewings fescue + 20% *P. trivialis* | 20 to 25 | 8 to 10 | 8 to 10 | 150 to 200 |

[a] Multiply pounds/1,000 square foot by 0.49 to obtain kilograms/100 square meter and pounds/acre by 1.12 to obtain kilograms/hectare.

ranges are exceeded, thin, open, and disease-susceptible turf can be expected.

### Overseeding Procedure

**Golf greens and tees.** Before overseeding, the steps listed in "Seedbed Preparation" should be implemented to provide a desirable seedbed. Following this, a uniform application of seed is needed to provide a smooth and uniform playing surface (Figure 9.61). Seed should be applied when the surface is dry. Drying can be hastened by dragging a hose across the turf surface. Wind speed should be less than 5 miles per hour (<8 km/hour) to minimize seed movement onto adjacent areas.

Spreaders should be carefully calibrated to deliver the appropriate amount of seed. Another method is to preweigh seed for a known area, such as a putting green, and carefully mete it out over the area. This usually requires numerous passes over the area to ensure uniform seed coverage, and at least two directions should be used when broadcasting the seed. Apply half of the seed in one direction and the other half while moving at right angles to the first pass.

Boundaries of the overseeded areas should be defined by using a drop spreader. The remaining seed is applied with a drop, or centrifugal, spreader within the outer fringe of the seeded area. Avoid seed spread to nontarget areas such as collars, fringes, and other nearby turf areas. Unwanted seed will reduce aesthetic value and create the need for additional maintenance. Clumps of ryegrass are most noticeable the spring following application and tend to remain visible into early summer (Figure 9.62). Control of these clumps, once they are established, is difficult and slow. When trying to reduce the amount of unwanted ryegrass drift, carpets and mats should be placed in the designated areas used to fill and empty spreaders. Mats also can be placed around the perimeter of the seeding area to minimize escaped seed. Workers' shoes and dragging equipment should be cleaned before and after entering the perimeter area surrounding the green. Also, a preemergence herbicide should be applied prior to overseeding. A short boom, backpack sprayer, or drop spreader should be used to treat areas adjacent to greens for a definitive look.

If non-fungicide-treated seed is used, a fungicide is needed for disease control. For best seed coverage, fungicides should

**FIGURE 9.61** Overseeding in multiple directions to help ensure even and complete seed coverage.

**FIGURE 9.62** Clumps of "renegade" ryegrass or escapes from the intended overseeded area.

be applied before the green is topdressed. Fresh seed also should be used, as good seedling vigor is necessary for plants to quickly develop past the susceptible seedling stage. Proper seeding rates should also be adhered to since higher rates should not be used because this could produce weak, succulent plants.

Once the seed has been applied, light topdressing of greens and tees at one-third to one-half of a cubic yard/1,000 square feet (0.27 to 0.41 $m^3/100\ m^2$) encourages desirable seed-to-soil contact and turf establishment. Seed and topdressing material are then incorporated by dragging a carpet across the seeded area. A steel mat may need to be placed on the carpet to provide sufficient weight. Topdressing should be dry before dragging to minimize seed pickup on shoes and equipment. A cover on the ground also should be used when entering and leaving overseeded areas to prevent unwanted seed movement.

**Fairways.** Following seedbed preparation, large areas such as fairways typically use a centrifugal spreader to disperse seed. As with golf greens, seed should be spread in at least two directions. Use a drop spreader for defining the perimeter of these areas. Fairways are seeded at 250 to 850 pounds/acre (280 to 952 kg/ha) with perennial ryegrass, or 250 to 500 pounds/acre (280 to 560 kg/ha) with annual ryegrass (Table 9.11). Ideally, fungicide-treated seed should be planted. Buying extra seed to repair worn, skipped, or poorly established areas is suggested.

Fairway seed should be matted into the soil surface following overseeding. Topdressing usually is not used on these large areas after seeding except for intensive traffic areas such as approaches. Matting of large areas can be achieved by dragging a rug or old carpet section over the seeded area. Another method to encourage good seed–soil contact is by operating a stiff-wire power sweeper in several directions over seeded areas. After dragging in the seed, topdress lightly (e.g., one-third to one-half of a cubic yard/1,000 square feet, 0.27 to 0.41 $m^3/100\ m^2$) over heavy play or traffic areas with a soil mix similar to the underlying soil.

Another means of seeding fairways is through a slit applicator. This machine cuts a thin slice through the thatch layer, drops the seed, and presses seed into the soil. Slit seeding helps overcome heavy thatch layering and generally provides good germination since seed is in direct contact with the soil and is less susceptible to drought. Slit seeding, however, requires a slit seeder, is slower to plant, is easy to have skips and gaps in the seeding pattern, and results in pronounced rows of seed until the ryegrass is mature enough to tiller. If thatch has become excessive, a vertical mowing will eventually be needed and/or additional seed used to compensate for losses through desiccation from the thatch layering.

The use of plant growth retardants (PGRs) and herbicides as overseeding aids has increased in recent years. The material is applied just prior to overseeding to retard the bermudagrass growth, reducing the competition between bermudagrass and the newly overseeded grasses. Only foliar-absorbed products should be used since root-absorbed materials can retard the germination and growth of the overseeded grasses. Trinexapac-ethyl 1 L applied two to four days prior to overseeding has worked well if the bermudagrass is still actively growing with little or no effect on overseeded ryegrass. If the area to be overseeded is cultivated by verticutting prior to seeding, the advantages of using PGRs are greatly diminished. The herbicides triclopyr (Turflon Ester 4 L) or pelargonic acid (Scythe) can also be used two to seven days prior to overseeding to retard bermudagrass growth.

To produce a sharp contrast line between the overseeded fairway and nonoverseeded rough, the outside perimeter of overseeded fairways may be treated with a preemergence herbicide prior to overseeding. This will help minimize "renegade ryegrass," *Poa annua*, and other winter weeds that give the overseeded area a ragged appearance.

**Pregerminated seed.** If bare areas result from excessive play or diseases, reseeding with ryegrass through early spring is possible. It is always a good idea to order an extra 10% of seed in the event of thinning. Ryegrass seed (5 to 10 pounds/1,000 square feet, 24 to 49 $g/m^2$) may be pregerminated by soaking it in water for 24 to 48 hours and mixing it with topdressing prior to dispersal. The soil in these areas should first be loosened by spiking or aerifying. Ryegrass seed also may be "cleated in" by simply broadcasting seed before play and letting the players' spikes push seed into the soil.

**Overseeding cool-season areas.** Areas established with cool-season grasses also require periodic overseeding to repair or thicken worn and damaged areas. The procedures of overseeding cool-season areas are similar to those for the warm-season areas, except for timing. Cool-season areas can be overseeded at different times of the year, but generally are seeded in late summer (best time) or early spring (second best time). Dormant seeding is sometimes used successfully in northern areas when seed is applied during late fall or winter when temperatures prevent germination. In spring, when thawing soils are often saturated and very difficult to get heavy machinery into, earlier seeding allows germination and establishment. Winter weather for dormant seeding should cooperate where temperatures stay consistently cold during winter to prevent premature germination and subsequent kill by a late cold snap. The areas should also remain under snow cover to maintain steady soil moisture with little rain to move seed. When using dormant seeding, seed mortality is high, requiring seeding rates up to 50% higher than normal.

In order to obtain good seed-to-soil contact, slit-seeding or slicing is often used when overseeding into existing cool-season areas (Figure 9.63). The sliced grooves should be at least 1/4 inch (6.4 mm) deep to provide this desired contact rather than merely scratching the surface. Several passes help ensure good coverage and less noticeable slits.

Generally, the turfgrass species chosen should closely match what is currently present unless a change in grass is desired or extensive turf damage has occurred. A Kentucky bluegrass fairway, for example, is generally overseeded with 100% bluegrass at a rate of 2 to 3 pounds/1,000 square feet (9.8 to 15 $g/m^2$). If extensive damage or exposed areas exist (25% to 50% bare soil), a mixture of 2 pounds bluegrass plus 5 pounds perennial ryegrass/1,000 square feet

**FIGURE 9.63** Slit-seeder designed for good seed-to-soil contact when seeding into an existing turf stand.

(1 kg + 2.4 kg/100 m$^2$) may be necessary for rapid cover. Pure perennial ryegrass areas are generally overseeded with ryegrass only.

### Postplanting Maintenance

**Irrigation.** Following seeding, irrigate lightly to carefully moisten the soil surface without puddling or washing the seed into surrounding areas. Three to four light daily irrigations may be needed until all seedlings establish. Once germination begins, the seed cannot be allowed to dry out or the stand will thin. If seed washes into concentrated drifts following intense rains or heavy irrigation, a stiff-bristled broom should be used to redistribute it. Once grass is established, gradually reduce watering frequency to decrease disease potential.

After seedling emergence (5 to 7 days for ryegrass, 10 to 14 days for bentgrass and for roughstalk bluegrass), apply a preventative fungicide to help protect against *Pythium* root rot and *Rhizoctonia* brown patch, which can destroy overseeding stands (Figure 9.64). Factors that encourage these diseases include:

1. Unseasonably warm weather
2. Using excessive seeding rates that produce young, succulent plants
3. Prolonged periods of high, free moisture on leaf and stem surfaces, such as extended foggy conditions
4. Bermudagrass aggressiveness

Proper timing and application rate of seeding, minimum use of N, efficient irrigation scheduling, and the use of pretreated fungicide seed are methods to reduce disease potential. Many areas often experience warm, foggy mornings in fall. Turf managers, especially, should be on the lookout for diseases during these conditions since they are ideal for rapid fungal reproduction and spread. To prevent development of resistant strains of *Pythium*, always follow the label and alternate between chemical groups.

**Mowing.** With overseeded greens, mow at 1/2 inch (1.3 cm) when the new stand reaches 2/3 to 3/4 inch (1.7 to 1.9 cm). Gradually lower the cutting height to 1/4 to 5/16 inch (6.1 to 8 mm) over a two- to three-week period at 1/32 inch (0.8 mm) increments and skip the "cleanup" mowing lap the first few

**FIGURE 9.64** Diseases such as *Pythium* (shown) can invade improperly timed or managed seeding events.

mowings to minimize traffic and wear on the tender seedlings. Continue this gradual reduction in height for four to six weeks until a 3/16 inch (4.8 mm) height is reached. Use a sharp mower that will not pull up seedlings. Once seedlings are well established, mowing heights gradually can be reduced to the desired height and the heavier triplex mowers can be used. Cups in greens and tee markers should be moved daily. On tees and fairways, initiate mowing when the grass reaches 3/4 to 1 inch (1.9 to 2.5 cm). This normally occurs 14 to 21 days after seeding and will allow time for seedlings to root. Wait until the morning dew is gone before mowing to prevent pulling of the seedlings from the ground. Tees and fairways usually are permanently mowed at 1/2 to 3/4 inch (1.3 to 1.9 cm).

**Fertilization.** Do not fertilize with N during overseeding since this may encourage excessive bermudagrass competition. Adequate levels of P and K, however, should be maintained for good plant growth. Begin to fertilize shortly after significant shoot emergence (two to three weeks after seeding for perennial ryegrass) and continue until cold weather halts bermudagrass growth. Normally, 1/4 to 1/2 pound N/1,000 square feet (1.2 to 2.4 g/m$^2$) every two to three weeks with a soluble N source (e.g., ammonium nitrate/sulfate), or 1 pound/1,000 square feet per month (4.9 g/m$^2$) with a slow-release N source (e.g., IBDU, Milorganite, sulfur-coated urea [SCU]), is adequate to promote desired growth without overstimulating growth and encouraging disease. Traffic during grass establishment should be minimized whenever possible.

### Winter Management

Maintaining an acceptable turf appearance during winter involves proper watering and fertilization practices, proper traffic control, and proper disease management. In addition, damaged areas that do not provide an acceptable stand after the initial seeding may require additional seed applications.

Irrigate regularly during the dry winter months to prevent plant desiccation. In addition, light, mid-day irrigations may be necessary if the overseeded grass begins to wilt. Do not overwater since this may promote algae and disease occurrence.

Traffic control in winter reduces injury to turf crowns and stolons. Ways to control this injury include frequently alternating or changing mowing patterns, regularly skipping "cleanup" laps, and changing pin placement daily to distribute traffic.

Nitrogen fertilizer influences the appearance of the overseeded grass and spring recovery of the permanent grass. Excessive N should be avoided to prevent unhealthy grass competition and to prevent succulent overseeding growth. Nitrogen applications every two to three weeks with 1/2 pound N/1,000 square feet (2.4 kg N/m$^2$) usually are sufficient. More frequent applications may be needed if the recovery time from traffic or weather damage is slow.

Applications of P, K, Mn, and Fe during winter should be considered. All of these provide desirable color without stimulating excessive shoot growth. In addition, K helps in carbohydrate formation. These, along with Fe, also prevent grass desiccation. Soil P and K levels and rates can be determined by soil testing. Iron is generally applied every three to four weeks as ferrous sulfate at 2 ounces/1,000 square feet (6.1 kg/ha). Iron sulfate or a chelated iron source usually can be tank-mixed with most fungicides. Manganese can be applied as manganese sulfate at 1/2 to 1 ounce (14 to 28 g) in 3 to 5 gallons water/1,000 square feet (12 to 20 L/100 m$^2$).

Once the overseeded grass becomes established, the chance of severe disease is reduced. Dollar spot usually develops when N levels are low or when *Poa trivialis* or bentgrass is used as an overseeded grass. Its occurrence usually is suppressed with sufficient N levels. Brown patch and *Pythium* blight generally are the exception and not the rule for today's overseeded grasses. However, greens that drain poorly or during continuous wet periods can trigger outbreaks of these diseases. Excessive amounts of soluble N can also trigger disease. Turf managers should constantly check the weather forecast and be ready to use a fungicide if extended warm, moist (foggy) conditions are forecasted.

In addition to these diseases, the overseeded grass and the nondormant bermudagrass base grass can develop *Helminthosporium* leaf spot or pink patch during the fall, winter, and spring months when temperatures slow grass growth. Leaf spot is similar to dollar spot where maintaining adequate N levels usually keeps the grass growing aggressively enough to outgrow the disease symptoms. Fungicides, however, may be required during extended periods of cool weather that prevent adequate bermudagrass shoot growth.

### Spring Transition

The main objective in spring is the gradual and smooth transition from overseeded grasses back to bermudagrass turf. Dormant bermudagrass shows signs of "green-up" when soil temperatures reach the 60°F (15.6°C) range. Some overseeded grasses, especially the new and aggressive heat-tolerant perennial ryegrasses, can successfully compete with bermudagrass through the spring. This results in a poor transition. Golf greens seeded with mixtures containing bentgrass often experience the same problem. However, mixtures high in *Poa trivialis* or fine fescue are difficult to maintain once temperatures reach the 80°F (26.7°C) range. During these times, these grasses become very sensitive to management practices designed to encourage bermudagrass recovery (Figure 9.65).

A general "rule of thumb" for maintaining a bermudagrass base is to provide at least 90 days of nonoverseeded turf during summer. If overseeding remains so that less than 90 days of noncompetitive growth conditions occur, the bermudagrass base will eventually weaken due to lack of sufficient carbohydrate reserves (or rhizomes). Permanent damage is likely as well as poor future overseeding stands due to an insufficient bermudagrass base.

As early spring approaches, a cultural program is initiated using lower mowing heights, brushes, topdressings, and other

**FIGURE 9.65** Thin spring transition turf from an area overseeded the previous fall (left) versus none (right).

reel implements such as grooved rollers. The following procedures have proven beneficial in encouraging bermudagrass at the expense of overseeded grasses with minimal disruption to the turf.

1. **Use the appropriate seed or mixture.** Reducing the amount of perennial ryegrass or bentgrass in an overseeding mixture tends to aid in spring transition. Intermediate ryegrass, fine fescues, and roughstalk bluegrass are less heat tolerant and therefore tend to transition earlier than perennial ryegrass or bentgrass.
2. **Time transition according to temperatures.** Temperatures (both day *and* night) are the most important criteria influencing overseeding. Unfortunately, turf managers have little direct control over this. However, management practices can be timed around traditional temperature patterns and short-term weather forecasts. Probably the most important temperature range to remember is that bermudagrass will not aggressively grow until night temperatures consistently reach the mid-60s°F (~18.3°C). Bermudagrass shoots will "green-up" much sooner when daytime temperatures reach the upper 50s°F (10°C), and slowly initiate growth when night temperatures reach 58°F (14.4°C). Members become excited with the first warm spell of the season and this puts added pressure on the superintendent to hasten transition. However, bermudagrass will not aggressively grow (especially laterally) until the mid-to-high 60s°F (~18°C to 21°C) are reached. Therefore, do not begin your transition steps until just before these temperatures are anticipated. If performed prematurely, bare areas may become exposed and will not recover until temperatures are high enough to favor bermudagrass growth (Figure 9.66). Also, if a late cool snap occurs and night temperatures drop into the 50s°F or low 60s°F (12.8°C to 16.7°C), it will require three to seven days of high 60s°F (~21°C) temperatures for the bermudagrass to resume active growth.
3. **Reduced mowing height.** Begin reducing the mowing height just before the expected spring transition period to reduce the negative shading effect of the overseeded grass on the bermudagrass. Begin in mid- to late winter by reducing the mowing height 1/32 inch (0.8 mm) every two weeks until a height of 5/32 to 1/8 inch (4 to 3 mm) is reached in late winter. Maintain a short mowing height until at least 50% of the overseeded grass has been removed (e.g., midspring). The mowing height might be raised back to 3/16 inch (4.8 mm) as the bermudagrass begins to recover, but should not be raised while the overseeded grass still dominates. A lower mowing height reduces the overseeding shading of developing bermudagrass, warms the soil, and inhibits the growth of the overseeded grasses. The use of grooved rollers and brushes also improves the putting surface and helps maintain the overseeded grass in an upright growth habit.
4. **Cultivation.** Spike approximately three weeks before bermudagrass normally begins to green-up and every week following to enhance soil warming and turf recovery. Spiking also reduces surface compaction and algae growth. Aerify several weeks before the expected spring green-up to promote bermudagrass growth by warming the soil and reduce the competition from the overseeded grass. Small (1/4 inch, 6.4 mm) tines should be used and good soil moisture should be present. The superintendent and club pro should coordinate spring tournaments around aerification times to minimize play disruption.
5. **Fertility.** Maintain low fertilizer application rates in late winter through early spring to reduce overseeded grass growth. Liquid iron will aid in maintaining desirable green color without an excessive flush of growth. When bermudagrass growth is apparent,

**FIGURE 9.66** Premature spring transition prior to the permanent turf reestablishing itself.

restore fertilizer applications. Approximately two weeks after the initiation of spiking, fertilize with 3/4 pound N/1,000 square feet (3.7 g N/m$^2$) to help stimulate new bermudagrass growth. Fertilize weekly at this rate until an adequate bermudagrass cover is achieved.

6. **Verticutting/grooming.** While overseeded grasses are still actively growing, initiate light and frequent (e.g., weekly) verticutting (grooming) to help maintain the overseeded grasses in an upright growth habit that allows increased sunlight and warmth to penetrate through to the soil and thus encourage an earlier and more rapid regrowth of the bermudagrass. Begin when daytime temperatures are consistently above 70°F (>21°C). Another method of judging when to initiate this process is the green-up of nonoverseeded adjacent fairways. Verticutting/grooming is generally no lower than 1/16 inch (1.6 mm) below the bed knife. It should cut above the soil surface and remove only surface leaves and not remove or damage bermudagrass stolons. Light verticutting/grooming, in addition to aiding the bermudagrass, also will improve the putting quality of greens. Light topdressings on a two- to three-week basis at approximately 1/8 cubic yard/1,000 square feet (0.1 m$^3$/100 m$^2$) also aids in maintaining a desirable putting surface. **Note:** Although traditional light, frequent verticutting/grooming and core aerification may promote the gradual transition to bermudagrass, visual turf quality may be reduced until the bermudagrass has had sufficient time to recover. This is especially true if medium to heavy verticutting is implemented. Typically, patches of thin turf form and remain unsightly until the bermudagrass greens-up and fills in.
7. **Maintain adequate soil moisture.** Reducing or withholding water in an attempt to encourage the overseeded grass to die from moisture stress is not recommended. Spring is when bermudagrass suffers from the natural decline of older roots and initiation of new roots. Withholding water during this root transformation may cause greater damage to the extremely shallow-rooted bermudagrass than to the deeper rooted winter overseeded species. Water deeply and infrequently to encourage deep bermudagrass rooting at the expense of the overseeded grass.
8. **Use of herbicides.** Selective herbicides have proven useful for gradual removal of overseeded grasses in spring. This allows the superintendent better control over transition timing. Herbicide use also provides an earlier indication on how well the bermudagrass wintered and allows more time if resprigging or sodding are required. Control of other weeds, such as *Poa annua*, also is possible with some of the herbicides.

Several products are available to assist in spring transition. Application rates and timings are extremely important. Products that require four to six weeks to control the overseeding include pronamide and various members of the sulfonylurea herbicides such as metsulfuron, trifloxysulfuron, and flazasulfuron. Results are desirably slow, depending on temperatures and rates. Warmer temperatures and the higher rates usually hasten this conversion. However, a weak stand of grass may result if greens are treated too soon in spring and if the weather remains cool (Figure 9.67). Thin turf will remain until temperatures are warm enough for bermudagrass to recover. Applications should be timed when nighttime temperatures are consistently in the mid-60s°F (~18°C to 21°C) and bermudagrass growth is evident. These products are typically applied in mid- to late spring, depending on the location.

Quicker-reacting products include rimsulfuron (TranXit) and foramsulfuron (Revolver). These typically control the overseeding in two to three weeks. These, therefore, are not typically applied until later in spring (e.g., May) so sufficient bermudagrass growth has resumed to provide the desired

**FIGURE 9.67** Using herbicides to obtain more desirable spring transition from overseeding to permanent turf.

smooth transition. Visual injury to ryegrass with this treatment normally lasts from one to three weeks.

Experience suggests turf managers wishing to use herbicides to enhance transition proceed with caution. Do not begin treatments until one to two weeks after bermudagrass resumes active growth. Lighter rates than listed may be wise if "weaker" overseeded grasses such as *Poa trivialis* or fine fescue are present. Sprayer calibration and application uniformity are extremely critical; once the overseeded grass begins to die from the herbicide, there is no turning back.

## Colorants, Dyes, and Paints

Coloring of golf courses for winter color instead of overseeding is gaining in popularity due to the disruption of overseeding in fall and problems with spring transition (Figure 9.68). Seed, fertilizers, pesticides, and labor costs are greatly reduced or eliminated by painting. Equipment wear-and-tear also is reduced. Improved putting quality results as painted greens tend to putt much more smoothly and farther (quicker) since overseeding grasses tend to disrupt or slow ball roll. To minimize complaints from excessive green speed, raise the mowing height for the last several regular mowing events to provide more leaf area, which restricts ball roll. Higher mowing heights in fall also benefit the bermudagrass by encouraging desirable carbohydrate production and storage prior to the dormancy period. Ball roll can also be slowed by heavily hand-watering just prior to a major tournament. Improvements in paints, their longevity, and application equipment/techniques also make painting more attractive than in previous situations. Maintenance is also eased by not overseeding. This allows the entire maintenance program to revolve around the needs of bermudagrass.

*Pigments* selectively absorb different wavelengths of light and generally are insoluble in solution, forming a suspension. When applied, they enter the treated plants. A *dye* is a colored substance that binds to the substrate applied to the solution and is soluble in it. A *paint* is composed of a pigment, binder, solvent, and various additives and, when applied as a thin layer, is converted to an opaque solid, covering (coating) the substrate. A *colorant* includes paints, pigments, and dyes and changes the color when applied to something else.

Coloring certainly has some challenges. Overseeded areas typically exhibit better wear tolerance as more leaf tissue is present and growth continues. Wear patterns and disease infestations are more easily seen in colored areas. Overseeding also helps mask imperfections in the turf stand, including off-type bermudagrasses, plugs, diseases, ball-marks, and grain.

**FIGURE 9.68** Using colorants, dyes, and paints for desirable winter color of warm-season turf.

A decent cover of turf is also needed for coloring to bind to, and sequential applications may be needed.

Coloring greens is still somewhat more of an art than science, though experience indicates that exceptional results are all about the application. Considerable practice on the nursery green, driving range, or out-of-play areas should be made prior to actually coloring greens. There is some debate as to whether a spray gun or spray boom should be used for paint applications. The advantage of a boom is the speed and ease of application. However, some streaking and small triangles may result on the collars or surrounds as the sprayer comes on and off greens. Using hand guns can provide a more uniform pattern with no triangles on the collars or surrounds. However, the quality of the painting is only as good as the talent of the applicator.

Paints tend to last four to six weeks, resulting in the need for reapplication; dyes last about two weeks. The colorant adheres most successfully to good, dense turf stands, while thin turf stands will require higher application rates. Worn, thin, bare areas or common bermudagrass patches will be less successful from coloring. Typical dilution rates are 5 gallons of paint per 100 gallons of water (19 L/379 L) with two passes made in different directions and a cleanup pass to keep the perimeters even in terms of color. Painted greens should still receive monthly aerification with hydrojets or solid tine to keep greens soft and provide soil $O_2$ and drying.

For best (most consistent) results and the need for less colorant, begin the process in fall when green turf is still present (about two weeks before the first anticipated frost) or during an actual frost event. Coloring brown, dormant turf requires considerably more product and streaking is more apparent. Raising the mowing height in fall also helps by providing more leaf tissue area for the colorant to adhere to. Lightly irrigating prior to applying a colorant also reduces the amount needed as the surface moisture reduces the wicking of the material into the leaf surface. Products are mostly applied using a diaphragm or roller pump as their seals are less affected by the viscous material compared to centrifugal pumps. Low-cost seals, however, are becoming more available for centrifugal pumps. Most colorants dry within an hour after application.

## SPECIAL MANAGEMENT: BUFFALOGRASS

Buffalograss characteristics and available cultivars are covered in Chapter 1 (Figure 9.69). Being a low-maintenance grass, buffalograss requires less intensive management than other grasses or the stand will weaken and thin.

**Establishment.** Propagation may be done vegetatively or by seed (burs). Vegetative propagation by sodding has been the main method of establishment primarily because of a seed shortage. If vegetatively propagated, female plants are generally used since male plants produce unsightly seedstalks. When viable seed (or burs) is available, seeding rates range from 1 to 5 pounds/1,000 square feet (48 to 245 kg/ha). The higher rate is needed if seeds are broadcasted and/or if irrigation is not available, or if a quicker establishment rate is desired. Lower rates (e.g., 10 to 20 pounds/acre, 11 to 22 kg/ha) may be used if the seeds are drilled in rows.

Plugs equal to or greater than 2 inches (5 cm) also may be used for establishment. These are spaced from 6 inches (15 cm) to 2 feet (0.6 m) apart, depending on the desired establishment rate. Soil should be well watered before and after planting for several weeks to aid survival. The establishment rate using seed or plugs is generally slower than for most other warm-season grasses.

**Maintenance.** Buffalograss is somewhat pest free in its native habitat but problems occur when it is grown under warmer, more humid, and wetter areas. Some authors even suggest the quickest way to thin a buffalograss stand is to start fertilizing and watering it. Therefore, only minimum maintenance practices should be used to keep a buffalograss stand. For golf course fairways, a cutting height from 3/4 to 2 inches (1.9 to 5 cm) is preferred. For golf course rough areas, buffalograss is mowed between 2 and 3 inches (5 to 13 cm) and only as needed or about once per growing month. Even at the higher heights, due to its open density, a golf ball can easily be found and played. The mowing interval is generally infrequent due to a slow vertical shoot growth rate.

**FIGURE 9.69** Buffalograss as a low-maintenance turf on a golf course.

Dense stands require minimum water, mowing, and weed control. Nitrogen requirements range from 0 to 1 pound N/1,000 square feet per year (0 to 48 kg N/ha). Higher fertilizer rates encourage bermudagrass encroachment.

Supplemental watering normally is not advisable since overwatering can produce undesirable effects. If watered, apply only in deep (e.g., 3/4 inch, 1.9 cm), infrequent (weekly or longer) cycles.

## SPECIAL MANAGEMENT: KIKUYUGRASS

For successful kikuyugrass management, proper mechanical, growth regulator, moisture, and fertility components must be integrated. By being integrated, these components help maintain a dense, less coarse, and less thatchy stand of kikuyugrass (Figure 9.70). Insufficient moisture restricts rooting, causing the grass to become puffy and more prone to scalping.

**Mowing, vertical mowing, and aerification.** Mowing heights for kikuyugrass range from 1/4 inch (6.4 mm) on tees and collars to more than 2 inches (5 cm) in roughs. Fairways are typically mowed from 1/2 to 5/8 inches (1.3 to 1.6 cm) three to five times weekly during active growth to minimize scalping. Unlike other fairway grasses, heavier, motor-driven reel mowing units are preferred to push reels into the turf canopy for a higher quality cut. Roughs are maintained between 1.5 and 2 inches (3.8 to 5 cm) and require twice weekly mowing to prevent scalping. Due to the tough shoots of kikuyugrass, added wear on mowing equipment generally reduces life expectancy by 10% or more.

Vertical mowing helps minimize thatch development and prevents formation of long, coarse, poorly rooted stolons. Thatch and long stolons increase scalping and excessive turf removal by mowing. Verticutting three times yearly (e.g., midspring, midsummer, and late summer at approximately 1/2 inch [1.3 cm]) deep seems necessary to achieve this standard for kikuyugrass fairways. Alternatively, light (e.g., 1/4 inch, 0.64 cm), monthly vertical mowings during periods of active growth help avoid the heavy debris generated by less frequent, heavier vertical mowings.

Vertical mowing is incorporated with deep tine aerification, which promotes deeper rooting by alleviating compaction. At least one deep-tine (e.g., 6 to 10 inches, 15 to 25 cm) aerification in spring followed by light verticutting in spring and two additional regular aerifications in summer help "tighten" the stand of kikuyugrass. The use of plant-growth regulators is also incorporated in this mechanical management. Gibberellic acid-inhibiting PGRs, such as trinexapac-ethyl (e.g., Primo), applied every two to three weeks, have dramatically improved kikuyugrass stands by reducing scalping and clipping production. PGRs cause a shorter, more compact growth habit, with finer leaf texture and denser stands. Treated plants also have a darker color. When combined with verticutting and aerification, PGRs allow mowing heights of 1/2 inch (1.3 cm) or less without scalping, excessive clipping production, or exposure of brown, senescing leaves underneath. Chelated iron should be tank-mixed with the PGRs, especially with the first seasonal application, as temporary discoloration often occurs then.

**Fertility.** Fertility rates also should be closely regulated so as not to be too heavy to encourage thatch or scalping of the kikuyugrass or underfertilized grass, which discourages aggressive growth. Apply 2 to 3 pounds N/1,000 square feet (9.8 to 147 g N/m$^2$) in late spring as a slow-release source or as 1 pound N/1,000 square feet applications (4.9 g N/m$^2$) in midspring plus early and late summer. Additional light (1/8 to 1/4 pound N/1,000 square feet, 0.6 to 1.2 g N/m$^2$) nitrogen (N) applications seem to promote good, but not too aggressive, kikuyugrass growth. Supplemental chelated iron (Fe) and manganese (Mn) also promote green color without promoting excessive growth. Phosphorus, potassium (K), and other nutrient needs should be based on annual soil and tissue tests.

**FIGURE 9.70** Kikuyugrass as a low-maintenance turf on a golf course.

**Pest problems.** Being a relatively new turfgrass, the full potential of pest problems on kikuyugrass probably has not been experienced. It is sensitive to most postemergence herbicides, especially MSMA, quinclorac, and triclopyr. Three-way mixtures of 2,4-D, MCPP, and dicamba can be used but only at one-half to one-fourth their normal recommended label rates. Clopyralid appears safe on kikuyugrass.

Brown patch and take-all patch are other known pest problems. Manganese sulfate at 1 pound product/1,000 square feet (4.9 g/100 $m^2$) helps control take-all patch without the use of fungicides. As the use range of kikuyugrass expands, additional pest problems will probably occur.

## BUNKER MAINTENANCE

Bunkers have, in a relatively short period, become the main focus of golfer complaints (Figure 9.71). Greens traditionally were the lightning rod of criticism on a golf course. However, industry advancements for greens maintenance have allowed superintendents to excel in putting green management, so the focus has shifted to bunkers.

Common bunker complaints from golfers concern the depth or inconsistent depth of sand, "fried-egg" lies (sand too soft), lack of consistency between bunkers, and varying moisture levels in the same and between bunkers (Figure 9.72). Many of these complaints stem from improper initial bunker design and construction. With complaints so rampant concerning bunkers, golf courses should dedicate adequate time and resources to their maintenance. Legitimate questions during the construction phase include:

- How many bunkers does the course really need to make it a challenging layout so that it will not become a maintenance nightmare?
- Are steep-faced bunkers, which easily and constantly wash following rainfall or even normal irrigation, needed?
- Do we need islands or fingers of green grass extending into the bunkers?
- These look interesting, but rarely come into play and significantly increase installation and maintenance costs. Can we afford the drainage systems and daily maintenance necessary for high-quality bunkers?

**FIGURE 9.71** Poorly draining bunker.

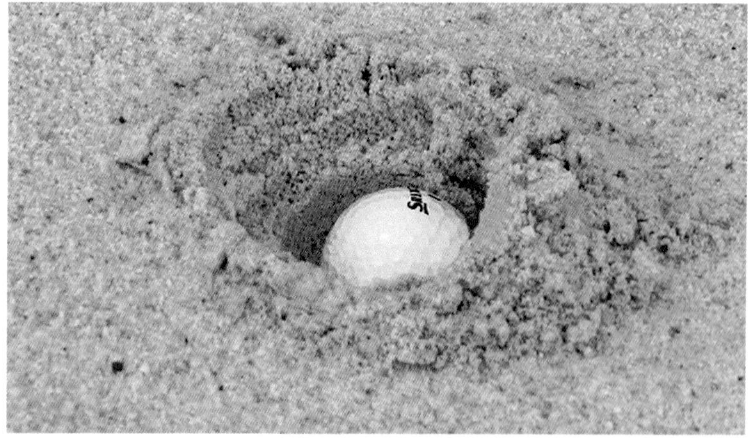

**FIGURE 9.72** Undesirable buried or "fried-egg" golf ball in soft bunker sand.

- Can some bunkers that seldom come into play be eliminated or at least converted to grass hollows?
- Routine practices associated with maintaining acceptable bunkers include:
  - Raking
  - Edging
  - Mowing and blowing
  - Firming sand faces
  - Checking sand depth and replacing lost sand
  - Pumping standing water

Bunker raking usually involves either a complete raking or "spot-raking" areas where play has occurred. This is typically performed via hand-raking or with mechanical equipment (Figure 9.73). The increased use of bunker liners has reduced washing, so hand-raking has become more popular. This helps firm sand loosened by washing and helps reduce snagging liners, but requires additional labor and is very costly. Situations when hand-raking may be required compared to mechanical means include:

- Bunkers too small to allow turning and maneuvering of mechanical rakes
- Bunkers with severe slopes
- Small pot bunkers with high vertical walls (Figure 9.74)
- Bunker sand that is too soft or loose to produce the desired firmness using mechanical rakes
- Preference for straight sand furrow patterns compared to circular patterns created by mechanical rakes

**FIGURE 9.73** Mechanical means of raking a bunker.

**FIGURE 9.74** "Pot" bunker with high vertical facing, which requires increased maintenance.

- Bunkers with irregularly shaped perimeters, numerous capes or fingers, narrow bays, or other severe design features
- Sand depths too shallow to allow mechanical rakes without snagging underlying soft liners
- Players' perception that hand-raking produces superior playing conditions

Edging preserves the sharp, well-defined perimeters of bunkers often seen on television. Weekly line trimming is performed during periods of active turfgrass growth while edging with a sidewalk-type edger is typically done once or twice monthly. Trimming debris then must be removed. PGRs can be applied to turf on bunker edges to reduce the frequency of trimming. Another means to reduce bunker maintenance is adapting the Aussie style of raking (Figure 9.75). With this, bunker perimeters are smoothed and firmed to reduce washing and buried lies while the bunker bottoms are hand-raked to provide a more traditional exit shot.

Mowing involves trimming grass-faced bunkers on a routine basis to keep a manicured appearance. Walk-behind mowers are used on more gently sloped bunker facings, while floating mowers (fly mowers) are used on more steeply sloped facings. These are performed weekly, followed by removing clippings from the bunker via blowing or raking. Additional blowing is needed to remove tree leaves, grass clippings, and other debris from bunkers. Frequency varies according to season, with daily blowing needed during major leaf drop events (fall, following rain events, etc.) and blowing being less common during slower growth periods. If the bunker sand is too soft, a short-term solution can be to incorporate finer particle sized sand or a crushed granite dust such as that used on clay tennis courts. However, if performed too often for too long, this technique will eventually cause the sand to become too firm and reduce drainage.

Checking, replacing, and packing sand in bunkers have become major considerations for routine bunker maintenance. Bunker sand depths are checked weekly with a goal of 1 to 2 inches (2.5 to 5 cm) minimum sand depth to prevent exposure or snagging of liners during play or routine maintenance. Hand-shoveling is used on smaller, routine depth corrections associated with normal rain events. Following shoveling, sand facings should be packed with hand tampers to prevent the "fried-egg" deep settling of incoming shots. This is performed once or twice monthly.

However, over time, additional sand is lost from play, wind, and routine maintenance practices. Also, often following a major rain event, a thin layer of silt and clay and organic debris accumulates at the bottom of the bunker. This layer should be removed to retain desirable bunker drainage. Ideally, smaller amounts (e.g., 1 to 2 inches, 2.5 to 5 cm) of sand should be added frequently to replenish bunkers compared to less frequent, larger amounts. The smaller amount helps prevent softer sand and packs better than deeper layers.

An undesirable site is seeing a bunker become more of a water hazard than a sand hazard (Figure 9.76). As bunkers age, their drainage slows due to compaction; introduction of silt, clay, and organic matter; and use of different sand-sized particles. Crews then use a hand- or powered pump to remove water and shovels to remove silt and clay. New sand is then added and hand-raking follows to restore the bunker to its original condition.

Applying a wetting agent every two weeks helps provide more consistent moisture in the bunker sand. This also helps firm or pack sands, especially those that have not had time to properly settle.

## PETROLEUM SPILLS

Turf managers periodically face various petroleum leaks or spills, causing serious, long-term turf damage. These include various fuels, oil, hydraulic and brake fluids, and grease (Figure 9.77). Extent of turf damage and recovery depends on the product type, volume applied, temperature of the product and atmosphere, and soil conditions.

Petroleum products typically contain carbon (~85%), N (~1%), $O_2$ (0% to 1.5%), S (~3%), and a trace amount of metals. A comparison of several commonly used fuels is provided

**FIGURE 9.75** Aussie style of bunker maintenance where perimeters are smoothed and firmed to reduce washing and buried lies while the centers are raked to provide a more traditional exit shot.

**FIGURE 9.76** A bunker being more of a water than sand hazard.

**FIGURE 9.77** Extended turf damage from a hydraulic leak from maintenance equipment.

in Table 9.12. In addition to various fuels, other lubricants and hydraulic fluids can be problematic as spills and leaks.

- **Motor oil.** Motor oils lubricate engine parts and are classified according to their viscosity with these mainly designed for temperature extremes.
- **Hydraulic fluid.** Hydraulic fluid is a common source of turf damage from machinery leaks, mainly from hose rupture or hose connection failure. It is designed for performing at high temperatures (e.g., 110°F to 130°F, 43°C to 54°C) and pressures (i.e., >3,000 psi, 204 atm). The base material used in

**TABLE 9.12**
**Comparing Several Fuels Used in Turfgrass Management**

| Fuel Type | Chemical Structure | Fuel Material | Flash Point | Ignition Temperature | Comments |
|---|---|---|---|---|---|
| Gasoline | $C_4$ to $C_{12}$ | Crude oil | –45°F (–43°C) | 495°F (272°C) | Insoluble in water |
| Diesel #2 | $C_8$ to $C_{25}$ | Crude oil | 165°F (74°C) | –600°F (–351°C) | – |
| Biodiesel | $C_{12}$ to $C_{22}$ | Fats and oil from animal fats, cooking oil, rapeseed, soybeans | 212°F to 338°F (100°C to 170°C) | –300°F (184°C) | Higher percentage blends may affect seals and hoses |
| Ethanol | $CH_3CH_2OH$ | Corn, small grains | 55°F (13°C) | 793°F (423°C) | Lubricants may be added; often contains ~5% water |

hydraulic fluid often includes refined mineral oil plus additives such as corrosion and oxidation inhibitors, defoaming agents, antiwear and antifriction agents, and detergents.

- **Brake fluid.** Brake fluid is a type of hydraulic fluid, either formed from mineral oil, silicon, or glycol; glycol is the most used. Various grades of brake fluid are available with dry boiling point generally increasing with the grade number. Most of these have a dry boiling point > 400°F (204°C).
- **Grease.** Grease is a lubricant typically containing a base oil, thickener, and various additives depending on the desired viscosity and use. Its thickness is often classified from zero to six with increasing firming with higher values.

If, how soon, and the extent of turf damage following a petroleum leak or spill depend on the turfgrass species, soil type, temperature of the material, and volume of spill. For example, at ambient air temperatures, synthetic hydraulic fluid spills heal more quickly on treated bermudagrass and seashore paspalum than either biodegradable vegetable/ester-based or petroleum/mineral-based hydraulic fluid. As the temperature and amount of fluid spilled increase, the amount and longevity of turf damage also increases.

Turf injury symptoms following a spill or leak vary and aids in recovery are summarized as

- **Gasoline.** Affected turf turns shiny and slightly oily, with a pungent smell. Within 30 minutes, turf starts to desiccate with rolled leaves and darker color. The turf turns completely brown after 16 hours (Figure 9.78). Applications of calcined clay, activated charcoal, detergent, or water drench have not had any positive effects or shortened turf recovery time.
- **Motor oil.** Affected turf is shiny and after about one day, leaf browning occurs and after two days, turf kill is evident. Detergent (anionic and nonionic granules) applied following motor oil exposure provides quickest recovery, shortening recovery time by half compared to no treatment.
- **Hydraulic fluid.** Turf damage resembles that of gasoline exposure by developing a dark brown color, eventually turning brown. Like motor oil, detergent applied following hydraulic fluid exposure provides quickest recovery, approximately twice as quickly compared to no treatment.
- **Brake fluid.** Within 30 minutes of exposure, turf appears shiny before beginning to roll its leaves, turning darker and desiccated. Turf may turn a pale grayish-green before turning completely brown within two days. Because it is water soluble, flushing with water following brake fluid exposure aids in recovery. Adding a detergent may hasten recovery slightly.
- **Grease.** Smearing of grease remains visible for several days, with turf dying in the thickest affected areas. Calcined clay, activated charcoal, detergent, or water drench does not have any positive effects on shortening turf recovery time of exposed turf.

## ORGANIC GOLF COURSE MANAGEMENT

A current trend in many areas is to provide "organic,"—or as near as possible to organic—golf course maintenance. This certainly is not impossible as golf has an almost 500-year history. For ~400 of these years, courses were maintained "organically" with little artificial inputs. However, with modern conditions and player demands, it has proved very elusive to maintain truly "organically" grown golf courses without major sacrifices in turf quality and playing conditions.

### DEFINITION

"Organic" has many definitions. Chemists consider carbon-containing compounds as organic. Another is "of, relating to, or derived from living organisms," while the original meaning was "of or pertaining to an organ or its function." A contemporary strict definition would be "no added synthetic pesticides or fertilizers." A more realistic definition could be "derived from plant materials, biological organisms, or mined

**FIGURE 9.78** Damaged turf from a gasoline spill.

from natural deposits." The U.S. Department of Agriculture (USDA) has a very detailed protocol concerning raising food crops and labeling them "organic." For additional information, refer to http://www.nal.usda.gov/afsic/pubs/ofp/ofp.shtml. However, a uniform definition or standards of managing golf courses "organically" currently does not exist.

### Management Considerations

The most consistent problems when golf courses go organic are disease infestations, followed by insects and weeds. This is severely compounded when proper construction criteria are not provided, improper grass selection occurs, improper or inadequate turf cultural practices are implemented, and golfers demand as high a quality of turf when grown organically compared with conventional maintenance.

**Construction considerations.** Everything possible must be done prior to going organic to minimize conditions that favor pests and weaken turf at the expense of healthy turf. Agronomically, this includes minimizing shade by adequately removing trees; providing excellent golf green and course drainage by building sand-based rootzones; installing extensive fairway, rough, and bunker tile drainage; and contouring the course for surface drainage. For cool-season grass greens grown in heat-stress environments, surface fans are required on all poor air circulation greens. Subsurface fans (pumps) for greens can remove excessive moisture and incorporate soil $O_2$. Grasses must be selected primarily on their ability to resist pest invasion and secondarily on their playing characteristics. Wells for monitoring along the perimeter of the course are needed to monitor ground water quality. Documenting water quality helps counter allegations the course is contaminating ground water. Water entering and exiting the site (e.g., rivers or streams) needs to be periodically monitored, again, to document that water quality is maintained on the property. Irrigation systems will require injection capability as many biological-control agents require daily exposure due to their short-term life expectancies. The latest and most complete cultural equipment and supplies need to be provided by grow-in. Extra trained personnel or "scouts" should be hired by grow-in to monitor and implement integrated pest management or nonchemical golf course management, regardless of whether major tournaments or other events are scheduled.

Unnecessary trees should be removed, as mowing is interrupted by extensive amounts of turning, leaf litter, and limb removal; this also eliminates pruning of branches and line trimming around bases of trees, shade and tree root competition, and eventual liability as larger limbs often fall as trees senesce. If trees are not removed, the area under the drip line should be mulched or a shade-tolerant ground cover planted. Another current trend by some courses is to eliminate traditional grass roughs and plant low-maintenance plants or ground covers, or convert them to waste bunkers.

Traffic to and from greens and tees should be dispersed to prevent bare soil paths, cart path curbing should be installed adjacent to these areas to avoid unwanted cart wandering, soil around concentrated traffic points such as water coolers and ball washers should be mulched or paved to prevent bare soil, and tee slopes should be softened to avoid steps as this concentrates traffic and opens possible liability issues from slippage. Also, avoid using gravel on areas with more than 12% slope or frequent regrading of washed-out stone will be needed.

The number and size of bunkers should be evaluated as they are expensive to construct and maintain. Remaining ones should be designed so they do not accept surface drainage, do not have high sand faces that are prone to washing, and avoid bright white sand that requires a liner to avoid soil contamination or losing color.

**Membership considerations.** Specialized equipment, monitoring techniques, and periodic less-than-ideal turf conditions will probably occur, especially during environmentally stressful times. Cultural practices, such as aerification and other disruptive practices, would be implemented as needed for turf health. Occasional renovation or replanting of damaged turf would be needed. At this point, whether or not replacement sod has been grown organically becomes an important consideration. Additional labor would also be needed to physically remove pests such as weeds, which will invariably increase under organically based maintenance. A competent animal trapper would be needed to capture and remove predatory animals such as raccoons, skunks, crows, armadillos, hogs, etc. as insect pressures increase. Finally, golfers would have to realize that turf quality and the game would change radically if organic maintenance is imposed. The cost of the game would also increase, especially if any semblance of modern playability is retained. Much is to be learned and discovered on how to successfully maintain "organic" golf courses. Research is often slow or lacking as turf is not viewed as a necessary food or fiber crop and thus is not eligible for federal research grants to address these issues.

# 10 Managing Greens and Tees

## BENTGRASS/*POA ANNUA* IN STRESSFUL ENVIRONMENTS

Bentgrass/*Poa annua* use on putting greens has expanded outside their natural region of adaptation into more hot and humid areas (Figure 10.1). This expansion frequently occurs in vacation or resort travel destinations where the facilities are operated primarily as daily fee courses. These players expect similar putting surfaces in heat-stress environments as those found in cooler regions. Advancements in innovative summer stress management practices and the availability of improved cultivars that possess better heat and drought tolerance have allowed bentgrass/*Poa* greens to be grown in these areas. These improved cultivars produce an extremely fine-textured, dense turf when closely mowed because of their extremely high shoot density. *Poa* greens are becoming more widespread primarily due to the lack of reliable selective control in bentgrass greens, heavy use of certain plant growth regulators (PGRs) that favor *Poa*, and the introduction of fungicides, which protect *Poa* from diseases. In many cases, superintendents have adjusted their management programs to maintain *Poa* greens.

### SUMMER DECLINE

Summer bentgrass decline is the term used to describe physiological stress decline along with disease complex (Figure 10.2). Symptoms begin as a gradual discoloration, loss in vigor, and decline in shoot density. Eventually large areas with slowed or stopped growth can be affected. In conjunction with shoot decline, root systems become extremely short (less than 2 inches, 5 cm, in many cases) and may eventually rot.

When organic layer temperatures exceed 86°F (30°C) for extended periods, soil $O_2$ levels drop, soil carbon dioxide ($CO_2$) levels increase, and these combine to weaken creeping bentgrass to the point at which diseases become deadly. A typical growth response of creeping bentgrass grown in stressful conditions that have chronic high organic layers is

- *Early summer*—plant respiration exceeds photosynthesis, stripping root carbohydrates. Root function then declines, leading to some root death.
- *Midsummer*—as roots degrade, a gel-like mass forms that reduces water infiltration and soil $O_2$ levels. Wet soil surfaces develop along with poor wear tolerance, causing plants to weaken and allowing disease activity to increase.
- *Late summer*—as roots continue to decline, shoots thin, allowing algae and moss to encroach. *Poa annua* eventually fills in bare or weak areas.

Summer decline usually begins in low-lying areas with poor or restricted air circulation and areas with compacted soils and/or poor surface and subsurface drainage (Figure 10.3). Various summer patch diseases may be confused with decline symptoms. However, patch diseases usually occur in isolated locations and have fairly distinguishable borders. Anthracnose, which primarily affects *Poa*, initially resembles the early development stages of localized dry spot. Leaves (or borders) in infected areas turn a brownish-bronze color and the plants eventually collapse in irregularly shaped patches. These patches, like *Pythium* blight, often are streaked from mowing equipment or water movement into long, thin, brown stripes.

Despite many advancements, considerable time, expertise, effort, and money are required to successfully maintain grasses grown outside their natural range of adaptation. The following information outlines management practices needed to successfully manage bentgrass grown in chronically stressful environments (Table 10.1).

## CULTURAL PRACTICES

### AIR MOVEMENT

In the absence of proper soil moisture and air movement, bentgrass turf can reach surface temperatures in excess of 125°F (52°C). At these temperatures, turf will quickly wilt and, if not addressed, substantial stand loss will occur. In an effort to reduce these lethal conditions, various surface and subsurface air movement techniques are available to complement other management practices used to maintain bentgrass.

Proper air circulation helps reduce the direct heat load on bentgrass surfaces. Air circulation also helps plants transpire during summer heat-stress periods, which naturally cools them. Newly constructed bentgrass greens should not be located in pocketed sites surrounded by anything preventing adequate wind or air movement. Additionally, established courses should remove these obstacles to encourage better air circulation and/or install fans around pocketed greens to improve surface air movement.

**Surface movement.** A cultural practice unique to bentgrass greens is to use surface area fans in warmer months to disperse the heat load and increase transpiration, which encourages evaporative cooling (Figure 10.4). Sufficient air movement is the most effective means of reducing stress on bentgrass greens grown in heat-stress environments. For example, a healthy bentgrass turf has a surface temperature of over 100°F (38°C), with hot spots being over 120°F (49°C) when air temperatures are 90°F (32°C). If a 3.5 mph (6 km/hour) breeze is present, the turf surface temperature drops to ~90°F (32°C). If misters are combined with the surface fans, the turf surface temperature drops to ~80°F (27°C). With fans,

**FIGURE 10.1** Bentgrass/*Poa annua* putting greens outside their region of adaptation into hotter, humid areas.

**FIGURE 10.2** Summer decline describes bentgrass physiological stress along with disease complex.

**FIGURE 10.3** Summer decline typically begins in poorly drained areas with restricted air circulation.

**TABLE 10.1**
**Management Techniques Necessary for Successfully Growing Creeping Bentgrass Golf Greens in Stressful Environments**

1. Sufficient air movement by trimming limbs and trees and installing fans at critical sites
2. Water management including proper greens/tees construction and sufficient funds for hand-watering
3. Thatch management including aggressive aerification, topdressing, plus summer venting
4. Strategic fungicide programs when certain diseases are expected and best means of control
5. Excellent water quality—if not available, excellent drainage and periodic flushing are required
6. Conservative mowing heights and grooming practices in summer to provide more leaf area and better rooting
7. Sufficient fertility to replenish depleted reserves
8. Using improved varieties as these typically have better stress and disease tolerance
9. Sufficient sunlight with morning sunlight generally most important
10. Minimizing physical turfgrass damage as extensive time will be required for recovery, allowing pests such as weeds and algae to invade

**FIGURE 10.4** Surface air movement disperses heat and increases transpiration, a natural cooling process.

**FIGURE 10.5** Fans placed too far away or above the turf canopy are considerably less effective.

however, care must be taken not to allow soil to excessively dry; with misters, increased disease incidence, especially dollar spot, may occur.

Fans work best when installed and used in late spring before heat and moisture stress start. Research indicates maximum benefits when fans provide 3.5 to 9.0 mph (6 to 14 km/hour) wind speed at the turf canopy across the whole green surface. Fans placed considerable distances away from and/or elevated above the turf surface seldom provide the necessary air movement at the turf canopy surface to reduce heat stress (Figure 10.5). Current trends involve placing oscillating fans near or on the collars of greens or placing larger fans just off the collars. Fans are operated 24/7 from late spring until late summer.

**Subsurface movement.** Another management tool in bentgrass management is providing subsurface air movement through drainage tiles (Figure 10.6). Blowers connected to the drainage tile force air through this system into the rootzone mixture or can create vacuum to remove excess soil moisture. This system should be considered on greens grown in especially hot areas, as the air movement displaces potentially toxic gasses like $CO_2$, methane, and hydrogen sulfide and provides deep soil oxygenation. These systems are also very beneficial by removing excess soil moisture following heavy rainfall.

**FIGURE 10.6** Subsurface air movement introduces oxygen by removing excessive soil moisture.

## Green Construction

Directly or indirectly, the primary cause of damage to creeping bentgrass in summer is excessive soil moisture. Without precise control, bentgrass can and will suffer from excessive or inadequate soil moisture that may lead to other conditions, such as diseases or soil compaction (Figure 10.7). Precise moisture control is best achieved with golf greens built as outlined in the green construction chapter. The intent is to

**FIGURE 10.7** Precise moisture control is necessary for bentgrass in hot, humid regions.

resist compaction, provide proper internal drainage, and prevent excessive prolonged soil wetness that may weaken bentgrass root systems. This is primarily due to root decline from anaerobic soil conditions such as "black layer" or attack from soil-borne diseases such as *Pythium* root rot or *Rhizoctonia* brown patch (Figure 10.8).

Poor root drainage in combination with hot summer temperatures also may lead to turf scald and wet wilt. *Scald* occurs when turfgrass is submerged in standing water after heavy rainfall and subsequent sunlight causes excessive (>100°F, >38°C) wet temperatures; this damages plants as heat is transferred downward to stems and roots (Figure 10.9). Standing surface water, therefore, must be removed before the sun reappears. *Wet wilt* occurs by a combination of saturated soils and high temperatures that prevents proper evapotranspirative cooling on the turf. Though adequate moisture is present, stems and roots are unable to absorb moisture and they collapse, becoming damaged. Wet wilt symptoms appear as brownish, water-soaked grass (Figure 10.10). Dead turf results in low-lying, poorly drained areas and greens with excessive thatch/mat layering that act like sponges, holding excessive moisture. Damage is exacerbated when mechanical practices, such as mowing or traffic, are allowed on the weakened turf.

**FIGURE 10.8** Excessive soil moisture often encourages black layer (shown) and/or soil-borne diseases.

**FIGURE 10.9** Turf scald from standing water during hot summer temperatures.

**FIGURE 10.10** Turf wet wilt from saturated soils during high temperatures.

Soil layering and excessive soil/surface compaction provide additional stress to the turf by further reducing proper surface drainage, maintaining proper soil $O_2$ exchange, and holding additional moisture. Flatter greens, with slower surface drainage, also tend to experience scald and wet wilt. Affected grass needs to be cooled but also have soil drying/aeration promoted by hand-syringing and sufficient coring and spiking.

Managing summer wet wilt on bentgrass greens includes:

1. Reduce aggressive management practice events.
2. Avoid mowing spongy-wet greens. Use solid rollers.
3. Raise mowing heights to 0.150 inch (3.8 mm) or at least 20% higher than currently mowed. Reduce weekly mowing frequency to four or five times when greens are weak and turf loss is occurring.
4. Stop all grooming practices aimed at increasing ball roll distances.
5. Hand-syringe to better control/regulate soil moisture.
6. Run fans continuously.
7. Aerify to increase soil $O_2$ and to reduce soil moisture.

### Irrigation/Water Management

Proper surface and internal soil drainage plus irrigation are key parameters in growing bentgrass/*Poa* in stressful areas

FIGURE 10.11 Excessive thatch acts as a water sponge.

FIGURE 10.12 Hand-watering to precisely apply appropriate moisture to various parts of greens.

(Figure 10.11). Three approaches are used in summer to water bentgrass: (1) overhead irrigation, (2) hand-watering, and (3) syringing.

**Overhead irrigation.** During summer stress, overhead irrigation should not be the main source of applying water to bentgrass greens. Daily overhead irrigation encourages shallow rooting, algae, traffic stress, ball marks, moss, and many diseases. The recommended philosophy is to use heavy and infrequent applications, allowing grass/soil to dry-down to near wilting between irrigations.

Most superintendents prefer to keep their greens slightly on the dry side to minimize disease occurrence, encourage deeper rooting, and limit ball marks. Drier greens also absorb summer thunderstorms and other rainfall events more quickly and easily than saturated greens. Furthermore, drier greens retain less heat, and cool off at night more efficiently than saturated greens. It is generally easier to apply water when needed rather than trying to remove excessive amounts. As a general rule, superintendents try to "stretch" the time between heavy, overhead irrigation cycles. Two to five days between the heavier overhead irrigation cycles are typically utilized to encourage deeper rooting and to allow the flushing of salts out of the surface. These deep, infrequent irrigation cycles are supplemented by hand-watering or syringing on localized areas displaying drought stress.

**Hand-watering.** Hand-watering allows deep watering where needed without overwatering areas with sufficient moisture. Hand-watering with a hose is often supplemented to normal sprinkler irrigation on "hot spots" throughout the summer heat-stress period (Figure 10.12). Constant (hourly) inspection from 10 a.m. to 6 p.m. daily, probing for moisture status, and supplemental hand-watering are necessary during peak summer stress. Hot-spot–prone areas often do not form typical morning dew patterns compared to normal turf. During cooler morning hours, these hot spots can be sufficiently watered. However, remaining hot spots during hot afternoon hours must receive a light mist or syringe (see later). Heavy hand-watering on hot spots during the heat of the day results in scald as discussed previously. To improve water penetration, hot spots are typically cored with a pitchfork or soil core probe to facilitate water infiltration through the thatch and propel water into the soil. Commercial water injection forks attached to irrigation hoses are available to help water penetrate thatch/mat to combat localized dry spots. Hand-watering is preferred to syringing because the former only treats stressed areas of the green, thereby reducing the chance for over-watering the entire green or its surrounds. Wetting agents should be used prior to hot-spot development. In general, one to three extra people per 18 holes will be needed to properly monitor and prevent wilt and lethal heat accumulation.

**Syringing.** Syringing is the process of periodically applying a very thin film of water, generally through the irrigation system or by hand, to gently wet the leaf blades and relieve heat stress via evaporative cooling, yet not be so heavy as to wet the soil (Figure 10.13). As this applied syringed water evaporates, it extracts heat from the turf surface, thereby cooling it. When water evaporates, it absorbs heat (e.g., 580 calories at 86°F or 30°C, for 1 g water) from the leaf and its environment. This is termed latent heat of vaporization. This principle works better when air moisture is at a lower relative humidity. A typical syringing program involves two to six daily cycles when temperatures exceed 85°F to 90°F

FIGURE 10.13 Combining hand-misting with surface fans to cool turf via evapotranspiration.

(29°C to 32°C). Syringing cycles generally are spaced 45 minutes to 1.5 hours apart and apply about 0.005 to 0.01 inch (0.13 to 0.29 mm) of water each. If the turf canopy does not dry between syringing cycles, the relative humidity may be too high to allow expedient ET to occur. Other possible side effects of constant, light syringing are the development of a shallow root system and excess salt accumulation. It is important to remember syringing is a supplementary cultural practice employed to manage severe summer stress. It is not a replacement for the regular, deep, infrequent irrigation cycles.

**Localized dry spots.** Another problem associated with bentgrass grown on sand-based greens is the occurrence of hydrophobic localized dry spots (referred to as LDSs, or isolated dry spots). In these areas, rootzones become hydrophobic, water is repelled from organic-coated sand particles, and bentgrass/*Poa* roots are unable to absorb water. LDSs form naturally as soil organic matter decomposes, coating sand particles with various acids that repel water. In *hydrophobic* (meaning water-hating) sands, shoots wilt and the color of shoots quickly change to a purplish- or bluish-gray. If allowed to go untreated, LDSs will eventually kill turf.

Detecting dry spots early is the key to turf survival. A reduction in morning dew formation serves as an early indicator of LDS. Another means of determining LDS severity is by pulling a soil core and placing water droplets every half inch (1.3 cm) or so along the length of the core. If these droplets do not penetrate the core within one minute (e.g., remain in droplet formation), this indicates the soil is hydrophobic (Figure 10.14). Handheld time-domain reflectometer (TDR) probes also provide soil moisture data. However, these devices do not indicate how much moisture is plant available; thus, they should be viewed as a supplement to dry spot detection. If the soil is dry, then the localized dry areas should be deeply spiked with a solid-tine aerator, soil probe, pitchfork, or other similar device. The area should then be deeply hand watered, and treated with a wetting agent to improve water penetration and retention. Overhead sprinklers should not be used in these situations because water is wasted on unneeded areas and also may encourage disease occurrence. LDSs often recur annually in the same areas. Therefore, identifying and mapping these chronically affected areas will help future management and allow preventative treatment with wetting agents. Refer to the chapter on irrigation for more information concerning LDSs and wetting agents.

**Salinity.** Another potential problem with growing bentgrass and regular shallow irrigation cycles is a gradual buildup of excess salinity. The light and frequent irrigation cycles often used to maintain these grasses, combined with the high summer evapotranspiration rates and applications of salt-based fertilizers, are conducive for salt accumulation. This is magnified when highly soluble fertilizer sources are used at relatively light but frequent rates and insufficient rainfall occurs to periodically leach salt below the rootzone. This is more common on newly constructed greens receiving light, frequent N and potassium (K) applications as well as light, frequent overhead irrigations to encourage rapid turf coverage. Under hot, dry, summer conditions, soil salinity levels as low as 600 ppm in the upper 1/2 to 1 inch (1.3 to 2.5 cm) of soil can begin to stress bentgrass. During cooler growing conditions, critical salinity levels often increase to ~1,000 to 1,200 ppm (1.6 to 1.9 dS/m).

## Thatch Management

The high water-holding capacities of the thatch/mat layer often prevent or minimize internal water and air movement into the rootzone, effectively sealing off the sand and roots below. It is not uncommon to have a wet thatch/mat layer while the sand below the thatch layer is powder dry. The thick, organic thatch/mat layer is much like a dishwashing sponge that holds significant quantities of water. This water can often be squeezed out of the thatch or mat layer for several days after irrigation or rainfall even though the sand mixture below is relatively dry (Figure 10.15). This environment is very favorable for fungi and algae development and discourages deep turf rooting. An excessively wet thatch/mat layer also retains significant quantities of heat during summer months, further stressing crowns and roots growing in this layer.

Without proper agronomic practices, bentgrass greens often develop a "puffy" surface as plant stems elongate, thatch

**FIGURE 10.14** Localized dry spots indicating hydrophobic soils.

**FIGURE 10.15** Thatch holding excessive moisture is the major cause of summer bentgrass decline.

swells, mowers sink into this layer, and scalping results. Aerial roots form seeking $O_2$ absent in the rootzone. Ways to prevent "puffy" bentgrass include:

1. Proper acrification
2. Spoon feeding N
3. Light, frequent topdressing
4. Properly timed light to moderate vertical mowings/grooming
5. Routine brushing during periods of active growth
6. Raising mowing heights

**Aerification.** Creeping bentgrass has poor tolerance to compacted soils and generally has a shallow (less than 3 inches, 7.6 cm) root system in summer. *Poa annua,* meanwhile, is better adapted to compacted soil conditions and thus has the competitive advantage over bentgrass when these conditions occur.

Soil cultivation should ideally be performed prior to a stress period when the turf is actively growing. This cannot be overemphasized, as once stress begins, the chances of successful aerification are reduced (Figure 10.16). The most popular method of soil cultivation on golf greens is hollow-tine aerification. These tines generally penetrate 3 inches (7.5 cm) deep, are 3/8 to 5/8 inch (9.5 to 16 mm) in diameter, and most often are spaced on 2-inch (5 cm) centers. Cores are removed if the existing rootzone media are unacceptable and sand topdressing mix is brushed into the holes. If the existing rootzone media are appropriate, cores are mechanically pulverized, thatch is removed, and rootzone materials brushed back into the holes.

Aerification frequency most often depends on the amount of traffic an area receives, generally three to four times yearly. A general goal is to impact 15% to 25% of the surface area yearly by core aerification. To obtain this, bentgrass greens are aerified from late summer through midspring when temperatures are between 50°F and 80°F (10°C and 27°C). Typically, at least three annual aerifications (e.g., early and late spring plus late summer) with relatively large (1/3 to 3/4 inch, 1.3 to 1.9 cm) tines are performed. In cooler regions, four annual core aerifications with smaller (3/8 inch, 1 mm) diameter tines typically are performed in midspring, early summer, late summer, and early fall. In addition, on older or native "push-up" greens, deep-tine aerification (e.g., 8 to 10 inches, 20 to 25 cm deep) is also annually performed. This helps improve internal drainage and breaks up any compacted "plow layers" that may have formed as a result of repetitive cultivation to the same depth year after year. Deep tine aerification should be timed in spring during active growth.

Core aerification during summer months is with small (e.g., 1/4 inch, 6.4 mm) hollow or solid tines. These small tines (often called "needle" or "pencil" tines) allow the benefits of aerification without extensive disruption to the playing surface (Figure 10.17) and the smaller holes heal more quickly. If hollow-tine aerification is practiced then, cores should be quickly removed from the surface via a backpack blower to prevent unwanted heat buildup. If the desire is to change or thicken the existing bentgrass stand, interseeding following each aerification with 1/4 pound seed/1,000 square feet (1.2 g/m$^2$) followed by topdressing may eventually help. During the stressful summer months, especially where high humidity accompanies heat, practices causing leaf abrasion should be avoided, including heavy sand topdressing, aggressive vertical mowing, or dragging-in aerification cores.

The least disruptive form of core cultivation is high-pressure water or air injection. These systems use small, highly pressurized jets of water or air that can penetrate more than 8 inches (20 cm). Regular use of hydroaerifiers dilutes (distributes) surface organic matter deeper into the rootzone, improves soil $O_2$, and reduces localized dry spot. Weekly aerification with these units during stressful summer months and every 21 days in fall, winter, and spring is often beneficial. Lastly, it is important to follow any core aerification event with a light rolling of the surface. This prevents mower scalping, which often occurs due to the uneven turf surface from the aerification event.

**Slicing and spiking.** Soil aerification should be supplemented with a variety of venting tools (Figure 10.18). Spiking,

**FIGURE 10.16** Healthier bentgrass (left) compared to summer decline bentgrass from insufficient aerification.

**FIGURE 10.17** Summer core aerification with small diameter tines to avoid extensive turf surface disruption.

**FIGURE 10.18** Spiking as a supplemental tool to aerification to provide additional soil venting.

slicing, and using a high-pressure water injection system are preferred in warmer months. Slicing involves thin, V-shaped blades that slice (or cut) into the top inch (2.5 cm) or so of turf. These thin slices help disrupt algae formation, divide stolons to encourage turf regrowth, and provide water and air entrance ports into thatch and soil. Slicing or spiking is normally performed weekly in two directions during summer and more frequently on problem areas. Many similarly shaped venting tools are available that are designed to minimize summer stress while providing the desired benefits.

**Verticutting and grooming.** Moderate to heavy verticutting is performed at least once in spring (March until early May) and repeated in early fall. Verticutting helps thin the turf, removes unwanted thatch and mat, and allows better soil $O_2$ diffusion and water penetration. Proper aerification coupled with sand topdressing and light verticutting also are needed to minimize development of the thatch/mat layer. A number of cultural practices designed to remove or relieve soil compaction and excessive thatch/mat development are available.

Verticutting is generally followed by topdressing, rolling, light fertilization, and moderate watering. Light vertical mowing (or grooming) at ~0.1 inch (3 mm) depth, enough to just stand the leaf blades up to promote an upright growth habit, also may be performed weekly in fall and early spring. This is followed each time by a light topdressing with the same soil mixture used in green construction.

**Topdressing.** Sand topdressing is the most effective means of preventing and diluting existing thatch/mat buildup. Typically, 40 to 50 cubic feet sand/1,000 square feet (1.2 to 1.5 $m^3$/100 $m^2$) yearly or about 4,500 pounds dry sand (2,200 kg/100 $m^2$) are needed to help reduce or dilute organic layering. Lower levels are possible if other mechanical means such as aerification and verticutting are incorporated to reduce thatch. Thatch control can be accomplished by light (1 to 3 cubic feet/1,000 square feet, 0.03 to 0.09 $m^3$/100 $m^2$) topdressing, which also reduces algae development and helps provide a smoother putting surface. This is performed weekly during spring, early summer, late summer, and early fall months. Less aggressive topdressing occurs during summer months due to potential damage to stressed plants from the topdressing and subsequent brushing. Turf injury from inappropriate topdressing during summer stress may not become apparent for three to five days.

A key to success is using a sand topdressing material that matches the particle-size distribution of the underlying rootzone. Additionally, all sand topdressing should be free of, or have very little, silt or clay. With repeated use, the presence of these finer diameter particles clogs the desirable macropores of a sand-based rootzone. Following topdressing, the sand is incorporated (removed from the turf surface), most often by light weight brushing. Hand-brushes have become popular to use to perform this due to their ease of handling and less negative impact on the turf leaf surface. During summer, hand-pushed rotary spreaders are preferred as are blowers that help work the sand into the canopy with minimum leaf abrasion. Also, syringe prior to topdressing if wilt is present.

**Black layer.** A distinctive black layer may develop in the rootzone when excessive moisture and/or soil compaction causes anaerobic (lacking $O_2$) soil conditions. Hydrogen and iron sulfide and methane gas concentrations rise and can reach toxic concentrations for bentgrass roots. If conditions persist, a layer of metal sulfides forms in the saturated zone, often in the top 4 inches (10 cm) of the soil surface. This anaerobic zone takes on a characteristic dark (or black) color due to the reduced metals present, as well as a characteristic rotten egg or sewer-type smell. Turf symptoms include leaves that turn yellow or reddish-brown and start to thin. It typically occurs in low, poorly drained sites. Since the problem is mainly due to anaerobic conditions, the primary solution is to improve soil $O_2$ diffusion. This involves prudent aerification, pH regulation, and water management practices to reduce the toxic effects of anaerobic conditions. Where the underlying soil or rootzone does not have adequate internal drainage, this problem will also need to be addressed.

### Mowing and Rolling

Rooting depths directly correlate with mowing height. As the mowing height is lowered, root systems become shorter (Figure 9.3, Chapter 9). The lowest mowing height for bentgrass should be from early fall through early spring. Greens during this time are maintained at heights generally between 0.125 and 0.156 inch (3.2 and 4 mm) and are mowed daily (Table 10.2). By midspring through early fall, the highest tolerable mowing height should be used to provide deeper and stronger rooting. As summer temperatures and stresses increase, raising the mowing height is one of the most effective management techniques. This and other mowing strategies to consider during summer on heat-stressed bentgrass greens include:

- Raise mowing heights > 0.140 (3.5 mm).
- Reduce mowing frequency to four to five times weekly, or replace a mowing event with rolling alone.
- Lightweight walk-behind mowers are preferred since "cleanup laps" with heavy triplex mowers compact and damage the turf.
- Double cutting should be avoided, and perimeter cleanup passes skipped to every other day.

**TABLE 10.2**
**Mowing Heights for Turfgrass Species Used for Golf Greens and Tees**

| Species | Mowing Height (inches)[a] | Mowings per Week |
|---|---|---|
| | **Greens** | |
| "Ultradwarf" bermudagrass | 1/8[b] to 3/16 | Five to seven |
| Tifdwarf bermudagrass | 5/32 to 3/16 | Five to seven |
| Tifgreen (328) bermudagrass | 3/16 to 1/4 | Five to seven |
| Creeping or colonial bentgrass/Poa | 1/10 to 3/16 | Five to seven |
| | **Tees** | |
| Creeping or colonial bentgrass | 1/4 to 1/2 | Three to five |
| "Ultrdwarf" bermudagrass | 1/4 to 1/2 | Three to five |
| Tifdwarf bermudagrass | 1/2 to 3/4 | Two to three |
| Kentucky bluegrass | 3/4 to 1 | Three to five |
| Kikuyugrass | 7/16 to 5/8 | Two to five |
| Perennial ryegrass, fine fescue | 3/8 to 1 | Three to five |
| Seashore paspalum | 5/16 to 1/2 | Three to five |
| Zoysiagrass | 7/16 to 5/8 | Two to five |

[a] Multiply inches by 25.4 to obtain millimeters or by 2.54 to obtain centimeters.

[b] Tifgreen and Tifdwarf bermudagrass mowing heights below 3/16 inch (0.48 cm) are recommended only for short-term durations such as during tournament play. Newer ultradwarf bermudagrass cultivars are more tolerant to longer term lower mowing heights.

- Avoid mowing greens when they are spongy-wet or have casual water present. Injury from this is long-lasting and typically occurs initially on cleanup laps. Algae and weeds often invade these areas as turf thins.
- Slow the traveling speed of the mower on cleanup laps.
- Use solid rollers (instead of grooved).
- Minimize excess water, which encourages compaction, scalping, and sliding (tearing) of mowers.
- When aerifying, double aerify cleanup laps, protect (e.g., cover) collars when turning mowers or rollers, and fertilize in the optimum range to promote recovery. Shade on these areas also slows turf recovery.
- Clippings are always collected on greens as they interfere with ball roll, reduce sunlight penetration, and may act as a source of inoculum for several diseases.
- Mowing direction should be changed or rotated daily to promote upright shoot growth and reduce grain.

Rolling of greens also has become an important cultural practice, especially in summer when mowing heights must be raised for grass survival. This helps smooth the putting surface, often improving ball roll distances—up to 8 to 12 inches (20 to 30 cm) as measured by a Stimpmeter. Optimal situations for rolling include following core aerification or spiking to avoid mower scalping and on days mowing is skipped or when mowing heights are increased. Rolling should not be performed when greens are wet or suffering from stress. With the demands for faster putting green speeds, this practice provides better ball roll without excessively low mowing during the summer-stress period. The major limiting factor to roller use is abrasion and mechanical injury occurring in the collars and chronically wet areas such as the low points of the green where water tends to settle. These areas should be avoided, especially during periods of high heat and humidity.

### Fertilizing

Devising and implementing a sound fertility program is one of the most important agronomic considerations when growing bentgrass. Superintendents should formulate strategies for each green according to their financial resources, traffic intensity, geographic location, microclimates, player/quality expectations, specialized rootzone construction method, and the age of the rootzone/turfgrass.

Regularly fertilizing bentgrass provides an aesthetically desirable and playable surface and greatly influences the recuperative potential and stress tolerance of the grass. However, overstimulation with N fertilizer can lead to short roots and lush, green top growth that is more susceptible to diseases and drought stress. Adequate K, meanwhile, is necessary for plant stress tolerance, and efforts to maintain sufficient soil K are an important management strategy for heat-stressed bentgrass.

**Rates.** The total amount of N used for bentgrass greens generally ranges from 3 to 8 pounds N/1,000 square feet (15 to 39 g N/m$^2$) annually for established putting greens, while slightly more (~25%) will be required for newly planted turf. Specifically, in heat-stress environments, research indicates 4 to 6 pounds N/1,000 square feet (20 to 29 g N/m$^2$) annually provides best bentgrass quality and healthy turf (Figure 10.19). Due to its leaching potential and high needs by

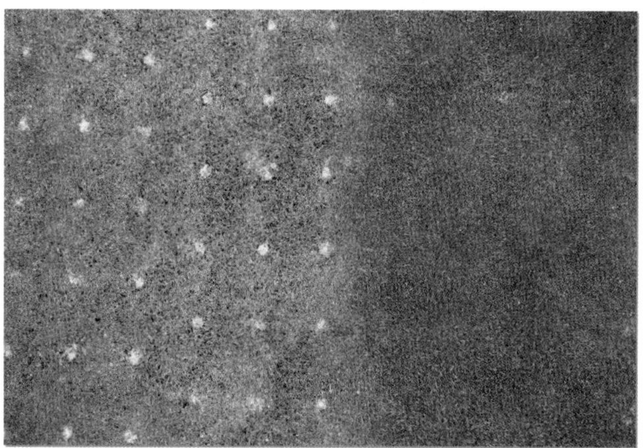

**FIGURE 10.19** Insufficiently fertilized bentgrass (left) to promote recovery from cultural practices and summer stress.

plants, K rates should equal N rates or even be slightly higher. Phosphorus (P) rates, in lieu of a soil test, should be approximately one-half the rate of N. An all-purpose fertilizer such as a 2-1-2 N-P-K ratio would be appropriate.

Research also indicates a 50:50 yearly granular-to-liquid-fertilizer source as best for building important root carbohydrates and providing adequate growth and shoot color. Typically, granular fertilizers are used during the fall, winter, and early spring months, especially for K and P needs. Liquid materials may be used during this time to enhance color. During the summer stress periods, many switch predominantly to liquid sources and "spoon feed" the greens as needed (e.g., when soil and tissue tests indicate nutrient needs, when color is needed, or when growth dramatically slows)—typically, every one to two weeks. Granular fertilizers are then used in fall after summer stress periods to promote recovery.

**Timing.** Timing for this 3 to 8 pounds N/1,000 square feet (15 to 39 g N/m$^2$) is based on the temperatures affecting shoot and root growth (Table 10.1). Table 10.3 lists a general rule of thumb for the percentages of this total N amount applied during the various seasons of the year.

**TABLE 10.3**
**Percentages of Nitrogen Fertilizer Applied to Bentgrass/*Poa* Golf Greens during the Year**

| Season | Annual N Application to Bentgrass/*Poa* Golf Greens (%) |
|---|---|
| Fall | 40 to 75 |
| Winter[a] | 0 to 30 |
| Spring | 10 to 20 |
| Summer | 10 to 20 |

[a] Nitrogen should not be applied to greens where soils regularly freeze during these months. 'Penncross' greens typically have a 40:30:20:10 annual N percentage use in fall:winter:spring:summer, while newer cultivars typically have a 40:30:10:20 annual N percentage use in fall:winter:spring:summer.

With improved bentgrass cultivars, superintendents are slightly altering their seasonal percentages from 40:30:20:10 fall:winter:spring:summer for 'Penncross' to 40:30:10:20 for newer cultivars. This alteration provides less spring fertilization, which typically produces a flush of undesirable top growth, in favor of more summer fertilization to promote grass recovery from summer decline, traffic damage, ball marks, and so forth on these newer cultivars.

A typical fertilization schedule based on Table 10.3 would include 3/4 to 1 pound N/1,000 square feet (3.7 to 4.9 g N/m$^2$) for each month of fall and early winter. Midwinter (depending on location) would probably receive slightly less (e.g., 3/4 pound N/1,000 square feet, 3.7 g N/m$^2$) to reduce succulent growth in the subsequent spring months. Late winter and early to midspring months might receive 1/4 to 1/2 pound N/1,000 square feet (1.2 to 2.5 g N/m$^2$) each, while summer months would receive minimal, but some, N (e.g., 1/4 pound N/1,000 square feet each, 1.2 g N/m$^2$) in an attempt to reduce heat, drought, and disease susceptibility. For the upper transition zone and in areas where the soil freezes in the winter, most of the annual N requirements should be applied during the late summer and autumn months, with little application during winter. This will avoid the spring flush of growth and minimize disease problems. Table 10.4 lists typical fertilizer programs for tees planted with various grasses.

**Micronutrients.** Iron (Fe), sulfur (S), manganese (Mn), and magnesium (Mg) nutrient deficiencies may appear, especially when soil pH and/or P levels are high and pure sand greens are used. Tissue testing indicates relative nutrient levels in plants. Light, frequent applications (e.g., 1 to 2 ounces of product/1,000 square feet, 3 to 6 L/ha, every three to four weeks) of one or more of these nutrients often produce desirable color response without undesirable succulent shoot growth. Research also suggests the lack of the micronutrients Mn and zinc (Zn), as well as the macronutrient P, can contribute significantly to typical summer decline symptoms. Tissue testing is the most accurate means of monitoring these and other elements in plants. Excessive use, however, can cause phytotoxicity and/or imbalances with other necessary nutrients. Humic acid or humates typically contain one or more micronutrients in chelated form plus some N, and therefore also add color when used.

**Biostimulants.** Biostimulants containing cytokinins and possibly gibberellic acid (GA), have proven beneficial, especially when sand-based soils are used. Cytokinins can help promote rooting, reduce water use, and provide an overall healthier plant. Biostimulants can also increase antioxidants such as beta-carotene and ascorbic acid, reducing oxidative damage to plant cell membranes exposed to heat stress. Biostimulant use should begin in spring prior to summer stress and continue according to label directions. Regular, continued use of biostimulants is necessary for these to provide the desired results. The use of GA-containing biostimulants, however, may negate the desirable effects of GA-inhibiting PGRs.

**TABLE 10.4**
**Typical Nitrogen Application Rates, Frequency, and Yearly Total Amounts for Golf Course Tees**

| | Tees | | |
|---|---|---|---|
| Grass | Single Rate (pounds N/1,000 square feet)[a] | Frequency (weeks)[b] | Yearly Total (pounds N/1,000 square feet) |
| Annual bluegrass | 0.25 to 0.5 | 2 to 4 | 2 to 8 |
| Bentgrass | 0.25 to 0.5 | 2 to 4 | 2 to 8 |
| Bermudagrass[c] | 0.5 to 1 | 2 to 4 | 4 to 12 |
| Fine-fescue | 0.25 to 0.5 | 2 to 4 | 2 to 8 |
| Kentucky bluegrass | 0.25 to 0.5 | 2 to 4 | 2 to 8 |
| Ryegrass | 0.25 to 0.5 | 2 to 4 | 2 to 8 |
| Seashore paspalum | 0.25 to 0.5 | 2 to 4 | 2 to 8 |
| Zoysiagrass[c] | 0.5 to 1 | 4 to 8 | 2 to 8 |

[a] Rates listed are typical for single applications. Single application rates, frequency, and yearly total N rates vary considerably between individual golf facilities, geographical regions, and desired results. Lower rates are acceptable for most golf facilities, especially for those turf areas not irrigated. Slow-release or controlled-release fertilizer sources should be considered with higher rates. Soluble sources should be used at lower rates and applied more frequently. Multiply pounds/1,000 square feet by 4.9 to obtain grams/square meter; multiply pounds/acre by 1.12 to obtain kilograms/hectare.

[b] Frequency represents fertilizing only during periods of active growth for each respective grass. The highest application rates should be reserved for optimum growth seasons of each grass. Typically, this is late summer through early spring months for cool-season grasses and late spring through late summer/early fall months for warm-season grasses. Fertilizer applications should be minimized or possibly eliminated during periods of natural slowed growth or where soils freeze.

[c] Refer to the chapter on cultural practices for fertilization recommendations of overseeded areas.

### Sunlight Requirements

$C_3$ (cool-season) plants such as bentgrass, in higher mowed environments, are more shade tolerant than most $C_4$ (warm-season) grasses. However, shady environments are often combined with reduced air movement and increased humidity that limits leaf-drying potential. Close mowing heights also reduce leaf surface area and eventually reduce rooting depths. The best growing and easiest to maintain bentgrass greens requiring fewest inputs tend to be those located in open, full-sun sites with no surrounding trees, buildings, or mounds to impede air movement. A rule of thumb is that greens should ideally receive 9 to 10 hours of sunlight in summer with half of this in the morning.

Greens receiving four hours of sunlight are able to survive if morning sun is available to help dry greens, but the more shade-tolerant *Poa* will eventually become the dominant species, ball roll distances will be shorter, and algae and certain weeds will be a constant threat. Mowing heights should be raised in shaded conditions. Avoid excessive surface and soil moisture by relying on hand-syringing, squeegeeing puddles, plus spiking and quadratining. Mowing should be avoided when the surface is wet. Promote air circulation by removing trees/brush and using surface fans. Less nitrogen (N) should be applied to shaded greens compared to greens growing in full sun. Use PGRs that discourage turf scalping and prune tree roots that compete with turf roots for nutrients and moisture.

## PEST MANAGEMENT

### Diseases

Several diseases can be chronic problems on bentgrass greens. Most diseases occur in summer, although some also occur in other seasons (Table 10.5). Mowing, fertilization, and seedhead control practices directly influence disease development, especially on *Poa* greens. For example, increasing mowing heights as little as 0.016 inch (0.4 mm) can decrease occurrence of certain diseases (e.g., anthracnose).

The most common diseases on bentgrass greens are root and crown rots from various species of *Pythium, Rhizoctonia, Colletotrichum* (or anthracnose), dollar spot (*Sclerotinia homoeocarpa*), and summer patch (*Magnaporthe poae*). Root and crown injury from these diseases stress bentgrass even more, beginning a cycle of stress–disease–stress. These diseases form a complex and, when combined with other stresses such as heavy traffic, low mowing heights, poor drainage, and

**TABLE 10.5**
**Common Diseases and Their Typical Occurrences for Bentgrass and/or *Poa annua* Golf Greens in Heat-Stress Environments**

| Disease | Winter | | | Spring | | | Summer | | | Fall | | |
|---|---|---|---|---|---|---|---|---|---|---|---|---|
| | 1 | 2 | 3 | 1 | 2 | 3 | 1 | 2 | 3 | 1 | 2 | 3 |
| Anthracnose (*Collectotrichum* sp.) | | | | | ◄ | — | — | — | — | ► | | |
| Brown patch (*Rhizoctonia solani*) | | | | | | ◄ | — | — | ► | | | |
| Dollar spot (*Sclerotinia homoeocarpa*) | | | ◄ | — | — | ► | | | ◄ | — | ► | |
| Fairy ring (*Lycoperdon* sp. + others) | | | ◄ | — | — | — | — | — | — | — | ► | |
| Moss/algae | | | | | | ◄ | — | — | ► | | | |
| *Pythium* diseases | | | | | | ◄ | — | — | ► | | | |
| Summer/take-all patch (*Magnaporthe* and *Gaeumannomyces* spp.) | | | | | | ◄ | — | — | ► | | | |
| Bacteria (*Xanthomonas* and *Acidovorax* spp.) | | | | | | ◄ | — | — | ► | | | |
| Yellow patch (*R. cerealis*) | ◄ | ► | | | | | | | | | ◄ | ► |

*Note:* Disease occurrence varies by regions, grass variety, microenvironment, management practices, etc. and this table is intended only as a guideline.

poor fertility, bentgrass can succumb to the summer bentgrass decline syndrome. Dollar spot is especially problematic during periods of warm weather and cool nights that produce heavy morning dews and during times when bentgrass is not aggressively growing. Some cultivars (e.g., Crenshaw, Century, Imperial, SR 1020, Grand Prix) are more susceptible than others, so check with a local turfgrass specialist regarding disease resistance. Anthracnose typically begins in spring and lasts until fall, while summer patch, brown patch, and *Pythium* diseases are most pronounced during summer. Refer to the turfgrass diseases chapter for considerably more information on these and other diseases.

**Fungicides.** Bentgrass management during summer often requires preventative disease control measures, since curative control options often are too late or work too slowly (Figure 10.20). Typically, this involves beginning a fungicide program in late spring when bentgrass-lethal temperatures approach those listed in Table 10.1. Normally, in early and late summer, dollar spot is the most prevalent disease. During hotter and drier summer months, disease strategies shift from dollar spot to brown patch, summer patch, *Pythium*, bacteria, and anthracnose. During this time, products providing brown patch control are tank-mixed with those providing *Pythium* root rot control since these diseases often occur simultaneously. For *Poa* greens, tank-mixing and rotating products are necessary for optimum anthracnose and summer patch control. Treatments in traditional heavy disease pressure areas should begin in spring prior to *Poa* seedhead production. Usually a 10- to 14-day spray interval is needed for this preventative measure. Superintendents should rotate between chemical classes to minimize the chance of disease resistance. Extra care should be taken when using the sterol inhibitor fungicides on bentgrass. Summer applications of these materials can reduce bentgrass growth and result in thin turf and increased algae.

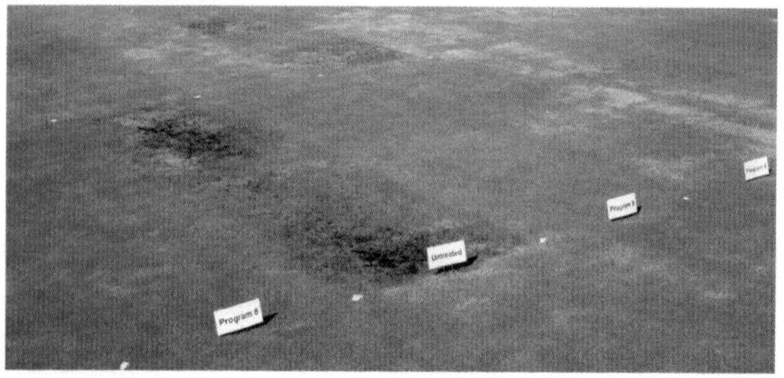

**FIGURE 10.20** A logical summer fungicide program is required for bentgrass survival.

One should remember, however, that successful bentgrass growth, especially through fungicide use, is still based on combining good science with the art of turfgrass management. Success with these and many other products on one golf course, in one year, or on one green on a golf course does not mean similar successes will universally occur. Too many unknown or poorly understood biological parameters influence this. Fungicide strategies constantly change; therefore, the reader should consult the extension turf specialist in his or her state for the latest available and recommended materials.

## Nematodes

Nematodes most often associated with bentgrass include sting (*Belonolaimus longicaudatus*), lance (*Hoplolaimus* species), and root-knot (*Meloidogyne* species). Activity is generally greatest on bentgrass in mid- to late spring, and again in early fall. These times follow active growth of turfgrass roots that subsequently supports higher soil nematode levels. The result of nematode feeding becomes most apparent when conditions become unfavorable (e.g., hot and dry) for the turfgrass. It may be prudent to sample your turf monthly during the growing season to determine the fluctuation of nematode populations. This will allow treatment before typical population peaks.

Analysis of soil and plant tissue is the only sure way to determine if, and what kinds of, nematodes are present and the level of risk they may pose to turf. Most states have an assay laboratory associated with the land-grant university. Other laboratories also provide nematode assay service. Consult these institutions about the services they offer, as well as gather instructions for preparing and submitting samples to them.

### Weed Control

**Herbicide use.** Using preemergence herbicides on bentgrass in stressful areas is a difficult choice. On the one hand, if not used, annual grasses such as goosegrass and crabgrass may invade. On the other hand, most preemergence herbicides act by preventing cell division when the growing point of the plant comes into contact with the chemical. Selective tolerance does not exist between weeds and desirable turfgrass roots trying to grow through this layer. Since most creeping bentgrass varieties have extremely shallow rooting depths during summer, herbicides applied on the soil surface may further inhibit this rooting. Read the label and consult with your state turfgrass specialist before putting any herbicide on bentgrass, especially during or just prior to summer heat stress. Postemergence herbicides are used only as a last resort, since turf tolerance to these during summer stress is poor. Refer to the chapter on weed management for additional herbicide use strategies on greens.

**Bermudagrass encroachment.** One of the hardest weeds to control in bentgrass is encroachment by bermudagrass (Figure 10.21). Bermudagrass from surrounding collars often invades weakly competitive bentgrass greens during summer. During construction, collars should be established on rootzone media identical to the putting greens. They should then be planted to bentgrass and/or perennial ryegrass to provide a buffer against invasive bermudagrass. Golf courses in warmer climates often sod zoysiagrass on the outside perimeter of the collar to provide an additional buffer against bermudagrass encroachment. Clubs also often heavily overseed bermudagrass collars with ryegrass to help suppress the bermudagrass throughout most of the year.

FIGURE 10.21 Bermudagrass is one of the weeds in bentgrass that is most difficult to control.

Bermudagrass control involves edging, mechanical handpicking, or physically removing the undesirable grass and resodding. Proper timing of certain herbicides such as siduron (Tupersan) or ethofumesate (Prograss) and plant growth retardants like flurprimidol (Cutless) helps suppress bermudagrass encroachment. Monthly applications of siduron beginning in spring with the resumed growth of bermudagrass work best. Summer applications do not appear to have much effect, but treatments should resume in late summer until expected heavy frost causes the bermudagrass to go dormant. Also available is an embedded steel track system that surrounds the green where a metal blade is run around the green periodically to sever encroaching bermudagrass stolons and rhizomes. Although this is effective, costs, installation disruptions, and diligence/dedication in use should be considered prior to purchase.

## ADDITIONAL CONSIDERATIONS

### Segregation and Patch Development

'Penncross' bentgrass, the most widely planted cultivar, is a polysynthetic from three parental clones that produces nine possible crosses. Only the first generation of this seed is certified. 'Penncross' greens will eventually (within five years) segregate into distinct mottled patches, which can sometimes give the false appearance of disease or grass contamination, and may add to surface grain (Figure 10.22). Lightly verticutting and reseeding each fall with 1 pound/1,000 square feet (5 g/m$^2$) will help mask this mottled appearance and help thicken weak areas thinned by summer stress. While 'Penncross' has frequently been observed to segregate, it appears the newer bentgrass cultivars also segregate after several years of use, although at a slower rate and with less intensity. It is worth noting that bentgrass seeds number about 4 million per pound and they are not all genetically identical.

### Reseeding/Interseeding

As mentioned, greens often experience loss shoot density following the summer-stress period and require seeding to regain density. Some superintendents are also interseeding newer

FIGURE 10.22 Winter bentgrass segregated patches.

improved cultivars into older ones in an attempt to increase the genetic pool and improve stress resistance. Seeding requires soil preparation to provide a suitable seedbed. Various intensities of aerification, verticutting, and topdressing are required for success, and it may take several years to successfully convert greens to the newer desired cultivar. Recent research on this topic indicates that only limited success (10% to 15%) in converting older cultivars to newer cultivars can be achieved while the green is still being used. The combination of close, frequent mowing and existing plant competition make it difficult for seedlings to establish and survive. The best chance of success would be to seed in late summer when the existing bentgrass is at its weakest. Competition from existing hardy *Poa* seedlings that are germinating at this time of year will also be a challenge. Spiking and seeding weekly at 0.25 pound of seed/1,000 square feet (1.2 g/m$^2$) during stressful summer months often take advantage of declining *Poa* growth, allowing increased chances of seedling establishment.

### Turf Nursery

An on-site turf nursery is highly recommended and practically a requirement when growing bentgrass/*Poa* in stressful areas. This allows repair of injured or dead turf on greens with the same grasses, and provides a test area for any new products or grasses being considered (Figure 10.23). The size of the nursery should be equal to or larger than the largest green on the property, or at least 10,000 square feet (929 m$^2$) in size. The rootzone media used for this nursery should be identical to what is used on the regular greens. Furthermore, this area should receive similar management practices as regular greens do.

### Collars

Bentgrass collars can easily decline/thin due to the compaction/traffic they receive from daily mechanical practices; excessive soil moisture, especially if native soil is used in their construction; more leaf surface area since collars are mowed at higher heights, which are more easily abraded; and abrasion from sand topdressing or from bunker splash accumulating from greens. Often, the most damaged areas are located between the edge of a green and another adjacent obstacle such as bunkers, water bodies, or steep drop-offs. Affected leaf tissue shows visible bruising and turf turns bronze-orange, purple, or yellow to lime green, especially if plant leaves are allowed to lie down. Leaf tips are initially most affected, but in later stages crowns are damaged, and turf turns a straw color, thins, and is invaded by algae or weeds. To avoid problems, reduce shade/tree root competition in collars. Topdress when temperatures are cooler and use blowers to move sand into the leaf canopy and off the leaves prior to mowing. Be careful not to overwater collars and avoid mowing when they are wet. Avoid turning equipment continuously around on collars. Use protective covers (e.g., plywood, plastic mats, lattice, etc.) when this is necessary (Figure 10.24). Mow collars at about 0.25 to 0.30 inch (6.2 to 7.6 mm) to reduce leaf surface area, which is more prone to mechanical damage and to encourage upright growth and increased plant density. Reducing cleanup laps to several days weekly also reduces mechanical stress. Incorporate extra yearly aerifications in collars or double aerify each time aerficiation is normally performed. On warm-season grasses, minimize exposure to certain PGRs (e.g., flurprimidol and paclobutrazol), especially during spring transition when they are vulnerable to stunting and thinning from repeated applications.

**FIGURE 10.23** An appropriately sized nursery green will be necessary for bentgrass in stressful environments.

**FIGURE 10.24** Avoiding collar damage by turning mowers on protective covers.

### Adequate Financial Resources

Needless to say, one of the most important prerequisites when growing bentgrass outside its naturally adapted region is adequate financial resources. Many of the cultural practices necessary to maintain bentgrass greens require significantly more labor, irrigation, fertilizer, mowing, chemical inputs, and specialized equipment than other greens. The major cost, however, will be associated with labor. When monitoring summer stress on greens at a bentgrass golf course, one to three additional employees will be needed from late spring through early fall. These employees will be primarily monitoring moisture conditions, hand-watering, and scouting for localized dry spots and disease symptoms. Additionally, they probably will be assisting in walk-mowing greens each morning.

Weekly preventative fungicide applications should also be budgeted during the summer stress period. Where air movement is an issue, fans are highly recommended on greens during summer. These have installation and power costs. Also, at least one additional employee is needed during the summer to repair damage from ball marks and to hand-pick weeds. The additional cost on an annual basis to maintain bentgrass greens is more than for bermudagrass putting greens. If the facility management/ownership choose to maintain bentgrass outside its region of adaptation, then a substantial financial commitment to the appropriate management practices needs to be made.

### Sympathetic Management and Members

Not only does a golf course need adequate financial resources, but it also needs management and/or members who are sympathetic to the dynamic seasonal life cycle and cultural requirements of bentgrass. As mentioned earlier, bentgrass will survive better during summer months outside its natural region of adaptation if the mowing height is raised. Uninformed members should not dictate lower mowing heights to increase ball roll and ultimately increase green speed at the expense of turf health. This may be difficult when a vocal group of low-handicap members or regular patrons demand fast greens. If this occurs, turf will decline. Also, chronic "hot" spots on greens have to be hand-watered quickly before the grass dies, regardless of play, which can be difficult when the golf course is full of players. Additionally, golfers hate nothing more than aerification, which can produce a temporarily bumpy putting surface. However, adequate aerification will be necessary to maintain drainage and ensure proper $O_2$ diffusion so turf roots survive. Fans may be annoying to many golfers, as some models are noisy, aesthetically disruptive, and sometimes hard to play around. Regardless, these have proven invaluable in heat stress environments.

Political communication skills are needed by the superintendent to educate the membership or golfers on bentgrass growth requirements. The superintendent must keep the channels of communication open. Greens committee members should be personally informed of vital cultural management practices and any situations affecting play. Providing members with a relevant article or research summary in a monthly newsletter has proven extremely beneficial and reduces confusion and misunderstandings by uninformed members. The scheduling of key tournaments should involve input from the superintendent along with the golf professional so that critical management practices like core aeration can be properly scheduled and the best playing conditions possible can be provided.

Experience, research, and trial by fire have proven the aforementioned principles as necessary when growing bentgrass outside its naturally adapted area (Table 10.6). Members and management should be fully informed and aware of the benefits and potential detriments of attempting this daunting task. Many successful superintendents have maintained high-quality bentgrass greens in very unfavorable environments. In the end, however, Mother Nature still has the final word on how easy or difficult any given particular year will be.

## BERMUDAGRASS

Bermudagrass has traditionally been the grass of choice for most golf courses in tropical and subtropical areas, as well as many other areas of the temperate growing zones (Figure 10.25). Bermudagrass provides a smooth putting surface and resists many pests to which other grasses, like bentgrass or *Poa*, are often susceptible. New dwarf bermudagrass cultivars offer similar short mowing height options that traditionally have been provided by bentgrass or *Poa*. In addition, bermudagrass is not as sensitive to heat, salinity, and moisture stress as bentgrass and is generally more tolerant to management practices such as aerification, topdressing, and vertical mowing. Soil pH for bermudagrass is between 5.5 and 7.0, although it often tolerates a wider range than this.

**Sunlight requirements.** One often unrecognized characteristic of bermudagrass is its inherently poor shade tolerance (Figure 10.26). Many facilities have trees that were small when the course was constructed. Over time, as trees enlarge, they cast larger and longer shadows that can interfere with normal bermudagrass growth and development. This is especially true in late fall and winter when the sun is lowest on the horizon and greens are overseeded. Therefore, in order to support bermudagrass greens, a minimum of eight hours of full sunlight (33 mol/m$^2$/day) must be provided on each green throughout the year. Morning sun should also not be neglected as this encourages early growth each day, helps dry the dew to minimize disease problems, and thaws frost so play can resume earlier. Morning sun is also important in summer as clouds typically build up by mid-day, severely limiting the quality and quantity of light reaching the green by early afternoon. The transition of some courses to zoysiagrass (as opposed to bermudagrass) has been most often because of excessive shade.

**Dormancy and winter-kill.** Bermudagrass begins going dormant when air temperatures approach 50°F (10°C). This dormancy often requires overseeding with a cool-season grass in late summer/early fall or painting to provide winter color (Figure 10.27). Overseeding, however, can be a disruptive process that often requires courses to close for several days to prepare the bermudagrass seedbed, perform the seeding, and provide postseeding management such as extensive watering (refer to Chapter 9 for an extensive discussion on overseeding and painting). Overseeding also goes through another transition period in spring, as the overseeded grass begins to weaken with the onset of heat stress and bermudagrass gradually greens up and fills in. Spring transition can be difficult when temperatures do not follow normal patterns,

**TABLE 10.6**
**Summary of Cultural Practices for Bentgrass/*Poa Annua* Golf Greens**

| Practice | Description |
|---|---|
| | **Fertilization** |
| Optimum timing for N fertilization | • Early fall through midspring |
| Damaging response to excessive N fertilizer | • During summer stress |
| Yearly N requirements | • 3 to 8 pounds N/1,000 square feet (15 to 39 g/m$^2$). Lower range for bentgrass, higher for *Poa* anthracnose management |
| Monthly N requirement | • 0.5 to 0.7 pound N/1,000 square feet (2.5 to 3.4 g/m$^2$) per growing month |
| Fertilizer application | • Slow-release granular (fall) + liquids for color |
| Liquid versus granular fertilizer use | • Yearly totals evenly split. 0.1 to 0.2 pound N/1,000 square feet (0.5 to 1.0 g N/m$^2$) every 7 to 14 days for liquid carrier during summer; 0.3 to 0.7 pound N/1,000 square feet (1.5 to 3.4 g N/m$^2$) every 3 to 4 weeks in fall and spring from slow-release granules |
| Acceptable pH range | • 6.0 to 6.5 (higher values encourage take-all patch) |
| Acceptable P levels | • Low to medium (excess levels encourage micronutrient deficiencies and *Poa annua*) |
| Optimum K levels | • Medium (improves drought and heat tolerance) |
| | **Cultural Requirements** |
| Overseeding | • In early fall, if needed at 1 pound seed/1,000 square feet; weekly during summer at 1/4 pound seed/1,000 square feet if new cultivar incorporation is the objective |
| Irrigation requirements | • Moisten full rootzone depth, rewetting just prior to wilting, plus heavy hand watering of hot (dry) spots; generally, more hand-watering needed for *Poa* greens |
| Air circulation requirement | • Required where natural air movement is restricted in heat-stressed environments |
| Mowing height | • 0.100 to 0.156 (1/10 to 5/32) inch (2.5 to 4 mm) |
| Mowing frequency | • Daily, reduced to three to five times weekly during stress |
| Mowing pattern | • Changed (rotated) daily |
| Mower type | • Walk-behind during stress periods, triplex during periods of active growth; solid rollers during stress |
| Clippings | • Removed |
| Thatch/mat control | • Frequent topdressing, lightly vertical mowing and grooming, regular aerification |
| Compaction control | • Using a sand-based rootzone, routine aerification, limit play (traffic) during wet conditions |
| Topdressing | • Lightly (0.2 cubic yard/1,000 square feet, 0.17 m$^3$/100 m$^2$) but frequently (weekly) during active growth, water in; monthly at 0.3 to 0.5 cubic yard/1,000 square feet (0.25 to 0.4 m$^3$/100 m$^2$) during slowed growth |
| Aerification (coring) | • Two to four times annually; minimum of midspring and late summer; at least one deep tine yearly; stop within 30 days of normal cold or hot weather |
| Regular (deeper) verticutting | • Once or twice yearly, once in late winter/early spring and another during active fall growth; often timed with core aerification |
| Grooming | • Performed weekly except during periods of extreme stress |
| Slicing | • Performed weekly in two directions to reduce surface compaction and algae and to facilitate soil air exchange and infiltration |
| Plant growth regulators | • Used to regulate turf growth; summer for bentgrass, almost year-round for *Poa* to regulate growth and seedhead development |
| Key diseases | • Bentgrass: dollar spot, *Pythium*, take-all patch<br>• Poa: summer patch, anthracnose |

often causing premature thinning of the overseeded grass before the bermudagrass has had the necessary temperatures to encourage its recovery.

Bermudagrass also has been susceptible to periodic “winter-kill” where freezes below 10°F to 15°F (–12°C to –15°C) can kill or thin the stand (Figure 10.28). This typically is more of a problem on fairways and roughs (see Chapter 2). Depending on the location, winter-kill is most severe in the northern and western areas of the transition zone, especially in shade- or wind-prone situations. Means to improve winter hardiness include:

1. Gradually raise the mowing height in fall to ~0.188 inch (4 mm) as night temperatures reach the low 50°F (10°C).
2. Minimize thatch layering as discussed later.
3. Maintain soil K levels in the medium range as deficient levels may increase turf susceptibility to winter damage.
4. Reduce shade and soil compaction as the worst winter damage often occurs in these situations.
5. Use covers if temperatures routinely fall below ~23°F (–5°C).

FIGURE 10.25 A desirable bermudagrass golf green.

FIGURE 10.26 Bermudagrass has inherently poor shade tolerance and will suffer under such conditions.

FIGURE 10.27 Overseeding dormant (brown) bermudagrass in winter to provide green color.

**Off-types.** In recent years, mutants or off-types of grasses in previously pure bermudagrass stands have occurred (Figure 10.29), presumably from the chance mutation of parent material or possibly contamination through mechanical means or encroachment from collars. Typically, off-types start to emerge about 7 years following planting, forcing many courses to regrass every 10 to 15 years. Off-types are not necessarily all bad, as these are often the basis of new, improved

FIGURE 10.28 Bermudagrass damage from a combination of winter shade and cold temperatures.

FIGURE 10.29 Patches of undesirable "off-type" bermudagrass.

grasses. However, they are different from the existing grass, most often in color and texture. As mowing heights have been lowered below the adapted heights for Tifgreen and Tifdwarf bermudagrass, mutations or biotypes have emerged that could survive these conditions. Tifdwarf, FloraDwarf, and Pee Dee 102 bermudagrasses are believed to be natural mutants from Tifgreen bermudagrass. Champion, Mini Verde, Reesegrass, MS Supreme, Classic dwarf, Sunday, and others are selections from previously planted Tifdwarf greens. TifEagle bermudagrass, being more genetically stable, has had fewer issues with off-types, presumably due to it being an induced mutant from cobalt radiation.

## CULTURAL PRACTICES

### Green Construction

Precise soil moisture control is not as critical for growing and maintaining bermudagrass as it is for bentgrass/*Poa*. However, the chances of growing successful bermudagrass increase with the quality of green construction. Many bermudagrass greens were initially constructed as "push-ups" from native soil or built with the fill resulting from pond construction or creek excavation. These native soils often contain high levels of silt and clay. "Push-up" greens have often proven unsatisfactory in today's market, where the pressure to open soon after rainfall events is greater and, due to increased play, soil compaction has also increased. Minimum construction specifications for bermudagrass greens include those described in the chapter dealing with putting green construction.

## Mowing

Bermudagrass greens are maintained at heights between 0.105 and 0.25 inch (2.7 and 6.4 mm) and are mowed five to seven times weekly. Older bermudagrass cultivars such as Tifgreen are mowed routinely between 3/16 (4.8 mm) and 1/4 (6.4 mm) inch; these are only mowed shorter (e.g., 5/32 inch, 4 mm) for short-term periods such as key tournaments. Tifdwarf is routinely mowed at approximately 3/16 inch (4.8 mm) while newer dwarf cultivars such as TifEagle, Champion, Mini Verde, FloraDwarf, and others tolerate routine mowing heights between 0.105 and 0.125 inch (2.7 and 3.2 mm). Mowing heights should be raised up to 25% higher during periods of reduced sunlight such as extended rainy periods or late summer as days become shorter. This cannot be overemphasized as continued low mowing during these stress periods typically results in thin, algae-prone greens that often suffer bermudagrass decline or mini-ring disease. Walk-behind mowers are often used, especially with mowing heights below 3/16 inch (4.8 mm), as it becomes difficult to set triplex mowers at these low heights without scalping and excessive compaction. Clippings are normally removed and the mower pattern changed or rotated daily to reduce grain. When reducing mowing heights during periods of active growth, avoid scalping by not decreasing heights more than 0.01 inch (0.25 mm) every four days.

## Fertilizing

To promote recovery from traffic and clipping removal, and to fulfill clientele color expectations, bermudagrass requires adequate nutrients. Furthermore, many courses are located in areas with extended growing seasons, high annual rainfall, and predominantly sandy soils that increase the need for precision of using the right amounts and ratios of nutrients.

**Timing.** Timing is partially based on the minimum and optimum temperatures necessary for turfgrass growth. Table 10.7 lists growth temperatures for cool- and warm-season turfgrasses. If temperatures are outside the growth range of the grass, fertilizer applications will be inefficiently utilized.

**Nitrogen rates.** A general yearly range of N needs for bermudagrass golf greens is from 6 to 18 pounds N/1,000 square feet or 1/2 to 1 1/2 pounds N/1,000 square feet per growing month (24 to 74 g N/m$^2$). Courses with sufficient resources, excessive traffic, sandy soils, older bermudagrass cultivars, and elevated demands from players would use the higher rate range. Courses interested in maintaining a less-intensive playing surface, or those that have ultradwarf cultivars or limited labor and financial resources, should use N rates in the lower range. Exceptions to these values may occur. For example, courses recovering from excessive traffic, pest or low-temperature stresses, or that are establishing new greens may require higher (~25% more) N rates than those listed until their grass is sufficiently reestablished. An additional 3/4 pound N/1,000 square feet around aggressive vertical mowing or aerification facilitates turf recovery. When applying granular products to greens, use a particle size guide number (SGN) of 100 or less to reduce mower pickup.

Spoon (also called foliar) feeding with liquid fertilizer programs is often used on high sand content areas to help regulate turf growth and color, provide a continuous supply of elements such as N or K that are often easily leached, and supply micronutrients that are needed in smaller quantities. N rates in a foliar feeding program typically range from 0.1 to 0.25 pound N/1,000 square feet (0.5 to 1.2 g N/m$^2$) on a 7- to 14-day interval. Phosphorus (as phosphoric acid or $P_2O_5$) is applied at about one-third the rate of N while K rates will be equal to one-half to three-fourths the rate of N (see following). Other elements, such as Mg or Fe, also may be added by foliar feeding to help regulate and maintain desirable turf color. Suggested seasonal percent N fertilizer applied to bermudagrass throughout the year is listed in Table 10.8.

**Overseeded bermudagrass greens.** Once established, overseeded greens should be fertilized every two to three weeks with 1/2 pound soluble N (plus K) per 1,000 square feet (2.4 g N/m$^2$) during fall and winter months. The objective is to provide enough N to maintain desirable color but not be so excessive as to weaken the overseeded grasses and promote premature growth of bermudagrass. In addition, highly soluble N used on overseeded grass often leads to excessive

**TABLE 10.7**
**Air Temperatures Affecting Turfgrass Shoot Growth and Soil Temperatures at 4 Inches (10 cm) Affecting Root Growth**

| Turfgrass | Shoot Growth | | Root Growth | |
|---|---|---|---|---|
| | Minimum | Optimum | Minimum | Optimum |
| | | °F[a] | | |
| Warm-season grasses | 55 | 80 to 95 | 50 to 60 | 75 to 85 |
| Cool-season grasses | 40 | 60 to 75 | 33 | 50 to 65 |

[a] °C = °F – 32 (5/9).

**TABLE 10.8**
**Percentages of Nitrogen Fertilizer Applied to Bermudagrass during the Year**

| Season | Overseeded | Nonoverseeded |
|---|---|---|
| | Percent | |
| Fall | 15 | 10 to 15 |
| Winter | 15 | 0 |
| Spring | 25 | 35 |
| Summer | 45 | 50 to 55 |

turf growth, slower putting speeds, and disease (e.g., brown patch and *Pythium*) occurrence. Many superintendents have discovered an application of Mn, and possibly Fe, can often substitute for a N application. Two to three ounces (56 to 84 g) of an Fe source (such as ferrous sulfate) or half an ounce (14 g) of manganese sulfate in 2 to 5 gallons of water applied per 1,000 square feet (8 to 20 L/100 m$^3$) provides 10 to 21 days of desirable dark-green color without an undesirable flush of growth. These elements are only foliarly absorbed by plants and provide relatively short color response time.

## Other Elements

**Potassium.** Potassium often is called the "health" element. Without a relatively available supply of K, turfgrasses will be more susceptible to environmental and pest stresses. Root growth also is related to K availability. Research and experience indicate N:K ratios on greens should be 1 part N to 0.5 or 0.75 part K. Excessive K may increase the soil salinity levels, especially if the leaching capability is limited, and also compete with the soil exchange sites at the expense of other essential elements such as Ca, Mg, and Fe.

**Phosphorus.** Soil phosphorus levels tend not to fluctuate as readily as N or K. Soil test results should be used to determine the amount needed for a particular course. Usually 1 to 4 pounds P/1,000 square feet (5 to 20 g P/m$^2$) are needed per year. Phosphorus is generally not very water soluble; therefore, if needed, its efficiency is increased if applications follow aerification. This allows the material to be placed more directly into the rootzone. Phosphorus levels can become limiting during a grow-in situation; an N:P ratio of one part N to one part P has proven best for bermudagrass sprig grow-in and is normally applied at 1 pound N/1,000 square feet (4.9 g N/m$^2$) every 7 to 14 days until complete coverage is achieved.

**Micronutrients.** Regular soil and tissue testing is the best preventative approach to solving many of the nutrient deficiency problems. Iron and Mn are two of the most common micronutrient deficiencies turf managers experience. Micronutrient deficiency symptoms can easily be confused with pest occurrence or other stresses. These problems, however, are usually more localized and appear as either irregular spots or in circular patterns. Refer to the fertilization chapter and sections on developing turf fertilizer programs and fertilizer calculations for information on applying micronutrients and chelates on greens.

## Aerification Strategies and Techniques

A key concept when dealing with ultradwarf cultivars is thatch/organic layering prevention rather than curative control. Practices to prevent excessive thatch layering include core aerification, vertical mowing, brushing, grooming, and topdressing.

**Frequency and Timing of Core Cultivation.** Frequency of core cultivation should be based on the intensity of traffic the turf is exposed to, soil makeup, hardness of the soil surface, drainage capability, and degree of compaction. Areas receiving intense, daily traffic such as golf greens, approaches, landing areas, aprons, and tees require a *minimum* of three annual core aerifications, often four. Typical timings are in midspring, summer, late summer, and late winter, which should be followed with medium to heavy topdressing with desirable sand. If followed for four to five consecutive years, this method will help relieve the undesirable condition of many native soil greens.

Another means to determine how much to aerify is based on the amount of turf surface impacted by aerification. A 15% to 25% surface area being impacted on an annual basis is a reasonable routine goal for many well-established golf courses. A percentage basis is sometimes easier for nonagronomists to understand instead of trying to comprehend tine diameters and spacing. Courses that have been neglected may need a more aggressive aerification program. During periods of active growth, tines up to 5/8 inch (16 mm) on 2- × 2-inch (5 cm) spacings are used to help reach the 15% to 25% annual surface area impacted goal. At other times, 1/4 to 3/8 inch (6.4 to 9.5 mm) diameter tines are used to improve ball roll and enhance turf recovery. As outlined in the chapter on cultural practices for golf courses, using various tine sizes and hole spacings will determine the surface area of turf impacted.

## Spiking and Slicing

Two supplemental cultural practices available to help relieve surface compaction, break up algae layers, and promote better water penetration and aeration are spiking and slicing. Since only minor disruptions of soil surfaces occur, spiking and slicing can be performed more often (e.g., every 7 to 14 days) than core aerification (e.g., every four to eight weeks) during periods of active turf growth.

## Vertical Mowing

Vertical mowing has been shown to be effective in reducing organic matter, mower scalping, and localized dry spot development (Figure 10.30). Light verticutting during periods of active growth alternating with topdressing helps reduce/dilute organic matter plus provide smoother ball roll. Depending

**FIGURE 10.30** Scalping due to excessive thatch accumulation.

on the depth of the blades, greens are often verticut in two directions.

**Depth.** Different objectives can be met with vertical mowing depending on the depth of penetration. Grain is reduced when knives are set shallow enough to just nick the turf's surface. This is often referred to as "grooming." Shallow vertical mowing also is used to break up cores following aerification, providing a topdressing effect. Deeper penetration of knives stimulates new growth and removes accumulated thatch when stolons and rhizomes are severed. This is often performed in conjunction with core aerification. Seedbed preparation prior to overseeding is also accomplished by deep vertical mowing.

When dethatching is the objective, thatch depth will determine the depth of blades. The bottom of the thatch layer should be reached by vertical mowing, and preferably the soil surface beneath the thatch layer should be sliced. However, there is a limit to the depth blades should be set or excessive removal of turf roots, rhizomes, stolons, and leaf surface may occur. For example, blades should be set at a depth to just cut stolons and no deeper if new growth stimulation is the objective. Vertical blade spacing for thatch removal should be between 1 and 2 inches (2.5 and 5 cm). This range provides maximum thatch removal with minimal turf damage.

Deep vertical mower penetration requires the use of a heavy-duty machine that can penetrate 1 1/2 to 3 inches (3.8 to 7.6 cm). Deep vertical mowing grooves the turf surface, so subsequent topdressing is often required to smooth the surface and cover exposed stolons. Shallow-rooted or immature turf can be severely damaged or torn out by deep vertical mowing. Preliminary testing at the site to be verticut should be performed by hand-pulling to measure if favorable rooting of the grass exists. Irrigation and topdressing should follow such deep vertical mowing to prevent quick desiccation of exposed roots, rhizomes, and stolons. This will also help smooth the turf surface and encourage turf recovery. Aggressive vertical mowing should only be performed during periods of active growth to prevent extended recovery time. Even during periods of active growth, three weeks or longer are often required for full recovery.

Soil and thatch should be dry when deep vertical mowing is performed or turfgrass injury will be more extensive, since moist conditions encourage excessive plant material to be removed. Following verticutting, debris should be removed and the area immediately irrigated. Approximately five to seven days following heavy vertical mowing, 1/2 to 1 pound N/1,000 square feet (2.5 to 4.9 g N/m$^2$) should be applied to encourage rapid recovery. Quick-release N sources are preferred at this time.

**Frequency.** The rate of thatch accumulation dictates the frequency of vertical mowing, which should begin once the thatch or mat layer on golf greens exceeds 1/4 to 1/2 inch (0.6 to 1.3 cm). Areas prone to thatch accumulation may require heavy vertical mowing several times per year. For bermudagrass, the first is during mid- to late spring when turf is actively growing. This removes thatch and encourages turf spread by slicing stolons and by warming the soil surface more quickly than if the thatch is allowed to remain. The second vertical mowing should be timed one to seven days before the anticipated fall overseeding. This discourages late-season bermudagrass growth that can compete with the overseeded grasses and exposes the soil surface so grass seed can reach the soil better and have optimum germination. However, fall vertical mowing will result in a degree of surface damage that may not mask until the overseeding has time to become established. Recovery time will be extended when the bermudagrass is not actively growing, such as late summer when days are shorter or during extended cloudy/rainy periods.

### Grooming and Conditioning

The grooming mower (also called turf conditioner) keeps greens smooth and fast by reducing grain and removing excessive top growth. On front of the reel-cutting unit of greens mowers is an attached miniature vertical mower (often referred to as vertical grooming), with blade spacing typically less than or equal to 1/4 inch (0.6 cm), that rotates through slots in the front slit roller. The depth of blades is typically 0.0625 to 0.125 inch (1.6 to 3.2 mm) below the mower bed knife. Each time turf is mowed with this unit, the turf is lightly vertically mowed (aka, groomed or conditioned). This unit improves the playing surface by standing up leaf blades before mowing, thereby removing much of the surface grain. New shoot development is also stimulated by slicing stolons and removing thatch near the surface. During growing months, daily grooming, along with timely topdressing and aerification, helps eliminate the need for traditionally performed turf renovation by severe vertical mowing. As grain worsens, double grooming is practiced where the unit makes two passes in opposite directions. Similarly to changing mowing direction daily, the direction of grooming should be changed each time it is performed. Other groomers use a rotary brush that rotates in the opposite direction of the mower blade. The brush stands up the grass prior to mowing, thereby reducing grain and providing a smooth surface.

### Double Cutting and/or Rolling

A popular practice "rediscovered" in the 1990s was double cutting and/or rolling greens to help provide smoother playing surfaces without having to mow at extremely low heights that often proved detrimental to Tifgreen and Tifdwarf greens. Double cutting, due to costs, is still used, but mostly only prior to and during tournament play. Rolling typically accompanies double cutting during tournaments, but more courses roll two to four times weekly as routine maintenance, especially when bermudagrass mowing heights are raised during periods of cloudy/rainy weather and/or when days become shorter (late summer). These practices, along with PGR use, help to maintain more desirable putting characteristics without overly damaging the grass. Courses, obviously, must weigh the benefits of double cutting and/or rolling greens compared to increased equipment needs, costs, and wear-and-tear, as well as the possibility of increased soil compaction or turf wear, especially on cleanup (perimeter) passes and on collars.

### PGR Use

Plant growth regulators are often used on bermudagrass to promote lateral growth (e.g., surface tightening) over vertical growth. PGRs also help to produce a consistent putting surface during periods of growth surges, provide more consistent putting throughout the day, promote a darker blue-green color, help control unwanted seedheads, and provide greens with "off-types" of bermudagrass patches with an illusion of having only one type of grass present. The most popular PGR product currently used for these purposes is trinexapac-ethyl. Using this product on a one- to two-week interval has proven beneficial during the growing months. PGRs, however, do not necessarily increase putting speeds (or distances), but rather help maintain those current playing conditions throughout the day and from one day to the next. A recently introduced product, prohexadione-Ca (trade name Anuew™) appears to also help evenly regulate the differential growth often associated with off-types, reducing scalping of these areas.

### Topdressing

Topdressing adds a thin layer of soil to the turf surface, which is then incorporated into the turf by dragging or brushing. Frequency and rate of topdressing depend on the objective. Following coring and heavy verticutting, moderate to heavy topdressing is used to help smooth the surface, fill coring holes, and cover exposed roots resulting from these two processes. Irregular playing surfaces or gradual soil profile modification will require frequent and relatively heavy topdressing. Rates ranging from 1/8 to 1/4 inch deep (3.2 to 6.4 mm) (0.4 to 0.8 cubic yard of soil/1,000 square feet, 0.33 to 0.66 m$^3$/100 m$^2$) are suggested, except if the turf has a limited capacity to absorb the material, since grass smothering would result. Topdressing, like grooming, should begin once a new green has formed a full cover of turf.

If the objective of topdressing is to change the characteristic of the underlying soil, then a heavy topdressing program following numerous core removal operations is required. However, even following a rigorous coring and topdressing program, adequate modification of underlying soil may take several years to accomplish. Deep-tine aerification and core removal should be performed as much as possible prior to topdressing.

If thatch control is the main objective of topdressing, the amount and frequency are governed by the rate of thatch accumulation. A suggested amount of a medium rate of topdressing when thatch is not excessively thick (1/4 to 1/2 inch, 0.6 to 1.2 cm) is approximately 1 cubic yard/5,000 square feet (0.17 m$^3$/100 m$^2$). If this relatively light rate does not adequately enhance the decomposition of the thatch layer, then the frequency of application and topdressing rate should be increased.

If the objective of topdressing is just to provide routine smoothing of the playing surface, then light, frequent topdressings are suggested. Matting or brushing the green following topdressing results in the material being dragged into low spots. Surface irregularities of greens are reduced and the area is somewhat leveled. Topdressing with 1/2 to 1 cubic yard/5,000 square feet (0.09 to 0.17 m$^3$/100 m$^2$) of green surface every 7 to 14 days provides a smoother, truer playing surface. Applying frequent, light topdressing on newly planted bermudagrass greens to cover stolons and to smooth the surface should be performed every two to four weeks until complete cover or the desired smoothness is achieved. Putting speed on a freshly topdressed green will initially be much slower but will increase in several days following irrigation, incorporation, and several mowings.

Using dry, bagged sand is becoming popular as a topdressing source or sand is stored in watertight sand "silos." These smaller amounts of dry topdressing material can be applied via walk-behind rotary spreaders to reduce heavy equipment frequently driving across greens. Dry sand is more easily incorporated into the turf canopy with less pickup by subsequent mowing.

### Irrigation/Water Management

Good water management is a key component of any golf course management plan. Maximizing the time between irrigations without producing significant moisture stress encourages a deeper root system. Fortunately, bermudagrass is not as sensitive to precise water management as bentgrass/*Poa*, since bermudagrass does not require routine hand-watering and extra labor to check for "hot spots" hourly during summer months. As stated, the recommended philosophy is to utilize heavy and infrequent applications, thereby allowing the grass/soil to dry-down between irrigation events.

**Water quality.** Another potential problem with all greens is a gradual buildup of salinity from the use of high pH or bicarbonate-containing irrigation water, effluent sources, or brackish water. This is magnified when highly soluble fertilizer sources are used at relatively light but frequent rates and a lack of frequent, heavy rainfall occurs to periodically

FIGURE 10.31 Turf damage from excessive salinity exposure.

FIGURE 10.32 Bermudagrass tees treated with a colorant to provide winter green color.

leach the salt below the rootzone (Figure 10.31). Greens also must drain relatively well so salts can be leached below the rootzone by heavy rainfall, flushing, or heavy hand-watering. Using low-salt–containing water sources also helps. Salt tolerance of ultradwarf bermudagrass is ~3.5 to 4.0 dS/m with 50% growth reduction at about 2.5 dS/m.

To combat potential salinity buildup, superintendents have modified their watering practices from light, frequent applications in summer to heavier, infrequent applications. Buffering the water pH from 6.0 to 6.5 and using periodic light applications of gypsum help reduce salinity stress. High-bicarbonate water is also commonly treated by artificially acidifying it, often by sulfur injection. This technology is rapidly evolving and dramatic effects can be realized.

## Winter Color

Overseeding can be performed in late summer through early fall to provide green color to the bermudagrass, which begins to turn brown when temperatures drop below 50°F (10°C) for an extended period. In addition, golf course fairways are overseeded to clearly mark suggested landing areas for golfers. Overseeding is an important economic aspect of golf because many resort courses enjoy their heaviest play from tourists during the fall, winter, and spring months.

Overseeding, however, requires various degrees of seedbed preparation in the form of aerifying, verticutting, brushing, and topdressing, each causing disruption to the playing surface. Extensive watering is needed until the overseeded grass becomes established, and *Poa annua* control is expensive and rarely 100%. Spring transition from the overseeded grass back to the bermudagrass also can be unpredictable and erratic, and can undesirably extend into summer. It is not recommended to overseed ultradwarf greens in the year of initial establishment. The chapter in this book on cultural practices for golf courses covers overseeding practices in detail.

Coloring of bermudagrass greens with various paints, dyes, and colorants for winter color instead of overseeding has become popular (Figure 10.32). Paints tend to last four to six weeks while dyes last about two weeks. A good turf base is also needed for the colorant to adhere to. Worn, thin, bare areas or common bermudagrass patches will be less successfully colored. Typical dilution rates are 5 gallons of paint per 100 gallons of water (19 L/379 L) with two passes made in two directions, with a cleanup pass to keep the perimeters even in terms of color. Painted greens should still receive monthly aerification with hydrojets or solid tine to keep greens soft and provide soil $O_2$ and drying.

For best (most consistent) results and the need for less colorant, begin the process in fall when green turf is still present (about two weeks before the first anticipated frost). Coloring brown, dormant turf requires considerably more product and streaking is more apparent. Raising the mowing height in fall also helps by providing more leaf tissue area for the colorant to adhere to. Lightly irrigating prior to applying a colorant also reduces the amount needed as the surface moisture reduces the wicking of the material into the leaf surface. Products are mostly applied using a diaphragm or roller pump as their seals are less affected by the viscous material compared to centrifugal pumps. Low-cost seals, however, are becoming more available for centrifugal pumps. Most colorants dry within an hour after application. Table 10.9 summarizes cultural practices for bermudagrass golf greens.

## Tees

Tees typically are not managed as intensively as greens (Figure 10.33). Mowing heights are 0.2 to 0.3 inch (5 to 7.6 mm). Nitrogen fertilization is 2 to 4 pounds N/1,000 square feet (1 to 2 kg N/100 m$^2$) annually with 3 to 6 pounds N/1,000 square feet (1.5 to 3.0 kg N/100 m$^2$) for sand-based tees. This is often supplemented by spoon feeding with fertigation or broadcast soluble N spray applications. Verticutting is performed two to four times yearly, depending on thatch layering and traffic patterns. PGRs such as trinexapac-ethyl and prohexadione-Ca are applied every 14 days for growth regulation and seedhead suppression.

**Mowing.** Mowing is performed three to seven times weekly, with higher quality turf requiring the daily mowing frequency. Higher quality courses not only mow more often but should also have a fast reel clip. Ten to 11 blade reels provide a cleaner clip of tough leaf blades of these turfgrasses.

Hydraulically driven units are now the norm to prevent turf ripping and slipping when mowing wet turf. Medium-weight five gang mowers are currently the industry norm. Heights typically range from 1/4 to 5/8 inch (13 to 16 mm) for routine play and 3/8 inch (9.5 mm) for tournament play. Higher heights (1 inch, 25 mm) are used for common-type bermudagrasses. In fall, raising heights to 3/4 inch (19 mm) encourages carbohydrate storage. As grasses resume growth in spring, heights are gradually reduced.

**Cultivation.** Due to lush growth, various cultivation practices are necessary to minimize thatch accumulation. At least one core aerification, plus verticutting, is extremely beneficial, with additional events implemented in areas with longer growing seasons (Figure 10.34). A variety of front and rear rollers, brushes, groomers, and vertical mowing accessories is also available to improve the quality of cut and to help reduce or prevent grain or thatch development. They help promote stand density and upright shoot growth to provide better ball lie. The organic debris produced can be removed with sweeping, blowing, and/or vacuuming operations. Aggressive verticutting should be performed in early to midsummer when the grass is actively growing. Waiting until late summer to perform such operations may not allow sufficient time for turf recovery before growth naturally slows in fall.

**TABLE 10.9**
**Summary of Cultural Practices for Bermudagrass Golf Greens**

| Practice | Description |
|---|---|
| | **Fertilization** |
| Optimum timing for N fertilization | • Late spring through late summer |
| Damaging response to excessive N fertilizer | • Late fall through early spring if turf is dormant |
| Yearly N requirements | • 6 to 18 pounds N/1,000 square feet (39 to 88 g/m$^2$) |
| Monthly N requirements | • 0.5 to 1.5 pounds N/1,000 square feet (2.4 to 7.3 g N/m$^2$) |
| Acceptable pH range | • 5.5 to 7.0 |
| Acceptable P levels | • Low to high |
| Optimum K levels | • Medium to high (improves low temperature tolerance); usually applied at 1/2 to 3/4 rate of N |
| | **Cultural Requirements** |
| Minimum full-sunlight hours required | • 8 hours daily (minimum 33 mol/m$^2$/day) |
| Overseeding/painting | • Necessary for winter color |
| Irrigation requirements | • Moisten full rootzone depth, rewetting just prior to wilting |
| Air circulation requirements | • Not normally required |
| Aerification (coring) | • Performed two to four times annually with one being deep tine; stop within 30 days of overseeding/expected frost |
| Additional venting operations (solid tine aerification, slicing, hydrojet, etc.) | • Monthly or more often |
| Cultivation timing optimum | • Midspring through late summer |
| Fertilizer application | • Slow-release (primarily) granular + liquids for color |
| Mowing height | • 0.105 to 0.25 inch (2.7 to 6.4 mm), higher height in late summer |
| Mowing frequency | • Five to seven times weekly |
| Mowing pattern | • Changed (rotated) daily |
| Clippings | • Removed |
| Thatch/mat control | • Frequent topdressing, lightly vertical mowing and grooming, regular aerification |
| Compaction control | • Use a high sand-based rootzone during construction (e.g., 90:10); routine aerification; limit play (traffic) during wet conditions |
| Topdressing | • Light (0.2 cubic yard/1,000 square feet, 0.17 m$^3$/100 m$^2$) but frequently (weekly) during active growth; monthly at 0.3 to 0.5 cubic yard/1,000 square feet (0.25 to 0.4 m$^3$/100 m$^2$) during slowed growth |
| Regular (deeper) verticutting | • Once or twice yearly; once in spring and another just prior to overseeding; often performed with core aerification |
| Grooming | • Performed frequently with ultradwarf cultivars |
| Rolling | • Performed less than or equal to three times weekly, daily during tournaments |
| Slicing | • Performed weekly in two directions to reduce surface compaction and algae and to facilitate soil air exchange and infiltration |
| Plant growth regulators (PGRs) | • Approximately weekly during periods of active shoot growth |
| Key diseases | • *Bipolaris*: midsummer through midfall<br>• *Pythium* blight: during slowed growth, most often in fall<br>• *Microdochium* (pink snow mold): spring<br>• *Rhizoctonia zeae*: mid- through late summer |

FIGURE 10.33 Sufficiently sized tees are necessary on heavy play holes.

FIGURE 10.34 Soil cores on tees following aerification.

Lighter (less aggressive) vertical cutting is also available by using mowers with vertical mower attachments. Though not designed for deep verticutting, these units help reduce complaints concerning dusty air and generate less organic debris at one time, reducing disposal issues.

In recent years, sand topdressing has been beneficial in diluting soil organic matter along with a drier, firmer, and smoother playing surface. A topdressing goal is to eventually develop a 4- to 6-inch (10 to 15 cm) homogenous combination of sand and organic matter. More frequent, lighter sand topdressing of 8 to 10 tons/acre (18 to 22 metric tons/ha) per application help provide this compared to less frequent, one to two heavier applications.

## SEASHORE PASPALUM

Seashore paspalum (*Paspalum vaginatum* O. Swartz.), commonly referred to as paspalum, is a halophytic, warm-season, perennial turfgrass species native to tropical and subtropical regions (Figure 10.35). Many of the turf-type cultivars originate from South Africa. In the United States, paspalum is found almost exclusively along the Atlantic coastal and southern California regions in marshy, brackish ecosystems.

FIGURE 10.35 Seashore paspalum use in a saline environment.

Paspalum is known for its aggressive growth habit, high salinity tolerance, and ability to be irrigated with saline and effluent water. It is generally established by vegetative propagation as paspalum has some self-incompatibility issues and possesses 20 chromosomes.

Paspalum spreads by seed, stolons, and rhizomes, with rooting occurring at internodes. Leaves of the grass are medium to coarse (3 to 8 mm leaf width), with leaf fineness increasing with decreasing mowing heights. Lowering mowing heights also often increases tillering, which causes greater competition between plants, thus reducing leaf blade width. Turf color during the growing season is a dark-green hue with striping capabilities resembling that of perennial ryegrass or Kentucky bluegrass. It does not produce heavy morning dew patterns like other turfgrasses and tends to retain green color longer in fall than bermudagrass. Inflorescence (seedhead) of paspalum is a V-shaped, spike-like raceme and it is notorious for its rapid inflorescence production (Figure 10.36).

FIGURE 10.36 V-shaped seedhead of seashore paspalum.

## CULTURAL PRACTICES

### Fertility of Mature Turf

Seashore paspalum evolved from sand dunes where ocean water and rainfall provided the majority of water and nutrients, forcing it to become efficient in uptake and use of these. High chloride ($Cl^-$) and $Mg^{+2}$ requirements can be attributed to the elevated levels of each found in sea water. Calcium and Mg are required nutrients for osmotic adjustments, from high levels of salts, and photosynthesis is allowed during salt stress, along with the basic nutrients needed for growth. Plant tissue should ideally contain a 2:1 ratio of Ca to Mg as a 1:1 ratio often results in turfgrass yellowing (Figure 10.37). A general rule of thumb is that as salinity increases, N fertility should decrease, while K, Ca, and micronutrients should increase. Potassium and Ca help paspalum offset salinity stress.

Excessive N rates in conjunction with saline water irrigation use on seashore paspalum can cause excessive thatch accumulation, scalping, disease pressure, and slower putting speeds. Mature stands of paspalum grown on sand-based soils generally require 4 to 6 pounds N/1,000 square feet (20 to 30 g N/$m^2$) annually while areas with longer growing seasons require 5 to 8 pounds N/1,000 square feet (24 to 39 g N/$m^2$) annually. Fertilizer schedules for paspalum often mirror cool-season turfgrass fertility regimes—for example, monthly high-rate applications in spring and fall (1/2 to 1 pound N/1,000 square feet, 2.4 to 4.9 g N/$m^2$) and minimal application rates in the summer months (1/4 to 1/2 pound N/1,000 square feet, 1.2 to 2.4 g N/$m^2$) every two weeks. About 50% of the annual fertilizer used should be granular, with the remaining 50% foliar applied. Some disease suppression has been noticed when supplemental granular Mg and zinc (Zn) are applied.

Fine-textured paspalums used on greens have 85% to 95% higher leaf tissue K content than more coarse-textured paspalums. Potassium should be applied at two to four times the N rate (higher rate for salt-stressed plants) and sites frequently leached. Tissue analysis sufficiency range of K should be 2.0% to 4.0% while soil cation exchange capacity (CEC) levels of 3% to 8% K are suggested. To improve winter hardiness in cooler grown areas, 1 to 2 pounds K/1,000 square feet (0.49 to 0.98 kg K/100 $m^2$) is applied in midfall. Phosphorus should be supplied at approximately 1 to 2 pounds P/1,000 square feet (0.49 to 0.98 kg P/100 $m^2$) annually at low rates throughout the growing season. Areas prone to high leaching should have P rates increased 25% to 50% to those listed.

**FIGURE 10.37** Seashore paspalum chlorosis (yellowing) from iron deficiency.

**TABLE 10.10**
**Suggested Tissue Cation Nutrient Ratios for Seashore Paspalum**

| Cations | Suggested Ratios |
|---|---|
| Na:K | 1:[2 to 4] |
| Ca:Mg | [3 to 8]:1 |
| Ca:K | [10 to 30]:1 |
| Mg:K | [2 to 10]:1 |
| Fe:Mn:Mg | 1:1:1 |
| Zn:Mn | 1:1 |
| N:$P_2O_5$:$K_2O$ | [2 to 3]:1:[4 to 8] |

For salinity management, a typical fertility program involves applying Ca at three to five times the yearly rate of N. Biweekly Ca and K applications are generally at 1/4 pound Ca and K/1,000 square feet (0.12 kg/100 $m^2$). Additional K applications at about 1/4 pound K/1,000 square feet (0.12 kg K/100 $m^2$) may also be applied as the quality of irrigation water is lowered.

For even color, the micronutrients Fe, Mn, and Mg are maintained at a tissue ratio of 1:1:1. However, at pH values > 8.0, Fe should be applied at a 3:1 ratio with Mn. Micronutrients are foliarly applied typically at 7- to 14-day intervals. Manganese and Zn may help with plant salt tolerance and appear to activate paspalum's growth. Key cation nutrient ratios are suggested in Table 10.10.

### Irrigation Strategies

Seashore paspalum has the ability to be irrigated with a wide array of water types including fresh, effluent, and pure seawater (if certain conditions are met). Heavier and less frequent application cycles are best. Because of its excellent drought tolerance, paspalum often requires less water input than other turfgrass species. In situations where water high in salt and sodium is used for irrigation, the most important issue becomes excellent drainage and periodic flushing of salts from the soil with fresh water. Along with flushing, frequent applications of sulfur and lime or gypsum become mandatory. Irrigation rates vary widely when using ocean water, depending on severity of salt stress, soil structure, and climate. Rates range from one-half inch (13 mm) daily on sites with good percolation to a total of 7 inches (18 cm) per week (Duncan and Carrow, 2005a). Again, proper drainage, sand-based soil media, and regular flushing with fresh water are imperative when irrigating with ocean water. A modern, state-of-the-art irrigation system plus technician also are needed to keep it in proper working order when using high-saline–content water. Following leaching events, additions of K and Mg may be needed.

Fresh or potable water can produce high-quality paspalum stands and greatly increase the plant's tolerance to other environmental stresses. Paspalum should be watered to a soil depth of 6 inches (15 cm) with each event and should be rewatered only when the soil profile has sufficiently dried from previous irrigation cycles. Typical irrigation rates with high-quality, fresh water are from 1 inch (2.5 cm) weekly for sandy soils in hot environments to 1 inch (2.5 cm) biweekly for heavier soils and/or cooler environments. Since paspalum requires low water inputs, it would be more beneficial to irrigate when drought symptoms appear, to encourage roots to exploit deeper soil moisture reserves, increasing drought tolerance.

Irrigation with recycled waters is also highly variable, depending on amounts of heavy metals, salts, pH, carbonates and bicarbonates, and other ingredients in it. Water should be monitored daily for very poor water quality, to every six months depending on the drainage characteristic of soils, climate (rainfall amounts), and typical water quality. Water quality testing includes sodium absorption ratio (SAR), total dissolved salts (TDS, $EC_w$), pH, and carbonate and bicarbonate levels (residual sodium carbonate; RSC). These determine the amount of amendments needed to counteract excess salinity, sodium, carbonates and bicarbonates. $EC_w$ values should ideally be below 0.75 dS/m (or mmhos/cm). Values between 0.75 and 3.0 indicate increasing problems and values above 3.0 dS/m should only be used with caution. RSC values indicate the relative amounts of carbonates and bicarbonates compared to Ca and Mg and are reported as milliequivalents/liter. Values below 1.25 meq/L are generally safe with readings between 1.25 and 2.5 indicating increasing problems and values above 2.5 being severe.

Management strategies to offset poor irrigation water use include excellent soil drainage, use of gypsum (calcium sulfate), sulfur, calcium nitrate, calcium chloride and other Ca sources, plus acid injection. Acid injection or sulfur generators are useful when RSC values are greater than 1.25 meq/L, plus high SAR values (>~2.0 dS/m) and soil exchangeable sodium percentage values greater than 15%. Imbalances of carbonate and bicarbonate compared to Ca and Mg can cause buildup of insoluble calcitic limestone in the soil and scale inside irrigation lines. Sodium then dominates soil CEC sites and reduces the availability of Ca, Mg, and K to plants. Acid injection involves using sulfuric, citric, or phosphoric acid.

### Aerification and Verticutting

As with any high-maintenance putting green, cultural practices to reduce excessive thatch buildup are necessary to prevent scalping and puffiness, provide proper air:soil gas exchange, and ensure adequate infiltration, percolation, and drainage. Due to paspalum's aggressive vegetative growth, core aerification followed by topdressing should be performed at least twice yearly in late spring or early summer and again in late summer with hollow tines 1/4 to 5/8 inch (13 to 16 mm) in diameter, typically on close center tine spacings. Under conditions of excessive thatch and soil compaction, three to four aerifications should be considered followed by sand topdressing with 10 to 12 cubic feet/1,000 square feet (0.31 to 0.37 $m^3$/100 $m^2$) to help fill the holes. In addition to these heavy topdressings in conjunction with core aeration, routine topdressing involves lighter rates (1/2 to 1 cubic foot/1,000 square feet, 0.015 to 0.03 $m^3$/100 $m^2$) using medium sand at least every two weeks, often weekly. Routine light topdressing should be performed in order to maintain acceptable ball roll speeds and smooth putting surfaces. Dry sand should be used to facilitate uniform application and ease of incorporation. Greens are typically lightly vertical mowed prior to topdressing to help with sand incorporation into the dense turf canopy.

Light but frequent vertical mowing, 6/10 to 1/4 inch (1.6 to 3.2 mm) below the mowing height, followed by topdressing every 14 days during periods of active growth, is ideal for routine maintenance. Triplex units equipped with standard or carbide-tipped blades are often used. Due to paspalum's predominant rhizomatous growth habit, aggressive verticutting is suggested only in conjunction with core aerifications. Grooming (or light vertical mowing) is performed four to six times weekly during peak growth periods. These are typically set 1/8 inch (3.2 mm) below the mowing height but may be raised during periods of high humidity to avoid unnecessary stress on the grass plant. Grooming helps stand grass leaf blades up (vertical) to achieve a truer cut, preventing grain and providing desirable putting surfaces. Grooming should be avoided the day following topdressing to prevent excessive turf damage (bruising). Brushing is performed daily during periods of active growth, typically with mower mounted units. Additional, regular venting with small-diameter solid tines is performed every two to four weeks during periods of limited growth. These aid in good moisture and soil gas exchange.

### Mowing

Low paspalum mowing heights of 0.075 (short term) to 0.125 inch (1.9 to 3.2 mm) can be achieved for putting greens if sharp blades are maintained and only small (e.g., 0.02 to 0.05 inch, 0.5 to 1.3 mm) daily increments are made to eventually achieve these heights. At these heights, one to two daily mowings are typically required for normal play with two to three mowings daily for tournament play. Greens are mowed six days per week while perimeter "cleanup" laps are performed only four or five times weekly to minimize soil compaction and turf shearing. Greens are rolled once daily for normal play and twice daily for tournament play. Electric greens mowers that have separate speed control for the reel and forward motion roller are popular in order to maintain high reel speeds to minimize scalping and turf gouging when turning the units around and to reduce neighborhood noises.

Scalping of seashore paspalum greens is typically associated with use of cultivars not suited to the mowing height, overapplication of N, drastic reduction in heights, waterlogged soils, or any combination of these. Four to six weeks may be

**FIGURE 10.38** Seashore paspalum scalping.

needed for full recovery following severe mower scalp. Reels and bed knives should be sharpened on a routine basis, as they quickly become dulled by the thick stems and high lignin content of paspalum. Other practices that increase scalping tendency are inadequate K levels, excessive irrigation or periods of rainfall, and improper reel mower blade maintenance (Figure 10.38). Mowers with dual back rollers and smooth (not grooved) front rollers are less likely to scalp. The proper use of PGRs also avoids excessive growth, which leads to scalping. If green speeds in the range of 10 feet (3 m) are required, double mowing is performed three to five times weekly along with daily rolling.

### Grain (Dysphasia) Management

Due to the predominant stoloniferous growth habits of paspalum, surface grain can occur where the leaf blades lie in varying directions, causing inferior putting surfaces and differing color hues. However, due to paspalum's more rhizomatous growth, these greens are less likely to develop grain. Altering mowing direction daily, and routine grooming, brushing, and frequent topdressing as discussed previously, help reduce this possibility while also providing a better putting surface.

## PEST CONTROL

### Disease and Nematode Management

Diseases of paspalum are similar to those of other warm-season grasses. These include leaf spots, *Pythium*, take-all patch, fusarium, anthracnose, yellow (winter) patch, dollar spot, *Rhizoctonia*, and fairy ring (Figure 10.39). Most diseases are minor unless the turf has suffered damage such as cold injury or scalping and improper nutrition levels are maintained. For example, high N levels encourage brown patch, take-all patch, *Pythium* blight, and yellow patch while low N levels favor anthracnose, leaf spot, and dollar spot (Table 10.11). Preventative fungicide programs are needed during the fall, winter, and spring when recovery potential is limited and play is at its heaviest.

Perhaps the most detrimental diseases on seashore paspalum are dollar spot, large or brown patch (*Rhizoctonia* spp.), and, to some extent, *Fusarium* blight. Dollar spot is a disease found commonly on many turf settings and is most closely related to, or induced by, low N levels. Preventative measures against dollar spot should include, but are not limited to, regular applications of N; decreasing leaf wetness; decreasing organic matter buildup; providing good soil moisture; avoiding irrigation with ocean water, which slows growth; and increasing air movement. Applying a broad-spectrum fungicide such as chlorothalonil also provides excellent dollar spot prevention. In the event of dollar spot occurrence, a low rate of liquid N can act as a corrective measure against this disease.

**FIGURE 10.39** Brown patch (*Rhizoctonia solani*) disease in insufficiently drained seashore paspalum.

**TABLE 10.11**
**Relationship between Nutrient Status of Paspalum and Diseases Often Favored by Them**

| Nutrient Status | Disease Favored |
|---|---|
| High N levels | • Brown patch, take-all patch, *Pythium* blight, yellow patch |
| Low N levels | • Anthracnose, leaf spot, dollar spot |
| Low P levels | • Take-all patch |
| Low K levels | • Take-all patch, crown and root rot, dollar spot |
| High K levels | • Brown patch |
| Low Ca levels | • *Fusarium* blight, *Pythium* blight |
| High Ca levels | • Take-all patch |
| Low S levels | • Take-all patch |
| Low Mn levels | • Take-all patch |
| High pH (>6.5) | • Take-all patch |

Large or brown patch, caused by *Rhizoctonia solani*, typically appears on paspalum in the spring and fall, causing rotting at the basal portion of the plant. Large patch is favored by prolonged leaf wetness, humid or cloudy weather, and poor air circulation and soil drainage. Preventative control methods begin by reducing the time during which leaves stay moist in the morning by mowing, rolling, or whipping; thatch management; reducing N inputs; and increasing air movement and drainage. Preventative fungicide applications for this pathogen should be applied one to three times at 14- to 28-day intervals in the fall.

*Fusarium* blight is especially problematic on seashore paspalum under hot (75°F to 90°F, 24°C to 32°C), and high-humidity conditions. Management practices that help reduce *Fusarium* blight include lower N inputs, relatively high P and K levels, controlling excessive thatch, increased air circulation, and reduced leaf wetness. Chemical control of this disease requires repeat applications at 7- to 14-day intervals and often involves formulations of thiophanate-methyl.

Nematodes associated with paspalum and suggested threshold values (number per 100 $cm^3$ soil) for each species include lance (*Hoplolaimus*), 40 to 120; sting (*Belonolaimus*), 10 to 25; sheathoid (*Hemicriconemoides*), 500 to 1,000; ring (*Mesocricomema*), 500 to 1,000; spiral (*Helicotylenchus, Peltamigratus*), 700 to 1,500 (150 to 300 for *Peltamigratus*); and root-knot (*Meloidogyne*), 80 to 300. Spiral nematodes appear more damaging to paspalum than the more traditional sting nematodes. Symptoms of damage are irregular patches of thin turf with poor, shallow roots.

## Weed Control

A major limitation to seashore paspalum is the lack of labeled selective herbicides as many commercial products are too phytotoxic. Weeds commonly associated with paspalum include crabgrass, goosegrass, annual bluegrass, purple and yellow nutsedge, kyllinga, bermudagrass, torpedograss, bahiagrass, dallisgrass, vaseygrass, foxtail, broadleaf signalgrass, barnyardgrass, and various broadleaf weeds (Figure 10.40). Preemergence herbicides currently safe on paspalum include benefin, benefin plus trifluralin, bensulide, dithiopyr, isoxaben, DCPA, metolachlor, granular oxadiazon, pendimethalin, prodiamine, and pronamide. “Marginally” tolerant preemergence herbicides include oryzalin (retards green-up), atrazine, simazine, and liquid formulations of oxadiazon. Tolerant postemergence herbicides include bentazon, carfentrazone, sulfentrazone, ethofumesate, imazaquin, quinclorac, metsulfuron-methyl, pronamide, and halosulfuron. Moderate postemergence herbicide tolerance includes three-way mixtures of 2,4-D plus dicamba plus MCPP, fluroxypyr, dicamba, metribuzin, sulfosulfuron, and 2,4-D amine. Lower rates should be used with these products, especially as temperatures increase. Herbicides with low tolerance by paspalum include asulam, arsenicals, foramsulfuron, clopyralid, triclopyr, trifloxysulfuron, diclofop-methyl, sethoxydim, fluazifop, clethodim, and fenoxaprop. As specific cultivars and environmental conditions often greatly influence herbicide tolerance, a small test area should always be available prior to widespread application.

**FIGURE 10.40** Cutting out patches of bermudagrass in seashore paspalum.

## Insects

Although not major problems on paspalum, several insects can periodically occur. Fall armyworms and tropical sod webworms cause damage similar to scalping, requiring extended periods for recovery. Isolated cases of billbugs, spittlebugs, mole crickets, bermudagrass mites, aphids or greenbugs, grubs, and frit flies occasionally occur.

## Plant Growth Regulators

Since salt is a natural growth regulator, PGR use rates and timings are adjusted according to irrigation water quality and cultivar salinity tolerance. The most popular PGR, trinexapac-ethyl (TE), has been used with success; however, rates and timings of applications of TE often vary widely. When used at low rates, TE can effectively suppress paspalum growth with minimal turf injury. Flurprimidol (Cutless) will slow paspalum growth and seedhead production when temperature exceeds 85°F (29°C). When temperatures are less than 85°F (29°C), paclobutrazol (Trimmit) and ethephon (Proxy) are combined to suppress seedheads. In order to take advantage of the growth surge that typically occurs following PGR use, some managers discontinue this just prior to aerification to promote quicker turf recovery.

## Overseeding and Painting

Where seashore paspalum enters dormancy each fall, managers in frost zones may need plans to maintain year-round green color during the winter. Two options for maintaining attractive color and playability include winter overseeding and use of turf colorants, or painting. The most popular grass species choices for overseeding include roughstalk bluegrass or perennial ryegrass, with less popular choices of bentgrass, fine fescue, or possibly a blend of two or more of these. Perennial ryegrass exhibits the fastest emergence and establishment of most cool-season species, while roughstalk bluegrass has been shown to be very cost effective and to have vigorous growth. Due to increased *Rhizoctonia* diseases associated with irrigation to overseeded turf, this practice is losing some of its importance.

Painting of greens is popular for those who do not want to face overseeding spring transition or increased diseases, or who lack adequate financial resources to overseed. Painting helps reduce disease susceptibility, weed pressure, cost, delayed green-ups, and the inability to remove overseeded grass. Painting increases turf quality color, ball roll speeds, and surface and soil temperatures. Painting should involve two to three applications of paint over the course of the winter. Many formulations of turf paint are available to superintendents today. It is suggested that managers experiment on nursery greens with a variety of paints, timings, and number of applications to determine which will be the best for their specific environmental and playing conditions. Mowing height should be raised slightly prior to painting in order to slow ball roll and help to promote paspalum winter survival. Table 10.12 summarizes cultural practices for seashore paspalum and zoysiagrass golf greens.

## ZOYSIAGRASS

A somewhat recent trend in the golf course industry is using various zoysiagrasses for golf greens and tees (Figure 10.41). In many cases, this need originated from excessive shade situations, which will not support successful long-term

**TABLE 10.12**
**Summary of Cultural Practices for Seashore Paspalum and Zoysiagrass Golf Greens**

| Practice | Description |
|---|---|
| | **Fertilization** |
| Optimum timing for N fertilization | • Midspring through summer |
| Damaging response to excessive N fertilizer | • Fall through early spring |
| Yearly N requirements | • 2 to 8 pounds N/1,000 square feet (9.8 to 39 g N/m$^2$) |
| Monthly N requirements | • 0.2 to 0.5 pound N/1,000 square feet (1.2 to 2.4 g N/m$^2$) |
| Acceptable pH range | • 5.5 to 8.0 |
| Acceptable P levels | • Medium to high |
| Optimum K levels | • High (improves salt and low-temperature tolerance); usually applied at two to four times the annual rate of N |
| | **Cultural Requirements** |
| Minimum full-sunlight hours required | • 6 to 8 hours daily; GA-inhibiting PGRs help |
| Coloring | • Necessary in temperate areas for winter color; overseeding not normally performed with these grasses |
| Irrigation requirement | • Moisten full rootzone depth especially to flush salts, rewetting just prior to wilting |
| Drainage requirement | • Excellent if using salt-laden irrigation water |
| Air circulation requirement | • Not normally required |
| Aerification (coring) | • Performed two to four times annually with at least one deep tine yearly |
| Additional venting operations (solid tine aerification, slicing, hydrojet, etc.) | • Monthly |
| Cultivation timing optimum | • Spring through summer |
| Fertilizer application | • Slow-release (primarily) granular + liquids for color |
| Mowing height (greens) | • 1/10 to 1/8 inch (2.5 to 3.2 mm) |
| Mowing frequency | • Daily; double cut during tournaments |
| Mowing pattern | • Changed (rotated) daily |
| Clippings | • Removed |
| Thatch/mat control | • Frequent topdressing, lightly vertical mowing and grooming, regular aerification |
| Compaction control | • Use a high sand-base rootzone during construction (e.g., 90:10); routine aerification; limit play (traffic) during wet conditions |
| Topdressing | • Light (0.2 cubic yard/1,000 square feet, 0.17 m$^3$/100 m$^2$) but frequently (weekly) during active growth; monthly at 0.3 to 0.5 cubic yard/1,000 square feet (0.25 to 0.4 m$^3$/100 m$^2$) during slowed growth |
| Regular (moderate) verticutting | • Once or twice yearly, typically with core aerification |
| Grooming | • Daily during growth, weekly otherwise |
| Rolling | • Performed daily during normal play, double during tournaments |
| Slicing | • Performed weekly in two directions to reduce surface compaction and algae and to facilitate soil air exchange and infiltration |
| Plant growth regulators (PGRs) | • Extensively throughout growing season; also in late summer for seedhead suppression |
| Key diseases | • Large patch: symptoms in spring, infections in fall; leaf spot; dollar spot |

**FIGURE 10.41** Zoysiagrass golf green in a shade-prone environment.

bermudagrass growth. In other situations, reasons for not using bermudagrass include diseases, in particular, spring dead spot; devastating winter damage; and, in some cases, just a desire to set oneself apart from others. Other possible advantages of using zoysiagrass include:

- Color contrast—bright green color versus darker or duller bermudagrass color/hue
- Extended fall turf color with zoysiagrass and earlier spring green-up color compared to bermudagrass
- Stiffer leaves, thus higher ball setup
- Little to no morning dew formation compared to bermudagrass
- Less divot removal compared to bermudagrass
- Lower N requirement
- Good cold, shade, and salt tolerance
- Slower growing
- Tolerates most soils
- Medium to fine texture
- Excellent wear resistance
- Excellent density, which helps prevents weeds
- Genetically stable, with few if any mutations reported to date

Of course, certain disadvantages exist with all grasses being considered. Some main ones associated with zoysiagrass include:

- Thatch/organic matter buildup
- Sensitive to scalping, requiring considerably more time to recover compared to other grasses
- Minor element (e.g., Fe) deficiency sensitive
- Drainage sensitive; roots easily rot in wet conditions
- Lower quality (e.g., bumpier) putting surface
- Mower maintenance—frequent blade sharpening and reel mowers are best
- Seedheads (typically with two yearly flushes, mid-spring and early to midfall)
- Slow recuperative capacity
- Disease susceptibility, especially brown patch and leaf spot
- Less tolerant of poor drainage and water-logged conditions
- Susceptible to bermudagrass invasion
- Susceptible to nematodes
- Poorer drought tolerance due to natural shallow rooting
- Mostly vegetatively (sodded) established, thus higher front-end costs
- Slower putting surfaces (approximately 9 feet, 2.74 m), in summer; faster during other times

Much of the information on managing zoysiagrass, especially on greens, is recent and still evolving. The following techniques will change and be updated as new information, experiences, machinery, and methods are developed.

#### Greens Management

**Mowing.** As with any situation, no two cultural systems are alike with zoysiagrass greens, as budgets, expectations, available resources, labor, etc. all influence management programs. In regards to mowing, trends are overall similar to those for bermudagrass. Heights typically range from 0.09 (spring/summer) to 0.15 inches (fall/winter) (2.3 to 3.8 mm) with a norm of 0.105 to 0.125 inch (2.7 to 3.2 mm). Mowing frequency is once daily for normal play, double cutting (or more) for tournaments. Greens are mowed five to seven days a week during growing months and grooved rollers are used at this time to help provide a higher quality cut and minimize grain formation. When mowing heights are lowered, this must be slowly performed to prevent scalping—no more than 0.03 inch (0.76 mm) daily. Additional backlapping of reels and sharpening of bed knives will be needed to provide a clean cut and not shred (fray) shoots.

Rolling is perfomed daily during growth periods to twice weekly for normal play during nongrowth months. Grooming occurs daily during periods of growth, typically at 25% below bed knife height. Also during periods of growth, greens are brushed daily. Plant growth regulator use is often personal preference or regionally based. Currently, trinexapac-ethyl 1 L is used at rates about double those for bermudagrass. Other PGRs are being investigated for possible use on zoysiagrass greens.

**Fertility.** In terms of fertility, N rates used on zoysiagrass greens are at least one-half those for bermudagrass, similar to those of bentgrass or seashore paspalum, typically ranging from 2 to 4 pounds N/1,000 square feet (9.8 to 20 g N/m$^2$) annually. Higher rates may be needed in areas with longer growing seasons, heavier rainfall, or extremely sandy soils. Nitrogen sources used are typically split evenly between granulars and liquids. To hasten green-up and growth in spring, 1/2 pound N/1,000 square feet (2.4 g N/m$^2$) is used as well as in conjunction with aerification. However, up to 70% of the N sources used in these products are slow release to prevent growth surges and resulting scalping. When using liquid N sources, these are applied about every 10 days at rates ranging from 0.1 to 0.15 pound N/1,000 square feet (0.49 to 0.78 g N/m$^2$).

Minor elements (especially Fe) are applied every 7 to 21 days for color.

Biostimulant use on zoysiagrass typically contains humic acid plus some minor elements and N sources. Therefore, greening often occurs following their use, providing an illusion these products are more beneficial than normal on zoysiagrass. In reality, they are just adding organic matter and various nutrients on a grass that is maintained at extremely low fertility rates to minimize thatch development and to help improve putting speeds.

**Thatch.** Thatch accumulation is a constant concern with all turfgrasses, but especially with zoysiagrass. With its elevated lignin content in leaves and stems, these typically decompose more slowly than most other turfgrasses. As mentioned, during the growing season, greens are groomed daily. This is supplemented by light verticutting 1/32 to 1/16 inch (0.8 to 1.6 mm) below mowing height every one to three weeks during active growth. Double verticutting is best, alternated with topdressing.

Greens are aerified one to three times yearly with three-eighths inch (9.5 mm) diameter hollow tines. These are timed in midspring, early summer, and late summer. Cores are removed, followed by sand topdressing. Aerification is supplemented with needle (pencil) or 1/4 inch (6.4 mm) solid tines. These are typically performed monthly, as is hydrojetting, if available. Routine topdressing is frequent (approximately every 7 to 14 days during growing months) and light using fine- to medium-sized sand using 1/8 to 1/4 cubic yard/1,000 square feet (0.1 to 0.21 $m^3/100\ m^2$). Heaviest topdressing timing should always coincide with grooming, verticutting, and aerifying.

**Irrigation.** Due to its naturally shallower rooting and relative intolerance to wet conditions, irrigating zoysiagrass requires closer observations and measurements than for many other grasses. Ideally, one would irrigate deeply, then skip as many days as possible before rewatering. This translates to two or three heavier weekly irrigation events at 1/4 inch (6.4 mm) each. Hand-watering should be reduced compared to bentgrass and bermudagrass, and only during the hottest, driest, windiest, low-relative-humidity conditions. As with most grasses, one should start using wetting agents early (late winter/early spring), using rates and timings as recommended by product labels. TDR probes are used once calibration for a specific soil type has been completed.

**Seedheads.** A characteristic with zoysiagrass greens is annual flushes of seedheads, typically midspring and late summer (Figure 10.42). The late summer flush tends to be the heavier of the two. Shorter days during these times presumeably trigger seedhead production, but these are accentuated when the plant is under additional stresses such as moisture, insufficient fertility, and cooler nights. As with any PGR program targeting seedhead production, these must be applied prior to emergence to be effective. If seedheads are expressed, mow the greens low and/or groom aggressively to remove seedheads prior to PGR application. Currently, products of choice include combinations of trinaxapac-ethyl plus ethephon or paclobutrazol. At the lower rates used, these are effective for about three weeks, with additional applications needed thereafter. Simazine also helps prevent seedheads. It is also thought that simazine in spring and fall helps prevent any seed from germination. All products should be tested on-site on a nursery green as rates and timings are often site or local environment specific.

**FIGURE 10.42** Zoysiagrass seedheads in a golf green.

**Winter management.** Winter green management issues are similar to those for most warm-season grasses. Compared to *Zoysia japonica*, *Zoysia matrella* is about 7°F less cold tolerant and should be treated accordingly where winter temperatures typically drop below about 18°F (–7.8°C). In terms of providing winter color, overseeding is not a normal practice on zoysiagrass greens due to the difficulty of obtaining a uniform stand with the thatch/organic matter accumulation. The extra fertilizer and water needed for overseeding also can trigger major *Rhizoctonia* disease outbreak. Like many bermudagrass greens, zoysiagrass is colored during winter in areas where green color is needed. Generally, greens are either painted two to three times each winter, or painted only once and then supplemented by pigments, or only pigments are used. Heavier traffic, smaller greens size, and wetter climates would need the higher range of paint applications. As with bermudagrass, colored zoysiagrass greens tend to show wear patterns more easily in winter and ball marks tend to be more noticeable. Cup hole rim stabilizers are often used as cups are typically changed two to three times weekly in summer and up to every two weeks in winter.

In areas where low-temperature damage (or "winter-kill") is a concern, several techniques are used for damage prevention. Mowing heights are raised prior to dormancy to encourage fall carbohydrate production. If resulting greens speeds are too slow, greens are rolled. If they become too fast, greens are fluffed by hand-brushing and/or watered. To encourage fall carbohydrate formation, the last major N application is made in late summer followed by K and Mg use several weeks later. PGR use is also discontinued. Cups are changed frequently enough to prevent collapsing of edges and to avoid showing wear patterns. This is performed typically once weekly or every two weeks. To help mask traffic patterns and/or ball marks, green topdressing sand can be used in winter. The top 2 inches (5 cm) of soil also should not be allowed to dry out in winter. This can easily happen when coloring is done as the discoloration normally associated with drought conditions is not as easily seen. Using a TDR probe is a means of tracking soil moisture levels. Covers are not normally needed for most areas where zoysiagrass greens are grown. Exceptions are shaded greens or areas that traditionally have frequent temperatures in the teens. It is suggested that covers be considered if temperatures drop quickly below ~23°F (–5°C) or slowly below 18°F (–8°C).

**Diseases.** The two most common damaging diseases of zoysiagrass greens are various leaf spots (Figure 10.43) and large patch (Figure 10.44). Other diseases such as dollar spot, spring dead spot, pink snow mold, and *Pythium*, can occur but usually are somewhat rare or are not as devastating as leaf spot and large patch. Of these diseases, large patch is the most commonly occurring and most noticeable. The causal agent, *Rhizoctonia solani*, attacks zoysiagrass in fall, with symptoms typically not seen until late winter or early spring when the grass begins to green up. Wet, cold fall conditions and excessive or late season N use are commonly associated with the worst outbreaks. Providing adequate drainage in chronically wet areas should be a long-term management strategy to discourage the disease. Thatch, which holds additional water and provides an ideal pathogen habitat, should also be managed. Higher mowing heights also appear to help prevent disease intensity since this presumably provides a stronger plant going into dormancy. For example, field observations indicate more disease incidence at 1/2 to 1 inch (13 to 25 mm) compared to 1 3/4 to 2 inches (4.4 to 5 cm).

**FIGURE 10.43** Leaf spot in zoysiagrass golf green.

**FIGURE 10.44** Brown patch (*Rhizoctonia solani*) disease in zoysiagrass.

Preventative fungicides for large patch should be applied in late summer and early fall prior to symptom expression when soil thatch temperatures are ~70°F (21°C). This also often corresponds with the onset of the first several cool nights. Once patches are obvious, fungicides provide little relief except possibly preventing future spread. An integrated pest management (IPM) strategy is to map patches from year to year so that these are priority treatment areas in subsequent years. Table 10.11 summarizes cultural practices for zoysiagrass golf greens.

## MANAGING PUTTING SPEEDS

Today's golf course superintendent is in the middle of a hotly debated issue concerning golf green putting speed (or distances). On one hand, club members see professional golfers playing every weekend on the best maintained golf courses in the world, where not a blade of grass is missing or out of place. Even drainage ditches and creek banks are completely covered with lush, dark-green turf. Greens putt like glass and professionals often complain they are too fast. Everything seems to be perfect and the members have trouble understanding why their home course cannot be maintained in a similar condition.

On the other hand, the superintendent wisely knows what is required in terms of money, time, labor, and resources to obtain "tournament" playing conditions. Courses can spend up to five years preparing for one tournament and spare no expense in achieving the best possible playing surface. Members do not realize the greens have been pampered for months leading up to the tournament, often not allowing membership play during this time. The finest equipment and supplies have been purchased or leased and used on the greens. Greens are built with the latest technology, which allows perfect soil water content control.

As with most things in life, a reasonable compromise must be struck to keep the majority of golf course members happy yet not be overly cost prohibitive. Before using putting speed as the sole criterion for judging the quality of a green, other components influencing putting speed need to be discussed (Figure 10.45). A high-quality green should be uniform in terms of density and coverage, deeply rooted, and free of disruptions from insects, diseases, or weeds. Individual leaves and tillers of the green should be oriented vertically to eliminate grain. The green does not necessarily need to be dark green in color and lush in growth. A diminishing quality putting surface can be expected if putting speed is emphasized long term over other components of a good golf green such as smoothness, density, and health. Other potential negative effects of emphasizing greens speed over healthy grass include:

**FIGURE 10.45** Many components, not just speed, ascertain the quality of a golf green.

- Fewer useable pin placements
- Flatter greens with less interesting contouring
- More putts, thus, slower rounds
- Slower recovery of ball marks and old hole plugs due to fewer carbohydrates from less fertilizer inputs and less rooting depth
- Increased turf stress from weaker (but faster) greens

Steps in managing greens speed with golfer expectation include:

1. Determine speeds using the Stimpmeter.
2. Survey golfers to gauge their expectations/satisfaction levels of speeds.
3. Evaluate maintenance practices to allow best management practices for your target range.
4. Communicate the results.

Golf course officials should first decide on the desired speed for both normal and tournament play. Table 10.13 lists guidelines for ball speeds based on the Stimpmeter. This decision should be based on the desires of the course members, as well as the amount of normal play received by the course, the superintendent's knowledge and experience, the course's budget and equipment, and other resources the club can make available to maintain the greens. Before unrealistic speeds are outlined by the club, reasonable expectations of the resources available to the superintendent should be discussed.

**THE STIMPMETER**

Edward Stimpson, an avid golfer and 1935 Massachusetts amateur champion, designed the Stimpmeter in the 1930s. It is an extruded 36-inch (0.9 m) aluminium bar with a V-shaped groove along its entire length (Figure 10.46). A golf ball is placed in a notch 30 inches (76 cm) from the end that rests on the ground and the bar is raised to an angle of ~20° until it is released. The distance the ball rolls is then measured and recorded. It is best to choose a flat surface for most accurate measurement.

The USGA suggests rolling three golf balls, one at a time, down the Stimpmeter bar, recording the distances, and then rerolling the balls in the opposite direction. The greens speed is then calculated from the average of the three ball rolls of the two directions.

## Grooming Practices

Surface smoothness is the most often determining parameter affecting greens speed and a variety of grooming practices are employed to maximize this quality. Grooming putting surfaces to maximize greens speed and quality involves the following (see Table 10.14):

1. Grass selection
2. Mowing practices
3. Fertilization
4. Aerification
5. Topdressing
6. Brushing/combing/grooming
7. Plant growth regulators
8. Water management
9. Rolling

## Suggested Sequence of Events

The following sequence of agronomic events is suggested to provide acceptable putting surfaces in terms of uniformity,

**TABLE 10.13**
**Putting Green Speeds as Determined by the Stimpmeter (USGA)**

| | Ball Roll Distance[a] | | | |
|---|---|---|---|---|
| | Normal Conditions | | Tournament Conditions | |
| **Putting Characteristic** | **Feet** | **Meters** | **Feet** | **Meters** |
| Slow | <7.5 | <2.3 | <8.5 | <2.59 |
| Medium | 7.5 to 9.0 | 2.3 to 2.7 | 8.5 to 10 | 2.6 to 3.1 |
| Fast | >9.0 | >2.7 | >10 | >3.1 |

[a] Based on an average of ball roll distance up and down a green's slope.

**FIGURE 10.46** Using a Stimpmeter to measure ball roll distances (speed) of a golf green.

density, and speed. If all of these are incorporated into a total management practice package, putting characteristics should be acceptable for the majority of participants.

**Three months prior to the tournament.** Begin a PGR program to help regulate growth throughout the day. It requires several applications of PGRs to "condition" the grass so as not to have any phytotoxicity. These are usually spaced one to three weeks apart depending on use rate.

**Six weeks prior to the tournament.** Fertilize with a 3-1-2 ratio fertilizer at the rate of 1 pound actual N/1,000 square feet (4.9 kg N/100 m$^2$). This fertilization will strengthen the grass for the upcoming aerification and aid recuperation.

**One month prior to the tournament.** Core aerify with relatively small tines (e.g., less than 3/8 inch, 9.5 mm) when the green has adequate moisture to minimize damage to the putting surface. Aerification is needed when greens are considered hard and do not hold approaching shots well, if the greens are not properly draining, or if localized dry spots are developing. If these problems are not present, then core aerification can be skipped. However, one month may be needed for the grass to recover and fill in holes left from coring. Individual cores should be removed following aerification. The exception to this is when a desirable soil mixture is currently present. The cores in this case should be incorporated back into the profile by lightly verticutting or dragging with a brush or piece of carpet or steel drag mat.

Following aerification, greens should be topdressed with a medium rate of desirable, clean soil. One to three cubic feet of material/1,000 square feet (0.03 to 0.09 m$^3$/100 m$^2$) is a common topdressing rate. This should be immediately incorporated by dragging. Care must be taken during this process to prevent excessive desiccation or turf damage. Irrigate heavily following aerification to help prevent exposed roots from drying.

One week following aerification, greens should receive the equivalent of 1 pound actual N/1,000 square feet (4.9 g N/m$^2$) as a quick-release N source (e.g., ammonium sulfate or ammonium nitrate). Rapid-release N will aid in recovery of aerifying holes and be timed in advance of the tournament so excessive growth will have subsided.

**Two weeks prior to the tournament.** All N fertilization should be completed by this time. If not, the grass will be growing too aggressively by tournament time and, consequently, the putting speed will be disappointingly reduced.

The greens should also be lightly vertical mowed in two directions at this time. Vertical mowing blades should just barely touch the plant crowns. This will remove some of the surface debris left over from aerifying and will reduce grain. Greens should be groomed daily at this time by using grooved rollers, brushing, and/or combing.

Between 10 and 14 days prior to the tournament, the greens should receive a light application of topdressing. One to two cubic yards/5,000 square feet (0.03 to 0.06 m$^3$/100 m$^2$) should be ample to provide the final touches in smoothing the putting

**TABLE 10.14**
**Individual Cultural Practices Influence on Putting Speeds (or Distances)**

| Practice | Comment |
|---|---|
| Grass selection | • Ultradwarf cultivars, due to their lower mowing height tolerance, and denser shoot counts and narrower leaves, generally provide best (farthest) putting speeds. |
| Mowing height | • A 1/8 inch (3.2 mm) decrease in height increases putting speeds 8 to 12 inches (20 to 30 cm). |
| Mower type | • Drum-type walk-behind mowers increase putting speeds 6 inches to 1 foot (15 to 30 cm) over floating head and triplex mowers. Floating heads increase speeds about 6 inches (15 cm) over triplex mowers. |
| Double cutting | • Double cutting starting two days prior to and continuing during a tournament increases speeds ~6 inches (15 cm). |
| Grooming | • Walk-behind mowers with groomers increase speeds ~2 inches (5 cm) versus nongroomed. |
| Fertilization | • For each pound of excessive N applied annually/1,000 square feet (4.9 g/m$^2$), ~4 inches (10 cm) decrease in speeds occur. |
| Aerification and topdressing | • Aerification not followed by topdressing decreases speeds ~5 inches (13 cm) for up to 28 days.<br>• Aerification followed by topdressing increases speeds 6 to 15 inches (15 to 38 cm) up to 29 days, depending on rate. The increases, however, are not normally seen until day eight following topdressing. |
| Rolling | • One pass increases speeds ~1 foot (0.3 m) that day and ~6 inches (15 cm) the next day. Interestingly, rolling can also help reduce disease activity such as dollar spot and anthracnose. Rolling has been shown not to be detrimental to the turf or to increase soil bulk density (compaction) if performed three times or less weekly. |
| Plant growth regulators (PGRs) | • PGRs have only limited impact on increasing ball roll distances but help maintain similar distances throughout the day. |

surface. The topdressing material should be immediately incorporated by dragging in several directions. If the green is under environmental stress, only one light dragging is suggested to reduce turf bruising.

**Five days prior to the tournament.** If the greens start to lose some color due to lack of N fertilization, application of liquid Fe sources should be considered. Two ounces of iron sulfate or a chelated iron source in several gallons of water should be applied per 1,000 square feet (6.4 L/ha). To prevent burn, this should not be applied during the heat of the day.

The mowing height can be reduced at this time from 3/16 inch (4.8 mm) to 5/32 or 1/8 inch (4 or 3 mm). This height, combined with the other management practices already outlined, should provide satisfactory putting conditions for the tournament.

**Three days prior to the tournament.** Begin double cutting greens at the tournament height so the maximum benefits from this practice are experienced. The cleanup mowing lap should be performed only once daily to prevent excessive tearing of the grass or soil compaction.

Water management during the weeks leading up to the tournament is critical. Excessive soil moisture should not be used as a method to soften the greens. Water should be applied to wet the rootzone and soil should be allowed to dry before reapplying it. Firm greens are often desired to challenge golfers for precise approach shots. However, with the added stresses of a lower mowing height and double cutting, constant inspection for wilting should occur and afternoon syringing should be considered.

**Tournament day (optional).** If green speeds are not acceptable, then the greens should be rolled. The greens should be allowed to drain (not wet) before this is attempted and routine mowing should be followed. One pass increases speed ~10% for that day but four passes are needed to increase speeds ~20%. Once the tournament is completed, the mowing height should be raised. Greens should then be fertilized and adequately watered to aid in recovery from the concentrated tournament play.

# Section VI

## Pest Management and Control Practices

# 11 Integrated Pest Management

## IPM HISTORY

Modern integrated pest management (IPM) concepts and practices began to develop in the late 1950s with alfalfa and apple production, and vastly expanded in the 1960s with cotton production. These concepts evolved from the mid-1940s when growers felt "silver-bullet" or ultimate specific weapons such as pesticides were needed to control all pest problems. Most traditional pest and plant ecological studies were then deemphasized and abandoned, along with nonchemical control alternative research efforts. This led to a new generation of producers and scientists who had little experience with nonchemical approaches to pest or plant management.

In recent years, turf managers, feeling dependent on pesticides, suffered from a lack of research and training in the pest management area. For example, in the early 1980s, two very effective and relatively inexpensive pesticides, EDB (ethylene dibromide) and chlordane, were banned from the turf market. EDB was a highly effective soil-injected nematicide. Since EDB was so effective and inexpensive, research on turf tolerance to higher nematode populations was essentially abandoned. Chlordane, an insecticide, was especially useful for mole cricket and grub control. With chlordane, managers did not worry about managing the turf to withstand higher pest populations or treating during the most susceptible point in the insect's life cycle. However, after losing these chemicals, nematodes, mole crickets, and grubs became very serious turf pests. Researchers are now trying to find alternative methods of pest management.

Plant breeding provides another method of manipulating grass for better pest resistance. In the past, turf breeders basically tried to satisfy the golfer by developing grasses that had finer texture, greater density, and the ability to survive at lower mowing heights. Today, turf breeders have redirected their efforts by including pest resistance in turf species lines. Genetically modified turfgrasses show great promise in providing innate ability to resist pest pressures and withstand extreme environmental conditions. More time and research will be necessary to solve problems that have basically been ignored for the last 40 years.

As noted, IPM practices have not been widely used in the turfgrass profession. One reason for this is the lack of adequate federal, state, or industry money available for turf researchers wishing to explore IPM approaches. Traditional "crops" such as cotton, soybeans, and corn have enjoyed the majority of available grants for this type of research and subsequently have made the greatest strides in IPM. Hopefully, turf researchers will soon receive similar considerations from granting agencies.

Only a few reports are available dealing with turf IPM programs. One experimental IPM program in selected urban Florida areas has resulted in an approximately 90% reduction in pesticide application without sacrificing visual quality. No differences were noted between lawns sprayed only when pests reached aesthetic thresholds and lawns sprayed preventatively. Results from a similar IPM scouting program in Maryland suggested 40% to 80% of the pest problems could be eliminated by simply substituting resistant ornamental varieties or eliminating pest-susceptible plants in residential lawns.

A country club in Massachusetts has used IPM practices to control Japanese beetle grubs, using sex tabs and floral lures instead of the traditional pesticide treatment approach to attract insects into traps. For example, during August, the club placed 47 traps in the rough and collected 160 gallons of beetles. The number of grubs were reduced from ~65 per square foot (6/m$^2$) to only one or two.

A golf club in Idaho traditionally had disease problems during the summer on its bentgrass greens. The superintendent, with some major reservations from club officials, decided that continued and frequent fungicide use only provided temporary masking of disease symptoms and began searching for the underlying disease causes. Poor drainage and soil layering were discovered on the most troublesome greens. Excessive nitrogen and water-use rates were applied in recent years and a distinct buildup of black layer occurred.

The club initially began implementing IPM practices by informing and soliciting support from the club members. After obtaining approval, greens were aggressively and frequently aerified deep enough to allow better and healthier root penetration. Natural organic fertilizers and a biostimulant were used to supplement and eventually replace a portion of the synthetic materials. Synthetic fungicide use was partially replaced by natural sea plant extract disease suppressants and a "compost soup" consisting of digested sewage sludge and wood wastes.

Other projects in New York have been able to reduce pesticide applications by about one-third. Diligent scouting and emergency pesticide applications were needed, but avoided the "calendar" spray schedules typical of many greens today. The IPM greens also usually had higher quality than regularly maintained greens. The IPM greens, however, had occasional pest outbreaks that caused poor conditions before control measures were implemented. Additional experience; evolving technology such as GPS/GIS, optical sensing, and more precise spray technology; and improved pest-resistant grasses will improve IPM results and acceptance.

South Carolina conducted a pilot project incorporating traditional IPM strategies into managing golf turf. Turfgrass Information and Pest Scouting (TIPS) was administered on seven golf courses. Scouting was performed and recommendations made to the superintendents about agronomic practices and judicious use of pesticides. Among this project's accomplishments was a 30% reduction in fungicides used by monitoring weather parameters and not applying chemicals until favorable conditions existed for disease development. Nitrogen use was also reduced 35% without sacrificing golf course quality by using judicious amounts timed to produce maximum benefits in relation to the plant's growth cycle and environment.

## STRATEGIES OF INTEGRATED PEST MANAGEMENT

Developing IPM strategies requires superintendents to have reliable information about the following:

1. Obtain information concerning the current site conditions regarding a golf course. This involves the total ecological situation involved with the particular turfgrass and pest. Identify the pest and know its biology and life cycles, as well as management practices that disrupt or influence these to reduce pest numbers. Understanding the strengths and weaknesses of the pest increases the chances of success. This allows one to know:
   a. When to expect the pest problem or when, in its life cycle, it is most susceptible for control.
   b. Where to expect their occurrence. Mapping and recordkeeping can help spot the "when" and "where" of infestation patterns.

   Site assessment also involves documenting current growing conditions such as:
   i. Soil type
   ii. Hydraulic conductivity of greens
   iii. Soil and irrigation water pH
   iv. Soil nutrient and CEC levels
   v. Soil pests such as nematodes and noted diseases
   vi. Surface air movement of each green
2. Developing IPM strategies includes correcting any major agronomic or environmental problems in growing as healthy turf as possible. Develop this in conjunction with club officials and submit to the general membership when completed.
3. Implement the plan including using a detailed monitoring system to carefully follow pest trends, which determines if a pesticide will be necessary, and if so, when and where it would be most effectively applied.
4. Maintain careful records to measure IPM effectiveness.
5. Periodically, evaluate the program, adjusting weaknesses as needed. Also inform membership and the general public these IPM practices are being implemented to demonstrate golf course managers are just as environmentally aware and sensitive as players are.

IPM control strategies can be subdivided into chemical and nonchemical strategies. Both are equally important in implementing a successful program.

### Nonchemical Strategies

The following contribute to the integration of nonchemical strategies for pest management: host-plant resistance, pest-free propagation (or sanitation), site preparation, cultural practices, and biological control.

**Host-plant resistance.** One of the oldest means of pest control is through careful selection and breeding of pest-resistant or pest-tolerant plants. People have traditionally selected those plants that grew best or had the highest yields, and then used them in subsequent years. Many turfgrass varieties are the result of this type of selection process, where a patch of turf grew better, was denser, or had better color than the surrounding grass. Genetic work has revealed certain genes for pest resistance and/or better turfgrass characteristics. This genetic resistance can be incorporated into a cultivated plant to provide more effective pest control. For example, "Floratam" St. Augustinegrass was the first commercially available variety resistant to chinch bugs (*Blissus* spp.). However, some plants are less resistant to certain pests and their use must be weighed against other desirable characteristics. For example, Ormond bermudagrass is highly susceptible to stunt mites (*Aceria* spp.) while Tifway bermudagrass is not. L-93 creeping bentgrass has good resistance to dollar spot while Crenshaw bentgrass does not.

Resistance usually involves one or a combination of the following:

1. Chemicals produced by the host repel the pest or prevent it from completing its life cycle.
2. The host is more vigorous or tolerant than other varieties and thus withstands higher pest populations and/or more pest damage.
3. The host has physical characteristics making it less vulnerable to attack.

As noted, turf-breeding efforts have traditionally been more concerned with improving the playability of grasses than with breeding for pest resistance. With the threat of losing many of today's effective pesticides, breeding for pest resistance has become a higher priority. Hopefully, as today's breeding and genetic engineering technologies evolve, better means of transferring genes will provide turf plants with desirable playing qualities and inherent pest resistance.

**Pest-free propagation.** One of the easiest, yet often overlooked, means of preventing pest establishment in turf is by using pest-free planting materials. Each state has established a seed certification program, one of the first regulatory measures designed to provide pest-free propagation. Each bag of seed, in order to meet certification, must provide information on seed purity and germination percentages. In addition, weed seed test date, origin, lot number, inert matter, and

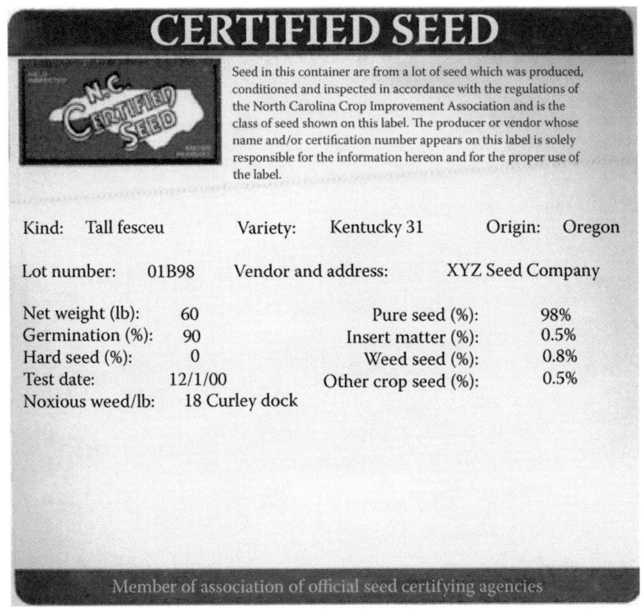

CERTIFIED SEED

N.C. CERTIFIED SEED

Seed in this container are from a lot of seed which was produced, conditioned and inspected in accordance with the regulations of the North Carolina Crop Improvement Association and is the class of seed shown on this label. The producer or vendor whose name and/or certification number appears on this label is solely responsible for the information hereon and for the proper use of the label.

Kind: Tall fesceu Variety: Kentucky 31 Origin: Oregon

Lot number: 01B98 Vendor and address: XYZ Seed Company

| | | | |
|---|---|---|---|
| Net weight (lb): | 60 | Pure seed (%): | 98% |
| Germination (%): | 90 | Insert matter (%): | 0.5% |
| Hard seed (%): | 0 | Weed seed (%): | 0.8% |
| Test date: | 12/1/00 | Other crop seed (%): | 0.5% |
| Noxious weed/lb: | 18 Curley dock | | |

Member of association of official seed certifying agencies

**FIGURE 11.1** Using certified seed, sprigs, and sod is the best assurance of receiving quality products.

variety/brand name are provided (Figure 11.1). No noxious weed seeds are allowed in the seed bags.

This same philosophy also should apply to planting in soil free of noxious pests. Soil fumigation to control many soil-borne pests such as nematodes and weeds should be considered in almost all cases before turf establishment. Likewise, soil added to an existing turf site via topdressing should be pest free. Do not accept pest-contaminated soil or plant material because of price.

**Site preparation.** Properly preparing the planting site is an important, yet often unrecognized, IPM step that involves planning and constructing highly utilized areas, such as putting greens or tees, with precise water management capabilities. Precise water management is the major key to successful turf maintenance under intense playing conditions. Disease and soil compaction are among the problems occurring if soil saturation is allowed during intense play. Adequate surface and subsurface drainage must be provided so play can quickly resume after a heavy rain without soil compaction. However, golf greens must be able to retain adequate levels of moisture and nutrients to avoid continuous reapplication of them.

Other considerations for site preparation and construction include (1) provision of proper sunlight and ventilation around golf greens, (2) adequate sizing of greens for even distribution of traffic, (3) proper drainage in fairways, and (4) provision of adequate quantity and quality of irrigation. All too often, these preconstruction criteria are neglected, causing even the most intense IPM programs to fail, leaving members dissatisfied with services rendered.

**Cultural practices.** Probably the best defense against pest invasion is providing a dense, healthy, competitive turf. This is achieved by providing cultural practices that favor turf growth over pest proliferation, which disrupts the normal relationship between the pest and its host plant, making the pest less likely to grow, survive, or reproduce. These practices include proper irrigation, fertilization, mowing, aerification, and topdressing (Figure 11.2). Prolonged use of incorrect cultural practices weakens the turf, promotes pest invasion and its spread, and often encourages excessive thatch development. Thatch not only harbors many insects and disease pathogens, but it also limits pesticide effectiveness while reducing the efficiency of a watering program.

**FIGURE 11.2** Discouraging pests such as weeds is best achieved by following sound agronomic practices.

**Biological control.** Biological pest control uses natural enemies, such as pathogens, predators, and parasites, to reduce pest populations to aesthetically acceptable levels (Figure 11.3). Criteria for a successful biological control agent include (1) the absence of nontarget effects on desirable plants or other organisms; (2) its ability to reproduce quickly to prevent the pest from attaining damage thresholds; (3) its persistence in the environment, even at nondamaging pest levels; and (4) its adaptability to the environment of the host. The biological control agent also should be free of its own predators, parasites, and pathogens.

**FIGURE 11.3** Scientists constantly seek viable biological control agents.

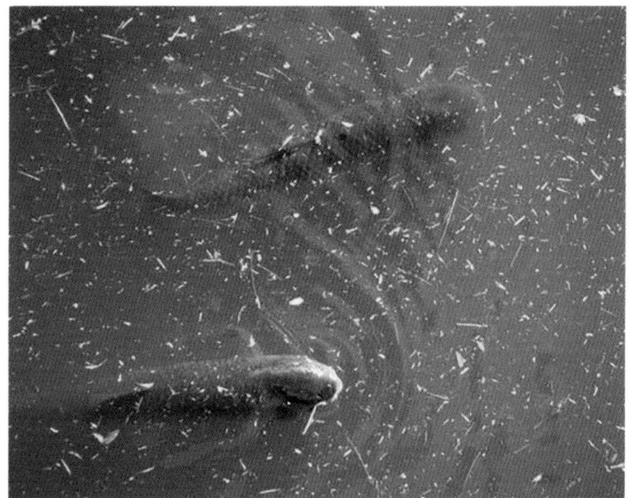

**FIGURE 11.4** White amur carp is a widely successful biological agent for certain aquatic weeds.

A few examples of biological control measures currently are being used in commercial turf production. *Bacillus popilliae* and *B. thurengensis,* bacteria commonly known to cause the milky spore disease, have been used to control Japanese beetle grubs. More effective strains are currently being developed. The white amur carp, a fish native to Southeast Asia, provides submerged aquatic weed control in golf course ponds (Figure 11.4). Extracts from various wood-decaying fungi are being reformulated as commercial fungicides. Other potential agents for biological control of turf pests include endophytic fungi for insect control, bacteria such as *Xanthomonas* species for annual bluegrass control, various rust (*Puccinia* spp.) fungi for nutsedge control, and several parasitic bacteria, nematodes, and fungi for the control of mole crickets and damaging nematodes in turf.

One underlying requirement of the use of biological control agents is the public's willingness to accept a certain level of pest damage. This level of pest pressure is necessary for the biological control agent to have a continual food source after the pest level has become acceptable. Thus, total pest elimination is not feasible when integrating biological control measures. There also is a time lag between the use of a biological agent and actual control. Club members must be educated to this fact and be willing to accept minor levels of pest pressure and time delay for long-term success.

### Chemical Pest Control

Not all pest problems can be solved by host-plant resistance, manipulating cultural practices in the plant environment, or by the use of biological control agents. In these cases, pesticides become the next line of defense. In the IPM scheme, pesticide use is not eliminated. However, indiscriminate spraying is eliminated and only practical pesticide use is employed so there is minimal damage to the natural biological control agents of the pest, as well as minimum damage to the environment. This requires extensive knowledge of the ecology and interrelation of the pest, the pesticide, the host plant, and the beneficial natural biological control agent. Several considerations for strategic pesticide use involve making management decisions concerning the following:

1. Locate and identify the pest using reliable monitoring techniques and establish an aesthetic threshold. Pest identification is a fundamental requirement when developing an IPM program. The task of identifying pests can be both frustrating and time consuming. Pest diagnosis often can be assisted by training and by having well-illustrated pest literature on site.

   Trained personnel, often referred to as **scouts**, examine or scout the golf course on a routine basis, monitoring pest levels (Figure 11.5). Monitoring ranges from visual inspection, sampling, and analyzing soil and plant tissue to the use of sophisticated "high-tech" detection techniques. Decisions based on what pests and symptoms are visible can immediately be made. When specialized tests must be run at public or private diagnostic laboratories (e.g., nematode assays, detection and accurate identification of certain viruses, species identification of some fungi), scouts must know how each type of sample should be taken and handled to provide the most useful and reliable information. They also must know how to specifically interpret the results of each test conducted.

   Once pests have been identified and their infestation levels recorded, a control action must be initiated at a predetermined pest threshold level. These threshold levels are known as economic, damage, or action thresholds in other IPM programs. For turf managers, economic and related threshold-level terms mean little since crop yield is not the ultimate goal, but rather an aesthetic level of turf quality is. An aesthetic threshold deals with the amount of visual damage a particular turf area can withstand before action is

**FIGURE 11.5** Trained personnel, often referred to as scouts, examine or scout a golf course for pests.

required. Highly maintained areas such as golf greens have a lower aesthetic threshold level than less maintained areas such as roughs, which can withstand a higher degree of pest damage before action is required. These thresholds also vary with the expectations for a particular golf course, the availability of financial resources, and available alternative control measures.

2. Attack a pest during the most susceptible point in its life cycle, which usually is during the early stages of development. For example, the most effective time for mole cricket control is during its first instars, which normally occur during late spring/early summer. However, prior scouting and mapping of mole cricket infested areas have proven most beneficial when pinpointing sites of future pesticide use. The same philosophy of attacking the most susceptible point in its life cycle also is true for most weeds. Young, actively growing weeds are usually the easiest to control. Once weeds begin to mature, they become more difficult and expensive to control.
3. If a pesticide is necessary, use the one that is most effective but the least toxic to nontarget organisms and the environment. Read the label and use the recommended rate, and spot treat, if possible, instead of applying "blanket" or "wall-to-wall" treatments. Obviously, this relies on effective scouting techniques and proper recording or mapping of pest outbreaks.

## IMPLEMENTING AN IPM PROGRAM

Developing and implementing an IPM program involves certain steps. However, each golfing facility differs and will require an IPM program tailored to its interest, level of expectation, and available budget. Pest problems are going to occur on any golf course; even the best management program cannot guarantee problems will not occur (Figure 11.6). The very nature of managing a golf course predisposes it to stress since the turf is maintained at its very edge of survival. For example, golf greens are generally mowed below 1/8 inch (3.2 mm), which is much lower than the natural adaptation of these grasses. This results in a precarious balance between the needs for grass maintenance and those for grass survival. The following steps have proven successful in developing an IPM program or mind-set and should provide a good starting point for golf course superintendents.

**FIGURE 11.6** Courses should plan for eventual pest infestations as well as natural and man-made disasters.

1. Define the roles and responsibilities of all persons who are involved in the pest management program. This includes establishing communication between club officials, players, and crew members, who must be aware of the new approach the superintendent is trying and that it is an ongoing experiment. They need to expect some successes and some setbacks. Assurance and understanding will be needed by all participants during the initial stages of development to prevent misunderstanding and provide ample time for desirable results to occur.
2. Determine management objectives for specific areas of the course and correct all practices that favor pest development or place undue stress on the turf. Obviously, highly maintained areas such as greens and tees receive a priority for pest control. Lower maintained grass, such as the driving range or roughs, is a lower priority. A thorough inspection should be conducted of each site on the course before implementing the IPM program. This will provide the groundwork from which all management decisions can be based and also will provide a record justifying the correction of problems made during course construction or from subsequent management. A field history form similar to Table 11.1 should be used to record data, including the current turf species, its area, mowing schedule, soil analysis, soil drainage, fertilizer programs, irrigation scheduling, and shade and traffic patterns. Be prepared to improve existing problems that weaken the turf, or the potential success of the IPM program will be greatly reduced.

   A weather monitoring system should also be installed, which will provide detailed, localized data on important variables such as rainfall patterns, soil temperature and moisture, wind movement, humidity, and sunlight indices. These climatic conditions usually play the most important role in specific turf growth patterns. Being able to track or pinpoint them enables the superintendent to modify cultural practices to supplement or offset its effects.
3. Set aesthetic or action thresholds and begin monitoring and recording pest levels. An aesthetic or action threshold is the point when pest populations or environmental conditions indicate some action must be taken to prevent intolerable damage. These thresholds will vary according to the location of the course, the specific pest being scouted, the level of use of the turf area, the expectations of club members, and budget constraints.

**TABLE 11.1**
**Field History Report Form Used for Golf Courses**

**Turf IPM Field History Report Form**

Club ____________ Superintendent ____________ Phone Number ____________ Date ____________
Hole Number ____________ Scout ____________ Phone Number ____________

| Site | Turf Species | Area | Mowing Schedule | Soil Analysis | Soil Drainage | Fertilization Amount (N/1,000 square feet) | Frequency | Irrigation Scheduling |
|---|---|---|---|---|---|---|---|---|
| | | | | pH P K | | Spring Summer Fall Winter | | |
| Green | | | | | | | | |
| Tee | | | | | | | | |
| Fairway | | | | | | | | |
| Rough | | | | | | | | |
| Driving range | | | | | | | | |
| Nursery green | | | | | | | | |
| Practice green | | | | | | | | |

*Note:* Comments on specific topics such as shade, overseeding blend, nitrogen carrier, topdressing mix, weather, irrigation salinity levels, and so forth.

The pest in question will partially determine its aesthetic threshold. For example, the number of mole crickets tolerated on an area basis is less than the number of sod webworms. Related to this threshold level is the site in which the pest is found. Golf greens have a much lower aesthetic threshold for mole crickets than a rough or out-of-play area. Unfortunately, exact threshold numbers have not been developed for every pest encountered in turf. Known threshold numbers for various turfgrass pests are listed in the pest control chapters.

Tools required for scouting vary with pest problems, scout training, and golf course budget. A good set of eyes and an inquisitive mind are essential, supported by a standard 10× hand or pocket lens, soil probe, soil profile probe, spade, cup cutter, pocket knife, tweezers, scalpel, collection vials and paper bags, digital camera, access to e-mail and the World Wide Web, and field identification guides. Soap and water also are necessary for insect monitoring.

More expensive but precise instruments may be used in a room designated as a diagnostic laboratory, including stereo- and compound microscopes, soil sieves, pH meter, conductivity meter, and elementary soil analysis kits. These need to be supplemented by ongoing scout training at short courses, formal classes, appropriate diagnostic guides, and opportunities to visit similar setups to exchange ideas.

Monitoring intensively maintained golf courses includes scouting greens, tees, fairways, roughs, ornamental plantings, and trees. Greens and tees generally require the greatest amount of attention and are monitored daily or every other day. Remaining areas are monitored less frequently, usually weekly. Monitoring frequency may require adjustment depending on climatic conditions and reports of nearby pest problems.

Greens and tees are scouted by simply walking around the area to observe insect and disease activity as well as other pest and noninfectious symptoms. Fairways and roughs are usually scouted from a golf cart or utility vehicle. This allows the scout closer examination if symptoms are observed.

The superintendent should provide the scout guidance for traditional pest problem areas or hot spots. This allows the scout to concentrate his or her efforts in these areas when conditions favor those pests. In order to minimize play disruptions and to better recognize specific pest damage such as disease symptoms and nocturnal insect feeding, early morning scouting is suggested.

Recording pest levels should be done on a form similar to the one in Table 11.2. This will allow the scout and superintendent to monitor pest trends and determine if these levels reach or exceed aesthetic thresholds. Maps developed for each golf hole should accompany these forms. Maps enable the scout to pinpoint pest problem areas, thereby allowing spot versus traditional blanket pesticide treatment. Over time, these maps can indicate where pest problems annually occur and

**TABLE 11.2**
**Field Infestation Report Form Used for Golf Courses**

**Turf IPM Field Infestation Report Form**

| Club ____________ | Superintendent ____________ | Phone Number ____________ | Date ____________ |
|---|---|---|---|
| Hole Number ____________ | Scout ____________ | Phone Number ____________ | |

| **Site (Turf Species)** | **Mowing Height** | **Soil Moisture** | **Weeds Species No. or %** | **Diseases Species No. or %** | **Insects Species No.** | **Nematodes Species No.** |
|---|---|---|---|---|---|---|
| Green: | | | | | | |
| Tee: | | | | | | |
| Fairway: | | | | | | |
| Rough: | | | | | | |
| Notes: | | | | | | |

| | | | |
|---|---|---|---|
| 1. Goosegrass | 1. Dollar spot | 1. Mole crickets | 1. Sting |
| 2. Crabgrass | 2. Leaf spot | 2. Sod webworms | 2. Lance |
| 3. Thin paspalum/dallisgrass | 3. *Pythium* blight | 3. Armyworms | 3. Stubby-root |
| 4. Broadleaves | 4. *Pythium* root rot | 4. Cutworms | 4. Root-knot |
| 5. Quackgrass | 5. Fairy ring | 5. White grubs | 5. Cyst |
| 6. Nutsedge (yellow, globe, purple, annual, kyllinga) | 6. Brown patch (*R. solani*) | 6. Fire ants | 6. Ring |
| 7. Nutsedge (purple) | 7. Rhizoctonia leaf/sheath blight (*R. zeae*) | 7. Mites | 7. Spiral |
| 8. *Poa annua* | 8. Bermudagrass decline | 8. Grass scales | 8. Sheath |
| 9. Broadleaves | 9. Algae/moss | 9. Billbugs | 9. Other |
| 10. Other | 10. Other | 10. Other | |

possibly allow superintendents to correct management or environmental variables influencing them. For example, mole cricket egg laying activity (tunneling) during April and May should be mapped and these areas treated in June, as mole crickets tend to deposit eggs in the same areas each year. Maps also provide area information for the superintendent and allow new crew members a visual aid in treating problem areas. Aerial photographs and digital cameras are valuable in identifying, tracking, and mapping problem areas.

4. Use pesticides correctly and only when threshold limits are reached. One of the goals of IPM is intelligent and prudent pesticide use. Once these thresholds are reached, the pesticide used should be the safest one available; spot treatments should be practiced, if possible; and all safety precautions should be followed. Pests should also be treated during the most vulnerable stage of their life cycle. In addition:
   - Use natural biological controls when possible.
   - Spray only when necessary and strictly control the location of pesticide application.
   - Select pesticides that are least toxic, less mobile, and have a short half-life.
   - Identify areas that may be particularly susceptible to ground/surface water contamination.
5. Evaluate the results of the habitat modification and pesticide treatments by periodically monitoring the site environment and pest populations. Keep written records of site pest management objectives, monitoring methods, and data collected. Also record the actions taken and the results obtained by the pest management methods. This will provide additional information for club members who do not understand the program but would understand results. This will also demonstrate that golf course superintendents are striving to reduce the chemical inputs in maintaining the course and obtain an ecological balance between man and nature.

### Additional Considerations

Pest management strategies for turf production are in their infancy but are nonetheless being developed and used. Strategies necessary for a successful IPM program have been outlined and should provide a starting point for golf course superintendents. No one recipe, however, will work uniformly across all golf courses. Individual programs will most likely evolve for each particular course depending on its demands, needs, acceptance of pest damage, and budgets to work with. Golf course members should realize that it will take time to implement and measure the success of IPM methods. Open communication between all crew members, club officials, players, and the superintendent is necessary for programs to be understood and to succeed. Read and study information on pests associated with golf courses in your state, their life cycles, the ecology surrounding their occurrence, and the pest management strategies to successfully suppress their numbers below threshold levels. More importantly, follow those management guidelines that encourage competitive turf growth and discourage pest establishment.

IPM represents a case of redirecting resources invested in turf maintenance, in which professional knowledge and judgment are partially substituted for chemicals. It may or may not be cheaper in up-front costs, but it does provide an excellent means of reducing unwelcome environmental and personnel risks without sacrificing turf health. IPM reduces pesticide use or "softens" the kinds of pesticides needed, as well as the potential liability for unforeseeable cleanups or other corrective actions for specific pesticide use or handling sites. IPM thus improves the image of the golf industry as good stewards of modern urban society.

## BMPs FOR PESTICIDES AND NUTRIENTS

### Fate of Pesticides and Nutrients

Environmentally introduced pesticides and nutrients experience numerous fates and routes of decomposition following application. Most pesticides and nutrients pose little, if any, threat when used according to the label. This is due to the numerous means by which they are absorbed and broken down by natural entities in the environment.

Turfgrasses are unique systems due to the extremely high density of plants with communities containing up to 2,500 plants/square foot (~27,000/m$^2$). Turfgrasses also produce a thatch layer high in organic matter that binds many pesticides before they can reach the soil. It has been estimated that a 150-acre (61 ha) golf course has the capacity to absorb 12 million gallons (45 million L) of water during a 3-inch (7.6 cm) downpour. This water infiltrates and is absorbed by the turf/thatch/soil profile before it moves horizontally as runoff. Thatch also harbors high populations of microorganisms that use pesticides and nutrients as energy sources as they break down these compounds into simple molecules. Turfgrasses also possess dense, fibrous root systems that are capable of adsorbing pesticides. Other avenues of pesticide and nutrient dissipation include (Figure 11.7):

- Volatilization and evaporation
- Photodegradation by ultraviolet light from the sun
- Microbial decay
- Hydrolysis
- Conversion to other compounds
- Plant uptake

Chemical and microbial degradation of pesticides and nutrients occurs through the processes of hydrolysis, oxidation, and reduction to transform the parent compounds into the production of carbon dioxide, water, and some inorganic products such as N, S, and P. Soil microbes can use pesticides

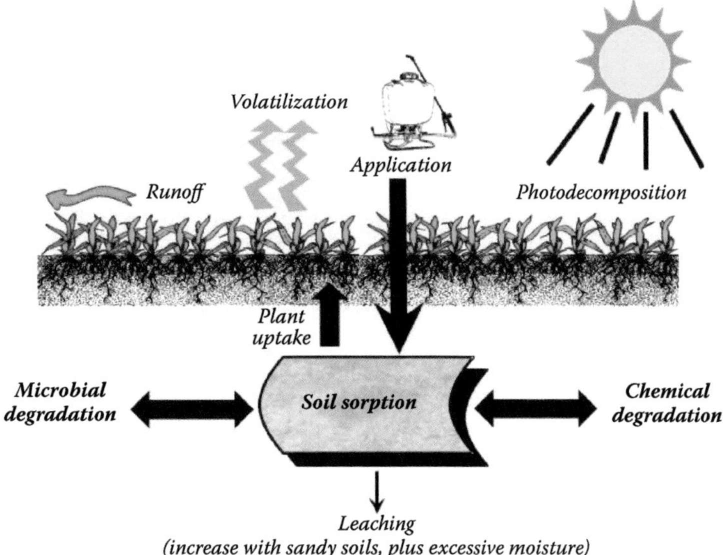


**FIGURE 11.7** A dense turf system minimizes negative impact of applied products.

as a food source; this rate of degradation is influenced by pesticide present, temperature, soil water content, pH, oxygen levels, organic matter, and prior pesticide use.

Pesticide persistence is often expressed as *half-life* (often indicated as $DT_{50}$), which is the time required for 50% of the original pesticide to break down into other products. For example, if a product has a $DT_{50}$ value of 10 days, the product is degraded to half of its original amount by 10 days after application, to one-fourth of its original amount by 20 days after application, to one-eighth of its original amount by 30 days after application, and so on (Figure 11.8). Half-life values change according to the location and site being treated; thus, they are guidelines instead of absolute values. Once a pesticide leaves the nozzle or spreader, there are four major places it can go: (1) air, (2) plants, (3) soil, and (4) water.

### Atmosphere

When using traditional applicators, the air is the first place a pesticide goes before it hits a target. Losses of pesticides to the air can occur. Here are some factors in these losses.

**Volatilization.** Volatilization and evaporation of pesticides involves the chemical transformation of a solid or liquid into a gas. This tendency to volatilize is expressed in units of vapor pressure. Vapor pressure (or volatility) is the measure of the intramolecular bonding forces of a compound and is usually expressed as mmHg (mercury) at 25°C (Table 11.3). Compounds possessing weak intramolecular forces readily volatilize and can easily change from a solid or liquid form to a gaseous form at room temperature.

Volatilization increases with the inherent vapor pressure of a chemical, with increasing temperature, with increasing

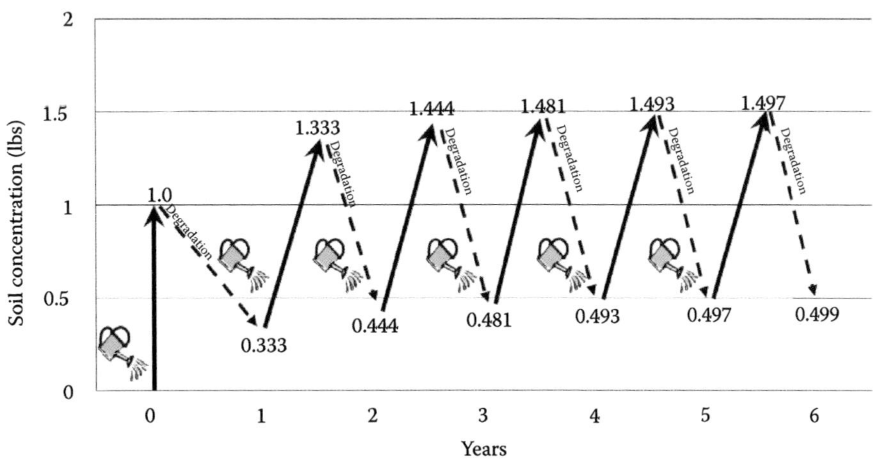


**FIGURE 11.8** A common misconception is that repeat pesticide use causes environmental "buildup". As illustrated, even through yearly use, pesticide is degraded once applied. (Modified from Ross, M.A. and C.A. Lembi, *Applied weed science* (3rd ed.), Prentice Hall, Upper Saddle River, NJ, 2009.)

**TABLE 11.3**
**Vapor Pressure as a Degree of Volatility**

| Volatility | Vapor Pressure Range at 20 to 30°C (mm Hg) |
|---|---|
| Very high volatility | $>10^{-3}$ |
| High volatility | $10^{-4}$ to $10^{-3}$ |
| Moderate volatility | $10^{-5}$ to $10^{-4}$ |
| Low volatility | $10^{-6}$ to $10^{-5}$ |
| Very low volatility | $10^{-7}$ to $10^{-6}$ |
| Extremely low volatility | $<10^{-7}$ |

air movement, and in the absence of incorporation. Losses to volatilization typically range from 5% to 20% but can be higher when a volatile compound is surface applied on a hot, windy day. Some of the factors in volatilization are

1. **Soil placement—**Pesticides incorporated by being irrigated-in or injected have a much lower rate of volatilization compared to those that are surface applied.
2. **Soil organic matter and clay content**—Higher soil organic matter or clay contents will increase the adsorption of a pesticide, which in turn reduces the amount available for volatilization.
3. **Soil moisture**—Higher moisture contents can work to slow adsorption to soil particles, which will keep more pesticides available for volatilization.
4. **Temperature**—Increasing temperatures can increase volatilization, but the relationship is not direct because changes in moisture usually accompany temperature changes. For example, dry conditions usually accompany hot weather periods.
5. **Temperature inversion**—If present, the unsettled, lightweight, and nearly invisible spray droplets will hover along the ground, much like a fog, instead of rising and mixing.
6. **Wind**—Increasing wind speed will increase volatile losses, although a rough surface such as turf can greatly reduce these effects.

As discussed, volatility is highly dependent on temperature, moisture, soil texture, and the properties of a specific pesticide, and is extremely important in determining how a particular pesticide can be used. For example, compounds with high volatility are applied and incorporated by watering in to prevent loss through volatilization to the air. Solids and liquids increase vaporization as the temperature increases. Pesticides formulated as esters also have a greater potential for volatility than do amine formulations.

**Drift.** Most pesticide applications are liquid sprays applied several feet (or meters) or more above the soil or turf. Drift is the movement of spray particles away from the target area. This may lead to several problems, including (a) damage to susceptible off-target sites; (b) a lower rate than intended, which can reduce the effectiveness of the material, wasting money; and (c) environmental contamination, such as water pollution. Several types of potential drift occur:

1. **Vapor drift**—Movement associated with the volatilization of pesticide molecules and their movement off-target, making it independent of the application. The only effective means of reducing drift when increasing spray volume is to increase the nozzle size rather than increasing pressure in the smaller nozzle size.
2. **Particle drift**—Off-target movement of spray particles formed during application that typically moves immediately off-target at the time of the application. Type of nozzle, pressure, height of boom, spray volume, and environmental conditions all affect the amount of spray drift that occurs.
3. **Atomization**—Breaking up of the spray solution as it passes through the nozzle orifice under pressure. Atomizing a known amount of spray solution into smaller droplets will increase the coverage; however, evaporation, drift potential, canopy penetration, and deposition characteristics must be taken into account.

This creates an opportunity for wind to carry a portion of the spray away from the target. Under worst-case conditions (high winds, small droplets), portions of the spray can be carried downwind. Several parameters that potentially influence this include.

1. **Wind**—The amount of drift is directly related to wind speed. Spraying in windy conditions (>10 mph, 16 km/hour, wind speed) may cause the spray to drift away from the target. Highly turbulent winds, which occur during inversions or on hot afternoons, are more likely to cause drift. Other environmental parameters influencing drift include temperature, humidity, and temperature inversion.
2. **Droplet size**—Droplet sizes are influenced by nozzle type and spray pressure. Nozzles that produce a significant number of small droplets (e.g., <100 μm) can significantly increase drift. There is always a trade-off between the improved coverage of small droplets and the risk of drift. Larger droplets (e.g., >200 μm) are least likely to drift but may reduce coverage. The only effective means of reducing drift by increasing spray volume is to increase the nozzle size rather than increasing pressure in the smaller nozzle size.
3. **Drift-control additives**—Tank additives have been shown to reduce drift primarily by reducing the amount of smaller droplets produced. Special adjuvants work well but only those intended for this purpose should be used.
4. **Boom height**—The farther the boom is from the intended target, the greater is the chance of drift. Nozzle manufacturers' charts will provide information for proper boom settings for different nozzle types. Wide-angle nozzles can be placed closer to

the ground than nozzles that produce narrow spray angles. Check the pattern uniformity of an applicator by spraying water across flat, black asphalt. The spray swath dries evenly when the spray pattern is uniform. If the swath dries in streaks, the pattern is uneven and the nozzles need checking.

**Best management practices for minimizing drift.** Ways of reducing drift include (a) selecting low or nonvolatile products; (b) following label directions on proper product use with minimal risk to the environment and using spray additives to increase droplet sizes and pesticide effectiveness; (c) using larger nozzle orifice sizes to produce larger drops to improve coverage and effectiveness; (d) avoiding high pressure, which creates finer droplets; (e) using drift-reduction nozzles, which produce larger droplets when operated at low pressures; (f) using wide-angle nozzles, low boom heights, and maintaining a stable boom; (g) avoiding applications when wind speeds are greater than 10 mph (16 km/hour), especially if treating upwind to sensitive crops, gardens, dwellings, livestock, or water sources; and (h) using shielded (or hooded) booms.

**Environmental impacts of pesticides in the atmosphere.** The most immediate impact of pesticide losses to the atmosphere is the reduction of pest control. If a certain rate of product is recommended to control a pest and half of it is lost to the atmosphere, the pest is probably not going to be controlled. The second possible impact is on adjacent property, where the pesticide could cause significant damage in the form of injured crops, residues on crops with no label for the pesticide, or negative wildlife impacts such as on fish or bees.

## Plants

The most important target of most pesticide applications is the turf itself. A number of fates await a pesticide once it lands on the turf:

**Photolysis.** The sun provides a wide spectrum of radiation beyond the narrow band called light. The higher energy ultraviolet radiation is sufficiently strong to cause some pesticides to break apart. This is the case for residues that are exposed on the plant (or soil) surface.

**Absorption.** Absorption of a pesticide by a plant can occur through either the leaf or the roots. Getting through the leaf is difficult due to leaf hairs and a waxy cuticle blocking entry. One way to improve coverage and penetration is by using an adjuvant or spray additive that allows the droplets to spread on the leaf surface without beading up.

**Effects of plants on pesticides in the environment.** Uptake and metabolism of pesticides by plants are important in a turf environment due to the high percentage of ground covered. Much of the applied pesticides still end up in the thatch and soil, however. The impacts of plants are still quite important. A turf cover significantly reduces drift through interception and increased surface "roughness," which slows wind. The canopy can reduce surface temperatures 10°F to 20°F, which slows volatilization. Turf also creates a more porous soil, allowing more water to infiltrate and slow runoff.

## Soil

A majority of applied pesticides end up in the thatch and soil where they are largely processed and disposed of. A very serious environmental contamination problem would exist if this did not occur. In the absence of a spill, pesticide fate in the soil/thatch region of the turfgrass is dictated by adsorption, chemical and biological degradation, and leaching.

**Adsorption.** Soil sorption is the affinity a chemical has to adhere to soils. The term "sorption" includes the processes of adsorption and absorption. Adsorption is the binding of a chemical onto the surface of a soil particle while absorption is the binding of a chemical into a soil particle. Adsorption is generally the more important means of binding a chemical to the soil surface. This is greatly influenced by the chemical and physical characteristics of the chemical (e.g., anionic, cationic, or nonionic charges), the soil characteristics or composition, and the nature of the soil solution. The major soil characteristics affecting pesticides are soil texture, permeability, and organic matter content. Coarse, sandy soil textures tend to have low cation exchange capacities (CECs) and high permeability rates; thus, they hold applied materials poorly. Increases in organic matter and certain clays tend to increase a soil's ability to hold applied materials. This is one reason organic matter should be considered in golf course construction. Clay soils are least likely to allow leaching, but are most likely to have runoff. Most insecticides readily adsorb to soil and thatch, which is why they are rarely detected in groundwater. The majority of fungicides and herbicides are also tightly bound by soil, but some do have the potential to leach to groundwater.

The constant used to measure the tendency of a chemical to sorb to the soil organic matter is called the *partition coefficient*. This is most often abbreviated as Koc but is sometimes reported as Kd. The relationship between Koc and Kd is

$$\text{Koc} = \text{Kd} \div \text{percent soil organic carbon content} \times 100$$

For an organic compound, the larger the Koc value is, the more strongly it will be sorbed to soils; therefore, it is less likely to leach or volatilize. A partition coefficient less than 300 to 500 is considered low and the chemical has a greater tendency to leach in low organic-matter–containing soil (Table 11.4).

Pesticides binding tightly to soil may be preferred from an environmental perspective since this reduces the movement of the pesticide. This can be a problem, however, because a pesticide may not be active if it is highly bound to soil. A good example is the herbicide glyphosate (Roundup Pro), which is inactivated in soil through tight binding to clays.

**Degradation.** Pesticides are often quickly degraded once they reach the thatch and soil layers. Sunlight can break them down through photolysis the same way as on plant surfaces. Acids commonly found in the soil can sever parts of the pesticide via hydrolysis. The most important factor, however, is the organisms living in soil that use them as food sources. Bacteria are the main degraders in soil and thatch. Most turf

**TABLE 11.4**
**Pesticide Characteristics Influencing the Potential for Groundwater and Surface Water Contamination**

| Chemical Characteristic | Range for Potential Contamination |
|---|---|
| Water solubility | >30 ppm |
| Kd | <5, usually <1 |
| Koc (ml/g) | <300 to 500 |
| Henry's law constant | $<10^{-2}$ atm $m^{-3}$ mol |
| Hydrolysis half-life | >175 days |
| Photolysis half-life | >7 days |
| Field dissipation half-life | >21 days |

*Source:* USEPA, 1988, as reported by Balogh, J. C., and W. J. Walker, eds. 1992. *Golf course management and construction: Environmental issues.* Chelsea, MI: Lewis Publishers.

pesticides will be degraded in thatch and soil in a matter of days or weeks. This is good if one wants to minimize the possibility of the pesticide reaching ground or surface water. Of course, this also means the effectiveness of the pesticide is reduced. Pesticide chemists developing new products are always confronted with this dilemma: A chemical that persists may be effective longer but may have an increased risk of contaminating water. The U.S.EPA will not allow a new chemical on the market if it poses an undue risk of contaminating water.

**Leaching.** A pesticide that is not bound to the soil and is not degraded can move through the soil with infiltrating water. This process depends on the chemical properties of the pesticide and the soil. As chemicals move downward in the soil, both the amount of adsorption and the rate of degradation rapidly decline. There is little organic matter or bacterial activity once the pesticide is past the rootzone. Fortunately, the adsorption and degradation processes handle all or nearly all of the pesticide before it can reach the lower layers.

*Solubility* is the extent a chemical will dissolve in water (Table 11.5). Generally, the higher a liquid's solubility is, the greater is the chance it may move from the site of application. Although solubility is normally a good indicator of the likelihood a chemical may be mobile, its sorption to soil also must be considered.

**TABLE 11.5**
**Solubility Rankings of a Compound Based on Water Solubilities**

| Relative Solubility | Water Solubility (ppm) |
|---|---|
| Very high solubility | >10,000 (1%) |
| High solubility | 1,000 to 10,000 |
| Moderate solubility | 100 to 1,000 |
| Low solubility | 10 to 100 |
| Very low solubility | 1 to 10 |
| Extremely low solubility | 0.1 to 1 |

## Water Quality

Much of the bad press pesticides have received is from detections in groundwater and surface water used for drinking water supplies. Surveys of ground and surface water have found pesticides in some areas. The extent of the contamination is becoming reasonably well defined, but the source or sources of contamination are often quite elusive. The sources and problems associated with groundwater and surface water contamination are quite different and will be dealt with separately in the following.

**Surface water.** Most pesticide contamination of streams, lakes, and estuaries occurs as runoff from agricultural and urban areas. Runoff carries with it a mix of suspended soil particles and any pesticides that were either attached to the particles or dissolved in surface moisture just before runoff began. The amount of pesticide loss to runoff is affected by the following factors:

1. **Rain intensity**—Heavy downpours result in minimal infiltration and maximum runoff.
2. **Surface conditions**—Recently tilled soil and soil with a good ground cover have the most resistance to runoff since water infiltrates relatively easily and the surface is "rough" enough to break up water flow. Maximum runoff is expected during the month after planting, since the soil is exposed and the turf has not grown large enough to intercept rain and reduce its impact energy.
3. **Magnitude and length of slope**—The steeper and longer the slope is, the greater is the chance of runoff picking up energy and soil.
4. **Method of application**—Pesticides tilled or injected into the soil are less likely to be lost in runoff, although the disturbance of the soil itself may increase soil (and attached pesticide) losses.
5. **Timing**—As previously mentioned, if a runoff event occurs soon after the pesticide is applied, substantial losses can occur.

Losses of pesticides to runoff generally are found at 1% to 5% of applied rates, depending on the various factors. Losses are usually the greatest in the one to two weeks after application and are highly dependent on storm events.

The effects of providing untreated grassed borders can be quite substantial, with reductions of pesticide movement into adjoining streams of 80% to 90% (Baird et al., 1997). The combination of infiltration, reduced overland flow rates, and adsorption in these zones can be quite effective in keeping pollutants from the fields from getting into the waterways (Figure 11.9). The role of forested buffers is likely quite similar.

It is important to emphasize that buffers function only when the waters they receive are spread across the strip. Runoff that moves through a buffer in a ditch or channel has little opportunity to degrade or adsorb before it intercepts surface water.

Once organic chemicals like pesticides enter surface water, their rate of degradation slows considerably compared to soil

FIGURE 11.9 Untreated grass borders avoid lateral movement of products.

degradation rates. A portion of the pesticide will partition onto the sediment and remain there until a flood event moves the sediment back into the moving water. This is thought to be the explanation for why low levels of pesticides can be detected long after the application season.

**Groundwater.** Groundwater is water located beneath the earth's surface. Groundwater is often erroneously perceived as occurring in vast underground lakes, rivers, or streams. In reality, groundwater usually occurs in aquifers composed of pore spaces and cracks in rock and soil. An *aquifer* may be defined as a formation containing sufficiently saturated permeable material to yield significant quantities of water to wells and springs. Aquifers, such as unconsolidated sands and gravels, can store and transmit water. They are quite extensive and may be confined above or below by a confining bed made by a relatively impermeable material such as clay or sandy clay.

Unconsolidated aquifers are sand and gravel deposits that can be relatively shallow, often only 10 to 20 feet (3 to 6 m) below the surface. The pores within these deposits are filled with water if they are below the water table. These aquifers are mostly located in coastal plain areas within 200 miles (322 km) of the coast. Recharge of these aquifers is usually local but there may be several layered on top of each other, with lower aquifers recharging some distance away. Wells drilled into the first or top layer, or superficial aquifer, are the most susceptible to contamination.

Groundwater moves very slowly through irregular spaces within otherwise solid rock or seeps between particles of sand, clay, and gravel. An exception is in limestone areas, where groundwater may, in fact, flow through large underground channels or caverns.

Groundwater is recharged (replaced) mostly from rain or snow that enters the soil. Water moving downward in the soil is absorbed by plants, is held in the upper layers of soil, or moves down through the rootzone until it reaches a zone saturated with water. This saturated zone is the uppermost layer of groundwater. The *water table* is the "dividing line" between the groundwater and the unsaturated rock or soil above it. Spring and fall generally are the seasons when the water table is closest to the soil surface.

The following largely determine whether a pesticide moving through the soil will reach the groundwater.

1. **Soil type—**Sandy soils leach more than loamy or heavier soils.
2. **Pesticide characteristics—**How tightly the pesticide binds to soil and thatch and how long it lasts in the soil. Pesticides with low Koc values and long half-lives have greater risk of groundwater contamination.
3. **Weather—**Cool, wet conditions can push the pesticide deep in the soil before it can degrade or bind.
4. **Depth to groundwater—**The deeper the groundwater is, the more time a pesticide has to break down before reaching it.

Most rural residents obtain their water from wells placed in groundwater tables. The water in those aquifers may come from the nearby area or many miles away, depending on the type of aquifer. The water may have been in the aquifer for several years or many decades.

### Sources of Contamination

**Point-source.** Environmental contamination originates either as point-source or non-point-source pollution. Point-source pollution comes from a specific, identifiable point. A pesticide spill into a well, sinkhole, or storm sewer exemplifies facilities identifiable point-source pollution. Sewage treatment plants and concentrated animal production also are potential point-sources. Other point-source pollution comes from improper disposal of pesticide containers and water from rinsates, leaks and spills at the site of pesticide storage facilities, and spills occurring while mixing and loading pesticides into application equipment. Groundwater can be contaminated directly in many ways (Figure 11.10). Some of the most serious include back-siphoning, surface water movement into wells, or drainage into limestone channels. These contamination problems can nearly always be prevented. Once they occur, however, the point of entry becomes a point-source for contamination.

FIGURE 11.10 Spills are a common point-source means of groundwater contamination. Shown is an appropriate storage bay of various fuels to avoid this.

A plume of contamination moves slowly away from the source and can spread to contaminate many wells down gradient.

**Nonpoint.** Nonpoint-source pollution generally involves contamination from a normal application over a wide surface area, which can occur if nutrients or pesticides leach through the soil and reach groundwater, or if they wash off in runoff and enter lakes or streams. Leaching through the soil profile and movement with runoff water are believed to be two major sources of non-point-source pollution. However, if a product is used according to the pesticide label, it is believed non-point-source pollution is actually very minor and contributes little to the overall pollution problem. Problems normally develop only when an intense rainfall occurs almost immediately after application, before the pesticide has been absorbed by the plant or soil.

Parameters influencing the potential of non-point-source pollution most often involve weather after application, chemical characteristics of pesticides or nutrients, depth of groundwater from the soil surface, and soil type. Pesticides should not be applied if standing water is on the treated surface, as this greatly increases the likelihood of downward movement. Applications prior to heavy rain or excessive irrigation also tend to move pesticides prematurely from the site of application.

In addition, chemical properties of pesticides influence the likelihood of groundwater contamination. These properties include solubility, adsorption and absorption capability, and persistence. Chemicals with high solubility, low adsorption capability, and extended persistence have more potential of becoming non-point-source polluters (Table 11.6).

Research indicates solid ground coverage by turf is probably the best "filter" in preventing both lateral and horizontal movement of applied pesticides, fertilizers, and soil sediments. Untreated turf buffer zones around sensitive areas such as ponds, lakes, or streams also provide the best assurance of preventing unwanted lateral pesticide movement. Turf plants also directly absorb pesticides and, when actively growing, can reduce the likelihood of water-soluble pesticides leaching.

**TABLE 11.6**
**Parameters That Minimize the Risk for Groundwater and Surface Water Contamination from Pesticide Use**

| Pesticide or Site Characteristic | Parameters that Minimize Off-Site Movement |
|---|---|
| Pesticide properties | • Low solubility |
| | • High soil adsorption (Koc) |
| | • Short half-life or little persistence |
| | • High volatility |
| Soil properties | • Finer textured soil |
| | • Higher organic matter content |
| Site characteristics | • Deep water table |
| | • Flat versus sloping land |
| | • Adequate distance from surface water, sink holes, or abandoned wells |
| | • Soil completely covered with turf, mulch, or ground cover |
| Management planning | • Adequate planning and consideration for impending weather events |
| | • Proper application and timing |
| | • Proper incorporation through irrigation following application |

## BEST ENVIRONMENTAL GOLF COURSE MANAGEMENT SUMMARY

Nutrient and pesticide movement or soil sediment erosion could negatively affect environmental quality if allowed to enter lakes or streams in an uncontrolled manner or in significant amounts. Best management practices are plans to reduce this possibility by (1) growing a solid turf stand that reduces the off-site transport of sediment, nutrients, and pesticides; (2) controlling the rate, method, and type of chemicals being applied; and (3) reducing the total chemical load by using BMPs (including economic thresholds, alternate pest control) and fertility testing. Following the labeling directions exactly is the best way to prevent contaminating groundwater. In addition, the following should be remembered:

1. **Avoid disturbing sensitive wildlife areas and wetlands during the initial design and construction of a golf course**—When designing and maintaining a golf course, wetlands and other sensitive natural areas should be incorporated into the design and not disturbed.
2. **Select plant species that are locally adapted and require minimum inputs**—Growing plants outside their naturally adapted range usually means increased inputs in terms of pesticides, water, nutrients, and other energy needed to maintain the plant. Native plants often are able to survive with fewer maintenance inputs such as nutrients, water, and pesticides, but are not always very attractive.
3. **During construction, all necessary steps of preventing soil erosion and managing stormwater runoff on disturbed areas should be implemented**—This step involves minimizing soil movement during construction by installing silt fencing, including providing a master plan with drainage for green construction, grading, clearing, and specific features such as trees, bunkers, mounding, ponds, and so forth.
4. **Maintain the turf to encourage a deep, actively growing root system**—Deep, active root systems are better able to recover applied nutrients and thereby prevent leaching. Deep root systems also allow efficient use of applied water as they can extract moisture from a greater area and require less irrigation. Cultural practices such as using the highest acceptable mowing height and proper frequency, minimizing soil compaction and layering, minimizing thatch layering, and eliminating root feeding

insects, diseases, and nematodes are means of promoting a healthy turf root system.

5. **When possible, select pesticides based on the following:**
   - Lowest toxicity to humans, mammals, fish, birds, and bees
   - Rapidity of degradation and lowest leaching potential; more persistent pesticides have a greater chance of off-site movement via leaching or lateral movement
   - Highest soil adsorption, as higher soil adsorbing pesticides are less likely to leach
   - Lowest volatility; pesticides with lower vapor pressures are less likely to volatilize and move off-site. This results in lower efficiency of pesticide use and could possibly pose problems to adjacent sensitive plants. Ester formulations of phenoxy herbicides, for example, may vaporize and drift when sprayed during warm temperatures, when winds are excessive, and when high soil moisture is present
6. **Using the appropriate pesticide and nutrient rates (not excessive, especially for N and P)**—An obvious management practice that is sometimes lost primarily due to pressure from club members or players is to maintain ideal year-round green grass. Higher nutrient rates usually mean lush, fast-growing turf, which is harder to mow and becomes more susceptible to certain pests and organic matter accumulation. Unusual circumstances, however, such as growing-in a newly established area or having to quickly recover from winter-kill, sometimes dictate using temporarily slightly higher than normal nutrient rates.
7. **Use the least amount of soluble nutrient sources commensurate with acceptable turfgrass quality**—Applying lighter but more frequent amounts of soluble nutrient sources has been shown to produce high-quality turf without groundwater leaching problems. Using slow-release nutrient sources in combination with soluble sources also helps minimize any negative impact. Water in soluble sources after application to reduce movement following a heavy rainfall event.
8. **Use the best application method to minimize any special risks**—For example, soil injection of some pesticides may not be wise when groundwater is close to the surface. However, in other cases slit injection may reduce exposure to wildlife. Liquid fertilizer application and fertigation are means of applying low amounts of nutrients on a frequent basis. These allow steady turf color and growth and minimize single application nutrient loads. Spot treat with pesticides whenever possible or practical.
9. **Prevent back-siphoning of the pesticide into your water source by keeping the end of the fill hose above the spray tank's water level**—The most pronounced means of contaminating groundwater with pesticides is during the mixing and loading process, especially when near a water source or well. Check valves should be installed to prevent back-siphoning, although these can fail. An air space is the best protection.
10. **Locate mix-load sites and equipment rinsing sites at least 100 feet (30 m) from surface water or from wells or sinkholes that have direct links to groundwater**—Dikes, sump pits, and containment pads may be necessary when these sites are closer than 100 feet (30 m). Also work on a flat, impermeable surface when mixing or loading pesticides and mix pesticides where a spill can be contained. Consider installing a closed system.
11. **Dispose all pesticide waste in accordance with local, state, and federal laws to prevent contaminating groundwater through improper disposal of unused pesticides, pesticide containers, and container rinse water**—Do not dispose of excess pesticide rinsates by dumping them on the ground, especially in a concentrated manner. Also, do not place them in a septic tank or sewer system unless allowed by law. The best means of disposing of excess rinsates is by applying them on a legal site or through an approved degradation pit. Rinse used containers three times and read and follow all label directions before use.
12. **Use ponds, basins, vegetation strips, riparian zones (e.g., bottomland hardwood, flood plain forests), and channels containing vegetation to filter or assimilate nutrients from drainage water**—The pond for irrigation reserve often serves to protect native surface waters. Created wetlands, which are less than 18 inches (46 cm) in depth, also are used to reduce runoff quantity and nutrient and pesticide discharge. Wetland flora create an environment in which bacteria facilitate the precipitation and degradation of pollutants. Algae, bacteria, sphagnum moss (*Sphagnum* sp.), cattails (*Typha* sp.), bulrushes (*Scirpus* sp.), and rushes (*Juncus* sp.) have been shown to remove nutrients, sulfate, and pesticides. Many of these and other wetland plants are available either locally or through companies. Fish and invertebrates should not be overlooked. While invertebrates will most likely colonize the wetland, small fish should be stocked. Fish serve as excellent mosquito control. Generally, nearby soil is used for the wetland substrate. A suggested ratio of wetland size to drainage area is 1 acre (0.4 ha) for a 200-acre (81 ha) drainage area. Water flow velocity for wetlands is between 0.1 and 1.0 feet (3 to 30 cm) per second. Average hydraulic retention should be about five days. Reuse drainage water as an irrigation source when possible.
13. **Drainage from greens and tees should be spread in grassed or natural areas**—This allows the nutrients and pesticides to be absorbed before the water

reaches a stream or pond. When this is not possible, subsurface drainage can be filtered through a sand–charcoal system to remove pollutants, or this drainage can be captured, stored in ponds, or stored in storage reserves for stormwater run-off collection, and reused as an irrigation source.

14. **Use a natural or constructed waterway or outlet maintained with vegetative cover to prevent soil erosion and to filter nutrients**—Dry ponds and golf course roughs serve in this capacity. Whenever possible, courses should be designed so irrigation and stormwater runoff move from the edges into the middle of the course. Drainage ditches should be bisected by small swales or natural or constructed wetlands. These geographic features slow down the water and allow for assimilation of nutrients and pesticides by the vegetation.
15. **Maintain an untreated (no spray) vegetative buffer zone of 10 to 50 feet (3 to 16 m) adjacent to all water sources to help filter nutrients or pesticides from runoff**—Research has repeatedly shown a vegetative buffer strip adjacent to water sources acts as a filter in preventing surface runoff of sediments, nutrients, and pesticides. Steeper sloping land adjacent to water sources or slopes with slowly permeable soils such as clays should consider using the wider buffer strip range.
16. **Manage a decline in dissolved $O_2$ levels in water by controlling algae and nutrient runoff, as well as providing adequate aeration**—When problems such as fish kills occur, pesticides are often erroneously blamed. Low dissolved oxygen (DO) levels from algal blooms or natural pond turnover from temperature stratification or weather changes is most often the reason for the problem. Large fish are more sensitive to DO fluctuations that typically occur in early morning hours. Limited fish stocking might be considered. A water quality monitoring program may be considered to track water pH, dissolved oxygen levels, and possibly nutrient and certain pesticide levels. If a fish kill occurs, the following steps and observations should immediately be taken to help pinpoint the cause:
    - Date and time of day
    - Location
    - Estimated time kill began
    - Water quality characteristics including dissolved oxygen, pH, temperature, color, odor, and salinity
    - Condition of fish seen (e.g., moribund, dead, or decaying)
    - Condition of other organisms in the area
    - Weather conditions of the day and previous day and night
    - Physical appearance of fish (e.g., gills flared, mouth agape, spinal curvature, excessive mucus, lesions)
    - Any other unusual characteristics, behaviors, or discolored vegetation
    - If effluent water is used, heavy metal concentration and other toxic materials tested
    - For possible organophosphate and carbamate exposure, fish samples can be tested for blood and brain cholinesterase, an enzyme responsible for normal nervous system function
17. **Test soil and tissues to optimize fertilization with the growth and use by turfgrasses to prevent excess use and possible runoff**—Use slow-release (e.g., coated) fertilizers when possible to minimize nitrogen losses through leaching. This book lists those optimum soil and tissue nutrient ranges for turfgrasses. These ranges should be checked frequently to ensure adequate but not excessive rates are available. Fertilizer use should also coincide with the growth needs of the turfgrass species.
18. **Prevent runoff by using irrigation water management so application rates do not exceed the infiltration capacity of the soil**—Watering according to evapotranspiration rates and soil moisture sensing devices, and using efficiently designed, installed, maintained, and frequently calibrated irrigation systems, not only saves precious water, but also minimizes surface and subsurface movement of any applied materials. Water to a depth just below the root system and water in early morning. Sloped areas should be irrigated in short, frequent intervals to minimize runoff. Excessive irrigation rates not only waste water, but also potential nutrient and pesticide leaching may occur. Using effluent water sources also helps conserve and recycle this precious commodity.
19. **Avoid pesticide or nutrient applications just prior to anticipated rainfall events**—Rainfall, especially as thunderstorms, can release large amounts of water in a relatively short period. Soils cannot always absorb this water quantity in a short time period; thus, surface runoff occurs. Nutrient and pesticide movement in this runoff may occur, reducing the effectiveness of the products and also potentially causing environmental problems. Applying nutrients or pesticides during the period of day (morning) that has the least chance of heavy rainfall, and irrigating them in immediately (if the label recommends this), are the best means of preventing runoff. Also, do not treat standing water puddles, as pesticides or nutrients may move with this water as it drains.
20. **Do not apply pesticides or nutrients to an area in excess of what can be quickly and efficiently watered in**—Many turf installations can irrigate only so many acres at one time. Therefore, if more area is treated with a product than the irrigation system can immediately water in, the chances of rainfall events or exposure problems increase, reducing the effectiveness of the product and potentially causing

problems. For example, if the irrigation system can only cover 10 acres (4 ha) at once, then only these 10 acres should be treated and irrigated-in before the next 10 acres are treated.

21. **Return grass clippings when practical**—Grass clippings are a source of most major nutrients and, by returning them after mowing, this aids in recycling. Certain areas, such as golf greens, do not return clippings, as they would interfere with the surface. These clippings can be recycled by mulching or composting them and applying them to landscape beds.
22. **Follow IPM practices as previously outlined in addition to the proper storage, mixing, application, and disposal of pesticides, their containers, and rinsates**—IPM practices help optimize the use of pesticides without discriminate applications. An IPM program includes hiring a knowledgeable and appropriately trained superintendent, devising a written IPM plan, defining pest thresholds for specific turf sites, implementing appropriate cultural practices, monitoring (scouting) pest levels and activity, and maintaining accurate records.
23. **Frequently check and calibrate all pesticide and fertilizer applicators**—A maximum of 10% deviation should be allowed before recalibrating or changing worn nozzles or other parts.
24. **Install a well-balanced landscape with a mix of shrubs, trees, grass areas, and water features that sustain and encourage wildlife and a diversity of plants**—Wildlife needs a variety of plants to feed on and to provide habitats. This can be supplemented by providing feeders and nesting sites such as birdhouses. As part of this landscape, conservation areas to be established throughout the golf course should be considered.
25. **Keep detailed, accurate records on nutrient and pesticide applications**—Record keeping becomes necessary for agronomic reasons and in case any environmental problems occur. Without accurate records, a golf course invites finger-pointing, as well as possible accusations and liability associated with its operation. Contact Audubon International for greater details on creating and maintaining natural conservative areas and wildlife habitats. Becoming a fully certified Audubon Cooperative Sanctuary should be a long-term goal of most golf facilities.
26. **Provide periodic pesticide handling, and use training and updating sessions for all employees**—Technology, trends, and laws dealing with pesticide and nutrient purchasing, storage, mixing, application, and disposal constantly change. All employees who handle or are exposed to these materials should routinely attend updates on these trends. Numerous turfgrass and pesticide seminars, workshops, and training sessions are available where continuing educational credits or units are provided for certification. Contact your local cooperative county agent or golf course superintendent's association for the latest information on training events in your area.

## ANIMAL PESTS

A number of animals can become problematic on fine turf installations. These include alligators, armadillos, bears, beavers, deer, fox, geese, hogs/pigs, moles, rabbits, raccoons, gophers, groundhogs/woodchucks, skunks, snakes, and other local problems. Rarely do these animals feed directly on the turf but rather, either destroy it searching for food or, through breeding and excretions, cause undesirable turf or playing conditions.

Sightings of wildlife, especially birds and mammals on turfgrass, give most people a sense of being close to nature. With increasing urbanization, it is literally impossible to live in most areas without seeing or hearing wildlife on a daily basis. Many of these experiences are enjoyable while others can be confrontational. Unpleasant encounters with animal pests can result in injury or fear of injury, property damage, or minor nuisances. Some of our frustrations with wildlife can be alleviated by simply learning why wildlife appear to be causing a problem. Others require more action-oriented prevention and control techniques.

### CANADA GEESE

Of the animal pests on golf courses, Canada geese (*Branta canadensis maxima*) in large numbers are among the most annoying and widespread (Figure 11.11). Resident Canada geese become aggressive during brood season, produce extensive fecal droppings, foul water quality, and shed feathers during molt. Geese also may represent a potential serious environmental threat or risk to human health and safety. Urban Canada geese may also fly to surrounding agricultural lands to feed on crops such as corn, soybeans, rice, winter wheat, and other grains.

Besides the direct impact of Canada geese residing in urban areas, they can act as decoys for migratory geese, causing periodic increases in urban goose populations. Geese in

**FIGURE 11.11** Canada geese are among the most annoying and widespread animal pests on golf courses.

urban areas are very aggressive around their nests or goslings and may attack or threaten pets, children, and adults.

Geese also trample grass in medium-heavy soils, which creates a surface "hard pan" that prevents vegetative growth. This causes erosion and loss of habitat for other species. Geese in high concentrations or even a smaller flock that remains in the same place for an extended period of time may overgraze the grass, creating large, dead spots of turf.

High concentrations of geese increase the likelihood that avian diseases will be transmitted, creating the potential for massive die-offs. Disease organisms originating from a single species of waterfowl can also spread to other species. Canada geese are suspected of transmitting salmonella to cattle. Transmission of disease or parasites from geese to humans has not been well documented, but the potential exists.

**Breeding, nesting, and feeding.** Canada geese have enormous reproductive potential as one pair can increase to more than 50 birds in five years. They usually begin nesting at three years of age. Pairs usually stay together for life unless one dies; in that case the remaining goose usually finds another mate, generally within the same breeding season.

Canada geese usually nest within 150 feet (46 m) of water, and most nests are surrounded by or are very close to water (Figure 11.12). Water provides access to food, a place to drink, aids in preening and bathing, and is an avenue of escape from predators. A typical Canada goose nest is bowl shaped, approximately 1.5 feet (46 cm) in diameter, and made from the surrounding vegetation lined with goose down plucked from the female's breast.

A good view of the surrounding area is always important in nest site selection. Nesting females tend to use the same immediate area year after year. Both males and females defend the nest site territory before incubation. Egg laying is initiated shortly after nest construction starts.

After the eggs are laid, the male does not incubate but will stand guard and defend the incubating female by striking at opponents with its wings or nipping with its beak. Average clutch size is slightly more than five eggs per nest; nests may contain 1 to 15 eggs. The average incubation period is 26 to 28 days. Incubation does not begin until all eggs are laid, so all goslings usually hatch the same day. If the nest is destroyed or the eggs are eaten or removed by predators, Canada geese may renest, usually in or near the first nest. Renesting is more common when nest failure occurs early in the egg-laying period.

FIGURE 11.12 Fecal deposits, aggressive behavior, feather molting, and destructive feeding habits are reasons Canada geese are major pests of fine turf.

Parents often move their broods to areas chosen for the presence of suitable food, visibility, and proximity to water. Canada geese are grazers and they prefer tender, well fertilized lawn grass in urban areas. As mentioned, they tend to choose open areas with few obstructions to give them views of potential predators.

**Molting.** Adult Canada geese undergo a complete replacement of flight feathers each summer, which takes about a month for most individuals. Nuisance goose management is very important during this period because all birds present are flightless and thus vulnerable to capture. Mown lawns, parks, and golf courses suit their habitat requirements. Adults with young will molt at the brood rearing area 10 to 20 days after the nonbreeding geese initiate their molt.

**Migration.** Canada geese nesting in the continental United States and southern Canadian provinces usually migrate only short distances, generally staying within their state of birth or flying to neighboring states. Although only a small portion of migratory geese use urban areas, they may have a startling impact on the environment when, within the span of a few days, a pond that previously had only a few geese on it suddenly supports a flock of several thousand. Migratory birds are usually much warier of human activities than resident geese.

**Regulations covering Canada geese management strategies.** Canada geese are migratory waterfowl and thus are protected by the Migratory Bird Treaty Act of 1918 (16 USC 703-711). This act made it illegal to harvest waterfowl or other migratory birds except during the hunting season or by permit. It prevented the unrestricted egg harvesting and commercial hunting for meat and feathers that was commonplace in the United States in the late 1800s and early 1900s. This treaty gave the United States and Canadian governments (U.S. Fish and Wildlife Service and Canadian Wildlife Service, respectively) the authority to set limits, make regulations, and issue permits to harvest or take waterfowl. In addition to federal permits, most states require permits anytime one intends to destroy eggs or nests, capture, translocate, disturb, or harvest Canada geese. Local laws or regulations may also affect the use of control techniques such as firearms, chemicals, and auditory and visual scaring devices.

**Control.** As with most wildlife control situations, solutions involve habitat modification, exclusion, deterrents, repellents, and population reduction. The following suggestions for controlling Canada geese populations have been met with success in many instances. Complete control, however, rarely follows only one avenue; thus, an integrated management approach that implements several means of control or deterrent is necessary.

1. **Do not feed geese.** Feeding Canada geese or other wild fowl promotes a dependency on humans and creates problems for them. Appropriate signage that explains why feeding is discouraged should be part

**FIGURE 11.13** Warning sign to avoid wildlife on a golf course.

of a campaign to encourage people to stop feeding and domesticating Canada geese and other animal pests (Figure 11.13).

2. **Habitat modification.** Golf courses provide an ideal habitat for Canada geese with water, lush grasses, and ample space. Modifying the habitat in which birds live, feed, and reproduce helps discourage them from becoming resident geese. A goal is to focus on eliminating as many nesting sites as possible and reducing water accessibility. Numerous ways and means of providing this are available. Some of the more popular include:

**Vegetative barriers.** Planting shrubs, hedges, or prairie plants around the water (heights > 30 inches, 76 cm, and widths at least 20 to 30 feet, 6 to 9 m), are most effective. These physically impede the movements of geese to and from the water, make fewer new shoots available to feed on, and block their line of site, making it more difficult for geese to see potential predators. Barrier plantings require protection from geese during establishment. Letting the grass grow longer and reducing fertilizer use will also help greatly. Elimination of mowing as a 20- to 30-foot (6 to 9 m) strip around water bodies discourages geese as they cannot see over the grass. Geese love the tender shoots of grass, especially Kentucky bluegrass, bromegrass, canarygrass, colonial bentgrass, perennial ryegrass, quackgrass, and red fescue. They tend not to prefer mature tall fescue, periwinkle, myrtle, pachysandra, English ivy, hosta or plantain lily, ground juniper, or switchgrass.

**Rock barriers.** Large boulders (greater than 2 feet, 0.6 m, diameter) placed along the shoreline may discourage goose use and access to grazing sites by making it difficult for them to exit the water. Their effectiveness is improved when used in conjunction with vegetative barriers.

**Fence barriers.** Fences can prevent geese from walking into an area. Fences should be at least 30 inches (76 cm) tall, and have openings no larger than 3 inches (7.6 cm) in diameter. Woven wire, chicken wire, picket fencing, erosion control fencing, plastic snow fencing, and construction fencing are examples of effective materials. The effectiveness of fence barriers may be enhanced when used in conjunction with landscaping modifications (vegetative barriers, rock barriers, etc.).

A popular fence that seems to be effective, especially for private areas, is a triple-strand electric fence. The wire should be strung 5, 10, and 15 inches (13, 25, and 38 cm) above the ground. The amperage required to exclude Canada geese is minimal and will not harm them. Two-strand portable electric fencing is also effective and economical in small areas. The lower strand is set up 8 inches (20 cm) off the ground and the upper strand at about 18 inches (46 cm). Low impedance energizers powered by batteries or plug-in outlets deliver short electrical bursts once a second that geese learn to avoid.

**Note:** To avoid accidentally shocking pedestrians, electric fences should be well marked with signs and not used in public-use areas.

**Tall trees.** Establishing tall trees around small ponds (less than half an acre, 0.2 ha) may prevent geese from landing since geese are large birds that require a relatively large open space to land and take off. However, shade provided by trees can also be attractive to geese.

**Steepen banks.** Canada geese prefer a gentle, grassy perimeter slope that enables them to easily walk into and out of the water to feed or rest. If access to the water is poor, adult geese may leave to raise their young elsewhere.

To steepen the shoreline, build a vertical seawall 3 feet (0.9 m) above the surface of the water or create a 60° to 65° angle from the water's edge. Allowing vegetation to grow tall along this slope will help protect it from erosion and keep the geese from walking up. Rip-rap, while ineffective on gentle slopes, is often effective on steeper ones.

**Allow water to freeze.** Aerating ponds is one of the reasons Canada geese have become year-round residents in colder climates. Allowing a pond to freeze over forces geese to seek alternative water sources and may encourage them to migrate. Concentrations of geese will maintain in open water even in below freezing temperatures. Harassment may be necessary to force the birds to leave long enough for ice to form.

**Swan decoys.** The premise is that aggressive swans will defend their territory, especially during the breeding season, and will exclude other waterfowl from the area. Place swan decoys out when the migratory season for geese has begun. Domestic waterfowl, including mute swans, act as decoys for Canada geese when they are flying over an area. Since native swans are difficult to acquire, non-native mute swans are commonly used instead. However, mute swans are much more tolerant of other waterfowl and may only defend the immediate area around their nests. It is not uncommon to find situations where mute swans and Canada geese peacefully share a site, adding to any fecal concerns that may already exist.

**Flagging.** Some have had success putting up plastic flags around 4 feet (1.2 m) off the ground. The 3 mm black plastic flags should be at least a 2- × 3-foot (0.6 × 0.9 m) wide sheet. These sheets blow in the wind, disturbing the geese. A "V"-shaped notch should be placed on the end of the flag to aid its being rustled by the wind.

**Overhead grid systems.** One of the most effective methods of exclusion is the installation of a grid system over the water surface. Grids work on a simple principle: Canada geese are large birds, requiring a long glide-slope to land, much like an airplane. A grid system above the water surface will be seen by the geese as a barrier between them and the water.

Grids work best on bodies of water less than 150 feet (46 m) across, but can be used on larger bodies up to 300 feet (91 m) across. Nearly any type of cord can be used to construct the grid, from cotton kite string to plastic-coated Kevlar cord. Twenty-pound clear monofilament line is a popular choice. Anchor points for the grid lines can be trees, wooden stakes. or "U" channel fence posts.

Grid system specifications are variable, but spacing the grid lines 20 feet (6 m) apart and suspending them at least 3 feet (0.9 m) above the water's surface should be sufficient to exclude geese, while allowing ducks, gulls, or other smaller birds access to the water.

Modifications can be made if water levels change or if geese penetrate the system. For example, geese may land on the shore and walk into the water under the grid. The solution would be to place a barrier around the water to keep them from entering under the grid. For example, place two strands of cord 6 inches (15 cm) and 12 inches (30 cm) above the ground running the length of the shore and attached to the anchor points. For a more permanent solution, plant a hedgerow or install a fence.

3. **Scarecrows.** This method can be done easily and without reservation. Scare techniques are mainly effective early in spring as adult geese are seeking secure, secluded places to nest. Noisemakers like sirens and natural gas exploders can haze geese, but the loud sounds are equally displeasing to people. Moreover, goose flocks can eventually get used to loud noises that are not accompanied by a real threat. One should be careful to have some limb of the scarecrow move, as European experience has shown that some motion is necessary to deter Canada geese.

   **Mylar.** Mylar helium balloons painted with eye spots can be tethered in fields. They scare geese because they look like large predators. Half-inch strips of flashing Mylar tape or ribbon can also persuade geese to move onto other nearby mown grass. Both the rattling sound and the light flashes frighten geese. Such tape is available through garden centers, feed co-ops and mail-order catalogs.

   **Reflective tape.** Mylar tape is a visual barrier used in conjunction with other exclusion methods. Mylar tape is 1/2 inch (13 mm) wide, red on one side and shiny on the other. To use as a fence, string one or two strands between two posts and twist the tape two or three times. When the wind blows, the tape rotates, creating a flash between the red and shiny sides. This unfamiliar flash acts as a visual barrier and makes the geese shy away from the area. It works best on bright days with breezes and before animal habituation begins.

4. **Chemical repellents.** Several chemical repellents are available with varying levels of success (Table 11.7). Application of dimethyl (DMA) and methyl anthranilate (MA) to areas surrounding ponds has decreased the number of geese and weight of fecal deposits.

   **Methyl anthranilate.** Products using the active ingredient methyl anthranilate (artificial grape flavoring), help change the birds' behavior. When applied to grass where geese feed, methyl anthranilate makes the grass unpalatable. Geese may still frequent the treated area, but they will not feed there. Methyl anthranilate does not readily wash off after a rain if allowed to dry first, but must be reapplied after mowing.

   **Anthraquinone.** Anthraquinone repels geese in two ways. First, geese experience a harmless "gut reaction" after eating the grass. Secondly, the grass appears unnatural and uninviting because the anthraquinone brings out the ultraviolet

**TABLE 11.7**
**Repellents for Various Vertebrate Pests[a]**

| Common Name of Ingredient(s) | Trade Name(s) |
|---|---|
| **Geese** | |
| Methyl Anthranilate | • Rejex-It Migrate, Fog Force, Goose-B-Gone, GooseChase |
| Pepper (capsaicin) | • Bobbex-G, Amorex |
| Anthraquinone | • Flight Control |
| **Rodents (Squirrels, Rabbits, Moles, Gophers, Chipmunks)** | |
| Castor oil | • Rejex-It Mole and Gopher Chaser, Scoot Mole, Mole Repellent, Get Away Mole Repellent, Mole Med |
| Pepper (capsaicin) | • Hot Pepper Wax Animal Repellent, Scoot Squirrel, Bobbex-R |
| Ammonium salts of fatty acids | • Professional Deer and Rabbit Repellent |
| Urine | • Predator urine |
| Strychnine bait | • Gopher Getter |
| Blood meal | • Plantskydd repellent |

[a] Products listed are not endorsed or guaranteed to work nor is criticism implied for those excluded.

spectrum when applied to turf. Combining the strange look of the grass with the intestinal reaction they experience, geese often look elsewhere to nest and feed. Anthraquinone does not readily wash off after a rain, but needs to be reapplied after mowing. Adding a growth regulator can keep the grass from growing as rapidly. This product is considered to be environmentally safe and does not produce long-term physical effects on the birds that ingest it.

5. **Dogs.** Using dogs to harass geese from an area has become one of the most popular and successful methods. While highly trained border collies are popular, just about any athletic, medium-large dog capable of obeying commands can be used (Figure 11.14). Control of the dog is vital because dogs used in this manner are legally considered an extension of your hand and must not be allowed to catch, injure, or kill a Canada goose. If geese are using an area for nighttime resting, disturbing them just before dark is more inconvenient to them than disturbing them during the day.

   Dogs are most effective in the autumn during migration by harassing the Canada geese and thereby discouraging geese from settling in for the winter. In spring, dogs are used to prevent Canada geese from developing nesting sites. Typically, a handler and a dog enter an area occupied by unwanted geese. On command, the dog is allowed to chase after the geese. Geese will likely seek refuge from the dog in a nearby body of water. If this is the case, the dog can be allowed to enter the water. To make this method more effective, use a boat or pyrotechnics to further harass the geese. Harassment should continue and be repeated until the geese leave the area permanently.

**FIGURE 11.14** Border collie dogs are popular means of discouraging geese on golf course property.

6. **Egg substitution and oral contraception.** There are at least five methods of intervening in the Canada goose reproductive cycle: egg addling, egg oiling, shaking, puncturing or freezing, or egg substitution (using wooden eggs). These methods should only be undertaken by a professional under contracted service, as approaching Canada geese is not without risk. Oral contraception is the newest option and remains untested but shows promise for the future. Nicarbazin (the active ingredient) is consumed by the birds, which then are unable to lay fertile eggs. This product may help minimize the growth of goose populations before their numbers become intolerable.

   **Egg shaking and coating.** This method requires eggs be shaken but not broken. Shaking the eggs kills the embryo, but breaking the eggs may just cause the mother goose to lay more. A dog or another person can help keep the geese away from the nest so the eggs can be shaken quickly and safely. Careful recordkeeping is important so you will know when to remove the eggs from the nest. If you do not, the goose may remain on the nest too long, deplete her food reserves, and suffer needlessly. Obviously, this technique does not help resolve a present problem but it will help reduce geese populations next year.

   As in shaking, egg coating (or addling) helps control geese damage by reducing future population growth. Coating the eggs with mineral or corn oil starves the embryonic goose of air, resulting in death. Typically, permits should be obtained in winter as nesting normally begins in early spring. Once proper permits are obtained, locate the nests with eggs. First check the incubation stage of the eggs.

If they are cool to the touch, the female is not finished laying eggs. Go back in one week. Once the eggs are warm, float several in a bucket of water. If the eggs sink, oil the entire egg clutch by treating the top two-thirds of the eggs. If the eggs float in water, they can be removed and buried. Floating eggs indicate they are in advanced stages of development and the female's egg follicles have dried up and cannot lay any additional ones.

7. **Hazing/harassment.** Noisemaking devices (cracker shells, propane cannons, whistles, or audible distress calls of different bird species as well as howls from coyotes and dogs, etc.) can help deter geese from an area when used in conjunction with habitat modification. If used alone, geese will become habituated to hazing. Many noisemaking devices often require a permit or license, especially within city limits. Projectiles are illegal to fire in some areas. Canada geese seek areas where they can go about their daily activities with minimum disturbance. If someone or something bothers them enough, they usually will find another area where they will not be disturbed. However, they often get used to some harassment techniques when they learn they will not be harmed.

   Harassment techniques usually will not stop damage once it has started. They are, however, useful in preventing damage before it begins. If Canada geese were raised on an area or have become accustomed to using it for feeding, they will be more difficult to move.

   **Dead geese effigies.** Nobody likes a dead goose, especially live geese. A lifelike goose decoy can be placed in an "agony" position, convincing live geese of nearby predators. Geese effigies can be used on land and water (water applications must be anchored so they will not float away). A minimum of two decoys must be used for flocking geese to believe predators will return to inflict more damage. Coyote and fox decoys have also been used with some success.

   **Laser harassment.** In certain situations, lasers have been used to quietly haze Canada geese. This technique is quiet and relatively easy to apply and appears to work on geese. As issues of safety, health, and property damage associated with wildlife populations increase, so too does public demand for nonlethal, noninjurious, and environmentally benign solutions to such problems. Relatively low-power, long-wavelength lasers (630–650 nm "red" beam) provide an effective means of dispersing some problem bird species under low-light conditions while presenting no threat to the animal or environment. Best success is typically between dusk and dawn or when extensive cloudy or overcast days occur. Do not look into the beam.

   **Pyrotechnics.** Although not all geese react to pyrotechnics, most do. Pyrotechnics are specially designed class C fireworks that are used to frighten wildlife. The types of pyrotechnics in this class include:
   - Screamers and bangers: large bottle-rocket type devices fired from a 15 mm starter's pistol that whistle loudly or explode
   - Shellcrackers: firecrackers fired from a 12-gauge shotgun

   The distance a particular pyrotechnic device will travel varies from 50 to several hundred yards (>46 m) depending on manufacturer and type. Check with the manufacturer to be sure a particular device fits your needs. They can be very effective when used as soon as the flock begins to use the pond or property. But if the flock become established, their effectiveness is lessened or may be only temporary. Individuals using pyrotechnics should be trained in their use and should wear eye and ear protection.

   Be cautious when using them in populated areas. These are most effective early and late in the day, which also avoids most golfers. *Note*: Check with local authorities for possible ordinances restricting the use of pyrotechnics before purchasing these devices.

   **Propane cannons.** Propane cannons are popular tools in use at numerous airports around the country. Many farmers also have used them with some success. They operate from the gas in a standard propane tank. On a timed basis, a small amount of propane is ignited, producing a loud report that can be heard more than a mile (1.6 km) away.

   The simplest models explode every 30 seconds to 30 minutes, based on the setting. More sophisticated models use computer chips to control the detonation more randomly, on a particular schedule or by remote control. Canada geese, like many other animals, have the ability to quickly adapt to propane cannons and sometimes quit responding without additional aversive conditioning. Propane cannons may not be suitable for large communities because the devices are loud and may be more of a nuisance than the geese.

   **Chasing.** Chasing geese on foot or in a golf cart is labor intensive, but in conjunction with other harassment methods, it can be successful if people are persistent. The idea is to chase geese long enough to cause them to go elsewhere, where they can live without being chased.

   Other techniques used to harass Canada geese include
   - High-pressure water sprayers
   - Air horns

- Beating pots and pans together
- Toy water craft, model airplanes, drones, and helicopters
- Paint ball and similar devices

When coupled with techniques mentioned, they encourage Canada geese to move from an area. The key is to be more persistent than the geese and to use at least two means of deterrent. As long as the geese are not physically harmed, these harassment techniques are legal.

8. **Capture and relocation.** Geese, like all waterfowl, molt all of their primary wing feathers at once and become flightless for a short period in late summer. You can "round up" the geese at this time and relocate them. This is a costly and time-consuming process best used when all other efforts have failed. Capturing and relocating geese requires federal and state permits. It is best to use preventative techniques BEFORE geese get established in an area. Once geese are established, it becomes very difficult to deter them from a given area, particularly after nesting has begun.

   Capture and relocation of adult geese is not viable as the birds imprint the area where they learn to fly and most will return to the capture site or a similar setting. Since giant Canada geese already occupy virtually all suitable habitat, there is limited opportunity to relocate juvenile geese without creating similar problems at release sites. Relocation is best for young juveniles because they imprint on the release area where they learn to fly rather than returning to the area where they were captured.

   In some instances, localized populations may be captured during the molting season and processed for human consumption through charitable organizations. This step may be taken only when other techniques have not been successful.

9. **Hunting.** Where allowed, hunting is an important tool for managing problems caused by Canada geese. Hunting helps to reduce the number of birds in an area, provides a repellent effect for geese not taken, and reinforces the use of nonlethal techniques, such as pyrotechnics. Early goose hunting opportunities are designed to harvest local giant Canada geese before the migrants arrive. Many areas with resident Canada geese prohibit the use of firearms. Check federal, state, and local regulations before hunting.
10. **Toxicants.** There are no toxicants registered with the Environmental Protection Agency for controlling Canada geese in the United States.

# 12 Turfgrass Diseases and Nematodes

## WHAT IS A DISEASE?

A **disease** is defined as a detrimental condition developing from a continuous interaction between a causal agent and a host plant. Diseases interfere with the normal function of some part of the plant, resulting in lower yields or reduced quality.

Diseases are caused by either living agents (*biotic agents*) or nonliving agents (*abiotic agents*). Abiotic agents can be environmental parameters, such as air pollutants, soil or water pollutants, lightning, inadequate moisture, light, air movement or nutrition, mechanical injury, etc. An important point to note is that abiotic causal agents induce disease that is noninfectious, meaning the disease does not spread from affected plants to nonaffected ones.

Biotic causal agents are living organisms capable of causing disease, including both microorganisms and parasitic plants. In contrast to abiotic agents, biotic agents cause disease that is infectious, whereby infected plants produce inoculum that is spread in various ways to nonaffected, susceptible plants. Biotic agents of disease include fungi, bacteria, phytoplasmas, nematodes, viruses and viroids, parasitic plants, and protozoa.

Infectious diseases in turfgrasses, as in other plants, originate from an interaction between a susceptible plant, a disease-producing organism (usually a fungus), and an environment favorable for the disease-causing organism to attack. Often the environment is unfavorable for ideal growth and development of the host plant. Diseases do not occur until all three parameters overlap (Figure 12.1). Information on all three factors should be gathered to obtain information for diagnosis of the problem. Turf managers should try to alter any or all of these three factors to combat the disease.

Turfgrass plants affected with disease give visible evidence that something is wrong. *Signs* involve directly seeing a pathogen or its parts and products on a plant. An example involves hyphae, a vegetative structure for most fungi, or mycelium, a mass of hyphae (Figure 12.2). The plant's visible reaction to a disease is referred to as its *symptoms* (Figure 12.3). Some symptoms of turfgrass disease may be leaf lesions, turfgrass thinning or discoloration, the development of small areas of dead plants within the turfgrass community, and many other symptoms characteristic of specific diseases that will be discussed later. Sometimes parts of the causal agent itself are directly visible with the unaided eye or with the assistance of a handheld magnifying lens. In this case, disease signs involve directly seeing a pathogen or its parts and products on a plant.

In most turfgrass situations, the environment is the key factor for disease development since the turfgrass host and turfgrass pathogens are virtually always present, especially in older turfgrass stands. While turfgrasses may be affected by diseases all year long, individual turf diseases are generally prominent for only a finite period each year, usually due to weather patterns and subsequent environmental effects. However, any stress (environmental or man-made) placed on turf will weaken it, making it more susceptible to disease development and less able to recover from disease once ideal conditions for disease cease (e.g., changes in weather or chemical treatments).

## BIOTIC ORGANISMS CAUSING TURFGRASS DISEASES

### FUNGI

Fungi are small, threadlike organisms composed of tiny filaments (hyphae) that cannot manufacture their own food. Although many fungi resemble plants without chlorophyll, they are not plants and are classified in their own kingdom (Fungi kingdom). Similar to fungi are the Stramenopiles. These fungus-like organisms include many algae but also *Pythium*, the downy mildews, and the Labyrinthulomycota, among the turf pathogens. Fungi and the Straminopiles are either parasitic or saprophytic (Table 12.1).

Most fungi produce spores (seedlike bodies), consisting of single or multiple cells, that are spread by wind, water, mechanical means such as mowers, or by infected plant materials such as grass clippings. Under moist conditions, spores germinate and grow by producing threadlike filaments, or hyphae. Collectively, a mass of hyphae is known as a mycelium (or mycelia for plural). Mycelia may function by absorbing nutrients from plant cells and also can become dormant, thereby providing some fungi a means to survive periods of inactive growth. Some fungi produce hardened, aggregated mycelial structures called sclerotia, which may aid in resisting unfavorable environmental conditions.

Fungi produce spores through sexual reproduction, by asexual means, or as a combination of both. The portion of a fungi's life cycle involving sexual reproduction is termed the perfect stage. The imperfect stage is the life cycle phase involving asexual reproduction. In some fungi, spores produced asexually in vast numbers may quickly colonize susceptible plants and produce epidemics.

When a fungal pathogen is not actively attacking the plant, it has not disappeared from the turfgrass area but is simply surviving in a dormant state or as a saprophyte living off organic materials in the thatch and soil layers in the environment. Fungicides generally do not eliminate turfgrass pathogens, but simply suppress the pathogen's activity on the plant while the environment is conducive for disease. Generally, they should be thought of as plant pharmaceuticals acting as plant protectants and/or chemotherapeutic agents.

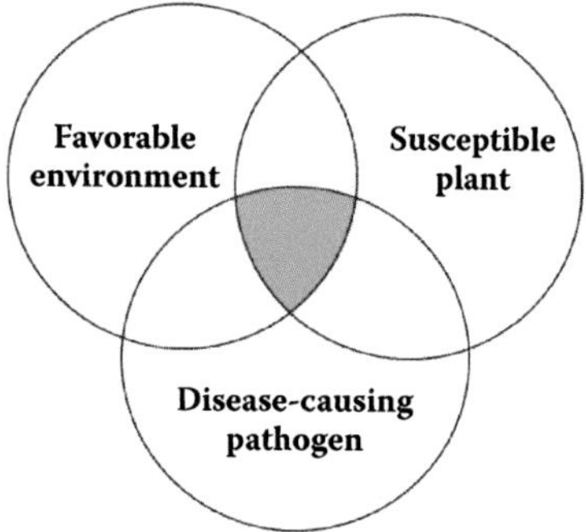


**FIGURE 12.1** For diseases to develop, a combination of favorable environment, susceptible host, and disease-causing pathogen is needed.

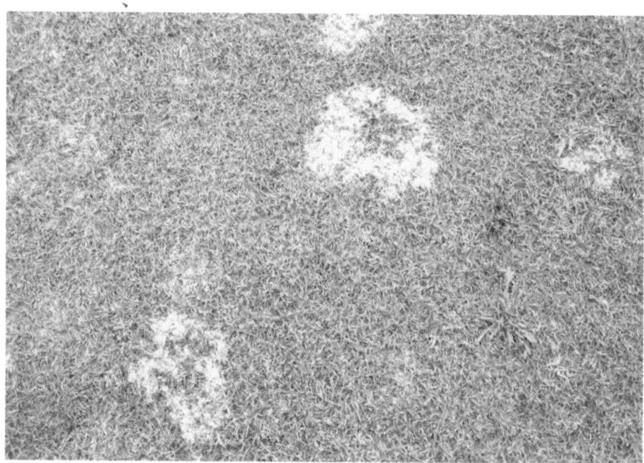

**FIGURE 12.2** Signs are visual pathogens or their parts and products, such as mycelium (shown).

**FIGURE 12.3** Symptoms are a plant's visible reaction to a disease, such as leaf lesions (shown).

## Bacteria

Bacteria are microscopic, single-celled organisms with rigid cell walls but no organized nucleus (termed *prokaryotes*). Fungi and Stramenopiles are *eukaryotes*, meaning they have organized nuclei within their cells. Bacteria reproduce by fission (or simple cell division), and can rapidly reproduce. They lack chlorophyll and overwinter, like fungi, in and on thatch, plants, seed, soil, and sometimes in insects. Bacterial pathogens enter plants only through wounds (which may be made by living organisms, such as insects) or natural openings, such as lenticels, hydathodes, or stomata. Although some bacteria are able to swim, splashing water, wind-blown water, insects, infected tools, or plant cuttings are the normal means of dispersal. Once inside the host, bacteria cause damage by producing toxins or by blocking vascular tissue. This blocking retards water movement and infected plants wilt. Practically all plant-infecting bacteria are rod shaped.

More than 80 bacteria species have been reported to cause plant diseases. Fortunately for turf managers, only a few damage finely maintained turf. The traditional economically important bacterial turf disease is *bacterial wilt,* which is caused by specific forms of the pathogen *Burkholderia (Xanthomonas) campestris* (= *Xanthomonas translucens* pv. *poae*). These forms attack a number of grass species but are found most often on annual bluegrass and the vegetatively propagated creeping bentgrass cultivars Toronto C-15, Nimisilla, and Cohansey. More recently, *Acidovorax avenae* has been shown to be pathogenic on creeping bentgrass under certain conditions of environmental stress. The term *bacterial decline* has been used to describe the symptoms and to differentiate the disease from the better known and studied bacterial wilt described earlier.

## Nematodes

Nematodes are small, eel-like worms that lack eyes. Those capable of causing disease in plants are mostly microscopic, with the largest approaching only 1/16 inch (1.6 mm) in length. Nematodes cause disease by infecting plant roots; however, some species infect leaves, stems, buds, and even replace embryos in the seeds of some plants. They possess a specialized feeding apparatus, a needlelike stylet, in their heads, which is utilized to tear and stab plant cells and extract their contents for nematode nutrition. Some, like root-knot and cyst nematodes, set up specialized feeding sites within plant roots, inducing giant cells and cell proliferation in root tissues, leading to gall formation in the case of root-knot nematodes. Nematodes as turfgrass pathogens are discussed later in this chapter.

## Phytoplasmas

Phytoplasmas (previously known as mycoplasma-like organisms) are also prokaryotes, but do not have rigid cell walls. They are also believed to be obligate parasites and are vectored by insects (many are transmitted by leafhoppers). Some diseases, called yellow diseases, that were once thought to be caused by viruses, have since been shown to be caused by phytoplasmas. Although diseases of turf in other parts of the world have reportedly been caused by phytoplasmas, such as yellow dwarf of bentgrass in Japan, none have been reported in the United States.

**TABLE 12.1**
**Parasitic Habits of Turfgrass Fungi and Straminopiles**

| Parasitic Habit | Turf Pathogen | Primary Infection Sites |
|---|---|---|
| **Facultative parasites**—parasitic fungi that can survive as saprophytes until conditions are favorable for infecting living plants, after which they become parasitic | *Colletotrichum, Curvularia, Xanthomonas* | Mostly shoots |
| | *Pythium, Typhula* | Mostly roots |
| **Facultative Saprophytes**—fungi that function primarily as parasites, but can temporarily subsist as saprophytes on dead organic matter | *Gaeumannonyces, Leptosphaeria, Magnaporthe, Pyricularia, Rhizoctonia, Sclerotinia,* | Roots |
| | *Microdochium* (*Fusarium*) | Shoots |
| | *Drechslera, Bipolaris* | Shoots and roots |
| **Obligate Parasites**—fungi that live only in association with a living host, feeding on it | *Erysiphe, Puccinia, Urocystis, Ustilago,* Panicum mosaic virus (SADV) | Shoots |
| | Parasitic nematodes | Roots |
| **Obligate Saprophytes**—fungi that only live with dead organic residues, such as thatch and mat, and do not infect living plants | *Fuligo, Mucilago, Physarum* | Shoots |
| | *Agaricus, Lepiota, Lycoperdon, Tricholoma* | Roots |

*Note:* **Saprophytes** are fungi that feed on dead organic matter such as thatch. **Parasites** are fungi that feed directly on plants to obtain some or all of their food.

### Viruses

Viruses are much smaller than bacteria and can only be viewed with the aid of electron microscopes. They are obligate parasites composed of nucleic acid (DNA or RNA) and a protective protein coat. Viruses are especially complex molecules, but are infectious in nature, infecting plants through wounds or by introduction into cells by means of a vector (insect, fungus, nematode), depending on the virus. For example, barley yellow dwarf virus is vectored by aphids, tomato spotted wilt virus is vectored by thrips, and wheat-soil–borne mosaic is vectored by a soil-borne fungus. Once inside the cell, a virus forces the plant's genetic material to make more viral particles. Being produced inside the plant, viruses cause systemic infections. They seldom kill their host plants, but do severely affect their quality and health.

The diseases, St. Augustinegrass decline (SAD) and centipede mosaic, are the only known economically important virus-induced diseases in turfgrasses. These are caused by various strains of panicum mosaic virus, which is easily mechanically transmitted, most often by mowers and possibly by insects. Symptoms first appear as a mild chlorotic mottling, blotching, and speckling or stippling of the leaf blades for the first several years. Initially, symptoms resemble Fe or Zn deficiency, mite damage, or yellow tuft (downy mildew) disease (Figure 12.4). Virus symptoms on infected leaf blades have a mosaic pattern with yellow streaking. Leaf nutrient deficiency symptoms, however, include continuous yellow stripes parallel to the veins. Symptoms intensify over time and infected plants may eventually die, leaving voids for weeds to fill.

### Parasitic Higher Plants

Parasitic plants, like mistletoe and dodder, and protozoa are also known to cause infectious diseases in some plants, but none are known to cause turfgrass diseases.

**FIGURE 12.4** Virus symptoms initially appear as mild chlorotic mottling and streaking of leaf blades. Shown is virus symptoms on Virginia buttonweed, a common turfgrass weed.

## THE DISEASE CYCLE

The nature of infectious disease is dynamic, meaning it is always changing. The dynamics involve the growth and reproduction of the host and the pathogen, as affected by environmental conditions. The host and the pathogen have definite life cycles, detailing their growth and reproduction. The disease cycle illustrates the phases of the interaction between the host and the pathogen as disease is developing and changing. A disease cycle might include the following phases: pathogen survival (overwintering or oversummering); pathogen dispersal (by wind, water, on plant material, or by vectors); inoculation (the pathogen coming in contact with the host); infection (penetration of host tissues); colonization of host tissues; reproduction (within tissues and by propagules such as spores); and, finally, pathogen dispersal once again.

The infection process begins when a fungus spore or mycelium comes into contact with a host plant. With germination, the fungus spore produces a germ tube (termed haustorium) that enters the plant through a wound or by penetrating the cuticle and epidermis of a leaf, sometimes through stomates, but also directly through the leaf surface. Depending on the pathogen, root and crown tissues may be attacked as well. Certain fungi, such as *Pythium*, directly disrupt cells by producing enzymes that cause plant cell walls to dissolve. The cell contents then leak into the surrounding tissue where the fungal hyphae can absorb nutrients.

Almost all fungi require moisture on leaves and a favorable temperature to start disease development. Because of this, turfgrass diseases are most common during rainy or foggy weather or when excessive moisture remains on leaves or in the soil for extended periods. Since most fungi do not infect turfgrass in the absence of moisture, proper irrigation and soil drainage are important keys to disease management.

Once inside the host plant, the pathogen begins to obtain nutrition directly from living cells or from products of cellular decomposition. This is followed by an incubation period where the pathogen produces spores, sclerotia, or other survival structures that can serve to infect other plants. Symptoms eventually develop and, with continuous infection cycles and favorable environmental conditions for the pathogen, an epidemic results, causing widespread loss of turf.

## DISEASE CONTROL STRATEGIES

### Correct Diagnosis

Turfgrass disease diagnosis involves a process of elimination. For accurate diagnosis, a number of steps are taken, the first being careful observation of symptoms. Leaf spots occur with various sizes and colors, most commonly with round or oval lesions on leaves with a distinct marginal border to the lesion. Gray leaf spot is an example of a fungal pathogen with leaf spots as a symptom. Foliar lesions are large, irregular areas on leaves with a distinct border such as is typical with brown patch. Stem lesions may occur with large irregular areas on leaf sheaths with a distinct border—symptoms typically seen with large patch. Foliar blight or dieback involves necrosis of leaves or entire tillers such as with *Pythium* blight. Crown rot symptoms involve the necrosis of crowns, rhizomes, or stolons as occurs with summer patch. Root rot involves the necrosis of roots, typical of "patch" diseases such as take-all patch.

Three main types of plant diseases occur on turf. Foliar diseases are generally easy to detect and diagnose and can be controlled curatively (Table 12.2). Stem and crown diseases are more difficult to detect and diagnose and control curatively. Root diseases can cause significant damage to root systems even before symptoms appear. Most root diseases cannot be controlled curatively.

As mentioned, most pathogens (fungi) causing turfgrass diseases are microscopic; therefore, individual spores cannot be seen by the naked eye. However, with some diseases, spores collect to sufficient levels that they become visible as threads, powders, or masses. When the causal fungus is visible, its appearance often becomes the most important clue for diagnosis.

True pathogenicity (ability to cause disease) can be proven only by reliable scientific methods, usually requiring a laboratory and greenhouse or growth chamber. Three disease factors (grass, pathogen, and environment) provide the sources of diagnostic information. Thus, the next step in identification is an examination of the environment during the onset of disease occurrence. For example, what were the temperature, light intensity, and moisture conditions prior to and during disease development? The nature of the disease site also is important. Air and water drainage, soil conditions, sun and shade, slope, and nearness of other plantings or buildings all may be important in the development of turfgrass diseases. Prior chemical applications to the site, including pesticides and fertilizers and even fungicides, may be important to make a correct diagnosis. Heavy thatch accumulation and poor mowing practices stressing the turf may trigger or amplify certain disease problems.

Finally, the submission of affected turf samples to a reputable laboratory or pathologist can assist in receiving a correct diagnosis and advice on corrective measures to take. This should be supplemented with close-up and head-high digital images of the affected turf, any noticeable symptoms or signs, and any unusual patterns.

### Cultural Control

Similarly to the human immune system, plants have a complex system for protection from disease. When stressed or

**TABLE 12.2**
**Major Turfgrass Diseases and Their Location (Occurrence Site) in Plants**

| Foliar Diseases | Stem and Crown Diseases | Root Diseases |
|---|---|---|
| Bacteria, *Bipolaris* and *Drechslera* leaf spots, brown patch, copper spot, dollar spot, gray leaf spot, pink patch, pink snow mold, powdery mildew, *Pythium* blight, red thread, rusts, southern blight, stripe smut, yellow tuft | Algae, anthracnose basal rot, dead spot, large patch, melting out, white patch | Bermudagrass decline, necrotic ring spot, *Pythium* root rot, spring dead spot, summer patch, take-all patch |

depleted of energy, this defense system does not work properly. Cultural practices should promote an environment that is not conducive for pathogen infection and disease development, but rather that promotes healthy, vigorous plant growth to better fight off an attack. This may be achieved by planting disease-resistant cultivars, if they are available; following sound principles of site preparation during turfgrass establishment; and managing an established turf through proper mowing, fertilization, irrigation, and cultivation practices.

**Disease-resistant grasses.** When establishing a new turf area or when renovating disease-damaged turf, it is important to select grasses known to be resistant to pests commonly occurring in the area (Table 12.3). Resistant cultivars can produce chemical toxins to the fungus, have morphological or chemical characteristics that prevent fungi entrance, or lack a nutrient the fungus needs. For example, certain Kentucky bluegrass varieties are resistant to melting out, stripe smut diseases, and spring leaf spot, diseases that devastate many bluegrass varieties. Crenshaw bentgrass, meanwhile, is highly susceptible to dollar spot disease (Table 12.4). Blends of various varieties or even turf species often provide a wider spectrum of pest resistance than planting a monoculture of one specific variety. If one cultivar loses its disease resistance, the entire turf stand is not ruined since other resistant cultivars are present. For example, blending Crenshaw and L-93 bentgrasses reduced dollar spot disease occurrence compared to Crenshaw-only stands.

**Proper turfgrass establishment.** Disease problems can often be avoided by implementing proper site preparation prior to planting. Buried debris, soil layering, and failure to incorporate amendments or construct drainage facilities where necessary can provide unfavorable turfgrass growing conditions.

**TABLE 12.3**
**Relative Disease Proneness of Turfgrass Species**

| Susceptibility Ranking | Turfgrass Species |
|---|---|
| Very high | Annual bluegrass |
| | "Colonial" bentgrass |
| | Creeping bentgrass |
| | Velvet bentgrass |
| | Roughstalk bluegrass |
| High | Kentucky bluegrass |
| | Perennial ryegrass |
| | Fine fescues |
| | Tall fescue |
| Moderate | Bermudagrass |
| | Seashore paspalum |
| | St. Augustinegrass |
| | Zoysiagrass |
| Low | Bahiagrass |
| | Carpetgrass |
| | Centipedegrass |
| | Buffalograss |
| | Kikuyugrass |

**TABLE 12.4**
**An Example of Natural Plant Resistance to Certain Diseases Is the Relative Susceptibility of Creeping Bentgrass Cultivars to Dollar Spot Disease**

| Most Tolerant | Moderately Tolerant | Moderately Susceptible | Most Susceptible |
|---|---|---|---|
| 007 | Authority | Alpha | Century |
| 13M | Cobra II | Backspin | Crenshaw |
| Barracuda | Crystal Blue Links | Bengal | Imperial |
| Benchmark SDR | L-93 | Grand Prix | Independence |
| Declaration | Penn A-1, A-2 | Mackenzie | Ninety-Six Two |
| Focus | Penncross | Penneagle II | 18th Green |
| Flagstick | Pennlinks | Penn A-4, G-1, G-2,G-6 | |
| Kingpin | Pin-up | Pin-Up | |
| Luminary | Proclamation | Princeville | |
| Memorial | Seaside | Providence | |
| Pennlinks II | Seaside II | Pure Distinction | |
| SR1150 | SR1119 | Putter | |
| | Tyee | Sandhill | |
| | V-8 | Shark | |
| | | Southshore | |
| | | T-1 | |

*Source:* NTEP, Purdue and Rutgers Universities.

The turf may then become weakened from stresses associated with site construction, causing the turf to become more susceptible to disease.

**Surface and subsurface moisture.** Moisture is critical for many fungal spores to germinate and develop. While irrigation is essential to prevent drought damage, the amount of water and the timing of its application can prevent or contribute to disease development. Dew (more importantly, the length of the dew or leaf wetness period)—a critical factor for disease development—is dependent on temperature and humidity. Extending the length of the dew (free water) period by irrigating in the evening before dew forms or in the morning after the dew evaporates extends the leaf wetness period. Therefore, irrigate when dew is already present, usually in the predawn hours. This will also dilute or remove the guttation fluid accumulating at the cut leaf tip, which may provide a food source for some pathogens. For many golf courses, eight hours or more may be needed to irrigate the entire course. Since the putting greens are most susceptible to diseases, irrigate them during the predawn hours, so they will dry more quickly. As more and more courses implement new computer-driven irrigation systems, this is often easier to accomplish. Mowing, whipping, and dragging; tree/shrub pruning and removal; and fan installation also help reduce leaf wetness duration.

Soil moisture levels also influence certain diseases. Adequate surface and subsurface drainage are two extremely important considerations prior to turf establishment that often dictate the level of turf health. Diseases encouraged by excessive soil moisture and low soil aeration include brown patch, *Pythium* blight and root rot, anthracnose basal rot, summer patch and take-all patch, spring dead spot, algae, and black layer. This is most acutely critical when trying to grow grasses outside their naturally adapted areas, such as growing bentgrass in hot, humid areas.

**Nutritional and soil pH status.** Many diseases are also influenced by the nutritional status of the grass, especially concerning N (Table 12.5). Achieving a perfect balance to maintain green, healthy, and growing plants should be the goal because both excessively high and low N fertility contributes to turfgrass diseases and recovery. Excessive N applications, for example, encourage *Rhizoctonia*, *Pythium*, snow molds, powdery mildew, and stripe smut. Low N levels, meanwhile, encourage dollar spot, anthracnose, rust diseases, and red thread (Figure 12.5). When a foliar disease is active, select a fertilizer blend with a high percentage of the slow-release component and a lower percentage of the rapid-release component. Adequate nutrition should be available to plants for continued growth, but the goal is to minimize luxuriant growth that induces greater susceptibility to pathogens.

**FIGURE 12.5** Proper nutrition often influences diseases such as insufficient nitrogen encouraging dollar spot disease (left).

**TABLE 12.5**
**Turfgrass Disease Occurrence as Influenced by Nutrient Levels**

| Diseases Favored by Excessive Levels | Diseases Favored by Inadequate Levels |
|---|---|
| **Nitrogen** | |
| Copper spot | Anthracnose |
| Gray leaf spot | Dollar spot |
| Gray snow mold (*Typhula* blight) | Leaf spots |
| Pink snow mold (*Fusarium* or *Microdochium* patch) | Melting out |
| *Pythium* diseases | Necrotic ring spot/summer patch |
| *Rhizoctonia* diseases | Red thread |
| Spring dead spot | Rust |
| Stripe smut | Mini ring |
| Take-all patch | |
| **Phosphorus** | |
| – | Pink snow mold |
| | *Pythium* diseases |
| | Take-all patch |
| **Potassium** | |
| Pink snow mold | Dollar spot |
| *Rhizoctonia* diseases | Leaf spots |
| | Red thread |
| | Spring dead spot |
| | Take-all patch |
| **Calcium** | |
| – | *Pythium* blight |
| | Red thread |
| **Sulfur** | |
| – | Pink snow mold |
| | Take-all patch |
| **Iron** | |
| – | Pink snow mold |
| **Manganese** | |
| – | Take-all patch and other ETRI diseases |
| **High Soil pH (>~6.5)** | |
| Bermudagrass decline | Pink snow mold; Summer patch |
| Necrotic ring spot | Spring dead spot; Take-all patch |

Potassium is an important component in disease prevention as it reduces plant stress. Again, a nonstressed plant is not as susceptible to diseases. This has probably best been documented with "*Helminthosporium*" leaf spot diseases. A N:K ratio of at least 1:1 is best, although a ratio of 1:2 may be necessary in certain circumstances. Soil and leaf tissue testing should be performed periodically to indicate plant availability of nutrients.

High soil pH (>6.5) also appears to aggravate certain ectotrophic root-infecting (ETRI) diseases such as spring dead spot, necrotic ring spot, summer patch, bermudagrass decline, and take-all patch (Table 12.5).

## CHEMICAL CONTROL PRACTICES

A primary misconception many turf managers have concerning fungicides is that these materials kill fungi. Fungicides do not eliminate pathogens from the turfgrass area, and should never be utilized with this strategy in mind. When a fungal pathogen is not actively attacking the plant, it has not disappeared from the turfgrass area but is simply surviving in the environment in a dormant state or as a saprophyte living off dead organic matter in the thatch and soil layers. Most fungicides act as fungistatic compounds, meaning they primarily suppress fungal growth to prevent plant infection during environmental conditions conducive for disease development. Fungicides are applied either prior to anticipated disease outbreaks (*preventative*) or following the appearance of disease symptoms (*curative* or *eradicants*). Thus, fungicides should be considered as prophylactic plant pharmaceuticals. Those with some eradicant activity still work as plant protectants, but may also behave as chemical therapeutic plant pharmaceuticals.

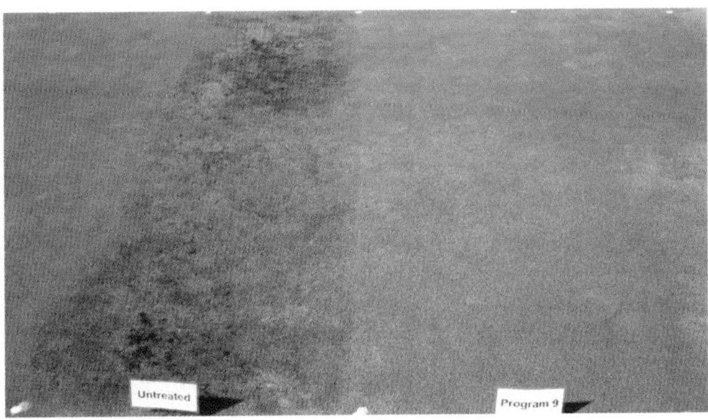


**FIGURE 12.6** Proper bentgrass fungicide program (right) compared to none in a hot, humid environment.

Preventative fungicides are needed when chronically damaging diseases occur. This is especially true on golf greens that have seasonal problems with certain diseases such as *Pythium* blight, anthracnose, gray leaf spot, summer decline, brown patch, and snow molds (Figure 12.6). Preventative use of fungicides often allows substantially lower rates than those required for curative treatment.

To be effective, a curative fungicide application strategy should be used at the onset of disease symptoms. Obviously, correct disease diagnosis is crucial for this strategy. Once significant turf damage has occurred, it is generally useless to apply curative fungicides. Many fungicides, such as mancozeb, ethazole, and others, act as contacts (nonplant penetrants). They are more economical to use and less likely to have pathogens develop disease resistance. However, they generally only provide short-duration (7- to 14-day) control under high disease pressure, and therefore, are used at high rates on short intervals.

## Fungicides

Fungicides can be classified in various ways, and managers should become familiar with all of the different classifications for knowledge on their mode of action, as well as strategies for minimizing undesirable effects and maximizing disease management effectiveness. Fungicides may be classified based on how they behave on or within turfgrass plants. Basically, there are two categories of their topical modes of action: plant penetrants or contact.

1. **Contact fungicides** remain on plant surfaces and are mostly curative, only affecting fungi present there and having no activity against root diseases. Good surface coverage is a key with contact fungicides.
2. **Penetrant fungicides** are absorbed into plant tissues due to concentration gradients and they generally restrict fungal growth within the plant. They protect against new infections and suppress existing ones. However, due to very specific modes of action, penetrants have a greater risk of resistance occurrence. Two types of penetrant fungicides exist: acropetal and localized.
    a. Acropetal penetrants are xylem mobile by having upward movement from the point of contact along a water-pressure gradient. Overall, acropetal penetrants are the best curative products and mixing them in a tank with a contact fungicide is often beneficial.
    b. Localized penetrants diffuse into and mostly remain in the plant's waxy cuticle.

For optimum activity for most fungicides, the following are required:

1. Fungicides are actually fungistatic (restrict growth) rather than fungicidal (complete control).
2. The pathogen must be active for fungicides to be effective.
3. Most fungicide residual activity is depleted over 14 to 21 days—thus, the possible need for reapplications.
4. Additional fungicides will be required to achieve adequate control under high disease pressure.

**Modes of action.** Within fungi, several known cellular targets are attacked by fungicides. These include:

- Fungal cell walls (provide fungal structure and protection)
- Cell membranes (regulate compounds that enter or exit the cell)
- The cytoskeleton (cell function and structure framework within the fungal cytoplasm)
- Nucleic acids (genetic transfer information)
- Cytoplasmic enzymes (regulate cell function and catalyze metabolic processes)
- Mitochondria (energy production via respiration, often affecting ATP formation)

Numerous specific sites of action may occur within each of these known cellular targets.

**Phosphonates (phosphorous acid salts).** Phosphonate-containing products have recently been linked with disease control and possible fertilizer benefits. Phosphonate is a compound containing a carbon-to-phosphorus bond. Salts of phosphorous acid are phosphates $\left(PO_4^{-3}\right)$ and phosphate products often contain a mixture of phosphorous acid and potassium hydroxide (KOH). From a fungicidal view, products made up of the salt and esters of phosphorous acid ($HPO(OH)_2$) typically provide the desired control. Phosphorous acid is a readily available solid substance that, when mixed with water, forms a strong acid referred to as phosphonic acid. In its pure form, phosphonic acid is so strong it causes excessive plant injury; thus, it is combined or diluted with other chemicals. This is often performed by mixing with an alkali salt such as potassium hydroxide (KOH) to form potassium phosphorous acid or potassium phosphite $\left(KPO_3^{-2}\right)$ and is the active ingredient in Alude, Appear, Magellan, Biophos, Resyst, Vital, and others. It is also the main ingredient in several phosphonate fertilizers such as K-Phite (0-29-26), Ele-Max Foliar Phosphite (0-28-26), and Nutri Phite P + K (0-28-26).

Other fungicides are produced by reacting phosphonic acid with ethanol to form ethyl-phosphonate with aluminum added to neutralize the ethyl-phosphonate ions and produce fosetyl-Al (also called aluminum tris (*o*-ethyl phosphonate)). Fosetyl-Al is found in Aliette WDG, Prodigy, and the Chipco Signature fungicides.

Although similar in structure, phosphonate-derived fungicides act differently in plants, soil, and plant pathogens than the traditional P-containing fertilizer, phosphate. In plants, phosphonate is transformed to phosphorous acid $\left(H_2PO_3^-\right)$, which does not appear to be involved in any P metabolism in plants and only very inefficiently in soils.

Phosphonate fungicides are unique in plants as they move both in the xylem and phloem, making them true systemics. Due to this mobility, especially to roots, phosphonate fungicides provide good control of root diseases caused by various *Pythium* spp., especially when applied preventatively. When combined with mancozeb or a patented blue pigment, phosphonate-containing fungicides also significantly improve turf quality and control or at least reduce the severity of bentgrass "summer decline" or bentgrass "summer stress complex." This blue inert pigment in mancozeb, copper phthalocyanine, is an insoluble organic pigment (#15) added to make the naturally yellow-colored mancozeb have a green hue. Chipco Signature is a combination of fosetyl-Al plus a similar pigment, chlorinated phthalocyanine. "StressGard" technology contains this pigment and is thought to add to turf health in addition to pathogen control.

The exact mode of action of phosphonate fungicides is not known but is suspected to stimulate a plant's natural chemical and physical defenses against disease, improve the plant's photosynthesis capability, interfere with phosphate metabolism in fungal cells (the mechanism with most evidence for *Pythium* control), or inhibit several key enzymes necessary for fungal growth and development. This is assumed to be why these fungicides work best when applied preventatively.

**Fungicide resistance.** Fungicide resistance by turf pathogens is an increasing problem, especially with systemic fungicides since the physiological processes these materials inhibit are genetically controlled by only one or a few genes. Being so specific in their mode of action, fungi may exist in nature at low frequencies that circumvent this, reducing the efficacy of the fungicide. Resistant fungi are then able to multiply as the susceptible fungi in the population are reduced by the fungicide. With repeat applications, resistant fungi multiply until the population is mostly fungicide resistant and disease control is severely reduced or completely lost. Usually, strains of fungi resistant to a fungicide will be resistant to all fungicides in the same chemical class (called *cross resistance*). Examples of diseases where resistance has been documented are listed in Table 12.6.

Strategies to minimize or delay resistance include:

1. The same fungicides or products with the same mode of action should not be used more than two consecutive applications before switching to products with different modes of action.
2. Combine the practice of mixing products in tanks with different modes of action along with the rotation strategy.
3. Rotate among contact and systemic products.
4. Proper fungicide timing with preventative applications typically is more effective than curative applications, especially with site-specific products.
5. Avoid at-risk, site-specific fungicides once the disease is active.
6. Always incorporate proper agronomic techniques to reduce disease severity.

**Mixing fungicides.** Fungicides are often mixed to increase the spectrum of disease control, to avoid resistance, and to improve turf quality. The most likely mixture contains a contact and a systemic fungicide. The contact fungicide provides a protective coating while the systemic fungicide protects root and crown pathogens and extends control of foliar pathogens. Common tank mixtures include a contact product such as chlorothalonil or mancozeb with a more systemic product such as azoxystrobin, fludioxonil, flutolanil, fosetyl-Al, iprodione, mefenoxam, myclobutanil, propamocarb, propiconazole, thiophanate-methyl, triadimefon, or trifloxystrobin. As previously mentioned, these mixtures are a common technique to delay or reduce the risk of resistant forms of pathogens from increasing and to provide curative and residual control from a single application. Use the low rate of each and apply on a 14- to 21-day interval.

A special consideration should be given to applications of fosetyl-aluminum (Signature). Due to the acidic and reactive

**TABLE 12.6**
**Examples of Resistance of Turfgrass Pathogens to Fungicides**

| Disease | Pathogen | Fungicide |
|---|---|---|
| Anthracnose | *Colletotrichum graminicola* | QoIs, benzimidozoles, DMIs |
| Dollar spot | *Sclerotinia homoeocarpa* | Benzimidazoles, dicarboximides, DMIs |
| Gray leaf spot | *Pyricularia grisea* | QoIs, benzimidazoles |
| Pink snow mold | *Microdochium nivale* | Dicarboximides |
| Powdery mildew | *Erysiphe graminis* | DMIs |
| *Pythium* blight | Pythium aphanidermatum | Acetanilides, QoIs |

nature of this chemical, care should be exercised in mixing Signature with certain other chemicals. Do not mix Aliette with either soluble fertilizers or flowable chlorothalonil or mancozeb formulations. In addition, do not add acidifying products to alkaline spray water when using Aliette. In a neutral spray water (pH = 7.0), Aliette will acidify the spray water to a pH of about 4.5. Similar precautions should be observed with other phosphonate fungicides.

**Spray volumes.** Application spray volumes for disease control are generally higher than for other pests such as weeds. The pathogen's location dictates application volumes, as foliar-based pathogens require excellent spray coverage. For example, crown-based pathogens require high spray volumes while root-based pathogens require watering in immediately following application. Delivery rates for most fungicides are in the 45- to 135-gallon/acre range (1 to 3 gallons/1,000 square feet)—specifically, 1 to 2 gallons/1,000 square feet (4 to 8 L/100 $m^2$) for fungicides targeting foliar and crown pathogens, and 2 to 3 gallons/1,000 square feet (8 to 12 L/100 $m^2$) for fungicides aimed at crown and root pathogens. Fungicide applications targeting root pathogens should be irrigated in immediately following application with 1/4 inch (6.3 mm) of water. For fairways, about 50 gallons/acre (935 L/ha) is standard for fungicide applications on dollar spot, but higher volumes should be utilized for root diseases such as summer patch. This high volume should be applied by regulating (slowing) sprayer speed and changing nozzle types instead of using higher spray pressures. A general rule of thumb is that spray pressure must be quadrupled to double sprayer output.

A broadcast boom fitted with flat fan or air-induction nozzles and delivering 30 to 60 psi (20 to 414 kPa) has provided very good results. Flooding nozzles have not been as effective for diseases of foliage or crowns. As pressure increases, droplet size decreases, spray volume and drift potential increase, and nozzles may wear more quickly.

#### Chemical Groups

Fungicides are also divided into groups based on their chemical properties (Table 12.7). To prevent fungicide resistance from developing, it is important to know which fungicides belong to the same chemical group or have the same biochemical mode of action. Biochemical mode of action refers to the way the fungicide inhibits fungi—whether at sites controlled genetically by one or a few genes or through multiple genetic control mechanisms. Fungicides should be periodically alternated or rotated, or used in mixtures with fungicides belonging to different chemical groups to prevent fungicide resistance. For example, alternating between propiconazole (Banner Maxx) and tebuconazole (Torque or Mirage) is not alternating between chemical groups, as both fungicides belong to the same chemical group, the demethylation inhibitors (DMIs). The occurrence of resistance has been problematic with plant-penetrant fungicides since their modes of action in controlling fungi are more genetically specific and the probability of resistant strains becoming selected is higher. In Table 12.7, fungicides are listed based on their chemical group.

#### Read Labels

You would not think of giving a family member any medication without reading the instructions first. Turfgrass fungicides deserve the same amount of respect. After all, they are plant pharmaceuticals. In addition to rates and intervals for application, labels provide information concerning the use or nonuse of surfactants with the material, compatibility between other pesticides or fertilizers, amount of water to use in the application process, posting or reentry restrictions, and so forth. Keep up to date with the labels. Take the time to read them completely at least once each year. Remember that labels are the law.

## DIAGNOSTIC FEATURES AND MANAGEMENT OF SPECIFIC TURFGRASS DISEASES

This section will discuss the primary diseases that occur on turfgrasses. Some are rare or their duration is so limited that they are relatively unimportant. Chemical control treatments are listed as the common name only.

#### Anthracnose Leaf Blight

The causal agent of anthracnose leaf blight is *Colletotrichum cereale* Manns sensu lato Crouch, Clarke, and Hillman.

**Occurrence.** Several manifestations of anthracnose diseases exist in turfgrasses. It occurs on many cool-season grasses as anthracnose leaf blight and occasionally as anthracnose basal rot. Bermudagrass and centipedegrass also may develop anthracnose leaf blight. On bermudagrass, the disease occurs in the fall as the growth of the grass slows. Anthracnose leaf blight occurs by itself and in combination with *Bipolaris sorokiniana* (or leaf spot) to severely thin annual bluegrass, fine-leaf fescue, perennial ryegrass, creeping bentgrass, and certain cultivars of Kentucky bluegrass. Anthracnose infection is associated with plant stress. It is favored during environmental periods of limited grass growth such as warm (80°F to 90°F, 27°C to 32°C) summer weather when the soil is dry, leaves are wet, and high atmospheric humidity occurs with cool-season grasses. Mechanically injured or winter desiccated bentgrass is most susceptible. Anthracnose leaf blight often selectively infects annual bluegrass in bentgrass/annual bluegrass stands in late spring, often when bluegrass is weakened or senescing (Figure 12.7). Although common, anthracnose leaf blight is usually of minor significance except on greens dependent on annual bluegrass as a component of the putting surface.

**Symptoms/signs.** The fungus overwinters either as mycelium or sclerotia, as a saprophyte in dead plant tissue in the thatch and on stems beneath the leaf sheath of live plants. It spreads from plant to plant via crescent-shaped, asexual reproductive spores called conidia when warm, high-humidity weather occurs along with high light intensity. The pathogen attacks grass shoots, stems, and roots. Initial symptoms are small (size of a dime) yellowing of the annual bluegrass turf, which turns bronze in color if warm weather

**TABLE 12.7**
**Turfgrass Fungicides Classified by Chemical Activity Group**

| Chemical Group (Activity) | Mode of Action | Common Name | Diseases Controlled |
|---|---|---|---|
| | | **Contact (Nonsystemic or Nonplant Penetrants)** | |
| **Aromatic hydrocarbons** (also called **substituted aromatics** or **substituted benzenes**) (protective) | Interferes with DNA synthesis, respiration and cell wall membranes needed for growth and reproduction; multisite action | Chloroneb, ethazol (etridiazole), PCNB (quintozene) | • *Pythium* spp. (chloroneb, ethazol, mancozeb, maneb, and propamocarb); brown patch (chloroneb, PCNB, thiram, mancozeb, and maneb); dollar spot (PCNB, mancozeb, and maneb); snow mold; "*Helminthosporium*" diseases (PCNB, mancozeb, and maneb); rust (mancozeb and maneb); melting out (PCNB, mancozeb); preventative blue-green algae (mancozeb). Little turf phytotoxicity or undesirable growth regulatory effects occur. Chloroneb and ethazol are short-lived (7 to 10 days) contacts. All have low resistance risk. |
| **Dithiocarbamates** or **carbamates** (protective) | Interferes with fungi cell respiration, enzyme activity, and cell membrane fluidity; multisite action | Mancozeb, maneb, propamocarb, thiram | |
| **Arylaminopyridine** | Dual mode of action, with uncoupling oxidative phosphorylation in mitochondria plus reactivity with thiols | Fluazinam | • Dollar spot, brown patch, "*Helminthosporium*" leaf spots, red thread, pink patch, anthracnose, pink and gray snow molds, algae. 2ee label for rapid blight. Low resistance risk. |
| **Benzonitriles** (also called **nitriles**) (protective) | Interferes with general cell constituents including DNA synthesis and nuclear division; multisite action | Chlorothalonil | • Dollar spot; copper spot; "*Helminthosporium*" diseases; brown patch; preventative anthracnose; gray leaf spot; snow mold; red thread; stem rust; preventative blue-green algae. Little turf phytotoxicity or undesirable growth regulatory effects occur. Low resistance risk. |
| **Oxidants** | Oxidation of cell membrane proteins and enzymes of microbial organisms. | Sodium carbonate, peroxyhydrate, hydrogen dioxide | • Predominantly algae. Low resistance risk. |
| **Pyrollnitrin** (or **phenylpyroles**) (protective); antifungal compound derived from the bacterium *Pseudomonas pyrocinia* | Interferes with fungal membrane transport and amino acid uptake; multisite action | Fludioxonil | • Brown patch; "*Helminthosporium*" leaf spot diseases, dead spot. Narrow control spectrum, short residual. Low to moderate resistance risk. |
| | | **Penetrants—Local-Systemic (or Local-Penetrant)** | |
| **Dicarboximides** (protective) | Inhibits DNA synthesis in cell division of spores, disrupts lipid metabolism in membranes, and blocks certain respiratory enzymes; multisite action | Iprodione, vinclozolin | • Dollar spot; "leaf spot" diseases; brown patch; copper spot; snow mold; melting out; red thread; *Curvularia* blight. Also combined with propiconazole for dollar spot control. Low root disease activity. Moderate resistance risk. |
| **Qi1** (protective) | Inhibits mitochondria functions. Specifically inhibits fungal spore germination by disrupting the Qi site of cytochrome bc1 in complex III of the fungal mitochondrial membrane | Cyazofamid | • *Pythium* blight, damping-off, and *Pythium* root dysfunction. Narrow spectrum of control. Moderate resistance risk. |

(*Continued*)

**TABLE 12.7 (CONTINUED)**
**Turfgrass Fungicides Classified by Chemical Activity Group**

| Chemical Group (Activity) | Mode of Action | Common Name | Diseases Controlled |
|---|---|---|---|
| **SDHI** or succinate dehydrogenase inhibitors (also called carboxamides, benzamides, or anilides) (curative and protective) | Inhibits mitochondria functions by inhibiting the respiratory enzyme succinate dehydrogenase, which is needed in the mitochondrial transport chain, specifically the complex II of the mitochondria; acropetal systemic; multisite action | Flutolanil | • Brown patch; large patch; red thread; southern blight; and basidiomycete (*Lycoperdon* spp.) fungi, which cause fairy rings. No control of nonbasidiomycete fungi such as dollar spot and anthracnose. Low to moderate resistance risk. |
| | | Boscalid | • Dollar spot. Narrow spectrum of control. |
| | | Fluxapyroxad, Penthiopyrad | • Dollar spot; brown patch; large patch; fairy ring; spring dead spot; "leaf spot" diseases. Moderate resistance risk. |
| | | Isofetamid | • Dollar spot; spring dead spot. |
| | | Fluopyram | • Sting and rootknot nematodes. |
| | | **Mobile Plant Penetrants (Systemics)** | |
| **Acetanilides** (also called **phenylamides**) (curative and protective) | Blocks RNA synthesis of mycelia and spore production. Does not inhibit spore germination or zoospore release. Acropetal systemics; specific site of action | Metalaxyl, mefanoxam, oxadixyl | • *Pythium* blight and damping-off; not always effective against *Pythium* root rot. Yellow tuft; downy mildew. Often combined with mancozeb for *Pythium* blight control. High resistance risk. |
| **Antibiotics** (or **polyoxins**). Antifungal compounds derived from *Streptomycin* (protective and curative) | Inhibits fungal cell walls by inhibiting chitin, a substance in fungal (and insect) cell walls; also inhibits spore germination and mycelium growth | Polyoxin D zinc salt | • Best for brown patch and gray leaf spot. Suppresses gray snow mold, *Rhizoctonia* damping-off, large patch, pink snow mold, leaf spot/melting out, snow mold, red thread, and yellow patch. Recently labeled for preventative anthracnose. Moderate resistance risk. |
| **Benzimidazoles** (curative and protective) | Inhibits microtubules via DNA synthesis, mitosis, and development of the fungal skeleton. Upon wetting, transform to methyl-2-benzimidazole carbamate (MBC); acropetal systemic | Thiophanate-methyl | • Dollar spot; brown patch; anthracnose; copper spot; snow molds; stripe smut; *Microdochium* patch, bermudagrass decline. High resistance risk. Root disease control is moderate to poor. |
| **Demethylation inhibitors** or **DMIs** including triazole, pyrimidine groups (curative and protective); also often referred to as **ergosterol biosynthesis inhibitors** (or **SBIs**) | Inhibits sterol synthesis in fungi cell membranes; acropetal systemic; specific site action | Fenarimol, myclobutanil, propiconazole, tebuconazole, triadimefon, triticonazole | • Curative against dollar spot; anthracnose; rust (cyproconazole, propiconazole, and triadimefon); snow mold; "*Helminthosporium*" diseases (propiconazole); copper spot (cyproconazole, triadimefon); powdery mildew; red thread; southern blight (cyproconazole, triadimefon); and gray leaf spot (cyproconazole, propiconazole). Protective against brown patch, especially if tank-mixed; summer patch; zoysia large patch; stripe smut. Undesirable turf growth regulatory effects may occur if used during stress (e.g., summer) periods. Moderate to high resistance risk. Excellent root disease control. |
| **Phosphonates** (curative and protective) | Increases plant resistance and blocks mycelial development and spore germination; specific site of action | Fosetyl-aluminum (Al), phosphite salts, potassium salts of phosphorous acid, sodium, K and ammonium phosphites | • *Pythium* spp.; better as a preventative rather than a curative. Yellow tuft. Often combined with metalaxyl, propamocarb, mancozeb, or chlorothalonil. Low resistance risk. |

(*Continued*)

**TABLE 12.7 (CONTINUED)**
**Turfgrass Fungicides Classified by Chemical Activity Group**

| Chemical Group (Activity) | Mode of Action | Common Name | Diseases Controlled |
|---|---|---|---|
| **Strobilurin** (or **Qols**) (curative and protective); antifungal compounds derived from various wood-decaying fungi such as *Strobilurus* and *Oudemansin* spp. | Inhibits spore germination and mycelial growth by blocking electron transport in the cytochrome bc complex, disrupting mitochondrial respiration, and ATP production. Acropetal systemic (azoystrobin), localized penetrant (trifloxystrobin); specific site action | Azoxystrobin, pyraclostrobin, trifloxystrobin, fluoxastrobin | • Brown patch; *Pythium* blight; summer patch; snow mold; anthracnose; red thread; "*Helminthosporium*" leaf spot; gray leaf spot; rust; take-all patch. Suppression of fairy ring caused by *Lycoperdon* spp. Minimum activity on dollar spot (Pyraclostrobin). Little turf phytotoxicity or undesirable growth regulatory effects occur. Broad spectrum of control. Long residual. Good root disease activity. Poor overall dollar spot activity. High resistance risk. |
| Biofungicides | One of: (1) antibiotic biosynthesis; (2) resource competition; or (3) hyperparasitism | *Bacillus licheniformis* SB3086, *Bacillus subtilis* GB03, *Bacillus* spp., *Burkholderia cepacia*, *Gliocladium catenulatum* J1446, *Pseudomonas aureofaciens* Tx-1, *P. chlororaphis* AFS009, *Streptomyces cacaoi* var. *asoensis*, *Streptomyces griseoveridis* (K61), *Streptomyces lydicus* WYEC 108, *Trichoderma harzianum* 1295-22, Rosemary oil, clove oil, thyme oil, wintergreen oil, butyl lactate, plus lecithin | • *Pseudomonas chlororaphis* bacteria are commercially available as a stand-alone or companion product with traditional fungicides for brown patch, *Pythium*, and anthracnose control. They interfere with the fungi life cycle, directly attack the fungi cell walls, and colonize their roots, competing for space and nutrient resources. They have a two-year shelf life, do not have to be refrigerated and are much less sensitive to UV light. |

*Note:* This is a partial list of trade names. Other fungicides are available alone and as mixtures under various trade names.

**FIGURE 12.7** Anthracnose leaf blight selectively attacking annual bluegrass in a creeping bentgrass green.

persists. Commonly, this occurs after flowering of annual bluegrass during heat stress. Lower leaves of plants initially show symptoms as elongated reddish-brown lesions that may enlarge and eventually encompass the entire leaf blade. The disease occurs as irregularly shaped patches. A water-soaked black rot of crown tissue may be evident if sheath tissue is removed (anthracnose basal rot). One-celled, crescent-shaped conidia are produced. The black fruiting bodies of the fungus (acervuli) can be seen growing in rows on infected leaves during hot weather. These acervuli have black spines called setae protruding from leaf tissue (Figure 12.8). Affected turf areas may thin and have a yellow- to orange- or reddish-brown cast to the overall area.

**Cultural controls.** Utilize cultural practices that minimize stress to cool-season grasses such as reducing soil compaction, heavy traffic, and provide adequate N fertility. Most damage occurs on annual bluegrass under heat stress. Irrigation based on 80% daily evapotranspiration rates provides best annual bluegrass health with least disease pressure. Begin a solid-tine aerification or slicing program in late spring, just prior to traditional heat-stress periods. Failure to aerify at this time may enhance anthracnose in summer. Raise the mowing height, water to prevent wilt, and topdress to help protect crowns. Double-cutting and rolling instead of relying on lower mowing heights for ball roll provides a healthier plant more easily able to withstand disease pressure. Adequate N fertility will help reduce the severity of anthracnose leaf blight in some areas during cool periods. For example, 0.1 pound soluble N/1,000 square feet (0.49 g N/m$^2$) every 7 days or 0.125 pound N/1,000 square feet (0.61 g N/m$^2$) every 14 days can reduce disease severity up to 25% compared to 28-day treatment intervals. Plant growth regulators do not appear to enhance disease occurrence. Fungicides will be needed in addition to light N applications during summer to prevent turf stand loss.

**FIGURE 12.8** Anthracnose black fruiting bodies called setae protruding from affected leaf tissue.

**Chemical controls.** Applying systemic fungicides preventatively on 21-day intervals starting at least 30 days prior to normal disease onset is much more effective than curative strategies. For best curative control, in a tank, mix a systemic fungicide with a contact and reapply in 14 to 21 days. Fungicides should be applied in 2 gallons of water per 1,000 square feet (82 mL/m$^2$) using nozzles that produce medium to coarse droplet sizes.

## Anthracnose Basal Rot

The causal agent of anthracnose basal rot is *Colletotrichum cereale* Manns sensu lato Crouch, Clarke, and Hillman.

**Occurrence.** Increasingly, anthracnose basal rot is occurring in stands of creeping bentgrass, especially 'Penncross,' 'Pennlinks,' and 'Dominant,' and/or annual bluegrass. Generally, anthracnose basal rot of affected annual bluegrass and bentgrass appears in summer, although cases have been noted in spring under conditions normally associated with good bentgrass growth. Presumably, the early appearance is from residual infected plant material from the preceding season. Usually, some stress factor is associated with the development of anthracnose basal rot. Stressful weather (heat, drought, or excessive rain) alone can spark the disease but is increasingly associated with putting greens under high management, which implies various combinations of traffic, low cutting heights, grooming, rolling, soil compaction, poor soil drainage, and low fertility. Mower cleanup laps often are the first areas to show disease symptoms. Mechanical damage from these practices typically intensifies the disease. In some cases, anthracnose basal rot has developed in bentgrass after it has been affected by localized dry spots. In addition to a stressed host, surface moisture film is needed for infection. Annual bluegrass is especially susceptible following seedhead production leaves the plant deficient in carbohydrates.

**Symptoms/signs.** The disease organism persists in turf as a saprophyte in thatch or infected plant material. The disease is initiated when a fungal spore penetrates a healthy grass plant, presumably in late spring into early summer. Hyphae then grow between plant cells at the crown region without symptom development as the fungus becomes semidormant until the plant is under stress. When stressed, the pathogen invades plant cells and starts killing them, triggering symptoms. At this time, initial symptoms are small (size of a dime) and chlorotic or even dark blue to purple. As it progresses, annual bluegrass turns yellowish with patches one to two inches (2.5 to 5 cm) in diameter. Bentgrass turf turns reddish-orange or bronze to

**FIGURE 12.9** Anthracnose basal rot on creeping bentgrass.

copper in color if warm weather persists (Figure 12.9). If left uncontrolled, spots coalesce, forming large, irregular areas of blighted turf. Crowns are affected, and leaf lesions are usually absent. Leaves die back from their tips, beginning with older leaves, turning orange or yellow. Healthy appearing, but infected, plants can easily be pulled up and their base or crown regions will be black and rotted, exhibiting a water-soaked appearance if the plants are only initially infected. It may occur in diffuse patches or in no particular pattern. Infection of leaves close to the soil surface typically produces "basal rot" symptoms where water-soaked, blackened tissue is easily pulled from the crown. As the disease worsens, the darkening (rotting) progresses up the stem (Figure 12.10). Usually dark, black masses of setae (black, hair-like projections seen with a 10× lens) and abundant sporulation in fruiting bodies called acervuli are noted. The spores from these acervuli spread the disease to neighboring plants. Eventually, irregularly shaped yellow to brown patches develop, ranging in size from 1/2 to 6 inches (1.3 to 21 cm) in diameter in annual bluegrass or up to 18 inches (46 cm) in creeping bentgrass. Healthy neighboring plants fill in bare areas, causing an uneven, unsightly appearance that disrupts ball roll. Basal rot infection inevitably leads to plant death.

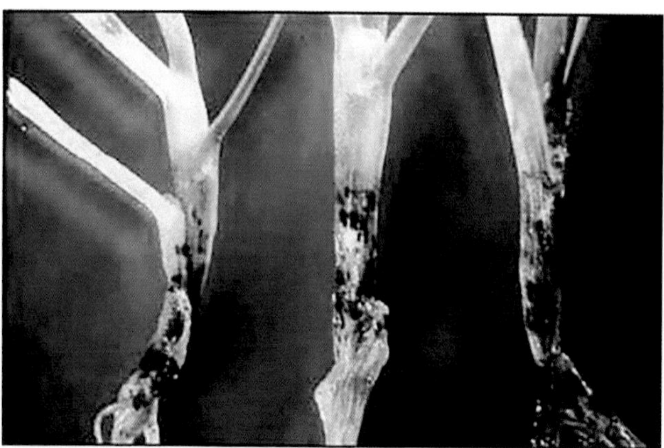

**FIGURE 12.10** Anthracnose basal rot symptoms of blackened tissue easily pulled from crowns.

**Cultural controls.** A major challenge of managing anthracnose basal rot is that infection occurs well before symptoms become apparent. Once the pathogen enters the plant, it remains in a semidormant state until the plant is stressed, when it then begins to kill plant cells. By the time symptoms develop, control of infected plants becomes extremely difficult. Therefore, using proper cultural practices to reduce stressful growing conditions for plants and minimizing excessive leaf moisture limits the time the pathogen attacks plants. Maintaining year-round tournament-like playing conditions is believed to be a major contributing factor to anthracnose basal rot outbreaks.

When the disease is active, avoid extremely low cutting heights and extremely low fertility; use floating head mowers and solid rollers, not grooved ones; and skip daily mowing cleanup laps. Research suggests sand topdressing in spring and summer reduces anthracnose, presumably by providing support to the susceptible crowns of annual bluegrass. Summer mowing heights should be raised to 1/2 inch (3.2 mm) or higher; sufficient yearly N (4 to 6 pounds N/1,000 square feet, 18 to 27 g N/m$^2$) and proper use of plant growth regulators are needed to produce a healthier, less disease-prone plant. Utilize cultural practices that minimize stress to cool-season grasses such as reducing soil compaction, heavy traffic, and moisture stress. To promote some growth and plant recovery, moderate applications of soluble ammonium sulfate or urea at 0.1 to 0.125 pound N/1,000 square feet (49 to 61 kg N/100 m$^2$) is suggested every 7 to 14 days as a tank mixture with fungicides. Irrigation should be provided to prevent wilt or development of localized dry spots. Some surfactants for prevention of localized dry spots may be beneficial. Do not mow when greens are excessively wet and change cups frequently to direct traffic from weakened areas. Syringe to prevent wilt and mow infected greens last to reduce disease spread. Also, avoid major club tournaments during peak disease period.

**Chemical controls.** Currently, using preventative and curative fungicide programs plus scouting for symptoms to catch the disease as early as possible appear best for management. Keep records on the specific environmental conditions under which the disease occurs, and then use this information as an application timing guide for subsequent years. Following first-year disease occurrence, areas infected the previous year can be targeted for preventative control. With anthracnose basal rot, affected plants generally die and recovery depends on surrounding, unaffected plants.

Apply effective fungicide products preventatively, typically from late spring through summer in areas of high disease potential. Under traditional heavy-pressure sites, preventative control should begin in spring prior to *Poa* seedhead formation. For best curative control, mix a systemic fungicide with a contact such as chlorothalonil in a tank, and reapply in 14 to 21 days. Other effective fungicides include polyoxin-D, tebuconazole, metconazole, and propiconazole. Resistance is common to QoIs, benzimidzoles, and DMI fungicides such as thiophanate methyl and the strobilurin fungicides. Using a multisite contact product helps reduce overall resistance potential and should be a component tank mix when high-risk, single-site products are used. Avoid using fungicides from the same chemical class in succession and be aware of the possible growth regulatory effects of sterol-inhibiting (DMI) fungicides.

## BACTERIAL DISEASES

### Bacterial Decline

The main causal agent of bacterial decline is believed to be *Acidovorax agrostis* subsp. *avenae*.

**Occurrence.** Reports almost exclusively concern intensively maintained turf—very high-profile golf courses with creeping bentgrass greens. Maintenance of those putting greens is most aggressive. Tolerance to damage—even cosmetic changes in color—is extremely low. Environmental stress is a significant component in all of the reports including heat, drought stress along with mechanical stress such as aggressive grooming, low mowing heights, and uncontrolled traffic.

**Symptoms/signs.** Numerous bacterial species are known to produce gibberellins in sufficient levels to cause etiolation (leaf elongation) (Figure 12.11). Etiolation leads to ragged appearing turf that easily scalps and is prone to wear and eventual stand decline. Leaf and internode etiolation often are followed by severe decline, leaf chlorosis, and weak, spindly growth, and seem to "melt" away as the disease progresses (Figure 12.12). Leaves turn yellow or a blotchy yellow-green and appear water soaked before dying (Figure 12.13). Bacterial streaming is observed in cut leaf tissue (Figure 12.14). Symptoms typically are first seen in summer, presumably due to heat and moisture excess. Disease patterns often follow a mower or topdressing pattern, presumably since these practices help transfer the disease and provide entry wounds into plants.

**Cultural control.** Control begins with reducing stress on bentgrass greens. Symptoms often occur in excessive or deficient fertility regimes, especially N. Raise the height of cut to

**FIGURE 12.11** Leaf elongation (or etiolation) due to gibberellins produced by bacterial diseases.

**FIGURE 12.12** Early symptoms of bacterial decline disease on creeping bentgrass.

**FIGURE 12.13** Creeping bentgrass thinning from bacterial decline.

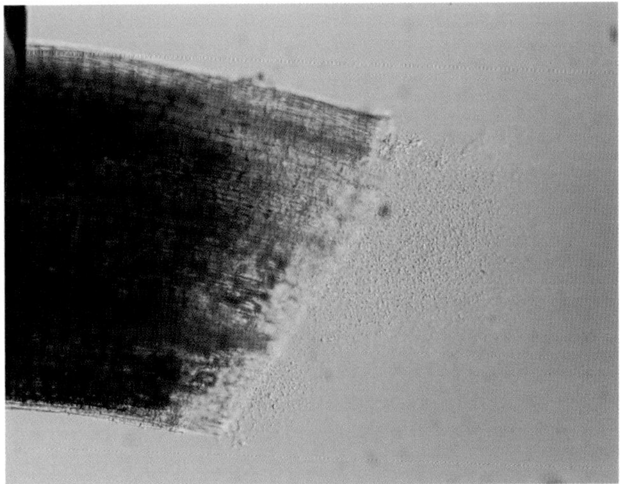

**FIGURE 12.14** Bacterial streaming from a cut leaf tissue. (Courtesy Bruce Martin.)

>0.15 inch (3.8 mm) to avoid scalping, switch to lightweight walk-behind mowers, reduce rolling frequency, mow when greens are dry, and use solid versus grooved rollers. Syringe turf to avoid soil moisture extremes. Avoid heavy topdressing during heat stress and disperse bunker sand blasted onto greens daily. Improve air circulation around greens. Plant growth regulator (PGR) applications help suppress leaf etiolation symptoms but this is controversial as some believe PGRs predispose plants to bacterial decline. Paclobutrazol and flurprimidol do not appear to be enhancing bacterial activity. Ammonium sulfate use also is often associated with disease symptoms. The A and G series of creeping bentgrass cultivars as well as L-93 have experienced the most bacterial damage.

**Chemical control.** Being bacterial in nature, fungicides have little effect on this disease. Symptom suppression may occur following applications of fosetyl-Al, chlorothalonil, or mancozeb.

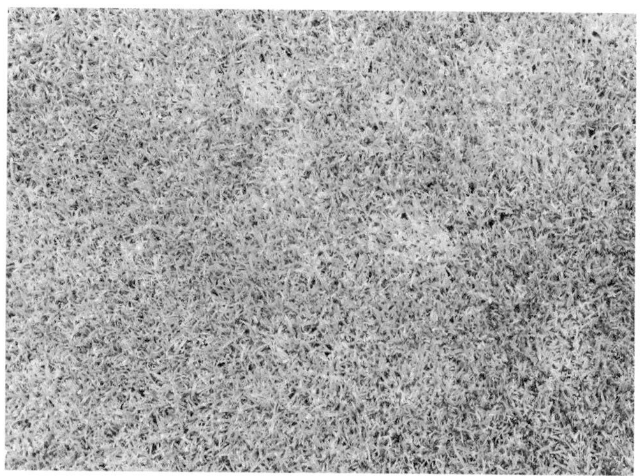

**FIGURE 12.15** Tissue etiolation and yellowing from bacterial decline.

## Bacterial Wilt

The causal agent of bacterial wilt is *Burkholderia* (*Xanthomonas*) *campestris* (= *Xanthomonas translucens* pv. *poae*).

**Occurrence.** These forms attack a number of grass species but are found most often on annual bluegrass and the vegetatively propagated creeping bentgrass cultivars Toronto C-15, Nimisilla, and Cohansey. The switch in the 1980s to sand topdressing programs was associated with several outbreaks of bacterial wilt in vegetatively propagated bentgrasses. The sand was thought to cause excessive wounds, allowing infection, or possibly the number of beneficial organisms in pure sand topdressing were too low to compete with the pathogen. Recently, bacterial wilt in annual bluegrass has been increasing in occurrence. Like anthracnose, plants severely stressed by shade, low mowing heights, aggressive grooming, soil compaction, underfertilization, and excessive wear are more susceptible to bacterial wilt. The disease is enhanced by extensive rainfall, followed by warm weather that induces wilting. Although not known why, bacterial wilt is often seen in spring following severe winter damage. Following a spring flush of disease, infected plants tend to recover until midsummer when temperatures allow optimum disease growth and increased plant stress. As temperatures decrease in fall, plants recover and the disease usually declines unless fall days are warm, causing disease resurgence.

**Symptoms/signs.** Bacterial wilt symptoms initially appear as tiny red- to copper-colored spots resembling dollar spot except for the unusual color. The disease grows best at about 86°F (30°C) but can be observed from midspring until early fall. It is most active in spring during periods of ample precipitation, though symptoms may not be seen until summer when infected plants are less likely to survive. Often etiolation (tissue yellowing or bleaching from the lack of photosynthesis) occurs, especially if plants are left unmowed for two or three days (Figure 12.15). It can be active whenever infected plants are growing. As the disease progresses, the spots become larger and portions of the green wilt as the bacteria clog the xylem tissue of the plants, given the green a speckled or pitted appearance. When cut and observed microscopically (≥100×), infected leaves ooze out streaming bacteria cells. Infected tillers of plants eventually die.

**Cultural controls.** Control of the disease in bentgrass involves switching from vegetatively planted turfgrass or annual bluegrass to seeded varieties with a wider genetic base for resistance. Avoiding plant injury through sand topdressing where the disease is present also helps as well as relieving other management stresses to the turf. It typically is most severe in mower cleanup (or perimeter) laps on greens. Increase mowing height and reduce its frequency. Mow when turf is dry, using lightweight machines. Avoid grooming, aeration, and topdressing during active disease symptoms. Spoon-feed N to encourage turf recovery.

**Chemical controls.** Being bacterial in nature, fungicides have little effect on this disease. The antibiotic oxytetracycline may provide temporary (four- to six-week) control. However, since bacteria enter plants by mower wounds, repeat treatments are necessary, which soon become cost prohibitive. In addition, the antibiotic can be very phytotoxic in warm weather and may cause more harm than good. Copper hydroxide, hydrogen dioxide, and mancozeb can reduce bacterial populations on leaves. However, since these products do not penetrate plants, only surface bacteria will be affected. These also need reapplication following mowing. If used preventatively, chlorothalonil + acylbenzolar-*S*-methyl and fosetyl-Al can reduce etiolation.

## Bermudagrass Decline/Take-All Root Rot

The causal agent of bermudagrass decline (aka take-all root rot) is *Gaeumannomyces graminis* (Sacc.) Arx and D. Olivier var. *graminis*. This disease organism occurs on many turfgrasses including bermudagrass, St. Augustinegrass, kikuyugrass, bluegrass, bentgrass, and seashore paspalum.

**Occurrence.** Bermudagrass decline is primarily observed during summer and early fall months when temperatures are above 75°F to 80°F (24°C to 27°C, including at night) with high humidity, cloudy skies, and frequent rainfall. This disease is mostly limited to putting greens due to the stress imposed by very low cutting heights combined with reduced sunlight from extensive cloudy weather, shorter days of late summer, cooler nights and, thus, slower bermudagrass growth and extensive rainy periods. Tees may also be affected if mowed excessively low. As a general rule, the outer margins of a golf green exhibit the disease symptoms first, presumably due to added pressure from turning mowers, but can spread slowly across an entire green (Figure 12.16). It occurs on all types of putting greens—old, new, poorly drained, and well drained. Tifdwarf bermudagrass often declines with continual mowing heights at 0.125 inch (3.2 mm) combined with the previously mentioned stress factors. The disease tends to be less severe with the ultradwarf bermudagrasses unless the mowing heights are below 0.125 inch (3.2 mm) for extended periods with the environmental stresses noted previously.

**Symptoms/signs.** This is a root rot disease; therefore, the plant is easily stressed for water and nutrients. By the time aboveground symptoms appear, pathogens have been active on roots for at least a few weeks and possibly months. Initial symptoms are irregular, off-color patches ranging in diameter from a few inches to a few feet. The fungus colonizes root cortex and vascular systems. Lower leaves will exhibit the aboveground symptoms first by becoming yellow and then brown (dead). Roots with initial symptoms will usually be thin and off-white in color with isolated black lesions. Eventually, roots will turn black and rot (Figure 12.17). Stolons and rhizomes may also have black lesions. Black strands of fungi (runner hyphae) are present on the outside of roots, as well as distinctive multilobed hyphopodia (specialized hyphae visible microscopically on root surfaces). Entire plants may die, resulting in an irregular thinning of grass and, if not controlled, bare patches may develop (Figure 12.18). This disease can be confused with *Pythium* root rot, which causes a general decline across the entire green or limited portions of the green. *Pythium* root rot does not usually result in plant death or develop black stands of fungi on the outside of roots.

**Cultural controls.** Bermudagrass decline is very difficult to control once it is established. Therefore, preventative measures that alleviate stress are the best methods for completely controlling the disease or at least decreasing the potential damage.

**FIGURE 12.16** Bermudagrass decline symptoms developing initially on outer margins of a green.

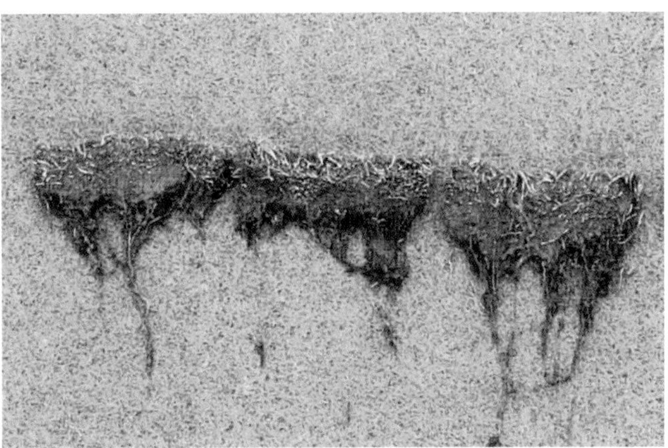

**FIGURE 12.17** Blackened roots from bermudagrass decline.

**FIGURE 12.18** Turf stand thinning from bermudagrass decline.

1. Raise the mowing height, especially during stressful growth periods during summer and early fall months. This is the most important preventative measure.
2. Aerify greens frequently to avoid compaction problems. Remove the cores.
3. Topdress after aerification with a topsoil mix containing at least 70% sand. More frequent topdressing may be necessary on putting greens where the disease has been previously observed. Topdressing also covers dying lower leaves, leaving the "appearance" of a healthy, green playing surface.
4. Balance N applications with equal amounts of K. On new sand greens, monitor other nutrients as well—especially P and micronutrients. Fertilize to encourage recovery. Remember that deeper roots are damaged; thus, shallower ones have to "work" for nutrients in the top few inches of soil. Nitrogen should be applied with K in a 1:1 or 1:2 ratio. Apply micronutrients, especially Fe and Mn, if they are in low supply or unavailable to the plant due to a high soil pH. Two to eight pounds Mn per acre (2.21 to 9.0 kg/ha) typically helps reduce symptoms. A readily available source of P may be useful. Foliar feeding of nutrients may be useful if the root system is severely damaged.

5. Do not raise soil pH by adding dolomitic lime to greens. If it is necessary to add Mg, use magnesium sulfate (Epsom salt) or, if Ca is needed, then consider calcium sulfate (gypsum). The disease is typically more severe in sites with neutral to alkaline soil pH, so the addition of lime may increase activity as the pH increases.

**Chemical controls.** Best control is achieved when fungicides are used preventatively, prior to symptom development. Do not use these excessively as they may have negative growth-regulating effects on bermudagrass. Treat in spring prior to infestations when daytime temperatures are <80°F (27°C). Apply in 2 gallons of water/1,000 square feet (8 L/100 $m^2$) and incorporate further with irrigation. Cultural control methods, especially raising the cutting height, should also be implemented in mid- to late summer before decline begins. These fungicides penetrate the plant, but are xylem-mobile (see fungicide section). The use of a contact fungicide such as mancozeb or chlorothalonil on the leaf tissue may also be useful to prevent secondary leaf infections from occurring and to prevent algae formation in thin areas of the green (see algae section). Fluxapyroxad plus pyrachlostrobin, tebuconazole alone, and trifloxystrobin plus triadimefon have shown suppression when used preventatively. Other options include mixing fosetyl-Al plus mancozeb in a tank, azoxystrobin alone and with difenoconazole. Tank mixtures are not watered in. If *Rhizoctonia zeae* is also present, do not use thiophanate methyl products, since it is not sensitive to this chemistry.

### Brown Patch (*Rhizoctonia* Blight or Large Patch)

The causal agent of brown patch is *Rhizoctonia solani* Kuhn.

**Occurrence.** Brown patch is the most widespread turfgrass disease. It affects all turfgrass species and is observed most often on creeping bentgrass, tall fescue, perennial ryegrass, and annual bluegrass during warm (>85°F, 29°C) weather when night temperatures are above 68°F (20°C) and foliage remains wet for extended (>10 hours) periods for several consecutive days. In cool-season grasses, infection is triggered by a rapid rise in air temperature combined with either rainfall or extended periods of high humidity, resulting in the leaf canopy being continuously wet for 48 hours or more. *Rhizoctonia* species are found in all soils and survive unfavorable environments as dormant, thick-walled mycelia or as compact masses of thick-walled cells (bulbils) resembling sclerotia in plant debris. There is some evidence inferred that infected but asymptomatic plants may become blighted when weather favors symptom expression and serves as primary inoculum for initiating new disease cycles. Strains of *R. solani* that cause brown patch in cool-season grasses are typically the subspecies groups AG-1 or AG 2-2 IIIB.

On warm-season grasses, large patch occurs in bermudagrass, zoysiagrass, seashore paspalum, St. Augustinegrass, and centipedegrass. The strain of *R. solani* (AG-2-2 subgroup 'LP') is most active as a pathogen in the spring or fall months when the temperatures are relatively cool and the turf surface stays continuously wet for several days due to extended rainfall or overirrigation. This disease in zoysia may be referred to as "zoysia patch."

**Symptoms/signs.** Symptoms of brown patch differ in cool-season grasses, depending on the height of cut. On putting greens, the disease begins as small, circular, light-green to dark-purplish patches that eventually turn yellow and then brown or straw colored (Figure 12.19). Patches may expand to several feet in diameter, and may or may not have green, relatively healthy appearing turf in the center of the patches. On lower heights of cut, a dark-gray, purplish, or dark-brown "smoke ring" may be visible, particularly under conditions of high relative humidity or when dew is on the green (Figure 12.20). This ring is composed of mycelium and freshly wilted, infected grass. These symptoms may change as the day progresses so that, by late morning, the patch appears more uniformly light brown, yellow-orange, or straw colored. On higher heights of cut, typical of roughs, leaf lesions are more apparent and a smoke ring symptom may not be visible, although mycelium may still be visible upon careful examination of affected leaves low in the turfgrass canopy. The turf may also appear "sunken." The later symptoms are typical in tall fescue (Figure 12.21). Brown patch is distinguished from summer patch in annual bluegrass by the characteristic leaf lesions on individual blades associated with brown patch. Lesions are irregularly shaped grayish-colored leaf spots with a dark-brown margin.

Large patch symptoms in warm-season grasses differ significantly from those observed in cool-season grasses. In warm-season grasses, discrete leaf lesions are not usually formed; rather, a soft, dark rot will occur on the lower portion of the leaf sheath (Figure 12.22). Whole leaf fascicles pull up easily due to this basal leaf sheath rot. Eventually, entire shoots will easily pull off the stolons. When disease is active, margins of patches exhibit yellowed shoots whose basal leaf sheaths have been recently infected and rotted. So, a yellow to light-brown band of more recently affected shoots surround the more brown patch of affected turf. Roots are not normally affected by this pathogen or those causing disease in

**FIGURE 12.19** Initial brown patch symptoms in creeping bentgrass.

**FIGURE 12.20** Dark-brown, "smoke ring" symptoms sometimes occur with brown patch disease.

**FIGURE 12.21** Sunken tall fescue from brown patch disease.

**FIGURE 12.22** Soft, dark rot on the lower leaf sheath from brown patch disease of warm-season turfgrasses.

cool-season turfgrasses. Roots may decline, however, as foliage is destroyed. Patches up to 20 feet (6 m) in diameter may develop on bermudagrass, zoysiagrass, seashore paspalum, St. Augustinegrass, or centipedegrass during cool, wet weather and shoot green-up in the spring (Figure 12.23). Leaf sheaths of plants at the perimeter of these patches are dark brown or black. As temperatures warm, turf vigor increases and eventually grows over diseased areas, but cultivars or species with lower turf recuperative potential (like Meyer zoysia, or centipede) may still exhibit symptoms into the summertime, and

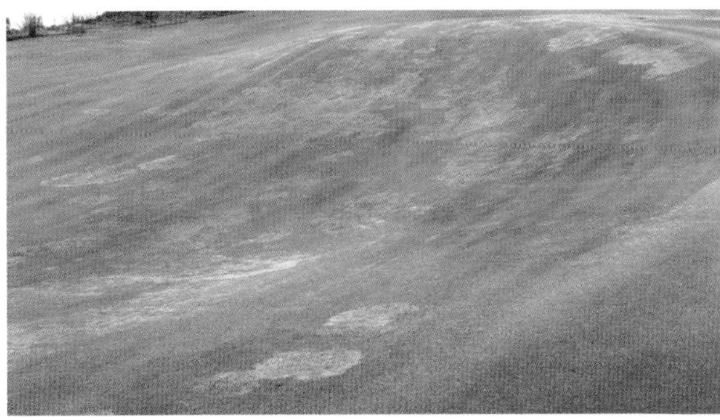

**FIGURE 12.23** Large patch symptoms on zoysiagrass during spring green-up.

weeds typically invade these weak areas of turf. These large patches are perennial in nature, and may reappear in the same location in following years, but expanded in size.

**Cultural controls.** For cool-season grasses, avoid excess N, especially readily available forms such as soluble liquids or quick-release N sources, just prior to hot, humid weather. For bentgrass greens, use very low rates and adopt a "spoon-feeding" approach to avoid stimulating succulent foliage, but still maintain the ability of bentgrass turf to recover from disease. Complete elimination of N fertility, even in the summertime for bentgrass greens, is not advised, as new root- and shoot growth is needed for plants to recover from disease and (hopefully) transient highly stressful high temperatures. Maintain adequate levels of P and K. Avoid excessive irrigation, and irrigate greens when dew is already present so leaves do not stay continuously wet. Dragging a hose across the turfgrass or whipping greens with a bamboo or fiberglass pole will also remove morning moisture. Increase air circulation by removing adjacent underbrush and consider tree removal to improve morning sunlight penetration to the green. The use of fans for bentgrass greens promotes healthier grass and reduces disease pressure. Remove clippings on infected areas or when conditions favor disease development. Use sharp mower blades to reduce turf stress and excessive wounding. Regularly aerify the core to increase soil drainage, improve soil oxygen status, and reduce thatch buildup. Increasing mowing height also encourages turf recovery.

For brown patch management in warm-season grasses, avoid early fall applications of excess N. Also, avoid early spring N applications until sustained spring turf growth occurs. Improve drainage, as it has been observed that large patch chronically appears in poorly drained areas. On golf course roughs where centipedegrass has been utilized, large patch may become severe as superintendents water to establish fall overseedings of ryegrass in fairways. This may be an argument against using centipedegrass for this purpose. Managing thatch accumulations will help to improve recuperative potential as well. Applications of a silica source such as calcium silicate or potassium silicate may reduce the incidence of brown patch (and dollar spot). Although not a stand-alone product under heavy disease pressure, silica additions may help lessen these diseases under mild infestations.

**Chemical controls.** Preventative control (such as fall application for warm-season grasses) in chronic disease areas helps prevent spring disease symptoms. For creeping bentgrass in summertime, preventative applications are necessary to allow conservation of healthy foliage for maintenance of root systems and hence the turf stand. During other times of the year, when favorable weather for epidemics is likely to be of short duration (spring, winter, and fall), curative approaches are successful. Apply on the appropriate schedule during hot, humid weather when night temperatures exceed 68°F (20°C). Waiting for symptoms to develop before chemical control, however, may result in dead or thinned areas of turf unable to recover until favorable turf-growing conditions resume in fall. In summer, algae development becomes problematic under these conditions.

### *Rhizoctonia* Leaf and Sheath Spot (Also Called "Mini Ring")

The causal agents of *Rhizoctonia* leaf and sheath spot are *Rhizoctonia zeae* Voorhees and *R. oryzae* Ryker and Gooch. Mini ring of bermudagrass is also thought to be caused by *Rhizoctonia zeae*.

**Occurrence.** *Rhizoctonia* leaf and sheath spot occurs primarily during the warm late spring, summer, and early fall months, especially when nighttime temperatures exceed 77°F (25°C) and humidity is high. However, it has also been observed in winter months after a sustained, high-temperature period (>80°F, 27°C) with high humidity. Another environmental factor associated with secondary infection is a dry rootzone as found with localized dry spots. The *Rhizoctonia* species infect by mycelial growth arising from infested plant debris or from sclerotia or thick-walled cells that survive in thatch or soil. These fungi rarely form spores, and therefore spread primarily by mycelial growth or movement through mechanical means.

**Symptoms/signs.** Lesions may be present on leaves and leaf sheaths. Normally, the disease appears as a total blight of leaf tissue, resulting in a reddish-yellow coloration of leaves. On creeping bentgrass greens, symptoms can include yellow rings similar to those of yellow patch disease, but also can resemble brown patch symptoms, including the formation of a faint smoke ring symptom in high humidity (Figure 12.24). Sometimes, a creamy visible foliar mycelium occurs as well. Individual infected leaves have gray or tan irregular lesions with tan or brown borders. Warm-season grasses rarely exhibit lesions on upper leaf blades as do cool-season grasses, but lower leaves are blighted. Roots are not normally affected by these pathogens. Two types of overall "patch" symptoms have been observed. One is large areas of chlorotic (yellow) turf that never turn brown. The symptoms most often observed are necrotic (brown) rings of all sizes (6 inches to 6 feet diameter, 0.2 to 1.8 m)—ranging from full circles to semicircles to quarter-circles. Rings often appear sunken, are never dark green as with true fairy rings, and they do not normally expand in size, only in number. On bentgrass greens, symptoms may resemble those of yellow patch except that yellow rings or arcs are present in summertime.

On bermudagrass, mini ring consists of frog-eye (centers still green) circular patches roughly 4 to 18 inches (10 to 46 cm) in diameter that typically develop during hot, humid days of late summer and early fall when bermudagrass growth has slowed (Figure 12.25). Centers of patches often have more intense green color than the perimeters. Disease patches often are on higher, drier parts of greens, lacking sufficient N.

**Cultural controls.** Control information is quite limited. As with all turfgrass diseases, maintain an adequate and balanced fertility regime, such as slow-release N sources. This is especially important for prevention and recovery of the mini-ring symptoms in bermudagrass. In fact, if fertility is kept low, experience has shown that fungicides are largely ineffective once symptoms occur. Irrigate only when dew is already present on the turf and irrigate the entire rootzone. Do not allow the turf to become drought stressed, especially putting greens. In other words, avoid having a dry root system when the leaf canopy is wet due to high humidity.

**Chemical controls.** Consistent disease control has been somewhat less successful with the fungicides currently registered for *R. solani* (brown patch/*Rhizoctonia* blight). Best control to date has been with strobilurin fungicides and SDHI fungicides, especially flutolanil. Polyoxin-D has been shown to have some efficacy although residual control is short. Some DMI fungicides, such as tebuconazole, have good activity, but on bermudagrass, use only low rates and long intervals of application due to growth-regulating tendencies. Do not

**FIGURE 12.24** Rhizoctonia leaf and sheath spot of creeping bentgrass. (Courtesy Bruce Martin.)

**FIGURE 12.25** Mini ring symptoms on ultradwarf bermudagrass putting green.

use fungicides containing benomyl or thiophanate methyl as *R. zeae* and *R. oryzae* are not sensitive to this group of fungicides.

## Yellow Patch

The causal agent of yellow patch is *Rhizoctonia cerealis* Van der Hoeven.

**Occurrence.** Yellow patch, also referred to as cool-season brown patch, is primarily a problem on cool-season grasses, especially bentgrass and annual bluegrass. It also occurs on Kentucky bluegrass, zoysiagrass, ryegrass, rough bluegrass overseedings, tall fescue, and bermudagrass. It is favored by cold (50°F to 68°F, 10°C to 20°C), wet conditions from fall to early spring.

**Symptoms/signs.** On bentgrass greens, irregular, narrow, yellow rings ranging from several inches to several feet in diameter appear in late fall through winter when cool, humid (or wet) conditions occur (Figure 12.26). Ring symptoms persist under these conditions while bentgrass has slow growth. Often patches appear sunken due to rapid decomposition of the thatch layer. Individual tillers appear yellow in color, with crowns and lower portions of leaves sometimes red in color. Leaf lesions are rare compared to brown patch (*R. solani*) on cool-season grasses. Once ring symptoms appear, fungicides become relatively ineffective. Preventative applications are needed before rings appear in order to control symptoms.

**Cultural controls.** Avoid excess N, especially readily available forms of N such as soluble liquids or quick-release N sources. Do not force grasses to grow in winter with excessive N. Irrigate greens when dew is already present so leaves do not stay continuously wet. Increase air circulation. Remove clippings and avoid thatch buildup. The disease survives periods of unfavorable growth as minute, dark-brown to black bulbils and as mycelium in the thatch and plant debris.

**Chemical controls.** Curative control has been difficult with this disease, and symptoms generally subside once good growing conditions resume with warmer weather. Propiconazole, flutolanil, azoxystrobin, and pyraclostrobin provide preventative control.

**FIGURE 12.26** Yellow patch on creeping bentgrass golf green. (Courtesy Bruce Martin.)

**FIGURE 12.27** Copper spot small reddish lesions.

## Copper Spot

The causal agent of copper spot is *Gloeocercospora sorghi* Bain and Edgerton ex Deighton.

**Occurrence.** Copper spot (also called zonate leaf spot) can be a serious disease during periods of moderately warm, wet weather. While not as widespread as most other turfgrass diseases, it can be destructive to creeping and velvet bentgrasses, and less frequently to bermudagrass and zoysiagrass. Disease development initiates when temperatures reach 69°F (21°C) and is favored by temperatures above 80°F (27°C).

**Symptoms/signs.** Leaves initially have small reddish lesions (Figure 12.27). The spots are salmon-pink to copper-red in appearance and range in size from 1 to 3 inches (2.5 to 7.6 cm). Lesions may spread rapidly to blight entire leaves. The disease can be easily confused with the early stages of *Pythium* blight if patches are copper colored. Overall symptoms resemble dollar spot disease except for the distinct copper- or pink-colored patches associated with copper spot. During wet weather, leaves may be covered with pustules of gelatinous, pink spore masses that turn bright orange when dry. Also, very tiny black sclerotia may be formed in necrotic portions of leaf lesions. When active, sclerotia germinate from sporodochia. Conidia (spores) are produced in the sporodochia and germinate to produce hyphae that quickly penetrate leaves in warm, moist weather. Conidia are spread by splashing, shoes, and turf equipment. The fungus survives in host debris as sclerotia and as thick-walled mycelium in thatch.

**Cultural controls.** Turf normally slowly recovers following cool, dry weather. To prevent spread, allow greens to dry before mowing. Dragging or poling greens hastens this drying. Avoid quickly available N sources when disease is active. Liming may help reduce disease severity if soil pH is low.

**Chemical Controls.** Chlorothalonil, iprodione, thiophanate, and triadimefon provide good control. Apply once or twice at 14- to 21-day intervals.

## Dollar Spot

The causal agent of dollar spot is *Sclerotinia homoeocarpa* F. T. Bennett.

**Occurrence.** This disease is favored by low soil N levels, close mowing, micronutrient deficiency, dry soil, heavy dews, soil compaction, and excessive thatch accumulation—characteristics of most modern golf greens. It occurs during periods of warm, humid weather when heavy dews occur. All turfgrasses are susceptible, with creeping bentgrass, ryegrass, Kentucky bluegrass, annual bluegrass, and roughstalk bluegrass (*Poa trivialis*) being most susceptible for cool-season grasses and bermudagrass and zoysiagrass for warm-season grasses. Activity begins at 55°F (13°C) and is optimum between 70°F and 84°F (21°C and 29°C), especially when free moisture (dew or >85% relative humidity) is present for 10 hours or more. A forecasting model is based on disease activity beginning when relative humidity levels are >70% for five days. Although it is a very persistent disease, dollar spot does not normally cause total kill of a turf area but, if left unchecked, it can spread to epidemic proportions, leaving a playing surface scarred and pitted.

**Symptoms/signs.** Small (2-inch, 5 cm, diameter), bleached patches of dead grass on low-cut turf such as golf greens will develop first (Figure 12.28). Irregular, light-tan lesions with distinct brown borders will be present on individual leaf tissue at the outside edge of the patch. Leaves may be girdled and collapse at the lesion even though leaf tips remain green. In contrast, lesions caused by *Pythium* fungi generally are water soaked in appearance, feel greasy to the touch, and do not have distinct borders around the bleached diseased leaf tissue. In taller turfgrass, 4- to 6-inch (10 to 15 cm) patches of mottled, straw-colored turf often occur. Lesions on taller mowed grass often die back from the tip and have straw-colored or bleached lesions shaped like an hourglass (Figure 12.29). White, cottony mycelium may be observed in early morning hours when dew is present and can easily be confused with early stages of *Pythium* blight or the webs of spiders. Spider webs, however, are flat, have a web pattern, and do not cause leaf lesions, while mycelium from dollar spot or *Pythium* blight is three dimensional.

Dollar spot fungi do not form spores; rather, they spread from mycelial growth and movement of infected plant parts, infested equipment, or traffic. Dollar spot fungi survive as dormant mycelium in plant parts and as thin flakes of fungal tissue on foliage or in soil. Disease often develops earlier in spring where it was not adequately controlled the previous fall. Spots in sod-forming grasses, such as bermudagrass and bentgrass, usually disappear once the disease is controlled; however, spots in bunch-type grasses, such as ryegrass, often remain due to the grass's inability to fill in damaged areas. Severity usually peaks in late spring and again in late summer on cool-season grasses when night temperatures are cool enough to allow heavy early morning dew formation and high humidity in the turf canopy. Dry soils also favor dollar spot. Leaf wetness periods of 10 or more hours are conducive for severe dollar spot outbreaks on putting greens cultured to cool-season grasses. On warm-season grasses, dollar spot occurs all season long on underfertilized turf.

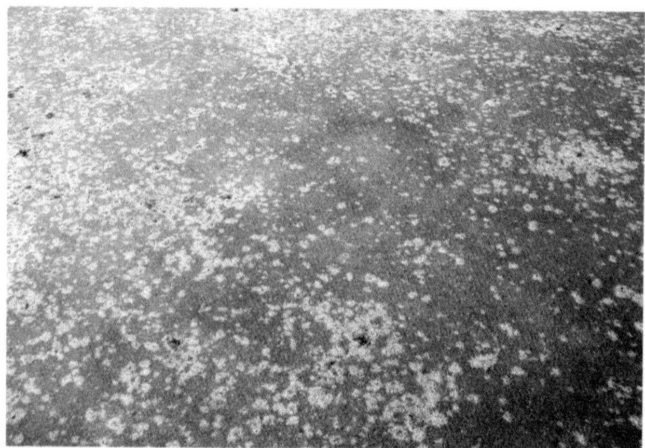

**FIGURE 12.28** Severe dollar spot disease on creeping bentgrass.

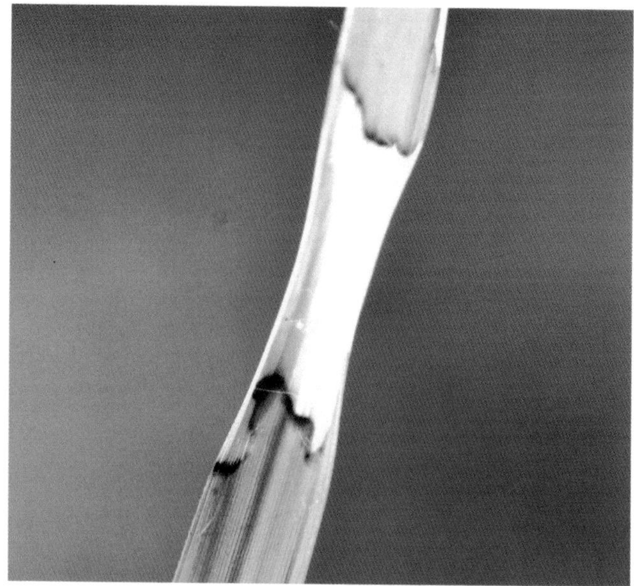

**FIGURE 12.29** Hourglass-shaped bleached lesion from dollar spot disease.

**Cultural controls.** Plant resistant cultivars, blends, and mixtures of various grasses whenever possible (Table 12.4). Avoid extreme N or soil moisture deficiencies (Table 12.5). If the disease develops, apply a quick-release source of N such as ammonium sulfate; symptoms will subside, although control is not sustained as long as with some fungicides. Do not use this approach with creeping bentgrass in heat-stress environments as quick-release fertilizers may cause excessive, succulent growth, promoting heat- and moisture-stress problems. Organic fertilizers may help delay dollar spot in early season, but do not adequately reduce its severity when disease pressure is high. Maintain adequate P, K, micronutrients, and lime. Irrigate during early morning hours to limit periods of high humidity and remove leaf moisture by rolling, dragging, or mowing. Remove trees and brush to increase sunlight and wind movement, which help dry the turf surface quicker. Avoid thatch buildup by aerifying, topdressing, and verticutting. Applications of a silica source such as calcium silicate or potassium silicate may reduce the incidence of dollar spot (and brown patch). Although not a stand-alone product under heavy disease pressure, silica additions may help lessen these diseases under mild infestations.

Prevent disease spread by removing clippings from infected areas, washing equipment before entering a noninfected area, and encouraging golfers to clean their shoes between rounds. Raising the mowing height also reduces disease severity on golf greens.

**Chemical controls.** Most fungicides labeled for control are effective, especially when used preventatively. Contact fungicides provide 10- to 14-day preventative control while effective systemics provide 14- to 28-day control. Use higher rates only for curative control situations. Resistance has been problematic for the benzimidazole class of fungicides (including thiophanate methyl) and the sterol biosynthesis inhibitors (propiconazole, triadimefon, cyproconazole, myclobutanil) (Table 12.7). Once induced, resistance to these fungicides appears to be long lasting. Resistance has also developed in response to overuse of dicarboxamides, which include iprodione, or vinclozolin, but the pathogen populations become sensitive to these fungicides if they are not used for several years before being used once again. Isofetamid is a recent product with very promising preventative and curative dollar spot control.

Rotating and tank-mixing fungicides in tanks are necessary for managing disease resistance. For resistance-prone fungicides, do not apply products from the same chemical class (or family) two or more consecutive times. Tank-mixing a low resistance-prone contact fungicide with a systemic (or penetrant), as previously discussed, also reduces the chances of resistance. Another way of reducing the risk of resistance is rotating different single-site inhibitors with the multisite fungicide with each treatment. Strains of the fungus *Trichoderma* and several formulations of antagonistic bacteria appear promising as biocontrol agents when disease pressure is low.

Preventative control treatments should start in spring when nightly low temperatures are between 50°F and 70°F (10°C and 21°C). Repeat applications on 14- to 21-day intervals, depending on the products and rates selected. In addition, flat fan nozzles work best due to their even coverage from fine to medium spray droplet sizes. Air-induction nozzles produce a larger spray droplet and are better for windy conditions. Spray application volume should also be 1.5 to 2.0 gallons water/1,000 square feet (6 to 8.2 L/100 m$^2$).

## Fairy Ring

The causal agents of fairy ring include *Chlorophyllum, Marasmius, Lepiota, Agaricus, Amanita, Lycoperdon, Calvatia, Vascellum,* and over 50 other basidiomycetes (mushroom-, toadstool-, or puffball-producing fungi). The name fairy ring is from English folklore where the rings were believed to be where fairies had danced. There are records of fairy rings in Europe over 100 years old and up to several hundred feet in diameter. These fungi do not infect turfgrass plants, but change the soil properties that affect the growth and/or appearance of the turf, making these very noticeable.

**Occurrence.** Fairy rings are most common in soils with high organic matter, often where old tree stumps, lumber, and other organic debris were buried during construction, or they may occur in soils of very poor fertility. Newly constructed putting greens, in which the rootzone mix consists of sand and an organic amendment (such as peat), may develop severe infestations after only a few years or even months. The fungi live by decomposing organic litter such as thatch and plant debris. Fairy rings caused by *Lycoperdon* (a common puffball) may develop on 100% sand-based greens as well. Fairy rings are most frequently observed during late spring through summer months, presumably due to warm, wet weather, which favors fungal growth. In some instances, organic acids associated with certain species of fairy ring may cause/promote hydrophobic localized dry spot (LDS). However, fairy ring can occur under any soil condition supporting turfgrass growth.

Usually growth stimulation of the turfgrass occurs in the form of darker green rings or arcs (Figure 12.30). Grass stimulation associated with fairy rings is due to the increased availability of N, from decomposition of organic complexes in soil by the fungi or from decomposition of the fungi themselves as the rings expand outward. Grass inside a fairy ring is usually in a state of decline and frequently infested with weeds or algae. This decline is thought to be from depletion of nutrients; lack of soil moisture due to an impervious mat of fungal tissue at or near the soil surface; a toxic agent, such as cyanide-containing compounds produced by the fungus; or a combination of two or more of these. Fairy rings caused by *Marasmius oreades* rarely cross each other as the fungi produce compounds that inhibit the growth of other fairy ring fungi. Thatch collapse is a condition where sunken patches form from the natural decaying of lignin associated with thatch by basidiomycete fungi. Other superficial fairy rings may form but do not cause soil hydrophobicity.

**Symptoms/signs.** Fairy rings have been classified in several ways, depending on the symptoms and conditions induced. One classification describes them as belonging to two basic types: edaphic or lectophilic. Edaphic fairy rings are those induced by soil-inhabiting fungi and they occur as rings or arcs of green stimulated turf, which may or may not be accompanied by adjacent areas of dead or declining grass. A mat of white to cream-colored fungal mycelium may be present at or just below the soil line. This mat becomes very evident when a plug of grass is incubated in a sealed

**FIGURE 12.30** Darker green turf rings or arcs associated with fairy ring disease.

**FIGURE 12.31** Whitish fungal mycelium mat following incubating turf infected with fairy ring disease.

bag for two or three days (Figure 12.31). Soil beneath rings may become very dry and difficult to wet during summer and autumn. Edaphic fairy rings are more common in lawns and pastures. Three types of symptom expression are described for edaphic fairy rings. It should be noted that the same fairy ring fungi can cause any of these symptoms at certain times of the season or under varying environmental conditions:

*Type I rings*—These have a zone of dead grass just inside a zone of dark-green grass (Figure 12.32). They are more prevalent on new greens than on established ones. The dead grass may form from mushroom mycelia accumulating below the soil surface and cause soil to become hydrophobic (water repelling). Type I rings generally cause the most damage due to this soil drying, which prevents water from reaching plant roots, killing the turf or severely stunting its growth.

*Type II rings*—These have only a circular band of dark-green, rapidly growing turf, with or without mushrooms present in the band (Figure 12.33). On frequently mowed areas (greens and tees), mature mushrooms may never be observed but the "button" stage may be present at ground level. Turf is normally not killed. When conditions turn hot and dry, Type II rings can evolve into type I rings.

**FIGURE 12.32** Type I fairy ring with a zone of dead grass inside a zone of dark-green grass.

**FIGURE 12.33** Type II fairy ring consisting of circular bands of dark-green, rapidly growing turf.

*Type III rings*—These do not exhibit a dead zone or a stimulated dark-green zone, but simply have a ring of mushrooms present (Figure 12.34). Mushrooms or puffballs often develop after rains or heavy irrigation during mild weather. Rings normally expand each year. The size and completeness (circular, semicircular, quarter-circles) of the bands vary considerably (e.g., 1 to 100 feet, 0.3 to 30 m).

Type II and III fairy rings are more prevalent during cool, wet springs whereas type I fairy ring is more common in hot, dry summers. Lectophilic fairy rings (sometimes called "superficial" fairy rings) are those inhabiting thatch and upper soil surfaces. Lectophilic fairy rings tend to be less problematic, although, on putting greens, they can be unsightly and may still induce hydrophobic conditions that require treatment.

**Cultural controls.** Remove type III mushrooms because some (e.g., *Amanita*, *Chlorophyllum*) are poisonous, but these pose little threat to the turf and their mushrooms are easily removed by mowing. However, other edaphic fairy rings are very difficult to control. If necessary for aesthetic

**FIGURE 12.34** Type III fairy ring consisting of a ring of mushrooms.

purposes, mask the dark-green ring symptoms with N fertilizers (especially for type II rings). However, do not overfertilize cool-season grasses in summer as this may stimulate other, more severe problems. Fumigation may be used where sod in affected areas is removed, and the top 6 inches (15 cm) of soil are then mixed with a fumigant such as metam sodium or dazomet. Once the fumigant has dissipated, the area is resodded. Affected areas may also be excavated by digging up all infested soil in the ring area and 2 feet (0.6 m) beyond. Soil in the affected area is then removed to a 12-inch (30 cm) depth. Uncontaminated soil is then replaced and the area reseeded or sodded.

Although it is possible to excavate and fumigate fairy ring sites, it is quite likely rings will return if the food source is still present underground. In some situations on putting greens, type I fairy rings are also associated with localized dry spots in which the fungi have produced humic and fulvic acids that coat sand particles. In those situations, it may be useful to spike the area and use wetting agents to increase water absorption. Also, aerify and remove soil cores to allow better nutrient and water penetration. Infected areas should be heavily hand-watered or water injected, being careful not to overwater adjacent, unaffected areas, which could result in other diseases and problems. Do not bury and plant over organic debris such as tree stumps, large roots, and lumber left over during the construction and establishment phases of turf areas.

**Chemical controls.** Removing afflicted sod, followed by soil fumigation, tilling, and replanting, helps reduce fairy ring occurrence. Best results occur when fungicides are applied following aerification and irrigation to move them into the zone of fungus infestation. Also, the use of a wetting agent program to combat hydrophobic soil conditions is helpful. However, wetting agents alone will not control fairy ring fungi. By inhibiting the fungi with fungicides and providing better water infiltration and soil movement with wetting agents, symptoms can sometimes be suppressed for months. Pressure water injection of the fungicides and wetting agent combined is also helpful. Preventative control is best and should start in early spring when soil temperatures are 55°F to 65°F (13°C to 18°C). At least one repeat application will probably be required and should be timed 30 days following the first.

## *MICRODOCHIUM* PATCH (*FUSARIUM* PATCH OR PINK SNOW MOLD)

The causal agent of *Microdochium* patch is *Microdochium nivale* (Fr.) Samuels and Hallett (formerly, *Fusarium nivale* Ces. ex Berl. and Voglino).

**Occurrence.** *Microdochium* patch is probably the most important disease in the Pacific Northwest, although it is also important in northern areas of the United States, Canada, and Europe. It occurs during periods of cool (32°F to 60°F, 0°C to 16°C), wet spring and fall weather with or without snow cover. It is referred to as *Microdochium* patch, in the absence of snow cover, and as pink snow mold with snow cover. Alternating thawing and snow cover, repeated frosts, cold fogs, and light rain are particularly favorable for leaf-to-leaf spread of the fungus. Creeping bentgrass, annual bluegrass, Kentucky bluegrass, ryegrass, and fine-leaf fescue are all susceptible, with annual bluegrass and bentgrass being most susceptible and Kentucky bluegrass, tall fescue, and red fescue least susceptible. Although warm-season grasses are sometimes susceptible, occurrence of *Microdochium* patch is uncommon, but may occur on bermudagrass at or near full dormancy.

**Symptoms/signs.** When *Microdochium* patch occurs without snow cover, the spots are 1 to 8 inches (2.5 to 20 cm) in diameter and reddish-brown in color (Figure 12.35). At this stage, the disease can be easily confused with *Pythium* blight or copper spot. When the disease occurs under snow, spots are usually from 2 to 3 inches (5 to 7.6 cm) up to 2 feet (0.6 m) in diameter, and range in color from tan to whitish-gray or reddish-brown. Immediately after the snow melts, spots will initially appear water soaked and may have pinkish, salmon-colored mycelium present in the margins (hence, the common name), later turning to light tan in color (Figure 12.36). In the center of patches, grass collapses and is frequently matted and water soaked. If severe, spots may coalesce to form large areas of diseased turf. Grass on the outer edges of patches generally appears water soaked, with profuse gray- or pink-colored mycelium of the fungus present. The disease may spread by mycelial growth or movement of spores that are produced in enormous amounts on diseased tissue in fruiting bodies called sporodochia. The white- or salmon-pink–colored

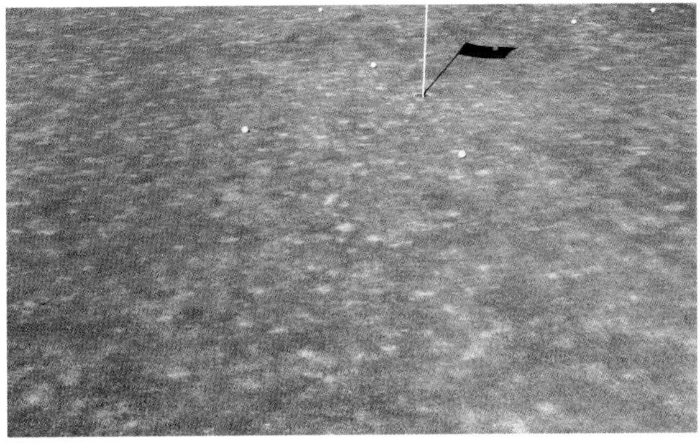

**FIGURE 12.35** *Microdochium* patch of semidormant bermudagrass.

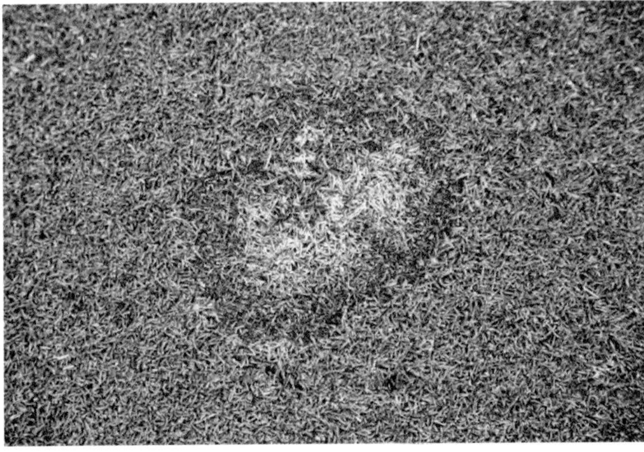

**FIGURE 12.36** Pink snow mold following snow melt.

sporodochia are very small and, with the aid of a hand lens, appear as flecks on dead plant tissue. These cannot be readily seen after the plant tissue dries. Spores are easily transported by water, machinery, animals, and foot traffic, and tracking patterns on greens are not uncommon. Conidia usually occur in orange-colored spore masses. It probably survives the summer months as resistant hyphae (mycelia) and spores in turfgrass debris. On dormant bermudagrass, under periods of high moisture, a grayish-colored patch can form. Permanent bermudagrass damage has not been seen. Unlike gray snow mold (*Typhula* spp.), pink snow mold does not produce sporocarps or sclerotia.

**Cultural controls.** Avoid late summer and early fall fertility that may lead to lush growth during cool, wet weather or under snow cover. Avoid excess spring N when *Microdochium* patch is a problem. Acidifying fertilizers such as ammonium nitrate and ammonium sulfate may aid in reducing soil pH and disease occurrence. Soil pH should be maintained below 7.0. Increase soil K in late fall to increase turf cold hardiness and to suppress the disease. Poor drainage and long leaf blades that mat down produce high humidity that favors disease development. Reduce shade and improve soil aeration and drainage. Remove tree leaves as the disease often develops under these, remaining on the turf for long periods during cold, wet weather. Use snow fences or plant landscape plants adjacent to golf greens to prevent excess snow accumulation. Avoid compaction during winter by preventing skiers and snowmobiles from being on greens and tees. If covers are used on greens for insulation, treat with a fungicide prior to installation of the covers.

**Chemical controls.** Contact fungicides generally provide the best control and should be applied before the first seasonal snowstorm. Other combination tank mixes with good suppression should be applied in late summer as a mix of a systemic fungicide and a strobilurin fungicide. This helps control *Microdochium* patch, dollar spot, anthracnose, and brown patch. Sequential treatments should be used in disease-prone areas during midwinter thaws. In warmer climates, a curative approach is generally successful due to the probability of more favorable weather for turf recovery in winter.

## Gray Leaf Spot (or Blast)

The causal agent of gray leaf spot is *Pyricularia grisea* (Cke.) Sacc. (= *Magnaporthe grisea* (Hebert) = *Pyricularia oryzae*).

**Occurrence.** Gray leaf spot (also called "blast") is a disease most often seen in St. Augustinegrass, tall fescue, and perennial ryegrass (Figure 12.37). Certain strains of the fungus also infect bermudagrass, kikuyugrass, crabgrass, foxtails, rice, corn, and barnyardgrass. It occurs mostly in late summer during prolonged hot (80°F to 90°F, 27°C to 32°C), rainy, humid conditions, and is generally more severe in newly established turf areas, shady locations, turf being heavily fertilized with N, and locations with poor air movement. Morning dew and high overcast (cloudy) conditions also are often associated with its occurrence, and higher mowed areas of perennial ryegrass, such as roughs and south-facing areas, seem most affected, presumably due to their increased canopy humidity. Six to 16 hours of wet periods favor infection. As temperature and humidity levels increase, a progressive buildup of disease occurs. Turf growth slowed by herbicides also becomes susceptible. The fungus forms spores (conidia) that are spread easily by wind and moisture or by maintenance equipment. Conidia germinate on host tissue when leaf surfaces are moist and daily air temperatures exceed 80°F (27°C). As temperatures and periods of high humidity increase, a progressive buildup of disease occurs. The disease overwinters as dormant mycelium in dead leaves; however, disease outbreaks tend to be less in years following colder winters. Conidia produced from these leaves serve as the primary inoculum early in the growing season and tend to initially occur in the same areas as the previous year. Conidia are then widely dispersed by maintenance equipment, golf carts, wind, and water. Therefore, in areas with typically cold winters, superintendents may manage the disease with fungicides. Courses located in warmer climates may need to convert to another turf species. However, thatch buildup and other diseases associated with these grass choices should carefully be weighed before use.

**FIGURE 12.37** Gray leaf spot of ryegrass.

**Symptoms/signs.** Infections may occur on all aboveground plant parts and begin as small olive-green to brown to tan water-soaked leaf spots (lesions) with a distinct brown to purple border or band surrounding the infected tissue (Figure 12.38). Lesions may become very numerous on leaves and individual spots may expand to consume leaves completely and to girdle stolons. Infected perennial ryegrass leaf blades

**FIGURE 12.38** Gray leaf spot lesions on St. Augustinegrass.

often are twisted or flagging and die back from the tips. Prolonged favorable weather for disease may leave turf with a scorched and ragged appearance that thins to resemble melting out or severe drought stress. However, unlike drought stress, a water-soaked and yellowing appearance of the leaf tips is first observed along with distinctive leaf spots. The circular spots may take on a grayish or grayish-brown appearance with purple to dark-brown borders and a yellowish halo. The lesions may resemble those caused by the *Drechslera*, *Bipolaris*, and *Curvularia* species. Advanced symptoms on perennial ryegrass commonly resemble *Rhizoctonia* leaf and sheath spot or *Pythium* blight where blighted leaves turn yellow, then brown, with characteristic twisted leaf tips. However, there are no foliar mycelium or smoke rings associated with these patches. Under conducive environmental conditions, such as 24 hours of continued leaf wetness and air temperatures above 85°F (29°C), spores germinate rapidly on the wet leaf surfaces and complete ryegrass stand loss can occur within 48 hours. This rapid decline causes confusion with *Pythium* diseases or may be misdiagnosed as a symptom of drought. Grass blades may also appear in early morning hours as felted or fuzzy, due to the massive production of spores on the lesions. It overwinters as spores and dormant mycelium in infected plants and in thatch. Disease is often along low-lying or poorly drained areas where prolonged leaf wetness occurs and on juvenile turf as late-summer seeded ryegrass. Spring use of ethofumesate herbicide to control annual bluegrass also has been associated with disease outbreaks.

**Cultural controls.** The rapid development and severe damage from this disease require appropriate monitoring and disease prevention steps. Avoid N fertility in wet summer months, which leads to lush growth. Modify irrigation practices to decrease the length of time leaves remain wet and to avoid drought stress. Improve air movement and light penetration in areas prone to chronic infections. Irrigate in early morning hours to promote maximum drying conditions during the day. Correct unfavorable turf management practices such as drought, compaction, and ill-timed herbicide use (e.g., spring applications of ethofumesate). Plant resistant cultivars when possible. Great strides have been made recently in this area. Replacing perennial ryegrass with another turf species such as bentgrass or Kentucky bluegrass mix is an alternative to fungicides.

**Chemical controls.** Azoxystrobin, chlorothalonil, propiconazole, trifloxystrobin, pyraclostrobin, and thiophanate methyl are used to control this disease. DMI fungicides (propiconazole and triadimefon) are not as effective, especially under heavy disease pressure. Resistance to strobilurin fungicides such as azoxystrobin, pyraclostrobin, and trifloxystrobin has been documented. An effective alternative is thiophanate methyl and combinations of thiophanate methyl and chlorothalonil or propiconazole and chlorothalonil. Apply as needed at 10- to 14-day intervals. Due to its potential explosive nature, early diagnosis is critical. If allowed to reach epidemic proportions, nearly all control strategies may fail. One strategy is to mix products with different modes of action in a tank during high disease pressure. Another strategy to reduce resistance potential is to rotate fungicides between different modes of action such as a strobilurin, then thiophanate methyl, and then a DMI fungicide plus chlorothalonil or mancozeb. Granular silica sources may also provide disease suppression. Like dollar spot disease, control of gray leaf spot faces fungicide expense, federal limitations on the use of chlorothalonil, and fungicide resistance.

## "*Helminthosporium*" Diseases—Leaf Spot and Melting Out

The causal agents of "*Helminthosporium*" diseases are primarily *Bipolaris* and *Drechslera* spp. (previously known as species of *Helminthosporium*). *Bipolaris* species tend to be more problematic on warm-season grasses while *Drechslera* species tend to occur on cool-season grasses (Table 12.8).

**Occurrence.** Leaf spot fungi previously belonged to the genus *Helminthosporium,* and they are still often referred to as "*Helminthosporium*" diseases. These fungi induce a variety of symptoms in many warm- and cool-season turfgrasses and attack all plant parts of turfgrasses. Thus, leaf spots, crown rots, and root, crown, rhizome, and stolon rots may occur, depending on the specific disease.

**TABLE 12.8**
**Common Names, Causal Pathogens, and Primary Turfgrass Hosts of Various "*Helminthosporium*" Diseases**

| Disease | Pathogen | Primary Turfgrass Hosts |
|---|---|---|
| Melting out | *Drechslera poae* (Baudys) Shoemaker | • Kentucky bluegrass, rough bluegrass, ryegrass, buffalograss |
| *Helminthosporium* leaf spot | *Bipolaris sorokiniana* (Sacc.) Shoemaker | • Kentucky bluegrass, bentgrasses, annual bluegrass, tall fescue, ryegrasses, bermudagrass, buffalograss |
| Red leaf spot | *Drechslera erythrospila* Paul and Parberry | • Bentgrasses |
| *Helminthosporium* blight (net blotch) | *Drechslera dictyoides* (Drechs.) Shoemaker | • Fescues, ryegrasses, Kentucky bluegrass |
| Zonate eyespot | *Drechslera gigantea* (Heald and Wolf) Ito | • Bermudagrasses, but also bentgrasses and Kentucky bluegrass |
| Stem and crown necrosis | *Bipolaris specifera* Nicot | • Bermudagrasses, zoysiagrasses |
| Leaf blotch | *Bipolaris cynodontis* Marignoni | • Bermudagrasses |

In general, leaf tissue as well as crowns, rhizomes, and roots may be affected. These diseases become severe under moderate temperatures in spring, fall, and summer under wet, humid conditions. Conidia (spores) of these fungi are abundantly produced in lesions and are dispersed by wind, water, and through dispersal of infested tissue removed by mowing. Crown and root rotting becomes more apparent in late spring and summer when stressful conditions eliminate severely infected plants, causing a "melting out" symptom common with *Drechslera poae* in Kentucky bluegrass. These fungi survive as dormant mycelia in thatch and leaf litter, on colonized plants, and as conidia.

**Symptoms/signs.** *Drechslera* species cause leaf spots during cool, humid conditions, especially following cloudy weather, with crown and root rot phases occurring during warm, dry weather or during wet periods following dry periods. Symptoms vary on different cool-season turfgrasses. Leaf lesions are generally distinct and begin when temperatures are between 70°F and 85°F (21°C and 27°C), and are tiny water-soaked areas that become dark brown to purplish-black (Figure 12.39). Lesions are usually surrounded by a yellow area of varying width that fades to the normal green of the leaf tissue. Older lesions may have a white or bleached area in their centers. Severely affected cool-season grasses may become almost entirely yellow in appearance. Leaf spots, crown rotting, and root rotting occur. On leaves, small, water-soaked lesions occur initially, are brown in color, or are a purplish color (Figure 12.40). Lesions on leaf sheaths similarly occur, but may encircle leaf sheaths and girdle leaf fascicles. When temperatures are above 85°F (27°C), lesions mature; the centers may turn tan or white in color, and yellow; and chlorotic tissue may surround the actual lesions. Microscopically, abundant, dark, multicellular spores borne on simple modified hyphae are associated with the lesion's necrotic tissue. Gradual browning and thinning occur over a period of weeks to months (Figure 12.41). As diseases progress, large, irregular areas turn yellow, then brown, and then thin out. Lesions on stolons, stems, crowns, rhizomes, and roots are dark purple to black. Crown/root rots will also occur at this time and infected plants lack vigor, often wilting even when adequate soil moisture is present.

**FIGURE 12.39** Leaf spot on zoysiagrass.

**FIGURE 12.40** Leaf spot lesions on ryegrass.

**FIGURE 12.41** Water-soaked lesions on leaves from leaf spot disease.

On closely mowed bentgrass, leaves from red leafspot turn reddish-brown or dark gray in irregularly shaped areas, causing a smoky blue appearance resembling drought stress. Lower (older) leaves tend to initially show symptoms, presumably to their slowed growth.

On bermudagrass, leaf spot is most pronounced in fall and early spring months when the grass is green but not actively growing due to cool temperatures. Leaf spot rarely causes permanent damage to bermudagrass but can cause streaking, browning, light-brown blotches, and stand thinning under severe infestations (Figure 12.42). Turf normally fully recovers when good grass growing temperatures return.

**Cultural controls.** Use resistant cultivars, blends, and mixtures of cool-season grasses whenever possible to alter the genetics of the host and prevent devastating epidemics. Balance N levels with K, and increase N rates to promote turf growth. Avoid drought stress and reduce leaf-surface moisture by deeply but infrequently watering. Avoid late afternoon and evening waterings and encourage good soil drainage. Encourage air movement and light penetration by removing shade sources and unneeded adjacent vegetation. Avoid thatch accumulation greater than 0.5 inch (1.3 cm). Raise mowing height between 1.5 and 2 inches (3.8 and 5 cm) for bluegrass to improve the survival of affected plants, but note this practice

**FIGURE 12.42** Browning and thinning of bermudagrass from leaf spot disease. (Courtesy Bruce Martin.)

may be dependent on the particular cultivars being used. Overuse of phenoxy herbicides (MCPP, 2,4-D, and dicamba) for broadleaf weed control may enhance disease development on cool-season grasses or if used when bermudagrass is not actively growing. Avoid using these pesticides or treat preventatively for "*Helminthosporium*" disease control prior to their use.

**Chemical controls.** Chronic problems with these diseases may be site specific, and preventative or curative control approaches should be used accordingly. Some disease enhancement has been noted with some of the sterol biosynthesis inhibitor fungicides and thiophanate-methyl. Preventative control is best. For cool-season grasses, begin applications in early spring after new growth is apparent and repeat at 7- to 21-day intervals until warm weather occurs. Curative control of the warm weather group such as leaf blight and melting out are often ineffective when symptoms are obvious since the fungus is well established and inaccessible in infected plant crowns, roots, rhizomes, and stolons. Fungicide applications in early to midfall are generally successful for managing leaf spot in bermudagrass putting greens.

## Necrotic Ring Spot

The causal agent of necrotic ring spot is *Ophiosphaerella (=Leptosphaeria) korrae* J. C. Walker and A. M. Smith.

**Occurrence.** A major disease of Kentucky bluegrass and annual bluegrass, along with summer patch, necrotic ring spot has a similar pattern of occurrence to yellow patch and summer patch. Undoubtedly, all of these distinct diseases have been confused due to the similarity of occurrence and symptoms in Kentucky bluegrass. Necrotic ring spot has been recorded on lawns and sports turf of Kentucky bluegrass, roughstalk bluegrass, annual bluegrass, and fine-leaf fescue. To date, necrotic ring spot occurs mostly in the northeastern, north central, mountainous western states, and Pacific Northwest.

**Symptoms/signs.** This is a root and crown rot caused by *Ophiosphaerella korrae*, one of the "ectotrophic" root-infecting fungi that destroy the plant's ability to absorb water, causing it to wilt and die more rapidly than healthy grass. This fungus colonizes roots and crown tissues, growing slowly and initially infecting cortical tissues, and eventually entire roots. Initial symptoms are thin, wilting, circular yellow to light-green patches of turf that range in size from a few inches up to a foot or more in diameter. Eventually, patches can expand up to 3 feet (1 m) in diameter, gradually turning brown or a straw color and die (Figure 12.43). Sunken, crater-like patches of this disease appear in spring or late summer in cool, wet weather. Patches may develop a "frog eye" appearance in which diseased turf occurs in circles or arcs, with surviving plants or weeds in the center. Individual plants exhibit brown to black, rotted roots and crowns, with varying degrees of chlorosis, necrosis, and tip die-back of shoots. Necrotic ring spot patches tend to be large (greater than 1 foot, 0.3 m, in diameter), often with the frog-eye symptoms, and mostly occur in spring and early fall (Figure 12.44). Summer patch tends to be smaller (less than 1 foot, 0.3 m) in diameter and occurs most often during summer-stress periods.

Distinctive leaf lesions are not associated with this disease, but leaves often die from the tip back. Plants at the edge of patches often have a bronze or copper color, while affected greens often have a spotted effect as infected annual bluegrass plants die out and adjacent bentgrass is unaffected. A dark-brown to black discoloration of lower shoots, stems, and roots is often associated with advanced stages of the disease and affected plants easily pull from soil.

These fungi survive as saprophytic mycelium and sclerotia in soil and as conidia. Sometimes, the sexual stage of the

**FIGURE 12.43** Necrotic ring spot of Kentucky bluegrass.

**FIGURE 12.44** Severe necrotic ring spot of Kentucky bluegrass with frog eye center symptoms.

fungus forms, a small, black, flask-shaped pseudothecium. They initiate in moist soil, thrive when temperatures are up to 80°F (27°C), and become more severe in higher temperatures and drought conditions. They are especially found in newly planted turf sights that were newly cleared woodlands. They are also found in compacted soils and areas high in N during spring and summer. Disease spreads by transporting affected roots, stems, crowns, and soil by maintenance equipment.

**Cultural controls.** As with most root diseases, control requires an integrated strategy involving cultural, genetic, and fungicidal use. Utilize watering practices to minimize drought stress and encourage deeper rooting. When disease has severely damaged the root system, it requires light and frequent (e.g., 0.1 inch/day, 25 mm) irrigation, most often during afternoon stress. Reduce soil compaction by aerifying and using lightweight maintenance equipment but do not aerify when the disease is active. Cutting heights can be raised to increase carbohydrate production and improve the chances of turf recovery. Utilize fertility practices to maintain a moderate soil pH of about 5.5 and utilize slow-release N sources. Using slow-release N forms consistently reduces disease symptoms in research trials. Differences also occur in susceptibility among Kentucky bluegrass cultivars; check with your local turf specialist for a recent listing of these. Plant a 75% to 85% blend of these and other resistant cultivars with 15% to 25% (by weight) of turf-type perennial ryegrass. Avoid herbicide use during hot weather.

**Chemical controls.** Preventative applications have shown efficacy if applied before symptoms develop (preventative) and repeated at 21- to 30-day intervals during the summer. A single midspring (e.g., April/May) application of one of the DMI fungicides has shown good control through late summer/early autumn. Adequate N and soil moisture are necessary or these fungicides are less effective.

## Pink Patch and Cream Leaf Blight

The causal agent of pink patch and cream leaf blight is *Limonomyces roseipellis* Stalpers and Loerakker.

**Occurrence.** Pink patch and cream leaf blight occur in cool- and warm-season grasses, including creeping bentgrass, fescues, ryegrasses, bluegrasses, and bermudagrass. Infections occur under nearly identical environmental conditions as red thread, and sometimes these can occur concurrently. Pink patch and cream leaf blight in bermudagrass putting greens occur in fall, winter, and early spring and infections appear to be largely superficial. The disease occurs, like red thread, under relatively cool, wet conditions on turf that is slowly growing or semidormant. However, abundant mycelium can develop on the lower leaves and cause unsightly pink to gray-white patches. It is often associated with the use of covers to prevent winter temperature damage. The causal organism is a basidiomycete fungus, and close microscopic examination reveals clamp connections (pink patch); however, those isolates associated with cream leaf blight symptoms lack clamp connections. Mycelium of the fungus usually occurs in "rhizomorph-like" aggregates of hyphae, giving a loose, ropy appearance when observed microscopically.

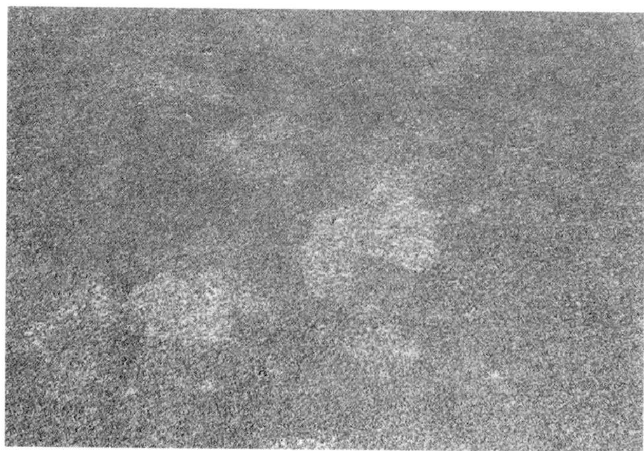

**FIGURE 12.45** Cream leaf blight on bermudagrass.

**Symptoms/signs.** Small, roughly circular patches appear in lower foliage of affected grasses (Figure 12.45). Patch coloration can be a reddish to tan color. The disease is slow moving; rapid spread of the disease upon observation of initial symptoms is rarely seen. Infected leaves appear chlorotic and tan in cool-season grasses. In bermudagrass, lower leaves are covered in mycelium and the turf may appear a darker green than surrounding unaffected turf. If bermudagrass goes dormant, patches have a distinct pink coloration.

**Cultural controls.** Fertilize turf adequately to promote even, normal growth. Maintain soil pH in the slightly acid range (pH 6 to 6.5). Prevent drought stress by supplemental irrigation when needed.

**Chemical controls.** Fungicides with good activity against basidiomycete fungi are recommended for pink patch control. Many fungicides labeled for red thread will also be labeled for pink patch. Because the infestations are in the turf canopy, spray as you would for any foliar disease and do not incorporate the treatments with irrigation.

## Powdery Mildew

The causal agent of powdery mildew is *Erysiphe graminis* DC.

**Occurrence.** Powdery mildew is a disease occurring under low-light intensities; therefore, it is mostly a problem on turf grown in shaded areas such as under trees or on the north sides of buildings. It can also occur in open areas, especially in fall and spring with consecutive cool, cloudy, humid days. Unlike most foliar blights or leaf-spot diseases, powdery mildew does not require wet foliage and usually does not cause severe damage. Cooler temperatures (55°F to 70°F, 13°C to 21°C) favor its development. Kentucky bluegrass is especially susceptible. *Erysiphe graminis* is an obligate parasite, obtaining nutrients from its host and only completing its life cycle in association with a living host.

**Symptoms/signs.** Turf appears to have a dull-white appearance and may be severely thinned. In severe cases the turf may take on a chlorotic (yellow) appearance or appear to have been dusted with ground limestone or flour. The fungus appears as a white to grayish-white cobwebby growth or

**FIGURE 12.46** Powdery mildew on Kentucky bluegrass.

**FIGURE 12.47** *Pythium* blight on overseeded perennial ryegrass.

powdery patch on upper surfaces of the leaves (Figure 12.46). The powdery appearance is from the production of numerous conidia (spores) in chains. Conidia are easily spread by wind, water, and turf-maintenance equipment. In older leaves, black cleistothecia (spore-bearing bodies) may be seen embedded in them.

The fungus overwinters in living, infected plants as ascospores in cleistothecia. Spores are produced in spring, become airborne, and only live for a few days, but can infect susceptible leaves over wide environmental conditions. Peak activity is generally when days are warm and nights are cool. Late summer and early fall are when the white covering of mycelium and spores occur.

**Cultural controls.** Use resistant cultivars and resolve shade issues, where possible. Increase air circulation and drying by removing underbrush, lower tree limbs and, if possible, selective trees. Plant shade-tolerant grasses such as creeping red fescue or other ground cover, increase the mowing height, and use a balanced N-P-K fertility program to reduce disease injury.

**Chemical controls.** When the turf stand has chlorotic plants and stand thinning, fungicides such as propiconazole and triadimefon may be justified. A single application usually is sufficient.

## *Pythium* Blight (Cottony Blight, Greasy Spot, Damping-Off, Seedling Blight)

The causal agents of *Pythium* blight are *Pythium aphanidermatum* (Edson) Fitzpatrick and other *Pythium* species such as *P. myriotylum* Drechs., *P. graminicola* Subrum., *P. arrhenomanes* Drechs., and *P. ultimum* Trow.

**Occurrence.** All turfgrasses are susceptible, with the cool-season grasses—creeping bentgrass, annual bluegrass, Kentucky bluegrass, and perennial ryegrass—being the most susceptible. Although this disease is frequently observed on bermudagrass putting greens, the disease is less severe when compared to *Pythium* blight on cool-season turfgrasses. With all cool-season grasses, disease can be especially severe on newly established stands (Figure 12.47). In cool-season grasses, it is most likely to occur when day and night temperatures exceed 85°F and 68°F (29°C and 20°C), respectively; when the relative humidity is high; and foliage wetness occurs more than 10 hours daily for several consecutive days. *Pythium* blight can be a highly destructive disease. Disease occurrence on bermudagrass greens is most commonly in fall, winter, and spring; it occurs when temperatures are relatively cool but when excessive moisture provides conditions favorable for outbreaks.

*Pythium* species are not true fungi but fungus-like water molds classified in the Stramenopiles; thus, they require sufficient surface moisture for development. All soils contain *Pythium* species unless the soil has recently been fumigated. *Pythium* species causing *Pythium* blight are facultative parasites, meaning they are saprophytes in thatch or soil and quickly become pathogenic when favorable conditions develop. The organisms produce thick-walled sexual spores that survive for long periods in the soil. The germination and "grow-in" periods of newly established bentgrass greens or overseeded cool-season grasses are ideal for disease development since fall temperatures can still be quite warm and the turf is being irrigated more frequently than normal. Extended periods of warm, foggy mornings also favor this disease, potentially destroying large areas in a short period of time on cool season grass hosts. On creeping bentgrass, conditions may be favorable for relatively long periods of time during summer, depending on location. Greens situated in low, protected areas with poor sunlight penetration and air movement are most at risk. Bentgrass is more susceptible when saline soil conditions (and presumably high-soluble salts from fertilizers) exist.

**Symptoms/signs.** Small, distinct reddish-brown patches of grass, usually about 1 to 6 inches (2.5 to 15 cm) in diameter, first appear dark and water soaked (slimy), but later shrivel and turn straw colored as humidity and/or temperature decreases. In some cases, after heavy rains and prolonged high relative humidity, the first sign of disease is a white, cottony mycelial growth (Figure 12.48). Turf grown in shaded, low-lying areas adjacent to water where air circulation is poor and humidity highest is likely to become diseased first. Water-soaked, blackened leaves often feel greasy or slimy. If conditions become less humid during the day, bleached lesions can be observed on partially damaged leaves, especially on bermudagrass and tall fescue. However, these lesions have no distinct border, which easily distinguishes them from dollar spot.

**FIGURE 12.48** White cottony mycelial growth associated with *Pythium* blight.

*Pythium* blight can initially be confused with both copper spot and *Microdochium* patch on creeping bentgrass in the fall. Patches may spread quickly in a "streak" pattern, usually following water drainage movement (Figure 12.49). *Pythium* fungi is spread by direct mycelial growth as well as through a spore or spore-containing sack called sporangia and small swimming spores released from sporangia called zoospores.

**FIGURE 12.49** Streak pattern of *Pythium* blight indicating surface water drainage patterns.

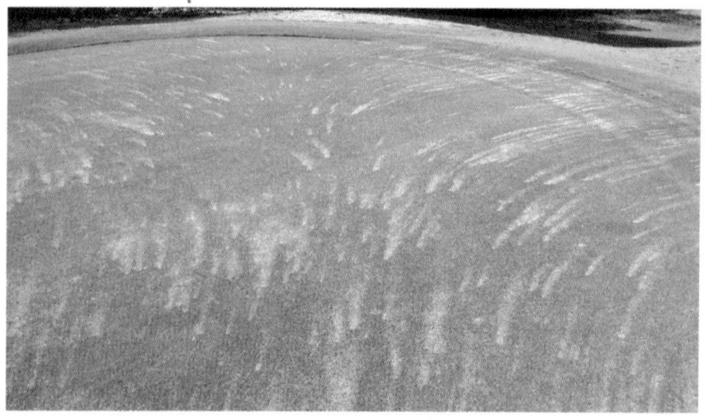

**FIGURE 12.50** *Pythium* blight on ultradwarf bermudagrass golf green. (Courtesy Bruce Martin.)

These fungi also produce thick-walled resting spores called oospores that may survive in soil and thatch for extended periods. Disease may spread rapidly when sporangia, zoospores, oospores, or infected plant parts are moved by water along drainage patterns, or by mowers or traffic. Often, two or more *Pythium* species may simultaneously occur, requiring growing the fungus on special laboratory media and microscopically examining the sporangia and oospores for specific identification. Being water molds, *Pythium* spores can survive in ponds used for irrigation, which makes control difficult as the turf is reinoculated with each watering. Although less common on bermudagrass, lower mowing heights and ultradwarf cultivars have seemingly led to an increase of *Pythium* on bermudagrass golf greens (Figure 12.50).

**Cultural controls.** Improve drainage, air circulation, and light penetration by removing shrubs, trees, and limbs and possibly add fans in pockets of poor air movement. Reduce soil and leaf moisture, as moisture control is a key to *Pythium* management. Therefore, early morning removal of dew and guttation water by poling or dragging a hose and utilization of early morning irrigation is beneficial. Plant fungicide-treated seed at the recommended amount, avoiding very high seeding rates. Avoid excessive use of N prior to warm weather. Minimize equipment or foot traffic across wet infected turf. Wash equipment before entering unaffected areas and encourage golfers to clean their shoes between rounds. Delay overseeding in late summer/early fall until the weather turns cool and dry. Alleviate soil compaction as this reduces turfgrass rooting, which then requires turf managers to irrigate lightly and frequently. Maintain a slightly acidic soil pH, using properly balanced fertilization and avoiding Ca deficiency. If growing in bermudagrass greens from sprigs, strive to complete the establishment during warm summer months, as *Pythium* can infect in fall if managers are still fertilizing and irrigating to "push" the sprigs to cover.

**Chemical controls.** Preventative, short-term (e.g., 5 to 10 days) control is provided by several contact fungicides. On cool-season grasses, apply during hot (>80°F, 27°C), humid (≥85%) weather when night temperatures exceed 65°F (18°C). Be careful using chloroneb and ethazole, as these are potentially phytotoxic to the turf. Preventative control is provided by systemic fungicides, and tends to provide longer residual (14 to 21 days) control. Whenever possible, plant fungicide-treated seed. Otherwise, apply a fungicide immediately after planting, and reapply approximately 7 to 14 days later (after the seed has germinated). Alternate between compounds to avoid development of fungicide-resistant strains of *Pythium*.

Due to the potential for rapid development of this disease and loss of large areas, turf managers growing cool-season grasses should consider a preventative fungicide program when hot, humid (e.g., foggy) weather is forecasted. Alternating between contact and preventative fungicides should also help reduce the risk of resistance problems. Use curative rates only when absolutely necessary to prevent resistance buildup. On ultradwarf bermudagrass, treat preventatively if extended warm, wet, cloudy conditions are forecasted, especially in fall and early winter months when days are shorter and grass not actively growing. For low infections, consider azoxystrobin, pyraclostrobin, and fosetyl-Al. For higher infections, mefenoxam, cyazofamid, propamocarb, and fluopicolide plus propamocarb have provided control.

### *Pythium* Root Rot/*Pythium* Root Dysfunction

The causal agents of *Pythium* root rot are *Pythium* spp. including *P. aristosporum* Vanterpool, *P. aphanidermatum* (Edson) Fitzpatrick, *P. arrhenomanes* Drechs., *P. graminicola* Subrum., *P. irregulare, P. myriotylum* Drechs., and *P. vanterpooli* Kouyeas and Kouyeas. In addition to these *Pythium* species, root dysfunction also involves *P. hyphae*, *P. volutum* Vanterpool and Truscott, and *P. torulosum* Coker and F. Patterson, with *P. volutum* appearing most often for root dysfunction.

**Occurrence.** These *Pythium* species are capable of infecting roots and crowns of cool-season grasses such as bentgrass and annual bluegrass year-round, causing stunting and yellowing (Figure 12.51). The newer ultradwarf bermudagrasses also are susceptible. *Pythium* root rot is still poorly understood for the different turfgrasses and different *Pythium* species. Symptoms may appear at any time of the year, but they will usually be associated with wet conditions—either from too much precipitation or too much irrigation. Poor drainage and soil compaction conditions will compound this problem and encourage algae development in areas where disease has weakened or killed the grass. Root damage from nematodes or *Gaeumannomyces* spp. also may contribute to this disease. It is thought most of the plant damage from the *Pythium* spp. is in the fall or spring when soil temperatures are between 54°F and 75°F (12°C and 24°C), yet symptoms are not expressed until the heat (>78°F, 26°C soil temperatures) and drought stress of summer. Relatively young stands of bentgrass (less than or equal to three years old) are most often affected, presumably due to less organic matter accumulation, which apparently helps suppress the disease. In addition, root dysfunction is often seen on exposed sites, especially on top of slopes. It occurs more during hot, dry conditions with extremely low N fertility.

**Symptoms/signs.** Symptoms are typically nonspecific declines in turf quality. However, circular patches in rings, arcs, or solid patches of a few inches up to about 3 feet (0.9 m) most often occur in summer under heat and drought stress on creeping bentgrass, typically on mounds, along slopes, or other drought-prone areas (Figure 12.52). However, drought-stricken greens can also show symptoms in fall and spring. Symptoms also are more prevalent on low N fed (annually at less than 4 pounds N/1,000 square feet, <20 g N/m$^2$) greens, especially sand-based rootzones. Damage is often initially observed in cleanup (perimeter) laps where mowers turn around. Small or large turf areas will become a general yellow, orange, or brown color and gradually begin to thin, often resembling drought stress. However, areas do not normally thin to bare soil. Symptoms may disappear following fertilization or irrigation, but often reappear when these again become limiting. Commonly, turf in aerification holes remains healthiest, indicating adequate soil $O_2$ is necessary to minimize the disease. Presumably due to injured roots, affected plants have less ability to absorb P and thus often take on a purplish color.

**FIGURE 12.51** *Pythium* root rot symptoms on bentgrass. (Courtesy Lane Tredway.)

**FIGURE 12.52** *Pythium* root dysfunction caused by *Pythium volutum* on bentgrass. (Courtesy Lane Tredway.)

*Pythium* root dysfunction describes root tissue that does not function properly but has no apparent rotting. Roots appear thin with few root hairs and have a general tan discoloration, but are not black and rotted as they are with other root-rotting diseases such as necrotic ring spot or bermudagrass decline. Infected roots also lack or have a loose cortical structure and have bulbous (swollen) root tips. Sand typically does not cling to infected roots as it does to healthy ones. Infected roots evidently have limited water and nutrient uptake capability, thus causing drought- or nutrient-stressed symptoms leading to eventual collapsing of the plant. Foliar mycelium is absent, often causing confusion with melting out disease, take-all patch, or possibly anthracnose. This disease cannot be diagnosed from field symptoms alone. Microscopic examination of affected roots and crowns is required to determine if *Pythium* spp. oospores and hyphae are associated with the symptoms. If affected turf is left untreated, death may occur in two to three weeks. Interestingly, symptoms often appear on higher elevations first, presumably due to the root dysfunction, which limits plant water uptake, whereas root rot develops in very wet or shady areas and often appears in more irregular patterns rather than distinct circles.

**Cultural controls.** The first line of defense against any *Pythium* disease is reducing management stress on the grass, as any stress on the grass tends to encourage the disease. Symptoms are often reported following a tournament, especially when mowing heights were lowered or greens allowed to dry out. Improve drainage, provide sufficient N, aerate, and reduce irrigation as *Pythium* spp. require very wet situations for disease development. Aerify in spring to improve surface drying, provide soil $O_2$, and promote rooting. When symptoms are present, irrigate only as needed to prevent permanent wilt and compensate for the damaged root system by hand-watering. To prevent the disease, avoid frequent shallow irrigations that constantly keep the turf wet. Reduce nematode populations, if justified by a nematode assay. Increase mowing height to >0.150 inch (3.8 mm) to reduce stress and promote root growth. Mow affected greens with walking mowers using solid rollers, skipping cleanup laps and possibly mowing on an alternate-day basis. Avoid mowing, grooming, and topdressing excessively wet greens. Wash mowers when used between infected and uninfected turf areas to minimize the spread of disease spores. Aerify and topdress to stimulate new root growth. Check for soluble salt levels in the upper soil and thatch, especially in the first year of turf establishment when excessive fertilizer is used for grow-in.

**Chemical controls.** Currently, a rotational treatment program appears to work best. Initially, apply pyraclostrobin or cyazofamid and water in with 1/8 inch (0.3 cm) of irrigation or apply in 5 to 10 gallons of water/1,000 square feet (20 to 40 L/100 $m^2$), which suppresses the disease for ~28 days. At least two applications will probably be required. Treat preventatively in fall and spring when soil temperatures are between 54°F and 75°F (12°C and 24°C) and continue at three-week intervals. Since different *Pythium* species may be involved, a spring application of cyazofamid with either propamocarb or fosetyl-Al is often beneficial. In summer, a drench application of ethazole followed in five days with fosetyl-Al plus mancozeb has been effective. Also, aerify in conjunction with fungicide use to increase soil $O_2$ while reducing soil moisture. If applied curatively, application intervals should be 14 to 28 days. This is due to inconsistent disease control. For root rot, curative control includes ethazole followed in two to three days with one of the preventative products, cyazofamid, fosetyl Al, or mefenoxam. For root dysfunction, preventative control is with pyraclostrobin or cyazofamid, which require watering in. Additional options include fosetyl-Al combined with either propamocarb or metalaxyl. Alternate between compounds to avoid development of fungicide-resistant strains of *Pythium*. Note that, except for pyraclostrobin, these fungicides are specific for *Pythium* spp. only. Treat preventatively as trying to "cure" bentgrass of this disease during summer heat and obtaining turf recovery requires weeks or even months.

## Red Thread

The causal agent of red thread is *Laetisaria fuciformis* (McAlphine) Burdsall (formerly, *Corticium fuciformis*).

**Occurrence.** Red thread can be a destructive disease on slow-growing turf. Perennial ryegrasses and fine leaf fescues are notably susceptible, especially in N-deficient turf. The disease is primarily a problem during winter or early spring months in cooler (60°F to 70°F, 16°C to 21°C), humid environments. The disease is fairly common in humid winter months in the upper elevation or northern U.S. regions.

**Symptoms/signs.** Red thread appears as irregularly shaped patches approximately 4 to 6 inches (10 to 15 cm) in diameter that are bleached-tan to reddish in color (resembling dollar spot). From a distance, the disease often appears similar to dollar spot and the turf often develops a ragged straw-brown appearance as if it had been mowed with a dull mower blade or may be suffering from drought stress (Figure 12.53). Decline of leaves is rapid in affected areas. Bright coral-pink to red antler-like "threads" or sclerotia of the fungi protruding from dying leaf tips that may have a split, torn appearance are easily seen with the naked eye (Figure 12.54). Threads are gelatinous in the early morning dew, but later dry and become thin and brittle. Threads occur most often during the cool, drizzly days of spring but can also occur throughout summer in higher elevations. The fungi overwinter on leaves and thatch as pink or red, gelatinous crusts of fungal threads or sclerotia. Red thread, like dollar spot, may increase with the current trend of using lower N rates.

**Cultural controls.** Increase N fertility in the spring and fall when possible. This increases plant vigor, which allows the plant

**FIGURE 12.53** Red thread on fine fescue.

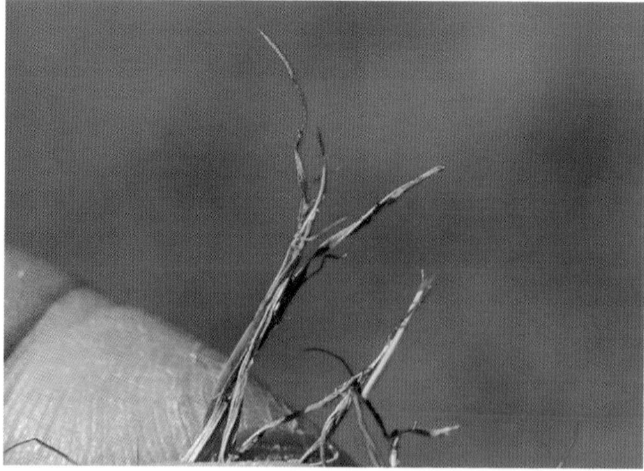

**FIGURE 12.54** Pink to red antler-like "threads" (sclerotia) associated with red thread disease.

to outgrow disease damage. Collect and destroy clippings to minimize spread of sclerotia. Potassium should also be applied with N. Water deeply but infrequently to prevent prolonged leaf wetness. Avoid watering in late afternoon and evening. Improve air circulation and reduce humidity by pruning trees and removing underbrush. Collect and dispose of clippings from infected areas and wash equipment before entering unaffected areas. Encourage golfers to clean their shoes between rounds.

**Chemical controls.** Fall and winter treatments help prevent spring disease development.

## Rust

The causal agents of rust are *Puccinia* spp. (such as *Puccinia graminis* Pers.:Pers., *P. striiformis* Westend., *P. coronata* Corda, *P. zoysiae* Dietel) or *Uromyces dactylidis* G. Otth.

**Occurrence.** Stem rust (*Puccinia graminis* f. sp. *poae*) has been a problem on slow-growing Kentucky bluegrass, especially in seed production fields. Crown rust (*P. coronata*) is an important problem on the new perennial ryegrasses, including their seed production fields. Rusts can also be a serious problem on zoysiagrass and fescue. The disease is most severe during cool weather in late summer and fall when conditions are less than adequate for good turfgrass growth. Nitrogen-deficient grasses grown in shade are especially affected.

Infected turfgrass foliage in areas with mild climates serves as the overwintering site for rust fungi spores or mycelium. Spore dispersal by wind can spread from warm regions in the country to cooler ones. Rust fungi are obligate parasites; thus, they must grow and reproduce only in living tissue.

**Symptoms/signs.** Overall symptoms of severe rust infection are a thinned, clumpy turf. Heavily infected turf areas will appear yellow and produce a cloud of orange dust (urediospores) when the foliage is disturbed. The orange powdery material easily rubs off on hands, machinery, shoes, and clothing. Turf will appear thin and weak, and is more susceptible to drought and winter injury (Figure 12.55). Oval to elongated pustules, yellow to orange or reddish-brown in color, are raised on the surface of leaves (Figure 12.56). These spores are dry, powdery, and spread easily by wind, machinery, shoes, and infected plants. Survival during periods of unfavorable growth is by dormant mycelium and urediospores in or on infected plants and equipment.

**FIGURE 12.55** Thinned, clumpy turf infected with rust disease on Kentucky bluegrass.

**FIGURE 12.56** Orange-colored pustules associated with rust disease on zoysiagrass.

**Cultural controls.** Management practices that provide steady grass growth during prolonged warm to hot periods when rust problems are most common should be followed. Maintain adequate N levels (1/4 to 1/2 pound N/1,000 square feet, 1.2 to 2.4 g/m$^2$) for sufficient growth to ensure weekly mowing will help reduce the severity of rust. If rust occurs in late fall, do not fertilize as the disease usually disappears during winter. Many cultivars of Kentucky bluegrass, perennial ryegrass, and tall fescue have moderate to good rust resistance. Planting blends of cultivars with different genetic backgrounds have limited disease occurrence. Remove clippings and remove shade sources, as low light intensity favors the disease. Wash equipment before entering an uninfected area. Water deeply but infrequently to encourage deep rooting and avoid drought stress.

**Chemical controls.** Zoysiagrass, due to its inherent slow-growing nature, and seed production fields of perennial ryegrass and tall fescue may require fungicide treatment. Triadimefon, triticonazole, and propiconazole control rusts when applied in spring or fall when the first signs of rust are visible. Repeat applications every 7 to 14 days are necessary while rust is present. Contact fungicides require applications on 7- to 10-day intervals and are not especially effective.

## Slime Mold

The causal agents of slime mold are *Mucilago, Physarum,* and *Fuligo* spp.

**Occurrence.** All grasses may be affected as slime molds are not parasites of turfgrass plants, but occur most often during warm, wet weather or after periods of heavy summer rainfall. These fungi move about as slow-moving protoplasmic amoebae and feed as saprophytes on bacteria, fungi, and decayed organic matter in soil. They do not feed on turfgrasses and survive as spores.

**Symptoms/signs.** White, gray, powdery fruiting bodies cover leaves in patches 6 to 12 inches (15 to 30 cm) in diameter during warm, wet weather. The protoplasmic forms of the fungi migrate to plant tips and produce the spore stage, which allows enhanced dispersal of the dark, powdery black spores by wind (Figure 12.57). During the day, ruptured fruiting bodies may cover grass leaves with the black, soot-like spores. Affected areas may appear a dull-gray color due to high populations of these spores.

**Cultural controls.** Remove the grayish fruiting bodies by brushing, mowing, or washing the turf. Slime molds are not considered harmful and usually disappear during dry weather.

**Chemical controls.** None are needed.

## Spring Dead Spot

The causal agents of spring dead spot are *Ophiosphaerella* (=*Leptosphaeria*) *korrae* J. Walker and A. M. Sm.; *Ophiosphaerella* (=*Leptosphaeria*) *narmari* J. Walker and A. M. Sm.; and/or *Ophiosphaerella herpotricha* (Fr.:Fr.) J. Walker.

**Occurrence.** Spring dead spot (SDS) is the most serious disease of bermudagrass in the United States, where it undergoes complete dormancy in winter. It also occurs sporadically in zoysiagrass. The highest maintained turf is generally most susceptible, such as hybrid bermudagrasses that tend to produce excessive thatch, while cold-hardy cultivars are more resistant. Late-summer N applications, abundant fall moisture, and low winter temperatures also predispose bermudagrass for spring dead spot development. Spots generally begin to appear after the turf is at least three to five years old; however, outbreaks have occurred on ultradwarf bermudagrasses within a year of establishment. Infected areas recover slowly and weeds frequently invade these areas during summer. Presumably, fungi causing spring dead spot infect in late summer and fall, and weaken the turf without visible symptoms. Parameters affecting winter hardiness and spring green-up also influence spring dead spot symptom development and turf recovery. All of the reported causal agents are slow-growing root ectotrophic fungi, similar to organisms causing take-all patch, necrotic ringspot, and summer patch in bluegrass and bentgrass. In fact, *Ophiosphaerella korrae* is known to cause both necrotic ringspot and spring dead spot. The dark-brown, black mycelium from ectotrophic-growing fungi directly penetrates roots, stolons, and rhizomes, filling vascular tissue with a brown substrate and dark, spindle-shaped sclerotia. *Gaeumannomyces graminis* var. *graminis* has been associated with spring dead spot symptoms as well; however, more recent research with molecular tools and results of successful field inoculations suggest *Ophiosphaerella korrae, O. herpotricha*, and *O. narmari* are the agents demonstrated to cause SDS symptoms in the field. Colonization and root infection are likely when soil temperatures range from 50°F to 75°F (12°C to 24°C) with maximum activity at approximately 60°F (15.6°C), giving the disease a competitive advantage as bermudagrass root growth considerably slows at this temperature.

**Symptoms/signs.** Dead, straw-colored spots 2 to 3 feet (0.3 to 0.6 m) in diameter appear in spring as affected bermudagrass begins to "green-up" and is often confused with winter-kill (Figure 12.58). Spring dead spot patches are sunken, generally well defined, and circular, in contrast to more diffuse dead areas caused by direct low-temperature injury. Roots and stolons of affected bermudagrass are severely rotted. Patches may enlarge over three to four years, develop into rings, and then disappear. Affected spots may also remain greener in late fall going into winter. Patches in overseeded affected bermudagrass may resemble brown

**FIGURE 12.57** Black, powdery spores of noninfectious slime mold.

**FIGURE 12.58** Spring dead spot disease of a bermudagrass fairway.

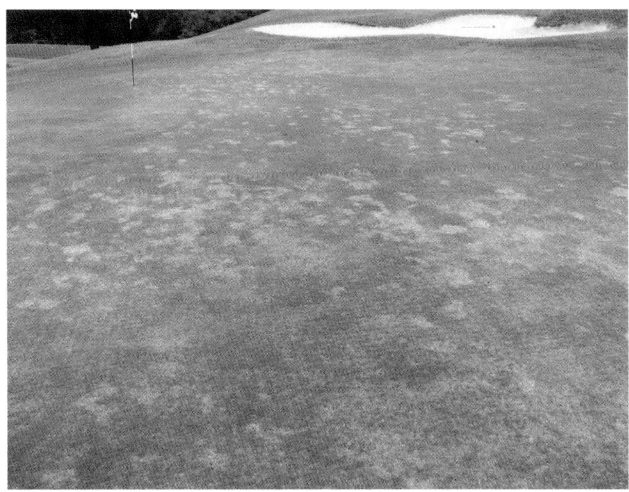

**FIGURE 12.59** Spring dead spot disease of a bermudagrass golf green.

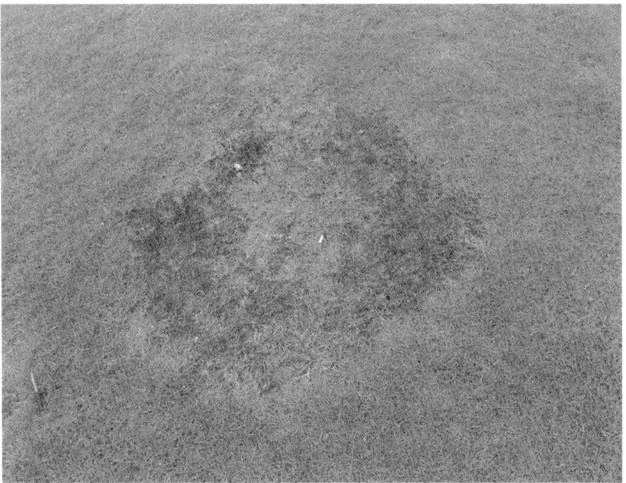

**FIGURE 12.60** Frog-eye symptom of spring dead spot disease of bermudagrass.

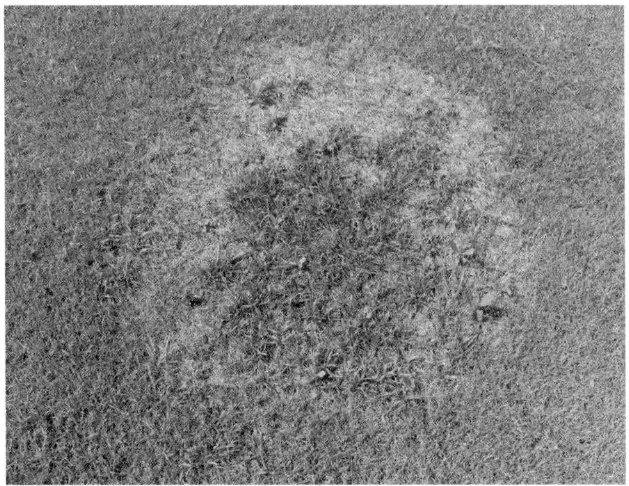

**FIGURE 12.61** Weeds invading the dead centers of spring dead spot disease of bermudagrass.

patch in spring. Patches are usually perennial in nature, often recurring in the same location over several years (Figure 12.59). After a year or two from their first occurrence, patches develop into doughnut or frog eye patterns with relatively nonsymptomatic bermudagrass in the centers (Figure 12.60). After several years, spring dead spot may entirely disappear from a site, although this phenomenon is not predictable.

Rhizomes and stolons from adjacent bermudagrass slowly fill in dead spots. This slow process allows summer annual weeds such as crabgrass to easily become established (Figure 12.61). Use of certain preemergence herbicides for summer annual grassy weed control may affect recovery from spring dead spot. Herbicides that inhibit cell division (the dinitroanaline group, as well as dithiopyr and indaziflam) may inhibit new stolons from colonizing patches and slow recovery. Oxadiazon, however, does not inhibit rooting and might be a better choice of a preemergence herbicide in SDS areas. Managers may also opt to use postemergence techniques for weed control in areas prone to severe spring dead spot.

**Cultural controls.** No single cultural or chemical practice provides complete control in one year. Spring dead spot is a disease of intensively managed mature bermudagrass. Use acidifying fertilizers such as ammonium nitrate and ammonium sulfate to help speed recovery and reduce disease severity. However, avoid excessive N fertilization and do not apply in the fall. Raise the mowing height and ensure adequate K levels in the fall. Reduce thatch by aerifying and pulverizing soil cores. Bermudagrass cultivars with varying resistance to SDS include Ashmore, Midiron, Midlawn, Patriot, Riviera, TifSport, Yukon, and U-3. Those with less SDS resistance include Arizona common, Mohawk, NuMex Sahara, Panama, Princess-77, Southern Star, Tifway, and Transcontinental.

**Chemical controls.** Fungicides work best when used in combination with appropriate maintenance practices. Some newer fungicides, including tebuconazole, fluxapyroxad + pyraclostrobin, penthiopyrad, difenoconazole + azoxystrobin, and propiconazole + azoxystrobin have provided good control of SDS in recent trials. Further research is needed but results have been encouraging since fenarimol, which had been the most efficacious fungicide for SDS control, has been withdrawn from the market. Applications should be made generally in late summer using a five-day average rootzone (2 to 4 inches, 5 to 10 cm), temperature between 60°F and 80°F (16°C and 27°C) with an optimum of about 75°F (24°C) for application timing, and a second application made about a month later. Sprays should be applied in higher volumes (greater than 5 gallons/1,000 square feet, 20 L/m$^2$) but can be applied in 2 gallons/1000 square feet (8 L/100 m$^2$) but then immediately incorporated with additional irrigation (approximately 1/4 inch, 6.4 mm) to move the fungicides to the rootzone. These fungicides can also be applied as drenches (5 gallons/1,000 square feet, 0.2 L/m$^2$) or at lower water volumes and incorporated with 1/4 inch (0.64 cm). Several years of consecutive use may be required on fairway height bermudagrass for complete control as the patches are typically reduced in size following each yearly fungicide use.

## Stripe Smut

The causal agents of stripe smut are *Ustilago striiformis* (Westend.) Niessl and *Urocystis* spp. such as *U. agropyri* (Preuss) Schrot (often called flag smut).

**Occurrence.** Stripe smut is an important disease of creeping bentgrass and Kentucky bluegrass, especially Kentucky bluegrass cultivars resistant to melting out. It also attacks older stands of tall fescue but not to the same extent as brown patch. It is a perennial systemic disease; once a plant is infected, it remains so for life. Daughter plants from stolons or rhizomes of infected mother plants will also be infected. Infected plants often are invaded by other organisms such as "*Helminthosporium*" diseases. Infected turf appears clumpy and patchy when viewed from a distance. Infected plants usually die during the heat- and drought-stress period of summer and are replaced by weeds. An additional smut, *Ustilago cynodontis* (Pass.) Henn., may infect common bermudagrass, replacing the flower heads with dark smut spores.

**Symptoms/signs.** Symptoms most commonly occur in spring and fall while experiencing cool (<70°F, 21°C), wet weather. Infected turf has a clumpy appearance due to the death of surrounding grass plants. Many "clumps" appear pale green in spring, but develop yellow streaks when disease is active, giving the appearance of N deficiency (Figure 12.62). Quackgrass and tall fescue often invade severely infected stripe smut turf areas in Kentucky bluegrass turf. Individual plants are stunted and upright in their growth habit. Disease first appears as dull-gray streaks in leaves. The epidermis above streaks is eventually ruptured, exposing black masses of stripe smut teliospores. These spores may be transported on seeds and turf-maintenance equipment. Leaves eventually split, shred, twist, and curl, exposing the black smut teliospores (Figure 12.63). Although plants are infected, they may appear amazingly healthy if properly maintained as weather warms during the summer. Turf stand loss mostly occurs during the hot, dry summer from heat and drought stress. By fall, typical symptoms may reoccur and, by winter, these develop a gray-brown desiccated appearance from the shredding of leaves by maturing fruiting bodies during fall. The fungus overwinters as dormant mycelium in plants and as teliospores in soil and plant debris.

**FIGURE 12.62** Strip smut symptoms of Kentucky bluegrass.

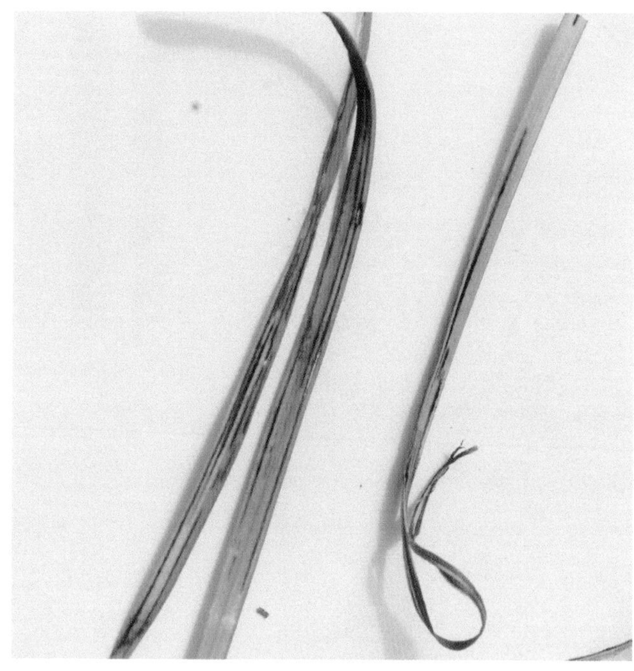

**FIGURE 12.63** Distorted leaves exposing black smut teliospores associated with strip smut disease.

**Cultural controls.** Use resistant cultivars in blends that have been propagated in smut-free environments and avoid heavy (over 1/2 pound N/1,000 square feet, 2.5 kg N/100 m$^2$, per month) N applications in late spring through summer. Prevent moisture stress on infected turf, use a balanced N-P-K fall fertilizer program, and raise the mowing height of infected turf.

**Chemical controls.** Some systemic fungicides, including triadimefon, isofetamid, penthiopyrad, tebuconazole, and propiconazole, are effective in slowing infection rates and in promoting healing of diseased turf stands when applied in spring (May) or fall (mid- to late October).

## Summer Patch

The causal agent of summer patch is *Magnaporthe poae* Landschoot and Jackson.

**Occurrence.** Summer patch is primarily a soil-borne disease of bluegrass caused by a slow-growing ectotrophic root-infecting fungus that spreads to adjacent plants by growing along roots. Being a soil-borne fungus, early disease diagnosis can be difficult and often goes undetected until plants begin to stress. It is easily confused with necrotic ring spot (*Ophiosphaerella korrae*), as their hosts and symptoms are very similar. It is a major disease of annual bluegrass, but also affects Kentucky bluegrass, fine fescues, and creeping bentgrass. Parameters increasing stress to these grasses also influence the development of summer patch. Summer patch primarily occurs in the northeastern and midwestern United States, and also in California and the Pacific Northwest. Colonization of new turfgrass roots, crowns, and rhizomes begins in spring when soil temperatures reach 65°F to 70°F (18°C to 21°C) for five consecutive days. Initially, plant damage is not noticeable as new root growth replaces diseased roots. However, as temperatures rise to 82°F to 86°F (28°C to 30°C), cool-season root growth slows while the pathogen's growth increases. Outbreaks occur during high summer temperature stress, especially when nighttime temperatures remain abovc 70°F (21°C) and during heavy rainfall. It is also favored by higher (>6.3) soil pH. Disease severity is directly related to infection severity and turfgrass vigor. Sunny areas and those adjacent to heat-stress areas such as sidewalks are most susceptible. Interestingly, mature turf is infected more than young immature stands. It can overwinter in infected host tissue, causing annual outbreak of disease.

**Symptoms/signs.** Patches of affected turf are circular, semicircular, or serpentine in shape from a few scattered plants ranging in size from a few inches up to about a foot (0.3 m) in diameter (Figure 12.64). Patches are generally smaller than those caused by necrotic ring spot. Initially, plants are wilted, gray-green in color, and may develop heat-stress banding (white bands across individual leaf blades). Affected plants wilt, die, and become matted, tan, or brown, leaving patterns of patches, rings, and arcs of symptomatic turf (Figure 12.65). Tufts of green grass or weeds may remain in the center, leaving a "frog eye" pattern. On golf greens with high populations of annual bluegrass, patches are initially small 2- to 3-inch (5 to 6.6 cm) reddish-brown patches. As stress increases, patches may coalesce. Leaf lesions are not found with summer patch. On turf roots, dark runner hyphae are visible and may infect crowns and rhizomes. Roots and crowns are rotted and are brown to black in color, similarly to necrotic ring spot. Due to the lack of leaf lesions, laboratory identification is necessary as summer patch resembles symptoms of many other diseases.

**Cultural controls.** Management of cultural practices that reduce plant stress is necessary in any summer patch control program. Once summer patch is established, it often reappears in the same area in successive years since it is a root disease. Utilize cultural practices to improve root systems, such as core aerification, proper watering (adequate moisture, avoiding even transient drought stress), and thatch management to eliminate plant stress. Core aerify prior to infection periods but not during an active disease outbreak. Fertilize with 1/3 to 1/2 pound N/1,000 square feet (1.5 to 2.4 g N/m$^2$) in conjunction with core aerification. Avoid heavy spring and summer N fertilization. Maintain a soil pH of 6.3 or less and thereby indiscriminate lime applications. Use ammonium sulfate or sulfur-coated urea to help keep soil pH acidic, which tends to help retard the disease. Raising cutting heights prior to and during anticipated periods of heat stress can be beneficial. Consider conversion of *Poa annua*-dependent greens to bentgrass, and other areas to perennial ryegrass or tall fescue, which are highly resistant.

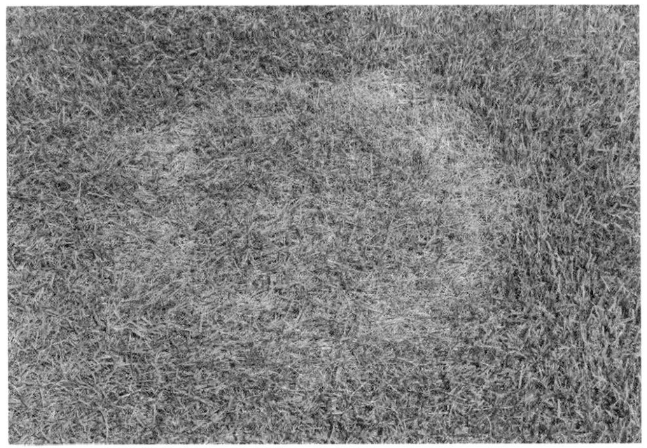

**FIGURE 12.64** Summer patch disease of Kentucky bluegrass.

**FIGURE 12.65** Summer patch disease showing arcs of symptomatic turf.

**Chemical controls.** Preventative fungicide applications have been effective. Apply fungicides with adequate water to move the active ingredients to the rootzone, which should have moderate moisture in the soil when the fungicides are applied. A key preventative control strategy is to treat before the pathogen colonizes and suppresses root growth. Since infections occur well in advance of symptom expression, apply preventatively in spring when soil temperatures at a depth of 2 inches (5 cm) reach 62°F (16.7°C), before symptoms develop, and repeat two to three times on 21- to 28-day intervals during summer. Use 2 gallons/1,000 square feet (8 L/100 m$^2$) when applying DMI and Qol fungicides and 4 to 5 gallons/1,000 square feet (16 to 20 L/m$^2$) with benzimidazoles, especially thiophanate-methyl.

## Take-All Patch (or *Ophiobolus* Patch)

The causal agent of take-all patch is *Gaeumannomyces graminis* (Sacc.) Arx and Olivier var. *avenae* (E. M. Turner) Dennis.

**Occurrence.** Take-all patch is a root disease of bentgrass and, to a lesser extent, annual bluegrass, ryegrass, and fescue. The disease occurs most frequently in cooler (40°F to 70°F, 4.4°C to 21°C) climates and in soils of high pH and excessive moisture. It appears mostly from late spring through summer months and survives as a saprophyte in grass debris and living plants. It may also occur in late fall and winter. Take-all patch is a problem in the northeastern, north central, and northwestern states of the United States; Europe; Japan; and Australia. It has also been a problem in the mid-Atlantic states and the mountains of North Carolina. Its occurrence elsewhere has not been officially confirmed or reported, although the disease is suspected to occur at times under special conditions, such as new (less than eight years old) golf courses or recently fumigated courses with moderate soil pH (>6.5) levels, recently limed sandy soils with low organic matter and low fertility, and conditions of soil compaction and high moisture during infection periods. As soil organisms recover from the fumigation process, disease severity declines, presumably due to the antagonistic effects of these soil organisms. During this time, turf rarely is killed but takes on a yellow, thin appearance (Figure 12.66). Plants grown in soils with low levels of plant-available Mn are especially susceptible to infection. Nitrate fertilizers also encourage the disease by raising soil pH.

**Symptoms/signs.** Roots, crowns, and stems are affected with no distinctive leaf spots or sheath lesions. Dead, yellow or reddish-brown to orange-tan, sunken circular patches ranging from several inches initially up to 3 to 4 feet (0.9 to 1.2 m) in diameter develop in late spring when soils tend to be cool and wet and may increase in size slowly over the summer months (Figure 12.67). Symptoms may fade with warmer soil temperatures and adequate moisture but reappear during drought during summer and fall. These may persist several years and increase in size each year. The fungus is active around the margins of patches and causes dark-brown to gray to black rotted tissue of affected roots and crowns. Dying bentgrass at advancing margins has a purplish tinge, and freshly infected plants may become "bronze" in color (Figure 12.68). Black strands of mycelium are visible under the base of leaf sheaths or on the surface of roots, rhizomes, and stolons. Black flask-shaped perithecia (fruiting bodies) may be visible with a hand lens on dead tissue. These contain septate, elongated ascospores. Weeds such as annual bluegrass, fescue, crabgrass, and various broadleaf weeds commonly invade the centers of patches. The fungus overseasons as a mycelium in plant debris, thatch, and on perennial plant parts.

**FIGURE 12.66** Take-all patch of creeping bentgrass.

**FIGURE 12.67** Severe infestation of take-all patch of creeping bentgrass.

**FIGURE 12.68** Bronze-colored bentgrass infected with take-all patch disease.

**Cultural controls.** Since take-all patch is favored by alkaline soils, fertilize with acidifying fertilizers such as ammonium nitrate and ammonium sulfate or use other acidifying compounds such as elemental S at 3 to 5 pounds/1,000 square feet (15 to 25 g/m$^2$). Split the total yearly amount of acidifying compound into several applications so as to maintain a soil pH of around 5.5. Minor elements such as Mn, Mg, and Zn should also be tested and soil applied if deficient. Soil (granular) applications of these nutrients are typically more effective than foliar sprays for take-all suppression. Either spring or fall applications of 2 pounds Mn/acre (5 kg/100 m$^2$) in heavy soils and 6 to 8 pounds/acre (30 to 40 kg/100 m$^2$) in sandy soils have proven beneficial. Patch suppression has followed use of manganese sulfate, ammonium chloride, and ammonium sulfate. Annual applications are usually needed. Acidic soils favor more Mn availability and thus tend to have better disease suppression.

Control thatch accumulation and promote rooting by aerifying, topdressing, verticutting, and deep, infrequent irrigation. In situations with chronic infections, replant affected areas with less susceptible grasses, blends, or mixtures of nonhost grasses.

**Chemical controls.** Preventative control is best and has been effective when applied in late fall or early winter. If take-all patch was prevalent the previous year, midspring applications when soil temperatures are between 40°F and 60°F (4.4°C and 15°C) may be beneficial but generally are best when applied in fall. Some curative control may be provided by azoxystrobin or thiophanate-methyl. At least 2 gallons/1,000 square feet (8 L/100 $m^2$) carrier volume should be used to help fungicides reach root tissue with 4 to 5 gallons/1,000 square feet (16 to 20 L/100 $m^2$) being more effective. Applications should be followed by 0.25-inch (6.4 cm) irrigation.

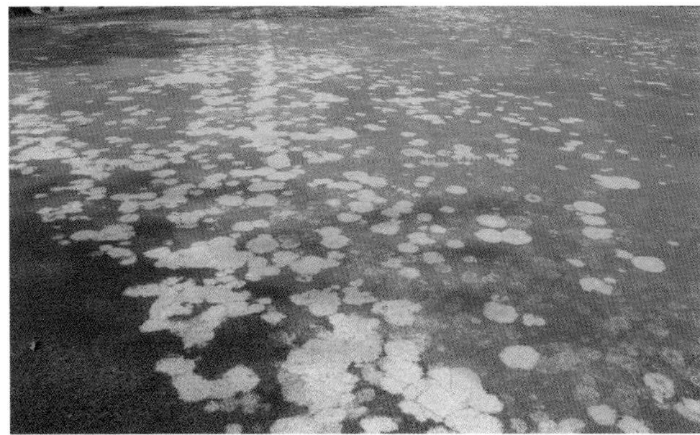

**FIGURE 12.69** *Typhula* blight (gray snow mold) spring symptoms on creeping bentgrass.

**FIGURE 12.70** Grayish-white fungal mycelium mat from *Typhula* blight (gray snow mold).

## *Typhula* Blight (Gray or Speckled Snow Mold)

The causal agents of *Typhula* blight are the basidiomycetes *Typhula incarnata* Lasch ex Fr. (gray snow mold) and *T. ishikariensis* Imai (speckled snow mold). These are psychrophilic (cold-loving) microorganisms. Other related diseases are *Coprinus* snow mold (*Coprinus psychromobidus* Redhead and J. A. Traquair); snow scald (*Myriosclerotinia borealis* (Bubák and Vleugel) L. M. Kohn (syn. *Sclerotinia borealis* Bubák and Vleugel); and snow rot or snow blight (various *Pythium* spp.).

**Occurrence.** *Typhula* blight, or gray snow mold, is a very destructive disease of turfgrass under snow cover. Optimum temperatures for development range from 32°F to 45°F (0°C to 7°C). Disease severity increases with duration of snow cover and often occurs under slushy conditions during thaws. Perennial ryegrass and creeping bentgrass are especially susceptible and severe destruction of this turf can occur with a subsequent invasion of annual bluegrass. *Typhula* blight is rare in the southern United States, occurring only in mountainous regions where persistent snow cover occurs.

**Symptoms/signs.** Snow mold severity is generally governed by fall and winter environmental conditions, depth and duration of snow cover, and level of disease inoculum. The fungi begin as saprophytes feeding on dead organic matter and invade turf when under snow. They survive summer as dark- to rust-colored sclerotia (or compact mass of hyphae) less than 1/8 inch (3 mm) in diameter. Sclerotia germinate in late fall when cool, moist weather occurs but before the soil is frozen. The fungi then proliferate under snow and spread in host tissue under dark, humid conditions. Foliar symptoms include light brown or red sclerotia embedded in leaves and crowns of infected plants. Later they dry and shrink to pinhead size and turn dark brown, giving a speckled appearance. Infected turf stands have straw-colored, roundish areas ranging in size from a few inches to 2 feet (0.6 m) in diameter (Figure 12.69). Spots may coalesce to form mats covering large areas. When spots are first observed in the margins of melting snow, they may contain fluffy white to bluish-gray fungal mycelium or a silvery crust that disappears as the turf dries (Figure 12.70). This gray-colored mycelium is where the disease gets its common name. Leaves in affected areas are matted together. Gray snow mold is more damaging under prolonged, deep snow coverage. Although the fungus can kill the turf when it infects crowns and roots, it is more common that only shoots are killed and new leaves appear in spring. Disease spots can be noticeable until late spring. The fungus survives as sclerotia during the summer.

Symptoms of gray snow mold caused by *T. ishikariensis* resemble diseases caused by *T. incarnata* except it is found when a snow cover remains for four months or more and it produces numerous small (<0.063 inch or 1.6 mm diameter) dark-colored sclerotia following snow melt. The sclerotia are so numerous the turf appears as if it had been sprinkled with pepper. It is found more often in the western portion of North America. The sclerotia for *Typhula incarnate*, meanwhile, are larger (0.125 inch or 3.2 mm) in diameter and reddish-brown in color (Figure 12.71).

**Cultural controls.** Avoid N fertilizing in fall, which leads to lush, succulent growth going into winter. If dormant N feeding is applied to promote early spring green-up and quicker recovery from snow mold damage, fungicides should also be applied. Erect snow fences or barriers to protect areas damaged yearly by snow drifts. Promote drying by providing good soil drainage, pruning limbs, trees, or shrubs to increase air

**FIGURE 12.71** Dark-colored sclerotia associated with snow mold.

circulation, and removing fallen leaves and clippings. The fine fescues are typically more resistant than bluegrass and bentgrass.

**Chemical controls.** Preventative control is provided by chloroneb, flutolanil, iprodione, PCNB, propiconazole, triticonazole, and triadimefon, but they provide little control in spring after sclerotia form. The combination of chlorothalonil and iprodione or a three-way combination of a systemic fungicide, a strobilurin fungicide, and either PCNB or chlorothalonil (e.g., fludioxonil + propiconazole + chlorothalonil or iprodione + fosetyl-Al + PCNB) applied in mid- to late October and repeated three to four weeks later appears to satisfactorily control both *Typhula* species as well as pink snow mold or *Fusarium* patch, which may occur at the same time and site. Apply before the first heavy snow or before cold, rainy weather, and repeat as needed during midwinter and early spring during snow melts.

## Algae

The causal agents of algal infestations are the blue-green cyanobacteria *Nostoc* and *Oscillatoria,* but also include the green and brown algae genera *Chlamydomonas, Hantzschia* spp., and others. Algae contain complex cells with nuclei, mitochondria, chloroplasts, an endoplasmic reticulum, and other organelles. Algae are considered plants since they contain chlorophyll. However, they are primitive plants since they lack roots, stems, and leaves and must be in or near aquatic environments to thrive.

In addition to true green algae, primitive bacteria called cyanobacteria, also called the blue-green algae, infest turf stands with dark, slimy colonies. Cyanobacteria (*Oscillatoria, Phormidium, Lynbya, Nostoc,* and *Anacystis* spp.) contain none of the complex eukaryotic cells as algae and are more closely related to other bacteria such as *E. coli* and *Xanthomonas.* They are single-cell and filamentous, reproducing by fission and their nuclei are not bound by a membrane. *Oscillatoria* is a small, primitive, photosynthetic microbe that produces dark crusts and slime layers on putting green soils and foliage. It produces energy by photosynthesis at low light intensities; thus, it invades closely mowed, shaded greens. Cyanobacteria can grow rapidly in water or wet soils with sufficient light, nutrients, and temperature and can move by sliding on their own produced mucilaginous (slimy) materials, which can seal the soil surface. Cyanobacteria are filamentous, producing trichomes, which are long chains of cells making up the black slime coating often seen on greens that can bind sand particles, plugging pore spaces. They typically move to plants from thatch at night or on shade greens and retract back to cooler thatch during hot temperatures. They also produce toxins that enter plants through mowing injury. Cyanobacteria are known to be the most ancient organism with photosynthetic capability and have the ability to synthesize and secrete large quantities of polysaccharides from cells. The polysaccharides provide a protective coating allowing the cyanobacteria to withstand moisture and heat stresses. Some cyanobacteria can survive deserts, hot springs, arctic conditions, and other environmental extremes. In addition, an antibiotic produced by *Oscillatoria* attacks photosystem II in plant cells. Due to lack of $O_2$ and drainage under cyanobacteria crusts, the plant may yellow and thin.

**Occurrence.** Algae are most noticeable on close-cut, poorly drained, low lying, and shaded areas on tees and putting greens. Overcast, rainy weather; soil layering; poor surface drainage; compaction; poor disease control; and other stresses can predispose greens and other highly managed sites to algae infestations. Excessive soil P and the use of damaging fungicides such as the DMIs during summer also encourage algae infestation as turf thins. Although erroneously thought to cause algae on golf courses, algae species found in irrigation ponds, lakes, and streams are different and do not appear to contribute to this problem.

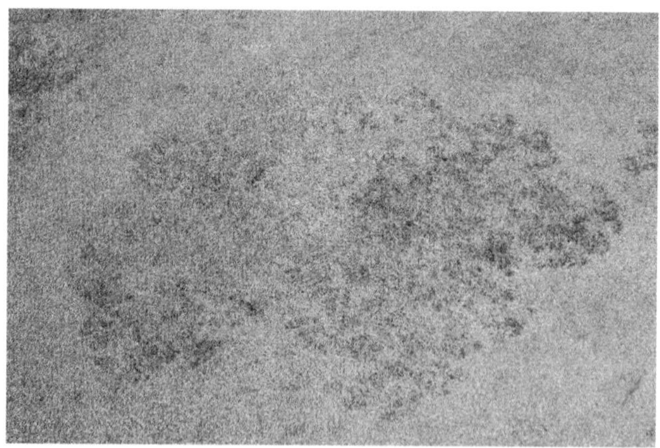

**FIGURE 12.72** Dark, slimy colonies of algae on a golf green.

**Symptoms/signs.** Symptoms begin slowly, often at the edges of greens. Turf areas in partially shaded, damp locations become weak and begin to thin, usually starting in early summer (Figure 12.72). Algae begin to predominate in these areas. These algae are commonly green or brown in color and can be sheetlike, leaflike, or cushion-like in appearance. Initially, these spots may be difficult to see, but as they enlarge, the turf begins to yellow and thin due to a lack of $O_2$ and drainage under the crust. Algae will develop on turf areas where grass is less dense than normal and surface soil moisture is high, resulting in a dark-color "scum" or "mat" forming on the soil surface (Figure 12.73). Frequently, on bentgrass greens affected by brown patch disease, algae invade the thinned areas. If left untreated, these areas will continue to "thin out" and expand in size until a large mat of algae forms, preventing turfgrass growth and penetration of irrigation water (Figure 12.74). Black algal scum development often occurs in summer following periods of rainy, overcast, warm days. Turf thinning from summer DMI fungicide use also may trigger algae invasion. Commonly in bentgrass greens cultured in heat stress summer environments, a "yellow spot" symptom occurs in mid- to late summer. Presumably this symptom is a result of algal toxins, and can be alleviated with the fungicides mancozeb or chlorothalonil. Other fungicides (DMIs and strobilurins) may increase the yellow spot symptoms although turf is rarely thinned.

**FIGURE 12.73** Algal "scum" or "mat" invading thin bentgrass.

**FIGURE 12.74** Crusted mat of algae forming following periods of rain, overcast, warm summer days.

**Cultural controls.** Due to their rapid development, preventative management is the best approach. Prevention begins by correcting those conditions that predispose the turf to algal growth. This involves reducing surface moisture by improving air circulation and light exposure by removing adjacent underbrush and selectively removing trees. Improve drainage and reduce irrigation frequency and amount. Reduce freely available N and P. Avoid organically based products if excessive P is present in them. Reduce irrigation and improve the growth of the turfgrass where algae are present so the turf can form a dense area. If the area occupied by algae is large, spiking, verticutting, and topdressing will help break up and dry the mat. Applying ground limestone or hydrated lime will help desiccate algae. Diluted bleach, copper sulfate, and chloride also may help reduce algae growth. However, these should not be used during hot temperatures, as they may cause varying levels of turf discoloration. Be judicious in the use of Cu-containing fungicides to avoid a buildup of Cu and potential Cu toxicity in soils. This would be more of a risk in sandy putting green mixes with low organic matter. Increase the mowing height, as low mowing aggravates the problem.

**Chemical controls.** Some fungicides help to prevent development of algae and their spread when a mat has already formed. They need periodic applications (e.g., every 7 to 14 days) and should be used **prior** to algae formation and continually as long as conditions remain favorable for occurrence. Control diseases, such as brown patch, to prevent thinning of the turf canopy. Some curative control is provided by hydrated lime or mancozeb + copper hydroxide. Being a desiccant, hydrated lime typically works quickly while the mancozeb + copper hydroxide takes several days. Once the algae are controlled, switch to a preventative program. In areas with high disease pressure and low turf recuperative potential, such as the transition zone, preventative disease control approaches are necessary in the summertime. Chlorothalonil, mancozeb, and various phosphite-containing fungicides are often used to manage algae.

## Moss

The causal agents of moss include *Selagimella, Byrum, Amblystegium, Brachythecium, Ceratodon, Hypnum, Polytrichum* spp., plus others. *Amblystegium trichopodium* and *Brachythecium* spp. are usually found in higher cut turf and are often referred to as "yard moss." *Bryum argetum*, referred to as silver thread moss, has a silvery appearance and is found more frequently on greens (Figure 12.75). Unlike most mosses, silvery thread moss grows well in both wet, shady environments and hot, dry sites in full sun.

**Occurrence.** Mosses are threadlike, branched, primitive (400 million years old) plant forms encompassing many species. They spread by spores disseminated by wind and water movement. Most moss species are **bryophytes**, meaning they do not form true roots, but rather form rhizoids, which are

**FIGURE 12.75** Silvery thread moss on a bentgrass putting green.

**FIGURE 12.76** Close-up of silvery thread moss patch on a bermudagrass putting green.

filamentous structures that provide anchoring. Consequently, they can survive on rocks, concrete, and masonry walls. Bryophytes also are nonvascular plants requiring constant contact with water to prevent drying. Others are able to absorb water through their rhizoids. Mosses are able to photosynthesize and fix N. Mosses are nonparasitic to the turf and spread by plant fragments (mainly)—less so as spores. They can also survive long periods of desiccation.

Moss is most noticeable on close-cut areas such as tees and putting greens that drain poorly (thus remaining continuously wet) and are heavily shaded. However, moss can rapidly fill a void if thin turf develops, sun or shade. Moss can survive weather extremes in a dormant state or by living symbiotically with blue-green algae. Algae, therefore, can be a precursor to moss encroachment and should be discouraged to prevent moss colonization. Silvery thread moss occurs first in weak turf areas such as ridges and mounds where grass is thin from scalping, drought, or nematode infestations. It is favored by acidic, infertile soils with thatch. It typically forms in summer following periods of rain or overcast, warm days. It is encouraged by (1) extremely low mowing heights, (2) minimal N fertilization, (3) increased use of sand growth medium, (4) intense topdressing with finer textured sands that slow drainage, and (5) loss of mercury-based fungicides. Moss appears sensitive to metal-contaminated soils, with the heavy metals, especially mercury, being most toxic. These metals participate in the destruction of chlorophyll molecules in moss, which is why many copper-based products are currently being used against it.

**Symptoms/signs.** Turf areas in partially shaded, damp locations become weak and begin to thin. Moss begins to predominate in these areas. Moss forms a tangled, thick, green mat occurring in patches over the soil surface (Figure 12.76). Moss will develop on turf areas where grass is less dense than normal and surface soil moisture is high. Acidic, infertile soils with excessive thatch also favor moss development. If left untreated, these areas will continue to "thin out" and expand in size until a large mat forms, preventing growth of the grass and penetration of irrigation water. Moss mats typically develop in summer following periods of rainy, overcast, warm days.

**Cultural controls.** Control involves a long-term, persistent program combining cultural and chemical control methods in realizing healthy turf is the only means to prevent and cure moss. Control begins by correcting those conditions that predispose turf to moss growth. This involves reducing surface moisture by improving air circulation and light exposure by removing adjacent underbrush and selectively removing trees. Improve surface and subsurface drainage and reduce irrigation frequency and amount. Reduce freely available N at the site. Reduce irrigation and improve turfgrass growth where moss is present so the turf can form a dense area. If the area occupied by moss is large, spiking, verticutting, and topdressing will help to break up and dry the mat. Moss turning orange-brown or golden brown in color indicates positive desiccation is occurring. However, mosses are adapted to survive periods of desiccation, so simple drying the moss will not cause it to die.

Several trends in fertility and moss development have been noted. For example, Ca-rich soil may encourage certain moss species while moss tends to be discouraged in K-adequate soils. Spike or rake the dehydrated moss layer to remove any remaining impervious layer. Products controlling moss that can be phytotoxic to turf include hydrogen peroxide, copper sulfate, copper soaps, and zinc sulfate. Baking soda applied lightly at 6 ounces/gallon of water (40 g/L) also helps desiccate moss.

**Chemical controls.** Chemical control is erratic and often unsuccessful, especially if agronomic practices are not corrected that favor moss growth and development. Ammonium sulfate at 1/10 to 1/8 pound N/1,000 square feet (0.5 to 6.1 g N/100 m$^2$) applied weekly is thought to help desiccate moss and encourage competitive turf growth. Applying ground limestone (75 to 100 pounds/1,000 square feet, 368 to 491 kg/100 m$^2$) or hydrated lime (2 to 3 pounds/1,000 square feet in 3 gallons of water, 9.8 to 15 kg/100 m$^2$ in 1,220 L) will help desiccate the moss and raise the soil to a pH level that favors competitive turf growth. Diluted bleach and dishwashing detergent at 4 ounces/gallon water (31 mL/L), applied as a drench; chloride; ferrous sulfate at 4 to 7 ounces per

1,000 square feet (13 to 22 L/ha); granular iron sulfate at up to 3 pounds/1,000 square feet (15 kg/100 $m^2$); or ferrous ammonium sulfate at 10 ounces/1,000 square feet (31 kg/100 $m^2$) also may help reduce moss growth. Five to seven treatments applied at two-week intervals are often necessary. However, these should not be used on greens during hot temperatures, as they may cause varying levels of turf discoloration. Applications should be performed on sunny days when temperatures are between 55°F and 80°F (13°C and 27°C). Iron (Fe)-containing products should be used if a copper (Cu)-containing product is also used.

Chlorothalonil applied for three consecutive weeks also has suppressed moss. This works best when air temperatures at the time of application are >80°F (27°C), preferably >85°F (29°C). In cooler weather, copper hydroxide alone (e.g., Kocide 2000) or copper hydroxide combined with mancozeb (e.g., Junction) at 0.1 to 0.15 pound Cu/1,000 square feet (0.1 to 0.17 kg Cu/ha) can be used every two weeks for a total of five to seven applications. Due to the potential of Cu buildup, limit applications to a total of 1 pound Cu/1,000 square feet (1.1 kg Cu/ha) yearly. Iron chlorosis may occur with copper use; therefore, apply Fe at 0.05 pound Fe/1,000 square feet (0.06 kg Fe/ha). Products containing K salts of fatty acids (e.g., DeMoss, HO2) applied weekly at 2 to 3 ounces/1,000 square feet (6.4 to 9.5 L/ha), or No-Mas (22% fatty acid) at 0.8 gallon/1,000 square feet (3.2 L/100 $m^2$) may be used to control moss in turfgrasses. These should be applied in high rates of water, greater than or equal to 6 gallons water/1,000 square feet (24 L/100 $m^2$). Two applications, two weeks apart, are typically needed. They control moss through a contact mode of action but should be carefully used. All label information should be closely followed. Again, if Fe chlorosis (turf yellowing) occurs, alternate the fatty acid applications with 0.05 pound Fe/1,000 square feet (0.06 kg Fe/ha). Mixing Fe with fatty acids in a tank often causes Fe precipitation, clogging spray nozzles. High soil or spray water pH may reduce control.

Carfentrazone (QuickSilver 1.9 pounds/gallon) also is used to help desiccate silvery thread moss. On greens, use 0.098 pound active ingredient/acre (1.1 kg/ha or ~6.7 ounces/acre, 0.5 L/ha) to control silvery thread moss and apply as often as every two weeks when temperatures are ≤85°F. Mix in a tank with a surfactant and chlorothalonil to improve control. Annual bluegrass can be damaged at rates greater than 2 ounces/acre (0.15 L/ha). Do not apply if bensulide has been applied within the previous 75 days. Before application, spike the moss and maintain a good N spoon-feeding program. Add bentgrass seed to improve the turf stand.

## Localized Dry Spots

At least one causal agent of localized dry spots is Basidiomycete fungi, probably *Lycoperdon* spp. There may be other reasons for development of localized dry spots not due to fungi, such as coating of sand particles by humic substances, initial poor mixing of the greens mix, buried rocks and other debris, and tree roots near the soil surface.

**Occurrence.** Localized dry spots are most noticeable on close-cut areas such as tees and putting greens, usually during warmer months where this water-repellent soil develops in the top 2 inches (5 cm) of the soil profile. Localized dry spots caused by fungi have been primarily observed on greens less than three to four years old, especially those aggressively topdressed with sand and during drought periods.

**Symptoms/signs.** Dry spots are several inches to several feet across and often irregularly or serpentine shaped (Figure 12.77). An affected area will appear drought stressed, despite daily irrigation or rainfall (Figure 12.78). "Puffball" mushrooms may be present throughout the dry area, but these signs are not always apparent. The fungus has colonized (covered) the sand particles in the rootzone mix. Due to this fungal covering, the sand is now hydrophobic and repels water, despite heavy rainfall or watering. Soil hydrophobicity primarily occurs in the upper 2 inches (5 cm) of the soil profile. It is thought that, as fungi mycelium decomposes, organic substances are released that coat and bind the coarse sand particles so tightly together they prevent water penetration. This organic matter decomposition is a natural process and cannot be stopped. Also, it has been recently observed that anthracnose basal rot of creeping bentgrass may be initiated by development of stress from localized dry spots. Localized dry spots can also be a symptom of nematode infestations in sand-based root zones.

**FIGURE 12.77** Localized dry spot on a bermudagrass golf green.

**FIGURE 12.78** Drought-stressed bentgrass from localized dry spot.

**FIGURE 12.79** A water drop test to ascertain water repellency of soil from localized dry spot area.

To determine the water repellency tendencies of a soil, the water drop test can be used (Figure 12.79). A soil core 1 inch (2.5 cm) in diameter and at least 6 inches (15 cm) in depth is extracted. Drops of water are place on soil at increments of 1/2 to 1 inch (1.3 to 2.5 cm), starting at the soil surface. If droplets remain intact on the surface for more than five seconds, the soil is hydrophobic. The longer droplets remain, the more water repellent the soil is.

**Cultural controls.** Since hydrophobic soils tend to be in the upper 2 inches (5 cm) of soil, management practices to encourage rooting beyond the 2-inch (5 cm) depth should be implemented. Hydrophobic sand must be broken up and wetted, which can be accomplished by spiking the dry patch every five to seven days or core aerifying. For a small area, a pitchfork or similar tool will accomplish this task. Irrigate dry patches by hand several times a day, in addition to any normal irrigation or rainfall.

The addition of a wetting agent to the water is also useful and should be watered in. Some programs of prevention by utilizing several applications of certain wetting agents have been successful. Ideally, it is best to treat the entire green with the specified wetting agent and then, if needed, go back and spot treat areas continuing to show localized dry spot. Wetting agents can last up to five months but are affected by organic matter content, cultural practices, and the degree of hydrophobicity. When excessive thatch, mat, or compacted soil is present in conjunction with LDS, core-cultivate prior to wetting agent application. Soil treated with wetting agents will not hold more water than its normal field capacity after treatment. However, the soil surface of sands may drain and dry out more quickly following wetting agent use. If excessive organic matter (thatch or mat) is present, moisture may be retained longer when treated with a wetting agent. Using high-pressure water injection cultivation in combination with a wetting agent also helps to alleviate symptoms.

**Chemical controls.** No chemical controls are currently registered as localized dry spot is from organic coating of sand particles and not directly from fungi. Therefore, fungicides rarely work to improve localized dry spot. If chronic infestations of fairy ring fungi such as *Lycoperdon* spp. are present, then control using flutolanil, azoxystrobin, or pyraclostrobin may be beneficial.

## ADDITIONAL LOCALLY IMPORTANT TURF DISEASES

### Brown Ring Patch

Brown ring patch, caused by *Waitea circinata* var. *circinata*, occurs on roughtstalk bluegrass (*Poa trivialis* L.), annual bluegrass, and less so on creeping bentgrass. Its asexual stage is *Rhizoctonia circinata* var. *circinata*. Affected plants develop a leaf and stem blight plus crown rot. Initial symptoms involve leaf yellowing followed by development of dark, water-soaked appearance of the whole plant. As the water-soaked tissue dries out, plants turf reddish brown, forming the thin, ring-like symptoms, approximately 4 to 8 inches (10 to 20 cm) in diameter, when late winter/early spring maximum daytime air temperatures reach 86°F to 95°F (30°C to 35°C). These can eventually reach several feet (>0.5 m) in diameter with green grass often remaining within the rings (Figure 12.80). Rings may become sunken and often appear as a series of small interconnected crescents rather than the circular style rings associated with other diseases. The pathogen survives in the lower plant canopy into the thatch layer. Disease severity is almost always higher in the presence of thatch and soil compaction. Less aggressive PGR use along with sufficient irrigation, N fertility, and fungicide applications help reduce disease epidemics. When applying PGRs in early spring to suppress *Poa* seedheads, mixing a preventative fungicide in a tank is suggested. Once suppressed, typically two to three weeks are needed for the disease symptoms to subside and turf to recover.

### Rapid Blight

The causal agent of rapid blight is believed to be a species of marine slime mold in the genus *Labyrinthula*, possibly *L. terrestris*. It occurs in arid climates or during periods of extended dry weather. Rapid blight has occurred on perennial ryegrass and rough bluegrass when used to overseed bermudagrass in certain western and southern U.S. states during fall, winter, and spring. It has also been a problem in annual bluegrass and creeping bentgrass putting greens in parts of California.

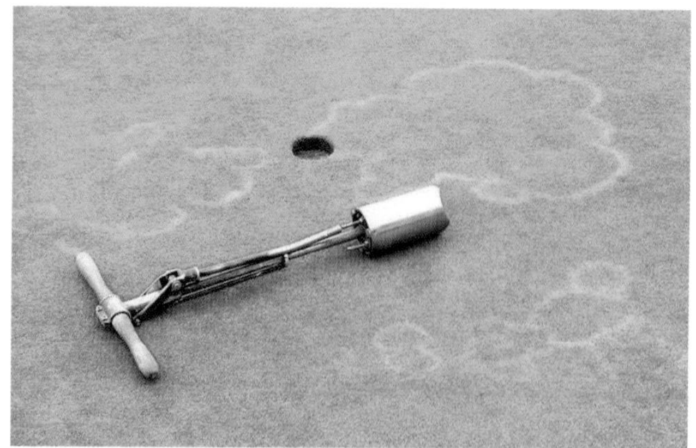

**FIGURE 12.80** Brown ring patch of annual bluegrass. (Courtesy Frank Wong.)

*Labyrinthula* spp. are known to occur in saline environments in terrestrial sites, and outbreaks of rapid blight have been problematic in golf courses with saline irrigation water and/or with soils with highly soluble salts. Damage to dormant and actively growing bermudagrass has not been observed.

Initial symptoms include small yellow, orange, or reddish-brown spots or patches that enlarge quickly to 1 foot (0.3 m) in diameter (Figure 12.81). Spots appear roughly circular with chlorotic turf in the center of patches, surrounded with a darker water-soaked (or greasy) appearance in the grass bordering the affected patches (Figure 12.82), but roots are typically unaffected. No association with mycelium of true fungi has been observed. Due to the remarkable speed the disease spreads, especially in immature seeded turf stands, the name "rapid blight" has been proposed.

Seedlings of overseeded bermudagrass turf appear to be the most susceptible. The disease has appeared in annual bluegrass and overseeded rough bluegrass, and perennial ryegrass greens closely following the first mowing. Slender creeping red fescues, creeping bentgrass, and alkaligrass have shown good resistance. Most severe cases include the wetter and lower areas on greens. The pathogen appears to be tracked by mowers across turf areas. The disease has been reported at sites with high irrigation water salinity, sodium, or bicarbonate levels, with sodium absorption ratios (SARs) ranging from 20 to 40 meq/L. Soil tests indicate high sodium-based saturation levels in the 10% to 35% range may also be a critical factor in the incidence of the disease. Courses that manage their water quality have fewer disease problems. In addition, it may be prudent to wash mowers after mowing affected greens. Minimize turf shoot abrasion or damage situations like mowing wet greens and using excessive sand topdressing rates that appear to increase the severity and spread of the disease.

**FIGURE 12.81** Initial symptoms of rapid blight disease of *Poa trivialis.*

**FIGURE 12.82** Advanced symptoms of rapid blight disease of *Poa trivialis.*

Traditional *Pythium* blight fungicides have so far been unsuccessful in controlling the disease. The strobilurin fungicides, trifloxystrobin and pyraclostrobin, appear promising. Applications of mancozeb alone or in combination with trifloxystrobin have also prevented the disease. Curative applications of mancozeb alone, trifloxystrobin, or pyraclostrobin may stop development in early stages. Copper-containing fungicides also provide control and have potential as a rotational treatment.

## Yellow Tuft or Downy Mildew

The causal agent of yellow tuft is *Sclerophthora macrospora* (Sacc.) Thirum., Shaw and Naras. All turfgrasses can be affected, but bentgrass greens and Kentucky bluegrass are most susceptible. Yellow tuft appears on golf greens as small yellow spots 1/4 to 2 inches (0.6 to 5 cm) in diameter (Figure 12.83). These spots have plants with prolific (e.g., 20 or more) tillers, giving them a tufted appearance (Figure 12.84). Infected plants have short root systems, are easily pulled up, and die during stress periods. Yellow tuft symptoms most often develop during cool, moist periods of spring and fall. Plants appear to recover in summer. The pathogen is an obligate parasite and a member of the oomycete fungal group. The fungus survives as dormant oospores and mycelium in plant tissue, thatch, and leaf litter. Oospores may germinate to produce a lemon-shaped fruiting body. Low areas that often puddle are most often infected since the fungus disperses from these fruiting bodies as swimming zoospores. Control begins with good surface drainage. Fosetyl-Al, metalaxyl, mefenoxam, and fludioxonil provide chemical control when applied preventatively in one or two applications in spring or fall when nighttime temperatures

**FIGURE 12.83** Yellow tuft (or downy mildew) of creeping bentgrass.

**FIGURE 12.84** Tufted bentgrass growth from yellow tuft disease.

are in the 50s°F (10s°C). Liquid Fe applications may help mask disease symptoms. Once tuft formation occurs, fungicides provide little cosmetic turf recovery.

### Yellow Spot (Formerly, *Curvularia* Blight)

Yellow spot is currently thought to be caused by cyanobacterium (a blue-green algae), either *Phormidium* or *Oscillartoria* species, but could also involve *Curvularia geniculata* (Tracy and Earle) Boedijn, *C. lunata* (Wakk.) Boedijn, and others. Bentgrasses and bluegrasses are most susceptible, especially during hot (≥85°F, 29°C) temperatures. Symptoms involve yellow, chlorotic spots of infected plants up to 4 inches (10 cm) in diameter and turf stands that thin in an irregular pattern (Figure 12.85). Brownish leaf lesions similar to leaf spot diseases may be present, particularly on older, senescent leaves. Plants damaged from heat, drought, and/or herbicide stress are most susceptible. Yellow spot is most common on newly established bentgrass greens, and generally becomes apparent in midsummer after abundant rainfall during the high-heat periods of mid- to late summer. Turf exposed to full sun, growing on south-facing slopes or adjacent to paved surfaces, is often most affected. Light, frequent watering generally is associated with more disease compared to deep and infrequent cycles. The disease rarely kills the grass, but symptoms of yellow patches of a few inches in diameter are unsightly. A mat of algae is consistently observed on plant crowns, leaves, and stems. The cyanobacteria climb from the mat up the turf surface, presumably seeking light. The yellowing is thought to be from toxins diffused from the cyanobacteria into the plant's vascular system. It has been observed that a good preventative fungicide program for brown patch and *Pythium* blight will help to prevent stress associated with yellow spot. Alleviating stress through soil aeration, good irrigation practices, increasing mowing height, and proper fertility helps plants resist infection. Fungicides that control leaf spot diseases also may be beneficial in controlling this disease. Chlorothalonil, mancozeb, and propiconazole appear to provide good disease suppression when applied in 2 gallons of carrier water per 1,000 square feet (8 L/100 m$^2$). Repeat applications 14 to 21 days apart may be needed.

### Dead Spot

The causal agent of dead spot is *Ophiosphaerella agrostis* Dernoeden, M.P.S. Camara, N.R. O'Neill, van Berkum et M.E. Palm. First identified in Maryland, the disease has been found on creeping bentgrass as far north as Michigan, west as far as Missouri, and along the eastern seaboard of the United States almost exclusively in newly seeded turf.

Environmental stress conditions appearing to enhance disease development include hot, dry weather, and low fertility. Symptoms begin as small, reddish-brown spots 1/2 to 1 inch (1.3 to 2.5 cm) in diameter and often are confused with dollar spot, copper spot, black cutworm damage, or golf ball injury (Figure 12.86). Spots enlarge at a slow rate, with dead tissue in the center of the spots turning tan, and leaves in the active, outer edge appearing bronze or red (Figure 12.87). Spots generally do not coalesce and usually enlarge to no more than 4 inches (10 cm) in diameter. Turf is slow to recover from dead spot damage, leaving depressions in the putting surface

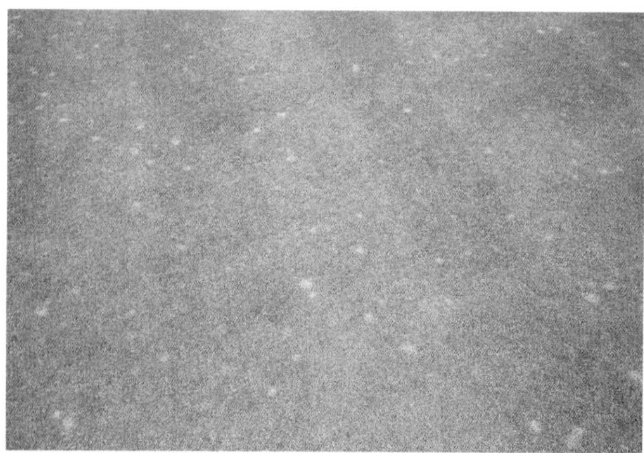

**FIGURE 12.85** Yellow spot (*Curvularia* blight) disease of creeping bentgrass.

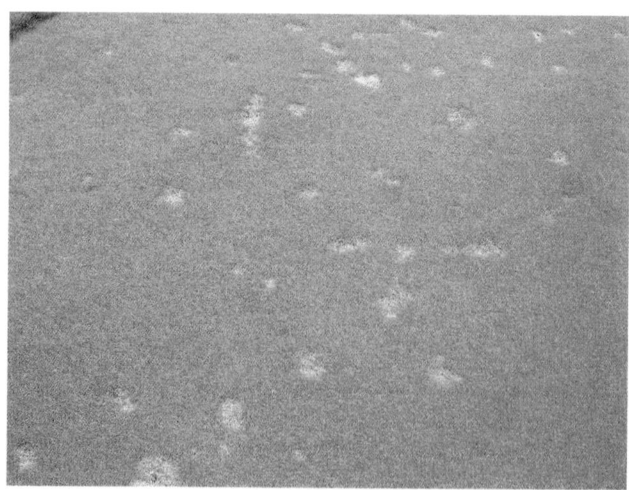

**FIGURE 12.86** Dead spot disease of creeping bentgrass.

**FIGURE 12.87** Tan leaves associated with dead spot disease of bentgrass.

for extended periods. Unlike other turfgrass pathogens within the genus, black, flask-shaped fruiting bodies (pseudothecia) often are found embedded in necrotic leaf, sheath, and stolons of infected plants. Mycelium is not observed in the field. Dark-brown to black hyphal masses and runner hyphae, however, are commonly found on the nodes of bentgrass stolons. Turf in the center of infected spots is killed and recovery from the disease is very slow.

The pathogen also affects hybrid bermudagrass greens overseeded with *Poa trivialis* in the southeastern United States. The disease is most commonly found within one to two years of seeding new sand-based greens or on older greens that have been fumigated with methyl bromide. Dead spot generally appears in areas receiving full sun and good air circulation, and infection centers first appear on ridges, mounds, or south-facing slopes of greens. On bentgrass, dead spot is most severe in midsummer. Disease symptoms on bermudagrass generally appear in early spring.

All major bentgrass species used on golf course putting greens are susceptible to the disease as no resistant cultivars are known. Limited information on cultural controls is available; however, applications of an acidifying fertilizer (e.g., ammonium sulfate) may reduce disease incidence and severity. Often, urea, nitrates, or lime can increase disease activity. If the disease becomes established, light fertilization, coupled with chemical control, aids in the healing of dead spots.

Dead spot is most successfully managed when treated prior to the onset of disease symptoms and is difficult to control once infection centers appear. When the disease is active, fungicides should be reapplied on a 7- to 10-day schedule. Although fungicide efficacy varies within each chemical class, most QoI and sterol-inhibiting fungicides have not provided adequate disease control to date. Recovery from existing damage is typically slow.

### White Patch

White patch (*Melanotus philiipsii* (Berk. and Br.) Singer) occurs on tall fescue (primarily), red fescue, chewings fescue, and creeping bentgrass. Symptoms appear as white circles or patches up to 1 foot (0.3 m) in diameter, and small white or tan mushrooms (called sporocarps) about 1/4 inch (5 mm) in diameter are produced directly on affected leaves (Figure 12.88). It tends to damage newly seeded areas, especially when excessive seeding rates result in a very dense stand. Unlike brown patch (*Rhizoctonia solani*), white patch is usually associated with low rather than high N fertility levels as well as insufficient soil moisture. It is most commonly seen during summer and early fall, especially when daytime temperatures are above 85°F (29°C), night temperatures higher than 70°F (21°C), and relative humidity above 70%.

**FIGURE 12.88** Whitish mushrooms of white patch disease of tall fescue.

### Additional Diseases

Many additional diseases can invade various turfgrasses, while many are present in the turfgrass environment but are not active. Most of these diseases are local in nature, often influenced by local microenvironments and agronomic growing conditions. Consult local turf specialists for proper identification and control options. Refer to the reference section of this text for further information on additional diseases.

## TURFGRASS NEMATODES

Plant parasitic nematodes, alone as well as in combination with other drought-related stresses, nutritional status, and fungal diseases, can cause serious damage to turfgrass stands. In subtropical and warm regions, parasitic nematodes are among the most important turfgrass pests. Symptoms of nematode damage to turfgrasses can be easily confused with the effects of nutritional deficiencies, water or heat stress, or other diseases. A good knowledge of nematode sampling, biology, and turfgrass management strategies for different environmental conditions is a prerequisite for successful management of turfgrass nematodes without heavy reliance on nematicides.

## NEMATODE DESCRIPTION

Nematodes (sometimes called eelworms) are probably the most abundant multicellular animals known. Fortunately, the overwhelming majority are nonparasitic (free living). Nematodes are microscopic (tiny), aquatic, nonsegmented roundworms measuring 0.1 to 3 mm (1/250 to 0.125 inch) in length. With the unaided eye, adults are barely visible. When viewed under magnification, they appear almost transparent but with definite body outlines (Figure 12.89). Most nematodes are slender (eel like) throughout their life stages, but a few become swollen as they mature. Although flowers, stems, leaves, and roots of some plants are attacked by certain plant parasitic nematodes, most nematode pests in turfgrasses are root parasites.

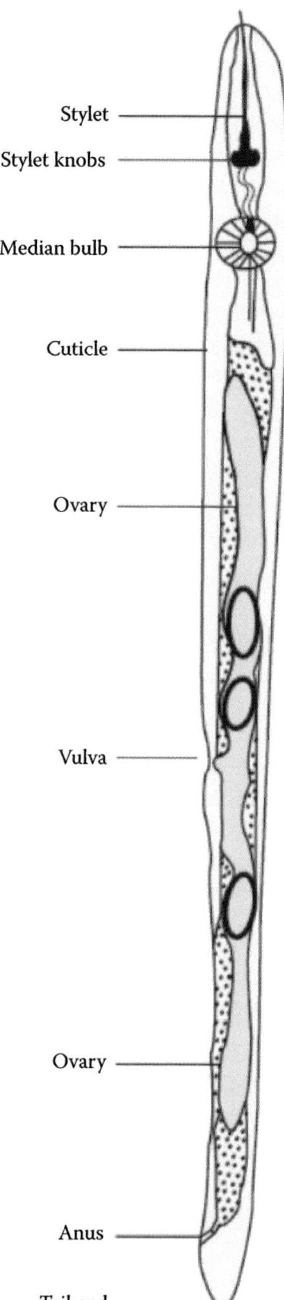


**FIGURE 12.89** Illustration of a plant parasitic nematode.

Most nematode parasites of turfgrasses remain entirely outside of roots with only their stylet (hollow, protrusible spear) thrust inside the root (these are termed *ectoparasites*). A few are *endoparasites,* where they spend part of their life cycle completely inside the roots. Endoparasitic nematode feeding often causes root swelling or galls, or rotting and lack of root-hair development. Some endoparasitic nematodes move freely within and out of the root for all developmental stages (termed *migratory endoparasites*), while other species remain in a permanent feeding position within the root tissue for most of their life cycle (termed *sedentary endoparasites*). Sting, ring, and stubby-root nematodes are examples of ectoparasites; lance and lesion nematodes are migratory endoparasites, while root-knot nematodes are sedentary endoparasites (Table 12.9).

Plant nematodes are aquatic animals living in the soil water film or in plant fluids. They are very well-adapted plant parasites. Females produce a few dozen to over 500 eggs each. When a host plant is unavailable, eggs of some species survive for years but hatch quickly when stimulated by exudates from plants. Their activity, growth, and reproduction increase as soil temperature rises from about 50°F to about 90°F (10°C to 32°C). Generation time is between three and six weeks for many nematodes.

### How Nematodes Cause Damage

Plant nematodes are obligate parasites, which means they must find and feed on a living plant host at some stage of their life cycle. Most species attack a wide range of plant material, and can survive on weeds or on overseeded grasses. Many also may survive short periods in frozen soils. Plant-parasitic nematodes damage plants directly by their feeding activities, which involve puncturing cells (mostly in the root system) with their stylet, injecting digestive juices into the cells, and then sucking the liquid contents out (Figure 12.90). A stylet is a hollow, needlelike organ in the head, adapted for feeding activities. Infected root tips typically stop growing and develop a "stubby" or abbreviated growth habit.

Although nematodes rarely kill their host, their feeding activities cause localized cell death, induce host reactions that change the integrity of root tissue, and provide entry points for secondary invasion by fungi, all of which result in abnormal root function. Furthermore, uptake of nutrients and water by the roots is adversely affected, eventually weakening plants and causing turf stand thinning. Aboveground symptoms often involve turf yellowing (chlorosis), wilting and decline (thinning), and a proliferation of weeds such as spurge or pusley (Figure 12.91). Since nematodes mostly damage turfgrass roots (and not aboveground plant parts), symptoms of injury often go unnoticed until soil water is limited (Figure 12.92). These symptoms are often confused with environmental stress symptoms or nutritional problems and may be difficult to diagnose. Fungi such as *Curvularia* and *Pythium* can feed on nematode-damaged roots and are often seen when soil is assayed for nematodes or disease.

**TABLE 12.9**
**Nematodes Affecting Turfgrasses and Approximate Damage Threshold Levels Typically Used to Justify Nematicide Application**

| Common Name (scientific name) | Most Susceptible Grasses | Threshold[a] (no./100 cm³ soil) |
|---|---|---|
| | **Endoparasitic** | |
| Root-knot (*Meloidogyne* spp.) | St. Augustinegrass | 80 |
| | Bentgrass | 80 to 100 |
| | Bermudagrass | 80 |
| | Zoysiagrass | 80 |
| | Seashore paspalum | 80 to 300 |
| Lesion (*Pratylenchus* spp.) | All | 150 |
| Cyst (*Heterodera* spp.) | St. Augustinegrass | na |
| | **Ectoparasitic** | |
| Sting (*Belonolaimus longicaudatus*) | St. Augustinegrass | 10 to 20 |
| | Bermudagrass | 10 to 20 |
| | Seashore paspalum | 10 to 25 |
| | Zoysiagrass | 10 to 20 |
| Lance (*Hoplolaimus* spp.) | St. Augustinegrass | 50 |
| | Bentgrass | 40 to 60 |
| | Bermudagrass | 80 |
| | Seashore paspalum | 40 to 120 |
| Ring (*Criconemella* spp.) | Centipedegrass | 150 to 300 |
| Stubby-root (*Nanidorus minor*) | All | >100 |
| Sheath (*Hemicycliophora* spp.) | All | 80 to 300 |
| Spiral (*Helicotylenchus* spp.) | All | 200 to 700 |
| Awl (*Dolichodorus heterocephalus*) | Turfgrass (especially bermudagrass) in wet locations | 80 |
| Dagger (*Xiphinema* spp.) | Rarely a turf pest | 150 to 300 |
| Stunt (*Tylenchorhynchus* spp.) | Rarely a turf pest | 100 to 400 |
| Root-gall (*Subanguina radicicola*) | *Poa annua* | N/A |

[a] Threshold level ranges commonly used in research by universities. Numbers are not absolute as the health of the turf and environmental conditions substantially influence these. N/A = data not available.

**FIGURE 12.90** Bermudagrass root damage (left) from sting nematodes compared to uninfected soil (right).

**FIGURE 12.91** Turf thinning, wilting, and decline from sting nematode infestation.

The extended growing season, warm climate, and sandy soils of tropical and subtropical areas favor development of very high nematode populations and also create conditions in which grasses are most susceptible to nematode damage. Moreover, many activities and management practices on golf courses promote the development of high numbers of nematodes and increase turfgrass susceptibility to nematodes. Nematode feeding alone rarely kills the turf but rather weakens it to where other stresses such as drought, heat, malnutrition, arthropods, pathogens, and weeds can reduce the turf's quality, color, and density.

**FIGURE 12.92** Nematode damage often resembles drought symptoms.

### Nematode Movement

Soils that are naturally sand in composition are often inherently high in plant parasitic nematodes and these readily attack newly planted sites, which are watered and fertilized regularly. Due to their minute size, nematodes are relatively sedimentary, moving only a few inches on their own during their life span. Nematodes, however, also are readily transported when contaminated soil and plants are moved. Infested sprigs and sod are common sources of nematode inoculum to new areas. They also move with infested soil adhering to equipment. Aerification equipment is a potential vectoring agent; thus, these should be cleaned when moving between sites, and areas known to be nematode infested should be aerified last.

Areas to be newly planted with turfgrasses are often fumigated to kill a wide array of pests, including nematodes. Seldom, however, does fumigation completely kill all nematodes and their populations eventually recover and increase.

## MAJOR NEMATODES

Several different kinds of nematodes can affect turf, with the most damaging described in the following pages. Table 12.9 lists damaged threshold numbers of each kind of nematode.

### Endoparasites (Enter Plants)

**Lance (*Hoplolaimus* spp.).** Lance is the most widely occurring nematode in the world. The stylet knobs of lance nematodes resemble a tulip, their tail length is <1.5 times the anal body diameter, they have a median bulb, two ovaries are present, and mature females are vermiform (bulb shaped). These nematodes are migratory endoparasites where they feed mostly ectoparasitically (outside the plant) but can enter and move through the root cortex, making them very difficult to control even with chemical nematicides. For this reason, lance nematodes frequently occur in high numbers more in older turfgrass areas, such as older putting greens where other nematodes were controlled over the years with various nematicides. Also, nematode soil assays often are not good indicators of their numbers, as many or most may be within turf roots and not in soil. In zones of heavy feeding, roots exhibit slight tissue swelling followed by necrosis and sloughing off of cortical tissues. Fewer feeder roots also tend to occur. Lance nematodes have a wide host range and attack all commonly grown turfgrasses (St. Augustinegrass, bahiagrass, creeping bentgrasses, bermudagrass, annual and Kentucky bluegrasses, annual ryegrass, and zoysiagrasses). It is the most important nematode pest of St. Augustinegrass and the hardest nematode to control/manage due to its migratory endoparasitic activity. Contact nematicides tend to be less effective compared to systemic ones.

**Root-knot (*Meloidogyne* spp.).** These nematodes enter and feed inside roots and are classified as sedentary endoparasites. The female enlarges into a pear-shaped mature form that remains embedded within the root. It possesses a knobbed stylet and median bulb. Second-stage larvae are worm shaped and are the infective stage found in soil. They eventually enter root cells with the aid of their stylet and migrate through the root cortex toward the vascular tissue, where they become stationary. Saliva secretion during feeding by the nematodes induces enlargement and accelerated division of root cells at certain points inside roots. Mature female nematodes enlarge, appear pearly white, and produce eggs in a gelatinous sac matrix that usually protrudes on the outside of the root surface. Symptoms on both fibrous and lateral roots appear as distinctively small, spherical, or elongated swellings having the same color as the adjacent root surface (Figure 12.93). Root swellings may be difficult to see without a hand lens. Root-knot nematodes are assumed to be injurious only when numerous, but their importance as a pest of turfgrasses has not been well assessed. In turfgrasses, they occur mostly in thatch and upper soil profiles. Root-knot nematodes have a very wide host range and can colonize all common turfgrasses, but they are most commonly found on zoysiagrass, St. Augustinegrass, bermudagrass, and bentgrasses. It has been observed that, similarly to lance nematodes, root-knot nematodes occur in large numbers in older greens. Their numbers may increase relative to ectoparasites, which may be differentially susceptible to certain nematicides. Single applications of contact nematicides tend not to be very effective as they only affect the second-stage juveniles in soil

**FIGURE 12.93** Root enlargement associated with root-knot nematodes. (Courtesy Bruce Martin.)

and not nematodes within roots. Root-knot nematodes tend to predominate in the top 2 inches (5 cm) of soil profiles so treatments that target this zone work best.

**Lesion nematodes (*Pratylenchus* spp.).** Lesion nematodes are occasional turfgrass pests. They are migratory endoparasites possessing a median bulb, a single ovary, and a short but very robust knobbed stylet that tunnels within turf roots. Root lesions are initially minute and brown. They progressively enlarge and promote secondary fungal invasion, and may eventually girdle the root. With high populations of lesion nematodes, the root system may appear severely pruned. Both larvae and adults can penetrate and move through and between root cells. There is a preference for feeding in the more mature cortical areas behind the root tips. Two common species, *P. brachyurus* and *P. penetrans,* favor soil temperatures of 80°F to 90°F (27°C to 32°C) and 70°F (21°C), respectively, and can complete their life cycles (egg to egg) in six to nine weeks under optimal conditions. Both warm- and cool-season grasses can be colonized by root lesion nematodes.

#### Ectoparasites (Remain in Soil)

**Sting (commonly *Belonolaimus longicaudatus*, also *B. gracilis*, and other species).** Sting nematodes are considered the most damaging nematode to turf. It is largest of the turf parasites in the southeastern United States, with adults reaching lengths of nearly 1/8 inch (3.2 mm). They occur naturally in dune sands and the sandy soils of sandhill regions. Sand-based rootzones in constructed putting greens are ideal habitats for sting nematodes as well as other species. The following combination of characteristics describes the genus *Belonolaimus:* possession of a very long stylet with knobs at the base, a median bulb (valvulated "pump"), vermiform (eel-like) adult, two ovaries, and a round tail end. Sting nematodes use their very long stylets to siphon plant sap from turfgrass roots. These nematodes are ectoparasitic; thus, they remain outside roots while feeding, with only their long stylets penetrated deep into the vascular tissue. Although lesions are evident throughout the root system, the most active feeding occurs at root tips, causing them to be cropped off below the thatch. In sufficient numbers (usually ≥20/100 mL soil), sting nematodes retard overall root development, making plants more sensitive to moisture stress. Turfgrass top growth is stunted, appears yellow, and thins out (Figure 12.94). Sting nematodes are generally found only in very sandy (>80% sand) soils. They tend to move deeper in soil during adverse weather conditions such as extreme hot or cold temperatures, drought, etc., but move back into the rootzone when conditions improve. They feed on all grasses grown for turf, but damage is most severe on bermudagrass, St. Augustinegrass, zoysiagrass, bentgrass, and ryegrasses. Many other grasses are hosts to sting nematodes including oats, centipedegrass, barley, and ryegrass. Centipedegrass in sandy soils can be severely damaged by sting nematodes. Weedy plants such as prostrate spurge, Florida pusley, knotweed, or bahiagrass may invade sting- and also ring-nematode–infested areas. Fortunately, sting nematodes are among the easiest to control, responding to both contact and systemic nematicides.

**FIGURE 12.94** Yellow, thin, poor growth nematode "hot spots" on a golf green.

**Ring (*Criconemella, Mesocriconemella* spp.).** Ring nematodes are characterized by a body with prominent rings, one ovary, vermiform adults, blunt tail, and a knobbed stylet. Ring nematodes feed ectoparasitically (from the outside the roots). Brown lesions appear along the roots and tips, and roots become stunted. When high numbers of ring nematodes are present, severe root rotting may occur. Ring nematodes are widely distributed on many turfgrasses (namely, Kentucky bluegrass, annual bluegrass, St. Augustinegrass, zoysiagrass, bermudagrass, and creeping bentgrass), but are considered to be a major pest on centipedegrass, especially in sandy soils. Being ectoparasites, contact or systemic nematicides can be effective on ring nematodes.

**Spiral (*Helicotylenchus* spp., *Peltamigratus christiei*).** Spiral nematodes are ectoparasites and are among the most frequently found nematodes on all common turfgrasses. They possess a median bulb, knobbed stylet, two ovaries, ventrally located tail tip, and curl in a characteristic spiral shape when relaxed. Spiral nematodes are rarely serious pests of turfgrasses when they are the dominant species encountered. However, they cannot be dismissed entirely as pathogens when they occur in high numbers or at lower populations in combination with other plant parasitic species. In high numbers, roots are poorly developed and discolored brown with premature sloughing of cortical tissues. Aboveground, the turfgrass stand becomes sparse and difficult to maintain because of reduced plant vigor. Leaf blades become narrower and appear pale to chlorotic. Symptoms are not evident during vigorous root-and-shoot growth in the moist months of spring and early summer even in the presence of very high numbers of spiral nematodes. The nematode populations peak in the advent of both higher temperatures and decreased availability of soil moisture in the top few inches of soil (which also coincides with plant decline). Nematode numbers decline during dormancy periods, but increase again in the cooler weeks of early fall when plant growth resumes. Both warm- and cool-season grasses are good hosts for spiral nematodes, with seashore paspalum and dwarf bermudagrass greens as the main hosts. Being ectoparasites, spiral nematodes can be managed with contact or systemic nematicides.

**Stubby-root (*NanidorusNanidorus, Trichodorus* spp.).** These ectoparasites are characterized by the possession of a distinctive short, curved spear without basal knobs and the absence of a median bulb. They mostly feed at growing root tips, which stop elongating, may become slightly swollen and discolored (or both), and often become very short and "stubby." Root symptoms resemble those induced by sting nematodes. Lesions are large, brownish, irregular in shape, and often extend deep into the root tissue. Aboveground symptoms appear as a pronounced chlorosis with reduced growth rates. Most warm-season grasses are hosts of stubby-root nematodes. Kentucky bluegrass, tall and red fescues, St. Augustinegrass, bermudagrass, centipedegrass, and zoysiagrasses are common hosts. Numbers can fluctuate in assays of bentgrass putting greens, and their importance as pathogens of bentgrass is not well understood. Knowledge to date has shown that *Trichodorus obtusus* is more pathogenic to bermudagrass, zoysiagrass, and St. Augustinegrass than is *Nanidorus* (*Paratrichodorus*) *minor.* Being ectoparasites, contact and systemic nematicides are generally effective control agents.

**Awl (commonly, *Dolichodorus heterocephalus*).** Awl nematodes are ectoparasites possessing a median bulb, long knobbed stylet, two ovaries, and female tail ending in a short terminal "awl-like" point. Awl nematodes are very damaging to turfgrasses in wet locations such as the low land areas near lakes, ponds, and canals. Turfgrass hosts include bentgrasses, centipedegrass, tall fescue, St. Augustinegrass, and especially bermudagrass.

**Stunt (*Tylenchorhynchus* spp.).** Stunt nematodes possess two ovaries, a median bulb, a round tail end, but a short stylet with round (not tulip-shaped) knobs. The tail is somewhat cylindrical and the tail end is round. They are an occasional turfgrass pest, but may reach very high populations on bentgrass and *Poa annua* in cooler regions in sand-based rootzones. Brown lesions may be evident on the roots, but definite lesions are not usually present (as a rule). Roots appear shriveled and severely shortened. *Tylenchorhynchus dubius* on bentgrasses feeds primarily on root hairs and areas immediately behind root tips with only the stylet penetrating the epidermal cell. Other species browse more and may group into clusters on epidermal cells that cause a mechanical breakdown of epidermal, cortical, and vascular tissue. Stunt nematodes can survive in the presence of adequate soil moisture for several months without a host. Both warm- and cool-season grasses are suitable hosts. Being ectoparasites, stunt nematodes can be managed by either contact or systemic nematicides.

**Dagger (*Xiphinema* spp.).** Dagger nematodes are long, slender, and shaped like a "dagger." The stylet is very long with basal flanges (not knobs). The guiding ring for the stylet is nearer to the base than it is to the apex of the stylet. A median bulb is absent. Dagger nematodes feed ectoparasitically and induce reddish-brown to black, slightly sunken root lesions. In very high numbers, extensive feeding reduces root growth. Warm-season grasses (especially zoysiagrasses) and perennial ryegrass are suitable hosts. Other hosts include bentgrasses and Kentucky bluegrass. Being ectoparasites, dagger nematodes can be managed by either contact or systemic nematicides.

**Other nematodes.** Other nematodes may damage turf, especially when numerous other pests, pathogens, or environmental conditions stress the turfgrass. Occasional turf pests include spiral, stunt, and dagger nematodes (already previously mentioned); sheath (*Hemicycliophora* spp.) and sheathoid (*Hemicriconemoides* spp.) nematodes; cyst nematodes (*Heterodera* and *Punctodera* spp.); and, root-gall nematodes (*Subanguina radicicola*). All, except the cyst and root-gall nematodes, feed ectoparasitically. *Hemicycliophora* is identified by the presence of a loose sheath around the body of the nematode. *Hemicriconemoides* do not have a loose sheath, but do have fairly pronounced annulations in the cuticle and a tail end that is more pointed than that of *Mesocriconemella* (ring nematode). Cyst nematodes are sedentary endoparasites, so only the second-stage juveniles are found in soil assays. The bodies of female cyst nematodes enlarge into lemon shapes that eventually become completely filled with eggs and the body forms a hard cyst that protects the eggs from unfavorable environmental conditions. Egg-filled cysts can be observed still attached to affected roots. St. Augustinegrass and creeping bentgrass are especially susceptible to cyst nematodes. An endoparasitic seed and gall nematode (*Subanguina radicicola*) can be found on *Poa annua* and bentgrass in coastal regions of Northern California. It has the ability to feed on crown tissue, causing gall formation, not just roots like most turf parasitic nematodes. Galls caused by these nematodes tend to be at ends of roots and sickle shaped compared to more rounded and located on the upper portion of roots with root-knot nematodes.

**Stem gall or Pacific shoot-gall (*Anguina pacificae*).** Unlike other turf-damaging nematodes that attack roots, stem gall nematodes infect stem and crown tissue of *Poa annua* (mainly) and bentgrass. It is limited in the United States to portions of coastal California and is also found in the UK and can be very damaging to *Poa* greens. These primarily are endoparasites as only the second-stage juveniles remain outside plant tissue. Juveniles move from the soil up to a shoot tip via a thin film of water. They penetrate plant tissue, inducing a cavity-containing gall formation that contains many nematodes (Figure 12.95).

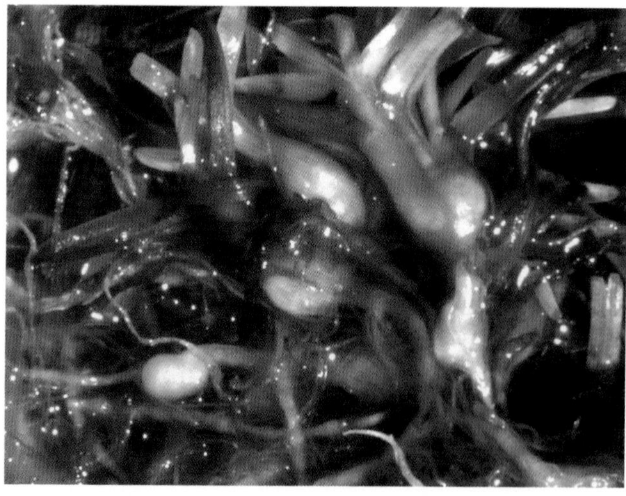

**FIGURE 12.95** Stem gall nematodes infecting *Poa annua* stem and crown tissue.

Over time, as galls decompose or are destroyed by mowing, juveniles are released and move to the next shoot tip via a film of water to continue the life cycle.

## DIAGNOSIS OF NEMATODE PROBLEMS

Although diagnosing nematode problems is often difficult, there are several clues that are used during the investigative process. These include the types of symptoms, pattern and timing of damage, previous history, nematode species present, and the results of nematode counts.

### Root Symptoms

Root symptoms include lesions; galls; stubby, swollen root tips; lateral root proliferation; and/or stunted shallow root systems with few feeder roots (Figure 12.96). Common symptoms associated with certain nematodes were previously described. The penetration and movement of endoparasitic nematodes within roots leave openings that allow root invasion by secondary microorganisms in the soil such as fungi. The result is accelerated rotting (blackening) of roots and proneness of plants to wilting. Heavily affected root systems have much less soil clinging to them when a plug is pulled from the turf compared to unaffected turfgrass stands. The root symptoms, however, are not unique to nematodes and should always be considered in conjunction with other observations when diagnosing nematode problems.

### Aboveground Symptoms

Aboveground symptoms include wilting, thinning or gradual decline, or the yellowing of leaves without lesions or deformities (Figure 12.97). Again, these symptoms are not unique to nematodes and can be caused by heat or drought stresses, nutrient deficiency, fungal diseases, insect feeding, soil compaction, prolonged saturation of soil with water, or chemical contamination. The turfgrass is weakened by the nematode damage and is unable to outcompete invading weeds (such as sedges, knotweed, Florida pusley, and spurges). Nematode-affected areas may appear weedier than other turf areas.

**FIGURE 12.96** Stunted, swollen turfgrass root tips indicating nematode feeding.

**FIGURE 12.97** Wilting, thinning, and gradual decline of nematode-infected bermudagrass fairway.

### Pattern of Damage

Nematodes do not cause uniform damage to an expanse of turfgrass (as occurs, for example, in rust diseases). Rust fungi produce millions of dry spores that are easily dispersed over long distances in open air by the wind. Nematodes, however, produce a mere 50 to 500 eggs (per female). These eggs are in the soil environment and are not immediately and easily dispersed from the source by an active agent (such as the wind). Nematodes do not migrate more than 1 meter in one growing season and must depend on movement in surface-water runoff, irrigation water, and soil clinging to equipment, sod, or plugs for long-range movement. Nematodes, therefore, show an irregular (somewhat patchy) horizontal and vertical distribution in the soil. Symptoms above ground also follow this irregular distribution, but nematode-affected areas usually do not show distinctly sharp boundaries. However, these symptoms resemble early stages of many fungal turfgrass diseases and could be misdiagnosed as nematode related.

### Timing of Damage

Plant parasitic nematodes are obligate parasites and feed most when the turfgrass roots are actively growing. They are therefore most numerous during mild weather, in late spring (May to June) and early fall (October to November) on warm-season grasses, and late spring (June to July) and again in fall (November to December) on cool-season grasses. The turf usually shows no aboveground symptoms of nematode damage until unfavorable environmental conditions prevail (for example, during hot, dry periods when soils are dry).

### Soil Sampling

Nematode counts are the surest way to determine whether a problem in the turf is indeed caused by parasitic nematodes. It is good practice to take soil and root samples monthly so changes in the populations of plant parasitic nematodes in the turfgrass stand can be monitored and kept below acceptable damage threshold levels. Given the irregular distribution of nematodes in the soil, it is imperative that adequate soil/root sampling be conducted in order to confirm the nematode problem to some degree of certainty. Golf course superintendents could be wasting precious time and thousands of dollars on fungicide applications if a problem is not correctly diagnosed as nematode related. The same waste of resources would occur if the problem is suspected to be caused by nematodes when the real cause is another stress.

To sample nematodes, composite samples of 10 to 20 soil cores are taken from suspected areas. If damage is visible, sample the margins of affected areas. If not, sample in a zigzag pattern. Soil cores should be at least 1/2 inch (1.3 cm) in diameter to 4 inches (10 cm) deep, placed into a clean plastic bag, and sealed. The laboratory will take a 100 $cm^3$ (about a cup) subsample for assay. If several areas show symptoms of nematode damage, separate the samples from each area and store these out of direct sunlight and at room temperature. Sample from the edge of an affected area, avoiding spots where the grass is already dead. If samples cannot be shipped to a lab within two days, place them in a refrigerator, but do not freeze them. Another common mistake is to place samples in a vehicle where they quickly heat to lethal temperatures. Also, sampling from an adjacent "good" turf site may help provide better insight into the nature of the problem. It is best to sample just prior to periods of root growth and less desirable to wait until damage symptoms appear.

## NEMATODE MANAGEMENT

Although no turfgrass cultivars are currently available that are resistant to all nematode species, significant differences exist among turfgrass varieties in terms of the numbers and species of nematodes feeding on them and their proneness to damage caused by feeding activities. Visual symptoms and even adequate soil/root sampling may sometimes not be enough to confirm a nematode problem. A positive growth response to an effective nematicide may sometimes be required for confirmation.

Turfgrasses tolerate some feeding by most nematodes; therefore, the most practical strategy for nematode control is often the promotion of vigorous root growth (using recommended cultural practices and timely nematicide applications).

1. **Cultural practices.** Certain cultural practices help minimize stresses that make the turfgrass more susceptible to nematodes. To facilitate deeper penetration of the soil by roots, irrigate deeply (but less frequently) instead of using shallow, daily watering. Also, raising mowing heights can improve nematode tolerance by increasing plant rooting. To achieve proper infiltration and adequate $O_2$ levels in soil, coring with narrow, hollow tines or spiking should be performed (in late spring and early summer). Cultivation should be performed at times of the year when the best turf recovery occurs (e.g., in late spring for warm-season turfgrasses and in midspring or early fall for cool-season turfgrasses).

   Excessive fertilization with water-soluble N must be avoided since nematode numbers increase rapidly on succulent roots and, during periods of environmental stress (for example, in summer), the roots are placed under an additional strain. Organic forms of N have been shown to be associated with lower nematode numbers than inorganic forms. However, judicious use of a balanced fertilizer is always advocated.

   Plant diseases, nutrient deficiencies, shade, and soil compaction (traffic) should be managed or minimized in order to decrease the impact of nematode diseases on turfgrass stands. Avoid mowing low to prevent additional stress to the nematode-infested turfgrass stand that is forced to survive with reduced shoot biomass.

   Certain soil amendments to turf grown in sandy soils are known to improve soil composition and reduce the impact of plant parasitic nematodes. Preplanting incorporation of colloidal phosphate and/or composted municipal sludge, or long-term use of the latter as a topdressing, has been shown to reduce nematode damage to turfgrass stands.
2. **Use of tolerant grasses.** Whenever possible, avoid planting species or cultivars that are the most susceptible to the nematode species deemed problematic in a given locality. By establishing turf initially with a tolerant variety, the impact of certain nematodes and the cost of nematode management will be reduced overall. Information on the relative tolerance of the hundreds of varieties of different turfgrasses is scarce. The bermudagrass variety Tifdwarf, however, appears to be more tolerant of stunt and ring nematodes than Tufcote, Tiffine, Continental, or U-3. Tifway is fairly tolerant of the sting nematode, while Tifdwarf, Tufcote, Tifgreen I, Tifgreen II, and Midiron are susceptible. Polyploid St. Augustinegrass varieties are typically more tolerant of sting nematodes compared to diploid varieties.
3. **Chemical control.** Because crop rotation, varietal resistance, biological control, and several other disease-management strategies are not always practical or effective for turfgrass nematode control, the use of chemical nematicides is currently the most reliable approach to reducing parasitic nematode levels in turfgrass stands. Chemical nematicides can be applied as preplant fumigants and as postplant nonfumigant contact chemicals. Fumigants are toxic to plants and are labeled for use only before establishment of the turfgrass stand. In established turfgrass

**FIGURE 12.98** Healthier bermudagrass (right) following nematicide use.

stands, contact nematicides are available as granular or spray formulations and are always watered in immediately after application (Figure 12.98). They also have some insecticidal activity. No single product is effective against all nematodes on a given turfgrass species.

### Soil Fumigation before Planting

Soil fumigants are chemicals applied as gases or liquids that readily vaporize. They are toxic to the turfgrass but may be used to treat soil prior to seeding or planting to reduce plant parasitic nematodes, weeds, fungal pathogens, and other soil-borne microorganisms. Turfgrasses established in fumigated soil show more uniform and vigorous growth. The fumigants used in turf are the gas methyl bromide, and the liquids 1,3-dichloropropene (Telone II), 1,3-dichloropropene-chloropicrin (Telone C-17), and metam-sodium (labeled as Vapam, Sectagon, or Busan 1020). All three fumigants are "restricted-use" pesticides requiring special equipment and application only by licensed professionals, especially when large areas are to be treated.

1. Methyl bromide is a very effective broad-spectrum biocide that has "served" the turf industry well. It is standard practice to fumigate new greens, tees, and areas being replanted with methyl bromide. Custom applicators often have their own name brand of methyl bromide, such as Terr-O-Gas. The commercial production of methyl bromide is being phased out except for sod production as it has been linked to ozone depletion. Methyl bromide will therefore not be available for nematode control after the phase-out period.
2. Telone C-17 is not without problems. It contains tear gas and is now under special review. Residues have been detected in the air near schools and residential areas.
3. Metam-sodium (Vapam) is a useful option, although not as effective as methyl bromide. It can be applied as a drench in water or by injection. After application, metam releases gases, especially methyl isothiocyanate, that provide control. Metam products do not, however, penetrate tough tubers, stolons, and rhizomes, as well as methyl bromide, and are more sensitive to soil temperature and moisture levels. Also, the fumes from metam sodium escape slowly from the soil, especially when the soil is cool, wet, or high in organic matter or clay content; thus, it should be used several weeks prior to an anticipated planting. Busan, Sectagon, and Vapam are examples of commercial formulations of metam sodium.
4. Dazomet (trade name Basamid) is a granular product that is applied preplanting, and can be incorporated with irrigation. The granules dissolve in water and release methyl isothiocyanate similarly to metam-sodium products. Treated areas can be tarped to improve efficacy.

### Nematicides for Established Commercial Turf

**1,3-Dichloropropene.** Until recently, the only synthetic fumigant nematicide available in the United States was 1,3-dichloropropene (trade name, Curfew). It can only be used

**TABLE 12.10**
**Relative Effectiveness of Nonfumigant Nematicides Used in Turfgrass Nematode Control**

| Nematode | 1,3-Dichloropropene (Curfew) | Abamectin (Divanem) | Fluopyram (Indemnify) | Fluensulfone (Nimitz Pro) | Furfural (MultiGuard) |
|---|---|---|---|---|---|
| Sting | Good | Good | Good | Good | Good |
| Awl | N/A | N/A | N/A | N/A | N/A |
| Spiral | Good | N/A | N/A | N/A | N/A |
| Ring | Good | Good | Good | N/A | N/A |
| Stubby-root | Good | N/A | N/A | N/A | N/A |
| Sheath, sheathoid | Good | N/A | N/A | N/A | N/A |
| Lance | Fair | N/A | N/A | Good | Fair |
| Lesion | N/A | N/A | N/A | Good | N/A |
| Root-knot | Good | Good | Good | Good | Fair |
| Root-gall | N/A | N/A | Good | N/A | N/A |

on commercial turf (including golf courses and sod farms) where the risks of exposure can be minimized. Depending on use rate, 1,3-dichloropropene has nematicidal, fungicidal, insecticidal, and herbicidal properties (Table 12.10). This material is injected 5 to 6 inches deep (13 to 15 cm) through chisels spaced 12 inches (30 cm) apart with a coulter, knife, and roller assembly. In the soil, the product turns from a liquid to a gas, killing nematodes as it disperses through the soil. The coulter slices the turf open at least 5 inches (13-cm) deep, with a 6-inch (15 cm) knife, and an attached tube that injects the material at a rate of 5 gallons of product per acre (9.4 L/ha), followed by rolling to seal the slit. With adequate fertility and irrigation, slits typically heal within two to three weeks (Figure 12.99). A 1/2 inch (1.3 cm) irrigation should follow injection to help "cap" the treatment, thereby reducing residual activity. The fumigant then diffuses throughout soil pores, killing nematodes on contact with 98% dissipation within 24 hours after treatment. Nematicide applications should be made in autumn or spring (before nematode populations' peak) during periods when soil temperatures are above 60°F (15.6°C), according to the product label. In transition zones where warm-season grasses go dormant or semidormant, Curfew should be applied in spring or early summer so that slits will have time to heal. Early fall applications would be acceptable on cool-season turfgrasses where healing of slits would occur relatively quickly. Since Curfew can only be used once in a season in states with 24(c) (special local need) labels, spring or early summer may still be the best timing for optimal results.

**FIGURE 12.99** Slits following 1,3-dichloropropene turf injection.

The reentry period of turf following injection is 24 hours unless full personal protective equipment (including suit, gloves, boots, and respirator) are worn. Injections cannot occur within 100 feet (30 m) of an occupied structure such as a residence or place of business. Obviously, being a soil injection, buried obstacles such as irrigation heads and drains must be flagged and avoided. Currently, it is available as a custom application only (Table 12.11).

Following treatment, turfgrass roots typically respond dramatically with greater root depths and mass. However, since the soil is not sterilized, nematode populations usually rebound over time. Even so, with improved rooting, higher nematode populations may be tolerated. The product is also effective on soil insects such as mole crickets.

**Abamectin.** Abamcectin (trade name Divanem) is a traditional insecticide and miticide recently developed for turf use. Two application schedule options include four applications of Abamectin 0.15 L at 70 g active ingredient/ha (12.2 fluid ounces/acre, 0.89 L/ha) at four-week intervals or eight applications of 35 g active ingredient/ha at two-week intervals for an annual limit of 280 g active ingredient/ha. Best results occur when the total sequence of applications is made and none skipped. Apply in 2 gallons water per 1,000 square feet (16 L/100 m$^2$) followed by 1/4 inch (6.4 mm) irrigation as the material is readily bound by soil organic matter, possesses low water solubility, and is quickly broken down by sunlight. Addition of a soil wetting agent may improve performance. Abamectin is best on root-knot nematodes, tending to be less effective on sting and ring nematodes due to its limited soil movement. It has good contact activity but is not plant systemic and has a relatively short half-life to 14 to

**TABLE 12.11**
**Characteristics of Various Commercial Nematicides Used in Turf**

| Characteristic | 1,3-Dichloropropene (Curfew) | *Bacillus firmus* (Nortica) | Abamectin (Divanem) | Fluopyram (Indemnify) | Fluensulfone (Nimitz Pro G) | Furfural (MultiGuard) |
|---|---|---|---|---|---|---|
| Best nematode activity | Sting | Root-knot, sting, lance | Root-knot, sting | Root-knot, sting | Sting, lance | Sting |
| Site of activity | Contact | Protectant | Contact | Contact and systemic | Contact and systemic | Contact |
| Longevity of activity | Short (~1 month) | Intermediate (~4 months) | Intermediate (~4 months) | Long (~8 months) | Intermediate (~4 months) | Intermediate (~4 months) |
| Signal word | Warning | Caution | Caution | Caution | Caution | Warning |
| Restricted-use product | Yes | No | Yes | No | No | No |
| Use sites | Golf, sod—custom application only | Golf, sports fields, cemeteries, sod, lawns | Golf greens/tees, fairways | Golf, sports fields, sod, lawns | Golf, sports fields, sod, lawns | Golf greens/tees, sod |

30 days. For lance nematodes, treating in early spring after most bermudagrass roots are naturally sloughed off, exposes the population for best control. For sting nematodes, timing is similar to that for lance, as in early spring, soils are cooler and these nematodes tend to migrate closer to the soil surface where abamectin tends to stay. Combinations of abamectin and a good fungicide such as azoxystrobin tend to reduce fungal infections associated with nematode feeding, promoting healthier turf. Abamectin is labeled for use on golf greens and tees and spot treatment of fairways. It is not labeled for use on athletic fields, lawns, sod, and other turf venues. It possesses little systemic activity and has a half-life ≥ 14 days.

**Furfural.** Furfural (trade nbame: Multiguard Protect 90EC), a by-product of sugar processing, is labeled for use on golf course greens, tees, practice greens, and sod farms. Apply at 1:9 dilution with water at a rate of 8 gpa (0.75 L/100 m$^2$) initially, followed by 5.5 to 8 gpa (0.5 to 0.75 L/100 m$^2$) in subsequent treatments 14 to 28 days apart. Best results are from four applications at three-week intervals. Due to its relative water insolubility and turf burn potential, incorporate with irrigation (1/4 to 1/2 inch, 6.4 to 13 mm water) within 15 minutes of application in sandy soil. It generally works best on nematodes nearer the soil surface. Furfural has a contact mode of action by cuticle disintegration and thus is more effective on ectoparasitic nematodes such as sting, but less effective on lance and rootknot nematodes. It is best applied with onset of turf growth in spring. Positive results often are not seen until after repeat applications are made. Its use requires appropriate personal protective equipment (PPE) and buffer zones. Golf courses must be closed during application, with a two-hour reentry interval to treated zones. See label for details.

**Fluensulfone.** Fluensulfone has the trade name "Nimitz Pro G" and demonstrates efficacy on sting and lance nematodes plus other genera. It is labeled for use on golf courses, athletic fields, sod farms, and lawns. It has both contact and systemic activity; thus, it is xylem-mobile (systemic in plants), leading to the higher probability of effectiveness for endoparasitic nematodes such as lance as well as the ectoparasitic species. Annual use rate is limited to 268 kg/ha (240 pounds/acre). Multiple applications at lower rates tend to work better than one or two heavier ones. Best results have been with four monthly applications of 60 pounds/acre (67 kg/ha) or three monthly applications of 80 pounds/acre (90 kg/ha). It should be incorporated with irrigation, similarly to abamectin. Fluensulfone moves more readily through thatch, and has a short half-life (less than two weeks); thus, control can be inconsistent. When it is applied, nematodes are almost immediately paralyzed by blocked cellular respiration, causing them to become stiff and straight. Being unable to feed, the nematode eventually dies. In contrast to other true nematicides, it has a caution label that makes it attractive in terms of low mammalian toxicity, not requiring courses to be closed following application. The addition of a good fungicide with fluensulfone tends to improve turf quality.

**Fluopyram.** Fluopyram is actually an SDHI (succinate dehydrogenase inhibitor) fungicide with nematicidal activity, a trade name "Indemnify" 3.34 L, and a relatively long half-life of up to two years. It demonstrates good activity on sting and root-knot nematodes, but is less effective on lance. It is a contact product and has limited mobility through thatch. The maximum allowed amount is 17.1 fluid ounces/acre/year (1.25 L/ha) for commercial turf. Curative spot treatments over smaller areas (no more than 10,000 square feet, 929 m$^2$, each) can be made at a maximum of 0.39 fluid ounce/1,000 square feet (1.24 L/ha) up to four times per calendar year. Apply in 2 gallons of water per 1,000 square feet (16 L/100 m$^2$) followed by 1/4 inch (6.4 mm) irrigation for sting nematodes but only lightly for root-knot. For shoot-gall nematodes, one to two applications at 0.2 to 0.35 ounce/1,000 square feet (6 to 12 g/100 m$^2$) provides season-long control. When applied, fluopyram almost immediately paralyzes nematodes by blocking cellular respiration, causing them to become stiff and straight. Being unable to feed, the nematode eventually dies. Fluopyram provides preventative and curative control and can be used on golf courses, athletic fields, sod, and lawns. It has a caution use label and does not require courses to be closed following application.

### Biological Control

There are several products on the market for management of plant parasitic nematodes using various natural products (Table 12.12). Examples of earlier biological control products include a mixture of chitin from shells and urea to suppress root-knot nematodes when the material is incorporated into soil. Microbes increase to enzymatically break down chitin (chitinases), which may, concomitantly, degrade nematode eggs in soil. ClandoSan is one commercial formulation of

**TABLE 12.12**
**Examples of "Biological" or "Organic" Products That Tout Nematode Suppression**

| Source | Contends or Mode of Action |
|---|---|
| Agroneem | Plant based |
| Bioblitz | Plant based |
| ClandoSan | Chitin + urea fertilizer |
| CMP | Mustard bran |
| Cyclewise Nema | Fungal |
| Dragonfire CPP | Plant based |
| EcoNem | Bionematicide (*Pasteuria* spp.) |
| Floradox | Stimulates plants' natural defenses |
| Keyplex 350DP | Stimulates plants' natural defenses |
| MustGro | Plant (mustard)-based fertilizer |
| Nemastop | Plant based |
| Nematac S | Plant based |
| Nematrol | Sesame extract |
| NeoTec S.O. | Plant based |
| Neotrol | Sesame extract |
| Nortica | Bionematicide (*Bacillus firmus* strain I-1582) |
| O2YS | Chitosan |
| Safe-T Green | Biologically derived |
| Vector MC | Predacious nematodes |

this. Another product utilizes preparations of sesame, which has been shown to be toxic to nematodes under some circumstances. Nematrol, Neotrol, and others are commercial formulations. Also, various bacteria have been shown to suppress nematodes, and various commercial preparations of bacteria or bacterial products have come on the market. Incorporation of effective quantities of these materials into existing turf is problematic, as are the relatively high amounts of N as urea used to break down chitin.

Biological nematode control is also receiving much attention. Two insect-parasitic nematodes (*Steinernema riobravis* and *S. carpocapsae*, trade name Vector MC and others) have been used for mole cricket control. Their effectiveness for certain plant-parasitic nematodes, however, has been somewhat erratic.

Although many of these materials may suppress nematodes for short time intervals, none have given results in the field to date that compare with the efficacy of chemical nematicides. However, the need for safe and effective control of nematodes for turfgrasses has never been greater. The bacteria genus *Pasteuria* are obligate parasites of nematodes, especially *P. usage*. The *Pasteuria* attach to the cuticle of the nematode, injecting themselves into the nematodes where they multiply, eventually killing them. The dead nematodes eventually rupture, releasing enormous additional bacteria, which can infect other nematodes. Sting nematodes appear sensitive to the bacteria, but when used in field conditions, control and/or positive turf response (rooting, color, density) are often inconsistent.

***Bacillus.*** Products containing various *Bacillus* spp. are perhaps currently the mostly widely used of these products. Nortica (*Bacillus firmus*, strain I-1582) is termed a "plant protectant" as it does not directly control nematodes but produces various "biometabolites" that do. This bacterium colonizes the root system of the turf and produces compounds that protect the root system from nematodes. Nortica is a wettable powder that disperses in water. It is sprayed onto the turf surface and then moved into the soil with irrigation. Scientists do not necessarily see large reductions in soil nematode populations following use, but the bacteria appear to "protect" roots against their feeding. Best results have been noted as preventative applications in early spring as bermudagrass turf resumes growth after winter dormancy rather than a "fix" of existing populations in midseason.

## Maximizing the Effectiveness of Nematicides

Neither fumigant nor nonfumigant nematicides completely eradicate plant parasitic nematodes. Some nematodes in deeper layers of soil and root tissue may escape exposure to lethal concentrations of the nematicide. Others are only temporarily paralyzed or disoriented by sublethal levels of the nematicide and will resume feeding when the chemical dissipates through diffusion, dilution, degradation, or leaching. Avoid the introduction of nematodes from other sources (for example, contaminated soil or sod). It is important to monitor the population levels of nematodes to know when nematicide treatments are needed.

Nematicides mainly affect nematodes and do not directly stimulate plant growth. Nematicide-treated turfgrass, therefore, needs time to grow new roots in order to support new foliage and recover from nematode-induced stresses. Factors limiting root growth must be taken care of immediately after nematicide applications in order to achieve complete recovery of turf affected by nematode parasites. Ensure good drainage, adequate irrigation and aerification, balanced soil fertility, control of other pests and diseases, and reduced pedestrian traffic, if possible. Aboveground plant responses after a nematicide application are usually slow or delayed.

Timing of applications is important. In the southeastern United States, a very good response of bermudagrass to nematicide application in sting- and ring-infested soil is obtained by a mid-April application, normally several weeks after spring green-up. Presumably, nematode populations are suppressed and allow new stolon and root development at the time of year when growth is maximized. Fall applications of nematicides to suppress damaging nematodes in bermudagrass turf may also be made, but overseeding establishment of cool-season grasses can be adversely affected if seeding and nematicide application coincide or if the interval is short. If a nematicide application is necessary in the fall, it should be done two to three weeks prior to the overseeding date.

Related to the timing of applications, the use of certain preemergence herbicides for crabgrass or goosegrass control, which act as inhibitors of cell division, may inhibit bermudagrass response to nematode suppression by nematicides. Although nematodes are suppressed by the nematicide application, turf may not respond because residual herbicides inhibit new stolons. Managers may opt to skip the preemergence herbicide application and use postemergence strategies instead, or use a material (e.g., oxadiazon, or Ronstar) that does not inhibit "tacking" of new stolons into treated areas. This becomes a problem when damage to bermudagrass is substantial and managers rely on new stolon development for recovery (e.g., rhizomes are absent or weakened).

Avoid overuse of any nematicide because soil microorganisms that can degrade the nematicide will build up to high population levels, decrease the efficacy and longevity of the chemical in subsequent applications, and consequently shorten the period of nematode control. Prolonged frequent use of a given pesticide also allows the buildup of one or more parasitic nematode species against which the chemical is less effective.

# 13 Turfgrass Insects

## INSECT BIOLOGY

**Arthropod** is a broad term that includes all insects, mites, and other relatives. The following are the defining characteristics of an arthropod:

1. Multiple pairs of jointed and articulated legs and antennae.
2. The presence of a chitinous exoskeleton (body wall, or outer covering or shell). The exoskeleton provides arthropods with structural integrity, but it also limits their growth. To grow, arthropods have to shed this exoskeleton in a process called molting.
3. A segmented body. Insects are segmented in head, thorax, and abdomen, whereas mites are segmented into prosoma (cephalothorax, equivalent to an insect's head and thorax) and opisthosoma (abdomen). The possession of wings is unique to insects.
4. A nervous system on the underside (ventral or belly) and a circulatory system on the upper side (dorsal or back).

Almost all arthropod pests of turfgrass are insects and mites, although other arthropod pests of turfgrass sites include spiders, fleas, ticks, centipedes, millipedes, sowbugs, and pillbugs.

**Metamorphosis.** The process by which insects and mites grow is called metamorphosis. During this process, the body of an insect grows, molts, and then becomes larger and more developed, and can sometimes appear different from the previous life stage. Insects undergo two types of metamorphosis:

1. **Incomplete** (or **gradual) metamorphosis:** Insects in this group are called hemimetabolous insects and include species such as chinch bug, mole cricket, leafhopper, and ground pearl. All mites are hemimetabolous. The juvenile form (called nymph) of these insects resembles the adult, but is smaller and lacks adult features such as wings and functional reproductive organs. The life cycle of a hemimetabolous insect or mite can be divided into three stages: egg, nymph, and adult. Within the nymphal stage, there may be several growth stages between molts (called instars). Many species of this group can damage turf as an adult and a nymph. Adults do not grow or molt.
2. **Complete metamorphosis:** Insects in this group are called holometabolous insects and include ants, armyworms, cutworms, billbugs, white grubs, and crane flies. The juvenile form (called larva) appears very different from the adult. Each group may have a specific name for their larvae: caterpillar (larva of a moth or butterfly), maggot (larva of a fly), or grub (larva of a beetle). The life cycle of a holometabolous insect can be divided into four stages: egg, larva, pupa, and adult. Several instars may also occur within the larval stage. Most holometabolous insects are damaging during their larval stage, but not during their adult stage.

**Insect feeding.** Insect and mite pests of turfgrass can be broadly categorized as

1. Foliage and stem feeders, which include the caterpillar (black cutworm, fall armyworm, and sod webworm), chinch bug, adult billbug, spittlebug, bermudagrass mite, rhodesgrass mealybug, greenbug, and scale insects
2. Root and rhizome feeders, which include mole crickets, white grubs, billbug grubs, and ground pearls
3. Nuisance pests, including fire ants, nuisance ants, ground-nesting bees and wasps, ticks and millipedes, which can invade homes, disrupt turf quality or playability, or have medical significance due to their stings or bites

Turf-damaging insects and mites either have *chewing* or *piercing and sucking* mouthparts. Chewing mouthparts are jaw-like arrangements that insects use to tear, chew, and grind plant tissue. Piercing and sucking mouthparts are needlelike beaks that insects or mites insert into plant tissue to remove internal fluids from plants.

## DETECTION AND IDENTIFICATION

The first step in formulating any pest management strategy is identifying the damage your turf is experiencing, especially before pest populations reach chronic levels. Insect pests attack grass plants either above- or belowground, usually either by feeding directly on roots (belowground) or by sucking sap or chewing grass leaves (aboveground). A few, such as southern mole crickets and fire ants, do not normally feed directly on the plants, but in their search for food or nest building, disrupt the turfgrass surface. This is compounded by animals, such as moles, pigs, birds, skunks, armadillos, and raccoons, digging through the turf as they search for these turfgrass insects as a food source.

### Surface (or Aboveground)-Feeding Insects

These insects feed on the leaves and stems of turfgrass plants. Many of them hide in the thatch during the day, and then surface and feed on leaves and stems at night. Other surface-feeding insects remain on the leaf surface and feed. The larvae (caterpillars) of various moths have chewing mouthparts and eat entire leaves and stems. Adult beetles also have chewing mouthparts.

**Indications of surface-feeding insects include:**

1. Birds feeding on the turf surface
2. Moths (mostly tan in color) flying zigzag patterns over the turf area, especially in late evening
3. Rapid turf color loss similar to fertilizer burn, mower scalp, or drought stress even though irrigation is provided
4. Notched leaves or residue of chewed grass (frass) at or near the soil surface
5. Fecal pellets, often green in color

Other surface-feeding insects have piercing and sucking mouthparts that tend to discolor the grass, resembling drought stress. These insects include aphids, chinch bugs, spittlebugs, and mealybugs.

### Subsurface (Root or Belowground)-Feeding Insects

These insects inhabit the thatch and rootzones, and can be very damaging due to their root and crown feeding. Due to their subterranean location, this group of insects is the most difficult to control. Pesticides must move through the leaf canopy, through the thatch layer, and into the soil. Thatch control, adequate incorporation through irrigation, and subsurface placement in the soil are means to increase the efficacy of control products. White grubs, billbug larvae, mole crickets, and ground pearls are the most damaging subsurface-feeding insects.

**Indications of subsurface-feeding insects include:**

1. Visible mounds of soil that are soft and cause excessive scalping injury or tunnels in soil or pulled up due to weak roots
2. Poor turf rooting where the grass can be rolled up like carpet
3. Damage from insect-feeding animals such as moles, pigs, birds, raccoons, and armadillos
4. Turf thinning
5. Yellow, chlorotic turf appearance

### Detection Techniques

Inspecting the turfgrass involves several common sampling techniques to identify the most common turfgrass insects. Early detection is important to head off undesirable and unacceptable aesthetic damage.

1. **Visual inspection**—Walking around the perimeter of the turfgrass area in question is the first step in detecting insect presence/damage. Pull up some grass stems and see if a particular feeding pattern is noticed. Certain insects and mites cause telltale feeding pattern/damage; for example, bermudagrass mites cause stunted terminals or "witch's broom" deformation on bermudagrass, caterpillars often remove or chew holes in the leaf blade, and spittlebugs cause yellowish discoloration on centipedegrass. Kneel on your hands and knees and, with a 10× handheld lens, see if tiny insects are detectable at the base of the turfgrass plant or on the soil surface. Physical disruption, such as tunneling from mole crickets, is relatively easy to detect when the insects are large nymphs or adults. However, remember the objective is to detect pest populations before major damage occurs. Surface-active insects feeding on aboveground plant portions are easiest to determine by visual inspection.
2. **Soil/thatch inspection**—Manual inspection of the top 2 to 3 inches (5 to 7.6 cm) of soil or turfgrass thatch involves severing three sides of a piece of turf with a shovel and peeling back the grass to expose the thatch and soil (Figure 13.1). White grubs, billbug grubs, and ground pearls are the principal insects found by this method. Several samples in an area should be taken to obtain a representative average.
3. **Soap flush**—Soap flushing helps indicate insects hiding in the thatch or those moving too rapidly to be seen or caught. Mix one to two fluid ounces (30 to 59 mL) of lemon-scented liquid dishwashing soap or one tablespoon (15 mL) of 1% pyrethrin in 2 to 3 gallons (7.6 to 11 L) of water and apply this to a 2- × 2-foot (0.6 × 0.6 m) area of turf. The solution interferes with insect respiration and, if present, they will soon emerge in an attempt to breathe (Figure 13.2). The area should be observed for at least two minutes to see if any insects emerge. Mole crickets, webworms, armyworms, cutworms, and beetles are detected by this method. Use of more than two fluid

**FIGURE 13.1** Soil inspection for subsurface insects such as grubs.

FIGURE 13.2 Topical soap flush for insects in the thatch such as mole crickets, webworms, and cutworms.

ounces (59 mL) of detergent may cause damage to the grass during hot and humid weather, especially in cool-season turfgrass.

4. **Water flotation**—This method is especially effective in sampling for chinch bugs. Water flotation involves inserting a cylinder, such as an open-ended, 3-pound (1.4 kg) coffee can or a 4-inch PVC pipe, through the turf 1 to 2 inches (2.5 to 5 cm) deep and filling with water (Figure 13.3). Do not add detergent to the water. Stirring the turf may help dislodge insects. Those present should float to the top of the water level within several minutes.
5. **Traps**—Several traps are used to detect turfgrass insect pests. Traps are effective, but some can be costly and require frequent maintenance. Pheromone traps are very pest specific, whereas other trap types may capture a large variety of nontarget insects.
   a. **Pitfall traps**—Pitfall traps are used to capture insects active on the soil surface and can be constructed of various materials. The simplest and cheapest variety is constructed by burying a plastic cup in the ground, with the rim of the cup flush with the soil surface. The cup is then filled to one-third full with 70% ethanol or undiluted propylene glycol (acts as killing liquid and preservative). A pitfall trap that captures insects in a larger area consists of 1-inch (2.5 cm) diameter PVC pipe with a portion removed from its side and placed face-up flush with the soil surface (Figure 13.4). A trap jar, one-third full of 70% ethanol or propylene glycol, is placed at one end of this pipe. As insects move across the soil surface, they fall in the open-sided PVC pipe, crawl down the pipe, and fall in the jar of killing liquid. To prevent dilution of the killing liquid by rainwater, a rain shelter constructed of a plastic plate can be installed over the pitfall traps. A turf manager or scout can pour the contents of the plastic cup or trap jar into a large, shallow container and examine them for insect species of interest.
   b. **Pheromone traps**—Females of many insect species, such as Japanese beetle, Oriental beetle, and various moth species (armyworms and cutworms), produce sex pheromones to attract males. To avoid mistaken mate identification, sex pheromone is very species specific. Pheromone traps consist of a lure that emits a synthetic sex pheromone over a long time, and a collecting device that is either a sticky card, a jar filled with killing liquid, or an insecticide-impregnated kill strip. A scout and turf manager can examine the catch and change the lure and collection device periodically. Pheromone traps are a good monitoring tool, especially for adult moth activity, but they are not effective in trapping and killing an entire insect population.
   c. **Light traps**—Night-flying insects are attracted to light, particularly ultraviolet (UV) or black light. Light traps use UV light to attract insects, which fall into a collection jar filled with a killing

FIGURE 13.3 Water flotation for sampling chinch bugs.

FIGURE 13.4 Pitfall trap to capture soil surface active insects.

liquid. Light traps attract a wide variety of both pest and nonpest insect species. If large numbers of insects are captured, determination of population density of a species of interest and species identification can be very time consuming.

**Scouts.** A professional scout, who may be employed by several nearby courses, may be used to identify and quantify pest populations. Since these scouts visit several courses, pest trends are more easily recognized, and useful information from one course can be used to assist others. A scout should typically hold a degree in agronomy, horticulture, entomology, or plant pathology with an emphasis in pest management.

Tools required for scouting vary with pest problems, scout training, and golf course budget. A good set of eyes and an inquisitive mind are essential. These are supported by a standard handheld or pocket 10× lens, soil probe, soil profile probe, spade, cup cutter, pocket knife, tweezers, scalpel, collection vials, paper bags, and field identification guides. Soap and water also are necessary for insect monitoring.

More expensive, but precise, instruments may be used in a room designated as a diagnostic laboratory. Included are stereo- and compound microscopes, soil sieves, pH meter, conductivity meter, and elementary soil analysis kits. These need to be supplemented by ongoing scout training at short courses, formal classes, appropriate diagnostic guides, and opportunities to visit similar facilities to exchange ideas.

**Monitoring or sampling plan.** The goals of a sampling plan are to estimate the pest population density or damage severity, and to inform turf managers of the potential of damage. Because of the uniqueness of each pest species, a sampling plan has to be developed for each pest species.

The fact that most insecticides are most effective against specific life stages of a pest requires a golf course superintendent to have a good knowledge of the seasonal life cycle of pests. Monitoring or sampling plans should be initiated before the appearance of a target life stage. For example, because white grubs are most effectively managed with long-residual preventative insecticides that target egg hatch, the sampling plan for scarab beetles (i.e., adults of white grubs) should begin before the adult flight.

**Considerations that aid in determining the timing of sampling plan initiation include:**

1. **Calendar date:** Insect pests often appear at approximately the same period each year; a golf course superintendent with prior experience can initiate sampling at about the same time each year. Although calendar-based sampling plans are the easiest to deploy, they are also the least accurate. Insect development depends on temperature and other environmental conditions, and may be accelerated or delayed for several weeks by favorable or adverse environmental conditions. The inaccuracy in sampling may cause additional damage or costs.
2. **Plant phenological indicators:** Both insect and plant development are tied to temperature. Therefore, plant phenological events can be used as a signal or indicator of the progression of insect development and events. Annual bluegrass weevils begin laying eggs when flowering dogwood and eastern redbud are in full bloom (use preventative larvicides), whereas larvae begin to appear when Catawba rhododendron hybrid is in full bloom (use curative larvicides). Each insect pest species may have a unique plant phenological indicator; a list could be developed based on a superintendent's experience and research.
3. **Degree-day models:** These models measure the total number of heat units (or degree-days) needed to achieve a particular life stage. Most models start accumulation of degree-days on January 1 and use a base temperature of 50°F. Most cooperative extension services, as well as a number of online sources, can provide degree-day information for free. Each insect has a unique degree-day model. Adult bluegrass billbugs first appear when 280 to 352 degree-days have accumulated, whereas the first generation of larger sod webworm adults appear when 846 to 882 degree-days have accumulated.

Monitoring frequency may require adjustment depending on climatic conditions (particularly temperature), reports of nearby pest problems, and focus areas. Greens and tees generally require the greatest amount of attention and are monitored daily or every other day. Fairways, roughs, ornamental plantings, and trees are usually monitored weekly.

Pest levels should be recorded on a form similar to the one in Chapter 11. This will allow the scout and superintendent to monitor pest trends and determine if these levels reach or exceed aesthetic or action thresholds. Pest maps should accompany these forms.

## Identification

Correct identification of insects is perhaps the most important step in developing an IPM program. Based on an incorrect identification or diagnosis of a pest or damage, a turf manager can make the wrong management decision, causing more damage, and wasting time, materials and labor.

Fortunately, only a small number of arthropod pests can cause damage to turf at any given time and area. A turf manager can often narrow down the list of likely culprits by considering the damage caused, where among the turf profile the insect is found, and the time of year. The following sections on individual pest groups will describe characteristics that can aid in identification. A turf manager can also seek help from the cooperative extension service (including county agents and extension specialists) or (public or private) plant problem diagnosis services. It is important to seek diagnosis assistance from reputable services staffed by trained and knowledgeable personnel.

## INSECT CONTROL STRATEGIES

### Set Threshold Levels

Once the insect and its damage have been positively identified and its life cycle understood, pest management involves setting threshold levels, recording pest levels, and possibly applying an appropriate control method. An aesthetic or action threshold is the point when pest populations or environmental conditions indicate some action must be taken to prevent intolerable damage. These thresholds will vary according to the location of the course, the specific pest being scouted, the turf area's level of use, club members' expectations, and budget constraints (Table 13.1).

### Biological Control Strategies

Much research and discussion has occurred on "organic," "nonsynthetic," "bio," or "biorational" pesticide control of pests. Biopesticides have little, if any, adverse effect on beneficial organisms or the environment. The best approach in pest control is to integrate all possible control and management strategies into a plan, since no single method of control is 100% reliable. These strategies include host-plant resistance, pest-free propagation (or sanitation), proper site preparation, cultural practices, and biological control. These are discussed in detail in the chapter on integrated pest management.

Biorational pesticides are derived from a variety of sources, including endophytes, bacteria, nematodes, fungi, other insects, and insect-derived pheromones and growth regulators. Biorational pesticides tend to be short lived. Therefore, they may require multiple applications, are often life-stage and insect specific, and require proper timing of application and positive insect identification.

**Endophytes.** A relatively new nonchemical control method of certain insects involves naturally occurring fungi called endophytes. Endophytes are fungi and other organisms that form symbiotic relationships with certain grasses but do not cause disease. Unlike most fungi, endophytes are not externally visible on plants. The endophyte fungus (*Acremonium coenophialum*) was initially discovered in forage production, as cattle and horses feeding on endophyte-containing tall fescue produced a syndrome referred to as "fescue toxicity." However, this fungus has been found to also produce ergot alkaloids (such as peramine, lolitrem B, ergovaline, paxilline, and others) that are toxic or incompatible to certain insects and have since been transferred to certain turfgrasses. Other fungal endophytes include *Neotyphodium lolli* in perennial ryegrass and *Neotyphodium coenophialum* in tall fescue, while two endophytes, *Neotyphodium typhinum* and *Epichloe typhina*, occur in the fine fescues. In addition to providing resistance to various insects, secondary effects of certain endophytes include providing dollar spot control and increasing plant tolerance to drought and other stresses.

Infected plants appear the same as endophyte-free plants, and laboratory examination is the only way to detect the presence of infection. Plants are not harmed by the fungus. In fact, the endophyte and the grass derive mutual positive benefits from their association. In addition to infested grass being more tolerant to insects, it also has more tolerance to nematodes and drought.

It appears the endophyte is spread only through infected seed. Therefore, cool-season grasses (fine fescue, perennial

**TABLE 13.1**
**Aesthetic or Action Levels for Several Common Turf Insects**

| Insect | Inspection Method | Approximate Aesthetic or Action Threshold (per square foot)[a] |
|---|---|---|
| Annual bluegrass weevil | Visual inspection of clipping (adult); soap flush (adult); soil core (grub) | • 30 to 80 grubs (spring); 20 to 40 grubs (summer); 5 to 10 grubs (preventative) |
| Armyworms | Visual and soap flush | • 1 to 2 caterpillars |
| Bermudagrass mite | Visual | • 4 to 8 tufts |
| Billbugs | Visual (adult); pitfall (adult); soil core (grub) | • 10 to 14 grubs |
| Black turfgrass *Ataenius* | Soil core (grub) | • 50 grubs |
| Cutworms | Pheromone trap (adult); visual and soap flush (caterpillar) | • 1 caterpillar |
| European crane fly (leatherjacket) | Visual and soil core | • 15 to 50 larvae |
| Green June beetle | Visual and soil core | • 5 to 7 grubs |
| Hairy chinch bug | Water float | • 15 to 20 bugs |
| Japanese beetle | Pheromone trap (adult); soil core (grub) | • 6 to 10 grubs |
| May beetles | Light trap (adult); soil core (grub) | • 4 to 5 grubs |
| Masked chafers | Light trap (adult); soil core (grub) | • 8 to 20 grubs |
| Oriental beetle | Pheromone trap (adult) and soil core (grub) | • 8 to 10 grubs |
| Temperate sod webworms | Visual and soap flush | • 2 to 6 caterpillars |
| Tropical sod webworm | Visual and soap flush | • 0.5 to 1 caterpillar |

[a] Smaller numbers represent threshold levels for highly maintained areas such as golf greens and tees. Larger numbers are for less intensively maintained areas such as fairways, roughs, athletic fields, and lawns. Multiply values by 0.09 to obtain numbers per square meter.

ryegrass, and tall fescue) currently have the most endophyte incorporated in them since most warm-season grasses are vegetatively established. Endophytes remain viable in storage as seed for only one or two years. They offer plant breeders and entomologists one of the most significant means for non-chemical control. However, market oversaturation, the fungi's fragility, and government regulations to define quantifiable levels have limited their commercial success. Hopefully, further research will determine how to transfer these fungi to additional turfgrasses, stabilize their shelf life, increase their efficacy, and possibly offer tolerance to other pests such as nematodes.

**Beneficial bacteria.** Several species of the soil bacterium *Bacillus* provide various levels of soil and forest insect control. The first, *Bacillus thuringiensis* (or Bt), was discovered in 1901 in Japan. Since then, over 30 subspecies and varieties of Bt have been identified (Table 13.2). These bacteria produce protein crystals (endotoxins) that must be consumed by the

**TABLE 13.2**
**Insecticides for Control of Turfgrass Insect and Mite Pests, Listed by Insecticide Resistance Action Committee (IRAC) Group Number, Mode of Action, Chemical Class, and Active Ingredient**

| IRAC | Chemical Classes | Mode of Action | Active Ingredient | Trade Name Examples[a] |
|---|---|---|---|---|
| 1A | Carbamates | Acetylcholine esterase inhibitors | Carbaryl | Sevin |
| 1B | Organophosphates | | Acephate | Orthene |
| | | | Chlorpyrifos | Dursban |
| | | | Trichlorfon | Dylox |
| 2B | Fipronil | GABA-gated chloride channel antagonists | Fipronil | TopChoice, Taurus |
| 3 | Pyrethroids | Sodium channel modulators | Bifenthrin | Allectus,[b] Aloft,[b] Onyx, Talstar, Bifen, etc. |
| | | | Cyfluthrin | Tempo |
| | | | Cypermethrin | Demon, Triple Crown T&O |
| | | | Deltamethrin | Deltagard |
| | | | Lambda-Cyhalothrin | Lambda, Battle, Demand, Scimitar, Tandem[b] |
| | | | Permethrin | Astro |
| 4A | Neonicotinoids | Nicotinic acetylcholine receptor agonists/antagonists | Clothianidin | Arena, Aloft[b] |
| | | | Dinotefuran | Zylam |
| | | | Imidacloprid | Allectus,[b] Imidacloprid, Merit, Mallet, etc. |
| | | | Thiamethoxam | Meridian, Tandem[b] |
| 5 | Spinosyns | Nicotinic acetylcholine allosteric activator | Spinosad | Conserve |
| 6 | Avermectins | Chloride channel activators | Abamectin | Avid, Award II fire ant bait |
| 7A | Junevile hormone | Juvenile hormone mimics | S-Methoprene | Firestrike,[b] Extinguish, Extinguish Plus[b] |
| 7C | Pyriproxyfen | | Pyriproxyfen | Distance Fire Ant Bait |
| 11A | *Bacillus thuringiensis* | Microbial disruptors of insect midgut membranes | *B.t.* subsp. *aizawai* | Xentari |
| | | | *B.t.* subsp. *galleriae* | grubGONE! G |
| | | | *B.t.* subsp. *kurstaki* | Biobit, Crymax, Dipel, Juvelin, Lepinox |
| 18A | Diacylhydrazines | Ecdysone agonists, molting disruptors | Halofenozide | Mach 2 |
| 20 | Hydramethylnon | Mitochondrial complex III electron transport inhibitors | Hydramethylnon | Amdro Firestrike,[b] Extinguish Plus,[b] SiegePro |
| 22 | Indoxacarb | Voltage-dependent sodium channel blockers | Indoxacarb | Advion fire ant bait, Provaunt |
| | Metaflumizone | | Metaflumizone | Siesta fire ant bait |
| 28 | Diamides | Ryanodine receptor modulator | Chlorantraniliprole | Acelypryn |
| | | | Cyantraniliprole | Ference |
| un | Azadirachtin | Unknown MOA | Azadirachtin | Azatin O, Azatrol, Molt-X, Ornazin |
| | Dicofol | | Dicofol | Dicofol 4E |
| uc | Bacteria | Unclassified: pathogens | *Bacillus popillae* | Milky spore powder |
| | | | *Chromobacterium subtsugae* | Grandevo PTO |
| | Nematodes | | *Steinernema* + *Heterorhabditis* spp. | Millenium, BioVector, NemaShield |
| | Fungi | | *Beauveria bassiana* | Botanigard, Naturalis |
| | | | *Metaehizium anisopliae* | Met52, Tick-Ex |

[a] Trade names are provided as examples only. No endorsement of products is intended, nor is criticism of unnamed products implied.
[b] Allectus = imidacloprid + bifenthrin; Aloft = clothianidin + bifenthrin; Tandem = thiamethoxam + lambda-cyhalothrin; Triple Crown T&O = zeta-cypermethrin + bifenthrin + imidacloprid; Amdro Firestrike and Extinguished Plus = s-methoprene + hydramethylnon.

insects. Other products produced by the bacteria also must be consumed to control certain insects. Once inside the insect's gut, the crystals dissolve if the proper acidity exists and bind to specific sites in the gut lining. Susceptible insect guts then become paralyzed and the insect stops feeding within a few hours and dies in several days.

Unlike many biological control agents, these bacteria generally are short lived in the environment, are readily degraded by direct sunlight, are slow to kill insects, do not reproduce in the insect host, are less effective on larger larvae, and are not spread from treated to nontreated sites. This has limited their acceptability in the commercial arena.

*Bacillus thuringiensis* varieties currently available include Bt *kurstake* and *aizawai* (best for caterpillars), Bt *israelensis* (for mosquitoes and European crane fly larvae), Bt *tenebrionis* and *san diego* (for potato beetle larvae), and, most recently, Bt *buibui* and *gallereiae* (for Japanese beetle, Oriental beetle, masked chafer, and green June beetle grubs). The newer strains have fewer of these negative attributes and appear to be more efficacious.

Another *Bacillus* species (*Bacillus popilliae*) was the first microbial agent registered as an insecticide in the United States and has been used to control Japanese beetle grubs. It is often referred to as the milky disease since the bacteria causes the insect's body fluids to turn a milky white color prior to grub death. The use of this bacterium has been limited due to the extended period of time needed for populations to build to infectious levels and become lethal (three to five years). Milky disease also is relatively expensive and can be extremely variable in its control. It is harmless to earthworms, wildlife, humans, and beneficial insects.

Recently, another bacterium, *Serratia entomophila,* has shown promise for the control of grass grubs in New Zealand. A new class of insecticides, Spinosad (called naturalytes), contains fermented-derived products from the bacterium *Saccharopolyspora spinosa.* It is a gut poison and must be eaten by the insect to be effective. It is used at low rates and has relatively short residual activity. Research continues on the commercial development of these and other beneficial bacteria and their products.

**Beneficial nematodes.** Research has led to the identification of several beneficial (or entomogenous) nematodes in the families Steinernematidae and Heterorhabditidae, which attack a specific host, yet will not attack plants or vertebrates. These nematodes lack a stylet, or the piercing mouthpart, characteristic of plant-parasitic nematodes; therefore, they do not feed on plants. In addition, these nematodes are relatively easy to mass-produce, can search out their target hosts, and can be applied with most standard pesticide application equipment—assuming the tank is ultraclean, coarse nozzles are used with no filter, and a carrier volume of at least 140 gallons per acre (1,200 L/ha) is used. The nematodes rapidly kill their host by entering the host's mouth or spiracles and then moving through the gut into the blood, where a colony of bacteria is released. The bacteria then multiply and produce toxins that kill the infected insect. The nematode continues to feed inside the infected host and eventually reproduces, producing thousands of new nematodes that emerge and search for new hosts.

Since beneficial nematodes are considered parasites, and not microbial insecticides, they are exempt from registration with the U.S.EPA. This greatly speeds up the registration process since these predators are exempt from long-term toxicological and environmental studies, which greatly increase the time and costs of bringing a new pesticide to the market.

Several beneficial nematodes are available, with additional ones currently being screened as potential control agents. *Steinernema scapterisci* was the first isolated and commercialized nematode, followed by *S. riobravis.* These are used for mole cricket control. Additional beneficial nematodes include *S. glaseri* (white grubs), *S. scarabaei* (white grubs), *S. carpocapsae* (for billbugs and caterpillars such as cutworms, webworms, and armyworms), and *Heterorhabditis bacterophora* and *H. zealandica* X1 (for white grubs).

**Beneficial fungi.** Various species of fungal pathogens have been discovered and screened as potential biological control agents. *Beauveria bassiana* and *Metarhizium anisopliae* are used against soil-dwelling and soft-bodied arthropods, such as chinch bugs, chiggers, and ticks. Host infection is initiated when the spores of the fungus adhere to the insect body. These spores germinate under the correct environmental conditions and grow into the insect, eventually penetrating its circulatory system. The fungi then rapidly reproduce and produce toxins that kill the insect. Additional fungal spores are then produced that can spread through the environment, infecting other insects. Similar to beneficial bacteria and nematodes, fungal products can be applied with existing spray equipment, provided the tanks are clean of fungicides, fertilizer, and other chemicals. Continued moisture and protection from ultraviolet light are required for the fungi to initiate infection, survive, and kill.

**Parasitic and predatory insects.** Insect and mite pests are constantly under attack by their predators and parasites. Big-eyed bugs can be found feeding on southern chinch bugs in the southern United States, and fall armyworms often are preyed upon by paper wasps and ground beetles. Many wasps parasitize turf insect pests. For example, *Tiphia* wasp is a common parasitoid of white grubs and *Larra bicolor* has reduced mole cricket densities at some locations in Florida and Georgia by 90%. These predators and parasites are endemic to turfgrass systems and can help to maintain pest populations at a low level. However, their populations and activities can be severely hampered by modification of the environment (such as building a monoculture) and the application of detrimental insecticides.

**Insect growth regulators (IGRs).** Insect growth regulators are a class of artificial compounds currently being developed that mimic the action of the natural hormone, ecdysone. IGRs interfere with the normal insect molting process, causing insect mortality, or they alter juvenile hormones, preventing insects from maturing. These also can inhibit the production of chitin, a polymer composite of the insect's exterior. The process is rate dependent and insect specific. Suboptimal application rates cause sublethal effects such as rapid maturation of insects to adult stages and deformities of larvae. As mentioned, certain IGRs are more effective on specific insects; therefore, the turf manager must identify the insect present for maximum effectiveness.

IGRs generally require ingestion for optimum activity; therefore, the insect must be actively feeding when the IGR is applied. Young larval stages are most susceptible to IGRs that attack chitin synthesis. Armyworms, cutworms, sod webworms, fire ants, and white grubs are most susceptible to current IGRs. Typically, they are slower acting than many conventional synthetic insecticides.

### Chemical Control

One of the goals of IPM is intelligent and prudent pesticide use. Once pest thresholds are reached, the pesticide used should be the safest one available; spot treatments should be practiced, if possible, and all safety precautions should be followed. Pests should also be treated during the most vulnerable stage of their life cycle. Refer to Figures 13.5 and 13.6 for typical life cycles and control timings for the major turf insect pests.

Evaluate the results of the habitat modification and pesticide treatments by periodically monitoring the site environment and pest populations. Keep written records of site pest management objectives, monitoring methods, and data collected. Also, record the actions taken and the results obtained by the pest management methods. This will provide additional information for club members who do not understand the program but would understand results. This also will aid in demonstrating that golf course superintendents are striving to reduce the chemical inputs in maintaining the course and obtain an ecological balance between man and nature.

**Resistance.** With an intensive and repeated use pattern, some pests have developed resistance to some of the most commonly used insecticides. A pest population is considered resistant to a particular chemical when a pesticide that was previously effective at controlling the pest has failed to reduce the pest population. Pesticide resistance develops when the same chemical class or mode of action is used repeatedly against the same pest population. Some annual bluegrass weevil populations in the northeastern United States are resistant to pyrethroids, whereas some southern chinch bug populations in Florida are resistant to carbamates, organophosphates, cyclodiene organochlorines, pyrethroids, and neonicotinoids. The most effective method to delay or prevent the development of pesticide resistance is to use pesticides of different modes of action against each pest generation. To aid in the design of a pesticide rotation program, the Insecticide Resistance Action Committee (IRAC) assigns each mode of action a unique group number. To avoid resistance development, pesticides of a mode of action or one IRAC group number should not follow another pesticide of the same mode of action or IRAC group number. For example, a sensible pesticide rotation program against a known pyrethroid-resistant chinch bug population will include a rotation of acephate (IRAC group no. 1), clothianidin (4A), and carbaryl (3). Table 13.2 lists current insecticides and their modes of action.

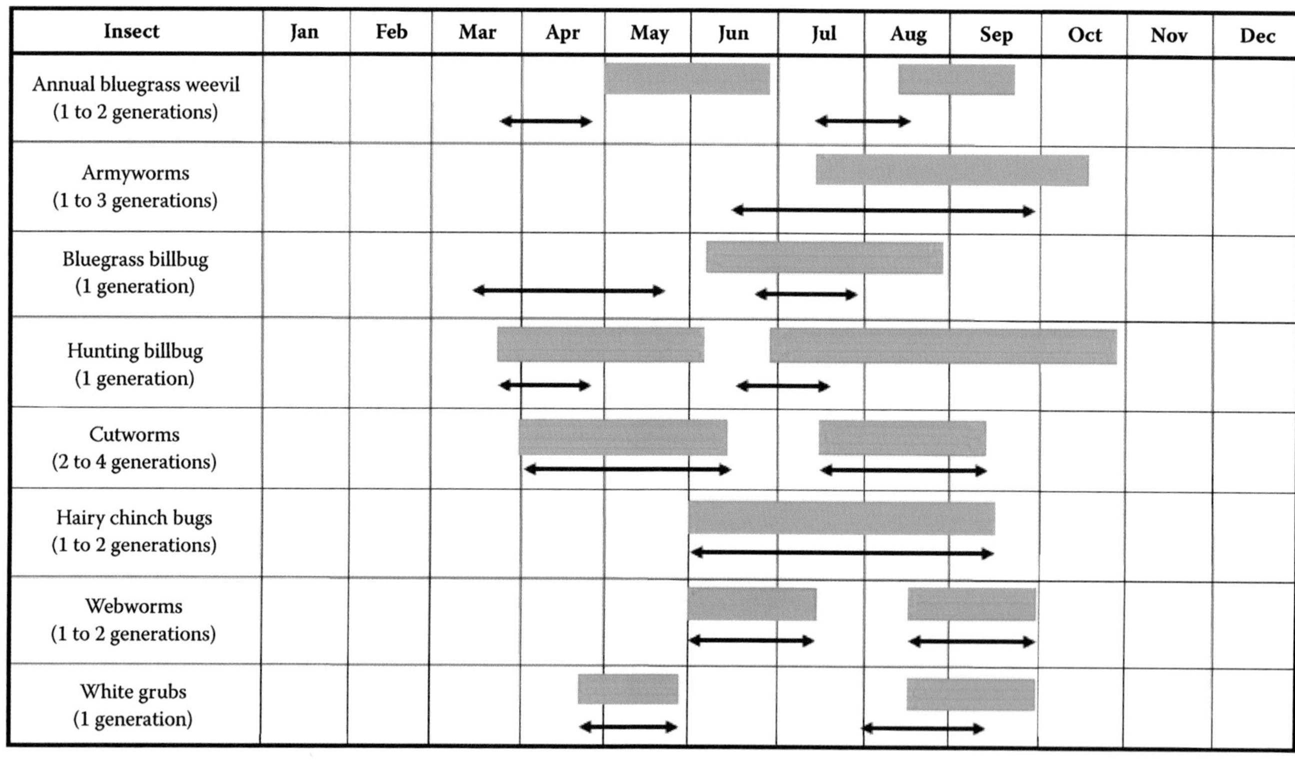


**FIGURE 13.5** Typical damage periods and control timings for insects in the northern United States.

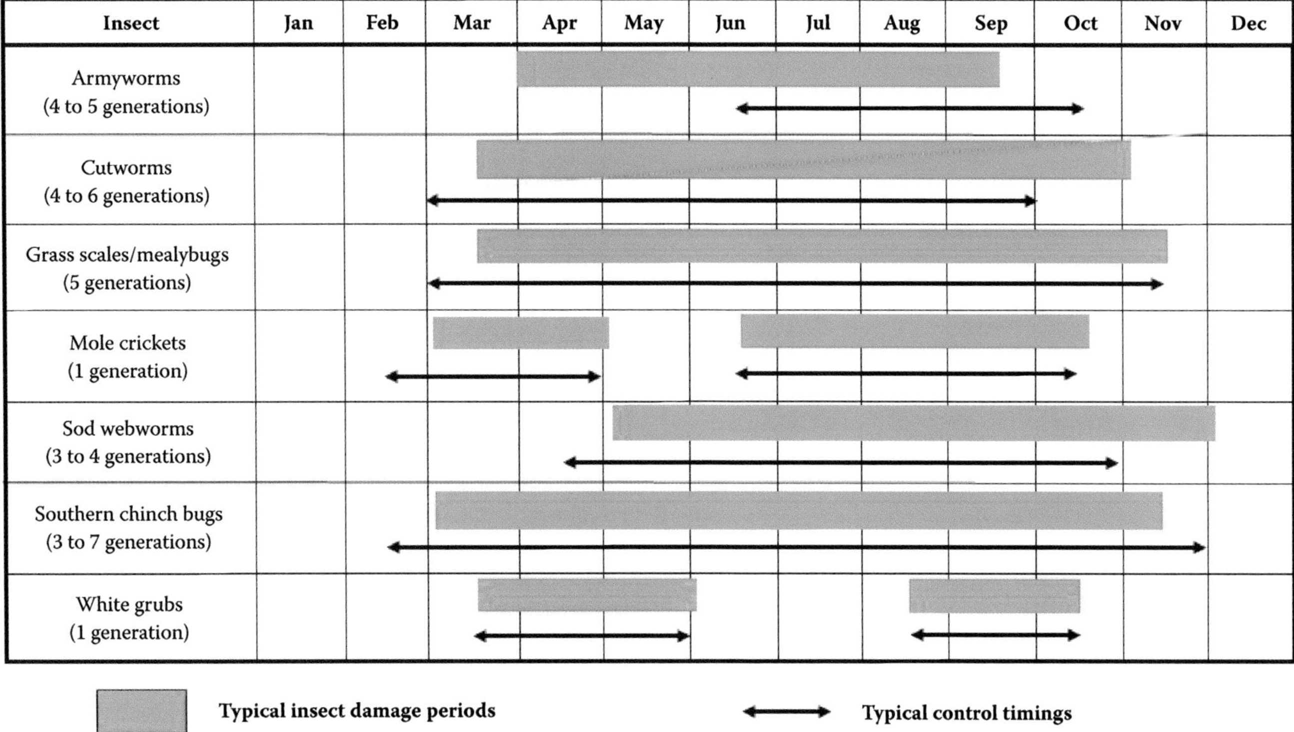


**FIGURE 13.6** Typical damage periods and control timings for insects in the southern United States.

Chemical control is not the total answer but is a contributing method in insect pest management. Influence of control is regulated by a number of inherent chemical properties and by environmental parameters interacting with the following:

**Thatch.** This partially decomposed layer of turfgrass leaves and stems is located just below the shoot tissue and right above the soil layer. Excessive thatch (more than half an inch, 1.3 cm) provides an ideal habitat (e.g., warmth and humidity) within which most insects can reside. Due to its high organic matter content, the thatch layer also tends to "tie up" or bind many applied pesticides. This lowers the effectiveness of these pesticides, especially when dealing with soil-inhabiting insects. Thatch control should be a routine practice for most golf courses, especially on greens and tees. One method of bypassing the thatch layer is application with a high-pressure injector or a slit granular applicator.

**Irrigation.** Irrigation water moves surface-applied insecticides into and through the soil profile. For insecticides applied against soil- or thatch-dwelling insects, such as white grubs and mole crickets, materials should be watered in with irrigation or rainfall soon after the application to allow penetration of the chemicals to the soil layer where the insects are feeding. When applied against surface- or leaf-feeding insects, however, irrigation should be withheld or rainfall should be avoided for at least 24 hours to allow contact of the insects with pesticide residue on the leaf surface. Read the label to determine if a specific product should be irrigated after application.

**Ultraviolet light (UV) degradation.** Exposure to sunlight often breaks chemical bonds that may make the compound inactive. Pyrethroids, insect growth regulators, bio (microbial) pesticides, and botanical insecticides are often susceptible to UV degradation. Applying susceptible products late in the evening and/or watering them off the leaf surface are two means turf managers can use to reduce UV degradation.

**Weather.** Weather conditions, especially temperature, often influence the frequency of insect generations and severity of insect infestations and damage. A rapid temperature drop to subfreezing temperatures often kills significant numbers of overwintering insects. Drought conditions often delay development or reduce pest reproduction. Generally, as temperatures increase, insects reproduce and develop faster.

**pH of spray tank and soil.** Most insecticides require a neutral (pH = 7) or slightly acidic pH in the tank mixture. When the pH in a tank mixture exceeds 7.0, the product begins to disassociate through hydrolysis into inactive or ineffective by-products. However, not all insecticides are sensitive to pH. After the tank-mix is made and agitated, the pH of the resulting solution should be taken. If a high pH is detected, a commercial buffering agent or acidifier should be added. Use the mixed pesticide solution immediately and never store overnight or leave the sprayer in open sunlight. Read the label and materials safety data sheet (MSDS) to determine if a particular pest control compound is sensitive to pH.

**Enhanced microbial degradation.** Microbes (usually bacteria and simple fungi) present in the soil often can use a portion of pesticides as a food source. During this process, complex compounds are broken down into those used for food. Certain continuously used pesticides can become victims of enhanced microbial degradation. Populations of aerobic microbes build up to levels that quickly break down pesticides within hours after being applied. Reduce the chance of enhanced microbial degradation by (1) rotating between classes of pesticides, (2) using one class of pesticides only once per year, and (3) applying only the recommended amounts of materials as excessive rates may favor microbial buildup.

**Longevity of control.** Modern synthetic insecticides classified as organophosphates, carbamates, and pyrethroids are generally very effective against the target pest if application timing and methods are followed as recommended. One drawback to these materials, however, is their relatively short control. The general range of control for most of these materials is 3 to 14 days, depending on the material, insect, and environmental conditions (as discussed earlier). Improper application and subsequent watering practices reduce the control longevity of these materials as do excessive thatch layering, sunlight (UV) exposure, and improper spray tank pH levels.

Control also is rarely 100% effective, even if ideal application parameters and environmental conditions exist. However, populations should be lowered to below aesthetic levels. Slow repopulation from survivors and subsequent hatching of eggs and migration of adults from nearby untreated areas may require retreatment. Newer insecticides, such as imidacloprid and fipronil, show good soil longevity, yet they are extremely environmentally safe.

## SURFACE- OR LEAF-FEEDING INSECTS

### Tropical Sod Webworm

**Species.** Tropical sod webworm, *Herpetogramma phaeopteralis* Guenée (order Lepidoptera: family Crambidae), feeds on warm-season turfgrasses in the southeastern United States from Florida to Texas, Oklahoma, Tennessee, and South Carolina. This species is also common in the Caribbean through Central America. The grass webworm [*Herpetogramma licarsisalis* (Walker)] is a turfgrass pest in Hawaii.

**Description.** Adults are dingy brown moths about 3/4 to 1 inch (1.9 to 2.5 cm) in length (Figure 13.7). Unlike the temperate sod webworms, adult tropical sod webworms hold their wings flat in a triangle shape and lack prominent snouts. Caterpillars have cream-colored bodies, yellowish brown heads, and brown spots along the length of their bodies (Figure 13.8). Body lengths are on average 1/21 inch (1.2 mm) when hatched and 1/2 inch (11.3 mm) when fully grown. Pupae are often buried in thatch, reddish brown, 35/100 inch (9 mm) long, and 1/10 inch (2.5 mm) wide.

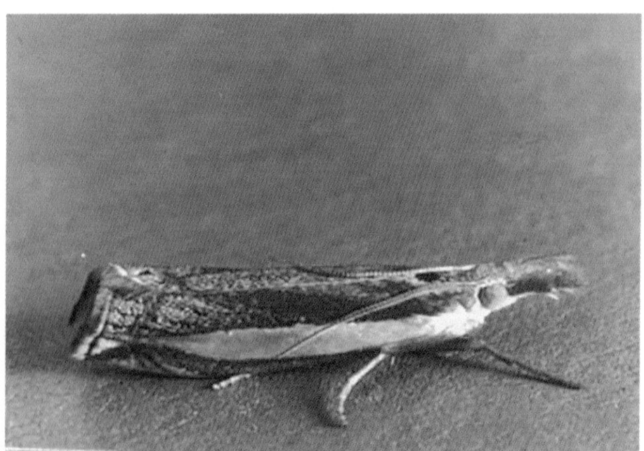

**FIGURE 13.7** Tropical sod webworm adult moth.

**Biology.** Adult moths prefer to rest in tall grass and shrubs during the day, and become active at dusk. Females live for 10 to 14 days and deposit several clusters of 10 to 35 eggs on the upper surface of grass blades. Eggs hatch in about 4 days, and the caterpillars develop through six larval instars in 21 to 47 days (depending on temperature) before reaching pupal and adult stages. The webworm name is derived from the insect's webbing together plant debris, soil particles, and fecal pellets in which they hide during the day. In southern Florida, adults are present year-round and there may be four or more generations per year. The population peaks in September through November, and declines in the winter before increasing again in March through May. It is likely that tropical sod webworms do not overwinter north of Florida. Moths migrating gradually from Florida in late spring and early summer may be the sources of infestations in other states. Even though adult moths do not feed on turf, noting their presence can serve as a monitoring tool. Adult moths fly in a zigzag pattern about a foot high (0.3 m) above turf at dusk.

**Damage symptoms/signs.** Tropical sod webworms feed on all warm-season turfgrasses but prefer St. Augustinegrass and bermudagrass. Caterpillars feed at night and hide in the thatch during the day, preferring to feed in dry, hot turf areas. Young caterpillars feed on the upper leaf surface, creating

**FIGURE 13.8** Tropical sod webworm larvae with cream-colored bodies, yellowish-brown heads, and brown body spots.

"windows" that can be easily overlooked. When mature caterpillars consume the entire leaf blade, a "ragged" or thinned turf often results (Figure 13.9). Damaged areas may appear as patches of thinned turf and exposed thatch that become increasingly large. Roots, stolons, and meristems of turfgrass are not affected by the infestation; therefore, with management and care, the turf can regrow and recover. In Florida, severe damage may occur in late fall when the moth population is the largest and the grass growth is slowest. Additional damage may be caused by birds and mammals digging for the caterpillars in the turf.

**Sampling and control.** Early detection of infestation is crucial to the management of tropical sod webworm. "Hot spots" where annual infestations repeatedly appear should be monitored every seven days through visual inspections and soap flush. Look for caterpillars and chewing damage in the periphery of thinned or poorly grown turf areas. Caterpillars are often found near the soil surface, and curl into a C-shape when touched. Also, look for the silk trails left behind by the night-feeding caterpillars. These trails are easier to see when dew is present. Caterpillars usually climb up the grass blades within five minutes of soap flush. The activities of birds and other mammals, and flying moths between dawn and dusk, also help pinpoint areas with potential infestations.

The action threshold of tropical sod webworm is 5 to 10 caterpillars per square meter (or about 1 caterpillar per square foot) in high-maintenance turf in dry, sunny areas. Insecticide applications 10 to 12 days after observing the adult moths can provide effective control of young caterpillars, typically in midsummer, and should be made before caterpillars burrow down to overwinter. Irrigation should be withheld for 24 hours after the application of contact insecticides, whereas post-treatment irrigation is needed to move systemic insecticides to the root zone. Granular formulations of contact insecticides may not be effective because the concentration of active ingredients on leaf blades, where the caterpillars are feeding, may not be high enough.

**FIGURE 13.9** Typical leaf notching from tropical sod webworm feeding.

Beneficial fungi (*Beauveria bassiana*) and nematodes (*Steinernema carpocapsae, Steinernema feltiae, Heterorhabditis bacteriophora,* and *Heterorhabditis indica*) can be used against the tropical sod webworm. A study in Florida reported a reduction up to 93% in webworm population after *S. carpocapsae* application. A large number of predatory and parasitic insects feed on the tropical sod webworm. An egg parasitoid, *Trichogramma fuentesi* (Torre), parasitizes >80% of eggs.

Avoid thatch buildup and stressful conditions by practicing fertilization, irrigation, and mowing that are appropriate for the turfgrass species and location in question. Clipping removal may help reduce webworm population by removing eggs deposited on the grass blades. 'Common' and 'FB-119' bermudagrass are less preferred for larval development, whereas 'Amerishade,' 'Floratine,' 'FX-10,' 'NUF-76,' and 'Winchester' St. Augustinegrass, and 'Cavalier,' 'DALZ8501,' and 'JZ-1' zoysiagrass are less preferred for oviposition and larval development.

### Temperate Sod Webworms

**Species.** A large number of sod webworms (order Lepidoptera, family Crambidae) feed on warm- and cool-season turfgrass in the temperate regions of the United States. Some of the most common species are

- Bluegrass sod webworm, *Parapediasis teterrellus* (Zincken), which is distributed in the eastern United States from Massachusetts to Connecticut and westward to Colorado and central Texas. This species is most prevalent in Kentucky and Tennessee where Kentucky bluegrass is dominant. Caterpillars also damage bentgrass greens, with symptoms similar to grub feeding.
- Striped sod webworm or changeable grass-veneer, *Fissicrambus mutabilis* (Clemens), which is found from Florida to New York and Ontario, and west to Illinois and Texas. This species is especially prevalent in Pennsylvania, Illinois, and Tennessee.
- Large sod webworm or greater sod webworm, *Pediasia trisecta* (Walker), which is distributed from North Carolina, Tennessee, Texas, New Mexico, and Colorado north to Washington state and southern Canada. This species is especially prevalent in Ohio and Iowa.
- Burrowing sod webworm, *Acrolophus popeanellus* (Clemens), which is sometimes a problem in the Midwest and the southeastern United States.
- Silver-striped webworm, *Crambus praefectellus* (Zincken), which is found along the Pacific Coast and areas west of the Rocky Mountains in the United States.

These species have similar biology and management; therefore, they are discussed as a group.

**Description.** Adult moths are small, dingy brown to almost white colored, with a wingspan of 3/4 inch (1.9 cm) and delicate fringes along the wing borders. Resting adult sod webworm moths have a very long, distinct, snout-like projection in front of their heads (giving this group a common name of "snout moths"), and roll their wings in a tube (different from the flat, triangular-shaped wings of the tropical sod webworm). Temperate sod webworm adults are identified to species by varying wing color patterns and male genitalia, while larvae are extremely difficult to separate into species. Sod webworm caterpillars have a dark, yellowish-brown head and greenish or beige, hairy body with numerous brown spots scattered over it (Figure 13.10). Mature larvae are 3/4 inch (1.9 cm) long. Most sod webworm larvae curl into a ball when disturbed.

**Biology.** Temperate sod webworms overwinter as larvae tightly coiled in silk-lined tunnels or cases. In spring, the larvae pupate within these tunnels or cases and the adults emerge, mate, and lay eggs. During the day, moths rest in shrubbery adjacent to turf areas. At dusk they fly in a zigzag pattern over the turf, depositing clusters of 6 to 15 eggs on grass leaves. Eggs hatch in about seven days when temperatures are at least 78°F (26°C) and larvae progress through seven instars, requiring 25 days to complete their development. When temperatures are lower (72°F or 23°C), the caterpillars develop through eight larval instars in 45 to 50 days. They pupate on the soil surface and emerge as adult moths in seven days. Life cycle from egg to adult requires five to six weeks at 78°F (26°C) and 12 weeks at 72°F (23°C).

In the southern United States, three to four generations may occur between late April and mid-October, with caterpillars overwintering from October to March. In the transition zone or middle region of the United States, two to three generations occur in May to September. In the northern United States, there may be one to two generations in May to September. Overwintering caterpillars can be found from October to April in the central to northern United States.

**Damage symptoms/signs.** Adult sod webworms do not feed, but larvae damage grass by chewing blades or severing blades above the thatch and pulling this into their silk-lined tunnels. When first hatched, they only rasp the surface or skeletonize the blades. Damaged areas appear grayish and usually are only 2 to 3 feet in diameter. Often, damage is first noticed adjacent to shrubbery and flower beds where moths are resting. Moths are attracted to dark-green, healthy turf, making golf greens a preferred feeding site. When larvae become larger, they notch the blades and grass becomes ragged in appearance. Injury initially appears as small, closely cropped grass patches as grass blades are clipped just above the crowns. Damaged areas then become larger and they often fuse. Continued feeding gives the turf a close-cropped yellowish, and then brownish, appearance from the exposed thatch. Webworms normally do not kill turf, and turf usually can recover if the webworm is controlled and proper fertilization and watering practices are used. Birds often pull out the paper-like white silk sacs where the caterpillars hide, leaving them lying on the grass.

FIGURE 13.10 Temperate sod webworm larvae with its hairy body and dark, yellowish-brown head.

**Sampling and control.** Approximately two to six caterpillars/square foot (0.2 to 0.6/m$^2$) are required to cause economic damage. Irregular brown spots resembling dollar spot disease are early signs of damage. The presence of larvae can be confirmed by parting the grass and observing the soil surface in suspected areas for frass, pellets, or curled-up resting larvae. They also can be flushed to the grass surface by using a soap solution. Flocks of birds frequently return to an infested turf area, especially in early morning. The presence of large numbers of moths flying over the turf area at dusk indicates the presence of sod webworms and egg-laying periods. The turf area should be monitored for webworms once a week when the moths are most active in an area. Damage is most evident in late summer when populations have increased and grass growth is slowed.

Insecticides may be applied after observing the adult moths or being informed of the presence of caterpillars by the sampling program. Irrigation should be withheld for 24 hours after the application of insecticides. Long residual control of sod webworm is difficult because moths are continually flying into the turf area and depositing eggs during the summer months. In addition, degradation of insecticide residues and constant growth of turfgrass also leave new grass unprotected, necessitating repeated applications during the peak activity of the sod webworms.

Control of early instar larvae is possible with various microbial insecticides containing *Bacillus thuringiensis* var *kurstaki*. Control is best on young larvae. Beneficial nematodes (*Steinernema carpocapsae* and *Heterorhabditis bacteriophora*) and fungi (*Beauveria bassiana*) are also available. Insecticides and microbial products should be applied late in the day to target caterpillars feeding at night. Endophyte-infested tall fescues and ryegrass may help reduce sod webworm damage. Several insect pathogenic nematodes are also being screened as possible control measures, as are some fungal endophytes and ichneumonid wasps.

### Fall Armyworm

**Species.** The fall armyworm, *Spodoptera frugiperda* (Smith), is a perennial pest problem of all turfgrass species in areas east of the Rocky Mountains, particularly in the southeastern United States. Other armyworm species, such as the yellow-striped armyworm [*Spodoptera ornithogalli* (Guenée)], the true or common armyworm [*Pseudaletia unipuncta* (Haworth)], and the lawn armyworm [*Spodoptera mauritia* (Boisduval)] (all order Lepidoptera, family Noctuidae), are occasional turfgrass pests in various parts of the United States, and can destroy a turf area during outbreak. Armyworms were described by American colonists as a pest that devoured crops like an "army of worms."

**Insect description.** Fall armyworm larvae are the injurious stage and are 1 1/2 inches (3.8 cm) long when mature. The caterpillars are greenish when small and dark brown when fully grown (Figure 13.11). They have a light dorsal stripe, darker bands on each side running the length of the body, and four black spots on the end of the body. The caterpillars also have a distinct inverted yellow or white "Y" shape on their heads (Figure 13.12). Adult moths are brownish with light and dark markings and a distinct white blotch near the tip of each front wing (Figure 13.13). They have a wing span of about 1 1/2 inches (3.8 cm). Common armyworms do not have the light-colored, Y-shaped mark but rather a brownish head with H-shaped, darker brown lines.

**FIGURE 13.11** Fall armyworm larvae with dorsal stripes.

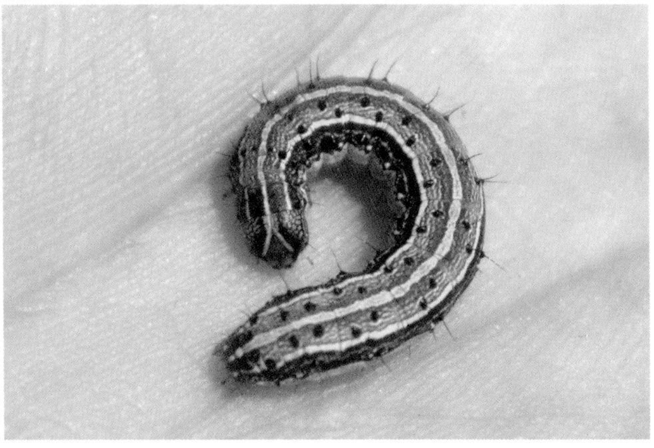

**FIGURE 13.12** Fall armyworm larvae with a distinct inverted yellow or white "Y" shape on their heads.

**FIGURE 13.13** Fall armyworm adult moth.

**Biology and distribution.** The life cycle of the fall armyworm varies considerably according to the region. Armyworms continuously reside in Central and South America and the West Indies. They also survive mild winters in southern Florida and Texas. They spread each spring from these areas into the eastern United States and into southern New Mexico, Arizona, and California, reaching the northern states in fall, providing their common name.

Female moths prefer to deposit eggs on lightly colored, erected objects near turf, such as flags, poles, posts, and fences. Each female is capable of producing 1,000 to 2,000 eggs over her lifetime of 7 to 21 days. Eggs are laid in clusters of 100 to 200, and are covered with grayish, fuzzy scales from the body of the female moth (Figure 13.14). Eggs hatch in

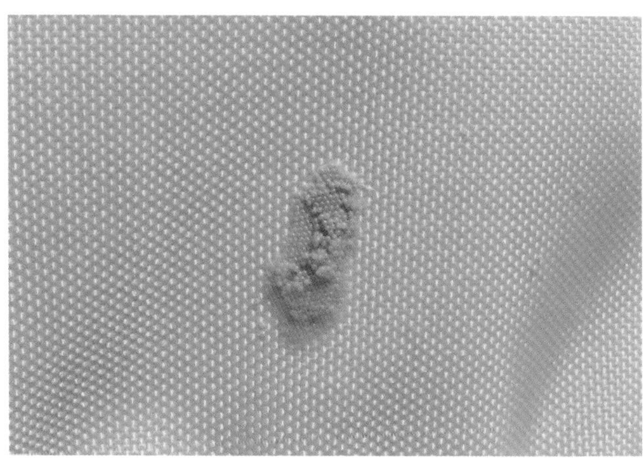

**FIGURE 13.14** Fall armyworm egg mass on a golf-green flag.

about three days, and caterpillars develop through six instars in 14 to 30 days, depending on the temperature, feeding on average for 28 days. Mature caterpillars burrow about 1 inch (2.5 cm) into the soil and pupate. A generation takes five to six weeks to complete. There may be four to five generations per year between May and November in the southeast. Except for those in southern Florida and Texas, all fall armyworm populations die each fall after freezing temperatures. The common armyworm is more tolerant of cold and can survive as larvae or pupae further north.

**Damage symptoms/signs.** Despite its name, the fall armyworm is capable of causing damage to turfgrass in early summer, especially following cool, wet springs, which may reduce populations of natural parasites. However, most damage occurs during late summer and early fall after populations have increased during the season. Larval feeding is similar to webworm feeding except it usually does not occur in patches, but in more uniform (sharp lines) and larger areas. Larvae feed day or night, but are most active early in the morning or late in the evening. Younger larvae feed on leaf margins, giving them a ragged look. Larger larvae eat all aboveground leaves and stems, resembling a "mowing" (Figure 13.15). Bentgrass, bermudagrass, fescue, bluegrass, seashore paspalum, ryegrass, and grain crops (especially sweet corn) are most often attacked.

**Inspection and control.** Moths typically lay eggs on structures such as buildings and fences and on plants such as shrubs and trees. Therefore, the infestations almost always begin along the margins or edge of the turfgrass area. They migrate or "march" when the larvae are about three weeks old and can cause extensive damage at this time (Figure 13.16). Soap flushing can bring the larvae to the soil surface. Feeding birds and the presence of green fecal pellets also indicate an armyworm presence. Adult moths are often attracted to lights at night during flight periods. Management using insecticides and microbial products should target younger caterpillars; therefore, early detection is crucial to the management of fall armyworm.

Threshold levels vary on lower maintenance turf, but control may be justified if one armyworm is found per square foot ($0.1/m^2$) on a green. Current strains of *Bacillus thuringiensis* offer inconsistent control. However, endophyte-containing

**FIGURE 13.15** Feeding damage from fall armyworms, resembling mower scalping or drought.

**FIGURE 13.16** Heavy fall armyworm infestation resembling "marching" as they migrate.

grasses such as ryegrasses or fescues are quite resistant to armyworms. Stoloniferous grasses, such as bermudagrass, generally recover from armyworm feeding since the pests do not destroy plant crowns. Nonstoloniferous grasses, such as fescue, may not fully recover.

The spinosyn toxins and parasitic *Steinernema* nematodes provide biorational control. As new Bt strains are produced, they also may provide acceptable biorational control. Most synthetic pesticides such as the pyrethroids, organophosphates, carbamates, indoxycarb, IGRs, neonicotinoids, and spinosad also provide effective control. In addition to treating obviously damaged areas, treat one or two boom widths outside the infested area to control the probable additional caterpillars feeding beyond.

### Cutworms

**Species.** There are several cutworm species (order Lepidoptera, family Noctuidae):

- Black cutworm, *Agrostis ipsilon* (Hufnagel), is the most important pest species on golf course turf throughout the United States, and the only cutworm species that requires regular management.
- Variegated cutworm, *Peridroma saucia* (Hubner), is an occasional pest of turfgrass.
- Granulate cutworm, *Feltia subterranea* (Fabricius), is found most often in southern regions.
- Bronzed cutworm, *Nephelodes minians* (Guenée), is an occasional pest of turfgrass and prefers bluegrass.

Cutworm species vary slightly in habits and appearance from each other, but their life histories are generally similar. Because of its economic importance, the following description is based on the black cutworm.

**Insect description.** Adult cutworm moths are generally stout-bodied, hairy, and have a wingspan of about 1.5 inches (3.8 cm). The forewings of adult moths are gray-black to dark brown, and mottled or streaked in color, whereas the hindwings are lightly colored and unmarked. Black cutworm adults have a distinctive black, dagger-shaped marking in the center of each forewing, which is lacking in other cutworm species. Resting cutworm moths hold their wings flat in a triangular position. Cutworm caterpillars are fat, hairless, dull gray or brown to nearly black colored on the back, light gray in the bottom, with a pale stripe on the middle of the back; they measure about 1.75 inches (4.4 cm) when fully grown (Figure 13.17). If disturbed, the larvae usually curl into a C-shaped position. They are about twice the size of webworms.

**Biology and distribution.** Cutworms are found throughout the United States. Adults and larvae are nocturnal and hide during the day, but may become active on cloudy days. Caterpillars often hide in silk-lined burrows, and may take advantage of existing coring holes. They overwinter in the soil either as pupae or mature larvae. In the spring, the hibernating larvae pupate and adults begin to appear in mid-March. Female moths deposit eggs singly or in clusters, often on tips of grass blades, and each female can lay as many as 1,200 to 1,600 eggs. Under optimum conditions, eggs hatch in three to five days, and larvae develop in three to four weeks, passing through six instars. Pupae mature in two weeks during summer but may require up to nine weeks in fall. As many as four generations occur each year.

**Damage symptoms/signs.** Adult moths are nondestructive, while cutworm caterpillars feed on the stems and leaves of young plants and often cut them off near the soil line, providing their common name (Figure 13.18). Many prefer wilted plant material and may climb ornamental plants and feed on unopened buds. Bentgrass is a major target of cutworms and they often reside in and eat extensively around aerification holes (Figure 13.19). Damage to greens from insect-feeding birds can also be substantial. Damage appears as 1- to 2-inch (2.5 to 5 cm) dead spots resembling ball marks on closely cut turf with a pencil-size hole in the middle.

**Inspection and control.** Adult flight activities can be monitored with commercially available pheromone traps. Once adults are trapped, visual detection for feeding damage and soap flush for caterpillars can begin. Since cutworms are mostly nocturnal, late afternoon is best for their detection. Examine the turf for damage. Initial feeding symptoms often

**FIGURE 13.17** Black cutworm larvae.

**FIGURE 13.18** Feeding damage from cutworm larvae.

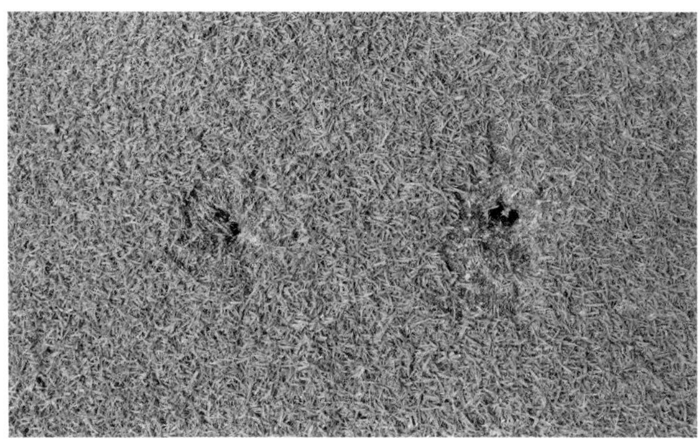

**FIGURE 13.19** Cutworm feeding adjacent an aerification hole where they reside during the day.

are mistaken for ball mark damage. Later symptoms often appear similar to dollar spot disease except a hole where the caterpillar resides can be found in the center of the damaged area. The soap or pyrethrum flush test will also aid in detection (use caution when using soap flush on bent greens in hot weather).

Threshold levels for golf greens are approximately one cutworm/square foot (11/$m^2$), while 5 to 10 larvae/square foot (54 to 108/$m^2$) are tolerable on fairways and lawns. Control is best when applied in late afternoon to early evening. Bt and *Steinernema* nematodes are effective on young (first to third instars) larvae. Many liquid insecticides provide good control. These should be applied late in the evening and **not** irrigated in, as the insecticide should be left on the leaf surface for the insect to feed on. Granular insecticides may not be as effective. Because cutworms can move long distances, a band 20- to 30-feet wide around golf greens should be treated to prevent the cutworms from moving into the greens. Repeated applications may be needed, especially for short-residual contact insecticides. Because moths from overlapping generations are always present, eggs may be continuously produced, and cutworms might migrate from nearby turf areas.

Mowing removes 80% to 90% of cutworm eggs deposited on the tips of the grass blades. However, as the clippings fall onto other areas, these eggs can still hatch and initiate new infestations. Clippings should be removed and not moved onto greens with the mowing equipment. Cutworms do not seem to be greatly affected by the endophytic toxins in ryegrass and fescues.

### Bermudagrass Mites

**Species.** The bermudagrass mite, *Eriophyes cynodoniensis* Sayed (order Acarina, family Eriophyidae), appears in all bermudagrass-growing states. It is also commonly referred to as the "bermudagrass stunt mite." Isolated cases of Eriophyid mites on zoysiagrass can also occur.

**Insect description.** The mites are extremely small, only about 0.008 inch (0.2 mm) long, yellowish-white, translucent, and somewhat wormlike in shape with only two pairs of short legs, as opposed to four pairs on typical mites and three pairs for most insects. A microscope with at least 30× magnification is needed to see them.

**Biology and distribution.** Bermudagrass mites are probably native to Australia but have spread to New Zealand, Africa, and America. They are found in all bermudagrass-growing regions in the United States. The biology of bermudagrass mites is poorly known. They multiply very rapidly, requiring only about seven days to complete their life cycle. This short life cycle allows for rapid buildup during late spring and summer. Each female is believed to produce 10 to 12 eggs in her lifetime. Eggs are deposited under the leaf sheath, and after hatching, mites molt twice before reaching adulthood. All stages are found under leaf sheaths. Mites appear well adapted to hot temperatures and become relatively inactive during cold temperatures. They spread through infected plants, clippings, machinery, mobile insects, and even by wind.

**Damage symptoms/signs.** Since the bermudagrass mite is so small, it remains hidden beneath the leaf sheath. Therefore, it can be identified more easily by symptoms of grass damage. The mites suck plant juices with their needlelike mouth parts that cause a characteristic type of damage. Grass blades turn light green and abnormally curl (Figure 13.20). Typical vigorous spring growth is noticeably absent. Internodes shorten, tissues swell, and the grass becomes tufted, so small clumps, often bushy in appearance, are observed. This is often called "witch's brooming." This characteristic growth is believed to be caused by a toxin injected into the developing grass node. The grass loses its vigor, thins out, and may die. Injury is more pronounced during dry weather and especially when grass is stressed due to poor maintenance. Since damage is often associated with drought stress, providing adequate moisture and nutrients helps grass outgrow mite damage. The witch's broom formation is permanent. Over time, turfgrass stems and stolons die, leaving voids in turf that may coalesce to form large bare patches.

**Inspection and control.** In most bermudagrass cultivars (such as 'Celebration'), the witch's broom symptom is very easily detected and a reliable diagnostic characteristic for the infestation. In other cultivars (such as 'Tifway'), however, the grass blades merely turn light green and the witch's broom symptom does not always appear. In this case, visual inspection of the light-green grass and the detection of the bermudagrass mite will be needed to confirm infestation (Figure 13.21). Inspection should focus on thinned turf, and symptoms may be most noticeable on taller mowed grass, such as rough areas, on slopes, around sand traps, and along canals and fence rows.

The following sampling plan and thresholds may be used. Construct a 3 × 4 foot (1 × 1.2 m) rectangular frame with PVC pipes. Inside the frame, thread string through holes drilled through the PVC pipes at 1-foot (0.3 m) intervals, creating a grid with 12 squares. Toss the frame onto an infested turf and count the total number of "witch's brooms" in 10 of the 12 squares. Once a month, take one sample every 50 feet (15 m) on fairways and roughs, four frame samples from each greens approach, and two samples from each tee bank. Apply miticide if on average four or more "witch's brooms" are found in each sample. If less than four damaged stems are found, then a cultural control program may be sufficient.

**FIGURE 13.20** Bermudagrass light-green color and abnormal curling from stunt mite feeding.

**FIGURE 13.21** Witch's broom symptoms from bermudagrass stunt mite feeding.

An alternative threshold for control is four to eight witch's brooms/square foot (2 to 4/5 $m^2$). These sampling plans and thresholds have not been rigorously verified in the field.

There is currently no effective pesticide against the bermudagrass mite. Azadirachtin, bifenthrin, chlorpyrifos, cyfluthrin, deltamethrin, dicofol, and lambda-cyhalothrin are currently registered for general mite control in turfgrass; however, most have been ineffective against bermudagrass mite. The most effective insecticide is diazinon, which is no longer registered for use in turfgrass. Abamectin, dicofol, and chlorpyrifos are alternatives to diazinon, but their efficacies are lower and repeated biweekly or monthly applications will be required. The addition of a surfactant that helps pesticide solution to penetrate into the leaf layer will increase efficacy.

Lower mowing height may help remove many infested stems. Scalping and vacuuming clippings may be effective in removing the most severely infested grass. A higher fertilization and irrigation rate after scalping may allow recovery of infected turf. It is important to clean the mowing equipment thoroughly because infested clippings can initiate new infestations in other areas. Maintain good soil moisture, as dry conditions tend to favor mite damage. Consider growing resistant or less susceptible bermudagrass cultivars such as Tifsport, Franklin, and Midiron.

**FIGURE 13.22** Grass scale or mealybug bodies covered with a white secretion, resembling tufts of cotton.

## Grass Scales or Mealybugs

**Species.** Over 37 species of mealybugs have been associated with grasses. Two major grass scales occur: (1) the rhodesgrass mealybug, also called rhodesgrass scale, *Antonia graminis* (Maskell); and (2) the bermudagrass scale, *Odonaspis ruthae* Kotinsky (order Homoptera, family Pseudococcidae and Diaspididae). These are not common pests but do occasionally occur in the United States from South Carolina to California, most often on bermudagrass and St. Augustinegrass, especially growing in shade. Rhodesgrass scale derives its name from feeding on its favorite host, rhodesgrass, a coarse-textured pasture grass. The buffalograss mealybug, *Tridiscus sporoboli* (Cockerell) and *Trionymus* sp., is often found on buffalograss but appears to cause little economic damage. Ground pearls also are scale insects and are discussed separately.

**Insect description.** The rhodesgrass mealybug body is round and dark brown but is covered with a white cottony secretion that appears like tufts of cotton on the grass (Figure 13.22). Male mealybugs resemble tiny gnats with a single pair of wings and three pairs of red eyes. Adult males are not considered harmful to turfgrasses. The bermudagrass scale is oval shaped, white, wingless, and approximately 1/15 inch (1.7 mm) in diameter. They prefer the taller grass in rough areas, especially in heavily thatched and shaded areas. They also are found around sand traps, along fence rows, and in other similar areas. When the scales hatch into the crawler stage, they migrate beneath the leaf sheath, usually at the nodes. Only the youngest, immature stages are mobile. Adults settle on the leaf or stem, insert their needlelike mouthparts, become immobile, and eventually start excreting a white, cottony, waxy covering.

**Biology and distribution.** The life cycles of both insects range from 60 to 70 days, and there are up to five generations per year in the southern United States. Continuous generations occur from Orlando, Florida, south.

**Damage symptoms/signs.** As mentioned, bermudagrass and St. Augustinegrass are the preferred hosts of mealybugs and scales. They infest the crown, nodes, or under leaf sheaths (not the leaves) and withdraw plant sap with their piercing/sucking mouthparts. Infested grass slowly loses vitality, discolors, and later appears to be suffering from drought (Figure 13.23). Stunting and thinning of the grass stand occur

**FIGURE 13.23** Grass-scale feeding damage on bermudagrass.

under high infestation levels. Under heavy infestations, plants are often covered with tiny masses of white, waxy secretions. Injury is most severe during extended hot, dry (stressful) periods. The rhodesgrass mealybug produces considerable honeydew, and other insects such as ants or bees may be present on heavily infested turfgrass.

**Inspection and control.** Plant leaves should be pulled away from the stem and sheaths examined for tiny, white, cottony masses. Ants feeding on the honeydew also can indicate mealybugs. Since these insects produce more damage during dry weather, keep the turf well irrigated and fertilized. Cultural control includes collecting grass clippings, which will contain some scales, and depositing off-site. Several insecticides provide control but are rarely needed. If needed, thorough spray coverage is necessary and a surfactant should be added.

### Annual Bluegrass (or *Hyperodes*) Weevils

**Species.** *Listronotus maculicollis* (Dietz) (order Coleoptera, family Curculionidae); formerly *Hyperodes* sp. near *anthracinus* (Dietz), is the annual bluegrass weevil.

**Insect description.** Larvae are C-shaped, legless, and from 1/32 to 3/16 inch (0.8 to 5 mm) long. Larvae are creamy white in color, with light brown to tan heads, and are difficult to distinguish from billbug grubs (Figure 13.24). Adults are small (1/8 inch, 3 mm, long) black to dark-gray weevils, about half the size of Kentucky bluegrass billbug, and can be distinguished from the billbugs by the position of their antennae. The antennae of the annual bluegrass weevil emerge from the end of the snout while the antennae of billbugs emerge from the snout closer to the head. The annual bluegrass weevil also has a broader snout than a billbug. Adults initially appear reddish in color when they first emerge from the pupal stage but turn black as their exoskeleton hardens. Damage is mostly to *Poa annua* and secondarily to creeping bentgrass.

**Biology and distribution.** Annual bluegrass weevils are native to the United States and occur in at least 40 states. It is most often encountered in the northeastern United States including New York, southwestern Connecticut, northern New Jersey, Pennsylvania, Maryland, Delaware, and all of the New England states into Ontario and Quebec. The insect has two to three generations annually in the United States (Figure 13.25). Adults are in the overwintering stage and often hibernate in the leaf litter under trees of golf course roughs, and become active in early spring, often corresponding with early plant flowering (e.g., between forsythia and flowering dogwood or eastern redbud full bloom). As adults are not capable fliers, they travel (walk) to playing surfaces to feed, mate, and deposit eggs. Eggs are deposited in midspring (early May) in chewed-out stems by adults. Larvae hatch in four to five days and burrow into stems. Five instar stages occur, each lasting 5 to 10 days, and they resemble each other except that they become larger in size, appearing as a grain of rice with a brown head. The first three larval instars feed on plant stems and eventually exit the stems; the fourth and fifth instars feed externally at the soil/thatch surface on plant crowns, eventually killing plants. Sawdust-like frass is evident from this feeding and is apparent by late spring. Mature larvae pupate near the soil surface in mid-June. Pupae are somewhat diamond shaped and creamy white. Young adults emerge in late June, lay eggs in mid-July, and the second generation of small larvae emerge in late July. Newly emerged adults are a rusty brown color, turning black in a few days. Most of these larvae pupate in late August. Since life cycles overlap, all stages of development can be found from late June through early September. This becomes a challenge as small larvae, large larvae, pupae, and adults can occur at the same time, making it difficult to determine the dynamics of the population. If, for example, the population is primarily pupae, no current insecticides will work satisfactorily. It may be prudent to wait a week or so to treat until most of the pupae have emerged as adults. If most of the sample is in the adult stage, treat immediately to minimize egg laying.

FIGURE 13.24 Annual bluegrass (or *Hyperodes*) weevil larvae are C-shaped, creamy-white, and legless.

**Damage symptoms/signs.** Damage is generally most severe to *Poa annua* in late spring (early June) and again in midsummer (late July) when other turf stresses, such as water, fertility, and mowing, weaken the turf. It appears only in short-mowed (≤0.5 inch, 1.3 cm) areas such as greens, tees, and fairways, often seen first in peripheral areas (Figure 13.26). Damage is from larvae feeding and begins as small yellow patches and brown spots that appear wilted or water soaked and increase into larger areas (Figure 13.27). Turf does not normally respond to watering. This is from larvae severing stems from the plant. The edges of fairways, near woods, tees, or collars are often initial sites of damage. Feeding damage resembles anthracnose symptoms, causing easy misdiagnosis. Severe damage resembles a water-soaked appearance from the hollowed grass stems. Reducing *Poa annua* populations reduces the presence of this insect as does the removal of fairway litter, where adults reside, in late fall or early spring.

**Inspection and control.** Suspected infested areas should have a physical inspection in spring (April to early May) for larvae densities. A cup cutter (or similar device) should be used to collect cores. Break up the cores and place the loosened soil and plant parts in a pan or cup filled with lukewarm water. This forces larvae and adults to crawl to the surface

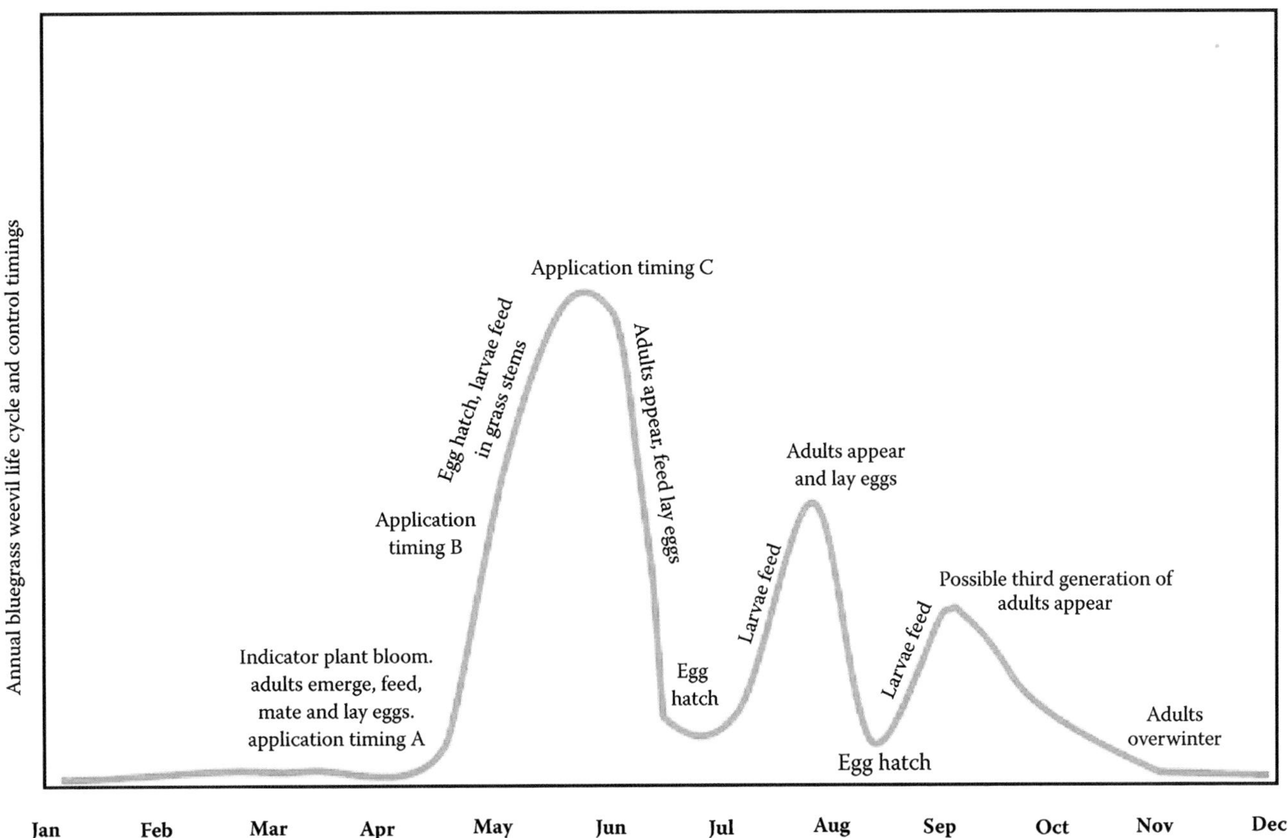


**FIGURE 13.25** Typical life cycle and application timings of annual bluegrass weevils in the northern United States.

**FIGURE 13.26** Annual bluegrass weevil damage on the peripheral areas of a golf green.

**FIGURE 13.27** Advanced damage from annual bluegrass weevil feeding.

within 10 minutes. A soapy flush can also be used in the field to indicate an adult presence. Another flush solution involves three-fourths cup (0.18 L) of table salt (NaCl) in a quart (0.9 L) of lukewarm water. If present, after about 5 minutes, all stages of weevils (except eggs) will float to the surface, although up to 45 minutes may be required. Another detection method involves inspecting the number of adults caught in mower clipping baskets or the newer vacuum or inverted leaf blower techniques. Since infestations occur in the same locations each year, keeping track of these helps pinpoint where future treatments will be needed.

Threshold levels range from 30/square foot (323/m$^2$) for golf greens and other areas in summer, and up to 50/square foot (540/m$^2$) for well-maintained fairways and other areas in spring. Lower maintained (nonirrigated) turf may have lower (e.g., 20 to 40 weevils/square foot, 215 to 430/m$^2$) threshold levels.

Insecticides should be used to target adults whenever possible as larvae are difficult to control. Timing for adults is in spring just prior to their beginning to lay eggs and can usually be timed when flowering shrubs and trees such as forsythia, wisteria, and dogwood bloom, typically in midspring (mid-April through early May). Due to its systemic action, chlorantraniliprole is very effective when applied to young larvae feeding inside the plant. It is applied in early spring when full bloom of dogwood or eastern redwood is occurring or just past full bloom (half yellow and green stage) of forsythia. This product also can provide season-long control of white grubs and caterpillars with the same application. Other larvicides are timed several weeks later including spinosad, trichlorfon, and indoxacarb. These must be washed into the soil following application. Where the second generation occurs, a follow-up insecticide application may be needed in early July. Adults at this time are detected in mower clippings, with soap flush or vacuuming in early summer (late June to early July). An adulticide such as spinosad, indoxacarb, or a pyrethroid should be considered. Always rotate products with different modes of action. Treatments should be lightly watered in. A third adult generation may occur in mid- to late summer (August) but rarely requires treatment. If needed, use one of the products listed that has not been used yet that year.

Chlorpyrifos and synthetic pyrethroids (bifenthrin, cyfluthrin, lambda-cyhalothrin, deltamethrin) are used as periphery sprays for adults in spring after forsythia full bloom, as are spinosad and a combination of bifenthrin and imidacloprid. A turf manager should rotate among insecticides of different modes of action because the annual bluegrass weevil has been reported to develop resistance to pyrethroids. All products typically require two to five annual applications with one or more of these as a fairway application.

The annual bluegrass weevil has few natural enemies. The beneficial nematode, *S. carpocapsae*, has been shown to reduce >70% of larvae feeding in the soil. Since the annual bluegrass weevil only damages annual bluegrass, a turf manager may consider reducing annual bluegrass in the turf area or growing a different grass species. Removal of fairway litter (where adults reside) in late fall or early spring also reduces the number of overwintering adults. Healthy turf is the first step against damage.

## SUBSURFACE- OR ROOT-FEEDING INSECTS

### Mole Crickets

**Species.** Mole crickets (order Orthoptera, family Gryllotalpidae) are subterranean insects and are considered the most serious turfgrass pest in sandy, coastal plain areas from North Carolina to Texas. Isolated outbreaks have also occurred in southern California and Arizona, causing severe damage to bermudagrass, bahiagrass, and centipedegrass. Several species exist, including:

- The southern mole cricket, *Scapteriscus borelli* Giglio-Tos
- The tawny mole cricket, *Scapteriscus vicinus* Scudder (previously called the changa or Puerto Rican mole cricket)
- The short-winged mole cricket, *Scapteriscus abbreviatus* Scudder (found in isolated areas in the United States)
- The northern mole cricket, *Gryllotalpa hexadactyla* Perty (the only species native to the United States—causes least damage)

The following discussion focuses on the two most damaging species on turfgrass, the tawny and southern mole crickets.

**Insect description.** Mole crickets are 1 to 1.5 inches (2.5 to 3.8 cm) long when mature and possess spadelike front legs that are well adapted for tunneling through soil (Figure 13.28). Nymphs resemble adults but are smaller and wingless. The color patterns of southern and tawny mole cricket are distinct. The tawny mole cricket is a lighter creamy brown, while the southern mole cricket is grayish to dark brown and usually has four distinct light spots on its prothorax. The two species also can be distinguished by their dactyls (digging claws). The southern mole cricket has a "U"-shaped space between its dactyls, while the tawny has a "V"-shaped space. A third species, the short-winged mole cricket, is abundant especially along the southeast and southwest coasts in Florida. It is similar in appearance to the tawny mole cricket, but has short wings and cannot fly, limiting its distribution. The southern, tawny, and short-winged species were introduced into the southeastern United States around 1900 as stowaways in sand used as the ballast material of South American ships. They are found throughout the Coastal Plain region of the southeast and northern Argentina, Uruguay, and Brazil. A fourth species, the northern mole cricket, is native to the United States, but is not considered to be a major pest. It primarily inhabits moist soil adjacent to water.

**Biology and distribution.** Mole crickets have a gradual metamorphosis life cycle where adults appear similar to nymphs except for the underdeveloped wings of the latter. In most locations, the tawny and southern mole crickets have one generation per year. The process of this life cycle begins in spring, when adults fly, mate, and lay eggs (Figure 13.29). In the spring, starting in February in Florida and peaking in March, tawny mole cricket adults are attracted to lights, with major flights during the full moon. Southern mole cricket flights begin in March in Florida and peak in early May. Flight periods of each cricket species are delayed about a month in more northern areas such as North Carolina. Flights also may be delayed during cool, wet, windy weather conditions.

**FIGURE 13.28** Adult mole cricket with its spade-like front legs used for tunneling through soil.

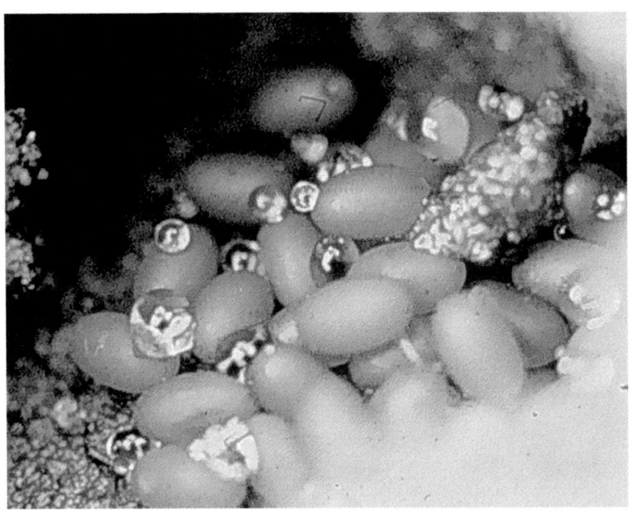

FIGURE 13.29 Mole cricket eggs, which turn tan-brown just prior to hatching.

The spring flight is the larger of the two flights. The second minor flight (the dispersal flight) is made in fall. This flight enables new-generation adults to reach previously uninfested areas. The flights may be up to 6 miles (9.7 km) per night.

Mole crickets mate and disperse via flying in the spring. Adult males attract females by using a harp-shaped area located on the wing between the two forelegs. This "harp" resonates to produce a mating call. Males construct trumpet-shaped chambers at the soil surface during the mating period to increase the intensity of the mating calls. This also helps ensure females are attracted to areas with good soil moisture to lay eggs. The hollow tops of these chambers are often visible in early spring.

Inspection by soap flushing should be performed throughout this period and females constantly inspected for egg development. Most of the eggs are laid within the first 12 inches (0.3 m) of soil, but cool and/or dry weather may cause these chambers to be constructed at a greater depth. Females often lay eggs near where they mate as this usually is moist soil, increasing the chances of egg and nymph survival. Females usually lay about four clutches of eggs per year, and average 35 eggs per clutch (Figure 13.30). Adult males die after mating while adult females die after depositing their eggs.

In most of the southern United States, oviposition (egg laying) begins in late March with a peak in May. Eggs hatch in 20 to 25 days, and their emergence is complete by mid-June. Extended drought conditions may delay oviposition and egg hatch. The small nymphs are easiest to control, and because they cause little visible damage, they are often ignored. Nymphs feed and mature throughout the summer, molting five to eight times, although their wing buds do not appear until the last two instars. As nymphs mature, it becomes more difficult to manage and control them. Adults begin to reappear in the fall. Tawny mole crickets overwinter mostly as adults, while southern mole crickets overwinter primarily as large nymphs. In most of the southeast, the southern and tawny mole crickets have only one distinct generation per year, with the southern mole cricket having a life cycle about two weeks later than the tawny.

Tawny mole crickets tend to tunnel deeper into the soil than the southern mole cricket, which tends to dig and feed near the soil surface. This is important as the tawny mole cricket may escape the lethal effects of some surface-applied insecticides. The tawny mole cricket also tends to dig two

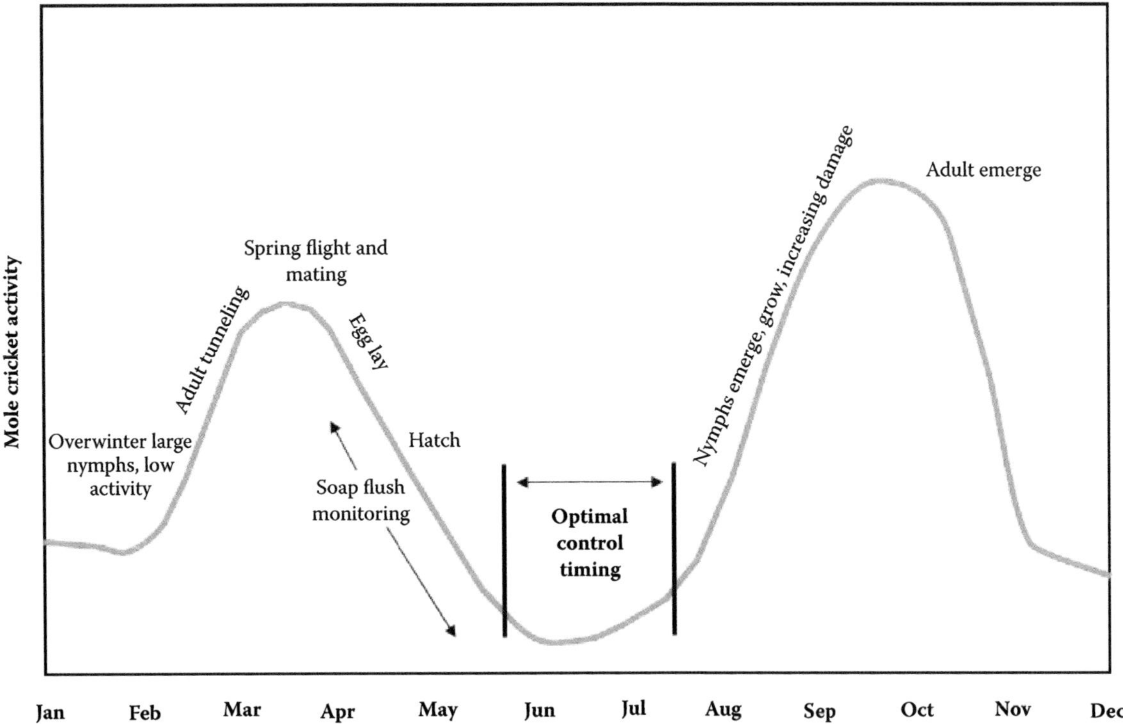


FIGURE 13.30 Typical life cycle and application timings of mole crickets in the southern United States.

V-shaped tunnels, presumably to provide an alternative route of escaping enemies.

**Damage symptoms/signs.** Mole crickets damage turf in several ways. Tawny mole crickets are the most damaging as they are herbivores, consuming all parts of the grass plant. The southern mole cricket is a carnivore and a scavenger, feeding on earthworms and insects, and is believed not to prefer plant material as food. Both species tunnel through the surface layer of the soil, causing considerable mechanical damage to grass roots. Tunneling also loosens the soil, so the grass is often uprooted, resulting in desiccation that can disrupt and break preemergence herbicide barriers, enabling weeds to germinate (Figure 13.31). The majority of turf damage occurs in early spring when adults are active, and in late summer and fall when nymphs are reaching maturity. As one walks across infested turf, the ground often feels spongy because of burrowing and displacement of soil near the surface. Counts of up to 100 nymphs/square yard (84/m$^2$) can occur in heavily infested areas.

Most mole cricket tunneling occurs at night, with highest activity occurring a few hours after dusk and again just before dawn. They are especially active after rain showers or after irrigation in warm weather. Most activity within the top 2 inches (5 cm) of soil occurs when night temperatures are above 60°F (15.6°C). Both nymphs and adults tunnel in the top inch (2.5 cm) of soil and come to the surface to feed when soil is moist and may tunnel up to 20 feet (6 m) per night in moist soils. Their feeding and tunneling are greatly reduced during cold weather or when soil is dry.

**Inspection and control.** In order to control mole crickets, a clear understanding of their life cycle and behavioral patterns is essential, followed by proper mapping of damaged areas, scouting for egg hatch, and timing of insecticides of small, newly hatched nymphs. A poor appreciation of what mole crickets are doing at any particular point in time is a recipe for failure (Figure 13.32). Without this appreciation, successful control is unrealistic. Maps of each golf hole are useful in scouting and should be made in fall (October or November) indicating heavily infested areas. Additional

**FIGURE 13.31** Tunneling in a bunker from mole crickets.

**FIGURE 13.32** Turf damage from mole cricket tunneling.

mapping should be made in late winter through spring when overwintering mole crickets become active. Tawny mole crickets typically infest the same sites yearly. Mapping can pinpoint these preferred sites, enabling spot treatment of the most heavily infested areas. Spring mapping during periods of adult activity provides knowledge on where the majority of nymphs reside and where to apply your insecticide during early summer. If these areas are not mapped and treatments applied shortly after hatch when nymphs are small, by the time one sees activity in mid- to late summer, insecticides become less effective.

The following critical steps are necessary for mole cricket control.

1. **Mapping.** Mole crickets tend to return to the same areas year after year. Mapping provides best indication where to scout later for nymphs to determine when the eggs are hatching. Mapping also pinpoints where spot treatment of insecticides is needed, eliminating wall-to-wall treatments. Map the infestation areas in spring, since adults usually lay eggs in the same areas where spring damage is seen. Mapping heavily infested areas in fall also indicates where damage is likely to occur in spring.
2. **Scouting.** Sample for the mole crickets with a soap flush. Mix three tablespoons (44 mL) of lemon-scented liquid dishwashing detergent in 2 gallons (7.6 L) of water. Apply the soap mixture over a 2- × 2-foot (0.6 × 0.6 m) area of infested turf using a 2-gallon (7.6 L) sprinkling can. Mole crickets present will surface in a few minutes. Flushing late in the afternoon or early in the morning, especially in moist soil, is best. As crickets mature, the soap flush becomes less effective. Sampling in the spring serves two purposes: (1) to determine the density and if the action threshold has been reached, and (2) to determine the life stage or size so that treatments can be applied when the nymphs are still small. When adults are collected, cut open the abdomen of a few females to determine if the eggs have matured (turned brown) and are ready to be deposited (Figure 13.29).

3. **Timing.** After the majority of egg hatch has occurred and before nymphs grow past 1/4 inch (6.4 mm) long, apply the appropriate control product. Waiting to treat past this growth stage will experience significantly less control with most commercial products. This is typically in late spring (late May to mid- to late June). Irrigate prior to application and lightly afterwards (according to label instructions) for best results. Do not irrigate afterwards if using baits. Continue to monitor the treated area with the soap flush and be prepared to spot-treat areas with unacceptable control.

The majority of turf damage occurs in late summer and fall when the nymphs are reaching maturity. Tunneling damage also occurs in late winter and spring from overwintering adult crickets. Damage subsides in May after eggs are deposited and most adults have died. It usually is mid-July before nymphs reach sufficient size to again cause noticeable turf damage.

By August, nymphs have grown considerably and become more difficult to control. Baits generally are still effective at this time, as are most curative insecticides including the pyrethroids and acephate (Orthene). Most commercial insecticides also are effective at this time. These should be applied when soils are moist to encourage cricket activity nearer the surface. Most insecticides (baits and Orthene are exceptions) need about a half inch (1.3 cm) irrigation following application to move the material into the soil and to encourage surface activity by the crickets. The residual activity of these products is relatively short; thus, they should be applied after the bulk of mole crickets have hatched.

Timing of the pesticide application is one of the most important aspects of successful mole cricket management. Mid- to late June usually is the optimum time to obtain maximum control with an insecticide application (especially long-residual ones). This is when nymphs are large enough to feed on or near the soil surface when insecticides are applied. Specifically, in the southeastern United States, treatment for tawny mole cricket nymphs is typically in late June. In comparison, southern mole crickets are treated in early to mid-July as tawny mole cricket eggs usually hatch about two weeks before those of the southern mole cricket. Yearly environmental variations in temperature and soil moisture may affect the timing of hatch as warm and moist conditions can result in earlier hatch than normal while cool, dry conditions may delay hatch.

In spring, feeding by adults declines as they prepare tunnels to lay eggs. Cutting open several fertilized females in spring helps better pinpoint when they will begin to lay eggs. Young eggs are white to almost clear in color, and turn tannish-brown just before being laid. When approximately 50% of the developing eggs inside the females become hard and BB-like, eggs will be laid in about one week and should begin to hatch in approximately one month. No registered insecticide provides outstanding control of adults in a single application at this time. Spot-treating the most damaged areas is currently the most economical means of control.

Materials with the longest soil residual (three to six weeks) should be used to control the nymphs at or even before this time and include fipronil (Choice) and imidacloprid (Merit). It is crucial when using these materials to apply them right at or soon after egg hatch (Table 13.3). Turf should be moist before insecticide treatments are made. Apply the pesticide as late in the day as possible, with dusk being the optimum time. Follow label directions explicitly regarding safety, dosage, application, and irrigation information. Later hatching nymphs should also be treated with another long-residual material in mid-July. Commercial baits also are effective for this mid-July application as crickets are then big enough to ingest enough bait to make it effective. Moist soil and dry turf foliage at the time of application provide the best control with baits. Irrigation after bait application should not occur for several days as this may degrade the material.

By fall, numerous adults are present. At this time, they cause extensive damage and are difficult to control. Feeding

**TABLE 13.3**
**Comparison of Insecticides for Mole Cricket Control in Turf**

| Active Ingredient | Trade Name | Residual | Comment |
|---|---|---|---|
| Fipronil | Chipco Choice[a] | Very long | Slit applications only; caution use label |
| Fipronil | Chipco TopChoice[a] | Long | Broadcast application; water in; 4 months' control |
| Imidacloprid | Merit | Intermediate | Several formulations for various sites |
| Imidacloprid | Advanced Lawn Season Long Grub Control | Intermediate | Best for nymphs; apply at egg hatch |
| Lambda-cyhalothrin | Battle/Scimitar/Demand | Intermediate | Best for nymphs; apply after egg hatch |
| Trichlorfon | Advanced Lawn 24-Hour Grub Control | Short | Low odor; controls nymphs and adults |
| Acephate | Orthene, Pinpoint, Velocity | Short | Standard for nymphs in summer |
| | | **Baits** | |
| Carbaryl | Sevin | Short | Good for mid- and late-season nymphs |
| Indoxycarb | Advion | Short | Quick acting; good for larger nymphs |
| Chlorpyrifos | Mole Cricket Bait, Dursban | Short | Do not water in; apply in evening |

[a] Not effective on white grubs.

continues throughout the fall until cold temperatures drive the crickets further down in the soil. Spot treatment generally is the best method of controlling adults. Timing for control is at dusk as feeding is optimum at this time.

Parasitic nematodes work best on adult crickets and may be considered in spring. The parasitic nematodes, *Steinernema scapterisci* and *S. riobravis,* and the red-eyed Brazilian fly, *Ormia depleta,* are being utilized as biological control agents against mole crickets. These natural enemies were imported from South America and are specific mole cricket parasites, but harmless to nontarget organisms. The nematodes enter the crickets through the mouth or spiracles. They penetrate the gut and enter the hemocoel where bacteria are released in the hemolymph. The mole cricket then dies from bacterial poisoning. The nematodes pass through several generations inside the dead mole cricket. About 100,000 nematodes can emerge from a single insect. The red-eyed fly, a tachinid, locates mole crickets by their singing and deposits live maggots on or near their host.

As with most biological control agents, the beneficial nematodes and the red-eyed fly are slower to control the mole cricket host and control is never 100%. However, as the populations of these beneficial organisms increase, control effectiveness also increases. The parasitic nematodes are most effective on adult crickets that are prevalent during early spring (e.g., February in Florida, March in North Carolina) and early fall (October to November before frost). Providing good soil moisture also is necessary for survival of these parasitic nematodes. Tawny mole crickets avoid contact with endopathogenic fungi (*Beauveria busiana* and *Metarhizium anidopliae*), suggesting these could potentially be used in mole cricket management.

## White Grubs

**Species.** White grubs (order Coleoptera, family Scarabaeidae), the larvae of scarab beetles, are among the most serious insect pests in the United States. Although over 1,500 different scarab beetles occur in North America, the most common in turf include (Figure 13.33):

- Japanese beetle, *Popillia japonica* Newman
- May beetle, *Phyllophaga* sp.
- Black turfgrass *Ataenius*, *Ataenius spretulus* Haldeman
- Green June beetle, *Cotinis nitida* (Linnaeus)
- Masked chafer, Cyclocephala borealis and C. lurida
- European chafer, *Rhizotrogus majalis* (Razoumowsky)
- Asiatic garden beetle (*Maladera castanea*)
- Oriental beetle (*Anomala orientalis*)

Native species to the United States include masked chafers, black *Ataenius*, and various May or June beetles. Introduced species include Japanese beetle, European chafer, Asiatic garden beetle, and Oriental beetles (*Anomala orientalis*).

**Insect description.** The grubs of different species are similar in appearance. They have white to cream-colored robust bodies with brown heads, three pairs of small legs, and a dark area at the rear of their 10-segmented abdomen (Figure 13.34). Depending on the species, they range from 3/8 to 2 inches (1 to 5 cm) long when mature and rest in a C-shaped position, especially when disturbed. Specific identification of larvae is difficult and is based on the form, shape, and arrangement of coarse hair, bristles, or spines (the rasters) on the end abdominal segments (Table 13.4).

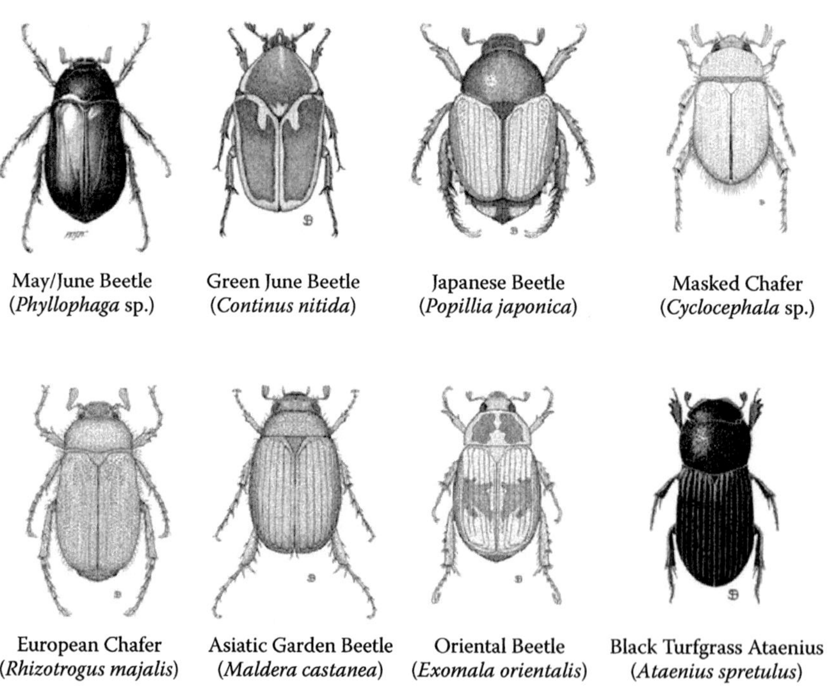


**FIGURE 13.33** Illustration of adult scarab beetles whose white grub larvae cause turf damage.

**FIGURE 13.34** White grub larvae with white to cream-colored robust bodes, brown heads, and three pairs of small legs.

Black turfgrass *Ataenius* (BTA) grubs are quite small—1/4 inch (0.6 cm)—compared to almost 2 inches (5 cm) for grubs of May and June beetles. They are sometimes mistaken for young grubs of Japanese beetles, masked chafers, or other larger species. BTA grubs can be distinguished by a pair of padlike bumps at the tip of the abdomen, just in front of the anal slit. These are fairly easy to discern with a handheld 10× lens. The main blood vessel, which runs longitudinally down the back, typically appears almost black in contrast to the gray-white body. The black *Ataenius* grubs tend to be a more serious problem in northern states, especially on bluegrass and bentgrass roots. Eggs of various species are similar in appearance.

Eggs are typically shiny and white, but vary in size depending on the species. The eggs of black turfgrass *Ataenius* are less than 1/32 inch (0.7 mm), whereas those of green June beetles are 1/16 to 1/8 inch (1.5 to 3 mm) long. They are roughly oval when first laid in the soil (at about 1 to 4 inches, 2.54 to 10 cm deep), and become more spherical as they absorb moisture from the soil.

**Biology and distribution.** Most scarab beetles, including the Asiatic garden beetle, the European chafer, the green June beetle, the Japanese beetle, the Oriental beetle, and the masked chafers, have a one-year life cycle. There are two generations of black turfgrass *Ataenius* per year, whereas each generation of May beetles takes two to four years to complete. May beetles and other species with an annual life cycle overwinter as large grubs. Adults of these species start flying in May or June, with peak flight activity in June in the transition zone. In the southern states, overwintering black turfgrass *Ataenius* adults begin flight in April, the adults from the first generation begin flight in July or August, and the adults from the second generation that are dispersing to overwintering sites begin flight in October or November.

Adults of many scarab beetle species, such as the Japanese and May beetles, feed on ornamental plants and other crops (Figure 13.35). Some species, such as the masked chafers, do not feed as adults. From May to August, females burrow into the soil and deposit eggs in the first 2 inches (5 cm) below the soil surface. During July through August, eggs hatch, and the young first instar larvae begin feeding on turfgrass roots. Grubs continue feeding on grass roots, molt twice, and grow larger until winter before working their way deeper into the soil, just below the frost line, to overwinter. The following spring they return to the rootzone and continue to feed on grass roots until April to June. They then pupate a few inches below the soil surface. Adult beetles emerge during late May through early July to mate and lay eggs. Turf damage from grubs with annual life cycles usually is most evident in late summer (late August through frost), with less damage apparent during the spring feeding season. May beetles spend the extra years in the grub stage, feeding throughout the growing season and moving deeper into the soil each winter.

**Damage symptoms/signs.** Grubs feed on all species of grass, although green June beetle grubs tend to feed mainly on decaying organic matter. They feed on the roots at, or just below, the soil–thatch interface and cause large patches of turf to die. Damage is most pronounced from mature grubs in late summer and early fall (August through October) and less so during spring (April and May). During heavy infestations, the soil surface may become very loose and spongy to walk on. In severe cases, roots are pruned so extensively the turf mat can be rolled back like a carpet exposing the C-shaped white grubs. Damage is ill-timed, being just prior to summer stress for cool-season grasses and just prior to dormancy for warm-season grasses. Symptoms of grub infestation include a gradual decline forming a yellow mosaic pattern, or consistently wilting grass in an area even though adequate water is available. Continued feeding causes larger patches of turf to thin and die, allowing weeds to invade (Figure 13.36). Additional damage may occur from predatory animals such as armadillos, birds, hogs, skunks, raccoons, moles, or opossums (Figure 13.37). Unlike the other white grubs, green June beetle grubs do not feed primarily on plant roots but rather uproot the grass and push up small mounds of soil. Adult scarab beetles do not feed on turfgrasses; however, some of these beetles (e.g., Japanese beetle) aggressively feed on ornamental plants and trees during June and July.

**Inspection and control.** Good turf management such as adequate moisture and fertilization helps the turf withstand moderate grub infestations. Adult beetles are often found in swimming pools, under lights, or slowly crawling across turf areas. To check for grubs, use a spade to cut three sides of a 1-square-foot (0.3 $m^2$) piece of sod. The cuts should be 2 inches (5 cm) deep at the edge of one of the off-color areas. Force the spade under the sod and lay it back. See if the grass roots are chewed off and sift through the soil looking for the larvae. Check several places in the turf area. As a rule of thumb, if an average of three to seven grubs is found per square foot, an insecticide should be applied. However, irrigated areas and lower maintenance areas can withstand 25 to 50 grubs/square foot (2.3 to 4.7/$m^2$), depending on the species and size.

The Japanese beetle is one of the easier species to manage, while oriental beetles and European chafers are more problematic. Biological control has been provided by strains of the bacteria, *Bacillus popilliae* and *B. thuringiensis* (Bt),

**TABLE 13.4**
**Identification of Adult Scarab Beetles and White Grubs**

| Species | Adult | White Grub | |
|---|---|---|---|
| | | Spine or Raster Pattern | Anal Slit Shape + Additional Characteristics |
| Asiatic garden beetle | • Body 5/16 to 7/16 inch (8 to 11 mm) long; 3/16 to 1/4 inch (5 to 6.4 mm) wide<br>• Dull chestnut brown; velvety or hairy<br>• Wing covers do not reach the tip of abdomen, exposing the last two abdominal segments<br>• Small erected hairs in the top of head | Tightly packed semicircle of spines just in front of the anal slit | • Anal slit branched<br>• A cream-colored tumor-like expansion on the sides of the mandibles on the head |
| Black turfgrass *Ataenius* | • Body 3/16 to 1/4 inch (3.6 to 5.5 mm) long; about half as wide as long<br>• Shiny black, with grooves running along the wing covers | No distinct pattern of spines is present | • Anal slit branched<br>• A pair of padlike bumps at the tips of the abdomen, just in front of the anal slit<br>• A distinguishable black-colored blood vessel running down the back |
| European chafer | • Body 9/16 inch (13 mm) long<br>• Light reddish-brown, with a slightly darker head and pronotum<br>• Wing covers have longitudinal grooves and minute punctures | Two rows, roughly parallel to each other, resembling a partly open zipper | • Anal slit branched |
| Green June beetle | • Body 3/4 to 1 inch (19 to 25 mm) long; 1/2 inch (12.5 mm) wide<br>• Upper side ranges from dull brown with lengthwise stripes of green, to velvety forest green; underside is shiny metallic green to gold | Two short rows or spines, parallel to each other; few additional spines are present | • Anal slit transverse;<br>• Legs are very small compare to the body |
| Japanese beetle | • Body 5/16 to 7/16 inch (8 to 11 mm) long; 1/4 inch (12.5 mm) wide<br>• Head and body are shiny metallic green; wing covers coppery-brown; legs dark green<br>• Each side of abdomen has patches of white hairs | Distinct V-shaped pattern of spines pointing toward the insect's head | • Anal slit transverse |
| May beetles | • Most species are 5/8 to 1 inch (11 to 25.4 mm) long<br>• Colors range from tan, light brown to reddish-brown or black<br>• Some species are hairless, but others are hairy<br>• Adult identification by experts is recommended | Two rows, roughly parallel to each other | • Anal slit V- or Y-shaped |
| Masked chafers | • Body 7/16 to 1/2 inch (11 to 12 mm) long; 1/4 inch (6 mm) wide<br>• Dull yellow-brown (northern masked chafer); shiny, reddish-brown (southern masked chafer)<br>• A chocolate-brown band across the head (between the eyes)<br>• Adult identification best performed by experts | No distinctive patter; 25–30 spines randomly scattered in the anal region; adults and grubs of southern and northern masked chafers are virtually indistinguishable | • Anal slit transverse |
| Oriental beetle | • Body 3/8 inch (9 to 10 mm) long<br>• Straw-colored to brownish-black; head is solidly brown<br>• Wing covers have variable black markings, and longitudinal grooves | With two parallel rows of 10 to 16 short inward-pointing spines | • Anal slit transverse |

**FIGURE 13.35** Adult Japanese beetles feeding on an ornamental plant.

**FIGURE 13.36** Turf damage resembling drought stress from white grub feeding.

**FIGURE 13.37** Turf damage from animals such as skunks, hogs, and armadillos feeding on white grubs.

and parasitic nematodes, *Heterorhadbitis bacteriophora* and *Steinernema* sp. The milky disease from this bacterium is most active on Japanese beetle grubs. Grubs ingest this while feeding and the bacteria causes the grub's body fluids to turn a milky white color (hence, the name) prior to death. Bt produces crystalline proteins that destroy an insect's gut lining. Bacteria population buildup and control require an extended time period, typically three to five years. The parasitic nematodes show promise but must be applied yearly within moist soil. A dark-colored, hairy wasp, *Scolia dubia,* is often seen hovering over turf in late August or September. Female wasps sting a grub to paralyze it and then deposit an egg; upon hatching, the wasp larva consumes the grub. These wasps are virtually harmless to humans unless picked up or stepped on with bare feet.

Commercial insecticides are partially (75% to 90%) effective. This is due, in part, to the subterranean habit of the larvae, which reduces the effectiveness of most surface-applied insecticides. As a result, watering in the products increases control. Timing is critical; application should be made just after egg laying for targeted species when using systemic insecticides. A contact insecticide such as trichlorfon or carbaryl also may be needed in spring if damage is excessive by overwintering second-instar grubs. Once grubs reach the third-instar stage in late spring, insecticides become less effective than when grubs are small and actively feeding. Larger grubs in fall are much more difficult to control and tend to go deeper in the soil when temperatures cool. These grubs emerge in spring and are also very difficult to control.

Two windows of opportunity exist for maximum grub control: pre-grub damage and post-grub damage. Pre-grub damage occurs during the time from pre-egg lay to second-instar grubs, typically early June through mid-July. Longer residual products such as chlorantraniliprole, imidacloprid, clothianidin, thiamethoxam, dinotefuran, and halofenozide typically provide best control then. Control is slow with these products, typically occurring 10 to 20 days after application. Although effective on small grubs, these materials are less efficacious on larger, more mature grubs.

Post-grub damage occurs when grubs are larger—late second to third instar, typically from late August through spring (March and April) of the following year. Turf damage from vertebrate pests also often occurs then. Trichlorfon, acephate, chlorpyrifos, and carbaryl provide quick knockdown as a rescue treatment during this time, but control is usually only about 50% to 60%. Control is fairly quick, within one to two days after application. These, however, have a very short residual (less than a week) and require multiple applications for extended effectiveness. One common problem of using post-grub damage products against some species such as green June beetle grubs is the presence of piles of dead, smelly grubs littering the soil surface that require a morning-after cleanup.

Control will be more effective if the soil is kept moist for several days before treatment to encourage grubs to come closer to the soil surface, and thereby more susceptible to insecticides. Apply as late in the afternoon as practical and irrigate immediately with 1/4 to 1/2 inch (0.6 to 1.3 cm) of water for maximum effectiveness. Thatch control also is extremely important as excessive thatch tends to bind or tie up most insecticides. Enhanced soil degradation of certain insecticides used for grub control also is currently suspected.

*Ataenius* control is best timed in spring (late May), when adults are laying eggs, using a longer residual product. If control methods are delayed until damage is evident, control becomes less effective since the grubs have finished feeding and are less affected by the insecticide. Insecticides should be lightly watered in to move them into the thatch layer where the insects are located. Alternative control timing is when larvae begin to hatch from late May on. These applications should be thoroughly watered in to move the materials through the thatch to the soil layer.

FIGURE 13.39 Legless billbug larvae with a white body and a tan head.

### Billbugs

**Species.** In cool-season grasses, the bluegrass billbug, *Sphenophorus parvulus* Gyllenhal; the lesser billbug, *S. minimus*; and the Denver or Rocky Mountain billbug, *S. cicatristriatus*, are most important. In warm-season grasses, the hunting billbug, *Sphenophorus venatus vestitus* Chittenden, and the Phoenix billbug, *S. phoeniciensis*, on bermudagrass are most important. Billbugs are in the insect order Coleoptera, family Curculionidae, and are not a true white grub.

**Insect description.** Adult billbugs are weevils or beetles with distinguishable snouts (or bills) (Figure 13.38). Adult beetles are about 3/8 inch (1 cm) long; typically weevil-like in appearance with a short, fairly broad recurved snout; and relatively broad in their shoulder regions. They have chewing mouthparts at the tip of their distinctive snouts. They are gray to black but often are covered with soil, giving the beetle a dirty appearance. Adults are slow moving and often play dead when disturbed. They are not capable fliers, so walking is their primary mode of movement. Eggs are generally white, bean shaped, and about 1/16 inch (1.6 mm) long. Grubs are 3/8 inch (1 cm) long when mature, their body is white with a tan head, and they are legless (Figure 13.39). Billbugs can be identified to species based on the wart patterns on their pronotum. However, life cycle and biology are similar among the species. Billbugs can have overlapping generations, where all life stages can be found throughout the year.

FIGURE 13.38 Billbug adult with its distinguishable snouts (or bills).

Billbugs are not a persistent problem in turfgrass. Damage from this pest usually is sporadic. Several years may elapse before infestations reoccur. Zoysiagrass and bermudagrass grown in the warm-season and transition areas are most damaged; however, bahiagrass, St. Augustinegrass, and centipedegrass also are attacked. The bluegrass billbug is a pest of bluegrass and other cool-season grasses in the north from Washington state across to the East Coast. Billbugs are native to North America.

**Biology and distribution.** Both larvae and adults injure turfgrasses by feeding on grass stems, crowns, and rhizomes. Adults feed by inserting their mouthparts into the center of the grass stem. However, adult billbug feeding is not considered very damaging. Adults begin to emerge from winter hibernation in mid- to late-spring (May to mid-June), when daytime temperatures consistently reach 65°F to 68°F (18°C to 20°C). They are active for four to six weeks, and often are observed climbing walls and windows, and over paved surfaces on their way to feed by chewing holes in grass stems and to deposit eggs on turf. In cool-season grasses, adult female billbugs bore cavities in grass stems near the crown in mid- to late spring (May and June) and deposit eggs in these cavities. Legless, grub-like larvae hatch in 3 to 10 days and feed inside the grass stem and crown area to the degree stems can easily be pulled out by hand by early summer (late June) (Figure 13.40). Eventually, the larvae destroy the crown and drop out the stem once they become too large to fit within the plant. They can destroy a plant's root system, causing the turf

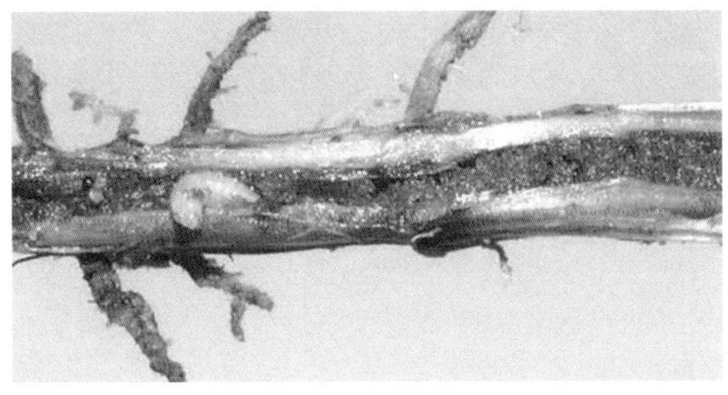

FIGURE 13.40 Billbug larvae feeding inside a grass stem.

to appear drought stressed. The larval stage is the most damaging stage. In early to midsummer (late June into early July), the mature (3/8 inch, 9.5 mm long) larvae dig 1 to 2 inches (2.5 to 5 cm) into the soil and form a pupal cell (Figure 13.41). In the cell or chamber, pupae gradually mature into adults over a two-week period. They overwinter as adults, often in the junction of turf and sidewalks as well as hedgerows. Normally, only one generation occurs per year. Peak feeding activity for adults is during July.

In warm-season grasses, females chew holes into stems (stolons) of bermudagrass and zoysiagrass and insert their eggs. After hatching, larvae feed on the stems, drop into the soil, and feed externally on grass parts. Adults first appear in spring (April and May) and again in late summer (September to October). However, larvae may still be present throughout much of the year.

Larvae commonly remain active through the winter, feeding on dormant stolons and crowns. Damage often is not evident until spring, when the bermudagrass or zoysiagrass starts to green-up. Damage is often mistaken for small spring dead spot disease or "delayed spring green-up syndrome."

Larvae mature from late winter into spring (February into early May), pupate, and lay eggs as adults throughout the summer. Since turf growth is rapid in summer, feeding damage often goes unnoticed.

**Damage symptoms/signs.** Most damage on cool-season turf is in early to midsummer (June and July). Damage resembles fertilizer burn, drought, or disease; however, the grass easily breaks off at the crown ("tug test") (Figure 13.42). Grab several affected stems and tug up. Damaged stems easily break off just below the thatch level and tan, sawdust-like frass is evident at the base (Figure 13.43). Lack of mobility of the legless larvae results in small irregular areas of dead grass, resembling dollar spot disease. When the stem is consumed, the larvae migrate downward and feed on roots. Damage to the turf from mature larvae resembles white grub damage with sloped, sunny areas often showing damage first. Larvae are found in the soil 1 to 3 inches (2.5 to 7.6 cm) deep, among roots and runners. The larval stage lasts three to five weeks. Pupation occurs in the soil. Turf injury is much more pronounced during extended dry weather than when ample rainfall or irrigation is available, and damage often resembles drought or heat stress. Billbugs overwinter in any stage in the south, but overwinter mostly as adults in the north.

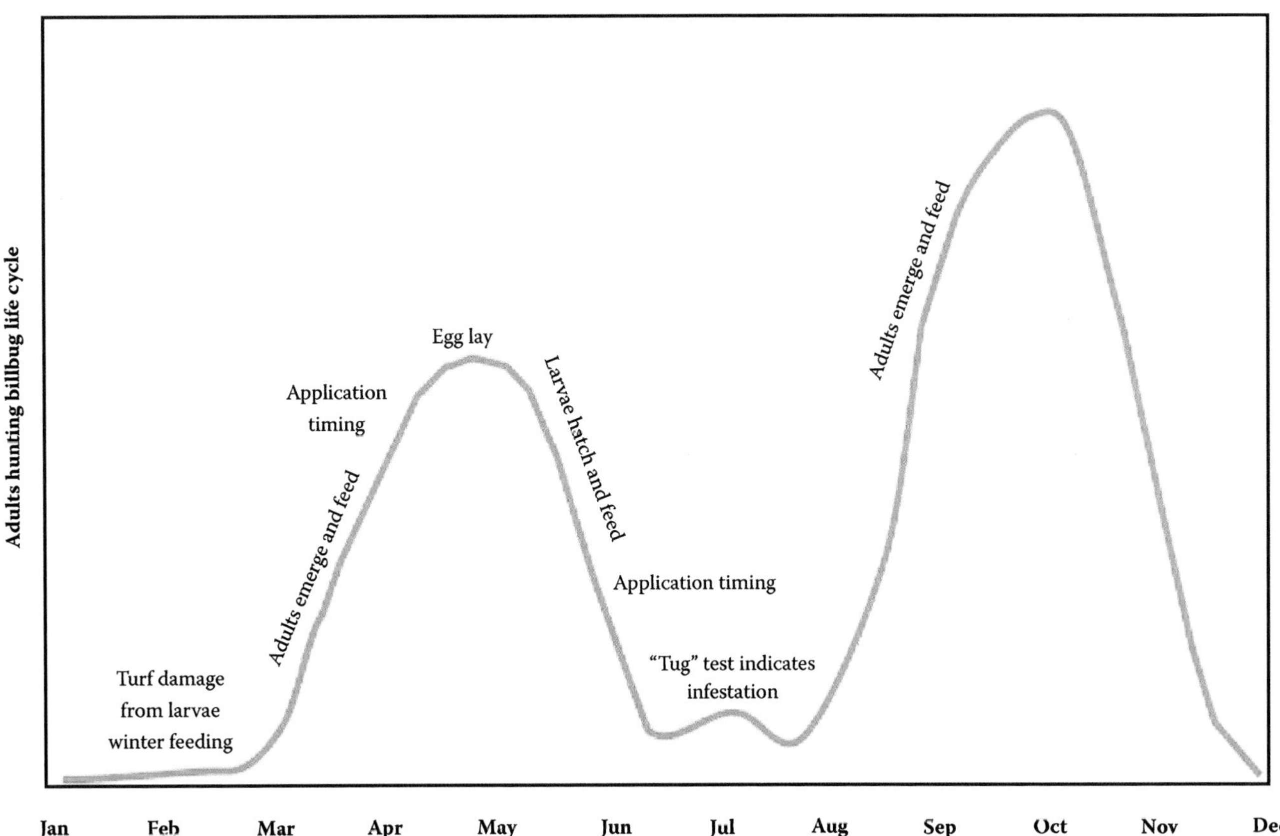


FIGURE 13.41 Typical billbug life cycle and application timing in the northern United States.

**FIGURE 13.42** Damaged turf from billbug feeding easily breaks off just below the thatch level.

**FIGURE 13.43** Turf stems, which easily break off from billbug feeding.

Adults tend to be a problem in spring while grubs cause damage in summer, especially when the turf suffers from moisture stress.

**Inspection and control.** To determine if billbugs are causing problems, inspect the rootzone, as with white grubs. Pitfall traps are used to monitor spring adult activity. Watching driveways, steep-walled sandtraps, and sidewalks for adult migration also aids in determining a pest presence. Damage usually occurs as spotty, brown patches, first along driveways and sidewalks. Control involves cultural (proper irrigation and fertilization), plant resistance (endophyte-enhanced ryegrass or fescue), and preventative and curative insecticides.

As a rule of thumb in cool season turf, if an average of 10 billbug larvae are found per square foot (107/m$^2$), an insecticide should be applied. Damage thresholds in more arid areas may be lower due to drier conditions, while they are higher in wetter areas. Damage from moderate infestations may be masked with light fertilization and adequate water. Beneficial fungus, *Beauveria* sp., and nematodes, *Steinernema* sp., show promise for biological control. Endophyte-containing ryegrasses and fescues often are resistant or tolerant to attack. Once billbug damage is noticeable, pesticide control often is unsatisfactory. Successful control, therefore, focuses on preventative treatment. Control with insecticides has been most beneficial when timed in spring from April to mid-May, when temperatures begin to approach 60°F (15.6°C), as adults emerge from winter and are searching for oviposition sites. If more than five adults are observed during a five-minute period, a pesticide application may be warranted. Treat newly mowed turf and lightly water in to move the insecticide into the thatch where grubs reside. Preventative control uses systemic insecticides, such as neonicotinoids (clothianidin, dinotefuran, imidacloprid, and thiamethoxam) and anthranilic diamides (chlorantraniliprole and cyantraniliprole), for better distribution within plants and long-residual control of larvae. Mixing a preventative insecticide with an organophosphate (chlorpyrifos) or pyrethroid (such as bifenthrin, deltamethrin, cyhalothrin and permethrin) in a tank during the time of adult activity may reduce adults and larvae. Curative control with carbamates and organophosphates is difficult to achieve; it is best attempted in mid- to late June into July as a soil drench. Water in the products following treatment.

## ADDITIONAL LOCALLY IMPORTANT TURFGRASS INSECTS

Other insects can occasionally become important pests of various turfgrasses. However, their occurrence and damage are usually regionalized and normally minor. Exceptional weather conditions (e.g., extremely dry) generally are associated with damage from these insects.

### Chinch Bugs

Chinch bugs include the southern chinch bug, *Blissus insularis* Barber; the common chinch bug, *Blissus leucopterus* Say; the western chinch bug, *Blissus occiduus* Barber; and the hairy chinch bug, *Blissus leucopterus hirtus* Montandon (order Hemiptera, family Lygaeidae). The most serious insect pest of St. Augustinegrass is the southern chinch bug, but it may also attack bermudagrass and zoysiagrass. The hairy chinch bug is a pest of bluegrass, ryegrass, zoysiagrass, bentgrass, and fescue in the north and upper midwestern United States.

Chinch bugs damage turf by piercing plant tissue and sucking sap with their needlelike mouthparts. They may also introduce toxic saliva into the plant during this feeding process that may block xylem and phloem tissue, causing wilting and discoloration of plant tissue. Chinch bug nymphs do the most damage. Affected areas appear as yellow spots or patches 3 to 10 feet (1 to 3 m) in diameter and often are noticed first along concrete or asphalt paved edges or in water-stressed areas where the grass is growing in full sun; this is often mistaken for drought stress (Figure 13.44). When xylem and phloem tissue is blocked, leaves wither as in drought and food from photosynthesis does not translocate to the roots. Plants eventually die.

**FIGURE 13.44** Turf damage resembling drought stress from chinch bug feeding.

Adults are about 1/4 inch (6.4 mm) long and black, with white patches on their wings (Figure 13.45). Nymphs range from 1/20 inch (1.3 mm) long to nearly adult size. There are five nymphal instars. The small nymphs are bright red with a white band across the back, but become black and white in color as they mature. Chinch bugs overwinter as inactive adults and become active in spring, where they feed and mate—two to three generations typically, while up to seven occur in south Florida. Eggs are laid in sheaths or pushed into soft soil and protected places. In summer, eggs hatch in 10 days to two weeks, and the young develop to adults in three weeks. Adults become active when daytime temperatures reach 70°F (21°C).

Detection of chinch bugs begins by parting grass on the edge of infested areas and observing the soil surface. All stages can be seen moving through the loose duff on the soil surface. In extremely heavy infestations, some can be seen crawling over grass blades, sidewalks, and outside walls of houses. Their presence can also be confirmed by using the flotation method.

Control begins by preventing drought stress, since damage symptoms are reduced when grass is well watered. Reduce summer fertilization with quick-release N sources because the resulting lush grass growth is more susceptible to damage. Spot spraying before extensive damage occurs is best and is directed where the highest feeding activity is occurring. Threshold levels of 15 to 20/square foot (160 to 215/m$^2$) may warrant control. Most insecticides for chinch bugs should not be watered in. To achieve greatest efficacy, the insecticides should be applied at high volume (i.e., 1.2 gallons/1,000 square feet, ~5 L/100 m$^2$) to deliver the active ingredients to the thatch layer where the insects are feeding. When using a granular product, watering-in should be done soon after an application to push and dissolve the product in the thatch layer. Repeated applications are often needed in areas with heavy infestation or damage. Use products of different modes of action during each application. Several chinch-bug-resistant St. Augustinegrass varieties—"Floratam," "Floralawn," and "Captiva"—are available. Beneficial fungi, *Beauveria bassiana*, can also be applied repeatedly to the infested turf with good results.

## Cicada Killers

Homeowners and golfers are often alarmed by the cicada killer, *Sphecius speciosus* (Drury), and other ground-nesting wasps (order Hymenoptera, various families). The cicada killer is a large wasp, about 1 5/8 inches (40 mm) long and somewhat resembling a giant hornet with bright colors; it appears fearsome but generally is docile (Figure 13.46). Females dig soil burrows 1/2 inch (1.2 cm) across and 6 to 9 inches (15 to 23 cm) deep that form a small U-shaped mound around the entrance on the turf surface. Sandier sites appear to be preferred. Often a "runway" leads to the entrance of the mound. There may be several mounds within the same area. Female cicada killers stock these mounds with cicadas as provisions for developing larvae. Adults feed only on plant nectar. They are widely distributed east of the Rocky Mountains. Although their activity causes fear and interferes with human activity, the cicada killers are actually beneficial insects. The mounds are unsightly and are often more noticeable in summer, but they disappear as fall approaches. Covering sandy areas with turf or mulch can discourage mound building by cicada killers. Often, chemical treatment is not needed; however, when needed, dusting the burrow entrance with carbaryl will kill the females as they enter and leave the nest.

**FIGURE 13.45** Adult chinch bugs with white patches on their wings.

**FIGURE 13.46** Cicada killer wasp dragging a stunned cicada back to its nest.

**FIGURE 13.47** Dull reddish-brown clover mite with eight orange legs.

## Clover Mites

The clover mite, *Bryobia praetiosa* Koch, is a minor pest of cool-season turfgrasses and is distributed throughout the United States. The clover mite is dull reddish-brown, with eight orange legs (Figure 13.47). It is most active in the spring and fall and becomes dormant in the summer (as adults) and winter (as eggs). They often become a nuisance pest when a large population, which builds up on turfgrass through the spring and fall, begins to move into homes and buildings in the summer and winter to seek hibernation sites. Damaged turf appears yellowish or silvery, and often fails to green-up in the spring. Although mites do not bite humans or pets, they can be annoying and leave behind red stains when squashed. Avoiding excessive fertilization that promotes lush plant growth, but also favors mite development and reproduction, can help control the population. Repeated insecticide applications may be needed to reduce mite populations on turfgrass.

## Crane Flies or Leatherjackets

The European crane fly, *Tipula paludosa* Meigen, and the common crane fly, *Tipula oleracea* L. (order Diptera, family Tipulidae), are pests most commonly found in Europe, the Pacific Northwest, and other cooler turf sites. Adults have very long legs and resemble large mosquitoes with their large bodies, approximately 1 inch (2.5 cm) long (Figure 13.48). Larvae initially are white, wormlike maggots and have four instars. Later, they become small, gray-brown, wormlike creatures, 1.2 to 1.6 inches (3 to 4 cm) in length, with a tough leathery-looking skin giving them another common name, "leatherjackets" (Figure 13.49). The third instar is the overwintering stage; the third and fourth instars do the most turf damage in spring. The larvae feed on turfgrass roots and crowns, causing browning when threshold numbers approach 20 to 25/square foot (215 to 269/m$^2$). Feeding by larvae progresses slowly in winter, with most damage to turf occurring in spring (April to May). Leatherjackets rest during the day in the soil and feed mostly at night. When adults are ready to emerge, the pupae wriggle to the surface; the pupal skin left on the turf surface is a telltale sign of infestation. Adults may congregate near houses, often exciting homeowners into thinking immediate control is necessary. However, they do not bite or sting and cause no damage to houses. Control may be necessary for sod operations to prevent accidental shipment of larvae to uninfested areas, as well as for golf greens that are expensive to replace if extensive damage occurs. Preventative insecticide treatment is best made in early fall (September to October) against eggs or small larvae. Larvae are sensitive to dry conditions; therefore, careful management of the timing and frequency of irrigation, as well as better drainage, can allow the turf area to dry and be less habitable to the crane fly larvae.

**FIGURE 13.48** Adult crane fly, which resembles a large mosquito with long legs.

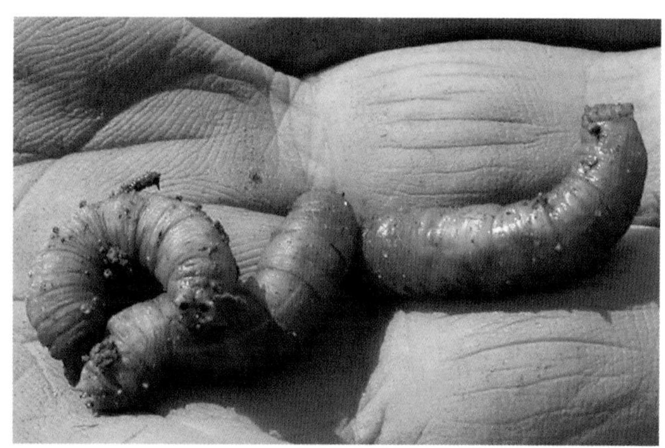

**FIGURE 13.49** Tough leathery looking wormlike larvae of crane flies or leatherjackets.

## Earthworms

Earthworms, also referred to as night crawlers, angleworms, fishing worms, and dew worms, are generally considered beneficial as their presence indicates ecologically healthy turfgrass growing conditions. Their two primary requirements are moist soil and an organic matter food source. Their feeding and burrowing habits initiate thatch decomposition, stimulate microorganism activity, increase the availability of certain soil nutrients, increase soil aeration, and generally improve overall soil quality. However, aside from these benefits, earthworm casting on golf course fairways is an extremely challenging turfgrass management issue. Casting occurs when earthworms ingest soil and leaf tissue to extract nutrients and then emerge from their burrows to deposit fecal matter (casts) as mounds of soil on the turfgrass surface (Figure 13.50). These castings interfere with maintenance practices, play, and overall turf aesthetics and are more prevalent on well-irrigated, shady sites. Turf beneath the leveled mounds is also smothered.

A major earthworm species in the United States is the common night crawler (*Lumbricus terrestris* Linnaeus). *Apporectodea calignosa* and *A. longa* are additional earthworms often found on golf courses. Peak casting is during the cool, wet weather in spring and late fall through winter. Night crawlers can live up to six to nine years, reoccurring in the same place. Earthworms dig burrows in the soil and migrate upward with fluctuations in moisture content, soil temperature, and availability of food such as clippings and soil organic matter. It has been shown, however, that neither clipping removal nor aerification reduces earthworm casting.

Green stinkworms (*Amynthas hupeiensis*) also are a periodic pest originating from eastern Asia that produces large, sandy, often sinuous casts on sand-based greens. Casts are especially heavy for several days after rain but are rarely seen in winter. Stinkworms are named for their foul-smelling body secretions.

No products specifically control earthworm casting. Any pesticide application in the United States specifically targeting earthworm control is illegal, increasing the need for alternative, nonchemical management strategies. Carbaryl and thiophanate-methyl have been shown to reduce earthworm castings but are not labeled for that purpose. For green stinkworms, combination products such as clothianidin + bifenthrin or zeta-permethrin + bifenthrin + imidacloprid applied in midspring at label rates for ant control provide up to eight weeks' control of castings. All products should be watered in following application with ~0.5 inch (1.3 cm). For other earthworms, products containing tea-seed meal, an organic fertilizer containing saponins, have been fairly successful. Other control attempts have involved changing soil pH (high or low), which does not influence night crawler activity. Other earthworm species, however, may respond differently to changes in soil pH. Since the outer skin (or cuticle) of earthworms is extremely sensitive, using abrasive substances such as sand may irritate and repel them. Control is currently focusing on the abrasiveness and desiccation ability of sand particles through aggressive topdressing. Due to the need for multiple applications, a long-term commitment of up to five years of topdressing is needed for success. Successful yearly total topdressing rates have been between 0.75 and 1.5 inches (1.9 to 3.8 cm). Due to the expense and labor required to perform this heavy rate of topdressing, it is suggested that infestation areas be mapped and a priority be placed on the most infested sites. Additional research continues to evaluate irritants such as various hot peppers as possible deterrents. Member education about the benefits of earthworms and long-term commitment for casting control also are needed. Because of the possibility of injuring turf roots and shoots from foot and maintenance equipment pressure, long-term use of sharply angular soil aggregates should be considered before use. Roots may die back, diseases become more prevalent, and turf wilting may occur more frequently.

## Grasshoppers

Numerous species (most often *Melanoplus* spp.) of grasshoppers (order Orthoptera, family Acrididae) are found throughout the world (Figure 13.51). They are turfgrass pests mostly during periods of exceptional drought in locations of low annual rainfall, such as the Great Plains. They feed by chewing the foliage of a wide range of plants. Well-maintained turfgrass rarely has problems unless the grasshoppers are present in high numbers on adjacent crops. Generally, one

**FIGURE 13.50** Surface mounds, or castings, of soil from earthworms.

**FIGURE 13.51** Grasshoppers occasionally are turf pests from foliage feeding.

or two generations occur yearly and they overwinter as eggs in the soil. If grasshoppers become a turfgrass problem, then they can be managed using insecticides such as azadirachtin, carbamates, indoxacarb, organophosphates, and pyrethroids.

### Greenbug

Greenbugs are a species of aphid, *Schizaphis graminum* (Rondani), that are a major pest of grain crops, but can occasionally become a pest in cool-season turfgrasses, particularly Kentucky bluegrass and ryegrass. Greenbugs are soft bodied, light green with a dark green stripe down the back, and a pair of black tailpipe-like cornicles at the end of the abdomen (Figure 13.52). They are sucking insects. They remove plant sap and cause yellowing and browning of turf; damage often appears as irregular patches of burnt orange, or yellowing in late spring (Figure 13.53). In turfgrass, damage often begins under shade trees. Populations are present year-round in the southern United States. There are many generations of greenbug each year, as each generation can be completed in seven days at typical summer temperatures. They overwinter as eggs, which hatch in the spring to initiate a new generation. Endophyte-enriched cool-season turfgrasses are resistant to greenbugs, which are very easy to control with contact and systemic insecticides applied as sprays. Control may be warranted by 10 to 25 aphids per leaf blade. Spot treatment, with an extended treatment band of 6 feet (2 m) outside the infested area, is sufficient to reduce localized infestations. Avoid irrigating the treated area for at least 24 hours after insecticide application.

**FIGURE 13.52** Greenbug aphids feeding on ryegrass.

**FIGURE 13.53** Ryegrass damage from greenbug feeding.

### Ground-Nesting Bees and Wasps

Many bee species nest in individual holes in the ground. These are of various sizes and colors (from reddish-brown to metallic blue). Females build nests 6 or more inches (15 cm) deep in the ground. During the process of excavation, the females build mounds of soil with a central hole for entrance. These nests often appear in large numbers in sunny, well-drained, sandy or thinned turf areas. The mounds often appear in the spring and can become a nuisance to turfgrass managers. Bees are pollinators and are not harmful to humans or pets. Chemical management of these ground-nesting bees is not recommended. Because bees do not like wet soil, heavy irrigation can discourage nest building. Tilling the soil while adding organic matter also makes the area less suitable for nest building. The most effective approach to controlling the bees is to establish a dense, healthy turf.

### Ground Pearls

*Margarodes meridionalis* Morrill (order Homoptera, family Margarodidae) are tiny scale insects that live in the soil and suck juices from grass roots. Ground pearls are most damaging to centipedegrass, but they will also feed on bermudagrass, St Augustinegrass, and zoysiagrass. The slow growth habit of centipedegrass makes it more susceptible to damage from ground pearls, which are distributed from North Carolina west to southern California. They are spherical and range in size from a grain of sand to about 1/8 inch (3 mm) in diameter. They are pearly white to yellowish in color and look very much like pearls (Figure 13.54). Adult females are soft bodied, pink, and about 1.16 inches (1.6 mm) long.

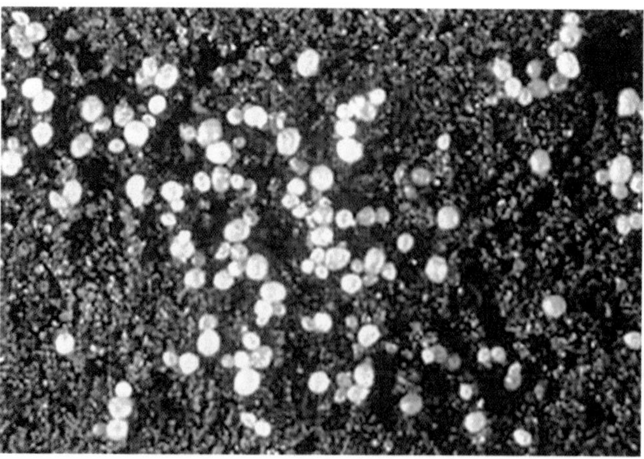

**FIGURE 13.54** Ground pearls' appearance in the soil.

Adult females emerge from cysts in the spring, and move around slowly in the thatch and soil. Eggs are laid in the soil from March to June. The first nymphal instar, called the crawler, locates a grass root, attaches itself, and begins to cover itself with a yellowish to light-pearl-colored hard wax coating. The nymphs continue developing inside this shell and overwinter in this stage. The life cycle from egg to adult requires one or possibly two years. Severely infested grass turns yellow, then brown, resembling drought stress occurring in irregular patches. Chemical control is currently unavailable for ground pearls since they are covered by a hard, waxy, practically impenetrable shell. In addition, ground pearls live at depths up to 10 inches (25 cm). Infested grass should be grown to follow best management practices such as raising the mowing height, providing good soil moisture, and providing adequate soil potassium to encourage rooting.

### Imported Fire Ants

Fire ants are small reddish-brown to nearly black ants in the genus *Solenopsis* (Figure 13.55). They range in size from about 1/8 to 3/8 inch (3 to 9.5 mm). Fire ant nests are easily recognized by their characteristic dome shape, size, and apparent lack of an entrance hole (Figure 13.56). Although they do not directly feed on turfgrasses, their mounds often desiccate turf areas, disrupt the turfgrass surface, cause mowing problems, smother closely mowed grass, clog machinery, and are a nuisance to golfers and maintenance personnel. In sandy soil, mounds may not maintain their shape.

Disturbance of a mound creates a characteristic aggressive "boiling" effect of ants coming out of the mound in a defensive action, attacking and stinging all intruders. Each nest may contain anywhere from a few hundred ants to 350,000 ants. It is not uncommon for infestations with 300 or more mounds per acre to occur.

Nests are usually located in open areas such as lawns, pastures, and golf courses. They frequently take advantage of protective structures such as rocks, pavement, stumps, rotten logs, and so forth. Fire ants also reduce the yield of some 40 different crops, cause damage to electrical units (a favorite nesting location), and also damage mowing machinery.

**FIGURE 13.55** Imported fire ant adults with reddish-brown bodies.

**FIGURE 13.56** Characteristic dome-shaped fire ant nests with an apparent lack of an entrance hole.

Wildlife is also greatly impacted by the fire ants' presence. Bobwhite quail populations have been reduced by competition for their primary food source—namely, small insects. Other species have fallen prey to the fire ants' omnivorous appetite.

**History.** Two major species of fire ants occur in the United States—*Solenopsis invicta,* the red imported fire ant, and *Solenopsis ricteri,* the black imported fire ant—as well as several native species of fire ants.

The two imported species accidentally entered through the port of Mobile, Alabama, in 1918. Their spread has continued at a rate of about 5 miles per year since. It is currently estimated that imported fire ants infest more than 300 million acres throughout the states of Alabama, Arkansas, California, Florida, Georgia, Louisiana, Mississippi, North Carolina, Oklahoma, South Carolina, Tennessee, Texas, and Puerto Rico. Spot infestations have occurred in Arizona, Delaware, Maryland, Nevada, and Virginia as well.

To limit the spread of imported fire ants, a federal quarantine restricts the movement of soil, potted plants, plants with soil attached, grass sod, hay, and used soil-moving equipment to uninfested areas of the United States. These items must be certified to be free from infestation.

**Biology.** An understanding of their biology is critical in understanding the management of fire ants. The following are some of the highlights of the fire ants' life history as it relates to matters of control.

**Establishing a colony.** When a colony reaches a critical size and resources are suitably abundant, the queen ant will begin to lay eggs that will develop into sexually reproductive winged ants called alates. Alates have a mating flight when the temperature is between 70°F and 95°F (21°C and 35°C), with high humidity, that usually occurs within 24 hours of a rain. Throughout much of the southeastern United States, these conditions can occur any time throughout the year and, indeed, mating flights have been recorded during 12 months of the year. The bulk of the mating flights takes place in the spring and the fall. It is important to note that reinfestation of a treated area can occur throughout the year.

Mating flights take place in the air. The males die immediately after mating, having served their only function within the colony. All of the other ants are female. The female usually lands within a mile or two of her mother colony, but with a tail wind some have been recorded to fly as much as 10 to 12 miles (16 to 19 km). This is important because it demonstrates that a treated area can be reinfested from fairly long distances.

After landing, the newly mated queen breaks her wings off and begins the search for a suitable nest site. This is a very dangerous time for the queen, as she is vulnerable to desiccation and numerous predators. Less than 1% of queens ever survive to establish a viable colony. This demonstrates the tremendous reproductive potential of the fire ant.

After finding a suitable site, the queen forms a waterproof chamber sealed with soil and saliva where she will lay her eggs. This chamber may be as deep as 6 to 8 inches (15 to 20 cm) below the surface of the soil, making it difficult to reach with pesticides. The queen will lay between 45 and 150 eggs in her founding brood. She will rear the larvae to adulthood, living off the energy provided by digesting her wing muscles. Once the young mature into adults, they begin to take on the work of the colony and take care of the queen. The queen, relieved of other maternal duties, begins producing eggs as her only duty. She may lay as many as 1,500 eggs per day or 350,000 eggs per year. Since she may live as long as seven years, it is imperative for any control measures to eliminate the queen.

**The caste system.** Fire ants are social insects. This means they have overlapping generations that care for the young and a caste system with specific jobs for each caste to perform. In general, the youngest ants serve as nursery workers and take care of the larvae and pupae (called brood). The middle-aged workers serve as reserves, providing help wherever needed. The oldest workers serve as foragers, venturing from the colony in search of food. Even the larvae have a job to do. Fire ants cannot eat solid food, so one stage of the larvae has a big "lip" called a bucal pouch. Solid food is placed on this "lip" and the larvae secrete digestive enzymes that liquefy the solid food. Nurse ants then stroke the larvae with their antennae and the larvae regurgitate an oily liquid. Other ants then repeat this process with the nurse ants and so on. This process of passing food from one individual to another within the colony is called **tropholaxis**, which serves as an efficient filtering system for food. Foods that prove toxic or unsuitable are discarded, as are any sick ants. The queen is among the last ants to get any food—so she can survive. Baits with toxicants used for control must act slowly enough so that all of the ants, including the queen, receive a lethal dose before any effects of the toxicant are revealed.

In the absence of larvae to digest food, adults can survive on liquid oils and sugars provided in the environment. Therefore, a mound can be treated and the queen killed, but a colony can still persist for up to six months.

**The mound.** In general, there is as much mound belowground as there is aboveground. However, the ants constantly change the mound—raising the mound to capture radiant heat when it is cold, and lowering the mound to reduce the effect of the sun's rays when it is too hot. The ants move up and down within the soil profile searching for the ideal temperatures of 70°F and 95°F (21°C and 35°C). Some colonies have been found as deep as 12 to 14 feet (3.6 to 4.3 m).

The fire ants can also regulate moisture and humidity levels as they do for heat. This is why fresh mounds are often seen after a rain even though the colony was always there. Each colony also contains deep tunnels to groundwater regardless of how deep that is.

Each colony has forage tunnels radiating out like the spokes of a wagon wheel in all directions. The tunnels have periodic openings that are used by the foragers. They will search for food around the opening, then return. If they do not find food, they continue down the tunnel and try at another opening. If they find food, they release a chemical called a pheromone that recruits other workers to help. Forage tunnels may reach as far as 100 yards (91 m) away from the colony.

**The sting.** Though most people refer to fire ant "bites," in reality this is a sting. The stinger is like a hypodermic needle delivering the venom below the skin that then burns "like fire." The stingers are used for both defense and subduing prey. The venom also contains 1% to 2% protein, which causes the allergic reaction many people experience.

When getting ready to sting, the fire ant will grab the victim with its mandibles, or jaws, and place the stinger precisely between the mandible and into the victim. At the same time, the ant releases an alarm pheromone, which is simply a chemical that tells the other ants to sting as well. Thus, the perception that the entire colony gets on you and they all sting at the same time is true.

**Management strategies for the fire ant.** A large number of products are labeled for fire ant control. Most are effective when used properly, but none are permanently effective. As a result, retreatment will be necessary. It is very easy to control individual mounds of fire ants, as there is no known pesticide resistance. Population levels, however, are difficult to reduce. The fire ant's ability to move vertically and horizontally within the soil profile, lack of ecological competition, high reproductive potential, ability to traverse relatively great distances, and biological behaviors make population reduction a difficult proposition. Three basic questions need to be addressed when choosing products: (1) speed of control, (2) duration of control, and (3) costs.

Most management strategies fall into one of four categories.

1. Bait applications
2. Individual mound treatments (IMTs)
3. Combination of bait and IMT—"two-step" program
4. General broadcast treatments

**Baits.** Most fire ant baits are formulated using defatted corn grit as the bait matrix and soybean oil as the carrier. The solid yellow material is defatted corn grit and the active ingredient is added to soybean oil, which is then added to the defatted corn grit. The soybean oil is what attracts the ants and is an important part of the bait. Over time, soybean oil can go rancid and become unattractive to the ants. Therefore, it is imperative that fresh bait be used. Best results occur when baits are applied to dry grass and soil and rain is not forecasted for one to two days. Apply when ants are actively foraging.

**Advantages of baits include:**

1. They have a low level of toxicity to nontarget organisms.
2. They are easy to find in most retail stores.
3. They can be quickly applied, requiring less labor.
4. They can kill mounds that cannot be seen or reached.
5. They work well for sensitive areas such as ponds or wells.
6. They are relatively inexpensive when properly used.
7. They kill the queen and the rest of the colony.
8. They kill colonies outside the treatment area.
9. Active ingredients do not persist in the environment.

**Disadvantages of baits include:**

1. If used as an individual mound treatment, they can be expensive.
2. They are often very slow acting.
3. A knowledge of fire ant biology is necessary for optimum control.
4. Control levels rarely exceed a 95% reduction.
5. Rebound of the population usually occurs in about three months.
6. Active ingredients break down quickly in sunlight and water.

The active ingredients used in baits can be grouped into two general categories. The first group includes the compounds that use a toxin as the active ingredient. The second group uses insect growth regulators (**IGRs**) as active ingredients.

In general, research demonstrates most baits provide a similar level of control—between 85% and 95% reduction in the fire ant population. Differences become evident when you start to look at how quickly a population reduction occurs and how long the reduction lasts.

The insect growth regulators take longer to see an effect, but they also tend to give control for an extended period of time when compared to products using toxins as the active ingredient.

Several things should be remembered to ensure success with any fire ant bait used. First, because the products break down quickly, they need to be applied when the fire ants are foraging, typically broadcasted in spring and/or fall when the surface soil temperature is between 70°F and 95°F (21°C and 35°C). The easiest way to tell if fire ants are foraging is to place a small amount of test bait or a potato, corn, or cheese chip in the treatment area and wait about 30 minutes. If fire ants are foraging, they will find the bait or food in that amount of time and bait can be applied. If fire ants are not foraging, do not apply the bait.

Secondly, the most effective method of bait application is to broadcast it. However, one of the greatest difficulties with broadcast treatment of fire ant bait formulations is the low rate. Most call for 1 to 1.5 pounds/acre (1.1 to 1.7 kg/ha). This is a small amount for a large area and overapplication is common. For best results:

- **Use fresh baits**—As previously stated, the baits can spoil and become unattractive to the fire ants. Buy only what you need for a single application.
- Apply baits on a dry surface and when rain is not expected for at least 12 hours.
- **Apply baits when ants are actively foraging**—These products break down quickly in sunlight and water. If the ants are actively foraging, they will pick up the bait before its effectiveness is reduced. If the fire ants are active, their ability to outcompete other ants will reduce the effect of the bait on nontarget ant species. Afternoon/evening applications often work best.
- **Do not mix baits with other materials such as fertilizer or seed**—Ants have a very good sense of smell and will not pick up bait tainted with undesirable odors. Make sure the spreader you are using is clean or, better yet, used only for spreading fire ant bait.
- Calibrate and measure properly.

**Individual mound treatment (IMT).** Products labeled for individual mound treatments are, as a whole, very effective in eliminating 98% or more of the mounds treated. The not-so-obvious disadvantage is that they kill only the mounds that are treated. Many newly established colonies, or colonies in intensively managed areas, are difficult to see and are therefore missed with this form of treatment.

**Advantages of IMT include:**

- High percentage of mounds treated are eliminated.
- Most act very quickly and eliminate mounds within a few hours.
- Most are inexpensive.
- Most are readily available.
- They kill the queen and the rest of the colony.
- They are very fast acting.

**Disadvantages of IMT include:**

- There is a greater concentration of toxins.
- This is not recommended for sensitive areas such as wells, ponds, playgrounds, and pet runs. (See reentry intervals listed on labels for these sites.)
- This is more expensive for large areas.
- It is more labor intensive.
- It is easy to miss some mounds that are difficult to see.
- Some products require premixing, or drenching.
- If the queen is not killed, then the mound may be relocated.

**The two-step method.** This method is a combination of the two methods previously discussed and is the most commonly recommended method by researchers and extension service personnel. All of the methods discussed for both baits and IMTs apply to this method of fire ant management. Broadcast fire ant baits while fire ants are foraging, wait 10 to 14 days, and then use individual mound treatments on mounds that continue to be a problem. The two-step process is recommended twice a year, usually in the April to May time frame and again in the September to October time frame.

Combining the use of baits and granular broadcast treatments also shows considerable promise. Strategies when using baits plus granular broadcast treatment include applying an area-wide bait treatment in spring and fall to outlying areas while using a granular broadcast treatment in early spring in critical areas. For example, on golf courses, roughs and out-of-play areas would be treated with baits in spring and fall, while fairways, tees, and greens would be considered critical areas and be treated in spring. Fipronil is a very effective, long-lasting product that does not require ants to be foraging when applied. Water is needed to activate it.

## Mound-Building Turfgrass Ants

Other than fire ants, most ant species do not pose a threat to humans, but can disrupt play and become a maintenance problem by their extensive mound construction on putting surfaces (Figure 13.57). In addition to the mounds, predatory birds may cause further damage by probing the nest openings. The most common ant species that causes these mounds is the turfgrass ant, *Lasius neoniger* (order Hymenoptera: family Formicidae). These social insects are light to medium brown and about 0.10 inch (2.5 mm) long. Each colony consists of numerous sterile female workers and usually only one reproductive queen. The workers feed and tend to the queen ant and her eggs and larvae. The worker ants often feed on small insects and insect eggs, seed, flower nectar, and the sugary honeydew produced by root aphids. Once food is found, worker ants deposit pheromone trails as they return to the nest for additional worker ants to follow.

Mounds are passageways for the workers, and interconnected chambers compose the underground living quarters, usually 10 to 15 inches (15 to 38 cm) below the soil surface. Each mound is a different nest entrance point. Seasonal mound-building patterns typically involve first appearing in late winter or early spring, peaking in early summer, and declining by late summer. Most ant mounds are located on the perimeter of greens while the main nests, with queen and brood, are located in native soil outside the green's collar. These secondary nests located in the perimeter of greens contain only worker ants and appear to expand the area from which the colony collects food. These sub- or secondary nests are connected via underground tunnels to main nests located just outside the collar. Mounds easily dry out and damage mower blades, reducing turf uniformity and quality.

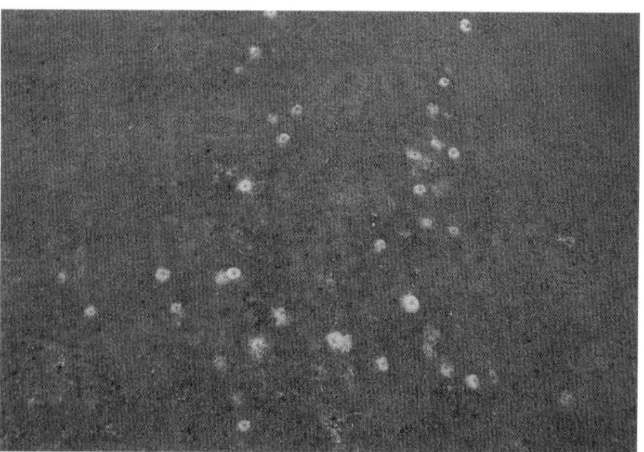

**FIGURE 13.57** Soil mounds on golf green from various turfgrass ants.

Treating with fast-acting products often only temporarily suppresses ant mounding as this usually fails to control the queen in her underground nest chamber. If contact products such as pyrethroids (bifenthrin, cyfluthrin, deltamethrin, and lambda-cyhalothrin) are used, the best time for application is early in the growing season, just after mounds appear. These applications typically provide 50% to 70% mound suppression for four to six weeks. A combination treatment of a pyrethroid along with a neonic insecticide like imadacloprid, thiamethoxam, or chlothianidin often provides longer term control, up to two to three months. Fipronil is another longer lasting product that has slow knockdown activity, allowing worker ants time to return to the nest and spread the product to other ants. However, it does not have a broad spectrum, does not kill all ant species, and must be used in fire ant regions in compliance with the label. Spot-treating with an insect bait such as hydramethylnon often provides good control if certain precautions are followed. The baits are typically slow acting, allowing worker ants time to feed it to the queen and her brood. Small amounts are sprinkled around mounds but must remain dry as ants are not attracted to wet bait. Once applied to dry turf, irrigation should be withheld for at least 12 hours after application to allow time for the worker ants to carry the bait to the queen. After the queen is eliminated, the colony cannot reproduce and dies out. About two days are needed to eliminate the colony. Baits are usually spot treatment, most often by shaker cans. Early spring is best for applications, since nests are small and thus the buildup of mounds that occurs in late spring and summer is avoided.

## Sugarcane Beetle

The sugarcane beetle [*Euetheola humilis* (Burmeister)] is a relatively new insect pest of turf, affecting bermudagrass and zoysiagrass. The insect is somewhat unique in that both juveniles and adults feed on and damage turf. Four stages occur: egg, larva, pupa, and adult. Juveniles are C-shaped white grubs, 1.25 inches (3.2 cm) long with a red head capsule and dark abdomen. Adults are dull black in color, 0.5 inch (1.25 cm) long with modified forelegs with coarse spines for digging (Figure 13.58). Adults emerge in spring (March until May), actively feed, and mate. Females lay eggs in moist soil. Larvae are present through late midsummer (late July into early August) and then pupate; adults remerge in late summer, causing another feeding period, late summer into fall (August through October). Adults are mostly active at night, being attracted to lights, though some emerge during the day. They feed on roots and crowns

**FIGURE 13.58** Dull, black-colored sugarcane beetle whose larvae and adults can damage turf.

while larvae feed mostly on organic matter. Adults damage turf as they dig into the soil and thatch and feed in the crowns of plants. Turf damage is often related to the fall populations as turfgrass is approaching dormancy and adult beetles are preparing to overwinter. Unlike most white grub species, which overwinter as grubs, the sugarcane beetle overwinters as adults. One generation occurs yearly. As with most white grubs, apply preventative treatments (clothiandin, dinotefuran, imidacloprid, thiamethoxam, chlorantraniliprole) when adults are flying.

## Two-Lined Spittlebug

Spittlebugs, *Prosapia bicincta* (Say) (order Homoptera, family Cercopidae) are an occasional pest of turfgrasses, especially centipedegrass and, to a lesser degree, St. Augustinegrass, bermudagrass, and ornamentals (especially holly) (Figure 13.59). Other grasses attacked by spittlebugs include bahiagrass, pangolagrass, and ryegrasses. They occur from Maine to Florida and westward to Iowa, Kansas, and Oklahoma. The nymphs are initially reddish, but turn white and live within a white frothy mass or "spittle" (Figure 13.60). Nymphs and adults feed by sucking juices from the grass through their needlelike mouthparts. Infested grass will have spittle masses present, and tips

**FIGURE 13.59** Adult two-lined spittlebug.

**FIGURE 13.60** Frothy mass or "spittle" where spittlebug nymphs reside.

of the grass will turn yellow, followed by browning and curling. Generally, two generations occur yearly (peaking in late spring and late summer). They overwinter as eggs in hollow stems or among plant debris. Spittlebugs prefer a thick thatch layer; therefore, the first step in management is to dethatch and topdress at appropriate times. Insecticide application targeting nymphs of the second generation (i.e., in July) may be most effective. Also avoid planting Japanese hollies near turf area as the hollies can serve as a host plant and attractant for the adult spittlebugs.

## Zoysiagrass Mite

The zoysiagrass mite, *Eriophyes zoysiae* Baker et al., is a close relative of the bermudagrass mite. This species is becoming more problematic in zoysiagrass turf across the southern United States. The appearance and life cycle of the zoysiagrass mite are similar to that of the bermudagrass mite. They are active mostly in the spring and summer. The zoysiagrass mite is specific to zoysiagrass, and does not attack any other grass species. Infested grass turns pale white or yellow, and is stunted. The upper leaf is rolled up and becomes caught within the older leaves, creating an arch; this type of damage is commonly termed "buggy whip" (Figure 13.61). The leaf sheath can be

**FIGURE 13.61** "Buggy whip" leaf response from zoysiagrass mite feeding. (Courtesy Jason Higingbottom.)

pulled back and examined for the presence of the zoysiagrass mites. The slow growth habit of zoysiagrass makes recovery from mite damage difficult. "Royal" and "Emerald" zoysiagrass are resistant to zoysiagrass mite damage. Management options for the zoysiagrass mite are poorly known, but are likely to be similar to those of the bermudagrass mite.

**Disclaimer.** All mentioned chemicals are for reference only and may not be available for turf use. They may be restricted by some state, province, or federal agencies; thus, be sure to check the current status of the pesticide being considered for use. Always read and follow the manufacturer's label as registered under the Federal Insecticide, Fungicide, and Rodenticide Act. Mention of a proprietary product does not constitute a guarantee or warranty of the product by the author or the publisher and does not imply approval to the exclusion of other products that also may be suitable.

# 14 Turfgrass Weeds and PGR Use

## DEVELOPING A WEED MANAGEMENT PROGRAM

A weed can be defined as a plant growing where it is not wanted or growing out of place (Figure 14.1). For example, tall fescue is considered a weed when grown in a pure stand of Kentucky bluegrass, but may be highly desirable when grown in a monoculture such as a golf course rough. In addition to being unsightly, weeds compete with turfgrasses for light, soil nutrients, soil moisture, and physical space. Weeds also are hosts for pests such as plant pathogens, nematodes, and insects. Certain weeds are irritants to humans when allergic reactions to pollen or chemicals occur.

Weeds often are the result, and not the cause, of weakened turf. Understanding this helps to explain the major reason for weed encroachment into a turf area (e.g., thin turf density and bare spots). Reasons for weak or bare turf areas are numerous, including but not limited to (1) improper turf selection of a species not adapted to environmental conditions; (2) damage from turfgrass pests such as diseases, insects, nematodes, and animals; (3) environmental stresses such as excessive shade, drought, heat, and cold; (4) improper turf management practices such as misuse of fertilizer and chemicals, improper mowing height, or improper mowing frequency and inadequate soil aeration; and (5) physical damage and compaction from concentrated or constant traffic. Unless the parameters contributing to the decline of a turf area are corrected, continued problems with weed infestations should be expected.

Weed management is an integrated process where good cultural practices are employed to encourage desirable turfgrass ground cover as well as the intelligent selection and use of herbicides. A successful weed management approach involves the following:

1. Proper weed identification
2. Prevention of weed introduction
3. Proper turfgrass management or cultural practices to encourage competitive turf growth against the weed
4. If necessary, the proper selection and use of an herbicide

### WEED IDENTIFICATION

The first step to successful weed management is proper identification. Turf managers should be able to identify each weed to the genus level—and preferably to species—in order to select the appropriate control technique. Weed identification also is the first step in understanding why weeds occur and how to control them. For instance, most sedges prefer moist, wet areas while sandspurs prefer drier sites.

Identification begins with classifying the weed type. **Broadleaves**, or dicotyledonous plants, have two seed cotyledons (young leaves) at emergence and have netlike veins in their true leaves (Table 14.1). **Grasses**, or monocotyledonous plants, only have one seed cotyledon present on an emerging seedling. Grasses also have hollow, rounded stems with nodes (joints) and parallel veins in their true leaves. Leaf sheaths are open. Most grass species have ligules, a projection at the inside junction of the leaf blade and collar. Ligules may be membrane-like, have a membrane with hairs on top, or membrane may be totally absent.

Although most nonbroadleaf weeds in turfgrasses are usually grasses, not all "grass-like" weeds in turf are true grasses and the two are often easily confused. **Sedges** and **rushes** have a closed leaf sheath, three-ranked leaf arrangements, mostly absent ligules, and stems that are triangular shaped and solid (**sedges**) (Figure 14.2), or round and solid (**rushes**). The distinguishing characteristics that help distinguish grasses from grass-like plants—sedges and rushes—are listed in Table 14.2.

Another group of turfgrass weeds are monocots, which are not grasses, sedges, or rushes. These include weeds such as various dayflower species, doveweed, spring starflower, wild garlic, onions, Star-of-Bethlehem, and others.

### WEED LIFE CYCLES

Weeds complete their life cycles in one growing season (**annuals**), two growing seasons (**biennials**), or three or more growing seasons (**perennials**). Annuals completing their life cycles from spring to fall are generally referred to as **summer annuals**, and those completing their life cycles from fall to spring are **winter annuals**. Summer annual grasses, as a class, are generally the most troublesome in turf.

### WEED PREVENTION

Prevention involves avoiding the introduction of weeds into an area. There are national, state, and local prevention efforts against the introduction and spread of weeds. A local preventative program is one of the best methods of avoiding future weed problems. Many of these methods are commonsense approaches that ensure sanitary conditions and minimize weed introduction. Some of these methods include use of weed-free turf seeds, stolons, sprigs, plugs, or sod. Washing or blowing equipment between mowings, maintaining weed-free fence lines and ditch banks, and the use of clean mulch and topdressing material are additional examples of preventative methods (Figure 14.3).

**FIGURE 14.1** Weeds are plants growing out of place. Shown is a bermudagrass-infested bentgrass golf green.

**FIGURE 14.2** Sedges are noted for their pitch-filled, triangular-shaped stems.

**TABLE 14.1**
**Distinguishing Characteristics of Monocots Compared to Dicots**

| Characteristic | Monocot | Dicot |
|---|---|---|
| Seedling cotyledons | • One | • Two |
| Leaf veination | • Parallel | • Netted |
| Leaf attachment | • Directly on stems | • On short stalks called petioles |
| Ligules | • Present, rarely absent | • Absent |
| Vascular bundles | • Scattered | • Distinct (arranged in a ring of bundles surrounding a central pith) |
| Vascular tissue growth | • Only primary | • Primary and secondary; thus, can become woody |
| Meristems | • Basal | • Terminal |
| Root system | • Fibrous without cambium layer | • Taproot with a cambium layer |
| Flowers | • Not showy | • Usually showy |
| Flower parts | • Group of 3s | • Usually groups of fours or fives |

### Cultural

Cultural practices promoting a vigorous, dense turf are perhaps the most important and least recognized means of preventing weed encroachment and establishment. Soil fertility, mowing schedule, aeration, and moisture levels should be maintained at an optimum level to promote turf cover. Since light is required for optimum germination of weeds such as crabgrass and goosegrass, cultural practices increasing turf density will prevent light from reaching the soil surface (Figure 14.4); this also delays spring germination of weed seeds requiring warmth because the soil surface is better insulated and remains cooler. Maintaining the highest cutting height possible and adequate fertility levels will help encourage a high shoot density and will also minimize light penetration to the soil surface.

High infestation levels of certain weeds also might indicate specific soil conditions that favor their presence. Table 14.3 lists some of these soil conditions and associated weeds. Continual weed problems can be expected until these growth conditions are corrected.

### Herbicide Classification and Nomenclature

Herbicides may be classified according to chemistry, method of application, timing of application, persistence, selectivity, and/or mode of action:

1. **Selective**—A selective herbicide controls or suppresses certain plant species without seriously affecting the growth of another plant species. Selectivity may be due to differential absorption, translocation, and morphological and/or physiological differences between turfgrasses and weeds. The majority of herbicides used in turfgrasses are selective. For example, 2,4-D (several trade names) is used for selective control of many broadleaf weeds, such as dandelion, without significant injury to turfgrasses.
2. **Nonselective**—Nonselective herbicides control plants regardless of species. These are primarily used to control all plants, as in the renovation or establishment of a new turf area, "spot treatments" in landscape beds, or for edging. Glyphosate (Roundup Pro, others), glufosinate (Finale), and diquat (Reward) are examples of nonselective herbicides. Herbicides such as atrazine (Aatrex) or MSMA (Bueno 6, others) can be nonselective at rates in excess of those used for selective control.
3. **Systemic**—Systemic herbicides are extensively translocated (moved) in the plant's vascular system. The vascular system translocates the nutrients, water, and organic materials necessary for normal

**TABLE 14.2**
**Distinguishing Characteristics between Grasses, Sedges, and Rushes**

| Characteristic | Grasses | Sedges | Rushes |
|---|---|---|---|
| Stem | • Usually hollow, round, or flattened | • Usually three-sided, pithy, rarely hollow | • Round and filled with sponge-like pith |
| Nodes | • Very noticeable | • Indistinct | • Indistinct |
| Leaf arrangement | • Two-ranked | • Three-ranked | • Three-ranked |
| Leaf sheath | • Usually split | • Usually closed | • Usually open |
| Leaf blade | • Flat, often folded, hairy, or smooth | • Flat, usually smooth | • Round or flat, usually smooth, often with visible partitions |
| Leaf margin | • Smooth, rough hairy, or sharp | • Usually rough | • Usually smooth |
| Collar | • Often a distinct band | • Indistinct | • Indistinct |
| Auricles | • Present or absent | • Absent | • Present or absent |
| Ligule | • Present, rarely absent | • Absent or only weakly developed | • Absent or only weakly developed |

**FIGURE 14.3** Washing equipment before entering a weed-free area is a key prevention practice.

**FIGURE 14.4** A densely growing turfgrass generally has fewer weed problems such as this hybrid bermudagrass patch within a goosegrass-infested common bermudagrass fairway.

growth and development. In contrast to the quick kill observed with contact herbicides, systemic herbicides require several days or even a few weeks to fully translocate throughout the plant's vascular system, and therefore require a longer period of time before kill. Systemic herbicides are also classified as selective or nonselective. Glyphosate is a nonselective, systemic herbicide; 2,4-D, dicamba (Vanquish), imazaquin (Image), and sethoxydim (Vantage) are examples of selective, systemic herbicides.

4. **Contact**—Contact herbicides only affect the portion of green plant tissue in contact with the herbicide spray. These herbicides are not translocated in the vascular system of plants—or only to a limited extent. Therefore, underground plant parts such as rhizomes or tubers are not usually controlled. Repeat applications are often needed with contact herbicides to kill regrowth from these underground plant parts. Adequate spray volumes and thorough coverage of the weed foliage are necessary for effective control. These herbicides kill plants quickly, often within a few hours of application. Contact herbicides may be classified as selective or nonselective. Bromoxynil (Buctril) and bentazon (Basagran TandO) are classified as selective, contact herbicides. Diquat (Reward) and glufosinate (Finale) are nonselective, contact herbicides.
5. **Chemical family and mode of action**—Herbicides from the same class of chemistry are grouped into families in much the same way plants are grouped into genus and species. These groupings are related to the chemical structure of the active ingredient. In general, members of an herbicide family are similarly absorbed and translocated and have a similar mode of action (Table 14.4). Mode of action refers to how an herbicide acts on (inhibits) a plant process.

### Timing of Herbicide Application

Herbicides are also classified by the time the chemical is applied in respect to turfgrass and/or weed seed germination—preplant, preemergence, and postemergence. Although the majority of herbicides may be classified into one category, atrazine (AAtrex), simazine (Princep), dithiopyr (Dimension), flumioxazin (SureGuard), and pronamide (Kerb) are notable

**TABLE 14.3**
**Weeds as Indicators of Specific Poor Soil Conditions**

| Soil Condition | Indicator Weed(s) |
|---|---|
| Acid soils | • Bentgrasses (*Agrostis* spp.)<br>• Red sorrel (*Rumex acetosella*) |
| Alkaline soils | • Plantains (*Plantago* spp.) |
| Compacted soils | • Annual bluegrass (*Poa annua*)<br>• Bermudagrass (*Cynodon dactylon*)<br>• Common chickweed (*Stellaria media*)<br>• Goosegrass (*Eleusine indica*)<br>• Knotweed (*Polygonum aviculare*)<br>• Mouse-ear chickweed (*Cerastium glomeratum*)<br>• Prostrate spurge (*Chamaesyce humistrata*)<br>• Rushes (*Juncus* spp.)<br>• Smutgrass (*Sporobulus* spp.) |
| Infertile/sandy soils | • Bahiagrass (*Paspalum notatum*)<br>• Black medic (*Medicago lupulina*)<br>• Broomsedge (*Andropogon* spp.)<br>• Carpetweed (*Mollugo verticillata*)<br>• Legumes (clover, lespedeza, medic)<br>• Red sorrel (*Rumex acetosella*)<br>• Poorjoe (*Diodia teres*)<br>• Sandbur (*Cenchrus pauciflorus*)<br>• Spurge (*Chamaesyce* spp.)<br>• Quackgrass (*Agropyron repens*)<br>• White clover (*Trifolium repens*)<br>• Yarrow (*Achillea millefolium*)<br>• Yellow woodsorrel (*Oxalis stricta*) |
| High-fertility soils | • Annual bluegrass (*Poa annua*)<br>• Bentgrasses (*Agrostis* spp.)<br>• Bermudagrass (*Cynodon dactylon*)<br>• Crabgrasses (*Digitaria* spp.)<br>• Henbit (*Lamium amplexicaule*)<br>• Purslane (*Portulacca oleracea*)<br>• Ryegrass (*Lolium* spp.)<br>• Yellow woodsorrel (*Oxalis stricta*) |
| High or infrequent mowing | • Chicory (*Cichorium intybus*)<br>• Clover (*Trifolium* spp.)<br>• Thistle (*Cirsium, Carduus* spp.) |
| Low mowing | • Algae<br>• Annual bluegrass (*Poa annua*)<br>• Chickweeds<br>• Crabgrass (*Digitaria* spp.)<br>• Moss (*Bryum* spp.)<br>• Pearlwort (*Sagina procumbens*) |
| Shaded soils | • Annual bluegrass (*Poa annua*)<br>• Common chickweed (*Stellaria media*)<br>• Ground ivy (*Glechoma hederacea*)<br>• Mouse-ear chickweed (*Cerastium glomeratum*)<br>• Nimblewill (*Muhlenbergia schreberi*)<br>• Roughstalk bluegrass (*Poa trivialis*)<br>• Violets (*Viola* spp.)<br>• Indian mock-strawberry (*Duchesnea indica*) |

(*Continued*)

**TABLE 14.3 (CONTINUED)**
**Weeds as Indicators of Specific Poor Soil Conditions**

| Soil Condition | Indicator Weed(s) |
|---|---|
| Wet soils | • Algae<br>• Alligatorweed (*Alternanthera philoxeroides*)<br>• Annual bluegrass (*Poa annua*)<br>• Barnyardgrass (*Echinochloa crusgalli*)<br>• Bentgrasses (*Agrostis* spp.)<br>• Common chickweed (*Stellaria media*)<br>• Ground ivy (*Glechoma hederacea*)<br>• Kyllingas (*Cyperus* and *Kyllinga* spp.)<br>• Mouseear chickweed (*Cerastium glomeratum*)<br>• Moss (*Bryum* spp.)<br>• Nutsedges (*Cyperus* spp.)<br>• Pennywort (*Hydrocotyle* spp.)<br>• Rushes (*Juncus* spp.)<br>• Speedwells (*Veronica* spp.)<br>• Torpedograss (*Panicum repens*)<br>• Violets (*Viola* spp.) |

exceptions. They are used as both preemergence and postemergence herbicides.

**Preplant herbicides.** Preplant herbicides are applied before turfgrass is established, usually to provide nonselective, complete control of all present weeds. Soil fumigants, such as metam sodium (Vapam), methyl bromide (Terr-O-Gas, Dowfume, Brom-O-Gas, others), and dazomet (Basamid), and nonselective herbicides, such as glyphosate (Roundup Pro, others), may be used as nonselective preplant herbicides.

**Preemergence herbicides.** Preemergence herbicides are applied to the turfgrass prior to weed seed germination and form a barrier at or just below the soil surface. Most preemergence herbicides prevent cell division during weed seed germination as the emerging seedling comes into contact with the herbicide (Figure 14.5). Weeds already emerged (visible) at the time of application are not controlled consistently by preemergence herbicides because their primary growing points escape treatment. Preemergence herbicides also do not control dormant weed seeds.

**Postemergence herbicides.** Postemergence herbicides are applied directly to emerged weeds. In contrast to preemergence herbicides, this group of herbicides provides little, if any, soil residual control of weeds. A complete chemical weed control program can be accomplished with postemergence herbicides, provided multiple applications are used throughout the year. However, due to the necessity of repeat applications and temporary turfgrass injury, most turfgrass managers use postemergence herbicides in conjunction with a preemergence herbicide. Postemergence herbicides are useful for controlling perennial grasses and broadleaf weeds not controlled by preemergence herbicides.

Postemergence herbicides enter the plant through microscopic cracks and crevices in the leaf cuticle. Herbicide uptake via plant stomates represents a tiny percentage of uptake and is not considered to be significant. The more active the plant growth is, the easier it is for the herbicide to penetrate the cracks and crevices in the leaf surface. Plants not actively growing generally have a thicker cuticle; thus, penetration via the cuticle is more difficult.

### Soil Fumigation for Nonselective Preplant Weed Control

Soil fumigants are volatile liquids or gases that control a wide range of soil-borne pests. Soil fumigants are also *highly toxic* to plants and other biological organisms, and are expensive. Their use is limited to small, high value crop acres such as tobacco, certain vegetables, fruits, bedding plants, and turf. The expense usually results from the impermeable cover necessary to trap the fumigant vapors in the soil. Fumigants control not only most weed species, but also many nematodes, fungi, and insects. Weed species possessing a hard, water-impermeable seed coat such as sicklepod, white clover, common mallow, redstem filaree, and morning glory are not effectively controlled with soil fumigants. Important considerations before choosing a particular soil fumigant include expense, soil moisture level, soil temperature, and time available before planting. Several compounds are, or have been, used as fumigants. Currently, the most used material in turf is **metham** or **metam sodium. Dazomet** also is available as a granular soil fumigant.

**Metham or metam sodium.** Metham (sodium methyldithiocarbamate) is a member of the thiocarbamate herbicide family. It is water soluble and, upon contact with moist soils, it breaks down to form the highly toxic and volatile chemical **methyl isothiocyanate**. Like methyl bromide, metham should be applied to moist soils with temperatures of at least 60°F (16°C). It is most effective when its vapors are confined with a cover; however, a water and soil-seal method may be used in the absence of a cover. With the water and soil-seal method,

**TABLE 14.4**
**Cross Reference of Herbicides, Their Use Timing, and Mode of Action to Help Aid in Managing Possible Herbicide Resistance**

| Timing | Mechanism of Action | Active Ingredient (Trade Name Example)[a] |
|---|---|---|
| Preemergence | Cellulose biosynthesis inhibition | Indaziflam (Specticle) |
| | | Isoxaben (Gallery) |
| | | Methiozolin (PoaCure) |
| | Mitotic (microtubule) inhibition | Benefin (Balan) |
| | | Dithiopyr (Dimension) |
| | | Oryzalin (Surflan) |
| | | Pendimethalin (Pendulum) |
| | | Prodiamine (Barricade) |
| | | Trifluralin (Treflan) |
| | Lipid biosynthesis inhibition | Bensulide (Bensumec) |
| | Long chain fatty acid biosynthesis inhibition | Metolachlor (Pennant) |
| | Protoporphyrinogen oxidase (PPO) inhibition | Oxadiazon (Ronstar) |
| Pre/Postemergence | Cellulose biosynthesis (grasses), synthetic auxin inhibition (broadleaves) | Quinclorac (Drive) |
| | Mitotic inhibition | Dimethenamid (Tower) |
| | | Pronamide (Kerb) |
| | Photosystem II inhibition | Atrazine (Aatrex) |
| | | Metribuzin (Sencor) |
| | | Simazine (Princep) |
| | Lipid biosynthesis inhibition | Ethofumesate (Prograss) |
| | Protoporphyrinogen oxidase (PPO or Protox) inhibition | Flumioxazin (SureGuard) |
| Postemergence | Acetyl CoA carboxylase (ACCase) enzyme (lipid biosynthesis) inhibition | Clethodim (Envoy) |
| | | Diclofop (Illoxan) |
| | | Fenoxaprop-ethyl (Acclaim Extra) |
| | | Fluazifop-P (Fusilade II) |
| | | Sethoxydim (Vantage) |
| | Acetolactate synthase (ALS) [aka, acetohydroxyacid synthase (AHAS)] enzyme inhibition | Bispyribac-sodium (Velocity) |
| | | Chlorsulfuron (Corsair) |
| | | Flazasulfuron (Katana) |
| | | Foramsulfuron (Revolver) |
| | | Imazaquin (Image) |
| | | Metsulfuron (Manor) |
| | | Rimsulfuron (TranXit) |
| | | Sulfosulfuron (Certainty) |
| | | Thiencarbazone-methyl |
| | | Trifloxysulfuron (Monument) |
| | Carotenoid biosynthesis (HPPD or 4-hydroxyphenyl-pyruvate dioxygenase) inhibition | Mesotrione (Tenacity) |
| | | Topramezone (Pylex) |
| | Cell membrane disruptors | Dilute sulfuric acid, herbicidal oils |
| | Dihydropteroate (DHP) synthase inhibition | Asulam (Asulox) |
| | Enolpyruvyl shikimate-3 phosphate synthase (EPSP) inhibition | Glyphosate (Roundup) |
| | Glutamine synthase inhibition | Glufosinate (Finale) |
| | Nucleic acid inhibition | MSMA, DSMA |
| | Photosystem I (lipid peroxidation) inhibition | Diquat (Reward) |
| | Photosystem II inhibition | Amicarbazone (Xonerate) |
| | | Bentazon (Basagran) |
| | | Siduron (Tupersan) |
| | Protoporphyrinogen oxidase (PPO or Protox) inhibition | Carfentrazone-ethyl (Quicksilver) |
| | | Pyraflufen-ethyl (Octane) |
| | | Sulfentrazone (Dismiss) |

*(Continued)*

**TABLE 14.4 (CONTINUED)**
**Cross Reference of Herbicides, Their Use Timing, and Mode of Action to Help Aid in Managing Possible Herbicide Resistance**

| Timing | Mechanism of Action | Active Ingredient (Trade Name Example)[a] |
|---|---|---|
| | Synthetic auxin inhibition | 2,4-D, dicamba, MCPP, 2,4-DP, MCPA |
| | | Clopyralid (Confront) |
| | | Fluroxypyr (Spotlight) |
| | | Triclopyr (Turflon) |

[a] Additional trade names are available for many products.

FIGURE 14.5 "Club roots" from turf stolons growing across a preemergence-treated soil.

the soil is cultivated, sealed with water, and kept moist for a week before treatment. Approximately seven days after treatment, the area should be cultivated to help release any residual gases. One to two weeks later (two to three weeks after initial application), the treated area may be planted. The longer waiting period before planting, and the lowered effectiveness in the absence of using a cover, are drawbacks to metham and should be considered before its use.

**Dazomet.** Dazomet also is a soil fumigant. Unlike methyl bromide and metham, dazomet is a granular product. Being a granular formulation, dazomet must be evenly applied and incorporated for maximum effectiveness. Its breakdown characteristics, application preparation, and effectiveness are closely associated to metham, as are its advantages and disadvantages.

### Selective Weed Control

Maintaining today's modern, multimillion dollar turf complexes at the desired level of aesthetics requires knowledge of specific weeds, their biology, and available control measures. The following sections discuss current selective weed control options turf managers have at their disposal. Weed control should be a carefully planned and coordinated program instead of being a hit-or-miss operation. Understanding how and why weeds are present on a site is more important than what control options are available once the weed is present.

When choosing an herbicide, the effectiveness at certain weed growth stages, tolerance or susceptibility of treated turf species, time required for its control, economics, and safety characteristics are important considerations when trying to choose among herbicides. The most effective herbicide is only as good as its application. Many variables influence successful herbicide application, including (1) proper equipment, (2) environmental factors at the time of application, (3) accurate and constant monitoring of calibration, and (4) adequate agitation. Most herbicide failures involve using the wrong chemical at the wrong date, or they are applied at an improper time or in an improper manner. It is most often not by failure of the herbicide itself.

## PREEMERGENCE HERBICIDES

Preemergence herbicides are the foundation of a chemical weed control program in turfgrasses and are used primarily to control annual grasses and certain annual broadleaf weeds. Preemergence weed control was first suggested in 1927. Some of the first chemicals evaluated for preemergence weed control included calcium cyanide, arsenate, and naphthylacetic acid. In 1959, the first true preemergence herbicide that provided consistent control became available for turf managers. DCPA (dimethyl tetrachloroterephthalate) (Dacthal) provided more consistent control with less damage to the turf than was previously available. With the subsequent release of dinitroaniline herbicides, the widespread acceptance of preemergence weed control in turfgrass was established.

### Turf Tolerance to Preemergence Herbicides

When considering any herbicide, one of the first questions is the tolerance of the desirable turfgrass species to the chemical in question. Table 14.5 lists the most widely used turfgrass species on golf courses and their tolerance to currently available preemergence herbicides. Herbicides such as bensulide (Betasan), dithiopyr (Dimension), indaziflam (Specticle), and members of the dinitroaniline herbicide family (e.g., benefin [Balan], oryzalin [Surflan], pendimethalin [Pre-M], prodiamine [Barricade], trifluralin [Treflan]) should be used only on well-established turfgrasses (Figure 14.5).

**TABLE 14.5**
**Preemergence Herbicides for Putting Greens[a]**

| Active Ingredients | Trade Names | Weeds Controlled | Comments | Bentgrass | *Poa*/ Bentgrass | Nonoverseeded Bermudagrass |
|---|---|---|---|---|---|---|
| Bensulide | • Bensumec<br>• Betasan<br>• Pre-San | • Annual grasses<br>• Select broadleaves | • PRE crabgrass, goosegrass and *Poa* control<br>• Only two applications yearly | Yes | No | No |
| Bensulide + oxadiazon | • Anderson's Goose/Crab Control | • Summer annual grasses | • PRE crabgrass and goosegrass<br>• Some yellowing may occur within 30 days of use | Yes | No | Yes |
| Dithiopyr | • Anderson's Golf Fertilizers with dithiopyr | • Summer annual grasses<br>• Select broadleaves | • Avoid stressed putting greens<br>• Use only on bentgrass with good root development<br>• Yellowing of *Poa* may occur during stressful conditions | Yes | No | Yes |
| Pendimethalin | • Pendulum AquaCap | • Annual grasses<br>• Select broadleaves | • Label neither allows nor restricts bermudagrass greens | No | No | Yes |
| Pronamide | • Kerb | • Annual grasses<br>• Select broadleaves | • Restricted use pesticide<br>• Label neither allows nor restricts bermudagrass greens; 2(ee) label is available | No | No | Yes |
| Siduron | • Tupersan 50WP | • Crabgrass<br>• Bermudagrass suppression | • Use on bentgrass for crabgrass control and bermudagrass suppression<br>• Band applications along bentgrass green perimeter to suppress bermudagrass stolons; apply prior to bermudagrass green-up; repeats on 30-day intervals | Yes | No | No |

*Source:* Modified from Brosnan, J. et al., 2014, *Golf Course Management* 82(8):92–98.

[a] Always consult and follow all label directions prior to use. Refer to the latest label version to confirm use on a particular grass or site.

### Effectiveness of Preemergence Herbicides

The effectiveness of preemergence herbicides varies because of many factors including application timing in relation to weed seed germination, soil type, environmental conditions (e.g., rainfall and temperature), target weed species and biotype, and cultural factors (e.g., aerification) that follow application. Preemergence herbicides generally are most effective for annual grass control, although some annual small seeded broadleaf weeds also are suppressed. Many websites are available which list the latest available products and their effectiveness. One such website is http://www.clemson.edu/extension/horticulture/turf/pest_guidelines/.

### Timing of Preemergence Herbicides

An important consideration in using preemergence herbicides is application timing. Most preemergence herbicides act as mitotic inhibitors, meaning they prevent cell division. Since the germinating shoot- and root tips are the two major sites of cell division, preemergence herbicides must contact these plant structures in the soil. Application should therefore be timed just prior to weed-seed germination since most preemergence herbicides are ineffective on emerged (visible) weeds. If applied too soon, natural herbicide degradation processes may reduce the herbicide concentration in the soil to a level resulting in ineffective or reduced control. If applied too late (e.g., weed seedlings are visible), weeds have been known to grow above the thin layer of preemergence herbicide located at the soil surface, resulting in the effectiveness of the materials being drastically reduced (Figure 14.6).

Crabgrass germinates from late winter through spring when soil temperatures at a 4-inch (10 cm) depth reach 53°F to 58°F (12°C to 14°C) for 24 consecutive hours. Alternating dry and wet conditions at the soil surface, as well as sunlight, greatly encourage crabgrass germination. Crabgrass germination often coincides with flowering of early spring plants such as redbuds, forsythia, pears, and cherry trees.

Goosegrass germinates in spring when soil temperatures at the 4-inch (10 cm) level reach 60°F to 65°F (16°C to 18°C) for 24 consecutive hours. Goosegrass also requires sunlight for optimum seed germination and is very competitive in compacted soils. Normally, because of higher temperature requirements for germination, goosegrass germinates two to four weeks later in spring than crabgrass. This often coincides with flowering of later plants such as dogwoods and azaleas. If herbicides are applied at the time for crabgrass control, the material will begin to break down in the soil and goosegrass

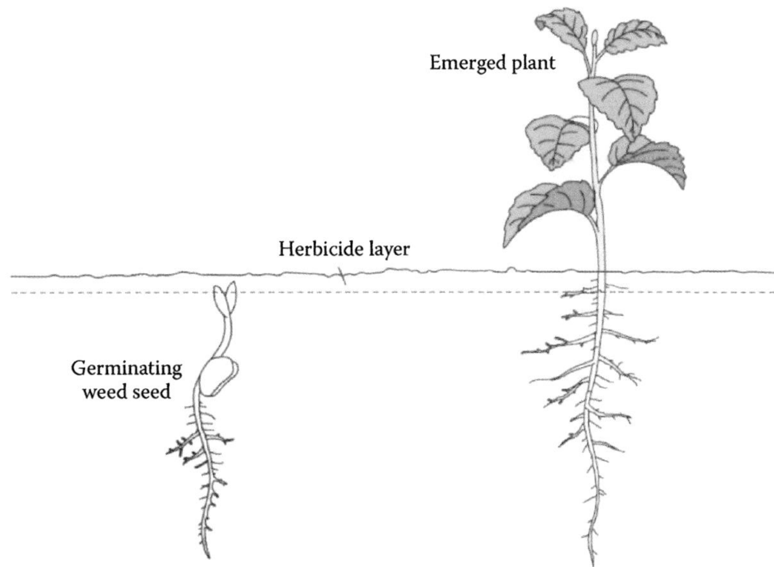


**FIGURE 14.6** Illustration of preemergence herbicides preventing germinating weed seeds from emerging.

control will be reduced. Therefore, when developing a goosegrass weed control program, delaying preemergence spring herbicide applications two to four weeks after the targeted date for crabgrass control is more efficient.

Annual bluegrass (*Poa annua*) begins to germinate in late summer through early fall when daytime temperatures consistently drop into the mid-70s°F (21s°C). On average, ~75% of the year's seed germination is in midfall, with the remaining occurring in winter and early spring when days are bright, air temperatures are in the 60s°F (16s°C), and night temperatures are cold (<35°F, 2°C). Thin turf areas, slightly shady areas, and excessively wet areas generally have the earliest and heaviest *Poa* germination.

## Sequential or Repeat Applications

Repeat applications of preemergence herbicides are generally necessary for full season control for crabgrass, goosegrass, and annual bluegrass. Most herbicides begin to degrade soon after application when exposed to the environment. Usually, the level of degradation occurring from 6 to 16 weeks after application reduces the herbicide concentration to the point where poor control of later germinating weed seeds, such as goosegrass, occurs. A general "rule of thumb" on whether sequential applications are needed is if the growing season of a particular region is more than 175 days long, repeat applications are needed. Repeat applications, therefore, are necessary between 60 and 75 days after the initial application for season-long preemergence weed control.

**Note:** On those areas to be established with turf, most preemergence herbicides should not be used two to four months prior to planting. Severe root damage and reduced turfgrass seed germination may result. Since oxadiazon has a different mode of action, it generally is the safest to use on newly sprigged or sodded turfgrasses.

## Core Aerification and Preemergence Herbicides

Core aeration has not traditionally been recommended or practiced following a preemergence herbicide application. This procedure was believed to disrupt the herbicide barrier in the soil, thereby allowing weed germination. Research, however, indicates that core aerification immediately prior to or one, two, three, or four months after application of many preemergence herbicides does not stimulate large crabgrass emergence. Aeration at one or two months after application also has little effect on large crabgrass cover. Core aeration at one, two, or three months after an application of a preemergence herbicide does not decrease goosegrass control on putting greens. However, greater amounts of crabgrass occur on greens aerified with the cores returned than nonaerified greens, or aerified greens with the cores removed.

## Fertilizer-Based Preemergence Herbicide

A growing trend in the turfgrass industry is the use of granular preemergence herbicides. The low cost of granular applicators and public perception of spraying has contributed to this trend. Additionally, many turfgrass managers are using dry fertilizer-based preemergence herbicides. Fertilizer–herbicide mixtures enable a "weed-and-feed" treatment in the same application or trip over the turfgrass. "Weed-and-feed" treatments may be convenient; however, certain factors must be considered prior to application. Depending upon the turfgrass, the time of year in which an herbicide should be applied may not coincide with the time of year a fertilizer is needed. For example, unless a slow-release nitrogen fertilizer is used, bermudagrass should not receive spring fertilization until the grass has greened-up approximately 40%, the chances of frost have passed, or the grass has recovered from winter dormancy. At the full green-up stage of bermudagrass, it is usually too late to apply most preemergence

**FIGURE 14.7** Fertilizer-based preemergence herbicides require sufficient coverage to work.

herbicides since many summer annual weeds have already emerged.

When using herbicides formulated on a fertilizer carrier, it is also important to determine if the manufacturers' recommended rate of the product supplies sufficient amounts of fertilizer needed by the turfgrass and herbicide required for weed control. Supplemental applications of fertilizer or herbicide may be required if the product does not supply enough of each to meet these needs. A spread rate of material when using "weed-and-feed" products should be at least 300 pounds of product/acre (34 g/m$^2$) to ensure adequate coverage (Figure 14.7). This is a general rule that applies to an average-size fertilizer particle. Smaller particles often used in more expensive premium products can provide adequate coverage with slightly lower product coverage rates. On greens, for example, an SGN size of 100 or less is recommended to minimize mower pickup.

### Corn Gluten for Preemergence Weed Control

The concept of using corn gluten meal as a natural herbicide originated in the 1980s. Accidentally discovered while using corn meal as a growing medium for *Pythium*, raw corn meal was applied in adjacent plots several weeks before seed bed planting. Secondary observations were made when germination of grass seed was inhibited by the raw corn meal. It was later found that an organic substance in the corn meal was destroyed when it was cultured with the fungal organism *Pythium*. Corn gluten is a by-product of the wet milling process of corn grain. Corn gluten contains 60% protein and 10% nitrogen (N) by weight, which may make an excellent fertilizer for plants with an established root system. The inhibitory substance of corn gluten prevents the formation of roots on germinating seedlings of a variety of grass and broadleaf plants. Five individual dipeptides (combinations of two amino acids) inhibit root formation of germinating seedlings. The material is applied in spring two to four weeks before the anticipated germination of summer annuals.

### Preemergence Herbicides for Golf Greens

Few preemergence herbicides are recommended for use by the manufacturer on golf greens due to the liability associated with these valuable areas. However, several are available for either bermudagrass or bentgrass greens. Several can be used on either grass; however, the user should check the latest herbicide label to ensure these are still available for putting green use.

## POSTEMERGENCE HERBICIDES

Postemergence herbicides are generally effective only on those weeds that have germinated and are visible. Most postemergence herbicides are relatively ineffective as preemergence herbicides. The timing of application should be when weeds are young (two- to four-leaf stage) (Figure 14.8) and actively growing. At this stage, herbicide uptake and translocation are favored, and turfgrasses are better able to fill in voids left by the dying weeds. The tolerances of different turfgrass species to postemergence herbicides are listed on the following website: http://www.clemson.edu/extension/horticulture/turf/pest_guidelines/.

### Broadleaf Weed Control

Broadleaf weeds in turf have traditionally been controlled with members of the phenoxy herbicide family (e.g., 2,4-D, dichlorprop, MCPA, and mecoprop) and benzoic acid herbicide family (e.g., dicamba). All are selective, systemic, foliar-applied herbicides. Only a very few broadleaf weeds, especially perennials, are controlled with just one of these materials. Usually, two- or three-way combinations of these herbicides and possible repeat applications are necessary for satisfactory weed control (Table 14.6). Special formulations of three-way type mixtures of 2,4-D, dichlorprop (2,4-DP), MCPP, MCPA, and dicamba are available. Manufacturers also are premixing a quicker acting product with these traditional two- and three-way mixtures to increase control and reduce the time for herbicide symptoms to show. Carfentrazone, sulfentrazone, and pyraflufen-ethyl quickly disrupt plant membranes, causing quicker herbicide symptoms. Sequential applications for all products should be

**FIGURE 14.8** Young weeds are much easier and cheaper to control with postemergence herbicides.

**TABLE 14.6**
**Postemergence Herbicides for Putting Greens[a]**

| Active Ingredients | Trade Names | Weeds Controlled | Comments | Bentgrass | *Poa*/ Bentgrass | Bermudagrass |
|---|---|---|---|---|---|---|
| 2,4-D + MCPP + dicamba | • Eliminate LO<br>• Threesome<br>• Trimec Classic/899/ Southern<br>• Triplet/Low Odor/ Hi-D<br>• TruPower2/3 | • Broadleaf weeds | • Rate limits on bentgrass greens<br>• Two applications/site/year<br>• May cause slight injury<br>• Do not apply above 85°F (29°C) | Yes | Yes | Yes |
| 2,4-D + MCPP + dicamba | • Trimec Bentgrass Formula | • Broadleaf weed | • Contains less 2,4-D than Trimec Classic<br>• May cause slight injury<br>• Two applications/site/year | Yes | No | No |
| 2,4-D + MCPP + dicamba + pyraflufen ethyl | • 4-Speed<br>• RedZone 2 | • Broadleaf weeds | • Avoid turf under heat or drought stress<br>• Mild yellowing for ~1 week<br>• Two applications/site/year | Yes | Yes | Yes |
| 2,4-D + triclopyr + dicamba + pyraflufen ethyl | • 4-Speed XT | • Broadleaf weeds | • Two applications/site/year | Yes | Yes | Yes |
| 2,4-D + dicamba + quinclorac | • 2DQ | • Broadleaf weeds | • Avoid temps. >90°F (32°C)<br>• Two applications/site/year<br>• Label neither allows nor restricts bermudagrass greens | Yes | No | Yes |
| Carfentrazone | • QuickSilver TandO | • Broadleaf weeds<br>• Silvery thread moss | • Moss: 6.7 fluid ounces/acre, every 2 weeks at ≤85°F (29°C)<br>• *Poa* damaged at >2.0 ounces/ acre (0.15 L/ha)<br>• Add NIS at 0.25% (v/v)<br>• Wait 75 days after bensulide use | Yes | Yes | Yes |
| Mecoprop (MCPP) | • Mecomec 2.5 + 4SL<br>• MCPP Amine | • Broadleaf weeds | • Avoid temps. >90°F (32°C) | Yes | Yes | No |
| MCPP + MCPA + dicamba | • Trimec Encore<br>• Tri-Power | • Broadleaf weeds | • Rate limit on bentgrass greens<br>• Avoid stressed bentgrass<br>• Avoid temps. >85°F (29°C)<br>• Mild yellowing may occur<br>• Label neither allows nor restricts bermudagrass greens | Yes | No | Yes |
| Foramsulfuron | • Revolver | • Goosegrass<br>• *Poa* | • Removes overseeding and *Poa* + certain broadleaf weeds | No | No | Yes |
| Trifloxysulfuron | • Monument 75WG | • *Poa*<br>• Sedge/kyllinga<br>• Select broadleaf weeds | • Label neither allows nor restricts bermudagrass greens<br>• Add NIS at 0.25% (v/v) | No | No | Yes |
| Rimsulfuron | • TranXit<br>• QP Rimsulfuron 25DF | • *Poa* and other cool-season species | • Removes overseeding + certain broadleaf weeds<br>• Add NIS at 0.25% (v/v) | No | No | Yes |
| Pronamide | • Kerb | • *Poa* and other cool-season species | • Removes overseeding + certain broadleaf weeds<br>• Add NIS at 0.25% (v/v)<br>• 2(ee) label is available | No | No | Yes |

*Source:* Modified from Brosnan, J. et al., 2014, *Golf Course Management* 82(8):92–98.

[a] Always consult and follow all label directions prior to use. Refer to the latest label version to confirm use on a particular grass or site.

spaced 10 to 14 days apart and only healthy-growing, nonstressed turf should be treated.

Clopyralid (Lontrel), triclopyr (Turflon), and various combinations with other herbicides also are alternatives to phenoxy herbicides for broadleaf weed control. Triclopyr belongs to the picolinic acid herbicide family. Compounds in this family are noted for their high degree of activity. These herbicides are up to 10 times more potent than 2,4-D on certain broadleaf weed species. They are rapidly absorbed by the roots and foliage of broadleaf plants, and are readily translocated throughout the plants via both xylem and phloem tissues. Problems with this herbicide family include its soil mobility and the extreme ornamental sensitivity. Clopyralid also is one of the newer members of this herbicide family. It is currently marketed in a mixture with triclopyr (Confront) for use on labeled cool- and warm-season turfgrasses. Clopyralid is especially effective on leguminous plants (nitrogen producing) such as black medic, clover, kudzu, lespedeza, and vetch. Penoxsulam (LockUp) is the most recently introduced broadleaf herbicide. It is noted for control of English daisy, Florida betony, and dollarweed and better turf tolerance. It is also formulated as a granule.

Metsulfuron (Manor, Blade) is a sulfonylurea herbicide for selective broadleaf and cool-season grass control in certain warm-season turfgrasses. Quinclorac also has postemergence broadleaf and crabgrass control. The following website lists the effectiveness of commonly used postemergence herbicides for broadleaf weed control: http://www.clemson.edu/extension/horticulture/turf/pest_guidelines/.

## Grass Weed Control

Traditionally, for tolerant turfgrass species, postemergence grass weed control was through single and repeat applications of the organic arsenicals (e.g., MSMA, DSMA, CMA). Two to four applications, spaced seven days apart, generally are required for complete control. The rate and number of applications necessary for weed control usually increases as weeds mature. On cool-season turfgrasses and zoysiagrass, organic arsenicals can be very phytotoxic, especially when used during high temperatures (>90°F, 32°C). Control also is reduced if rainfall occurs within 24 hours of treatment. Recently, new herbicide releases have provided alternatives to the arsenicals for postemergence grass weed control in cool-season turfgrasses. Decreased phytotoxicity as well as a reduced number of applications are often associated with these herbicides. Products include fuazifop-methyl, fenoxaprop-ethyl, quinclorac, sethoxydim, mesotrione, and topramezone. For postemergence crabgrass and goosegrass control in bentgrass, fenoxaprop-ethyl applied every 14 to 18 days beginning in midspring, when crabgrass has one to two leaves, is generally safe and effective. On larger plants, quinclorac can be applied twice, 14 days apart, but often causes objectionable yellow-green mottling of turf for about 30 days. Tank mixing with chelated iron and N may help mask this injury. Mesotrione provides postemergence goosegrass control in certain cool-season grasses. The following website lists the effectiveness of commonly used postemergence herbicides for grass control: http://www.clemson.edu/extension/horticulture/turf/pest_guidelines/.

## Nutsedge and Kyllinga

Sedges generally thrive in soils that remain wet for extended periods of time due to poor drainage or excessive irrigation. The first step in sedge weed control is, therefore, to correct the cause of continuously wet soils. Do not overirrigate an area; if necessary, provide surface and subsurface drainage.

Yellow and purple nutsedge are low-growing perennials resembling grasses. Sedges, in general, are yellow-green to dark green, with triangular stems bearing three-ranked leaves—unlike the two-ranked leaves of the grass family. Yellow and purple nutsedge have fibrous root systems with deep-rooted tubers or nutlets and rhizomes for reproduction and spread (Figure 14.9). Seedhead color is often used to distinguish between these two major nutsedges. Yellow nutsedge has a yellowish- to straw-colored inflorescence, while purple nutsedge has a reddish to purple inflorescence (Figure 14.10).

**FIGURE 14.9** White rhizome of purple nutsedge.

**FIGURE 14.10** Comparison of yellow (left) and purple (right) nutsedge inflorescences.

**TABLE 14.7**
**Relative Sedge Control of and Turf Tolerance to Various Herbicides**

| | Sedge Control | | | | | Turf Tolerance (Excluding Greens) | | | | | | | | | | | |
|---|---|---|---|---|---|---|---|---|---|---|---|---|---|---|---|---|---|
| Herbicide (trade names)[a] | Annual Sedge | Purple Nutsedge | Yellow Nutsedge | Annual Kyllinga | Perennial Kyllinga | Bahiagrass | Bermudagrass | Bentgrass | Bluegrass, Ryegrass | Buffalograss | Centipedegrass | Fine Fescue | Kikuyugrass | St. Augustinegrass | Tall Fescue | Zoysiagrass |
| **Preemergence Control** | | | | | | | | | | | | | | | | | |
| Dimethenamid (Tower) | G[b] | F-G | G | G | F-G | NR | S | NR | NR | S | NR | NR | NR | NR | S | S |
| Metolachlor (Pennant Magnum) | G | P | G | F-G | P | S | S[c] | NR | NR | NR | S | NR | NR | S | NR | S |
| **Postemergence Control** | | | | | | | | | | | | | | | | | |
| Bentazon (Basagran TandO) | G | P | G | F-G | F-G | S | S | S-I | S | S | S | S | NR | S | S | S |
| Flazasulfuron (Katana) | G | F | F-G | G | G | NR | S | NR | NR | S | I-S | NR | NR | NR | NR | S |
| Imazaquin (Image) | G | G | F | G | G | NR | I-S | NR | NR | NR | I | NR | NR | I | NR | S |
| Imazosulfuron (Celero) | G | G-E | G-E | G | F | NR | S | S | S | NR | S | S | NR | S | S | S |
| Halosulfuron (Sedgehammer) | G | G-E | G-E | G | F-G | S | S | S | S | S | S | S | D | S | S | S |
| Mesotrian (Tenacity, TRIONE) | P | P | G | P | P | NR | I | NR | S | NR | S-I | S-I | NR | S-I | S-I | NR |
| MSMA/DSMA/CMA | G | P-F | F | G | G | NR | S-I | I | NR | I | NR | I | NR | NR | I-S | S-I |
| Image + MSMA/DSMA | G | G | G | G | G | NR | S-I | NR | NR | NR | NR | NR | NR | NR | NR | S-I |
| Sulfentrazone (Dismiss) | G | P-F | F | F | F | S | S | S | I-S | S | S | S | S | NR | S | S |
| Sulfentrazone + imazethapyr (Dismiss South) | G | G | G | G | G | S | S | NR | NR | S | S | NR | S | NR | NR | S |
| Sulfosulfuron (Certainty) | G | G | G | G | G | S | S | NR | NR | S | S | NR | S | S | NR | S |
| Trifloxysulfuron (Monument) | G | G | G | G | G | NR | S | NR | NR | S-I | NR | NR | NR | NR | NR | S |

*Notes:* Refer to herbicide label for specific turf species use listing. These are relative rankings and depend on many factors such as environmental conditions, turfgrass vigor or health, application timing, etc. and are intended only as a guide.

[a] Repeat applications are necessary for complete control from all herbicides.

[b] E = excellent (>89%) control; F = fair to good (70% to 89%) (good control sometimes with high rates; however, a repeat treatment one to three weeks later each at the standard or reduced rate is usually more effective); P = poor (<70%) control in most cases.

[c] S = safe at labeled rates; I = intermediate safety, use at reduced rates; NR = not registered for use on and/or damages this turfgrass; D = dormant turf only.

Leaf-tip shape is another distinguishing method. Leaf tips of purple nutsedge are generally thicker and more rounded than yellow nutsedge leaf tips, which are very narrow, ultimately forming a needlelike end. Yellow and purple nutsedge have a great capacity to reproduce and spread due in part to their massive underground tuber and rhizome systems. They are not believed to produce viable seed.

Selective yellow nutsedge control is available with bentazon (Basagran TandO) and sulfentrazone (Dismiss) with minimum turf damage (Table 14.7). Control of most other sedges, except purple nutsedge, also will result from bentazon treatments. Bentazon is a contact material and will control only those portions of the weeds contacted by the spray. Complete coverage of the weeds is necessary for the greatest bentazon activity. Even with good herbicide coverage, regrowth will normally occur from the roots and tubers and repeat applications will be necessary. Sulfentrazone does not provide complete control of purple nutsedge. It has relatively rapid burndown properties and is safe on most warm- and cool-season turfgrasses. Being root absorbed, rainfall or irrigation will be necessary following application.

Purple and yellow nutsedge, as well as several other sedges, can be suppressed with imazaquin (Image). Selective broadleaf weeds, such as pennywort, are also controlled with imazaquin. As with bentazon, repeat applications—possibly over several years—will be required to control all the underground reproductive parts with imazaquin. The addition of MSMA increases the activity of either bentazon or imazaquin and broadens the range of controlled weeds. However, this tank mix should be used only on actively growing bermudagrass or zoysiagrass.

Halosulfuron (Sedgehammer) and imazosulfuron (Celero) also have good control on most sedges and good warm- and cool-season turf tolerance. Both products are somewhat slow, requiring two to three weeks for control. As with other sedge control materials, repeat applications three to four weeks after the initial application are necessary, especially when trying to control perennial sedges. In most cool-season grasses, halosulfuron, sulfentrazone, bentazon, and imazosulfuron are the key nutsedge control choices.

Trifloxysulfuron (Monument) and sulfosulfuron (Certainty) control all major sedges and kyllinga species in bermudagrass and zoysiagrass. Two applications four weeks apart will be necessary for satisfactory control of perennial species. Cool-season grasses and other warm-season grasses should not be treated with these products. The following website lists the effectiveness of commonly used postemergence herbicides for nutsedge and kyllinga control: http://www.clemson.edu/extension/horticulture/turf/pest_guidelines/.

## SPECIAL WEED MANAGEMENT SITUATIONS

### Herbicide-Resistant Weeds

Herbicide-resistant weeds have plagued the row-crop sector of weed science for the past 15 or so years. However, more recently, turfgrass herbicide resistance has become a major issue for golf courses. Resistance problems generally begin to show up when a class of compounds with a similar mode of action is continuously used over an approximately 7- to 10-year period. These herbicides selectively control susceptible weed biotypes, thereby gradually allowing the resistant biotypes to spread and increase over time. Just as turf managers should rotate between fungicide groups to prevent disease resistance, herbicides with different modes of action should also be rotated to prevent weed resistance. Herbicide-resistant weeds in turf, such as *Poa annua*, spurges, goosegrass, nutsedges, and crabgrass, are becoming more prevalent. Fortunately, this can be contained if prudent action is taken. Table 14.4 summarizes the main herbicides used in turf including their timing (pre- versus postemergence), their mode of action within plant (how they control them), and the various active ingredients. Rotating between and mixing herbicides with different mechanisms of action in tanks are keys to delaying or preventing herbicide-resistant weeds from dominating a population.

### *Poa annua*

Annual bluegrass (*Poa annua* L.) is the most troublesome winter annual grass weed on golf courses. Its low growth habit and unique ability to thrive in moist conditions and compacted areas make it difficult to control with management practices alone. Annual bluegrass has a lighter green color than most grass species used to overseed golf greens, and produces numerous seedheads that disrupt the playing surface (Figure 14.11). Also, due to low heat tolerance, annual bluegrass quickly dies in warm weather, leaving areas bare until desirable grass species have time to fill in.

Chemical control of annual bluegrass is difficult to achieve due to (1) the inability of most preemergence herbicides to prevent annual bluegrass germination selectively while allowing the desirable grass to become established, and (2) most postemergence herbicides effective on annual bluegrass also damage the desirable grass species, especially cool-season grasses.

In recent years, annual bluegrass has quickly reached epidemic proportions for many golf courses. The reasons for this

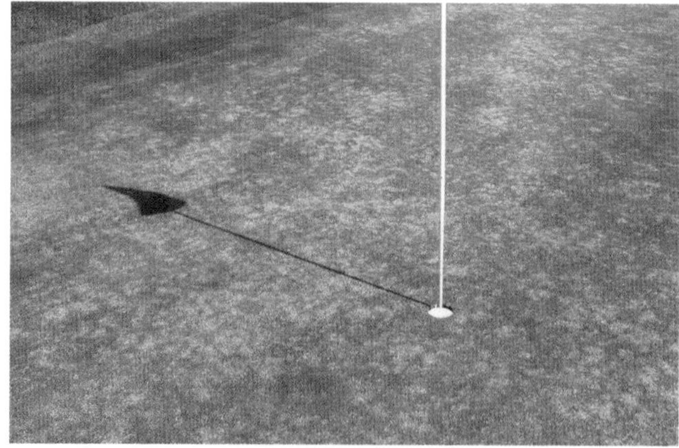

**FIGURE 14.11** Annual bluegrass (*Poa annua*) infestation on a golf course green.

**TABLE 14.8**
**Growth Responses of Annual Bluegrass (*Poa annua*) to Varying Temperatures**

| Growth Response | Temperature |
|---|---|
| Optimum germination | • Mid-70s°F (21s°C), daytime<br>• Mid-50s°F (10s°C), nighttime |
| Germination RANGE | • 35°F to 102°F (1.7°C to 39°C) |
| Maximum growth | • 60°F to 70°F (15.5°C to 21°C) |
| Maximum root growth | • 65°F to 70°F (18°C to 21°C) (soil temperatures) |
| Seedhead development | • ≈70°F (26.7°C) |
| Maximum heat tolerance | • 85°F to 95°F (29°C to 35°C) |
| Minimum growth | • 50°F (10°C) (soil temperatures) |
| Lethal cold temperature | • ≈5°F (–15°C) |

are numerous: (1) a variety of biotypes are present, (2) the switch of many golf courses to the similar *Poa trivialis* for overseeding greens reduces herbicide options, (3) increased overseeding of fairways generally increases *Poa annua* populations, and (4) the occurrence of herbicide-resistant plants.

***Poa annua* primer.** The annual *Poa* biotype (*Poa annua* ssp. *annua* L.) has a nonstoloniferous bunch-type growth habit and generally a light-green color. It is a true annual and begins to germinate when daytime temperatures consistently drop into the mid-70s°F (~21°C) and nighttime temperatures are in the mid-50s°F (~10°C) for several consecutive days in late summer and early fall (Table 14.8). Maximum seed germination occurs when full sunlight is present; thus, thin, weak turf stands often have earliest germination. Areas remaining cooler, including shaded areas and continuous wet areas, also have earlier germination. Another flush of germination typically occurs in early winter when daytime temperatures are warm and nighttime temperatures are cold. This alternating warm/cold temperature scarifies additional seed; thus, it results in another flush of germination. Many herbicides applied in late summer for the initial fall flush of germination will not satisfactorily control this second germination; thus, repeat applications may be necessary.

After germination, *Poa* grows and then tillers (mostly unnoticeable) throughout the fall and early winter months. Once late winter arrives, *Poa* begins to shift its growth from vegetative toward reproductive by forming numerous seedheads that can literally turn a turf stand snow-white (Figure 14.12). Since *Poa* has both male and female flowers on each plant, they are capable of either self-pollinating or cross-fertilizing nearby plants. Each small plant can produce several hundred viable seeds. From the time *Poa* seed germinates in the soil, 44 to 149 days are required before seedheads form. However, the timing of seedhead formation is influenced by many factors including weather, geography, and, most importantly, the biotypes of *Poa* present. The annual biotype reduces its growth in late spring and dies when daytime temperatures reach the mid-80s°F to mid-90s°F (~29°C to 35°C) for several consecutive days. Seed can remain viable in the soil for more than six years, thereby ensuring a continual supply.

**FIGURE 14.12** Whitish seedheads from a heavy annual bluegrass infestation.

Several cultural factors favor *Poa* growth and occurrence. The first step, therefore, in a total *Poa* management program would be to shift the following to those favoring turfgrass growth:

1. **Continuous wet and compacted soils**—*Poa* thrives under wet and/or compacted soil conditions. Due to its shallow root system, *Poa* can tolerate lower soil oxygen levels while turfgrass stands will begin to thin. Use appropriate soil mixtures to reduce soil compaction. Soils should be frequently aerified, and greens should be mowed with walking mowers. Greens also should be spiked frequently to reduce surface compaction and to sever bermudagrass stolons to encourage a thicker turf stand. Also, do not overwater.
2. **Excessive nitrogen rates**—A high available N supply will profusely encourage *Poa* occurrence, growth, and tillering.
3. **Excessive soil phosphorus levels**—*Poa* has been shown to be favored by excessive phosphorus (P) levels and will outcompete turfgrasses under these conditions. Supply sufficient phosphorus as indicated by soil tests for the turf but do not use excessive rates.
4. **Leaving clippings**—Due to its tolerance to low mowing height and abundant seed production, leaving clippings only helps spread *Poa* seed, thereby slowly increasing its occurrence.
5. **Overseeding**—Overseeding bermudagrass provides winter and early spring turf color for golfers. However, overseeding greatly encourages *Poa* invasion since prepping bermudagrass via aerifying and verticutting helps thin the turf stand. This opens the turf to more sunlight, which encourages *Poa* to germinate. The increased watering need for overseeding establishment also encourages *Poa*. Overseeding weakens the bermudagrass in spring, extending the time needed for recovery and allowing *Poa* to encroach slowly. In addition, overseeding restricts the number of herbicide options available, as controlling one cool-season grass in another is very difficult, especially if the overseeded grass is *Poa trivialis*, a relative of *Poa annua*.

A buffer zone of overseeding around the approach helps intercept *Poa* seed from golfers and equipment. Courses with epidemic *Poa* levels may opt to skip overseeding fairways until the population is brought under control. Others may overseed fairways but skip approaches as more control options are available for nonoverseeded areas.

**Herbicide-resistant *Poa annua*.** Besides those listed management practices that encourage *Poa annua* stands, the recent occurrence of herbicide-resistant *Poa* biotypes has further reduced the number of effective control options. Resistant biotypes to glyphosate, simazine and atrazine, the aryloxyphenoxy-propionates, the sulfonylurea herbicides, the dinitroaniline herbicides, and ethofumesate have been noted.

**Preemergence control—overseeded bermudagrass.** Preemergence annual bluegrass control is currently achieved with several herbicides. Each has its own precautions before use, and if these are not followed, unsatisfactory results may occur.

Pronamide (Kerb) is applied before annual bluegrass germination and planting of overseeding grass. Thirty to 45 days is the minimum recommended period between application and overseeding. It also is recommended that applications not be made where drainage flows onto areas planted with cool-season grasses, or onto bermudagrass golf greens.

Superintendents who have to apply pronamide closer than 30 days before overseeding can offset the problems of reduced ryegrass seed germination by applying a thin layer of charcoal. The charcoal helps bind the pronamide and prevents it from damaging the overseeded ryegrass. However, the mess and dark color associated with charcoal applications and the risk of it not working must be considered before use. Activated charcoal should be applied at 2 to 4 pounds/1,000 square feet (10 to 20 kg/100 m$^2$), and reseeding should occur no sooner than seven days following charcoal use.

Dithiopyr (Dimension) provides good control of annual bluegrass when applied six to eight weeks prior to overseeding bermudagrass with perennial ryegrass. Dithiopyr is available in various formulations; however, best control often accompanies the use of a granular formulation. Due to its soil persistence characteristics, a six- to eight-week waiting period is necessary to prevent dithiopyr from affecting perennial ryegrass establishment.

Prodiamine (Barricade) has a long soil half-life but can be applied before overseeding bermudagrass with perennial ryegrass for annual bluegrass control. Similarly to dithiopyr, prodiamine is available in various formulations. It is applied six to eight weeks in advance of overseeding with only slight effects on perennial ryegrass establishment.

Foramsulfuron (Revolver), trifloxysulfuron (Monument), and rimsulfuron (TranXit GTA) control annual bluegrass in bermudagrass scheduled to be overseeded with perennial ryegrass. These herbicides have short half-lives (2 to 14 days) in the soil and can be applied 10 to 14 days in advance of overseeding. When used in this manner, very young annual bluegrass plants that have germinated in the early fall months are effectively controlled. The best *Poa* control strategy with these products is to delay overseeding as late as possible to allow maximum *Poa* germination, and then apply prior to overseeding according to the reseeding interval listed on the label.

**Postemergence control.** Ethofumesate (Prograss) provides preemergence and early postemergence annual bluegrass control in bermudagrass. However, to prevent undesirable turfgrass injury, the application rate, timing, and frequency are important. If applied in fall before bermudagrass dormancy, an immediate cessation of bermudagrass growth occurs. A delay in spring transition from ryegrass to bermudagrass also occurs with early fall application. Spring green-up of bermudagrass can be severely retarded with February applications. Therefore, ethofumesate should only be used where sufficient cold weather occurs for complete bermudagrass dormancy, and then applied during late fall through early winter. Ethofumesate is not labeled for golf greens, zoysiagrass, or fine fescues. Another selective postemergence herbicide designed to selectively control *Poa* in cool-season grasses is amicarbazone (Xonerate). It is applied in late winter when daytime air temperatures are between 55°F and 75°F (13°C and 24°C), twice, 14 to 21 days apart.

**Plant growth retardants.** Certain plant growth retardants (PGRs) also help suppress the seedheads of *Poa annua* in overseeded bermudagrass fairways. Timing is critical with these, and the club must be willing to accept some degree of short-term turfgrass discoloration (or phytotoxicity).

Paclobutrazol (Trimmit 2SC) is applied in late winter to early spring after the growth of desired grasses has resumed and one or two mowings have occurred. Do not apply after late winter to avoid delaying green-up of bermudagrass. Paclobutrazol is root absorbed, and 1/4 inch (0.64 cm) of rainfall or irrigation water should be applied within 24 hours of application. Repeat applications may be made three to four weeks apart. It is not recommended for use with ultradwarf bermudagrass golf greens. Do not use if *Poa annua* populations exceed 70%. Other products include combinations of flurprimidol, paclobutrazol, and trinexapac-ethyl.

**Preemergence control—nonoverseeded bermudagrass.** *Poa* control in nonoverseeded bermudagrass turf is generally much easier and cheaper to achieve as more effective herbicide options are available. Control is available with most commonly used preemergence herbicides such as benefin (Balan), dithiopyr (Dimension), indaziflam (Specticle), oryzalin (Surflan), oxadiazon (Ronstar), pendimethalin (Pre-M, Pendulum), prodiamine (Barricade), and various combinations such as Team (benefin + trifluralin) and XL (benefin + oryzalin). Approximately 80% to 90% control is the best level of control expected from a single preemergence herbicide application. For season-long control, some preemergence products will need a repeat application approximately 60 days following the first.

**Postemergence control—nonoverseeded bermudagrass.** Postemergence control options for *Poa* in nonoverseeded bermudagrass are numerous. However, timing and selective placement are very important to prevent damage to the adjacent overseeded stand or permanent bermudagrass.

Pronamide (Kerb) can be applied either as preemergence or early postemergence for excellent *Poa* control.

Applications should be timed for midfall, with a possible repeat in midwinter. If applications are made later in spring, control will drop and the time required for control significantly increases. Pronamide also will help control perennial (or renegade) ryegrass before it tillers and clumps. Pronamide, however, should *not* be used upslope of desirable overseeding or bentgrass as it can and will move with water and run across these areas, causing death of the overseeding.

Rimsulfuron (TranXit), trifloxysulfuron (Monument), and foramsulfuron (Revolver) are members of the sulfonylurea herbicide family that provide good to excellent postemergence control of *Poa* in nonoverseeded bermudagrass or zoysiagrass. Control is slow, at two to six weeks. Like Kerb, neither TranXit, Monument nor Revolver should be used upslope of desirable overseeding or bentgrass, as these can and will move with water and run across these areas, causing death of the overseeding (Figure 14.13).

Simazine and atrazine (e.g., Princep TandO, Aatrex), like Kerb, provide good selective control of *Poa* if appropriately timed. Timing of these is very similar to that for Kerb, with midfall and early winter follow-up applications best. Simazine and atrazine also provide good to excellent control of many winter annual broadleaf weeds such as spurweed, chickweed, and henbit. These materials, however, should not be used during or just before normal spring green-up of bermudagrass as they can significantly delay this process. As with pronamide, do not use simazine or atrazine upslope of desirable cool-season grasses.

Nonselective control of *Poa* and other winter weeds is available with several herbicides if these are appropriately timed. Glyphosate (Roundup Pro), glufosinate (Finale), diquat (Reward), and pelargonic acid (Scythe) provide varying levels of *Poa* control (Figure 14.14). These must be timed when the bermudagrass is fully dormant, with no green leaves or stolons present. This situation generally occurs in midwinter after several consecutive hard frosts. If applied at other times, delayed green-up of the bermudagrass can be expected with possible permanent damage. Subtropical areas such as Florida, southern California, south Texas, and other Gulf Coast areas do not typically experience sufficient cold weather to have completely dormant bermudagrass. Nonselective *Poa* control, therefore, is not recommended for these areas.

**FIGURE 14.13** Turf damage from lateral herbicide movement.

**FIGURE 14.14** Postemergence control of annual bluegrass.

These herbicides should be applied on relatively warm days (>50°F, 10°C) or extended periods will be required for control. Diquat, glufosinate, and pelargonic acid provide relatively quick burn of winter weeds while glyphosate provides better long-term control, but with much slower results. Glyphosate also provides better control when applied with low application volumes. High volumes, in excess of 40 GPA (374 L/ha), should be avoided. Do not apply to dormant zoysiagrass or centipedegrass as significant turf injury may occur.

**Bentgrass.** Bentgrass is sensitive to most postemergence grass herbicides. Additionally, most of these products are ineffective against annual bluegrass. As a result, preemergence herbicides are the most common means of controlling this weed on bentgrass golf greens. Erratic control of annual bluegrass, however, often occurs. The presence of perennial biotypes of the species may contribute to this erratic control. Low-growing, creeping perennial types become dominant over the annual biotype under frequent close mowing. Moist soil conditions and high soil nutrient levels, which are conditions normally maintained with creeping bentgrass golf greens, also contribute to creeping perennial biotype dominance. As the annual biotype is controlled by herbicides, the perennial biotype begins to dominate the green.

Limited research on preemergence control of the perennial biotype suggests poor short-term control. Repeat applications over multiple years (minimum of four years) are necessary for significant reduction of the perennial biotypes in bentgrass golf greens. Multiple-year treatments during February and March are considered superior for control and turf safety to August or September treatments. Currently available preemergence herbicides labeled for bentgrass golf greens include bensulide (Betasan, Weed Grass Preventer) and bensulide plus oxadiazon (Goosegrass/Crabgrass Control) (Table 14.5).

Selective postemergence control of annual bluegrass in bentgrass golf greens is highly variable. Products with activity include ethofumesate (Prograss) and amicarbazone (Exonerate). In the future, methiozolin (PoaCure) may become available. Ethofumesate and amicarbazone can work but also can be erratic, depending on temperatures, sunlight, drainage, and *Poa* biotypes present. All products should be viewed as *Poa* reduction tools and not a "silver bullet" that completely eradicates *Poa*.

### Best Management Practices for Controlling *Poa annua* in Golf Greens

1. Fumigate all soil mix before planting.
2. Begin with and retain good drainage to prevent soil compaction and excessive soil moisture that favor the *Poa*. Irrigate deeply and infrequently. Moisture stress discourages *Poa*.
3. Use certified seed, sprigs, or sod free of *Poa* when planting.
4. Obtain and maintain good turf density to reduce *Poa* invasion.
5. Improve sunlight for better bentgrass growth.
6. Use lightweight mowers to help reduce compaction. Remove clippings when *Poa* is seeding.
7. Aerify consistently to relieve soil compaction. Perform before peak *Poa* germination period.
8. Use fumigated sand/soil when topdressing.
9. Use preemergence herbicides in spring and fall.
10. Use PGRs in spring and fall to reduce *Poa* competition and seedhead development.
11. Hand-pick or -wick nonselective herbicide (e.g., glyphosate) on small (e.g., 1 inch, 2.5 cm diameter) *Poa* plants.
12. Plug larger spots with *Poa*-free turf.
13. Control *Poa* in green surrounds and in other areas of the golf course to minimize seed transfer to greens by players, water movement, and maintenance equipment.

**PGRs.** With the realization that annual bluegrass elimination in golf greens is rarely achievable with current herbicide technology, the industry has focused on suppressing its growth and seedhead production (Table 14.9). The turf growth regulators (TGRs) ethephon (Proxy), paclobutrazol (Trimmit), and flurprimidol (Cutless) currently are available for annual bluegrass suppression in bentgrass golf greens. Combining these PGRs with other products such as the sterol inhibitor (DMI) fungicides may improve suppression. In a typical program, paclobutrazol, flurprimidol, and ethephon alone, or in combination with trinexapac-ethyl, are applied to actively growing bentgrass two to three times in fall as well as in early spring when the turf is actively growing (50°F to 70°F, 10°C to 21°C). Thirty-day intervals should occur between applications. DMI fungicides applied two weeks following each PGR treatment increase *Poa* control. Repeating applications during these time periods over several years is necessary to gradually eliminate the perennial biotype. Treatments should not be made during periods of heat, moisture, or cold stress to the bentgrass. Treated turf also may appear more "grainy" with a wider leaf texture, while treated *Poa* plants often have noticeable discoloration in terms of a lighter green to yellow color.

Differential species' susceptibility to the herbicide is attributed to a greater uptake of the TGR by the shallower rooted annual bluegrass when compared to the deeper rooted bentgrass. Paclobutrazol and flurprimidol are considered type II growth regulators since they inhibit gibberellin biosynthesis and suppress internode elongation (Figure 14.15). These materials are root absorbed (xylem-mobile) and work by reducing the competitive ability of the annual bluegrass for three to eight weeks after application. This allows the creeping bentgrass to outcompete the weed. These materials should only be applied during periods of active bentgrass root growth. Prevention of annual bluegrass seedhead formation may be inconsistent with these materials but they do prevent seedhead stalk elongation, which may result in improved turf uniformity and appearance.

***Poa annua* control in fairways.** For creeping bentgrass, Kentucky bluegrass, and tall fescue fairways and roughs, preemergence *Poa annua* control is provided by those preemergence herbicides previously discussed. Early postemergence control is provided by ethofumesate (Prograss). Young (one- to five-leaf stage) *Poa* should be treated in fall with a sequential application three weeks following the initial application.

**TABLE 14.9**
**Typical *Poa annua* Suppression Program in Creeping Bentgrass Using Plant Growth Regulators**

| Timing | *Poa annua* Control in Bentgrass | Bentgrass/*Poa annua* Management |
|---|---|---|
| Spring | Ethephon + Paclobutrazol—monthly | Ethephon + Trinexapac-ethyl—monthly |
| | Ethephon + Paclobutrazol—monthly | Ethephon + Trinexapac-ethyl—monthly |
| | Ethephon + Paclobutrazol—monthly | Ethephon + Trinexapac-ethyl—monthly |
| Summer | Paclobutrazol (1/3 rate)—14 day | Trinexapac-ethyl (1/2 rate)—14 day |
| | Paclobutrazol (1/8 rate)—14 day | Trinexapac-ethyl (1/3 rate)—14 day |
| | Paclobutrazol (1/8 rate)—14 day | Trinexapac-ethyl (1/3 rate)—14 day |
| Fall | Core + seed: | Core + seed: |
| | Paclobutrazol (3/4 rate)—monthly | – |
| | Paclobutrazol (full rate)—monthly | – |

*Note:* PGR rates used are dependent on many variables, including environmental extremes, health of bentgrass, level of suppression desired, regrowth needs, etc.

**FIGURE 14.15** Bermudagrass stunting (left) following use of a gibberellin-inhibiting plant growth retardant.

Turf should be at least eight weeks old before treatment with ethofumesate. Control of perennial biotypes with ethofumesate is erratic. Turf growing in shaded, wet, and/or low areas may have increased damage from ethofumesate.

Amicarbazone (Xonerate) is applied in later winter or early sping just prior to seedhead formation when daytime/nighttime temperatures are 70°F/50°F (10°C/21°C) at the time of application and several days later. A repeat application will be needed two to three weeks following the initial one.

### Bentgrass Control in Most Cool-Season Turfgrasses

Mesotrione plus triclopyr ester are highly effective and relatively safe on most cool-season grasses such as tall fescue, Kentucky bluegrass, and perennial ryegrass. It will require at least three applications at 14- to 21-day intervals.

### Moss

*Byrum argentum* is referred to as silvery thread moss due to its silvery appearance and is found frequently on greens. Mosses are threadlike, branched, primitive nonvascular plant forms encompassing many species. They are not parasitic and they spread by spores disseminated by wind and water movement. Mosses are able to photosynthesize and fix nitrogen. Moss is most noticeable on close-cut areas such as tees and putting greens that are poorly drained (and thus remain continuously wet) and heavily shaded. Moss can survive weather extremes in a dormant state or by living symbiotically with blue-green algae. Algae, therefore, can be a precursor to moss encroachment and should be discouraged to prevent moss colonization. Moss mats typically develop in summer following periods of rainy, overcast, warm days.

**Cultural controls.** Control involves a long-term, persistent program combining cultural and chemical control methods realizing healthy turf is the only means to cure and prevent moss occurrence. Control begins by correcting those conditions that predispose the turf to moss growth. This involves reducing surface moisture by improving air circulation and light exposure by removing adjacent underbrush and selectively removing trees. Improve surface and subsurface drainage and reduce irrigation frequency and amount. Reduce freely available nitrogen at the site. Reduce irrigation and improve growth of the turfgrass where the moss is present so the turf can form a dense area. If the area occupied by moss is large, spiking, verticutting, and topdressing will help to break up and dry the mat. Moss turning orange-brown or golden brown in color indicates positive desiccation is occurring. Several trends in fertility and moss development have been noted. For example, calcium-rich soil may encourage certain moss species, but moss tends to be discouraged soils adequate in potassium. Ammonium sulfate at 1/10 to 1/8 pound N/1,000 square feet (0.49 to 0.61 g N/m$^2$) applied weekly is thought to help desiccate moss and encourage competitive turf growth. Use only when air temperatures are below 80°F (27°C) and adequate moisture is present. Applying ground limestone (75 to 100 pounds/1,000 square feet, 366 to 488 g/m$^2$), baking soda (6 ounces/gallon water to drench, 45 g/L), hydrogen peroxide, or hydrated lime (2 to 3 pounds/1,000 square feet, 9.8 to 15 g/m$^2$, in 3 gallons, 11 L, of water) will help desiccate the moss and raise the soil pH level, which favors competitive turf growth. Diluted bleach and dishwashing detergent, chloride, ferrous sulfate at 4 to 7 ounces/1,000 square feet (1.2 to 2.1 g/m$^2$), granular iron sulfate at up to 3 pounds/1,000 square feet (14.6 g/m$^2$), or ferrous ammonium sulfate at 10 ounces/1,000 square feet (3 g/m$^2$) also may help reduce moss growth. However, these should not be used on greens during hot temperatures, as they may cause varying levels of turf discoloration. Increase the mowing height. as low mowing aggravates the problem. Spike or rake the dehydrated moss layer to remove any remaining impervious layer.

**Chemical controls.** Chemical control is erratic and often unsuccessful, especially if agronomic practices that favor moss growth and development are not corrected. Carfentrazone (Quicksilver 1.9 L) at 2.1 to 6.7 ounces/acre (6.7 to 21 L/ha) has provided good moss control. Treat when air temperatures are <85°F (29°C) and use at 100 GPA (935 L/ha) and repeat in two to three weeks. Products containing potassium salts of fatty acids (e.g., DeMoss) applied weekly at 2 to 3 ounces/1,000 square feet (6.2 to 9.5 L/ha) or formaldehyde control moss through a contact mode of action, but should be carefully used and all label information followed closely. Chlorothalonil may be used but only during summer. Chlorothalonil should be applied every seven days for three consecutive weeks in 5 gallons water per 1,000 square feet (20 L/100 m$^2$). For success air temperatures should be above 80°F (preferably, >85°F, 27°C to 29°C) at the time of application. Being nonvascular plants, high gallonage is needed for complete coverage.

### Precautionary Statements

When using any postemergence herbicide, certain precautions should be followed to minimize any problems. Treat weeds when they are young (e.g., two- to four-leaf stage). Larger weeds require repeat applications, which will result in an increased chance of phytotoxicity and increased labor costs with added wear-and-tear on equipment. Treat when weeds, and preferably the turf, are actively growing and good soil moisture is present. Treating when the weed is actively growing results in better herbicide uptake and translocation, resulting in better efficacy. If weeds are treated after they begin to flower or produce seedheads, herbicide activity will be reduced and repeat applications will be necessary. If seedheads or flowers are present, mow the weeds as low as possible, wait several days until new regrowth is evident, and then make the herbicide application. Allowing weeds to produce seedheads may add to the soil's weed-seed reserve; therefore, mowing or herbicide treatments should be in advance of seedhead development.

An adjuvant (surfactant, wetting agent, or crop oil concentrate) is generally needed by most postemergence herbicides. The label should be consulted, however, as many postemergence herbicides already contain them. If other pesticides are to be mixed in tanks with herbicides, always conduct a compatibility test unless the specific tank mix is recommended on the herbicide label. Indiscriminate tank mixing can lead to chemical compatibility problems (e.g., flakes, gels, precipitates) in the spray tank and may result in excessive turfgrass injury. As application volumes and pre- or post-treatment irrigation recommendations dramatically vary between herbicides, fungicides, and insecticides, it usually is not advisable to mix the various types of pesticides. Compatibility tests should also be conducted when mixing herbicides with liquid fertilizers.

## TURFGRASS PLANT GROWTH REGULATORS

Plant growth regulators (PGRs) or inhibitors are increasingly being used on golf courses to suppress seedheads and vegetative growth of desirable turfgrasses, enhance turfgrass quality, and as briefly discussed, manage annual bluegrass (*Poa annua*) growth and development. Depending upon the turfgrass and situation, PGRs may reduce mowing costs, prevent scalping, increase turf density, and decrease the need to mow steep slopes (Figure 14.16).

**FIGURE 14.16** Plant growth regulators reduce mowing requirements for hard-to-reach areas.

### PGR Uses on Golf Courses

1. Reduce clipping and mowing requirements.
2. Improve and maintain consistent ball roll.
3. Manage summer stress.
4. Increase carbohydrate reserves.
5. Improve turf quality in shade.
6. Improve ball lie.
7. Suppress *Poa annua* and its seedheads.

Fine-turfgrass PGRs suppress vertical top growth, but usually do not radically influence the lateral or horizontal spread of stolons. The most noticeable effect is usually a reduction in the number of clippings, and a reduction in mowing frequency. On tee boxes and fairway landing areas, turf recovery from golf club divots and other injuries occurs while vertical top growth remains suppressed. Depending upon the product, fine-turfgrass PGRs also enable superintendents to reduce mowing frequency on fairways, suppress annual bluegrass in creeping bentgrass greens, improve ball lie or playability, and suppress the growth of bermudagrass during overseeding with a cool-season turfgrass.

## PLANT GROWTH RETARDANT CLASSIFICATION

Similarly to herbicides, PGRs are placed into groups based on mode of action, or the way they inhibit turfgrass growth. Classification schemes can vary; however, three distinct groups of PGRs exist (Table 14.10).

### Site of Absorption

Plant growth regulators are absorbed, or enter the turfgrass plant, by roots, foliage (or shoots), or, with some products, both roots and foliage (Table 14.11). Root-absorbed PGRs, such as paclobutrazol and flurprimidol, require irrigation or rainfall after application to move the material into the turfgrass rootzone. In contrast, trinexapac-ethyl is rapidly absorbed by turfgrass foliage, and irrigation after application is not necessary. Compared to cell division inhibitors, there is less likelihood of leaf burn due to improper spray pattern overlaps with the gibberellin biosynthesis inhibitor PGRs. Most foliar-absorbed materials (e.g., maleic hydrazide, and herbicides) require uniform, even coverage to prevent phytotoxicity and must be absorbed by turfgrass leaves before irrigation or rainfall occurs.

### Growth Suppression

Cell division inhibitor PGRs quickly (five to seven days) suppress vegetative growth, but usually provide a shorter period of growth suppression than gibberellin biosynthesis inhibitors (three to six weeks). However, unlike the latter, cell division inhibitors are highly effective in suppressing seedhead development. The growth suppression activity of gibberellin biosynthesis inhibitors is often not immediately evident. Compared to cell division inhibitors, paclobutrazol and flurprimidol are slower (10 to 14 days) in suppressing turfgrass

**TABLE 14.10**
**Plant Growth Regulators (PGRs) for Various Turfgrass Species**

| Active Ingredient (Trade Name Example) | Turfgrass Uses | | | | | | | | | | | | Site of Uptake | | Specific Uses | | | |
|---|---|---|---|---|---|---|---|---|---|---|---|---|---|---|---|---|---|---|
| | Bahiagrass | Bermudagrass | Centipedegrass | Creeping Bentgrass | Fine Fescues | Ky. Bluegrass | Kikuyugrass | Perennial Ryegrass | *Poa annua* | St. Augustinegrass | Tall Fescue | Zoysiagrass | Root | Foliar | Overseeding Aid | Golf Greens | Seedhead Suppression | Mode of Action |
| Ethephon (Proxy) | – | – | – | Y | Y | Y | – | Y | – | – | Y | | – | Y | – | – | Y | Promotes ethylene, which reduces cell elongation |
| Flurprimidol (Cutless) | – | Y | – | Y | – | Y | – | Y | – | Y | – | Y | Y | – | – | Y | – | Type II GA inhibitor of cell elongation |
| Gibberellic acid (RyzUp) | – | Y | – | – | – | – | – | – | – | – | – | – | Y | Y | – | – | – | Promotes gibberellin synthesis |
| Indolebutyric acid + gibberellic acid | – | Y | Y | Y | Y | Y | Y | Y | – | Y | Y | Y | – | Y | – | Y | – | Enhance root growth and plant vigor |
| Maleic hydrazide (Slo Gro) | Y | Y | – | – | Y | Y | – | Y | – | – | Y | – | – | Y | Y | – | Y | Type I growth and seedhead inhibitor |
| Mefluidide (Embark 2S) | – | Y | Y | Y | Y | Y | Y | Y | Y | Y | Y | Y | – | Y | Y | – | Y | Type I growth and seedhead inhibitor |
| Paclobutrazol (Trimmit/TGR) | – | Y | – | Y | Y | Y | – | Y | – | Y | Y | – | Y | – | Y | Y | P | Type II GA inhibitor of cell elongation |
| Trinexapac-ethyl (Primo) | Y | Y | Y | Y | Y | Y | Y | Y | Y | Y | Y | Y | – | Y | Y | Y | P | Type II GA inhibitor of cell elongation |
| Prohexadione-Ca (Anuew 27.5WP) | – | Y | – | Y | – | Y | – | Y | – | – | – | – | – | Y | – | Y | P | Type II GA inhibitor of cell elongation |
| Amidochlor (Limit) | – | – | – | – | – | Y | – | Y | – | – | Y | – | Y | – | – | – | Y | Type I cell division inhibitor |

*Note:* Y = yes; P = partial.

**TABLE 14.11**
**Characteristics of Various Plant Growth Regulators Used in Turfgrass Management**

| Common Name | Trade Name | Absorption Site | Mode of Action | Comments |
|---|---|---|---|---|
| Ethephon | Proxy | Foliar | Converted to ethylene, a growth-inhibiting hormone | • Often tank-mixed with trinexapac-ethyl for *Poa annua* seedhead suppression in bentgrass golf greens. Apply in spring at *Poa* "boot stage" to suppress seedheads. Excessive use may cause stem elongation, followed by scalping. |
| Flurprimidol | Cutless | Root | Early gibberellic acid inhibitor | • Water is required for activation (for *Poa* seedhead suppression). Multiple applications in fall and spring suppress *Poa* more than bentgrass. Treated bentgrass may turn bluish-green. Treated *Poa* may turn yellow or brownish-green. Do not overseed within 4 to 8 weeks of application. Use with GA biostimulant may reduce its effects and cause turf chlorosis. |
| Glyphosate | Roundup, others | Foliar | Inhibit amino acid biosynthesis | • Seedhead/growth suppression in low-maintenance turf. |
| Imazapic | Plateau | Foliar and root | Inhibit amino acid biosynthesis | • Seedhead/growth suppression in low-maintenance turf. |
| Paclobutrazol | Trimmit, Turf Enhancer | Root | Early gibberellic acid inhibitor | • Water required for activation. Multiple applications in fall and spring for *Poa* seedhead suppression. Treated bentgrass may turn bluish-green. Treated *Poa* may turn yellow or brownish-green. Do not overseed within 4 to 8 weeks of application. Avoid using with DMI fungicide or GA biostimulant, as these may enhance/reduce its effects and cause turf chlorosis. Reapplications based on 350 GDDs. |
| Prohexadione-Ca | Anuew | Foliar | Late gibberellic acid inhibitor | • Reduces clippings, improves putting surfaces and ball roll distances. Reapplications based on 280 to 350 GDDs. |
| Trinexapac-ethyl[a] | Primo, others | Foliar | Late gibberellic acid inhibitor | • Often tank-mixed with other PGRs to improve turf quality and reduce mower clippings. Improves *Poa*'s ability to tolerate summer stress. Improves turf quality when grown in shade plus enhances turf color and density. Improves putting surfaces and ball roll distances. Reapplications in cool-season turfgrasses based on 200 GDDs. |
| **Chemicals for Growth and Color Promotion** | | | | |
| Gibberellic acid | Gibgrow, ProGib, RyzUp 4% | Foliar | Promotes gibberellin biosynthesis | • Apply 10 g active ingredient/acre (10 fluid ounces/acre) weekly or 25 g active ingredient/acre biweekly in 25 to 100 gallons/acre (234 to 935 L/ha) to promote growth and prevent discoloration (e.g., purpling) during cold stress and light frosts on ultradwarf bermudagrasses. Do not apply when night temperatures exceed 65°F (18°C). |

[a] Trinexapac-ethyl is commonly mixed with flurprimidol or paclobutrazole to take advantage of two sites of absorption, thus yielding more rapid growth suppression with less turf discoloration and bronzing. Combinations also provide more consistent foliar growth suppression and decrease clipping production.

**FIGURE 14.17** Plant growth regulators reduce mowing clippings, which can smother turf.

growth, but their duration of activity is usually longer, lasting from four to eight weeks, depending on the application rate. Selective suppression of *Poa annua* in creeping bentgrass with paclobutrazol is from bentgrass being able to metabolize the PGR more quickly than *Poa*; thus, it has less suppression of the bentgrass. Trinexapac-ethyl has been shown to reduce common and hybrid bermudagrass clipping weights 50% at seven days after application (Figure 14.17). Depending upon the application rates and schedules, trinexapac-ethyl also provides long-term (four to eight weeks) growth suppression. As temperatures increase, however, the longevity of trinexapac-ethyl in plants is reduced, requiring more frequent application to provide comparable growth suppression as in cooler environments. Another key difference is that while gibberellin biosynthesis inhibitors decrease seedhead stalk height, they have little effect on the actual formation of seedheads. However, GA-inhibiting PGRs redirect plant growth from shoots to tiller production, encouraging thicker turf stands. If GA-inhibiting PGR use is suddenly stopped, many plants experience a "rebound" effect, where a surge of growth in terms of clipping production results. Continuing to apply a PGR suppresses leaf elongation and the rebound effect.

### Application Timing

Timing of application with PGRs is critical to achieve desired results. When used for seedhead suppression, the PGR must be applied before seedhead formation and emergence. Applications made after seedhead emergence will not be effective. For bahiagrass, mow the area as seedheads initially emerge (usually late spring) to provide a uniform, even appearance to the site. For tall fescue or bluegrass, mow the area in early spring. The PGR treatment should be applied about 7 to 10 days following mowing or just prior to new seedhead appearance. Additional applications six to eight weeks later may be required if new seedheads begin to emerge.

If PGRs are being used on creeping bentgrass golf greens, applications should be made during periods of active root growth. In most areas, this would be during the midfall and spring months. Applications should not be made during stressful midsummer and midwinter months. On warm-season turfgrasses, such as bermudagrass, the appropriate PGR should be applied to actively growing turfgrasses after full spring green-up and several mowings. Applications can be repeated during the summer months if additional growth regulation is needed.

### Weed Control Considerations

An integrated weed management program must accompany any PGR use as PGRs usually do not suppress weed growth, particularly that of broadleaf weeds. In addition, after the PGR has been applied, annual and perennial weeds can become a problem, as PGR-treated turfgrass often does not compete well with weeds. On high-maintenance turfgrasses, it usually is advisable to continue preemergence herbicide use to control annual grass weeds. For postemergence control, 2,4-D, dicamba, or various two- and three-way herbicide mixtures are normally used to control broadleaf weeds. Other postemergence herbicides such as MSMA, used for annual grass weed control, or nutsedge control herbicides may also be needed in some situations. Postemergence herbicides often cause temporary phytotoxicity to turfgrasses. Postemergence herbicides can be mixed with PGRs in a tank; however, turfgrass injury is often greater than when either type of product is used alone. Therefore, on high-maintenance turfgrasses, where color and appearance may be of utmost importance, it is advisable not to mix postemergence herbicides with PGRs in tanks. Additionally, if a postemergence herbicide has injured the turfgrass, PGR application should be delayed until the turfgrass has fully recovered. The PGR label and personal experience provide the best guide to determine the suitability of mixing PGRs and postemergence herbicides.

## PLANT GROWTH PROMOTERS

An available plant growth promoter is gibberellic acid (GA), which encourages cell division and elongation. When used, GA helps initiate or maintain growth and prevent color changes (e.g., purpling) during periods of cold stress and light frosts on bermudagrass such as Tifdwarf and Tifgreen (Figure 14.18). Oftentimes, fall golf tournaments may experience an early light frost before the overseeding has become established. GA helps the turf recover from this discoloration. A plant growth promoter also

**FIGURE 14.18** Plant growth promoters help overcome bermudagrass discoloration following light frost.

may be used for anecdotal purposes if excessive rates of a gibberellic acid inhibitory PGR, like trinexapac-ethyl, prohexadione-Ca, paclobutrazol, or flurprimidol, are applied. GA-containing products should be used carefully and only where necessary because unacceptable yellowing of turf can occur.

## GROWING DEGREE DAYS

A growing degree day (GDD) model is a mathematical method to track the heat units that have accumulated and are needed for plant growth and development. For example, accumulated GDDs can be used to determine when weeds might germinate or flower or when grasses might produce a seedhead. GDDs also are used to help optimize application timing, such as pre-emergence crabgrass application, when various insects emerge, breed, and feed; when optimal reapplications of plant growth regulators should be made; or for the selection of amines or esters for spring broadleaf applications. When accumulated GDDs are tracked, superintendents have a window of the best predicted time to apply these applications or when these events might occur. For example, the PGR trinexapac-ethyl is known to last a shorter time during warmer portions of the year, as its half-life is 6.4 days at 64°F (18°C) but only 3.1 days at 86°F (30°C). To mathematically track this, it has been determined in cool-season turfgrasses that 200 GDDs is when trinexapac-ethyl reapplication intervals are maximized to sustain clipping yield suppression during summer without allowing the "rebound" effect to occur. Conversely, it has been suggested that 350 GDDs are needed to predict optimal reapplication timing for paclobutrazol. To calculate GDDs, the high and low air temperatures are averaged, subtracted from a base temperature where metabolism is minimal, and added to values from previous days. The formula for calculating GDDs is

$$\text{GDD} = \left[\frac{\text{max temperature} + \text{min temperature}}{2}\right] - \text{base temperature}$$

In the case of trinexapac-ethyl, the base temperature of 32°F (0°C) has been used. For example, if the high today was 75°F, the low was 51°F, and the base temperature of 32°F was used, the calculation would be

$$\text{GDD} = \left[\frac{75 + 51}{2}\right] - 32 \\ = 31$$

One would then simply add the GDDs calculated each day and, in the case of trinexapac-ethyl, when the sum reaches 200 GDDs, then a sequential application would be made.

Much research is still needed to verify GDDs in different environments, when using various rates, or when mixing products with different modes of action. Most GDD base temperatures are targeted for cool-season turfgrass. For example, when treating bermudagrass fairways with trinexapac-ethyl, 400 to 600 GDDs are suggested using a base temperature of 32°F (0°C). For warm-season turfgrass, a base temperature of 50°F or 60°F (10°C or 15.6°C) may be more appropriate since these grasses grow progressively less at these and lower temperatures. Obviously, additional research is needed to identify the appropriate base temperature for various turfgrasses and products being considered.

**Disclaimer.** All mentioned chemicals are for reference only and may not be available for turf use. They may be restricted by some state, province, or federal agencies; thus, be sure to check the current status of the pesticide being considered for use. Always read and follow the manufacturer's label as registered under the Federal Insecticide, Fungicide, and Rodenticide Act. Mention of a proprietary product does not constitute a guarantee or warranty of the product by the author or the publisher and does not imply approval to the exclusion of other products that also may be suitable.

# 15 Pesticide Use and Calibration

## PESTICIDE NOMENCLATURE

A pesticide is defined as any substance or mixture of substances intended for preventing, destroying, repelling, or mitigating any pest, and any substance or mixture of substances intended for use as a plant regulator, defoliant, or desiccant. Three types of names are normally associated with a pesticide. The *chemical name* describes the chemistry of the compound, which is usually technical and lengthy. The *common name* is a generic term assigned to the chemical that is often a simpler version of the chemical name. Chemical and common names must be approved by an appropriate authority. The *trade name* is used by the chemical company for marketing purposes to promote a specific product's sale. It often is the most recognizable pesticide name. A pesticide with one common name can have a number of trade names. For example, Fungo, Cleary 3336, and SysTec 1998 are trade names for the fungicide thiophanate methyl. Thiophanate methyl is the common name, while the chemical name is dimethyl 4,4′-*o*-phenylenebis[3-thioallophanate]. Due to the number of trade names and their constant change, most scientific journals and university publications refer to pesticides by their common name.

## PESTICIDE REGISTRATION

Registering a new pesticide is a time-consuming, complicated, and expensive venture. Only one successful pesticide reaches the market out of every 140,000 compounds tested (Figure 15.1). Costs associated with discovering, developing, and registering a new pesticide are nearing $300 million, with an additional $50 million to $100 million required to build a production plant. Nine to 10 years of testing normally are required before a compound reaches the market. Since a patent protects a compound for 20 years, about 10 years of exclusive marketing can be expected to protect the investment of development and continued research.

Much of the money needed for pesticide development is used to generate required data from extensive and rigorous testing. Testing for crop safety, pest control efficacy, environmental hazards, and a wide array of toxicological issues is performed (Figure 15.2). These increasing costs and regulations have significantly reduced the number of companies developing new products.

Data are submitted to the Environmental Protection Agency (EPA) for review and possible registration. The EPA has the responsibility of ensuring that the pesticide poses no undue environmental and health hazards when used as instructed. It is not the responsibility of the EPA to document efficacy. A label providing sufficient use instructions and warnings on the safe and proper handling must accompany each product.

The U.S. congressional law regulating the registration, manufacture, transportation, and use of pesticides is the Federal Insecticide, Fungicide and Rodenticide Act (FIFRA). A legally registered pesticide will have an EPA registration number on its label. The EPA registers pesticides; it does not "approve" them. The law provides the following:

1. All pesticides must be used only as directed on the label.
2. All pesticides must be classified as "general" or "restricted-use" pesticides.
3. Persons who buy or use "restricted-use" pesticides must be certified as competent pesticide applicators, or must be supervised directly by a certified applicator.
4. Persons who do not obey the law will be subject to civil and/or criminal penalties including fines and jail terms.

### PESTICIDE CLASSIFICATION

**General-use pesticides.** General-use pesticides do not cause unreasonable adverse effects to the environment and may be purchased and applied by the general public when used according to label directions (Table 15.1). However, in some states, anyone applying any pesticide for monetary compensation or to public property (including golf courses) must have a license or be directly supervised by someone with a license. *Restricted-use pesticides* are pesticides that pose some risk to the environment or human health even when used according to the label. These pesticides must be applied by certified applicators, or persons under their direct supervision, who have shown the ability to use these materials safely and properly. Persons handling restricted use pesticides also must wear approved protective clothing. A third category, mixed use, covers pesticides that are classified as "general" for certain purposes and "restricted" for others.

A state may be allowed to register additional uses for a federally registered (EPA) pesticide under certain circumstances. This is called a *special local needs* (SLN) *registration* or *24(C) registration.* These registrations often involve adding application sites, pests, or alternate techniques to those already listed on the federally registered pesticide label. The EPA registration number will have the initials "SLN" and the two-letter abbreviation for the state that issues the special registration. The applicator must have a copy of an SLN label in

**FIGURE 15.1** Registering a new pesticide is a time-consuming, complicated, and expensive venture.

**FIGURE 15.2** Much of the expense of registering a new pesticide is from the extensive testing required.

his or her possession in order to apply the pesticide for the purpose listed on it.

A *section 18* is another special temporary registration allowed under *emergency* situations where an unregistered pesticide may be used. Such conditions must be deemed where no alternate control method is available and significant economic or health problems would exist if the pest is left uncontrolled. Nonfood or nonfiber commodities such as turf and ornamentals rarely are granted section 18 exemptions.

An *experimental use permit* (EUP) is often granted by the EPA for products during the final stages of development to allow companies to expand their database under actual field conditions and commercial applications before full registration. Only a limited amount of the product can be used or sold in specific geographical locations under an EUP. These are also time limited.

A *section 2(ee)* use recommendation allow pesticides be applied under certain limited conditions that are not specified on a pesticide label. Registrants sometimes issue section 2(ee) use recommendations for their products in the form of written bulletins, which are considered "labeling" and must be consistent with FIFRA requirements and the section 3 label. Section 2(ee) labels allow a pesticide to be

- Applied at any dosage, concentration, or frequency less than that specified on the label
- Applied against any target pest not specified on the label
- Applied by any method not prohibited by the label; examples include aerial, ground and airblast
- Mixed with another pesticide or a fertilizer

Section 2(ee) use recommendations are subject to the following restrictions:

- A 2(ee) cannot add a crop/site, or expand an existing crop/site.

**TABLE 15.1**
**Table of Pesticide Toxicity Categories**

| | Categories | | | |
|---|---|---|---|---|
| | I | II | III | IV |
| Pesticide classification | Restricted use | Some restricted use, some general use | Mostly general use | General use |
| Oral $LD_{50}$ (mg/kg) | 0 to 50 | 50 to 500 | 500 to 5,000 | >5,000 |
| Inhalation $LC_{50}$ (mg/L) | 0 to 0.2 | 0.2 to 2 | 2 to 20 | >20 |
| Dermal $LD_{50}$ (mg/kg) | 0 to 200 | 200 to 2,000 | 2,000 to 20,000 | >20,000 |
| Eye effects | Corneal opacity not reversible within 7 days | Corneal opacity reversible within 7 days | No corneal opacity; irritation reversible within 7 days | No irritation |
| Skin effects | Corrosive | Severe irritation at 72 hours | Moderate irritation at 72 hours | Mild or slight irritation at 72 hours |
| Signal words/symbols | DANGER/POISON with skull and crossbones symbol | WARNING | CAUTION | CAUTION |
| Description | Very highly toxic | Highly toxic | Moderately toxic | Low toxicity |
| Oral dosage lethal to human adults | Few drops to 1 teaspoon | 1 teaspoon to 2 tablespoons | 1 ounce to 1 pint | >1 pint |

- A 2(ee) cannot increase the dosage, concentration, or frequency of application.
- A 2(ee) cannot change the timing or type of application.

Certified pesticide applicators for golf courses in most states are regulated by the state's pesticide regulatory agency—Department of Agriculture, Department of Pesticide Regulation, or a similar department. Two tests must be completed before a person can obtain a restricted-use pesticide applicator license. The first test queries a person's knowledge of general areas of safe pesticide use and handling (core test), while the second covers specific information concerning pesticide use for turfgrass and ornamental pest control (category test). These tests are administered by the state regulatory agency at various locations or online. Study guides for the tests can be obtained from the state pesticide agency or cooperative extension service. Once the license has been obtained, it must be renewed by accumulating credits, such as continuing education units (CEUs) or continuing certification credits (CCCs), over a prescribed period of years. Lapsed licenses usually require retaking tests. Programs that provide recertification credits are offered by a number of agencies, trade organizations, or private entities throughout the year.

**Restricted-use pesticides.** As previously indicated, restricted-use pesticides must be applied by a licensed, certified pesticide applicator, or by persons under the direct supervision of a licensed, certified pesticide applicator. "Direct supervision" is legally defined under federal and state law. If an unlicensed worker is applying the pesticide, the licensed supervisor may be required by law to tell or provide written or printed information to unlicensed workers:

- Exact location of area to be treated
- Safety procedures to be followed as given on the label
- Safety clothing and equipment to be used
- Common symptoms of pesticide poisoning
- Dangers of eating, drinking, smoking, or toileting while handling pesticides
- How to contact the supervising certified applicator and where to obtain emergency medical treatment

The licensed certified pesticide applicator must maintain records of the use of restricted-use pesticides. This information also should be maintained for general-use pesticides. The records must usually be kept for a state-prescribed length of time and usually include (exact requirements are state specific):

- Date and time of treatment
- Specific location of property
- Pests to be controlled
- Size of area or number of items or animals treated
- Owner or person authorizing application
- Name of applicator
- Pesticide used, including trade name and/or common name and/or active ingredient and/or EPA registration number, amount of product or active ingredient

Simplifying recordkeeping makes the process less burdensome. For example, superintendents can use a computer to print out customized forms with blanks to be filled in for each pesticide application. Another simplification of keeping is the use of a rubber stamp that can be made at local office supply stores. Stamp the book pages and fill in the blanks accordingly. While pesticide recordkeeping is required by law, good records can help you duplicate successful applications and trouble-shoot pest control failures as well as manage your pesticide purchases and inventory. An example of the minimum amount of information recorded for each pesticide application is shown on the following Pesticide Application Record form.

**Pesticide Application Record**

| | | | | |
|---|---|---|---|---|
| Company Name: | Commercial Applicator: | | | License No.: |
| License Category: | Trade Name: | | | Active Ingredient and Formulation: |
| % AI: | Manufacturer: | Lot No.: | | EPA Registration No.: |
| Restricted-entry Interval (REI): | | PPE Needed/Worn: | | |

**Application Information**

| | | | | |
|---|---|---|---|---|
| Application Start Time: | Treated Site Location: | | | Type of Area Treated: |
| Target Pest(s): | Total Treated Area: | | Application Rate (per acre or 1,000 square feet): | |
| Timing: | Amount of Pesticide Mixed: | Per | Gallon | Gallon per Acre (GPA): |
| Additives (Surfactant/Wetting Agent/ Crop Oil, etc.) | | | | Rate: |
| Additives (Surfactant/Wetting Agent/ Crop Oil, etc.) | | | | Rate: |

**Weather Conditions**

Air Temperature (F): Relative Humidity (%): Dew Present (Y/N) Wind Velocity (MPH)

Wind Direction: First Hour: Second Hour: Soil Temperature at 4 in (F):

Soil Moisture (%): Cloud Cover (%:) Rainfall/Irrigation after application (date/time/amount):

**Application Equipment**

Method of Application: Speed (mph): Motor Speed (RPM): Nozzle Height (in):

Nozzle Type and No.: Spacing: Boom Width: Spray Pressure (PSI):

Nontarget Plant, Animal, or Human Exposure: Yes No (If yes, identify and list corrective or emergency action taken):

**Other comments:**

Signature Date

## PESTICIDE FORMULATIONS AND CARRIERS

Pesticides are not sold to end-users as pure chemicals, but active ingredients are formulated or combined with appropriate solvents, diluents, or adjuvants to make formulations. The primary function of formulating a pesticide is to permit uniform application. However, formulations also extend the stability and storage life of pesticides, enhance pesticide activity, allow pesticides to be packaged in convenient containers, and allow for safer use. They also may vary in their effectiveness on pests and the tolerance of turf and ornamentals to the herbicide. Some formulations cost more than others, and the ease of application and compatibility with your application equipment can vary according to the formulation. Pesticides are available in a variety of formulations, and often the same pesticide is sold as several different formulations (Table 15.2). A *material safety data sheet* (MSDS) for each formulation of a pesticide should be obtained and cataloged.

Pesticides are applied to the target site with the use of a carrier. A carrier is a gas, liquid, or solid substance used to propel, dilute, or suspend a pesticide during its application. Water is the most commonly used liquid carrier, although fluid fertilizers also may be used. Granules and pellets consisting of clay, corn cobs, ground nut hulls, sand, or dry fertilizer serve as carriers for dry pesticide formulations.

### Sprayable Formulations

Sprayable formulations are applied with liquid carriers, usually water. The amount of liquid carrier required to uniformly

**TABLE 15.2**
**Comparisons of Commonly Used Pesticide Formulations[a]**

| Formulation (abbreviation) | Mixing/Loading Hazards | Plant Phytotoxicity | Effect on Application Equipment | Agitation Required | Visible Residues | Compatible with Other Formulations |
|---|---|---|---|---|---|---|
| Dry flowables/water-dispersible granules (DF or WDG) | Minimum | Safe | Abrasive | Yes | Yes | Good |
| Emulsifiable concentrates (EC) | Spills and splashes | Maybe | May affect rubber pump parts | Yes | No | Fair |
| Flowables (F) | Spills and splashes | Maybe | May affect rubber pump parts; also, abrasive | Yes | Yes | Fair |
| Dusts (D) | Severe inhalation | Safe | – | Yes | Yes | – |
| Granules (G) and pellets (P or Ps) | Minimum | Safe | – | No | No | – |
| Microencapsulated (M) | Spills and splashes | Safe | None | Yes | – | Fair |
| Solutions (S) | Spills and splashes | Safe | Nonabrasive | No | No | Fair |
| Soluble powders (SP) | Dust inhalation | Safe | Nonabrasive | No | Some | Fair |
| Wettable powders (WP) | Dust inhalation | Safe | Abrasive | Yes | Yes | Highly |

[a] Read individual labels for product-specific information.

cover the turfgrass will be indicated on the label. Use the label recommendation for each pesticide you apply, because amounts above the label rate are illegal and those below the label rate may be ineffective.

**Aerosols (A).** Aerosols contain one or more active ingredients and a solvent. Most aerosols contain a very low percentage of active ingredient. There are two types of aerosol formulations: the ready-to-use type and those made for use in smoke or fog generators. Insecticides are the pesticides most often used as aerosols.

**Water-soluble liquids (S or SL).** A water-soluble pesticide formulation typically consists of the pesticide, water as solvent for the pesticide, selected surfactants to improve wetting and penetration, and possibly an antifreeze. These form true solutions (completely dissolved) when mixed with water; they are nonabrasive and do not plug screens or nozzles. The resulting solution can be clear or colored and is stable, requiring no agitation once initially mixed. Most water-soluble formulations contain from 2 to 4 pounds active ingredient/gallon (240 to 480 g/L).

**Water-soluble powders (SP or WSP).** Water-soluble powders are finely divided, dry solids that look like wettable powders; however, they completely dissolve in water to form true solutions requiring no agitation once initially mixed. Wettable powders, however, do not form true solutions and must be constantly agitated. Water-soluble powders possess all the advantages of wettable powders with none of the disadvantages except inhalation hazard during mixing. SP may also refer to soluble packets, where the powder is available in water-soluble packets added directly to the spray tank.

**Emulsifiable concentrates (E or EC).** Emulsifiable concentrates are oily (or nonpolar) liquids forming emulsions (droplets of oil surrounded by water) in water (polar) instead of forming true solutions. The *emulsifying agent*, a type of surfactant, acts as a binder-coupler between the oil–water surface, reducing interfacial tension and allowing the tiny droplets of oil to remain in suspension. This allows water-insoluble pesticides to be uniformly dispersed in water, even though each maintains its original identity, similarly to oil and vinegar salad dressing. After EC compounds are added to water, the resulting emulsions are milky colored and require mild agitation to keep the pesticide uniformly suspended in the spray tank. Since the emulsifier is a surfactant, additional surfactant is usually not required for foliar applications. Each EC gallon usually contains 25% to 75% (2 to 8 pounds, 0.24 to 0.96 kg/L) of active ingredient. Emulsions present few problems in mixing, pumping, or spraying. They are not abrasive, do not plug screens or nozzles, and leave little visible residue. **Gels** (GL) are thickened emulsifiable concentrates available in water-soluble packets added directly to the spray tank.

**Wettable powders (W or WP).** Wettable powders are finely ground solids resembling dusts, and consist of a dry diluent (usually a hydrophilic clay such as bentonite or attapulgite) plus the pesticide and perhaps adjuvants. Usually, pesticides make up 50% to 80% (by weight) of a wettable powder formulation. The various adjuvants in the formulation prevent lumpiness or flocculation of the finely ground materials and improve mixing in the spray tank. Wettable powders do not dissolve in water but rather form unstable suspensions giving a cloudy appearance. Vigorous agitation is required to prevent settling of the suspended particles. Inhalation hazards also exist when pouring and mixing the powder. Wettable powder formulations cause rapid nozzle wear, often clog nozzles and screens, and may leave visible residues.

**Dispersible granules or dry flowable granules (DG or DF or WDG).** These finely ground solids are formulated as water-dispersible granules and form a suspension in water. These are similar to wettable powder formulations, except they are granule-sized particles and thus less dusty. Once in solution, the granules break apart into a fine powder. Agitation is required to prevent settling of the suspended particles, and nozzle wear is similar to flowables. Granules are made up of finely ground solids combined with suspending and dispersing agents. Their chief advantage over wettable powders and flowables is the ease of measurement and handling (pourable) and that they are less susceptible to inhalation and wind blowing. **Note:** These formulations are always applied with a liquid (water or fluid fertilizer) carrier.

**Flowables or aqueous suspensions (F or AS).** Also designated as *liquids* (L) or *water-dispersible liquids* (WDL), these highly viscous (not easily poured) liquids contain finely ground solids suspended in a liquid system (usually water). The particles are smaller than those of the wettable powders. Wetting agents are usually added to keep the solid surfaces wetted as they tend to be hydrophobic. These formulations form a suspension in water and require agitation to remain distributed. These also settle out when in storage; therefore, they require vigorous shaking before use but have fewer of the mixing and dust exposure problems typical of wettable powders. They may leave a visible residue. Nozzle wear is intermediate between WP and EC, but more similar to WP.

**Water-soluble bags/packs.** Some pesticides are being sold with a premeasured amount of the pesticide formulation (usually a wettable powder or dry flowable) packaged inside a water-soluble bag. This bag will dissolve when placed in a tank of water, thereby releasing the pesticide. This eliminates the need to directly handle the pesticide, reducing worker exposure to it. These bags are not to be broken before being placed in the tank. Exact sprayer calibration is necessary since parts or pieces of a bag cannot be used.

**Microencapsulated (M or ME).** These particles of pesticides (liquid or dry) are surrounded by a plastic-coating barrier layer. The formulated product is mixed with water and applied as a spray. Once applied, the capsule slowly releases the pesticide. Advantages of microencapsulated pesticides include increased safety to the applicator, ease of mixing and handling, and controlled release, thereby extending the period of pest control. Since microencapsulation substantially reduces the amount of liquid carrier required, this formulation tends to be more concentrated and lower in cost per unit of active ingredient. Constant agitation is necessary to maintain the solution.

### Dry Formulations

Dry pesticide formulations are not applied with liquid carriers but are applied as purchased. Normally the pesticide is formulated in relatively low concentrations on the dry carrier to aid in uniform distribution.

**Granules (G).** Small granular particles (<10 mm$^3$) are applied in the dry state. These consist of the pesticide plus a dry carrier such as clay, vermiculite, walnut shells, sand, or corn cobs. The active material either coats the outside of the granules or is absorbed into them. Pesticide concentrations typically range from 2% to 20%. Advantages of granules include: (1) ready to use—water is not needed for application; (2) generally quicker to apply than liquid applications; (3) can be combined with fertilizer, thereby reducing costs; and (4) public perception is usually more favorable using granules compared to liquid applications. Granular formulations require slightly more rainfall for activation than sprayable formulations and tend to be more expensive. Uniform application can also be a problem, and application equipment is harder to calibrate. When combined with fertilizer, proper timing of pest control and appropriate plant fertilization timing needs should coincide.

**Pellets (P).** Pellets are similar to granules except the particles are usually larger (≥100 mm$^3$). Pellets are frequently used for spot applications, and are applied either "by hand" from shaker cans or with hand spreaders. These usually contain low active ingredient concentrations of approximately 5% to 20%.

**Dusts (D).** Dusts are popular homeowner formulations of insecticides and fungicides requiring simple equipment and are effective in hard-to-reach areas. Most formulations are ready to use and contain between 0.5% and 10% active ingredient plus a fine, dry, inert carrier such as talc, chalk, clay, or ash. Due to drift hazards, few herbicides are currently formulated as dusts.

### Adjuvants

An **adjuvant** is a spray additive enhancing the performance, safety, or handling characteristics of a pesticide. "Adjuvant" is a broad term and includes surfactants, crop oils, crop oil concentrates, antifoaming agents, drift control agents, pH modifiers, and compatibility agents. These help modify the surface properties of liquids by enhancing and facilitating emulsifying, dispersing, wetting, spreading, sticking, and penetrating of liquids into plants and soil. Surfactants, crop oils, and crop oil concentrates are added according to label directions since indiscriminate use may cause severe turfgrass injury or decreased pesticide performance. Some pesticides, such as postemergence herbicides and a few fungicides, have surfactants included in their formulation; therefore, additional surfactant is unnecessary. Always read the pesticide label before adding any adjuvants. Look for recommendations as to the type of adjuvant to add. Use only the recommended rates as too much of some adjuvants can cause an unsprayable tank-mix or excessive turf phytotoxicity. Surfactants are most often used in liquid (soluble, emulsifiable) and dry (wettable powders, others) formulations applied in aqueous sprays.

**Surfactants.** Substances without affinity for each other (such as water and leaf wax) tend to repel. To "bind" the two surfaces, surfactants with a lipophilic (oil-loving) portion and a hydrophilic (water-loving) portion on the same molecule are used. The term surfactant is an acronym for surface-active agents to indicate the changes they produce at surfaces. At low concentrations, surfactants reduce surface tension between spray droplets and the waxy leaf surface, allowing the spray droplets to spread out and contact a greater portion of the leaf. This aids in penetration and helps prevent droplets from rolling off the leaf. At higher concentrations, surfactants help dissolve the wax in the leaf cuticles, allowing easier penetration of the leaf by the pesticide. However, this also accounts for undesirable phytotoxicity to the turfgrass.

Three major types of surfactants include *emulsifiers*, *wetting agents*, and *stickers*. Emulsifiers stabilize the dispersal of oil-soluble pesticides in water so the pesticide will not settle out. These allow petroleum-based formulations such as emulsifiable concentrates to mix with water. These usually are added by the chemical company during the pesticide formulation process. Invert emulsifiers allow water-based pesticides to mix with petroleum-based carriers. Stickers (or adhesives) cause the spray droplet to adhere to the leaf surface and reduce spray runoff during application and wash-off by rain or irrigation. Stickers often are combined with wetting agents (spreader-stickers) to increase adhesion and spray droplet coverage.

**Dispersing agents** also are surfactants that enhance the dispersal of a powder in a solid–liquid suspension. As with wetting agents, the lipophilic (oil-loving) end of the dispersing agent molecule partitions into the suspended particle, while the hydrophilic (water-loving) end partitions into the surrounding water medium, reducing the water tension and enabling the particle to repel other particles and remain in suspension. As a result, they help prevent clumping of the pesticide.

**Wetting agents** help the spray droplet to spread over the leaf surface by reducing the interfacial tension between the leaf surface and spray droplets. Wetting agents also allow wettable powders to mix with water. The three types of wetting agents (*anionic, cationic,* and *nonionic*) are classified based on how they ionize or separate into charged particles in water. Nonionic surfactants do not ionize; thus, they remain uncharged. They are the most commonly used type of surfactant and are compatible with most pesticides. They are unaffected by water containing high levels of calcium, magnesium, or ferric ions. They also can be used in strong acid solutions. Anionic surfactants ionize with water to form a negative charge, while cationic surfactants ionize with water to form a positive charge. These are only occasionally used. A pesticide mixed with an anionic surfactant will stick to the leaf tissue but will not be absorbed by the plant. These should be used with pesticides that remain on the plant surface (contact pesticides).

A pesticide mixed with a nonionic surfactant will help a pesticide penetrate plant cuticles. These are best used with systemic pesticides that need to be absorbed by the plant to be effective. Cationic surfactants are extremely phytotoxic. Do not use them unless it is specifically stated on a pesticide label.

**Crop oils. Crop oils** and **crop oil concentrates** are non-phytotoxic light oils containing varying percentages of surfactants and primarily emulsifiers. These are phytobland petroleum or vegetable oils increasing pesticide absorption through leaf cuticles (or waxy layer). Crop oils contain 1% to 10% surfactant and commonly are used at concentrations of 1 gallon/acre (9.4 L/ha). Crop oil concentrates contain 17% to 20% surfactant and are generally used at concentrations of 1 quart/acre (2.3 L/ha). Crop oil concentrates have replaced crop oils since reduced amounts of the adjuvant are required.

**Methylated seed oil.** An MSO is a methylated seed oil, such as soybean oil, produced by reacting methanol with whole soybean oil in a process called "transesterification" that changes soybean triglyceride into soybean methyl ester. MSOs are especially effective in postemergence weed control as the MSO has a solvency action on the protective wax cuticles formed by weeds. Control can often be achieved at a rate of 1 to 2 pints/acre (1.12 to 2.3 L/ha), compared to a quart or more for other types of adjuvants per acre (2.3 L/ha).

**Miscellaneous adjuvants.** The use of *antifoaming* agents (or defoamers) minimizes air entrapment during agitation and may be necessary if excessive foaming occurs in the spray tank. *Drift control agents* (or thickeners) reduce spray droplet drift by reducing the percentage of very fine spray particles in the spray mist. *Stickers* reduce losses of spray droplets from leaf surfaces by increasing the viscosity of spray droplets and reducing the interfacial tension between the spray droplet and the leaf surface. The droplets then resist washing off and, at the same time, spread out on the leaf surface. *Compatibility agents* are added to fluid fertilizer and pesticide mixtures to prevent these individual components from separating or clumping. Follow label directions closely for mixing compatibility agents. However, before adding any such mixture to a spray tank, *test* the mixture in a small jar to ensure there will be no clumping or separation (see the procedure in the following "Pesticide Compatibility" section).

**Penetrants** allow pesticides to enter the outer surface of plants while spreaders allow a uniform coating layer over the treated surface.

Some pesticides will be inactivated if the pH of the spray water is too high or too low. *Modifiers* are compounds, either buffers or acidifiers, available to adjust the pH of the water to be used as the pesticide carrier. Buffers change the pH to a desired level, and then keep it relatively constant. Acidifiers neutralize alkaline solutions (lower the pH) but will not maintain the pH at this level as well as buffers do. Lower solution pH occurs with an acidifier after the pesticide is added to the tank. Ask your sales representative to provide you with the manufacturer's recommendation for the pH of the carrier.

## Pesticide Compatibility

Two or more pesticides, or one that can be mixed with fertilizer, are compatible if no adverse effects occur as a result of the mixture. Possible effects of mixing incompatible chemicals include:

- Effectiveness of one or both compounds may be reduced.
- A precipitate may form in the tank, clogging screens and nozzles of application equipment (Figure 15.3).
- Plant phytotoxicity, stunting, or reduced seed germination and production may occur.
- Excessive residues may occur.
- Excessive runoff may be present.

**Compatibility test.** A compatibility test (the jar test) should be made well before mixing chemicals for application. Some pesticide labels give the procedure for this, but a general procedure is

1. Measure 1 pint (0.47 L) of carrier water in a clear quart (0.95 L) jar that is not used for any other purpose.
2. Add ingredients in the proper mixing order (Table 15.3), stirring each time a new chemical is added. Check for the formation of foam, flakes, sludge, gels, oil films, scum, or precipitates after adding each ingredient. It is sometimes necessary to premix (slurry) some chemicals (some wettable powder (WP), dry flowable (DF), water-dispersing

**FIGURE 15.3** A precipitate from incompatibility of two pesticides. (Courtesy John Boyd.)

**TABLE 15.3**
**Proper Mixing Procedures for Mixing Chemicals in Tank and Amount of Each Chemical Needed to Perform a Jar Test**

| Order of Addition | Chemical | Amount for Jar Test (per 40 to 50 gpa of spray volume) |
|---|---|---|
| 1 | Water conditioning agents and activators | 1 teaspoon for each pint |
| 2 | Wettable powders and dry flowables | 1 tablespoon for each pound |
| 3 | Water soluble concentrates or solutions | 1 teaspoon for each pint |
| 4 | Soluble powders | 1 teaspoon for each pound |
| 5 | Emulsifiable concentrates | 1 teaspoon for each pint |
| 6 | Surfactants and oils | 1 teaspoon for each pint |
| 7 | Fertilizers | proportional |

*Notes:* Agitate (vigorously mix or shake) mixture following each step. One teaspoon = ~5 mL; one tablespoon = ~15 mL.

granule (WDG), or liquid flowable formulations as indicated on the labels) before adding to the jar. *Do not mix the chemicals together without dilution before adding to the jar or spray tank.*

3. Let the mixture sit for 15 minutes. Check for foam, scum and precipitates, and other unexpected results or appearance (for example, wettable powders will not dissolve) (Figure 15.4). Feel the side of the jar to gauge temperature. If it is warm, let the jar sit and recheck in another 15 minutes.

**Mixing compatible chemicals.** Mixing some pesticides requires premixing, making a slurry, in a smaller, separate container or tank.

**FIGURE 15.4** Insoluble wettable powder formulation.

1. Always add a wettable powder first. Make a slurry with it in a separate container by adding a small amount of water until it forms a gravy-like consistency. Slowly add this slurry to the tank with the spray tank agitator running.
2. Dry flowable or water-dispersible granules are added second. Flowables should be premixed (one part flowable to one part water) and poured slowly into the tank (Table 15.3).
3. Liquid flowables should be added third, after dry flowables. Liquids should also be premixed (one part liquid chemical to two parts water or liquid fertilizer) before blending in the tank. Many labels provide the proper mixing sequence.
4. Emulsifiable concentrates and water-soluble liquids should be added last.

Proper order for mixing various pesticide formulations in tanks:

$$\text{WP} \rightarrow \text{DF} \rightarrow \text{F} \rightarrow \text{EC} \rightarrow \text{S}$$

### pH Problems

The measure of pH (how alkaline [or basic] or acidic a solution is) can greatly influence how pesticide and other products perform. Most waters are slightly basic because of the presence of dissolved carbonate and bicarbonate salts. The water pH greatly affects the breakdown (or hydrolysis) of pesticides. In general, high pH water conditions (basic) cause a shorter pesticide chemical half-life (the time it takes for half of the pesticide to break down). Insecticides are especially affected by spray-tank water pH. For example, acephate (Orthene) has a half-life of 65 days at a pH of 3 (acidic), and a half-life of 16 days at pH 9 (basic). Carbaryl (Sevin) has a half-life of 100 to 150 days at pH 6, but only 24 hours at pH 9.

The pH of spray-tank water should be adjusted with buffers or acidifiers to within a certain range for adequate usage (Table 15.4).

To test and adjust the pH of water to be used for mixing pesticides, do the following:

1. Test the water by using a clean container to obtain a 1-pint (0.47 L) sample of water to be used. Check the pH using a pH meter, test kit, or test paper and determine if the pH needs to be adjusted.
2. Adjust the pH by using a standard eyedropper to add three drops of buffer or acidifier to the measured pint of water. Stir well with a clean glass rod or clean, nonporous utensil. Check pH as before and, if further adjustments are needed, repeat the previous steps until proper pH is obtained, recording the number of times three drops of buffer or acidifier were added.
3. Correct the pH in the spray tank by filling the tank with water and your pesticide. Add 2 ounces (60 mL) of buffer or acidifier for each time three drops were

**TABLE 15.4**
**Spray-Tank pH Levels and Precautionary Statements**

| pH Range | Comments |
|---|---|
| 3.5 to 6.0 | Satisfactory for spraying and short-term storage of most spray mixtures in the spray tank |
| 6.1 to 7.0 | Adequate for immediate spraying of most pesticides; *do not allow mixture to sit over ~2 hours to prevent loss of effectiveness* |
| ≥7.1 | Should add buffer or acidifier; do not allow to sit in direct sunlight, and use immediately following mixing |

used in the previous jar test, in every 100 gallons (380 L) of water in the spray tank. Recheck the pH of the tank mix and adjust, if necessary.

## HANDLING PESTICIDES SAFELY

Pesticide "handling" includes virtually all transport, storage, mixing, loading, applying, and disposal activities.

### Pesticide Storage

Proper storage of pesticides not only helps protect against accidental spills and leaks, but can also influence the shelf life and efficacy of relied-upon products. Factors such as air temperature, humidity, sunlight, and ventilation should be carefully considered. The following are some storage suggestions; however, the best source of specific product storage and spill control information is the product MSDS.

### Storage Facility and Site (Figure 15.5)

- Keep the amount of storage space to a minimum to discourage storing unneeded pesticides, but large enough to handle what might reasonably be stored. This should include not only newly purchased chemicals, but also opened containers; empty, clean containers; and waste pesticides held for proper disposal.
- Locate the storage site in a safe location: a place not flooded by rivers, ditches, runoff, or tides.
- To prevent moisture accumulation, elevate the storage facility above the immediate ground level by at least 12 inches (0.3 m). Free moisture will rust metal containers; disintegrate paper or cardboard packaging; make labels unreadable; cause labels to detach; cause dry formulation to clump, cake, break down, or dissolve; and possibly cause pesticides to spread or move from the storage area.
- Use a portable storage building for best storage. These can be repositioned easily in case of flooding or change in the area use patterns.
- Use tie-downs on portable storage buildings to prevent tipping, rolling, or movement due to water or wind.
- Select a site with as little runoff as possible to prevent contamination of surface waters in case of a leak or spill.
- Locate the storage facility at least 100 feet (30 m) from surface waters and wells, animal feeding stations or shelters, and food or feed storage and dwellings.
- Determine the direction of prevailing winds and consider what is downwind from your storage site. This may be important in the event of spills or fires.

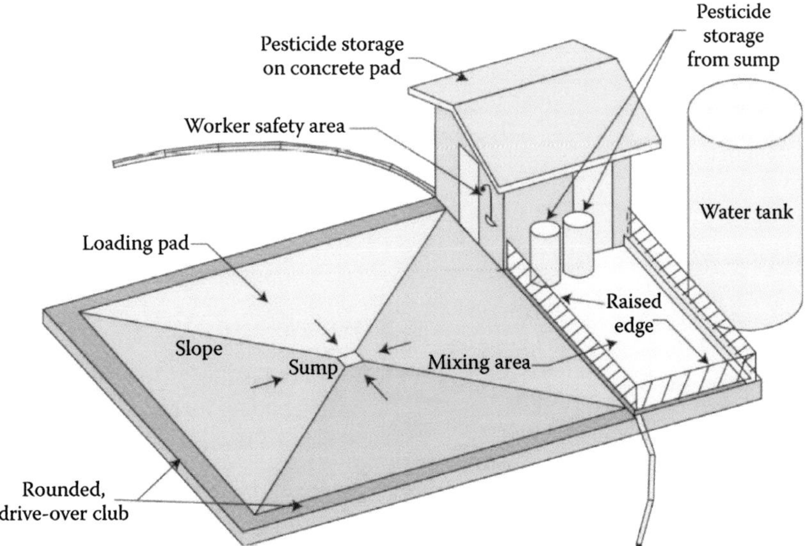


**FIGURE 15.5** Diagram of a possible pesticide storage facility and use site.

- Locate your storage site so it is easily accessible by vehicles for pesticide delivery and pickup, as well as emergency vehicles.
- Determine and comply with any applicable local zoning and building codes.
- Consider using barriers, such as posts, to prevent damage to the storage unit by vehicles delivering pesticides or picking them up for applications.

## Physical Construction of the Storage Unit

- Use a separate storage unit made of nonflammable materials to reduce fire hazard.
- The best storage is a detached structure positioned far enough away from other structures that could threaten the storage if they should burn.
- Use sealed floors—metal; sealed concrete; epoxy-coated metal, wood, or concrete; no-wax sheet flooring; or other easily cleaned, nonabsorbent material. Dirt or unsealed wood flooring should not be used.
- Use nonabsorbing materials throughout. The best shelving is metal with a lip. Consider leak-proof plastic trays on shelves.
- Have a built-in sump, or drain to a sump. Locate any external sump adjacent to the building instead of under it for easy access if you need to remove spilled materials. Protect sumps from filling with water from rain or runoff.
- Have a continuous internal lip or curb 2 to 4 inches (5 to 10 cm) high to prevent spills from overflowing and going outside the building.
- If possible, provide electrical power to the storage unit. This allows interior lighting, as well as an exhaust fan and heater. It also allows exterior security lighting and alarms. Explosion-proof wiring and switches are best. The light/fan switch should be on the outside of the storage unit and weatherproof.
- Have good lighting; explosion-proof lighting is best. Good lighting allows:
  - Reading labels easily
  - Exposure of leaky and damaged containers
  - Proper cleanup of spills
  - Recording of inventory

## Storage Environment (Figure 15.6)

- Keep the storage unit dry. Keep doors and windows to the outside closed and locked, unless windows are needed for ventilation.
- Keep the unit well ventilated by passive ventilation, mechanical ventilation, or both.
- Locate a louvered vent or exhaust fan high at one end (back) of the unit and an air makeup louvered vent low at the other end (e.g., in the lower part of the door). This allows vapors to flow away from anyone entering or inside the storage unit.

**FIGURE 15.6** Example of a pesticide storage facility.

- One recommendation for mechanical ventilation is an exhaust fan capable of exchanging air in the unit at least once every six minutes; increase to every three minutes when pesticide handlers are in the unit. For best operation, wire the fan to the light switch so that the fan is always on or increases speed whenever pesticide handlers are in the unit. If possible, also wire the fan to a thermostat set between 75°F and 85°F (24°C and 29°C).
- Keep pesticides from freezing and extremely high temperatures. Most pesticides should be stored above 40°F and below 90°F (4.4°C and 32°C).
- Freezing can cause containers to burst or develop intermittent leaks. Freezing may cause formulations to separate. Many pesticide labels say, "Store in a cool, dry place." Some must be stored at temperatures below 90°F (32°C).
- High temperatures can cause plastic to melt or become brittle, or glass to burst. Pressure caused by high temperatures may cause intermittent leaks, swelling, or a spill when the product is opened. High temperatures can break down some chemicals or cause some chemicals to volatilize. Heat can also cause explosion or fire.
- Insulate the unit to prevent freezing or overheating.
- Install a heater—ideally, an explosion-proof one.
- Keep containers out of direct sunlight. Do not put containers, especially glass or aerosol containers, in windows, even temporarily.

## Storage Security

- Store pesticides in a separate location, preferably in a separate, locked storage building dedicated solely for this use.
- Always lock pesticide storage cabinets, closets, rooms, and buildings.
- Besides locking a storage building, it is a good idea to contain the building inside a fence with the ability to lock a gate.

- Limit access to pesticide storage—allow access to only essential persons. Take necessary steps to keep out any unauthorized persons—children, workers who do not use pesticides, visitors, and so forth.
- Consider installing security lighting and possibly an alarm system.
- If pesticide storage is located within a larger structure, have storage access through a separate, outside door.
- Post signs on the door, building, or fence that indicate pesticide storage and tell people to stay out: "Danger—Pesticides—Unauthorized Persons KEEP OUT." Consider writing this information in a second, additional language, such as Spanish.
- Signs should have at least two emergency phone numbers. One person should not be the sole contact in an accident, as that person could be the victim of an accidental exposure. At least one phone number should be for emergency response (fire, rescue, etc.). Poison Control Center phone numbers are a good additional choice.
- Indicate the location of the nearest accessible telephone.
- Post NO SMOKING signs and do not allow smoking in or near your storage area or facility.

### Reusing Stored Pesticides

Stored pesticides should be examined before they are used to indicate if they have deteriorated (Table 15.5). Containers should be checked for leaks, cracks, tears, erosion, seal failures, or the development of gases. Some formulations, such as those containing emulsifiers, may stratify at low temperatures and should be stirred or shaken and allowed to sit for 30 minutes. These should then be rechecked to see if the formulation stays in solution. If not, they should be properly disposed and not used.

Resuspending crystals that have separated out is accomplished by heating the mixture to 100°F to 140°F (38°C to 60°C) using electrically heated warming blankets, hot water baths, or steam, but never by using an open flame. This should be performed by experienced pesticide handlers and not by the general public. If the pesticide appears normal but there are doubts about its effectiveness, test a small amount according to label instructions. Product-specific information on reuse may be found on the product MSDS and the label. If a pesticide exhibits any of the signs of deterioration listed in Table 15.5 and cannot be restored to usable condition, it should be properly disposed of.

### Pesticide Containment Area

A pesticide containment area is designed to keep storing, mixing, and loading isolated from other operations. It usually consists of a concrete pad on which a storage building is built and where a drainage system can collect spills and washwater into a holding tank. An example is shown in Figure 15.5. The key concept is that spills and equipment washwater can be easily contained, temporarily stored, and then added back to the spray tank as mix water to be sprayed according to the label.

The concrete pads must be sufficiently reinforced and thick enough to accommodate the gross weight of any item or combination of items to be washed without any structural damage. The area also should exclude runoff in the case of excessive rainfall. State and local laws often govern the proper design and construction of pesticide storage and containment areas and should be consulted when construction is planned on such a facility. Pesticide containment and the wash pad site should be concrete. This prevents the accumulation of rainwater and overflow.

### Worker Protection Standard

The Worker Protection Standard (WPS) applies to agricultural workers performing tasks related to the cultivation and harvesting of agricultural plants on farms, in nurseries, in greenhouses, and in forests. The law also applies to employees who handle (mix, load, apply, repair application equipment, etc.) pesticides in these work environments. The WPS mandates specific restricted entry intervals, personal protective equipment (PPE), emergency assistance, employee pesticide safety education, and worker access to displayed information. Pesticides requiring compliance with the WPS can be identified by WPS reference in the "Agricultural Use" section of a product's "Directions for Use."

The WPS specifically excludes many turf uses of pesticides (e.g., golf courses and recreational areas) but not others (e.g., turf [sod] farms). In either case, follow all precautions

**TABLE 15.5**
**General Signs of Deterioration in Pesticide Formulations**

| Formulation | General Signs of Deterioration |
|---|---|
| Dusts and granulars | Excessive lumping |
| Emulsifiable | Milky coloration does not occur with the addition of concentrate to water; sludge or separation of components is evident |
| Oil sprays | Milky coloration does not occur by addition of water |
| Solutions | Crystallization, precipitation |
| Wettable powders | Lumping occurs and powder will not suspend in water |

**FIGURE 15.7** Personal protective clothing and equipment worn when handling and using regulated products.

possible to protect employees from exposure and keep up with the latest changes in laws and regulations.

### Personal Protection

To determine the specific personal protective clothing and equipment required for a particular product, you must refer to the instructions on the product label (Figure 15.7). These instructions carry the weight of law. The toxicity level of the chemical determines the correct body protection. The pesticide label should list a toxicity class, or so-called signal word, with class I ("danger") being the most toxic, followed by class II ("warning") for moderately toxic, and class III and IV ("caution") for the least toxic chemicals. The material safety data sheet (MSDS) provides additional information in helping to determine personal protection.

### Mixing and Loading

Opening pesticide containers, connecting application equipment, or transferring pesticides to another container for application all entail the possibility of exposure. Having an appropriate pesticide mixing center provides a place where the operator can perform all mixing and loading duties without spills escaping into the environment (Figure 15.5).

### Disposal of Pesticide Wastes

Proper pesticide waste disposal is an important part of responsible pesticide use by the applicator. Improper disposal can lead to contamination of soil and ground and surface water, causing serious liability problems for the pesticide user. Federal and state laws, including the Federal Insecticide Fungicide and Rodenticide Act (FIFRA), regulate the disposal of pesticide waste.

Pesticide wastes include (1) empty containers, (2) excess mixture, (3) rinse water from containers and application equipment, and (4) material generated from the cleanup of spills and leaks. These types of pesticide wastes are classified as either hazardous waste or solid waste. Pesticides classified as hazardous wastes are regulated by the Federal Resource Conservation and Recovery Act (RCRA). This waste must be disposed of properly, usually by a licensed hazardous waste contractor. Properly rinsed empty containers are solid waste.

Pesticides not classified as hazardous wastes may be disposed of as regular solid waste or trash. This is regulated under state law and must be disposed of in a careful manner according to label instructions. Ask your local pesticide regulatory agency about the specific laws and regulations affecting your area.

### Emergency Planning and Community Right-to-Know

In 1986, in a response to the 1984 toxic gas disaster in Bhopal, India, Title III of the Superfund Amendment and Reauthorization Act (SARA) mandated a federal program subtitled the Emergency Planning and Community Right-to-Know Act (EPCRA). The intent of EPCRA is to assure information regarding hazardous chemicals is made available to emergency response agencies and the general public. The responsibilities of these efforts are assigned to the U.S.EPA and individual states. Four separate categories are covered by this legislation:

1. Emergency planning involving extremely hazardous substances
2. Reporting spills and leaks of extremely hazardous substances
3. Reporting haza[illegible]ances in the workplace
4. Reporting toxic chemical releases

Contact your local emergency planning committee or law enforcement agencies for more information on local and state requirements involving compliance.

### Equipment Cleanup

After each day and/or pesticide use, application equipment should be flushed with water inside and out to prevent chemical accumulation. The cleaning area should be chosen with care to prevent cleaning water from contaminating water supplies and streams, or injuring plants. The cleaning water should be stored and reused to dilute the next batch of spray solution containing the same chemical(s). When changing chemicals or finishing spraying for an extended time, clean the sprayer thoroughly inside and out by following these steps:

1. Completely hose down the inside of the tank, filling it about 10% full of water, and then flush it out through the nozzles by operating the sprayer for at least 30 seconds. Repeat this step twice more.
2. Remove nozzle tips and screens and check for wear. Clean them in kerosene or detergent solution using a soft brush. Do not use a knife, wire, or other hard material to clean nozzle tips because the finely machined tip surfaces can be easily damaged, causing spray pattern distortion and an increased application rate.
3. Fill the tank half full of water and add about a pound of detergent for every 50 gallons (2.4 kg/L) of water.

4. Operate the pump to circulate the detergent solution through the sprayer for about 30 minutes, then flush it out through the bottom.
5. If 2,4-D or an organophosphate insecticide has been used, follow these steps, in addition to the final flush using water:

   - Replace the screens and nozzle tips.
   - Fill the tank half full of water and add 1 pint of household ammonia for every 25 gallons water (5 mL/L).
   - Operate the pump to circulate the ammonia solution through to the sprayer for about five minutes, and discharge a small amount through the boom and nozzles.
   - Keep the remaining solution in the sprayer overnight.
   - The next day, agitate the system and flush out all ammonia solution through the nozzles by operating the sprayer.
   - Finally, fill the tank about half full of clean water while hosing down both the inside and outside, then flush out through the boom. When finished with the sprayer for an extended period, remove and store the nozzle tips, strainers, and screens in light oil. Store the sprayer in a clean, dry shed. If the pump cannot be completely drained, store it where it will not freeze, or run antifreeze through the pump.

## PESTICIDE LABELS

### Material Safety Data Sheets

Each pesticide product has a material safety data sheet (MSDS) written by the manufacturer (Figure 15.8). These sheets provide information on

1. Chemical product/company identification
2. Composition/information on ingredients
3. Hazardous identification
4. First aid measures
5. Firefighting measures
6. Accidental release measures
7. Handling and storage
8. Exposure controls/personal protection
9. Physical and chemical properties
10. Stability and reactivity
11. Toxicological information
12. Ecological information
13. Disposal considerations
14. Transportation information
15. Regulatory information
16. Other information

MSDS sheets for each pesticide formulation must be kept readily available for workers to read prior to handling the pesticide and to refer to in emergency situations.

### Technical Information Bulletins

It has become quite common for some golf courses to submit pesticide use plans to water management districts or other state agencies. In most cases, the information required may either be found on the label or on the MSDS for the pesticide. However, some information may only be found in technical bulletins. These also are written by the company manufacturing the product, but normally are ***not*** provided to the pesticide user. If a specific piece of information cannot be located on the label or MSDS, contact the company representative (salesperson, sales manager, R&D representative, etc.) and request a technical bulletin.

### Labeling Is the Law

It is extremely important to remember that the pesticide label is the *law*. Pesticides may not be used in a manner not permitted by the labeling. Pesticide uses inconsistent with the label include:

1. Applying pesticides to plants, animals, or sites not specified in the directions for use; if the label does not state it is for use on turfgrass, then it is not legal to use on turfgrass
2. Using higher dosages, higher concentrations, or more frequent applications than specified on the label
3. Not following the directions for use, safety, diluting, storage, and disposal; this also includes any restrictions on course reentry, not only for employees but for golfers as well

The law does allow you to

1. Apply pesticides at dosages, concentrations, and frequencies less than those listed on the label if you obtain expert opinion or have data to justify the lower rate.
2. Apply a pesticide against any target pest not listed on the label *if* the application is to a crop/plant, animal, or site listed on the label. In other words, if a new weed suddenly appears, it is legal to use an herbicide for control as long as turfgrass is listed on the label and you know the material will control the weed.
3. Mix a pesticide with a fertilizer if the mixture is not prohibited by the label.
4. Mix two or more pesticides together if all the dosages are at or below the labeled rate(s), and the mixture is not prohibited by any of the labels.

# MATERIAL SAFETY DATA SHEET

**Date Prepared:** May 3, 2006 **Infotrac**......................................(800) 535-5053
**Prepared By:** Stewart Holyk **U.C.I. Poison Control**................(714) 634-5988

## GENERAL INFORMATION

**Product**
GOLDEN BIO EXTRA

**Manufacturer**
Golden Bio
1200 N. Jefferson Street
Anaheim, CA 92807
Phone (714) 414-1021
Fax (714) 630-4807

NFPA RATING
0 = Non-Hazardous
1 = Slight Hazard
2 = Hazardous
3 = Extreme Hazard
4 = Deadly

FIRE 1
HEALTH 2
REACTIVITY 0
SPECIAL

**Product Number**
Special

**Chemical Family**
Bacterial Adjunct

**DOT Classification**
Non Regulated

## SECTION 2 - HAZARDOUS INGREDIENTS

| Chemical Name | CAS Number | % By Vol. | Exposure Limits in Air ACGIH (TLV) OSHA (PEL) |
|---|---|---|---|

All ingredients contained within Golden Bio Extra are generally considered Non Hazardous, Non-Toxic and Non-Rerstricted

## SECTION 3 - HEALTH HAZARD DATA

### Effects of Overexposure

**Skin:** When the concentrated form of this product comes in contact with the skin it may cause slight irritation and / or irritation to the skin. There is also the possibility of defatting of the skin due to the removal of the skins natural oils.

**Eyes:** When this product comes in contact with the eyes or eyelids it may cause severe irritation and / or damage to the eyes and immediate areas.

**Ingestion:** When concentrated solutions of this product are swallowed severe discomfort and / or damage of the stomach and intestinal tract may occur.

**Inhalation:** Fumes from this product are generally considered non - toxic, and pose no safety or health threat.

### First Aid Procedures

**Skin:** If skin comes in contact with product: Flush exposed area with cool water for 15 minutes. Consult a physician if irritation persists.

**Eyes:** If this product comes in contact with eyes; flush eyes with luke warm water for 30 minutes and consult a physician if irritation persists.

**Ingestion:** **DO NOT INDUCE VOMITING.** Have patient drink large amounts of milk or plain water. Consult a physician immediately.

**Inhalation:** Remove to clean atmosphere and consult physician immediately

### Carcinogens

**The hazardous ingredients contained in this product are NOT listed on the following regulatory lists of suspected carcinogens: Federal OSHA, National Toxicology Program (NTP), International Agency for Reasearch on Cancer (IARC), State of California - OSHA**

## SECTION 4 - SPILL, LEAK, AND DISPOSAL PROCEDURES

### Spill and Leak Control

**Small Spill / Leaks:** Flush area with water to an industrial sewer line.

**Large Spill / Leaks:** Contain spill with dikes of absorbant mateirals such as clay, sand, or vermiculate. This material is considered Non Hazardous in concentrated form.

### Waste Disposal

All hazardous waste must be solidified and disposed of in an EPA approved class one disposal facility. When disposing of chemicals contact local, state, and federal environmental agencies to fully understand the necessary regulations governing the disposal of chemical wastes.

**FIGURE 15.8** A material safety data sheet describing the characteristics and precautionary use of a product.

Read the entire label of any pesticide before you buy, mix, apply, store, or dispose of it. If you have questions on how to use a pesticide, it is quite likely other applicators have the same questions. Be a good consumer and tell the manufacturer your concerns. The company may not realize there are problems or questions with the label directions. The label must contain the following items:

1. Trade name
2. Ingredient statement
   a. Active ingredient (chemical name; common name may be present)
   b. Inert ingredient(s)
3. Type of pesticide (herbicide, insecticide, nematicide, fungicide, etc.)
4. Net contents
5. Name and address of manufacturer—establishment number
6. EPA registration number—indicates the label is approved by the EPA
7. Signal words and symbols
   - Danger: highly toxic; some products may also carry the word "Poison" printed in red plus the skull and crossbones symbol (category I)
   - Warning: moderately toxic (category II)
   - Caution: slightly toxic (categories III and IV)
8. Precautionary statements
   - Route of entry (to the body) statements
   - Specific action statements (to prevent poisoning accidents)
   - Protective clothing and equipment statements
   - Other statements may be listed in regards to precautions to take while handling the product
9. Statement of practical treatment in case of poisoning
10. Environment hazards
    - Special toxicity statements (e.g., toxic to bees, fish, etc.)
    - General environmental statements
11. Physical or chemical hazards
12. Classification statement: general- or restricted-use pesticide
13. Reentry statement
14. Storage and disposal
15. Directions for use

## PREPARING PESTICIDE STORAGE FACILITIES FOR A MAJOR DISASTER

Hurricanes, floods, tornadoes, and other severe storms can seriously damage agricultural chemical storage facilities and the chemicals they contain. Storm-damaged facilities may adversely affect the environment and people. The following are steps to consider if a storm approaches and damages an agrochemical storage facility.

### INVENTORY

Do an inventory of what pesticides and other chemicals you have on hand. Such an inventory will be useful for insurance purposes, or in the event of necessary pesticide or chemical cleanups. Include product and active ingredient names and container sizes. Receipts for the purchase of these materials are useful for this, or in some cases may suffice themselves. Put the inventory in a safe location. In the case of large-scale storms, it may be useful to make a copy of your inventory and mail or fax it to a friend or business associate living outside the potentially affected area.

### INSURANCE

Know where your insurance policy is and know exactly what kind of coverage you have. Does it cover your chemical inventory or the damage it could cause? Find out now because if you need to know later, your insurance agent will be very busy.

### CHEMICAL USE

At this point, consider not using or making applications of agricultural chemicals, or at least holding off until the potential of the impending severe weather event is resolved. Delay purchase or delivery of additional chemicals to your operation until after the impending storm risk is past. If deliveries are scheduled for the coming week, it may be best to cancel them.

### CHEMICAL STORAGE AND SECURITY

Secure all chemicals including fertilizers, pesticides, solvents, fuels, and so forth. Close and secure container lids, moving containers and application equipment to the most secure location. Raise chemicals from the floor or cover materials potentially damaged by water. Do what you can to protect product labels and labeling. Doors, windows, and other points of access to storage locations should be secured and locked. Consider boarding up the pesticide and other chemical storage areas. Do not leave chemicals in vehicles or application equipment. Be sure all of the stored items are compatible. Do not, for instance, put pesticides and fuels in the same building with animals, or animal feeds.

## ACTIVATED CHARCOAL TO DEACTIVATE PESTICIDES

Activated charcoal (also called activated carbon) is often used to adsorb or deactivate organic chemicals such as pesticides. Activated charcoal has been used for many years to remove organic contaminants from wastewaters and in water purification systems. Since most pesticides are organic chemicals, activated charcoal can effectively be used to deactivate or "tie up" these products in soil. Once the pesticide has been adsorbed onto activated charcoal, it is biologically inactive and cannot cause injury to the turfgrass. Therefore, this product can be beneficial to turfgrass managers in the case of an accidental pesticide spill or where a pesticide needs to be inactivated for seeding or sprigging of turfgrasses. Due to its dark color and its consequent ability to absorb heat, activated charcoal is also used to artificially warm the soil to minimize the effects of light frosts or to allow earlier seeding of an area.

Charcoal is a porous, soft, black substance made by heating, in a restricted amount of air, substances containing carbon such as material from hardwood trees and coconut shells. Powdered activated charcoal is made up of very small carbon particles with a high affinity for organic chemicals such as pesticides. Activated charcoal has a large surface area and can absorb 100 to 200 times its own weight.

The amount of activated charcoal to apply to a pesticide-contaminated area varies with the chemical characteristics of the particular pesticide. Rates generally range from about 100 to 400 pounds activated charcoal per acre, or 110 to 450 kg/ha (2.3 to 9.2 pounds/1,000 square feet, 11 to 45 g/m$^2$) for each pound of active ingredient of a pesticide applied per acre (1.12 kg/ha). A general rule is to apply about 200 pounds activated charcoal per acre, 224 kg/ha (4.6 pounds/1,000 square feet, 20 g/m$^2$) for each pound of pesticide active ingredient per acre (Table 15.6).

### Example

Suppose Balan 2.5G was inadvertently applied at 2 pounds active ingredient/acre (2.24 kg active ingredient/ha) to an area designated to be seeded with a turfgrass:

$$\frac{\text{200 pounds activated charcoal}}{\text{pounds active ingredient in product per acre}} \times \frac{\text{2 pounds active ingredient of product applied}}{\text{acre}} = \frac{\text{400 pounds activated charcoal}}{\text{acre}}$$

When using a general rate of 200 pounds activated charcoal/ pound active ingredient applied/acre to completely inactivate this herbicide, the total activated charcoal required is 400 pounds/acre [450 kg/ha (or 9.2 pounds/1,000 square feet, 45 g/m$^2$)].

### Application Methods

Activated charcoal can be applied by various methods. It can be applied in the dry form with a drop spreader. However, activated charcoal particles are easily moved by wind, so it may be difficult to distribute the charcoal evenly when applied in the dry form. The easiest method is to suspend the charcoal in water and apply it by hand with a watering can (for small areas) or a power sprayer. Because activated charcoal does not mix easily with water, a 0.5% solution of a nonionic surfactant (equivalent to 1 quart per 50 gallons, 1.3 mL/L) will enhance its suspension in water. Note that charcoal particles are very abrasive and can damage spray equipment (particularly rotary type pumps). Therefore, if a sprayer is used to apply activated charcoal, care should be taken to thoroughly clean the equipment when finished.

When deactivating a pesticide in a seedbed, the activated charcoal should be incorporated with a rotary tiller or other appropriate equipment so the charcoal is placed in the upper few inches of soil. The objective is to get the activated charcoal in the same proximity as the pesticide. Uniform application of activated charcoal followed by thorough mixing is the key to inactivating a pesticide-contaminated area. If the pesticide is on the turf, in the thatch layer, or on the uppermost surface of the soil (for instance, if the pesticide has not been watered in), the pesticide can be inactivated by simply applying the charcoal to the area and thoroughly watering once charcoal application is complete. Again, the objective is to place the charcoal in the same proximity as the pesticide. If activated carbon is applied and either incorporated or watered correctly, inactivation of the pesticide will be successfully accomplished. For application convenience, it is recommended that activated charcoal be applied as a water slurry. To minimize dusting, always add activated charcoal to water slowly, keeping the bag as close to the water surface as possible. The following steps are suggested when mixing and applying charcoal.

**TABLE 15.6**
**Rates of Activated Charcoal Used for Spills and Deactivating Turf Pesticides**

| Application | Recommendation | Comments |
|---|---|---|
| Spills | For reducing the effects from spills of organic pesticides, some petroleum products, and hydraulic fluids | • Use 100 to 400 pounds activated charcoal to every pound active material spilled per acre, or 110 to 450 kg/ha (2.3 to 9.2 pounds/1,000 square feet, 11 to 45 g/m$^2$). If the active material has not been diluted with water at the time of spill, apply the charcoal directly as a dry powder. If the active material has been diluted with water, apply the activated charcoal in a slurry with a sprinkle can or common sprayer equipment. The charcoal must be incorporated into the contaminated soil, preferably to a depth of 6 inches (15 cm). With severe spills, some of the contaminated soils may need removal prior to activated charcoal application. |
| "Deactivating" turf herbicides and soil warming | Turf areas treated with preemergence herbicides can be reseeded earlier than normal by treating with activated charcoal | • Whenever it is desirable to terminate a preemergence herbicide, apply charcoal slurry at a rate of 2 to 4 pounds/1,000 square feet (9 to 19 g/m$^2$). Water the slurry into the soil. Make sure the grass is washed free of heavy charcoal deposits. Where possible, it is desirable to thoroughly rake the charcoal into the soil. The area can be seeded 24 hours after treatment. |

### Spray Application

1. Make sure spray equipment, tubing, and nozzles are completely clean. Screens should be removed, if practical.
2. The final spray mixture should contain 1 to 2 pounds of charcoal per gallon of water.
3. Add sufficient water to begin moderate agitation. Simultaneously add the balance of required water and charcoal. Continue agitation until a uniform mixture is obtained.
4. Maintain moderate agitation while spraying.

It is important to understand situations where activated charcoal will not work. If an herbicide has been applied for several weeks and rainfall has occurred and/or irrigation water has been applied, the herbicide is most likely past the thatch layer and probably in the upper inch or so of the soil. In this case, activated charcoal would have to be physically incorporated with a rotary tiller or other implement to place the charcoal in contact with the herbicide because activated charcoal will not leach through soil. If activated charcoal is applied to the soil surface and watered, the charcoal will remain on top of the soil and will not inactivate the herbicide below the soil surface. Activated charcoal also is ineffective for inorganic pesticides such as arsenates, lead compounds, sodium chlorate, sulfur, borax, and so forth, and water-soluble organic pesticides such as, but not limited to, MSMA (monosodium methyl arsenate) and DSMA (disodium methyl arsonate).

It is a good idea to keep several bags of activated carbon on hand so it can be applied immediately instead of having to wait for delivery. Several different brands and formulations are on the market. There appears to be little, if any, difference in the effectiveness of the different brands. However, some may be easier to apply than others, depending on the particular situations where they are to be used.

**FIGURE 15.9** Turf damage from using wrong spray nozzles or improper boom heights.

## SPRAYER AND SPREADER CALIBRATION

Proper sprayer/spreader calibration is extremely important to assure successful application of pesticides, fertilizers, lime, and seed. Applying excessive product wastes materials and money, may injure the turf, and can potentially cause unnecessary environmental contamination. Applying inadequate pesticide may result in poor control of the pest, resulting in undesirable turf and lost revenue. Constant observation of applicator operating conditions and frequent calibration of the equipment are necessary to avoid these problems (Figure 15.9). Helpful pesticide calibration formulas and metric conversion factors are listed at the end of the book.

## SPRAYERS

Spraying is the most common method for applying pesticides. Liquid sprays can be applied uniformly and accurately, as long as the sprayer has been properly designed, calibrated, and operated, and the mixing of the formulation in the spray tank has been performed correctly (Figure 15.10).

A typical sprayer consists of the following components:

- *Tank* to hold the spray mixture
- *Pump* to deliver the mixture through the sprayer
- *Pressure regulator* to maintain constant pressure
- *Bypass line* to return excess spray to the tank (so the pressure regulator can function)
- *Pressure gauge* to provide pressure readings
- *Boom with nozzles* or a *handgun* to deliver the spray to the application site
- *Shutoff valve* located upstream from the boom (or in the handgun) to control the flow of the spray mixture
- Mechanical or hydraulic *agitator* to keep suspended particles or oil droplets uniformly dispersed within the spray tank
- Strainers and/or *screens* to prevent large particles from clogging the sprayer
- *Suction line* to convey the spray mixture from the tank to the pump
- Various *hoses*, *fittings*, *clamps*, and other devices—perhaps including governors and controllers—that enable or enhance the overall functionality of the sprayer

### Spray Tanks

Sprayer tanks can be constructed from a variety of materials, including stainless steel, aluminum, aluminized steel, galvanized steel, fiberglass, and plastic. Stainless steel is the most adaptable, but also the most expensive. Galvanized steel should be avoided, as it tends to react with some pesticides and is susceptible to rust, generating rust flakes that can clog sprayers. Some pesticides can also react with aluminum and aluminized steel. Fiberglass and plastic resist chemical reactions with pesticides and do not rust; they are lighter in weight than metal tanks but are more susceptible to breakage and possible degradation from long-term exposure to sunlight. Tanks mounted on turf utility vehicles typically range in size

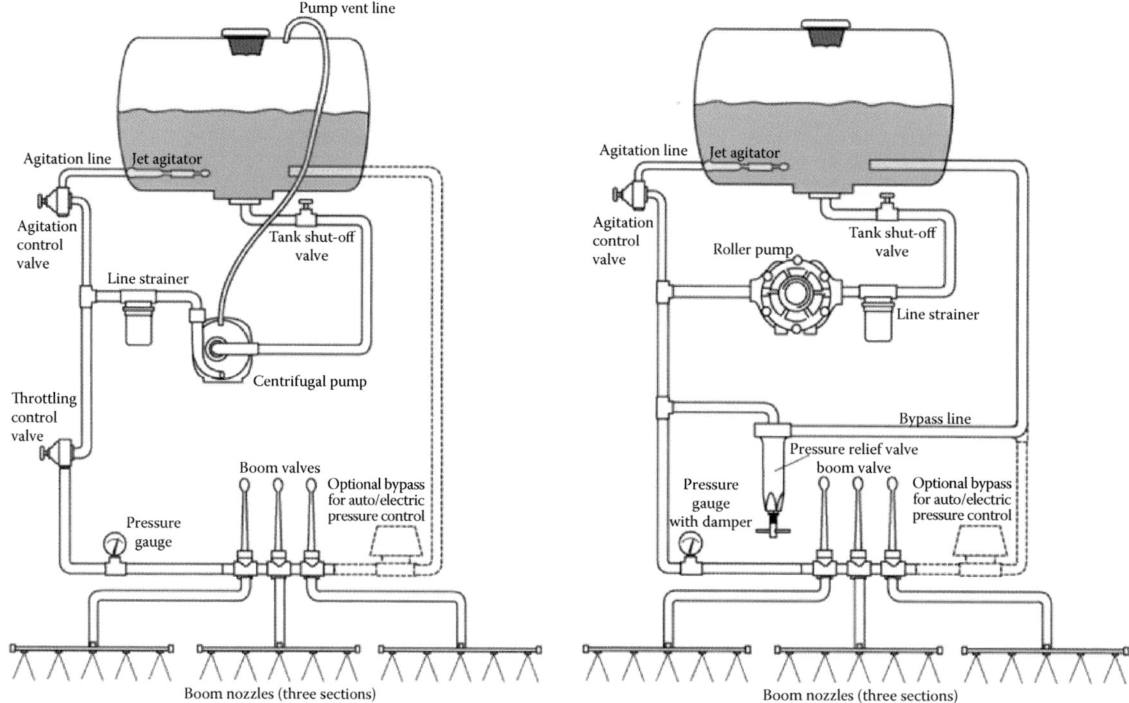


**FIGURE 15.10** Illustration of a sprayer using a centrifugal (left) compared to a roller (right) pump.

from 100 to 300 gallons (379 to 1,136 L). The largest tanks usually have baffles to reduce liquid shifting during operation.

## Sprayer Pumps

Pumps used on spray equipment create the required amount of pressure to force spray solution through the nozzles. Pumps also provide agitation if the sprayer has a hydraulic agitation system. The three major types of low-pressure pumps include (Table 15.7) *roller, centrifugal* (or *turbine*), and *diaphragm*. *Piston pumps* are also used in some special situations when high pressure is needed (Figure 15.11).

Pumps are also classified as positive or nonpositive displacement. With positive displacement, the pump output is independent of pressure. A constant volume of liquid is displaced with each pump stroke or revolution. Positive type pump sprayers require a pressure relief valve, or some sort of pressure control device to recirculate liquid not being used for spraying back to the nurse tank. These pumps also are self-priming; therefore, they do not need to be mounted below the tank level in order to be primed or start pumping. Positive displacement pumps include the piston pump, diaphragm pump, and the roller pump operating at low pressures.

Nonpositive displacement pumps (primarily centrifugal) have output that is dependent upon pressure; thus, output decreases if pressure increases. They do not require bypass relief valves and are not self-priming. Nonpositive pumps generally have longer life spans than positive displacement pumps, which require close housing of rotary parts and easily wear, especially from abrasive suspensions. The power to drive both types of pumps includes tractor power take-off (PTO), ground wheel traction drive, or separate fuel-driven

**TABLE 15.7**
**Characteristics of Various Sprayer Pump Types**

| Pump Type | Pressure Range (psi) | Flow Rate (gpm) | Type Displacement | Comments |
|---|---|---|---|---|
| Roller | 5 to 300 | 1 to 35 | Positive | • Low costs, ease of service, self-priming, operates at PTO speeds, medium volume, short life if WPs used |
| Centrifugal (or turbine) | 5 to 100 | 20 to 120 | Nonpositive | • Medium costs, handles all materials, high volume, long life, low pressure, not self-priming, requires high PTO speed |
| Piston | 400 to 1,000 | 5 to 60 | Positive | • High costs, handles all materials, wear resistant, self-priming, high pressure, needs surge tank, long life |
| Diaphragm | 75 to 100 | 15 to 60 | Positive | • High costs, long life, resistant to most chemicals, self-priming, needs surge tank |

*Note:* Pound-force per square inch (psi) × 6.89 = kilopascals (kPa); gallons per minute (gpm) × 3.785 = liters/minute.

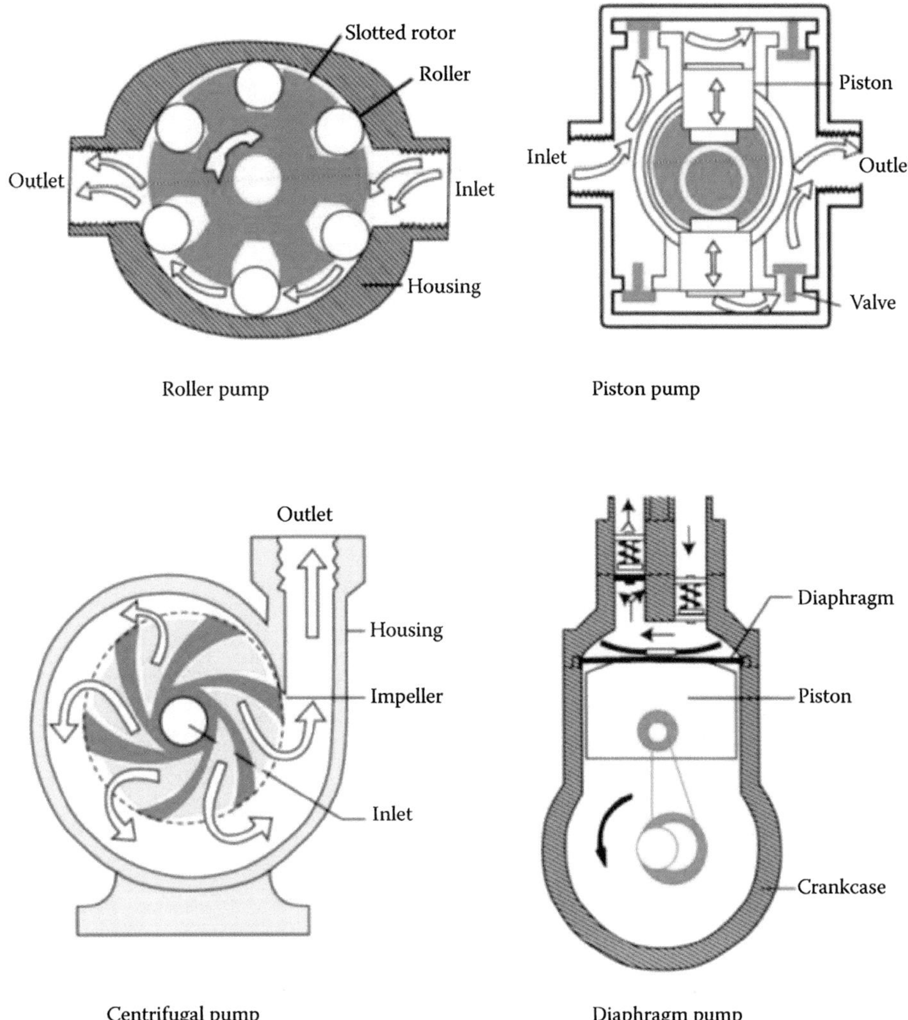


**FIGURE 15.11** Illustration comparing roller, piston, centrifugal, and diaphragm sprayer pumps.

engines or electric motors. Nonpositive displacement pumps include the centrifugal pump and the roller pump when operated at higher pressures.

**Roller pumps.** Roller pumps have elliptically shaped inner walls of the pump housing with a slotted rotor containing rollers in the slots. Spray solution accumulates in these slots since more space is located between the rotor and pump housing on the intake side. As the rotor turns toward the outlet side, less space occurs between the rotor and housing, creating pressure. As the roller pushes back into the slot, the spray solution is forced out. They are often preferred for their low-pressure operation (40 to 280 pounds per square inch [psi]) and deliver 5 to 40 gallons per acre.

Roller pumps are considered positive displacement, but do experience some decline in output as pressure is increased. They require a pressure relief or control device to divert unsprayed solution back to the tank. They have the advantages of being low in initial costs and in maintenance. They are compact in size in relation to their capacity and operate at tractor PTO speeds. They prime easily and have relatively long life expectancies when properly operated. Their chief disadvantage is being sensitive to abrasive pesticide formulations (e.g., WP, DF, F, sand, and grit). Using rubber rollers helps extend the pump's life span if abrasive materials are used often. Roller pumps are widely used for turfgrass spray operations.

**Centrifugal pumps.** Centrifugal pumps create pressure by spinning a fly wheel within the pump housing; this centrifugal force creates pressure. The spray material feeds into the center of a high-speed impeller and is thrown outward, creating pressure in the outlet line. Centrifugal pumps are considered nonpositive displacement since their output decreases as pressure increases; thus, they are not self-priming. Their chief advantages include being able to handle suspensions well, being resistant to wear, and having relatively long operating lives. They also have a high capacity (70 to 130 gallons per minute [gpm] at 30 to 40 psi), simple maintenance, and no need for pressure relief valves. They are considered high-volume, low-pressure pumps.

Disadvantages include the limited low pressure, usually 50 psi or less. They tend to be hard to prime, requiring positive inlet pressure, or they are placed below the tank, using a small vent at the top of the pump to allow trapped air to escape to prevent vapor lock. Normal tractor speed also may be too slow, requiring a step-up gear box or pulley arrangement. Turbine pumps are similar to centrifugal pumps but can operate as a direct drive from the PTO due to their low operating speeds; thus, they do not require step-up mechanisms.

**Diaphragm pumps.** Diaphragm pumps work on a piston action where a flexible diaphragm moves up and down in a sealed chamber, creating suction and pressure. As the diaphragm moves up, suction is created and the spray material is drawn into the chamber. As the diaphragm moves down, pressure is created and the spray material is forced out of the chamber through a discharge valve. They are self-priming positive displacement pumps and require less horsepower than comparable pumps with similar flow and pressure ratings. Medium- to high-pressure models (75 to 100 psi, 517 to 689 kPa, and higher) are available, creating flow rates ranging from 15 to 50 gpm (57 to 189 L/minute). Abrasive materials can successfully be used with diaphragm pumps, although the diaphragm should be resistant to other caustic materials.

**Piston pumps.** Piston pumps work on the basic principle of pistons in a motor where the intake stroke draws the liquid in through one valve, and the output stroke forces the liquid out through another valve. Piston pumps are low-volume, high-pressure capacity pumps. They are true positive displacement, meaning the output is independent of pressure. They tend to handle suspensions well, are resistant to wear, and have relatively long life spans. They also prime easily and handle high pressures (up to 600 psi, 4.13 mPa, or more). Disadvantages include a higher purchase price and the need for a surge tank, and they are generally too large to mount directly on most tractor PTOs.

### Pressure Regulator

A pressure regulator is needed in systems with positive displacement pumps (e.g., roller, piston, and diaphragm pumps) to control operating pressure. The valve opens when pressure increases to some adjustable level, relieving the pressure and conducting some of the flow back to the spray tank. In some systems, the same *bypass line* is used for hydraulic agitation of the tank spray mixture. The disadvantage of linking pressure relief and hydraulic agitation, however, is that the flow through the bypass line can fluctuate due to changes in the system's pressure requirements.

### Pressure Gauge

A pressure gauge is used to measure liquid pressure. The range should be about twice the maximum anticipated pressure encountered in the operation of the sprayer. Liquid-filled pressure gauges are preferable to dry gauges, as they dampen vibrations caused by pressure spikes in the system. Gauges should be installed in a position easily observed by the sprayer operator.

### Booms

Spray booms include those that are mounted onto the sprayer and walking booms that are attached to the sprayer by a long flexible hose. Booms can also be wet or dry; wet booms actually carry the spray mixture through a pipe—usually stainless steel—to the nozzles, while dry booms carry the spray mixture through flexible tubing to the nozzles, which are attached to a rigid frame. Mounted booms are usually 15 to 20 feet (4.6 to 6.1 m) in length and often come in three sections, with each section operating independently. The lateral sections can be folded up for transport. Walking booms measure up to 80 inches (2 m), sometimes with fold-up lateral sections. These are used where compaction and wear from vehicles precludes using mounted booms, such as on golf greens or newly established areas.

### Nozzles

Nozzles mounted on a spray boom convert the spray mixture into deliverable spray droplets (Figure 15.12). Nozzles

**FIGURE 15.12** Examples of various spray nozzles available for boom application.

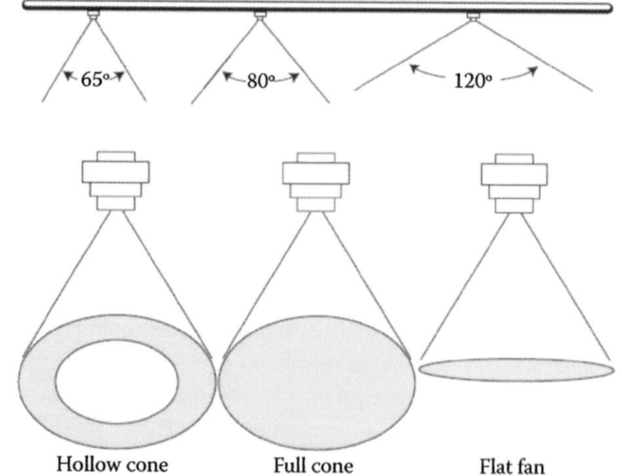


**FIGURE 15.13** Illustration of the angles and spray patterns of three popular spray nozzles.

typically consist of four parts: body, strainer, tip, and cap. The cap screws onto the body and holds the tip and strainer in place. The strainer is placed immediately behind the nozzle tip to filter the liquid and prevent nozzle clogging. The nozzle helps determine the amount of spray applied to an area, uniformity and coverage of application, and drift potential.

Nozzle spray patterns vary according to the intended pest target or application equipment. The main spray patterns include hollow cone, solid or full cone, and regular flat fan (Figure 15.13). Flat fan nozzles form a narrow, elliptical, inverted "V" pattern. Spray output is heaviest at the center of the pattern and dissipates toward the outer edge. A uniform distribution pattern is achieved when the spray boom height and nozzle spacing are optimized for proper spray pattern overlap of adjacent nozzles. A 30% overlap between nozzles is recommended for most flat-fan spray patterns. A one-to-one relationship between nozzle spacing and nozzle height above the turf is best. Variations of the flat fan include:

1. An extended-range flat fan for broadcast spraying over a wider range of spray pressures allows lower operating pressure for drift control. These nozzles are available with 80° to 110° spray angles, produce a fine to coarse droplet size, and are rated as excellent for contact products and good for systemic products (Table 15.8). Since extended range nozzles have an excellent spray distribution over a wide range of pressure 15 to 60 psi), they can be used on sprayers equipped with flow controllers.
2. Flood flat fans for broadcast spraying with a wide (120°) angle flat pattern using larger droplets.
3. Even flat fan spray can be used for band spraying, which provides even coverage without overlapping. These have 40° to 110° spray angles.

In addition, a venture air-induction flat fan nozzle is available that has two orifices with a venture or air aspirator between them that mixes air with the spray. These produce a coarse to very coarse droplet and are excellent for use with systemic products and good with contact products.

Hollow-cone nozzles form a round ring pattern that helps provide complete coverage by creating a finely atomized spray pattern. These are available with 65° to 80° spray angles. The hollow-cone nozzle is more difficult to set up for uniform application. Full-cone nozzles form a round, full pattern for specialty applications. They typically produce a 120° spray angle. Flat fans are most often used in turfgrass situations due to crop uniformity. Hollow cones, however, are sometimes used to apply fungicides or insecticides.

Commonly used nozzles have spray angles ranging from 65° to 120° (Figure 15.13). Wide-angle types can be mounted lower to the ground or spaced further apart on the boom. Narrow-angle types produce a more penetrating pattern, and tend to better force droplets through trash or foliage.

Most normal spraying operations in turf are easily and efficiently performed using regular flat fan nozzles. Nozzles break liquids into droplets that are measured in microns (1/1,000 mL). As a comparison, a human hair is about 100 microns. Nozzles also are classified according to color and nomenclature (Table 15.9). Color indicates nozzle flow capacity at a particular pressure while the numbering system indicates the spray angle and output (in gallons per minute) for the particular nozzle at 40 psi (276 kPa). Spraying Systems, for example, code their regular flat fans with two sets of numbers. The first two or three indicate the spray angle and the second two indicate the output. For example, 8002 tips have 80° angles and apply 0.2 gallon per minute at 40 psi. An XR8008 nozzle is white and has an output of

**TABLE 15.9**
**Sprayer Nozzle Droplet Size Classification**

| Droplet Category | Symbol | Color Code | Approximate Median Drop Size (microns) |
|---|---|---|---|
| Very fine | VF | Red | <145 |
| Fine | F | Orange | 145–225 |
| Medium | M | Yellow | 226–325 |
| Coarse | C | Blue | 326–400 |
| Very coarse | VC | Green | 401–500 |
| Extremely coarse | XC | White | >500 |

**TABLE 15.8**
**Efficacy of Common Spray Nozzles with Several Pesticides**

| | Herbicides | | | | | | |
|---|---|---|---|---|---|---|---|
| | | Postemergence | | Fungicides | | Insecticides | |
| Nozzle | Preemergence | Contact | Systemic | Contact | Systemic | Contact | Systemic |
| Extended range flat fan | – | Excellent | Good | Excellent | Good | Excellent | Good |
| • at <30 psi | Good | Good | Very good | Good | Very good | Good | Very good |
| Wide angle flat spray | – | Very good | Very good | Very good | Very good | Very good | Very good |
| • at <30psi | Good | Good | Excellent | Good | Excellent | Good | Excellent |
| Venturi air-induction flat spray | Very good | Good | Excellent | Good | Excellent | Good | Excellent |
| Hollow cone | – | Excellent | Good | Excellent | Good | Excellent | Good |
| Wide angle flood fan | Excellent | – | Very good | – | Very good | – | Very good |

0.80 gpm (3 L/minute) at an 80° spray angle, while a 11004 nozzle is red with an output of 0.40 gpm (1.5 L/minute) at a 110° spray angle. Delavan, another nozzle company, indicates the same nozzle as Delavan 80° LF2, meaning an 80° spray angle that applies 0.2 gpm at 40 psi. The LF indicates a regular flat fan. Manufacturer specifications should be consulted on nozzle types available and their required nozzle spacings, operating pressures, and application heights.

Select the spray nozzle most applicable for the pesticide being used. If you only have one sprayer and do not want to change nozzles for each type of pesticide applied, flat fan nozzles are the best overall choice. These also are the best choice for broadcast spraying. When coverage is critical, such as with contact products, nozzles that produce fine to medium droplets help ensure thorough leaf surface coverage. For products that need to reach the lower plant canopy or soil surface, nozzles that provide coarse to very coarse droplet size are needed. These products are either applied at higher carrier volume or are watered in to help move products down. Drift reduction nozzle technology is available and includes air induction and preorifice nozzles, which create pressure drop resulting in larger droplet size.

Nozzles are made of aluminum, brass, ceramic, plastic, stainless steel, or tungsten carbide. Brass tips are the most often used due to their lower costs, but tend to wear faster. Stainless-steel and tungsten carbide tips last up to six times longer than brass and are more resistant to abrasive powders, but are more expensive. Plastic materials with stainless steel or ceramic inserts cost less than stainless steel and last two to three times longer than brass. Ceramic tips last 20 to 50 times longer than brass. No one nozzle material provides combined cost effectiveness with corrosion resistance. Check valves also are available that help eliminate "dribble" once the sprayer is turned off. A rule of thumb for replacing nozzles is to do so when the measured flow rate is 10% higher than the intended rate. Should a nozzle become clogged, blow out the debris or use a soft-bristle brush such as a toothbrush. Using a hard object such as wire or a nail can easily damage the nozzle's orifice.

### Spray Markers

Important concerns in sprayer operations are skips and excessive overlaps. Skips occur where no spray has been received, often resulting in poor pest control and the necessity to make supplemental applications. Some overlap is necessary for ensuring that the entire treated area receives the proper rate of the pesticide. Depending on nozzle selection, this may involve overlapping by 15% to 50% of the last nozzle's spray width; however, overlapping by more than these percentages means the pesticide application rate will be greater than that intended, possibly resulting in phytotoxicity to nontarget turfgrasses or ornamental plants. A sprayer accessory that can be helpful for minimizing these problems is a foam or dye marker system that clearly shows the edge of the spray swath (Figure 15.14). The spray operator is then aided in determining exactly where the next pass should be made to avoid skips and excessive overlaps. An alternative would be to include a turf colorant to the spray mixture to show previously treated areas.

**FIGURE 15.14** A foam marker system indicating where products were previously applied.

### System Monitors and Controllers

Historically, sprayer-mounted vehicles were operated mostly by feel, with the pressure regulator set and the throttle or accelerator positioned to maintain the proper traveling speed. Some spraying systems have employed ground-speed governors to maintain constant ground speed over variable terrain in an attempt to provide consistent chemical application rates. Today, highly sophisticated systems can sense and continuously display on a monitor various operating conditions, including pressure, travel speed, spray volume, field capacity (in acres or hectares per hour), amount applied, amount remaining in the tank, area covered, and distance traveled. Accurate monitoring offers the possibility of achieving greater effectiveness and efficiency (e.g., better pest control and reduced pesticide use) from spraying operations. Some systems employ speed-compensating controllers that sense the ground speed of the vehicle and adjust the flow rate from the sprayer to ensure a consistent application rate. Geographic positioning systems (GPS) can be used to acquire data from satellites to accurately determine the vehicle's locations on the ground, and then plot the areas treated in spatial displays and relay application data to a digital storage unit for automated recordkeeping. Where GPS is linked to a geographic information system (GIS) containing site-specific information from soil maps and/or soil test results that can be spatially displayed on a monitor, an option available is variable rate technology (VRT), in which the application rate of specific materials (e.g., preemergence herbicides, fertilizer nutrients) is based, in part, on variable field conditions.

Another technology that shows promise is direct injection (DI) of chemicals into the water stream moving toward the nozzles, thus providing the capability of choosing materials on a site-specific basis while avoiding the problem of spray-tank chemical residues afterwards. Linking GPS, GIS, and

DI, along with specialized sensors, could provide a truly site-specific approach to management, in that the materials applied could be based not only on historical records, but also on direct and remote (satellite-based) observations of field conditions. In choosing from an increasing array of rapidly evolving technologies, one must balance the presumed advantages against the acquisition and maintenance costs of these technologies, along with the technical complexities of their use. Finally, no technology, no matter how sophisticated, can replace the proper selection and timing of pesticide application, accurate sprayer calibration, and common sense.

## SPRAYER CALIBRATION

In order to deliver the appropriate rate of any applied pesticide, the sprayer calibration (delivery volume per unit area) must be correctly determined and maintained. The amount of chemical solution applied depends upon (1) the sprayer's forward speed, (2) the pressure at the nozzle, (3) the size of the nozzles, and (4) the spacing of the nozzles on the boom. The boom height also influences spray coverage (Figure 15.15). A change in any one of these will change the application rate and coverage.

### Sprayer Speed

To measure travel speed, drive at the speed used for spraying (Figure 15.16). In addition, drive on a turfgrass area rather than a concrete or asphalt road, so the travel "test" conditions reflect "actual" conditions. Even if sprayers are equipped with digital speedometers, it is necessary to confirm ground speed using another method. The following formula may be used to

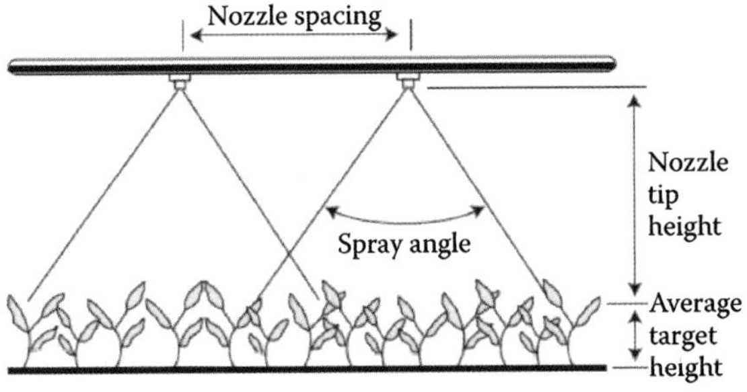


**FIGURE 15.15** Nozzle spacing, boom height, and spray angle all influence which nozzles should be selected.

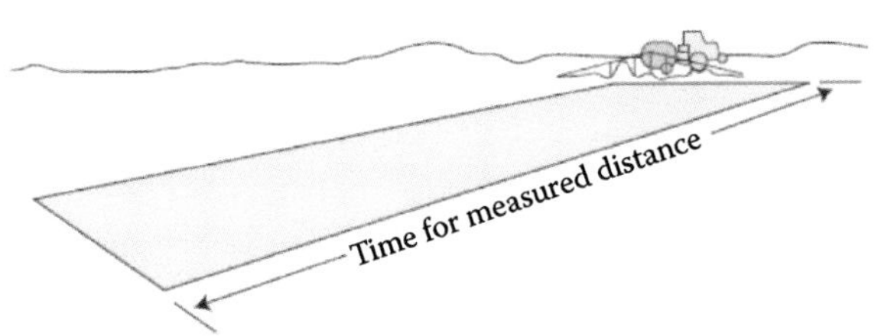


**FIGURE 15.16** To calibrate a sprayer, forward speed, spray width, and sprayer output are needed.

determine miles per hour when the time necessary to travel a certain distance is known. Table 15.10 is a chart that converts the time to drive 300 feet (91 m) to the speed in miles per hour (mph).

$$\text{Speed (mph)} = \frac{\text{distance travelled (feet)}}{\text{time (seconds)}} \times \frac{1\text{ mile}}{5{,}280\text{ feet}} \times \frac{3{,}600\text{ seconds}}{\text{hour}}$$

or

$$= \frac{\text{distance traveled (feet)} \times 0.682}{\text{time (seconds) to travel distance}}$$

$$\text{Speed (km/h)} = \frac{\text{distance traveled (m)} \times 3.6}{\text{time (seconds)}}$$

#### Example

A superintendent fills his sprayer and marks off a 100-foot (30.5 m) strip. He rides the course several times and determines an average of 20 seconds is needed to cover this area. What is the sprayer's speed, in miles per hour?

Simply insert the distance and time values into the following formula:

$$\text{Speed (mph)} = \frac{\text{distance traveled (feet)}}{\text{time (seconds)}} \times \frac{1\text{ mile}}{5{,}280\text{ feet}} \times \frac{3{,}600\text{ seconds}}{\text{hour}}$$

$$= \frac{100\text{ feet}}{20\text{ seconds}} \times \frac{1\text{ mile}}{5{,}280\text{ feet}} \times \frac{3{,}600\text{ seconds}}{\text{hour}}$$

$$= 3.4\text{ mph}$$

$$\text{or } \frac{\text{distance traveled (feet)} \times 0.682}{\text{time (seconds) to travel distance}}$$

$$= \frac{100 \times 0.682}{20} = 3.4\text{ mph}$$

$$\text{Speed (km/h)} = \frac{\text{distance traveled (m)} \times 3.6}{\text{time (seconds)}} = \frac{30.5 \times 3.6}{20} = 5.5\text{ km/h}$$

### Pressure and Nozzle Selection

The type of pesticide and nozzle being used often determines the pressure needed for spraying. The pesticide label usually recommends the sprayer pressure. Low pressures of

**TABLE 15.10**
**Time (Seconds) Required to Cover a Specific Distance to Obtain a Desired Speed (mph)**

| | | Time Required (seconds) to Travel a Distance of | | |
|---|---|---|---|---|
| **Desired mph** | **Feet per Minute** | **100 feet** | **200 feet** | **300 feet** |
| 2.0 | 176 | 34 | 68 | 102 |
| 2.5 | 220 | 27 | 54 | 81 |
| 3.0 | 264 | 23 | 45 | 68 |
| 3.5 | 308 | 20 | 39 | 58 |
| 4.0 | 352 | 17 | 33 | 51 |
| 4.5 | 395 | 15 | 30 | 45 |
| 5.0 | 440 | 14 | 27 | 41 |
| 6.0 | 528 | – | 23 | 34 |
| 7.0 | 616 | – | 19 | 29 |
| 8.0 | 704 | – | 17 | 26 |
| 9.0 | 792 | – | 15 | 23 |

*Note:* Miles per hour (mph) × 1.61 = kilometers per hour; meters = feet × 0.3048.

**TABLE 15.11**
**General Operating Pressure Ranges for Various Nozzle Types**

| **Nozzles** | **Pressure (psi)** |
|---|---|
| Flat fan | 15 to 40 |
| Even flat fan | 15 to 40 |
| Whirl chamber | 5 to 25 |
| Flooding nozzle | 5 to 20 |

*Note:* Pound-force per square inch (psi) × 6.89 = kilopascals (kPa).

30 to 60 psi (207 to 414 kPa) are typical for most turfgrass pesticides. Spray nozzles are designed to be operated within a certain pressure range (Table 15.11). Higher than recommended pressure increases delivery rate, reduces droplet size, distorts spray patterns, increases nozzle wear, and increases drift potential. Lower pressures reduce delivery rate and prevent a full spray width pattern, resulting in "skips." Always follow the nozzle manufacturer's pressure recommendations as explained in its product catalog. Determine pressure at the nozzles or boom and not at the control valve.

Nozzle manufacturers help applicators choose the right tip for each job by providing detailed performance charts. The pesticide applicator matches the specific needs of the job to the information on the chart to determine the tips and strainers to use. The charts include factors to be considered in order to choose appropriate nozzles—pressure, equipment speed, and spray volume. Charts showing spray volume in terms of both gallons per acre (gpa) and gallons per minute (gpm) allow the nozzle choice without further calculations. Some manufacturers, however, list only gallons per minute. Desired gallons per acre must be converted to gallons per minute as shown here:

Gallons per minute (per nozzle)

$$= \frac{\text{gallon}}{\text{acre}} \times \frac{\text{mile}}{\text{hour}} \times \frac{\text{inch}}{\text{nozzle}} \times \frac{1\text{ hour}}{60\text{ minutes}} \times \frac{5{,}280\text{ feet}}{\text{mile}} \times \frac{1\text{ foot}}{12\text{ inches}} \times \frac{\text{acre}}{43{,}560\text{ square feet}}$$

$$= \frac{\text{gallons per acre} \times \text{miles per hour} \times \text{nozzle spacing (inches)}}{5{,}940}$$

### Example

If the pesticide label recommends a spray volume of 10 gallons per acre, the sprayer travels at 4 mph and the spray boom has a 20-inch (51 cm) spacing between nozzles, find the output per nozzle.

Gallons per minute (per nozzle)

$$= \frac{\text{gallons}}{\text{acre}} \times \frac{\text{mile}}{\text{hour}} \times \frac{\text{inches}}{\text{nozzle}} \times \frac{1\text{ hour}}{60\text{ minutes}} \times \frac{5,280\text{ feet}}{\text{mile}} \times \frac{1\text{ foot}}{12\text{ inches}} \times \frac{\text{acre}}{43,560\text{ square feet}}$$

$$= \frac{10\text{ gallons}}{\text{acre}} \times \frac{4\text{ miles}}{\text{hour}} \times \frac{20\text{ inches}}{\text{nozzle}} \times \frac{1\text{ hour}}{60\text{ minutes}} \times \frac{5,280\text{ feet}}{\text{mile}} \times \frac{1\text{ foot}}{12\text{ inches}} \times \frac{\text{acre}}{43,560\text{ square feet}}$$

$= 0.13$ gallon per minute (per nozzle)

**or**

$$\frac{\text{gallons per acre} \times \text{miles per hour} \times \text{nozzle spacing (inches)}}{5,940}$$

$$= \frac{10\text{ gpa} \times 4\text{ mph} \times 20\text{ inches}}{5,940} = 0.13\text{ gpm, } 0.5\text{ L (per nozzle)}$$

Nozzle spacing usually is predetermined based on the type or make of the boom purchased (Figure 15.15). Nozzles usually are spaced 20, 30, or 40 inches (0.5, 0.76, or 1 m) apart along the boom. To obtain a uniform spray pattern with proper overlap, nozzles must be mounted at the proper height (e.g., boom height) above the target. Boom height varies depending on the nozzle spacing on the boom. For example, a 20-inch (0.5 m) nozzle spacing allows nozzles to be mounted closer to the spray target than a 30-inch (0.76 m) nozzle spacing does.

Each nozzle on a sprayer should apply nearly equal amounts of pesticide. For any one nozzle, the discharge should not vary by more than 10% below or above the average. Under normal operating pressure, collect the discharge from each nozzle for 30 seconds (Figure 15.17). Determine the average output and compare the output for each nozzle with the average.

**FIGURE 15.17** Capturing the output of a nozzle when calibrating a sprayer.

### Example

To evaluate the nozzles of a sprayer, the following test should be conducted:

**Step 1:** Measure and average the amount of water applied with the sprayer running for 60 seconds. For example, for individual nozzle readings, the fluid ounces collected in 60 seconds are shown:

| | |
|---|---|
| Nozzle 1: | 28 ounces (0.82 L) |
| Nozzle 2: | 36 ounces (1.06 L) |
| Nozzle 3: | 34 ounces (1.00 L) |
| Nozzle 4: | 30 ounces (0.88 L) |
| Nozzle 5: | 32 ounces (0.95 L) |
| Total | 160 ounces (4.73 L) |

Average = 32 ounces per nozzle (160 ounces ÷ 5) (or 0.95 L)

**Step 2:** A 10% allowance for the nozzles is determined:

10% of 32 ounces is 3.2 ounces (32 ounces × 0.10)

32 ounces + 3.2 ounces (10% greater) = 35.2 ounces (1.04 L)

32 ounces − 3.2 ounces (10% less) = 28.8 ounces (0.85 L)

The range of 28.8 to 35.2 ounces (0.85 to 1.04 L) allows for a ±10% variation for these five nozzles. Therefore, nozzles 1 and 2 fall outside the ±10% variation and should be replaced.

To convert output per nozzle from ounces to gallons per minute (gpm), the following equation is used:

Gallons per minute

$$= \frac{\text{ounces}}{\text{nozzle}} \times \frac{\text{run}}{60\text{ seconds}} \times \frac{1\text{ gallon}}{128\text{ ounces}} \times \frac{60\text{ seconds}}{\text{minutes}}$$

$$= \frac{\text{ounces}}{\text{nozzle}} \times \frac{1\text{ gallon}}{128\text{ ounces}}$$

Use the average discharge per nozzle as previously calculated.

### Example

A broad-spectrum fungicide is needed on your bentgrass greens for brown patch control. You decide to use Fung-No-Mo and remembered that flat fan nozzles spaced 20 inches (51 cm) apart on the boom were the best choice for this application. The label suggests this material be applied in a volume of 20 to 30 gpa, with water as the carrier, and at a pressure of 30 to 40 psi. Decide on the best nozzle type for this application.

**Step 1:** First, determine the output required per nozzle at each recommended spray pressure:

| For 20 gpa | For 30 gpa |
|---|---|
| $gpm = \frac{GPA}{5{,}940} \times mph \times w$ | $gpm = \frac{GPA}{5{,}940} \times mph \times w$ |
| $= \frac{20}{5{,}940} \times 5 \times 20$ | $= \frac{30}{5{,}940} \times 5 \times 20$ |
| $= 0.34$ | $= 0.51$ |

**Step 2:** Next, refer to the nozzle catalog to select a size of flat fan nozzle that delivers between 0.34 and 0.51 gpm at pressures of 30 to 40 psi.

### Sprayer Pressure and Speed Changes

If only minor changes are needed in the output of a sprayer, it is possible to accomplish this by adjusting pressure. However, major changes in output should be adjusted by other means as all types of nozzles have a certain pressure range within which they should be operated. The following formula is used to determine the effects of pressure on output.

Flow rate (as influenced by pressure):

$$\frac{gpm_1}{gpm_2} = \frac{\sqrt{psi_1}}{\sqrt{psi_2}}$$

or

$$gpa_2 = gpa_1 \times \frac{\sqrt{psi_2}}{\sqrt{psi_1}}$$

or

$$psi_2 = psi_1 \times \frac{(gpa_2)^2}{(gpa_1)^2}$$

To double output, pressure must be increased fourfold. To cut output in half, pressure must be decreased fourfold.

For any change in travel speed (mph), calculate the resulting $gpa_2$ by:

$$gpa_2 = \frac{gpa_1 \times mph_1}{mph_2}$$

or

$$\frac{gpa_1}{gpa_2} = \frac{mph_2}{mph_1}$$

or

$$mph_2 = \frac{gpa_1 \times mph_1}{gpa_2}$$

By doubling travel speed, sprayer output per acre is cut in half. By reducing travel speed to one-half of the original speed, sprayer output per acre is doubled.

#### Example

A superintendent finds it takes seven minutes, 15 seconds to empty a 40-gallon tank operated at 25 psi. However, he wishes to extend this time to seven minutes, 45 seconds. How is this done?

**Step 1:** First, find the gallons per minute for each time period:

$$gpm_1 = \frac{40 \text{ gallons}}{7.25 \text{ minutes}} = 5.52$$

$$gpm_2 = \frac{40 \text{ gallons}}{7.75 \text{ minutes}} = 5.16$$

**Step 2:** Now, insert these data into the following:

$$psi_2 = psi_1 \times \frac{(gpa_2)^2}{(gpa_1)^2} = 25 \times \frac{(5.16)^2}{(5.52)^2} = 21.89 \text{ psi}$$

Now the superintendent wants to adjust the sprayer speed. If it applies 20 gpa at 4 mph, what will be the output if speed is increased to 8 mph?

**Step 1:** Given: $gpa_1 = 20$; $mph_1 = 4$; $mph_2 = 8$.

**FIGURE 15.18** Visually determining the uniformity of a spray rig over an asphalt surface.

**Step 2:** Again, set up the formula to determine this:

$$\frac{gpa_1}{gpa_2} = \frac{mph_2}{mph_1}$$

$$\frac{20}{gpa_2} = \frac{8}{4}$$

$$gpa_2 = 20 \times \frac{8}{4}$$

$$gpa_2 = 10$$

## Determining Sprayer Output

Once the sprayer speed (mph) and nozzle discharge (gpm) are determined as previously described, sprayer output in gallons per acre (gpa) is calculated.

Calculate gallons per acre using the following formula. If the resulting output appears to be too high or too low, remember that altering speed or adjusting pressure will alter output. However, do not deviate beyond the manufacturer's recommendations. The other alternative is to reevaluate your nozzle selection to determine if you selected the proper one based on expected nozzle output and pressure (Figure 15.18).

$$\text{Gallons per acre} = \frac{\text{gallons}}{\text{minutes}} \times \frac{\text{hours}}{\text{miles}} \times \frac{\text{nozzles}}{\text{inches}} \times \frac{60 \text{ minutes}}{1 \text{ hour}} \times \frac{\text{miles}}{5{,}280 \text{ feet}} \times \frac{12 \text{ inches}}{\text{feet}} \times \frac{43{,}560 \text{ square feet}}{\text{acre}}$$

**or**

$$= \frac{\text{gallons per minute per nozzle} \times 5{,}940}{\text{miles per hour} \times \text{nozzle spacing (inches)}}$$

To determine gallons per 1,000 square feet, divide the gallons per acre value by 43.56.

### Example

A superintendent wishes to apply 1 pound active ingredient/acre of 2,4-D amine at 30 gpa and at 25 psi. The formulation of 2,4-D purchased has 4 pounds active ingredient/gallon. A total area of 338,000 square feet is to be treated. Nozzle spacing on the boom is 19 inches and the spray tank holds 35 gallons.

1. You check the sprayer speed on a fairway and find it takes 34 seconds to travel 200 feet. What is your speed in mph?

$$\text{Speed (mph)} = \frac{\text{distance traveled (feet)}}{\text{time (seconds)}} \times \frac{1 \text{ mile}}{5{,}280 \text{ feet}} \times \frac{3{,}600 \text{ seconds}}{\text{hours}}$$

$$= \frac{200 \text{ feet}}{34 \text{ seconds}} \times \frac{1 \text{ mile}}{5{,}280 \text{ feet}} \times \frac{3{,}600 \text{ seconds}}{\text{hours}}$$

$$= 4.0 \text{ mph}$$

**or** $\frac{\text{distance traveled (feet)} \times 0.682}{\text{time (seconds) to travel distance}}$

$$= \frac{200 \times 0.682}{34}$$

$$= 4.0 \text{ mph}$$

2. You catch the following output of the nozzles when operated at 25 psi. Should any need to be replaced?

| Nozzle # | Ounces/30 seconds |
|---|---|
| 1 | 29 |
| 2 | 29 |
| 3 | 30 |
| 4 | 31 |
| 5 | 31 |
| 6 | 30 |
| 7 | 31 |
| 8 | 29 |

The average output of these nozzles is 30 ounces per 30 seconds. If a 10% error term is allowed, then any nozzles with an output plus or minus 3 ounces per 30 seconds higher or lower than 30 ounces per 30 seconds would need replacing. Since this range would then be 27 to 33 ounces per 30 seconds and all current nozzles fall within this range, none need replacing.

3. What is your sprayer output (gpa)?

You first need to determine the average gallons per minute for each nozzle. From the previous step you know the average output is 30 ounces per 30 seconds. This equals 60 ounces per minute. Since 128 ounces are in each gallon, the following is calculated:

$$\text{Gallons per acre} = \frac{60 \text{ ounces}}{\text{minutes}} \times \frac{1 \text{ gallon}}{128 \text{ ounces}} = \frac{0.47 \text{ gallon}}{\text{minutes}}$$

Gallons per acre

$$= \frac{\text{gallons}}{\text{minute}} \times \frac{\text{hour}}{\text{mile}} \times \frac{\text{nozzle}}{\text{inch}} \times \frac{60 \text{ minutes}}{1 \text{ hour}} \times \frac{\text{mile}}{5{,}280 \text{ feet}} \times \frac{12 \text{ inches}}{\text{foot}} \times \frac{43{,}560 \text{ square feet}}{\text{acre}}$$

$$= \frac{0.47 \text{ gallon}}{\text{minute}} \times \frac{\text{hour}}{4 \text{ miles}} \times \frac{\text{nozzle}}{19 \text{ inches}} \times \frac{60 \text{ minutes}}{1 \text{ hour}} \times \frac{\text{mile}}{5{,}280 \text{ feet}} \times \frac{12 \text{ inches}}{\text{foot}} \times \frac{43{,}560 \text{ square feet}}{\text{acre}}$$

$$= 36.7 \text{ or } 37 \text{ gpa}$$

**or** $$\frac{\text{gallons per minute per nozzle} \times 5{,}940}{\text{miles per hour} \times \text{nozzle spacing (inches)}}$$

$$= \frac{0.47 \times 5{,}940}{4 \times 19}$$

$$= 36.7 \text{ or } 37 \text{ gpa}$$

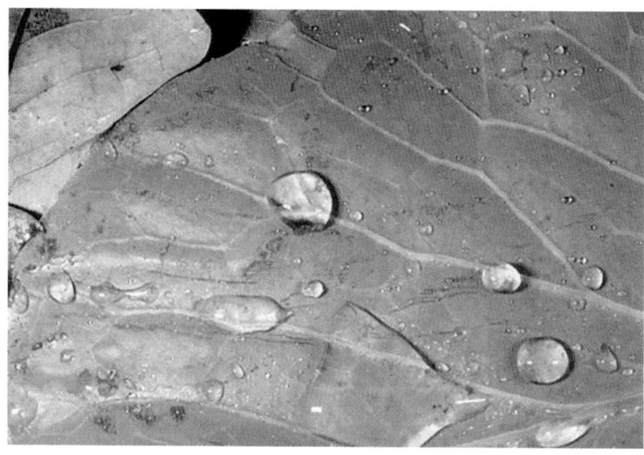

**FIGURE 15.19** Various adjuvants are available to improve adhering of products to waxy surfaces such as leaves.

4. How much herbicide would you need to buy to treat the 338,000 square feet?
   **Step 1:** Determine the number of acres in 338,000 square feet:

$$\text{Acres} = 338{,}000 \text{ square feet} \times \frac{1 \text{ acre}}{43{,}560 \text{ square feet}} = 7.8$$

   **Step 2:** Determine the amount of formulated herbicide needed per acre:

$$\frac{1 \text{ pound active ingredient}}{\text{acre}} \times \frac{1 \text{ gallon}}{4 \text{ pounds active ingredient}} = 0.25 \text{ gallon (or 1 quart)}$$

   **Step 3:** Combine these two previous steps to find the total amount of herbicide needed for the total area:

$$0.25 \text{ gallon/acre} \times 7.8 \text{ acres total} = \sim 2 \text{ gallons total}$$

5. How many ounces of this herbicide are needed to make a total spray volume of 35 gallons?
   First, find the number of gallons per tank needed for the appropriate herbicide application rate:

$$\frac{35 \text{ gallons}}{\text{tank}} \times \frac{\text{acres}}{34 \text{ gallons}} \times \frac{0.25 \text{ gallon}}{\text{acre}} = \frac{0.26 \text{ gallon}}{\text{tank}}$$

   Next, convert this value to ounces (oz):

$$\frac{0.26 \text{ gallon}}{\text{tank}} \times \frac{128 \text{ ounces}}{\text{gallon}} = \frac{33 \text{ ounces}}{\text{tank}}$$

   To convert to pints (pt):

$$\frac{0.26 \text{ gallon}}{\text{tank}} \times \frac{8 \text{ pints}}{\text{gallon}} = \frac{2.08 \text{ pints}}{\text{tank}}$$

   To convert to quarts (qt):

$$\frac{0.26 \text{ gallon}}{\text{tank}} \times \frac{4 \text{ quarts}}{\text{gallon}} = \frac{1.04 \text{ quart}}{\text{tank}}$$

   To convert to milliliters (mL):

$$\frac{0.26 \text{ gallon}}{\text{tank}} \times \frac{3{,}785 \text{ mL}}{\text{gallon}} = \frac{984 \text{ mL}}{\text{tank}}$$

6. Lastly, the number of acres that can be treated with a full tank of mixture must be determined.
   Use the following formula to determine this.

$$\text{Acre per tank} = \frac{\text{tank volume (gallons)}}{\text{gallons per acre}} = \frac{35 \text{ gallons}}{\text{tank}} \times \frac{\text{acre}}{34 \text{ gallons}} = 1.03 \text{ acre per tank}$$

Another way to determine how much pesticide to add per tank is to set up a ratio:

$$\frac{\text{Pesticide rate}}{\text{gallons per acre}} = \frac{X \text{ amount pesticide}}{\text{tank size (gallons)}}$$

In the previous example:

$$\frac{1 \text{ quart pesticide/acre}}{34 \text{ gpa}} = \frac{X \text{ quart pesticide}}{35\text{-gallon tank}}$$

$$X \text{ quart pesticide} = \frac{(1 \text{ quart})(35 \text{ gallons})}{34 \text{ gpa}}$$

$$X = \frac{1.03 \text{ quarts pesticide}}{35\text{-gallon tank}}$$

## Adding Adjuvants

Many pesticides require that various adjuvants (crop oils, surfactants, wetting agents, etc.) be added to the spray solution (Figure 15.19). Most of these are added on a volume basis. To determine this, go through the same types of calculations that determined how much pesticide needed to be added per tank.

### Example

An insecticide is being applied and the label suggests a crop oil concentrate be added at 1% volume of crop oil per volume of spray tank. The sprayer is calibrated to apply 20 gpa and the tank holds 100 gallons. For a full tank, how much crop oil concentrate is needed?

$$\text{Amount/100-gallon tank} = \frac{100 \text{ gallons}}{\text{tank}} \times 1\%$$
$$= \frac{1 \text{ gallon crop oil}}{100\text{-gallon tank}}$$

To add 0.25% nonionic surfactant per tank, the following is calculated:

$$\text{Amount/100-gallon tank} = \frac{100 \text{ gallons}}{\text{tank}} \times 0.25\%$$
$$= \frac{0.25 \text{ gallon NIS}}{100\text{-gallon tank}}$$

## Dilutions

Many turf products, including pesticides, are formulated as concentrated liquids for ease of handling and to save costs. Typically, these concentrated liquids must be diluted before being applied. Hose-end applicators often are calibrated where the concentrate is diluted by siphoning in a ratio to the volume of water (also called the diluent) that passes through the hose. Use the following formula for dilution problems:

Concentrate needed

$$= \frac{\text{percent active ingredient desired final concentration} \times \text{tank}}{\text{percent active ingredient initial product concentration}}$$

### Example

Five gallons of a 15% Clorox solution is desired to sterilize hand tools. How much Clorox is needed? (**Note:** Clorox is bought as a 100% concentrate.)

Concentrate needed

$$= \frac{\text{percent active ingredient desired final concentration} \times \text{tank size}}{\text{percent active ingredient initial product concentration}}$$
$$= \frac{15\% \times 5 \text{ gallons}}{100\%}$$
$$= 0.75 \text{ gallon (or 96 ounces)}$$

Add 96 ounces of Clorox to the tank and bring the final solution volume to 5 gallons.

### Example

Two gallons of 15% herbicide solution are desired to spot-treat weeds. How much herbicide (4 pounds active ingredient/gallon formulation) is needed? (Assume this herbicide is bought as a 50% concentrate.)

$$\text{Amount concentrate needed} = \frac{0.15 \times 2 \text{ gallons}}{0.5}$$
$$= 0.6 \text{ gallon (or 4.8 pints)}$$

Add 4.8 pints to the tank and bring the final solution volume to 2 gallons.

### Example

A hose-end sprayer applied 2.5 gallons of spray while siphoning 2 quarts out of a container.

1. If a fungicide is to be applied at a rate of 3 ounces/gallon of water, how should the stock solution be mixed?

$$\text{Concentrate needed} = \frac{\text{desired final concentrate}}{\text{initial product concentrate}} \times \text{tank size}$$
$$= \frac{3 \text{ ounces/gallon}}{0.5 \text{ gallon}} \times 2.5 \text{ gallons}$$
$$= 15 \text{ ounces/gallon}$$

Therefore, add 15 ounces of the fungicide and add water to bring the final solution volume to 1 gallon.

2. What is the parts per million of this final solution?

One gallon of water weighs 133 ounces. Therefore, insert the following:

$$\text{Parts per million} = \frac{\text{weight material used}}{\text{weight of tank mixture}} \times 1{,}000{,}000$$
$$= \frac{15 \text{ ounces}}{133 \text{ ounces}} \times 1{,}000{,}000$$
$$= 112{,}782$$

### Converting between Active Ingredients and Formulated Product

Pesticide recommendations are often given in pounds of active ingredient per acre because several different trade names and formulations may accompany a particular chemical. For example, mancozeb is the common name of a fungicide; however, trade names include Fore, Dithane, Formec, and Protect. Each product may have a different formulation (or strength); thus, pounds active ingredient/acre provides the basic rate regardless of formulation. Therefore, turf managers should be able to make the necessary calculations from pounds or gallons or pints of product to pounds of active ingredient and vice versa. The following formula determines liquid conversion:

$$\text{Liquid conversion} = \frac{\text{pounds active ingredient recommended}}{\text{acre}} \times \frac{\text{gallons}}{\text{pounds active ingredient}} = \frac{\text{gallons of material needed}}{\text{acre}}$$

Wettable powders (WPs), soluble powders (SPs), dry flowables (DFs), and water-dispersible granules (WDGs) are formulated in percent active ingredient and require the following for conversion:

$$\text{Powder (pounds) needed per acre} = \frac{\text{pounds active ingredient recommended}}{\text{acre}} \times \frac{\text{product}}{\text{percent active ingredient of material}}$$

#### Example

To control a certain insect, the superintendent tells you to apply 0.5 pound active ingredient/acre of Trumpet 4E liquid insecticide over 54 acres. How much Trumpet 4E should be bought?

**Step 1:** Determine the total pounds of active ingredient needed to treat 54 acres,

$$54 \text{ acres to treat} \times 0.5 \text{ pounds active ingredient/acre} = 27 \text{ pounds active ingredient needed}$$

**Step 2:** To find the total amount of the 4E formulation of Trumpet insecticide needed,

$$27 \text{ pounds active ingredient needed} \div 4 \text{ pounds active ingredient/gallon} = 6.75 \text{ gallons of Trumpet 4 needed to treat 54 acres at 0.5 pound active ingredient/acre}$$

If you find the 4E formulation of Trumpet is no longer available, but instead a 75DF formulation is, then the following is calculated (the 75DF formulation contains 75% active ingredient).

$$27 \text{ pounds active ingredient needed} \div 0.75 \text{ active} = 36 \text{ pounds of Trumpet 75DF needed to treat 54 acres at 0.5 pound active ingredient/acre}$$

### Acid Equivalents of Salts and Esters

Some pesticides, particularly herbicides, are formulated as salts, esters, or both. Recommendations are often based on pounds of the acid equivalent of the active ingredient per acre to provide standardization, since salt and ester forms vary in weight. The acid equivalent of a salt or ester is the portion of the molecule that is the parent acid form of the molecule.

Generally, the salt or ester portion of a pesticide is inactive; thus, the acid portion of this salt or ester molecule is the active portion. When determining the acid equivalent rate of a salt or ester form of a pesticide, it is necessary to adjust the rate to account for the heavier weight of the salt or ester. This is usually reported in terms of pounds of acid equivalent per gallon.

## SPREADERS AND THEIR CALIBRATION

Dry pesticide or fertilizer application is only effective if uniform coverage is ensured. These can be applied with either a drop (gravity) spreader or a rotary (centrifugal) spreader (Figure 15.20).

A drop spreader has the advantage of applying a fairly exact pattern since this is limited to the distance between the wheels. This method also allows a "tight" pattern (line) to be cut but requires each pass to exactly meet with the previous pass or skips will occur. Wide (more than 6 feet, 2 m) drop spreaders can become cumbersome in the landscape by limiting access around trees and shrubs and getting through gates.

**FIGURE 15.20** Centrifugal (left) and drop spreaders are used to applied granular products.

The agitator in the bottom of the drop spreader's hopper also may break the coating of some slow-release fertilizers.

The cyclone (also known as rotary or centrifugal) spreader generally has a wider pattern of distribution compared to a drop spreader and can cover a larger area in a short time. The application pattern of the cyclone spreader also gradually diminishes away from the machine, reducing the probability of an application skip. The uneven, wide pattern of the cyclone spreader is initially harder to calibrate, and heavier particles tend to sling farther away from the machine. However, proper calibration and experience minimize these incidents.

A recent improvement in spreader technology is the use of air to apply the material to the turf. This technique produces a fairly wide pattern (like the cyclone spreader) that is somewhat exact (like the drop spreader) without damaging the granules or slinging heavier particles farther. Wind and rain effects also are reduced using the technology but initial equipment expense and application expertise needed are higher.

Spreader calibration involves measurement of the granular output as the spreader is operated over a known area (Figure 15.21). One way to ensure uniform application of material is to divide the material into two equal portions. Use a spreader calibration that will deliver one-half the correct amount of material. Make an application over the entire area, turn the spreader direction 90° from the initial application, and make a second application, thereby eliminating skips in the coverage. Accordingly, calibration of the spreader should be based on one-half the desired application rates. A flat surface, a method of collecting the material, and a scale for weighing the material are needed for calibration. The following sequence of steps will aid in calibrating a fertilizer spreader.

**FIGURE 15.21** Turf damage from nonuniform application with a granular spreader.

**FIGURE 15.22** Attached catch pan when calibrating the output of a drop spreader.

### Calibrating a Drop-Type (Gravity) Spreader

To calibrate your drop-type (gravity) spreader, follow these steps, in order:

1. Check the spreader to make certain all the parts are functioning properly.
2. Mark off an area, multiplying it by the width of the spreader to give 100 square feet of area. For example, the length required for 1.5-, 2-, and 3-foot spreaders is 66.33, 50, and 33.33 feet, respectively.
3. Fill the spreader with the material you wish to apply (fertilizer, seed, pesticide, lime, other). Fill the hopper only to the level you will have when the material will actually be applied.
4. Make several trial runs over the area and practice opening the spreader as you cross the starting line and closing it at the finish line. Opening the spreader before it is in motion will result in nonuniform distribution. Walk at a pace that will be used when actually applying the material. Open and close the spreader gradually, not in a fast, jerky motion.
5. The weight of the material applied by the spreader must be determined. It can be swept up from a hard surface or caught on a large piece of paper or plastic. The easiest method is to attach a catch pan (cardboard works nicely) under the spreader openings; the material in the catch pan during the test run can determine how much was applied (Figure 15.22).
6. Begin calibration at the lowest setting and proceed at progressively higher settings (larger openings). The more trials there are at a given setting, the better will be the average rate of application. Three trials at a given setting are usually enough to obtain a reliable application rate. Weigh the material and record the information on each trial run for future use.
7. One of the calibrated settings will approximate the correct rate of material. Settings are not necessarily linear; therefore, half of a particular application rate may not necessarily be obtained by using a setting number of half the original.

#### Example

You want to calibrate a spreader to apply onr pound N/1,000 square feet using a 10-10-10 fertilizer. This calculates to 10 pounds fertilizer per 1,000 square feet since the material is 10% N (10% × 10 pounds = 1 pound N).

Since the area for calibration trials is in only 100 square feet, apply 1/10 of 10 pounds, or 1 pound fertilizer, per 100 square feet. The spreader setting should be 11 for this example if you obtain the following results shown from your calibration trials with your spreader. If the desired application rate was 1/2 pound N (5 pounds material per 1,000 square feet, or 1/2 pound/100 square feet), a setting of 7 should be used.

| Setting | Output (ounces) |
|---|---|
| 1 | 2 |
| 3 | 3 |
| 5 | 6 |
| 7 | 8 |
| 9 | 10 |
| 11 | 16 |

8. The same calibration procedure is used for any material you want to apply. Since the quantity applied depends upon the physical properties of the material, the same settings cannot be used for different materials, even if the ratios are the same. Once the spreader is calibrated and set for the proper rate, any size area can be accurately treated.

### Example

The application rate for a granular insecticide is 3 pounds product/1,000 square feet. Your applicator is 4 feet wide and you decide to catch the output when pushed for 200 feet. How much should be caught to calibrate the spreader to provide 3 pounds/1,000 square feet?

**Step 1:** First, determine the amount of total material needed to be caught in the area used for calibration.

$$\text{Four-foot-wide spreader} \times 200 \text{ feet long}$$
$$= 800 \text{ square feet of total area being treated}$$

**Step 2:** Next, set up a ratio involving the rate needed on a 1,000 square foot basis to the actual area being treated:

$$\frac{3 \text{ pounds}}{1{,}000 \text{ square feet}} = \frac{X \text{ pounds}}{800 \text{ square feet}}$$
$$X = \frac{(3)(800)}{1{,}000}$$
$$= 2.4 \text{ pounds}$$

Now use a catch basin to collect the output during the 200 feet. The output is then weighed and the spreader setting is adjusted to give more or less output as needed. Repeat this step until the unit is set to apply 2.4 pounds in 200 feet traveled with a spreader 4 feet wide.

### Calibrating a Rotary (Centrifugal) or Pendulum Spreader

It is important that the "effective" width of application be determined first (Figure 15.23). To calibrate your rotary (centrifugal) spreader, follow these steps, in order:

1. Check the spreader to make certain that all parts are operating properly.
2. Fill the hopper about half full with the material you plan to apply and run it with the spreader setting about half open (medium setting). Make the application on bare ground or a hard surface where the width of surface covered by the material can be measured.
3. Rotary spreaders do not apply a constant amount of material across the entire width of application. More material is applied toward the center and less at the edges. For this reason, the width of application is accurate for a constant application rate only at about two-thirds (60% to 70%) of the actual width measured.

**FIGURE 15.23** Using catch pans to determine the spread width and uniformity of a centrifugal spreader.

**FIGURE 15.24** Using a catch box to measure the output of a centrifugal spreader. (Courtesy John Boyd.)

### Example

If the application width is 12 feet, only about 8 feet (or 4 feet across both sides of the spreader), within the band of application, is receiving approximately the same application rate. The other 2 feet on each edge, respectively, receive much less material than the center area. Once this "effective" width is determined, calibration is fairly simple.

**Step 1:** Mark off a test distance that, when multiplied by the effective width, provides a 1,000 square-foot area. For this example, assume the "effective" width is 10 feet. The test strip will then be 100 feet long since width times length is 10 × 100 or 1,000 square feet. **Note:** This calculation is based on "effective" width of application and not the total width.

**Step 2:** Determine the amount of material to be applied.

### Example

To apply 1 pound N/1,000 square feet using a 16-4-8 fertilizer, 6.25 pounds of material should be applied per 1,000 square feet.

**Step 1:** Fill the hopper with a known weight of fertilizer and adjust the spreader to the lowest setting that will allow the material to flow. Push the spreader down the center of the test area, opening the hopper at the starting line and closing it at the finish. Weigh the material left in the spreader and subtract the amount from the starting weight to determine the amount used per 1,000 square feet. The beginning weight minus the ending weight tells how much material was applied per 1,000 square feet.

**Step 2:** Repeat the preceding step at successively greater settings (openings) and record the material applied at each setting (Figure 15.24).

**Step 3:** Select the spreader setting that most closely applies the desired rate of material, set the spreader accordingly, and use it on any size area. To obtain uniform spread of material, remember to set the spreader at half the desired rate of application and make two passes at 90° to each other. Strive for proper spread overlap during application.

If the "effective" width is 10 feet, after each pass, move the spreader over 10 feet from the center of the tire tracks. This will give a fairly constant rate of application over the entire area.

A simpler way to use a centrifugal spreader is to determine the amount of material to apply on a given area and then set the spreader at a low setting and apply this amount in several directions.

### Example

A rate of granular fungicide is 2.5 pounds/1,000 square feet. The nursery green to be treated is 40 feet by 150 feet. How much fungicide should go on the area?

**Step 1:** First, find the total amount of area to be treated:

$$40 \text{ feet} \times 150 \text{ feet} = 6{,}000 \text{ square feet to treat}$$

**Step 2:** Next, set up a ratio to convert the amount needed on a 1,000 square feet basis to the actual 6,000 square feet green area:

$$\frac{2.5 \text{ pounds}}{1{,}000 \text{ square feet}} = \frac{X \text{ pounds}}{6{,}000 \text{ square feet}}$$

$$X = \frac{(2.5)(6{,}000)}{1{,}000}$$

$$= 15 \text{ pounds}$$

Finally, place the 15 pounds of fungicide into the spreader hopper and set the opening in the bottom of the applicator to give a low rate of material flow. Push the spreader over the area to be treated until all of the fungicide has been applied. To improve the uniformity of application with this type of spreader, you should always go over the area in at least two directions.

# Section VII

## Budgeting Practices

# 16 Budgets and Personnel Management

## BASIC BUSINESS TERMS AND PRINCIPLES

Just as the layout and expectations vary from one golf course to another, so do costs, personnel requirements, equipment and supply needs, and budgets. Careful planning and communication are essential to the financial success of the operation, no matter what the decision makers agree to spend on the facility, as the owner and manager use the budget process to plan, project, and control the course's financial resources. Successful superintendents learn to develop and implement budgets and personnel issues and how to communicate these needs to the owners or managers (Figure 16.1).

Budgets are organized systematically in stages. This includes long-term improvements of the course, daily expectations of course conditions and personnel and budget needs to obtain these, and cost projections to meet the desired goals. Comparison with similar facilities provides a starting point for this, as do accounting firms or banks experienced in dealing with golf courses.

Accounting involves the design of a recordkeeping system, preparation of summarized reports based on the recorded data, and interpretation of those reports. Superintendents need a basic understanding of accounting and business terms and principles to intelligently talk with club officials, office managers, bookkeepers, accountants, loan institutions, etc. Without a basic understanding of these principles, the superintendent will probably not receive full consideration of financial requests, budget adjustments, or equipment purchases.

**Cash versus accrual basis.** These are the two primary means for recording financial transactions. The cash basis records revenue (or income) when cash is received and records expenses when cash is paid out. The cash basis is easiest for most to understand as personal checkbooks are based on this method. However, for larger businesses, the accrual method is generally more appropriate. The accrual basis records revenues in the period when earned, such as when services are performed or goods delivered, regardless of when the cash is actually received. Similarly, expenses are recorded when incurred (or created), regardless of when the expense is paid.

### Example

If a golf course has a membership drive and sells a membership for the next year on October 1 of the current year, the cash basis would record the cost of the membership as received revenue for October of the current year. However, if the accrual method for financial statement reporting is being used, the revenue would not be recorded for the current year as with the cash method, since income had not yet been earned. Instead, an unearned revenues account would be credited for the amount since the club "owes" the customer the cost of the year's membership until a service or product is delivered.

**Transactions** are an occurrence or event that must be recorded.

**Ledgers** are the complete set of accounts for a business. The classical account has two sides—left and right—separated by a vertical line, and the account is topped by a horizontal line with the account's name written on it (often referred to as a "T" format).

**Assets** are tangible or intangible properties owned by a business.

**Equities** are the rights or claims to those assets.

**Liabilities** reflect equities that are debts of a business (or rights of creditors). *Owner's equity* reflects the rights of the owners (often as capital stocks and retained earnings), or the cumulative total of annual net income and net losses since the business was first incorporated. The list of assets provides a description of the various business properties, while the list of liabilities and equities indicates the funding source of those assets. These form the accounting equation: assets = liabilities + owner equity.

**Depreciation.** This is a business expense reflecting a decrease in usefulness over an asset's lifetime. This decline in usefulness is due to physical wear and tear or technical obsolescence. For example, a $20,000 mower with a projected useful life expectancy of 10 years would indicate an annual expense deduction (or depreciation) of $2,000 instead of a one-time expense of $20,000.

One of the major concerns of accounting is the preparation of summarized reports of the recorded data. Income and balance sheets summarize the financial results and positions of a business. The income sheets summarize the revenue (income) and expenses of a business during a specific period of time. If total revenues exceed total expenses, the result is net income (or net profit)—often referred to as a business's "bottom line."

The balance sheet lists the assets, liabilities, and retained earnings (also called owner's equity including capital stocks) as of a specific date (Table 16.1). Assets are usually listed first, followed by a list of liabilities and a section detailing owner's equity. Readers of the balance sheet often seek the current liquidity and ratio of debt to equity in the business. Current liquidity refers to the amount of cash and cash equivalents relative to total assets and the ability to pay off debt due within the operating period. The ratio of current assets to total current liabilities indicates the comparable levels of asset creation produced by debt versus that produced by earnings and stockholder investment. This also provides a rough determination of the amount of working capital available.

**FIGURE 16.1** Successful golf course operations require sound agronomic practices, a qualified crew, and a strong financial understanding of budgeting.

## BUDGETS

Budgets for golf courses come in all shapes and sizes depending on the quality of the initial course design and construction; the size and location of the course; the proximity of available resources, such as water, topdressing materials, irrigation supplies, and fertilizers; the equipment age and preventative maintenance; membership makeup; player demand; play; financial resources; and desired playing conditions for the individual course. There is no standard cost per round, or cost per acre, for managing a golf course. Rather, golf course owners, managers, and members determine the conditions they desire and the money they are willing to spend to achieve their goal.

### Budget Purpose

The purposes of a budget are for planning, projecting, and controlling. Planning involves determining the labor needs and resources for maintaining the course at the desired level of quality and within its financial resources. Projecting the budget over a period of time allows the golf course manager and owner to determine how much money will be needed in the short term to cover bills, and how much money should be held in reserve for future expenditures. The budget can also be a tool for controlling expenditures, based on income or previous projections, by measuring the plan and the projection in comparison to the actual financial activity. In summary, the budget is a process of communicating current and future financial needs and of tracking the actual financial activity of the course.

**TABLE 16.1**
**A Sample Balance Sheet Businesses Typically Use**

SAMPLE COMPANY
BALANCE SHEET
DATE

| | |
|---|---|
| **Current assets:** | **Current liabilities:** |
| • Cash | • Accounts payable |
| • Accounts receivable | • Notes payable (includes current portion of long-term debt) |
| • Notes receivable | • Other payables |
| • Marketable securities | **Total current liabilities** |
| • Inventory | |
| • Prepaid expenses | |
| **Total current assets** | |
| **Long-term assets:** | **Long-term liabilities:** |
| • Land | • Notes payable (net of current portion) |
| • Office equipment | • Bonds payable |
| • Maintenance equipment | **Total long-term liabilities** |
| **Total long-term assets** | **Total liabilities** |
| **Other assets:** | **Owner's equity:** |
| • Goodwill | • Capital stock |
| • Intangible assets | • Retained earnings |
| **Total other assets** | **Total owner's equity** |
| **Total assets** | **Total liabilities and owner's equity** |

| | | | |
|---|---|---|---|
| Working capital | = | current assets | – current liabilities |
| Working capital ratio | = | $\frac{\text{current assets}}{\text{current liabilities}}$ | where: <1 = negative working capital<br>1.2 to 2.0 = sufficient capital ("net" working capital),<br>>2 = excess assets |

## Income

Income (or revenue) sources for a golf course are multifaceted. Revenue sources vary depending upon the type of golf course and services provided. Private facilities charge membership dues and fees. Semiprivate, public, or "for a fee" courses charge a fee for each round played. Other revenue categories are typically associated with the game or are generated by additional amenities such as food and beverage services and can generate substantial revenues. These are generally separated into two groups: (1) fees, and (2) golf-related services (Table 16.2). Other means of funding a club include member assessment, borrowing, issuing bonds, and municipal funding.

## Expenses

Costs for operating a golf course facility include capital improvements, capital expenses, fixed costs, and variable (or maintenance or operating) costs:

**Capital improvements.** Capital improvements are those expenditures (assets) that improve the value of the property. This includes permanent constructions such as a pump house, shelters, bridges, cart paths, parking lot, tree planting, tennis courts, permanent drainage, swimming pools, new green construction, and so forth. These items normally are prime objectives of long-range improvements and are often funded from the sale of property and initiation fees, as well as windfall profits.

Long-range planning is a must for any forward-thinking organization and, for golf courses, it typically involves a minimum planning period of two to three years with a maximum of five years. Failure to plan for large-ticket capital improvement items on an annual basis may lead to an excessive, one-time expense to replace equipment or buildings. Without long-range goals, operations become outdated, efficiency is lost, and employee turnover is often higher.

**Capital expenses.** Capital expenses or investments are nonhuman durable inputs of production, composed mostly of equipment, tools, and office furniture. Capital expenses are often the prime objective of short-range improvements. Equipment becomes worn, outdated, or even dangerous. Player demands and increasing labor costs dictate the need to purchase labor-saving devices that provide equal or better playing conditions than previously achieved. Planning for new or additional capital expenses typically is a continuous three- to five-year process. Equipment leasing and contract maintenance are two relatively recent means through which courses are maximizing their capital dollars. Capital items are allocated to the capital budget and are usually subject to tax depreciation deductions.

**Fixed costs.** Fixed costs include those required to maintain capital ownership and employ resources regardless of play, including land rent, insurance, taxes, depreciation, interest charges, and other costs.

**Variable (or maintenance) costs.** Variable or maintenance costs are resources allocated for routine golf course maintenance. Revenues from golf course fees generally fund the expenses in the golf course operating budget. Typical golf course maintenance (or variable) expenses are listed in Table 16.3 and do not include capital improvements, capital expenses, or fixed costs. Variable expenses are recorded to the operating budget and are fully tax deductible against revenue.

Table 16.3 does not, of course, cover all typical expenses that golf courses incur. Other maintenance considerations involve golf equipment such as pins, flags, cups, tee markers, benches, towels, and rope; repair to fences, shelters, and bridges; repair and upkeep of the maintenance building, tennis courts, and swimming pool; and miscellaneous items such as employee uniform laundry, office supplies, and vehicle licenses.

Floods, freezes, droughts, accidental spills or misapplied products, tornadoes, hurricanes, fire, vandalism, and other unusual expenses do not frequently occur, but most courses usually experience one or more of these at some point (Figure 16.2). Proper insurance coverage, continued employee training, and preventative measures should be planned for these and other unplanned expenses. Tables 16.4 and 16.5 offer examples of forecasting maintenance budgets and actual budgets for a typical 18-hole facility. Table 16.6 provides an example of a form listing approved, projected, and proposed operating budgets.

Tables 16.7 through 16.10 list employee salaries, equipment inventories, equipment purchases, and various supplies purchased. Much of the mundane recordkeeping for a modern golf course maintenance unit can be managed by a number of available database, spreadsheet, and/or specialized computer programs targeted for golf course maintenance and record keeping.

Due to escalating labor costs, courses often contract certain labor needs or hire part-time laborers. Courses also shop

**TABLE 16.2**
**Means and Approximate Percentages of Each from That a Typical Golf Facility Generates Income**

| Source | ~% of Total Income |
|---|---|
| **Fees** | |
| • Membership dues | 15 to 20 |
| • Golf cart fees | 15 |
| • Golf green/guest fees | 15 |
| • Initiation fees | 5 |
| **Services** | |
| • Food sales | 10 to 15 |
| • Merchandise sales | 5 to 8 |
| • Caddie services | ~5 |
| • Beverage sales | ~4 |
| • Tournament operations | ~2 |
| • Golf range | ~2 |
| • Golf instruction | ~1 |
| • Club rental | <1 |
| • Club repair | <1 |
| • Handicapping service | <1 |
| • Golf bag storage | <1 |
| • Locker fees | <1 |
| • Other (e.g., cart storage, catering, facility rental, raffles, etc.) | ~5 |

**TABLE 16.3**
**Typical Expenses and Their Approximate Percentages of the Total Maintenance Budget for a Moderately Budgeted 18-Hole Golf Course**

| Expense | ~Total Maintenance Budget (%) |
|---|---|
| • Salaries and wages (including benefits such as retirement, group health insurance, unemployment insurance, etc.) | 45 to 65 |
| • Fertilizer | 5 to 10 |
| • Chemicals | 5 to 10 |
| • Irrigation repair and drainage | 2 to 10 (location dependent) |
| • Irrigation water | Varies (~0 to 20) |
| • Parts, service, tools | 1 |
| • Utilities (electrical, heat, AC, wireless, phone) | 3 to 5 |
| • Gas, oil, and lubricants | 2 |
| • Landscaping (annuals, perennials, mulch, etc.) | 1 to 3 |
| • Topdressing sand and gravel | 1 to 2 |
| • Seed, sod, sprigs | 5 |
| • Tree service (where needed) | 3 |
| • Travel and association dues/professional development | 1 to 3 |
| • Equipment parts and repairs | 5 to 10 |
| • Cart path repairs | 1 to 3 |
| • Contract maintenance | 1 to 5 |
| • Office supplies | <1 |
| • Auto/vehicle allowance | 1 to 2 |
| • Soil, water, and disease analysis (testing) | <1 |
| • First aid | <1 |
| • Temporary labor | 1 to 5 |
| • Uniforms | 1 to 2 |
| • Equipment lease | 1 to 5 |
| • Equipment rental | <1 |
| • Cleaning supplies | <1 |
| • Golf maintenance supplies (e.g., cups, flags, ropes, stakes, coolers, paint, signs, etc.) | <1 |
| • Aquatic weed control and plant material | <1 |
| • Special projects | 1 to 5 |
| • Consultants | 1 to 2 |
| • Contingency | 1 to 5 |

insurance plans annually for best rates and are starting to adopt insurance plans that cater to healthier employee lifestyles and, thus, receive more favorable premiums. This ranges from gym memberships, no-smoking policies, healthier diets, minimum daily exercise, adequate sleep, routine checkups, etc.

Courses also realize it generally is cheaper to retain current employees compared to constant rehiring. Incentives to retain employees include providing a bonus structure, continued training and promotions, facility privileges, and other means to make employees feel appreciated.

**FIGURE 16.2** Disasters occur infrequently but facilities should be prepared for one.

**Tournament costs.** Local courses often desire playing conditions similar to those that professional golfers experience at major tournaments. Superintendents and their course conditions are often unfairly compared to the latest televised professional tournament with little regard to the time, labor, equipment, and financial resources required to achieve those conditions. The goal of golf courses hosting professional events is to have the course prepped to test participants' skills and have each player encounter similar conditions each day of the tournament.

Additional labor is a primary cost consideration for hosting a major tournament. For example, a recent U.S. Open Tournament required an additional $100,000 the two years preceding the event for enhanced course preparation—not including capital investments for irrigation additions, greens construction, tee additions, tee leveling, or bunker renovation. Five times the normal labor pool was required during the tournament to provide acceptable playing conditions. Club officials determined that maintaining the golf course under continuous tournament conditions would require an additional $1 million each year. This is in addition to the normal labor allocation budget of $750,000 for "routine" playing conditions. Maintaining tournament conditions for long periods also often sacrifices grass quality since the lower mowing heights and increased stresses in terms of lean nitrogen and less water often combine to eventually damage the grass.

## LONG-TERM PLANNING/GOLF COURSE STANDARDS

It is highly recommended that superintendents develop, along with the club professional/manager, greens committee, and board of directors/owner, long-term plans for the enhancement and care of the facility. This should include desired playing conditions (or expectations), anticipated facility upgrades, equipment replacement, capital improvements, and labor changes (Figure 16.3). The complex nature and expectations of modern golf course maintenance and play conditions are becoming increasingly evident. Finely conditioned courses do not happen by accident,

**TABLE 16.4**
**Example of a Form Used for Forecasting Budget Expenses**

FORECASTED BUDGET EXPENSES Year: ____________

| | Monthly Expenses—Estimated | | | | | | | | | | | | |
|---|---|---|---|---|---|---|---|---|---|---|---|---|---|
| **Item** | **J** | **F** | **M** | **A** | **M** | **J** | **J** | **A** | **S** | **O** | **N** | **D** | **Total** |
| Salaries and wages | | | | | | | | | | | | | |
| Chemicals | | | | | | | | | | | | | |
| Irrigation/drainage repair | | | | | | | | | | | | | |
| Irrigation water | | | | | | | | | | | | | |
| Parts, service, tools | | | | | | | | | | | | | |
| Utilities | | | | | | | | | | | | | |
| Gas and oil | | | | | | | | | | | | | |
| Landscaping | | | | | | | | | | | | | |
| Sand and gravel | | | | | | | | | | | | | |
| Seed, sod, sprigs | | | | | | | | | | | | | |
| Trees service | | | | | | | | | | | | | |
| Professional development | | | | | | | | | | | | | |
| Equipment parts/repairs | | | | | | | | | | | | | |
| Cart path repairs | | | | | | | | | | | | | |
| Contract maintenance | | | | | | | | | | | | | |
| Office supplies | | | | | | | | | | | | | |
| Auto/vehicle allowance | | | | | | | | | | | | | |
| Soil, water, and disease tests | | | | | | | | | | | | | |
| First aid | | | | | | | | | | | | | |
| Temporary labor | | | | | | | | | | | | | |
| Uniforms | | | | | | | | | | | | | |
| Equipment lease | | | | | | | | | | | | | |
| Equipment rental | | | | | | | | | | | | | |
| Cleaning supplies | | | | | | | | | | | | | |
| Golf supplies | | | | | | | | | | | | | |
| Aquatic weed control | | | | | | | | | | | | | |
| Special projects | | | | | | | | | | | | | |
| Consultants | | | | | | | | | | | | | |
| Other | | | | | | | | | | | | | |
| Contingency | | | | | | | | | | | | | |
| | | | | | | | | | **Grand total** | | | | $ |

but rather are the planned result of many contributing factors. Members often desire to take their course to the "next level," and written maintenance standards help determine what this level is and what is needed to achieve it. Each segment of the course and its facilities should be included. This involves long-term goals and means of achieving these for each hole in terms of greens, tees, bunkers, fairways, cart paths, landscaping, roughs, out-of-play areas, and surrounds. Equipment, labor, and facilities goals also should be included (Figure 16.4).

Plans for course goals, play expectations, capital investments, and expenses need to be discussed, approved, and documented by all affected parties. The board of directors and greens committees continually change, often with little regard to previous committee or board wishes. Complete club membership turnover often occurs every 15 or so years. Problems and unnecessary expenses are avoided if a long-range plan is documented and on file, to be passed on from retiring board members to newly elected board members. This planning

**TABLE 16.5**
**Example of a Form Used for Actual Budget Expenses**

ACTUAL BUDGET EXPENSES Year: ____________

| Item | Monthly Expenses—Actual | | | | | | | | | | | | |
|---|---|---|---|---|---|---|---|---|---|---|---|---|---|
| | J | F | M | A | M | J | J | A | S | O | N | D | Total |
| Salaries and wages | | | | | | | | | | | | | |
| Chemicals | | | | | | | | | | | | | |
| Irrigation/drainage repair | | | | | | | | | | | | | |
| Irrigation water | | | | | | | | | | | | | |
| Parts, service, tools | | | | | | | | | | | | | |
| Utilities | | | | | | | | | | | | | |
| Gas and oil | | | | | | | | | | | | | |
| Landscaping | | | | | | | | | | | | | |
| Sand and gravel | | | | | | | | | | | | | |
| Seed, sod, sprigs | | | | | | | | | | | | | |
| Trees service | | | | | | | | | | | | | |
| Professional development | | | | | | | | | | | | | |
| Equipment parts/repairs | | | | | | | | | | | | | |
| Cart path repairs | | | | | | | | | | | | | |
| Contract maintenance | | | | | | | | | | | | | |
| Office supplies | | | | | | | | | | | | | |
| Auto/vehicle allowance | | | | | | | | | | | | | |
| Soil, water, disease tests | | | | | | | | | | | | | |
| Dues and subscriptions | | | | | | | | | | | | | |
| First aid | | | | | | | | | | | | | |
| Temporary labor | | | | | | | | | | | | | |
| Uniforms | | | | | | | | | | | | | |
| Equipment lease | | | | | | | | | | | | | |
| Equipment rental | | | | | | | | | | | | | |
| Cleaning supplies | | | | | | | | | | | | | |
| Golf supplies | | | | | | | | | | | | | |
| Aquatic weed control | | | | | | | | | | | | | |
| Special projects | | | | | | | | | | | | | |
| Consultants | | | | | | | | | | | | | |
| Other | | | | | | | | | | | | | |
| Contingency | | | | | | | | | | | | | |
| | | | | | | | | | Grand total | | | | $ |

**TABLE 16.6**
**Example of a Form Used for Approved, Projected, and Proposed Operating Budgets**

| APPROVED, PROJECTED, AND PROPOSED OPERATING BUDGETS | | | | | |
|---|---|---|---|---|---|
| **Item** | **20_–_ Approved Budget** | **20_–_ Projected Budget** | **20_–_ Proposed Budget** | **$ Increase from Projected to Proposed Budgets** | **% Increase from Projected to Proposed Budgets** |
| Salaries and wages | | | | | |
| Chemicals | | | | | |
| Irrigation/drainage repair | | | | | |
| Irrigation water | | | | | |
| Parts, service, tools | | | | | |
| Utilities | | | | | |
| Gas, oil, lubricants | | | | | |
| Landscaping | | | | | |
| Sand and gravel | | | | | |
| Seed, sod, sprigs | | | | | |
| Trees service | | | | | |
| Professional development | | | | | |
| Equipment parts/repairs | | | | | |
| Cart path repairs | | | | | |
| Contract maintenance | | | | | |
| Office supplies | | | | | |
| Auto/vehicle allowance | | | | | |
| Soil, water, and disease tests | | | | | |
| Dues and subscriptions | | | | | |
| First aid | | | | | |
| Temporary labor | | | | | |
| Uniforms | | | | | |
| Equipment lease | | | | | |
| Equipment rental | | | | | |
| Cleaning supplies | | | | | |
| Golf supplies | | | | | |
| Aquatic weed control | | | | | |
| Special projects | | | | | |
| Consultants | | | | | |
| Other | | | | | |
| Contingency | | | | | |
| Totals | | | | | |

**TABLE 16.7**
**Example of a Form Used for Tracking Overall Salaries and Wages**

| OVERALL SALARIES AND WAGES | | | | | | | Year: ________ | | | | | | |
|---|---|---|---|---|---|---|---|---|---|---|---|---|---|
| | Monthly Expenses | | | | | | | | | | | | |
| Position | J | F | M | A | M | J | J | A | S | O | N | D | Total |
| Superintendent | | | | | | | | | | | | | |
| First assistant | | | | | | | | | | | | | |
| Second assistant | | | | | | | | | | | | | |
| Irrigation technician | | | | | | | | | | | | | |
| Spray technician | | | | | | | | | | | | | |
| Equipment manager | | | | | | | | | | | | | |
| Equipment operator 1 | | | | | | | | | | | | | |
| Equipment operator 2 | | | | | | | | | | | | | |
| Equipment operator 3 | | | | | | | | | | | | | |
| Equipment operator 4 | | | | | | | | | | | | | |
| Horticulturist | | | | | | | | | | | | | |
| Office assistant | | | | | | | | | | | | | |
| Summer intern 1 | | | | | | | | | | | | | |
| Summer intern 2 | | | | | | | | | | | | | |
| Monthly totals | | | | | | | | | | | | | |
| | | | | | | | | | Grand total | | | | $ |

**TABLE 16.8**
**Example of a Form Used for Recording Equipment Inventory**

| EQUIPMENT INVENTORY | | | Year: ________ | |
|---|---|---|---|---|
| Item | Serial Number | Total Cost | Depreciation, Years | Yearly Depreciation, $ |
| | | | | |
| | | | | |

**TABLE 16.9**
**Example of a Form Used for Recording New Equipment Purchases**

| NEW EQUIPMENT PURCHASES | | | | | Year: ________ | | |
|---|---|---|---|---|---|---|---|
| Date | Amount | Item | Make | Serial Number | Supplier | Cost | Expected Life, Years |
| | | | | | | | |
| | | | | | | | |
| | | | | | Total cost | | $ |

**TABLE 16.10**
**Example of a Form Used for Recording Supplies and Materials Purchased**

| SUPPLIES, MATERIALS PURCHASED | | | Year: ________ | |
|---|---|---|---|---|
| Date | Amount | Item | Supplier | Cost |
| | | | | |
| | | | | |
| | | | Total costs | $ |

would allow continuity and consistency into the total organizational operation and could result in long-term savings by not "doing and undoing." The most unwelcome news to the decision makers is a crisis situation and the unexpected request for large sums of money to cope with it. Table 16.11 provides an abbreviated example of basic objectives, maintenance standards, and known limiting factors that prevent the desired standards from being reached for a golf course. Also, developing a year-end letter recapping the previous year's goals, successes, and failures plus the goals for the upcoming season is an excellent business tactic that is welcomed by most decision makers.

**FIGURE 16.3** Visual demonstration of golf condition expectations for a course.

**FIGURE 16.4** Long-term planning is necessary for all successful courses.

## MAINTENANCE EQUIPMENT NEEDS

Maintaining today's golf courses requires adequate equipment, supplies, facilities, and personnel (Figure 16.5). As labor, salaries, and other costs such as workers' compensation rise, the manufacturing industry will continue to produce quicker, more efficient machinery that requires less labor to operate and maintain. However, golf equipment also becomes more specialized, requiring larger operating budgets to buy and use machinery (Figure 16.6). Table 16.12 has been compiled by the U.S. Golf Association as a suggested list for maintaining the majority of 18-hole courses. Courses with more demanding players or unusual environmental requirements may need additional or other specialized equipment.

### Miscellaneous Equipment

Additional miscellaneous equipment that is needed on golf courses includes axes, brooms, buckets, crowbars, cup cutters, dew whips, funnels, gas cans, hoses, ladders, picks, pruners, pumps, rakes, scales, shovels, soil probes, traps for vertebrate pests, tree saws, and a wheelbarrow. Other equipment may also be needed (Figure 16.7).

**General repair tools.** General tools are needed for repairs on all power equipment (trucks, tractors, mowers, etc.), plumbing, drains, electrical equipment, sewers, roadways, fences, cement and concrete, water systems, painting and woodworking repairs, and so forth.

**Shop equipment.** Shop equipment includes a backlapping machine, air compressor, steam cleaner, bench grinder, bed knife grinder, reel grinder, pipe threader, paint sprayer, welding equipment, drill press, vises, work benches, and an equipment lift or hoist (Figure 16.8).

**Irrigation equipment.** Irrigation equipment includes a pumping station, field satellite controllers, centralized computer control system and irrigation heads, hoses, trenching shovels, wire locators (metal detectors), pipe and wire pullers, pipe, fittings, and specialized repair tools.

**Maintenance facility.** Having a spacious, modern turf care center (maintenance facility) is one of the best ways to improve professionalism, increase worker safety, improve staff morale, and improve overall efficiency. Of course, no "one-size-fits-all" maintenance facility exists due to affordability, size availability, layout, and location (Figure 16.9). Ideally, a maintenance facility should provide the following:

- Administrative office for the superintendent, assistant superintendent, spray technician, irrigation technician, and secretary, including equipment such as desks, file cabinets, phones, computers, etc.
- Toilet facilities, showers, and locker room
- Break room/meeting room with audio-video equipment for training purposes
- Adequate heating, cooling, and ventilation
- Paint-spraying room
- Reel-grinding room
- Pesticide storage facility
- Covered wash rack area

**TABLE 16.11**
**Example of Basic Objectives, Maintenance Standards and Any Known Limiting Factors That Influence Maintenance Goals or Standards of Golf Course Maintenance Areas**

| Area | Objectives | Maintenance Standards | Limiting Factors |
|---|---|---|---|
| **Roughs, wood/ natural areas** | • Maintain turf at 1.25 to 1.75 inches (3.2 to 4.4 cm) and appropriate density so golfers can readily find errant shots and not delay play.<br>• Maintain the turf with few weeds and provide a uniform height of cut.<br>• Maintain wooded areas with as few leaves as possible to speed play and provide a neat appearance.<br>• Provide nonmaintained areas away from play to promote wildlife. | • Mow weekly at proper height.<br>• Increase mowing frequency during periods of rapid growth, to maintain consistent height and minimize clippings.<br>• Provide pre- and postemergent herbicides to reduce weeds.<br>• Treat the perimeter of trees and mulched areas with glyphosate to reduce weed growth.<br>• Develop a sound agronomic program for fertility, aeration, water management, and pest control. | • Extreme weather conditions that affect growth<br>• During drought, supplemental water will not be applied as others areas have priority |
| **Driving range/ practice tee** | • Maintain a quality turf area similar to fairway quality.<br>• Maintain a neat and tidy appearance in practice area. | • Regularly fill divots.<br>• Overseed tee annually for winter play.<br>• Aerate three times annually followed by topdressing.<br>• Mow three times weekly between 3/8 and 5/8 inches (9.5 and 16 mm).<br>• Move ropes daily to control usage and location of teeing area.<br>• Blow and topdress area as needed to keep a smooth playing surface.<br>• Fertilize for quickest turf recovery. | • Excessive heavy practice on our relatively small (7,000 square feet, 650 $m^2$) facility will have periods of heavy wear, especially in winter and early spring when bermudagrass is not actively growing.<br>• Expanding the practice teeing area is a primary long-term capital improvement project scheduled to be completed by__. |
| **Bunkers** | • Maintain hazards to enhance the golf course and make exacting shots to targets without modifying the game or unduly penalizing the player.<br>• Bunkers should be firm with 4 to 6 inches (10 to 15 cm) of sand. | • Replenish bunker sand as needed.<br>• Monitor and record of sand depth monthly.<br>• Rake bunkers three times weekly; waste bunkers twice weekly.<br>• Edge bunkers as needed.<br>• Maintain drains in working order. | • Heavy rain (>0.25 inch, 6.4 mm) typically cause bunkers to wash. These will be repaired as soon as possible following such weather events.<br>• A long-term goal is to rebuild five bunkers yearly utilizing the latest technology and techniques to reduce washing. |

*(Continued)*

**TABLE 16.11 (CONTINUED)**
**Example of Basic Objectives, Maintenance Standards and Any Known Limiting Factors That Influence Maintenance Goals or Standards of Golf Course Maintenance Areas**

| | | | |
|---|---|---|---|
| **Tees** | • Produce a teeing area of adequate size and alignment with a level and firm surface suitable for golfers.<br>• Maintain a dense, weed-free playing surface.<br>• Maintain a mowing height between 1/3 and 3/4 inch (8.4 and 19 mm).<br>• Develop a basic agronomic program for all tee boxes, including aeration, vertical mowing, soil fertility, drainage, pest control, and traffic wear strategies.<br>• Provide adequate sunlight (>6 hours full sunlight daily) for teeing areas.<br>• Maintain all benches, yardage plates, ball washers, trash receptacles, and tee markers. | • Overseeding versus painting teeing areas will be discussed yearly but is at the discretion of the superintendent.<br>• Topdressing sufficiently during growing season to maintain a desirable smooth, level playing surface and minimize thatch accumulation.<br>• Conduct aeration and vertical mowing during growing season.<br>• Mow three times weekly.<br>• Use growth regulators during growing season.<br>• Proper fertility is necessary to maintain healthy, aggressively growing turf.<br>• Divots should be repaired weekly. | • Extreme weather conditions (e.g., low-temperature stress or extensive drought) may alter long-term survival strategies of turf.<br>• Playing size area for some par-3 tees will become inadequate as rounds increase.<br>• Conducted continued negotiation with homeowners to remove problematic trees from golf course property.<br>• Install championship tees if the course hosts regional or state-wide tournaments.<br>• Incomplete cart path paving increases turf damage to adjacent area. |
| **Fairways** | • Maintain quality fairways at mowing heights of 5/8 inch (16 mm) from April until July, 3/4 inch (19 mm) thereafter.<br>• Maintain the architectural characteristics of the hole by mowing outlines.<br>• Develop sound agronomic programs for fertility, aeration, watering, and pest control.<br>• Implementation of traffic rules during excessively wet periods to protect the turf.<br>• No intermediate rough-cut around fairways. | • Annually core aerify to smooth fairways, and reduce soil compaction, thus enhancing soil structure and surface drainage.<br>• Maintain moderate fertility just to encourage reasonable divot recovery; apply additional treatments to weaker fairways.<br>• Refrain from golf cart use anytime the surface is soft. "Cart-paths only" decisions will be at the discretion of the golf course superintendent.<br>• Maintain low soil moisture that allows some turf browning, placing highest priority on playability; aesthetics is secondary.<br>• Overseeding is at the discretion of the golf course superintendent.<br>• Use PGRs in summer to enhance turf quality, reduce mowing and clippings.<br>• Apply annual pre- and postemergent herbicides for weed control. | • Wet conditions are possible, especially following a >0.25 inch (6.4 mm) rainfall event.<br>• Different turf species are present (common and 419 hybrid bermudagrass); thus some variation in texture, growth, color and spring green-up often occur.<br>• Some shading, especially in winter, may cause thin turf or low-temperature damage following severe winter.<br>• A consumptive water use permit restricts water available during periods of extreme drought, with greens and tees having priority.<br>• Periods of slow growth can occur. |

*(Continued)*

**TABLE 16.11 (CONTINUED)**
**Example of Basic Objectives, Maintenance Standards and Any Known Limiting Factors That Influence Maintenance Goals or Standards of Golf Course Maintenance Areas**

| | | | |
|---|---|---|---|
| **Putting greens** | • Maintain a smooth and uniform surface with a medium to fast speed for putting.<br>• Maintain all putting surfaces with fine-textured bentgrass with minimum thatch and grain.<br>• Maintain the original character of the architectural design of the putting surface outline.<br>• Maintain a dry putting surface for firmness.<br>• Develop a basic agronomic program that includes aeration, soil-fertility, drainage, pest-control, and traffic-wear strategies.<br>• Provide adequate sunlight and air circulation for bentgrass growth and survival.<br>• Maintain a high-density, weed-free turf stand surrounding all putting complexes. | • Monitor the rootzone of each green daily to maintain the putting surface as dry as possible without affecting turf health. Hand-watering will be an integral part of daily water management.<br>• Light topdressing should be done every 7 to 14 days during the growing season.<br>• Aeration (cores removed, followed by topdressing to fill holes) at least four times yearly (typically, March, May, September, and November), supplemented with hydrojetting, solid-tine aeration, slicing, spiking, and other means as needed to maintain bentgrass health and survival. A minimum of 25% of the surface area of the putting green will be impacted by these practices yearly.<br>• Greens should be mowed between 0.125 and 0.187 inch (3.2 and 4.7 mm), depending on weather conditions and time of year.<br>• Greens should be mowed three to six times weekly, often rotated with rolling to minimize stress.<br>• Conduct cleanup mowing lap three to four times weekly.<br>• Pins should be moved daily to distribute traffic and minimize surface wear. | • Surface fans are not available for all greens due to lack of nearby power.<br>• Trees and brush not on the golf course property limit proper air movement and sunlight on certain greens.<br>• Several greens are inadequately sized; thus, limited pin placements may cause thinning of turf, especially in late summer.<br>• Although greens currently drain sufficiently, they are constructed with river-bottom sand, which may require rebuilding in the future and is a primary long-term capital improvement goal.<br>• Damaging nematode population levels exist on several greens. However, due to the expense, only temporary suppression, and limited availability, the club has decided against synthetic control, thus, is willing to periodically experience temporary unthrifty turf. |
| **Golf course landscape** | • Maintain all landscape beds to enhance the quality of the facility. | • Irrigate as needed for plant survival.<br>• Mulch twice yearly.<br>• Remove and replace plant material as needed to maintain original design.<br>• Fertilize, hand-weed, prune, and apply nonselective herbicides as needed to keep beds neat and weed free.<br>• Use preemergence herbicides yearly. | • Unusual weather patterns such as hot, dry periods or excessively cold weather may damage/kill sensitive plants. |
| **Cart paths** | • Maintain all cart paths in an edged, acceptable fashion void of grass clippings and debris. | • Edge monthly during turf growth.<br>• Blow off weekly. | • Tree roots heave asphalt paths.<br>• New cart paths will be concrete, and at least 8 feet (2 m) wide for dual cart passing plus maintenance equipment access. |

**FIGURE 16.5** Adequate budgets are needed for course condition expectations.

**FIGURE 16.6** Balancing equipment updates with budgets is a prime responsibility of superintendents.

- Fertilizer and seed storage area
- Storage for oil, gasoline, and diesel fuel as well as a covered fuel island
- Adequate storage for all equipment, parts, supplies, and tools
- Adequate maintenance work area
- First aid supplies, including an automated external defibrillator (AED)
- Soil bins for topdressing material, bunker sand, and soil amendments, which should be kept in a dry area, free of contamination, and preferably covered

Having logistically located work zones improves efficiency, reduces wasted time, and increases worker safety. It is important to be cognizant of daily workflow patterns from employee parking, clocking in, receiving daily duties, equipment storage, and proceding to the course.

### Golf Equipment Maintenance Training

With the vast investment in equipment, supplies, and a modern maintenance facility, courses should develop an employee manual and a safety program. These provide proper training on the maintenance, use, and storage of various machines and other equipment necessary for a facility to safely operate. More importantly, if designed and implemented properly, it should sustainably reduce workplace accidents or miscommunications. Although state and local requirements vary,

**TABLE 16.12**
**Typical Equipment Needs for an 18-Hole Golf Course**

| Equipment | Quantity Needed |
|---|---|
| **Greens and Tees** | |
| Putting green mowers, walk-behind, **or** | 6 to 8, **or** |
| Triplex mowers (two for greens, one for collars) | 3 |
| Triplex vertical mowing reel set | 1 |
| Brushes or groomers for mowing equipment | 2 |
| Three-gang tee, and fringe mowers | 1 to 2 |
| Power aerifiers | 2 |
| Topdressing machines | 1 to 2 |
| Core harvesting machines | 1 |
| Power dethatcher | 1 |
| Power spiker (or slicer) or spiker attachment | 1 |
| Power sprayer for pesticide and fertilizer application | 1 |
| Rotary fertilizer spreader/seeder | 2 |
| Dragmats, brushes, or brooms to incorporate topdressing material | 2 |
| Blowers | 1 to 5 |
| Roller, tournament speed | 1 |
| Hydroject water inject aerifier | 1 |
| **Fairways and Roughs** | |
| Lightweight, five-gang fairway mowers, **or** | 2 to 3, **or** |
| Seven-gang hydraulic-powered mowers | 2 |
| Fairway aerifier | 1 |
| Fairway dethatcher/vertical mower | 1 |
| Power sprayer (200- to 300-gallon capacity) | 1 |
| Large volume fertilizer spreader | 1 |
| Five-gang rough mower, **or** | 1, **or** |
| Multideck rotary rough mowers | 2 |
| Triplex trim mowers | 2 to 3 |
| Tractor-mounted blower | 1 |
| Fairway sweeper/blower | 1 |
| **General Equipment** | |
| Transportation vehicles for the staff | 4 to 5 |
| General construction PTO tractors | 2 to 3 |
| Dump trailers | 1 to 2 |
| Pickup truck | 1 |
| Dump truck (5-ton minimum) | 1 |
| Front-end loader and backhoe with forks (this can be rented as needed) | 1 |
| Forklift or front-end loader | 1 |
| Trencher (this can be rented as needed) | 1 |
| Slicer/seeder | 1 |
| Snow plow, where necessary | 1 |
| Power sod cutter | 1 |
| Rotary trim mowers | 4 to 5 |
| Power edgers | 4 to 5 |
| Backpack blowers | 4 to 5 |
| String trimmers | 4 to 5 |

FIGURE 16.7 Golf course maintenance requires sufficient miscellaneous equipment.

FIGURE 16.8 A well-organized and equipped maintenance shop increases efficiency and worker safety.

FIGURE 16.9 Like their courses, maintenance facilities vary in size, complexity, layouts, and locations.

the following are the typical components in developing a working and effective safety program:

- Safety policy statement
- Employer information
- Designation of safety responsibilities
- Employee reporting and conduct
- Inspection procedures
- Written hazard communication and correction procedures
- Accident investigation
- Disciplinary procedures
- Employee training
- Recordkeeping
- Inventory
- Signatures

## Justifying an Asset

Justifying a needed asset and justifying needed personnel are two of the biggest budgetary challenges facing superintendents. Governing boards at golf courses are typically composed of bankers, accountants, lawyers, small-business owners, engineers, and doctors—professionals accustomed to justifying their own needs within their organizations. To those familiar with the golf maintenance business, justifying a piece of turf equipment or extra personnel seems obvious. However, when trying to balance a club's total budget, while keeping expenses in check, superintendents need to have a plan of justification before success in acquiring an asset can be expected. When trying to explain the actual use of a piece of equipment, it often is helpful to relate this in everyday, understandable terms. For example, explaining to a committee that a piece of equipment has 4,000 hours probably means little to them. However, if it is explained in terms of miles per hour, it might make more of an impact. For example, if the equipment has 4,000 hours, multiplying this by 60 miles per hour equals approximately 240,000 miles on a vehicle. This commonly used mile per hour value probably is more impressive than little known or used units like hours alone. The following outline is often successfully used when an asset that board members are familiar with and typically expect is needed:

1. **Description of the asset.** The description lists the name of the asset and model number, with all accessories, plus the name and address of the seller. It is helpful to attach a color picture of the asset from a manufacturer's brochure to help the committee visualize the asset.
2. **Justification of the asset.** The justification explains what the asset does and why the club needs the asset (for example, to replace a worn, broken, or outdated piece of equipment). Describe any safety issues or deficiencies the current asset has or is lacking. Attach a performance specification sheet from the manufacturer to explain features, accessories, and expected performance of the proposed asset.
3. **Selection of the new asset.** Explain the various assets available on the market, the results of any in-house testing or equipment demonstrations, and why a certain model or manufacturer was chosen.
4. **Availability and delivery of the asset and part replacement.** Explain the time frame in which the

new asset can be made available and if replacement parts are easily accessible.

5. **Asset maintenance plan.** Explain how routine maintenance of the new asset would be handled. Explain preventative maintenance procedures, schedule, and parts inventory needs. Cover service plans for on-site versus off-site equipment by a dealer technician.
6. **Use life of the asset.** List a reasonable life expectancy of the new asset, assuming proper maintenance is followed. Sales or technical brochures may be helpful in this task.
7. **List alternative(s) if this proposal is rejected.** Explain how expensive it would be to repair the existing machine or how much additional labor would be needed if the asset proposal is rejected (Figure 16.10). Keep an equipment log on each equipment piece to record repair costs and eventually to help justify when purchasing new equipment is cheaper than maintaining an existing piece.
8. **Fate of the asset being replaced.** Explain what is planned for the asset being replaced. Can any useful parts be salvaged? Can it be traded in? Can it be sold?

**FIGURE 16.10** Having an equipment replacement plan reduces labor costs and improves playing conditions.

### Example

A golf course superintendent determines a new mechanical bunker rake is needed but feels the new club manager will have difficulty agreeing with the request. The following capital equipment purchase request is prepared to justify this purchase:

1. **Description of the asset.** Sand Trap Master model #12345 with a 16 hp v-twin gas engine, three-wheel drive, midmount cultivator, manual control front blade, rear rakes, and spike attachment, to be purchased from Great Turf Equipment of Anywhere, U.S. (attach a picture).
2. **Justification of the asset.** This machine is a basic component of our equipment inventory, used to maintain our sand bunkers on the golf course. The front blade is needed for pushing sand up into washouts after heavy rains and to distribute newly placed sand. It works ~25% quicker than comparable machines. Additionally, this machine will be used to do small grading projects, spike greens, and drag in topdressing (attach performance sheet).
   This machine is needed to replace our current 10-year-old worn-out Sand Trap Mutilator. This machine has provided 10 years of commendable service; however, it is currently in poor condition. At present, the machine is not running due to a blown engine. In addition, the frame is broken, the brakes are nonresponsive (dangerous), and the hydraulic system needs replacing. Due to liability and mechanical issues, we have parked the machine, feeling it is not worth repairing. Since the machine is inoperable, we are having difficulties accomplishing routine bunker maintenance to the level outlined in our long-term play performance goals.
3. **Selection of the new asset.** Recently, several new bunker rakes have been introduced into the marketplace that are more time effective and provide a better quality product than our current model. Several models have been demonstrated on our course with our personnel. In addition, several manufacturers left demo models on the course for further evaluation. All machines did a commendable job, but the Sand Trap Master model #12345 would best fit our needs for the following reasons:
   a. This machine was most cost effective with the added accessories.
   b. If the operator prefers a piece of equipment, he/she will take better care of it and appreciate the job being performed.
   c. Maintenance is lower due to a superior steering system and simpler hydraulic system.
   d. Transport speed is superior, reducing the total man-hours needed to perform the total job.
   e. The mounting of the front blade is such that a wider turn can be made in the bunkers without digging into steep bunker faces.
4. **Availability and delivery of the asset and part replacement.** Upon the receipt of our order, Great Turf Equipment of Anywhere, U.S., can deliver the machine within 10 to 14 days. Repair parts are usually available overnight while service is available within one week from Great Turf Equipment.
5. **List alternative(s) if this proposal is rejected.** If the proposal is rejected, our options are limited. We would have to spend several thousand dollars repairing a machine that is worn out and beyond its usual life expectancy. The money to do so is not in our operations budget, nor do we have the time to undertake a rebuilding job of this magnitude at this point of the season.
6. **Asset maintenance plan.** Maintenance will be handled by the staff mechanic. A preventative maintenance schedule will be established and computer tracked. An inventory of frequently used

parts, of which there are very few, will be established. Many parts are currently inventoried, as this machine shares similar parts with other equipment. The dealer has agreed to two annual visits by its technicians to ensure proper maintenance is being performed. In turn, the dealer will extend the equipment warrantee for an additional year.

7. **Use life of the asset.** With proper maintenance, this machine will have an active and useful life of 8 to 10 years.
8. **Fate of the asset being replaced.** Due to safety issues, we feel the current machine should not be sold. The unit will be stripped of any useful parts and sold for scrap metal. A notification will be forwarded to accounting to remove it from the books.

### Example

Two mowers are being considered. Mower A's specs include: 62-inch (157 cm) wide cut, 5 mph (8.1 km/hour) operating speed and cost of $15,000. Mower B's specs are 56-inch (142 cm) wide cut, operating at 4 mph (6.4 km/hour) and cost of $12,000. The mowed area is 20 acres, averaging 45 mowings yearly, and payroll currently is at $10/hour. Assuming fuel usage, longevity, depreciation, and maintenance costs are similar for both machines, what is the "better" buy short term (1 year) and long term (5 years)?

**Step 1:** Determine total area (square feet) being mowed:

$$20\text{ acres}\times\frac{43{,}560\text{ square feet}}{\text{acre}}=871{,}200\text{ square feet (8.2 ha)}$$

**Step 2:** Convert mower width into feet:

**Mower A:**

$$62\text{-inch cut}\times\frac{\text{feet}}{12\text{ inches}}=5.17\text{ feet (1.57 m)}$$

**Mower B:**

$$56\text{-inch cut}\times\frac{\text{feet}}{12\text{ inches}}=4.7\text{ feet (1.42 m)}$$

**Step 3:** Convert total mowed area into miles for each mower:

**Mower A:**

$$871{,}200\text{ square feet}\times\frac{\text{cut}}{5.17\text{ feet}}\times\frac{\text{mile}}{5{,}280\text{ linear feet}}$$

$$=31.9\text{ miles (51.4 km)}$$

**Mower B:**

$$871{,}200\text{ square feet}\times\frac{\text{cut}}{4.7\text{ feet}}\times\frac{\text{mile}}{5{,}280\text{ linear feet}}$$

$$=35.1\text{ miles (56.5 km)}$$

**Step 4:** Determine total yearly operating hours for each mower:

**Mower A:**

$$31.9\text{ miles}\times\frac{\text{hour}}{5\text{ miles}}\times\frac{45\text{ mowings}}{\text{year}}=287.1\text{ hours}$$

**Mower B:**

$$35.1\text{ miles}\times\frac{\text{hour}}{4\text{ miles}}\times\frac{45\text{ mowings}}{\text{year}}=394.9\text{ hours}$$

**Step 5:** Determine total yearly payroll costs:

**Mower A:**

$$\frac{287.1\text{ hours}}{\text{year}}\times\frac{\$10}{\text{hour}}=\frac{\$2{,}871}{\text{year}}$$

**Mower B:**

$$394.9\text{ miles}\times\frac{\$10}{\text{hour}}=\frac{\$3{,}949}{\text{year}}$$

**Step 6:** Determine total operation costs for 1 year:

**Mower A:**

$$\frac{\$2{,}871}{\text{year}}+\$15{,}000\text{ mower cost}=\frac{\$17{,}871}{\text{year}}$$

**Mower B:**

$$\frac{\$3{,}949}{\text{year}}+\$12{,}000\text{ mower cost}=\frac{\$15{,}949}{\text{year}}$$

**Step 7:** Determine total operation costs over 5 years:

**Mower A:**

$$\frac{\$2{,}871}{\text{year}}\times 5\text{ years}+\$15{,}000\text{ mower cost}=\$29{,}355$$

**Mower B:**

$$\frac{\$3{,}949}{\text{year}}\times 5\text{ years}+\$12{,}000\text{ mower cost}=\$31{,}745$$

Therefore, Mower A's one-year costs are $17,871, compared to $15,949 for Mower B. However, over five years, Mower A's costs are $29,355, while costs for Mower B are $31,745. This results in approximately $2,400 savings, even though Mower A's initial purchase cost was higher ($15,000 vs. $12,000). Mower speed and width are key considerations in calculating total operating costs.

## EQUIPMENT PURCHASING, RENTING, AND LEASING

Turf equipment can be financed by several means, including long-term financed purchasing, renting, or leasing (Figure 16.11). Custom contracting services also are available in many areas.

**FIGURE 16.11** Financing turf equipment includes long-term financed purchasing, renting, and leasing.

Questions one should consider before entering into any equipment financing plan include:

1. Is the value of depreciation on the equipment worth more than the ability to make payments and keep debt off the balance sheets?
2. An operation that owns a piece of equipment can depreciate deductions, while if an operation leases, it takes an expense deduction. Table 16.13 provides a brief comparison of these means of obtaining equipment.

### Leasing

Leasing is a form of asset financing where a predetermined fee (often monthly) is paid for the right to use ("borrow") equipment. The title stays with the owner of the asset (e.g., dealer or lessor) during the lease term and the lessee (user of the asset) acquires no *equity* during the lease term. Typically, one- to three-year leases are agreed where the lessee pays a predetermined fee for the use of new equipment. Once the lease expires, the equipment is replaced with newer models, assuming the lease is renewed.

Several types of leasing programs or options exist. Two main programs for golf course equipment are

1. *Operating* (or true) lease, also called fair market value (FMV). This option is more like renting for relatively short durations:
   a. Users enter into an operating lease for a period for up to 75% of the equipment's life.
   b. The financing (leasing or lessor) company retains ownership of the equipment so no asset is exchanged, but also retains responsibility for property taxes (if any), insurance, and maintenance.
   c. All payments are treated as an operating expense (just like fuel, electricity, insurance, etc.) and appear in the budget as a rental or operational fee but not debt; rather, they are considered a monthly operating expense.
   d. The equipment is not depreciated as no asset transfer occurs; thus lessees lose the depreciation deduction (i.e., "residual" value at the end of the lease remains with the lessor). However, lease payments are tax deductible as operating expenses.
   f. This is the most common form of leasing today.
2. *Finance* or *capital lease*, also called conditional sales contract (CSC). This leasing option is more like rent to own (or "dollar buyout") and of relatively long duration:
   a. This involves a traditional financing lease wherein a near or full payout of the purchased equipment occurs, and the ownership of the leased assets is then transferred to the lessee (golf course) at the end of the lease term.
   b. The lessee (golf course) often assumes the responsibility of property taxes (if any), insurance, and maintenance.
   c. This option is often noncancelable or fairly costly to cancel.
   d. It is often referred to as "dollar buyout," as end-of-lease purchase prices are set prior to the lease going into effect, similarly to rent to own.
   e. The club shows the purchase as an asset and a lease liability, rather than a purchase, on the balance sheet.
   f. A finance or capital lease is considered a capital expense and thus depreciable. At fiscal year's end, the club can deduct the amount depreciated during the year against any tax liability.

In a lease, the lessee is typically allowed a set number of hour usages per year. If this amount is exceeded, a set dollar amount per hour "penalty" is charged. This amount is usually high as the financing company owns the asset and assumes a greater liability for it if the contracted usage limit is exceeded. However, if the lessee overestimates the number of hours expected to use the equipment to avoid such penalties, a higher overall lease payment will be imposed. To avoid high payments, the lessee has to carefully consider the equipment's historical hourly usage and negotiate a lower limit.

As a general rule of thumb, leasing is usually more advantageous than purchase financing when the asset significantly decreases in value soon after being purchased. In leasing, one pays for the decrease in value of the equipment over the lease term; the leasing company utilizes the depreciation or credit incentive, passing on savings in the form of lower equipment leasing costs. In addition, the interest factor built into the monthly payment will include the time value of money, commissions to the dealer, and a desired profit for the lessor. Table 16.14 compares obtaining a five-gang mower and overall costs to the club by making a cash purchase, financed purchase, or lease.

Some final thoughts on leasing include:

- A true lease is more expensive in the long term (more than three years) than purchasing.
- A true lease is usually better when one has
  - Frequent equipment turnover
  - Lease equipment that rapidly declines in value

**TABLE 16.13**
**Comparing Equipment Purchases by Renting, Purchasing, Leasing, or Using Contracting Services**

| Means | Advantages | Best for | Disadvantages |
|---|---|---|---|
| **Renting** | • The club does not have to buy, maintain, store, or haul equipment.<br>• This involves only a one-time fee. | • When used short term (~1 month or less)<br>• Equipment not used on a daily basis, only for special needs/projects | • Time and money needed for training operator or hire additional personnel<br>• Weather or use needs disrupt schedule, costs increase with additional rental times<br>• Desired model may be unavailable<br>• Additional pickup/drop-off or delivery fees |
| **Purchasing** | • You own the equipment.<br>• You have the ability to trade it in.<br>• There is a flexible use schedule in case of delays.<br>• Maintenance is low with infrequent use.<br>• Depreciation counts as tax write-off.<br>• It can free up money after the loan is paid for. | • Long-term equipment<br>• Equipment not used on a daily basis<br>• Equipment not requiring frequent or precision maintenance<br>• Inexpensive equipment<br>• Equipment that holds value | • Higher up-front costs<br>• Regular maintenance and storage needs<br>• Depreciation<br>• Increased maintenance as equipment ages<br>• Stuck with outdated technology |
| **Leasing** | • Asset management keeps debt (credit) lines open for working capital, rather than tied up in capital expenditures (no big down payment).<br>• Keep up-to-date equipment; warranty starts over each time equipment is replaced.<br>• Equipment disposal is not your responsibility.<br>• Leases do not show up as debt or liability.<br>• There are customizable lease programs (e.g., pay seasonal, monthly, annual, etc.).<br>• Predictable monthly payments are usually lower.<br>• It is easier to obtain than third-party financing.<br>• There is immediate write-off with no depreciation.<br>• There are options (e.g., buy) at the end of a lease.<br>• Agreement is easy.<br>• Operating leases are not considered a purchase and can be a tax-deductible overhead expense.<br>• There can be positive employee response to new equipment.<br>• There is continuous demand of customer service from the equipment company. | • When used long term (>1 year)<br>• Equipment used on a daily basis<br>• Equipment that loses its value quickly<br>• Equipment that wears quickly<br>• Equipment that affects the physical quality of the course<br>• Equipment that requires precision maintenance on a regular basis | • User does not build equity.<br>• This is costly for long-term use.<br>• There are equipment condition requirements prior to return.<br>• Equipment use (hour) is limited.<br>• It may lock you into a perpetual leasing program.<br>• Equipment is not owned once lease expires and thus does not build equity.<br>• Financial requirements are usually tougher than payments; after all, with leasing, payments are made in advance and the decision to lease is based on the customer's ability to pay, not the value of the equipment. |
| **Contracting services** | • One-time fee<br>• Little to no maintenance costs or efforts<br>• No time or money spent on training operators | • Specific, infrequent tasks such as aerification or verticutting<br>• Highly technical tasks<br>• Use of specialized and expensive equipment<br>• Freeing labor for other tasks | • Location of contractor may increase costs.<br>• Scheduling conflicts may make contractor unavailable when desired.<br>• Unexpected additional contracting dates may outweigh costs of leasing or ownership. |

**TABLE 16.14**
**Comparison of Overall Costs of Obtaining a Five-Gang Mower as Cash, Financed, or Leased**

| Five-Gang Mower | | Cash Purchase | Financed Purchase ($1,323/ Month) | Lease ($800/ Month) |
|---|---|---|---|---|
| Cash flows | Year 1 | $41,000 | $15,875 | $9,600 |
| | 2 | | $15,875 | $9,600 |
| | 3 | | $15,875 | $9,600 |
| Net of asset value | | $41,000 | $47,625 | $28,800 |
| Fair market value | | $(22,000) | $(22,000) | |
| Payments deductible | | | | $(8,064) |
| Depreciation ($23,070) | | $(6,460) | $(6,460) | |
| Interest ($6,626) | | | $(1,856) | |
| | | $12,540 | $17,309 | $19,936 |
| | Tax on sale (28%) | $1,140 | $1,140 | |
| | | $13,680 | $18,449 | $19,936 |
| Net of time value of money (5% rate) | | $13,680 | $18,449 | $19,936 |
| Lost earnings | | $6,620 | $3,644 | $2,203 |
| **Bottom-line cost** | | **$20,300** | **$22,093** | **$22,139** |

- Cash is always cheaper in the long term, while bank (third-party) financing is next cheapest.
- Leasing is usually less costly in the short term than ownership is.
- Leasing is the quickest and easiest way to obtain new equipment since preapproved loans are not needed.

## PERSONNEL REQUIREMENTS

Staffing a golf course maintenance crew is a never-ending challenge (Figure 16.12). Labor costs are skyrocketing, forcing some clubs to cut back on personnel numbers and/or benefits. However, golf courses require an extensive staff that is dependable and continuously trained to handle complicated machinery and certain chemicals. It is much cheaper in the long run to pay qualified people good salaries to minimize turnover and to ensure highly skilled labor.

**FIGURE 16.12** A major challenge for most courses is attracting and retaining a quality, dependable work force.

Labor costs are generally the largest line item in the operating budget and include salaries, wages, taxes, and benefits. In order to effectively plan and project labor costs, the owner and superintendent must agree upon the number of employees needed to maintain the golf course, how many will be full- and part-time employees, the wage and salary scale, what benefits are offered, and so forth.

A simple spreadsheet will help plan and project wages and salaries (Table 16.15). The number of weeks and hours per week, including overtime hours, is needed to add additional costs from unemployment insurance and employer's liability, Medicare and worker's compensation insurance, and any owner-sponsored benefit plans offered (Table 16.16).

Having a written expected standard level is another helpful piece of information when justifying these additions (Table 16.11). When labor needs do not match the expected standard of a course, then the superintendent has the information needed to justify hiring additional help. Table 16.17 lists a method for documenting the total hours needed to perform each job task. Hours necessary to perform these tasks typically vary for different months or seasons. Therefore, separate subcategories may be necessary to reflect these seasonal differences.

Trying to list the personnel requirements for an 18-hole facility is difficult, as each course has varying needs, financial resources, and player demands. A detailed manpower analysis by a professional group is suggested as a specific method to determine staff size based on the amount of labor and times performed for every golf course function. In the absence of a manpower analysis, Table 16.18 is a suggested working list of personnel for a typical facility.

### Labor Justification

As mentioned, justifying additional labor and equipment are two of the biggest challenges superintendents face. For this to be successful, detailed records on the time required to perform certain tasks on a particular sized area will be needed. The following formula can be used to determine the amount of labor required to perform certain tasks; if the number of personnel available does not meet this value, then several options are available. These include reducing the expected playing standards of the course, reducing the area being maintained at this level, increasing the labor pool, or, if available, purchasing equipment that reduces labor needs.

$$\text{Time (hours) per unit area} = \frac{\text{total hours per year to perform a given job}}{\text{yearly times a job is performed} \times [\text{area(square feet or acres), or hours needed}]}$$

**TABLE 16.15**
**Example of Tracking and Recording Various Tasks Needed to Maintain a Course for Budgetary or Personnel Increase Justification**

| Job/Task | No. Employees/Task | Days Performed/Week | Total Hours Weekly | Total Hours Monthly | Total Hours Annually |
|---|---|---|---|---|---|
| Mow greens (walk) | | | | | |
| Mow collars (walk) | | | | | |
| Mow tees (ride) | | | | | |
| Mow approaches | | | | | |
| Mow fairways | | | | | |
| Collect clippings | | | | | |
| Mow roughs | | | | | |
| Drag roughs | | | | | |
| Mow bunker edges | | | | | |
| Fly mow bunkers | | | | | |
| Roll greens | | | | | |
| Bunkers | | | | | |
| • Rake (daily) | | | | | |
| • Edge (monthly) | | | | | |
| • Blow out | | | | | |
| Cups/tee markers/ball washers/trash | | | | | |
| Divots and broken tees | | | | | |
| • Fairway | | | | | |
| • Tees | | | | | |
| Marking hazards | | | | | |
| Trimming | | | | | |
| Hand-water greens | | | | | |
| Edging (monthly) | | | | | |
| • Sprinkler heads | | | | | |
| • Cart paths | | | | | |
| • Yardage blocks | | | | | |
| • Catch basins | | | | | |
| • Pond rock walls | | | | | |
| Blow cart paths | | | | | |
| Greens aerification | | | | | |
| Tees Aerification | | | | | |
| Plug removal | | | | | |
| Solid-tine greens | | | | | |
| Greens topdressing | | | | | |
| Mow natural areas (3× yearly) | | | | | |
| Pick up sticks/cones | | | | | |
| Waste material removal (monthly) | | | | | |
| Flood/rain debris removal | | | | | |
| Vandalism (4× yearly) | | | | | |

*(Continued)*

**TABLE 16.15 (CONTINUED)**
**Example of Tracking and Recording Various Tasks Needed to Maintain a Course for Budgetary or Personnel Increase Justification**

| Job/Task | No. Employees/Task | Days Performed/Week | Total Hours Weekly | Total Hours Monthly | Total Hours Annually |
|---|---|---|---|---|---|
| Shop cleanup | | | | | |
| Arborist visit (annually) | | | | | |
| Chipping | | | | | |
| Drainage projects (annually) | | | | | |
| Misc. jobs (monthly) | | | | | |
| **Club grounds** | | | | | |
| Pruning-shrub cleaning | | | | | |
| Trash pickup | | | | | |
| Blowing parking lot | | | | | |
| Leaf debris removal | | | | | |
| Mowing grounds | | | | | |
| Fertilizing (annually) | | | | | |
| Weeding beds | | | | | |
| Thinning ornamental grasses | | | | | |
| Bed prep and planting | | | | | |
| Seasonal planting | | | | | |
| Mulching beds | | | | | |
| Pool landscaping | | | | | |
| Tennis court landscaping | | | | | |

**TABLE 16.16**
**Example of Tracking Work Hours for Employees and Corresponding Salaries and Wages**

| DETAILED SALARIES AND WAGES | | | | | Year: ________ | | | | |
|---|---|---|---|---|---|---|---|---|---|
| Employee | Hourly Rate | Nonovertime Hours | No. Weeks Worked | Total | Overtime Rate | Overtime Hours Weekly | No. Weeks | Overtime Total | Total Salary |
| 1 | $9.44 | 40 | 52 | $19,635.20 | $14.16 | 5 | 26 | $1,840.80 | $21,476.00 |
| 2 | 8.74 | 40 | 52 | 18,179.20 | 13.11 | 5 | 26 | 1,704.30 | 19,883.50 |
| 3 | 7.00 | 40 | 52 | 14,560.00 | 10.50 | 5 | 26 | 1,365.00 | 15,925.00 |
| 4 | 7.00 | 40 | 52 | 14,560.00 | 10.50 | 5 | 26 | 1,365.00 | 15,925.00 |
| 5 | 7.00 | 40 | 52 | 14,560.00 | 10.50 | 5 | 26 | 1,365.00 | 15,925.00 |
| 6 | 7.00 | 40 | 52 | 14,560.00 | 10.50 | 5 | 26 | 1,365.00 | 15,925.00 |
| 7 | 7.00 | 40 | 52 | 14,560.00 | 10.50 | 5 | 26 | 1,365.00 | 15,925.00 |
| 8 | 7.00 | 40 | 52 | 14,560.00 | 10.50 | 5 | 4 | 210.00 | 14,770.00 |
| 9 | 7.00 | 40 | 52 | 14,560.00 | 10.50 | 5 | 4 | 210.00 | 14,770.00 |
| 10 | 7.00 | 40 | 16 | 4,480.00 | 10.50 | 5 | 4 | 210.00 | 4,690.00 |
| 11 | 8.50 | 40 | 16 | 5,440.00 | 12.75 | 5 | 16 | 1,020.00 | 6,460.00 |
| 12 | 7.00 | 40 | 16 | 4,480.00 | 10.50 | 5 | 4 | 210.00 | 4,690.00 |
| 13 | 7.00 | 40 | 16 | 4,480.00 | 10.50 | 5 | 4 | 210.00 | 4,690.00 |
| **Totals** | | | | **$156,614.40** | | | | **$12,440.10** | **$171,054.00** |

**TABLE 16.17**
**Example of Documenting Hours Necessary to Perform Anticipated Tasks on a Golf Course Green during a Specific Time of Year**

| Task | Employees/ Task | Hours/ Person/Task | Total Hours/Day | Days/Week | Total Hours/ Week | Weeks/Category | Total Hours/ Category |
|---|---|---|---|---|---|---|---|
| | | | Greens (June 1–August 31) | | | | |
| Mowing | 2 | 3 | 6 | 7 | 42 | 13.1 | 550.2 |
| Rolling | 1 | 7 | 7 | 1 | 7 | 13.1 | 91.7 |
| Blowing | 1 | 2 | 2 | 1 | 2 | 3 | 6 |
| Grinding | 2 | 6 | 12 | 2 | 24 | 9 | 216 |
| Cup changing | 1 | 4.5 | 4.5 | 7 | 31.5 | 13.1 | 412.7 |
| Hand-watering | 2 | 5 | 10 | 6 | 60 | 13.1 | 786 |
| Verticutting | 0 | 0 | 0 | 0 | 0 | 0 | 0 |
| Aeration | 1 | 6 | 6 | 2 | 12 | 3 | 36 |
| Topdressing | 1 | 8 | 8 | 2 | 16 | 9 | 144 |
| Debris removal | 2 | 5 | 10 | 2 | 20 | 3 | 60 |
| Spraying | 1 | 5 | 5 | 2 | 10 | 13.1 | 131 |
| Hand-weeding | 2 | 8 | 16 | 1 | 16 | 1 | 16 |
| | | | | | | **Total** | **2449.6** |

**TABLE 16.18**
**Suggested Maintenance Personnel Needed for a Typical 18-Hole Golf Course**

| Position Description | Number |
|---|---|
| Superintendent | 1 |
| Assistant superintendent or foreman | 1 |
| Secretary (full- or part-time) | 1 |
| Equipment manager (mechanic) | 1 |
| Irrigation technician | 1 |
| Spray technician | 1 |
| Section people/greenskeepers/syringers | 3 |
| Equipment operators | 4 |
| Utility people | 2 |
| Landscaper/horticulturist | 1 |
| **Total** | **16** |

## Example

Through detailed records, the following labor hours have been determined as necessary each year. Each employee works, on average, 45 hours per week, and the main season is 40 weeks long.

| Task | Hours Required (Yearly) | Percentage of Hours |
|---|---|---|
| Greens and collars | 5,245 | 19 |
| Tees and nursery | 2,834 | 10 |
| Fairways | 3,100 | 11 |
| Approaches | 752 | 2.7 |
| Roughs | 5,543 | 20 |
| Clubhouse grounds and pool | 1,589 | 5.7 |
| Sand traps | 2,238 | 8 |
| Topdressing | 303 | 1 |
| Nursery | 165 | 0.5 |
| Grinding/rebuilding mowers | 1,056 | 3.8 |
| Pesticide application | 660 | 2.4 |
| Weed trimming | 824 | 3 |
| Repairs/maintenance | 2,062 | 7.4 |
| Miscellaneous | 1,499 | 5.4 |
| Total | 27,870 | 99.9 |

To determine the "ideal" number of employees to perform these tasks:

$$\text{Time (hours) per unit area} = \frac{\text{total hours per year to perform a given job}}{\text{yearly times a job is performed} \times [\text{area(square feet or acres), or hours needed}]}$$

$$= \frac{27{,}870}{40 \text{ weeks} \times 45 \text{ hours/work}}$$

$$= 15.5 \text{ employees}$$

If the total area is known, then the time (minutes or hours) needed to perform the particular operation can be determined. For example, if it requires two minutes to walk-mow each 1,000 square-foot area of greens and this is performed 225 times yearly, then it requires 7.5 hours (or 450 minutes) to perform this task. Multiply this number by the amount of greens area to determine the total hours necessary to mow greens each year. If one has 19 greens (counting the practice green) averaging 5,000 square feet each, then a total of 95,000 square feet (or 2.2 acres) needs mowing. Based on 7.5 hours needed yearly to mow 1,000 square feet, it requires 713 hours yearly to mow all greens a total of 225 times.

Obviously, conditions and playing area at every golf course can vary significantly and this formula cannot account for all duties performed or varying conditions throughout a season or year. However, the formula can be a guide to compare operating practices and possibly justify additional personnel and/or faster, more efficient equipment.

## Job Descriptions

In today's labor market and legal atmosphere, companies are requiring formal job descriptions for their employees. Job descriptions typically provide an overall statement of what is expected of the position followed by a detailed breakdown and percentages of each expected task. The following provides an example of a job description plus the necessary qualifications for an assistant golf course superintendent, plus a yearly evaluation.

### Example

#### ASSISTANT GOLF COURSE SUPERINTENDENT

##### Job Purpose

Under direct supervision of the golf course superintendent (GCS), the assistant GCS directs and participates in the maintenance of the golf course greens, tees, fairways, roughs, and cart paths; supervises the maintenance and repair of motorized and other mechanical equipment; coordinates pesticide and fertilizer application; oversees operational irrigation systems; and does related work as required. The assistant GCS may serve in the golf course superintendent's capacity during his or her absence.

| Job Tasks—Assistant Golf Course Superintendent | Percentage of Time |
|---|---|
| • Assists the golf course superintendent in planning and supervising the maintenance of greens, tees, and fairways; schedules work and supervises employees and use of the equipment | 60 |
| • Instructs equipment operators on the proper use and care of mowing and other equipment; supervises pesticide applications and/or operates and calibrates pesticide application equipment; supervises and participates in the operation and maintenance of pumps and in the maintenance of irrigation and drainage systems | 20 |
| • Assists in personnel management and evaluation, employee safety, and personnel discipline; may also modify the daily work schedule on professional interpretation | 15 |
| • Miscellaneous duties as required | 10 |

##### Minimum Training and Experience Requirements

AAS in horticulture, agronomy, or closely related field and a minimum of two years' experience in professional turfgrass/golf course management and construction. BS degree preferred. Must possess, maintain, and be willing to use a state pesticide license in turfgrass and ornamentals. Must be able to pick up 50 pounds.

##### Preferred Qualifications

BS degree and three years' experience as previously described. Formal training in Spanish, irrigation technology, and personnel management is preferred.

##### What Knowledge, Skills, and Abilities Are Needed by an Employee upon Entry to This Job?

The job requires knowledge of the goals and objectives of turfgrass maintenance programs and safety practices relevant to turf maintenance, including knowledge and/or experience with managing bentgrass greens in heat stress and overseeding bermudagrass with ryegrass, *Poa trivialis*, fine fescue, and so forth, for winter color. Ability to plan, coordinate, and conduct daily work schedules for the maintenance crew and plan capital improvement projects using modern construction methods and techniques. Must also coordinate and oversee the functional legal operation of pesticide mixing/loading sites, rinsate handling, and proper storage and inventory of pesticides, fertilizers, and other hazardous materials. Confers with and makes suggestions to the golf course superintendent on daily and long-term planning strategies, evaluating personnel, and handling minor personnel disputes/discipline.

##### Describe the Guidelines and Supervision an Employee Receives in Order to Do This Job, Including the Employee's Independence and Discretion

After conferring with the golf course superintendent, the assistant GCS must work independently and make informed decisions independently of the supervisor, with minimal supervision.

Other brief examples of job descriptions for positions on the golf course maintenance staff include the following examples.

### Example

#### EQUIPMENT MANAGER

##### Job Purpose

Under direct supervision of the golf course superintendent and/or assistant GCS, the equipment manager supervises a comprehensive preventative maintenance program on the golf course equipment. This includes the repair of failing or failed equipment, maintaining records of parts and labor required to perform this duty, and placing orders for necessary parts and supplies. The equipment manager communicates any needs or problems relating to equipment maintenance or repair. Safety is a top priority, and the equipment manager is responsible for maintaining a clean, organized service area and maintenance building. Basic hand tools normally associated with routine equipment maintenance are the responsibility of the equipment manager. Specialized tools necessary for additional equipment maintenance will be the responsibility of the facility.

| Job Tasks—Equipment Manager | Percentage of Time |
|---|---|
| • Inspects, diagnoses, and repairs mechanical defects/failures of maintenance equipment including diesel-, electric-, and gasoline-powered automobiles, trucks, mowers, rollers, and other mechanical equipment used on the golf course, including hydraulics and routine mower-blade sharpening | 60 |
| • Instructs equipment operators on the proper use and care of mowing and other equipment, preventative maintenance procedures, and proper cleaning, storing, and use of equipment | 15 |
| • Keeps a complete set of records for equipment and inventory purchases and needs, equipment condition, and costs of repairs; purchases repair parts and replacement supplies | 10 |
| • Maintains a clean, organized shop maintenance facility | 5 |
| • Miscellaneous duties as required | 10 |

#### Minimum Training and Experience Requirements

The position requires a working knowledge of light and heavy maintenance equipment, hydraulics, basic electrical circuitry, internal combustion engines, mowing equipment, and automotive pieces. This includes the ability to diagnose mechanical problems and determine appropriate maintenance needs. Requires skill in the use of various equipment repair tools with the ability to purchase appropriate tools and supplies within budget to provide the necessary working order of equipment. Requires a working knowledge of the safety precautions of the profession to maintain an OSHA-approved workplace. Must possess a valid driver's license and possibly a commercial driver's license (CDL). Must be able to move at least 75 pounds.

#### Preferred Qualifications

Formal mechanical training and three years' experience as previously described. Understands the needs of modern golf maintenance shops and playing conditions and how these affect the golf course.

### Example

#### IRRIGATION TECHNICIAN/SPECIALIST

#### Job Purpose

Under direct supervision of the golf course superintendent and/or assistant GCS, the irrigation technician/specialist is responsible for the programming, operation, and maintenance of the golf course irrigation system(s). This includes repair of failing or failed irrigation equipment, maintaining records of parts and labor required to perform this duty, and placing orders for necessary parts and supplies. The irrigation specialist communicates any needs or problems relating to irrigation equipment maintenance or repair. If water-permitting/-licensing becomes required, the irrigation technician shall help develop water usage and other materials/information necessary to conform or apply for this.

| Job Tasks—Irrigation Technician/Specialist | Percentage of Time |
|---|---|
| • Coordinates necessary watering for the entire golf course, including supplemental hand-watering/-syringing; water quality issues/adjustments; monitoring ponds, lakes, and streams for water quality and wildlife habitats | 50 |
| • Inspects, diagnoses, coordinates, and repairs defects/failures of irrigation equipment including controllers, pumps, motors, valves, heads, leaks, and other related irrigation situations on the golf course | 20 |
| • Keeps a complete set of records for irrigation equipment and inventory purchases and needs, equipment condition, and costs of repairs; purchases repair parts and replacement supplies | 10 |
| • Coordinates and supervises labor for special projects such as repairing greens with seeds or plugs of new grass when needed | 10 |
| • Miscellaneous duties as required | 10 |

#### Minimum Training and Experience Requirements

The position requires working knowledge of basic electricity and irrigation hydraulics including automatic valves and controllers, as well as various types of pumps and pumping systems. Requires knowledge of basic tests and data interpretation necessary to monitor water quality and ways to adjust or modify water to suitable levels. Requires understanding of soil and turfgrass science and the rules of golf, ability to operate light motorized equipment, and ability to follow oral and written instructions.

#### Preferred Qualifications

AAS or BS in turfgrass science, irrigation technology, or a closely related field. Training in water quality is desirable. Requires a valid driver's license.

## Example

### CHEMICAL TECHNICIAN/IPM SPECIALIST

#### Job Purpose

Under direct supervision of the golf course superintendent and/or assistant GCS, the chemical (spray) technician/IPM specialist is responsible for following all integrated pest management (IPM) strategies involving scouting, detection, and threshold determination. This person also oversees the application of all chemicals and fertilizers on golf course properties, including proper sprayer/spreader calibration, right-to-know laws, and appropriate applicator apparel. The chemical technician is responsible for reading, understanding, and keeping current with pesticide and fertilizer storage information, as well as application and disposal rules and regulations. This includes maintaining up-to-date pesticide labels and material safety data sheets (MSDSs). Records indicating pesticide usage on the golf course will also be provided to the golf course superintendent.

| Job Tasks—Chemical Technician/IPM Specialist | Percentage of Time |
|---|---|
| • Coordinates necessary pesticide and fertilizer application for the entire golf course, including proper sprayer calibration and application, applicator safety apparel, scouting for pests, and proper disposal of rinsates | 60 |
| • Keeps a complete set of records for pesticide and fertilizer inventory, purchases, application equipment, and repairs; conducts periodic training of the golf course crew on pesticide and fertilizer safety and handling issues | 20 |
| • Maintains chemical and fertilizer storage and mixing areas in compliance with county, state, and federal regulations and is responsible for the safety and direction of chemical handlers and/or helpers | 10 |
| • Miscellaneous duties as required | 10 |

#### Minimum Training and Experience Requirements

This position requires complete knowledge of chemical and fertilizer application equipment including hand-operated spreaders, spray equipment, and pressurized sprayers. Requires current state certification or licensing as a restricted-use pesticide applicator, American Red Cross first-aid training, and ability to follow oral and written instructions.

#### Preferred Qualifications

AAS or BS in turfgrass science, pest management, entomology, plant pathology, or closely related field. Requires a valid driver's license and the ability to move 50 pounds. Prefer ability to communicate in Spanish.

## Example

### EQUIPMENT OPERATOR

#### Job Purpose

Under direct supervision of the golf course superintendent and/or assistant GCS, the equipment operator works using large equipment, including tractors, loaders, trenchers, backhoes, trucks, and other maintenance equipment. This person also assists the equipment manager on routine preventative maintenance of such equipment.

| Job Tasks—Equipment Operator | Percentage of Time |
|---|---|
| • Mows tees, fairways, and roughs with gang units; operates walk-behind and motorized greens mowers | 50 |
| • Operates other large equipment such as turf vac, skid-loader, fairway and rough aerifier, and boom sprayer in fairways and rough | 20 |
| • Checks and properly maintains equipment cooling system, fuel, and oil levels; cleans equipment daily; immediately reports equipment problems or failures to the equipment manager | 10 |
| • Under the direction of the equipment manager, provides basic preventative maintenance on equipment and returns it to the designated location at the maintenance facility | 10 |
| • Miscellaneous duties as required | 10 |

#### Minimum Training and Experience Requirements

This position requires knowledge of safe, efficient mechanical operation of tractors and other motorized equipment, mechanical aptitude, and ability to follow oral and written directions.

#### Preferred Qualifications

One year's experience with similar equipment at a comparable golf course facility. Requires a valid driver's license and must randomly submit to a drug test.

## Example

### OFFICE ASSISTANT

#### Job Purpose

Under direct supervision of the golf course superintendent, the office assistant provides administrative, clerical, and office management support for the golf course maintenance operation.

| Job Tasks—Office Assistant | Percentage of Time |
|---|---|
| • Answers and directs all incoming phone calls and e-mails, screens the golf course superintendent's calls; greets and directs all visitors; distributes all incoming mail; and oversees outgoing mail, shipping operations, and services | 50 |
| • Maintains the golf course superintendent's calendar of meetings, and so forth Dispatches employees' daily tasks, posting assignments, tracking appointments and workflow; records hourly employees' work time | 20 |
| • Maintains records and files for the golf course maintenance operation, including current and accurate job position descriptions; job applicant database and file of applications received; inventory of building keys and credit card accounts; golf course maintenance staff payroll information, including regular and overtime hours, sick time, vacation time, and so forth; and schedules annual performance evaluations and other appointments | 20 |
| • Miscellaneous duties as required | 10 |

#### Minimum Training and Experience Requirements

This position requires telephone experience and the ability to organize an office, calls, appointments, job applications, and applicants. Some bookkeeping experience would be helpful.

#### Preferred Qualifications

High school diploma or equivalent and two years' postsecondary work in office administration, office machines, and recordkeeping. Requires typing skills of 50 words per minute and knowledge of word processing, spreadsheet, and database management software. Requires excellent organizational skills and sensitivity with confidential information. Specific knowledge of specialized materials and equipment necessary for golf course maintenance would be beneficial.

## Example

### LANDSCAPER/HORTICULTURIST

#### Job Purpose

Under direct supervision of the golf course superintendent, the landscaper/horticulturist manages the greenhouse and nursery to produce and install ornamental landscaping for the golf course and clubhouse properties. Routine maintenance of such plantings in terms of pruning, weeding, mulching, fertilizing, irrigation, and others also is included.

| Job Tasks—Landscaper/Horticulturist | Percentage of Time |
|---|---|
| • Manages the greenhouse and ornamental nursery facilities; selects, propagates, and grows plant materials (except trees and larger shrubs) for pleasing ornamental use on and around the golf course, clubhouse, and other buildings; applies mulch and coordinates hand weed control as needed | 60 |
| • Designs flower beds; cultivates shrubs, and flowers; trims and prunes hedges and shrubbery; advises the golf course superintendent of the need for tree-trimming services; | 10 |
| • Designs, installs, and maintains low-pressure irrigation systems for landscaping | 10 |
| • Applies fertilizer, pesticides, and other chemicals to ornamental plantings, keeping the appropriate records and following all local, state, and federal laws in providing this | 10 |
| • Miscellaneous duties as required | 10 |

#### Minimum Training and Experience Requirements

This position requires knowledge of locally adapted plants for various uses on the golf course property. Two years of such experience with similar climatically adapted plants would be helpful. Experience in golf course maintenance and the game of golf is helpful.

#### Preferred Qualifications

AAS or BS in horticulture or closely related field. Requires knowledge of the characteristics, growth habits, and proper cultivation of locally adapted flowering bulbs and annual, biennial, and perennial plant materials, including herbaceous and woody plants. Requires knowledge of various fertilizers, soil amendments, and pest control options for landscape plantings. Requires current state licensing as a pesticide applicator for indoor and outdoor landscape plantings.

## EMPLOYMENT CONTRACTS

To protect both the golf course and the superintendent, a memorandum of agreement or contract on the exact job descriptions and duties, extent of authority, specific responsibilities, immediate supervisor, salary, and list of privileges or benefits should be drawn up and agreed upon (Figure 16.13). This should also include a fair severance clause and possibly renewal (rollover) dates and timing plus a force majeure clause (occurrences out of one's control such as floods, ice storms, hurricanes, etc.). It should also include an indemnification clause, which protects one against lawsuits. This formal written contract is valid and enforceable while oral agreements probably are not. This is important as many members of the board of directors and/or greens committee constantly change; thus, the contract helps protect the superintendent with the different personalities and agendas of each board.

Many times, potential employees focus only on the base salary. However, in the golf business, additional considerations are often available to supplement a compensation package. These include:

- Signing bonus
- Performance bonus that is identifiable, realistic, and attainable. Schedule annual reviews when they are most beneficial, such as when the course is in the best condition, when one is needed the most, and after budget numbers are complete. Other bonus criteria that can be in one's favor include having a communication program with employers, quantifiable safety records, environmental compliance, budget development, and management, as well as possessing a good attitude, including being identifiable as a team player.
- Deferred compensation, which is an employer contribution
- Retirement plan including employer contribution
- Vacation
- Sick days
- Family leave
- Bereavement
- Holidays

Other possible benefits include special outing bonuses, low-interest loans, golf ball concession, and percentage of guest fees, cart fees, and range fees. Insurance considerations include disability, medical, dental, vision, dependent coverage, including employer–employee contributions. Relocation packages and continuing education are additional possible benefits. Table 16.19 provides a simple contract on what may be included. Time and consultation are needed in the contract negotiating and writing process so both parties understand what is expected.

**FIGURE 16.13** Good superintendents are organized, delegate responsibilities, and empower their employees.

## COMMUNICATING WITH GOLFERS/EMPLOYERS

Running a golf maintenance facility is like most businesses, where building a rapport with the appropriate people generally increases the relationships needed for long-term success. If employers sense one is extremely responsible with the budget provided and responsive to their needs (however mundane or unconventional they may be), chances are their confidence in one and one's abilities will entice them to better support one's efforts. The first step in this endeavor is opening and solidifying a line of communication. Ways to do this include:

- Being accessible at all times—home and away
- Providing copies of turf-related trade journals
- Reading the journals that employers read
- Attending greens committee meetings
- Inviting the greens committee members to one's building and demonstrating some of the equipment
- Writing thank-you and congratulatory notes as appropriate
- Learning the names and interests of the golfers
- Developing a website or blog, soliciting comments
- Writing articles for the club newsletter highlighting the upcoming maintenance calendar or extraordinary events such as unusual weather experienced (Figure 16.14)
- Always documenting changes on the course by taking before, during, and after photographs
- Playing a reasonable game and playing with key management personnel
- Having periodic open houses at the turf care center (maintenance shop)
- Getting to know the local media and governing bodies such as the town council and mayor
- Not arguing—understanding the hierarchy and one's place within it

**TABLE 16.19**
**Example of a Simple Contract between a Golf Course Superintendent and Golf Course**

LETTER OF UNDERSTANDING AND AGREEMENT BETWEEN

______________________________________________

Representing ABC Golf Course/Country Club (the Employer) and

______________________________________________

The Superintendent (the Employee)
is hereby employed as golf course superintendent of the ABC Golf Course for a period between ________, 20 and ________, 20 ____. Annual restitution will be at the rate of $__________________ subject to annual review and renewal. The mutual objective between the Superintendent and Club is to bring about and maintain a quality golf course at a reasonable cost.

**Golf Course Superintendent's Responsibilities:**

1. The Superintendent will be responsible only to the Chair of the Green Committee, the Club's Manager or______________.
2. The Superintendent shall be solely responsible for hiring, firing, and supervising all grounds and green maintenance personnel; the salaries of such persons shall be approved by the Green Committee, Club Manager, or ________________.
3. All purchases/leases for maintenance of the golf course grounds/greens shall be the responsibility of the Superintendent after previous approval by the Green Committee, the Club's Manager or______________.
4. The Superintendent shall follow all local, state, and national laws and policies concerning all pesticide/fertilizer/fuel and similar hazardous materials, purchases, storage and use; maintain the maintenance facility according to insurance and/or local codes, such as OSHA or similar, and follow club policy on hiring, firing, promotions, and employee reprimand.
5. The golf course shall be maintained in accordance to the performance standards agreed upon by the Green Committee and Superintendent with reasonable allowances for necessary routine maintenance, construction or renovation, and/or acts of God.

**Privileges:**

1. The Superintendent shall enjoy all privileges offered by the course and clubhouse, including guests, in good taste, in accordance with the rules of the Club.
2. Lunch shall be available to the Superintendent at his/her discretion. Supper will be available when duties require the Superintendent to be at the Club during evening hours.
3. Vacation, with pay, will be__________days annually, with timing approved by the Chair of the Green Committee, Club Manager, or________.
4. The Club shall provide a vehicle (usually a truck), fuel, insurance, and routine maintenance for the Superintendent to use during and after work not to exceed__________annual miles for personnel use.
5. As the superintendent is a key representative of the golf club, he/she is expected to maintain acceptable dress/attire in accordance to Club policy. An annual clothing allowance not to exceed $__________will be provided.
6. Reasonable expenses for educational meetings/training/materials will be provided by the Employer, not to exceed $________annually.

**Terms:**

These terms may be renewed or extended annually by the agreement of both parties. If either the Employer or Superintendent does not desire to renew or extend this Agreement, written notice will be provided to the other party at least 90 days prior to the expiration of this Agreement by certified mail. The Employer may terminate this Agreement for cause, effective the date the Employer so notifies the Superintendent in writing. The standard minimum two weeks' courtesy notice shall be invoked by either party prior to employment termination.

______________________________ Date______________

ABC Golf Club President

______________________________ Date______________

Superintendent

- Not getting defensive if feeling threatened and not counterattacking
- Being the "go-to guy" by offering solutions to problems

In the golf business, superintendents must be flexible in order to move up and, frankly, if/when the current job is terminated. Many variables beyond the superintendent's control influence this, including yearly elections of club presidents or greens chair, ownership changes, hiring a new director of golf, contracting with a management company, unusual weather patterns, or simply the perception that the superintendent's salary is excessive. It is prudent to negotiate a severance package, preferably when hired and less so when termination occurs. Items to negotiate include any severance pay plus benefits and duration of benefits, ways both parties can maintain a positive reputation, and the speed of management transition.

**DON GARRET, CGCS** **SUPERINTENDENT'S REPORT**

**June Status**

It's been a busy june as we have had lots of play. 2 weeks of golf camp and acrified all collars, approaches, fairways and greens.

We backed off a little on our fairway acrification, using a machine that punches fewer holes and we did not verticut. I hope to slice the fairways and scalp mow them in early August to help compensate. We were able to topdress lower 9 fairway in our continuing efforts to improve the drainage of this area. We planned to topdress #1 as well but were rained out and have it rescheduled for early July.

We also had the project on holes 14 and 15 going on the last two weeks of the month. Rain the first week slowed things down a bit, but we were able to complete everything except for a little cleanup and much work by months end. These areas will be marked as ground under repair until the sod establishes.

Doing all this aerification and project work has us a little bit behind on some of our trim work but we will be working hard to catch up on this in early July. Weeds in mulch beds and along the perimeter of the course have been especially bad this year. I am guessing that the mild winter. followed by a wet April and May have caused the weeks to flourish.

The newly reisodded white tee on #1 was reopened for play on June 22 and looks great. We hope to early it soon to break up any layering between the sod and the underlying soil.

**Greens**

The greens were aerified as scheduled on Monday June 26-27. We used a ½ inch coring time and topdressed to fill the holes, then the greens were rolled. Full recovery will take 21 days. An application of quick release nitrogen was made 7 days prior to aerification to promote recovery. This fertilizer will provide rapid growth then be gone allowing us to regain control of the growth rate and the green speed. Following recovery, we will be spoon feeding small quantities of plant food, applying growth regulator, ligthly and frequently verticutting, brushing, and topdressing.

I am also considering the possibility of some type of aerification in early August. I am just not sure that one aerification is enough. Rest assured it wont't be anything as aggressive as what we just did. It will likely be a ¼ inch solid tine, quad coring tine, or a sand injection machine.

**FIGURE 16.14** Example of communicating maintenance schedules to members through a newsletter.

# Useful Mathematics and Metric Conversion

**Acres covered/hour:** $= \text{mph} \times \text{swath (feet)} \times 0.1212$ or $\frac{\text{mph} \times \text{swath (feet)}}{8.25}$

**Gallons per acre (gpa):**

$$= \frac{\text{gpm (whole boom)} \times 495}{\text{mph} \times \text{swath (feet)}} \quad \text{or} \quad \frac{\text{gpm per nozzle} \times 495}{\text{mph} \times \text{nozzle spacing (feet)}}$$

$$= \frac{\text{gpm per nozzle} \times 5940}{\text{mph} \times \text{nozzle spacing (inches)}} \quad \text{or} \quad \frac{\text{gpm per nozzle} \times 5940}{\text{mph} \times \text{width of nozzle spray (inches)}}$$

$$= \frac{\text{fluid ounces collected per nozzle in 100 feet} \times 40.8375}{\text{nozzle spacing (inches)}} \quad \text{or} \quad \frac{\text{fluid ounces collected per nozzle} \times 4084}{\text{feet traveled} \times \text{nozzle spacing (inches)}}$$

$$= \frac{\text{gallons collected per nozzle} \times \text{no. nozzles} \times 43560}{\text{feet traveled} \times \text{swath (feet)}} \quad \text{or} \quad \frac{\text{gallons per 1,000 square feet}}{0.023}$$

**Gallons per 1,000 ft²** $= 0.023 \times \text{gpa}$

**Ounces per 1,000 ft²** $= 2.94 \times \text{gpa}$

**Gallons per minute (gpm):**

$$= \frac{\text{gpa} \times \text{mph} \times \text{swath (ft)}}{495} \quad \text{or} \quad \frac{\text{fl oz per minute}}{128}$$

$$= \frac{\text{gpa} \times \text{mph} \times \text{nozzle spacing (inches)} \times \text{no. nozzles}}{5940}$$

**Gallons per minute/nozzle:**

$$= \frac{\text{gpa} \times \text{mph} \times \text{nozzle spacing (inches)}}{5940} \quad \text{or} \quad \frac{\text{gpa} \times \text{mph} \times \text{nozzle spacing (ft)}}{495}$$

$$= \frac{\text{test jar fl oz} \times 0.46875}{\text{seconds to fill test jar}} \qquad \frac{7.5}{\text{seconds to fill 1 pint (16 fl oz)}}$$

$$= \frac{15}{\text{seconds to fill 1 quart (32 fl oz.)}}$$

**Minutes/acre:** $= \frac{495}{\text{mph} \times \text{swath (ft)}}$

Acres covered per tank: $= \frac{\text{gallons per tank}}{\text{gpa}}$

**Minutes/load:** $= \frac{\text{gallons/load} \times 495}{\text{mph} \times \text{gpa} \times \text{swath (ft)}}$

Material needed per tank $= \frac{\text{rate/acre} \times \text{gallons/tank}}{\text{gpa}}$

**Travel speed (miles per hour, mph)** $= \frac{\text{distance traveled (ft)} \times 0.68}{\text{time (seconds) to travel distance}}$

**Flow rate** (as influenced by pressure):

$$\frac{GPA_1}{GPA_2} = \frac{\sqrt{PSI_1}}{\sqrt{PSI_2}} \quad \text{or} \quad GPA_2 = GPA_1 \times \sqrt{\frac{PSI_2}{PSI_1}} \quad \text{or} \quad PSI_2 = PSI_1 \times \left(\frac{GPA_2}{GPA_1}\right)^2$$

For any change in travel speed (mph), calculate the resulting $GPA_2$ by

$$GPA_2 = \frac{GPA_1 \times MPH_1}{MPH_2} \quad \text{or} \quad \frac{GPA_1}{GPA_2} = \frac{MPH_2}{MPH_1} \quad \text{or} \quad MPH_2 = \frac{GPA_1 \times MPH_1}{GPA_2}$$

## Fluid Application

**ppm** $= \frac{1{,}000{,}000 \times \text{lb ai used}}{\text{gal/tank} \times 8.34}$ or $\frac{\text{wt. of material to be used (lb)} \times 1{,}000{,}000}{\text{wt. of tank mixture (lb)}}$

| | | | |
|---|---|---|---|
| | $= \frac{1,000,000 \times \text{oz commercial material used} \times \text{percent ai (decimal)}}{\text{gal/tank} \times 8.34 \times 16}$ | or | $\frac{1,000,000 \times \text{fl oz used} \times \text{lb ai/gal}}{\text{gal/tank} \times 8.34 \times 128}$ |
| **Pounds nutrients applied/acre** | = ppm of the element in the water × acre-inches water applied × 0.226464 | | |
| **Pounds ai to use per tank** | $= \frac{\text{ppm desired} \times \text{gal/tank} \times 8.34}{1,000,000}$ | or | $\frac{\text{ppm desired} \times \text{gal/tank} \times 8.34}{1,000,000 \times \text{percent ai}}$ |
| **Pounds commercial material to use per tank** | $= \frac{\text{ppm desired} \times \text{gal/tank} \times 8.34}{1,000,000 \times \text{percent ai (decimal)}}$ | or | $\frac{\text{ppm desired} \times \text{gal/tank} \times 8.34}{\text{percent ai (decimal)}}$ |
| **Fluid ounces to use per tank** | $= \frac{\text{ppm desired} \times \text{gal/tank} \times 8.34 \times 128}{1,000,000 \times \text{ai per gal}}$ | | |
| **Gallons commercial material to use per tank** | $= \frac{\text{ai (decimal)} \times 8.34 \times \text{gal/tank}}{\text{ai per gal} \times 100}$ | | |
| **Percent ai in a spray mix** | $= \frac{\text{lb commercial material used} \times \text{percent ai (decimal)}}{\text{gal/tank} \times 8.34}$ | | |
| **Gallons commercial material for total treated acres** | $= \frac{\text{ppm desired} \times \text{GPA} \times \text{acres} \times 8.34}{1,000,000 \times \text{lb ai/gal}}$ | | |

### Active Ingredients (ai)

| | | | |
|---|---|---|---|
| **Pounds commercial material/acre** | $= \frac{\text{lb ai to be applied per acre}}{\text{\% ai of material}}$ | **Gallons commercial material/tank** | $= \frac{\text{gallon/tank} \times \text{lb ai to be applied per acre}}{\text{gallons/acre} \times \text{lb ai per gallon}}$ |
| **Gallons commercial material/acre** | $= \frac{\text{lb ai to be applied per acre}}{\text{lb ai per gallon}}$ | | |

## Time (Seconds) Required to Cover a Specific Distance to Obtain a Desired Speed (mph)

| Desired mph | Feet per Minute | Time Required (seconds) to Travel a Distance of 100 ft | 200 ft | 300 ft |
|---|---|---|---|---|
| 2.0 | 176 | 34 | 68 | 102 |
| 2.5 | 220 | 27 | 54 | 81 |
| 3.0 | 264 | 23 | 45 | 68 |
| 3.5 | 308 | 20 | 39 | 58 |
| 4.0 | 352 | 17 | 43 | 51 |
| 4.5 | 395 | 15 | 30 | 45 |
| 5.0 | 440 | 14 | 27 | 41 |
| 6.0 | 528 | – | 23 | 34 |
| 7.0 | 616 | – | 19 | 29 |
| 8.0 | 704 | – | 17 | 26 |
| 9.0 | 792 | – | 15 | 23 |

## Approximate Rates of Application Equivalents

**Weights**

1 oz/ft² = 2722.5 lb/A
1 oz/yd² = 302.5 lb/A
1 oz/100 ft² = 27.2 lb/A
1 oz/1000 ft² = 43.46 oz/A = 2.72 lb/A
1 lb/A = 1 oz/2733 ft² = 8.5 g/1000 ft²

**Liquid**

1 oz/1000 ft² = 43.56 oz/A = 1.4 qt/A
1 pt/1000 ft² = 5.4 gal/A
100 gal/A = 2.3 gal/1000 ft² = 1 qt/100 ft²

100 lb/A = 2.5 lb/1000 $ft^2$
1 $yd^3$ sand = 1.3 to 1.5 tons
1 bushel = 1¼ $ft^3$ = 0.046 $yd^3$

When determining surface areas, the following calculations and formulas are helpful:

**Calculations and Formulas for Various Shapes**

| | | | |
|---|---|---|---|
| **Rectangle, square, or parallelogram:** | Area | = | length (L) × width (W) |
| **Trapezoid:** | Area | = | [a + (b × h)] ÷ 2 |
| **Circle:** | Area | = | radius $(r)^2$ × 3.1416 (or π) = diameter $(d)^2$ × 0.7854 |
| | Radius | = | d ÷ 2 |
| | Diameter | = | r × 2 |
| | Circumference | = | π × d |
| **Sphere:** | Volume | = | $r^3 \times 4.1888 = d^3 \times 0.5236$ |
| **Triangle:** | Area | = | ½(b × h) |
| **Cylinder:** | Volume | = | $r^2\pi L$ |
| **Cone:** | Area | = | ⅓($\pi r^2 h$) |
| **Cube:** | Volume | = | length × L × L |

## Example

From Figure A.1, determine the area ($ft^2$) of the tee, dogleg fairway, and green.

**Step 1:** Segment the total area into common geometric figures, including parallelograms, triangles, rectangles, trapezoids, circles, and ovals.
**Step 2:** Determine the area of each geometric figure:

*Tee*

$$\text{Area of a square} = \text{length} \times \text{width} = 30\,\text{ft} \times 30\,\text{ft} = 900\,\text{ft}^2$$

*Fairway*

$$\text{Area of a parallelogram} = \text{length} \times \text{width} = 85\,\text{ft} \times 200\,\text{ft} = 17{,}000\,\text{ft}^2$$

$$\text{Area of a triangle} = (\text{width} \times \text{height}) \div 2 = 20\,\text{ft} \times 200\,\text{ft} \div 2 = 2{,}000$$

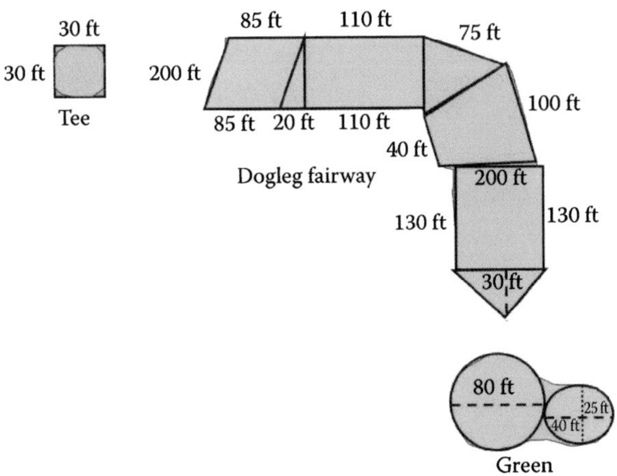


**FIGURE A.1** Determining the surface areas of a dogleg golf hole by segmenting each area into smaller units.

$$\text{Area of a rectangle} = \text{length} \times \text{width} = 110\,\text{ft} \times 200\,\text{ft} = 22{,}000\,\text{ft}^2$$

$$\text{Area of a triangle} = (\text{width} \times \text{height}) \div 2 = 75\,\text{ft} \times 200\,\text{ft} \div 2 = 7{,}500\,\text{ft}^2$$

$$\text{Area of a trapezoid} = \left[a + (b \times \text{height})\right] \div 2 = \left[40\,\text{ft} + (100\,\text{ft} \times 200\,\text{ft})\right] \div 2 = 10{,}020\,\text{ft}^2$$

$$\text{Area of a rectangle} = \text{length} \times \text{width} = 130\,\text{ft} \times 200\,\text{ft} = 26{,}000\,\text{ft}^2$$

$$\text{Area of a triangle} = (\text{width} \times \text{height}) \div 2 = (30\,\text{ft} \times 200\,\text{ft}) \div 2 = 3{,}000\,\text{ft}^2$$

$$\text{Total area of the fairway} = 17{,}000 + 2{,}000 + 22{,}000 + 7{,}500 + 10{,}020 + 26{,}000 + 3{,}000 = 87{,}520\,\text{ft}^2 = 87{,}520\,\text{ft}^2 \div 43{,}560\,\text{ft}^2/\text{acre} = 2 \text{ total acres}$$

*Green*

$$\text{Area of a circle} = \text{diameter (d)} \times 0.7854 = (80\,\text{ft})^2 \times 0.7854 = 5{,}027\,\text{ft}^2$$

$$\text{Area of an oval} = (\text{radius} \times \text{diameter}) \times 0.7854 = (25\,\text{ft} \times 40\,\text{ft}) \times 0.7854 = 785\,\text{ft}^2$$

$$\text{Total area of the green} = 5{,}027\,\text{ft}^2 + 785\,\text{ft}^2 = 5{,}812\,\text{ft}^2$$

## OFFSET METHOD

Another way to measure irregularly shaped areas is the offset method. Large areas are broken down into a series of smaller trapezoids equally spaced along a center line. The precision is increased by decreasing the spacing between lines. Each 90° offset line is added together, and this sum is multiplied by the spacing(s) of each line.

### Example

Determine the area of the following sand trap (Figure A.2).

**Step 1:** Establish a center line and divide it into equally spaced offset lines. The center line in this example is 80 feet long and is divided into seven offset lines spaced 10 feet apart.
**Step 2:** Add up the lengths of the seven offset lines:

1 = 12 ft 2 = 11 ft 3 = 10 ft 4 = 5 ft 5 = 7 ft 6 = 10 ft 7 = 9 ft

total = 64 ft

**Step 3:** Multiply the total lengths of the offset lines by the spacing(s) of each line to obtain total area in square feet:

$$\text{Area} = 64\,\text{ft} \times 10\,\text{ft} = 640\,\text{ft}^2$$

## AVERAGE RADIUS METHOD

A variation of the offset method to determine an irregularly shaped circular area is the **average radius method.** The average radius method is most often used for irregularly shaped greens (Figure A.3).

**Step 1:** From the center of the irregularly shaped circular area, such as a green, divide it into equally spaced 10° or 20° pie-shaped increments. For greater precision, use the 10° increments.
**Step 2:** Measure the length of each radial from the center to the edge of the green. This totals 36 measurements for 10° increments and 18 for 20° increments.
**Step 3:** Total these measurements and divide by the number of measurements to achieve an average radius length.
**Step 4:** Calculate the area of the green using the formula:

$$\text{Area} = 3.14 \times (\text{average radius})$$

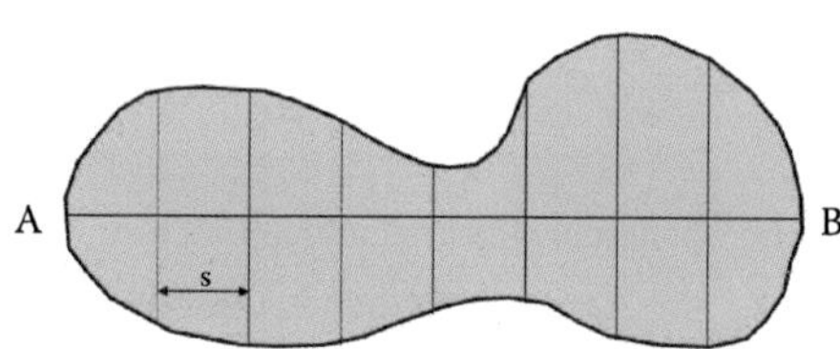


**FIGURE A.2** Using the offset method to determine the size of irregularly shaped areas. Large areas are segmented into a series of smaller trapezoids equally spaced along a centerline. Each 90° offset line is added together and this sum is multiplied by the spacing of each line.

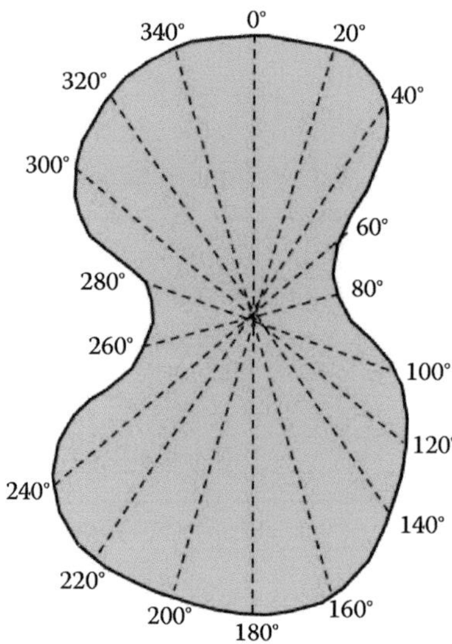


**FIGURE A.3** Using the average radius method to determine an irregularly shaped circular area. From the center of the circular area, it is divided into equally spaced 10 or 20° pie-shaped increments. The length of each radial from the center to the edge of the circular area is measured and are totaled and divided by the number of measurements to achieve an average radius length. The area is then calculated by using the formula: area = 3.14 × (average radius)$^2$.

### Example

From the previous illustration, determine the area of the green.

**Step 1:** Divide the green into equally spaced degree increments. For this example, 20° increments are used.
**Step 2:** M Measure the length of each radial from the center of the green to the edge. Add these and divide by the total number of measurements (18 in this example).

0° = 53 ft 60° = 36 ft 120° = 46 ft 180° = 52 ft 240° = 37 ft 300° = 47 ft
20° = 50 ft 80° = 34 ft 140° = 48 ft 200° = 48 ft 260° = 34 ft 320° = 50 ft
40° = 47 ft 100° = 37 ft 160° = 50 ft 220° = 46 ft 280° = 36 ft 340° = 49 ft

800 ft ÷ 18 = 44.4 ft

**Sum** = 800 ft

Step 3: Calculate the surface area using the formula:

$$\begin{aligned}\text{Area} &= 3.14 \times (\text{average radius})\\ &= 3.14 \times (44.4)^2\\ &= 6.190\,\text{ft}^2\end{aligned}$$

## MODIFIED OFFSET METHOD

When measuring large areas, such as ponds, a modification of the offset method can be used.

**Step 1:** The pond is first bracketed by a large rectangle so all of the pond is included (Figure A.4).
**Step 2:** Offset lines are marked at even spacings along the long edge of the rectangle and their lengths are recorded.
**Step 3:** Add the two lines for each spacing and subtract this amount from the total width of the rectangle.
**Step 4:** Total these values and multiply this total by the spacing between the offset lines.

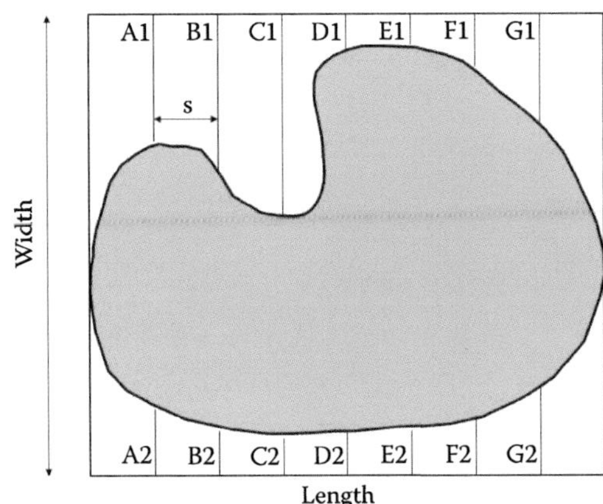


**FIGURE A-4** Using the modified offset method to determine the size of large areas, such as ponds. The pond is first bracketed by a large rectangle so all of it is included. Offset lines are marked at even spacings along the long edge of the rectangle and their lengths recorded. The two lines for each spacing are then added and subtracted from the total width of the rectangle. These values are totaled and multiplied by the spacing between the offset lines.

### Example

Determine the area of the previously illustrated pond.

**Step 1:** Draw and measure a rectangle around the area. The length is 360 feet long and the width is 320 feet. Divide the length into equally spaced offset lines.
**Step 2:** Measure and total the lengths of each equally spaced offset line.

| A | B | C | D | E | F | G |
|---|---|---|---|---|---|---|
| A1 = 40 ft | B1 = 50 ft | C1 = 60 ft | D1 = 20 ft | E1 = 16 ft | F1 = 24 ft | G1 = 24 ft |
| A2 = 20 ft | B2 = 24 ft | C2 = 18 ft | D2 = 20 ft | E2 = 22 ft | F2 = 26 ft | G2 = 22 ft |
| 60 ft | 74 ft | 78 ft | 40 ft | 38 ft | 50 ft | 46 ft |

**Step 3:** Subtract each offset line total from the width of the rectangle and then total these values.

| A: 160 ft | B: 160 ft | C: 160 ft | D: 160 ft | E: 160 ft | F: 160 ft | G: 160 f |
|---|---|---|---|---|---|---|
| 60 ft | 74 ft | 78 ft | 40 ft | 38 ft | 50 ft | 46 ft |
| 100 ft | 86 ft | 82 ft | 120 ft | 122 ft | 110 ft | 114 ft |

**Sum** = 734 ft

**Step 4:** Multiply the total by the spacing distance:

$$\begin{aligned}\text{Area} &= \text{total area} \times \text{spacing distance}\\ &= 734\,\text{ft} \times 80\,\text{ft}\\ &= 58{,}720\,\text{ft}^2\end{aligned}$$

To convert square feet to acres, divide the square feet by 43,560:

$$58{,}720 \text{ square feet} \times \frac{\text{acre}}{43{,}560 \text{ square feet}} = 1.35 \text{ acre}$$

If the average depth of the pond is known, the volume of water in the pond can be determined by multiplying the surface area by the average depth and converting to gallons, with the knowledge that 7.48 gallons can be held per cubic foot. For example, if the previous pond has an average depth of 50 feet, then:

$$\begin{aligned}\text{Volume} &= \text{surface area} \times \text{average depth} \times \frac{7.48 \text{ gallons}}{\text{cubic feet}} \\ &= 58{,}720 \text{ square feet} \times 50 \text{ feet} \times \frac{7.48 \text{ gallons}}{\text{cubic feet}} \\ &= 21{,}961{,}280 \text{ gallons}\end{aligned}$$

### Finding Tank Capacity (gallons)

| | | | |
|---|---|---|---|
| Cylindrical tanks: | (inches) | = | $L \times d^2 \times 0.0034$ |
| | (feet) | = | $L \times d^2 \times 5.875$ |
| Rectangle tanks: | (inches) | = | L × W × height × 0.004329 |
| | (feet) | = | L × W × height × 7.48 |
| Elliptical tanks: | (inches) | = | L × short diameter (sd) × long diameter (ld) × 0.0034 |
| | (feet) | = | L × sd × ld × 5.875 |

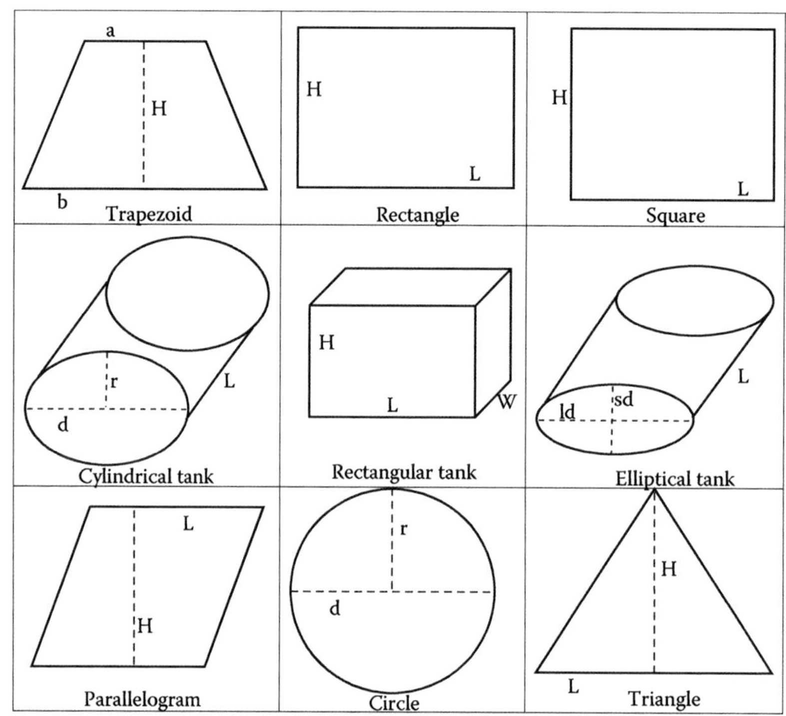


### Peat Moss Coverage

| Depth (inches) | Coverage (square feet) | |
|---|---|---|
| | 5.6 Cubic feet bale (Compressed) Covers | 4.0 Cubic feet bale (Compressed) Covers |
| 0.25 | 480 | 346 |
| 0.50 | 240 | 173 |
| 1.00 | 120 | 86 |
| 2.00 | 60 | 43 |
| 3.00 | 40 | 29 |
| 4.00 | 30 | 22 |
| 6.00 | 20 | 14 |

## Conversions for Determining Turfgrass Irrigation Needs

| | | | | | | |
|---|---|---|---|---|---|---|
| One acre-inch | = | 27,154 gallons | = | 43,560 cubic inches | = | 3,630 cubic feet |
| One inch 1,000 $ft^{-1}$ | = | 620 gallons | = | 83 cubic feet | | |
| One gallon | = | 0.134 cubic foot | = | 8.34 pounds | = | 231 cubic inches |
| One million gallons | = | 3.07 acre-feet | | | | |
| 7.5 gallons | = | 1 cubic foot | | | | |
| One acre-foot | = | 325,851 gallons | = | 43,560 cubic feet | | |
| One pound of water | = | 0.1199 gallon | | | | |
| Precipitation rate (inches/hour) | = | $\frac{\text{gpm} \times 96.3}{\text{area (ft}^2)}$ | | | | |

## Slopes

| | |
|---|---|
| 10% = 6E = 10:1 | 33% = 18E = 3:1 |
| 18% = 10E = 6:1 | 50% = 26E = 2:1 |
| 25% = 14E = 4:1 | 100% = 45E = 1:1 |

## Approximate Weight of Dry Soil

| Type | Bulk Density (g $cm^{-3}$) | Weight | | |
|---|---|---|---|---|
| | | Pounds $ft^{-3}$ | Kilograms $m^{-2}$ | Pounds $acre^{-1}$ (6 inches deep) |
| Sand | 1.6 | 100 (or 2,700 lb $yd^{-3}$) | 1,623 | 2,143,000 |
| Loam | 1.3–1.55 | 80–95 | 1,299–1,542 | 1,714,000 |
| Clay or silt | 1.0–1.30 | 65–80 | 1,055–1,299 | 1,286,000 |
| Muck | 0.65 | 40 | 649 | 860,000 |
| Peat (compact) | 0.325 | 20 | 325 | 430,000 |

Sand weights (tons) = cubic yards × 1.3

Gravel weights (tons) = cubic feet × 110

–0.5- to 1-inch diameter gravel = 2,700 lb/$yd^3$

–0.25- to 0.375-inch diameter gravel = 3,000 lb/$yd^3$

## Approximate Organic Materials for 6-Inch Depth per 1,000 Square Feet[a]

| Organic Material Volume in Mix | Approximate Thickness Applied to Soil Surfaces | | Organic Material Needed | |
|---|---|---|---|---|
| Percent | Inches | Centimeters | Cubic yards/1,000 square feet | Cubic meters/100 $m^2$ |
| 5 | 0.33 | 0.84 | 1.0 | 0.83 |
| 10 | 0.67 | 1.70 | 2.0 | 1.70 |
| 15 | 1.00 | 2.54 | 3.0 | 2.48 |
| 20 | 1.33 | 3.38 | 4.0 | 3.30 |
| 25 | 1.67 | 4.24 | 5.0 | 4.16 |
| 30 | 2.00 | 5.08 | 6.0 | 4.95 |

**Example:** If 10% organic materials is incorporated into the top 6 inches of a 1,000-square-foot area, the organic material is applied to a depth of 0.67 inch and 2.0 cubic yards will be needed (1.7 cm and 1.7 $m^3$ 100 $m^2$).

[a] Weight variance in materials may occur.

## Surface Area Impacted and Topdressing Sand Needed to Fill Aerification Holes

| Spacing (inches) | Tine Diameter (inches) | Tine Diameter (mm) | Holes $ft^{-2}$ (no.) | Surface Area Impacted (%) | Dry Sand to Fill Holes 3-inch depth | |
|---|---|---|---|---|---|---|
| | | | | | ~$ft^3$ 1,000 $ft^{-2}$ | ~lb 1,000 $ft^{-2}$ |
| 1.0 × 1.0 | 0.250 | 6.350 | 144 | 4.91 | 12.3 | 1,227 |
| | 0.375 | 9.525 | 144 | 11.04 | 27.6 | 2,761 |
| | 0.500 | 12.700 | 144 | 19.63 | 49.1 | 4,909 |
| | 0.625 | 15.875 | 144 | 30.68 | 76.7 | 7,670 |

| | | | | | | |
|---|---|---|---|---|---|---|
| | 0.750 | 19.050 | 144 | 44.16 | 110.4 | 11,040 |
| | 1.000 | 25.400 | 144 | 78.50 | 196.4 | 19,640 |
| 1.0 × 2.0 | 0.250 | 6.350 | 72 | 2.45 | 6.1 | 614 |
| | 0.375 | 9.525 | 72 | 5.52 | 13.8 | 1,381 |
| | 0.500 | 12.700 | 72 | 9.82 | 24.5 | 2,454 |
| | 0.625 | 15.875 | 72 | 15.34 | 38.4 | 3,855 |
| | 0.750 | 19.050 | 72 | 22.09 | 55.2 | 5,520 |
| | 1.000 | 25.400 | 72 | 39.27 | 98.2 | 9,820 |
| 1.5 × 1.5 | 0.250 | 6.350 | 64 | 2.18 | 5.5 | 550 |
| | 0.375 | 9.525 | 64 | 4.91 | 12.3 | 1,230 |
| | 0.500 | 12.700 | 64 | 8.72 | 21.8 | 2,180 |
| | 0.625 | 15.875 | 64 | 13.63 | 34.1 | 3,410 |
| | 0.750 | 19.050 | 64 | 19.63 | 49.1 | 4,910 |
| | 1.000 | 25.400 | 64 | 34.89 | 87.3 | 8,730 |
| 2.0 × 2.0 | 0.250 | 6.350 | 36 | 1.23 | 3.1 | 307 |
| | 0.375 | 9.525 | 36 | 2.76 | 6.9 | 690 |
| | 0.500 | 12.700 | 36 | 4.91 | 12.3 | 1,227 |
| | 0.625 | 15.875 | 36 | 7.67 | 19.2 | 1,917 |
| | 0.750 | 19.050 | 36 | 11.04 | 27.6 | 2,760 |
| | 1.000 | 25.400 | 36 | 19.63 | 49.1 | 4,910 |
| 7.0 × 7.0 | 0.750 | 19.050 | 2.94 | 0.90 | 2.3 | 2,30 |
| (drill and fill) | 1.000 | 25.400 | 2.94 | 1.60 | 4.0 | 400 |

## METRIC SYSTEM CONVERSION

**Metric Prefix Definitions (Basic Metric Unit = 1)**

| | | | | | |
|---|---|---|---|---|---|
| tera | = | $10^{12}$ | deci | = | $10^{-1}$ |
| giga | = | $10^{9}$ | centi | = | $10^{-2}$ |
| mega | = | $10^{6}$ | milli | = | $10^{-3}$ |
| kilo | = | $10^{3}$ | micro | = | $10^{-6}$ |
| hecto | = | $10^{2}$ | nano | = | $10^{-9}$ |
| deca | = | $10^{1}$ | pico | = | $10^{-12}$ |

**Metric Prefix Example (weight)**

| | | | | | | | | |
|---|---|---|---|---|---|---|---|---|
| 1 kg | = | $10^{3}$ g | = | $10^{6}$ mg | = | $10^{9}$ μg | = | $10^{12}$ ng |
| 1 g | = | $10^{-3}$ kg | = | $10^{3}$ mg | = | $10^{6}$ μg | = | $10^{9}$ ng |
| 1 mg | = | $10^{-6}$ kg | = | $10^{-3}$ g | = | $10^{3}$ μg | = | $10^{6}$ ng |
| 1 μg | = | $10^{-9}$ kg | = | $10^{-6}$ g | = | $10^{-3}$ mg | = | $10^{3}$ ng |
| 1 ng | = | $10^{12}$ kg | = | $10^{-9}$ g | = | $10^{-6}$ mg | = | $10^{-3}$ μg |

**Metric Prefix Example (volume)**

| | | | | |
|---|---|---|---|---|
| 1 L | = | $10^{3}$ mL | = | $10^{6}$ μL |
| 1 mL | = | $10^{-3}$ L | = | $10^{-6}$ μL |
| 1 $^{9}$μL | = | $10^{-6}$ L | = | $10^{-3}$ mL |

**Area Equivalents**

| | | | | | | | | | | | | |
|---|---|---|---|---|---|---|---|---|---|---|---|---|
| 1 acre | = | 43,560 $ft^2$ | = | 4840 $yd^2$ | = | 0.4047 hectares (ha) | = | 160 $rods^2$ | = | 4047 $m^2$ | = | 0.0016 $mi^2$ |
| 1 ha | = | 10,000 $m^2$ | = | 100 acres | = | 2.471 acres | = | 107,639 $ft^2$ | | | | |
| 1 $yd^2$ | = | 9 $ft^2$ | = | 0.836 $m^2$ | | | | 1 $yd^3$ | = | 27 $ft^3$ | = | 0.765 $m^3$ |
| 1 $ft^2$ | = | 144 $in.^2$ | = | 929.03 $cm^2$ | = | 0.09290 $m^2$ | | 1 $m^2$ | = | 10,000 $cm^2$ | | |
| 1 $ft^3$ | = | 1728 $in.^3$ | = | 0.037 $yd^3$ | = | 0.02832 $m^3$ | = | 28,320 $cm^3$ | | | | |
| 1 acre-inch | = | 102.8 $m^3$ | = | 27,154 gal | = | 3630 $ft^3$ | | | | | | |

**Liquid Equivalents**

| | | | | | | | | | | | | | | | | |
|---|---|---|---|---|---|---|---|---|---|---|---|---|---|---|---|---|
| 1 gal | = | 4 qt | = | 8 pt | = | 16 cups | = | 128 fl oz | = | 8.337 lb | | | | 1 barrel | = | 42 gal |
| | = | 231 in$^3$ | = | 256 tbsp | = | 0.134 ft$^3$ | = | 3.785 L | = | 3,785 mL | | | | | | |
| 1 qt | = | 0.9463 L | = | 2 pt | = | 4 cups | = | 32 fl oz | = | 64 tbsp | = | 57.75 in$^3$ | = | 946.4 mL | | |
| 1 L | = | 2.113 pt | = | 1,000 mL | = | 1.057 qt | = | 33.8 fl oz | = | 0.26 gal | = | 0.0001 m$^2$ | = | 1,000 cm$^3$ | | |
| 1 pt | = | 16 fl oz | = | 2 cups | = | 473.2 ml | = | 32 tbsp | = | 0.125 gal | = | 0.5 qt | | | | |
| 1 cup | = | 8 fl oz | = | 0.5 pt | = | 16 tbsp | = | 236.6 mL | | 1 tbsp | = | 14.8 mL | = | 3 tsp | = | 0.5 fl oz |
| 1 floz | = | 29.57 mL | = | 2 tbsp | = | 6 tsp | = | 0.0313 qt | | 1 tsp | = | 4.93 mL | = | 0.1667 fl oz | = | 80 drops |
| 1 ft$^3$ of water | = | 7.5 gal | = | 62.4 lb | = | 28.3 L | | 1 mL | = | 1 cm$^3$ | = | 0.034 fl oz | = | 0.002 pt | | |

**Pressure Equivalents**

| | | | | | | | | | | | | | | |
|---|---|---|---|---|---|---|---|---|---|---|---|---|---|---|
| 1 mmHg | = | 133.32 Pa | = | 0.133 kPa | = | 133,333 mPa | | | | | | | | |
| 1Pa | = | $10^{-3}$ kPa | = | $10^{-6}$ mPa | | | | | | | | | | |
| 1 psi | = | 6.9 kPa | = | 2.31 ft-head | | | | | | | | | | |
| 1mPa | = | $10^3$ kPa | = | $10^6$ Pa | = | 10 bar | = | 10.2 kg cm$^{-2}$ | = | 100 N cm$^{-2}$ | | | | |
| 1 atm | = | 760 mmHg | = | 29.92 in. Hg | = | $1.013 \times 10^5$ Pa | = | 1.013 bar | = | 14.69 psi | = | 33.89 ft water | | |
| 1 kPa | = | 0.001 mPa | = | 10 cm $H_2O$ | = | 10 mbar | = | 0.01 bar | = | 1J kg$^{-1}$ | = | 0.0099 atm | = | 0.145 psi |

**Temperature Equivalents**

| | | | | |
|---|---|---|---|---|
| °C | = | (°F – 32) | × | 5/9 |
| °F | = | (°C × 9/5) | + | 32 |

**Length Equivalents**

| | | | | | | | | | | |
|---|---|---|---|---|---|---|---|---|---|---|
| km | = | 0.621 statute mile | = | 1,000 m | = | 100,000 cm | = | 3,281 ft | = | 39,370 in. |
| m | = | 3.28 ft | = | 39.4 in. | = | 100 cm | = | 1.094 yd | = | 1,000 mm |
| cm | = | 0.3937 in. | = | 0.01 m | = | 0.03281 ft | | | | |
| In. | = | 2.54 cm | = | 25.4 mm | = | 0.0254 m | = | 0.08333 ft | | |
| ft | = | 0.3048 m | = | 30.48 cm | = | 12 in. | | | | |
| yd | = | 0.9144 m | = | 3 ft | = | 91.44 cm | | | | |
| statute mile | = | 1,760 yd | = | 5,280 ft | = | 1.61 km | = | 1,609 m | | |

**Mixture Ratios**

| | | | | |
|---|---|---|---|---|
| 1 mg g$^{-1}$ | = | 1000 ppm | = | 1 fl oz gal$^{-1}$ |
| 1 fl oz 100 gal$^{-1}$ | = | 75 ppm | = | 1 qt 100 gal$^{-1}$ |
| 1 pt 100 gal$^{-1}$ | = | 1 tsp gal$^{-1}$ | | |

**Flow**

| | | | | | | |
|---|---|---|---|---|---|---|
| 7,490 ppm | = | 1 gpm | = | 0.134 ft$^3$ min$^{-1}$ | = | 0.06308 L sec$^{-1}$ |
| 2 tbsp gal$^{-1}$ | = | 1 ft$^3$ min$^{-1}$ | = | 448.83 gal hr$^{-1}$ | = | 7.481 gal min$^{-1}$ |
| 1 ft$^3$sec$^{-1}$ | = | 448.83 gal min$^{-1}$ | | | | |

**Weight Equivalents**

| | | | | | | | | | | | | | | |
|---|---|---|---|---|---|---|---|---|---|---|---|---|---|---|
| 1 ton (US) | = | 2,000 lb | = | 0.907 metric tons | = | 907.2 kg | | 1 metric ton | = | $10^6$ g | = | 1,000 kg | = | 2,205 lb |
| 1 lb | = | 16 oz | = | 453.6 g | = | 0.4536 kg | | | | 1 oz (wt) | = | 28.35 g | = | 0.0625 lb |
| 1 g | = | 1,000 mg | = | 0.0353 oz | = | 0.001 kg | = | 0.002205 lb | | 1 mg | = | 0.001 g | | |
| 1 kg | = | 1,000 g | = | 35.3 oz | = | 2.205 lb | | | | 1 µg | = | $10^{-6}$ g | = | 0.001 mg |
| ng | = | $10^{-9}$ g | = | 0.001 micrograms (µg) | | | | | | picogram | = | $10^{-12}$ g | | |
| 1% (v/v) | = | 1.28 fl oz gal$^{-1}$ | = | 1 gal 100 gal$^{-1}$ | = | 10,000 ppm | = | 10g L$^{-1}$ | = | 1 g 100 mL$^{-1}$ | = | 1.33 oz (wt) gal$^{-1}$ | = | 8.34 lb 100 gal$^{-1}$ |
| 1 ppm | = | 0.0001% | = | 1 mg kg$^{-1}$ | = | 1 mg L$^{-1}$ | = | 1 µg g$^{-1}$ | = | 1 µL L$^{-1}$ | = | 1µg mL$^{-1}$ | | |
| | = | 0.379 g 100 gal$^{-1}$ | = | $8.34 \times 10^{-6}$ lb gal$^{-1}$ | = | 0.013 fl oz 100 gal$^{-1}$ | | 10 ppm | = | 0.001% | = | 10 mg L$^{-1}$ | | |
| 100 ppm | = | 0.01% | = | 100 mg L$^{-1}$ | | | | 1,000 ppm | = | 1 mg g$^{-1}$ | = | 0.1% | = | 1,000 mg L$^{-1}$ |
| 1 ppb | = | 1 µg kg$^{-1}$ | = | 1 µg L$^{-1}$ | = | 1 ng mL$^{-1}$ | = | 1 ng 1,000,000,000$^{-1}$ | | 1 ppt | = | 1 picogram g$^{-1}$ | | |

**Water and Soil Calculations**

| | | | | | | | | | | |
|---|---|---|---|---|---|---|---|---|---|---|
| 1 mmhos $cm^{-1}$ | = | 1,000 μmhos $cm^{-1}$ | = | 1 dS $m^{-1}$ | = | 0.1 S $m^{-1}$ | = | 1 mS $cm^{-1}$ | = | 10 meq $L^{-1}$ |
| 1 meq $L^{-1}$ | = | 1 mmol $L^{-1}$ | = | 1 mol $m^{-3}$ | | | | | | |
| 1 meq 100 $g^{-1}$ | = | 1 mmol 100 $g^{-1}$ | = | cmol $kg^{-1}$ | | | | | | |
| Electrical conductivity (mmhos $cm^{-1}$ or dS $m^{-1}$) | | | | × 640 | = | total dissolved salts (mg $L^{-1}$ or ppm) | | | | |
| Total dissolved salts (mg $L^{-1}$ or ppm) | | | | × 0.0016 | = | electrical conductivity (mmhos $cm^{-1}$ or ds $m^{-1}$) | | | | |

**Energy**

| | | |
|---|---|---|
| 1 calorie (cal) | = | 4.184 J |
| Joule (J) | = | 1 kg $m^2$ $s^{-2}$ |
| 1 kcal | = | 4.184 kJ |

**Decimal and Millimeter Length Equivalents**

| Fraction (inch) | Decimals (inch) | Millimeters |
|---|---|---|
| **1** | **1.00** | **25.4** |
| 15/16 | 0.9375 | 23.812 |
| 7/8 | 0.875 | 22.225 |
| 13/16 | 0.8125 | 20.638 |
| **3/4** | **0.75** | **19.05** |
| 11/16 | 0.6875 | 17.462 |
| 5/8 | 0.625 | 15.875 |
| 9/16 | 0.5625 | 14.288 |
| **1/2** | **0.5** | **12.70** |
| 7/16 | 0.4375 | 11.112 |
| 3/8 | 0.3750 | 9.525 |
| 11/32 | 0.34375 | 8.731 |
| 5/16 | 0.3125 | 7.938 |
| 9/32 | 0.28125 | 7.144 |
| **1/4** | **0.25** | **6.350** |
| 15/64 | 0.234375 | 5.953 |
| 7/32 | 0.21875 | 5.556 |
| 13/64 | 0.203125 | 5.159 |
| 1/5 | 0.200 | 5.08 |
| 3/16 | 0.1875 | 4.762 |
| 23/128 | 0.1797 | 4.564 |
| 11/64 | 0.171875 | 4.366 |
| 1/6 | 0.167 | 4.242 |
| 21/128 | 0.1641 | 4.168 |
| 5/32 | 0.15625 | 3.969 |
| 1/7 | 0.143 | 3.633 |
| 19/128 | 0.1484 | 3.769 |
| 9/64 | 0.140625 | 3.572 |
| **1/8** | **0.1250** | **3.175** |
| 7/64 | 0.109375 | 2.778 |
| 1/10 | 0.100 | 2.540 |
| 3/32 | 0.09375 | 2.381 |
| 5/64 | 0.078125 | 1.984 |
| 1/16 | 0.0625 | 1.588 |
| 3/64 | 0.046875 | 1.191 |
| 1/32 | 0.03125 | 0.794 |
| 1/64 | 0.015625 | 0.397 |

## Metric Conversion

| To Convert | Multiply by | To Obtain |
|---|---|---|
| acres (ac) | 0.4047 | hectares (ha) |
| acres | 43,560 | sq. feet ($ft^2$) |
| acres | 0.00405 | sq. kilometers ($km^2$) |
| acres | 4047 | sq. meter ($m^2$) |
| acres | 4840 | sq. yards ($yd^2$) |
| acre-feet | 325,851 | sq. feet ($ft^2$) |
| acre-feet | 43560 | cu. feet ($ft^3$) |
| acre-feet | 1233.5 | cu. meter ($m^3$) |
| acre-inches | 102.8 | $m^3$ |
| bars | 14.5 | pounds (lb)/in.$^2$ |
| bars | 1019.7 | gram (g)/$cm^3$ |
| bars | 29.53 | inches Hg @ 0°C |
| bars | 75 | cm Hg @ 0°C |
| bars | 0.001 | J/kg |
| bars | 100 | kPa |
| bushels—dry | 0.03524 | $m^2$ |
| bushels | 1.245 | $ft^3$ |
| calories (cal) | 4.184 | joules (J) |
| centimeters (cm) | 0.03281 | feet (ft) |
| cm | 0.3937 | inches (in.) |
| cm | 0.1094 | yards (yd) |
| cm | 0.01 | meters (m) |
| cm | 10 | millimeters (mm) |
| cm/sec = cm $sec^{-1}$ = cm per sec | 1.9685 | ft/min |
| cm/sec | 0.0223694 | miles per hour (mph) |
| $cm^2$ (square centimeters) | 0.001076 | $ft^2$ |
| $cm^2$ | 0.1550 | in.$^2$ |
| $cm^2$ | 0.01 | sq. decimeter |
| $cm^3$ (cubic centimeters) | 0.0610237 | in.$^3$ |
| $cm^3$ | 0.0338 | fl oz (fluid ounces) |
| $cm^3$ | 0.001057 | $qt^3$ |
| $cm^3$ | 0.000264172 | gal (gallon) |
| $cm^3$ | 0.001 | cu. decimeter |
| cup | 8 | fl oz |
| cup | 236.6 | $cm^3$ |
| feet (ft) | 30.48 | cm |
| ft | 0.3048 | m |
| ft | 305 | mm |
| $ft^2$ (square feet) | 929 | $cm^2$ |
| $ft^2$ | 0.0929 | $m^2$ |
| $ft^2$ | $9.294 \times 10^{-6}$ | hectares (ha) |
| $ft^2$ | 144 | $in^2$ |
| $ft^3$ (cubic feet) | 0.0283 | $m^3$ |
| $ft^3$ | 7.4805 | gallons |
| $ft^3$ | 1728 | cubic inches (in.$^3$) |
| $ft^3$ | 0.037 | cubic yards ($yd^3$) |
| $ft^3$ | 28.32 | liters (L) |
| $ft^3$/1,000 $ft^2$ | 0.030463 | $m^3$/100 $m^2$ |
| feet per minute | 0.01136 | mph |
| Feet-head of water | 0.433 | psi (pounds/in.$^2$) |
| Foot-candle | 10.764 | lux |
| gallons (gal) | 3.785 | liters |

| | | |
|---|---|---|
| gal | 3785 | mL (milliliters) |
| gal | 128 | ounces (liquid) |
| gal | 0.13368 | $ft^3$ |
| gal | 231 | $in.^3$ |
| gal | 3,785 | $cm^3$ |
| gallons per acre (gpa) | 9.354 | L/ha |
| gpa | 0.09354 | L/100 $m^2$ |
| gpa | 2.938 | ounces (oz)/1,000 $ft^2$ (liquid) |
| gal/1,000 $ft^2$ | 4.0746 | L/100 $m^2$ |
| gal/minute | $2.228 \times 10^{-3}$ | $ft^3$/sec |
| gal/min | 0.06309 | L/sec |
| gal/min | 0.227125 | $m^3$/hr (hour) |
| grams (g) | 0.002205 | lb |
| gram | 0.035274 | oz |
| g/$cm^3$ | 0.036127 | lb/$in.^3$ |
| g/$cm^3$ | 62.428 | lb/$ft^3$ |
| g/$ft^2$ | 96 | lb/acre |
| g/ha | 0.000893 | lb/acre |
| g/ha | 0.014275 | oz/acre |
| g/kg | 0.10 | percent (%) |
| g/liter | 1000 | parts per million (ppm) |
| g/liter | 10 | % |
| g/liter | 0.00834595 | lb/gal |
| g/liter | 0.13351 | oz/gal |
| g/$m^2$ | 0.00020481 | lb/$ft^2$ |
| g/$m^2$ | 0.20481 | lb/1,000 $ft^2$ |
| hectares (ha) | 2.471 | acres |
| ha | 107,639 | $ft^2$ |
| ha | 107.64 | 1,000 $ft^2$ |
| horsepower (electrical or mechanical) | 746 | watts |
| hp | 550 | ft-lb/sec |
| hp | 33,000 | ft-lb/min |
| hp | 1.014 | metric horsepower |
| hp | 33,000 | ft-lb/min |
| inches (in.) | 2.540 | cm |
| in. | 0.0254 | m |
| in. | 25.40 | ml |
| inches of mercury | 3.4 | kilopascals (kPa) |
| in./acre | 6.28 | cm/ha |
| in./ft | 0.083 | mm/mm |
| $in.^2$ | 6.4516 | $cm^2$ |
| $in.^3$ | 16.3871 | $cm^3$ |
| $in.^3$ | 0.55411 | fl oz |
| $in.^3$ | 0.01732 | quart (qt) |
| $in.^3$/hr | 0.00434 | gal/hr |
| joules per kilogram (J/kg) | 1 | kPa |
| kilograms (kg) | 2.2046 | lb |
| kg/hectare | 0.892 | lb/acre |
| kg/ha | 0.02048 | lb/1,000 $ft^2$ |
| kg/100 $m^2$ | 2.048 | lb/1,000 $ft^2$ |
| kg/L | 8.3454 | lb/gal |
| kilometers (km) | 100,000 | cm |
| kilometers | 3281 | ft |
| kilometers | 1,000 | m |
| kilometers | 0.6214 | miles |
| kilometers | 1094 | yd |
| km/hr | 0.62137 | mph |
| km/hr | 54.6807 | ft/min |

| | | |
|---|---|---|
| kilopascals (kPa) | 0.145 | lb/in.$^2$ (psi) |
| kPa | 1 | 0.01 bar |
| kPa | 1 | J/kg |
| kPa | 0.01 | bar |
| liters (L) | 0.2642 | gal |
| L | 33.814 | fl oz |
| L | 2.113 | pint (pt) |
| L | 1.057 | qt |
| L | 0.035315 | ft$^3$ |
| L/m$^2$ | 3.2808 | ft$^3$/1,000 ft$^2$ |
| L/100 m$^2$ | 0.2454 | gal/1,000 ft$^2$ |
| L/100 m$^2$ | 1.9634 | pt/1,000 ft$^2$ |
| L/ha | 0.107 | gal/ac |
| L/ha | 0.0025 | gal/1,000 ft$^2$ |
| L/ha | 0.314 | oz/1,000 ft$^2$ |
| L/ha | 0.855 | pt/ac |
| L/min | 15.85 | gal/hr |
| meters (m) | 3.281 | ft |
| m | 39.37 | in. |
| m | 1.094 | yd |
| m | 100 | cm |
| m | 0.001 | km |
| m | 1000 | mm |
| m/sec | 2.2369 | mph |
| m$^2$ (square meters) | 10.764 | ft$^2$ |
| m$^2$ | 1,550 | in.$^2$ |
| m$^2$ | 1.196 | yd$^2$ |
| m$^3$ (cubic meters) | 35.3147 | ft$^3$ |
| m$^3$ | 1.30795 | yd$^3$ |
| m$^3$ | 1,000 | L |
| m$^3$/ha | 14.29 | ft$^3$/ac |
| m$^3$/ha | 0.0122 | yd$^3$/1,000 ft$^2$ |
| m$^3$/ha | 0.328 | ft$^3$/1,000 ft$^2$ |
| mil | 0.001 | in. |
| mil | 0.0254 | mm |
| miles (nautical) | 1.1508 | miles (statute) |
| miles (nautical) | 6,076 | ft |
| miles (statute) | 160,900 | cm |
| miles | 5280 | ft |
| miles | 1.609 | km |
| miles | 1760 | yards |
| miles per hour (mph) | 1.467 | ft/sec |
| mph | 88 | ft/min |
| mph | 1.61 | km/hr |
| mph | 0.447 | m/sec |
| mg/kg | 1 | ppm |
| milliequivalents per liter (meq/L) | 1 | millimoles per liter (mmol/L) |
| milliequivalents per 100 g (meq/100 g) | eq. wt. × 10 | ppm |
| millimhos per centimeter (mmhos/cm) | 1 | decisiemens per meter (dS/m) |
| mmhos/cm | 1,000 | micromhos per centimeter (μmhos/cm) |
| milliliters (mL) | 0.0338 | oz (fluid) |
| mL | 0.0002642 | gal |
| mL/m$^2$ | 3.14 | oz/1,000 ft$^2$ |
| mL/L | 0.12793 | oz/gal |
| mL/10,000 L | 0.0128 | fl oz/1,000 gal |
| millimeters (mm) | 0.03937 | in. |

| | | |
|---|---|---|
| 1 mmHg @ 0°C | 0.13332 | kPa |
| 1 mmHg | 133333.3 | mPa |
| ounces (fluid) (oz) | 0.02957 | L |
| oz (fluid) | 29.573 | mL |
| oz (fluid) | 0.03125 | qt |
| oz (fluid)/gal | 7.81 | mL/L |
| oz (fluid)/acre | 0.0731 | L/ha |
| oz (fluid)/acre | 73.1 | mL/ha |
| oz (fluid)/1,000 $ft^2$ | 3.18 | L/ha |
| oz (weight) | 28.35 | g |
| oz (weight) | 0.0625 | lb |
| oz (weight)/acre | 0.07 | kg/ha |
| oz (weight)/acre | 70 | g/ha |
| oz (weight)/1,000 $ft^2$ | 3.05 | kg/ha |
| oz (weight)/$ft^2$ | 305.15 | g/$m^2$ |
| oz (weight)/gal | 7.5 | g/L |
| oz (weight)/1,000 $ft^2$ | 0.305 | g/$m^2$ |
| percent (%) | 10 | g/kg |
| pint (liquid) (pt) | 0.473 | liter |
| pt/ac | 1.1692 | L/ha |
| pt/ac | 0.3673 | oz/1,000 $ft^2$ |
| pt/1,000 $ft^2$ | 0.50932 | L/100 $m^2$ |
| parts per million (ppm) | 2.719 | lb active ingredient (ai)/acre-foot of water |
| ppm | 2.0 | lb/acre slice 7-in. deep |
| ppm | 2.25 | kg/ha slice 7-in. deep |
| ppm | 0.001 | g/L |
| ppm | 8.34 | lb/million gal |
| ppm | 1 | mg/kg |
| ppm | 0.013 | oz/100 gal of water |
| ppm | 0.3295 | gal/acre-foot of water |
| ppm | 8.2897 | lb/million gal of water |
| pounds (lb) | 0.4536 | kilograms (kg) |
| lb | 453.6 | g |
| lb/acre | 1,120 | g/ha |
| lb/acre | 1.12 | kg/ha |
| lb/acre | 1.0413 | g/100 $ft^2$ |
| lb/acre | 0.02296 | lb/1,000 $ft^2$ |
| lb/acre | 0.112 | g/$m^2$ |
| lb/acre-foot | 0.3682 | g/$m^3$ |
| lb/acre-foot | 0.0003682 | kg/$m^3$ |
| lb/$ft^2$ | 4883 | g/$m^2$ |
| lb/$ft^3$ | 16.02 | kg/$m^3$ |
| lb/1,000 $ft^2$ | 4.88 | g/$m^2$ |
| lb/1,000 $ft^2$ | 48.83 | kg/ha |
| lb/1,000 $ft^2$ | 43.5597 | lb/ac |
| lb/1,000 $ft^2$ | 488 | g/100 $m^2$ |
| lb/1,000 $ft^2$ | 0.4883 | kg/100 $m^2$ |
| lb/1,000 $ft^2$ | 0.91 | lb/100 $yd^2$ |
| lb/1,000 $ft^2$ | 1.1 | lb/1,000 $ft^2$ |
| lb/$yd^3$ | 0.0005937 | g/$cm^3$ |
| lb/$yd^3$ | 594 | g/$m^3$ |
| lb/$yd^3$ | 0.5932 | kg/$m^3$ |

| | | |
|---|---|---|
| lb/gallon | 0.12 | kg/liter |
| lb/1,000 gal | 0.12 | g/1,000 L |
| pounds per square inch (psi) | 6.89 | kPa |
| psi | 0.06895 | bar |
| psi | 0.068046 | atmosphere (atm) |
| psi | 2.31 | feet-head of water |
| quarts (qt) | 0.9463 | L |
| qt | 946 | mL |
| qt/acre | 2.3385 | L/ha |
| qt/acre | 0.7346 | oz/1,000 $ft^2$ |
| qt/100 gal | 2.5 | mL/L |
| temperature, °C + 17.98 | 1.8 | temperature, °F |
| temperature, °F – 32 | 0.5555 | temperature, °C |
| ton (2,000 lb) | 907 | kg |
| ton (2,000 lb)/acre | 2,240 | kg/ha |
| ton (2,000 lb) | 0.907 | ton (metric) |
| ton (2,000 lb)/acre | 2.241 | ton (metric)/ha |
| ton (metric) | 2,205 | lb |
| ton (metric) | 1,000 | kg |
| ton (metric) | 1.102 | ton (2,000 lb) |
| yards (yd) | 91.44 | cm |
| yd | 0.9144 | m |
| yd | 914.4 | mm |
| square yards ($yd^2$) | 0.836 | $m^2$ |
| $yd^2$ | 9 | $ft^2$ |
| $yd^2$ | 1,296 | $in.^2$ |
| cubic yards ($yd^3$) | 27 | $ft^3$ |
| $yd^3$ | 46,656 | $in.^3$ |
| $yd^3$ | 0.7645 | $m^3$ |
| $yd^3$ | 765 | L |
| $yd^3$/1,000 $ft^2$ | 0.825 | $m^3$/100 $m^2$ |
| $P_2O_5$ | 0.437 | P |
| $K_2O$ | 0.830 | K |
| CaO | 0.715 | Ca |
| MgO | 0.602 | Mg |
| meq $Ca^{+2}$/100 g soil | 400 | lb $Ca^{+2}$ per acre furrow slice |
| meq $K^+$/100 g soil | 780 | lb $K^+$ per acre furrow slice |
| meq $Na^+$/100 g soil | 460 | lb $Na^+$ per acre furrow slice |
| meq $Mg^{+2}$/100 g soil | 109 | lb $Mg^{+2}$ per acre furrow slice |
| meq $Fe^{+3}$/100 g soil | 372 | lb $Fe^{+3}$ per acre furrow slice |
| meq $Zn^{+2}$/100 g soil | 654 | lb $Zn^{+2}$ per acre furrow slice |
| meq $H^+$/100 g soil | 20 | lb $H^+$ per acre furrow slice |
| meq $Al^{+3}$/100 g soil | 180 | lb $Al^{+3}$ per acre furrow slice |
| meq $Ca^{+2}$/100 g soil | 9.2 | lb $Ca^{+2}$ per 1,000 $ft^2$ furrow slice |
| meq $K^+$/100 g soil | 18 | lb $K^+$ per 1,000 $ft^2$ furrow slice |
| meq $Na^+$/100 g soil | 10.6 | lb $Na^+$ per 1,000 $ft^2$ furrow slice |
| meq $Mg^{+2}$/100 g soil | 2.5 | lb $Mg^{+2}$ per 1,000 $ft^2$ furrow slice |
| meq $Fe^{+3}$/100 g soil | 8.5 | lb $Fe^{+3}$ per 1,000 $ft^2$ furrow slice |
| meq $Zn^{+2}$/100 g soil | 15 | lb $Zn^{+2}$ per 1,000 $ft^2$ furrow slice |
| meq $H^+$/100 g soil | 0.46 | lb $H^+$ per 1,000 $ft^2$ furrow slice |
| meq $Al^{+3}$/100 g soil | 4.1 | lb $Al^{+3}$ per 1,000 $ft^2$ furrow slice |

# References


Ayers, R. S., and D. W. Westcot. 1985. Water quality for agriculture. Irrigation and drainage paper 29. Rome, Italy: Food and Agriculture Organization of the United Nations. http://www.fao.org.

Baird, J. H., N. T. Basta, R. I. Huhnke, M. E. Payton, G. V. Johnson, D. E. Storm, and M. D. Smolen. 1997. Influence of buffer length and mowing height on surface runoff of pesticides and nutrients from bermudagrass turf. *Agronomy Abstracts* 89:130.

Balogh, J. C., and W. J. Walker, eds. 1992. *Golf course management and construction: Environmental issues*. Chelsea, MI: Lewis Publishers.

Beard, J. B. 1985. An assessment of water use by turfgrasses. In *Turfgrass water conversation. Proc. Symposium ASPA*, San Antonio, TX. 15–16 Feb. 1983, ed. V. B. Youngner and S. T. Cockerham, 47–60. Oakland, CA: Cooperative Extension University of California.

Brosnan, J., G. K. Breeden, and A. J. Patton. 2014. Weed management options on golf course putting greens. *Golf Course Management* 82(8):92–98.

Brown, K. W., and J. C. Thomas. 1986. Bunker sand selection. *Golf Course Management* 54:64–70.

Buchen T. 2017. Assessing infrastructure. *Golf Course Management* 85(8):48–56.

Carrow, R. N., D. V. Waddington, and P. E. Rieke. 2001. *Turfgrass soil fertility and chemical properties*. Chelsea, MI: Ann Arbor Press.

Duncan, R. R., and R. N. Carrow. 2005a. Managing seashore paspalum greens. *Golf Course Management* 73(2):114–118.

Green, R. L. 2005. Golf course water use and regulation in California. *Golf Course Management* 73(10):67–72.

Gross, C. M., J. S. Angle, R. L. Hill, and M. S. Welterlen. 1991. Runoff and sediment losses from tall fescue under simulated rainfall. *Journal of Environmental Quality* 20:604–607.

Habeck, J., and N. Christians. 2000. Time alters greens' key characteristics. *Golf Course Management* 68(5):54–60.

Harivandi, M. A. 2007. Using recycled water on golf courses. *Golf Course Management*. 75(6):98–108.

Harivandi, M. A., J. D. Butler, and L. Wu. 1992. Salinity and turf culture. In *Turfgrass*, ed. D. V. Waddington, R. N. Carrow, and R. C. Shearman, 208–230. Agronomy Monograph Series 32. Madison, WI: ASA, CSSA, and SSSA.

Huck, M. 2000. Does your irrigation system make the grade? *USGA Green Section Record* 39(5):1–5.

Jones, J. B., Jr., B. Wolf, and H. A. Mills. 1991. *Plant analysis handbook*. Athens, GA: Micro-Macro Publishing, Inc.

Kavanagh, T., and R. M. Jelley. 1981. Soil atmospheric studies in relation to compaction. In *Proceedings of the 4th International Turfgrass Research Conference*, ed. R. W. Sheard, 181–188. Ontario, Canada: University of Guelph.

Kopec, D. M. 1996. Managing turf with effluent water. *Golf Course Irrigation* 4(3):14–16.

Li, D. 2004. How to categorize organic materials in turfgrass root zones. *Turfgrass Trends* 60(4):58–61.

McCarty, L. B., ed. 2011. *Best golf course management practices* (3rd ed.). Upper Saddle River, NJ: Prentice Hall, Inc.

McCarty, L. B., L. R. Hubbard, Jr., and V. Quisenberry. 2016. *Applied soil physical properties, drainage, and irrigation strategies*. Springer Inter. Publ. Switzerland.

Ross, M. A. and C. A. Lembi. 2009. *Applied weed science* (3rd ed.). Upper Saddle River, NJ: Prentice Hall.

Zuberer, D. 2012. Soil microbes—Some practical perspectives for turfgrass systems. *USGA Green Record* 50(15):5 pp.


## ADDITIONAL READING


Anderson, J., C. Taliaferro, and D. Martin. 2002. Freeze tolerance of bermudagrasses: Vegetatively propagated cultivars intended for fairway and putting green use, and seed-propagated cultivars. Crop Science 42:975–977.

Anonymous. 1972. *Water quality criteria. Agricultural use of water.* Washington, DC: National Academy of Science—National Academy of Engineering, sec. V, 323–353.

Anonymous. 1982a. Lime and fertilizer recommendations based on soil-test results. Clemson University Extension Service, circular 476.

Anonymous. 1982b. *USGA refining the green section specification for putting green construction*. Far Hills, NJ: Golf House.

Anonymous. 1992. *Guidelines for water reuse*. EPA/625/R-92/004. Cincinnati, OH: U.S. EPA, Center for Environmental Research Information.

Anonymous. 1993. USGA recommendations for a method of putting green construction. *USGA Green Section Record* 31(2):1–3.

Anonymous. 1995. *Best management practices for golf course maintenance departments*. Tallahassee, FL: Florida Department of Environmental Protection.

Anonymous. 1996. Environmental principles for golf courses in the United States. Salt Lake City, UT: The Center for Resource Management, 1–15.

Anonymous. 2002. *Western fertilizer handbook* (9th ed.). Long Grove, IL: Waveland Press, Inc.

Anonymous. 2009. Golf course environmental profile, vol. II. *Water Use and Conservation Practices on US Golf Courses*. GCSAA. Lawrence, KS, 49 pp. www.eifg.org.

Asano, T., R. G. Smith, and G. Tschobanoglous. 1984. Municipal wastewater: Treatment and reclaimed water characteristics. In *Irrigation with reclaimed municipal wastewater—A guidance manual*, Report no. 84-1, ed. G. S. Pettygrove and T. Asano, 2:1–2:26. Sacramento: California State Water Resources Control Board.

Augustin, B. J., and G. H. Snyder. 1984. Moisture sensor-controlled irrigation for maintaining bermudagrass turf. *Agronomy Journal* 76:848–850.

Backman, P. A., E. D. Miltner, G. K. Stahnke, and T. W. Cook. 2002. Worming your way out of a turf situation. *USGA Green Section Record* 40(4):7–8.

Baldwin, L. B., and D. A. Comer. 1986. Utilizing treated sewage for irrigation of urban landscapes. Florida Cooperative Extension Service, circular 714.

Beard, J. B. 1973. *Turfgrass: Science and culture*. Upper Saddle River, NJ: Prentice Hall, Inc.

Beard, J. B. 2002. *Turf management for golf courses* (2nd ed.). Chelsea, MI: Ann Arbor Press.

Beard, J. B., and R. L. Green. 1994. The role of turfgrass in environmental protection and their benefits to humans. *Journal of Environmental Quality* 23:452–460.

Brandenburg, R. L. 2003. Cricket control. *Grounds Maintenance* 38(2):22–24, 33–35.

Bremner, J. M., and A. M. Blackmer. 1982. Composition of soil atmosphere. In *Methods of soil analysis*, part 2, ed. A. L. Page, 873–901. Agronomy Monographs 9. Madison, WI: ASA, CSSA, SSSA. American Society of Agronomy, Crop Science Society of America, Soil Science Society of America.

Brilman, L. A. 2003. PVP: What does it mean? *Golf Course Management* 71(3):116–119.

Bruneau, A. H., J. E. Watkins, and R. L. Brandenburg. 1992. Integrated pest management. In *Turfgrass*, ed. D. V. Waddington, R. N. Carrow, and R. C. Shearman, 653–688. Agronomy Monograph 32. Madison, WI: ASA, CSSA, SSSA.

Buchanan, B. B., W. Gruissem, and R. L. Jones. 2000. *Biochemistry and molecular biology of plants*. Rockville, MD: American Society of Plant Physiologists.

Butler, J. D., P. E. Rieke, and D. D. Minner. 1985. Influence of water quality on turfgrass. In *Turfgrass water conversation. Proceedings Symposium ASPA*, San Antonio, TX. 15–16 Feb. 1983. ed. V. B. Youngner and S. T. Cockerham, 71–84. Oakland, CA: Cooperative Extension, University of California.

Callahan, L. M., R. S. Freeland, R. D. Von Bernuth, D. P. Shepard, J. M. Parham, and J. M. Garrison. 1997. Geotextiles as substitutes for choker layer sand in USGA greens. I. Water infiltration rates and water retention. *International Turfgrass Society Research Journal* 8:65–74.

Carrow, R. N. 1985. Soil/water relationships in turfgrass. In *Turfgrass water conversation. Proceedings Symposium ASPA*, San Antonio, TX. 15–16 Feb. 1983. ed. V. B. Youngner and S. T. Cockerham, 85–102. Oakland, CA: Cooperative Extension, University of California.

Carrow, R. N. 1995a. Drought resistance aspects of turfgrasses in the Southeast: Evapotranspiration and crop coefficients. *Crop Science* 35:1685–1690.

Carrow, R. N. 1995b. Soil testing for fertilizer recommendations. *Golf Course Management* 63:61–68.

Carrow, R. N., L. Stowell, W. Gelernter, S. Davis, R. R. Duncan, and J. Skorulski. 2004a. Clarifying soil testing: II. Choosing SLAN extractants for macronutrients. *Golf Course Management* 72(1):189–193.

Carrow, R. N., L. Stowell, W. Gelernter, S. Davis, R. R. Duncan, and J. Skorulski. 2004b. Clarifying soil testing: III. SLAN sufficiency ranges and recommendations. *Golf Course Management* 72(1):194–197.

Carrow, R. N., M. Huck, and R. R. Duncan. 2000. Leaching for salinity management on turfgrass sites. *USGA Greens Record.* 38(6):15–24.

Carrow, R. N., R. R. Duncan, and D. Wienecke. 2005. BMPs approach to water conservation on golf courses. *Golf Course Management* 73(7):73–76.

Christians, N. 1993. The fundamentals of soil testing. *Golf Course Management* 61(6):88–99.

Christians, N. 1999. Why inject acid into irrigation water? *Golf Course Management* 67(6):52–56.

Chong, J.-H. 2013. A witch's brew of troubles with the bermudagrass mite. *USGA Green Section Record* 51(14):1–4.

Colbaugh, P. 2002. Algae—Crusty foes for golf greens. *Grounds Maintenance* 37(1):6–8.

Cooper, R. J. 1996. Soils teem with tiny organisms. *Golf Course Management* 64:63–67.

Couch, H. B. 1995. *Diseases of turfgrasses* (3rd ed.). Malabar, FL: Krieger Publishing Company, 421.

Crow, W. T., J. O. Becker, and J. H. Baird. 2017. New golf course nematicides. *Golf Course Management* 85(7):66–71.

Davis, W. B., J. L. Paul, and D. Bowman. 1990. The sand putting green—Construction and management. University of California Division of Agriculture and Natural Resources Publication 21448.

Dernoeden, P. H. 1995. Turfgrass diseases and their management. In *Managing turfgrass pests*, ed. T. L. Watschke, P. H. Dernoeden, and D. J. Shetlar, 87–170. Boca Raton, FL: Lewis Publishers.

Dernoeden, P. H. 2013. *Creeping bentgrass management* (2nd ed.). Boca Raton, FL: CRC Press, 362 pp.

Dowling, E., and B. Meentemeyer. 2017. Ten ways to mitigate summer stress on putting greens. *USGA Green Section Record* 55(11):1–6. TGIF record no: 284875.

Dubey, R. S., and M. Pessarakli. 1995. Physiological mechanisms of nitrogen absorption and assimilation in plants under stressful conditions. In *Handbook of plant and plant and crop physiology*, ed. M. Pessarakli, New York: Marcel Dekker, Inc., 605–626.

Duncan, R. R., and R. N. Carrow. 2005b. Preventing failure of seashore paspalum greens. *Golf Course Management* 73(3):99–102.

Engelstad, O. P., ed. 1985. *Fertilizer technology and use*. Madison, WI: Soil Science Society of America.

Ennis, J., and K. Bilawa. 2000. Golf course ponds are complete ecosystems. *Golf Course Management* 68(5):61–64.

Eskelson, D. 1992. Implementing IPM strategies. *Golf Course Management* 60:68–75.

Farnham, D. S., R. F. Hasek, and J. L. Paul. 1985. Water quality: Its effects on ornamental plants. University of California Cooperative Extension publication 2995, 15.

Fidanza, M. 2002. Figuring out fairy rings. *Grounds Maintenance* 37(3):49–54.

Frank, L. 2002. Executing irrigation inspections. *Sportsturf* 18(4):14–15.

Frank, M. J. 2000. Pointers for perfect *Poa*. *Golf Course Management* 68(10):106–112.

Fry, J., and B. Juang. 2004. *Applied turfgrass science and physiology*. Hoboken, NJ: Wiley & Sons.

Gee, G. W., and J. W. Bauder. 1986. Particle-size analysis. In *Method of soil analysis*, part 1, ed. A. Klute, 383–411. Madison, WI: ASA and SSSA.

Gelernter, W., and L. Stowell. 2001. Learning to love kikuyugrass. *Golf Course Management* 69:55–59.

Gelernter, W., and L. Stowell. 2017. National trends in land use and environmental practices on US golf courses, part 1. *Golf Course Management* 85(4):72–82.

Gross, P. J. 2003. Looking kindly at kikuyugrass. *USGA Green Section Record* 41(4):2–8.

Gross, P. J. 2008. A step-by-step guide for using recycled water. *USGA Green Section Record* 46(2):1–8.

Guertal, E. 2000. Nitrogen: Finding the form that fits. *Grounds Maintenance* 35(9):15–18.

Guertal, E., and C. Waltz, Jr. 2008. Adding inorganic amendments to a poorly performing green. *Golf Course Management* 76(5):133–137.

Handreck, K. A., and N. D. Black. 1986. *Growing media for ornamental plants and turf.* Kensington NSW, Australia: New South Wales University Press, 124–142.

Hanks, R. J., and G. L. Ashcroft. 1980. *Applied soil physics*. New York: Springer–Verlag, 159.

Huang, B., and Hongwen, G. 2000. Growth and metabolism of creeping bentgrass cultivars in response to increasing temperatures. *Crop Science* 40:1115–1120.

Hummel, N. W., Jr. 1998. Which root-zone recipe makes the best green? *Golf Course Management* 66(12):49–51.

Hummel, N. W., Jr. 2000. What goes best with sand: Peat, soil, or compost? *Golf Course Management* 68(4):57–60.

Hurdzan, M. J. 1996. *Golf course architecture*. Chelsea, MI: Sleeping Bear Press.

Hurley, R. H. 1990. Best turfgrasses for southern winter overseeding. *Grounds Maintenance* 26(1).

Johnson, P. G., and T. P. Riordan. 1999. Buffalograss: Home on the range. *Golf Course Management* 67(6):66–70.

Jones, J. B., Jr. 2001. *Laboratory guide for conducting soil tests and plant analysis*. Boca Raton, FL: CRC Press.

Karnock, K. 2000. Promises, promises: Can biostimulants deliver? *Golf Course Management* 68(8):67–71.

Karnock, K., and K. Tucker. 2002. Water repellent soils part 1: Where are we now? *Golf Course Management* 70(6):59–62.

Kneebone, W. R., D. M. Kopec, and C. F. Mancino. 1992. Water requirements and irrigation. In *Turfgrass*, ed. D. V. Waddington, R. N. Carrow, and R. C. Shearman, 441–472. Agronomy Monograph Series 32. Madison, WI: ASA, CSSA, and SSSA.

Knight, J. 2012. De-goosing golf courses. *Golf Course Management* 80(2):132–137.

Kopec, D. M., J. L. Walworth, J. J. Gilbert, G. M. Sower, and M. Pessarakli. 2007. 'SeaIsle 2000' paspalum putting surface response to mowing height and nitrogen fertilizer. *Agronomy Journal* 99:133–140.

Koppenhofer, A. M., S. R. Alm, R. S. Cowles, B. A. McGraw, S. Swier, and P. J. Vittum. 2012. Controlling annual bluegrass weevil: Optimal insecticide timing and rates. *Golf Course Management* 80(3):98104.

Krenisky, E. C., M. J. Carroll, R. L. Hill, and J. M. Krouse. 1998. Runoff and sediment losses from natural and man-made erosion control materials. *Journal of Soil Water Conservation* 52:96–102.

Landschoot, P., and J. Cook. 2005. Sorting out the phosphonate products. *Golf Course Management* 73(10):73–77.

Latin, R. 2013. Fungicides for turf disease control. *International Turf Society Research Journal* 12:61–68.

Lea, P. J., and R. C. Leegood. 1999. *Plant biochemistry and molecular biology*. Chichester, UK: John Wiley & Sons, Inc.

Leinauer, B., T. Barrick, M. Serena, M. Schiavon, and B. Maier. 2012. Physical water conditioners for managing turfgrasses. *Golf Course Management* 70(11):90–97.

Leinauer, B., T. Barrick, M. Serena, M. Schiavon, B. Maier, and C. Robertson. 2012. Usefulness of nonchemical water conditioners for managing turfgrasses and their rootzones. Online. *Applied Turfgrass Science* doi:10.1094/ATS-2012-0924-01-RS.

Lewis, J. D., R. E. Gaussoi, R. C. Shearman, M. Mamo, and C. S. Wortmann. 2010. Soil physical properties of aging golf course putting greens. *Crop Science* 50:2084–2091.

Liu, H., L. B. McCarty, C. M. Baldwin, W. G. Sarvis, and S. H. Long. 2007. Painting dormant bermudagrass putting greens. *Golf Course Management* 75(11):98–102.

Love, W. R. 1992. *An environmental approach to golf course development*. Chicago, IL: American Society of Golf Course Architects.

Lowe, T., and B. Vavrek. 2015. Managing bunkers. *Green Section Record* 53(10):8 pp.

Lubin, T. 1995. Controlling soil pH with irrigation water. *Golf Course Management* 63:56–60.

Martin, B., L. Stowell, and W. Gelernter. 2002. Rough bluegrass, annual bluegrass, and perennial ryegrass hit by new disease. *Golf Course Management* 70(4):61–65.

McCarty, L. B., and A. Canegallo. 2005. Tips for managing ultradwarf bermudagrass greens. *Golf Course Management* 73(6):90–95.

McCarty, L. B., and J. P. Kerns. 2016. Best management practices for Carolinas golf courses. Liberty, SC: CGCSA, 296 pp.

McCarty, L. B., and G. L. Miller. 2002. *Managing bermudagrass turf*. Chelsea, MI: Ann Arbor Press.

McCloud, D. E. 1955. Water requirements of field crops in Florida as influenced by climate. *Proceedings Soil Science Society Florida* 15:165–172.

McGuirk, S., and A. Harivandi. 1998. Irrigation systems can offer alternative uses. *Ground Maintenance* 33:28–32, 86.

McIntyre, K., and B. Jakobsen. 2000. *Practical drainage for golf, sportsturf and horticulture*. Chelsea, MI: Ann Arbor Press.

McNeely, W. H., and W. C. Morgan. 1968a. Review of soil amendments. Classification and development. *Turf-Grass Times* 3:23.

McNeely, W. H., and W. C. Morgan. 1968b. Soil amendments. *Turf-Grass Times*. 3:3, 4.

Meyer, F. P., and L. A. Barclay. 1990. Field manual for the investigation of fish kills. Washington, DC: Department of the Interior Fish and Wildlife Service, resource publication 177.

Miller, R. W., and D. T. Gardiner. 1998. *Soils in our environment* (8th ed.). Upper Saddle River, NJ: Prentice Hall, Inc.

Mitsch, W. J., and J. G. Gosselink. 1993. *Wetlands*. New York: Van Nostrand Reinhold.

Moeller A., and T. Lowe. 2016 Managing organic matter in putting greens. *USGA Green Section Record* 54(21):7 pp.

Moore, J. F. 2008. The money pit. *USGA Green Section Record* 46(4):1–6.

Musser, H. B., and A. T. Perkins. 1969. Guide to planting. In *Turfgrass science*, ed. A. A. Hanson and F. V. Juska, 474–490. Agronomy Monograph 14. Madison, WI: ASA.

Nelson, M. 1998. The microbial world. *USGA Green Section Record* 36(4):1–5.

Paire, J. 2017. The Masters economy. *Superintendent Magazine* 3:12–17.

Pettyman, G. W., and E. L. McCoy. 2002. Effect of profile layering, rootzone texture, and slope on putting-green drainage rates. *Agronomy Journal* 94:358–364.

Pier, J., and D. Barlow (eds.). 2012. *Western fertilizer handbook—3rd horticulture edition*. Sacramento, CA: Western Plant Health Association.

Potter, D. A., and R. M. Maier. 2006. Turfgrass ants: Biology dictates strategies for control. *Golf Course Management* 74(11): 87–90.

Potter, D. A., C. T. Redmond, and A. Saeed. 2017. Invasion of the green stinkworms: Managing *Amynthas hupeiensis* on greens. *Golf Course Management* M 85(2)76–82.

Pound, C. E. 1973. *Wastewater treatment and reuse by land application* (vol.1). Washington, DC: EPA-660/2-73-006a.

Raymer, P. L. 2006. Salt tolerance in seashore paspalum. *Turfgrass Trends* Sept. pp. 1 3.

Reasor, E. H., J. T. Brosnan, M. E. Staton, T. Lane, R. N. Trigiano, P. A. Wadl, J. A. Conner, and B. M. Schwartz. 2018. Genotypic and phenotypic evaluation of off-type grasses in hybrid bermudagrass [*Cynodon dactylon* (L.) x *C. transvaalensis* Burtt-Davy] putting greens using genotyping-by-sequencing and morphological characterization. *Hereditas* 155:8.

Salisbury, F. B., and C. W. Ross. 1992. *Plant physiology* (4th ed.). Belmont, CA: Wadsworth Publishing Company.

Samples, T., J. Sorochan, and A. Thomas. 2013. Identifying and managing petroleum spills and leaks on turf. *SportsTurf* 29(10):24–26.

Shaw, J. W., and R. D. Andrews. 2001. Cation exchange capacity affects greens' turf growth. *Golf Course Management* 69(3):185–189.

Shearman, R. C. 1985. Turfgrass culture and water use. In *Turfgrass water conservation. Proceedings Symposium ASPA*, San Antonio, TX. 15–16 Feb. 1983, ed. V. B. Youngner and S. T. Cockerham, 61–70. Oakland, CA: Cooperative Extension, University of California.

Shetlar, D. J. 1995. Turfgrass insect and mite management. In *Managing turfgrass pests*, ed. T. L. Watschke, P. H. Dernoeden, and D. J. Shetlar, 171–343. Boca Raton, FL: Lewis Publishers.

Smiley, R. W., P. H. Dernoeden, and B. B. Clarke. 2005. *Compendium of turfgrass diseases* (3rd ed.). St. Paul, MN: American Phytopathological Society Press.

Snow, J. T. 1993. USGA explains its new greens specifications. *Ground Maintenance* 28(1):20–22.

Snyder, G. H. 1979. Fertigation for managing turf nitrogen nutrition. *Proceedings GCSAA 50th International Turf Conference and Show*, 163–167. Lawrence, KS: GCSAA.

Snyder, G. H., and B. J. Augustin. 1986. Managing micronutrient application on Florida turfgrass. In *Advances in turfgrass fertility*, ed. B. J. Joyner, 149–179. Columbus, OH: ChemLawn Corp.

Stawn, D. G., J. L. Bohn, and G. A. O'Connor. 2015. *Soil chemistry* (4th ed.). Hoboken, NJ: Wiley Blackwell.

Sullivan, D. L. 1970. Wastewater for golf course irrigation. *Water and Sewage Work* 117(5):153–159.

Taiz, L., and E. Zeiger. 2002. *Plant physiology* (3rd ed.). Sunderland, MA: Sinauer Associates, Inc.

Taylor, D. H., G. R. Blake, and D. B. White. 1987. Athletic field construction and maintenance. Minnesota Extension Service AG-BU-3125, 16.

Thom, W. O., and A. J. Powell, Jr. Sampling soils with turfgrass. In *Soil sampling procedures for the southern region of the United States*, ed. W. O. Thom, and W. Sabbe, 31–34. Southern Cooperative series, bulletin no. 377.

Throssell, C. S. 1985. Management practices affecting bentgrass putting green speed. *Oklahoma Turf*: 3(1 & 2).

Tisdale, S. L., W. L. Nelson, and J. D. Beaton. 1985. *Soil fertility and fertilizers*. New York: Macmillan Publishing Company.

Tucker, B. 1984. The use of gypsum on turf. *Oklahoma Turf* 2(1):1–3.

Turgeon, A. J. 1996. *Turfgrass management* (4th ed.). Upper Saddle River, NJ: Prentice Hall, Inc.

Turner, T. R., and N. W. Hummel, Jr. 1992. Nutritional requirements and fertilization. In *Turfgrass*, ed. D. V. Waddington, R. N. Carrow, and R. C. Shearman, 382–439. Agronomy Monograph Series 32. Madison, WI: ASA, CSSA, and SSSA.

Vargas, J. M., Jr. 1994. *Management of turfgrass diseases*. Boca Raton, FL: Lewis Publishers.

Vincelli, P., and E. Dixon. 2002. Update: Fungicide failure against gray leaf spot. *Golf Course Management* 70(1):53–55.

Vittum, P. J., M. G. Villani, and H. Tashiro. 1999. *Turfgrass insects of the United States and Canada* (2nd ed.). Ithaca, NY: Cornell University.

Voigt, T. 2004. Ornamental grasses. *Golf Course Management* 72(2):137–142.

Volterrani, M., S. Miele, S. Magni, M. Gaetani, and G. Pardini. 2001. Bermudagrass and seashore paspalum winter overseeded with seven cool-season turfgrasses. *International Turfgrass Research Journal* 9:957–961.

Waddington, D. V. 1992. Soils, soil mixtures, and soil amendments. In *Turfgrass*, ed. D. V. Waddington, R. N. Carrow, and R. C. Shearman, 331–383. Agronomy Monograph 32. Madison, WI: ASA, CSSA, SSSA.

Ward, C. Y. 1969. Climate and adaptation. In *Turfgrass science*, ed. A. A. Hanson and F. V. Juska, 27–29. Agronomy Monograph 14. Madison, WI: American Society of Agronomy.

Watson, J. R. 1985. Water resources in the United States. In *Turfgrass water conversation. Proceedings Symposium ASPA*, San Antonio, TX. 15–16 Feb. 1983, ed. V. B. Youngner and S. T. Cockerham, 19–36. Oakland, CA: Cooperative Extension University of California.

Weil, R. R., and N. C. Brady. 2017. *The nature and properties of soils* (15th ed.). Upper Saddle River, NJ: Pearson.

Went, F. W. 1926. On growth accelerating substances in the coleoptile of Avena sativa. *Proceedings of the Section of Sciences, Koninklijke Akademie van Wetenschappen te Amsterdam* 30:10–19.

Wildmon, J. 1991. Converting golf courses to effluent irrigation. *Florida Turf Digest* 8(5):17–18.

Williamson, R. C. 1999. Biorational pesticides: What are they anyway? *Golf Course Management* 67:61–64.

Youngner, V. B. 1985. Physiology of water use and water stress. In *Turfgrass water conversation. Proceedings Symposium ASPA*, San Antonio, TX. 15–16 Feb. 1983, ed. V. B. Youngner and S. T. Cockerham, 37–43. Oakland, CA: Cooperative Extension, University of California.

# Glossary

## A

**abaxial:** Located away from the axis side; for example, the underside of a leaf.

**AB-DTPA extractant:** Acronym for an extraction reagent of ammonium bicarbonate (AB) in diethylenetriamine-penta acetic acid (DTPA) with a pH of 7.6 used to extract P, K, Na, Fe, Mn, and Zn from alkaline soils.

**abiotic:** Nonliving environmental elements such as rainfall, temperature, wind, lightning, and minerals.

**abscisic acid:** Plant hormone that, among other things, promotes bud dormancy, maintains seed dormancy, and causes stomatal closing.

**absorption:** In plants, passing of a compound from one system into another, such as water movement from soil into roots; in soils, binding of a chemical into a soil particle.

**acclimation:** Physical and physiological processes in plants that prepare them for temperature extremes, such as winter.

**acervulus:** Erumpent, cushion-like mass of black spore-bearing hyphae, with prominent black spines (or setae), often correlated with *Colletotrichum graminicola*, the causal pathogen of anthracnose.

**acid equivalent (ae):** Theoretical amount of parent acid from the active ingredient content found in a pesticide formulation.

**acid injection:** Adding acidic materials such as sulfur or phosphoric acid to irrigation water with excessive pH, bicarbonate, and carbonate contents. This lowers the pH and converts bicarbonate and carbonate into carbon dioxide and water.

**acid rain:** Rainwater with a pH below 5.6.

**acidity:** pH of a soil or solution with a value $<7.0$.

**acropetal:** Toward the top of a plant.

**actinomycetes:** Microorganisms that are intermediate between bacteria and fungi.

**activated sludge:** Aerated sludge subjected to bacterial processes and used as a fertilizer or soil amendment.

**active acidity:** Acidity of a soil solution from free hydrogen ions measured as pH; also called potential acidity and buffer capacity.

**active ingredient (ai):** Actual amount of active concentration in a formulation; for example, Aatrex 4L contains 4 pounds of active ingredient per gallon of product, and Ronstar 2G has 2% active ingredient per pound of product.

**activity index:** Cold water insoluble nitrogen (CWIN) percentage of a fertilizer that is soluble in hot (100°C) water, of not less than 40%.

**acute:** Sharply pointed.

**acute toxicity:** Ability of a substance to cause injury or death shortly following exposure, usually due to a relatively large dose.

**Adam–Evans buffer:** Buffer solution used for measuring $H^+$ ions and to determine lime requirements for acid, low CEC soils.

**adaptation:** Unique structure, physiology, or process that aids an organism to fit into its environment.

**adaxial:** Located toward the axis side; for example, the upper side of a leaf.

**adenosine triphosphate (ATP):** Nucleotide consisting of adenine, ribose sugar, and three phosphate groups that, upon metabolism, provide usable chemical energy. On hydrolysis, energy is released when ATP loses one phosphate group to become adenosine diphosphate (ADP).

**adhesion:** Molecular attraction and contact between the surfaces of two unlike substances or objects.

**adjuvant:** Substance in a formulation that enhances its effectiveness. Includes surfactants, crop oils, crop oil concentrates, antifoaming agents, drift control agents, pH modifiers, and compatibility agents.

**adsorption:** Bonding or adhering of ions or compounds to the surface of soil particles or plant parts.

**adventitious bud:** Bud produced in an unusual or unexpected place; for example, near a point of stem injury or on a leaf or a root.

**adventitious root:** Root originating in an unusual or unexpected place, such as from stem or leaf tissue instead of from another root.

**aerification:** In turf, a method of cultivation where hollow or solid tines or spoons are inserted into and removed from the turf to control soil compaction and increase water and fertilizer penetration. *Hollow-tine* aerification involves hollow tines that remove soil plugs; *solid-tine* aerification does not remove plugs when holes are made; *deep-drill* aerification involves removing soil via long drill bits, and *hydro-*, *hydraulic-*, or *water-injection* aerification uses fine streams of high-pressure water to penetrate the soil surface.

**aerobic:** Requiring oxygen or having oxygen present in the environment.

**aggregate:** To collect together in tufts, groups, or bunches, such as soil clods.

**agronomy:** Science of crops and soils.

**aleurone layer:** Layer of high-protein cells in seeds surrounding the storage cells of endosperm. It secretes enzymes for digesting food reserves in the endosperm during seed germination.

**alga (plural: algae):** Single- to multiple-celled plant found in damp habitats containing chlorophyll and having no true roots, stems, or leaves.

**algal bloom:** Population explosion of algae that is often followed by green- or red-colored water and low oxygen levels. It is commonly stimulated by phosphorus and/or nitrogen enrichment.

**alkalinity:** pH of a soil or solution >7.0.

**allelopathy:** Plant chemical product that inhibits the growth of adjacent plants.

**alternate:** In botany, an arrangement of a single leaf, bud, or branch attached singly at different points on the stem, thereby appearing to alternate.

**amendment:** Any material, such as sand, sawdust, gypsum, diatomaceous earth, peat, or calcined clay, that is added to soil to alter its chemical and/or physical characteristics.

**amino acids:** Nitrogen-containing organic acids that are building blocks to form proteins. Amino acids contain one or more amino ($-NH_2$) groups, a carboxyl group (–COOH), and possibly sulfur.

**ammoniacal nitrogen ($NH_3$-N):** With fertilizer, its ammonium- or ammonia-containing amount.

**ammonification:** Conversion of amino acids and other nitrogen-containing organic compounds into ammonia ($NH_3$) and ammonium ions $\left(NH_4^+\right)$.

**amylase:** Enzyme that converts starch into smaller sugars, especially in seed germination.

**anaerobic:** Not requiring oxygen, or the absence of oxygen in the environment.

**anion:** Negatively charged ion that is attracted to a positively charged cation.

**anion exchange capacity (AEC):** Sum of exchangeable anions a soil can adsorb, expressed as centimoles of charge per kilogram of soil ($cmol_c$/kg). AEC only occurs to any appreciable extent when soil pH is <5.

**annual:** A plant starting from seed and completing its life cycle, including death, within one year; *summer annuals* germinate in spring, grow through summer, and die in fall; *winter annuals* germinate in fall, live through winter, and die the following spring or summer.

**antagonism:** General term for counteraction between organisms or chemical mixtures, usually making them less effective than when used together.

**anther:** Sac-like portion of the male part of a flower (stamen) that bears pollen.

**anthesis:** Opening of the flower bud when pollination occurs.

**anthocyanin:** Water-soluble blue or red pigment that aids in light harvesting during photosynthesis.

**anthracnose:** Crown and/or leaf disease caused by fungi producing asexual spores in an acervulus.

**antibiosis:** In biology, an association between two organisms where one is adversely affected.

**antibiotic:** Chemical substance that inhibits the growth of or destroys microorganisms.

**antibody:** Chemical substance in a host that opposes the action of parasites, their products, or other foreign materials.

**antigen:** Substance introduced into animal tissue that stimulates the production of an antibody.

**antioxidants:** Compounds that react with toxic oxygen and free radicals to produce nontoxic water and molecular oxygen.

**apex:** Tip of a stem, root, or leaf.

**apical:** At or near the apex or tip.

**apical dominance:** Ability of a terminal plant bud to inhibit the sprouting and growth of adjacent lateral buds.

**apomixis:** Seed development without sexual fusion of the egg and sperm cells.

**apoplast:** Nonliving, interconnecting cells of xylem tissue that primarily move substances upward in plants.

**approach:** In golf, the fairway area nearest a golf green that is often mowed at lower heights to improve turf playability.

**apron:** In golf, area between the fairway and collar of the putting green that is usually mowed at lower heights than the fairway, but higher than the collar and green; often called the *approach*.

**aquatic:** Related to water; a plant or organism that grows in or on water.

**aquifer:** Reservoir for underground water, which is often a water source for wells and springs.

**artificial turf:** Synthetic turf surface.

**as-built drawing:** Drawing showing installed location of all elements of a construction project.

**ascending:** Sloping or growing upward or outward.

**ascomycetes:** Group (phylum) of fungi that produce asci, usually containing eight ascospores.

**ascospores:** Fungal spores produced in a sac-like structure called an ascus.

**ascus (plural asci):** Specialized cells that nuclei fuse, undergo meiosis, and are subsequently incorporated into ascospores.

**aseptate:** Lacking cross walls; usually refers to fungus mycelium.

**ASTM:** American Society for Testing and Materials, an association that establishes safety and performance standards for a number of products, including turf construction materials.

**asymmetrical:** Lopsided; having two sides different in shape or area.

**atom:** Smallest unit of a chemical element that retains its characteristic properties.

**atomic number:** Number of protons in the nucleus of an atom.

**atomic weight:** Weight of an atom of an element relative to the weight of an atom of carbon $^{12}C$, which has been assigned the value 12.

**ATP:** *See* adenosine triphosphate.

**auricle:** Small, ear-shaped lobes or appendages at the junction of the leaf sheath and blade in grasses, or at the leaf base of broadleaf plants.

**autoecious:** Requiring only one host on which to complete a life cycle; often refers to some rust fungi.

**autotrophic:** Living on inorganic materials as nutrients.

**auxin:** Plant growth regulating hormone that promotes cell elongation.

**available water:** Portion of soil water readily absorbed by plant roots; often considered to be the water held in soil against a pressure of –33 kPa to approximately –1,500 kPa (or –15 bars).

**awn:** Slender or stiff bristle, usually extending from a grass floret, specifically on the glumes or lemma.
**axil:** Angle between the leaf and stem.
**axillary bud:** Bud located in the leaf axil or internal fold.
**axis:** Main stem of an inflorescence; a panicle is an example.

## B

**bacillus (plural bacilli):** Rod-shaped bacterium.
**backflow preventer:** In irrigation, an in-line specialty valve which prevents flow reversal of contaminated water into the water source. Also called *antisiphon valve.*
**backlapping:** Backward turning of mower reel blades against a bed knife with a corresponding addition of fluid-dispersed grinding compound to hone the blades.
**bacteria (singular: bacterium):** Microscopic, single-celled, nongreen organisms with rigid cell walls that reproduce by cell division. Bacteria are prokaryotes.
**ball mark:** Depression or tear in the turf surface from the impact of a ball.
**ball roll distance:** Distance a golf ball moves following a putting stroke, often measured with a stimpmeter.
**bar:** In science, a unit of pressure used to express water potential (1 atmosphere = 1.013 bars).
**basal rosette:** Cluster of leaves radiating at the base of a plant at ground level.
**base saturation:** Relative amount of the basic cations, $Ca^{+2}$, $Mg^{+2}$, $Na^{+}$, and $K^{+}$, in a soil compared to acidic cations, $H^{+}$ and $Al^{+3}$. Mathematically, base saturation is a sum of the basic cations divided by the total cations present:

$$\text{base saturation} = \frac{Ca^{+2} + Mg^{+2} + Na^{+} + K^{+}}{CEC}$$

**basiomycetes:** Fungi that produce sexual spores (basidiospores) on a club-shaped structure called a basidium.
**basipetal:** Toward the base of a plant.
**beak:** Hard point or projection, seen frequently on seeds and fruits.
**bearded:** Having long hairs.
**bed knife:** Stationary lower blade on a reel mower where the rotating reel blade makes contact resulting in cutting.
**bench mark:** Fixed reference point used in surveying to measure elevations.
**bench setting:** Mowing height as determined from a hard, level surface to the blade. Also termed *bench height.*
**best management practices (BMPs):** In turf, management or cultural practices that minimize inputs and undesirable effects yet maximize outputs and aesthetic value.
**bicarbonates ($HCO_3^-$):** Ion in irrigation water that combines with calcium and magnesium to form insoluble lime (or calcite).
**biennial:** Plant completing its life cycle in two years. In the first year, seeds germinate and form vegetative growth; in the second year, flowering, seed set, and death occur.
**bioassay:** Using indicator plants or other biological organisms to determine the levels of a substance present.
**biochar:** Charcoal-based material used to sequester carbon and as a soil amendment.
**biodegradation:** Breakdown (digestion) by biological organisms.
**biological control:** Using any biological agent to control a pest.
**biological oxygen demand (BOD):** Quantity of dissolved oxygen in water used in the oxidation of organic matter. High BOD usually means anaerobic water.
**biostimulant:** Organic material applied in small quantities that enhances plant growth and development.
**biotechnology:** Branch of technology utilizing living organisms or biological systems to improve health and modify the environment.
**biotype:** Population within a species possessing a distinct genetic variation.
**bipinnate:** Two rows of lateral branches along an axis again divided into two rows; feather-like.
**bisexual:** Flowers with male (stamens) and female (pistil) elements.
**blade:** In biology, the expanded, usually flat, portion of a leaf.
**blend:** In turf cultivars, a combination of two or more cultivars of a single turfgrass species. With granular fertilizers, a fertilizer composed of individual prills of different fertilizer material.
**blight:** Plant disease symptom characterized by general and rapid killing of leaves, flowers, and stems, resembling heat injury.
**blotch:** Plant disease symptom characterized by large, irregularly shaped spots or blots on leaves, shoots, and stems.
**boat-shaped:** Leaf tips shaped like the front (or bow) of a boat such as *Poa* species.
**boot growth stage:** Grass growth stage when the upper leaf sheath swells due to the developing seedhead.
**Bordeau mixture:** Fungicide mixture of copper sulfate and lime.
**brackish:** Water description with a high content of soluble salts; salty.
**bract:** Modified, usually reduced leaf associated with a flower or flower cluster.
**branch:** A lateral stem.
**Bray P1 and P2 extraction reagents:** Soil extraction reagents of ammonium fluoride in hydrochloric acid use for determining soil-extractable P in acid soils of moderate CEC.
**brine:** Salt residue remaining following water desalination.
**bristle:** Short, coarse, stiff, hair-like part.
**brushing:** In turf, use of brushes in front of mower blades to upright the leaf for a cleaner cut.
**bryophytes:** Members of the phyla of nonvascular plants; the mosses, hornworts, and liverworts.
**bud:** Tightly bunched, undeveloped shoot or flower usually located at the tip of a stem or branch (called *apical bud*) or in the axil or internal fold of a leaf sheath (called *axillary bud*).
**buffer:** Substance that resists any change in pH.

**buffer capacity:** Ability of soils to resist chemical change caused by the high cation exchange capacity and, in some cases, free calcium carbonate.

**buffer pH:** Substance used to determine buffering capacity of a soil for lime requirements.

**bulb:** Underground short, thickened shoot where food is stored, such as in wild onion.

**bulk blending:** Physical mixing of two or more dry fertilizer materials (generally, >50-pound bags, 23 kg).

**bulk density:** Mass (or weight) per unit volume of soil. Higher bulk density indicates harder or more compact soil.

**bulk flow:** Overall movement of a liquid such as water, induced by gravity, pressure, or an interplay of both.

**bunchgrass:** Intravaginal tilling at the crown without stolon or rhizome formation.

**bunker:** In turf, a hazard, usually a depression filled with sand or grass, that penalizes an errant golf shot. Also called a *sand trap.*

**bur:** Structure with spines or prickles frequently hooked or barbed; sandspur is an example.

## C

**$C_3$ plants:** Plants using only the Calvin (or $C_3$) cycle or pathway in the fixation of $CO_2$; the first stable product is the three-carbon compound, 3-phosphoglycerate (3-PGA); hence, the designation $C_3$.

**$C_4$ cycle (Hatch–Slack) or pathway:** Reactions where carbon dioxide is fixed to phosphoenolpyruvate (or PEP) to yield a four-carbon compound, oxaloacetate (or OAA); hence, the designation $C_4$.

**$C_4$ plants:** Plants in which the first product of $CO_2$ fixation is a four-carbon compound, oxaloacetate (or OAA), where both the Calvin (or $C_3$) cycle and the $C_4$ pathway are used.

**calcareous soil:** Soil containing 10 to 1,000 g/kg of $CaCO_3$ (calcium carbonate, lime) equivalent.

**calcined clay:** Granular soil modification amendment of highly fired or heated (*calcining*) clay minerals to ~1,200°F (~650°C), such as montmorillonite and attapulgite, that are absorbent and stable.

**calcined diatomaceous earth:** Fired, single-celled ocean organisms called *diatoms* that are absorbent and stable; often used as a granular soil modification amendment.

**calcite:** Crystalline form of calcium carbonate ($CaCO_3$) often formed when high bicarbonate-containing water is used for irrigation; also called *calcitic limestone.*

**calcium carbonate ($CaCO_3$):** An insoluble form of calcium combined with carbonate that remains insoluble until reacted with an acid. Occurs in nature as limestone, marble, chalk, marl, shells, and similar substances.

**calcium carbonate equivalent (CCE):** Relative measurement of the purity of a liming material compared to 100 for calcium carbonate ($CaCO_3$).

**calcium chloride ($CaCl_2$):** Very soluble calcium salt dissolved in irrigation water to lower the sodium adsorption ratio (SAR) or increase the water's electrical conductivity (ECw).

**calcium/magnesium carbonate ($CaCO_3 \cdot MgCO_3$):** Insoluble calcium/magnesium combination precipitated from water high in calcium, magnesium, bicarbonates, and carbonates. It remains insoluble until reacted with an acid.

**calcium nitrate ($CaNO_3$):** Highly soluble source of calcium and nitrogen that can be dissolved in irrigation water to lower the sodium adsorption ratio (SAR) or increase the electrical conductivity (ECw).

**calcium oxide (CaO):** A compound composed of Ca and $O_2$ formed by heating calcium carbonate ($CaCO_3$) to remove carbon dioxide. Also known as *quick lime, unslaked lime, burnt lime*, or *caustic lime*; it does not occur in nature.

**calibration:** Process of adjusting equipment so the desired rate of material application occurs.

**Calvin cycle:** Series of reactions during photosynthesis where carbon dioxide is reduced to 3-phosphoglyceraldehyde (or 3PGA) and the carbon dioxide acceptor, ribulose 1,5-bisphosphate (RuBP), is regenerated.

**capillary fringe:** Zone immediately above the water table that is nearly saturated.

**capillary porosity:** Percentage of soil volume occupied by water due to capillary forces following gravity drainage.

**capillary water:** Water remaining in soil pores following gravity drainage.

**carbohydrates:** Plant food sources including sugars and starches containing carbon with hydrogen and oxygen in a 2:1 ratio, as in water ($H_2O$).

**carbon:nitrogen (C:N) ratio:** Ratio of organic carbon weight to total nitrogen weight in soil or organic material.

**carbonates $\left(CO_3^{-2}\right)$:** Ion in irrigation water that combines with calcium and magnesium to form insoluble lime (or calcite).

**carotenoid:** Yellow or red plant pigment.

**carpel:** Unit of the pistil in a flower; a simple pistil is formed from a single carpel, while two or more carpels compose a compound pistil.

**caryopsis:** The grass fruit, normally dry at maturity, consisting of a single seed within the ovary.

**catch basin:** "Junction box" for a drainage system, which also often allows surface water to flow in.

**cation:** Positively charged ion that is attracted to a negatively charged anion.

**cation exchange capacity (CEC):** Sum of exchangeable cations a soil can adsorb and retain against leaching; expressed as centimoles per kilogram of soil.

**cellulose:** Unbranched sugar that is the chief substance composing cell walls or woody parts of plants.

**centrifugal spreader:** Machine that spreads granules as they drop onto a spinning disc or blade beneath the hopper.

**certified seed:** Seed from registered or foundation plants produced under an officially designated system of maintaining genetic identity and purity. It is typically identified by a blue tag.

**character:** Observable feature that distinguishes between organisms.

**chelates (chelating or sequestering agents):** Chemical formulation that binds a metal atom (such as iron) with an organic component to improve its soil availability and uptake by plants.

**chilling injury:** Plant damage from nonfreezing, low temperatures, usually 35°F to 60°F (1.7°C to 16°C).

**chlorophyll:** Green photosynthetic pigment found in the chloroplasts of all photosynthetic plants that enables them to capture solar energy to make food.

**chloroplast:** Sac-like structure in plant cells containing chlorophyll; the site of photosynthesis.

**chlorosis:** Yellowing of normally green plant tissue due to chlorophyll loss.

**chromosome:** Structure that carries the genes during cell division.

**chronic toxicity:** Ability of a substance to cause injury or death after long-term exposure, usually to small doses.

**ciliate:** Fringed with hairs on the margin; hairy.

**clasping:** Type of leaf attachment where the leaf base partly or completely encircles the stem.

**clay:** Soil particles 0.002 mm in diameter; also indicates a soil mixture containing more than 40% clay.

**clear liquid solution:** One or more plant nutrients in solution with no suspended particles.

**Clegg impact hammer:** Device used to measure surface firmness by dropping a weight and measuring its deceleration speed.

**clipping height:** Distance above the soil line at which grasses are clipped or mowed.

**clippings:** Leaf and stem portions of plants severed by mowing.

**clump-forming or tufted:** Plant that grows in a compact cluster.

**coated fertilizer:** Water-soluble fertilizer granule covered with a durable, permeable covering.

**coefficient of uniformity (design uniformity):** In irrigation, the efficiency of a sprinkler system based on precipitation rates at various points.

**cohesion (surface tension):** Mutual attraction of molecules of the same substance.

**cold hardiness:** Ability of plants to survive extreme cold and dry weather.

**cold water insoluble nitrogen (CWIN):** Insoluble nitrogen fertilizer that forms in cold (77°F or 25°C) water.

**cold water soluble nitrogen (CWSN):** Soluble nitrogen fertilizer that forms in cold (77°F or 25°C) water.

**coleoptile:** Protective sheath covering the shoot tip and leaves of emerging grass seedlings.

**coleorhiza:** Transitory protective sheath covering the root apex (tip) of emerging seedlings.

**collar:** In botany, outer side of a grass leaf at the junction of blade and sheath; in golf, area surrounding the green or fringe (if present), normally maintained between the green and fairway.

**colloidal phosphate:** Soil amendment used to increase cation-exchange capacity and water-holding capacity of sandy soils, which is a by-product from phosphate washing.

**colloids:** Soil particles (inorganic or organic) 0.005 to 0.20 µm in diameter characterized by high ion exchange.

**colorant:** Paint-like dye or pigment used to create a green color when the turf is discolored or dormant.

**combing:** Mowing technique utilizing a series of metal teeth or flexible tines (comb) in front of a mower to lift shoots for a more desirable cut.

**common name:** In pesticides, a name applied to a pesticide's active ingredient, usually agreed upon by the American National Standards Institute and the International Organization for Standardization. For example, oryzalin is the common name of the herbicide with the trade name Surflan. In plants, layman or regional names associated with a particular plant; for example, annual bluegrass is a common name of the plant with the scientific name *Poa annua*.

**compaction:** An unfavorable increase in soil bulk density (grams/cubic meter) and corresponding decrease in soil porosity from mechanical pressures to the soil.

**complete fertilizer:** Fertilizer containing all three primary fertilizer nutrients (N, P, and K).

**composite:** Member of the Compositae or Asteraceae plant family that has a dense inflorescence, usually composed of florets, a receptacle, and bracts.

**compost:** Residues from organic matter and soil piles allowed to undergo biological decomposition.

**compost tea:** Generic term for actively aerated compost tea (AACT) where oxygen is added to water to promote aerobic conditions for increased microorganism populations, generally from compost.

**compound leaf:** Type of leaf composed of two or more distinct, similar parts often called leaflets.

**conduction:** Heat transfer through a conducting medium due to temperature differences.

**conductivity:** Measure of electrical resistance of a soil-water extract, or irrigation water, to determine the level of ions in solution. Reported as mhos/cm (micro- or milli-) or decisiemans/meter (dS/m).

**contact fungicide:** Product that protects aboveground treated plant parts from fungal infection.

**contours:** Outline of a figure, body, or mass; contour lines connect the points on a land surface with the same elevation.

**convection:** Heat transfer within the atmosphere due to temperature differences.

**cool-season grass:** Grass with optimum growth in the temperature range of 60°F to 75°F (15.5°C to 24°C). Cool-season grasses are $C_3$ plants.

**core aerification:** *See* aerification.

**core cultivation:** *See* aerification.

**coring:** *See* aerification.

**corm:** Stout, short, vertical, bulblike underground food storage stem.

**corolla:** Flower petals that surround the stamens and pistil.

**cotyledon:** Seed leaf of the embryo; most often the storage sites of reserve food used by germinating seedlings.

**crop coefficient:** Mathematical expression of differences in moisture needs of various plants.

**crop oil concentrate:** Nonphytotoxic petroleum-based light oil adjuvant containing surfactants used to enhance the effectiveness of liquid solutions applied to plant foliage.

**crown:** In plants, a meristematic growing point at or just below the ground where stems and roots join and new shoots emerge.

**culm:** Flowering stem of a grass plant not including the leaves.

**cultipacking:** Seeding technique utilizing a seeder with a larger, ridged, front roller to prepare a shallow seedbed, with an offset, smaller, rear roller to firm the soil around the planted seed.

**cultivar:** Form, type, or variety of a cultivated plant.

**cultivation:** In turf, the disturbance of soil and/or thatch layer without destroying the turf (e.g., aerifying, slicing, spiking, slitting, etc.).

**cup cutter:** Hollow cylinder used to cut a 4.25-inch (10.8 cm) diameter hole at least 4 inches (10.8 cm) deep for the cup (or pin) in a green or to replace damaged turf.

**curative:** Fungicides applied following the appearance of disease symptoms; also called eradicants.

**cut and fill:** In construction, soil is removed ("cut") or added ("fill") to achieve the desired grade.

**cuticle:** Waxy outer layer of a leaf or stem.

**cutin:** Waxy substance found on the surface of certain seeds or leaves to conserve water.

**cutting height:** Height above the plane of travel and the parallel plane of cut on a mower.

**cytochrome:** Heme proteins serving as electron carriers in photosynthesis and respiration.

**cytokinins:** Group of plant growth regulating hormones that regulate cell division.

## D

**$D_{15}$:** Particle diameter of gravel whereby 15% of the gravel particles are finer and 85% are coarser. Used to determine if a drainage stone meets bridging requirements with a rootzone mix.

**$D_{85}$:** Particle diameter below which 85% of the sand particles are finer and 15% are coarser. Used to determine in turf if a rootzone mix meets bridging requirements with a coarser diameter gravel.

**damping-off:** Disease (*Pythium* spp.) that causes seedlings to rot at the soil surface and fall over.

**Darcy's law:** Law describing saturated water flow through a porous media.

**daylight:** In drainage, where water flows downward and out of a collector pipe.

**day-neutral plants:** Plants with no day length requirement for floral initiation.

**decumbent:** Lying on the ground, but rising at the tip.

**deflocculation (dispersion):** To cause particles to separate into individual components and disperse by chemical and/or physical means; often used to describe sodic soil in which poor structure is exhibited.

**defoliant:** Chemical that causes a plant to lose its leaves.

**denitrification:** Biological conversion of nitrate or nitrite to gaseous nitrogen as either $N_2$ or $N_2O$.

**deoxyribonucleic acid (DNA):** Double-stranded helix chain made up of nucleotides containing the genetic instructions for the development and function of all living organisms.

**desalination:** Process of separating a saline solution into pure fresh water and brine.

**desiccation:** Plant tissue withering due to acute lack of moisture.

**dethatching:** Removal of excessive turf thatch and/or mat using stiff rakes or a series of vertically mounted knives or tines.

**dew:** Moisture formation when water vapor condenses.

**dew point:** Temperature at which air becomes saturated (100% relative humidity), forming fog or dew. Higher dew points mean higher moisture content of air at a given temperature. As dew point becomes lower, the chances of frost formation increase.

**diagnosis:** Process for determining and identifying the cause of a problem.

**diatomaceous earth:** Geologic deposit of siliceous skeleton material of diatoms (algae); used as an inorganic soil amendment.

**diatoms:** Algae with siliceous cell walls that persist as a skeleton following death.

**dicotyledon, dicot:** Broadleaf plants with two seed embryos (leaves) or cotyledons when they emerge from the soil; these also have netted leaf veins, showy flowers, flower parts in fours or fives, and often a cambium for secondary growth.

**diffuse:** Loose and widely spreading.

**digitate:** Branches arising from a common point, resembling the fingers of a human hand.

**dikaryotic:** Possessing two haploid nuclei per cell.

**dioecious:** Separate male (stamens) and female (pistil) plants; unisexual.

**diploid:** Containing a double ($2n$) number of chromosomes.

**disease:** Disturbance that interferes with normal plant structure, function, or economic value.

**disease triangle:** Concept where disease development requires concurrent presence of a susceptible host, a virulent pathogen, and conducive environmental conditions.

**disinfectant:** Chemical that eliminates plant pathogens from plant surfaces, seed, or inanimate objects such as tools, pots, or benches.

**dispersible granule (DG):** Dry granular formulation that separates (or disperses) when added to water to form a suspension.

**dissected:** Divided into numerous narrow segments or lobes.

**dissolved oxygen:** Atmospheric oxygen held in solution within water.

**distribution uniformity (DU):** Uniformity (evenness) in irrigation water application within an irrigated area. Commonly determined by dividing average lowest one-fourth amount of irrigation by average of the entire irrigation zone.

**ditch:** Deeper than wide valley-like excavation designed to conduct water.

**divot:** Removal of a small turf piece from twisting cleated foot traffic or a golf swing, often exposing soil. *Divot resistance* is ability of the turf to resist divoting when a golf club displaces turf and soil while striking the ball; *divot recovery* is the turf recovery rate after divots have been made.

**DNA sequencing:** Process of determining the nucleotide order of a given DNA fragment.

**dogleg hole:** Right or left bend in a golf hole resembling a dog's leg.

**dolomite:** Magnesium carbonate ($MgCO_3$) containing limestone.

**dolomitic limestone:** Limestone containing 10% to 49% dolomite, 50% to 90% calcite, and 4% to 23% $MgCO_3$.

**dormancy:** Arrested plant growth during periods of unfavorable environmental conditions.

**dormant seeding:** Planting in fall or winter when it is too cold for germination until the following spring.

**dormant sodding:** Sodding in fall or winter when it is too cold for active growth.

**double green:** Large putting surface servicing two different golf holes with two separate flagsticks and holes.

**drain line:** Underground pipe that collects and removes excessive soil moisture.

**drought:** Prolonged water stress that limits or prevents plant growth.

**drought avoidance:** Plant's ability to sustain internal water levels through morphological and physical growth features such as more efficient and deeper root systems, effective stomata closure, and thicker leaf cuticles.

**drought resistance:** Plant's ability to withstand drought conditions by combining drought avoidance and drought tolerance.

**drought tolerance:** Plant's ability to sustain internal water content through biochemical and physiological processes.

**dry flowables (DF):** Dissolvable dry granular pesticide formulation.

## E

**ecology:** Study of plant life in relation to its environment.

**ecosystem:** Interacting system of living organisms and their physical environment.

**ecotype:** Strain or selection within a given species adapted to a particular environment.

**ectotrophic:** Root-infecting fungal hyphae that develop mainly on the root surface.

**edaphic:** Pertaining to soil.

**effective calcium carbonate (ECC):** The quality of limestone determined by multiplying its fineness by its purity (or calcium carbonate equivalent).

**effluent:** Water that has undergone one cycle of human use and is partially or completely treated to make it suited for limited use. Also referred to as *recycled, reclaimed, wastewater*, and *treated sewage water.*

**electrical conductivity (EC):** A measure of salinity using electrical conductance expressed as millimhos per centimeter (mmho/cm) or decisiemens per meter (dS/m). $EC_e$ designates the electrical conductivity of soil from a saturated paste, while $EC_w$ designates the electrical conductivity of water.

**electrodialysis:** Water desalination method where ions pass through a semipermeable membrane toward an electrical field, leaving purer water behind.

**electron:** A negatively charged subatomic particle that orbits the atom's positively charged nucleus, determining the atom's chemical properties.

**electron transport system ("Z" scheme):** Movement of electrons down a series of electron-carrier molecules with subsequent release of energy. This is an energy-harvesting procedure during photosynthesis.

**element:** Substance composed of only one kind of atom. These combine to compose all materials.

**elliptic:** Narrow shape with relatively rounded ends that is widest at the middle.

**elongate:** To make narrow and long.

**embryo:** Seed portion that develops into a juvenile plant.

**emergence:** Visible protrusion of the shoots of a newly germinated seed.

**emulsifiable concentrate (EC):** Single phase liquid formulation that forms an emulsion when added to water.

**emulsifier:** Substance that promotes the suspension of one liquid in another. Often a type of surfactant.

**emulsifying agent:** Colloidal substance forming a film around immiscible particles to obtain a suspension.

**emulsion:** System of oil dispersed as minute globules in water or system of water dispersed in oil; not a true solution.

**endocarp:** Inner layer of the pericarp (fruit wall).

**endodermis:** Layer of cells forming a sheath around the vascular region in roots and some stems.

**endophyte:** Plant or fungus that develops and lives inside another plant.

**endosperm:** Seed portion containing food reserves.

**entire:** In plants, a type of leaf margin without teeth, lobes, or divisions; smoothed edge.

**environment:** Conditions, influences, or forces affecting living forms.

**enzyme:** Protein that speeds up, or catalyzes, cellular reactions without being altered in the process.

**epicotyl:** Young stem of a seedling or embryo just above the cotyledon(s).

**epidemic:** Disease occurrence increase in a population.

**epidermis:** Outer cellular layer of plants that helps prevent drying and mechanical injury.

**epinasty:** Bending or curling of a plant organ (especially leaf) when more rapid growth occurs on the upper side of the organ.

**Epsom salt:** Common name for magnesium sulfate ($MgSO_4$).

**equivalent:** Amount of material that reacts with or provides 1 g formula weight of hydrogen.

**equivalent weight (moles of ion charge):** Amounts of substances equivalent to each other in chemical reactions—determined in an acid as the weight of substance furnishing 1 mol of hydrogen ($H^+$) ions, while in a base as the weight furnishing 1 mol of hydroxide ($OH^-$) ions. Also measured as the change in oxidation (valence) atoms undergo in a chemical reaction:

$$\text{equivalent weight} = \frac{\text{molecular weight}}{\text{no. } H^+ \text{ or } OH^- \text{ per molecule}} \text{ or } \frac{\text{molecular weight}}{\text{valence no.}}$$

**eradicant:** Product that destroys a pathogen at its source.

**erosion:** Wearing of land by running water, wind, or other geological agents.

**ester:** In turf, an oil-soluble chemical formulation better able to penetrate waxy leaf surfaces but generally more volatile and precipitate out in "hard" water.

**ethylene:** Hydrocarbon hormone gas ($C_2H_4$) whose effect on plants is most prominent in fruit ripening, preventing seedhead (floral) expression, and suppressing root growth.

**eukaryote:** Organism with a membrane-bound nucleus and organelles whose DNA is associated with proteins.

**eutrophication:** Water condition with excess dissolved nutrients and insufficient dissolved oxygen.

**evaporation:** Process of water evaporation as vapor from land, water, and vegetation surfaces.

**evapotranspiration (ET):** Combined loss of water from an area by evaporation from the soil surface and transpiration from plants; expressed as inches/week and millimeters/day.

**exchangeable acidity:** Measure of hydrogen ions adsorbed to colloidal surface. Also known as *potential acidity*.

**exchangeable ions:** Replaceable ions on the soil complex by other ions of like charge.

**exchangeable sodium percentage (ESP):** A measure of excessive sodium hazard in the soil as the ratio (as percent) of exchangeable sodium to the remaining exchangeable cations (Mg, Ca, and K).

**exergonic:** Energy-yielding chemical reaction.

**exodermis:** Outer layer of cells of root cortex.

**extraction reagent:** Reagent of pure water or a mixture of acids or buffered salts used to extract nutrients from soil or plant tissue.

**extravaginal:** Growth from stem penetration through the basal leaf sheath as seen in rhizomes and stolons.

## F

**$F_1$:** First generation offspring from a cross; $F_2$ and $F_3$ are the second and third generations resulting from such a cross.

**facultative parasite:** Mostly saprophytic organisms that may become parasitic when environmental conditions are favorable.

**facultative saprophyte:** Mostly parasitic organisms that may become saprophytic when environmental conditions are favorable.

**fairway:** Primary playing area on a golf course between the tees and putting green.

**family:** In the plant and animal kingdom, a taxonomic group between order and genus in rank. A family contains one or more genera and each family belongs to an order.

**fat:** Molecule composed of glycerol and three fatty acid molecules, with its proportion of oxygen to carbon being less in fats than in carbohydrates. Liquid fats are called oils.

**fatty acid:** Organic compound of carbon, hydrogen, and oxygen that combines with glycerol to form a fat.

**ferredoxin (Fd):** Electron-transferring proteins of high iron content, often involved in electron transport during photosynthesis and respiration.

**fertigation:** Fertilizing through irrigation systems.

**fertilizer:** Any material, except lime, supplying essential elements.

**fertilizer analysis:** Percentage of composition of a fertilizer as total nitrogen (N), phosphoric acid ($P_2O_5$), and potash ($K_2O$).

**fertilizer burn:** Plant injury from dehydration due to contact with materials containing salts.

**fertilizer formula:** Quantity or grade of materials used in a fertilizer mixture.

**fertilizer grade:** Guaranteed minimum analysis of the major nutrients of a fertilizer.

**fibrous roots:** Slender, branched roots of similar size arising from a similar point.

**field capacity:** Water content a soil retains against the action of gravity; typically the water remaining in a soil two or three days after the soil is saturated and free drainage has occurred; often estimated at –33 kPa water potential.

**filament:** Anther-bearing stalk of a flower's stamen (male part); thread.

**filiform:** Threadlike, long, and very slender.

**fission:** Asexual reproduction involving the division of a single-celled individual into two new single-celled individuals of equal size.

**flaccid:** Without rigidity; limp or weak.

**flail mower:** Mower with a series of retractable, free hanging T-shaped blades that extend when a shaft rotates at high speeds. It cuts and recuts until clippings are sufficiently pulverized to escape the mower housing; also called a *hammerknife* mower.

**flocculated soil:** Soil with a well aggregated structure.

**flocculation:** To aggregate or clump together, especially clay, into larger clumps or aggregates.

**floret:** Small flower or one of individual, closely clustered, small flowers (having both pistil and stamens) enclosed by bracts (lemma and palea).

**flowable (F):** Two-phase formulation containing a solid substance suspended in liquid that forms a suspension when mixed with water.
**flow-by:** Sprinkler head leakage from a worn seal.
**folded:** In botany, lengthwise folding arrangement of the youngest leaf in a plant's shoot.
**foliar burn:** *See* fertilizer burn.
**foliar feeding:** Light liquid fertilizer application to plant foliage.
**footprinting:** Extended turf indentation from surface traffic due to the lack of leaf turgidity (moisture).
**forking:** Turf cultivation technique of using a spading fork to punch holes in the turf and soil.
**formulation:** Form of a product available to the consumer, including both the active and inert ingredients; examples include concentrates, emulsifiables, flowables, granulars, and powders.
**foundation seed:** Progeny (seed) from breeder stock; used as planting stock for registered and certified seed.
**fraze mowing:** Mechanical removal of the top layer of soil, thatch, and organic matter with a rotary spindle using specialized tines or blades. This is preformed to (a) smooth the surface, (b) stimulate new growth from rhizomes and stolons, (c) remove thatch and organic matter, (d) remove weed seed, and (e) remove overseeding.
**French drain:** *See* slit drain.
**frequency of clip:** Distance of travel between successive cuts of mower blades.
**fringe:** Optional area, 2 to 4 feet wide (0.6 to 1.2 m), between the golf green surface and collar, which is maintained between the green and collar but is not considered part of the putting surface.
**frost:** Ice crystal formation when water vapor condenses below freezing (32°F or 0°C). Forms when surfaces are cooled below the dew point of the air and below freezing.
**fruit:** Matured ovary with its enclosed seeds; a ripened pistil.
**fulvic acid:** Soluble organic substance remaining after a soil alkali extract has been acidified.
**fumigation:** Process of applying a fumigant that disinfests an area from various pests.
**fungicide:** Compound toxic to fungi.
**fungistatic:** Compound preventing fungus growth without killing it.
**fungus (plural: fungi):** An undifferentiated threadlike organism without chlorophyll and conductive tissue. Fungi are eukaryotes.

## G

**gene:** Locatable segment of a DNA molecule that controls a hereditary characteristic.
**general-use pesticide:** Pesticide not causing unreasonable adverse effects on the environment that is safe for application by the general public without requiring special training.
**genetic engineering:** Adding genetic material to the genome of an organism.
**genome:** Entirety of an organism's hereditary information.
**genotype:** Hereditary makeup of a plant (or variety) that determines its inheritance.
**genus (plural: genera):** Taxonomic group of related species ranked between family and species.
**germination:** Initiation of plant growth from seed.
**gibberellic acid (GA):** Growth-promoting substances that regulate many growth responses, such as increased plant stem elongation, and in seed priming to speed germination.
**glabrous:** Smooth, without hairs or bristles.
**glandular hair:** Small hair terminated in a small pinhead-like gland, frequently secreting resin, wax, or other substances.
**glaucous:** Covered with a waxy coating resulting in a whitish to blue-green color.
**glucose:** Common six-carbon sugar ($C_6H_{12}O_6$).
**glume:** Pair of bracts at the base of a grass spikelet that does not enclose flowers.
**glycolysis:** Anaerobic enzymatic breakdown of complex organic molecules with the release of energy.
**gradation index:** Expression of particle size uniformity of a sand sample in the middle 80% of the range based on $D_{90}/D_{10}$. Lower values indicate more uniform particle size and less compaction potential.
**grade stakes:** Stakes driven into the ground that indicate the desired elevation at that point.
**grain:** In turf, mostly undesirable horizontal lying of grass blades in one direction, usually caused by repeated mowing in the same direction, which tends to deflect a rolling ball from a true course.
**Gram-negative bacteria:** Negative reaction to the Gram's stain, wherein the purple dye washes out and the cells stain red.
**Gram-positive bacteria:** Positive reaction to the Gram's stain, wherein the cells retain the purple dye.
**granular (G):** Dry formulation of particles often $<10$ mm$^3$ in diameter applied without a liquid carrier.
**gravitational water:** Water movement through soil due to gravity.
**green:** Putting surface of a golf course.
**grooving:** Turf cultivation method utilizing vertical, rotating blades that cut shallow slits through the turf and into the soil.
**groundwater:** Subsurface water in the zone of saturation that moves freely, and often horizontally.
**growth regulator:** Synthetic compound that controls plant growth responses.
**grub:** In biology, an insect larva; typically associated with the Coleoptera and Hymenoptera insect families.
**grubbing:** In construction, mechanical means of gathering and removing rocks, stumps, roots, sticks, and other unwanted material prior to grading and planting.
**guano:** Decomposed dried bat droppings used for fertilizer purposes.

**guard cells:** Pairs of specialized epidermal cells surrounding a pore or stomata that open and close in response to changes in turgor pressure.
**guttation:** Exudation of liquid from leaf tips caused by root pressure.
**gypsum:** Common name for calcium sulfate ($CaSO_4$). Also called *landplaster.*

## H

**habit:** In botany, growth form of the plant.
**habitat:** Natural environment where an organism grows.
**half-life:** Time required for half of a substance to be inactivated.
**halophyte:** Plant that grows in salty soil.
**haploid:** Having only one set of chromosomes (*n*) compared to two in diploids (2*n*).
**hard water:** Water containing calcium, magnesium, or ferrous ions, which forms a precipitate with soap or crusting. Often determined by [(Ca × 2.5) + (Mg × 4.1)] and expressed as milligrams/liter.
**haustorium (plural haustoria):** Projection of fungal hypha that penetrates and absorbs content from the host tissue.
**hazard:** Penalty area on a golf course that does not allow normal play, such as a bunker or water hazard.
**head:** Dense cluster of stalkless flowers, as in dandelion.
**heat of fusion:** Heat released (80 calories/gram of water frozen) when water freezes to solid ice.
**heat of vaporization:** Heat released (540 calories for water at 100°C) when water vaporizes.
**heavy metals:** Metals with densities of 5.0 mg/m or greater (atomic weights >55), including cadmium, cobalt, chromium, copper, iron, lead, mercury, manganese, molybdenum, nickel, and zinc.
**herbaceous:** Nonwoody plant that may die back to the ground in winter.
**herbicide:** Weed (plant)-controlling compound.
**heredity:** Transfer of characteristics from parents to offspring through gametes.
**hole:** Final target on a golf green 4.25 inches (10.8 cm) in diameter, and at least 4 inches (100 mm) deep.
**homogenize:** With fertilizer, a process where nutrients are a slurry and used directly as a liquid fertilizer or undergo heating and drying to form granulars, possibly coated to extend release rates, and then screened for desired prill size.
**homogenous fertilizer:** Fertilizer where each granule or pellet has the same analysis.
**hormone:** Chemical substance produced in one part of a plant and used in minute quantities to induce a growth response in another part of the plant; often refers particularly to auxins.
**host:** An organism invaded by a plant parasite, which obtains its nutrients and reproduces upon it.
**hot water insoluble nitrogen (HWIN):** Insoluble nitrogen fertilizer formed in hot (212°F or 100°C) water.
**humate (humin):** Portion of soil organic matter that is insoluble in dilute alkali.
**humic acid:** Portion of humus that is water insoluble and is extracted from soil with dilute alkali and precipitated upon acidification.
**humidity:** *See* relative humidity.
**humus:** Relatively stable, dark-colored colloidal soil organic matter containing no recognizable plant parts.
**hybrid:** Cross between two species.
**hydrated lime:** Calcium hydroxide from reacting burnt lime (CaO) with water.
**hydraulic conductivity:** Rate of water flow in soil as imposed by a hydraulic head.
**hydrolysis:** Splitting of one molecule by adding water.
**hydrophilic:** Water-loving; attracted to water.
**hydrophobic:** Water-hating; not attracted to water.
**hydroscopic:** Substance capable of absorbing moisture from the air.
**hydroscopic water:** Unavailable soil water tightly held by bonding (absorption) to soil particles.
**hydroseeding:** Planting of turf seed through a hose mixed with water and possibly fertilizer and mulch.
**hydroxyl group:** An $OH^-$ group formed by the dissociation of a water molecule.
**hypha (plural: hyphae):** Filaments of fungal cells.
**hypocotyl:** Stem part below the cotyledons of a seedling.

## I

**IAA:** *See* indole-3-acetic acid.
**immersed:** Growing under water; submerged.
**immobilization:** Conversion of an element from the inorganic to organic form in microbial or plant tissues; often used to describe the conversion of nitrate or ammonium into organic forms by soil microorganisms.
**imperfect flower:** Flowers lacking either male or female parts; unisexual flowers.
**imperfect fungi:** Fungus that does not produce sexual spores.
**imperfect stage:** Asexual sporulating stage of a fungus.
**incidence:** Proportion of a population affected by a phenomenon such as a disease.
**indole-3-acetic acid (IAA):** Naturally occurring auxin, a kind of plant hormone.
**inductively coupled plasma emission spectrometry (ICP, ICAP, or ICP-AES):** Using an inductively coupled plasma as an excitation source to determine elements in water, soil extracts, and plant tissue based on the level of light emission of each element.
**inert ingredients:** Portion of a formulation not active (or is inert) but aids in dispersal, shipping, and longevity such as water, wetting agents, emulsifiers, buffers, and spreading agents.
**infect:** To invade or penetrate.
**infiltration:** Downward entry and movement of water into and through soil.
**infiltration rate:** Quantity (inch or centimeter) of water entering soil per unit time (hour).

**infiltrometer:** Device used for measuring the rate of water's entry into a soil.

**inflorescence:** Flowering portion of a plant.

**injury:** Plant damage from a biotic, physical, or chemical agent.

**inoculum:** Infectious bodies that initiate infection in a host organism.

**inorganic:** Non-carbon-containing matter, excluding carbonates.

**insecticide:** Insect-killing compound.

**instar:** Stages between insect molts or shedding of the exoskeleton.

**integrated pest management (IPM):** Combining methods to control pests, such as resistant plant varieties, chemical and natural or biological pesticides, pest exclusion, and plant health management techniques.

**intercalary meristem:** Meristematic area at the base of each internode that accounts for stem elongation in grasses.

**intermediate rough:** Optional narrow (4 to 10 ft, 1.2 to 3 m) strip of turf parallel to the sides of the fairway maintained in height between the fairway and primary rough.

**internode:** Section of stem between two successive nodes or joints.

**interpacking:** Process of compacting sand containing various sized particles.

**interseeding:** Seeding into an established stand of grass.

**interspecific hybrid:** Cross between two different species; [e.g., Tifway (419) bermudagrass is an interspecific hybrid from a cross between a tetraploid *Cynodon dactylon* and diploid *C. transvaalensis*].

**intraspecific hybrid:** Cross between two plants of the same genus and species; (e.g., Princess-77 bermudagrass is an intraspecific hybrid from crossing two *Cynodon dactylon* bermudagrasses).

**intravaginal growth:** Shoot growth in plants that does not penetrate through the enclosing sheath.

**invert emulsion:** Suspension of minute water droplets in a continuous oil phase.

**ions:** Electrically charged atoms resulting from the loss of electrons (cations) or gain of electrons (anions).

**irrigation:** Supplying water to a stand of plants.

**isomer:** Compounds with identical atomic composition but that differ in structural arrangement; for example, glucose and fructose.

**isomorphous substitution:** Replacement of similarly sized atoms in a soil crystal lattice.

## J

**joint:** In turf, node of a grass stem.

## K

**keel:** Prominent ridge, often comprising, for example, tissue on both sides of a glume's midrib or leaf blade, that has grown together.

**kinetin:** Purine that acts as a cytokinin in plants.

**kingdom:** One of seven chief taxonomic categories; for example, Fungi or Plantae.

**Kranz anatomy:** In $C_4$ plants, a wreathlike arrangement of mesophyll cells around large bundle-sheath cells, forming two concentric layers around the vascular bundle, that aids in concentrating carbon dioxide needed for photosynthesis and subsequent carbohydrate formation.

## L

**lamella:** Layer of photosynthetic, chlorophyll-containing plant membranes.

**lamina:** Extended flattened portion of a leaf or petal.

**lanceolate:** Shape longer than wide, broadest below the middle; lance shaped.

**lapping, mower:** *See* backlapping.

**larva:** Immature stage of young insects called a caterpillar, slug, maggot, or grub.

**lateral bud:** Bud originating in the leaf axil, on the side of the stem.

**lateral movement:** Herbicide movement, mostly in a horizontal plane, from the original site of application.

**lateral shoot:** Shoot originating from vegetative buds in the axil of leaves or from the nodes of stems, rhizomes, or stolons.

**layering:** In soil, alternating stratification of different soil texture that affects water movement and soil aeration.

**$LD_{50}$:** Measure of toxicity where the dose of product causes mortality in 50% of the treated test animals. Generally, the lower or smaller the $LD_{50}$ value is, the more acutely toxic the product is.

**LDS:** *See* localized dry spot.

**leaching:** Downward movement of soluble materials in a soil.

**leaching requirement (LR):** Water amount needed to maintain soil salinity below a phytotoxic threshold.

**leaf axil:** Upper angle formed between the axis of a stem and another structure, such as a leaf.

**leaf bud:** Emerging grass blade.

**leaflet:** One of the divisions of a compound leaf.

**legume:** Member of the pea or bean family having a dry fruit that splits open along two longitudinal sutures; pods.

**lemma:** Lowermost of the two bracts enclosing a grass flower; the other bract is a palea.

**lesion:** Wound or delimited diseased area.

**level:** In surveying, an instrument used in combination with an elevation rod to read differences in grade.

**lignin:** Strengthening or deposition material in plant cells, along with cellulose, that tends to make them hard. Chemically, lignin has both phenolic and alcoholic characteristics that are very resistant to decomposition by soil organisms and constitute much of the soil's residual humus.

**ligule:** Projection at the inside junction of the grass leaf blade and collar, which may be membranelike or a row of hairs.

**lime:** Calcium oxide (CaO) and/or a variety of acid-neutralizing materials containing calcium or calcium and magnesium.

**limestone:** Sedimentary rock composed of more than half calcium carbonate ($CaCO_3$).

**linear:** Long and narrow shape with parallel margins.

**liquid fertilization:** Applying nutrients as dissolved fertilizer in solution.

**lipophylic:** Nonpolar molecules that are soluble in fat or oil.

**littoral region:** Shoreline area of a body of water usually composed of a ring of plants.

**loam:** Soil composed of 7% to 27% clay, 28% to 50% silt, and less than 52% sand.

**lobe:** Segment of a simple leaf cut rather deeply into curved or angular segments.

**localized dry spot (LDS):** Soil that resists rewetting associated with thatch and/or organic acid coating of sand particles, buried debris, fairy ring fungi, or insufficient irrigation.

**long-day plants:** Plants that initiate flowering under long-day (short-night) regimes.

**luxury consumption:** Plant uptake of an essential nutrient in amounts higher than needed.

## M

**macronutrient:** Nutrients needed in the largest amounts (usually 50 mg/kg) for plant growth (e.g., C, N, O, K, Mg, Ca, S, and H).

**macropores:** Large soil pores that drain freely from gravity.

**magnesium carbonate ($MgCO_3$):** Insoluble form of magnesium precipitated by water high in magnesium, bicarbonates, and carbonates. It remains insoluble until reacted with an acid.

**marcelling:** Wave or washboard turf pattern from mower misadjustment, excessive mower speed, or too slow cutting frequency for the cutting height; also called *rippling*.

**marl:** Freshwater lake deposit consisting of soft calcium carbonate mixed with clay, silt, or other impurities.

**mat:** Tan- to brown-colored, tightly intermingled layer of thatch intermixed with soil.

**material safety data sheet (MSDS):** Information on first aid measures, fire fighting, spill or leak handling, storage procedures, worker protection standards, etc. about a product (e.g., pesticide).

**matric (or capillary) potential:** How tightly water is held (or adsorbed) in soil; carries a negative (minus) sign.

**matting:** Process of dragging or brushing the turf surface to incorporate topdressing or disperse cores lifted from aerifying.

**Mehlich I and III extraction reagents:** Extraction reagents for removing P, K, Ca, Mg, Na, and Zn from acid to neutral pH soils.

**meiosis:** Sexual cell division where the nucleus is divided into four genetically dissimilar nuclei due to chromosome recombination.

**membranous:** Thin, transparent, and flexible; membranelike.

**meristem:** Cluster of dividing cells at the root and stem apices or tips.

**mesophyll:** Leaf area between the upper and lower epidermis; the large, thin-walled cell parenchyma.

**metamorphosis:** In biology, a series of changes where an insect passes from an egg to an adult.

**methylated seed oil:** Process of reacting methanol with crop seed oil to increase plant entry of applied products.

**microclimate:** Environmental conditions in the immediate vicinity of plants or planting groups such as a golf green; also referred to as *microenvironment*.

**microencapsulated (ME):** Pesticide formulation where small particles are surrounded by a barrier layer and concentrated in a liquid carrier.

**micronutrient:** Elements needed in only small (usually less than 100 mg/kg) amounts for plant growth (e.g., boron, chlorine, copper, iron, manganese, and zinc).

**midrib:** Main or central vein or rib of a leaf or leaflet.

**midvein:** The primary vein.

**milliequivalent (meq):** Measurement of charge concentration as 1/1000 of an equivalent. One equivalent is 1 g hydrogen in 1 L water, while 1 meq is 0.001 g (or 1 mg) hydrogen in 1 L water. Milliequivalents per liter (meq/L) is the unit from dividing parts per million by equivalent weight.

**mineralization:** Conversion of an organic form of an element to an inorganic form (e.g., conversion of organic nitrogen to ammonium nitrogen by microbial decomposition).

**mitochondrion (plural mitochondria):** Double-membrane-bounded organelle in eukaryotic cells containing the enzymes of the Krebs cycle and the electron-transport chain.

**mitosis:** Asexual cell division where the nucleus is divided into two genetically similar nuclei.

**mixture:** In turf, seed combination of two or more turfgrass species.

**mold:** Any profuse or woolly fungus growth on damp or decaying matter or on plant tissue.

**mole:** In chemistry, the number of particles in 1 mol of any substance; always equal to Avogadro's number: $6.022 \times 10^{23}$.

**molecular weight:** Sum of relative weights of atoms in a molecule, with carbon atoms being the reference of 12.

**molecule:** Smallest possible unit of a compound, consisting of two or more atoms.

**monocotyledon, monocot:** Grass and grass-like plants in which embryos (seedlings) have one cotyledon (seed leaf), parallel-veined leaves, inconspicuous flowers, flower parts in multiples of three, and no secondary growth.

**monoecious:** Situation of the staminate (male) and pistillate (female) flowers being in separate inflorescences but occurring on the same plant, such as corn.

**monostand:** Area composed of one species of plants.

**morphological:** Plant form or structure.

**mosaic:** Disease symptoms composed of mixed green, light-green, and yellow patches.

**moss:** Small, threadlike, branched primitive plants often growing in shaded, wet turf areas.
**moss peat:** Nutrient-deprived, partially decomposed peat located directly under the moss layer in a bog.
**mottle:** Irregular pattern of indistinct light and dark areas; often used to characterize disease symptoms.
**mower:** Machine used to uniformly cut the upper shoot growth of turfgrasses.
**mowing frequency:** Time (usually in days) between mowing events.
**mRNA:** Single-stranded molecule of "messenger" ribonucleic acid that directs protein production (translation).
**muck:** *See* peat.
**muck soil:** Soil containing 20% to 50% of well-decomposed organic matter.
**mulch:** Layer of plant residues, leaves, sand, plastic, or paper on the soil surface to level the soil surface and protect it from the effects of raindrops, freezing, evaporation, and so forth.
**mutation:** Changes in DNA that produce new characteristics; often artificially induced by exposure to radioactive radiation or certain chemicals.
**mycelium (plural: mycelia):** Cottony strands of individual hyphae that form the body of a fungus.
**mycoplasmas:** *See* phytoplasmas.
**mycorrhiza:** Symbiotic association of a fungus with plant roots.

## N

**native:** In biology, species of plants or animals indigenous to a specific region.
**native soil:** Unamended soil normally found in a particular area.
**naturalized species:** Introduced species that has adapted to a specific region.
**necrosis:** Plant death, usually accompanied by darkening or discoloration.
**nematicide:** Nematode-controlling compound.
**nematode:** Microscopic, nonsegmented worm-like animals that may parasitize plants or animals.
**nitrification:** Microbial oxidation of ammonium nitrogen to nitrites and eventually to nitrates.
**node:** Point or level of a stem at which one or more leaves and roots are attached.
**noncapillary (or aeration) porosity:** Percentage of a soil occupied by air at field capacity (or free drainage).
**nonionic surfactant:** Surfactant that does not ionize (become charged) in aqueous solutions and therefore is unaffected by hard water and can be readily used in strong acid solutions and cold water.
**non-point-source pollution:** Contamination from a general area and not from a specific site.
**noxious plants:** Weedy plants that are usually difficult to control or eradicate.
**nucleic acid:** Single- or double-stranded polynucleotide containing either DNA or RNA linked by phosphodiester bonds.
**nursery:** In turf, designated area of turf often maintained similarly to a golf green used to repair or replace damaged grass.

## O

**obligate parasite:** Parasite that requires a living organism on which to grow and multiply.
**off-site mixing:** Soil and amendment mixing away from the planting site.
**opposite:** Arrangement of paired leaves attached oppositely from each other at the same node.
**oral toxicity:** Degree of toxicity of a compound when ingested.
**order:** In biology, a category of classification between the rank of class and family; classes contain one or more orders, and orders comprise one or more families.
**organic:** Carbon-containing compounds other than inorganic carbonates.
**organic fertilizer:** Fertilizer derived from natural organic materials.
**organic matter:** Residual decomposition of plant or animal content in soil.
**organic soil:** Soil containing >15% organic matter.
**osmosis:** Diffusion of water, or any solvent, from a region of greater water potential to one of lesser water potential across a selectively permeable membrane.
**osmotic (solute) potential:** Change in chemical potential (or free energy) of water produced by the addition of solutes; carries a negative (minus) sign. Also called *solute potential.*
**ovary:** Lower part of the pistil containing the ovules or, later, the seed.
**overseeding:** Planting or seeding onto an existing turf; performed primarily to provide a temporary turf cover for green color.
**oxidation:** Loss of an electron by an atom or molecule; the atom or molecule that loses the electron is referred to as being *oxidized.* This is called oxidation since oxygen, which strongly attracts electrons, is often the electron acceptor.
**oxidation-reduction reactions (or redox):** Passing of electrons from one atom or molecule to another during chemical reactions.
**oxidative phosphorylation:** Formation of ATP from ADP and inorganic phosphate.

## P

**palea:** Uppermost (inner) of the two bracts enclosing the flower of a grass floret; this bract is enclosed by the largest lower bract (lemma) and is located on the side opposite the embryo.
**palmate:** Type of leaf where leaflets or lobes originate from a common point, and diverge like the fingers from the palm of the hand.

**panicle:** Inflorescence composed of several branches and sub-branches.
**parasite:** Organism that obtains its food from another.
**particle density:** Mass (or weight) of dry soil per unit volume of soil solids (excluding pores).
**particle-size distribution:** Fractions of various soil separates in a soil sample.
**parts per million (ppm):** Parts by weight of a compound in 1 million parts of the final mixture (ppm = mg/L).
**pathogen:** Any organism capable of causing disease by obtaining nutrition from its diseased host.
**pathogenicity:** Ability to cause disease.
**pea gravel:** Uniformly sized gravel 1/8 to 3/8 inch (3.2 to 9.5 mm) in diameter.
**peat:** Partially decomposed organic matter accumulating under wet conditions. *Peat* refers to partially decomposed deposits while *muck* includes highly decomposed materials.
**peat moss:** Dried peat from various plants that is slow to decompose.
**peat soil:** Organic soil containing >50% organic matter.
**pedicel:** Stalk of a simple flower or spikelet.
**peletized lime:** Lime produced by binding or compressing fine lime particles into large granules or pellets.
**penetrant:** Pesticide class absorbed into plants following application.
**perched water table:** Saturated zone of fine-textured soil over an underlying coarser-textured soil. Water in the saturated zone will not move into the coarse-textured soil interface until sufficient water potential (or weight) builds to overcome the attraction between water and the fine-textured soil.
**percolation rate:** Downward movement of water through soil, especially saturated or near-saturated soil.
**percolation test:** Rate of water percolation in a soil profile.
**perennial:** Plant that normally lives for more than two years.
**perfect flower:** Flower with both functional pistils (female) and stamens (male); bisexual.
**perfect fungus:** Fungus that produces sexual spores.
**pericarp:** Outermost wall of a matured ovary when it becomes a fruit or the outerlayer (or "seed coat") of the caryopsis in grass seeds.
**perlite:** Light, expanded soil amendment from firing volcanic rock that resists weathering but may be fragile and is used to increase soil porosity and available soil water.
**permanent wilting point:** Soil water content at which plants wilt and do not recover; often considered to be the soil water content at –1.5 MPa (–15 bars) water potential.
**permeability:** In soil, ease in which gas, liquids, or plant roots penetrate or pass through.
**pesticide:** Compound designed to kill pests or render them harmless.
**petal:** Inner floral leaf that makes up a flower's corolla, which is generally colored or white.
**petiole:** Stalk or stem of a leaf.
**pH:** Degree of acidity or alkalinity; defined as the negative logarithm of hydrogen ion activity. A scale of 0 to 14 is used where 7 is neutral, <7 is increasingly acidic, and >7 is increasingly basic (or alkaline).
**phenolics:** Compounds with a hydroxyl group (–OH) attached to an aromatic ring (a ring of six carbons containing three double bonds); includes flavonoids, lignins, salicylic acid, and tannins.
**phloem:** Food-conducting tissue of plants.
**phosphate:** Phosphorus oxide ($P_2O_5$).
**photodegradation:** Breakdown from sunlight exposure.
**photorespiration:** Consumption of oxygen and release of carbon dioxide when the enzyme RuBisCO binds oxygen instead of carbon dioxide.
**photosynthesis:** Process of converting water and carbon dioxide by chlorophyll into sugars, starches, and oxygen, using energy from sunlight.
**phototropism:** Plant growth response to light.
**phytochrome:** Bluish photoreversible protein pigment responsible for the photoperiodic control of flowering and seed germination.
**phytoplasmas:** Microorganisms that lack an organized and bounded nucleus (like bacteria) but also lack a true cell wall or the ability to form one (unlike bacteria); formally called *mycoplasmas.*
**phytotoxicity:** Plant injury from a chemical.
**pistil:** Female flower composed of stigma, style, and ovary, and formed from one or more carpels.
**pith:** The central soft tissue of a stem.
**pitot tube:** Gauge used to measure operation pressure at a sprinkler head.
**plant growth regulator (PGR):** Applied substance that controls or modifies plant growth.
**plugging:** Vegetative propagation of turf using plugs or small sod pieces.
**point-source pollution:** Pollution from a known, identifiable source such as a spill, smokestack, or discharge from a pipe.
**poling:** In turf, using a long pole to mechanically break up dew on plants, clipping clumps, or earthworm castings.
**polyacrylamide gels:** Synthetic polymer used as a soil amendment due to its ability to hold water.
**polyploid:** Containing more than two complete sets of chromosomes.
**polystand:** Area composed of more than one plant species.
**pore space (soil porosity; void space):** Pore portion of soil bulk volume occupied by air or water.
**porosity:** Ratio percentage of pores (or voids) to total volume of solids.
**postemergence (POST):** Weed control after seedling emergence.
**potash:** Potassium oxide ($K_2O$).
**potential acidity:** *See* exchangeable acidity.
**preemergence (PRE):** Weed control before seedling emergence.
**pregermination:** Process of soaking seed to promote partial germination before planting.
**pressure potential:** Pressure that develops in cell walls due to its contents; also called *turgor potential.*

**preventative:** Fungicide applied prior to anticipated disease incidence.
**primary root:** First root of plants; the taproot.
**primary rough:** Rough area immediately adjacent to the intermediate rough (if present) or fairway where errant balls often land that is maintained between the levels of the fairway and more distant secondary rough.
**prokaryote:** Simple organism without nuclear material enclosed in a nuclear membrane such as actinomycetes, bacteria, and cyanobacteria.
**prostrate:** Parallel to or lying flat on the ground.
**protein:** Large complex molecule of one or more chains or amino acids that are the cell's main building materials and do most of a cell's work.
**protoplasm:** Living substance in plant or animal cells.
**pubescent:** Covered with hairs.
**puffiness:** Sponge-like characteristic of turf, usually from excessive thatch, causing inferior smoothness and scalping.
**pure live seed (PLS):** Percentage of germinable pure seed in a seed lot. Mathematically, it is the percent germination multiplied by the percent pure seed, which is then divided by 100.

## R

**raceme:** Elongated inflorescence with each flower on individual stalks.
**radiation:** Energy radiated or transmitted as rays or waves (e.g., sunlight).
**reduction (redox):** Gain of an electron by an atom or molecule.
**reel mower:** Mower with a vertically mounted rotating reel (or blade) that cuts against a stationary bed knife.
**registered seed:** Class of certified seed produced from foundation seed and planted to produce certified (or blue-tag) seed. It is identified by a purple tag.
**relative humidity (RH):** Ratio (expressed as a percentage) of water vapor in the atmosphere to the maximum possible quantity at the same temperature.
**remediation:** Correcting, restoring, or eliminating a contaminated or polluted soil.
**renovation:** To renew; make over; repair.
**residual sodium carbonate (RSC):** Measurement of irrigation water to determine whether sodium in it will cause soil structure problems from the potential precipitation of calcium and magnesium ions.
**resistance:** Ability of an organism to overcome a normally phytotoxic material or pathogen.
**respiration:** Intracellular process in which molecules are oxidized with the release of energy.
**restricted-use pesticide:** Pesticide that can be purchased and used only by certified applicators to avoid adverse health or environmental effects.
**rhizome:** Creeping, horizontal underground stem, producing shoots above ground and roots below; distinguished from a root by the presence of nodes, buds, or scale-like leaves; may originate from the main stem or from tillers.
**rhizosphere:** Soil immediately adjacent to plant roots containing microorganisms that may differ from those in the general bulk soil.
**ribonucleic acid (RNA):** Type of nucleic acid formed on chromosomal DNA and involved in protein synthesis.
**rippling:** Wave or washboard turf pattern from mower misadjustment, excessive mower speed, or too slow cutting frequency for the cutting height; Also called *marcelling.*
**riprap:** Placing broken rock, cobbles, or boulders on slopes to prevent soil erosion.
**roguing:** In turf, hand-removing of undesirable plants.
**rolled:** Cylindrical arrangement of the youngest leaf in the bud shoot.
**rolling:** In golf, use of a mechanical roller on greens to smooth the playing surface or increase ball roll.
**root:** Subterranean portion of a plant used for anchorage, aeration, and absorption.
**rootzone:** Soil layer where plant roots grow.
**rosette:** Circular cluster of leaves usually appressed to or located near the ground level.
**rotary mower:** Mower with a horizontal, rapidly rotating impact blade.
**rough:** In golf, the area surrounding fairways maintained with taller grass. Consists of intermediate rough (if present), primary rough, and secondary rough.
**row sprigging:** Planting turf sprigs in rows or furrows.
**RSC:** *See* residual sodium carbonate.
**rudimentary:** Small, often incompletely developed.
**runner:** A slender stolon (or horizontal stem).
**rust:** In turf, orange- or rust-colored disease, caused by one of the Uredinales (rust fungi).

## S

**saline-sodic soil:** Soil with enough soluble salts (>4 dS/m) and exchangeable sodium (>15%) to impair its productivity.
**saline soil:** Soil with enough soluble salts (>4 dS/m) to impair its productivity.
**salt:** Compound containing positive ions from a base and negative ions from an acid, or that results from direct combination of metal and nonmetal (e.g., NaCl dissociates into $Na^+$ and $Cl^-$ in water).
**salt index:** Numerical comparison of fertilizer compounds to 100 for sodium nitrate.
**sand:** Soil textural class consisting of particles between 0.05 and 2.0 millimeters in diameter.
**saprophyte:** Organism feeding on dead or decaying organic matter.
**SAR:** *See* sodium adsorption ratio.
**saturate:** Completely soak.
**scald:** Collapse of turfgrass when exposed to conditions of standing water and intense light.
**scalping:** In turf, excessive low mowing of plant leaf tissue, leaving brown stems, clippings, and exposed soil.

**scum:** In turf, a thin, slippery layer of algae on the turf surface, usually from excessive moisture and insufficient light and drainage that impedes healthy turf growth.

**scutellum:** Shield-shaped organ of grass embryos.

**secondary rough:** Rough farthest away from the fairway, normally maintained at the tallest heights.

**seed:** Ripened ovule, often used to establish a turf sward.

**seedbed:** Soil preparation that supports seed germination and seedling growth.

**seedhead:** Collection of flowers clustered upon a main stem; the inflorescence of grasses, sedges, and rushes.

**seeding:** To sow (plant) with seeds.

**seedling:** A young plant.

**selective herbicide:** Chemical that controls one plant species within another.

**semiarid turfgrass:** Turfgrass adapted to low-rainfall, semi-arid regions; includes buffalograss, bluegrama, wheatgrass, and sideoats grama.

**seminal root:** Root arising from the base of the hypocotyl.

**sensible heat loss:** Loss of heat from leaf surfaces due to air circulation around the leaf (conduction and convection) when leaf surface temperatures are higher than that of air.

**sepal:** Part of a flower that is usually petal-like in appearance and green in color.

**serrate:** Type of leaf margin with sharp teeth pointing forward; saw-toothed.

**sessile:** Without a petiole, stem, or stalk; usually refers to a leaf being attached directly on the axis or stem of a plant.

**seta (plural, setae):** Prominent black spines present on the spore-bearing bodies (acervuli) of *Colletotrichum graminicola*, the anthracnose fungus.

**sewage sludge:** Settled sewage solids removed by screening, sedimentation, chemical precipitation, or bacterial digestion.

**shattering:** Turf cultivation method involving fragmentation of a rigid or brittle soil mass, usually by a vibrating tine or mole device.

**sheath:** Lower portion of a leaf that encircles the stem.

**shoot:** In botany, a general term for the aboveground portion of a plant.

**short-day plants:** Plants that initiate flowering best under short-day (long-night) regimes.

**sign:** In biology, visible pathogen or its parts or products on a host plant.

**silt:** Soil textural class consisting of fine particles between 0.05 and 0.002 mm in diameter.

**simple:** Type of leaf consisting of a blade not divided into individual leaflets; unbranched.

**slag:** By-product material of smelting, containing mostly silicates, often used as a nutrient source.

**slicing:** Turf cultivation procedure where vertically positioned blades rotate into the turf, providing a slicing or cutting action of stolons, rhizomes, and thatch.

**slime molds:** Fungi in the class *Myxomycetes*; also, superficial diseases caused by these fungi on low-lying plants.

**slit drain:** Narrow trench backfilled with a porous medium, such as sand or gravel, used to intercept surface or lateral subsurface drainage water; also called *French drain*.

**slit seeding:** Process of using vertically mounted knives or blades to cut slits in the turf, followed by seeding and rolling to increase seed-to-soil contact with minimum surface disruption.

**slow-release (water-insoluble) fertilizer:** Fertilizer source designed to control its nutrient release rate.

**smooth:** Lacking hairs, divisions, or teeth; not rough to the touch.

**smut:** Black, sooty-colored disease caused by the smut fungi (Ustilaginales).

**sod:** Vegetative planting material consisting of turfgrass strips with or without adhering soil.

**sod strength:** Ability of sod to resist tearing during harvesting and handling.

**sodding:** Process of laying sod.

**sodic soil:** Nonsaline soil with sufficient sodium (exchangeable sodium percentage of ≥15 or sodium adsorption ratio of ≥13), a pH from 8.5 to 10, and dispersed soil colloids to reduce permeability.

**sodium adsorption ratio (SAR):** Relative hazard of irrigation water from its sodium content relative to its amount of calcium and magnesium; measured in milliequivalents per liter (meq/L):

$$SAR = \frac{Na^{+}}{\sqrt{\frac{Ca^{+2} + Mg^{+2}}{2}}}$$

**sodium sulfate ($Na_2SO_4$):** Soluble salt formed when gypsum (calcium sulfate) is used to treat soils high in sodium content.

**soil:** Upper layer of earth surface used as the natural medium for plant growth.

**soil conditioner:** Any material added to a soil to improve its physical properties.

**soil mix:** Prepared mixture of soil or sand plus amendments used as a growth medium.

**soil modification:** Artificial altering of soil by adding amendments to improve its physical conditions.

**soil test:** Chemical, physical, or microbiological property of a soil.

**soil texture:** Relative soil coarseness or fineness as determined by its proportions of sand, silt, and clay.

**soluble concentrate (SC):** Liquid formulation that forms a solution when added to water.

**soluble granule (SG):** Coarser textured dry formulation that forms a solution when added to water.

**soluble liquid (SL):** Liquid formulation in which the active ingredients are dissolved in water.

**soluble powder (SP):** Finer textured dry formulation that forms a solution when added to water.

**soluble salts:** Total soluble ions in a soil measured as electrical conductivity (EC).

**solute:** Molecule dissolved in a solution.
**solute potential:** *See* osmotic potential.
**species:** Group of individuals having certain distinctive characteristics in common.
**sphagnum peat:** Layer below moss in a bog that is partially decomposed.
**spike:** In plants, an unbranched inflorescence with the spikelets sessile (stalkless) on a rachis.
**spikelet:** Basic individual unit of the spike of grasses and some sedges; composed of one or more flowers and their subtending bracts (two glumes and one or more florets).
**spiking:** Turf cultivation procedure where solid tines or spoons penetrate into the turf and soil surface.
**spore:** Special structure, similar to a seed, produced by fungi for spread, survival in unfavorable environments, and reproduction and multiplication by the fungus.
**spray drift:** Movement of airborne particles from the intended site of application.
**sprig:** Single turfgrass unit containing viable stolons, rhizomes, or tillers and roots, used for establishment.
**sprigging:** Vegetative establishment by placing stolons, rhizomes, or tillers (with roots) into furrows.
**spring green-up:** Initial growth of green shoots in spring.
**stalk:** Any slender supporting structure such as a petiole for a leaf, a peduncle for an inflorescence, or a pedicel for a flower.
**stamen:** Male or pollen-bearing organ of flowering plants consisting of the filament (stalk) and the anther.
**starch:** Complex insoluble principal food storage substance of plants consisting of glucose units forming polysaccharides.
**stem:** Plant organ for support, leaf production, food storage, and limited food production.
**sterile:** Without seeds or pollen.
**stigma:** Upper feathery part of the female flower pistil that receives pollen.
**stimpmeter:** Device used on golf greens to measure ball roll distance and putting surface uniformity.
**stipule:** Bract-like appendages at the base of some leaves.
**Stokes's law:** Equation used to determine soil texture from the settling rates of suspended sands, silts, and clay based on the diameter of the particle and viscosity of the suspension medium.
**stolon:** Creeping, aboveground stem that roots at the nodes.
**stolonize:** Vegetative establishment method of broadcasting stolons over a prepared soil and covering with topdressing or press rolling.
**stomate (plural: stomata):** Small pore-like opening in the leaf or stem epidermis through which gas exchange occurs.
**strain:** Descendents of an isolated organism; biotype; race.
**stroma:** In plants, soluble environment of chloroplasts; in fungi, a mass of hyphae adhering together.
**style:** Stalk that connects the stigma to the ovary in flowers.
**stylet:** Hollow, spear-like device parasitic nematodes use to pierce and feed on grass roots.
**subgrade:** Soil elevation beneath the final grade that allows subsequent placement of topsoil to achieve the final grade.
**subsoil:** Soil below the plow layer.
**subspecies:** Subdivision of a species.
**succulent:** Soft and fleshy.
**sucker:** Plant shoot arising from an adventitious bud on a root.
**sulfuric acid ($H_2SO_4$):** Acid often injected into irrigation water to reduce pH or an acid that forms in soil when acidifying amendments or fertilizers are used such as soil sulfur (S), ammonium sulfate, etc.
**sulfurous ($SO_3$) generator:** Equipment that heats elemental sulfur to produce sulfurous gas that, when combined with water, becomes sulfuric acid, which converts bicarbonates and carbonates to carbon dioxide and, in the process, reduces water pH. Also known as a *sulfur burner*.
**summer annual:** Plant that germinates in spring, grows and flowers in summer, and sets seed in fall, after which it dies.
**summer dormancy:** Cessation of growth of perennial plants due to heat and/or moisture stress.
**surfactant:** Material that improves the emulsifying, dispersing, spreading, or wetting properties of liquids. Acronym for **SURF**ace-**ACT**ive **A**ge**NT**.
**surveying:** Determining boundaries and elevations of an area by measuring angles, distances, and elevations.
**suspension (S):** Mixture containing finely divided particles dispersed in a solid, liquid, or gas.
**suspension fertilizer:** Fluid containing dissolved and undissolved plant nutrients.
**swale:** Shallow, valley-like excavation designed to conduct surface water.
**swing joint:** Piping configuration that flexes to protect irrigation heads and lines from traffic.
**symptom:** Visual reaction by a plant indicating presence of disease.
**synergism:** When the combination of two factors is greater than either factor used separately.
**syringing:** In turf, the light application of moisture to reduce surface temperatures, prevent wilt, and remove dew, frost, and exudates.
**systemic:** Compounds that enter and move (translocate) throughout plants.

## T

**taproot:** Single enlarged vertical main root lacking major divisions.
**tee:** In golf, the area on each hole where play begins.
**tee-marker:** Movable device that indicates where golf play begins on each hole.
**tensiometer:** Device used to measure the tension water is held at in soil.
**terminal bud:** Bud located at the end or apex of a stem or branch.
**texture:** In turf, leaf blade width and arrangement.

**thatch:** Brown- to black-colored layer of dead leaves, stems, rhizomes, crowns, and stolons between the green vegetation and soil surface.

**thermal stratification:** Layering in bodies of water due to temperature differences. *Epilimnion* is an upper layer of warm water while *hypolimnion* is a lower layer of cold or cooler water. *Thermocline* is the dividing line between the two layers.

**three-ranked:** Diverging from the stem in three directions, as in the sedge family.

**thrust blocks:** Concrete slabs used to protect the integrity of an irrigation pipe connection.

**thylakoid:** Sac-like membranous structures in chloroplasts that contain chlorophyll and stack to form grana.

**tiller:** Grass shoot, usually erect, originating intravaginally (growing upward within the enclosing leaf sheath) from axillary buds in the axis of a leaf or in the unelongated crown portion of a stem.

**tipburn:** Leaf-tip necrosis from desiccation, mower, temperature, or pesticide damage.

**tissue testing:** Chemical analysis of plant leaves to ascertain nutrient levels.

**toothed:** Sawtooth-like projections (or "teeth") on the margins of leaves.

**topdressing:** In turf, addition of a thin layer of soil, sand, or fertilizer following plant establishment.

**topsoil:** Uppermost layer of soil.

**total dissolved salts (TDS):** Sum of all dissolved solids in a water or wastewater, normally reported as parts per million (ppm).

**transition:** In turf, a gradual conversion of a grass sward from one turfgrass to another without thin or bare areas developing (e.g., spring transition from overseeded ryegrass back to the permanent bermudagrass base).

**transition zone:** In turf, the area or zone between warm (subtropical) and cool (temperate) regions where both warm-season and cool-season turf species grow although it is not optimal for either.

**transpiration:** Water vapor loss from plants, primarily through leaf stomata.

**trifoliate:** Type of compound leaf composed of three leaflets.

**triploid:** Having three complete chromosome sets per cell ($3n$).

**TruFirm turf firmness meter:** Instrument utilizing a hemisphere-shaped impact hammer dropped from a consistent height to determine surface firmness based on maximum turf penetration. The lower the penetration value is, the firmer the turf surface is.

**tuber:** Thickened storage portion of a rhizome or stolon (stem), bearing nodes and buds.

**tufted:** In compact clusters, forming clumps.

**turf:** Uniform ground covering of mowed vegetation, usually a turfgrass.

**turfgrass:** Grass species or cultivar maintained as a uniform mowed vegetation.

**turgor pressure:** Cell pressure from the movement of water into it.

**turning point:** In surveying, a hub used in hilly terrain to allow surveying to continue when the instrument is moved and a new height of the instrument can be established.

**two-ranked:** In two vertical rows on opposite sides of a stem or an axis, as in the grass family.

## U

**ubiquitous:** Occurring everywhere.

**uniformity coefficient (CU):** Numerical expression when evaluating sands for the particle size uniformity based on $D_{60}/D_{10}$ with an optimum range of two to three. The higher the value is, the less uniform is the sand and therefore the greater is potential for particle packing. Values less than two may pack insufficiently, resulting in unstable surfaces. For gravel, CU is defined as $D_{90}/D_{15}$ and should be ≤2.5.

## V

**vapor pressure:** Pressure exerted above a liquid from the tendency of vapor to escape from the surface.

**variety:** Distinguishing, stable subdivision of a plant species.

**vein:** Ribs of a leaf; one of the vascular bundles of a leaf.

**verdure:** Green plant tissue remaining following mowing.

**vermiculite:** Two-to-one clay mineral noted for its high nutrient-holding capacity and low shrink/swell capacity.

**vernalization:** Flowering induction from cold treatment or exposure.

**vernation:** Arrangement of the youngest leaf in the bud shoot; either rolled or folded.

**vertebrate:** Organism with a backbone or spinal column.

**vertical mowing:** Slicing turf with a series of vertically mounted blades rotating on a shaft, usually to reduce thatch; also called *verticutting.*

**virus:** Infectious microscopic agents consisting of a protein sheath surrounding a nucleic acid core that is totally dependent on living cells (obligate parasite).

**vitrified clay:** Clay heated at ~2000°F (1090°C) becoming hard and stable. Used as a soil amendment.

**volatilization:** Changing of a liquid to vapor.

## W

**warm-season grass:** $C_4$ grass with optimum growth between temperatures of 80°F and 95°F (27°C to 35°C).

**washboard effect:** *See* rippling.

**waste area:** Large sand area not routinely raked and/or maintained as sand bunkers. Waste areas are not considered hazards; therefore, rules of play are not the same as for regularly maintained bunkers. Also called *waste bunkers.*

**water hammer:** Shock waves in pipelines from water surges.

**water potential:** The potential (or gradient) energy of water measured as the sum of matric, osmotic, and pressure potential; it is a negative value, and as the value is more negative, lower water potential occurs.

**watershed:** Portion of a landscape that contributes water to a single discharge location. A watershed is defined by its boundary or divide, across which no runoff occurs.
**water table:** Top level of permanent groundwater zone.
**weed:** Plant growing out of place or where it is not wanted.
**weed-and-feed:** Combination product of fertilizer and herbicide.
**wetland:** Land area that floods for part of the year, supports vegetation adapted to wet conditions (called *hydrophytes*), has unique soil conditions that differ from adjacent uplands, and forms a transition zone between aquatic and terrestrial systems; generally includes swamps, marshes, bogs, and similar areas.
**wettable powder (WP):** Finely divided dry formulation that readily suspends in water.
**wetting agent:** Substance that reduces surface tension and causes liquids to make better contact with treated surfaces. Commonly separated into four major groups, based on their ionization in water: (1) anionic, (2) cationic, (3) nonionic, and (4) amphoteric.
**wet wilt:** Turf wilting in the presence of water when evaporation exceeds root water uptake.
**white grub:** C-shaped, whitish larvae of insects in the family Scarabaeidae.
**whorl:** In botany, a cluster of leaves around a stem.
**wilt:** Plant collapse, usually when evapotranspiration exceeds water uptake by roots.
**winter annual:** Plant that germinates in late summer, grows vegetatively during winter, and flowers and sets seed in late spring to early summer, after which it dies.
**winter desiccation:** Plant death from drying during winter dormancy.
**winter feeding (or fertilization):** Late autumn application of fertilizer to cool-season grasses to maintain green color without adverse effects.
**winter-kill:** Injury to plants during winter; most often from desiccation, low mowing height, or shade plus low temperatures.
**witches' broom:** Abnormal brush- or broom-like growth of weak, tightly clustered plant shoots.
**woody:** Consisting or composed of wood or wood-like tissue.

## X

**xanthophyll:** Yellow chloroplast-located pigment that is a member of the carotenoids.
**xerophytes:** Plants that grow in extremely dry habitats.
**xylem:** Water-conducting tissue of a plant's vascular system.

## Z

**zeatin:** Natural cytokinin plant hormone isolated from maize.
**zeolite:** Mined soil mineral used as a soil amendment to improve water- and nutrient-holding capacity; also used in water softening by replacing calcium ions with sodium.
**zoospore:** Reproductive spore capable of moving (swimming) in water.
**zygote:** Fertilized egg.

# Index

Page numbers followed by f and t indicate figures and tables, respectively.

## C

## D

## E

## F

## G

## H

## Q

## T

## U



## V



## W

## X



## Y



## Z